Graph Theory and Interconnection Networks

Graph Theory and Interconnection Networks

Lih-Hsing Hsu and Cheng-Kuan Lin

CRC Press
Taylor & Francis Group
Boca Raton London New York

CRC Press is an imprint of the
Taylor & Francis Group, an **informa** business

CRC Press
Taylor & Francis Group
6000 Broken Sound Parkway NW, Suite 300
Boca Raton, FL 33487-2742

Printed in the United States of America on acid-free paper
10 9 8 7 6 5 4 3 2 1

International Standard Book Number-13: 978-1-4200-4481-2 (Hardcover)

Library of Congress Cataloging-in-Publication Data

Hsu, Lih-Hsing.
Graph theory and interconnection networks / Lih-Hsing Hsu, Cheng-Kuan Lin.
p. cm.
Includes bibliographical references and index.
ISBN 978-1-4200-4481-2 (hardback : acid-free paper)
1. Graph theory. I. Lin, Cheng-Kuan. II. Title.

QA166.H78 2008
511'.5--dc22 2008019917

Visit the Taylor & Francis Web site at
http://www.taylorandfrancis.com

and the CRC Press Web site at
http://www.crcpress.com

Contents

Preface

Since the origin of graph theory with Euler, people have liked graph theory, because it is a delightful playground for the exploration of proof techniques without much previous background needed. With the application of graph theory to many areas of computing, social, and natural sciences, the importance of graph theory has been recognized. For this reason, many good graph theoretic results have been developed in recent years. It is almost impossible to write a book that covers all these good graph theories.

The architecture of an interconnection network is always represented by a graph, where vertices represent processors and edges represent links between processors. It is almost impossible to design a network that is optimum from all aspects. One has to design a suitable network depending on its properties and requirements. Thus, many graphs are proposed as possible interconnection network topologies. Thus, these graphs can be referred to as good graphs. For this reason, the theory of interconnection networks is referred to as *good-graph theory*.

Thus, we need some basic background in graph theory to study interconnection networks. The first 10 chapters cover those materials presented in most graph theory texts and add some concepts of interconnection networks. Later, we discuss some interesting properties of interconnection networks. Sometimes, these properties are too good to be true—so the beginner finds them hard to believe. Such properties can be observed from interconnection networks, diameter, connectivity, Hamiltonian properties, and diagnosis properties. We can also study these properties in general graphs.

Ends of proofs are marked with the symbol □. This symbol can be found directly following a formal assertion. It means that the proof should be clear after what has been said. There are also some theorems that are stated, without proof, via background information. In this case, such theorems are stated without proofs and the symbol □.

The main purpose of this book is to share our research experience. Our experience tells us that much background is not needed for doing research. Looking for a proper theme is a difficult problem for research. We are just lucky in finding problems. For this reason, we put some of our results at the end of each chapter but we do not put any exercises in this book. We observed that all the material, contents, and exercises in every book are exercises of previous books. For this reason, we hope that the readers can find their own exercises.

It is a great pleasure to acknowledge the work of Professor Jimmy J.M. Tan. We have worked together for a long time. Yet, at least one-third of this book is his contribution. We also thank all the members of the Computer Theory Laboratory of National Chiao-Tung University: T.Y. Sung, T. Liang, H.M. Huang, Chiuyuan Chen, Y.C. Chuang, S. S. Kao, Y.C. Chang, T.Y. Ho, R.S. Chou, C.N. Hung, J.J. Wang, S.Y. Wang, C.Y. Lin, C.H. Tsai, J.Y. Hwang, C.P. Chang, J.N. Wang, J.J. Sheu, C.H. Chang, T.K. Li, W.T. Huang, Y.C. Chen,

H.C. Hsu, K.Y. Liang, Y.L. Hsieh, M.C. Yang, P.L. Lai, L.C. Chiang, Y.H. Teng, C.M. Sun, and K.M. Hsu.

We also thank CRC Press for offering us the opportunity of publishing this book. Lih-Hsing Hsu also thanks his advisor, Professor Joel Spencer, for entering the area of graph theory. We also thank Professor Frank Hsu for strongly recommending to us the area of interconnection networks. Finally, we thank all of our teachers; without their encouragement, this book could never have been written.

Authors

Lih-Hsing Hsu received his BS in mathematics from Chung Yuan Christian University, Taiwan, Republic of China, in 1975, and his PhD in mathematics from the State University of New York at Stony Brook in 1981. From 1981 to 1985, he was an associate professor at the Department of Applied Mathematics at National Chiao Tung University in Taiwan. From 1985 to 2006, he was a professor at National Chiao Tung University. From 1985 to 1988, he was the chairman of the Department of Applied Mathematics at National Chiao Tung University. After 1988, he joined the Department of Computer and Information Science of National Chiao Tung University. In 2004, he retired from National Chiao Tung University, holding a title as an honorary scholar of that university. He is currently the chairman of the Department of Computer Science and Information Engineering, Providence University, Taichung, Taiwan. His research interests include interconnection networks, algorithms, graph theory, and VLSI layout.

Cheng-Kuan Lin received his BS in applied mathematics from Chinese Culture University, Taiwan, Republic of China, in 2000, and his MS in mathematics from the National Central University in Taiwan in 2002. His research interests include interconnection networks, algorithms, and graph theory.

1 Fundamental Concepts

1.1 GRAPHS AND SIMPLE GRAPHS

A *graph* G with n vertices and m edges consists of the *vertex set* $V(G) = \{v_1, v_2, \ldots, v_n\}$ and *edge set* $E(G) = \{e_1, e_2, \ldots, e_m\}$, where each edge consists of two (possibly equal) vertices called, *endpoints*. An element in $V(G)$ is called a *vertex* of G. An element in $E(G)$ is called an *edge* of G. Because graph theory has a variety of applications, we may also use a *node* for a vertex and a *link* for an edge to fit the common terminology in that area. We use the *unordered pair* (u, v) for an edge $e = \{u, v\}$. If $(u, v) \in E(G)$, then u and v are *adjacent*. We write $u \leftrightarrow v$ to mean *u is adjacent to v*. A *loop* is an edge whose endpoints are equal. *Parallel edges* or *multiple edges* are edges that have the same pair of endpoints. A *simple graph* is a graph having no loops or multiple edges. For a graph $G = (V, E)$, the *underlying simple graph* UG is the simple graph with vertex V and $(x, y) \in E(UG)$ if and only if $x \neq y$ and $(x, y) \in E$. A graph is *finite* if its vertex set and edge set are finite. *We adopt the convention that every graph mentioned is finite.*

We illustrate a graph on paper by assigning a point to each vertex and drawing a curve for each edge between the points representing its endpoints, sometimes omitting the name of the vertices or edges. In Figure 1.1a, e is a loop and f and g are parallel edges. Figures 1.1b and 1.1c illustrate two ways of drawing one simple graph.

The *order* of a graph G, written $n(G)$, is the number of vertices in G. An *n-vertex graph* is a graph of order n. The *size* of a graph G, written $e(G)$, is the number of edges in G, even though we also use e by itself to denote an edge. The *degree* of a vertex v in a graph, written $\deg_G(v)$, or $\deg(v)$, is the number of nonloop edges containing v plus twice the number of loops containing v. The *maximum degree* of G, denoted by $\Delta(G)$, is $\max\{\deg(v) \mid v \in V(G)\}$ and the *minimum degree* of G, denoted by $\delta(G)$, is $\min\{\deg(v) \mid v \in V(G)\}$. A graph G is *regular* if $\Delta(G) = \delta(G)$, and *k-regular* if $\Delta(G) = \delta(G) = k$. A vertex of degree k is *k-valent*. The *neighborhood* of v, written $N_G(v)$ or $N(v)$, is $\{x \in V(G) \mid x \leftrightarrow v\}$; x is a *neighbor* of v if $x \in N(v)$. An *isolated vertex* has a degree of 0.

A *directed graph* or *digraph* G consists of a *vertex set* $V(G)$ and an *edge set* $E(G)$, where each edge is an ordered pair of vertices. We use the *ordered pair* (u, v) for the edge uv, where u is the *tail* and v is the *head*. (Note that the notation (u, v) is used both as an unordered pair in an unordered graph and as an ordered pair in a directed graph. However, it is easy to distinguish each from the other from the context.) Sometimes, we may use the term *arc* for an edge of a digraph. We write $u \rightarrow v$ when $(u, v) \in E(G)$, meaning *there is an edge from u to v*. A *simple digraph* is a digraph in which each ordered pair of vertices occur at most once as an edge. For a digraph $G = (V, E)$, the

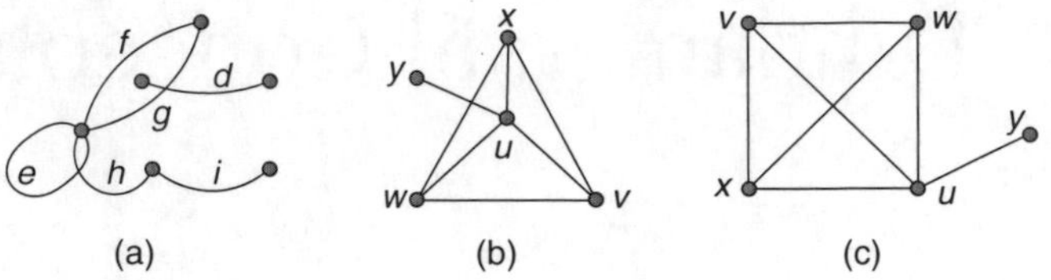

FIGURE 1.1 Examples of graphs.

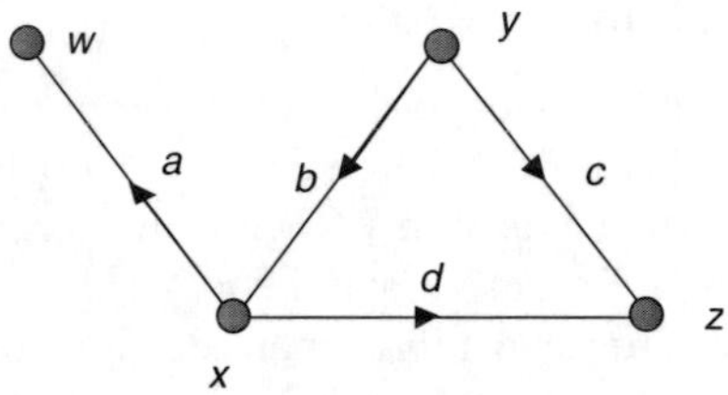

FIGURE 1.2 A digraph.

underlying graph UG is the simple graph with the vertex set V and $(x, y) \in E(UG)$ if and only if $x \neq y$ and either $(x, y) \in E(G)$ or $(y, x) \in E(G)$.

The choice of head and tail assigns a *direction* to an edge, which we illustrate by assigning edges as arrows. See Figure 1.2 for an example of a digraph. Sometimes, weights are assigned to the edges of graphs or digraphs. Thus, we have *weighted graphs*, *weighted digraphs*, *(unweighted) graphs*, and *(unweighted) digraphs*.

Graphs arise in many settings, which suggests useful concepts and terminologies about the structure of graphs.

Example 1.1

A famous brain-teaser asks: Does every set of six people have three mutual acquaintances or three mutual strangers?

Because an *acquaintance* is symmetric, we can model it by a simple graph having a vertex for each person and an edge for each acquainted pair. The *nonacquaintance* relation on the same set yields another graph. The *complement* $\overline{G}$ of a simple graph G with the same vertex set $V(G)$ is defined by $(u, v) \in E(\overline{G})$ if and only if $(u, v) \notin E(G)$. In Figure 1.3, we draw a graph and its complement.

Let two graphs equal $G = (V(G), E(G))$ and $H = (V(H), E(H))$. We say that H is a *subgraph* of G or G is a *supergraph* of H if $V(H) \subseteq V(G)$ and $E(H) \subseteq E(G)$; we write $H \subseteq G$ and say that G *contains* H. In particular, H is called a *spanning subgraph* of G or G is a *spanning supergraph* of H if H is a subgraph of G and $V(H) = V(G)$.

Let $G = (V, E)$ be a graph and S be a subset of V. The subgraph of G *induced* by S, denoted by $G[S]$, is the subgraph with vertex set S and those edges of G with both ends in S. In particular, the graph $G[V - S]$ is denoted by $G - S$. Let v be a vertex of G. We use $G - v$ to denote $G - \{v\}$. Let F be a subset of E. The subgraph generated

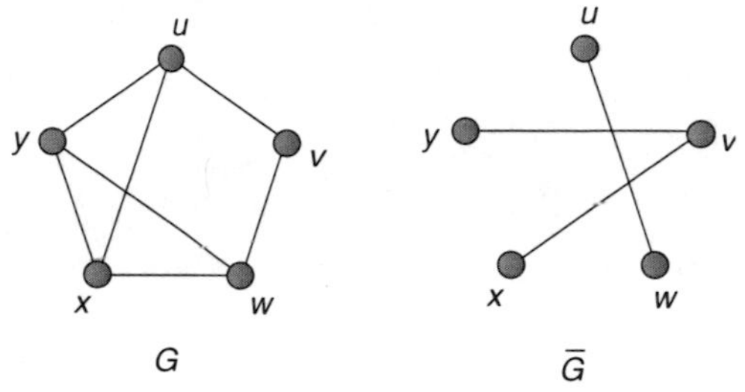

FIGURE 1.3 A graph and its complement.

by F, denoted by G_F, is the subgraph of G with F as its edge set and $\{v \mid v$ is incident with some edge of $F\}$ as its vertex set. The subgraph $G-F$ denotes the subgraph of G with V as its vertex set and $E-F$ as its edge set. Let e be an edge of E. The graph $G-\{e\}$ is denoted by $G-e$.

A *complete graph* or *clique* is a simple graph in which every pair of vertices is an edge. A complete graph has many subgraphs that are not cliques, but every induced subgraph of a complete graph is a clique. The complement of a complete graph has no edges.

An *independent set* in a graph G is a vertex set $S \subseteq V(G)$ that contains no edge of G (that is, $G[S]$ has no edges). In the graph G shown in Figure 1.3, the largest clique and the largest independent set have cardinalities three and two, respectively. These values reverse in the complement $\overline{G}$, because cliques become independent sets (and vice versa) under complementation. Our six-person brain-teaser asks whether every six-vertex graph has a clique or an independent set of size 3. It is worthwhile to verify this statement. The generalization of the six-person brain-teaser leads the Ramsey Theory [128].

Example 1.2

Suppose that we have m jobs and n people, and each person can do some of the jobs. Can we make assignments to fill the jobs? We model the available assignments by a graph H having a vertex for each job and each person, putting job j adjacent to person p if p can do j.

A graph G is *bipartite* if $V(G)$ is the union of two disjoint sets such that each edge consists of one vertex from each set. The graph H of available assignments is bipartite, with the two sets being the people and the jobs. Because a person can do only one job and we assign a job to only one person, we seek m pairwise disjoint edges in H.

A *complete bipartite graph*, illustrated in Figure 1.4, is a bipartite graph whose edge set consists of all pairs having a vertex set from each of two disjoint sets covering the vertices. When the assignment graph H is a complete bipartite graph, pairing up vertices is easy, so we seek the *best* way to do it. We can assign numerical weights on the edges to measure desirability. The best way to match up the vertices is often the one where the selected edges have maximum total weight.

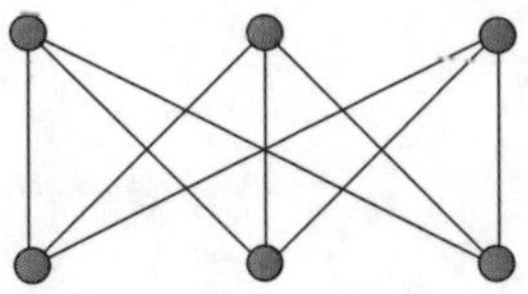

FIGURE 1.4 A complete bipartite graph.

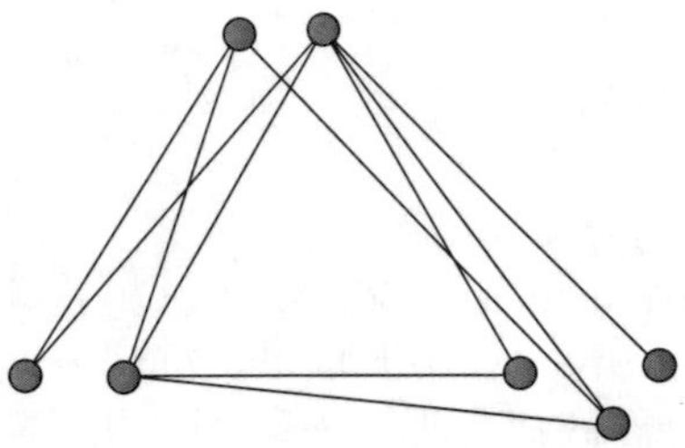

FIGURE 1.5 A 3-partite graph.

EXAMPLE 1.3

Suppose that we must schedule for Senate committee meetings into designated weekly time periods. We cannot assign two committees to the same time if they have a common member. How many different time periods do we need?

We create a vertex for each committee, with two vertices adjacent when their committees have a common member. We must assign labels (time periods) to vertices so that the endpoints of each edge receive different labels. The labels have no numerical value, so we call them *colors*, and the vertices receiving a particular label form a *color class*. The minimum number of colors needed is the *chromatic number* of G, written $\chi(G)$. Because vertices of the same color must form an independent set, $\chi(G)$ equals the minimum number of independent sets that partition $V(G)$. This generalizes bipartite graphs. A graph G is *k-partite* if $V(G)$ is the union of k (possibly empty) independent sets. When they are pairwise disjoint, these sets that partition $V(G)$ are called *partite sets* (or *color classes*). The graph in Figure 1.5 has chromatic number 3 and is 3-partite (there could be graphs that are also 4-partite, 5-partite, etc.). We will study the chromatic number of graphs in Chapter 10.

The most (in)famous problem in graph theory involves coloring. A *map* is a partition of the plane into connected regions. Can we color the regions of every map, using at most four colors, so that neighboring regions have different colors? In each map, we introduce a vertex for each region and an edge for regions sharing a boundary. We obtain a planar graph. A graph is *planar* if it can be drawn in the plane without line crossings. In other words, the map coloring problem asks whether each planar graph has a chromatic number of at most 4. We will study planar graphs in Chapter 12.

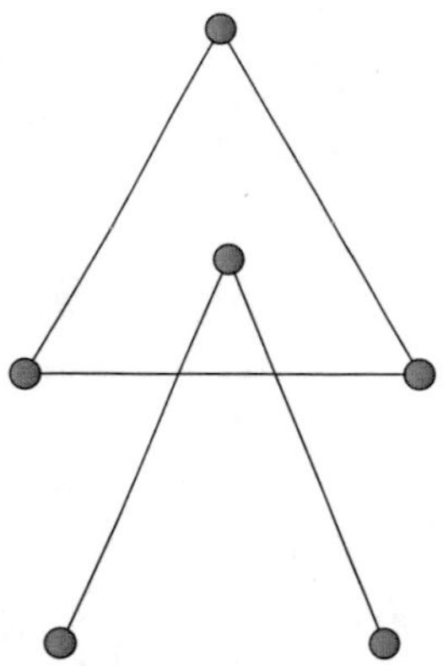

FIGURE 1.6 A disconnected graph.

Example 1.4

We can model a road network by a graph having edges that correspond to road segments between intersections. We can assign edge weights to measure distance or travel time. We may want to know the shortest route from x to y.

Informally, we think of a *path* in a graph as a single vertex or an ordered list of distinct vertices $v_1, v_2, \ldots, v_n$ such that (v_{i-1}, v_i) is an edge for $2 \leq i \leq n$. The first and the last vertices of a path are its *endpoints*. A path from u to v is a path with endpoints u and v. Similarly, we think of a *cycle* as an ordered list $v_1, v_2, \ldots, v_n$ such that all (v_{i-1}, v_i) for $2 \leq i \leq n$ and also (v_n, v_1) are edges.

To find the shortest route from x to y, we want to find the path from x to y having the least total weight among all paths from x to y in G. Similarly, in a network of n cities, we may want to visit all the cities and return home at the minimum total cost. Using the cost that we weight on the edges of the complete graph, we seek the n-vertex cycle with minimum total cost. This is the *Traveling Salesman Problem*.

In a road network or communication network, every site should be reachable from each other. A graph G is *connected* if it has a path from u to v for each pair $u, v \in V(G)$. A graph is *disconnected* if it is not connected. The graph in Figure 1.6 is disconnected.

1.2 MATRICES AND ISOMORPHISMS

How can we specify a graph? We can list the vertices and edges, but there are other useful ways to encode this information.

Given a graph or digraph G with vertices indexed as $V(G) = \{v_1, v_2, \ldots, v_n\}$, the *adjacency matrix* of G, written $A(G)$, is the matrix in which entry a_{ij} is the number of copies of the edges (v_i, v_j) in G. If vertex v belongs to edge e, then v and e are

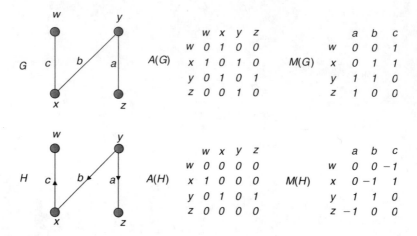

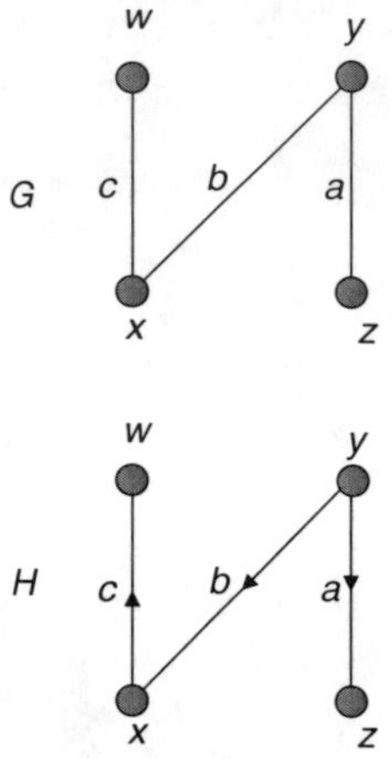

$$A(G) = \begin{array}{c|cccc} & w & x & y & z \\ \hline w & 0 & 1 & 0 & 0 \\ x & 1 & 0 & 1 & 0 \\ y & 0 & 1 & 0 & 1 \\ z & 0 & 0 & 1 & 0 \end{array} \qquad M(G) = \begin{array}{c|ccc} & a & b & c \\ \hline w & 0 & 0 & 1 \\ x & 0 & 1 & 1 \\ y & 1 & 1 & 0 \\ z & 1 & 0 & 0 \end{array}$$

$$A(H) = \begin{array}{c|cccc} & w & x & y & z \\ \hline w & 0 & 0 & 0 & 0 \\ x & 1 & 0 & 0 & 0 \\ y & 0 & 1 & 0 & 1 \\ z & 0 & 0 & 0 & 0 \end{array} \qquad M(H) = \begin{array}{c|ccc} & a & b & c \\ \hline w & 0 & 0 & -1 \\ x & 0 & -1 & 1 \\ y & 1 & 1 & 0 \\ z & -1 & 0 & 0 \end{array}$$

FIGURE 1.7 Adjacency matrices and incidence matrices.

incident. The *incidence matrix* $M(G)$ of a loopless graph G has rows indexed by $V(G)$ and columns indexed by $E(G)$, with $m_{ij} = 1$ if vertex v_i belongs to e_j; otherwise $m_{ij} = 0$. For a loopless digraph, $m_{ij} = +1$ if v_i is the tail of e_j, $m_{ij} = -1$ if v_i is the head of e_j, and $m_{ij} = 0$ if otherwise.

A graph may have many adjacency matrices, depending on the ordering of the vertices. Each ordering of the vertices determines an adjacency matrix. The adjacency matrix of G is symmetric if G is a graph (not a digraph). Moreover, every entry in $A(G)$ is 0 or 1, with 0's on the diagonal if G is a simple graph.

Example 1.5

In Figure 1.7, we draw a simple graph G and a digraph H, together with the adjacency matrices and incidence matrices that result from the vertex ordering w, x, y, z and the edges ordering a, b, c. The adjacency matrix for the graph having two copies of each of these edges would obtained by changing each 1 to 2.

When we present an adjacency matrix for a graph, we are implicitly naming the vertices by the order of its rows. The ith vertex corresponds to the ith row and column. This provides names for the vertices. We cannot store a graph in a computer without naming the vertices. Nevertheless, we want to study properties (like *chromatic number* or *connected*) that do not depend on the names of the vertices. If we can find a one-to-one correspondence (*bijection*) between $V(G)$ and $V(H)$ that preserves the adjacency relation, then G and H have the same structural properties.

An *isomorphism* from G to H is a bijection $f : V(G) \rightarrow V(H)$ such that $(u, v) \in E(G)$ if and only if $(f(u), f(v)) \in E(H)$. We say "G is isomorphic to H," written $G \cong H$, if there is an isomorphism from G to H.

When G is isomorphic to H and H is also isomorphic to G, we may say *G and H are isomorphic* (to each other). Because an adjacency matrix encodes the adjacency

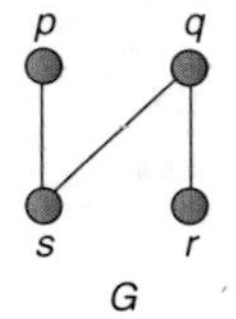

A(G)

	p	r	q	s
p	0	0	0	1
r	0	0	1	0
q	0	1	0	1
s	1	0	1	0

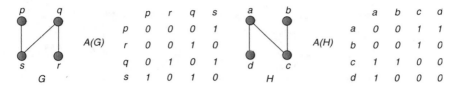

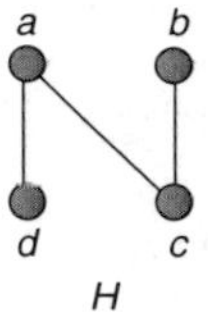

A(H)

	a	b	c	d
a	0	0	1	1
b	0	0	1	0
c	1	1	0	0
d	1	0	0	0

FIGURE 1.8 Two isomorphic graphs.

relation, we can also describe isomorphism using adjacency matrices. The graphs G and H are isomorphic if and only if we can apply a permutation to the rows of $A(G)$ and the same permutation to the columns of $A(G)$ to obtain $A(H)$.

Example 1.6

The graphs G and H in Figure 1.8 are 4-vertex paths. They are isomorphic by an isomorphism that maps p, q, r, s to b, a, d, c, respectively. Rewriting $A(G)$ by placing the rows in the order p, q, r, s and the columns also in order yields $A(H)$. Another isomorphism maps s, r, q, p to a, b, c, d, respectively.

The set of the pair (G, H) such that G is isomorphic to H is the *isomorphism relation* on the set of graphs. Obviously, isomorphism is an equivalence relation. An *isomorphism class* of graphs is an equivalence class of graphs under the isomorphism relation.

When discussing the structure of a graph G, we consider a fixed vertex set, but our comments apply to every graph isomorphic to G. Therefore, we sometimes use the informal expression *unlabeled graph* to mean an isomorphic class of graphs. When we draw a graph, its vertices are named by their physical locations, even if we give them no other names. We often use the name *graph* for the drawing of a graph. When we redraw a graph to display some structural aspect, we have chosen a more convenient member of the isomorphism class.

We often discuss two isomorphic graphs using the same name. Our statement applies to all graphs in their isomorphism class. Hence, we usually write $G = H$ instead of $G \cong H$. Similarly, when we say that *H is a subgraph of G*, we mean technically that H is isomorphic to a subgraph of G, or G contains a *copy* of H.

This treatment of isomorphism classes leads us to using the notation K_n, P_n, and C_n, respectively, to denote a clique, a path, or a cycle with n vertices. Similarly, $K_{r,s}$ denotes the complete bipartite graph with partite sets of cardinalities r and s.

Example 1.7

Suppose that X is a set of size n. Obviously, X contains $C(n; 2) = n(n-1)/2$ unordered pairs of (u, v) with $u \neq v$. We may include or omit each pair as an edge. Thus, there are $2^{C(n;2)}$ simple graphs with vertex set X. For example, there are sixty four simple graphs having four vertices. However, these fall into eleven isomorphism classes as illustrated in Figure 1.9 in complementary pairs. Only P_4 is isomorphic to its complement. Note that

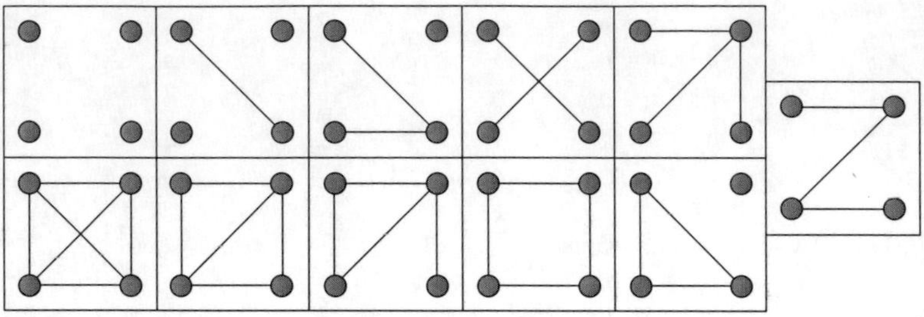

FIGURE 1.9 Graphs with four vertices.

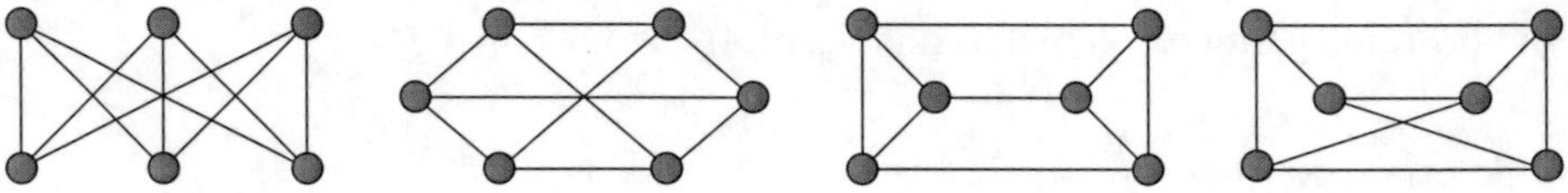

FIGURE 1.10 Graphs for Example 1.8.

the isomorphism classes have different sizes. We cannot count the isomorphism classes of n-vertex simple graphs, though, by dividing $2^{C(n;2)}$ by the size of a class.

An isomorphism from G into H preserves the adjacency relation. As structural properties of graphs are determined by their adjacency relations, we can prove that each G and H are not isomorphic by finding some structural properties that are not true of the other. If they have different vertex degrees, different sizes of the largest clique or smallest cycle, and so on, then they cannot be isomorphic, because these properties are preserved by isomorphism. On the other hand, no known list of common structural properties implies that $G \cong H$. We must present a bijection $f : V(G) \rightarrow V(H)$ that preserves the adjacency relation.

EXAMPLE 1.8

In Figure 1.10, we have four graphs with each vertex having a degree of 3. However, these graphs are not pairwise isomorphic. Several are drawings of $K_{3,3}$, as shown by their exhibited isomorphisms. One contains C_3 and hence cannot be a drawing of $K_{3,3}$.

Consider the graphs in Figure 1.11. Because they have many edges, we prefer to consider their complements. Note that graphs G and H are isomorphic if and only if $\overline{G}$ and $\overline{H}$ are isomorphic. It is observed that the complement of one of these graphs is connected, but the complement of the other is not connected. These two graphs cannot be isomorphic.

An *automorphism* of G is a permutation of $V(G)$, that is, an isomorphism from G into G. A graph G is *vertex-transitive* if for every pair $u, v \in V(G)$ there is an automorphism that maps u to v. A graph G is *edge-transitive* if for every pair $e, f \in E(G)$ there is an automorphism that maps e to f. Obviously, every cycle is not only vertex-transitive but also edge-transitive. Suppose that $n > 2$. Then P_n is not

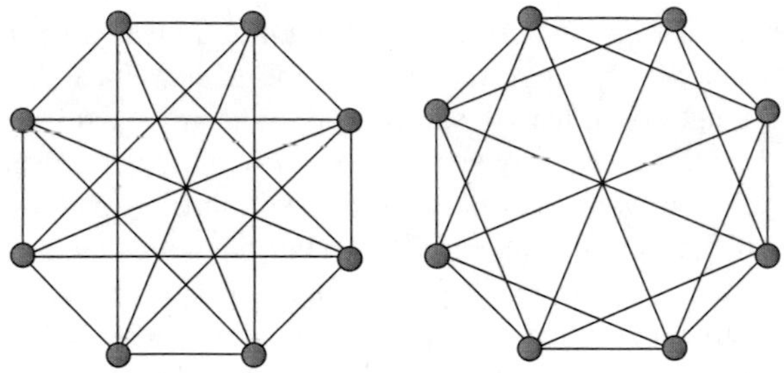

FIGURE 1.11 Two nonisomorphic graphs.

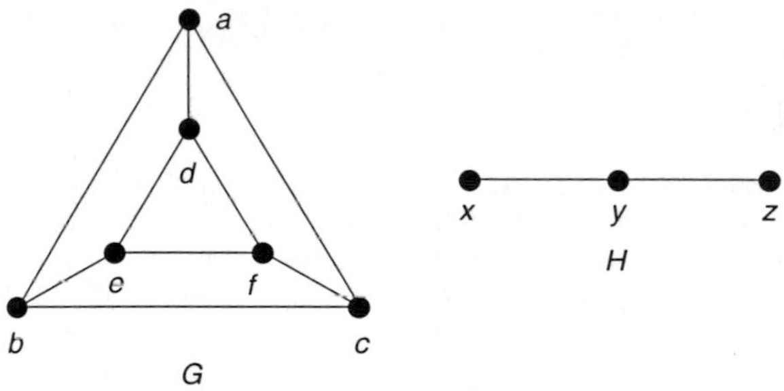

FIGURE 1.12 A vertex-transitive graph and an edge-transitive graph.

vertex-transitive, because no automorphism can map a vertex of degree 1 into a vertex of degree 2.

Example 1.9

Let G be the path with a vertex set $\{1, 2, 3, 4\}$ and an edge set $\{(1, 2), (2, 3), (3, 4)\}$. This graph has two isomorphisms: an identity permutation and the permutation that switches 1 with 4 and switches 2 with 3. Interchanging vertices 1 and 2 is not an automorphism of G, although G is isomorphic to the graph H with the vertex set $\{1, 2, 3, 4\}$ and the edge set $\{(1, 2), (1, 3), (3, 4)\}$.

The automorphisms of G can be viewed as the permutations that can be applied simultaneously to the rows and columns of $A(G)$ without changing $A(G)$. For example, permuting the vertices of one independent set of $K_{r,s}$ does not change the adjacency matrix. Thus, $K_{r,s}$ has $r!s!$ automorphism if $r \neq s$. Moreover, $K_{t,t}$ has $2(t!)^2$ automorphisms. Thus, the complete bipartite graph $K_{r,s}$ is vertex-transitive if and only if $r = s$.

The graph G in Figure 1.12 is vertex-transitive but not edge-transitive. Yet the graph H in Figure 1.12 is edge-transitive but not vertex-transitive.

1.3 PATHS AND CYCLES

A *walk* of *length* k is a sequence $v_0, e_1, v_1, e_2, \ldots, e_k, v_k$ of vertices such that $e_i = (v_{i-1}, v_i)$ for all i. The length of a walk W is denoted by $l(W)$. A *trail* is a walk with no repeated edge. A *path* is a walk with no repeated vertex. A *walk from u to v* has first vertex u and last vertex v; these are its *endpoints*. A walk (or trail) is *closed* if it has a length of at least 1 and its endpoints are equal. A *cycle* is a closed trail in which *first* = *last* is the only vertex repetition (a loop is a cycle of length 1).

Paths and trails are walks and hence have endpoints. A path or trail may have length 0, but a cycle or closed walk cannot. We use the words *path* and *cycle* in three closely related contexts: as a graph or subgraph, as the special case of a walk, and as a set of edges. In a simple graph, a walk is completely specified by its sequence of vertices. Hence, we usually describe a path or a cycle (or walk) in a simple graph by its ordered list of vertices. Although we may list only vertices, the walk still consists of both vertices and edges. We may start a cycle at any vertex. To emphasize the cyclic aspect of the ordering, we often list each vertex only once when naming a cycle.

Example 1.10

The graph in Figure 1.13 has a closed walk of length 11 that visits vertices in the order $\langle a, x, a, x, u, y, c, d, y, v, b, a\rangle$. Omitting the first two steps yields a closed trail (no edge repetition). The edge set of this trail is the the union of the edge sets of three pairwise edge-disjoint cycles. The trail $\langle u, y, c, d, y, v\rangle$ contains the edge of the path $\langle u, y, v\rangle$.

With the previous example, we have the following observation: suppose that u and v are distinct vertices in G. Obviously, every walk from u to v in G contains a path from u to v. Moreover, every closed odd walk contains an odd cycle.

The definitions and remarks hold for digraphs, with each e_i being an *ordered* pair (v_{i-1}, v_i). In a path or cycle of a directed graph, successive edges *follow the arrows*. For example, the digraph consisting only of the edge (x, y) has a path from x to y but no path from y to x.

A graph G is *connected* if it has a path from u to v for each pair $u, v \in V(G)$. Otherwise, it is *disconnected*. The vertex u is *connected to* the vertex v in G if G has a path from u to v. The *connection relation* in a graph consists of the vertex pairs (u, v) such that u is connected to v.

Connectedness is a property of graphs. However, the connection relation on vertices and the phrase *u is connected to v* are convenient when writing proofs. To specify the stronger condition $(u, v) \in E(G)$, we say *u and v are adjacent* or *u and v are joined by an edge*. We use *connected to* for the existence of a path from u to v, and we use *joined to* for the existence of an edge (u, v).

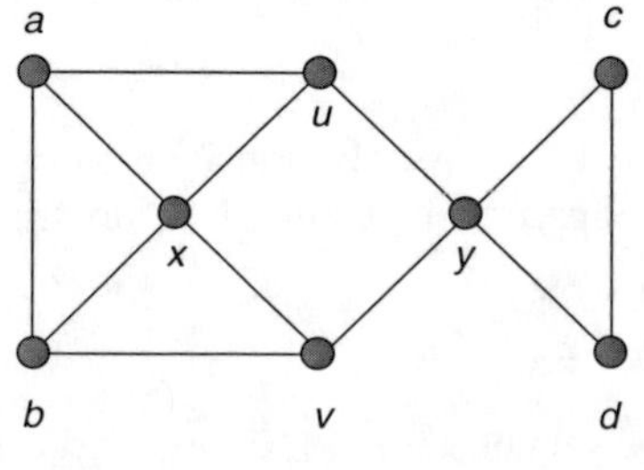

FIGURE 1.13 Graph for Example 1.10.

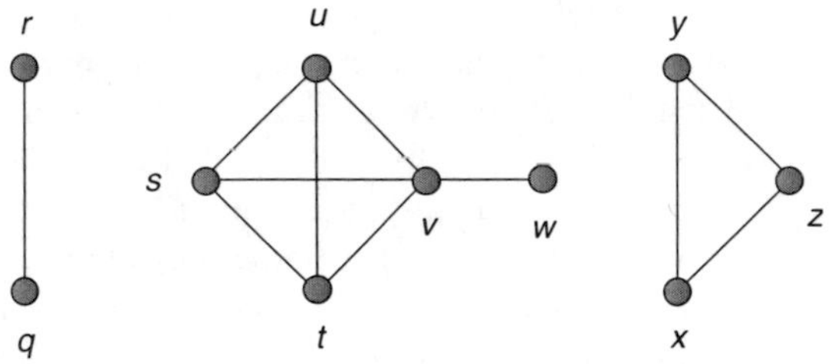

FIGURE 1.14 Graph for Example 1.11.

The *components* of a graph G are its maximal connected subgraphs. A component (or graph) is *nontrivial* if it contains an edge. A *cut vertex* of a graph is a vertex whose deletion increases the number of components. Similarly, a *cut edge* of a graph is an edge whose deletion increases the number of components.

Example 1.11

The graph in Figure 1.14 has three components. The vertex sets of components are $\{q, r\}$, $\{s, t, u, v, w\}$, and $\{x, y, z\}$. The vertex v is a cut vertex. The edges (q, r) and (v, w) are the cut edges.

The connection relation is an equivalence relation. The equivalence classes of this relation are the vertex sets of the components.

PROPOSITION 1.1 *A graph is connected if and only if there is an edge with endpoints in both sets for every partition of its vertices into two nonempty sets.*

Proof: Suppose that G is connected. Let S and T be any partition of $V(G)$; that is, $V(G) = S \cup T$, $S \cap T = \emptyset$, $\min\{|S|, |T|\} \neq 0$. We choose $u \in S$ and $v \in T$. Because G is connected, G has a path P from u to v. After its last vertex in S, P has an edge from S to T.

Suppose that G is disconnected. Let H be a component of G. Then no edge has exactly one endpoint in H. This means that the partition S and T with $S = V(H)$ has no edge with endpoints in both sets. We have proved that if G is disconnected, then the partition condition fails. By the contrapositive, the partition condition implies that G is connected.

Adding an edge e to G reduces the number of components by at most one. Similarly, deleting a cut edge increases the number of components by exactly one. □

THEOREM 1.1 An edge of a graph is a cut edge if and only if it belongs to no cycle.

Proof: Suppose that edge $e = (x, y)$ belongs to a component H of G. Note that the deletion of e affects no other components. It suffices to prove that $H - e$ is connected if and only if e belongs to a cycle.

Suppose that $H - e$ is connected. Then $H - e$ contains a path P from x to y. Obviously, $P \cup \{e\}$ forms a cycle containing e.

Conversely, suppose that e belongs to a cycle C. Let u and v be any two different vertices in $V(H)$. Because H is connected, H has a path P from u to v. Suppose that P does not contain e. Then P exists in $H-e$. Thus, $H-e$ has a path from u to v. Suppose that P contains e. By the symmetry role of x and y, we can assume that x is between u and y on P. Since $H-e$ contains a path from u to x along P, a path from x to y along C, and a path from y to v along P. The transitivity of the connection relation implies that $H-e$ has a path from u to v. Since u and v were chosen arbitrarily from $V(G)$, we have proved that $H-e$ is connected. □

PROPOSITION 1.2 *Let G be a graph with* $V(G)=\{v_1,v_2,\ldots,v_n\}$ *with* $n\geq 3$. *Suppose that at least two of the subgraphs* $G-v_1, G-v_2,\ldots,G-v_n$ *are connected. Then G is connected.*

Proof: Suppose that G is not connected and has components $H_1,H_2,\ldots,H_k$. When we delete a vertex from H_i, we still have at least k components unless $H_i=K_1$, in which case $k-1$ components remain. Thus, having at least two of $G-v_1, G-v_2,\ldots, G-v_n$ be connected requires that $k=2$ and that each of these two components is a single vertex. This contradicts the hypothesis that G has at least three vertices. □

COROLLARY 1.1 *Every nontrivial graph has at least two vertices that are not cut-vertices.*

THEOREM 1.2 A graph is bipartite if and only if it has no odd cycle.

Proof: Suppose that G is a bipartite graph. Every walk in G alternates between the two sets of bipartition of G. Thus, every return to the original class (including the original vertex) happens after an even number of steps. Hence, G has no odd cycle.

Suppose that graph G has no odd cycle. We prove that G is bipartite by constructing a bipartition of each nontrivial component. Let u be a vertex in a nontrivial component H of G. For each $v\in V(G)$, we claim that all walks from u to v have the same parity. If not, the concatenation of two walks from u to v of different parity (reversing the second) is a closed walk of odd length. Then G has an odd cycle. This contradicts the assumption of no odd cycle.

We can thus partition $V(G)$ into disjoint sets X, Y by letting $X=\{v\in V(G)\mid \text{walks from } u \text{ to } v \text{ have even length}\}$ and $Y=\{v\in V(G)\mid \text{walks from } u \text{ to } v \text{ have odd length}\}$. Because an edge joining two vertices in X will create a closed odd walk, X is an independent set. Similarly, Y is also an independent set. Thus, G is a bipartite graph. □

1.4 VERTEX DEGREES

We begin with an essential tool of graph theory, sometimes called the First Theorem of Graph Theory or the Handshaking Lemma.

THEOREM 1.3 **(Degree-Sum Formula).** If G is a graph, then $\sum_{v\in V(G)}\deg(v)=2e(G)$.

Proof: Summing the degrees counts each edge twice, as each edge has two ends and contributes to the degree at each endpoint. □

By the degree-sum formula, the average vertex degree is $2e(G)/n(G)$; hence $\delta(G) \leq 2e(G)/n(G) \leq \Delta(G)$. We list two other immediate corollaries.

COROLLARY 1.2 *Every graph has an even number of vertices of odd degree. No graph of odd order is regular with odd degree.*

COROLLARY 1.3 *A k-regular graph with n vertices has $nk/2$ edges.*

EXAMPLE 1.12

The Petersen graph is drawn in Figure 1.15. The graph is useful enough to have an entire book devoted to it (Holten and Sheehan [156]). The Petersen graph has a simple description using the set S of all two-element subsets of a 5-element set. Let G be the graph with $V(G) = S$ such that two vertices join an edge if and only if they are disjoint as sets. Hence, the Petersen graph is vertex-transitive. Obviously, the Petersen graph is a 3-regular graph with 10 vertices. Thus, it has $15 = 3 \times 10/2$ edges.

EXAMPLE 1.13

Let S be the set of all n-tuples in which each position is 0 or 1. Obviously, S contains 2^n elements. The n-dimensional hypercube is the graph Q_n with vertex set S in which two n-tuples are adjacent if and only if they differ in exactly one position. In Figure 1.16, we illustrate Q_3. The hypercube is an architecture for parallel computers [220]. Processing units can communicate directly if they correspond to adjacent vertices in Q_n.

The Hamming weight of a 0, 1-vector is the number of 1's. Every edge of Q_n consists of a vector of even weight and a vertex of odd weight. Hence, the vectors of all even weights form an independent set, and the vectors of all odd weights form an independent

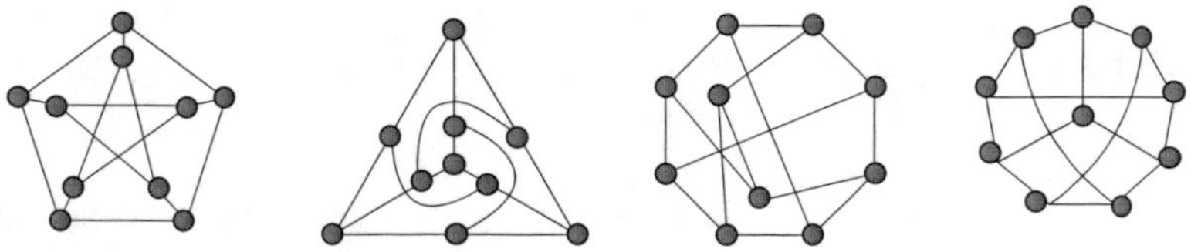

FIGURE 1.15 Some drawings of the Petersen graph.

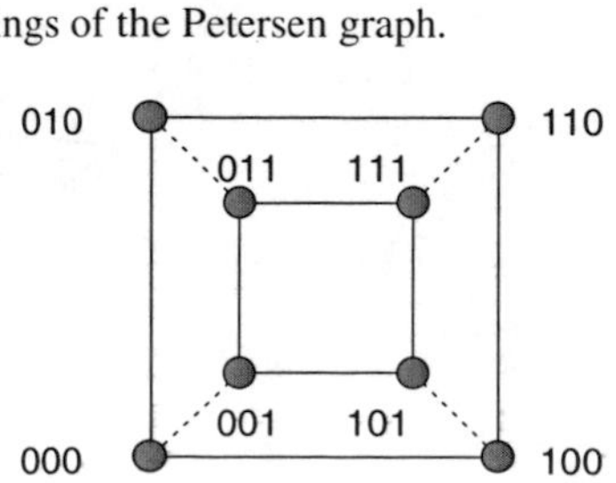

FIGURE 1.16 Q_3.

set. Thus, Q_n is bipartite. Because each vector can be changed in n places, Q_n is n-regular. By Corollary 3.3, Q_n has $n2^{n-1}$ edges.

Deleting the dashed edges in the illustration leaves $2Q_2$. This suggests an inductive description of Q_n. The basis is the 1-vertex graph Q_0, which consists of the unique binary vector of length 0. Given Q_{n-1}, we construct Q_n in two steps: (1) Take two disjoint copies of Q_{n-1}. Call them Q^0 and Q^1; (2) For each $v \in V(Q_{n-1})$, append a 0 to the end of the vector for v in Q^0 and a 1 to the end of the vector for v in Q^1, and add an edge consisting of these two vertices.

1.5 GRAPH OPERATIONS

We need more methods to generate graphs and discuss their properties. For this reason, we present some graph operations.

Let $G_1 = (V_1, E_1)$ and $G_2 = (V_2, E_2)$ be two graphs. The *union* of G_1 and G_2, written $G_1 \cup G_2$, has vertex set $V_1 \cup V_2$ and edge set $E_1 \cup E_2$. To specific the disjoint union with $V_1 \cap V_2 = \emptyset$, we write $G_1 + G_2$. More generally, mG is the graph consisting of m pairwise disjoint copies of G. The *join* of G_1 and G_2, written $G_1 \vee G_2$, is obtained from $G_1 + G_2$ by adding the edges $\{(x, y) \mid x \in V_1, y \in V_2\}$. We illustrate $C_3 + C_4$ in Figure 1.17a and $C_3 \vee C_4$ in Figure 1.17b.

Note that the adjacency matrix of $G_1 + G_2$ is the direct sum of the adjacency matrix of G_1 and the adjacency matrix of G_2.

Let $G_1 = (V_1, E_1)$ and $G_2 = (V_2, E_2)$ be two graphs. The *Cartesian product* of G_1 and G_2, denoted by $G_1 \times G_2$, is the graph with vertex set $V_1 \times V_2$ such that (u_1, v_1) is joined to (u_2, v_2) k times if and only if either $u_1 = u_2$ and v_1 is joined to v_2 k times in G_2 or $v_1 = v_2$ and u_1 is joined to u_2 k times in G_1. We illustrate the Cartesian product of G_1 in Figure 1.18a and G_2 in Figure 1.18b as the graph in Figure 1.18c.

Many interconnection networks are constructed by Cartesian product [6,220]. An *n-dimensional mesh* (abbreviated as *mesh*) $M(m_1, m_2, \ldots, m_n)$ is defined as the Cartesian product of n paths $P_{m_1} \times P_{m_2} \times \cdots \times P_{m_n}$, where P_{m_i} is the path graph with m_i vertices, and an *n-dimensional torus* (abbreviated as *torus*) $C(m_1, m_2, \ldots, m_n)$ is defined as the Cartesian product of n cycles $C_{m_1} \times C_{m_2} \times \cdots \times C_{m_n}$ where C_{m_i} is the cycle graph with m_i vertices. Meshes and Tori are widely used computer architectures [220]. In particular, the hypercube Q_n is $M(m_1, m_2, \ldots, m_n)$ where $m_i = 2$ for $1 \le i \le n$.

We introduce another family of such interconnection networks based on the Cartesian product. The Petersen graph is a 3-regular graph with ten vertices such that the

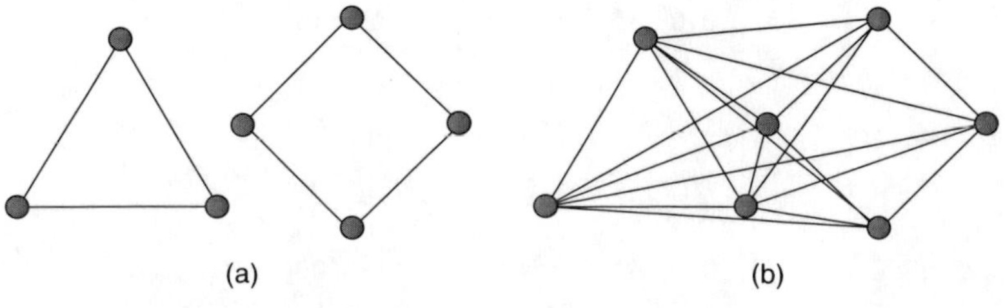

FIGURE 1.17 (a) $C_3 + C_4$ and (b) $C_3 \vee C_3$.

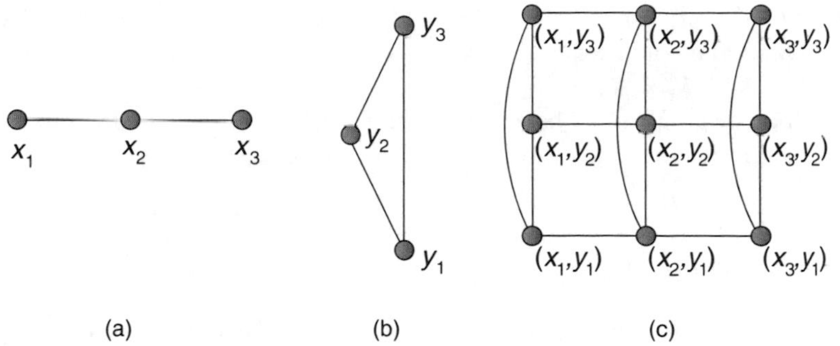

FIGURE 1.18 (a) G_1, (b) G_2, and (c) $G_1 \times G_2$.

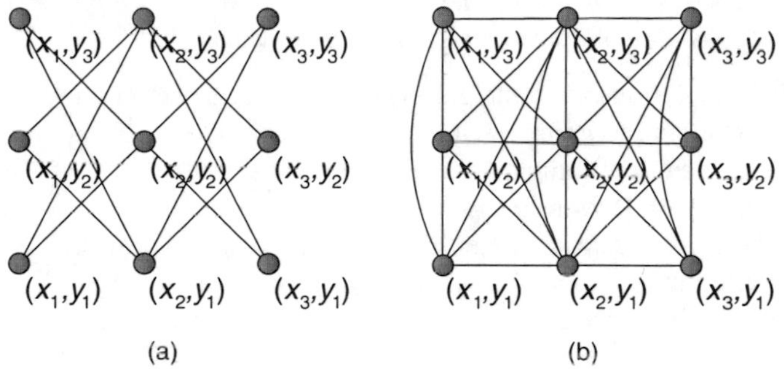

FIGURE 1.19 (a) $G_1 \cdot G_2$ and (b) $G_1 * G_2$.

length of the shortest path between any two vertices is at most 2. Compared to this graph, the three-dimensional hypercube is a 3-regular graph with eight vertices such that the length of the shortest path between any two vertices is at most 3. It has more vertices as compared to the three-dimensional hypercube and a shorter distance between any two vertices. We call it the simple Petersen graph. As an extension, Öhring and Das [260] introduce the k-dimensional folded Petersen graph, FP_k, to be P^k. It is observed that FP_k possesses qualities of a good network topology for distributed systems with a large number of sites, because it accommodates 10^k vertices and is a symmetric, $3k$-regular graph such that the distance between any two vertices is at most $2k$. Being an iterative Cartesian product on the Petersen graph, it is scalable. Moreover, Öhring and Das [260] define the folded Petersen cube networks $FPQ_{n,k}$ as $Q_n \times P^k$. It turns out that $FPQ_{n,k}$ and its special derivations $FPQ_{0,k} = FP_k$ and $FPQ_{n-3,1} = HP_n$, called the n-dimensional hyper Petersen network (originally proposed by Das and Banerjee [79]), are better than the comparable-size hypercubes and several other networks with respect to the usual metrics of a multicomputer architecture.

Let $G_1 = (V_1, E_1)$ and $G_2 = (V_2, E_2)$ be two graphs. The *tensor product* of G_1 and G_2, denoted by $G_1 \cdot G_2$, is the graph with vertex set $V_1 \times V_2$ such that (u_1, v_1) is joined to (u_2, v_2) k times if and only if u_1 is joined to u_2 m times in G_1 and v_1 is joined to v_2 n times in G_2 with $k = mn$. The *strong product* of G_1 and G_2, denoted by $G_1 * G_2$, is the graph $(G_1 \times G_2) \cup (G_1 \cdot G_2)$. We illustrate $G_1 \cdot G_2$ in Figure 1.19a and $G_1 * G_2$ in Figure 1.19b where G_1 and G_2 are shown in Figure 1.18.

Note that the adjacency matrix of $G_1 \cdot G_2$ is the tensor product of the adjacency matrix of G_1 and the adjacency matrix of G_2. Let m and n be positive integers. It is apparent that $K_m * K_n = K_{mn}$.

Let G be any graph and k be any positive integer. We use ${}_kG$ to denote the graph obtained by duplicating k times to each edge in G. Obviously, ${}_1G = G$. The graph ${}_2K_2$ is denoted by C_2.

1.6 SOME BASIC TECHNIQUES

We can count a set by finding a one-to-one correspondence between it and a set of known size.

PROPOSITION 1.3 *For $n \geq 2$, there are $2^{C(n-1;2)}$ simple graphs with the vertex set $\{v_1, v_2, \ldots, v_n\}$ such that every vertex degree is even.*

Proof: Let A be the set of simple graphs with the vertex set $\{v_1, v_2, \ldots, v_{n-1}\}$ and let B be the set of simple graphs with the vertex set $\{v_1, v_2, \ldots, v_n\}$ such that every vertex degree is even. In Example 1.7, we observed that $|A| = 2^{C(n-1;2)}$. To prove this proposition, we set a bijection from A to B as follows.

Let G be a simple graph with vertices $v_1, v_2, \ldots, v_{n-1}$. We form a new graph G' by adding a vertex v_n and making it adjacent to each vertex that has an odd degree in G, as illustrated in Figure 1.20.

The vertices with an odd degree in G have an even degree in G'. Also, v_n itself has even degree because the number of vertices of odd degree in G is even. Conversely, deleting the vertex v_n from any graph on $\{v_1, v_2, \ldots, v_n\}$ with even degrees produces a graph on $\{v_1, v_2, \ldots, v_{n-1}\}$, and this is the inverse of the first procedure. We have established a one-to-one correspondence between the sets. Hence, $|B| = 2^{C(n-1;2)}$. □

The pigeonhole principle is a simple notion that leads to elegant proofs and may reduce case analysis.

PROPOSITION 1.4 *Every simple graph with at least two vertices has two vertices of equal degree.*

Proof: In a simple graph with n vertices, every vertex degree belongs to the set $\{0, 1, \ldots, n-1\}$. Suppose that fewer than n values occur. By the pigeonhole principle, the proposition holds. Otherwise, both $(n-1)$ and 0 occur as vertex degrees. However,

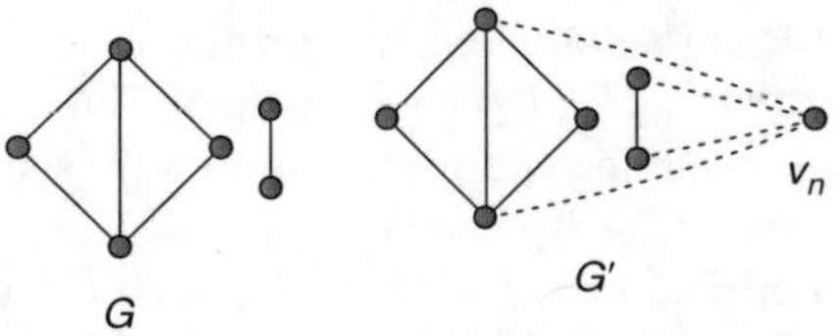

FIGURE 1.20 Illustration of Proposition 1.3.

this is impossible, because the vertex of degree $(n-1)$ is adjacent to all others, but the vertex of degree 0 is an isolated vertex. We get a contradiction. □

PROPOSITION 1.5 *Suppose that G is a simple n-vertex graph with* $\delta(G) \geq (n-1)/2$. *Then G is connected. Moreover, the bound* $(n-1)/2$ *is tight.*

Proof: Let G be a simple n-vertex graph with $\delta(G) \geq (n-1)/2$. Suppose that u and v are any two different vertices of G. Suppose that u is not adjacent with v. Because $\delta(G) \geq (n-1)/2$, at least $(n-1)$ edges join $\{u, v\}$ to the remaining vertices. However, there are $(n-2)$ other vertices. By the pigeonhole principle, one of them receives two of these edges. Because G is simple, this vertex is a common neighbor of u and v. In other words, every pair of vertices is either adjacent or has a common neighbor. Therefore, G is connected.

The graph $K_{\lfloor n/2 \rfloor} + K_{\lceil n/2 \rceil}$ has two components. Note that it has a minimum degree of $\lfloor \frac{n}{2} \rfloor - 1$ and is yet disconnected. Thus, the bound is tight. □

We can prove that something exists by building it. Such proofs can be implemented as computer algorithms. A constructive proof requires more than stating an algorithm. We must also prove that the algorithm terminates and yields result.

THEOREM 1.4 Every loopless graph G has a bipartite subgraph with at least $e(G)/2$ edges.

Proof: We start with an arbitrary partition of $V(G)$ into two sets, X and Y. By including the edges having one endpoint in each set, we obtain a bipartite subgraph H with bipartition X and Y. Suppose that H contains fewer than half the edges of G incident to a vertex v. Then v has more neighbors in its own class than in the other class, as illustrated in Figure 1.21. By moving v to the other class, we gain more edges of G than we lose.

We make such a local switch in the bipartition as long as the current bipartition subgraph has a vertex that contributes fewer than half its edges. Each such switch increases the number of edges in the subgraph. Obviously, the process must terminate. When it terminates, we have $\deg_H(v) \geq \deg_G(v)/2$ for every $v \in V(G)$. Hence, $e(H) \geq e(G)/2$ by the degree-sum formula. □

EXAMPLE 1.14

The algorithm used in Theorem 3.4 does not necessarily produce a bipartite subgraph with the most edges—only a bipartite graph with at least half the edges. For example, the graph in Figure 1.22 is 5-regular with 8 vertices. Hence, it has 20 edges. The bipartition $X = \{a, b, c, d\}$ and $Y = \{e, f, g, h\}$ yields a bipartite subgraph with 12 edges. In this

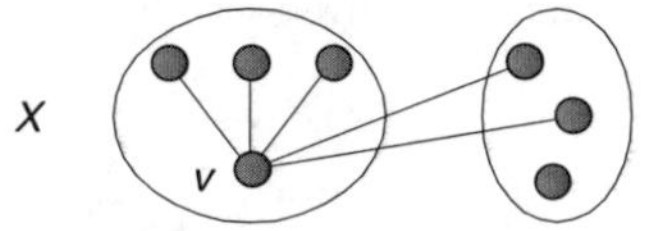

FIGURE 1.21 Illustration of Theorem 1.4.

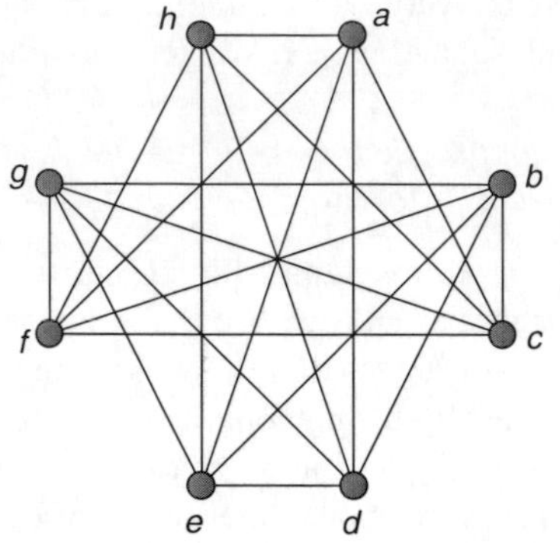

FIGURE 1.22 Illustration of Example 1.14.

bipartite graph, each vertex has degree 3. The algorithm terminates here. Any switching of one vertex would pick up two edges but lose three. Nevertheless, the bipartition $X = \{a, b, g, h\}$ and $Y = \{c, d, e, f\}$ produces a 4-regular bipartite subgraph with 16 edges. Thus, an algorithm seeking a maximum example by local changes. It may get stuck in a local maximum.

The proof in Theorem 1.4 illustrates one way to prove the existence of the desired configuration. Define a sequence of changes to an arbitrary configuration that must terminate but can terminate only when the desired property occurs.

1.7 DEGREE SEQUENCES

The *degree sequence* of a graph is the list of vertices' degrees, usually written in nonincreasing order, as $d_1 \geq d_2 \geq \cdots \geq d_n$. Some applications use nondecreasing order. Let $d_1, d_2, \ldots, d_n$ be a sequence of nonnegative integers. Can we determine whether there is a graph with the required sequence?

PROPOSITION 1.6 *The nonnegative integers $d_1, d_2, \ldots, d_n$ are the vertex degrees of some graph if and only if $\sum_{i=1}^{n} d_i$ is even.*

Proof: Obviously, the degree-sum formula makes the condition necessary. Conversely, suppose that $\sum_{i=1}^{n} d_i$ is even. We need to construct a graph with the vertex set $v_1, v_2, \ldots, v_n$ and $\deg(v_i) = d_i$ for all i.

Because $\sum_{i=1}^{n} d_i$ is even, the number of odd values is even. We can form an arbitrary pairing of $\{v_i \mid d_i \text{ is odd}\}$ and establish an edge for each pair. Now the remaining degree needed at each vertex is even and nonnegative. To satisfy this demand for each i, we put $\lfloor \frac{d_i}{2} \rfloor$ loops at v_i. We find the required graph. □

The availability of loops makes the construction easy. If a loop is forbidden, $(2, 0, 0)$ is not realizable. Thus, that the condition $\sum_{i=1}^{n} d_i$ is even is no longer sufficient.

A *graphic sequence* is a list of nonnegative numbers that is the degree sequence of a certain simple graph. A simple graph with degree sequence d *realizes* d.

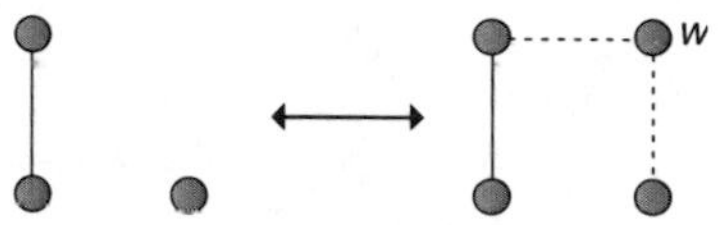

FIGURE 1.23 Graphs with degree sequences 1, 0, 1 and 2, 2, 1, 1.

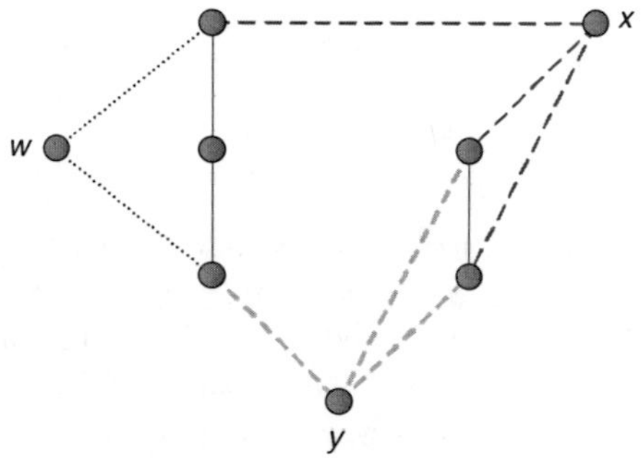

FIGURE 1.24 Graph with degree sequence 33333322.

EXAMPLE 1.15

The list 2, 0, 0 is not graphic. However, the list 2, 2, 1, 1 and the list 1, 0, 1 are graphic. Obviously, the graph $K_2 + K_1$ realizes 1, 0, 1. Suppose that we add a new vertex adjacent to the isolated vertex and to one vertex of degree 1. We obtain a graph with degree sequence 2, 2, 1, 1 as illustrated in Figure 1.23. Conversely, if we have a graph realizing the list 2, 2, 1, 1 in which some vertex w of maximum degree is adjacent to vertices of degrees 2 and 1, we can delete w to obtain a graph with degree list 1, 0, 1.

EXAMPLE 1.16

These observations suggest a recursive test for graphic sequences. To test the sequence 33333322, we can seek a realization with a vertex y of degree 3 that has three neighbors of degree 3. Such graph exists if and only if 2223322 is graphic by deleting y. We rearrange this as 3322222 and seek a realization having a vertex x of degree 3 with one neighbor of degree 3 and two neighbors of degree 2. Such a graph exists if and only if 211222 is graphic by deleting x. We rearrange this as 222211 and seek a realization having a vertex w of degree 2, with neighbors of degree 2. Such a graph exists if and only if 11211 is graphic. Perhaps you recognize that this is indeed graphic. Beginning with a realization 11211, we can insert w, y, x with the properties desired to obtain a realization of the original sequence 33333322 as shown in Figure 1.24. The realization is not unique.

THEOREM 1.5 **(Havel [141], Hakimi [135])** For $n > 1$, the nonnegative integer list d of size n is graphic if and only if d' is graphic, where d' is the list of size $n-1$ obtained from d by deleting its largest element Δ and subtracting 1 from its Δ next largest elements. The only 1-element graph is sequence is $d_1 = 0$.

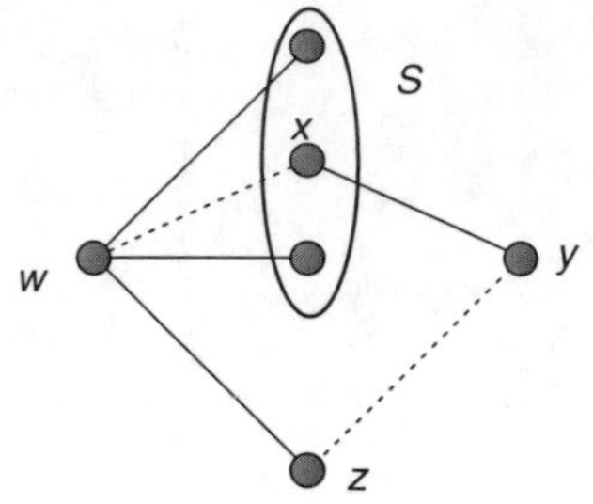

FIGURE 1.25 Illustration of Theorem 1.5.

Proof: The statement is trivial for $n = 1$. Assume that $n > 1$. We first prove that the condition is sufficient. Let d be the sequence $d_1 \geq d_2 \geq \cdots \geq d_n$, and G' be a simple graph with degree sequence d'. We can add a new vertex adjacent to vertices in G' having degrees $d_2 - 1, d_3 - 1, \ldots, d_{\Delta+1} - 1$. Note that these d_i are the Δ largest elements of d after (only copy of) Δ itself. However, the numbers $d_2 - 1, d_3 - 1, \ldots, d_{\Delta+1} - 1$ need not be the largest numbers in d'.

To prove necessity, let G be a simple graph with degree sequence d. We want to produce a simple graph G' realizing d'. Let w be a vertex of degree Δ in G. Let S be a set of Δ vertices in G having the desired degrees $d_2, d_3, \ldots, d_{\Delta+1}$.

Suppose that $N(w) = S$. We can delete w to obtain G'. Suppose that $N(w) \neq S$. Some vertex of S is missing from $N(w)$. In this case, we will modify G to increase $|N(w) \cap S|$ without changing the degree of any vertex. Because $N(w) \cap S$ can increase at most Δ times, repeating this procedure converts an arbitrary G that realizes d into a graph G^* that realizes d and has $N(w) = S$. From G^*, we then delete w to obtain the desired graph G' realizing d'.

The modification of G that increases $|N(w) \cap S|$ without changing the degree of any vertex is described as follows: because $\deg(w) = \Delta = |S|$, there exists a vertex $x \in S$ and another vertex $z \notin S$ such that w is adjacent to z and w is not adjacent to x. By the choice of S, $\deg(x) \geq \deg(z)$. Thus, there exists a vertex $y \notin \{x, z, w\}$ such that y is adjacent to x, but y is not adjacent to z. Now, we modify G by adding the edge set $\{(w, x), (z, y)\}$ and deleting the edge set $\{(w, z), (x, y)\}$ (see Figure 1.25 for illustration). The desired modification is obtained. □

The vertex degree notation for digraphs incorporates the distinction between heads and tails of edges. Let v be a vertex in a directed graph. The *out-neighborhood* or *successor set* $N^+(v)$ is $\{x \in V(G) \mid v \to x\}$. The *in-neighborhood* or *predecessor set* $N^-(v)$ is $\{x \in V(G) \mid x \to v\}$. The *out-degree* of v, $\deg^+(v)$, is the cardinality of $N^+(v)$. Similarly, the *in-degree* of v, $\deg^-(v)$, is the cardinality of $N^-(v)$.

2 Applications on Graph Isomorphisms

The main goal of this book is to share our research with the readers. We believe that anyone can do well in research with very little background. So it is time to see what we can do with the limited background provided in Chapter 1.

2.1 GENERALIZED HONEYCOMB TORI

Stojmenovic [288] introduced three different honeycomb tori by adding wraparound edges on honeycomb meshes; namely, the honeycomb rectangular torus, the honeycomb rhombic torus, and the honeycomb hexagonal torus. Recently, these honeycomb tori have been recognized as an attractive alternative to existing torus interconnection networks in parallel and distributed applications.

We use the brick drawing [266,288] to define the honeycomb rectangular torus. Assume that m and n are positive even integers. The *honeycomb rectangular torus* $\text{HReT}(m,n)$ is the graph with $V(\text{HReT}(m,n)) = \{(i,j) \mid 0 \leq i < m, 0 \leq j < n\}$ such that (i,j) and (k,l) are adjacent if they satisfy one of the following conditions:

1. $i = k$ and $j = l \pm 1 \ (\mathbf{mod}\ n)$
2. $j = l$ and $k = i - 1 \ (\mathbf{mod}\ m)$ if $i + j$ is even

Assume that m and n are positive integers, where n is even. The *honeycomb rhombic torus* $\text{HRoT}(m,n)$ is the graph with $V(\text{HRoT}(m,n)) = \{(i,j) \mid 0 \leq i < m, 0 \leq j - i < n\}$ such that (i,j) and (k,l) are adjacent if they satisfy one of the following conditions:

1. $i = k$ and $j = l \pm 1 \ (\mathbf{mod}\ n)$
2. $j = l$ and $k = i - 1$ if $i + j$ is even
3. $i = 0$, $k = m - 1$, and $l = j + m$ if j is even

Assume that n is a positive integer. The *honeycomb hexagonal mesh* $\text{HM}(n)$ is the graph with $V(\text{HM}(n)) = \{(x_1, x_2, x_3) \mid -n+1 \leq x_1, x_2, x_3 \leq n \text{ and } 1 \leq x_1 + x_2 + x_3 \leq 2\}$. Two vertices (x_1^1, x_2^1, x_3^1) and (x_1^2, x_2^2, x_3^2) are adjacent if and only $|x_1^1 - x_1^2| + |x_2^1 - x_2^2| + |x_3^1 - x_3^2| = 1$. The *honeycomb hexagonal torus* $\text{HT}(n)$ is the graph with the same vertex set as $\text{HM}(n)$. The edge set is the union of $E(\text{HM}(n))$ and the wraparound edge set

$$\begin{aligned} &\{((i, n-i+1, 1-n), (i-n, 1-i, n)) \mid 1 \leq i \leq n\} \\ \cup\ &\{((1-n, i, n-i+1), (n, i-n, 1-i)) \mid 1 \leq i \leq n\} \\ \cup\ &\{((i, 1-n, n-i+1), (i-n, n, 1-i)) \mid 1 \leq i \leq n\} \end{aligned}$$

Chou and Hsu [62] unify these three families of honeycomb torus into generalized honeycomb torus. Assume that m and n are positive integers, where n is even. Let d be any integer such that $(m-d)$ is an even number. The *generalized honeycomb rectangular torus* $\text{GHT}(m,n,d)$ is the graph with $V(\text{GHT}(m,n,d)) = \{(i,j) \mid 0 \le i < m, 0 \le j < n\}$ such that (i,j) and (k,l) are adjacent if they satisfy one of the following conditions:

1. $i=k$ and $j=l \pm 1 \ (\textbf{mod}\ n)$
2. $j=l$ and $k=i-1$ if $i+j$ is even
3. $i=0$, $k=m-1$, and $l=j+d \ (\textbf{mod}\ n)$ if j is even

See Figure 2.1 for various honeycomb tori. Obviously, any $\text{GHT}(m,n,d)$ is a 3-regular bipartite graph. We can label those vertices (i,j) white when $i+j$ is even or black if otherwise.

It is easy to confirm that the honeycomb rectangular torus $\text{HReT}(m,n)$ is isomorphic to $\text{GHT}(m,n,0)$ and the honeycomb rhombic torus $\text{HRoT}(m,n)$ is isomorphic to $\text{GHT}(m,n,m \ (\textbf{mod}\ n))$. With the following theorem, the honeycomb hexagonal torus $\text{HT}(n)$ is isomorphic to $\text{GHT}(n,6n,3n)$.

THEOREM 2.1 $\text{HT}(n)$ is isomorphic to $\text{GHT}(n,6n,3n)$.

Proof: Let h be the function from the vertex set of $\text{HT}(n)$ into the vertex set of $\text{GHT}(n,6n,3n)$ by setting $h(x_1,x_2,x_3)=(x_3,x_1-x_2+2n)$ if $0 \le x_3 < n$, $h(x_1,x_2,x_3)=(0,x_1-x_2+5n \ (\textbf{mod}\ 6n))$ if $x_3=n$, and $h(x_1,x_2,x_3)=(x_3+n,x_1-x_2+5n \ (\textbf{mod}\ 6n))$ if otherwise. We need to check whether f is an isomorphism.

For any $1-n \le c \le n$, we use X_c to denote the set of those vertices (x_1,x_2,x_3) in $\text{HT}(n)$ with $x_3=c$. We use Y_c to denote the set of vertices (i,j) in $\text{GHT}(n,6n,3n)$ where (1) $i=c+n$ and $j \in \{k \mid 4n-c-3 < k < 6n\} \cup \{k \mid 0 \le k < n+c\}$ if $c<0$, (2) $i=0$ and $j \in \{1 \le j < 4n\}$ if $c=0$, (3) $i=c$ and $\{j \mid c \le j \le 4n-c\}$ if $0<c<n$, and (4) $i=0$ and $j \in \{k \mid 4n \le k < 6n\} \cup \{0\}$ if $c=n$. Let h_c denote the function of h induced by X_c. It is easy to confirm that h_c is a one-to-one function from X_c onto Y_c. Thus, h is one-to-one and onto.

We need to check that h preserves the adjacency. Let $e=((x_1,x_2,x_3),(x_1',x_2',x_3'))$ be an edge of $\text{HT}(n)$. Without loss of generality, we assume that $x_1+x_2+x_3=2$ and $x_1'+x_2'+x_3'=1$.

Assume that e is an edge of $\text{HM}(n)$. Then either $x_3=x_3'$ or $x_3-x_3'=\pm 1$.

Case 1. $x_3=x_3'$. Obviously, either $(x_1',x_2',x_3')=(x_1-1,x_2,x_3)$ or $(x_1',x_2',x_3')=(x_1,x_2-1,x_3)$ holds.

Assume that $0 \le x_3 < n$. Then $h(x_1,x_2,x_3)=(x_3,x_1-x_2+2n)$. Moreover, $h(x_1',x_2',x_3')=(x_3,x_1-x_2-1+2n)$ if $(x_1',x_2',x_3')=(x_1-1,x_2,x_3)$ and $h(x_1',x_2',x_3')=(x_3,x_1-x_2+1+2n)$ if $(x_1',x_2',x_3')=(x_1,x_2-1,x_3)$. Assume that $x_3=n$. Then $h(x_1,x_2,x_3)=(0,x_1-x_2+5n \ (\textbf{mod}\ 6n))$. Moreover, $h(x_1',x_2',x_3')=(x_3,x_1-x_2-1+5n \ (\textbf{mod}\ 6n))$ if $(x_1',x_2',x_3')=(x_1-1,x_2,x_3)$ and $h(x_1',x_2',x_3')=(x_3,x_1-x_2+1+5n \ (\textbf{mod}\ 6n))$ if $(x_1',x_2',x_3')=(x_1,x_2-1,x_3)$. Assume that $x_3<0$. Then $h(x_1,x_2,x_3)=(x_3+n,x_1-x_2+5n \ (\textbf{mod}\ 6n))$. Moreover, $h(x_1',x_2',x_3')=(x_3,x_1-x_2-1+5n \ (\textbf{mod}\ 6n))$ if $(x_1',x_2',x_3')=(x_1-1,x_2,x_3)$ and $h(x_1',x_2',x_3')=(x_3,x_1-x_2+1+5n$

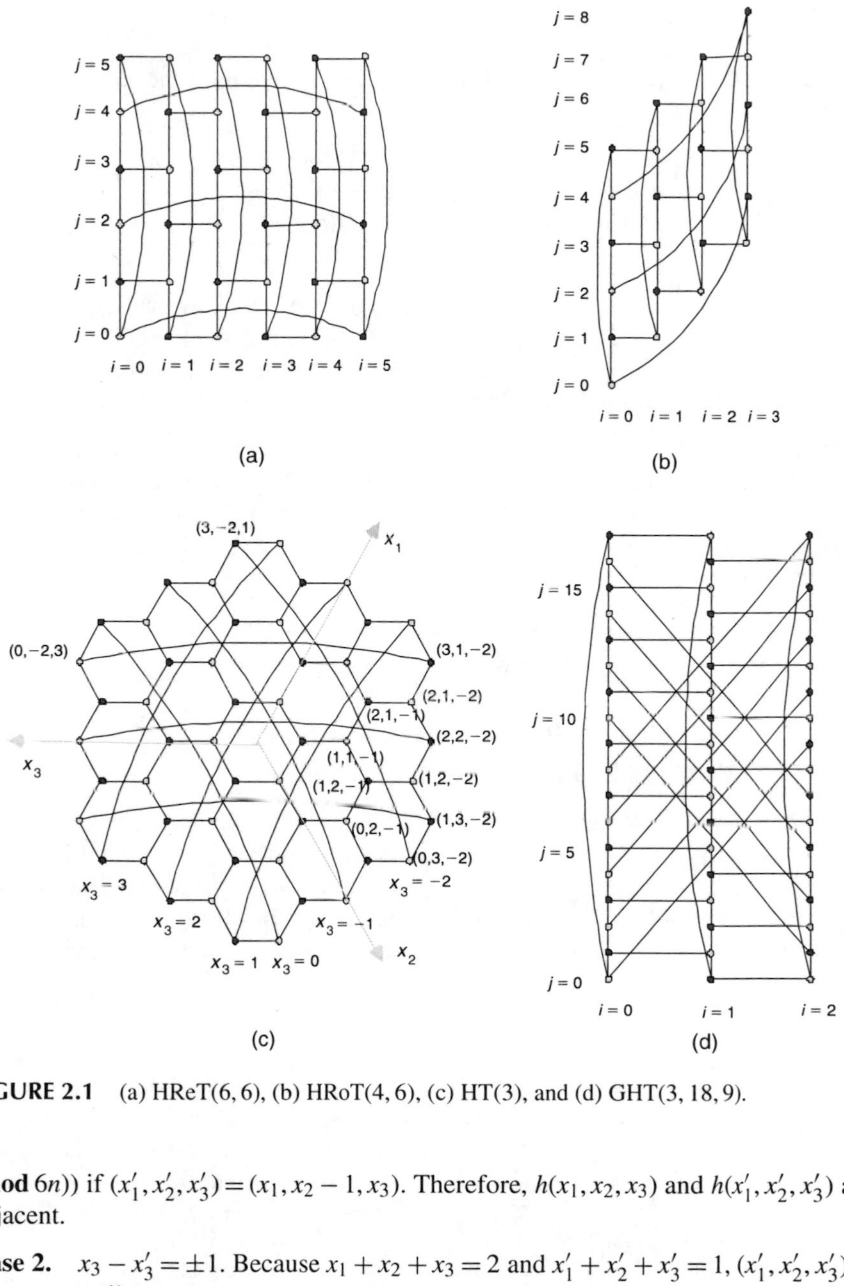

FIGURE 2.1 (a) HReT(6, 6), (b) HRoT(4, 6), (c) HT(3), and (d) GHT(3, 18, 9).

(**mod** $6n$)) if $(x_1', x_2', x_3') = (x_1, x_2 - 1, x_3)$. Therefore, $h(x_1, x_2, x_3)$ and $h(x_1', x_2', x_3')$ are adjacent.

Case 2. $x_3 - x_3' = \pm 1$. Because $x_1 + x_2 + x_3 = 2$ and $x_1' + x_2' + x_3' = 1$, $(x_1', x_2', x_3') = (x_1, x_2, x_3 - 1)$.

Assume that $1 \leq x_3 < n$. Then $h(x_1, x_2, x_3) = (x_3, x_1 - x_2 + 2n)$ and $h(x_1', x_2', x_3') = (x_3 - 1, x_1 - x_2 + 2n)$. Assume that $x_3 = 0$. Then $h(x_1, x_2, x_3) = (0, x_1 - x_2 + 2n)$ and $h(x_1', x_2', x_3') = (n - 1, x_1 - x_2 + 5n \pmod{6n})$. Assume that $x_3 = n$. Then $h(x_1, x_2, x_3) = (0, x_1 - x_2 + 5n \pmod{6n})$ and $h(x_1', x_2', x_3') = (n - 1, x_1 - x_2 + 2n)$. Assume that $2 - n \leq x_3 \leq -1$. Then $h(x_1, x_2, x_3) = (x_3 + n, x_1 - x_2 + 5n \pmod{6n})$

and $h(x_1', x_2', x_3') = (x_3 + n - 1, x_1 - x_2 + 5n \pmod{6n})$. Therefore, $h(x_1, x_2, x_3)$ and $h(x_1', x_2', x_3')$ are adjacent.

Assume that e is an wraparound edge of HM(n). We have the following three cases:

Case 3. $e \in \{((i, n-i+1, 1-n), (i-n, 1-i, n)) \mid 1 \le i \le n\}$. Obviously, $(x_1, x_2, x_3) = (i, n-i+1, 1-n)$ and $(x_1', x_2', x_3') = (i-n, 1-i, n)$. Thus, $h(x_1, x_2, x_3)$ is $(1, 4n + 2i - 1)$ and $h(x_1', x_2', x_3')$ is $(0, 4n + 2i - 1)$. Therefore, $h(x_1, x_2, x_3)$ and $h(x_1', x_2', x_3')$ are adjacent.

Case 4. $e \in \{((1-n, i, n-i+1), (n, i-n, 1-i)) \mid 1 \le i \le n\}$. Obviously, $(x_1, x_2, x_3) = (1-n, i, n-i+1)$ and $(x_1', x_2', x_3') = (n, i-n, 1-i)$. Thus, $h(x_1, x_2, x_3)$ is $(0, 4n)$ if $i = 1$ and $(n-i+1, n-i+1)$ if $1 < i \le n$. Again, $h(x_1', x_2', x_3')$ is $(0, 4n-1)$ if $i = 1$ and $(n-i+1, n-i)$ if $1 < i \le n$. Therefore, $h(x_1, x_2, x_3)$ and $h(x_1', x_2', x_3')$ are adjacent.

Case 5. $e \in \{((i, 1-n, n-i+1), (i-n, n, 1-i)) \mid 1 \le i \le n\}$. Obviously, $(x_1, x_2, x_3) = (i, 1-n, n-i+1)$ and $(x_1', x_2', x_3') = (i-n, n, 1-i)$. Thus, $h(x_1, x_2, x_3)$ is $(0, 0)$ if $i = 1$ and $(n-i+1, 3n+i-1)$ if $1 < i \le n$. Again, $h(x_1', x_2', x_3')$ is $(0, 1)$ if $i = 1$ and $(n-i+1, 3n+i)$ if $1 < i \le n$. Hence, $h(x_1, x_2, x_3)$ and $h(x_1', x_2', x_3')$ are adjacent.

Thus, the theorem is proved. □

For example, the honeycomb torus illustrated in Figure 2.1c is actually isomorphic to the generalized honeycomb torus illustrated in Figure 2.1d.

2.2 ISOMORPHISM BETWEEN CYCLIC-CUBES AND WRAPPED BUTTERFLY NETWORKS

The *wrapped-around butterfly network* $WB(n, k)$ has $n \cdot k^n$ vertices, and each vertex is represented by $(n+1)$-bit vectors $a_0 a_1 \ldots a_{n-1} i$ where $0 \le i \le n-1$ and $1 \le a_j \le k$ for all $0 \le j \le n-1$. Two vertices $a_0 a_1 \ldots a_{n-1} i$ and $b_0 b_1 \ldots b_{n-1} j$ are adjacent in $WB(n, k)$ if and only if $j - i = 1 \pmod{n}$ and $a_t = b_t$ for all $0 \le t \ne j \le n-1$. The graph in Figure 2.2 is $WB(3, 2)$. The $WB(n, k)$ is widely used in communication networks [220].

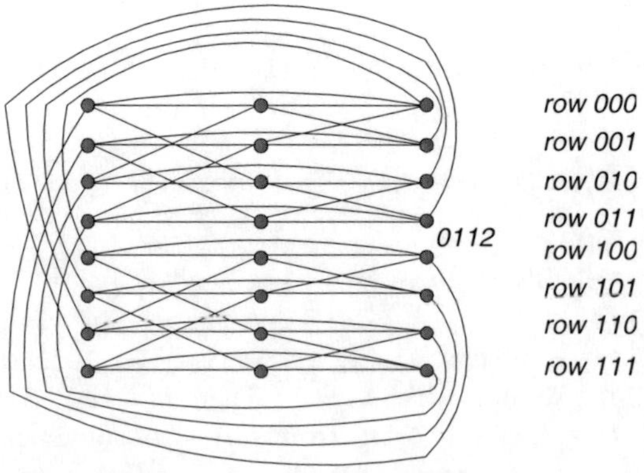

FIGURE 2.2 The graph $WB(3, 2)$.

In 1998, Fu and Chau [118] proposed a family of graphs called cyclic-cubes, which have even fixed degrees. They show that this family of graphs has many useful properties, so that it can serve as an interconnection network topology. Let G_n^k denote k-ary n-dimensional cyclic-cubes. In 2000, Hung et al. [184] proved that this family of graphs is indeed isomorphic to $WB(n,k)$.

To define G_n^k, let $t_1, t_2, \ldots, t_n$ be n distinct symbols with ordering $t_1 > t_2 \cdots > t_n$. Each symbol t_j is assigned a rank i for $1 \le i \le k$, and this ranked symbol is denoted by t_j^i. The graph G_n^k has $n \cdot k^n$ vertices, and each vertex of G_n^k is represented by an n-bit vector, which is a circular permutation of $t_1^{i_1} t_2^{i_2} \ldots t_n^{i_n}$ for $1 \le i_1, i_2, \ldots, i_n \le k$. For example, in G_4^2 $t_3^1 t_4^2 t_1^1 t_2^2$ is a vertex and $t_3^2 t_2^1 t_4^2 t_1^1$ is not. In other words,

$$V\left(G_n^k\right) = \left\{ t_j^{i_j} t_{j+1}^{i_{j+1}} \ldots t_n^{i_n} t_1^{i_1} \ldots t_{j-1}^{i_{j-1}} \;\middle|\; \text{for } 1 \le j \le n \quad \text{and} \quad 1 \le i_1, i_2, \ldots, i_n \le k \right\}$$

To define edges in G_n^k, we first define the function f_s for every $1 \le s \le k$, mapping $V(G_n^k)$ onto itself as follows:

$$f_s\left(t_j^{i_j} t_{j+1}^{i_{j+1}} \ldots t_n^{i_n} t_1^{i_1} \ldots t_{j-1}^{i_{j-1}}\right) = t_{j+1}^{i_{j+1}} \ldots t_n^{i_n} t_1^{i_1} \ldots t_{j-1}^{i_{j-1}} t_j^s \quad \text{for any } 1 \le s \le k$$

Note that all f_j are bijective functions. Each vertex $x \in V(G_n^k)$ is linked to exactly $2k$ vertices $f_j(x)$ and $f_j^{-1}(x)$ for all $1 \le j \le k$. For example, in G_4^2 the vertex $t_3^1 t_4^1 t_1^1 t_2^2$ is linked to $t_4^1 t_1^1 t_2^2 t_3^1$, $t_4^1 t_1^1 t_2^2 t_3^2$, $t_2^1 t_3^1 t_4^1 t_1^1$, and $t_2^2 t_3^1 t_4^1 t_1^1$.

THEOREM 2.2 G_n^k is isomorphic to $WB(n,k)$.

Proof: For each vertex $a_0 a_1 \ldots a_{n-1} i$ in $WB(n,k)$, we define a function π mapping $V(WB(n,k))$ to $V(G_n^k)$ as follows:

$$\pi\left(a_0 a_1 \ldots a_{n-1} i\right) = t_i^{a_{i-1}+1} t_{i+1}^{a_i+1} \ldots t_n^{a_{n-1}+1} t_1^{a_0+1} t_2^{a_1+1} \ldots t_{i-1}^{a_{i-2}+1}$$

For example, $n=3$, $k=2$, $f(0000) = t_1^1 t_2^1 t_3^1$, $f(0001) = t_2^1 t_3^1 t_1^1$, and $f(0112) = t_3^2 t_1^1 t_2^2$. Obviously, the function π is bijective.

Let $u = a_0 a_1 \ldots a_{n-1} i$ and $v = b_0 b_1 \ldots b_{n-1} j$ be two distinct vertices in $WB(n,k)$. Then $\pi(u)$ and $\pi(v)$ are two distinct vertices in G_n^k given as follows:

$$\begin{aligned}
\pi(u) &= t_i^{a_{i-1}+1} t_{i+1}^{a_i+1} \ldots t_n^{a_{n-1}+1} t_1^{a_0+1} t_2^{a_1+1} \ldots t_{i-1}^{a_{i-2}+1} \\
\pi(v) &= t_j^{b_{j-1}+1} t_{j+1}^{b_j+1} \ldots t_n^{b_{n-1}+1} t_1^{b_0+1} t_2^{b_1+1} \ldots t_{j-1}^{b_{j-2}+1}
\end{aligned}$$

Assume that u and v are adjacent in $WB(n,k)$. Without loss of generality, we may assume that $j = i-1 \pmod n$. Thus, $a_t = b_t$ for all $0 \le t \ne j \le n-1$; that is, $v = a_0 a_1 \ldots a_{i-2} b_{i-1} a_i \ldots a_{n-1}(i-1)$. Therefore,

$$\pi(v) = t_{i+1}^{a_i+1} t_{i+2}^{a_{i+1}+1} \cdots t_n^{a_{n-1}+1} t_1^{a_0+1} t_2^{a_1+1} \cdots t_{i-1}^{a_{i-2}+1} t_i^{b_{i-1}+1} = f_{b_{i-1}}(\pi(u))$$

Thus, $\pi(u)$ and $\pi(v)$ are adjacent in G_n^k.

On the other hand, suppose that $\pi(u)$ and $\pi(v)$ are adjacent in G_n^k. Then $\pi(v)$ can be $f_s(\pi(u))$ or $f_s^{-1}(\pi(u))$ for some $1 \leq s \leq k$. Suppose that $\pi(v) = f_s(\pi(u))$ for some $1 \leq s \leq k$. Then $\pi(v) = t_{i+1}^{a_i+1} t_{i+2}^{a_{i+1}+1} \ldots t_n^{a_{n-1}+1} t_1^{a_0+1} t_2^{a_1+1} \ldots t_{i-1}^{a_{i-2}-1} t_i^{s+1}$ and $v = a_0 a_1 \ldots a_{i-1} s a_i \ldots a_{n-1}(i-1)$. Therefore, $(u, v) \in E(WB(n, k))$. Similarly, $\pi(v) = f_s^{-1}(\pi(u))$ also implies $(u, v) \in E(WB(n, k))$.

This theorem is proved. □

2.3 1-EDGE FAULT-TOLERANT DESIGN FOR MESHES

Motivated by the study of computer and communication networks that can tolerate failure of their components, Harary and Hayes [139] formulated the concept of edge fault tolerance in graphs. Let G be a graph of order n and k be a positive integer. An n-vertex graph G^* is said to be *k-edge fault-tolerant*, or k-EFT, with respect to G, if every graph obtained by removing any k-edges from G^* contains G. For brevity, we refer to G^* as k-EFT(G) graph. A k-EFT(G) graph G^* is *optimal* if it contains the least number of edges among all k-EFT(G) graphs. Let $eft_k(G)$ be the difference between the number of edges in an optimal k-EFT(G) graph and that in G.

LEMMA 2.1 Assume that G^* is a k-EFT(G). Then $\delta(G^*) \geq \delta(G) + k$.

Proof: Suppose that the lemma is false. Then there exists a k-EFT(G) graph G^* with $\delta(G^*) < \delta(G) + k$. Let v be a vertex of G^* with $\deg_{G^*}(v) = \delta(G^*)$. We delete k edges that are incident with v from G^* to obtain a graph G'. Obviously, $\delta(G') < \delta(G)$. Hence, G' is not a subgraph of G. We get a contradiction, because G^* is a k-EFT(G). Hence, the lemma is proved. □

It is easy to see that ${}_{k+1}G$ is a k-EFT(G) for any graph G. Moreover, C_n is an optimal 1-EFT(P_n). It is easy to confirm that the torus graph $C(m_1, m_2, \ldots, m_n)$ is a 1-edge fault-tolerant graph for $M(m_1, m_2, \ldots, m_n)$ where $m_i \leq 2$ with $1 \leq i \leq n$. In Ref. 139, Harary and Hayes conjectured that $C(m_1, m_2, \ldots, m_n)$ is optimal 1-EFT $(M(m_1, m_2, \ldots, m_n))$ for every $m_i \geq 2$ with $1 \leq i \leq n$. The two-dimensional mesh $M(3, 4)$ is illustrated in Figure 2.3a. The graph $C(3, 4)$ with a faulty (dashed) edge $F = (x_{2,3}, x_{2,4})$ is illustrated in Figure 2.3b. A recovery from a fault affecting the dashed edge $F = (x_{2,3}, x_{2,4})$ is shown in Figure 2.3c.

Chou and Hsu [64] disprove the conjecture by proposing the following counterexample. We assume that the vertices of $M(m_1, m_2, \ldots, m_n)$ are labeled canonically. Thus, $x_{i_1,i_2,\ldots,i_n}$ is a vertex of $M(m_1, m_2, \ldots, m_n)$ if and only if $1 \leq i_j \leq m_j$ for $1 \leq j \leq n$.

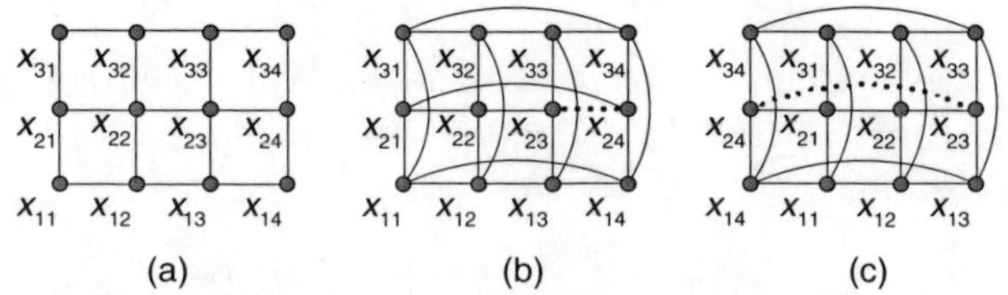

FIGURE 2.3 (a) The graph $M(3, 4)$, (b) the graph $C(3, 4)$ with a faulty (dashed) edge $F = (x_{2,3}, x_{2,4})$, and (c) a recovery from a fault affecting the dashed edge $F = (x_{2,3}, x_{2,4})$.

Moreover, $x_{i_1,i_2,\ldots,i_n}$ is adjacent to another vertex $x_{j_1,j_2,\ldots,j_n}$ if all indices t with $1 \le t \le n$ but one index k satisfy $i_t = j_t$; and the index k is such that $i_k \ne j_k$ satisfies $|i_k - j_k| = 1$. Let $V_p = \{x_{i_1,i_2,\ldots,i_n} \mid i_k = 1 \text{ or } m_k \text{ for some } 1 \le k \le n\}$ be the set of *peripheral vertices*. Assume that $x_{i_1,i_2,\ldots,i_n}$ is a vertex in V_p. The *antipodal vertex* of $x_{i_1,i_2,\ldots,i_n}$ is $x_{j_1,j_2,\ldots,j_n}$, with $j_k = m_k - i_k + 1$, which is another vertex in V_p. It is easy to confirm that every vertex in V_p has exactly one antipodal. In $M(m_1, m_2, \ldots, m_n)$, we add the edges joining each vertex in V_p to its antipodal counterpart to form a new graph $P(m_1, m_2, \ldots, m_n)$. We call these $P(m_1, m_2, \ldots, m_n)$ *projective-plane graphs* because their construction is similar to that of the projective plane when $n = 2$.

THEOREM 2.3 $P(m_1, m_2, \ldots, m_n)$ is 1-EFT($M(m_1, m_2, \ldots, m_n)$) and it contains fewer edges than that of $C(m_1, m_2, \ldots, m_n)$.

Proof: Suppose that F is the faulty edge of $P(m_1, m_2, \ldots, m_n)$ joining $x_{i_1,i_2,\ldots,i_n}$ to $x_{i'_1,i'_2,\ldots,i'_n}$ where $i'_s = i_s$ if $s \ne k$ and $i'_k = i_k + 1$. Now, we reconfigure $M(m_1, m_2, \ldots, m_n)$ through the following. First we delete all the edges joining $x_{j_1,j_2,\ldots,j_n}$ to $x_{j'_1,j'_2,\ldots,j'_n}$ where $j'_s = j_s$ for every $s \ne k$, $j_k = i_k$, and $j'_k = i_k + 1$. Let $A = \{x_{j_1,j_2,\ldots,j_n} \mid j_k \le i_k\}$ and $B = \{x_{j_1,j_2,\ldots,j_n} \mid j_k \ge j'_k + 1\}$. In $M(m_1, m_2, \ldots, m_n)$, the sets A and B induce two submeshes. These two submeshes are connected by edges including the set $E^* = \{(x_{j_1,j_2,\ldots,j_n}, x_{j'_1,j'_2,\ldots,j'_n}) \mid j_s = m_s - j'_s + 1 \text{ for every } s \ne k,\ j_k = 1, \text{ and } j'_k = m_k\}$. Therefore, the subgraphs of $P(m_1, m_2, \ldots, m_n)$ generated by A, B, and E^* form a mesh M' isomorphic to $M(m_1, m_2, \ldots, m_n)$. For example, the recovery of $M(3,4)$ for the faulty edge $F = (x_{2,3}, x_{2,4})$ in $P(3,4)$ is illustrated in Figure 2.4.

Obviously, $|E(P(m_1, m_2, \ldots, m_n))| \le |E(C(m_1, m_2, \ldots, m_n))|$. Hence, the theorem is proved. □

COROLLARY 2.1 $eft_1(M(m_1, m_2, \ldots, m_n)) \le 1/2 \left(\prod_{i=1}^{n} m_i - \prod_{i=1}^{n}(m_i - 2)\right)$.

The projective-plane graphs are optimal for some cases, but not for all. Note that every n-dimensional hypercube can be viewed as the mesh $M(2, 2, \ldots, 2)$. Our $P(2, 2, \ldots, 2)$ is actually the same 1-EFT graph as that proposed in Refs 34, 139, 319, and 347. Thus, $P(2, 2, \ldots, 2)$ is an optimal 1-EFT graph. It is easy to prove that the graph in Figure 2.5a is a 1-EFT($M(3,2)$) and the graph in Figure 2.5b is a 1-EFT($M(4,2)$). With these two examples, we know that the projective-plane graphs may not be optimal for some cases. Furthermore, the problem of finding the optimal 1-EFT for all n-dimensional meshes remains unsolved.

In Ref. 67, Chuang et al. study the 1-EFT graphs for $M(k,2)$ with $k \ge 2$. For simplicity, the kth *ladder graph* L_k is defined to be $M(k,2)$. The vertices of L_k can be

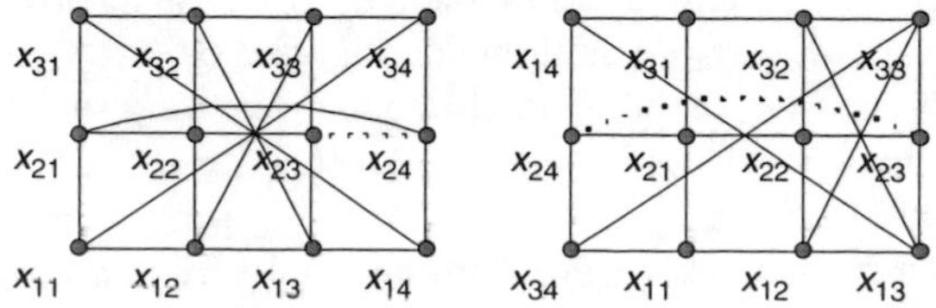

FIGURE 2.4 The recovery of $M(3,4)$ for the faulty edge $F = (x_{2,3}, x_{2,4})$ in $P(3,4)$.

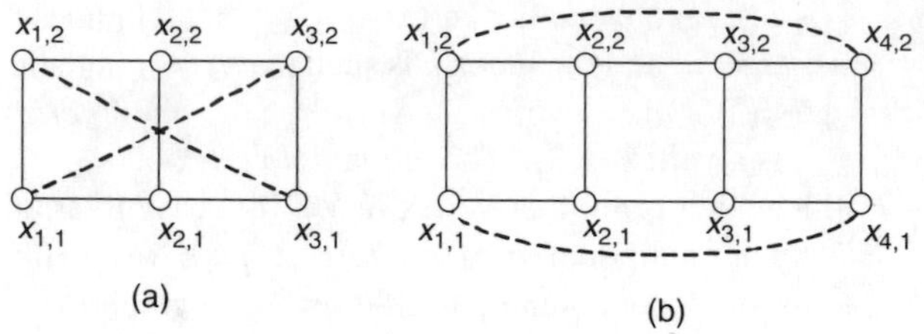

FIGURE 2.5 (a) A 1-EFT($M(3,2)$), L_3^* and (b) a 1-EFT($M(4,2)$), L_4^*.

labeled by $x_{i,j}$ with $1 \leq i \leq k$ and $1 \leq j \leq 2$ canonically. The vertices $x_{1,1}$, $x_{k,1}$, $x_{1,2}$, and $x_{k,2}$ are called the *corner vertices* of L_k. We have the following theorem.

THEOREM 2.4 Assume that L_k^* is a 1-EFT(L_k) graph. Then (*i*) $\deg_{L_k^*}(x) \geq 3$ for any vertex x of L_k^* and (*ii*) $eft_1(L_k) \geq 2$.

Proof: Assume that some vertex x with $\deg_{L_k^*}(x) = 2$. Let e be any edge incident with x. Then $\deg_{L_k^*-e}(x) = 1$. Because $\deg_{L_k}(x) \geq 2$ for any vertex x of L_k, L_k is not a subgraph of $L_k^* - e$. We find a contradiction to L_k^* being a 1-EFT(L_k) graph. Hence, $\deg_{L_k^*}(x) \geq 3$. Because there are exactly four corner vertices in every L_k, we have $eft_1(L_k) \geq 2$. □

COROLLARY 2.2 *$eft_1(L_k) > 2$ if $k > 4$.*

Proof: It is easy to check whether there are exactly three different ways of joining the four corner vertices in L_k with two edges; namely, $\{(x_{1,1}, x_{1,2}), (x_{k,1}, x_{k,2})\}$, $\{(x_{1,1}, x_{k,1}), (x_{1,2}, x_{k,2})\}$, and $\{(x_{1,1}, x_{k,2}), (x_{1,2}, x_{k,1})\}$. It is observed that none of the graphs obtained by joining two edges to the corner vertices of L_k with $k > 4$ is 1-EFT(L_k). Hence, $eft_1(L_k) > 2$ if $k > 4$. □

Let L_2^* (L_3^*, and L_4^*, respectively) be the graph $P(2,2)$ (the graph in Figures 2.5a, and 2.5b, respectively). From the previous discussion, L_k^* is 1-EFT(L_k) for $k = 2, 3$, and 4. Note that there are exactly two edges added to L_k with $k = 2, 3$, and 4. By Theorem 2.4, these graphs are optimal. By checking all three cases joining two edges to the corner vertices of L_k, the optimal 1-EFT(L_k) is unique for $k = 2, 3$, and 4. We obtain the following theorem.

THEOREM 2.5 *$eft_1(L_k) = 2$ for $k = 2, 3$, and 4.*

Let us consider the spanning supergraph L_5^* of L_5 given by $E(L_5^*) = E(L_5) \cup \{(x_{1,1}, x_{5,2}), (x_{1,2}, x_{4,2}), (x_{2,1}, x_{5,1})\}$ as shown in Figure 2.6.

Obviously, edges of L_5 can be divided into the following seven classes: namely, $A = \{(x_{1,1}, x_{1,2})\}$, $B = \{(x_{i,1}, x_{i,2}) \mid 2 \leq i \leq 4\}$, $C = \{(x_{5,1}, x_{5,2})\}$, $D = \{(x_{1,1}, x_{2,1}), (x_{1,2}, x_{2,2})\}$, $E = \{(x_{2,1}, x_{3,1}), (x_{2,2}, x_{3,2})\}$, $F = \{(x_{3,1}, x_{4,1}), (x_{3,2}, x_{4,2})\}$, and $G = \{(x_{4,1}, x_{5,1}), (x_{4,2}, x_{5,2})\}$. We can reconfigure L_5 in L_5^* for any faulty edge e in A, B, C, D, E, F, and G, respectively, as shown in Figures 2.7a through 2.7g, respectively. Thus, L_5^* is 1-EFT(L_5). By Corollary 2.2, we have the following theorem.

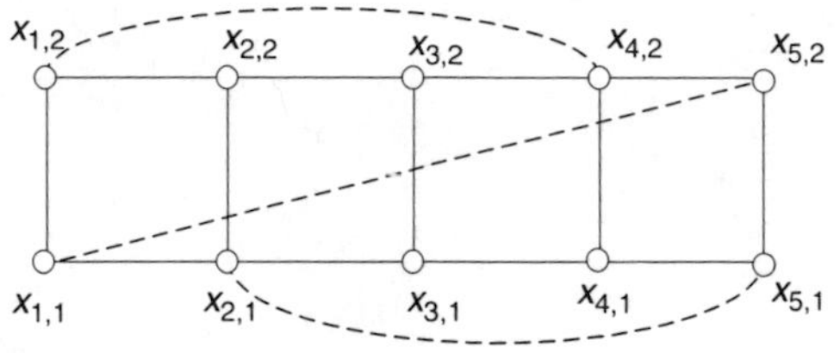

FIGURE 2.6 A 1-EFT($M(5,2)$), L_5^*.

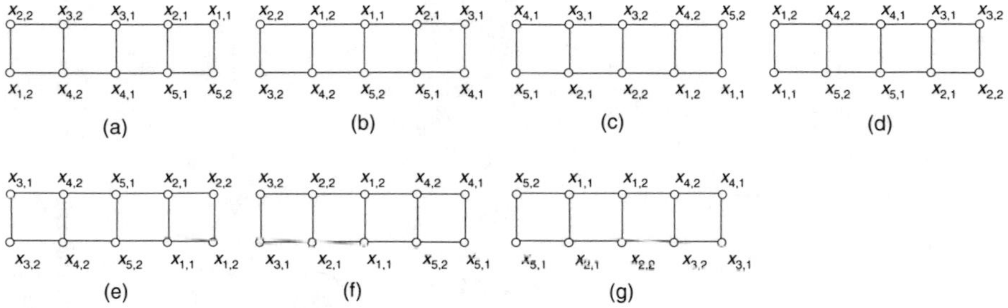

FIGURE 2.7 A 1-EFT($M(5,2)$), L_5^*.

THEOREM 2.6 $eft_1(L_5) = 3$.

Let k be an even integer with $k \geq 4$. The spanning supergraph L_k^* of L_k is the graph that adds $E' = \{(x_{i,j}, x_{k-i+1,j}) \mid 1 \leq i < k/2,\ j = 1,2\}$ to $E(L_k)$ as shown in Figure 2.8a. The graph in Figure 2.8a is actually isomorphic to $M(k/2, 2, 2)$, as shown in Figure 2.8b. We can reconfigure L_k in L_k^* as shown in Figure 2.8c for any faulty edge of the form $(x_{i,1}, x_{i,2})$ or as shown in Figure 2.8d for any faulty edge of the form $(x_{i,1}, x_{i+1,1})$ or $(x_{i,2}, x_{i+1,2})$. Thus, $M(k/2, 2, 2)$ is a 1-EFT(L_k). We obtain the following theorem.

THEOREM 2.7 Assume that k is an even integer with $k \geq 4$. Then $eft_1(L_k) \leq k - 2$.

Let k be an odd integer with $k \geq 7$. Construct the spanning supergraph L_k^* of L_k by adding $E' = \{(x_{1,2}, x_{4,2}),\ (x_{3,2}, x_{6,2}),\ (x_{2,1}, x_{5,1}),\ (x_{4,1}, x_{7,1}),\ (x_{1,1}, x_{5,2}),\ (x_{3,1}, x_{7,2})\} \cup \{(x_{2i,j}, x_{2i+3,j}) \mid 3 \leq i \leq (k-3)/2,\ j = 1,2\}$ as shown in Figure 2.9.

Obviously, edges of L_k can be divided into the following seven classes:

$$A = \{(x_{i,1}, x_{i,2}) \mid i = 1,2\} \cup \{(x_{2i,j}, x_{2i+1,j}) \mid 4 \leq i \leq (k-3)/2,\ j = 1,2\}$$
$$B = \{(x_{i,1}, x_{i,2}) \mid i = 3,4\} \cup \{(x_{2i-1,j}, x_{2i,j}) \mid 4 \leq i \leq (k-1)/2,\ j = 1,2\}$$
$$C = \{(x_{5,1}, x_{5,2})\} \cup \{(x_{i,1}, x_{i,2}) \mid 4 \leq i \leq (k-1)/2\}$$
$$D = \{(x_{i,1}, x_{i,2}) \mid i = 6,7\} \cup \{(x_{i,1}, x_{i,2}) \mid i = k, k-1\}$$
$$E = \{(x_{1,j}, x_{2,j}) \mid j = 1,2\} \cup \{(x_{3,j}, x_{4,j}) \mid j = 1,2\}$$
$$F = \{(x_{2,j}, x_{3,j}) \mid j = 1,2\} \cup \{(x_{5,j}, x_{6,j}) \mid j = 1,2\}$$
$$G = \{(x_{4,j}, x_{5,j}) \mid j = 1,2\} \cup \{(x_{6,j}, x_{7,j}) \mid j = 1,2\} \cup \{(x_{k-1,j}, x_{k,j}) \mid j = 1,2\}$$

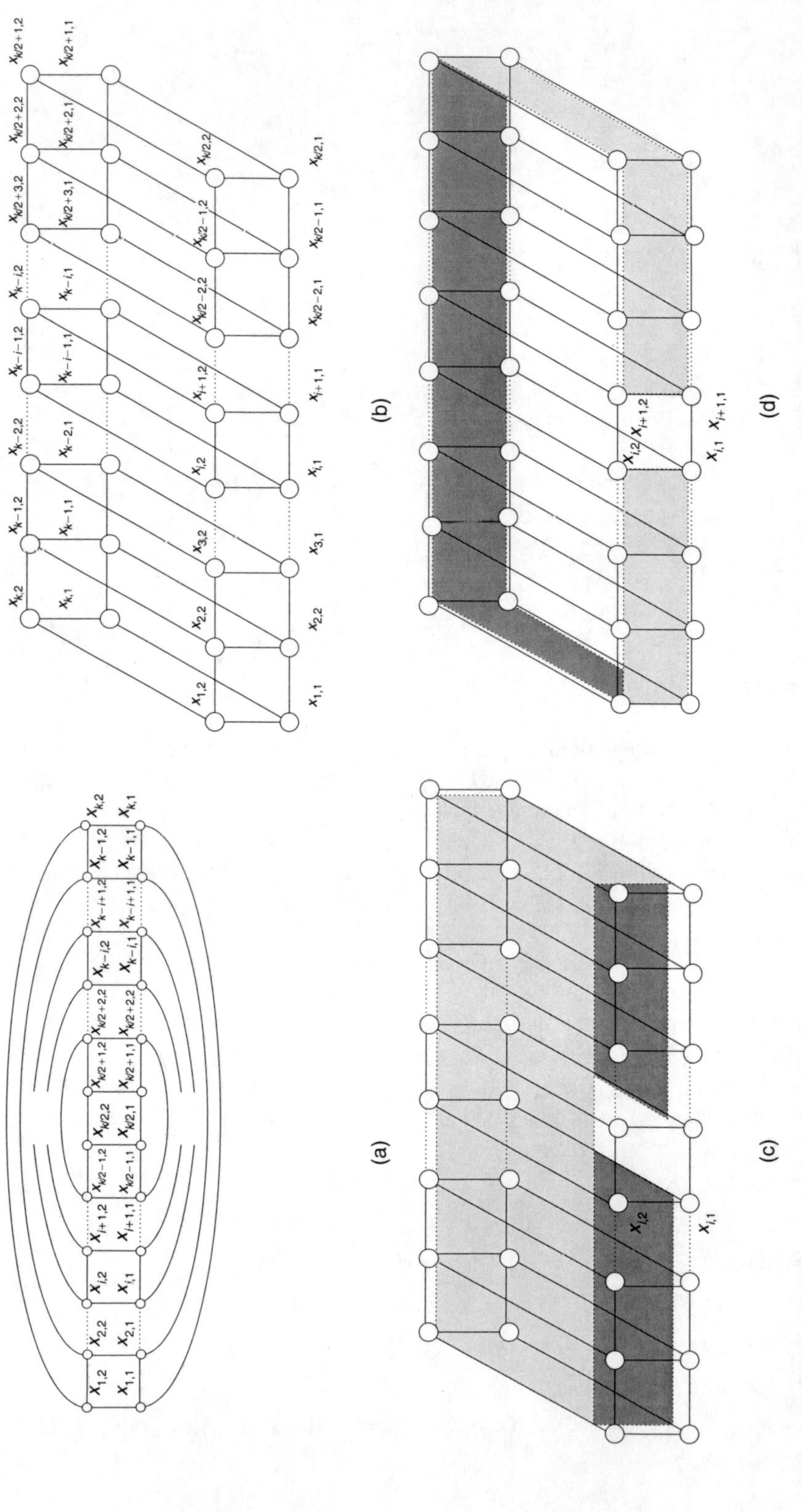

FIGURE 2.8 (a) L_k^*, a 1-EFT(L_k) where k is even and $k \geq 4$, (b) the three-dimensional mesh $M(k/2, 2, 2)$, (c) reconfigured L_k for any faulty edge of the form $(x_{i,1}, x_{i,2})$, and (d) reconfigured L_k for any faulty edge of the form $(x_{i,1}, x_{i+1,1})$ or $(x_{i,2}, x_{i+1,2})$.

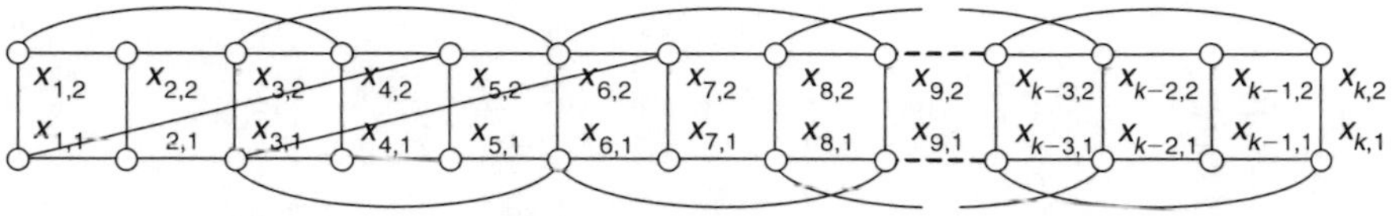

FIGURE 2.9 A 1-EFT(L_k) where k is odd and $k \geq 7$.

We can reconfigure L_k in L_k^* for any faulty edge e in A, B, C, D, E, F, and G, respectively, as shown in Figures 2.10a through 2.10g, respectively. Thus, L_k^* is 1-EFT(L_k). We have the following theorem.

THEOREM 2.8 Assume that k is an odd integer with $k \geq 7$. Then $eft_1(L_k) \leq k - 1$.

We conjecture that L_k^* is optimal 1-EFT(L_k) for $k \geq 6$.

2.4 FAITHFUL 1-EDGE FAULT-TOLERANT GRAPHS

Families of k-EFT graphs with respect to some graphs have been studied [64,139,153,284,293]. It is observed that there is no general approach to the construction of edge fault tolerant graphs. However, we note that meshes, tori, and hypercubes can be expressed as Cartesian products of several primal graphs. Wang et al. [332] aim at providing a scheme for constructing one-edge fault-tolerant graphs with respect to some graph products. Once we can find certain 1-EFT graphs with respect to these primal graphs having some desired properties, this scheme enables us to construct a 1-EFT graph with respect to the graph product. In particular, we apply this scheme to construct a 1-EFT($C(m_1, m_2, \ldots, m_n)$) and show it is optimal, where $m_1, m_2, \ldots, m_n$ are positive even integers with each $m_i \geq 4$.

Because multiple edges are allowed in graphs, all set operations in this section are defined with multisets; for example, $\{a, b\} \uplus \{a\} = \{a, a, b\}$ and $\{a, a, b\} - \{a, c\} = \{a, b\}$, where $\uplus$ denotes the sum operation of two multisets. We call (G^*, G) a *graph pair* if G^* is a spanning supergraph of G. Moreover, (G^*, G) is a *1-EFT pair* if G^* is a 1-EFT(G). Throughout this section, let (G_i^*, G_i) be a graph pair for all i with $G_i^* = (V_i, E_i^*)$ and $G_i = (V_i, E_i)$.

We use $(G_1^*, G_1) \oplus (G_2^*, G_2)$ to denote the graph with $V_1 \times V_2$ as its vertex set and $E(G_1 \times G_2) \uplus E((G_1^* - E_1) \cdot (G_2^* - E_2))$ as its edge set, where $\times$ is the Cartesian product of graphs and $\cdot$ is the tensor product of graphs. Obviously, $G_1 \times G_2$ is a spanning subgraph of $(G_1^*, G_1) \oplus (G_2^*, G_2)$. Now, we define an operator $\otimes$ on two graph pairs (G_1^*, G_1) and (G_2^*, G_2), denoted by $(G_1^*, G_1) \otimes (G_2^*, G_2)$, as the graph pair $((G_1^*, G_1) \oplus (G_2^*, G_2), G_1 \times G_2)$. For example, let $G_1 = C_6$, $G_2 = C_4$, G_1^* be the graph in Figure 2.11a, and G_2^* be the graph in Figure 2.11b. In Figures 2.11a and 2.11b, dashed lines represent edges $G_1^* - E_1$ and $G_2^* - E_2$. We illustrate $G_1 \times G_2$, $(G_1^* - E_1) \cdot (G_2^* - E_2)$, and $(G_1^*, G_1) \oplus (G_2^*, G_2)$ in Figures 2.11c through 2.11e, respectively. Note that $\times$ and $\cdot$ are commutative and associative. The following theorem can easily be obtained.

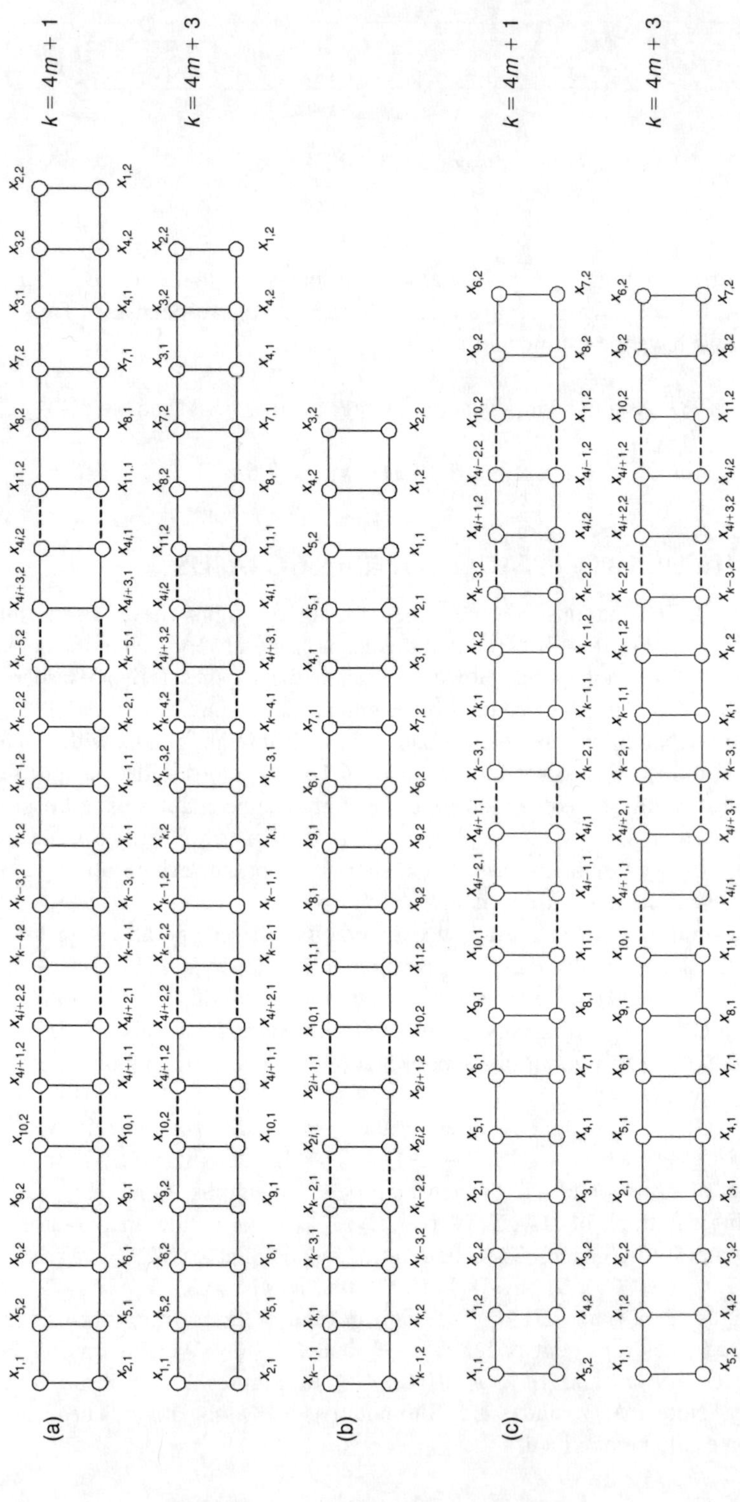

FIGURE 2.10 Reconfigures of L_k in L_k^* where k is odd and $k \geq 7$ for any faulty edge in (a)–(g), respectively.

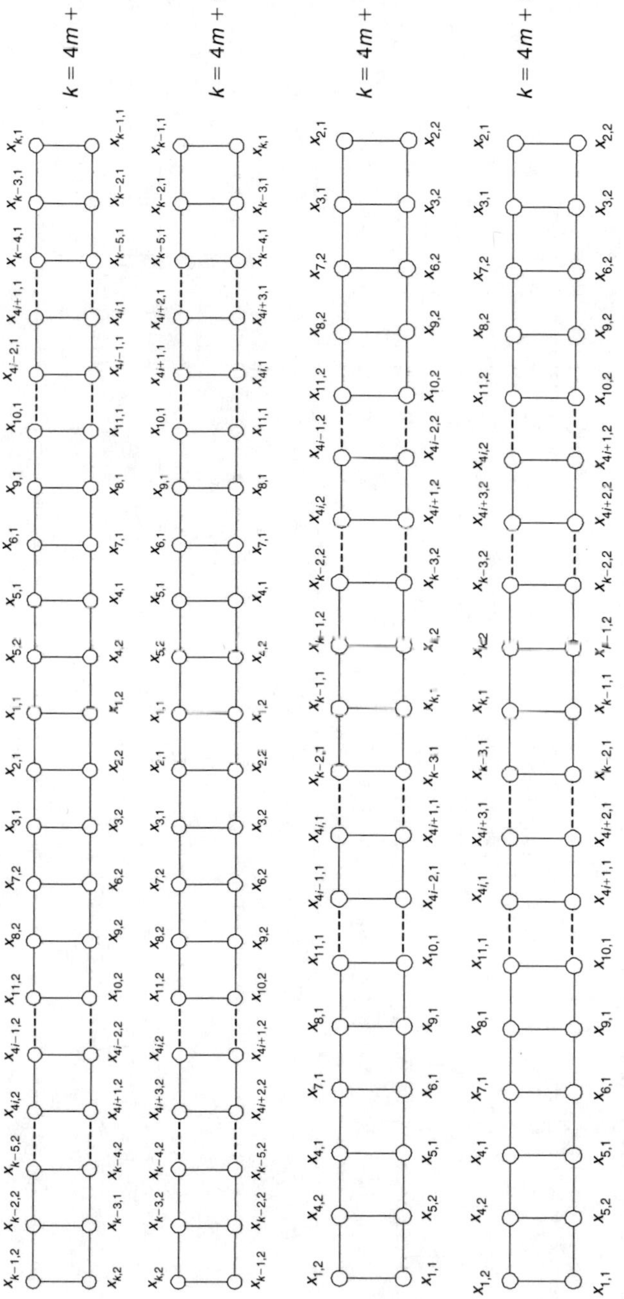

FIGURE 2.10 (continued)

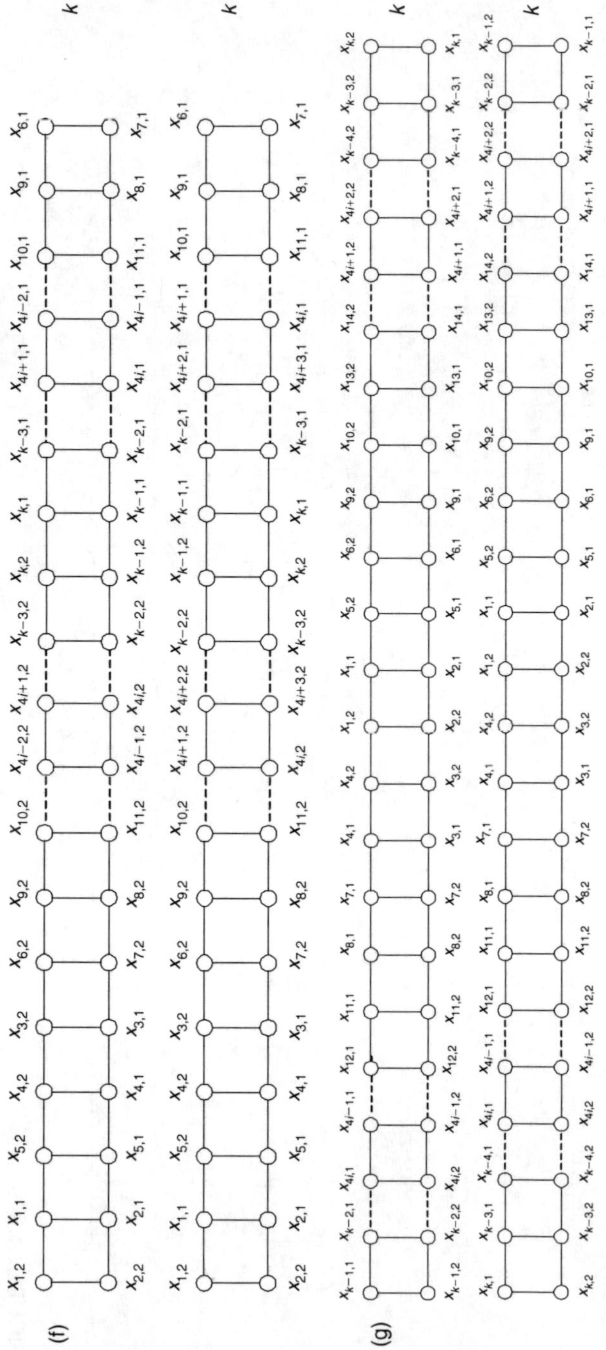

FIGURE 2.10 (continued)

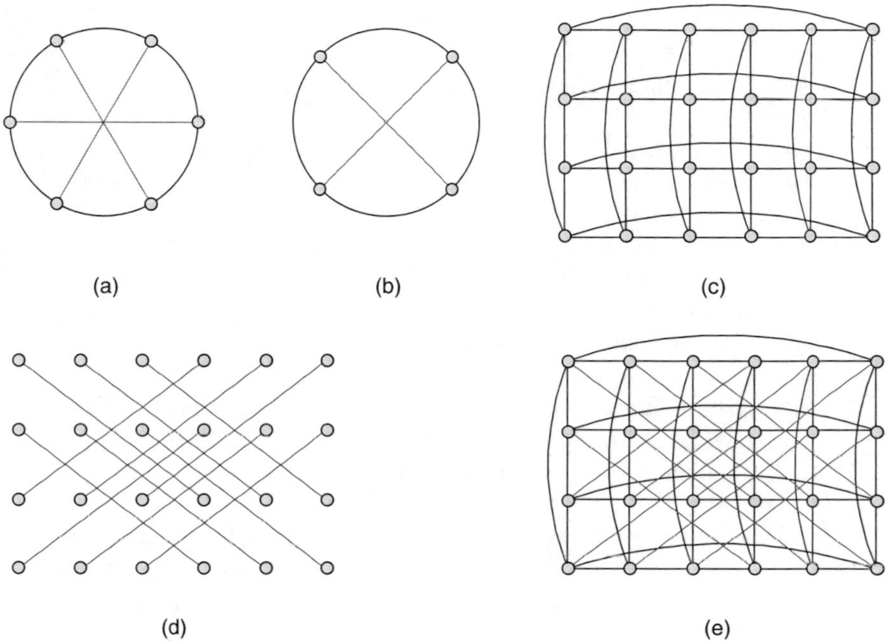

FIGURE 2.11 (a) The graph pair (G_1^*, G_1), (b) the graph pair (G_2^*, G_2), (c) the graph $G_1 \times G_2$, (d) the graph $(G_1^* - E_1) \cdot (G_2^* - E_2)$, and (e) the graph $(G_1^*, G_1) \oplus (G_2^*, G_2)$.

THEOREM 2.9 $(G_1^*, G_1) \otimes (G_2^*, G_2) = (G_2^*, G_2) \otimes (G_1^*, G_1)$; and $((G_1^*, G_1) \otimes (G_2^*, G_2)) \otimes (G_3^*, G_3) = (G_1^*, G_1) \otimes ((G_2^*, G_2) \otimes (G_3^*, G_3))$.

Recursively, we define $(G_1^*, G_1) \otimes (G_2^*, G_2) \otimes \cdots \otimes (G_n^*, G_n)$ as $((G_1^*, G_1) \otimes (G_2^*, G_2) \otimes \cdots \otimes (G_{n-1}^*, G_{n-1})) \otimes (G_n^*, G_n)$. Again, we define $(G_1^*, G_1) \oplus (G_2^*, G_2) \oplus \cdots \oplus (G_n^*, G_n)$ as $((G_1^*, G_1) \otimes (G_2^*, G_2) \otimes \cdots \otimes (G_{n-1}^*, G_{n-1})) \oplus (G_n^*, G_n)$. We have the following corollary.

COROLLARY 2.3 *Let π be any permutation on the set $\{1, 2, \ldots, n\}$. We have*

1. $(G_1^*, G_1) \otimes (G_2^*, G_2) \otimes \cdots \otimes (G_n^*, G_n) = (G_{\pi(1)}^*, G_{\pi(1)}) \otimes (G_{\pi(2)}^*, G_{\pi(2)}) \otimes \cdots \otimes (G_{\pi(n)}^*, G_{\pi(n)})$
2. $(G_1^*, G_1) \oplus (G_2^*, G_2) \oplus \cdots \oplus (G_n^*, G_n) = (G_{\pi(1)}^*, G_{\pi(1)}) \oplus (G_{\pi(2)}^*, G_{\pi(2)}) \oplus \cdots \oplus (G_{\pi(n)}^*, G_{\pi(n)})$

For $1 \le i \le n$, the ith *projection* of $V_1 \times V_2 \times \cdots \times V_n$ is defined as the function $p_i : V_1 \times V_2 \times \cdots \times V_n \to V_i$ given by $p_i((x_1, x_2, \ldots, x_n)) = x_i$ where $x_j \in V_j$ for $1 \le j \le n$.

Assume that G_i^* is a 1-EFT(G_i) graph for $i = 1, 2$. It is easy to verify that $G_1^* \times G_2^*$ is a 1-EFT($G_1 \times G_2$) graph. However, $G_1^* \times G_2^*$ may contain many more edges than that of optimal 1-EFT($G_1 \times G_2$). For example, let $G_1 = C_6$ and $G_2 = C_4$. The graphs G_1^* shown in Figure 2.11a and G_2^* shown in Figure 2.11b are 1-EFT(G_1) and 1-EFT(G_2), respectively. Then the graph $G_1^* \times G_2^*$ is 1-EFT($G_1 \times G_2$). It can be verified that the graph $(G_1^*, G_1) \oplus (G_2^*, G_2)$ in Figure 2.11e is also 1-EFT($G_1 \times G_2$). Note that the number of edges in $(G_1^*, G_1) \oplus (G_2^*, G_2)$ is less than that in $G_1^* \times G_2^*$. Hence $G_1^* \times G_2^*$ is not an optimal 1-EFT($G_1 \times G_2$) graph.

Let K_2 be the complete graph on two vertices z_1 and z_2. We refer to C_2 as ${}_2K_2$. Obviously, C_2 is 1-EFT(K_2). Assume that $G=(V(G), E(G))$ is a graph with $V(G)=\{x_1, x_2, \ldots, x_n\}$, and $G^*=(V, E^*)$ is a spanning supergraph of G. Then $G \times K_2$ is a spanning subgraph of $(G^*, G) \oplus (C_2, K_2)$. Any edge in $G \times K_2$ is of the form either $((x_i, z_1), (x_i, z_2))$ for some $x_i \in V(G)$ or $((x_i, z_k), (x_j, z_k))$ for some $(x_i, x_j) \in E(G)$ and $k=1, 2$. Let X be a set of edges given by $X=\{((x_i, z_1), (x_i, z_2)) \mid \text{for every } x_i \in V(G)\}$. For an edge $e=(x_i, y_j)$ in G, let $Y_e=\{((x_i, z_1), (x_j, z_1)), ((x_i, z_2), (x_j, z_2))\}$. The graph G^* is said to be *faithful* or a *faithful graph* with respect to G, denoted by FG(G), if it satisfies the following conditions:

1. There exists a function σ from $V(G)$ into itself such that the function $h: V(G \times K_2) \to V(G \times K_2)$ given by $h((x_i, z_1))=(x_i, z_1)$ and $h((x_i, z_2))=(\sigma(x_i), z_2)$ induces an isomorphism from $G \times K_2$ into a subgraph of $(G^*, G) \oplus (C_2, K_2) - X$.
2. For any edge $e=(x_i, x_j)$ in G, there exists an isomorphism f_e from $G \times K_2$ into a subgraph of $((G^*, G) \oplus (C_2, K_2)) - Y_e$ satisfying $p_1(f_e((x_i, z_1)))=p_1(f_e((x_i, z_2)))$ for every $x_i \in V(G)$ where p_1 is the first projection of the specified vertex.

REMARK: Suppose that function σ satisfies condition 1. Then σ is an automorphism on G. We call such σ an *inversion* of G^*.

Let P_n be the path graph of n vertices with $V(P_n)=\{x_0, x_1, \ldots, x_{n-1}\}$ and $E(P_n)=\{(x_i, x_{i+1}) \mid 0 \le i \le n-2\}$. We consider a spanning supergraph P_n^* of P_n defined by $E(P_n^*)=E(P_n) \cup \{(x_i, x_{n-i-1}) \mid 0 \le i \le \lfloor \frac{n}{2} \rfloor - 1\}$. We illustrate P_4^* and $(P_4^*, P_4) \oplus (C_2, K_2)$ in Figure 2.12. Let σ be a function from $V(P_n)$ into $V(P_n)$ defined by $\sigma(x_i)=x_{n-i+1}$ for every i. It can be easily verified that σ satisfies condition (1). We use $(P_5^*, P_5) \oplus (C_2, K_2)$ for illustration. Figures 2.13a and 2.13b show P_5^* and $(P_5^*, P_5) \oplus (C_2, K_2)$. In Figure 2.13c, we illustrate $(P_5^*, P_5) \oplus (C_2, K_2)$, where the vertices are labeled according to the function h, and the the graph isomorphic to $P_5 \times K_2$ is shown by thick lines.

Let $e=(x_i, x_{i+1})$ be an edge of P_n. Let f_e be the function defined by $f_e((x_j, z_k))=(x_{n-i+j-1}, z_k)$ for $1 \le j \le i$ and $f_e((x_j, z_k))=(x_{j-i-1}, z_{3-k})$ for $i < j \le n$.

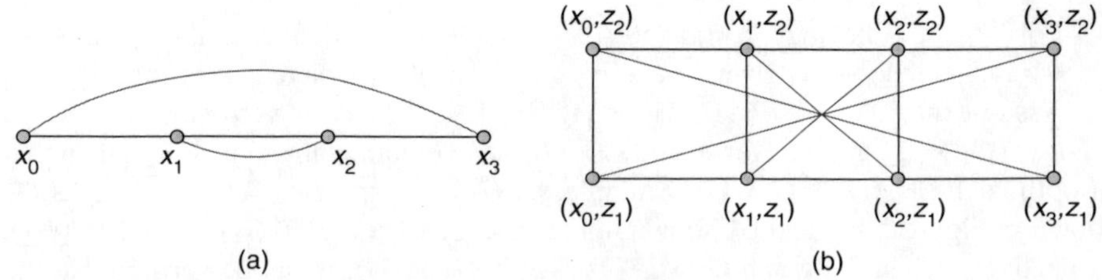

FIGURE 2.12 (a) The graph P_4^* and (b) the graph $(P_4^*, P_4) \oplus (C_2, K_2)$.

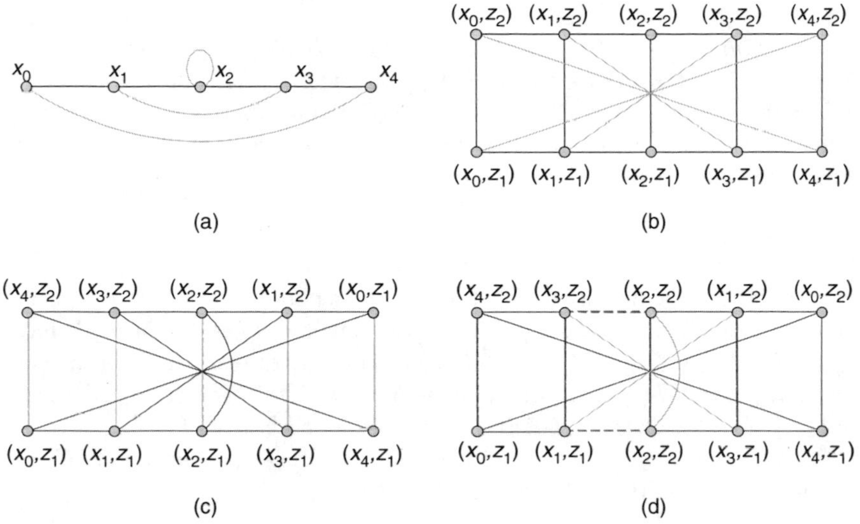

FIGURE 2.13 (a) The graph P_5^*, (b) the graph $(P_5^*, P_5) \oplus (C_2, K_2)$, (c) the function h satisfies condition 1, and (d) the function f_e satisfies condition 2 where $e = (x_1, x_2)$.

It is observed that the function f_e is an isomorphism from $P_n \times K_2$ into a subgraph of $((P_n^*, P_n) \oplus (C_2, K_2)) - Y_e$ such that $p_1(f((x_i, z_1))) = p_1(f((x_i, z_2)))$ for every $x_i \in V(P_n)$. Thus, the function f_e satisfies condition 2. In Figure 2.13d, we illustrate the case of $n = 5$ and $e = (x_1, x_2)$ where the vertices are labeled according to the function f_e, and the dark lines represent the graph isomorphic to $P_5 \times K_2$. Hence, P_n^* is FG(P_n), which is stated in the following lemma.

LEMMA 2.2 P_n^* *is FG*(P_n).

Let us consider the example illustrated in Figure 2.13d. Identifying $f_e(x_i, z_1)$ and $f_e(x_i, z_2)$ specified in condition 2, we obtain a graph that can tolerate fault on edge (x_1, x_2). It is not surprising that the faithful graph P_n^* is 1-EFT(P_n). To generalize this result, let G be an arbitrary graph, and G^* be FG(G). Then $(G^*, G) \oplus (C_2, K_2)$ is 1-EFT($G \times K_2$). Obviously, the number of edges in $(G^*, G) \oplus (C_2, K_2) - X$ is at most $|E(G \times K_2)| + 2(|E(G^*)| - |E(G)|) - |V(G)|$. Because the function h is an isomorphism from $G \times K_2$ into a subgraph of $(G^*, G) \oplus (C_2, K_2) - X$, it follows that $2(|E(G^*)| - |E(G)|) - |V(G)| \geq 0$, that is, $|E(G^*)| - |E(G)| \geq \lceil \frac{|V(G)|}{2} \rceil$. Let f_e be a function satisfying condition 2. The function $g_e : V(G) \to V(G)$ defined by $g_e(x_i) = p_1(f_e((x_i, z_1)))$ is called an *e-rotation* of G^*. Obviously, g_e induces an isomorphism from G into $G^* - e$. As a summary, we have the following lemma.

LEMMA 2.3 Any faithful graph G^* with respect to G is 1-EFT(G). Moreover, $|E(G^*)| - |E(G)| \geq \lceil \frac{|V(G)|}{2} \rceil$ holds for any faithful graph G^* with respect to G.

Because any faithful graph G^* with respect to G is also 1-EFT(G), we call G^* a *faithful 1-EFT(G)*. Let G^* be the spanning supergraph of G with $E(G^*) = E({}_2G) \uplus \{(x_i, x_i) \mid x_i \in V(G)\}$. Let σ be the identity function defined on $V(G)$, and f_e be the identity function defined on $V(G \times K_2)$ for every $e \in E(G)$. With these functions, it is easy to verify that G^* is a faithful graph with respect to G. We have the following lemma.

LEMMA 2.4 Any graph has a faithful supergraph.

The question whether any 1-EFT(G) is FG(G) naturally arises. Let us consider G to be the n-path graph P_n. The n-cycle C_n is a spanning supergraph of P_n defined by $E(C_n) = E(P_n) \cup \{(x_1, x_n)\}$. Harary and Hayes [145] point out that C_n is an optimal 1-EFT(P_n) graph. By Lemma 2.3, C_n is not FG(P_n) if $n \geq 3$. Therefore, C_n is FG(P_n) if and only if $n = 2$. Thus any 1-EFT(G) graph is not necessarily FG(G).

Assume that n is a positive even integer. Let us construct a supergraph C_n^* of C_n as follows: $E(C_n^*) = E(C_n) \cup \{(x_i, x_{i+\frac{n}{2}}) \mid 1 \leq i \leq n/2\}$.

LEMMA 2.5 Assume that n is a positive even integer n. C_n^* is FG(C_n) if and only if $n \geq 4$.

Proof: Note that $C_2 = {}_2K_2$ and $C_2^* = {}_3K_2$. Thus, there are two parallel edges in $C_2 \times K_2$, but there is no parallel edge in $(C_2^*, C_2) \oplus (C_2, K_2) - Y_e$ for any $e \in E(C_2)$. Hence, there is no f_e that satisfies condition (2). Thus, C_2^* is not FG(C_2).

Now, we discuss the case $n \geq 4$. Let σ be the function from $V(C_n)$ into $V(C_n)$ defined by $\sigma(x_i) = x_{i+\frac{n}{2}(\mathbf{mod}\, n)}$ for every i. It can be observed that σ satisfies condition 1. (See Figure 2.14a for the case $n = 4$; the vertices are labeled according to the function h.) Choose an arbitrary edge e from C_n; say, $e = (x_{\frac{n}{2}}, x_{\frac{n}{2}+1})$. Consider the function $f_e : C_n \times K_2 \to (C_n^*, C_n) \oplus (C_2, K_2)$ given by $f_e((x_j, z_k)) = (x_j, z_k)$ if $1 \leq j \leq n/2$, and $f_e((x_j, z_k)) = (x_{\frac{3n}{2}-j-1}, z_{3-k})$ otherwise. It follows that f_e satisfies condition 2. (See Figure 2.14b for the case $n = 4$ with $e = (x_2, x_3)$; the vertices are labeled according to f_e.) Since $(C_n^*, C_n) \oplus (C_2, K_2)$ is vertex-transitive, we can always find $f_{e'}$ for every edge $e' \in E(C_n)$ that satisfies condition 2. Hence, C_n^* is FG(C_n). □

THEOREM 2.10 Assume that G_i^* is FG(G_i) for $i = 1, 2$. The graph $(G_1^*, G_1) \oplus (G_2^*, G_2)$ is FG($G_1 \times G_2$). In other words, let $W = \{(G^*, G) \mid G^* \text{ is FG}(G)\}$. Then W is closed under the operation $\otimes$.

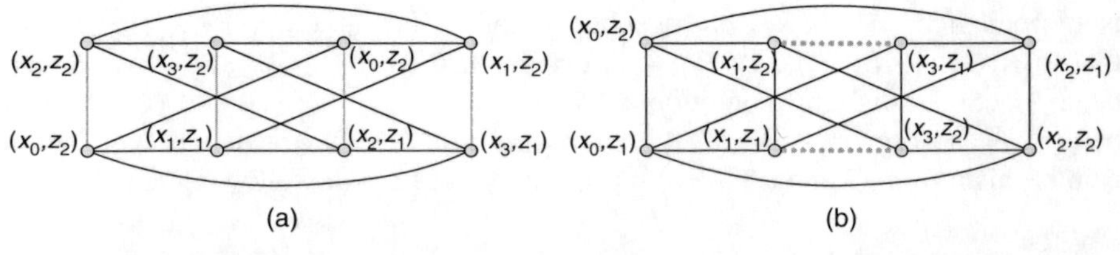

FIGURE 2.14 The graph $(C_4^*, C_4) \oplus (C_2, K_2)$. (a) The function h satisfies condition 1 and (b) the function f_e satisfies condition 2, where $e = (x_1, x_2)$.

Proof: Let $V_1=\{x_1,x_2,\ldots,x_m\}$ and $V_2=\{y_1,y_2,\ldots,y_n\}$. Since G_i^* is FG(G_i), G_i^* has an inversion σ_i for $i=1,2$. Let the function $\sigma : V(G_1 \times G_2)\to V(G_1\times G_2)$ defined by $\sigma((x_r,y_s))=(\sigma_1(x_r),\sigma_2(y_s))$. Obviously, σ is a one-to-one mapping of $V(G_1\times G_2)$. Since σ_i is an automorphism on G_i for $i=1,2$, it follows that $(x_i,x_j)\in E(G_1)$ implies $(\sigma_1(x_i),\sigma_1(x_j))\in E(G_1)$ and that $(y_k,y_l)\in E(G_2)$ implies $(\sigma_2(y_k),\sigma_2(y_l))\in E(G_2)$. Suppose that $((x_i,y_k),(x_i,y_l))\in E(G_1\times G_2)$. It follows that $(\sigma((x_i,y_k)),\sigma((x_i,y_l)))\in E(G_1\times G_2)$. Similarly, we have $(\sigma((x_i,y_k)),\sigma((x_j,y_k)))\in E(G_1\times G_2)$ if $((x_i,y_k),(x_j,y_k))\in E(G_1\times G_2)$. Thus, σ is an automorphism on $G_1\times G_2$ and satisfies condition 1.

Let e be an edge of $G_1\times G_2$. Without loss of generality, we assume that $e=((x_i,y_j),(x_k,y_j))$ where $e'=(x_i,x_k)$ is an edge of G_1. Let $f_{e'}$ be a function from $G_1\times K_2$ into a subgraph of $((G_1^*,G_1)\oplus(C_2,K_2))-Y_{e'}$ such that $p_1(f_{e'}((x_i,z_1)))=p_1(f_{e'}((x_i,z_2)))$ for every $x_i\in V_1$. We define $f_e : V_1\times V_2\times V(K_2)\to V_1\times V_2\times V(K_2)$ as follows: $f_e((x_r,y_s,z_t))=(x_u,y_v,z_w)$ where $(x_u,z_w)=f_{e'}(x_r,z_t)$, and $y_v=y_s$ if $z_w=z_t$ and $y_v=\sigma_2(y_s)$ otherwise.

For every $y_s\in V_2$, the image of f_e for $V_1\times\{y_s\}\times V(K_2)$ is either

$$\{(x_u,y_v,z_w)\mid (x_u,z_w)=f_{e'}(x_r,z_t),\ y_v=y_s\in V_2,\text{and } x_r\in V_1, z_t\in V(K_2), z_w=z_t\}\text{ or}$$
$$\{(x_u,y_v,z_w)\mid (x_u,z_w)=f_{e'}(x_r,z_t),\ y_v=\sigma_2(y_s),\text{ and } x_r\in V_1, z_t\in V(K_2), z_w\neq z_t\}$$

Since the function $f_{e'}$ satisfies condition 2, f_e induces an isomorphism from $V_1\times V(K_2)$ into its image in both cases.

For every $(x_r,z_t)\in V_1\times V(K_2)$ the image of f_e for $\{x_r\}\times V_2\times\{z_t\}$ is either

$$\{(x_u,y_v,z_w)\mid (x_u,z_w)=f_{e'}(x_r,z_t), y_v=y_s,\text{ and } y_s\in V_2, z_w=z_t\}\quad\text{or}$$
$$\{(x_u,y_v,z_w)\mid (x_u,z_w)=f_{e'}(x_r,z_t), y_v=\sigma_2(y_s),\text{ and } y_s\in V_2, z_w\neq z_t\}$$

Since the function σ_2 is an automorphism on V_2, f_e induces an isomorphism from V_2 into its image in both cases.

From this discussion, f_e induces an isomorphism from $V_1\times V_2\times V(K_2)$ into a subgraph of $(G_1^*,G_1)\oplus(G_2^*,G_2)\oplus(C_2,K_2)-\{((x_r,y_s,z_1),(x_r,y_s,z_2))\mid x_r\in V_1,y_s\in V_2\}$. Furthermore, $p_{1,2}(f_e((x_r,y_s,z_1)))=p_{1,2}(f_e((x_r,y_s,z_2)))$ where $p_{1,2}(x,y,z)=(x,y)$. Hence, f_e satisfies condition 2. Thus, $(G_1^*,G_1)\oplus(G_2^*,G_2)$ is FG($G_1\times G_2$). □

COROLLARY 2.4 *Assume that G_i^* is a faithful graph of G_i for $i=1,2$. The graph $(G_1^*,G_1)\oplus(G_2^*,G_2)$ is 1-EFT($G_1\times G_2$). Furthermore, $eft_1(G_1\times G_2)\le 2(|E(G_1^*)|-|E(G_1)|)(|E(G_2^*)|-|E(G_2)|)$.*

Proof: By Theorem 2.10 and Lemma 2.3, $(G_1^*,G_1)\oplus(G_2^*,G_2)$ is 1-EFT($G_1\times G_2$). Moreover, $|E((G_1^*,G_1)\oplus(G_2^*,G_2))|-|E(G_1\times G_2)|=2(|E(G_1^*)-E(G_1)|)(|E(G_2^*)|-|E(G_2)|)$. The corollary is proved. □

COROLLARY 2.5 *Assume that G_i^* is FG(G_i) for $i=1,2$, and $e'=(x_i,x_k)$ is any edge of G_1. Let f: $V(G_1\times G_2)\to V(G_1\times G_2)$ be a function given by $f((x_r,y_s))=(x_u,y_v)$ where $x_u=g_{e'}(x_r)$ (i.e., e'-rotation of G_1^*), and $y_v=y_s$ if $p_2(f_{e'}((x_r,z_t)))=z_t$,* *and* $y_v = \sigma_2(y_s)$ *otherwise. Then the function f induces an isomorphism from* $G_1 \times G_2$ *into a subgraph of* $(G_1^*, G_1) \oplus (G_2^*, G_2) - e$ *for any edge* $e = ((x_i, y_j), (x_k, y_j))$ *with* $y_j \in V(G_2)$.

COROLLARY 2.6 *Assume that* $m_i \geq 4$ *is a positive even integer for all i. Then the graph* $C^*(m_1, m_2, \ldots, m_n)$ *is an optimal 1-EFT graph with respect to* $C(m_1, m_2, \ldots, m_n)$. *Moreover,* $eft_1(C(m_1, m_2, \ldots, m_n)) = 1/2 \prod_{i=1}^{n} m_i$.

It would be interesting to know $eft_1(C(m_1, m_2, \ldots, m_n))$ with some m_i being odd integer.

2.5 *k*-EDGE FAULT-TOLERANT DESIGNS FOR HYPERCUBES

The hypercube Q_n can be viewed as a vector space over the field with two elements 0 and 1. In this case, any vertex of Q_n can be viewed as an n-dimensional binary vector. Let $u = (u_1, u_2, \ldots, u_n)$ be a vertex in the hypercube Q_n. The *Hamming weight* of u is the cardinality of $\{i \mid u_i = 1\}$. Let $u = (u_1, u_2, \ldots, u_n)$ and $v = (v_1, v_2, \ldots, v_n)$ be two vertices of Q_n. The *Hamming distance* between u and v, denoted by $h(u, v)$, is defined as the cardinality of $\{i \mid u_i \neq v_i\}$.

An optimal 1-EFT(Q_n) graph G^* has been proposed, which is given by adding to Q_n the set of edges $\{(u, v) \mid h(u, v) = n\}$. This 1-EFT($Q_n$) graph is called a *folded hypercube* in Ref. 96. Then the question whether k-EFT(Q_n) graphs for $k \geq 2$ can be derived naturally arises.

Yamada, Yamamoto, and Ueno [347] used a vector-space approach to develop k-EFT(Q_n) graphs for $k \geq 1$. Assume that B is any $m \times n$ matrix over the Galois field $GF(2)$ and R is a proper subset of $\{1, 2, \ldots, m\}$. We use $B(\overline{R})$ to denote the matrix obtained from B by deleting those rows with indices in R. Let k be any positive integer. Suppose that B is an $m \times n$ matrix such that the rank of $B(\overline{R})$ is n for any $|R| \leq k$. With this matrix B, we can build a graph $G_B = (V_n, E_B)$ where $V_n = V(Q_n)$ and any vertex $v \in V_n$ is joined with u if and only if $u = v + r^i$ where r^i is the ith row vector of B. Obviously, the degree of any vertex v in G_B is m. We call the edge joining v to $v + r^i$ be of class i. Assume that E' is a subset of E_B with $|E'| \leq k$. Let $R = \{i \mid e$ is an edge in E' and e is of class $i\}$. Then $|R| \leq k$. Hence, the rank of $B(\overline{R})$ is n. Therefore, we can choose n linearly independent rows from $B(\overline{R})$. Then all the edges of classes in $B(\overline{R})$ induce a graph isomorphic to Q_n. Hence, G_B is a k-EFT(Q_n). We call the corresponding graph G_B a *linear k-EFT*(Q_n).

Let $EFT_L(n, k)$ be the set of matrix B such that G_B is a linear k-EFT(Q_n). Obviously, the matrix $B \in EFT_L(n, k)$ with the smallest number of rows will derive a linear k-EFT(Q_n) graph with the least number of edges among all linear k-EFT(Q_n) graph. Thus, we say a matrix $B \in EFT_L(n, k)$ with the smallest number of rows is an *optimum linear* k-EFT(Q_n) and we use $eft_L(n, k)$ to denote $(m - n)$ where m is the number of rows in any optimum linear k-EFT(Q_n).

The concept of linear k-EFT(Q_n) had already been used before the formulation proposed by Yamada et al. [347]. Bruck et al. [34] used this approach to construct 1-EFT(Q_n). Assume that B is a matrix in $EFT_L(n, k)$. Obviously, the rank of B is n. By changing coordinates, we can transform B into $\left[\begin{smallmatrix} I_n \\ D \end{smallmatrix}\right] = \left[\begin{smallmatrix} I_n \\ d^1, d^2, \ldots, d^n \end{smallmatrix}\right]$. Shih and Batcher

[284] proved that any such B in $EFT_L(n,k)$ satisfies the following two conditions: (1) $w(d^j) \geq k$ for every $1 \leq j \leq n$; that is, the Hamming weight of each d^i is at least k and (2) $w(d^i + d^j) \geq k-1$ for every $1 \leq i < j \leq n$; that is, the Hamming distance $h(d^i, d^j)$ between d^i and d^j is at least $k-1$. Then they employed an *ad hoc* program to verify edge fault tolerance and thus generate optimal linear k-EFT graphs for $k = 2, 3$ and $n \leq 26$.

Sung et al. [293] show that these two conditions are actually the necessary and sufficient conditions for matrix $B \in EFT_L(n,k)$ with $k=2$. They also show that $eft_L(n,3) = eft_L(n,2)+1$ and present a construction scheme for optimal linear k-EFT(Q_n) for $k = 2, 3$. They also conjectured that $eft_L(n,5) = eft_L(n,4)+1$. Ho et al. [153] extend the idea in Ref. 347 to present some necessary and sufficient conditions for matrix B in $EFT_L(n,k)$. With this result, Ho et al. [153] prove that $eft_L(n,k+1) \geq eft_L(n,k)+1$ and $eft_L(n,k+1) = eft_L(n,k)+1$ if k is even.

Assume that B is any $m \times n$ matrix and R is any proper subset of $\{1, 2, \ldots, m\}$. Using column vectors, we can write B as $[c^1, c^2, \ldots, c^n]$ and $B(\overline{R})$ as $[c_{\overline{R}}^1, c_{\overline{R}}^2, \ldots, c_{\overline{R}}^n]$. Let h be a column or row vector. We use h^{tr} to denote the transpose of h.

THEOREM 2.11 A matrix B is in $EFT_L(n,k)$ if and only if the Hamming weight of any column vector—that is, a summation of t different columns of B with $1 \leq t \leq n$—is greater than k; that is, $w(c^{i_1} + c^{i_2} + \cdots + c^{i_t}) > k$ for any $1 \leq i_1 < i_2 < \cdots < i_t \leq n$ and $1 \leq t \leq n$.

Proof: Assume that B is a matrix such that the Hamming weight of some column vector—that is, a summation of t different columns of B with $1 \leq t \leq n$—is at most k. Then there exist some $1 < i_1 < i_2 < \cdots < i_t \leq n$ and $1 \leq t \leq n$ satisfying $w(c^{i_1} + c^{i_2} + \cdots + c^{i_t}) \leq k$. Let $h = (h_1, h_2, \ldots, h_m)^{tr} = c^{i_1} + c^{i_2} + \cdots + c^{i_t}$. Let $R = \{i \mid h_i = 1\}$. Obviously, $|R| = w(h) \leq k$ and $c_{\overline{R}}^{i_1} + c_{\overline{R}}^{i_2} + \cdots + c_{\overline{R}}^{i_t} = 0$. Therefore, $\{c_{\overline{R}}^{i_1}, c_{\overline{R}}^{i_2}, \ldots, c_{\overline{R}}^{i_t}\}$ is linearly dependent. Hence, the rank of $B(\overline{R})$ is less than n. Therefore, $B \notin EFT_L(n,k)$.

On the other hand, assume that B is a matrix such that the Hamming weight of any column vector—that is, a summation of t different columns of B with $1 \leq t \leq n$—is greater than k. Let R be any subset of $\{1, 2, \ldots, m\}$ with $|R| \leq k$. Thus, any nontrivial linear combination of at most n different columns of $B(\overline{R})$ is not zero. Hence, the rank of $B(\overline{R})$ is n. Therefore, $B \in EFT_L(n,k)$. □

THEOREM 2.12 $eft_L(n,k+1) \geq eft_L(n,k)+1$. Moreover, $eft_L(n,k+1) = eft_L(n,k)+1$ if k is even.

Proof: Assume that B^* is a matrix in $EFT_L(n,k+1)$. Let B be any matrix obtained by deleting any row from B^*. Obviously, B is a matrix in $EFT_L(n,k)$. Hence, $eft_L(n,k+1) \geq eft_L(n,k)+1$.

Let k be an even integer. Assume that $B = (b_{i,j})$ is any $m \times n$ matrix in $EFT_L(n,k)$. Form a new matrix $B' = (b'_{ij})$ from B by adding a new row $(b'_{m+1,1}, b'_{m+1,2}, \ldots, b'_{m+1,n})$ where $b'_{m+1,j} = \sum_{i=1}^{m} b_{i,j}$. In other words, the new row is the even parity check row of B. Thus, the Hamming weight of any column in B' is even. Therefore, the Hamming weight of any linear combination of column vectors of B' is even. Let

$h = (h_1, h_2, \ldots, h_m, h_{m+1})^{\text{tr}}$ be a summation of t columns of B' with $1 \le t \le n$. We set $h' = (h_1, h_2, \ldots, h_m)^{\text{tr}}$. Obviously, $w(h) \ge w(h')$. By Theorem 2.11, $w(h') > k$. Since both $w(h)$ and k are even integers, $w(h) > k + 1$. Thus, $B' \in EFT_L(n, k+1)$ follows from Theorem 2.11. Therefore, $eft_L(n, k+1) = eft_L(n, k) + 1$ if k is even. □

Since the rank of any matrix is an invariant on changing coordinates, we can find an optimal linear k-EFT(Q_n) among all the matrices of the form $\left[\begin{smallmatrix} I_n \\ D \end{smallmatrix}\right]$.

THEOREM 2.13 Let $B = \left[\begin{smallmatrix} I_n \\ D \end{smallmatrix}\right] = \left[\begin{smallmatrix} & I_n & \\ d^1, d^2, & \ldots, & d^n \end{smallmatrix}\right] = [c^1, c^2, \ldots, c^n] = \begin{bmatrix} r^1 \\ r^2 \\ \vdots \\ r^m \end{bmatrix}$. Then B is a matrix in $EFT_L(n, k)$ if and only if the Hamming weight of any column vector that is a summation of t different columns of D with $1 \le t \le k$ is greater than $k - t$.

Proof: It is observed that $w(c^{i_1} + c^{i_2} + \cdots + c^{i_j}) = w(d^{i_1} + d^{i_2} + \cdots + d^{i_j}) + j$ for any $1 \le j \le n$ and $1 \le i_1 < i_2 < \cdots < i_j \le n$.

Suppose that the Hamming weight of some column vector—that is, a summation of t different columns of D with $1 \le t \le k$—is at most $k - t$. Then there exist some $1 \le i_1 < i_2 < \cdots < i_t \le n$ and $1 \le t \le k$ satisfying $w(c^{i_1} + c^{i_2} + \cdots + c^{i_j}) \le k$. By Theorem 2.11, $B \notin EFT_L(n, k)$.

On the other hand, assume that the Hamming weight of any column vector—that is, a summation of t different columns of D with $1 \le t \le k$—is greater than $k - t$. Assume that R is any subset of $\{1, 2, \ldots, m\}$ with $|R| \le k$. Let $I_R = \{t \in R \mid t \le n\}$ and $|I_R| = s$. Then $s \le \min\{k, n\}$. Let $R^* = R - I_R$.

Assume that $s = 0$. Then $r^1, r^2, \ldots, r^n$ form n independent rows in $B(\overline{R})$. Thus, the rank of $B(\overline{R})$ is n. Assume that $0 < s \le n$. Let B^* be the submatrix of B formed by those columns with their indices not in I_R. By our assumption, any column vector that is a summation of t different columns of B with $1 \le t \le s$ is greater than k. Therefore, any column vector that is a summation of t different columns of $B(\overline{R})$ with $1 \le t \le s$ and with indices not in I_R is greater than 0. Thus, any nontrivial linear combination of at most s different columns of $B^*(\overline{R^*})$ is not a zero vector. Therefore, the rank of $B(\overline{R^*})$ is s. Note that the row rank of any matrix equals its column rank. We can find s independent rows, $r^{i_1}, r^{i_2}, \ldots, r^{i_s}$, that span all the row vectors of $B^*(\overline{R^*})$. Obviously, the rows of $\{r^{i_1}, r^{i_2}, \ldots, r^{i_s}\} \cup \{r^t \mid t \notin R \text{ and } 1 \le t \le n\}$ in $B(\overline{R})$ forms n independent row vectors. Hence, the rank of $B(\overline{R})$ is n. Therefore, B is in $EFT_L(n, k)$. □

Obviously, I_n is an optimal linear 0-EFT(Q_n). With Theorem 2.12, we obtain an optimal linear 1-EFT(Q_n). By Theorem 2.13, any $B_{m \times n}$ in $EFT_L(n, 2)$ of the form $\left[\begin{smallmatrix} I_n \\ D \end{smallmatrix}\right]$ satisfies $\binom{m-n}{2} + \binom{m-n}{3} + \cdots + \binom{m-n}{m-n} \ge n$. Assume that m and n satisfy this inequality. We can choose any n different $1 \times (m - n)$ columns with Hamming weights of at least 2 to form the matrix D. Again, by Theorem 2.13, the matrix B is in $EFT_L(n, 2)$. Therefore, $eft_L(n, 2)$ is the smallest integer r that satisfies $\binom{r}{2} + \binom{r}{3} + \cdots + \binom{r}{r} \ge n$. By Theorem 2.12, we obtain an optimal linear 3-EFT(Q_n). However, we have difficulty in constructing the optimal linear k-EFT(Q_n) with $k \ge 4$.

3 Distance and Diameter

3.1 INTRODUCTION

Let u and v be two vertices of a graph G. Suppose that there exists a path from u to v in G. The *distance* from u to v, written $d_G(u, v)$ or simply $d(u, v)$, is the least length among all paths from u to v. Suppose that there exists no path from u to v in G. Then $d(u, v) = \infty$.

THEOREM 3.1 Assume that G is a graph. Then

1. $d(u, v) \geq 0$ and $d(u, v) = 0$ if and only if $u = v$
2. $d(u, v) = d(v, u)$ for all $u, v \in V(G)$
3. $d(u, v) \leq d(u, w) + d(w, v)$ for all $u, v, w \in V(G)$ (*the triangle inequality*)

Proof: The proof of (1) and (2) are obvious. We need to prove the third statement. Let u, v, and w be vertices of G. Let P be the shortest path from u to w and Q the shortest path from w to v in G. Then P followed by Q is a path W from u to v. Thus, $d(u, v) \leq d(u, w) + d(w, v)$. □

The *eccentricity* of a vertex u, written $\epsilon(u)$, is the maximum of its distances to other vertices; that is, $\epsilon(u) = \max\{d(x, u) \mid x \in V(G) \text{ and } x \neq u\}$. In the graph G, the *diameter* $D(G)$ and the *radius* $r(G)$ are the maximum and minimum of the vertex eccentricities, respectively. In other words, $D(G) = \max\{\epsilon(u) \mid u \in V(G)\}$ and $r(G) = \min\{\epsilon(u) \mid u \in V(G)\}$. The *center* of G is the subgraph of G induced by the vertices of minimum eccentricity.

COROLLARY 3.1 $r(G) \leq D(G) \leq 2r(G)$ *for every graph* G.

Proof: By the definitions of radius and diameter, $r(G) \leq D(G)$. To prove $D(G) \leq 2r(G)$, let x and y be two vertices of G such that $d(x, y) = D(G)$. Let z be any vertex of G such that $\epsilon(z) = r(G)$. By the triangle inequality, we have $D(G) = d(x, y) \leq d(x, z) + d(z, y) \leq 2r(G)$. □

The distance of any two vertices of a graph (or digraph) can be computed by the breadth-first search algorithm that is discussed in Chapter 4. With this method, we can determine the radius and the diameter of a graph. However, we can more easily determine these parameters in some particular graphs.

3.2 DIAMETER FOR SOME INTERCONNECTION NETWORKS

Network topology is a crucial factor for an interconnection network, as it determines the performance of the network. Many interconnection network topologies have been proposed for the purpose of connecting hundreds or thousands of processing elements

[220]. Network topology is always represented by a graph, where vertices represent processors and edges represent links between processors. The diameter is an important parameter for interconnection networks. In this section, we discuss about the diameter of some interconnection networks.

Among all network topologies, the binary hypercube, Q_n, is one of the most popular one. Let $\mathbf{u} = u_{n-1}u_{n-2}\ldots u_1u_0$ and $\mathbf{v} = v_{n-1}v_{n-2}\ldots v_1v_0$ be two n-bit strings. Let $H(\mathbf{u}, \mathbf{v})$ denote the indices set $\{i \mid u_i \neq v_i\}$. The *Hamming distance* between $\mathbf{u}$ and $\mathbf{v}$, $h(\mathbf{u}, \mathbf{v})$, is $|H(\mathbf{u}, \mathbf{v})|$. For example, $h(11000, 11110) = 2$. The *n-dimensional hypercube* consists of all n-bit strings as its vertices, and two vertices $\mathbf{u}$ and $\mathbf{v}$ are adjacent if and only if $h(\mathbf{u}, \mathbf{v}) = 1$. $(\mathbf{u}, \mathbf{v})$ is an edge in $E(Q_n)$ of *dimension i* if the ith bit of $\mathbf{u}$ is different from that of $\mathbf{v}$. It is easy to check whether $d_{Q_n}(\mathbf{u}, \mathbf{v}) = h(\mathbf{u}, \mathbf{v})$.

Actually, we can easily find a path of length $h(\mathbf{u}, \mathbf{v})$ by applying the following algorithm: route $\mathbf{u}$ to $\mathbf{v}$ by changing the rightmost differing bit iteratively. For example, 11000, 11010, 11110 forms a path of length 2 joining 11000 to 11110. Hence, we can easily conclude that $D(Q_n) = n$.

However, Q_n does not make the best use of its hardware in the following sense: given $N = 2^n$ vertices and $nN/2$ links, it is possible to fashion networks with smaller diameters than the hypercube's diameter n. There are many variants proposed to lower the diameter, such as the twisted cube TQ_n [1,146], the crossed cube CQ_n [92–94], and the Möbius cube MQ_n [76]. The diameters of TQ_n, CQ_n, and MQ_n are around $n/2$. In the following discussion, we describe the twisted cubes.

The n-dimensional twisted cube, denoted by TQ_n, is a variant of an n-dimensional hypercube Q_n. TQ_n has the same number of vertices and edges as Q_n. We restrict the following discussion on TQ_n to the cases where n is odd. Assume that $n = 2m + 1$. To form the twisted cube, we remove some links from the hypercube and replace them with links that span two dimensions in such a manner that the total number of links $(nN/2)$ is conserved. To be precise, let $\mathbf{u} = u_{n-1}u_{n-2}\ldots u_1u_0$ be any vertex in TQ_n. We define the parity function $P_i(\mathbf{u}) = u_i \oplus u_{i-1} \oplus \cdots \oplus u_0$, where $\oplus$ is the exclusive-or operation. Suppose that $P_{2j-2}(\mathbf{u}) = 0$ for some $1 \leq j \leq m$. We divert the edge on the $(2j-1)$th dimension to vertex $\mathbf{v}$ such that $v_{2j}v_{2j-1} = \bar{u}_{2j}\bar{u}_{2j-1}$ and $v_i = u_i$ for $i \neq 2j$ or $2j-1$. Such diverted edges are called *twisted edges*. TQ_3 and TQ_5 are shown Figure 3.1.

We may formally define the term of *twisted cube* recursively as follows: a twisted 1-cube, TQ_1, is a complete graph with two vertices, 0 and 1. Assume that n is an odd integer and $n \geq 3$. We decompose vertices of TQ_n into four sets $S^{0,0}, S^{0,1}, S^{1,0}$ and $S^{1,1}$ where $S^{i,j}$ consists of those vertices u with $u_{n-1} = i$ and $u_{n-2} = j$. For each $(i, j) \in \{(0,0), (0,1), (1,0), (1,1)\}$, the induced subgraph of $S^{i,j}$ in TQ_n is isomorphic to TQ_{n-2}. Edges that connect these four subtwisted cubes are described as follows: any vertex $u_{n-1}u_{n-2}\ldots u_1u_0$ with $P_{n-3}(\mathbf{u}) = 0$ is connected to $\bar{u}_{n-1}\bar{u}_{n-2}u_{n-3}\ldots u_0$ and $\bar{u}_{n-1}u_{n-2}u_{n-3}\ldots u_0$; and $u_{n-1}\bar{u}_{n-2}u_{n-3}\ldots u_0$ and $\bar{u}_{n-1}u_{n-2}u_{n-3}\ldots u_0$, if $P_{n-3}(\mathbf{u}) = 1$.

We define the *0th double bit* of vertex address $\mathbf{u}$ as the single bit u_0, and the *jth double bit* as $u_{2j}u_{2j-1}$. Let $\mathbf{u}$ and $\mathbf{v}$ be any two vertices of TQ_n. The *double Hamming distance* of $\mathbf{u}$ and $\mathbf{v}$, denoted by $h_2(\mathbf{u}, \mathbf{v})$, is the number of different double bits between $\mathbf{u}$ and $\mathbf{v}$. Obviously, $d_{TQ_n}(\mathbf{u}, \mathbf{v}) \geq h_2(\mathbf{u}, \mathbf{v})$.

We can find the shortest path between any two vertices using the algorithm proposed in Ref. 1. Let $\mathbf{u}$ and $\mathbf{v}$ be two vertices of TQ_n. Let $\mathbf{z} = \mathbf{u}$. The basic strategy of

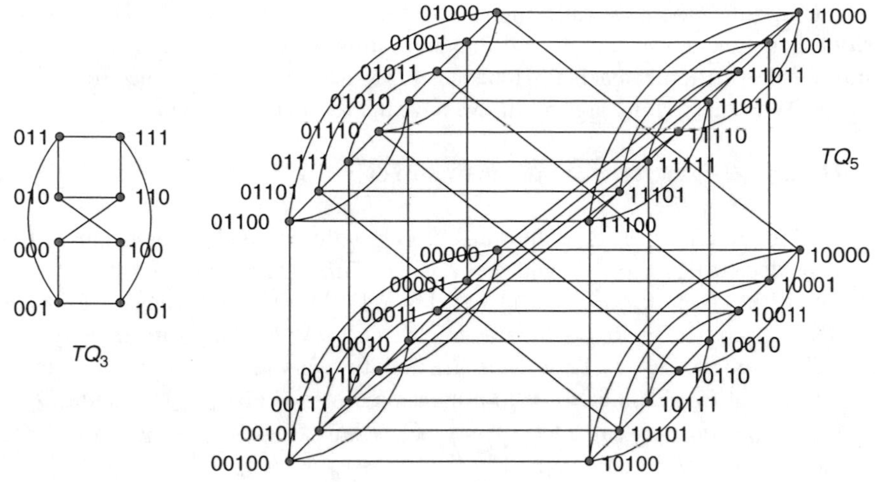

FIGURE 3.1 TQ_3 and TQ_5.

the algorithm is to find recursively a neighborhood **w** of **z** that reduces $h_2(\mathbf{w}, \mathbf{v})$. More precisely, the strategy is described as follows:

ALGORITHM H

1. If $\mathbf{z} = \mathbf{v}$, then the path is determined.
2. Assume that there exist neighbors **w** of **z** such that $h_2(\mathbf{w}, \mathbf{v}) = h_2(\mathbf{z}, \mathbf{v}) - 1$. Let $\mathbf{w}'$ be such **w** that differs from **z** with the largest double bit. We reset **z** to be $\mathbf{w}'$.
3. Assume that all the neighbors **w** of **z** satisfy $h_2(\mathbf{w}, \mathbf{v}) \geq h_2(\mathbf{z}, \mathbf{v})$. Let j be the smallest index of double bits that **z** differs from **v**. Choose $\mathbf{w}'$ to be the neighbor of **z** that differs from **z** in the $2j$th bit. We reset **z** to be $\mathbf{w}'$.

We use $P_H(\mathbf{u}, \mathbf{v})$ to denote the path joining from **u** to **v** obtained by the previous algorithm. Following are two examples.

Example 3.1

Find $P_H(000100101, 111000010)$ and $P_H(000100101, 111000011)$. We list the paths move by move:

P_H(000100101, 111000010)	P_H(000100101, 111000011)
⟨000100101 →	⟨000100101 →
001000101 →	001000101 →
001000100 →	001000001 →
111000100 →	111000001 →
111000010⟩	111000011⟩

□

Note that the rightmost differing double bit is selected in step 3. The resulting parity change guarantees that all subsequent routing for the message will be by step 2 until the destination is reached. Hence, step 3 is executed at most once for a given message. With this routing algorithm, we have the following theorems.

THEOREM 3.2 The diameter of the twisted cube TQ_n is $\lceil \frac{n+1}{2} \rceil$.

Next, we briefly give the definitions of crossed cubes and Möbius cubes. Two 2-digit binary strings $\mathbf{x} = x_1x_0$ and $\mathbf{y} = y_1y_0$ are *pair-related*, denoted by $\mathbf{x} \sim \mathbf{y}$, if and only if $(\mathbf{x}, \mathbf{y}) \in \{(00, 00), (10, 10), (01, 11), (11, 01)\}$. The n-dimensional *crossed cube* CQ_n, proposed in Ref. 93, is a graph $CQ_n = (V, E)$ that is recursively constructed as follows: CQ_1 is a complete graph with two vertices labeled by 0 and 1. CQ_n consists of two identical $(n-1)$-dimension crossed cubes, CQ^0_{n-1} and CQ^1_{n-1}. The vertex $\mathbf{u} = 0u_{n-2}\ldots u_0 \in V(CQ^0_{n-1})$ and vertex $\mathbf{v} = 1v_{n-2}\ldots v_0 \in V(CQ^1_{n-1})$ are adjacent in CQ_n if and only if (1) $u_{n-2} = v_{n-2}$, if n is even; and (2) for $0 \le i < \lfloor \frac{n-1}{2} \rfloor$, $u_{2i+1}u_{2i} \sim v_{2i+1}v_{2i}$. The crossed cubes CQ_3 and CQ_4 are illustrated in Figure 3.2.

The n-dimensional Möbius cube MQ_n, proposed in Ref. 76, has 2^n vertices. Each vertex is labeled by a unique n-bit binary string as its address and has connections to n other distinct vertices. The vertex with address $\mathbf{x} = x_{n-1}x_{n-2}\ldots x_0$ connects to n other vertices $\mathbf{y_i}$, $0 \le i \le n-1$, where the address of $\mathbf{y_i}$ satisfies (1) $\mathbf{y_i} = (x_{n-1}\ldots x_{i+1}\overline{x_i}\ldots x_0)$, if $x_{i+1} = 0$; or (2) $\mathbf{y_i} = (x_{n-1}\ldots x_{i+1}\overline{x_i\ldots x_0})$ if $x_{i+1} = 1$.

From the previous definition, $\mathbf{x}$ connects to $\mathbf{y_i}$ by complementing the bit x_i if $x_{i+1} = 0$, or by complementing all bits of $x_i \ldots x_0$ if $x_{i+1} = 1$. For the connection between $\mathbf{x}$ and $\mathbf{y_{n-1}}$, we can assume that the unspecified x_n is either 0 or 1, which gives slightly different topologies. If x_n is 0, we call the network generated the *0-Möbius cube*, denoted by 0-MQ_n; and if x_n is 1, we call the network generated the *1-Möbius*

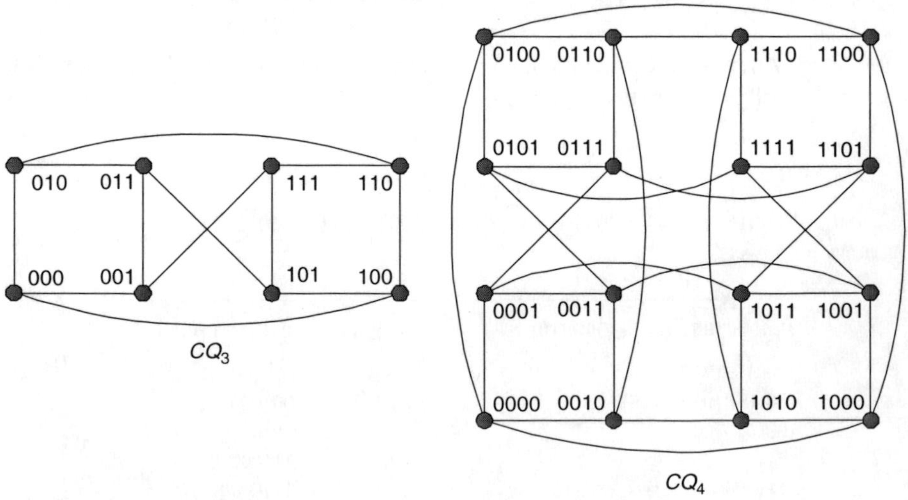

FIGURE 3.2 CQ_3 and CQ_4.

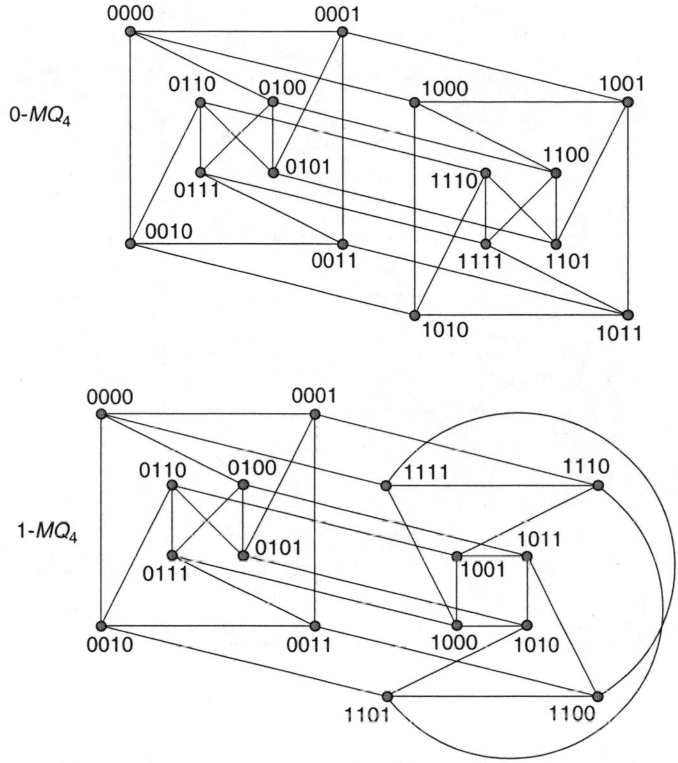

FIGURE 3.3 0-MQ_4 and 1-MQ_4.

cube, denoted by 1-MQ_n. The Möbius cubes 0-MQ_4 and 1-MQ_4 are illustrated in Figure 3.3.

3.3 SHUFFLE-CUBES

Li et al. [224] present a variant of hypercubes called the *shuffle-cube*, SQ_n. SQ_n is obtained from Q_n by changing some links of Q_n. It has a diameter of around $n/4$.

Let $\mathbf{u} = u_{n-1}u_{n-2}\ldots u_1u_0$ be an n-bit string. We use $p_j(\mathbf{u})$ to denote the *j-prefix* of $\mathbf{u}$; that is, $p_j(\mathbf{u}) = u_{n-1}u_{n-2}\ldots u_{n-j}$, and $s_i(\mathbf{u})$ the *i-suffix* of $\mathbf{u}$; that is, $s_i(\mathbf{u}) = u_{i-1}u_{i-2}\ldots u_1u_0$. Let $\mathbf{u}$ and $\mathbf{v}$ be two vertices. Let $\oplus$ denote an addition with modulo 2. We define the following four sets: $V_{00} = \{1111, 0001, 0010, 0011\}$, $V_{01} = \{0100, 0101, 0110, 0111\}$, $V_{10} = \{1000, 1001, 1010, 1011\}$, and $V_{11} = \{1100, 1101, 1110, 1111\}$. For ease of exposition, we limit our discussion to $n = 4k + 2$ for $k \geq 0$.

We recursively defined the *n-dimensional shuffle-cube*, SQ_n, as follows: SQ_2 is Q_2. For $n \geq 3$, SQ_n consists of 16 subcubes $SQ_{n-4}^{i_1i_2i_3i_4}$, where $i_j \in \{0, 1\}$ for $1 \leq j \leq 4$ and $p_4(u) = i_1i_2i_3i_4$ for all vertices u in $SQ_{n-4}^{i_1i_2i_3i_4}$. The vertices $\mathbf{u} = u_{n-1}u_{n-2}\ldots u_1u_0$

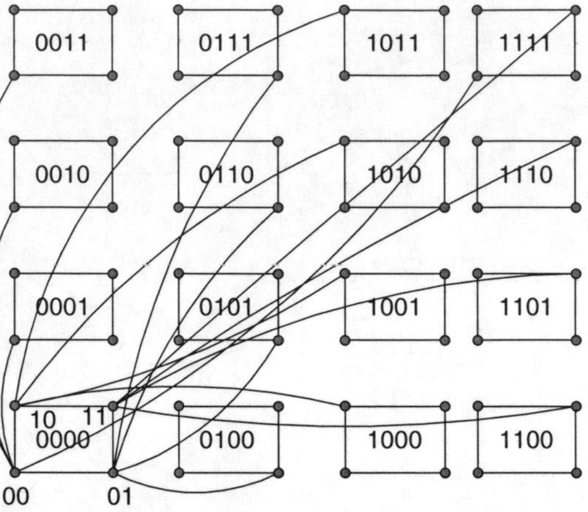

FIGURE 3.4 SQ_6.

and $\mathbf{v} = v_{n-1}v_{n-2}\ldots v_1v_0$ in different subcubes of dimension $(n-4)$ are adjacent in SQ_n if and only if

1. $s_{n-4}(\mathbf{u}) = s_{n-4}(\mathbf{v})$
2. $p_4(\mathbf{u}) \oplus p_4(\mathbf{v}) \in V_{s_2(\mathbf{u})}$

For example, the vertex 111101 in SQ_6 is linked to the following vertices in different subcubes of dimension 2: 101101, 101001, 100101, and 100001. We illustrate SQ_6 in Figure 3.4 by showing only edges incident at vertices in SQ_2^{0000} and omitting the others. It is easy to see that the degree of each vertex of SQ_n is n and the number of vertices (edges, respectively) is the same as that of Q_n.

For $1 \le j \le k$, the *jth 4-bit* of $\mathbf{u}$, denoted by $\mathbf{u}_4^j$, is defined as $\mathbf{u}_4^j = u_{4j+1}u_{4j}u_{4j-1}u_{4j-2}$. In particular, the *0*th *4-bit* of $\mathbf{u}$, $\mathbf{u}_4^0$, is defined as $\mathbf{u}_4^0 = u_1u_0$. Thus, $\mathbf{u}_4^j = \mathbf{v}_4^j$ if and only if $u_{4j+i} = v_{4j+i}$ for $-2 \ge i \ge 1$. Similar to the Hamming distance, we define the *4-bit Hamming distance* between $\mathbf{u}$ and $\mathbf{v}$, denoted by $h_4(\mathbf{u}, \mathbf{v})$, as the number of 4-bits $\mathbf{u}_4^j$ with $0 \le j \le k$ such that $\mathbf{u}_4^j \ne \mathbf{v}_4^j$, that is, $h_4(\mathbf{u}, \mathbf{v}) = |\{j \mid \mathbf{u}_4^j \ne \mathbf{v}_4^j \text{ for } 0 \le j \le k\}|$.

With the notion of $h_4(\mathbf{u}, \mathbf{v})$, we can redefine SQ_n as follows: the vertex $\mathbf{u}$ and the vertex $\mathbf{v}$ are adjacent if and only if one of the following conditions holds:

1. $\mathbf{u}_4^{j^*} \oplus \mathbf{v}_4^{j^*} \in V_{\mathbf{u}_4^0}$ for exactly one j^* satisfying $1 \le j^* \le k$ and $\mathbf{u}_4^j = \mathbf{v}_4^j$ for all $0 \le j \ne j^* \le k$
2. $\mathbf{u}_4^0 \oplus \mathbf{v}_4^0 \in \{01, 10\}$ and $\mathbf{u}_4^j = \mathbf{v}_4^j$ for all $1 \le j \le k$

For example, the ten neighbors of 1011000010 in SQ_{10} are given by 0011000010, 0010000010, 0001000010, 0000000010, 1011100010, 1011100110, 1011101010, 1011101110, 1011000000, and 1011000011. In other words, the vertex $\mathbf{u}$ is adjacent to the vertex $\mathbf{v}$ only if $h_4(\mathbf{u}, \mathbf{v}) = 1$. However, the converse is not necessarily true.

For example, $\mathbf{u} = 0000000000$ is not adjacent to $\mathbf{v} = 0000000011$, though $h_4(\mathbf{u}, \mathbf{v}) = 1$. Thus, $d(\mathbf{u}, \mathbf{v}) \geq h_4(\mathbf{u}, \mathbf{v})$ for any two vertices $\mathbf{u}, \mathbf{v}$ of SQ_n.

LEMMA 3.1 Assume that $n = 4k + 2$ with $k \geq 4$. Then $D(SQ_n) \geq \lceil \frac{n}{4} \rceil + 3$.

Proof: Let P be any path of SQ_n from $\mathbf{u}$ to $\mathbf{v}$. We can view P as a sequence of four bits changing from $\mathbf{u}$ to $\mathbf{v}$. Assume that $\mathbf{u} = u_{n-1}u_{n-2}\ldots u_1u_0 = \mathbf{u}_4^k\mathbf{u}_4^{k-1}\ldots\mathbf{u}_4^0$ with $\mathbf{u}_4^0 = 00$, $\mathbf{u}_4^1 = 1100$, $\mathbf{u}_4^2 = 1000$, $\mathbf{u}_4^3 = 0100$, and $\mathbf{u}_4^j = 0001$ if $4 \leq j \leq k$. Assume that $\mathbf{v} = v_{n-1}v_{n-2}\ldots v_1v_0$ with $v_j = 0$ for $0 \leq j < n$.

Note that the four bits 0001, 0100, 1000, and 1100 are only in V_{00}, V_{01}, V_{10}, and V_{11}, respectively. We can change any 4-bit 0001, 0100, 1000, or 1100 into 0000 in one step, only if the 0th 4-bit is 00, 01, 10, or 11, respectively. Therefore, $d(\mathbf{u}, \mathbf{v}) \geq \lceil \frac{n}{4} \rceil + 3$. Moreover, $D(SQ_n) \geq \lceil \frac{n}{4} \rceil + 3$. □

Next, we propose a routing algorithm on SQ_n. Assume that $\mathbf{u}$ and $\mathbf{v}$ are two vertices of SQ_n. We use $h_4^*(\mathbf{u}, \mathbf{v})$ to denote the number of $\mathbf{u}_4^j$ for $1 \leq j \leq k$ such that $\mathbf{u}_4^j \neq \mathbf{v}_4^j$.

3.3.1 ROUTE1(U, V)

1. Suppose that $\mathbf{u} = \mathbf{v}$. Then accept the message.
2. Find a neighbor $\mathbf{w}$ of $\mathbf{u}$ such that $h_4^*(\mathbf{w}, \mathbf{v}) = h_4^*(\mathbf{u}, \mathbf{v}) - 1$ if $\mathbf{w}$ exists. Then route into $\mathbf{w}$.
3. Suppose that there is no neighbor $\mathbf{w}$ of $\mathbf{u}$ such that $h_4^*(\mathbf{w}, \mathbf{v}) = h_4^*(\mathbf{u}, \mathbf{v}) - 1$. Then route into the neighbor $\mathbf{w}$ of $\mathbf{u}$ that changes u_1u_0 in a cyclic manner with respect to 00, 01, 11, and 10. For example, $\mathbf{w} = p_{n-2}(u)00$ if $u_1u_0 = 10$.

EXAMPLE 3.2

Let $\mathbf{u} = 0001000101001000110000$ and $\mathbf{v} = 0000000000000000000011$ be two vertices of SQ_{18}. The path obtained from **Route1(u, v)** is

0001000101001000110000,	$\underline{0000}000101001000110000$,	$0000\underline{0000}01001000110000$,
$000000000100100011000\underline{1}$,	$00000000\underline{0000}1000110001$,	$00000000000010001100\underline{1}1$,
$0000000000001000\underline{0000}11$,	$000000000000100000001\underline{0}$,	$000000000000\underline{0000}000010$,
$00000000000000000000\underline{0}0$,	$000000000000000000000\underline{1}$,	$00000000000000000000\underline{1}1$.

Note that this path is not the shortest path.

Applying the previous algorithm to any two vertices $\mathbf{u}$ and $\mathbf{v}$ on SQ_n, it is observed that we may apply step 3 at most three times to obtain a vertex $\mathbf{w}$ such that $h_4^*(\mathbf{w}, \mathbf{v}) = 0$. Hence, the algorithm will find a path—not necessarily the shortest path—of length at most $h_4^*(\mathbf{u}, \mathbf{v}) + 6$ that joins $\mathbf{u}$ to $\mathbf{v}$. Thus, $D(SQ_n) \leq \lceil \frac{n}{4} \rceil + 5$. The actual diameter of SQ_n and its generalization is discussed in Chapter 9. Actually, for any positive integer g, Li et al. [224] construct an n-dimensional generalized shuffle-cube with 2^n vertices that is n-regular and of a diameter about n/g if we consider g as a constant.

3.4 MOORE BOUND

We have seen several variations of cubes. Obviously, we prefer those variations with a lower diameter. The famous (n, d, D) problem [25] can be stated as follows: what is the largest number of vertices n that a graph (or digraph), each of whose vertices with degree (outdegree in digraph) at most d and of diameter D, can have? The following theorem gives a bound on the diameter.

THEOREM 3.3 **(Moore Bound)** Suppose that G is a graph with diameter D and maximum degree d. Then

$$n(G) \leq 1 + \frac{[(d-1)^D - 1]d}{d-2}$$

Proof: Let x_0 be a center of G. The number of vertices at distance 1 from x_0 is at most d. The number of vertices at distance 2 from x_0 is at most $d(d-1)$. The number of vertices at distance i with $i \leq D$ from x_0 is at most $d(d-1)^{i-1}$. Hence,

$$n(G) \leq 1 + d + d(d-1) + \cdots + d(d-1)^{D-1} = 1 + \frac{[(d-1)^D - 1]d}{d-2}$$

The theorem is proved. □

We have a similar result for a digraph.

THEOREM 3.4 **(Moore Bound)** Suppose that G is a digraph with diameter D and maximum degree d. Then

$$n(G) \leq \frac{d^{D+1} - 1}{d-1}$$

The (n, d, D) problem is one of the most difficult problems in graph theory [68]. It is known that only very few graphs satisfy the Moore bound. In the following discussion, we present two families of graphs (digraphs). The orders of the diameter of these families of digraphs meet the Moore bound.

Let d be an integer with $d \geq 2$. Let $X(t)$ denote the set $\{(x_1, x_2, \ldots, x_t) | x_i \in \{0, 1, \ldots, d-1\}\}$. Obviously, $|X(t)| = d^t$. The *de Bruijn digraph*, proposed in Ref. 85, $BRU(d, t)$, is the digraph with vertex set $X(t)$ such that an edge exists from $(x_1, x_2, \ldots, x_t)$ to $(y_1, y_2, \ldots, y_t)$ if and only if $x_j = y_{j-1}$ for $2 \leq j \leq t$. In Figure 3.5, we illustrate $BRU(2, 2)$ and $BRU(2, 3)$. We use $UBRU(d, t)$ to denote the underlying simple graph of $BRU(d, t)$.

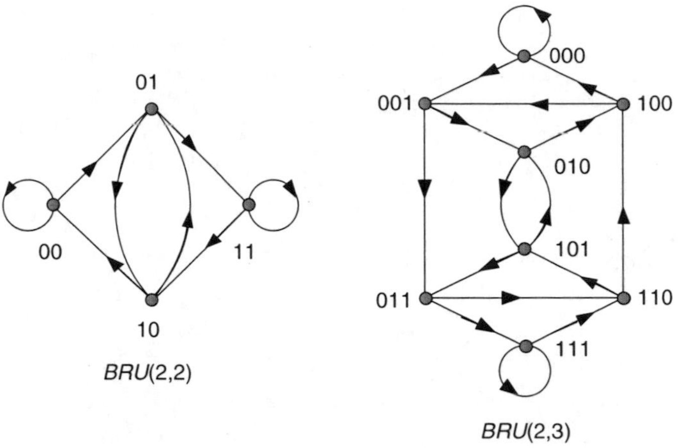

FIGURE 3.5 $BRU(2,2)$ and $BRU(2,3)$.

LEMMA 3.2 There exists a path of length t joining any two vertices x and y in $BRU(d,t)$.

Proof: Suppose that $x=(x_1,x_2,\ldots,x_t)$ and $y=(y_1,y_2,\ldots,y_t)$ are two vertices of $BRU(d,t)$. Obviously, $\langle p_0=x=(x_1,x_2,\ldots,x_t),\ p_1=(x_2,x_3,\ldots,x_t,y_1),\ldots,p_i=(x_{i+1},\ldots,x_t,y_1,\ldots,y_{i-1}).\ldots,p_t=y=(y_1,y_2,\ldots,y_t)\rangle$ forms a path of length t joining x to y. □

COROLLARY 3.2 $D(BRU(d,t))=D(UBRU(d,t))=t$.

Proof: From Lemma 3.2, we know that $D(BRU(d,t))\leq t$. However, it is obvious that the distance between $00\cdots00$ and $11\cdots11$ in $BRU(d,t)$ is t. Hence, $D(BRU(d,t))=t$. Similarly, we have $D(UBRU(d,t))=t$. □

Let d be an integer with $d\geq 2$. Let $Y(t)$ denote the set $\{(u_1,u_2,\ldots,u_t)\mid u_i\in\{0,1,\ldots,d\}, u_i\neq u_{i+1}\text{ for }1\leq i<t\}$. Obviously, $|Y(t)|=(d+1)d^{t-1}$. The *Kautz digraph*, proposed in [195], *Kautz* (d,t), is the digraph with vertex set $Y(t)$ such that an edge exists from $(x_1,x_2,\ldots,x_t)$ to $(y_1,y_2,\ldots,y_t)$ if and only if $x_j=y_{j-1}$ for $2\leq j\leq t$. In Figure 3.6, we illustrate *Kautz* (2, 3). We use *UKautz* (d,t) to denote the underlying simple graph of *Kautz* (d,t).

We have a similar result for Kautz digraphs.

LEMMA 3.3 There exists a path of length at most t joining any two vertices x and y in $Kautz(d,t)$. Hence, $D(Kautz(d,t))=D(UKautz(d,t))=t$.

Proof: Let $x=(x_1,x_2,\ldots,x_t)$ and $y=(y_1,y_2,\ldots,y_t)$ be two vertices of $Kautz(d,t)$.

Suppose that $x_t\neq y_1$. Then $\langle p_0=x=(x_1,x_2,\ldots,x_t),\ p_1=(x_2,x_3,\ldots,x_t,y_1),\ldots,$ $p_i=(x_{i+1},\ldots,x_t,y_1,\ldots,y_{i-1}),\ldots,\ p_t=y=(y_1,y_2,\ldots,y_t)\rangle$ forms a path of length t joining x to y.

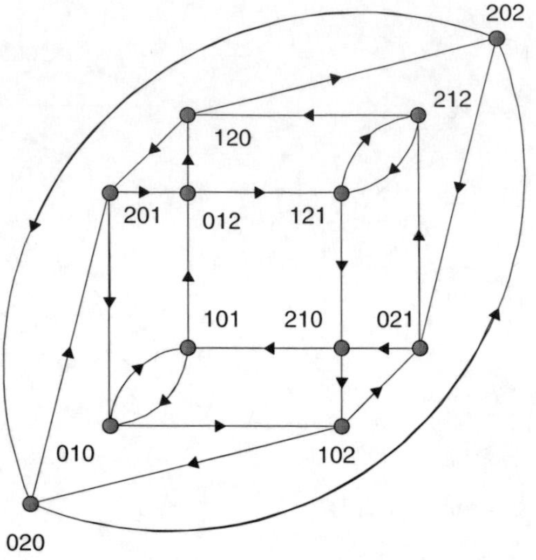

FIGURE 3.6 *Kautz*(2, 3).

Suppose that $x_t = y_1$. Then $\langle p_0 = x = (x_1, x_2, \ldots, x_t), p_1 = (x_2, x_3, \ldots, x_t, y_2), \ldots, p_i = (x_{i+1}, \ldots, x_t, y_1, \ldots, y_i), \ldots, p_{t-1} = y = (y_1, y_2, \ldots, y_t)\rangle$ forms a path of length $(t-1)$ joining x to y.

Similar to the de Bruijn digraphs, $D(Kautz(d,t)) = D(UKautz(d,t)) = t$. □

With Corollary 3.4, we notice that the diameter of the de Bruijn graphs and the diameter of the Kautz digraphs are closed to the Moore bound.

3.5 STAR GRAPHS AND PANCAKE GRAPHS

Here, we introduce two important families of interconnection networks. Assume that $n \geq 2$. We use $\langle n \rangle$ to denote the set $\{1, 2, \ldots, n\}$, where n is a positive integer. A *permutation* of $\langle n \rangle$ is a sequence of n distinct element of $u_i \in \langle n \rangle$, $u_1 u_2 \ldots\ldots u_n$. An *inversion* of $u_1 u_2 \ldots\ldots u_n$ is a pair of (i,j) such that $u_i < u_j$ and $i > j$. An *even permutation* is a permutation with an even number of inversions, and an *odd permutation* is a permutation with an odd number of inversions. The *n-dimensional star network*, denoted by S_n, is a graph with the vertex set $V(S_n) = \{u_1 u_2 \ldots u_n \mid u_i \in \langle n \rangle$ and $u_i \neq u_j$ for $i \neq j\}$. The edges are specified as follows: $u_1 u_2 \ldots u_i \ldots u_n$ is adjacent to $v_1 v_2 \ldots v_i \ldots v_n$ by an edge in dimension i with $2 \leq i \leq n$ if $v_j = u_j$ for $j \notin \{1, i\}$, $v_1 = u_i$ and $v_i = u_1$. The star graphs S_2, S_3, and S_4 are illustrated in Figure 3.7.

By definition, S_n is an $(n-1)$-regular graph with $n!$ vertices. Moreover, it is vertex-transitive and edge-transitive. Furthermore, S_n is a bipartite graph with one partite set containing those vertices corresponding to odd permutations and the other partite set containing those vertices corresponding to even permutations. The star graphs are an important family of interconnection networks proposed by Akers and Krishnameurthy [3,4].

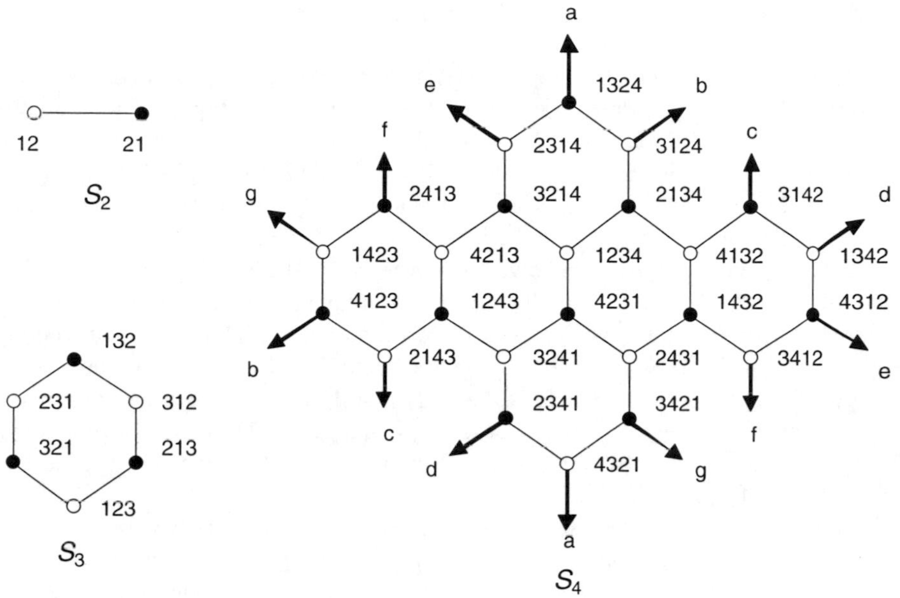

FIGURE 3.7 The star graphs S_2, S_3, and S_4.

The identity permutation is denoted by $I=(1,2,3,4,\ldots)$. Because the star graph is vertex-transitive, we can always view the distance between any two arbitrary vertices as the distance between the source vertex and the identity permutation by suitably renaming the symbols representing the permutations. Hence, the destination vertex is always assumed to be the identity vertex I without loss of generality.

It is known that any permutation of n elements can also be specified in terms of its unique cycle structure with respect to the identity permutation I. For example, $3,4,5,2,1,6=(1,3,5)(2,4)(6)$. The maximum number of cycles in a permutation of n elements is n and the minimum number is 1. When a cycle has only one symbol, the symbol is in its correct position in the permutation with respect to the identity permutation. For the rest of the work, we omit the singletons if the number of symbols in a permutation with respect to the identity permutation is understood from the context. Thus, $(1,3,5)(2,4)(6)=(1,3,5)(2,4)$. The length of a cycle is defined as the number of symbols present in the cycle.

Let $\Pi(u)=\pi_1\pi_2\ldots\pi_k$ be the cycle representation of the (vertex) permutation u where π_i's are the individual cycles in the vertex. Without loss of generality, we assume that $1=\pi_1^1$. In Ref. 4, a general algorithm to find the shortest path to the destination vertex (the identity permutation) is proposed. The algorithm has shown that in general this shortest path is not unique; that is, there may be multiple shortest paths to the destination from the given vertex u.

ALGORITHM In order for the path to be minimal in length, the moves must be restricted to two possible choices: for any vertex $u=\pi_1\pi_2\ldots\pi_k$ with $1=\pi_1^1$, either (1) exchange π_1^1 with π_1^2, thereby reducing the length of the cycle π_1 by 1, or (2) exchange

π_1^1 with π_j^l (where $2 \le j \le k$ and $1 \le l \le |\pi_j|$), thereby merging the two cycles π_1 and π_j into one cycle.

A path from a vertex u to the destination is said to follow the *shortest routing scheme* if the moves are restricted to these two types. For example, $3,4,5,2,1,6 = (1,3,5)(2,4)(6)$. Applying the algorithm, we obtain $(1,3)(1,5) \times (1,2)(1,4)(1,2)$. It is easy to confirm that $(1,3,5)(2,4)(6)(1,3)(1,5)(1,2)(1,4)(1,2) = 1,2,3,4,5,6$. Thus, the product $(1,3)(1,5)(1,2)(1,4)(1,2)$ corresponds to a shortest path from $3,4,5,2,1,6$ to the identity $1,2,3,4,5,6$.

Let $d(u)$ denote the distance between the vertex u to the identity permutation. Let c be the number of cycles of length at least 2 in any permutation u and m be the total number of symbols in those c cycles of the permutation u. With the previous routing algorithm, $D(u)$ is $d(u) = c + m$ if 1 is the first symbol, and $c + m - 2$ if otherwise.

For example, let $u = 3,2,5,4,1,6 = (1,3,5)(2)(4)(6) = (1,3,5)$. Then $c = 1$, $m = 3$, and $d(u) = c + m - 2 = 2$. Let $u = 1,4,3,2,5,6 = (1)(2,4)(3)(5)(6) = (2,4)$. Then $c = 1$, $m = 2$, and $d(u) = 1 + 2 = 3$. With this observation, it can be proved that the diameter of S_n is $\lfloor \frac{3(n-1)}{2} \rfloor$.

The *n-dimensional pancake graph*, denoted by P_n, is a graph with the vertex set $V(P_n) = \{p_1p_2 \ldots p_n \mid p_i \in \langle n \rangle \text{ and } p_i \ne p_j \text{ for } i \ne j\}$. The adjacency is defined as follows: $p_1p_2 \ldots p_i \ldots p_n$ is adjacency to $p_ip_{i-1} \ldots p_1, p_{i+1} \ldots p_n$ through an edge of dimension i, where $2 \le i \le n$ (prefix substring reversal). The pancake graphs P_2, P_3, and P_4 are shown in Figure 3.8. Again, P_n is an $(n-1)$-regular graph with $n!$ vertices. Note that P_n is bipartite if and only if $n = 2$ and 3.

The star graph and the pancake graph [3,4] are attractive alternatives to the n-cube. They have many significant advantages over the n-cube, such as a lower degree and a

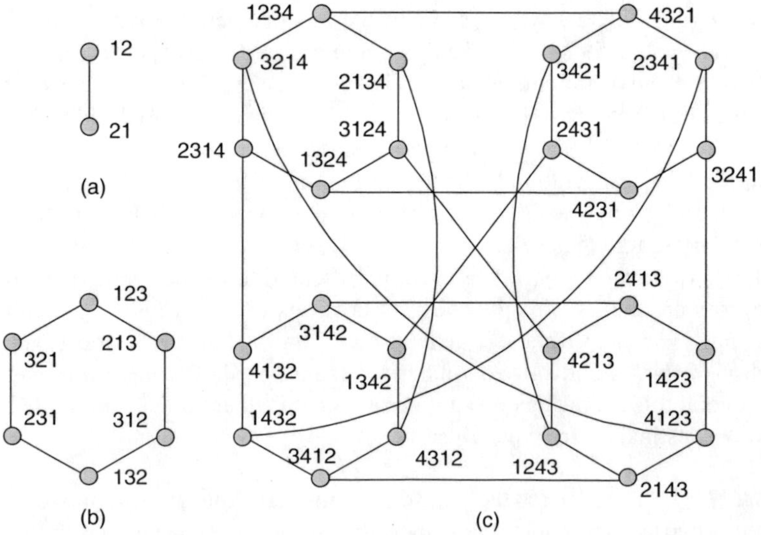

FIGURE 3.8 (a) P_2, (b) P_3, and (c) P_4.

smaller diameter. There are a lot of studies on the properties of these two families of graphs. We note that Gates and Papadimitriou [121] began the study on the diameter of the pancake graphs. At this point, the diameter of the pancake graphs is still unsolved [144].

The concrete values of $D(P_n)$ for small n are studied in Refs. 73, 144, and 205. Collective results are listed in the following table:

n	1	2	3	4	5	6	7	8	9	10	11	12	13	14	15
$D(P_n)$	0	1	3	4	5	7	8	9	10	11	13	14	15	16	17

3.6 EDGE CONGESTION AND BISECTION WIDTH

We can advance our study on the shortest paths in a interconnection network. Let us treat a path as a directed route from a source to a destination. To distinguish different orientations of an edge (u, v), we write $[u, v]$ and $[v, u]$ as traversing from u to v and from v to u, respectively. A path $P = \langle x = x^0, x^1, \ldots, x^m = y \rangle$ is treated as a directed path from x to y consisting of $[x^i, x^{i+1}]$ for $0 \le i \le m-1$. For convenience, we also treat a path as a set of edges and write $[x^i, x^{i+1}] \in P$ to mean that $[x^i, x^{i+1}]$ is in P. We say that an edge $e = (u, v)$ is *incident* on P if $[u, v] \in P$ or $[v, u] \in P$, that is, $u = x^i, v = x^{i+1}$ or $v = x^i, u = x^{i+1}$ for some i. When considering edge congestion, we can treat any routing algorithm A for a network $G = (V, E)$ as a function assigning each $(x, y) \in V \times V$ to only one path from x to y, denoted by $P_A(x, y)$. Then we define the *edge congestion of an edge* $e \subset E$ under the routing algorithm A, denoted by $c_A(e)$, as the number of (x, y)-pairs such that e is incident on $P_A(x, y)$; that is,

$$c_A(e) = |\{(x, y) \mid x, y \in V, \ \ e \text{ is incident on } P_A(x, y)\}|$$

Furthermore, we define the *edge congestion of the network* G *under the routing algorithm* A, denoted by $c_A(G)$, as

$$c_A(G) = \max\{c_A(e) \mid \text{for all } e \in E\}$$

and the *edge congestion of the network* G, denoted by $c(G)$, as

$$c(G) = \min\{c_A(G) \mid \ \text{for all routing algorithms } A \text{ for } G\}$$

The following lemma is proved by Fiduccia and Hedrick [109].

LEMMA 3.4 Assume that G is an n-vertex graph such that the removal of w edges partitions G into two not necessarily connected subgraphs of k vertices and $n-k$ vertices, respectively. Then $c(G) \ge \lceil \frac{2k(n-k)}{w} \rceil$. Moreover, this bound is tight.

Proof: Assume that A and B are the subgraphs of the partition and let π be any shortest path routing of G. Note that π contains a shortest path $\pi(u, v)$ for each (u, v). It contains $k(n-k)$ paths from A to B and same $k(n-k)$ paths from B to A that cross the w edges in the cut. Thus, some edge of the cut must crossed at least $\lceil \frac{2k(n-k)}{w} \rceil$ times. As this is true for every shortest path routing π of G, $c(G) \geq \lceil \frac{2k(n-k)}{w} \rceil$. Because a shortest path may cross the cut many times, this is not an upper bound. However, the bound is tight for G if G is the complete graph K_n. □

The *bisection width* of a graph G, $bw(G)$, is the least number of edges to be removed to partition G into two disconnected components with roughly equal number of vertices. From the previous lemma, we have the following corollary.

COROLLARY 3.3 *Assume that G is an n-vertices graph with edge congestion $c(G)$ and bisection width $bw(G)$. Then $c(G) \geq \lceil \frac{n^2}{2bw(G)} \rceil$ if n is even and $c(G) \geq \lceil \frac{n^2-1}{2bw(G)} \rceil$ if n is odd.*

Now, we compute the edge congestion and the bisection width of the hypercube Q_n.

THEOREM 3.5 $c(Q_n) = 2^n$ and $bw(Q_n) = 2^{n-1}$.

Proof: Let u be any vertex in Q_n. It is known that exactly $\binom{n}{i}$ vertices are at distance i from u. Thus, for any routing algorithm A, we have

$$\sum_{e \in E(Q_n)} c_A(e) = \sum_{(x,y) \in V \times V} |P_A(x, y)| \geq \sum_{x \in V} \sum_{i=0}^{n} i \binom{n}{i} = 2^n \sum_{i=0}^{n} i \binom{n}{i} = 2^n n 2^{n-1}$$

Note that there are exactly $n\,2^{n-1}$ edges in $E(Q_n)$. We have

$$c_A(Q_n) = \max\{c_A(e) \mid e \in E(Q_n)\} \geq 2^n$$

Thus, $c(Q_n) \geq 2^n$.

On the other hand, suppose that C is the routing strategy that routes x to y by changing the rightmost differing bit iteratively. To be precise, assume that x differs from y at k bits, say, the l_1-, l_2-, ..., l_k-th bits with $0 \leq l_1 < l_2 < \cdots < l_k \leq n-1$. Then $P_C(x, y)$ is given by $\langle x = x^0, x^1, \ldots, x^k = y \rangle$, where x^d differs from x^{d+1} at the l_{d+1}-th bit. Let $e = (u, v)$ be an edge of Q_n with u and v differing at the jth bit such that (u, v) is incident on $P_C(x, y)$. That is, either $[u, v] \in P_C(x, y)$ or $[v, u] \in P_C(x, y)$. Let us consider $[u, v] \in P_C(x, y)$. It follows from algorithm C that $x_{n-1} \ldots x_j = u_{n-1} \ldots u_j$ and $y_j \ldots y_1 y_0 = v_j \ldots v_1 v_0$. Therefore,

$$|\{(x, y) \mid x, y \in V(Q_n), [u, v] \in P_C(x, y)\}| = 2^{n-j-1} 2^j = 2^{n-1}$$

Similarly,

$$|\{(x, y) \mid x, y \in V(Q_n), [v, u] \in P_C(x, y)\}| = 2^{n-j-1} 2^j = 2^{n-1}$$

Hence, $c(Q_n) = 2^{n-1} + 2^{n-1} = 2^n$.

Now, we prove that $bw(Q_n)=2^{n-1}$. By Corollary 3.3, $2^n=c(Q_n)\geq\left\lceil\frac{2^{2n}}{2bw(Q_n)}\right\rceil$. Thus, $bw(Q_n)\geq 2^{n-1}$. Let $V_0=\{\mathbf{u}\mid\mathbf{u}=u_{n-1}u_{n-2}\ldots u_1u_0\in V(Q_n)$ with $u_{n-1}=0\}$ and $V_1=\{\mathbf{u}\mid\mathbf{u}=u_{n-1}u_{n-2}\ldots u_1u_0\in V(Q_n)$ with $u_{n-1}=1\}$. Let X denote the edges joining vertices between V_0 and V_1. Obviously, the deletion of X from Q_n partitions Q_n into V_0 and V_1. It is observed that $|V_0|=|V_1|$ and $|X|=2^{n-1}$. Hence, $bw(Q_n)=2^{n-1}$. □

In Section 3.2, we proposed a shortest path routing algorithm, Algorithm H, for a twisted cube. However, Algorithm H does not lead to optimum edge congestion of twisted cubes. Li et al. [223] propose another shortest path routing algorithm for twisted cubes. With this routing algorithm, it is proved that $c(TQ_n)=2^n$ and $bw(TQ_n)=2^{n-1}$ [223]. In Ref. 46, Chang et al. discuss the edge congestion and the bisection width of cross cubes.

3.7 TRANSMITTING PROBLEM

Assume that $G=(V,E)$ is a graph representing the topology for a network. Let v_0 be a special vertex outside G, called the *host processor* (abbreviated as *host*), which is connected to each vertex of V. The host processor is the sender or source of the message to be transmitted to all of the vertices in G. Each time unit, the host may send its message to any single vertex of the graph G, according to its choice. At the same time, each processor that has already received the message can send the message to all of its neighbors in one unit of time.

For $u, v\in V$, let $d(u, v)$ denote the shortest distance from u to v. Suppose that in the ith time unit the host sends its message to processor p_i. Then after the kth time unit, $k>i$, all of the processors v satisfying $d(p_i, v)\leq k-i$ can receive the message. The objective is to minimize the number of time units such that all of the vertices in G can receive the message. This minimum number of time units for graph G is called the *optimal transmitting time* for G, denoted by $t(G)$.

For an n-dimensional hypercube Q_n, Alon [8] proves that $t(Q_n)=\left\lceil\frac{n}{2}\right\rceil+1$ and gives a simple procedure to achieve this goal. The host processor simply sends its message to an arbitrary processor p_1 in the first time unit, then to the antipodal of p_1—that is, the vertex having the Hamming distance n from p_1—in the second time unit. Afterward, the host just waits.

This transmitting scheme gives the host a rather light workload. We note that even when the host sends the message to a processor in each time unit, the transmitting time remains the same. Thus, we can consider the transmitting problem as a problem to not only determine $t(G)$ but also to minimize the workload of the host when $t(G)$ is achieved. The workload of the host is defined as the number of time units in which the host sends the message to processors in G.

Now, we give a formal description of the transmitting problem. We are given a graph $G=(V,E)$ and a special vertex v_0 as introduced before. For $v\in V$ and a nonnegative integer r, the *r-neighborhood* of v is defined as $N_r(v)=\{u\in V\mid d(u,v)\leq r\}$. By convention, $N_0(v)=\{v\}$. Let t be a positive integer, and $V'=V\cup\{v_0\}$. We use $[t]$ to denote the set $\{1,2,\ldots,t\}$. Let f be a function from $[t]$ into V' defined as follows: $f(i) = v \neq v_0$ if the host sends its message to v at time i, and $f(i) = v_0$ if the host is idle at time i. Let $s = |\{i \mid f(i) \in V\}|$, which is called the *workload of the host*. Obviously, $s \leq t$. The function f is called a (t,s)-*transmitting scheme* if $\cup_{f(i)\in V} N_{t-i}(f(i)) = V$; that is, all vertices in V can receive the message in t time units.

In a (t,s)-transmitting scheme, each processor in G receives the message either from the host or from its neighbor within t time units, while the workload of the host is given by s. Such a t is called a *feasible transmitting time*. The *cost* of a (t,s)-transmitting scheme f, denoted by $c(f)$, is defined as the ordered pair (t,s). We use $TS(G)$ to denote the set of all possible transmitting schemes for G. Let $f, g \in TS(G)$ be two transmitting schemes with $c(f) = (t_1, s_1)$ and $c(g) = (t_2, s_2)$. We say $f \leq g$ if $c(f) \leq c(g)$ lexicographically; that is, $t_1 < t_2$ or $t_1 = t_2$ and $s_1 \leq s_2$.

A scheme $f^* \in TS(G)$ is an *optimal transmitting scheme* for G if $f^* \leq g$ for all $g \in TS(G)$. The *optimal transmitting cost* of G, denoted by $c(G)$, is defined as $c(G) = c(f^*) = (t^*, s^*)$, where $t^* = t(G)$ is the *optimal transmitting time*, and $s^* = s(G)$ is the *optimal workload of the host*.

Obviously, we can always obtain a $(D+1, 1)$-transmitting scheme, where D is the diameter of the underlying network. An optimal transmitting scheme is what we seek. The problem of designing an optimal transmitting scheme for a graph G is important to the communication of interconnection networks. Alon [8] proposes an optimal $\left(\lceil \frac{n}{2} \rceil + 1, 2\right)$-transmitting scheme for n-dimensional hypercube Q_n.

In the following sections, we study the transmitting problem of some graphs.

THEOREM 3.6 **[149]** $c(P_n) = (\lceil \sqrt{n} \rceil, \lceil \sqrt{n} \rceil)$.

Proof: Without loss of generality, we assume that $n = m^2$ for some positive integer m. Assume that f is an optimal transmitting scheme with $c(f) = (t,s)$. P_n can be easily partitioned into m nonoverlapping intervals $I_1, I_2, \ldots, I_m$ with $|I_i| = 2i - 1$ vertices for $1 \leq i \leq m$. An (m,m)-transmitting scheme can be easily obtained by setting $f(i)$ to the center of I_{m-i+1}. Thus, $m \geq t$.

On the other hand, it is observed that $|N_r(v)| \leq 2r + 1$ for any vertex v and nonnegative integer r. Because f is a transmitting scheme,

$$m^2 = |\cup_{f(i)\in V} N_{t-i}(f(i))| \leq \sum_{r=0}^{t-1} (2r+1) = 1 + 3 + \cdots + (2t-1) = t^2 \quad (3.1)$$

Thus, $t^2 \geq m^2$; that is, $t \geq m$. Hence, $t = m$.

Suppose that $s \leq m - 1 = t - 1$. Then

$$|\cup_{f(i)\in V} N_{t-i}(f(i))| \leq \sum_{r=1}^{t-1} (2r+1) = 3 + \cdots + (2t-1) = t^2 - 1 = m^2 - 1$$

leads to a contradiction. Consequently, the proposed (m,m)-transmitting scheme is optimal; moreover, the theorem follows. □

The optimal transmitting cost for C_n can be easily obtained as a corollary.

COROLLARY 3.4 **[149]** $c(C_n) = (\lceil\sqrt{n}\,\rceil, \lceil\sqrt{n}\,\rceil)$.

We can also study the transmitting problem on digraphs.

THEOREM 3.7 The optimal transmitting cost of $BRU(d,n)$ is $(1,1)$ if $d=1$, $(2,1)$ if $n=1$, $(n,2)$ if $d=2$, and $(n+1,1)$ if $d \geq 3$.

Proof: It is trivial to check the cases of $d=1$ or $n=1$. Now we consider the cases of $d \geq 2$ and $n \geq 2$. Because $BRU(d,n)$ has outdegree d, it follows that for any vertex v and a nonnegative integer $r \leq n$, we have $|N_r(v)| \leq 1 + d + d^2 + \cdots + d^r = \frac{d^{r+1}-1}{d-1}$. Therefore,

$$|\cup_{f(i)\in V} N_{t-i}(f(i))| \leq \sum_{r=0}^{t-1} \frac{d^{r+1}-1}{d-1} = \frac{d(d^t-1)}{(d-1)^2} - \frac{t}{d-1} \tag{3.2}$$

Now we consider $d \geq 3$. Because for any vertex v we have $N_n(v) = V(BRU(d,n))$, we can give an $(n+1,1)$-transmitting scheme as follows: the host sends a message to an arbitrary vertex in $BRU(d,n)$ and then idles. After $(n+1)$ units of time, all processors in $BRU(d,n)$ can receive the message. Assume that the optimal transmitting time is $t \leq n$. It follows that

$$|\cup_{f(i)\in V} N_{t-i}(f(i))| \leq \frac{d(d^n-1)}{(d-1)^2} - \frac{n}{d-1} < d^n$$

It means that not all of the vertices in $BRU(d,n)$ can receive the message in n time units, which is a contradiction. Thus, the $(n+1,1)$-transmitting scheme is optimal for $BRU(d,n)$ when $d \geq 3$.

Let $d=2$. Assume that (t^*, s^*) is the optimal transmitting cost for $BRU(2,n)$. For all (t,s)-transmitting schemes, it follows from (3.2) that $|\cup_{f(i)\in V} N_{t-i}(f(i))| \leq 2^n - n - 1 < 2^n$ if $t \leq n-1$. Thus we have $t \geq n$, and in particular, $t^* \geq n$. Suppose that $t^* = n$ and $s^* = 1$. It follows that $|\cup_{f(i)\in V} N_{t-i}(f(i))| = |N_{n-1}(f(1))| \leq \sum_{i=1}^{n} 2^{i-1} = 2^n - 1 < 2^n$, which is a contradiction. Thus, if $t^* = n$, we must have $s^* \geq 2$.

We give an $(n,2)$-transmitting scheme f as follows: $f(1) = 1 = (0,0,\ldots,0,1)$, $f(k) = 0 = (0,0,\ldots,0)$ for an arbitrary k with $2 \leq k \leq n$, and $f(i) = v_0$ for all $i \neq 1, k$. In other words, the host sends a message to vertex 1—that is, $(0,\ldots,0,1)$, at the first time unit, and then to vertex 0—that is, $(0,\ldots,0)$ at the kth unit of time, $2 \leq k \leq n$.

For any vertex $v = (v_{n-1}, v_{n-2}, \ldots, v_0)$, $2 \leq v \leq 2^n - 1$, it is obvious that $d(1,v) \leq n-1$, as $v_i = 1$ for some $1 \leq i \leq n-1$. Thus, $N_{n-1}(1) = V(BRU(d,n)) - \{0\}$. Therefore, $N_{n-1}(1) \cup f(k) = N_{n-1}(1) \cup \{0\} = V(BRU(d,n))$. Hence, the proposed $(n,2)$-transmitting scheme is optimal. □

4 Trees

4.1 BASIC PROPERTIES

Trees as graphs have many applications, especially in data storage and communication.

A graph having no cycle is *acyclic*. A *forest* is an acyclic graph. A *tree* is a connected acyclic graph. A *leaf* (or *pendant vertex*) is a vertex of degree 1. A *spanning subgraph* of G is a subgraph with vertex set $V(G)$. A *spanning tree* is a spanning subgraph that is a tree.

We first prove that deleting a leaf from a tree yields a smaller tree. This implies that every tree with more than one vertex can be grown from a smaller tree by adding a vertex of degree 1. This facilitates inductive proofs for trees by allowing the induction step to grow an $(n+1)$-vertex tree by adding a vertex to an arbitrary n-vertex tree without falling into the induction trap.

LEMMA 4.1 Every tree with at least two vertices has at least two leaves. Moreover, deleting a leaf from a tree with n vertices produces a tree with $(n-1)$ vertices.

Proof: Assume that T is a tree with at least two vertices. Obviously, it has an edge e because T is connected. Let P be any maximal path containing e. Obviously, the two endpoints x and y of P have only one neighbor on the path. Thus, $\deg_T(x) = \deg_T(y) = 1$. Hence, x and y are leaves.

Let v be a leaf of a tree T and let $T' = T - v$. Suppose that u and w are any two vertices of T'. Let P be a path of T from u to w. Obviously, $v \notin P$ because $\deg_T(v) = 1$. Thus, P is also present in T'. Hence, T' is connected. Because a vertex deletion cannot create a cycle, T' is also acyclic. Thus, T' is a tree with $(n-1)$ vertices. □

Trees have many equivalent characterizations, any of which could be taken as the definition. Such characterizations are useful because we need only to verify a graph that satisfies any one of them to prove that it is a tree. Then we can use all the other properties.

THEOREM 4.1 Let T be an n-vertex graph with $n \geq 1$. The following statements are equivalent:

1. T is connected and has no cycles.
2. T is connected and has $(n-1)$ edges.
3. T has $(n-1)$ edges and no cycles.
4. T has exactly one path from u to v for every u and v in $V(T)$.

Proof: We first prove that statement 1 implies statements 2 and 3. Assume that T is a connected n-vertex graph with no cycles. We use induction on n to prove statements 2 and 3. Because any acyclic 1-vertex graph has no edge, the implication holds for $n = 1$. For the induction step, we assume that the implication holds for those m-vertex graphs with $1 \leq m < n$. Assume that T is a connected graph with no cycles. By Lemma 4.1, there exists a leaf v such that $T' = T - v$ is acyclic and connected. By induction hypothesis, $e(T') = n - 2$. Hence, $e(T) = n - 1$.

We prove that statement 2 implies statements 1 and 3. Assume that T is a connected n-vertex graph with $(n-1)$ edges. Let us delete edges from cycles of T one by one until the resulting graph T' is acyclic. As no edge of a cycle is a cut edge, T' remains connected. By the previous paragraph, T' has $(n-1)$ edges. In other words, $T = T'$. Thus, T is acyclic.

We prove that statement 3 implies statements 1 and 2. Assume that T is an n-vertex graph with $(n-1)$ edges and no cycles. Suppose that T has k components with orders $n_1, n_2, \ldots, n_k$. Because T has no cycles, each component has $(n_i - 1)$ edges. Summing this over all components yields $e(T) = \sum_{i=1}^{k} (n_i - 1) = n - k$. Because $e(T) = n - 1$, $k = 1$. Thus, T is connected.

We prove that statement 1 implies statement 4. Assume that T is a connected n-vertex graph with no cycles. Let u and v be any two vertices of $V(T)$. Because T is connected, T has at least one path from u to v. Suppose that G has two distinct paths P and Q from u to v. Obviously, some edge $e = (x, y)$ appears in exactly one of them. There exists a closed walk containing e. We can delete e from T, which yields a walk W from x to y, not including e. Hence, W contains a path R from x to y. Obviously, $R \cup \{e\}$ forms a cycle of T, which contradicts the hypothesis that T is acyclic. Hence, T has exactly one path from u to v.

Finally, we prove that statement 4 implies statement 1. Assume that T has exactly one path from u to v for every u and v in $V(T)$. Obviously, T is connected. Suppose that T has a cycle C. Then T has two paths between any pair of vertices on C. Thus, T has no cycles. □

With Theorem 4.1, it is easy to see that every connected graph contains a spanning tree. Moreover, every n-vertex graph with $(n-k)$ edges has at least k components. With m components, there must be at least $(n-m)$ edges. The next result illustrates induction using deletion of a leaf.

PROPOSITION 4.1 *Assume that T is a tree with k edges and G is a simple graph with $\delta(G) \geq k$. Then T is a subgraph of G.*

Proof: We prove this proposition by induction on k. Obviously, every simple graph contains K_1. The proposition holds for $k = 0$. Suppose that the proposition holds for trees with fewer than m edges with $0 \leq m < k$. Because $k > 0$, by Lemma 4.1 we can choose a leaf v with neighbor u in T. Let $T' = T - v$. T' is a tree with $(k-1)$ edges. Because $\delta(G) \geq k > k - 1$, by the induction hypothesis, G contains a copy of T' as a subgraph. Let x be the vertex in this copy of T' that corresponds to u (see Figure 4.1 for illustration). As T' has only $(k-1)$ vertices other than u, x has a neighbor y in G that does not appear in this copy of T'. Obviously, we can add the edge (x, y) on T' to obtain a copy of T in G. □

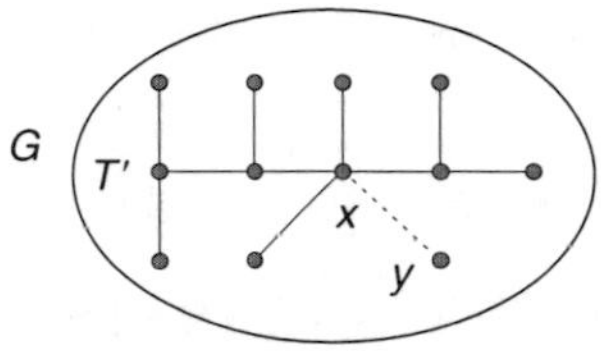

FIGURE 4.1 Illustration of Proposition 4.1.

Because a tree has no cycle, every edge of a tree is a cut edge. Because a tree has a unique path linking each pair of vertices, adding any edge creates exactly one cycle. We apply these observations to prove results that are used by algorithms for finding spanning trees in weighted graphs. We use subtraction and addition as operations involving sets to indicate the deletion and inclusion of edges.

PROPOSITION 4.2 *Suppose that T and T' are two spanning trees of a connected graph G such that $e \in E(T) - E(T')$. Then there is an edge $e' \in E(T') - E(T)$ such that $T - e + e'$ is a spanning tree of G.*

Proof: Since e is a cut edge of T, $T - e$ is disconnected. Let U and $\overline{U}$ be the vertex sets of the component $T - e$. Because T' is connected, it has an edge e' with endpoints in U and $\overline{U}$. Now $T - e + e'$ is a spanning tree. □

PROPOSITION 4.3 *Suppose that T and T' are two spanning trees of a connected graph G such that $e \in E(T) - E(T')$. Then there is an edge $e' \in E(T') - E(T)$ such that $T' + e - e'$ is a spanning tree of G.*

Proof: Obviously, the graph $T' + e$ contains a unique cycle C. Since T is acyclic, there exists an edge $e' \in E(C) - E(T)$. Since C is the only cycle in $T' + e$, $T' + e - e'$ is connected and acyclic. By Theorem 4.1, $T' + e - e'$ is a spanning tree of G. □

The diameter of a tree is the length of its longest path, because every path in a tree is the shortest path between its endpoints. We next describe the centers of trees. The proof uses deletion of all leaves to obtain a subtree; this sometimes provides a cleaner inductive proof than deleting one leaf.

THEOREM 4.2 **(Jordan [187])** The center of a tree is one vertex or one edge.

Proof: We use induction on the number of vertices. The statement is trivial for trees with one or two vertices. Assume that the statement holds for trees with fewer than n vertices with $n > 2$. Let T be an arbitrary n-vertex tree. Form T' by deleting every leaf of T. Since T contains n vertices and $(n - 1)$ edges, T' is a tree with at least one vertex.

Let u be a vertex in T'. It is observed that the vertex at maximum distance from u is a leaf. Since all the leaves have been removed, $\epsilon_{T'}(u) = \epsilon_T(u) - 1$ for every $u \in V(T')$. Moreover, the eccentricity of a leaf in T is greater than the eccentricity of its neighbor in T. Hence, the vertices minimizing $\epsilon_T(u)$ are the same vertices minimizing

$\epsilon_{T'}(u)$. By the induction hypothesis, for T' they consist of a single vertex or two adjacent vertices. □

4.2 BREADTH-FIRST SEARCH AND DEPTH-FIRST SEARCH

Many graph algorithms require a systematic method of visiting the vertices of a graph (or *traversing the graph*). In this section, we describe two types of spanning tree (or forest) that can be used to produce some particularly efficient algorithms. The *breadth-first search* (*BFS*) and the *depth-first search* (*DFS*) are such methods. To discuss these two search algorithms, we introduce two basic data structures: stacks and queues.

A *stack* is a list in which insertions and deletions are always made at one end, called the *top*. The *top item* in a stack is the one that was most recently inserted. Stacks are sometimes referred to as LIFO lists, meaning *last in, first out*. A *queue* is a list in which all insertions are made at one end, called the *tail*, and all deletions are made at the other end, called the *head*. Queues, then, are *first in, first out* lists (FIFO lists).

ALGORITHM: (Breadth-first search.)

Input. An unweighted graph (or digraph) and a start vertex u.

Idea. Maintain a set R of vertices that have been reached but not searched and a set S of vertices that have been searched. The set R is maintained as a FIFO list (queue) so that the first vertices found are the first vertices explored.

Initialization. $R = \{u\}$, $S = \emptyset$, $d(u, v) = 0$.

Iteration. As long as $R \neq \emptyset$, we search the head vertex v of R. The neighbors of v not in R or S are added to the tail of R and assigned distance $d(u, v) + 1$, and then v is removed from the head of R and placed in S.

It can be proved by induction that a BFS correctly computes the distance between any vertex v to the start vertex u. The longest distance from u to another vertex is $\epsilon(u)$. Hence, we can compute the diameter of a graph G by running a BFS from each vertex.

In a DFS, we explore always from the most recently discovered vertex that has an unexplored edge (this is also called *backtracking*). In contrast, a BFS explores from the oldest vertex, so the difference between DFS and BFS is that, in a DFS, the set R is maintained as a LIFO *stack* rather than a queue.

ALGORITHM: (Depth-first search.)

Input. An unweighted graph (or digraph) and a start vertex u.

Idea. Maintain a set R of vertices that have been reached but not searched and a set S of vertices that have been searched. The set R is maintained as a LIFO list (stack) so that the first vertices found are the last vertices explored.

Initialization. $R = \{u\}$ and $S = \emptyset$.

Iteration. As long as $R \neq \emptyset$, we remove the top element v of R and search v. The neighbors of v not in R or S are added to the top of R.

A DFS that traces only one path at a time is much easier to do by hand or to program. Further, in cases where we need to find only one of the possible solutions, it pays to search all the way down a path for a solution rather than to take a long time building a large number of partial paths, only one of which in the end will actually be used. On the other hand, when we want a solution involving a shortest path or when there may be very long dead-end paths (while solution paths tend to be relatively short), then the BFS is better.

DFS and BFS are common techniques for problem solving. Try to use these techniques for the following puzzles.

Example 4.1

Suppose that we are given three pitchers of water, of sizes 10, 7, and 4 quarts. Initially, the 10 quart pitcher is full and the other two empty. We can pour water from one pitcher into another, pouring until the receiving pitcher is full or the pouring pitcher is empty. Is there a way to pour among pitchers to obtain exactly 2 quarts in the 7 or 4 quart pitcher? If so, find a minimum sequence of pourings to get 2 quarts.

Example 4.2

Three jealous wives and their husbands come to a river. The party must cross the river (from near shore to far shore) in a boat that can hold at most two people. Find a sequence of boat trips that will get the six people across the river without ever letting any husband be alone (without his wife) in the presence of another wife.

Example 4.3

Find all ways to place eight nontaking queens on an 8×8 chessboard. Recall that one queen can capture another queen if they are both in the same row or the same column or a common diagonal.

4.3 ROOTED TREES

In this section, we change our emphasis from trees to directed trees. A *directed tree* is an orientation digraph whose underlying graph is a tree. A directed tree T is called a *rooted tree* if there exists a vertex r of T, called the *root*, such that for every vertex v of T, there is a path from r to v in T.

Let T be a rooted tree. It is customary to draw T with root r at the top, at level 1. The vertices adjacent from r are placed one level below, at level 2. Any vertex adjacent from a vertex at level 2 is at level 3, and so on. See Figure 4.2 for illustration. In general, every vertex at level $i > 1$ is adjacent from exactly one vertex at level $(i - 1)$. More formally, a vertex x in a rooted tree with root r is at level i if and only if the path from r to x in T has length $(i - 1)$. Each arc is directed from a vertex at some level j to a vertex at level $(j + 1)$. The largest integer h for which there is a vertex at level h in a rooted tree is called its *height*.

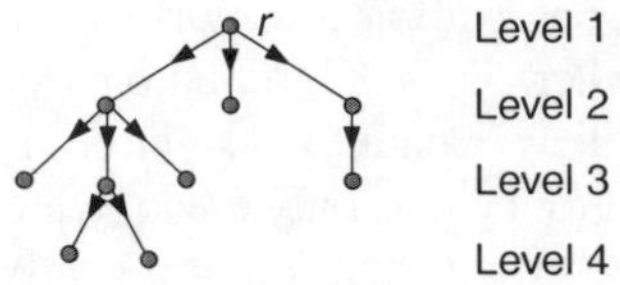

FIGURE 4.2 A rooted tree.

THEOREM 4.3 A directed tree T is a rooted tree if and only if T contains a vertex r with $\deg^-(r)=0$, and $\deg^-(v)=1$ for all other vertices v of T.

Proof: Let T be a rooted tree with root r. Then $\deg^-(r)=0$. Let v be any nonroot vertex of T. Suppose that v is at level i. Obviously, $i>1$. Thus, v is adjacent from exactly one vertex at level $(i-1)$. We have $\deg^-(v)=1$.

Conversely, let G be a directed graph containing a vertex r with $\deg^-(r)=0$ such that $\deg^-(v)=1$ for all vertices with $v\neq r$. Let u_1 be any vertex of T with $u_1\neq r$. Since $\deg^-(u_1)=1$, there is a vertex u_2 such that u_2 is adjacent to u_1. If $u_2\neq r$, then there exists a vertex u_3 adjacent to u_2. These vertices u_1, u_2, and u_3 are the beginning of a sequence.

We continue this sequence as far as possible. First, we claim that this sequence contains only distinct vertices. Suppose that $u_i=u_j$ with $i<j$ and the vertices $u_i, u_{i+1}, \ldots, u_{j-1}$ are distinct. Then in the underlying tree, $u_i, u_{i+1}, \ldots, u_{j-1}, u_j=u_i$ is a cycle, which is impossible. Thus, the sequence is finite; say, $u_1, u_2, \ldots, u_n$. Since only $\deg^-(r)=0$, $u_n=r$. Therefore, $\langle r=u_n, u_{n-1}, \ldots, u_2, u_1\rangle$ is a path from r to u_1. Hence, G is a rooted tree with root r. □

With Theorem 4.3, every rooted tree contains a unique root. Since it is common to draw the root r of a rooted tree T at the top and the remaining vertices at the appropriate level i $(i=2,3,\ldots)$ below r, all arcs are directed downward. Hence, the directions indicated in such a drawing are superfluous. Consequently, it is customary to draw a rooted tree in this manner, without arrows on the edges.

Let T be a rooted tree. Suppose that vertex v of T is adjacent to u lies in the level below v. We say that u is called a *child* of v, and v is the *parent* of u. Suppose that there is a path from v to w in T such that w lies below v. We say that w is a *descendant* of v and v is an *ancestor* of w. Let v be a vertex of T. We consider the subtree of a rooted tree T that induces by a vertex v and all of its descendants as a rooted tree with root v. This subtree is called the *maximal subtree* of T rooted at v. A vertex with no children is called a *leaf*. All other vertices are called *internal vertices*.

A *rooted plane tree* is a rooted tree with a left-to-right ordering specified for the children of each vertex. A *binary tree* is a rooted plane tree in which each vertex has at most two children, and each child is designated as its *left child* or *right child*. The subtree rooted at a children of the root are the *left subtree* and the *right subtree* of the tree. A *k-ary tree* allows each vertex up to k children. A rooted plane tree is called a *full m-ary tree* if every vertex of T has m children or no children. Thus, in a *full binary tree* every vertex has two children or no children.

THEOREM 4.4 A full m-ary tree with i internal vertices has order $(mi+1)$.

Proof: Let T be a full m-ary tree with n vertices having i internal vertices. Every internal vertex of T has m children, and every child has a unique parent. Since only the root is not a child of other vertices, $n = mi + 1$. □

COROLLARY 4.1 *Every full binary tree with i internal vertices has* $(i+1)$ *leaves.*

Proof: Suppose that T is a full binary tree with i internal vertices. By Theorem 4.4, T has $(2i+1)$ vertices. Since T has i internal vertices, the remaining $(i+1)$ vertices are leaves. □

THEOREM 4.5 Suppose that T is a binary tree of height h and order n. Then

$$h \leq n \leq 2^h - 1$$

Proof: For $1 \leq k \leq h$, let p_k be the number of vertices of T at level k. Obviously, $\sum_{k=1}^{h} p_k = n$, $p_k \geq 1$ for each k, and $p_k \leq 2p_{k-1}$ for $1 < k \leq h$. By induction, we can prove $p_k \leq 2^{k-1}$. Using the fact that

$$\sum_{k=1}^{h} 2^{k-1} = 1 + 2 + \cdots + 2^{h-1} = 2^h - 1$$

we conclude that

$$h = \sum_{k=1}^{h} 1 \leq n = \sum_{k=1}^{h} p_k \leq \sum_{k=1}^{h} 2^{k-1} = 2^h - 1$$

which gives the desired result. □

COROLLARY 4.2 *Suppose that T is a binary tree of height h and order n. Then*

$$h \geq \lceil \log_2((n+1)/2) \rceil - 1$$

An *hth complete m-ary tree* is a full m-ary tree with $(m^h - 1)/(m-1)$ vertices and of height h. In particular, the complete binary tree of height h, denoted by BT_h, is a rooted tree given by $V(BT_h) = \{1, 2, \ldots, 2^h - 1\}$, and $(i, j) \in E(BT_h)$ if and only if $\lceil j/2 \rceil = i$. The vertex 1 is the *root* of BT_h. The level of a vertex i is $\lceil \log_2 i \rceil + 1$.

The binary tree is a commonly used data structure. Three common methods are used to recover information from its inherent relationships; namely, inorder, preorder, and postorder traversal. These three algorithms are described as follows:

1. **Procedure inorder**(r)
   a. if $T \neq \emptyset$, then
   b. inorder(left child of v)

c. visit the present vertex
d. inorder(right child of v)

2. **Procedure preorder**(r)

a. if $T \neq \emptyset$, then
b. visit the present vertex
c. preorder(left child of v)
d. preorder(right child of v)

3. **Procedure postorder**(r)

a. if $T \neq \emptyset$, then
b. postorder(left child of v)
c. postorder(right child of v)
d. visit the present vertex

It is worthwhile to give at least one application for each tree traversal.

4.4 COUNTING TREES

There are $2^{C(n;2)}$ simple graphs with vertex set $\{1, 2, \ldots, n\}$. In this section, we count the trees with this vertex set and extend this to count spanning trees in arbitrary graphs. Cayley proved that there are n^{n-2} trees with a fixed set of n vertices. Suppose that $n = 1$ or 2. It is easy to check whether there is only one tree with n vertices. Suppose that $n = 3$. We can check whether there is still only one isomorphic class; a path of three vertices. However, there are three trees, considering the vertex in the center. Suppose that $n = 4$. There are 4 stars and 12 paths. We have 16 trees in total.

We use $\tau(G)$ to denote the number of spanning trees in G. Let K_n be the complete graph defined on the set $\langle n \rangle$, where $\langle n \rangle$ denotes the set $\{1, 2, \ldots, n\}$.

THEOREM 4.6 **(Cayley's Formula [37])** $\tau(K_n) = n^{n-2}$.

Proof: **(Prüfer [274])** Let A be the set of spanning trees in the complete graph K_n defined on the set $\langle n \rangle$. Let B be the set of all sequences of length $(n-2)$ with entries from $\langle n \rangle$. Obviously, $|B| = n^{n-2}$. To prove $|A| = n^{n-2}$, we establish a bijection between A and B by computing the Prüfer sequence $f(T)$ for a labeled tree T as follows.

We iteratively delete the leaf with smallest label and append the label of its *neighbor* to the sequence. After $(n-2)$ iterations, a single edge remains and we have produced a sequence $f(T)$ of length $(n-2)$. The sequence for the tree in Figure 4.3 is 144771, leaving the edge (1,8).

Next, we define a function g that produces a tree from each sequence a. Later, we prove $g = f^{-1}$. We begin with a forest consisting of n isolated vertices $1, 2, \ldots, n$. At the ith step, let x be the label in position i of a. Let y be the smallest label that does not appear in later positions and has not been marked *finished*. Add the edge (y, x) and mark y *finished*. After exhausting a in $(n-2)$ steps, join the two

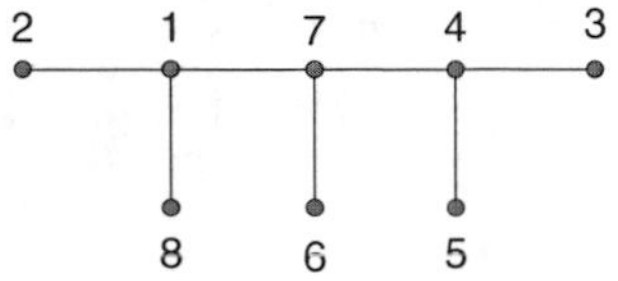

FIGURE 4.3 Illustration of the Prüfer sequence.

remaining unfinished vertices by an edge. For example, $g(144771)$ is the tree in Figure 4.3.

Vertices cannot be marked finished until they no longer appear in the sequence, and we finish one at each step. Hence, we start the ith step with $(n-i+1)$ unfinished vertices and $(n-i)$ remaining entries in a. This implies that y can be chosen as described and the algorithm does create a graph with $(n-1)$ edges. We show that it is a tree by showing that it is connected. Each step joins two unfinished vertices and marks one of them finished. With the basis remarked previously, it follows inductively that after i steps the graph has $(n-i)$ components, each containing one unfinished vertex. The edge added at the last step, therefore, connects the graph.

Hence, $g(a)$ is a tree. We want to prove that $g=f^{-1}$. At each step in computing $f(T)$, we can mark the deleted leaf as *finished*. The labels that do not appear in the remainder of the sequence we generated are the unfinished vertices that are now leaves. Because the next leaf deleted is the least, the edge deleted at each stage in computing $f(T)$ from T is precisely the edge added when computing $g(f(T))$. Therefore, $g(f(T))=T$. □

COROLLARY 4.3 *The number of trees with vertex set $\langle n \rangle$ in which vertices $1,2,\ldots,n$ have degrees $d_1,d_2,\ldots,d_n$, respectively, is $(n-2)!/\prod_{i=1}^{n}(d_i-1)!$.*

Proof: As we delete the vertex x from T while constructing the Prüfer sequence, all neighbors of x except one have already been deleted. We recorded x once for each such neighbor. After the deletion of x, x never appears again. Suppose that x remains at the end. One edge incident to x also remains. Thus, x appears $(\deg(x)-1)$ times in the sequence.

Therefore, we can count trees with each i having degree d_i by counting sequences of length $(n-2)$ having (d_i-1) copies of i for each i. Suppose that we assign subscripts to the copies of each i to distinguish them. There are $(n-2)!$ sequences. However, the copies of each i are in fact indistinguishable. We have counted each desired arrangement $\prod_{i=1}^{n}(d_i-1)!$ times, once for each way to order the subscripts on each type of label. □

EXAMPLE 4.4

Let us count those trees with vertices $\{1,2,3,4,5,6,7\}$ that have degrees (3, 1, 2, 3, 1, 1, 1), respectively. We compute $(n-2)!/\prod_{i=1}^{n}(d_i-1)!=30$. These trees are suggested in

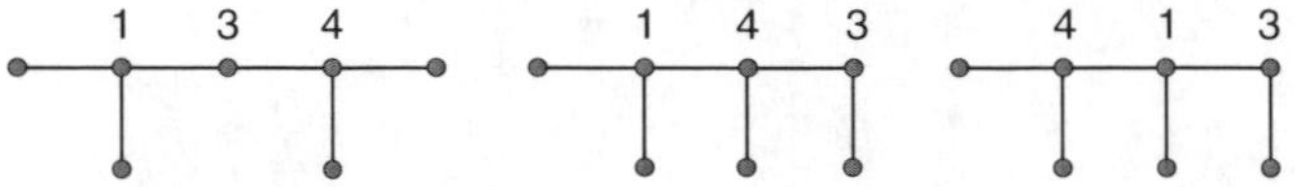

FIGURE 4.4 Illustration of Example 4.4.

Figure 4.4. For the first figure, it corresponds to six trees by picking from the remaining four vertices where two are adjacent to vertex 1. For each of the remaining figures, it corresponds to twelve trees by picking the neighbor of vertex 3 from the remaining four and then picking the neighbor of the central vertex from the remaining three.

There is a recursive way to count spanning trees by separately counting those that contain a particular edge e and those that omit e.

Let $e=(u, v)$ be an edge of G. The *contraction* of e is the operation of replacing u and v by a single vertex whose incident edges are the edges other than e that were incident to u or v. The resulting graph, denoted $G \cdot e$, has one fewer edge than G.

Visually, we think of contracting e as shrinking e to a single point. Contraction may produce multiple edges. To count spanning trees correctly, we must keep the multiple edges. However, in other applications of contraction, the multiple edges may be irrelevant. Since no spanning tree contains a loop, we may freely discard loops that arise in contraction. The following recurrence applies for all graphs.

PROPOSITION 4.4 $\tau(G)=\tau(G-e)+\tau(G \cdot e)$.

Proof: The spanning trees of G that omit e are precisely the spanning trees of $G-e$. Since there is a natural bijection between the spanning tree of $G \cdot e$ and spanning trees of G contain e, the number of spanning trees that contain e is $\tau(G \cdot e)$. Thus, $\tau(G)=\tau(G-e)+\tau(G \cdot e)$. □

EXAMPLE 4.5

It is easy to confirm that $\tau(G-e)=\tau(G \cdot e)=4$ in Figure 4.5. By Proposition 4.4, $\tau(G)=8$.

Computation using the recurrence in Proposition 4.4 requires initial conditions for graphs with no edges. Suppose that G is a graph with one vertex and no edge. Then $\tau(G)=1$. Suppose that G has more than one vertex and no edge. Then $\tau(G)=0$. However, the recursive computation is impractical for large graphs. We will introduce the following Matrix Tree Theorem to compute $\tau(G)$ using a determinant. However, we omit the proof, because it is rather difficult.

Let G be a loopless graph G with the vertices set $\{v_1, v_2, \ldots, v_n\}$. Let a_{ij} be the number of edges of the form (v_i, v_j). The *Kirchoff matrix* associated with G is the matrix $K(G)$ with entry (i,j) being $-a_{ij}$ when $i \neq j$ and $d(v_i)$ when $i=j$.

THEOREM 4.7 **(Matrix Tree Theorem, Kirchoff [201])** Let K^* be the matrix obtained by deleting any row s and column t of $K(G)$. Then $\tau(G)=(-1)^{s+t}\det(K^*)$.

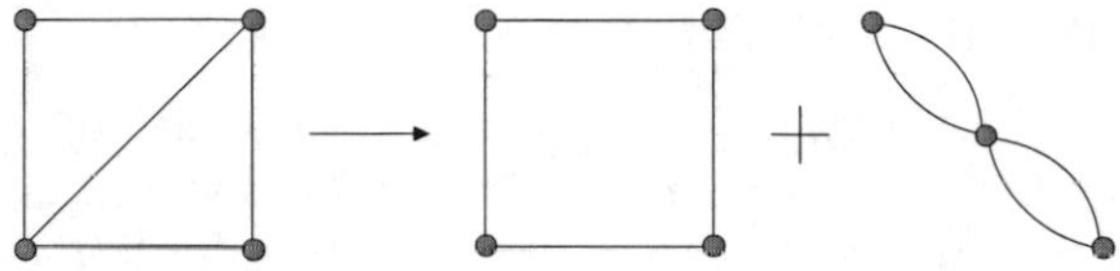

FIGURE 4.5 Illustration of Proposition 4.4.

$$K(D)^+ = \begin{bmatrix} 2 & -1 & -1 \\ 0 & 0 & 0 \\ 0 & -1 & 1 \end{bmatrix}$$

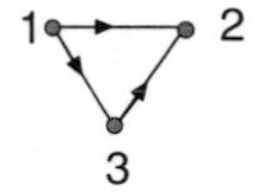

$$K(D)^- = \begin{bmatrix} 0 & -1 & -1 \\ 0 & 2 & 0 \\ 0 & -1 & 1 \end{bmatrix}$$

FIGURE 4.6 Illustration of Theorem 4.8.

EXAMPLE 4.6

Let G be the graph in Example 4.5. The vertex degrees are 3,3,2,2. We can form the associated Kirchoff matrix on the left below. Take the determinant of the matrix in the middle. We obtain $\tau(G) = 8$.

$$\begin{pmatrix} 3 & -1 & -1 & -1 \\ -1 & 3 & -1 & -1 \\ -1 & -1 & 2 & 0 \\ -1 & -1 & 0 & 2 \end{pmatrix} \to \begin{pmatrix} 3 & -1 & -1 \\ -1 & 2 & 0 \\ -1 & 0 & 2 \end{pmatrix} \to 8$$

Tutte [314] extends this theorem to directed graphs.

An *out-tree* is an orientation of a tree having a root of in-degree 0 and all other vertices of in-degree 1. An *in-tree* is an out-tree with its edges reversed.

Let D be a loopless digraph with the vertices set $\{v_1, v_2, \ldots, v_n\}$. Let D^-, D^+ be the diagonal matrices of in-degrees and out-degrees in G. Let A' be the matrix with i and j entries as the number of edges from v_i to v_j. Then the *Kirchoff in-matrix* $K^-(D)$ is $D^- - A'$ and the *Kirchoff out-matrix* $K^+(D)$ is $D^+ - A'$.

THEOREM 4.8 **(Directed Matrix Tree Theorem, Tutte [314])** The number of in-trees rooted at v_i is the value of any cofactor in the *ith* row of $K(D)^+$. Similarly, the number of out-trees rooted at v_i is the value of any cofactor in the *ith* column of $K(D)^-$.

EXAMPLE 4.7

The digraph in Figure 4.6 has two out-trees rooted at 1 and two in-trees rooted at 2. We can check these results using determinants.

4.5 COUNTING BINARY TREES

In the following discussion, we are going to count the number of binary trees with n vertices. However, we first study the set of well-formed sequences of parentheses. The set of *well-formed sequences of parentheses* is defined by the following recursive definition:

1. The empty sequence is well-defined.
2. If A and B are well-formed sequences, so is AB (the concatenation of A and B).
3. If A is well-formed, so is (A).
4. There are no other well-formed sequences.

For example, (( ) (( ))) is well-formed; (( ))) (( ) is not. The following lemma can easily be proved by induction.

LEMMA 4.2 A sequence of (left and right) parentheses is well-formed if and only if it contains an even number of parentheses, half of which are left and the other half are right, and we read the sequence from left to right, the number of right parentheses never exceeds the number of left parentheses.

Let B_n denote the set of all binary tree with n nodes, W_n denote the set of all well-formed parentheses. In the following discussion, we will prove that $|B_n| = |W_n| = \frac{1}{1+n}\binom{2n}{n}$.

LEMMA 4.3 $|B_n| = |W_n|$.

Proof: We are going to set a one-to-one correspondence between B_n and W_n. For any binary tree T in B_n, we use the following algorithm to transform T into a sequence $f(T)$ in W_n.

Procedure transform(T)

1. if $T \neq \emptyset$, then
2. write "("
3. transform(left child of v)
4. write ")"
5. transform(right child of v)

It can be easily proved by induction that the function f has one-to-one correspondence. Hence, $|B_n| = |W_n|$. □

THEOREM 4.9 $|B_n| = |W_n| = \frac{1}{1+n}\binom{2n}{n}$.

Proof: We first show a one-to-one correspondence between the non-well-formed sequences of n left and n right parentheses, and all sequences of $(n-1)$ left parentheses and $(n+1)$ right parentheses.

Let $p_1p_2 \ldots p_{2n}$ be a sequence of n left and n right parentheses that is not well-formed. By Lemma 4.2, there is a prefix of it that contains more right parentheses than left. Let j be the least integer such that the number of right parentheses exceeds the number of left parentheses in the subsequence $p_1p_2 \ldots p_j$. Clearly, the number of right parentheses is then one larger than the number of left parentheses, or j is not the least index to satisfy the condition. Now, invert all p_i's for $i > j$ from left parentheses to right parentheses, and from right parentheses to left parentheses. Clearly, the number of left parentheses is now $(n-1)$ and the number of the right parentheses is now $(n+1)$.

Conversely, given any sequence $p_1p_2 \ldots p_{2n}$ of $(n-1)$ left parentheses and $(n+1)$ right parentheses, let j be the first index such that $p_1p_2 \ldots p_j$ contains one more right parenthesis than left parentheses. If we now invert all parentheses in the section $p_{j+1}p_{j+2} \ldots p_{2n}$ from left to right and from right to left, we get a sequence of n left and n right parentheses that is not well-formed. This transformation is the inverse of the one of the previous paragraph. Thus, the one-to-one correspondence is established.

The number of sequence of $(n-1)$ left and $(n+1)$ right parentheses is

$$\binom{2n}{n-1}$$

We can choose the places for the left parentheses, and the remaining places will have right parentheses. Thus, the number of well-formed sequences of length n is

$$\binom{2n}{n} - \binom{2n}{n-1} = \frac{1}{1+n}\binom{2n}{n}$$

Hence, the theorem is proved. □

These number are called *Catalan numbers*. There is another method for computing the number of binary trees of n vertices through the use of a generating function as follows. Let b_n denote the number of binary trees with n nodes.

1. Show that $b_0 = 1$ and that $n \geq 1$, $b_n = \sum_{k=0}^{n-1} b_k b_{n-1-k}$.
2. Let $B(x)$ be the generating function $B(x) = \sum_{n=0}^{\infty} b_n x^n$. Show that $B(x) = xB(x)^2 + 1$ and hence $B(x) = \frac{1}{2x}(1 - \sqrt{1-4x})$.
3. Using the Taylor's expansion, show that $b_n = \frac{1}{1+n}\binom{2n}{n}$.

4.6 NUMBER OF SPANNING TREES CONTAINS A CERTAIN EDGE

The concept of *all-terminal network reliability*, proposed by Kel'mans [198], is defined as follows: Let G be a graph that represents the topology of a network. Assume that the vertices are perfectly reliable, but the edges operate independently with the same probability, p. The network is operational if the underlying probabilistic graph is connected. Colbourn's monograph [74] is an excellent survey of the work of this problem.

In computing, the all-terminal reliability of a network appears to be intractable. When the edges are very unreliable, maximizing the number of spanning trees is critical to maximizing reliability. Maximizing the number of spanning trees over all graphs with the same numbers of edges and vertices does not guarantee a network that is most reliable for all values of p. There are some studies concentrated on the problem of determining the graph on n vertices and m edges with the maximum number of spanning trees [27,55,197,199,308,338]. However, in many application, the graph topology is already given. For this situation, Tsen et al. [306] consider the problem that determines the effect of edge deletion with respect to the number of spanning trees.

The digraph D *associated with* an undirected graph G is constructed by replacing each edge (u, v) of G with the arcs (u, v) and (v, u). The following lemma can easily be obtained.

LEMMA 4.4 Assume that G is an undirected graph and D is the digraph associated with G. There is a one-to-one correspondence between the spanning trees of G and the out-tree rooted (in-tree) at any vertex of D.

From the one-to-one correspondence stated in Lemma 4.4, the number of spanning trees of an undirected graph G containing a particular edge (u, v) equals the number of out-trees in the associated digraph containing (u, v) plus the number of out-trees in the associated digraph containing (v, u).

To avoid treating the arcs that are incident to the root vertex r of D as a special case, we augment the original digraph D with a new vertex u, along with a new arc (u, r). Again, there is a one-to-one correspondence between the out-tree of D and those of the augmented digraph. Correspondingly, the *augmented Kirchoff* matrix A for the rooted digraph D is defined as the Kirchoff matrix K with one added to the diagonal entry for the root vertex (or equivalently, the matrix of the augmented digraph with *the row and the column corresponding to the new root* removed). We have the following corollary.

COROLLARY 4.4 *The number of out-tree of D rooted r equals the determinant of the corresponding augmented Kirchoff matrix.*

Because exactly one arc enters each nonroot vertex in an out-tree, there is a one-to-one correspondence between the out-trees of D that contain arc a and those $D \circ a$, which is obtained from D by removing all arcs, except a, that enter the terminus of a. Thus, the number of spanning trees containing an arc can be computed by evaluating the determinant of the augmented Kirchoff matrix for $D \circ a$.

Suppose that we first compute the inverse of the augmented Kirchoiff matrix. The determinant of the augmented Kirchoff matrix for $D \circ a$ can be found in constant time. The *classical adjoint matrix* of a matrix A, $adj(A) = [b_{ij}]$, is $\det(A) \cdot A^{-1}$. The value of b_{ij} is $(-1)^{i+j} \cdot \det(A_{ji})$, where A_{ji} is the matrix obtained from A by deleting row j and column i. The determinant of $A = [\alpha_{ij}]$ is $\sum_{j=1}^{n} \alpha_{ij} \cdot (-1)^{i+j} \cdot \det(A_{ij})$. This is *expansion by cofactors across row i*. Thus, the determinant of A can be computed by cross-multiplying row i of A and column i of the matrix $adj(A)$. The crucial observation is that column i of $adj(A)$ contains the cofactors of A obtained by deleting row i of A (along with one of the columns). Hence, any change to row i of A has no effect on the column i of $adj(A)$.

Let e_i be the vector which is 1 in the position i but is 0 otherwise. The augmented Kirchoff matrix of $D \circ a$ is obtained from A by changing row j to $(e_j - e_i)$. Therefore, from the previous observations, the number of out-tree containing a is $b_{jj} - b_{ij}$. We have the following theorem.

THEOREM 4.10 Assume that A is the augmented Kirchoff matrix of a digraph G. For each arc (i, j), the number of out-trees containing this arc is $b_{jj} - b_{ij}$. Alternatively, for the undirected case, the number of trees containing edge (i, j) is: $b_{jj} - b_{ij} + b_{ii} - b_{ji}$.

Example 4.8

Let G be the graph shown in Figure 4.7a. The corresponding Kirchoff matrix of G is the matrix A in Figure 4.7d. The associated digraph D is shown in Figure 4.7b. Obviously, the corresponding Kirchoff matrix of D is again the matrix A. With Lemma 4.4, the number of spanning trees of G is the number of rooted trees rooted at the vertex 1. We add to the digraph D a root u and joining u to the vertex 1 to obtain a directed graph D' as shown in Figure 4.7c. Obviously, the number of spanning rooted trees of D rooted at 1 is the number of spanning rooted roots of D' rooted at u. Thus, the number of spanning rooted trees of D rooted at 1 is the determinant of the augmented Kirchoff matrix B shown in Figure 4.7e. Obviously, the number of spanning trees of G containing the edge (6, 2) is the sum of the number of spanning trees of D' containing the arc (6, 2) and the number of spanning trees of D' containing the arc (2, 6). Let T be any spanning rooted tree of D rooted at 1 containing the arc (6, 2). Note that the in-degree of the vertex 2 in T is 1. Obviously, T does not have any arcs in $\{(1, 2), (3, 2), (4, 2), (5, 2)\}$. Thus, the number

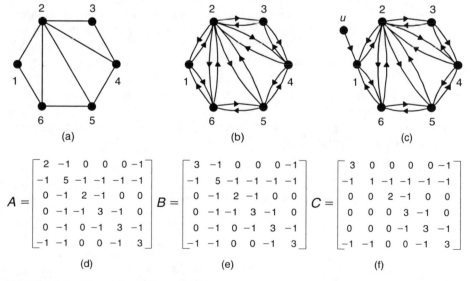

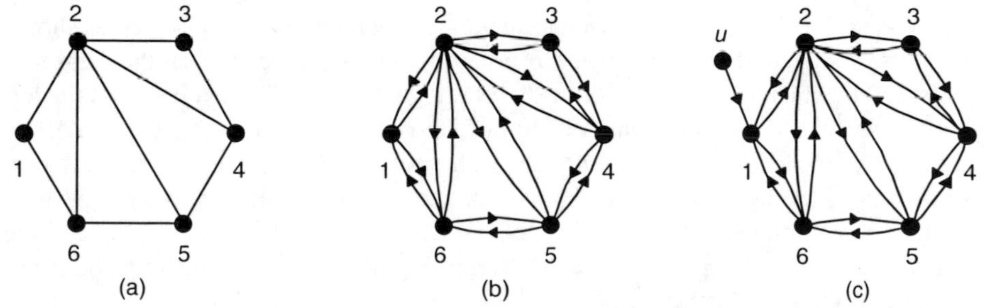

$$A = \begin{bmatrix} 2 & -1 & 0 & 0 & 0 & -1 \\ -1 & 5 & -1 & -1 & -1 & -1 \\ 0 & -1 & 2 & -1 & 0 & 0 \\ 0 & -1 & -1 & 3 & -1 & 0 \\ 0 & -1 & 0 & -1 & 3 & -1 \\ -1 & -1 & 0 & 0 & -1 & 3 \end{bmatrix}$$

(d)

$$B = \begin{bmatrix} 3 & -1 & 0 & 0 & 0 & -1 \\ -1 & 5 & -1 & -1 & -1 & -1 \\ 0 & -1 & 2 & -1 & 0 & 0 \\ 0 & -1 & -1 & 3 & -1 & 0 \\ 0 & -1 & 0 & -1 & 3 & -1 \\ -1 & -1 & 0 & 0 & -1 & 3 \end{bmatrix}$$

(e)

$$C = \begin{bmatrix} 3 & 0 & 0 & 0 & 0 & -1 \\ -1 & 1 & -1 & -1 & -1 & -1 \\ 0 & 0 & 2 & -1 & 0 & 0 \\ 0 & 0 & 0 & 3 & -1 & 0 \\ 0 & 0 & 0 & -1 & 3 & -1 \\ -1 & -1 & 0 & 0 & -1 & 3 \end{bmatrix}$$

(f)

FIGURE 4.7 Illustration of Example 4.8.

of spanning rooted trees of D rooted at 1 containing the arc (6, 2) is the determinant of the matrix C shown in Figure 4.7f.

4.7 EMBEDDING PROBLEM

Both the complete binary tree BT_n and the hypercube Q_n are widely used in computer architecture. We would like to know whether BT_n is a subgraph of Q_n. Obviously, BT_1 is a subgraph of Q_1 and BT_2 is a subgraph of Q_2.

THEOREM 4.11 BT_n is a subgraph of Q_n if and only if $n = 1, 2$.

Proof: We note that both BT_n and Q_n are connected bipartite graphs. The bipartition of any connected graph is unique. It is observed that both the sizes of the bipartition for Q_n are 2^{n-1}. However, one of the sizes of the bipartition of BT_n is at least $2^{n-1} + 2^{n-3}$ if $n \geq 3$. By the pigeonhole principle, it is easy to see that BT_n is not a subgraph of Q_n if $n \geq 3$. □

Despite the preceeding argument, BT_n is almost a subgraph of Q_n. In particular, a very similar graph, the 2^n-node double-rooted complete binary tree, denoted $DRCB(n)$, is a subgraph of Q_n. $DRCB(n)$ is a complete binary tree with the root replaced by a path of length 2. For example, the graph $DRCB(4)$ is shown in Figure 4.8.

THEOREM 4.12 $DRCB(n)$ is a subgraph of Q_n for $n \geq 1$.

Proof: We prove this theorem by induction. Obviously, $DRCB(1)$ is a subgraph of Q_1. Assume that $DRCB(k)$ is a subgraph of Q_k for $1 \leq k < n$. We partition Q_n into two Q_{n-1}. Each contains 2^{n-1} nodes $DRCB(n-1)$ trees by induction. When embedding the two $DRCB(n-1)$ trees, we need to make sure that the trees are oriented as shown in Figure 4.9. In particular, we want to orient the two subcubes so that the right root of the left tree is matched to the left root of the right tree so that the left root of the left tree is matched to the son of the left root of the right tree, and so that the son of the right root of the left tree is matched to the right root of the right tree. By the symmetry properties of the hypercubes, we know that such an orientation is always possible. We can now complete the embedding of the $DRCB(n)$ tree by adding the matching edges and cross-linking the trees as shown in Figure 4.9. This completes the proof that $DRCB(n)$ is a subgraph of Q_n. □

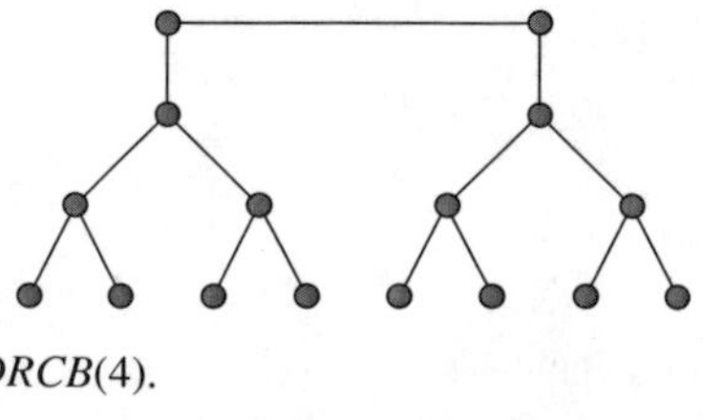

FIGURE 4.8 The graph $DRCB(4)$.

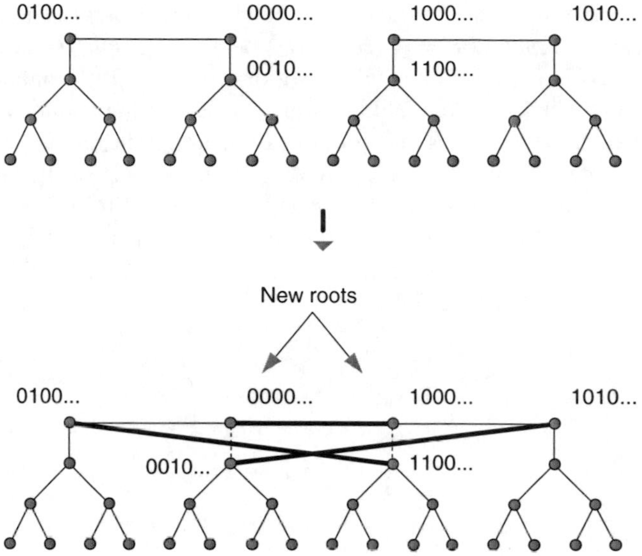

FIGURE 4.9 Embedding $DRCB(n)$ into Q_n.

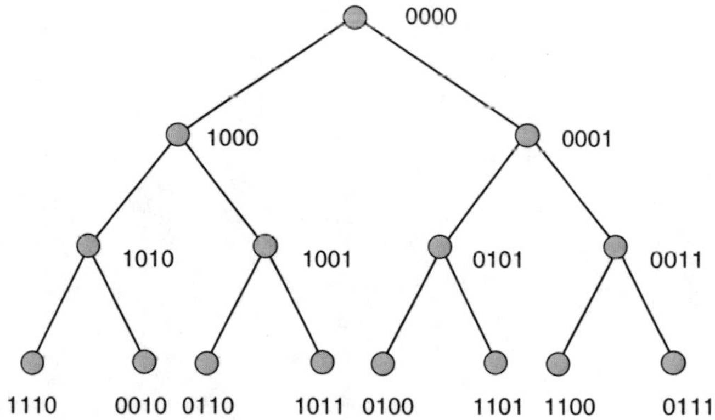

FIGURE 4.10 An embedding of BT_4 into FQ_4.

Similarly, we would like to know whether BT_n is a subgraph of the folded hypercube FQ_n. Note that FQ_n is a bipartite graph if n is odd. Using an argument similar to that in Theorem 4.11, we have the following theorem.

THEOREM 4.13 BT_n is not a subgraph of FQ_n if n is odd and $n \geq 3$.

So, the remaining cases are that n is even and $n \geq 4$. Figure 4.10 shows an embedding of BT_4 into FQ_4. In Ref. 65, Choudum and Nahdini proved that BT_n is a subgraph of FQ_n if n is even.

More generally, we can consider the following terminologies. Let $G=(V(G), E(G))$ be a *guest graph* and $H=(V(H), E(H))$ be a *host graph*. A one-to-one function ϕ from $V(G)$ into $V(H)$ is an *embedding* of G into H. Two common measures of quality of an embedding are the dilation that measures the communication delay and the expansion that measures the processor utilization. The *dilation* of the embedding ϕ, *dilation*(ϕ), is defined as $\max\{d_H(\phi(x), \phi(y)) \mid (x, y) \in E(G)\}$. The *expansion* of the embedding ϕ, *expansion*(ϕ), is defined as $|V(H)|/|V(G)|$. For example, the complete binary tree BT_n with n being even can be embedded into FQ_n with dilation 1 and expansion $2^n/(2^n-1)$. There are some studies on the embedding problems [17,32,106,159–162,222,253,267,270].

5 Eulerian Graphs and Digraphs

5.1 EULERIAN GRAPHS

Some say that graph theory was born in the city of Königsberg, located on the banks of the Pregel. The river surrounded the island of Kneiphof, and there were seven bridges linking the four land masses of the city, as illustrated in Figure 5.1a. It seems that the residents wanted to know whether it was possible to take a stroll from home, cross every bridge exactly once, and return home. The problem reduces to traversing the graph in Figure 5.1b, where the vertices represent land masses and the edges represent bridges. We want to know when a graph contains a single closed trail traversing all its edges.

The Swiss mathematician Leonhard Euler observed [100] that Königsberg had no such traversal. Because a closed trail contributes twice to the degree of a vertex for each visit, the degree of each traversed vertex is even. Moreover, the edge set generates a connected component. All vertex degrees of the graph in Figure 5.1b are odd. Thus, it fails the necessary condition. Euler stated that the condition is also sufficient. In honor of Euler, we say that a graph whose edges constitute a single closed trail is *Eulerian*. In fact, Euler's paper, which appeared in 1741, gave no proof that the obvious necessary condition is sufficient. Hierholzer [145] gave the first published proof. The diagram in Figure 5.1b did not appear until 1894 (see [342] for a discussion of the historical record).

We use the term *circuit* as another name for *closed trail.* A circuit containing every edge of G is a *Eulerian circuit*. We also use *odd vertex* or *even vertex* to indicate the parity of a vertex degree. Moreover, we say that a graph is *even* if all its vertices are even. Recall that a *nontrivial* graph is a graph having at least one edge. A graph is *Eulerian* if it has a Eulerian circuit.

LEMMA 5.1 Nontrivial maximal trails in even graphs are closed.

Proof: Assume that T is a nontrivial maximal trail in an even graph G. Since T is maximal, T includes all edges of G incident to its final vertex v. Moreover, T has an odd number of edges incident to v if T is not closed. □

THEOREM 5.1 A graph G is Eulerian if and only if the degree of each vertex is even and its edge set generates a connected component.

Proof: Assume that graph G is Eulerian. From the previous discussion, we know that the degree of each vertex is even and its edge set generates a connected component. Conversely, suppose that the degree of each vertex of G is even and the edge set of G generates a connected component. Let T be a trail in G of maximum length. Obviously,

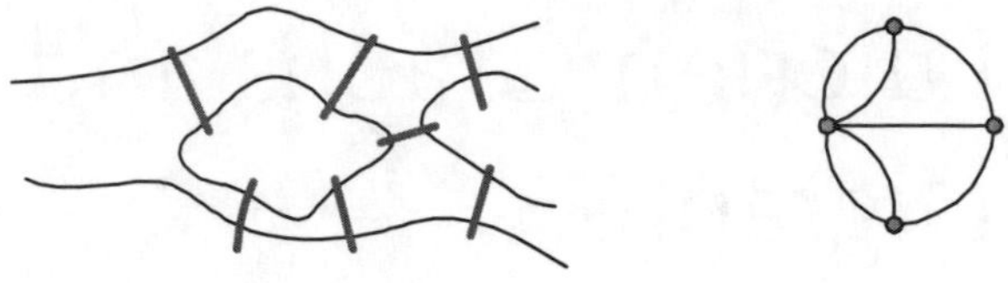

FIGURE 5.1 The Königsberg seven-bridge problem.

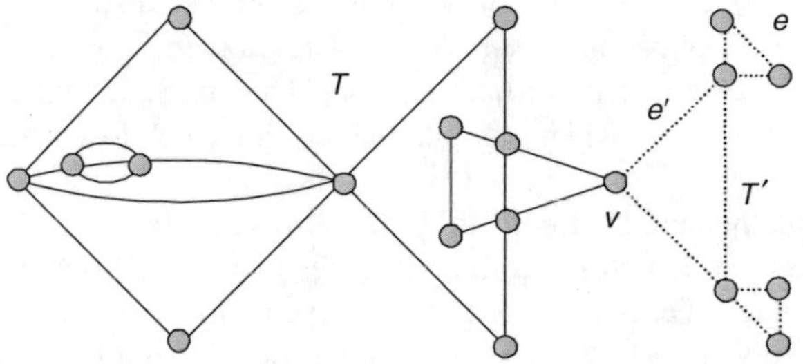

FIGURE 5.2 Illustration of Theorem 5.1.

T is a maximal trail. By Lemma 5.1, T is closed. Let $G' = G - E(T)$. Since the closed trail T has an even degree at every vertex, G' is an even graph. We will claim that $E(G') = \emptyset$. Obviously, T traverses every edge of G and is a Eulerian circuit once our claim holds.

Suppose that $E(G') \neq \emptyset$. Since G has one nontrivial component, G has a path from e to a vertex of T. Thus, some edge e' of G' is incident to some vertex v of T. Let T' be any maximal trail in G' beginning from v along e'. See Figure 5.2 for illustration. By Lemma 5.1, T' is closed. Hence, we can detour from T along T' when we reach v, and then complete T to obtain a longer trail than T. However, T is the longest trail in our hypothesis. Thus, $E(G') = \emptyset$. □

This proof and the resulting algorithm are basically those of Hierholzer [145]. Before discussing constructive aspects, we generalize to all connected graphs.

Given a figure G to be drawn on paper, how many times must we pick up and move the pen (or plotter) to draw? This is the minimum number of pairwise edge-disjoint trails whose union is $E(G)$. Since the number of trails needed to draw G is the sum of the number needed to draw each component, we may reduce the problem to connected graphs. Recall that every circuit is a closed trail. The graph G in Figure 5.3 has four odd vertices and we can decompose it into two trails. Adding dashed edges forms G to G', as in Figure 5.3, making it Eulerian.

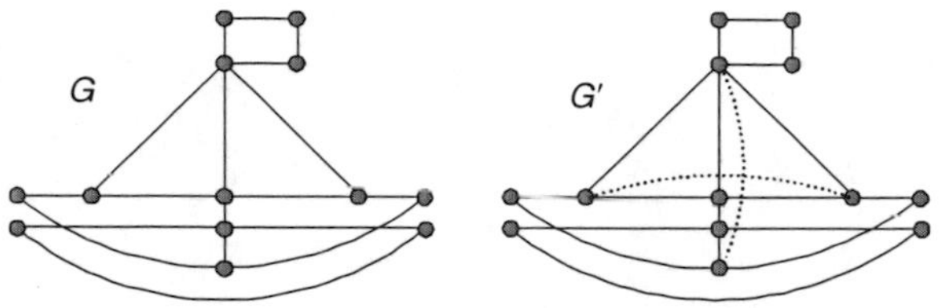

FIGURE 5.3 The graphs G and G'.

THEOREM 5.2 Let G be a connected nontrivial graph with $2k$ odd vertices. The minimum number of pairwise edge-disjoint trails covering the edges is $max\{k, 1\}$.

Proof: A trail contributes an even degree to every vertex, except that a nonclosed trail contributes an odd degree to its endpoints. Therefore, a partition of the edges into trails must have some trail ending at each odd vertex. Note that each trail has (at most) two ends. It requires at least k trails. Since $e(G) > 0$, it also requires at least one trail. Thus, one trail suffices if $k = 0$.

To prove that k trails suffice when $k > 0$, we can pick up the odd vertices in G arbitrarily and form G' by adding a copy of each pair as an edge, as illustrated in Figure 5.3. Obviously, the resulting G' is connected and even. By Theorem 5.1, G' has an Eulerian circuit. As we traverse the circuit, we start a new trail in G each time we traverse an edge of $G' - E(G)$. This yields k-disjoint trails partitioning $E(G)$. □

The proof of Theorem 5.1 provides an algorithm that constructs an Eulerian circuit. It iteratively merges circuits into the current circuit until all edges are absorbed. In the following, we will introduce an algorithm that approaches the *draw the figure* problem more directly, building a circuit one edge at a time without backtracking. An Eulerian circuit can start anywhere. However, an Eulerian trail must start at an odd vertex. We will not traverse an edge whose deletion cuts the remaining graph into two nontrivial components, because it could not return to pick up the stranded edges. Once we avoid such an edge, we can complete the traversal.

ALGORITHM: (Fleury's Algorithm—constructing Eulerian trails.)

Input. A graph G with one nontrivial component and at most two odd vertices.
Initialization. Start at a vertex that has an odd degree unless G is even, in which case start at any vertex.
Iteration. From the current vertex, traverse any remaining edge whose deletion from the remaining graph does not leave a graph with two nontrivial components. Stop when all edges have been traversed.

THEOREM 5.3 Suppose that G has one nontrivial component and at most two odd vertices. Then Fleury's algorithm constructs an Eulerian trail.

Proof: We use induction on $e(G)$. The claim is immediate for $e(G) = 1$. Assume that $e(G) > 1$, and assume that the claim holds for graphs with $e(G) - 1$ edges. Suppose

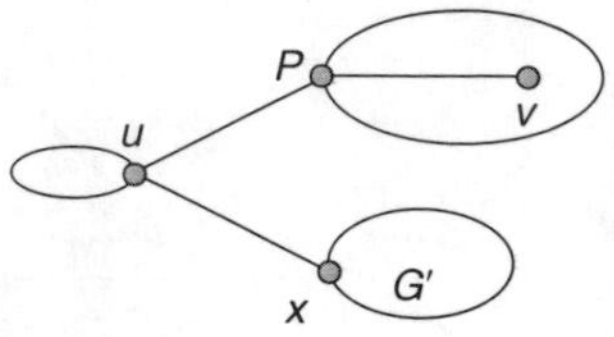

FIGURE 5.4 Illustration of Theorem 5.3.

that G is even. Then every edge of G is in a cycle. Thus, G has no cut edge. Hence, starting with some vertex v along an edge (u, v) leaves a graph with two odd vertices. By induction hypothesis, the algorithm completes the circuit of G by completing a trail from u to v in $G-(u, v)$.

Suppose that G has two odd vertices u and v. Assume that $d(u)=1$. Then traversing (u,x) leaves one nontrivial component. Thus, we assume that $d(u)>1$. Let P be a path from u to v and (u,x) be an edge incident to u but not on P. Suppose that (u,x) is a cut edge. Since u and v are connected in $G-(u,x)$, x would be the only odd vertex in its component of $G-(u,x)$ (see Figure 5.4). Thus, (u,x) is not a cut edge. Therefore, u has an incident edge (u,x) whose traversal leaves a graph $G-(u,x)$ having at most one nontrivial component and at most two odd vertices.

Suppose that $x=v$. Then $G-(u,x)$ is even. Suppose that $x \neq v$. Then the vertices of $G-(u,x)$ are even except for x and v. By induction, the algorithm completes an Eulerian trail of G from u to v by traversing an Eulerian trail of $G-(u,x)$ from x to v. □

5.2 EULERIAN DIGRAPHS

A *circuit* in a digraph traverses edges from tail to head. An *Eulerian* circuit traverses every edge. A digraph is *Eulerian* if it has an Eulerian circuit. Now, we study the necessary and sufficient condition for Eulerian digraphs.

A digraph G is *strongly connected* or *strong* if there is a path from u to v in G for every ordered pair $u, v \in V(G)$.

Each entrance to a vertex is followed by a departure, so an Eulerian digraph satisfies $\deg^+(u)=\deg^-(v)$ for all $u \in V(G)$. This is also sufficient, if each edge of G is reachable from every other. The proofs are similar to those of undirected graphs. We present a constructive proof, analogous to Fleury's algorithm. It needs only one search computation to create a tree at the beginning. After that, the information needed to construct the circuit is present in the tree.

ALGORITHM: (Directed Eulerian circuit.)

Input. A digraph G satisfies $\deg^+(u)=\deg^-(u)$ for all $u \in V(G)$.

Step 1. We choose any vertex $v \in V(G)$. Let G' be the digraph obtained from G by reversing direction on each edge. Using BFS and DEF on G' to construct a tree T' consisting of paths from v to all other vertices.

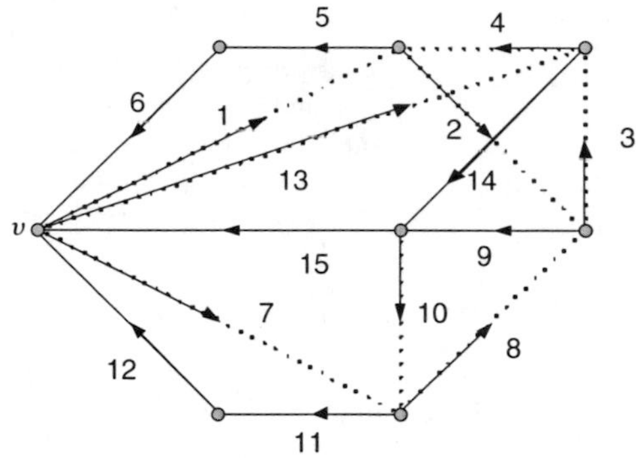

FIGURE 5.5 The digraph for Example 5.1.

Step 2. Let T be the reversal of T'. Thus, T contains a path from u to v in G for each $u \in V(G)$. We specify an arbitrary ordering of the edges that leaves each vertex u, except that $u \neq v$ and the edge leaving u in T must come last.

Step 3. Construct an Eulerian circuit from u as follows: whenever u is the current vertex, exit along the next unused edge in the order specified for each edge leaving u. □

Example 5.1

In the digraph in Figure 5.5, the solid edges indicate an in-tree T of paths into v. We follow the Eulerian circuit starting with edge 1. Then, we avoid using edges of T until we have no choice. Assuming that the ordering at v places 1 before 7 before 13, the algorithm traverses the edges in the order indicated.

THEOREM 5.4 Assume that G is a digraph with one nontrivial component and $\deg^+(u) = \deg^-(u)$ for all $u \in V(G)$. Then the previous algorithm constructs a Eulerian circuit of G.

Proof: First, we construct T'. We must prove that T forms a spanning tree of the subgraph generated by the set of edges in G. The search (BFS) from v reaches a new vertex at each step. Otherwise, all edges between the current reached set R and the remaining vertices enter R. Note that each edge within R contributes once to the in-degree and once to the out-degree of vertices in R. Moreover, each edge entering R contributes only to the in-degree. Hence, the sum of in-degrees in R exceeds the sum of out-degrees, which contradicts our hypothesis that $\deg^+(u) = \deg^-(u)$ for each vertex u.

Now we use T to build a trail as directed. We need to prove the trail obtained by the algorithm is indeed an Euler circuit. Since $\deg^+(u) = \deg^-(v)$, there exists an edge leaving v once we enter a vertex $u \neq v$. Since we cannot use an edge of T until it is the only remaining edge leaving its tail, we cannot use all edges entering v until we have finished all the other vertices. Since T contains a path from each vertex to v, the trail can end only at v. Moreover, we end at v only if we have used all edges leaving

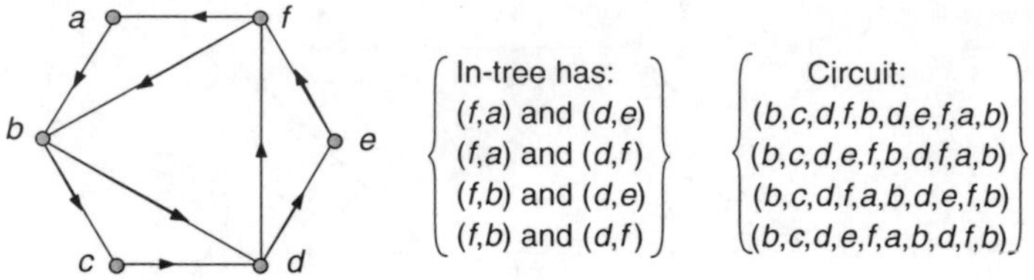

FIGURE 5.6 The digraph for Example 5.2.

v, and hence also all edges entering v. Thus, the trail obtained by the algorithm is an Euler circuit. □

EXAMPLE 5.2

Consider digraph in Figure 5.6. Every in-tree to b contains the edges $(a, b), (e, f), (c, d)$, exactly one of $\{(f, a), (f, b)\}$, and exactly one of $\{(d, e), (d, f)\}$. Thus, there are exactly four in-trees to b. It is observed that the number of legal sets of orderings of the edges leaving the vertices for each in-tree is $\prod (d_i - 1)! = (0!)^3(1!)^3 = 1$. Hence, we can generate one Eulerian circuit for each in-tree, starting along the edge (b, c) from b.

Two Eulerian circuits are the same if the successive pairs of edges are the same. From each in-tree to v, the previous algorithm can generate $\prod_{u \in V(G)} (\deg^+(u) - 1)!$ different Eulerian circuits. The last out-edge is fixed by the tree for vertices other than v. Since we consider only the cyclic order of the edges, we may choose any particular edge e to start the ordering of edges leaving v. Any change in the exit orderings at vertices specifics at some point different choices for the next edge. Thus, these circuits are distinct. Similarly, the circuits obtained from distinct Ts are distinct. Therefore, we have $c \prod_{u \in V(G)} (\deg^+(u) - 1)!$ distinct Eulerian circuits where c is the number of in-trees to v.

In fact, these are all Eulerian circuits. This provides a combinatorial proof that the number of in-trees to each vertex of an Eulerian digraph is the same. Since the graph obtained by reversing all the edges has the same number of Eulerian circuits, the number of out-trees from any vertex also has this same value, c. The value c can be computed using the Directed Matrix Tree Theorem.

THEOREM 5.5 **(van Aardenne-Ehrenfest and de Bruijn [323], Tutte and Smith [316]).** Assume that G is a Euler digraph. Let $d_i = \deg^+(v_i) = \deg^-(v_i)$. Then the number of Eulerian circuits is $c \prod_{i=1}^{n} (d_i - 1)!$ where c is the number of in-trees or the number of out-trees from any vertex.

Proof: We have observed that the previous algorithm generates these many distinct Eulerian circuits using in-trees to v with e being the first edge in the exit ordering at v. We need only to show this produces all Eulerian circuits.

To find the tree and ordering for an Eulerian circuit C, we can follow C from e and record the exit ordering of edges leaving each vertex. For each vertex w other than v, we collect in a set T the last edge leaving w. Since the last edge leaving a vertex occurs in C after all edges entering it, each edge in T extends to a path in T

that reaches v. With $(n-1)$ edges, T thus forms an in-tree to v. Furthermore, C is the circuit obtained from T and these exit orderings by the previous algorithm. □

By Theorem 5.5, we can compute the number of Euler circuits for directed graphs. However, we do not know how to compute the number of Euler circuits for undirected graphs.

5.3 APPLICATIONS

5.3.1 Chinese Postman Problem

Suppose that a mail carrier traverses all edges in a road network, starting and ending at the same vertex. The edges have nonnegative weights representing distance or time. We seek a closed walk of minimum total length that uses all the edges. This is called the *Chinese Postman Problem*, in honor of the Chinese mathematician Guan Meigu [131], who proposed it.

Suppose that every vertex is even. The graph G is Eulerian and the answer is the sum of the weights. Otherwise, we must repeat edges. Every traversal is an Eulerian circuit of a graph obtained by duplicating edges of G. Finding the shortest traversal is equivalent to finding the minimum total weight of edges whose duplication will make all vertices even. We say *duplication* because if we need not use more times in making all vertices even, then deleting two of those copies will leave all vertices even. There may be many ways to choose the duplicated edges.

Example 5.3

Let us consider the graph shown in Figure 5.7. The eight outer vertices are odd. If we match them around the outside to make the degrees even, the extra cost is $4+4+4+4=16$ or $2+6+6+2=16$. We can do better by using all the vertical edges, which total only 12. □

The example illustrates that adding an edge from an odd vertex to an even vertex makes the even vertex odd. We must continue adding edges until we complete a trail to an odd vertex. The duplicated edges must consist of a collection of trails that pair

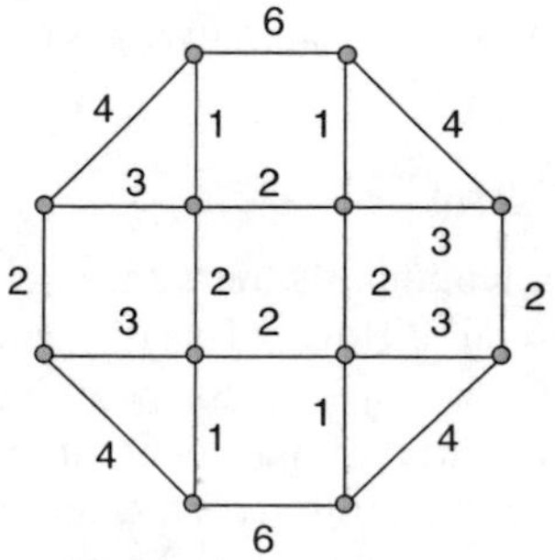

FIGURE 5.7 The graph for Example 5.3.

up the odd vertices. We may restrict our attention to paths pairing up the odd vertices, though the paths may need to intersect.

Edmonds and Johnson [91] described a way to solve the Chinese Postman Problem. Suppose that there are only two odd vertices. We can find the shortest path between them and solve the problem. Suppose that there are $2k$ odd vertices. Then we can find the shortest paths connecting each pair of odd vertices. We use these lengths of weights as the edges of K_{2k}. Then our problem is equivalent to finding the minimum total weight of k edges that pair up these $2k$ vertices. An exposition appears in Ref. 126.

5.3.2 Street-Sweeping Problem

In traversing a road network, it may be important to consider the direction of travel along segments of road, such as when sweeping the curb. The curb must be traversed in the direction that traffic flows, so a two-way street contributes a pair of opposite directed edges, and a one-way street contributes two edges in the same direction. We consider a simple version of the street-sweeping problem, discussed in more detail in Ref. 276, following the results of Tucker and Bodin [309].

In New York City, parking is prohibited from some street curbs each weekday to allow for street sweeping. This defines a *sweep subgraph* G of the full digraph of curbs. G consists of the edges available for sweeping, directed to follow traffic. The question is how to sweep G while minimizing the *deadheading time*; that is, the time when no sweeping is being done. Suppose that the in-degree equals the out-degree at each vertex of the subgraph. Obviously, no deadheading is needed. Otherwise, we must duplicate edges or add edges from the full digraph to obtain a Eulerian superdigraph of the sweep digraph. Each edge e in the full graph H has a deadheading time $t(e)$.

Let X be the set of vertices with in-degrees exceeding out-degrees. We set $\sigma(x) = \deg_G^-(x) - \deg_G^+(x)$ for $x \in X$. Let Y be the set of vertices with out-degrees exceeding in-degrees. We set $\rho(y) = \deg_G^+(y) - \deg_G^-(y)$ for $y \in Y$. Note that $\sum_{x \in X} \sigma(x) = \sum_{y \in Y} \rho(y)$. The Eulerian superdigraph must add $\sigma(x)$ edges with tails at $x \in X$ and $\rho(y)$ edges with heads at $y \in Y$. Since we must finish with a supergraph having degrees in balance, we can think of the additions as paths from X to Y. The cost of adding a path from x to y is the distance from x to y in H.

We now have a *Transportation Problem*. Given supplies $\sigma(x)$ for each $x \in X$ and demand $\rho(y)$ for $y \in Y$ and cost $c((x, y))$ for sending a unit from x to y, with $\sum \sigma(x) = \sum \rho(y)$, the problem is to satisfy the demands with the least total cost. The problem is discussed at length in Ford and Fulkerson [110].

5.3.3 Drum Design Problem

There are 2^n binary strings of length n. Is there a cyclic arrangement of 2^n strings of n where consecutive digits are all distinct? This can be used to test the position of a rotating drum, as observed by Good [127]. Suppose that the drum has 2^n rotational positions, and a strip around the surface is partitioned into 2^n portions that can be coded 0 or 1. Taps reading n consecutive portions can determine the position of the drum if the specific cyclic arrangement exists. For $n = 4$ we can use (0000111101100101), shown in Figure 5.8.

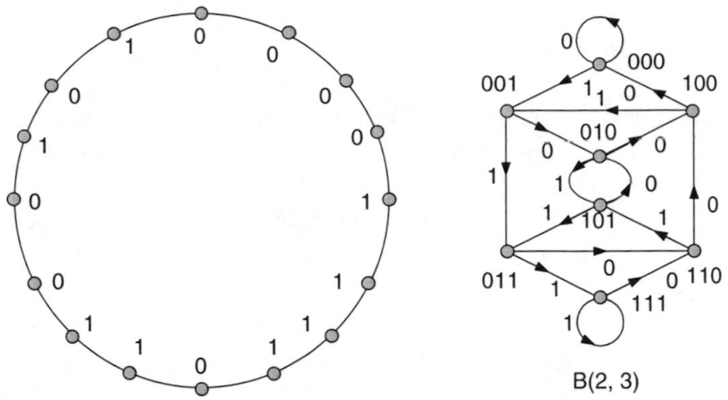

FIGURE 5.8 The drum design.

We can construct such an arrangement by finding an Eulerian circuit of the de Bruijn digraph $BRU(2, n-1)$. Note that the vertex set of $BRU(2, n-1)$ is the set of all $(n-1)$-digit binary sequences. Place an edge from sequence a to sequence b if the last $(n-2)$ digits of a agree with the first $(n-2)$ digits of b. Label the edge with the final digit of b. For each sequence a, there are two digits that we can append to obtain a new sequence, and hence there are two edges leaving each vertex. Similarly, because there are two digits that we could drop from a sequence to obtain a, there are two edges entering each vertex. Hence, $BRU(2, n-1)$ is Eulerian and has 2^n edges. The graph $BRU(2, 3)$ and the corresponding Euler circuit are shown in Figure 5.8.

THEOREM 5.6 The edges labels in any Eulerian circuit of $BRU(2, n-1)$ form a cyclic arrangement in which the consecutive segments of length n are the 2^n distinct binary vectors.

Proof: Let us traverse an Euler circuit C in $BRU(2, n-1)$. Suppose that we arrive at a vertex whose sequence $a = a_1 a_2 \ldots a_{n-1}$. The previous edge $(n-1)$ labels, looking backward, must be $a_{n-1} a_{n-2} \ldots a_1$, because the label on an edge entering a vertex agrees with the last digit of the sequence at the vertex. If C next traverses an edge with label a_n, then the subsequence ending there is $a_1, a_2, \ldots, a_n$. As the 2^{n-1} vertex labels are distinct and the two edges leaving each vertex have distinct labels, we have 2^n distinct subsequences determined from C. □

5.3.4 Functional Cell Layout Problem

A *series–parallel network* (*network,* for short) N of *type* $t \in \{L, S, P\}$ (which represent *leaf, series,* and *parallel,* respectively) defined on W is recursively constructed as follows:

1. N is a network of type L if $|W| = 1$.
2. Suppose that $|W| > 1$. Then N is a network of either type P or type S and consists of $k \geq 2$ networks $N_1, \ldots, N_k$ as *child subnetworks* parallel or series

connected together where each N_i is defined on a set W_i of type t_i with $t_i \neq t$ and the collection of W_is forms a partition of W.

A network can be expressed by a tree structure. Networks are useful in practice because they correspond to Boolean formulae with series connection (denoted by S) implementing logical-AND and parallel connection (denoted by P) implementing logical-OR. For example, the Boolean function $\mathbf{e} \wedge (\mathbf{a} \vee \mathbf{b}) \wedge (\mathbf{c} \vee \mathbf{d})$ can be represented by the network shown in Figure 5.9.

Obviously, networks can be used as a model for electrical circuits. For example, we can use the tree structure shown in Figure 5.9 to represent the network corresponding to the electrical circuit shown in Figure 5.10. In the tree representation of N, every node together with all of its descendants forms a *subnetwork* of N. A node together with some children and their descendants forms a *partial subnetwork*. The subnetwork of N formed by a child of the root is called a *child subnetwork* of N. The leaf node is labeled x if it is a subnetwork of type L defined on $\{x\}$. Every *internal node* is labeled by S or P according to the type of subnetwork it represents. Note that the order of the subtrees in the tree representation is immaterial, because different orders lead to the same Boolean formula.

On the other hand, every network can also be represented by a *series–parallel graph* (*s.p. graph*, for short), which is an edge-labeled graph with two given

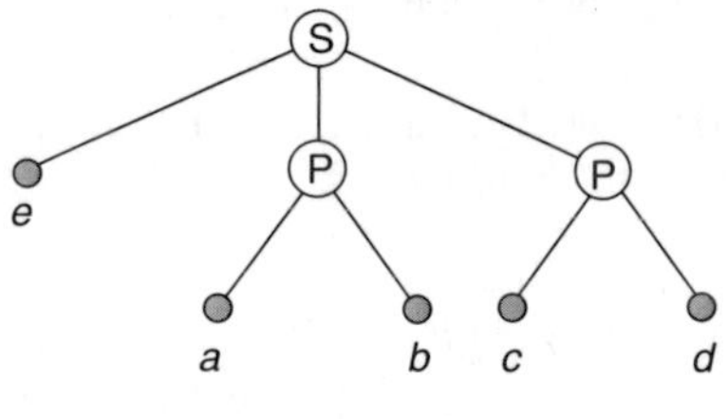

FIGURE 5.9 Tree representation of a series–parallel network.

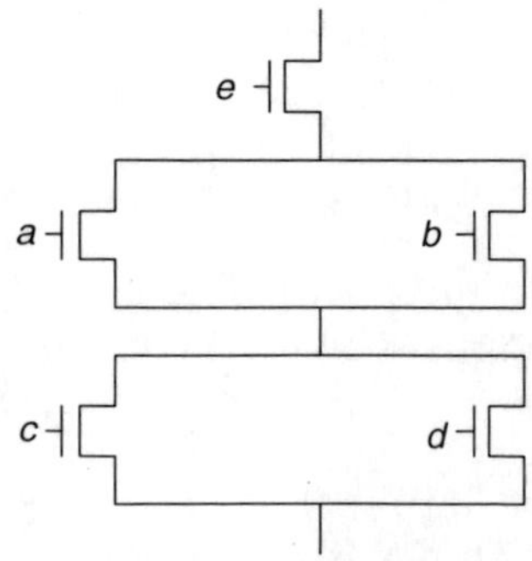

FIGURE 5.10 Electrical circuit corresponding to the network shown in Figure 5.9.

distinguished vertices denoted by s and t. We recursively construct an s.p. graph to represent a network as described:

1. Every network N defined on $W = \{x\}$ of type L is represented by an edge-labeled graph $G[N]$ having only one edge labeled x and having the two endpoints of this edge as distinguished vertices.
2. Assume that N is a network having child subnetworks $N_1, \ldots, N_k$. Let $G[N_i]$ be an s.p. graph representing N_i with the distinguished vertices s_i, t_i for every i. We identify t_i with s_{i+1} for $1 \leq i \leq k-1$. The resulting graph $G[N]$ with the distinguished vertices s_1, t_k represents the network N if N is of type S. We identify all s_i's to obtain a new vertex s and identify all t_i's to obtain a new vertex t. The resulting graph $G[N]$ with distinguished vertices s, t represents the network N if N is of type P.

The subgraph G_i of G induced by the child subnetwork N_i of N is called a *child s.p. subgraph* of G. Note that a graph representation for a network is not unique, because we can vary the order of the subnetworks and the order of the two distinguished vertices to obtain different graph representations. For example, both the nonisomorphic s.p. graphs shown in Figure 5.11 represent the network in Figure 5.9.

Let N be a network defined on set W. We define its *dual network* N' on set W by interchanging the types S or P of each node. For example, the network in Figure 5.12a is the dual network of the network in Figure 5.9. Figures 5.12b and 12c show a graph representation and the corresponding circuit, respectively, of the dual network. We observe that the Boolean formula corresponding to N' is the dual of the Boolean formula that corresponds to N. It is obvious that $(N')' = N$. If two s.p. graphs G, G' represent some network N and its dual, respectively, we say that (G, G') is an *s.p. graph pair*.

Let N be a network. Let $G[N]$ and $G[N']$ be graph representations for the network N and its dual network N', respectively. A sequence of edges $e_1, \ldots, e_m$ is said to form a *common trail* in $(G[N], G[N'])$ if $L = \langle v_0, e_1, v_1, e_2, \ldots, e_m, v_m \rangle$ and $L' = \langle v'_0, e_1, v'_1, e_2, \ldots, e_m, v'_m \rangle$ are trails in $G[N]$ and $G[N']$, respectively. We call (L, L') a *trail pair*. (We use L' to denote a trail for the dual network.) Sometimes,

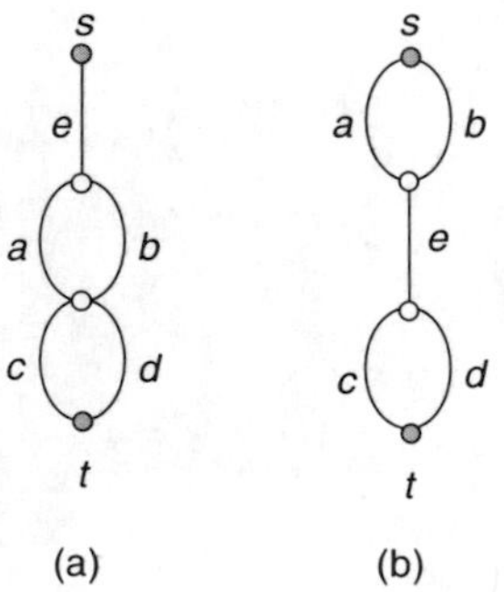

FIGURE 5.11 Two nonisomorphic series–parallel graphs representing the network in Figure 5.9.

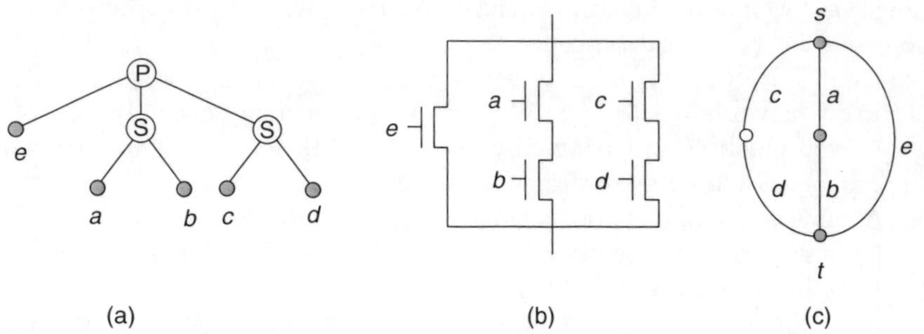

FIGURE 5.12 (a) The dual network of the network in Figure 5.9, (b) electrical circuit corresponding to the network in (a), and (c) a graph representation to the network in (a).

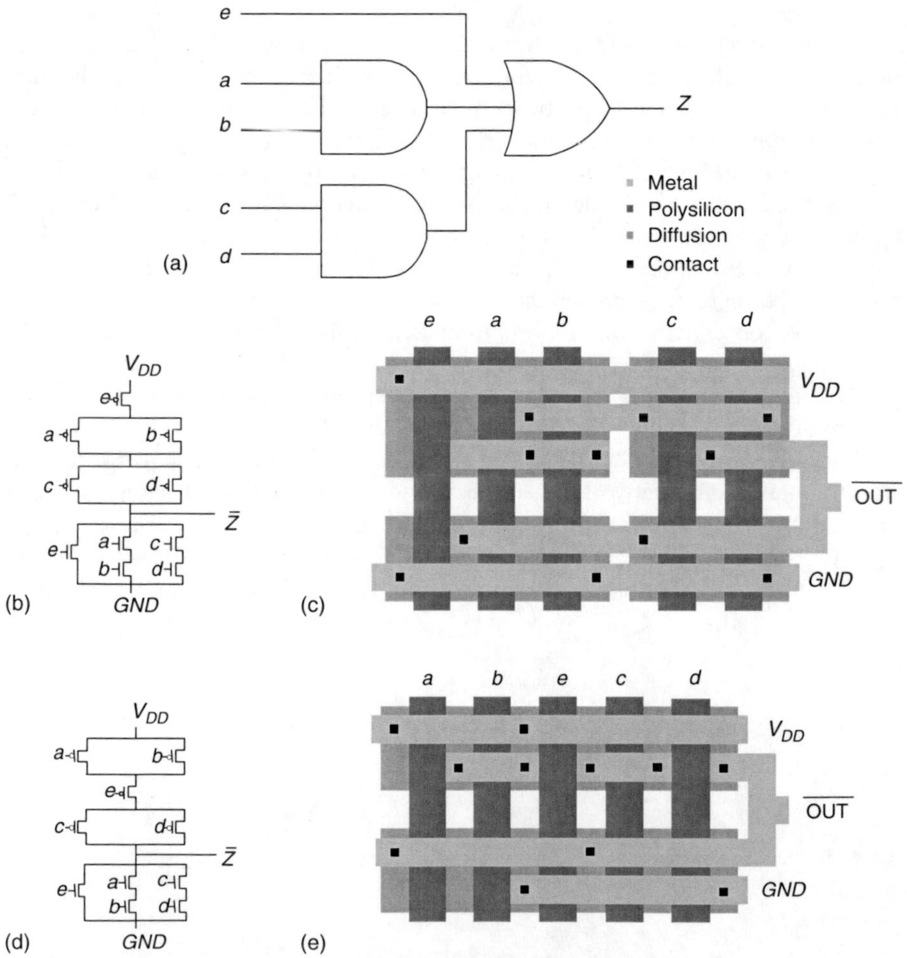

FIGURE 5.13 CMOS functional cell: (a) Gate-level scheme, (b) electric-level scheme, (c) geometric layout corresponding to the scheme in (b), (d) another electric-level scheme, and (e) geometric layout corresponding to the scheme in (d).

we write (L, L') as L for short. Suppose that $e_1, e_2, \ldots, e_m$ are all the edges in both $G[N]$ and $G[N']$. We say that $e_1, e_2, \ldots, e_m$ forms a *common Euler trail* (*circuit*) in $(G[N], G[N'])$. A network N has a *double Euler trail* (DET) if there is a common Euler trail for both some $G[N]$ of N and some $G[N']$ of the dual network N'. We say that $(G[N], G[N'])$ *realizes a DET pair* (L, L') for N and N', where L and L' are the corresponding Euler trails in $G[N]$ and $G[N']$, respectively. We say that a network N is DET if N possesses a DET. For example, the s.p. graphs in Figures 5.11a and 5.12c do not have a common Euler trail, but the ones in Figures 5.11b and 5.12c have a common Euler trail **abecd**. Thus, the network in Figure 5.9 is DET.

Actually, the problem of DET networks arises from a more general problem called $DCT(N)$. Let $DCT(G[N], G[N'])$ be the minimum number of disjoint common trails that cover all of the edges in $G[N]$ and $G[N']$. We define $DCT(N)$ as the minimum of $DCT(G[N], G[N'])$ among all possible graph representations $G[N]$ and $G[N']$. Uehara and vanCleemput [318] proposed a solution method for the layout of cells in the style shown in Figure 5.13. We assume that the height of each cell is fixed by technological considerations. The width of the cell, and therefore the area of the cell, can be minimized by ordering the transistors in the layout so that chains of transistors can share a common diffusion region. Uehara and vanCleemput [318] defined a graph model for functional cells on two dual multigraphs $(G[N], G[N'])$ and proposed a heuristic method for finding a small number of common trails that cover the given $(G[N], G[N'])$. Later, Maziasz and Hayes [249] gave a linear time algorithm for solving $DCT(G[N], G[N'])$ and an exponential algorithm for finding the $DCT(N)$. Several other papers have also explored the use of graph models to find solutions for layout [35, 249, 318]. Ho et al. [154] presented a linear time algorithm to recognize DET networks. We hope that the $DCT(N)$ problem can be solved in the near future.

6 Matchings and Factors

6.1 MATCHINGS

Many applications of graphs involve pairings of vertices. For example, among a set of people, some pairs are compatible as roommates. We want to know under what conditions we can pair them all up. In Chapter 1, we considered the case of jobs and applicants, asking whether all the jobs can be competently filled. Bipartite graphs have a natural vertex partition into two sets, and we want to know whether the edge pair can pair up the two sets. However, the graph need not be bipartite in the roommate problem.

A *matching* of *size* k in a graph G is a set of k pairwise disjoint edges. The vertices belonging to the edges of a matching are *saturated* by the matching. The others are *unsaturated*. A *perfect matching* or *1-factor* is a matching that saturates every vertex of G.

To find a large matching, we could iteratively select an edge disjoint from those previously selected. This yields a *maximal matching*. A maximal matching cannot be enlarged, because edges are incident to all others. A *maximum matching* is a matching of maximum size.

EXAMPLE 6.1

(Maximal $\neq$ maximum.) The smallest graph having a maximal matching that is not a maximum matching is P_4. Obviously, the middle edge forms a maximal edge of size 1. But the two pendant edges form a larger matching.

EXAMPLE 6.2

(Problem of the truncated chessboard.) Consider an 8×8 chessboard whose upper-left and lower-right corner squares have been removed. (See Figure 6.1a.) We have 31 dominoes, each domino covering exactly two adjacent squares of the chessboard. Can we cover the 62 squares of the chessboard with the 31 dominoes?

The problem is equivalent to finding a maximum matching in a graph whose vertices correspond to the squares of the truncated chessboard. In this graph, two vertices are adjacent if they represent adjacent squares in the chessboard. (See Figure 6.1b.) It is easy to see that any matching in Figure 6.1b is not perfect. A simple argument that does not use matching theory shows that no perfect matching is possible. Color the squares of the chessboard black and white (as usual). Note that the truncated chessboard does not have the same number of black and white squares, as the two missing squares must have the same color. Clearly, each arrangement of the dominoes covers the same number of black and white squares. Hence, no perfect matching can exist.

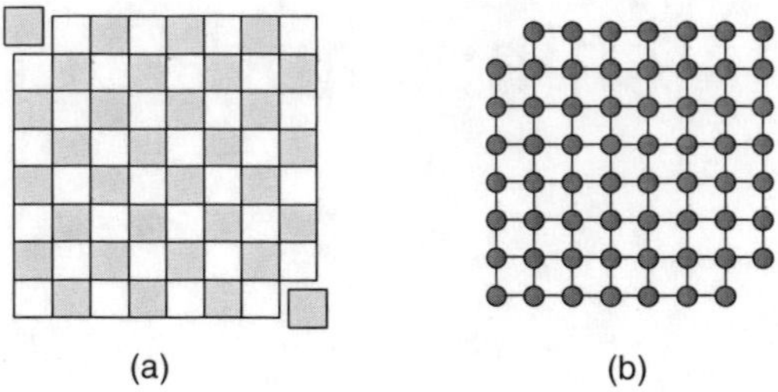

FIGURE 6.1 (a) A truncated chessboard and (b) the graph corresponding to (a).

EXAMPLE 6.3

(Dating problem.) In a coeducational college, each girl has k boyfriends, and each boy has k girlfriends. Is it possible to have a dance in which all students simultaneously dance with one of their friends? Later, it will be shown that this is possible if $k > 0$.

Let M be a matching for graph G. An *M-alternating path* is a path that alternates between edges in M and edges not in M. An M-alternating path P that begins and ends at M-unsaturated vertices is an *M-augmenting path.* We can replace $M \cap E(P)$ by $E(P) - M$ to produce a new matching M' of G with one more edge than M.

The absence of augmenting paths in G is used to characterize a maximum matching. We will prove this by checking the subgraph formed from the union of two matchings and deleting the common edges. We define this operation for any two graphs with the same vertex sets.

Assume that G and H are graphs with vertex set V. The *symmetric deference* $G \Delta H$ is the graph with vertex set V whose edges are all those edges appearing in exactly one G and H. We also use this notation for the set of edges. In particular, if M and M' are matchings, then $M \Delta M' = (M \cup M') - (M \cap M')$.

THEOREM 6.1 **(Berge [22])** A matching M in a graph G is a maximum matching in G if and only if G has no M-augmenting path.

Proof: We prove the contrapositive of each direction. We have noted that an M-augmenting path produces a larger matching. For the converse, suppose that G has a matching M' larger that M. To complete the proof, we need to find an M-augmenting path. Let $F = M \Delta M'$. Because M and M' are matchings, every vertex has at most one incident edge from each of them. Thus, F can be viewed as a subgraph of G having maximum degree at most 2.

Since $\Delta(F) \leq 2$, F consists of disjoint paths and cycles. Furthermore, every path or cycle in F alternates between edges of M and edges of M'. This implies that each cycle in F has even length. Since $|M'| \geq |M|$, F has a component with more edges of M' than of M. See Figure 6.2 for illustration. Such a component can only be a path that starts and ends with an edge of M'. Every such path is an M-augmenting path in G. □

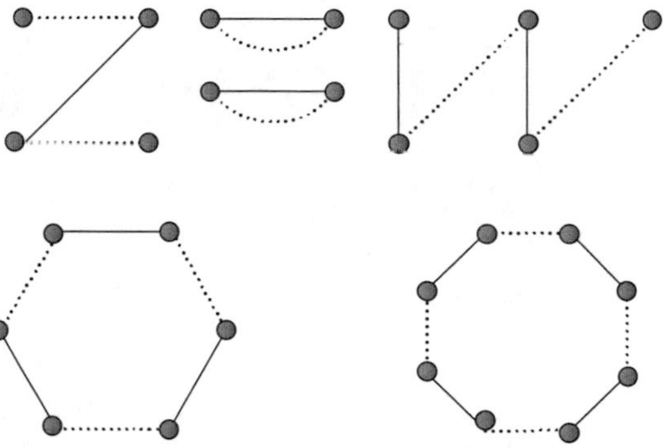

FIGURE 6.2 Illustration of Theorem 6.1.

6.2 BIPARTITE MATCHING

In the example of jobs and applicants, there may be many more applicants than jobs. In this case, we want to fill the jobs without using all the applicants. Hence, when G is a bipartite graph with bipartition X and Y, we may ask whether G has a matching that saturates X. We call such matching a *matching of X into Y*.

Let S be any subset of X. We use $N_G(S)$ or simply $N(S)$ to denote the set of vertices having a neighbor in S.

Suppose that M saturates X. Then $|N(S)| \geq |S|$, because the vertices matches to S are in $N(S)$. Thus, $|N(S)| \geq |S|$ is a necessary condition for M saturating X. Hall proved that this obvious necessary condition is also sufficient.

THEOREM 6.2 **(Hall [137])** Suppose that G is a bipartite graph with bipartition X and Y. Then G has a matching of X into Y if and only if $|N(S)| \geq |S|$ for all $S \subseteq X$.

Proof: Necessity was observed previously. For sufficiency, suppose that G satisfies the condition $|N(S)| \geq |S|$ for all $S \subseteq X$. Let M be a maximum matching of G. Suppose that M does not saturate S. We want to find a set X that violates the hypothesis. Suppose that $u \in X$ is unsaturated. Let S and T be the subsets of X and Y, respectively, that are reachable from u by M-alternating paths.

We claim that M matches T with $S - \{u\}$. The M-alternating paths from u reach Y along edges not in M and reach X along edges in M. Hence, every vertex of $S - \{u\}$ is reached along an edge in M from a vertex in T. Since there is no M-augmenting path, every vertex of T is saturated. Thus, an M-alternating path reaching $y \in T$ extends via M to a vertex of S. Hence, these edges of M establish a bijection between T and $S - \{u\}$. Thus, $|T| = |S - \{u\}|$.

The matching between T and $S - \{u\}$ implies that $T \subseteq N(S)$. In fact, $T = N(S)$. (See Figure 6.3.) An edge between S and a vertex $y \in Y - T$ is an edge not in M. Then we can find an M-alternating path to y that contradicts $y \notin T$. With $T = N(S)$, we have $|N(S)| = |T| = |S| - 1 < |S|$. This contradicts the hypothesis of the theorem. □

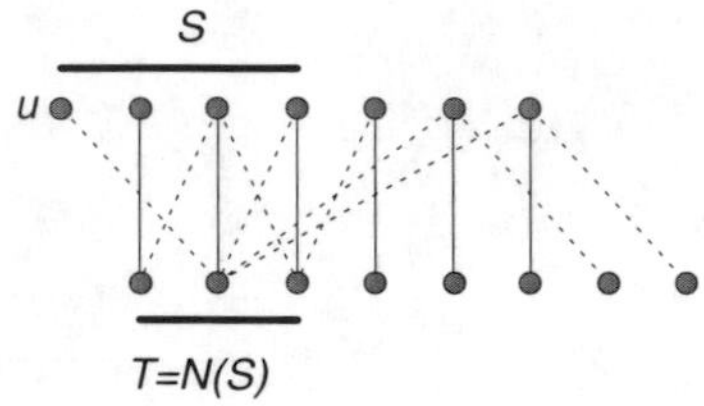

FIGURE 6.3 Illustration of Theorem 6.2.

When the sets of bipartition have the same size, Hall's Theorem is known as the *Marriage Theorem*. The name arises from the scenario of a symmetric compatibility relation between a set of n men and a set of n women. If every man is also compatible with k women and every woman is compatible with k men, then there must be a perfect matching using compatible pairs. The proof applies also to bipartite graphs with multiple edges, which enlarges the scope of applications.

COROLLARY 6.1 *For $k > 0$, every k-regular bipartite graph has a perfect matching.*

Proof: Suppose that the bipartition is X and Y. Counting the edges by endpoints in X and endpoints in Y shows that $k|X| = k|Y|$. Since $k > 0$, $|X| = |Y|$. Suppose that Hall's condition holds. By Theorem 6.2, a matching saturating X is obtained. As $|X| = |Y|$, this matching is a perfect matching.

Now, we check Hall's condition. Consider $S \subseteq X$, and suppose that there are m edges between S and $N(S)$. Since G is k-regular, we have $m = k|S|$. Since these m edges are adjacent to $N(S)$, we have $m < k|N(S)|$. Hence, $k|S| \leq k|N(S)|$, which yields $|N(S)| \geq |S|$. Having chosen $S \subseteq X$ arbitrarily, we have established Hall's condition.

Thus, the corollary is proved. □

With Corollary 6.1, we solved the dating problem in Example 6.3.

Let $\mathcal{A} = \{A_1, A_2, \ldots, A_m\}$ be a collection of sets such that $\cup_{i=1}^{m} A_i = X$. A set $\{x_1, x_2, \ldots, x_m\}$ is a *system of distinct representatives* (SDR) if $x_i \neq x_j$ if $i \neq j$ and $x_i \in A_i$ for $1 \leq i \leq m$.

THEOREM 6.3 Let $\mathcal{A} = \{A_1, A_2, \ldots, A_m\}$ be a collection of sets such that $\cup_{i=1}^{m} A_i = X$. $\mathcal{A}$ has an SDR if and only if $|\cup_{i \in J} A_i| \geq |I|$ for all $I \subseteq [m]$.

Proof: We build a bipartite graph with one partite set corresponding to $\{A_1, A_2, \ldots, A_m\}$ and the other partite set corresponding to the set X. We join an edge between $x \in X$ and A_j if and only if $x \in A_j$. It is easy to see the condition $|\cup_{i \in J} A_i| \geq |I|$ for all $I \subseteq [m]$ is actually Hall's condition. Hence, the theorem is proved. □

Suppose that a graph G does not have a perfect matching. We can prove that a matching M is a maximum matching by proving that G has no M-augmenting path. However, exploring all M-alternating paths to seek an augmentation would take a long time. Instead, we find an explicit structure in G that forbids a matching larger than M. A *dual* optimization problem provides a short proof that the answer is optimal.

A *vertex cover* of a graph $G = (V, E)$ is a subset S of V such that S contains at least one endpoint of every edge in E. The vertices in S *cover* the edges of G.

A road network can be modeled into a graph. The problem of finding a minimum vertex cover of a graph is the problem of installing the minimum number of policemen that can watch the entire road network.

Since no edges of a matching can be covered by a single vertex, the size of every vertex cover is at least the size of every matching. Therefore, exhibiting a matching and a vertex cover of the same size *proves* that each is optimal. Actually, such equality holds for every bipartite graph. However, the equality may not be true for general graphs.

Example 6.4

(Matchings and vertex covers.) Let G and H be the graph shown in Figure 6.4. The graph G has a matching and a vertex cover of size 2. The vertex cover of size 2 prohibits matchings with more than two edges. Moreover, the matching of size 2 prohibits vertex covers with fewer than 2 vertices. In graph H, these sizes differ by one for an odd cycle. The difference can be arbitrarily large in general graphs.

A *min–max relation* is a theorem stating equality between the answers to a minimization problem and a maximization problem over a class of instances. The König-Egerváry Theorem is such a relation for matching and vertex covering in bipartite graphs:

THEOREM 6.4 **(König [203], Egerváry [95])** Assume that G is a bipartite graph. Then the maximum size of a matching in G equals the minimum size of a vertex cover of G (see Figure 6.5).

Proof: Suppose that G has bipartition X and Y. Since distinct vertices must be used to cover the edges of a matching, we have $|U| \geq |M|$ whenever U is a vertex cover and M is a matching in G. Let U be a minimum vertex cover U of G. To prove that equality can always be achieved, we need to construct a matching of size $|U|$.

Let $R = U \cap X$ and $T = U \cap Y$. Moreover, let H and H' be the subgraphs of G induced by $R \cup (Y - T)$ and $T \cup (X - R)$. We use Hall's Theorem to show that H has a matching of R into $Y - T$ and H' has a matching of T into $X - R$. Since H and H' are disjoint, the two matchings together form a matching of size $|U|$ in G.

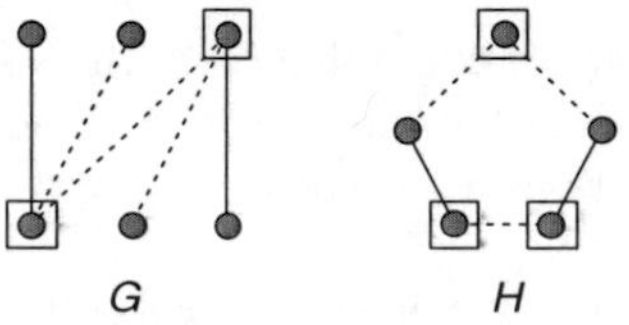

FIGURE 6.4 Graphs for Example 6.4.

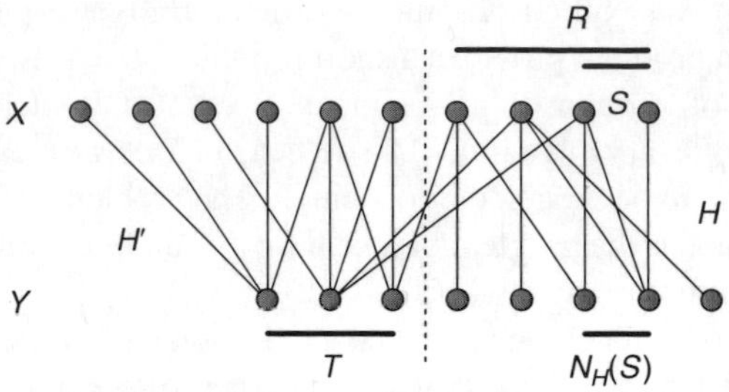

FIGURE 6.5 Illustration of Theorem 6.4.

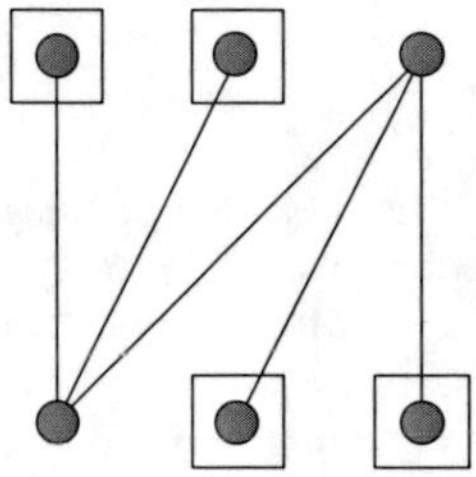

FIGURE 6.6 The independent number of a bipartite graph does not always equal the size of a partite set.

Since $R \cup T$ is a vertex cover, G has no edge from $Y - T$ to $X - R$. Let S be a subset of R. We consider $N_H(S) \subseteq Y - T$. Suppose that $|N_H(S)| < |S|$. Since $N_H(S)$ covers all edges incident to S that are not covered by T, we can substitute $N_H(S)$ for S in U to obtain a smaller vertex cover of G. We get a contradiction. Thus, the minimality of U yields Hall's condition in H. Hence, H has a matching of R into $Y - T$. Applying the same argument to H' yields the rest of the matching. □

We now change the study of independent sets of edges into independent sets of vertices. The *independence number* of a graph is the maximum size of an independent set of vertices. From the graph in Figure 6.6, it is observed that the independent number of a bipartite graph is not always equal to the size of a partite set.

6.3 EDGE COVER

Just as no vertex can cover two edges of a matching, no edge can contain two vertices of an independent set. Again, we have a dual covering problem for an independent number.

An *edge cover* of G is a set of edges that cover the vertices of G. Obviously, only graphs without isolated vertices have edge covers. For the independence and covering problems, the following notations are used:

maximum size of independent set	$\alpha(G)$
maximum size of matching	$\alpha'(G)$

minimum size of vertex cover	$\beta(G)$
minimum size of edge cover	$\beta'(G)$

In this notation, the König–Egerváry Theorem states that $\alpha'(G) = \beta(G)$ for every bipartite graph G. We will prove that $\alpha(G) = \beta'(G)$ for bipartite graphs without isolated vertices. We have already observed that $\beta'(G) \geq \alpha(G)$.

LEMMA 6.1 In a graph G, $S \subseteq V(G)$ is an independent set if and only if $\overline{S}$ is a vertex cover. Hence $\alpha(G) + \beta(G) = n(G)$.

Proof: Suppose that S is an independent set. Then there are no edges within S. Hence, every edge is incident to at least one vertex of $\overline{S}$. Thus, $\overline{S}$ is a vertex cover. Conversely, suppose that $\overline{S}$ is a vertex cover. Then there are no edges between vertices of S. Hence any maximum independent set is the complement of a minimum vertex cover. Moreover, $\alpha(G) + \beta(G) = n(G)$. □

Since the edges of G omitted by a matching need not form an edge cover of G, the relationship between matchings and edge covering is more complicated. Nevertheless, we have a similar formula.

THEOREM 6.5 **(Galli [120])** Assume that G is a graph without isolated vertices. We have $\alpha'(G) + \beta'(G) = n(G)$ (see Figure 6.7).

Proof: Suppose that we use a maximum matching M to construct an edge cover of size $n(G) - M$. Then the smallest edge cover is no bigger than $n(G) - M$. We conclude that $\beta'(G) \leq n(G) - \alpha'(G)$. On the other hand, suppose that we use a minimum edge cover L to construct a matching of size $n(G) - L$. Then the largest matching is no smaller than $n(G) - L$. We conclude that $\alpha'(G) \geq n(G) - \beta'(G)$. These two inequalities complete the proof.

Let M be a maximum matching in G. We can construct an edge cover of G by using M and adding one edge incident to each unsaturated vertex. The total number of edges used is $n(G) - |M|$, as desired. Thus, we save $|M|$ edges because each edge of the matching covers two vertices instead one.

For the other inequality, let L be the minimum edge cover. Suppose that the two endpoints of some edge e in L belong to other edges in L. Then $L - e$ is also an edge cover. Hence, L is not a minimum edge cover. Therefore, L consists of k disjoint stars, for some k. (Note that *stars* are the trees with a diameter of 2). Since L has one edge for each vertex that is not a center of its stars, we have $|L| = n(G) - k$. We can form a matching of size $k = n(G) - |L|$ by choosing one edge arbitrarily from each star in L. □

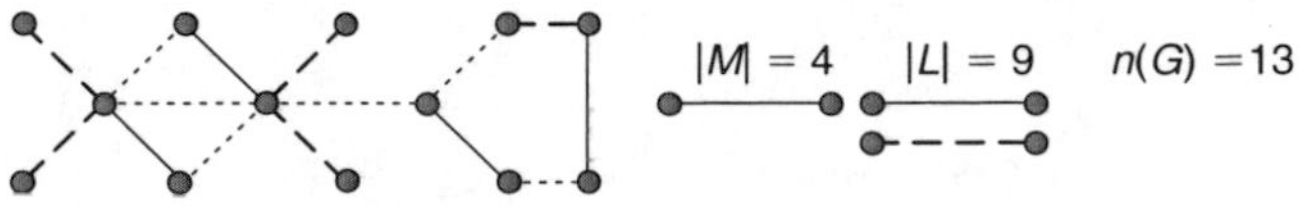

FIGURE 6.7 Illustration of Theorem 6.5.

COROLLARY 6.2 **(König [204])** *Assume that G is a bipartite graph with no isolated vertices. Then* $\alpha(G) = \beta'(G)$. *Thus,* max |*independent set*| = min |*edge cover*|.

Proof: The two proceeding results imply $\alpha(G) + \beta(G) = \alpha'(G) + \beta'(G)$. By subtracting the König-Egerváry min–max relation $\alpha'(G) = \beta(G)$, we obtain $\alpha(G) = \beta'(G)$. □

6.4 PERFECT MATCHING

A *factor* of G is a spanning subgraph of G. A *k-factor* is a spanning k-regular subgraph (a perfect matching is a 1-factor). An *odd component* of a graph is a component of odd order. The number of odd components of H is denoted by $o(H)$.

Tutte found a necessary and sufficient condition for an arbitrary graph to have a 1-factor. Suppose that G has a 1-factor. Let S be a subset of $V(G)$. Obviously, every odd component of $G - S$ has a vertex matched to something outside it, which can only belong to S. See Figure 6.8 for illustration.

As these must be distinct vertices of S, we conclude that $o(G - S) \leq |S|$. Tutte proved that this obvious necessary condition is also sufficient. Many proofs of this have appeared. We present the proof by Lovász using the ideas of symmetric difference and extremality.

THEOREM 6.6 **(Tutte [315])** A graph G has a 1-factor if and only if $o(G - S) \leq |S|$ for every $S \subseteq V(G)$.

Proof: **(Lovász [242])** As noted previously, Tutte's condition is necessary. We need to prove only that Tutte's condition is also sufficient.

We note that Tutte's condition is preserved by addition of edges. Suppose that $G' = G + e$. Let $S \subseteq V(G)$. Because the addition of e combines two components of $G - S$ into one, the number of components that have an odd order does not increase. Thus, $o(G' - S) \leq o(G - S)$.

Therefore, it suffices to consider a simple graph G such that G satisfies Tutte's condition, G has no 1-factor, and adding any edge to G creates a 1-factor. We will obtain a contradiction in every case by constructing a 1-factor in G. This implies that every graph satisfying Tutte's condition has a 1-factor.

Let $S = \emptyset$. Then $o(G) = 0$. Then $n(G)$ is even, because a graph of odd order must have a component of odd order. Let $U = \{x \in V(G) \mid N(x) = V(G) - \{x\}\}$.

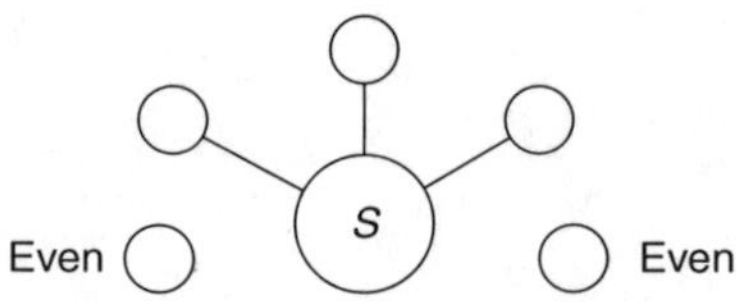

FIGURE 6.8 Illustration of $G - S$.

Suppose that $G-U$ consists of disjoint complete graphs. We build a 1-factor for such a G as follows. The vertices in each component of $G-U$ can be paired up arbitrarily, with one left over in the odd components. As $o(G-U) \le |U|$ and each vertex of U is adjacent to all of $G-U$, we can match these leftover vertices arbitrarily to vertices of U to complete a 1-factor (see Figure 6.9).

This leaves the case where $G-U$ is not a disjoint union of cliques. Here $G-U$ contains two nonadjacent vertices x, z, and a common neighbor y. Since $y \notin U$, y is not adjacent to some vertex w in $G-U$. By the maximality of G, adding any edge to G produces a 1-factor. Let M_1 and M_2 be 1-factors in $G+(x,z)$ and $G+(y,w)$. Using $M_1 \cup M_2$, we will find a 1-factor avoiding (x,z) and (y,w). This contradicts the fact that G has no 1-factor.

Let F be the set of edges that belong to exactly one of M_1 and M_2. Obviously, F contains (x,z) and (y,w). Since every vertex of G has degree 1 in each of M_1 and M_2, every vertex of G has degree 0 or 2 in F. Hence, F is a collection of disjoint even cycles and isolated vertices.

Let C be the cycle of F containing (x,z). Suppose that C does not also contain (y,w). Let $M^* = \{e \mid e \in M_2 \cap C\} \cup \{e \mid e \in M_1 - C\}$. Obviously, M^* is a 1-factor of G. Thus, we consider the case that C contains both (y,w) and (x,z). We can write C as $\langle y, P_1, x, z, P_2, y\rangle$.

Suppose that $d_C(y,x)$ is even, as shown in Figure 6.10a. Obviously, (y,w) is an edge in P_2. Let $M^* = \{e \mid e \in M_1 \cap P_2\} \cup \{(x,z)\} \cup \{e \mid e \in M_1 \cap P_1$ or $e \in M_2 - C\}$. Obviously, M^* forms a 1-factor of G.

Suppose that $d_C(y,x)$ is odd, as shown in Figure 6.10b. Obviously, (y,w) is an edge in P_1. Let $M^* = \{e \mid e \in M_1 \cap P_1\} \cup \{(y,x)\} \cup \{e \mid e \in M_2 \cap P_2$ or $e \in M_2 - C\}$. Obviously, M^* forms a perfect matching of G. Again, we have a 1-factor of G. □

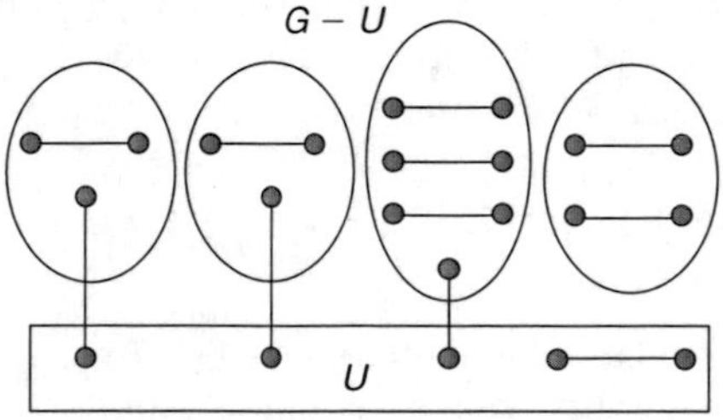

FIGURE 6.9 Illustration of $G-U$.

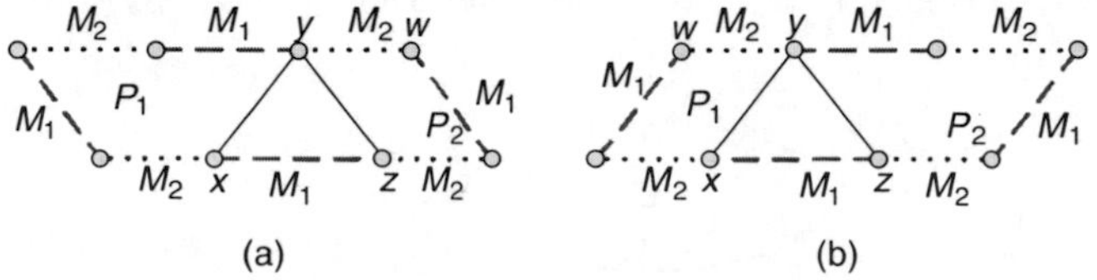

FIGURE 6.10 Illustration of M_1 and M_2.

REMARK: For any graph G and $S \subseteq V(G)$, the difference $o(G-S) - |S|$ has the same parity as $n(G)$. Hence, $o(G-S)$ exceeds $|S|$ by at least two for some S if $n(G)$ is even and G has no 1-factor.

COROLLARY 6.3 **(Berge [21])** *Let $d(S) = o(G-S) - |S|$. Then the largest number of vertices in a matching in G is*

$$\min_{S \subseteq V(G)} \{n - d(S)\}.$$

Proof: Let S be a subset of $V(G)$. Obviously, at most $|S|$ edges can match vertices of S to vertices in odd components of $G-S$. Thus, every matching has at least $o(G-S) - |S|$ unsaturated vertices. We want to achieve this bound.

Let $d = \max\{o(G-S) - |S| \mid S \subseteq V(G)\}$. The case $S = \emptyset$ implies $d \geq 0$. Let $G' = G \vee K_d$. Since $d(S)$ has the same parity as $n(G)$ for each S, we know that $n(G')$ is even. Suppose that G' satisfies Tutte's condition. Deleting the d-added vertices eliminates edges that saturate at most d vertices of G, and so we can obtain a matching of the desired size in G from a perfect matching in G' (see Figure 6.11).

The condition $o(G'-S') \leq |S'|$ holds for $S' = \emptyset$, because $n(G')$ is even. Suppose that S' is nonempty but does not contain all of K_d. Then $G' - S'$ has only one component. We have $1 = o(G'-S') \leq |S'|$. Suppose that $K_d \subseteq S'$. Let $S = S' - V(K_d)$. We have $G' - S' = G - S$, so $o(G'-S') = o(G-S) \leq |S| + d = |S'|$. Thus, G' indeed satisfies Tutte's condition. □

This corollary guarantees that there is a *proof* that a maximum matching is maximum by exhibiting a vertex set S whose deletion leaves the appropriate number of odd components. This is a short proof, but finding S may be hard.

Most applications of Tutte's Theorem involving showing that some other condition implies Tutte's condition and hence guarantees a 1-factor. Some applications were proved by other means long before Tutte's Theorem.

COROLLARY 6.4 **(Petersen [272])** *Every 3-regular graph with no cut edge has a 1-factor.*

Proof: Let G be a 3-regular graph with no cut edge. To prove this corollary, we can prove that each edge set $S \subseteq V(G)$ satisfies Tutte's condition.

Let us count the edges between S and the odd component of $G-S$. Since G is 3-regular, each vertex of S is incident to at most three edges between S and the odd component of $G-S$. Suppose that each odd component H of $G-S$ is incident to at least three such edges. Then, $3o(G-S) \leq 3|S|$. Hence, $o(G-S) \leq |S|$. Since G has

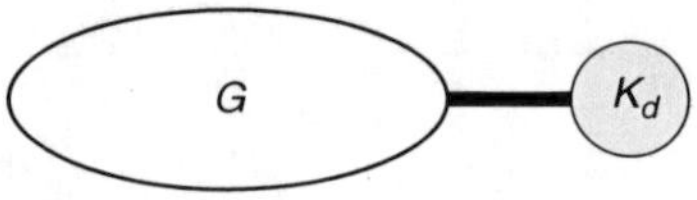

FIGURE 6.11 The graph G'.

no cut edge, the number of edges between S and H cannot be 1. Moreover, the number of edges between S and H cannot be even, because then the sum of degrees in H would be odd. Hence, there are at least three edges from H to S, as desired. □

6.5 FACTORS

Petersen also found a sufficient condition for 2-factors, which can be proved using only Eulerian circuits and bipartite matching. Note that every connected $2k$-regular graph is Eulerian. Thus, the edge set can be partitioned into edge-disjoint cycles. Yet, we prove a stronger result that these cycles can be organized into 2-factors.

THEOREM 6.7 **(Petersen [272])** Every regular graph of even degree has a 2-factor.

Proof: Let G be a $2k$-regular with vertices $v_1, v_2, \ldots, v_n$. Obviously, every component of G has an Eulerian circuit C. For each component, define a bipartite graph H with vertices $u_1, u_2, \ldots, u_n$ and $w_1, w_2, \ldots, w_n$ by putting $u_i \leftrightarrow w_j$ if v_j immediately follows v_i somewhere on C. Because C leaves and enters each vertex k times, H is k-regular.

Every regular bipartite graph has a 1-factor. A 1-factor in H designates one edge *leaving* v_i (incident to u_i in H) and one *entering* v_i (incident to w_i in H). Combining these edges for each component of G yields a 2-regular spanning subgraph of G. □

EXAMPLE 6.5

Let G be the graph K_5. Obviously, $\langle 1, 2, 3, 4, 5, 3, 1, 4, 2, 5, 1\rangle$ forms an Euler circuit of G. The corresponding bipartite graph H is shown in Figure 6.12. Obviously, $\{(1, 2), (2, 3), (3, 4), (4, 5), (5, 1)\}$ forms a 1-factor of H. The corresponding 2-factor in G is the cycle $\langle 1, 2, 3, 4, 5\rangle$. The remaining edges of H form another 1-factor. Again, it corresponds to the remaining 2-factor $\langle 1, 3, 5, 2, 4, 1\rangle$ in G.

Note that a factor is an arbitrary spanning subgraph of G. We would like to ask about the existence of factors of special types. For example, a k-factor is a k-regular factor. We have studied 1-factor and 2-factors. More generally, we can specify the degree at each vertex. Let f be a function from $V(G)$ into $\mathcal{N} \cup \{0\}$. We ask whether G has a subgraph H such that $d_H(v) = f(v)$ for all $v \in V(G)$. Such a subgraph H is an *f-factor* of G.

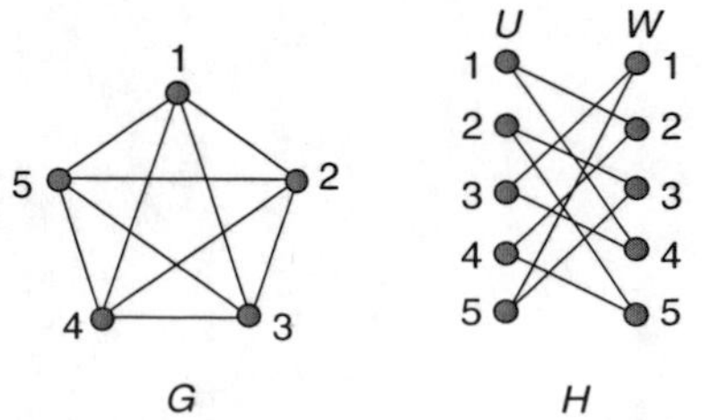

FIGURE 6.12 Illustration of Example 6.5.

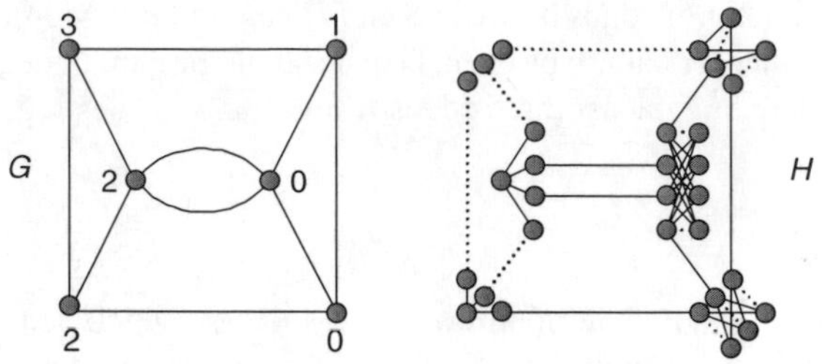

FIGURE 6.13 Graphs G and H.

Multiple edges don't affect the existence of 1-factors. However, they do affect the existence of an f-factor. Tutte [315] proved a necessary and sufficient condition for a graph to have an f-factor. The original proof was quite difficult. Later Tutte reduced it to checking for a 1-factor in a related graph. We describe the construction of this related graph. This is a beautiful example of transforming a graph problem into a previously solved problem.

We assume that $f(w) \leq d(w)$ for all w. Otherwise, immediately G has no f-factor. We construct a graph H that has a 1-factor if and only if G has an f-factor as follows.

Let $e(w) = d(w) - f(w) \geq 0$ be the *excess* degree at w. To construct H, we replace each vertex v by a bipartite graph graph $K_{d(v),e(v)}$, with partite sets $A(v)$ of size $d(v)$ and $B(v)$ of size $e(v)$. For each edge (v, w) in G, we join one vertex of $A(v)$ to one vertex of $A(w)$ so that each A-vertex participates in one such edge.

In Figure 6.13, we illustrate a graph G with the indicated function f and the corresponding simple graph H. The dashed edges in H form a 1-factor that translates back into an f-factor of G.

THEOREM 6.8 A graph G has an f-factor if and only if the graph H constructed previously from the pairs G,f has a 1-factor.

Proof: Suppose that G has an f-factor. Then the corresponding edges in H leave $e(u)$ vertices of $A(v)$ unmatched. We match them arbitrarily to the vertices of $B(v)$ to obtain a 1-factor of H. Similarly, we first delete the matched edges involving B-vertices in a 1-factor of H. The remaining matched edges of H will correspond to the edges of 1-factor of G. Hence, H has a 1-factor if and only if G has an f-factor. □

7 Connectivity

7.1 CUT AND CONNECTIVITY

A good communications network is hard to disrupt. We want to preserve network service by ensuring that the graph (or digraph) of possible transmissions remains connected, even when some vertices or edges fail. When communication links are expensive, we want to achieve these goals with few edges. Loops are irrelevant for connection, so we may assume that the graphs and digraphs of this chapter have no loops.

A *separating set* or *vertex cut* of a graph G is a set $S \subseteq V(G)$ such that $G - S$ has more than one component. A graph is *k-connected* if every vertex cut has at least k vertices. The *connectivity* of G, written $\kappa(G)$, is the minimum size of a vertex cut. In other words, $\kappa(G)$ is the maximum k such that G is k-connected. Hence, a graph has connectivity 0 if and only if it is disconnected. A clique has no separating set. We adopt the convention that $\kappa(K_n) = n - 1$, so that most results about connectivity will extend to cliques. Obviously, $\kappa(G) \leq n(G) - 1$ for all G.

EXAMPLE 7.1

(Connectivity of the n-dimensional cubes Q_n.) The cube Q_n is n-regular. We can cut Q_n by deleting the neighbors of a vertex, so $\kappa(Q_n) \leq n$. To prove that $\kappa(Q_n) = n$, we show that every vertex cut of Q_n has at least n vertices.

We use induction on n. Obviously, Q_n is a clique with connectivity n for $n = 0, 1$. This completes the basis. Suppose that $\kappa(Q_{n-1}) = n - 1$ with $n \geq 2$. We know that Q_n is obtained from two copies Q, Q' of Q_{n-1} by adding a matching joining corresponding vertices in Q and Q'. Let S be an arbitrary vertex cut in Q_n.

Suppose that $Q - S$ is connected and $Q' - S$ is connected. Then $Q_n - S$ is also connected unless S deletes at least one endpoint of every matched pair. This requires that $|S| \geq 2^{n-1}$. However, $2^{n-1} \geq n$ for $n \geq 2$. Hence, we may assume that $Q - S$ is disconnected. By induction hypothesis, S contains at least $n - 1$ vertices in Q. Suppose that S contains no vertices of Q'. Then $Q' - S$ is connected and all vertices of $Q - S$ have neighbors in $Q' - S$. Thus, $Q_n - S$ is connected. Hence, S must also contain a vertex of Q'. Therefore, $|S| \geq n$. □

Suppose that G is not a clique. Deleting the neighbors of a vertex separates G. Thus, $\kappa(G) \leq \delta(G)$. Equality need not hold. For example, $2K_m$ has a minimum degree of $m - 1$ but connectivity 0. Let G be a k-connected graph. From the previous discussion, $\delta(G) \geq k$. Thus, G contains at least $\lceil kn/2 \rceil$ edges. This is achievable.

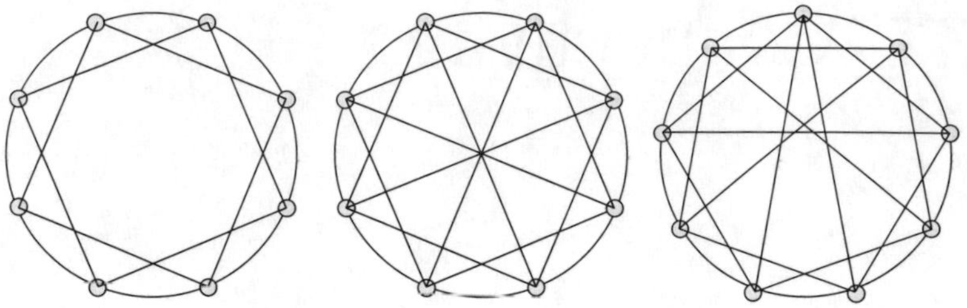

FIGURE 7.1 The Harary graphs $H_{4,8}, H_{5,8}$, and $H_{5,9}$.

EXAMPLE 7.2

(The k-connected Harary graph $H_{k,n}$.) Let $k < n$. Place n vertices, $\{x_0, x_1, \ldots, x_{n-1}\}$, in circular order. Suppose that $k = 2r$. Form $H_{k,n}$ by making each vertex adjacent to the nearest r vertices in each direction around the circle. Suppose that $k = 2r + 1$ and n is even. Form $H_{k,n}$ by making each vertex adjacent to the nearest r vertices in each direction and to the vertex opposite it on the circle. In each case, $H_{k,n}$ is k-regular. Suppose that $k = 2r + 1$ and n is odd. Construct $H_{k,n}$ from $H_{2r,n}$ by adding the edges $x_i \leftrightarrow x_{i+(n+1)/2}$ for $0 \leq i \leq (n-1)/2$. The graphs $H_{4,8}, H_{5,8}$, and $H_{5,9}$ are shown in Figure 7.1.

THEOREM 7.1 **(Harary [138])** $\kappa(H_{k,n}) = k$. Hence, the minimum number of edges in a k-connected graph on n vertices is $\lceil kn/2 \rceil$.

Proof: We prove only the case that $k = 2r$. Let $G = H_{2r,n}$. From the symmetric of G, it suffices to prove that x_0 and x_j, $j = 1, 2, \ldots, n-1$, cannot be disconnected by less than $2k$ vertices. Suppose that x_0 and x_j can be disconnected by $2k - 1$ vertices, $x_{i_1}, x_{i_2}, \ldots, x_{i_{2k-1}}$, where $1 \leq i_1 < i_2 < \cdots < i_{2k-1} \leq n-1$. One of the two intervals $[0, j]$ and $[j, n]$ contains at most $(k-1)$ of these indices. Suppose it is interval $[0, j]$. Note that the difference between their indices is less than $(k-1)$. Two consecutive vertices of the sequence obtained by removing $x_{i_1}, x_{i_2}, \ldots$ from $x_0, x_1, \ldots, x_j$, are joined by an edge. Consequently, there is a path from x_0 to x_j, which is a contradiction.

The remaining cases can be proved similarly. □

Perhaps our transmitters are secure and never fail. However, our communication links are subject to noise or other disruptions. In this situation, we want to make it hard to disconnect our graph by deleting edges.

A *disconnecting set* of edges is a set $F \subseteq E(G)$ such that $G - F$ has more than one component. A graph is *k-edge-connected* if every disconnecting set has at least k edges. The *edge-connectivity* of G, written $\kappa'(G)$, is the minimum size of a disconnecting set. In other words, $\kappa'(G)$ is the maximum k such that G is k-edge-connected.

Let S and T be two subsets of $V(G)$. We use $[S, T]$ to denote the edge set having one endpoint in S and the other in T. An *edge cut* is an edge set of the form $[S, \overline{S}]$ where S is a nonempty proper subset of $V(G)$.

The notation for edge-connectivity continues our convention of using a *prime* for an edge parameter analogous to a vertex parameter. Using the same base letter emphasizes the analogy and avoids the confusion of using many different letters.

Since there is no path of $G-[S,\overline{S}]$ from S to $\overline{S}$, every edge cut is a disconnecting set. However, the converse is false. In K_3, the set of all edges is a disconnecting set but is not an edge cut. In $K_{3,3}$, every set of seven edges is a disconnecting set, but none of them is an edge cut.

Suppose that $n(G)>1$. Then every minimal disconnecting set of edges is an edge cut. Suppose that $G-F$ has more than one component for some $F\subseteq E(G)$. Then we can delete all edges having one endpoint in some component H of $G-F$. Thus, F contains the edge cut $[V(H),\overline{V(H)}]$. Therefore, F is not a minimal disconnecting set unless $F=[V(H),\overline{V(H)}]$.

Deleting one endpoint of each edge in an edge cut F will delete every edge of F. Hence, we expect that $\kappa(G)\le\kappa'(G)$ always holds. To prove this, we must be careful not to delete the only vertex of a component of $G-F$ and thereby produce a connected subgraph. The inequality $\kappa(K_n)\le\kappa'(K_n)$ holds by our convention that $\kappa(K_n)=n-1$.

THEOREM 7.2 **(Whitney [340])** $\kappa(G)\le\kappa'(G)\le\delta(G)$.

Proof: The edges incident to a vertex v of minimum degree form a disconnecting set. Hence, $\kappa'(G)\le\delta(G)$. It remains to show $\kappa(G)\le\kappa'(G)$. Let G be a graph with $n(G)>1$ and $[S,\overline{S}]$ be a minimum edge cut having size $k=\kappa'(G)$.

Suppose that every vertex of S is adjacent to every vertex of $\overline{S}$. Obviously, $k\ge|S||\overline{S}|\ge n(G)-1$. Thus, the inequality holds.

Hence, we assume that there exists $x\in S$ and $y\in\overline{S}$ such that x is not adjacent to y. Now, let T be the vertex set consisting of all neighbors of x in $\overline{S}$ and all vertices of $S-\{x\}$ that have neighbors in $\overline{S}$ (illustration in Figure 7.2). Since x and y belong to different components of $G-T$, T is a separating set. Since T consists of one endpoint of each edge in $[S,\overline{S}]$, we have $|T|\le|[S,\overline{S}]|=k$. Hence, $\kappa(G)\le\kappa'(G)$. □

Because connectivity equals minimum degree for cliques, complete bipartite graphs, hypercubes, and Harary graphs, the inequality of the previous theorem implies also that edge-connectivity equals minimum degree for these graphs. Although the set of edges incident to a vertex of minimum degree is always an edge cut, it need not be a minimum edge cut. The situation $\kappa'(G)<\delta(G)$ is precisely the situation in which there is no minimum edge cut that disconnects a single vertex from the rest of the graph.

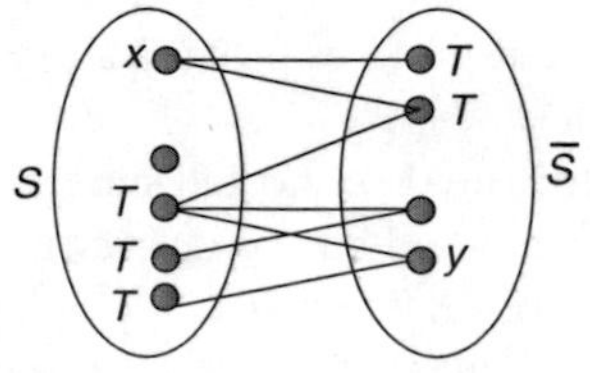

FIGURE 7.2 Illustration of Theorem 7.2.

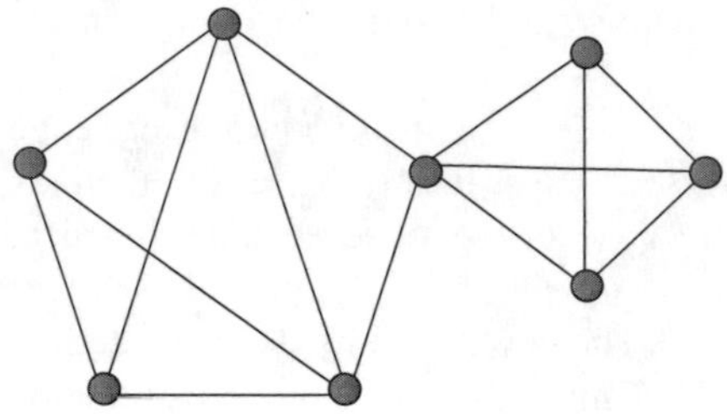

FIGURE 7.3 A graph G with $\kappa(G) = 1, \kappa'(G) = 2$, and $\delta(G) = 3$.

EXAMPLE 7.3

(Possibility of $\kappa < \kappa' < \delta$.) Let G be the graph in Figure 7.3. Obviously, $\kappa(G) = 1$, $\kappa'(G) = 2$, and $\delta(G) = 3$.

An edge cut may contain another edge cut as a subset. For example, $K_{1,2}$ has three edge cuts. However, one of them contains the other two. The minimal nonempty edge cut of a graph have useful structural properties and a special name: a *bond* is a minimal nonempty edge cut. Here *minimal* means that no proper nonempty subset of these edges is also an edge cut. Bonds in connected graphs are the edge cuts that leave only two components.

PROPOSITION 7.1 *Suppose that G is a connected graph and S is a nonempty proper subset of $V(G)$. Then the edge cut $F = [S, \overline{S}]$ is a bond if and only if $G - F$ has two components.*

Proof: Suppose that $G - F$ has two components. Let F' be a proper subset of F. Obviously, $G - F'$ contains the two components of $G - F$ plus at least one edge between the two components. Hence, $G - F'$ is connected. Thus, F is a bond.

Suppose that $G - F$ has more than two components. By symmetric, we may assume that $S = A \cup B$ with no edges between A and B. Now the edge cuts $[A, \overline{A}]$ and $[B, \overline{B}]$ are a proper subset of F. Thus, F is not a bond. □

In Chapter 1, we introduced the terms *cut vertex* and *cut edge* for vertex cuts and edge cuts of size 1. Some authors have used *articulation point* to mean cut vertex, and some authors have used *isthmus* or *bridge* to mean cut edge. Every graph with connectivity 1 other than K_2 has a cut vertex. These graphs are sometimes called *separable graphs*. Obviously, K_1 and K_2 have no cut vertex. Assume that G is a connected graph with no cut vertex and G is neither K_1 nor K_2. Then G is 2-connected. Connected subgraphs without cut-vertices provide a useful decomposition of a graph into subgraphs.

A *block* of a graph G is a maximal connected subgraph of G that has no cut vertex. Suppose that G itself is connected and has no cut vertex. Then G is a block.

Suppose that H is a block of G. Then H is a graph has no cut vertex. However, H may contain vertices that are cut vertices of G. For example, the graph drawn in Figure 7.4 has seven blocks.

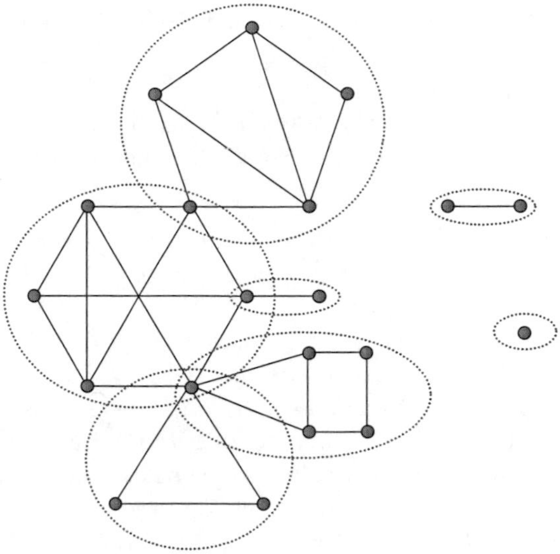

FIGURE 7.4 A graph with seven blocks.

An edge of a cycle cannot itself be a block, because it belongs to a larger subgraph having no cut vertex. Hence, an edge is a block of G if and only if it is a cut edge of G. The blocks of a tree are its edges. Suppose that a block has more than two vertices. It is 2-connected. The blocks of a graph are its isolated vertices, its cut edges, and its maximal 2-connected subgraphs.

PROPOSITION 7.2 *Two blocks in a graph share at most one vertex.*

Proof: Suppose that the statement is not true. There are two blocks B_1 and B_2 of some graph G sharing two vertices. Let x be any vertex in $V(B_1 \cup B_2)$. Note that deleting one vertex cannot disconnect a block. Deleting x leaves a path within B_i from every vertex remaining in B_i to each vertex of $(B_1 \cap B_2) - \{x\}$. Since $B_1 \cap B_2 - \{x\} \neq \emptyset$, $(B_1 \cup B_2) - \{x\}$ is connected. Therefore, $B_1 \cup B_2$ is a subgraph with no cut vertex. This contradicts the maximality of B_1 and B_2. □

7.2 2-CONNECTED GRAPHS

It is believed that a communication network is fault-tolerant if it has alternative paths between vertices. The more disjoint paths exist, the more reliable the network is. In this section, we prove that this alternative measure of connection is essential the same as k-connectedness. Obviously, a graph G is 1-edge-connected if and only if every pair of vertices is connected by a path. In the following, we generalize this characterization to k-edge-connected graphs and to k-connected graphs. We begin by characterizing 2-connected graphs. Let $P = \langle x_1, x_2, \ldots, x_k \rangle$ be a path of a graph G joining x_1 to x_k. We use $V(P)$ to denote the set $\{x_1, x_2, \ldots, x_k\}$ and $I(P)$ to denote the set $V(P) - \{x_1, x_k\}$.

Let P_1 and P_2 be two paths of a graph G. We say that P_1 and P_2 are *internally disjoint* if $I(P_1) \cap I(P_2) = \emptyset$.

THEOREM 7.3 **(Whitney [340])** A graph G having at least three vertices is 2-connected if and only if each pair u and v in $V(G)$ is connected by a pair of internally disjoint paths from u to v in G.

Proof: Suppose that each pair $u, v \in V(G)$ is connected by a pair of internally disjoint paths from u to v in G. Thus, we cannot separate u from v by deleting one vertex of G. Thus, G is 2-connected.

For the converse, suppose that G is 2-connected. We prove by induction on $d(u, v)$ that G has two internally disjoint paths from u to v.

Suppose that $d(u, v) = 1$. As $\kappa'(G) \geq \kappa(G) = 2$, $G - (u, v)$ is connected. There exists a path from u to v P_1 in $G - (u, v)$ joining u to v. Let P_2 be the path $\langle u, v \rangle$. Obviously, P_1 and P_2 forms two internally disjoint paths from u to v.

Assume that G has internally disjoint paths from x to y whenever $1 \leq d(u, v) < k$. Assume that u and v are two vertices with $d(u, v) = k$. Let w be the vertex before v on a shortest path from u to v. Thus, $d(u, w) = k - 1$. By the induction hypothesis, G has internally disjoint paths P and Q from u to w. Since $G - w$ is connected, $G - w$ contains a path R from u to v. Let z be the last vertex of R belonging to $P \cup Q$. By symmetry, we may assume that $z \in P$. We combine the subpath from u to z of P with the subpath from z to v of R to obtain a path from u to v internally disjoint from $Q \cup \{(w, v)\}$. See Figure 7.5 for illustration. □

LEMMA 7.1 **(Expansion Lemma.)** Suppose that G is a k-connected graph. Let G' be a graph obtained from G by adding a new vertex y adjacent to at least k vertices of G. Then G' is k-connected.

Proof: Suppose that S is a separating set of G'. Suppose that $y \in S$. Then $S - \{y\}$ separate G. Thus, $|S| \geq k + 1$. Suppose that $y \notin S$ and $N(y) \subseteq S$. Then $|S| \geq k$. Otherwise, S must separate G. Again, $|S| \geq k$. □

THEOREM 7.4 Suppose that G is a graph with at least three vertices. Then the following statements are equivalent (and characterize 2-connected graphs):

A. G is connected and has no cut vertex.
B. There exist two internally disjoint paths between any two vertices x and y of G.

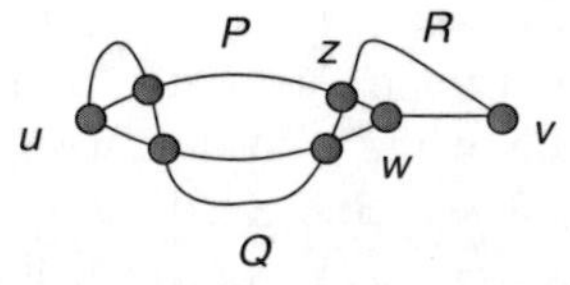

FIGURE 7.5 Illustration of Theorem 7.3.

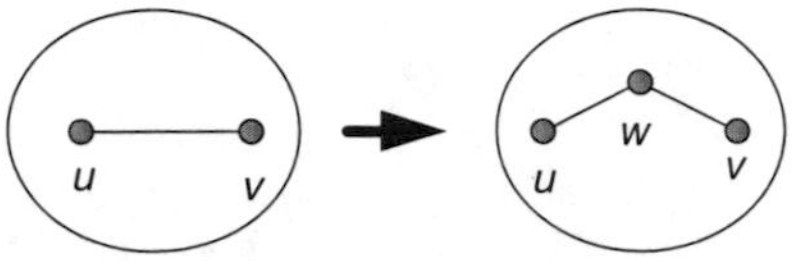

FIGURE 7.6 Subdividing the edge (u, v).

C. There exists a cycle through any two vertices x and y of G.
D. $\delta(G) \geq 1$ and every pair of edges in G lies on a common cycle.

Proof: By Theorem 7.3, A is equivalent to B. The existence of a cycle between x and y is equivalent to the existence of two internally disjoint paths from x to y. Thus, B is equivalent to C.

$D \Rightarrow C$: Suppose that G is a graph such that $\delta(G) \geq 1$ and every pair of edges in G lies on a common cycle. Let x and y be any two vertices of G. Since $\delta(G) \geq 1$, there exists an edge e_1 incident to x. Similarly, there exists an edge e_2 incident to y. By assumption, e_1 and e_2 lie on a common cycle C. Obviously, C is a cycle through x and y.

$A, C \Rightarrow D$: Suppose that G is a 2-connected graph and (u, v) and (x, y) are two edges of G. Add to G the vertices w with neighborhood $\{u, v\}$ and z with neighborhood $\{x, y\}$. By Lemma 7.1, the resulting graph G' is 2-connected. Hence, w and z lie on a common cycle C in G'. Obviously, $\deg_{G'}(w) = \deg_{G'}(z) = 2$. The cycle must contain the paths $\langle u, w, x\rangle$ and $\langle x, z, y\rangle$ but not $\langle u, v\rangle$ or $\langle x, y\rangle$. Replace the paths $\langle u, w, v\rangle$ and $\langle x, z, y\rangle$ in C by the edges (u, v) and (x, y) to obtain the desired cycle in C. □

Subdividing an edge (u, v) of a graph G is the operation of deleting (u, v) and adding a path $\langle u, w, v\rangle$ through a new vertex w. See Figure 7.6.

COROLLARY 7.1 *Suppose that G is 2-connected. Then the graph G' obtained by subdividing an edge of G is 2-connected.*

Proof: Suppose that G' is formed from G by adding w to subdivide (u, v). We will use condition D in Theorem 7.4 to prove that G' is 2-connected. Thus, we need to find a cycle through the arbitrary edges e and f of G'.

Suppose that e and f are edges in G. By condition D in Theorem 7.4, there exists a cycle C through these edges. If this cycle uses the edge (u, v), we will modify this cycle to pass through w between u and v. Suppose that $e \in E(G)$ and $f \in \{(u, w), (w, v)\}$. We modify a cycle passing through e and (u, v). Suppose that $\{e, f\} = \{(u, w), (w, v)\}$. We modify a cycle through (u, v). □

7.3 MENGER THEOREM

We have introduced two measures for connectivity: invulnerability to deletions and multiplicity of alternative communication paths. Here, we will extend Whitney's Theorem by showing that the two notions of k-connected graphs are the same. We have similar results for k-edge-connected and digraphs. We need some definitions.

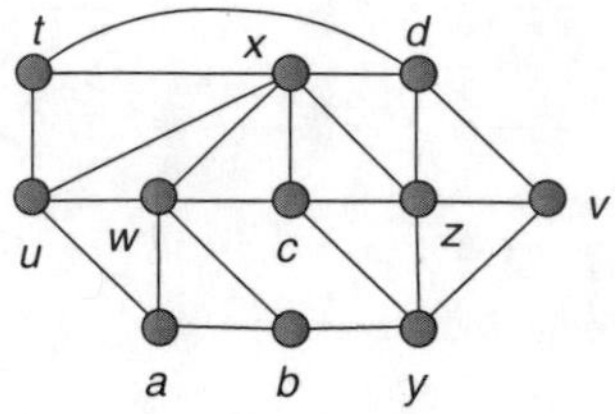

FIGURE 7.7 The graph for Example 7.4.

Let x and y be two vertices of G. A set $S \subseteq V(G) - \{x, y\}$ is an *x,y-separator* or *x,y-cut* if $G - S$ has no path from x to y. Let $\kappa(x, y)$ be the minimum size of an x, y-cut. Let $\lambda(x, y)$ be the maximum size of a set of pairwise internally disjoint paths from x to y. Let $\lambda(G)$ be the largest k such that $\lambda(x, y) \geq k$ for all $x, y \in V(G)$. Let X and Y be two subsets of $V(G)$. An *X,Y-path* is a path having its first vertex in X and its last vertex in Y.

Example 7.4

The graph in Figure 7.7 has four pairwise internally disjoint paths from x to y. Thus, $\lambda(x, y) \geq 4$. It can be confirmed that $S = \{b, c, z, d\}$ is an x, y-cut of size 4. Thus, $\kappa(x, y) \leq 4$. Note that every path from x to y intersects S and every x, y-cut has a vertex from each path. We have $\kappa(x, y) = \lambda(x, y) = 4$. We can further check that $\kappa(a, z) = \lambda(a, z) = 3$. Moreover, $\{y, u, w\}$ is an a, z-cut. Actually, this graph is 3-connected. We can find three pairwise internally disjoint paths for every pair $u, v \in V(G)$.

Yet we find five pairwise edge-disjoint paths from w to z and a set $\{(w, a), (w, b), (w, c), (w, u), (w, x)\}$ of five edges whose deletion breaks all paths from w to z.

Menger's original theorem states the local equality $\kappa(x, y) = \lambda(x, y)$. The global equality $\kappa(G) = \lambda(G)$ and the analogous results for edge-connectivity and digraphs were observed by others. All are considered forms of Menger's Theorem.

THEOREM 7.5 **(Menger [252])** Let x, y be any two vertices of a graph G with $(x, y) \notin E(G)$. Then the minimum size of an x, y-cut equals the maximum number of pairwise internally disjoint paths from x to y.

Proof: It is observed that any x, y-cut must contain an internal vertex from each path in a set of pairwise internally disjoint paths from x to y. These vertices must be distinct. Thus, $\kappa(x, y) \geq \lambda(x, y)$.

To prove equality, we use induction on $n(G)$. Suppose that $n = 2$. Then $(x, y) \notin E(G)$ implies $\kappa(x, y) = \lambda(x, y) = 0$. The theorem holds for $n = 2$. Assume that the theorem holds for any graph with less than n vertices with $n > 2$. Let G be any graph containing two vertices x and y with $k = \kappa_G(x, y)$. We need to construct k pairwise internally disjoint paths from x to y in G.

Case 1. *G has a minimum x, y-cut S such that* $S - (N(x) \cup N(y)) \neq \emptyset$. Let V_1 be the set of vertices on x, S-paths and let V_2 be the set of vertices on S, y-paths.

We claim that $S = V_1 \cap V_2$. As S is a minimal x, y-cut, every vertex of S lies on a path from x to y. Hence $S \subseteq V_1 \cap V_2$. Suppose $S \neq V_1 \cap V_2$. There exists a vertex $v \in (V_1 \cap V_2) - S$. The portion from x to v of some x, S-path followed by the portion from v to y of some S, y-path yields a path from x to y that avoids the x, y-cut S. This is impossible. Thus, $S = V_1 \cap V_2$.

Now, we claim that $V_1 \cap (N(y) - S) = \emptyset$. Suppose that there exists a vertex $v \in V_1 \cap (N(y) - S)$. The portion from x to v of some x, S-path followed by $\langle v, y\rangle$ yields a path from x to y that avoids the x, y-cut S. This is impossible. Thus, $V_1 \cap (N(y) - S) = \emptyset$. Similarly, $V_2 \cap (N(x) - S) = \emptyset$.

Now, we splice the graph G into H_1 and H_2 as follows. Form H_1 by adding to the induced subgraph $G[V_1]$ a vertex y' with each edge from S. Form H_2 by adding to the induced subgraph $G[V_2]$ a vertex x' with each edge to S.

Every path from x to y in G that starts with an x, S-path is contained in H_1. Thus, every x, y'-cut in H_1 is an x, y-cut in G. Hence, $\kappa_{H_1}(x, y') = k$. Similarly, $\kappa_{H_2}(x', y) = k$. As $V_1 \cap (N(y) - S) = \emptyset$ and $V_2 \cap (N(x) - S) = \emptyset$, both H_1 and H_2 have fewer vertices than G. By induction, $\lambda_{H_1}(x, y') = k = \lambda_{H_2}(x', y)$. Since $V_1 \cap V_2 = S$, deleting y' from the k-paths in H_1 and deleting x' from the k-paths in H_2 yields the desired x, S-paths and the S, y-paths in G. Combining these paths, we obtain k pairwise internally disjoint paths from x to y in G. See Figure 7.8 for an illustration.

Case 2. *Every minimum x, y-cut S of G satisfies $S - (N(x) \cup N(y)) = \emptyset$.* Thus, $S \subseteq N(x) \cup N(y)$. Again, we need to construct k pairwise internally disjoint paths from x to y in G.

Suppose that G has a vertex $v \notin \{x, y\} \cup N(x) \cup N(y)$. Then v is not in any minimum x, y-cut. Hence, $\kappa_{G-v}(x, y) = k$. By induction, $G - v$ contains k internally disjoint paths from x to y.

Suppose that $N(x) \cap N(y) \neq \emptyset$. Let v be any vertex in $N(x) \cap N(y)$. Obviously, v appears in every x, y-cut. Hence, $\kappa_{G-v}(x, y) = k - 1$. By induction, $G - v$ contains $(k - 1)$ internally disjoint paths from x to y. Adding the path $\langle x, v, y\rangle$, we obtain k internally disjoint paths from x to y of G.

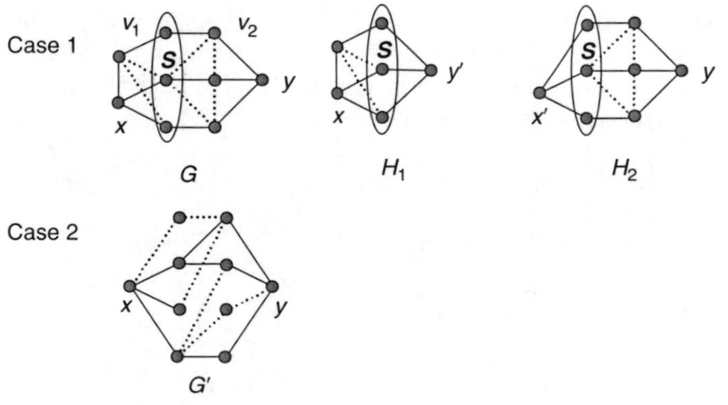

FIGURE 7.8 Illustration of Theorem 7.5.

Thus, we assume that $N(x)$ and $N(y)$ are disjoint and exhaust $V(G) - \{x, y\}$. Let G' be the bipartite graph with bipartition $N(x), N(y)$ and edge set $[N(x), N(y)]$. Every path from x to y in G uses some edge from $N(x)$ to $N(y)$. Therefore, x, y-cuts in G are precisely the vertex covers of G'. Hence, $\beta(G') = k$. By the König–Egerváry Theorem, there is a matching of size k from $N(x)$ to $N(y)$. These edges yield k pairwise internally disjoint paths from x to y of length 3. See Figure 7.8 for an illustration. □

The following global version for k-connected graphs, observed by Whitney [340], is also referred to as the Menger Theorem.

THEOREM 7.6 The connectivity of G equals the maximum k such that $\lambda(x, y) \geq k$ for all $x, y \in V(G)$.

Proof: By Theorem 7.5, $\kappa(x, y) = \lambda(x, y)$ when $(x, y) \notin E(G)$. By definition, $\kappa(G) = \min\{\kappa(x, y) \mid (x, y) \notin E(G)\}$. Thus, we need to show that $\lambda(x, y) \geq \kappa(G)$ if $(x, y) \in E(G)$.

We first claim that $\kappa(G) \leq \kappa(G - (x, y)) + 1$. Suppose that G is the complete graph K_n. Obviously, $\kappa(G - (x, y)) = n - 2 = \kappa(G) - 1$. Suppose that G is not a clique. Then $\kappa(G) \leq n - 2$. Let S be a minimum separating set of $G - (x, y)$. Suppose that S is also a disconnecting set of G. Then $\kappa(G) = \kappa(G - (x, y)) = |S|$. Thus, we assume that S is not a minimum separating set of G. Since $G \neq K_n$, $|S| < n - 2$. Moreover, $G - (x, y) - S$ consists of two components: one contains x and the other contains y. Thus, one of these components has at least two vertices. Without loss of generality, we assume that x is in a component with at least two vertices. Obviously, $S \cup \{x\}$ is a separating set of G. Thus, $\kappa(G) \leq \kappa(G - (x, y)) + 1$.

By definition, $\lambda_G(x, y) = 1 + \lambda_{G-(x,y)}(x, y)$. Thus,

$$\lambda_G(x, y) = 1 + \lambda_{G-(x,y)}(x, y) = 1 + \kappa_{G-(x,y)}(x, y) \geq 1 + \kappa(G - (x, y)) \geq \kappa(G)$$

The theorem is proved. □

The *line graph* of a graph G, written $L(G)$, is the graph whose vertices are edges of G, with $ef \in E(L(G))$ if $e = (u, v)$ and $f = (v, w)$ in G. For graphs in Figure 7.9, e and f share a vertex. We use the notation $\lambda'(x, y)$ to be the maximum size of a set of pairwise edge-disjoint paths from x to y, and $\kappa'(x, y)$ to be the minimum number of edges whose deletion makes y unreachable from x.

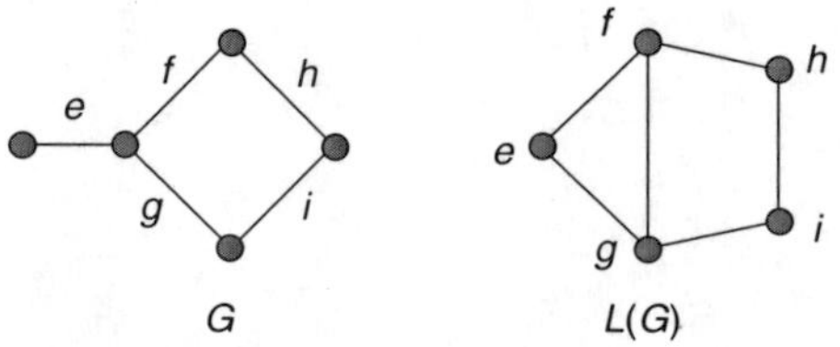

FIGURE 7.9 The graph and its line graph.

THEOREM 7.7 Suppose that x and y are distinct vertices of a graph G. Then the minimum size of an x,y-disconnecting set of edges equals the maximum number of pairwise edge-disjoint paths from x to y; that is, $\kappa'(x,y)=\lambda'(x,y)$.

Proof: We modify G by adding two new vertices s,t and two new edges (s,x) and (y,t) to obtain G'. It is easy to see that $\kappa'(x,y)=\kappa'_{G'}(x,y)$ and $\lambda'(x,y)=\lambda'_{G'}(x,y)$. We extend each path from x to y by starting from the edge (s,x) and ending with the edge (y,t). A set of edges disconnects y from x in G if and only if the corresponding vertices of $L(G')$ form an $(s,x),(y,t)$-cut. Similarly, edge-disjoint paths from x to y in G become internally disjoint paths from (s,x) to (y,t) in $L(G')$, and vice versa. Since $x\neq y$, $((s,x),(y,t))\notin E(L(G'))$. By Menger's Theorem, $\kappa'_G(x,y)=\kappa_{L(G')}((s,x),(y,t))=\lambda_{L(G')}((s,x),(y,t))=\lambda'_G(x,y)$. □

There are similar Menger-type results for edge connectivity and for digraphs. Dirac extended Menger's theorem to other families of paths. An (x,U)-fan is a set of (x,U)-paths such that any two of them share only the vertex x.

THEOREM 7.8 **(Fan Lemma, Dirac [87])** A graph is k-connected if and only if it has at least $k+1$ vertices and there is an (x,U)-fan for every choice of x and U with $|U|\leq k$ and $x\notin U$.

Proof: Suppose that G is k-connected. We construct G' from G by adding a new vertex y to all of U. By Lemma 7.1, G' also is k-connected. By Menger's Theorem, there exist k pairwise internally disjoint paths from x to y in G'. Deleting y from these paths produces an (x,U)-fan of size k in G.

Conversely, suppose that G has at least $k+1$ vertices and there is an (x,U)-fan for every choice of x and U with $|U|\leq k$ and $x\notin U$. Obviously, the statement holds for for every choice of x and U with $|U|=k$ and $x\notin U$. Thus, $\deg(x)\geq k$ for every vertex in G. Let x and y be any two different vertices of G. Let $U'=\{y_1,y_2,\ldots,y_{k-1}\}$ be a set of $(k-1)$ neighbors of y not containing x. We set $U=U'\cup\{y\}$. Obviously, we can extend the (x,U)-fan by adding the edges $\{(y_i,y)\mid y_i\in U'\}$ to obtain k internally disjoint paths from x to y. Hence, G is k-connected. □

Using a similar augment, we have the following theorem.

THEOREM 7.9 A graph $G=(V,E)$ is k-connected if and only if it has at least $(k+1)$ vertices and there exist k disjoint path joining any two vertex sets A and B with $|A|=|B|=k$.

7.4 AN APPLICATION—MAKING A ROAD SYSTEM ONE-WAY

Given a road system, how can it be converted to one-way operation so that traffic may flow as smoothly as possible? This is clearly a problem of the orientations of graphs. We consider, for example, the two graphs G_1 and G_2, representing road networks, in Figures 7.10a and 7.10b.

No matter how G_1 is oriented, the resulting orientation cannot be strongly connected, so that the network will not be able to flow freely through the system. The

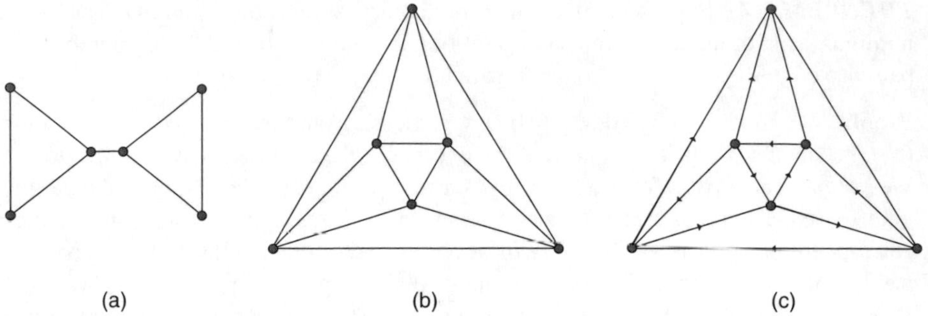

FIGURE 7.10 (a) G_1, (b) G_2, and (c) D_2.

trouble is that G_1 has a cut edge. On the other hand, G_2 has the *balanced* orientation D_2, shown in Figure 7.10c, in which each vertex is reachable from each other vertex in at most two steps; in particular, D_2 is strongly connected.

Certainly, a necessary condition for G to have a strongly connected orientation is that G is 2-edge-connected. Robbins [275] showed that this condition is also sufficient.

THEOREM 7.10 If G is 2-edge-connected, then G has a strong orientation.

Proof: Let G be a 2-edge-connected graph. Then G contains a cycle G_1. We define inductively a sequence $G_1, G_2, \ldots$ of connected subgraphs of G as follows. Suppose that G_i $(i = 1, 2, \ldots)$ is not a spanning subgraph of G. Let v_i be a vertex of G not in G. Then there exist edge-disjoint paths P_i and Q_i from v_i to G_i. Define $G_{i+1} = G_i \cup P_i \cup Q_i$. As the number of vertices of G_{i+1} is larger than that of G_i, this sequence must terminate in a spanning subgraph G_n of G.

We now orient G_n by orient G_1 as a directed cycle, each path P_i as a directed path with origin v_i, and each path Q_i as a directed path as terminus v_i. Suppose that $G_n \neq G$. All the edges of $G - G_n$ are oriented arbitrarily. Clearly every G_i—and hence in particular G_n—is thereby given a strong orientation. Since G_n is a spanning subgraph of G, G has a strong orientation. □

Nash-Williams [259] has generalized Robbins' theorem by showing that every $2k$-edge-connected graph G has a k-arc-connected orientation. Although the proof of this theorem is difficult, the special case in which G has an Euler trail carries a simple proof.

THEOREM 7.11 Let G be a $2k$-edge-connected graph with an Euler trail. Then G has a k-arc-connected orientation.

Proof: Let $v_0 e_1 v_1, \ldots e_m, v_m$ be an Euler trail of G. Orient G into a digraph D by converting the edge e_i with ends v_{i-1} and v_i to an arc a_i with tail v_{i-1} and head v_i for $1 \leq i \leq m$. Let $[S, \overline{S}]$ be an r-edge cut of G. The number of times that the directed trail $(v_0, a_1, v_1, \ldots, a_m, v_m)$ crosses from S to $\overline{S}$ differs from the number of times it crosses from $\overline{S}$ by at most one. Since it includes all arcs of D, both $[S,\overline{S}]$ and $[\overline{S},S]$ contain at least $\lfloor r/2 \rfloor$ arcs. The result follows. □

7.5 CONNECTIVITY OF SOME INTERCONNECTION NETWORKS

An interesting problem in network designs is as follows. Let n and d be positive integers. We want to find a graph (or digraph) with n vertices, each of which has degree (or outdegree) at most d, to minimize the diameter and to maximize the connectivity. In Example 7.1, we prove that $\kappa(Q_n)=n$. In Chapter 3, we introduced twisted cubes TQ_n. We also proved that the diameter of TQ_n is $\lceil \frac{n+1}{2} \rceil$ in Chapter 3. It is proved in [1] that $\kappa(TQ_n)=n$. In Chapter 3, we also introduced another variation of cubes, namely, shuffle cubes SQ_n. The diameter of SQ_n is proved to be about $n/4$. Here, we prove that the connectivity of SQ_n is still n.

THEOREM 7.12 SQ_n is n-connected.

Proof: We prove this theorem by induction. Since $SQ_2=Q_2$, SQ_2 is 2-connected. Since SQ_n is n-regular, it suffices to show that after removing arbitrary f vertices from SQ_n for $1 \le f \le n-1$, the remaining graph is still connected. Assume that F is an arbitrary set of f vertices removed from SQ_n.

Now consider $n=6$. By definition, SQ_6 consists of 16 SQ_2 subcubes. We decompose SQ_6 into two subgraphs H_1 and H_2, where H_1 consists of those SQ_2 subcubes containing vertices in F, and H_2 consists of the remaining SQ_2 subcubes. Thus, H_2 is connected and H_1-F is not necessarily connected. We distinguish the following two cases:

Case 1. Each SQ_2 subcube has at most one vertex in F. It follows that each subcube of SQ_6-F is still connected and has at least three vertices. Furthermore, H_1 contains at most five subcubes, since $|F| \le 5$. Let Q' be a subcube in H_1-F. Note that Q' contains three vertices. Q' has 12 edges connected with another 11 or 12 subcubes in SQ_6-F. Since there are at most five subcubes in H_1-F, Q' is connected to some subcubes in H_2. Since H_2 is connected and each subcube in H_1-F is also connected to H_2, SQ_6-F is connected.

Case 2. There is a subcube containing at least two vertices of F. Suppose that v is a vertex in H_1-F. Then v is connected to other four subcubes. Since $|F| \le 5$, H_1 contains at most four subcubes. Therefore, v is connected to a subcube in H_2. Since H_2 is connected, it follows that each vertex in H_1-F is connected to some vertices in H_2. Therefore, SQ_6-F is connected.

Hence, SQ_6 is 6-connected.

Now, we assume that SQ_{4k-2} is $(4k-2)$-connected for $k \ge 2$. Consider SQ_n for $n=4k+2$ and $k \ge 2$, and there are at most $4k+1$ vertices in F. Each subcube of SQ_{4k+2} is an SQ_{4k-2}. We distinguish the following two cases of F:

Case 1. Each subcube contains at most $4k-3$ vertices in F. By induction hypothesis, each subcube is still connected. Consider two arbitrary subcubes $SQ_{4k-2}^{i_1i_2i_3i_4}$ and $SQ_{4k-2}^{j_1j_2j_3j_4}$. The edges (u,v) between $SQ_{4k-2}^{i_1i_2i_3i_4}$ and $SQ_{4k-2}^{j_1j_2j_3j_4}$ satisfy

$s_{4k-2}(u) = s_{4k-2}(v)$, and $p_4(u) \oplus p_4(v) = i_1i_2i_3i_4 \oplus j_1j_2j_3j_4 \in V_{s_2(u)}$. Therefore, the number of edges in SQ_n between $SQ_{4k-2}^{i_1i_2i_3i_4}$ and $SQ_{4k-2}^{j_1j_2j_3j_4}$ is 2^{4k-4}, which is greater than $|F|$. As a consequence, each subcube $SQ_{4k-2}^{i_1i_2i_3i_4} - F$ is connected to every subcube $SQ_{4k-2}^{j_1j_2j_3j_4} - F$. Furthermore, it follows from the induction hypothesis that each subcube $SQ_{4k-2}^{i_1i_2i_3i_4} - F$ is connected. Hence, $SQ_{4k+2} - F$ is connected.

Case 2. There is a subcube containing at least $4k-2$ vertices in F. It follows that H_1 contains at most four subcubes. The proof is similar to **Case 2** for SQ_6.

Hence, SQ_{4k+2} is $4k+2$ connected. The theorem follows. □

Li et al. [224] also generalize shuffle cubes into generalized shuffle cubes by lowering the diameter and keeping both the connectivity and the number of edges.

The problem of finding a graph (or digraph) with n vertices each of which has degree (or outdegree) at most d to minimize the diameter and to maximize the connectivity by giving n and d is a multiobjective optimization problem. Usually, for such a problem, a solution is selected based on a tradeoff between two objective functions. However, for this problem, it is different; that is, there exist solutions that are nearly optimal to both. In the following, we present some solutions to this problem.

As discussed in Chapter 3 the de Bruijn digraph $BRU(d,t)$ contains d^t vertices of diameter t and with the outdegree of each vertex being d. The diameter is very closed to the Moore bound. It is proved that the connectivity of $BRU(d,t)$ is $d-1$, which is one less than the degree bound d. The Kautz digraph $Kautz(d,t)$ contains $(d+1)d^{t-1}$ vertices of diameter t and with the outdegree of each vertex being d. The diameter is very closed to the Moore bound. It is proved that the connectivity of $Kautz(d,t)$ is d, which is the best possible connectivity.

The maximum degree of the undirected de Bruijn graph $UBRU(d,t)$ is $2d$. Hence, the diameter of $UBRU(d,t)$ is bounded by a constant times the Moore bound. The connectivity of $UBRU(d,t)$ is shown to be $2d-2$. Similarly, the maximum degree of the undirected Kautz graph $UKautz(d,t)$ is $2d$. Hence, its diameter is again bounded by a constant times the Moore bound. The connectivity of $UKautz(d,t)$ is shown to be $2d-1$.

7.6 WIDE DIAMETERS AND FAULT DIAMETERS

Assume that $G=(V,E)$ is a graph with $\kappa(G)=\kappa$. For any pair of vertices, say u and v, by the Menger Theorem we can find κ internally disjoint paths such that the longest path length of κ disjoint paths is minimum, denoted by $d_\kappa(u,v)$, among all possible choice of κ internally disjoint paths. The wide diameter is defined as the maximum of $d_\kappa(u,v)$ over all $u, v \in V$. A small wide diameter is preferred, since it enables fast multipath communication. Fault diameter estimates the impact on diameter when faults occur; that is, the removal of vertices from G. For a pair of vertices u and v, we find the maximum of shortest path length between u and v over all possible $(\kappa-1)$ faults, denoted by $d_{\kappa-1}^f(u,v)$. The $(\kappa-1)$ fault diameter is the maximum of $d_{\kappa-1}^f(u,v)$ for all $u, v \in V$; that is, the maximum transmission delay of $(\kappa-1)$ faults. A small $(\kappa-1)$ fault diameter is also desirable to obtain a small communication delay when a fault occurs.

We here formally define the wide diameter and the fault diameter of an underlying network $G = (V, E)$. Let u and v be two distinct vertices in G, and let $\kappa(G) = \kappa$. Let $C(u, v)$ denote the set of all α internally disjoint paths between u and v. An element in $C(u, v)$ is called a *container* between u and v. Each element i of $C(u, v)$ consists of α internally disjoint paths, and the longest length among these α paths is denoted by $l_i(u, v)$. The number of elements in $C(u, v)$ is denoted by $|C(u, v)|$. We define $d_\alpha(u, v)$ as the minimum over all l_i; that is, $d_\alpha(u, v) = \min_{1 \le i \le |C(u,v)|}\{l_i(u, v)\}$. We write $d_1(u, v)$ as $d(u, v)$, which means the shortest distance between u and v. $D_\alpha(G)$ is called the *α-diameter* of G and is given by

$$D_\alpha(G) = \max_{i, v \in V} \{d_\alpha(u, v)\}$$

By the Menger Theorem, $D_\alpha(G) = \infty$ if $\alpha \ge \kappa + 1$. We usually write $D_1(G)$ as $D(G)$ and call $D(G)$ the *diameter* of G. In particular, the *wide diameter* of G is defined as $D_\kappa(G)$. The wide diameter, proposed by Hsu [164], measures the performance of multipath communication.

For a positive integer β, $d^f_\beta(u, v)$, is given by

$$d^f_\beta(u, v) = \max_{F \subset V, |F| = \beta} \{d_{G-F}(u, v) \mid u, v \notin F\}$$

The *β-fault diameter*, denoted by $D^f_\beta(G)$, is given by

$$D^f_\beta(G) = \max_{u, v \in V} \left\{d^f_\beta(u, v)\right\}$$

By the Menger Theorem, $D^f_\beta(G) = \infty$ if $\beta \ge \kappa$. In particular, the *fault diameter* of G is defined as $D^f_{\kappa-1}(G)$. The fault diameter, proposed by Krishnamoorthy and Krishnamurthy [206], estimates the impact of the diameter when a fault occurs. Fault diameters and wide diameters are important measures for interconnection networks.

Obviously, we have $D(G) \le D^f_{\kappa-1}(G) \le D_\kappa(G)$.

THEOREM 7.13 **(Saad and Shultz [279])** $D_n(Q_n) = D^f_{n-1}(Q_n) = n + 1$.

Proof: Each vertex $\mathbf{u}$ of Q_n can be expressed as an n-bit string $\mathbf{u} = u_{n-1}u_{n-2}\ldots u_0$. We use $\mathbf{u}^k$ to denote the string $u_{n-1}u_{n-2}\ldots u_{k+1}\overline{u}_k u_{k-1}\ldots u_0$.

Assume that $\mathbf{u}$ and $\mathbf{v}$ are two different nodes in the n-dimensional hypercube. We use $H(\mathbf{u}, \mathbf{v})$ to denote the set of indices such that $u_i \ne v_i$. Obviously, $|H(\mathbf{u}, \mathbf{v})| = h(\mathbf{u}, \mathbf{v})$. Let $\{\alpha_i\}_{i=0}^{h(u,v)-1}$ be the decreasing sequence of indices in $H(\mathbf{u}, \mathbf{v})$ and $\{\alpha_j\}_{j=h(u,v)}^{n-1}$ be the decreasing sequence of indices such that $u_{\alpha_j} = v_{\alpha_j}$. For $1 \le i \le h(\mathbf{u}, \mathbf{v})$, we set $P_i = \langle \mathbf{u},\ \mathbf{x}_{i,1} = \mathbf{u}^{\alpha_{0+i}},\ \mathbf{x}_{i,2} = \mathbf{x}_{i,1}^{\alpha_{1+i}}, \ldots, \mathbf{x}_{i,h(u,v)} = \mathbf{x}_{i,h(\mathbf{u},\mathbf{v})-1}^{\alpha_{h(\mathbf{u},\mathbf{v})-1+i}} = \mathbf{v}\rangle$ with the addition of superscripts being performed under modulo $h(\mathbf{u}, \mathbf{v})$; and for $h(\mathbf{u}, \mathbf{v}) \le j \le n-1$, we set $P_j = \langle \mathbf{u}, \mathbf{u}^{\alpha_j}, \mathbf{x}_{0,1}^{\alpha_j}, \mathbf{x}_{0,2}^{\alpha_j}, \ldots, \mathbf{x}_{0,h(u,v)-1}^{\alpha_j}, \mathbf{v}^{\alpha_j}, \mathbf{v}\rangle$.

For example, let $n = 5$, $\mathbf{u} = 00000$, and $\mathbf{v} = 00111$. Then $P_1 = \langle 00000, 00100, 00110, 00111\rangle$, $P_2 = \langle 00000, 00010, 00011, 00111\rangle$, $P_3 = \langle 00000, 00001, 00101,$

00111⟩, $P_4 = \langle 00000, 10000, 10100, 10110, 10111, 00111\rangle$, and $P_5 = \langle 00000, 01000, 01100, 01110, 01111, 00111\rangle$.

Obviously, $P_1, P_2, \ldots, P_n$ form n disjoint paths joining $\mathbf{u}$ to $\mathbf{v}$. With this construction, $d_n(\mathbf{u},\mathbf{v}) \le 2 + h(\mathbf{u},\mathbf{v})$ if $h(\mathbf{u},\mathbf{v}) \ne n$ and $d_n(\mathbf{u},\mathbf{v}) \le n$ if $h(\mathbf{u},\mathbf{v}) = n$. Hence $D_n(Q_n) \le n+1$.

Let $\mathbf{u} = \overbrace{00\ldots0}^{n}$ and $\mathbf{v} = 0\overbrace{11\ldots1}^{n-1}$. Assume that $F = \{\mathbf{u}^i \mid 0 \le i \le n-2\}$. Obviously, $\mathbf{u}^{n-1}$ is the only fault-free node that is adjacent to $\mathbf{u}$. Let $P = \langle \mathbf{u} = \mathbf{x}_0, \mathbf{x}_1, \ldots, \mathbf{x}_k = \mathbf{v}\rangle$ be any path in $Q_n - F$ joining $\mathbf{u}$ to $\mathbf{v}$. Obviously, $\mathbf{x}_1 = \mathbf{u}^n$. Since $h(\mathbf{x}_1, \mathbf{v}) = n$, the length of P is at least $n+1$. Therefore, $D^f_{\kappa-1}(Q_n) \ge n+1$.

Since $D(G) \le D^f_{\kappa-1}(G) \le D_\kappa(G)$ holds for any graph G, $D_n(Q_n) = D^f_{n-1}(Q_n) = n+1$ is obtained. □

It is interesting to study the wide diameter and the fault diameter of other interconnection networks. There are some studies on the fault diameter and wide diameter for the other interconnection networks [130,217].

7.7 SUPERCONNECTIVITY AND SUPER-EDGE-CONNECTIVITY

Assume that G is a k-regular graph with connectivity κ and edge-connectivity λ. We say that G is *maximum-connected* if $\kappa = k$; and G is *super-connected* if it is a complete graph, or it is maximum-connected and every minimum vertex cut is $N_G(v)$ for some vertex v. We say that G is *maximum-edge-connected* if $\lambda = k$; and G is *super-edge-connected* if it is maximum-edge-connected and every minimum edge disconnecting cut is $NE_G(v)$ for some vertex v where $NE_G(v) = \{(v,x) \mid x \in E\}$.

It has been shown that a network is more reliable if it is super-connected [59,61, 351]. Some important families of interconnection networks have been proven to be super-connected [59,61,351]. Here, we present three schemes, proposed in Ref. 52, to construct super-connected and super-edge-connected graphs.

Assume that t is a positive integer. Suppose that $G_0 = (V_0, E_0)$ and $G_1 = (V_1, E_1)$ are two disjoint graphs with t vertices. A *1-1 connection* between G_0 and G_1 is defined as an edge set $E_r = \{(v, \phi(v)) \mid v \in V_0, \phi(v) \in V_1$ and $\phi : V_0 \to V_1$ is a bijection$\}$. We use $G_0 \oplus G_1$ to denote the graph $G = (V_0 \cup V_1, E_0 \cup E_1 \cup E_r)$. The operation $\oplus$ may generate different graphs depending on the bijection. Sometimes, G can also be expressed as $G(G_0, G_1, M)$ where M is the perfect matching generated by ϕ. See Figure 7.11, for an illustration of $G_0 \oplus G_1$. We note that the Cartesian product of a graph H and a complete graph K_2 can be viewed as a $H \oplus H$. Let x be any vertex of $G_0 \oplus G_1$. We use $\bar{x}$ as the vertex matched under ϕ.

THEOREM 7.14 Assume that t is a positive integer. Suppose that G_0 and G_1 are two k-regular maximum-connected graphs with t vertices, and ϕ is a 1-1 connection between G_0 and G_1. Then, $G_0 \oplus G_1$ is $(k+1)$-regular super-connected if and only if (1) $t > k+1$ or (2) $t = k+1$ with $k = 0, 1, 2$.

Proof: Since G_0 and G_1 are k-regular-connected graphs, $t \ge k+1$. By definition, $G_0 \oplus G_1$ is a $(k+1)$-regular graph. In order to prove that $G_0 \oplus G_1$ is super-connected,

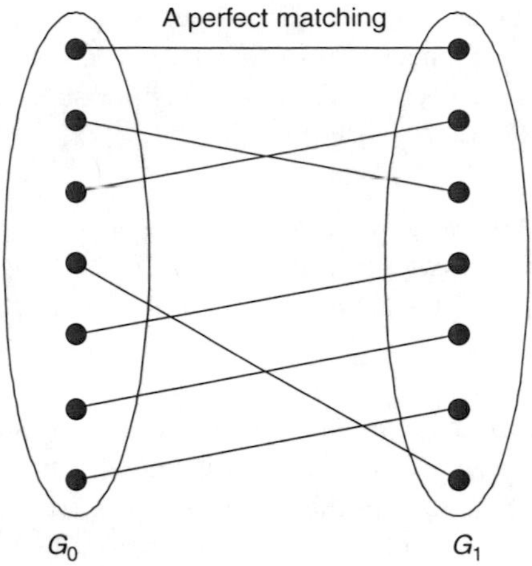

FIGURE 7.11 Illustration of $G_0 \oplus G_1$.

we need to check whether $G_0 \oplus G_1 - F$ is connected for any $F \subset V(G_0 \oplus G_1)$ such that $|F| = k+1$ and $F \neq N_{G_0 \oplus G_1}(v)$ for any vertex $v \in V(G_0 \oplus G_1)$.

Suppose that $t = k+1$. Then, G_0 and G_1 are isomorphic to the complete graph K_{k+1}. Moreover, $G_0 \oplus G_1$ is isomorphic to the Cartesian product of K_{k+1} and K_2. Without loss of generality, we assume that $V(G_0) = \{a_0, a_1, \ldots, a_k\}$ and $V(G_1) = \{\overline{a}_0, \overline{a}_1, \ldots, \overline{a}_k\}$. By brute force, we can check whether $G_0 \oplus G_1$ is super-connected for $k = 0, 1, 2$. Suppose that $k \geq 3$. We set $F = \{a_0, a_1\} \cup \{\overline{a}_i \mid 2 \leq i \leq k\}$. It is easy to see that $|F| = k+1$, $F \neq N_{G_0 \oplus G_1}(v)$, and F is a vertex cut of $G_0 \oplus G_1$. Therefore, $G_0 \oplus G_1$ is not super-connected.

Now, assume that $t > k+1$. Note that K_1 is the only connected 0-regular graph and K_2 is the only connected 1-regular graph. Let $k \geq 2$. We set $X_0 = F \cap V(G_0)$ and $X_1 = F \cap V(G_1)$.

Case 1. $|X_0| < k$ and $|X_1| < k$. Thus, both $G_0 - X_0$ and $G_1 - X_1$ are connected. Since $t = |M| > k+1$ and $|F| = k+1$, there exists $a \in V(G_0) - F$ and $\overline{a} \in V(G_1) - F$. Therefore, $G_0 \oplus G_1 - F$ is connected.

Case 2. Either $k \leq |X_0| \leq k+1$ or $k \leq |X_1| \leq k+1$. Without loss of generality, we assume that $k \leq |X_0| \leq k+1$. Hence, $|X_1| \leq 1$. Since $k \geq 2$, $G_1 - X_1$ is connected. Let C be any connected component of $G_0 - X_0$. We will claim that there exists $a \in C$ and $\overline{a} \in V(G_1) - F$. With this claim, $G_0 \oplus G_1 - F$ is connected.

First, suppose that C consists of only one vertex a. Then $N_{G_0}(a) \subset F$. Since $F \neq N_{G_0 \oplus G_1}(a)$, $\overline{a} \in V(G_1) - F$. Thus, the claim holds. Now, suppose that C contains at least two vertices a and b. Since at most one vertex of $\{\overline{a}, \overline{b}\}$ is in F, we may assume that $\overline{a} \notin F$. Thus, our claim holds.

Therefore, the theorem is proved. □

A similar argument leads to the following theorem for super-edge-connected graphs, and the following corollary.

THEOREM 7.15 Assume that t is a positive integer. Suppose that G_0 and G_1 are two k-regular maximum-edge-connected graphs with t vertices, and ϕ is a 1-1 connection between G_0 and G_1. Then, $G_0 \oplus G_1$ is $(k+1)$-regular super-edge-connected if and only if (1) $t > k+1$ or (2) $t = k+1$ with $k = 0$.

COROLLARY 7.2 *Assume that t is a positive integer. Suppose that* G_0 *and* G_1 *are two k-connected and k′-edge-connected graphs with t vertices, and* ϕ *is a* 1-1 *connection between* G_0 *and* G_1*. Then,* $G_0 \oplus G_1$ *is* $(k+1)$*-connected and* $(k'+1)$*-edge-connected.*

As noted in Chapter 3, the hypercube does not have the smallest diameter for its resources. Various networks are proposed by twisting some pairs of links in hypercubes [1,76,92,93]. Because of the lack of the unified perspective on these variants, results of one topology are hard to extend to others. To make a unified study of these variants, Vaidya et al. introduced the class of hypercube-like graphs [322]. We denote their graphs as H'-graphs. The class of H'-graphs, consisting of simple, connected, and undirected graphs, contains most of the hypercube variants.

Now, we can define the set of n-dimensional H'-graph, H'_n, as follows:

1. $H'_1 = \{K_2\}$, where K_2 is the complete graph with two vertices.
2. Assume that $G_0, G_1 \in H'_n$. Then $G_0 \oplus G_1$ is a graph in H'_{n+1}, where N is any perfect matching between $V(G_0)$ to $V(G_1)$.

COROLLARY 7.3 *Every graph in* H'_n *is both super-connected and super-edge-connected if* $n \geq 2$.

Assume that n is a positive integer. The *alternating graph* A_n [59] is another attractive interconnection graph topology. $V(A_n) = \{p \mid p = p_1p_2 \ldots p_{n-2}$ with $p_i \in \{1, 2, \ldots, n\}$ for $1 \leq i \leq n-2$ and $p_i \neq p_j$ if $i \neq j\}$ and $E(A_n) = \{(p, q) \mid$ there exists a unique $i \in \langle n-2 \rangle$ such that $p_i \neq q_i\}$. In Ref. 59, *split-star* S_n^2 is proposed as an attractive interconnection network. $V(S_n^2) = \{p \mid p = p_0p_1p_2 \ldots p_{n-2}$ with $p_0 \in \{0, 1\}$, $p_i \in \{1, 2, \ldots, n\}$ for $1 \leq i \leq n-2$, and $p_i \neq p_j$ if $i \neq j \neq 0\}$ and $E(S_n^2) = \{(p, q) \mid$ there exists a unique i with $0 \leq i \leq n-2$ such that $p_i \neq q_i\}$.

It is pointed out in Ref. 59 that S_n^2 can be viewed as $A_n \oplus A_n$. Moreover, it is proved that A_n is super-connected unless $n = 4$ and is super-edge-connected for any n. By Theorems 7.1 and 7.2, we can easily prove the result in Ref. 59 that S_n^2 is super-connected for any n and super-edge-connected unless $n = 3$.

Suppose that r and t are positive integers with $r \geq 3$. Assume that $G_0, G_1, \ldots, G_{r-1}$ are graphs with $|V(G_i)| = t$ for $0 \leq i \leq r-1$. We define $H = G(G_0, G_1, \ldots, G_{r-1}; \mathcal{M})$ as the graph with $V(H) = V(G_0) \cup V(G_1) \cup \cdots \cup V(G_{r-1})$ such that every subgraph of H induced by $V(G_i) \cup V(G_{i+1 \bmod r})$ with $0 \leq i < r$ can be viewed as $G_i \oplus G_{i+1 \bmod r}$ for some 1-1 connection. See Figure 7.12 for an illustration of $G(G_0, G_1, \ldots, G_{r-1}; \mathcal{M})$.

The following results are proved in Ref. 52.

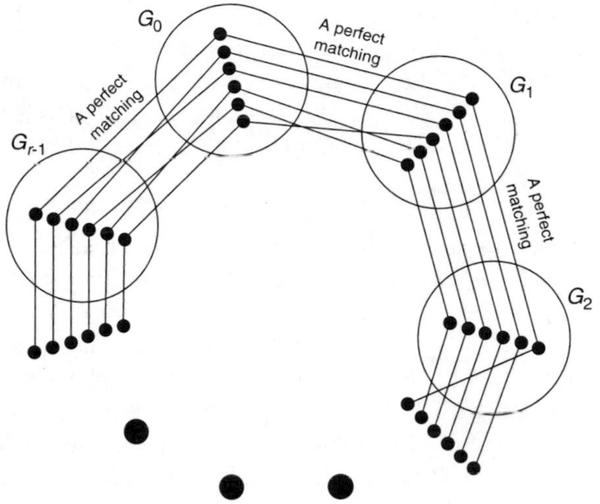

FIGURE 7.12 Illustration of $G(G_0, G_1, \ldots, G_{r-1}; \mathcal{M})$.

THEOREM 7.16 Suppose that r and t are positive integers with $r \geq 3$. Assume that $G_0, G_1, \ldots, G_{r-1}$ are k-regular maximum-connected graphs with $|V(G_i)| = t$ for $0 \leq i \leq r-1$. Then, $H = G(G_0, G_1, \ldots, G_{r-1}; \mathcal{M})$ is $(k+2)$-regular super-connected if and only if (1) $k \geq 1$ or (2) $k = 0$ and $r = 3, 4, 5$.

THEOREM 7.17 Suppose that r and t are positive integers with $r \geq 3$. Assume that $G_0, G_1, \ldots, G_{r-1}$ are k-regular maximum-edge-connected graphs with $|V(G_i)| = t$ for $0 \leq i \leq r-1$. Then, $H = G(G_0, G_1, \ldots, G_{r-1}; \mathcal{M})$ is $(k+2)$-regular super-edge-connected if and only if (1) $k \geq 2$, (2) $k = 1$ and $r \neq 3$, or (3) $k = 0$ and $r = 3$.

COROLLARY 7.4 *Suppose that r and t are positive integers with $r \geq 3$. Assume that $G_0, G_1, \ldots, G_{k-1}$ are k-connected and k'-edge-connected graphs with $|V(G_i)| = t$ for $0 \leq i \leq r-1$. Then, $H = G(G_0, G_1, \ldots, G_{r-1}; \mathcal{M})$ is $(k+2)$-connected and $(k'+2)$-edge-connected.*

Let t be an integer with $t \geq 2$. Let G_1 and G_2 be two graphs with t vertices such that $V(G_1) = \{a_i \mid 0 \leq i < t\}$ and $V(G_2) = \{b_i \mid 0 \leq i < t\}$. Let $\mathcal{C}$ be a set of edges given by $\mathcal{C} = \{(a_i, b_i) \mid 0 \leq i < t\} \cup \{(b_i, a_{i+1 \textbf{ (mod } t)}) \mid 0 \leq i < t\}$. The graph $G(G_1, G_2; \mathcal{C})$ is defined as the graph with the vertex set $V(G(G_1, G_2; \mathcal{C})) = V(G_1) \cup V(G_2)$, and edge set $E(G(G_1, G_2; \mathcal{C})) = E(G_1) \cup E(G_2) \cup \mathcal{C}$.

THEOREM 7.18 Let t be an integer with $t \geq 2$. Let G_1 and G_2 be two k-regular maximum-connected graphs with t vertices such that $V(G_1) = \{a_i \mid 0 \leq i < t\}$ and $V(G_2) = \{b_i \mid 0 \leq i < t\}$. Let $\mathcal{C}$ be a set of edges given by $\mathcal{C} = \{(a_i, b_i) \mid 0 \leq i < t\} \cup \{(b_i, a_{i+1 \textbf{ (mod } t)}) \mid 0 \leq i < t\}$. Then, $G(G_1, G_2; \mathcal{C})$ is $(k+2)$-regular super-connected if (1) $t > k+1$ with $k \geq 3$ or (2) $t = k+1$ with $k = 1, 2, 3$.

THEOREM 7.19 Let t be an integer with $t \geq 2$. Let G_1 and G_2 be two k-regular maximum-edge-connected graphs with t vertices such that $V(G_1) = \{a_i \mid 0 \leq i < t\}$ and $V(G_2) = \{b_i \mid 0 \leq i < t\}$. Let $\mathcal{C}$ be a set of edges given by $\mathcal{C} = \{(a_i, b_i) \mid 0 \leq i < t\} \cup \{(b_i, a_{i+1 \, (\textbf{mod}\, t)}) \mid 0 \leq i < t\}$. Then, $G(G_1, G_2; \mathcal{C})$ is $(k+2)$-regular super-edge-connected if (1) $t > k+1$ with $k \geq 2$ or (2) $t = k+1$ with $k \geq 1$.

Let c, d, r be integers with $r \geq 0$, $d > 1$, and $1 \leq c < d$. The *recursive circulant graph* $RC(c, d, r)$, defined as the circulant graph $G(cd^r; \{1, d, \ldots, d^{\lceil \log_d cd^r \rceil - 1}\})$, is proposed in Ref. 269. For $0 \leq i < d$, let V_i^r denote the set $\{j \mid 0 \leq j < cd^r, j = i(\textbf{mod}\ d)\}$. We use $RC_i(c, d, r)$ to denote the subgraph of $RC(c, d, r)$ induced by V_i^r. For a positive integer N, let Z_N be the additive group of residue classes modulo N. We can recursively describe $RC(c, d, r)$ as follows: Suppose that $r = 0$. Then $RC(c, d, 0)$ is the graph with $V(RC(c, d, 0)) = \{0\}$ and $E(RC(c, d, 0)) = \emptyset$ if $c = 1$, $V(RC(c, d, 0)) = \{0, 1\}$ and $E(RC(c, d, 0)) = \{(0, 1)\}$ if $c = 2$, and $V(RC(c, d, 0)) = Z_c$ and $E(RC(c, d, 0)) = \{(i, i+1) \mid i \in Z_c\}$ if $c \geq 3$. Suppose that $r \geq 1$. The induced subgraph $RC_i(c, d, r)$ is isomorphic to $RC(c, d, r-1)$ for $0 \leq i < d$. More precisely, let f_i^r be the function from $Z_{cd^{r-1}}$ into V_i^r defined by $f_i^r(x) = dx + i$. Then, f_i^r induces an isomorphism from $RC(c, d, r-1)$ into $RC_i(c, d, r)$. Let $H(c, d, r)$ denote the set of edges of $RC(c, d, r)$ not in $\cup_{i=0}^{d-1}(E(RC_i(c, d, r)))$. Then, $H(c, d, r) = \{(i, i+1) \mid i \in Z_{cd^r}\}$.

Suppose $d \geq 3$. Obviously, $RC(c, d, r)$ can be expressed as $G(H, H, \ldots, H; \mathcal{M})$, with d Hs, where H is $RC(c, d, r-1)$ for some $\mathcal{M} = \cup_{i=0}^{d-1} M_{i,i+1\,(\textbf{mod}\ d)}$. Suppose $d = 2$. Obviously, $RC(c, d, r)$ can be expressed as $G(H, H; \mathcal{C})$, where H is $RC(c, d, r-1)$. Recursively applying Theorems 7.3 through 7.6, we can easily prove that $RC(c, d, r)$ is super-connected unless (1) $r = 0$ and $c \geq 5$ or (2) $r = 1$, $c = 1$, and $d \geq 5$; and super-edge-connected unless (1) $r = 0$ and $c \geq 4$, (2) $r = 1$, $c = 1$, and $d \geq 4$, or (3) $r = 1$, $c = 2$, and $d = 3$.

COROLLARY 7.5 *Suppose that t is an integer with $t \geq 2$. Assume that G_1 and G_2 are two k-connected and k'-edge-connected graphs with t vertices. Then, $G(G_1, G_2; \mathcal{C})$ is $(k+2)$-connected and $(k'+2)$-edge-connected.*

8 Graph Coloring

8.1 VERTEX COLORINGS AND BOUNDS

Coloring problems arise in many contexts. The committee-scheduling example in Chapter 1 has other settings: Assume that G is the graph with the set of courses given in a university as a vertex set. We join an edge between two courses if some students take both courses. The chromatic number is the minimum number of time periods needed to schedule examinations without conflicts. The problem of 4-coloring the regions of maps such that regions with common boundaries receive different colors is another example.

A *k-coloring* of G is a labeling $f: V(G) \to \{1, 2, \ldots, k\}$. The labels are *colors*. The vertices with color i are a *color class*. A k-coloring f is *proper* if $x \leftrightarrow y$ implies $f(x) \neq f(y)$. A graph G is *k-colorable* if it has a proper k-coloring. The *chromatic number* $\chi(G)$ is the minimum k such that G is k-colorable. A graph G is *k-chromatic* if $\chi(G) = k$. Suppose that $\chi(G) = k$ but $\chi(H) < k$ for every proper subgraph H of G. We say G is *color-critical* or *k-critical*.

We call the labels colors because their numerical value is usually unimportant. We may interpret them as elements of any set. Discarding loops and copies of multiple edges does not change chromatic number. Therefore, *in this chapter—except for the final section—all graphs mentioned are simple graphs*, even though we say only *graph*.

Example 8.1

Each color class in a proper coloring is an independent set. Thus, G is k-colorable if and only if G is k-partite. In Figure 8.1, we illustrate optimal colorings of the 5-cycle and the Petersen graph. These two graphs have chromatic number 3.

Example 8.2

Every k-colorable graph is a k-partite graph. Thus, K_1 and K_2 are the only 1-critical and 2-critical graphs. By the characterization of bipartite graphs, the 3-critical graphs are odd cycles. We can test 2-colorability by using breadth-first search (BFS) from a vertex in each component. A graph is bipartite if and only if $G[X]$ and $G[Y]$ are independent sets where $X = \{u \in V(G) \mid d(u,x) \text{ is even}\}$ and $Y = \{u \in V(G) \mid d(u,x) \text{ is odd}\}$. Beyond this, we have no characterization of 4-critical graphs and no good algorithm to test 3-colorability.

Let $\omega(G)$ denote the order of the largest complete subgraph of G. Obviously, $\chi(G) \geq \omega(G)$. Since each color class is an independent set, $\chi(G) \geq n(G)/\alpha(G)$. With Example 8.1, we have examples with $\chi(G) > \omega(G)$.

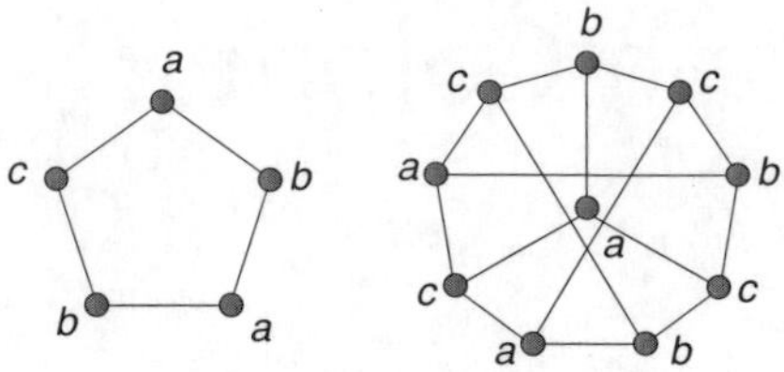

FIGURE 8.1 Optimal colorings of the 5-cycle and the Petersen graph.

We may study the behavior of the chromatic number under various graph operations. It is easy to obtain the following theorem.

THEOREM 8.1

1. $\chi(G+H) = \max\{\chi(G), \chi(H)\}$
2. $\chi(G \vee H) = \chi(G) + \chi(H)$
3. $\chi(G \times H) = \max\{\chi(G), \chi(H)\}$
4. $\chi(G \cdot H) \leq \min\{\chi(G), \chi(H)\}$

It is interesting to point out that Hedetniemi conjectured that the equality holds for all graphs in the equation 4 in 1966. This conjecture remains open.

CONJECTURE 8.1 (Hedetniemi [143]) $\chi(G \cdot H) = \min\{\chi(G), \chi(H)\}$.

8.2 PROPERTIES OF k-CRITICAL GRAPHS

It is helpful when dealing with colorings to study the properties of k-critical graphs. An easy consequence of the definition is that every k-critical graph is connected.

LEMMA 8.1 Suppose that H is a k-critical graph. Then $\delta(H) \geq k-1$.

Proof: Suppose that H is a k-critical graph and x is a vertex of H. Since H is k-critical, $H - x$ is $(k-1)$-colorable. Suppose that $d_H(x) < k-1$. The $(k-1)$ colors used on $H - x$ do not all appear on $N(x)$. We can assign a missing color to x to extend the coloring to H. This contradicts our hypothesis that H has no proper $(k-1)$-coloring. Hence every vertex of H has a degree of at least $k-1$. □

COROLLARY 8.1 *Every k-chromatic graph has at least k vertices of degree at least* $k-1$.

Proof: Suppose that G is a k-chromatic graph. Let H be a k-critical subgraph of G. By Lemma 8.1, each vertex has a degree of at least $(k-1)$ in H. Obviously, these vertices have a degree of at least $(k-1)$ in G. Since H contains at least k vertices, G contains at least k vertices of degree of at least $(k-1)$. □

Suppose that S is a vertex cut of a connected graph H. Let $V_1, V_2, \ldots, V_n$ be the vertex set of the components of $H - S$. The subgraphs $H_i = G[V_i \cup S]$ are called the *S-components* of H. See Figure 8.2 for illustration. We say that the colorings of $H_1, H_2, \ldots, H_n$, *agree* on S if vertex v is assigned the same color in each coloring for every $v \in S$.

THEOREM 8.2 No vertex cut is a clique in a critical graph.

Proof: Assume that the statement is false. Let G be a k-critical graph such that G has a vertex cut S that is a clique. Let $G_1, G_2, \ldots, G_n$ be the S-components of G. Since G is k-critical, each G_i is $(k-1)$-colorable. Since S is a clique, the vertices in S must receive distinct colors in any $(k-1)$-coloring of G_i. It follows that there are $(k-1)$-colorings of $G_1, G_2, \ldots, G_n$ that agree on S. But these colorings together yield a $(k-1)$ coloring of G. We get a contradiction. Hence, the theorem is proved. □

COROLLARY 8.2 *Every critical graph is a block.*

Proof: Suppose that v is a cut vertex. Then $\{v\}$ is a clique. It follows from Theorem 8.2 that no critical graph has a cut vertex. Thus, every critical graph is a block. □

Let G be a k-critical graph G with a 2-vertex cut $\{u, v\}$. By Theorem 8.2, u and v cannot be adjacent. A $\{u, v\}$-component G_i of G is of *type 1* if every $(k-1)$-coloring of G_i assigns the same color to u and v. A $\{u, v\}$-component G_i of G is of *type 2* if every $(k-1)$-coloring of G_i assigns different colors to u and v. See Figure 8.3 for illustration.

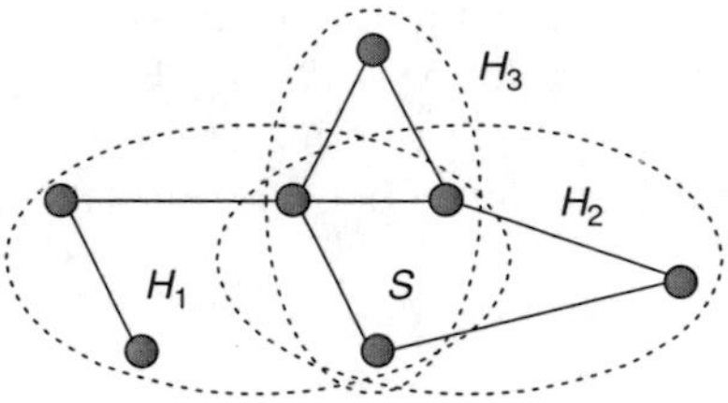

FIGURE 8.2 Illustration of S-component.

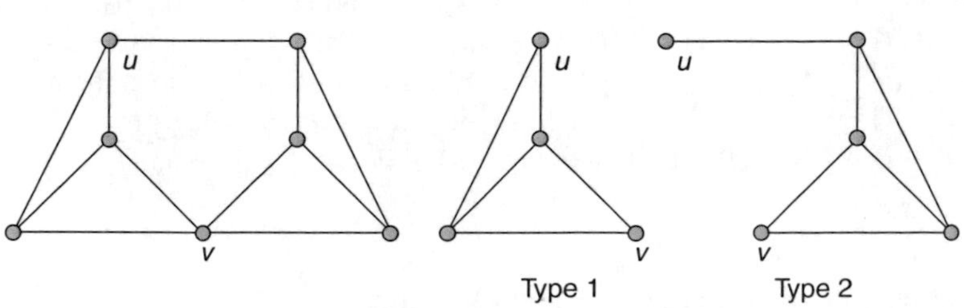

FIGURE 8.3 A 4-critical graph with a 2-vertex cut $\{u, v\}$ and its two $\{u, v\}$-components.

THEOREM 8.3 **(Dirac [89])** Let G be a k-critical graph with a 2-vertex cut $\{u, v\}$. Then

1. $G = G_1 \cup G_2$, where G_i is a $\{u, v\}$-component of type i, $(i = 1, 2)$
2. Both $G_1 + (u, v)$ and $G_2 \cdot (u, v)$ are k-critical

Proof:

1. Since G is critical, each $\{u, v\}$-component of G is $(k-1)$-colorable. There cannot exist $(k-1)$-colorings of these $\{u, v\}$-components all of which agree on $\{u, v\}$, because such colorings would together yield a $(k-1)$-coloring of G. Therefore there are two $\{u, v\}$-components G_1 and G_2 such that no $(k-1)$-coloring of G_1 agrees with any $(k-1)$-coloring of G_2. Clearly one, say G_1, must be of type 1 and the other, G_2, of type 2. Since G_1 and G_2 are of different types, the subgraph $G_1 \cup G_2$ of G is not $(k-1)$-colorable. Since G is critical, there are exactly two $\{u, v\}$-component of G. Hence $G = G_1 \cup G_2$.
2. Let $H_1 = G_1 + (u, v)$. Since G_1 is of type 1, H_1 is k-chromatic. We shall prove that H_1 is critical by showing that $H_1 - e$ is $(k-1)$-colorable for every edge e of H_1. Suppose that $e = (u, v)$. Since $H_1 - e = G_1$, $H_1 - e$ is $(k-1)$-colorable. Suppose that e is some other edge of H_1. Since G_2 is a subgraph of $G - e$, the vertices u and v must receive different colors in any $(k-1)$-coloring of $G - e$. The restriction of such coloring to the vertices of G_1 is a $(k-1)$-coloring of $H_1 - e$. Thus, $G_1 + (u, v)$ is k-critical.
3. Let $H_2 = G_2 \cdot (u, v)$. Since $\chi(G_2) = k - 1$, H_2 is k-chromatic. We shall prove that H_2 is critical by showing that $H_2 - e$ is $(k-1)$-colorable for every edge e of H_1. Since G_1 is a subgraph of $G - e$, the vertices u and v must receive same colors in any $(k-1)$-coloring of $G - e$. The restriction of such coloring to the vertices of G_2 is a $(k-1)$-coloring of $G_2 - e$ such that the vertices u and v receive same colors. Thus, $H_2 - e$ is $(k-1)$-colorable for every edge e of H_1. □

COROLLARY 8.3 *Assume that G is a k-critical graph with a 2-vertex cut* $\{u, v\}$. *Then* $\deg_G(u) + \deg_G(v) \geq 3k - 5$.

Proof: Let G_1 be the $\{u, v\}$-component of type 1 and G_2 be the $\{u, v\}$-component of type 2. Set $H_1 = G_1 + (u, v)$ and $H_2 = G_2 \cdot (u, v)$. By Theorem 8.3 and Lemma 8.1, we have $\deg_{H_1}(u) + \deg_{H_2}(v) \geq 2k - 2$ and $\deg_{H_2}(w) \geq k - 1$ where w is the new vertex obtained by identifying u and v. It follows that $\deg_{G_1}(u) + \deg_{G_1}(v) \geq 2k - 4$ and $\deg_{G_2}(u) + \deg_{G_2}(v) \geq k - 1$. Hence, $\deg_G(u) + \deg_G(v) \geq 3k - 5$. □

8.3 BOUND FOR CHROMATIC NUMBERS

There are some upper bounds on $\chi(G)$ coming from the following coloring algorithm.

ALGORITHM: *(Greedy coloring.)*

The *greedy coloring* with respect to a vertex ordering $v_1, v_2, \ldots, v_n$ of $V(G)$ is obtained by coloring vertices in the order $v_1, v_2, \ldots, v_n$ by the following rule: use the smallest index color not already used on its lower-indexed neighbors to color v_i.

PROPOSITION 8.1 $\chi(G) \leq \Delta(G) + 1$.

Proof: In greedy coloring, each vertex has at most $\Delta(G)$ earlier neighbors. Thus, the greedy coloring cannot be forced to use more than $\Delta(G) + 1$ colors. This proves $\chi(G) \leq \Delta(G) + 1$. □

PROPOSITION 8.2 (Welsh and Powell [337]) *Suppose that G is a graph with degree sequence* $d_1 \geq d_2 \geq \cdots \geq d_n$. *Then* $\chi(G) \leq 1 + \max\{\min\{d_i, i-1\} \mid 1 \leq i \leq n\}$.

Proof: We apply the greedy coloring with vertices in nonincreasing order of degree. As we color the ith vertices, there at most min $\{d_i, i-1\}$ of its neighbors already have colors. Thus, the index of its color is at most $1 + \min\{d_i, i-1\}$. Maximizing this over i yields the upper bound. □

THEOREM 8.4 **(Szekeres and Wilf [296])** $\chi(G) \leq 1 + \max_{H \subseteq G} \delta(H)$ for any graph G.

Proof: Suppose that $k = \chi(G)$ and H' is a k-critical subgraph of G. By Lemma 8.1, $\chi(G) - 1 \leq \delta(H') \leq \max_{H \subseteq G} \delta(H)$. □

The next bound involves orientations.

THEOREM 8.5 **(Gallai [119], Roy [278], Vitaver [325])** Suppose that D is an orientation of G. We use $l(D)$ to denote its longest path. Then $\chi(G) \leq 1 + l(D)$. Furthermore, there exists an orientation of G to make the equality hold. (See Figure 8.4).

Proof: Suppose that D is an orientation of G. Let D' be a maximal acyclic subdigraph of D. Obviously, D' is an orientation of a spanning subgraph H of G. Color $V(G)$ by assigning $f(v)$ to be the length of the longest path in D' that ends at v plus 1. Since D' has no cycle, f is strictly increasing along each path in D'.

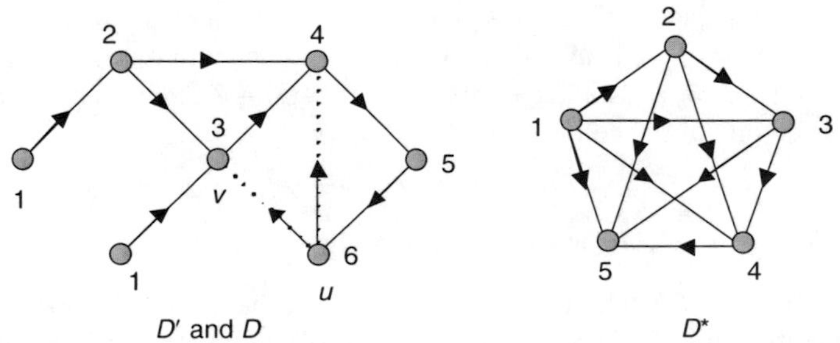

FIGURE 8.4 Illustration of Theorem 8.5.

The coloring f uses 1 through $1+l(D')$ on $V(D)=V(G)$. We claim that f is a proper coloring of G. Suppose that (u, v) is in D'. Obviously, $f(u) \neq f(v)$. Suppose (u, v) is not in D'. Then the addition of (u, v) to D' creates a cycle. Again, $f(u) \neq f(v)$. Thus, f is a proper coloring of G.

To proof the second statement, we construct an orientation D^* such that $l(D^*) \leq \chi(G)-1$. Let f be an optimal coloring of G. For each edge (u, v) in G, we orient the edge as $u \rightarrow v$ in D^* if and only if $f(u) < f(v)$. Since f is a proper coloring, this defines an orientation. Since the labels used by f increase along each path in D^* and there are only $\chi(G)$ label in f, we have $l(D^*) \leq \chi(G)-1$. □

The bound $\chi(G) \leq 1+\Delta(G)$ is tight for cliques and odd cycles. Using the greedy algorithm with a proper vertex ordering, we can show that these are essentially the only graphs with $\chi(G) > \Delta(G)$. With this result, the Petersen graph is 3-colorable. To avoid unimportant complications, we phrase the statement only for connected graphs. It can be extended to all graphs, because the chromatic number of a graph is the maximum chromatic number of its components. We present the proof by Lovász [242].

THEOREM 8.6 **(Brooks [33])** Suppose that G is a connected graph other than a clique or an odd cycle. Then $\chi(G) \leq \Delta(G)$.

Proof: Suppose that G is a k-chromatic connected graph other than a clique or an odd cycle. Without loss of generality, we may assume that G is k-critical. By Corollary 8.2, G is a block. Since 1-critical and 2-critical graphs are complete and 3-critical graphs are odd cycles, we have $k \geq 4$.

Suppose that G has a 2-vertex cut $\{u, v\}$. By Corollary 8.3, $2\Delta(G) \geq \deg_G(u) + \deg_G(v) \geq 3k-5 \geq 2k-1$. Since $2\Delta(G)$ is even, $\chi(G)=k \leq \Delta(G)$.

Assume that G is 3-connected. Since G is not complete, there are three vertices u, v, and w in G such that $(u, v) \in E(G)$ and $(v, w) \in E(G)$ but $(u, w) \notin E(G)$. Set $u=v_1$ and $w=v_2$ and let $v_3, v_4, \ldots, v_{n(G)}=v$ be any ordering of the vertices of $G-\{u, w\}$ such that each v_i is adjacent to some v_j with $j>i$. This can be achieved by arranging the vertices of $G-\{u, w\}$ in nonincreasing order of their distance from v; using, say, BFS.

We can now describe a $\Delta(G)$-coloring of G. Assign color 1 to $v_1=u$ and $v_2=w$. Then successively color $v_3, v_4, \ldots, v_{n(G)}$, each with the first available color in the list $1, 2, \ldots, \Delta(G)$. By the construction of the sequence $v_1, v_2, \ldots, v_{n(G)}$, each vertex v_i, $1 \leq i \leq n(G)-1$, is adjacent to some vertex v_j with $j>i$. Therefore, v_i is adjacent to at most $\Delta(G)-1$ vertices v_j with $j<i$. It follows that, when its turn comes to be colored, v_i is adjacent to at most $\Delta(G)-1$ colors, and, thus, that one of the colors $1, 2, \ldots, \Delta(G)$ will be available. As $v_{n(G)}$ is adjacent to two vertices of color 1 (namely v_1 and v_2), $v_{n(G)}$ is adjacent to at most $\Delta(G)-2$ other colors. Hence $v_{n(G)}$ can be assigned one of the colors $2, 3, \ldots, \Delta(G)$. □

8.4 GIRTH AND CHROMATIC NUMBER

The bound $\chi(G) \geq \omega(G)$ can be tight, but it can also be arbitrarily bad. There have been many constructions of graphs having arbitrarily large chromatic numbers, even though they do not contain K_3.

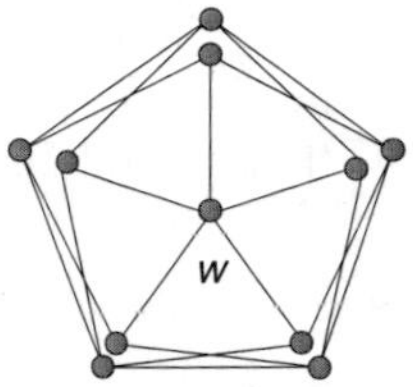

FIGURE 8.5 The Grötzsch graph.

EXAMPLE 8.3

(Mycielski's construction.) Mycielski [257] found a construction scheme that builds from any k-chromatic triangle-free graph G a $(k+1)$-chromatic triangle-free supergraph G'. Let G be a k-chromatic triangle-free graph with vertex set $V=\{v_1, v_2, \ldots, v_n\}$. Let $U=\{u_1, u_2, \ldots, u_n\}$. Then $V(G')=V\cup U\cup\{w\}$. Let the subgraph of G' induced by V, $G'[V]$, is G. We add edges to make u_i adjacent to all of $N_G(v_i)$, and then make $N(w)=U$. The resultant graph is G'.

Note that U is an independent set in G'. From the 2-chromatic graph K_2, the first iteration of Mycielski's construction yields the 3-chromatic C_5. The second iteration yields the 4-chromatic Grötzsch graph, shown in Figure 8.5. These graphs are the triangle-free k-chromatic graphs with fewest vertices for $k=2,3,4$.

THEOREM 8.7 Mycielski's construction produces a $(k+1)$-chromatic triangle-free graph from a k-chromatic triangle-free graph.

Proof: Suppose that $V(G)=\{v_1, v_2, \ldots, v_n\}$ and $V(G')=\{v_i \mid 1\leq i\leq n\}\cup\{u_i \mid 1\leq i\leq n\}\cup\{w\}$ as described previously. Since $\{u_i \mid 1\leq i\leq n\}$ is independent in G', the other two vertices—say, x and y— of any triangle containing u_i belong to $V(G)$. These two vertices are neighbors of v_i. Thus, $\{x, y, v_i\}$ forms a triangle in G. Since G is triangle-free, G' is also triangle-free.

We can extend a proper k-coloring f of G to a proper $(k+1)$-coloring of G' by setting $f(u_i)=f(v_i)$ and $f(w)=k+1$. Thus, $\chi(G')\leq\chi(G)+1$. We prove that $\chi(G)<\chi(G')$ to obtain equality.

Suppose that G' has a proper k-coloring g. We may assume that $g(w)=k$. Thus, $g(U)\subseteq\{1,2,\ldots,k-1\}$. Suppose that $g(V)\subseteq\{1,2,\ldots,k-1\}$. Suppose that $A=\{v_i \mid g(v_i)=k\}\neq\emptyset$. For each $v_i\in A$, we change the color of v_i to $g(u_i)$. We need to check whether it remains a proper coloring. Since A is contained in one color class of g, A is an independent set. We need only to check edges of the form (v_i, v') with $v'\in V(G)-A$. Since $(v_i, v')\in E(G')$, then the construction of G' yields $(u_i, v')\in E(G)$. Thus, $g(v')\neq g(u_i)$. Hence, our alternation does not violate edges with G. Thus, we obtain a proper coloring of G' with $f(w)=k$ and $A=\emptyset$. We now delete $U\cup\{w\}$ to obtain a proper $(k-1)$-coloring of G. We got a contradiction. Thus, $\chi(G')=\chi(G)+1$. □

8.5 HAJÓS' CONJECTURE

Hajós [134] conjectured that every k-chromatic graph contains a subdivision of K_k (a graph obtained from K_k by a sequence of edge subdivisions). The statement is trivial for $k \leq 3$. For $k = 2$, it states that every 2-chromatic graph contains a nontrivial path. For $k = 3$, it states that every 3-chromatic graph contains a cycle. Dirac [86] proved that the conjecture holds for $k = 4$. If F is a subdivision of H, then we call F is an *H-subdivision*.

THEOREM 8.8 **(Dirac [86])** Every graph with chromatic number of at least 4 contains a K_4-subdivision.

Proof: Let G be a 4-chromatic graph. Without loss of generality, we may assume that G is critical. Hence, G is a block and $\delta(G) \geq 3$. We proceed by induction on $n(G)$. Suppose that $n(G) = 4$. Then G is K_4 and the theorem holds trivially. Assume that the theorem is true for all 4-chromatic graphs with fewer than $n(G)$ vertices.

Suppose that G has a 2-vertex cut $\{u, v\}$. By Theorem 8.3, G has two $\{u, v\}$-components G_1 and G_2, where $G_1 + (u, v)$ is 4-critical. Since $n(G_1 + (u, v)) < n(G)$, by induction $G_1 + (u, v)$ contains a subdivision of K_4. Let P be a path from u to v in G_2. Then $G_1 \cup P$ contains a subdivision of K_4. Hence, G contains a subdivision of K_4.

Now, suppose that G is 3-connected. Since $\delta \geq 3$, G has a cycle C of length at least four. Let u and v be nonconsecutive vertices on C. Since $G - \{u, v\}$ is connected, there is a path P in $G - \{u, v\}$ connecting the two portions of $C - \{u, v\}$. We may assume that the origin x and the terminus y are the only vertices of P on C. Similarly, there is a path Q in $G - \{x, y\}$.

Suppose that P and Q have no vertex in common. Then $C \cup P \cup Q$ is a subdivision of K_4. See Figure 8.6a for illustration. Otherwise, let w be the first vertex of P on Q and let P' denote the section from x to w of P. Then $C \cup P' \cup Q$ is a subdivision of K_4. See Figure 8.6b for illustration. Hence, in both cases, G contains a subdivision of K_4. □

Catlin [36] proved that Hajós' conjecture fails for $k \geq 7$. Hadwiger [133] proposed a weaker conjecture: every k-chromatic graph contains a subgraph contractible to K_k. It means that a subgraph of a k-chromatic graph becomes K_k after a sequence of edge contractions. This is a weaker conjecture because a K_k-subdivision is a special type of subgraph contractible to K_k. Hadwiger's conjecture is still open.

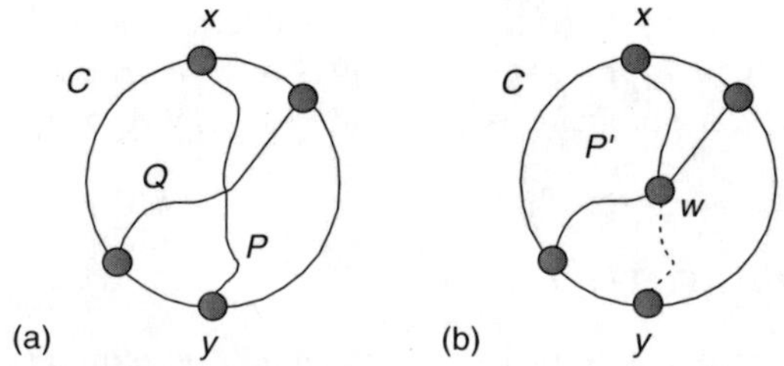

FIGURE 8.6 Illustration of Theorem 8.8.

8.6 ENUMERATIVE ASPECTS

Sometimes we can shed light on a difficult problem by considering a more general problem. Up to now, there has been no good algorithm to compute the minimum k such that G has a proper k-coloring. However, we define $\chi(G;k)$ as the number of proper k-colorings of G. Suppose that we know $\chi(G,k)$ for all k. We can compute $\chi(G)$ by finding the minimum k such that $\chi(G,k)$ is positive. Birkhoff [26] introduced this function as a possible way to attack the Four Color Problem.

The function $\chi(G;k)$ is defined to be the number of functions $f: V(G) \to [k]$ that properly color G from the set $[k] = \{1, 2, \ldots, k\}$. In this definition, the k colors need not all be used. Moreover, permuting the colors used produces a different coloring.

EXAMPLE 8.4

(Elementary examples.) Suppose that we are coloring the vertices of an independent set. We can independently choose one of the k colors at each vertex. Each of the k^n functions to $[k]$ is a proper coloring. Hence, $\chi(\overline{K}_n;k) = k^n$.

We can color the vertices of K_n in some order. The colors chosen earlier cannot be used on the ith vertex. There remain $(k-i+1)$ choices available for the ith vertex, no matter how the earlier colors were chosen. Hence, $\chi(K_n;k) = k(k-1)\cdots(k-n+1)$. Thus, $\chi(K_n;k) = 0$ if $k < n$, as it should be, because K_n is k-chromatic.

Suppose that we want color a tree with k coloring. We can choose any vertex as a root. We can color this vertex in k ways. Along with a coloring, we grow the tree from the root. At every stage, only the color of the parent is forbidden. Thus, we have $(k-1)$ choices for the color of the new vertex. Furthermore, we can see inductively that every proper k-coloring arises in this way by deleting a leaf. Hence, $\chi(T;k) = k(k-1)^{n-1}$ for every n-vertex tree.

All examples of $\chi(G,k)$ are polynomials in k of degree $n(G)$. This holds for every graph. For this reason, we call $\chi(G;k)$ the *chromatic polynomial* of G.

PROPOSITION 8.3 *Let $x_{(r)} = x(x-1)\cdots(x-r+1)$. Let $p_r(G)$ denote the number of partitions of $V(G)$ into exactly r independent sets. Then $\chi(G;k) = \sum_{r=1}^{n(G)} p_r(G)k_{(r)}$, which is a polynomial of degree $n(G)$.*

Proof: Suppose that r colors are actually used in a proper coloring. The coloring classes partition $V(G)$ into exactly r independent sets. Suppose that k colors are available. The number of ways to assign colors to a partition when exactly r colors are used is $k_{(r)}$. This accounts for all colorings. Thus, the formula for $\chi(G;k)$ holds. Since $k_{(r)}$ is a polynomial in k and $p_r(G)$ is a constant for each r, this formula implies that $\chi(G;k)$ is a polynomial function of k. Suppose that G has n vertices. There is exactly one partition of G into n independent sets and no partition using more sets. The leading term is k^n. □

Computing the chromatic polynomial using $\chi(G;k) = \sum_{r=1}^{n(G)} p_r(G)k_{(r)}$ is not feasible, because listing partitions into independent sets is no easier than finding the smallest such partition (chromatic number). There is also a recurrence, which is similar to counting the numbers of spanning trees. Again $G \cdot e$ denotes the graph obtained by contracting the edge e in G. However, since we are counting colorings, we can discard multiple copies of edges that arise from the contraction.

THEOREM 8.9 **(Chromatic recurrence.)** Suppose that G is a simple graph and e is an edge in G. Then $\chi(G;k) = \chi(G-e;k) - \chi(G \cdot e;k)$.

Proof: Let $e = (u, v)$. Obviously, every proper k-coloring of G is a proper k-coloring of $G-e$. Moreover, any proper k-coloring of $G-e$ is a proper k-coloring of G if and only if it gives distinct colors to the endpoints u and v of e. Hence, we can count the proper k-colorings of G by substracting from $\chi(G-e;k)$ the number of proper k-colorings of $G-e$ that give u and v the same color. The colorings of G that give u and v the same color correspond directly to proper k-colorings of $G \cdot e$, in which the color of the contracted vertex is the common color of u and v. □

Example 8.5

Deleting an edge from C_4 produces P_4. Contracting an edge of C_4 produces K_3. Since P_4 is a tree, $\chi(P_4;k) = k(k-1)^3$. Since K_3 is a clique, $\chi(K_3;k) = k(k-1)(k-2)$. Using the chromatic recurrence, we obtain $\chi(C_4;k) = \chi(P_4;k) - \chi(K_3;k) = k(k-1)(k^2-3k+3)$.

Since both $G-e$ and $G \cdot e$ have fewer edges than G, we can use the chromatic recurrence inductively to compute $\chi(G;e)$. The initial conditions are graphs with no edges. We have already computed $\chi(\overline{K}_n;k) = k^n$.

Example 8.6

(Near-complete graphs.) Suppose that a graph G has many edges. We may compute its chromatic polynomial by working up to cliques instead of down to their components. For example, we want to compute $\chi(K_n - e;k)$. We can write $\chi(G-e;k) = \chi(G;k) + \chi(G \cdot e;k)$ instead of $\chi(G;k) = \chi(G-e;k) - \chi(G \cdot e;k)$. We let G be K_n in this alternative formula and obtained

$$\chi(K_n - e) = \chi(K_n;k) + \chi(K_{n-1};k) = (k-n+2)^2 \prod_{i=0}^{n-3}(k-i)$$

THEOREM 8.10 $\chi(G;k)$ is a polynomial in k of degree $n(G)$ with integer coefficients. Moreover, the leading coefficient is 1, next coefficient $-e(G)$, and its coefficients alternate in sign.

Proof: We proof this theorem by induction on $e(G)$. These claims hold trivially when $e(G) = 0$, because $\chi(\overline{K}_n;k) = k^n$. Suppose that G is an n-vertex graph with $e(G) \geq 1$. Obviously, $G-e$ has $e(G)-1$ edges and n vertices and $G \cdot e$ has $(n-1)$ vertices. By the induction hypothesis, there are nonnegative integers $\{a_i\}$ and $\{b_i\}$

such that $\chi(G-e;k)=\sum_{i=0}^{n}(-1)^i a_i k^{n-i}$ and $\chi(G\cdot e;k)=\sum_{i=0}^{n-1} b_i k^{n-1-i}$. By the chromatic recurrence,

$$\begin{array}{rl} \chi(G-e;k): & k^n-[e(G)-1]k^{n-1}+a_2k^{n-2}-\cdots+(-1)^i a_i k^{n-i}\cdots \\ -\chi(G\cdot e;k): & -(k^{n-1}-b_1k^{n-2}+\cdots+(-1)^{i-1}b_{i-1}k^{n-i}\cdots) \\ \hline =\chi(G;e): & k^n-e(G)k^{n-1}+(a_2+b_1)k^{n-2}-\cdots+(-1)^i(a_i+b_{i-1})k^{n-i}\cdots \end{array}$$

Hence, $\chi(G;k)$ is a polynomial with leading coefficient $a_0=1$ and next coefficient $-e(G)$, and its coefficients alternate in sign. □

The following result with regard to $\chi(G;k)$ is very interesting, because $\chi(G;k)$ has meaning when evaluated at negative integers.

THEOREM 8.11 **(Stanley [286])** $\chi(G;-1)$ is $(-1)^{n(G)}$ times the number of acyclic orientations of G.

Proof: We use induction on $e(G)$. Let $a(G)$ be the number of acyclic orientations of G. Suppose that G has no edges. Obviously, $a(G)=1$ and $\chi(G;-1)=(-1)^{n(G)}$. So the claim holds. We will prove that $a(G)=a(G-e)+a(G\cdot e)$ for $e\in E(G)$. Suppose this recurrence, $a(G)=a(G-e)+a(G\cdot e)$ for $e\in E(G)$, holds. Combined with the induction hypothesis,

$$a(G)=(-1)^{n(G)}\chi(G-e;-1)+(-1)^{n(G)-1}\chi(G\cdot e;-1)=(-1)^{n(G)}\chi(G;-1)$$

Now, we discuss the recurrence. Let $e=(u,v)$ and D be an acyclic orientation of $G-e$. We can use $u\to v$ or $v\to u$ to obtain an acyclic orientation of G.

Suppose that D has no path from u to v. We can choose the orientation $v\to u$ to get an acyclic orientation of G. Suppose that D has no path from v to u. We can choose the orientation $u\to v$ to get an acyclic orientation of G. It is possible to use both orientation $u\to v$ and $v\to u$. In this case, D has two possible ways to obtain an acyclic orientation of G. Since D is acyclic, D cannot have both a path from u to v and a path from v to u. Hence, at least one possible orientation for e is to extend D into an acyclic orientation of G. Thus, $a(G)$ equals $a(G-e)$ plus the number of orientations that extend in both ways.

Thus, those extending in both ways are the acyclic orientations of $G-e$ with no path from u to v and no path from v to u. Since a path from u to v or a path from v to u in the orientation of $G-e$ becomes a cycle in $G\cdot e$, there are exactly $a(G\cdot e)$ such orientations.

Thus, the theorem is proved. □

EXAMPLE 8.7

There are exactly 4 edges in C_4. Thus, there are 16 orientations for C_4. Of these, 14 are acyclic. We know that $\chi(C_4;k)=k(k-1)(k^2-3k+3)$. Obviously, $\chi(C_4;-1)=14$.

8.7 HOMOMORPHISM FUNCTIONS

Let $G=(X,E)$ and $H=(Y,F)$ be two graphs. A function ϕ from X into Y is a *homomorphism* from G into H if $(u,v)\in E$ implies $(\phi(u),\phi(v))\in F$. For a fixed graph G, we can define the function h_G from $\mathcal{G}$ into $\mathcal{R}$ by setting $h_G(H)$ to be the number of homomorphisms from G into H. It can be checked that $\chi(G;k)=h_G(K_k)$ for any graph G. Thus, a homomorphism is a generalization of coloring. Recently, there are a lot of studies on homomorphism. The following theorem is proved in Refs 168 and 169.

THEOREM 8.12 Assume that G, H, and K are graphs. Then $h_G(H\cdot K)=h_G(H)\times h_G(K)$. Moreover, $h_G(H+K)=h_G(H)+h_G(K)$ if G is connected. Furthermore, $h_G(H)\leq h_G(K)$ is H is a subgraph of K.

Proof: We only prove that $h_G(H\cdot K)=h_G(H)\times h_G(K)$. The statement that $h_G(H+K)=h_G(H)+h_G(K)$ if G is connected follows from the statement that the homomorphic image of a connected graph is still connected. The statement that $h_G(H)\leq h_G(K)$ is H is a subgraph of K follows from the definition of homomorphism.

Let $\alpha: G\to H$ and $\beta: G\to K$ be homomorphisms. We can define $\gamma: G\to H\cdot K$ by letting $\gamma(v)=(\alpha(v),\beta(v))$ for and $v\in V(G)$. Let (v_1,v_2) be an edge of G. We have $(\alpha(v_1),\alpha(v_2))\in E(H)$ and $(\beta(v_1),\beta(v_2))\in E(K)$. Thus, γ is a homomorphism from G into $H\cdot K$.

Conversely, let γ be a homomorphism from G into $H\cdot K$. We can define $\gamma_H: G\to H$ and $\gamma_K: G\to K$ by $\gamma_H(v)=a$ and $\gamma_K(v)=b$ if and only if $\gamma(v)=(a,b)$. Thus, $\gamma(v)=(\gamma_H(v),\gamma_K(v))$.

Hence, $h_G(H\cdot K)=h_G(H)\times h_G(K)$. □

A function f from the set of natural numbers $\mathcal{N}$ into the set of real numbers $\mathcal{R}$ is *additive* if $f(m+n)=f(m)+f(n)$ for all $m,n\in\mathcal{N}$, f is *multiplicative* if $f(m\cdot n)=f(m)\cdot f(n)$ for all $m,n\in\mathcal{N}$, and f is *increasing* if $f(m)\leq f(n)$ whenever $m\leq n$. The following theorem can easily be obtained.

THEOREM 8.13 If f is a multiplicative increasing on $\mathcal{N}$, then either $f(n)=0$ for all $n\in\mathcal{N}$ or $f(n)=n^\alpha$ for some $\alpha\geq 0$.

From the mathematical point of view, the previous theorem is very good. It classifies all multiplicative increasing functions on $\mathcal{N}$. We also observed that all multiplicative increasing functions are generated by additive multiplicative increasing functions. It is very natural to study similar results on other algebraic systems.

Let f be a real-valued function defined on the set of all graphs, $\mathcal{G}$. The function f is *additive* if $f(G+H)=f(G)+f(H)$ for any $G,H\in\mathcal{G}$. The function f is *multiplicative* if $f(G\cdot H)=f(G)\times f(H)$ for any $G,H\in\mathcal{G}$. The function f is *increasing* if $f(G)\leq f(H)$ when G is a subgraph of H. A graph function f is *MI* if it is multiplicative and increasing. A graph function f is *AMI* if it is additive, multiplicative, and increasing. Note that *MI* is closed, undertaking the nonnegative power, finite product, and pointwise convergence. Let $S\subseteq MI$. We use $\langle S\rangle$ to denote the set of functions obtained by taking nonnegative power, finite product, and pointwise convergence from elements of S. In other words, the following functions are elements in $\langle S \rangle$:

1. f^{α}, $\alpha \geq 0$, and $f \in S$
2. $\prod_{i=1}^{k} f_i^{\alpha_i}$, $\alpha_i \geq 0$, and $f_i \in S$
3. $\lim_{m\to\infty} f_m$, where f_m is of type 1 or 2

Lovász observed this fact in Theorem 8.12 and asked if the set of multiplicative increasing functions (with respect to a weak product) is generated by $\{h_G \mid G$ is a graph$\}$; that is, it is conjectured that $MI = \langle \{h_G \mid G \text{ is a graph}\} \rangle$. Although the conjecture is disproved by Hsu [168,169], we present some interesting examples of multiplicative increasing graph functions.

Let $f_1(G)$ be defined as the number of vertices of G. Obviously, $f_1 = h_{K_1}$. Moreover, it is proved in Ref. 168 that $f = h_{K_1}^{\alpha}$ for some $\alpha \geq 0$ if f is an MI function with $f(K_1) \neq 0$. Let $f_2(G)$ be defined as $2|E(G)|$. It can be checked that $f_2 = h_{K_2}$. Let $f_3(G)$ be defined as $\max\{\deg(v) \mid v \in G\}$. It is proved in Ref. 168 that $f_3 = \lim_{m\to\infty} h_{K_{1,m}}^{1/m}$. Let $f_4(G)$ be defined as $\max\{|\lambda| \mid \lambda$ is an eigenvalue of $A(G)\}$. It is proved in Ref. 168 that $f_4 = \lim_{m\to\infty} h_{P_m}^{1/m}$. Let $f_5(G)$ be defined as $f_5(G) = \lim_{n\to\infty} \alpha'(G^n)^{1/n}$, where $\alpha'(G)$ is the maximum size of matching in G. It is proved in Ref. 49 that f_5 is an MI function but f_5 is not in $\langle \{h_G \mid G \text{ is a graph}\} \rangle$.

Some interesting multiplicative increasing graph functions can be found in [39, 49,165–171]. The classification of all multiplicative increasing functions on graph is still unsolved [39].

8.8 AN APPLICATION—TESTING ON PRINTED CIRCUIT BOARDS

Let $G = (V, E)$ be a graph. For subsets S and S' of V, we denote by $[S, S']$ the set of edges with one end in S and the other end in S'. A *cut* of G is a subset of E of the form $[S, \overline{S}]$, where S is a nonempty proper subset of V and $\overline{S} = V - S$. A *cut cover* of G is a family of cuts, $\mathcal{F} = \{[S_i, \overline{S}_i] \mid 1 \leq i \leq m\}$, such that each edge e in E belongs to $[S_i, \overline{S}_i]$ for some i. We use $cc(G)$ to denote the minimum cardinality of all possible cut covers of G. Loulou [241] formulates the problem of minimizing the test for printed circuit boards into finding $cc(G)$ for some graph G.

THEOREM 8.14 **(Ho and Hsu [147])** $cc(G) = \lceil \log_2 \chi(G) \rceil$.

Proof: Assume that $cc(G) = r$, $2^r = s$, and $\chi(G) = t$. For any integer k, we use k_i to denote the ith bit for the binary representation of k.

Assume that $\mathcal{F} = \{[S_i, \overline{S}_i] \mid i = 1, 2, \ldots, r\}$ is a cut cover of G. We define an s-coloring of G, $\pi : V \to \{1, 2, \ldots, s\}$ by assigning $\pi(v) = k$ such that for every i $(1 \leq i \leq r)$, $k_i = 0$ if $v \in S_i$ and $k_i = 1$ otherwise. Let (u, v) be any edge of G. Since $\mathcal{F}$ is a cut cover, $e \in [S_j, \overline{S}_j]$ for some j. We have $\pi(u)_j \neq \pi(v)_j$ and hence $\pi(u) \neq \pi(v)$. Thus, π is a proper s-coloring of G. We have $s \geq t$ and $cc(G) \geq \lceil \log_2 \chi(G) \rceil$.

On the other hand, suppose that $\pi : \{1,2,\ldots,t\}$ is a proper t-coloring of G. For $1 \le i \le \lceil \log_2 t \rceil$, let $S_i = \{v \in V \mid \pi(v)_i = 0\}$. Then $\mathcal{F} = \{[S_i, \overline{S}_i] \mid 1 \le i \le \lceil \log_2 t \rceil\}$ forms a cut cover of G. We have $cc(G) \le \lceil \log_2 t \rceil = \lceil \log_2 \chi(G) \rceil$.

The theorem is proved. □

8.9 EDGE-COLORINGS

We want to schedule games for a league with $2n$ teams so that each pair of teams plays each other. However, each team plays at most once a week. Since each team must play with $(2n-1)$ other teams, the season lasts at least $(2n-1)$ weeks. The games of each week must form a matching. We can schedule the season in $(2n-1)$ weeks if and only if we can partition $E(K_{2n})$ into $(2n-1)$ disjoint matchings. Since K_{2n} is $(2n-1)$ regular, the games for each week forms a perfect matching.

We use Figure 8.7 to describe a solution. Arrange $(2n-1)$ vertices cyclically. The solid line indicates a matching for a week. Rotating the picture as indicated by the dashed matching yields another matching. The $(2n-1)$ rotations of the figure yield the desired matchings.

A *k-edge-coloring* of G is a labeling $f : E(G) \to [k]$. The labels are colors, and the set of edges with one color is a *color class*. A k-edge-coloring is *proper* if edges sharing a vertex have different colors. Thus, each color class is a matching. A graph is *k-edge-colorable* if it has a proper k-edge-coloring. The *edge-chromatic number* $\chi'(G)$ of a loopless graph G is the least k such that G is k-edge-colorable. The *multiplicity* of an edge is the number of times its vertex pair appears in the edge set.

Chromatic index is another name used for $\chi'(G)$. In contrast to $\chi(G)$, multiple edges greatly affect $\chi'(G)$. Obviously, a graph with a loop has no proper edge-coloring. The word *loopless* excludes loops but allows multiple edges.

The bound $\chi'(G) \le 2\Delta(G) - 1$ follows easily for all loopless graphs. We use the greedy algorithm to color the edges of G in some order. Always assign the current edge the least-indexed color different from those already appearing on edges incident to it. Since no edge is incident to more than $2\Delta(G) - 1$ colors, we have $\chi'(G) \le 2\Delta(G) - 1$. Since any proper edge-coloring assigns different colors for the edges incident to any vertex, $\chi'(G) \ge \Delta(G)$. For bipartite graphs, $\chi'(G)$ achieves the trivial lower bound.

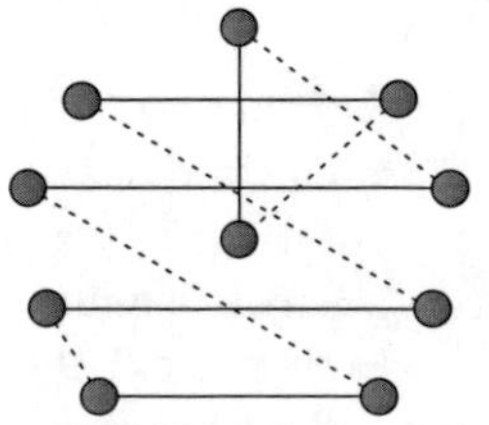

FIGURE 8.7 Illustration of edge-coloring of K_{2n}.

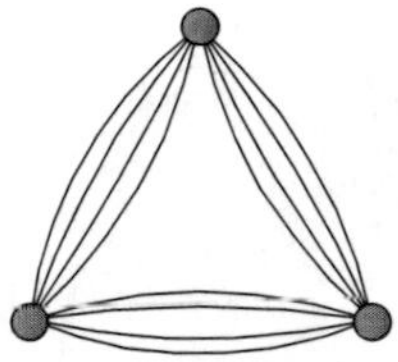

FIGURE 8.8 A multigraph with $\chi'(G) = 3\Delta(G)/2$.

THEOREM 8.15 **(König [204])** Let G be a bipartite graph with multiple edges allowed. We have $\chi'(G) = \Delta(G)$.

Proof: By Corollary 6.1, every regular bipartite graph H has a 1-factor. By induction on $\Delta(H)$, this yields a proper $\Delta(H)$-edge-coloring. It therefore suffices to show that every bipartite graph G with maximum degree k has a k-regular bipartite supergraph H.

We construct such a supergraph. Add vertices to the smaller partite set of G, if necessary, to equalize the sizes. If the resulting G' is not regular, then each partite set has a vertex with a degree of less than $\Delta(G') = \Delta(G)$. Add an edge consisting of this pair. Continue adding such edges until the graph becomes regular. □

Vizing proved that $\chi' \leq \Delta(G) + 1$ for any loopless simple graph G. The proof is omitted.

THEOREM 8.16 **(Vizing [326,327])** Every simple graph with maximum degree Δ has a proper $\Delta(G) + 1$-edge-coloring.

For loopless graphs with multiple edges, $\chi'(G)$ may exceed $\Delta(G) + 1$. Shannon [283] proved that the maximum of $\chi'(G)$ in terms of $\Delta(G)$ alone is $3\Delta(G)/2$. Vizing proved that $\chi'(G) + \mu(G)$, where $\mu(G)$ is the maximum edge multiplicity. The edges of the graph in Figure 8.8 are pairwise-intersecting and require distinct colors: $\chi'(G) = 3\Delta(G)/2 = \Delta(G) + \mu(G)$.

9 Hamiltonian Cycles

9.1 HAMILTONIAN GRAPHS

A *Hamiltonian cycle* is a spanning cycle in a graph (a cycle through every vertex). Introduced by Kirkman in 1855, Hamiltonian cycles are named for Sir William Hamilton, who described a game on the graph of the dodecahedron, in which one player would specify a 5-vertex path and the other would have to extend it to a cycle. The game was marketed as the *Traveler's Dodecahedron*, a wooden version, in which the vertices were named for 20 important cities. Here, we use $a, b, c, \ldots, t$ to represent these vertices as illustrated in Figure 9.1. It was not commercially successful.

Another variation of the Hamiltonian cycle problem is a Hamiltonian path problem. A *Hamiltonian path* is a spanning path. Every graph with a spanning cycle has a spanning path, but P_n shows that the converse is not true.

Until the 1970s, the interest in Hamiltonian cycles had been centered on their relationship to the 4-color problem. More recently, the study of Hamiltonian cycles in general graphs has been fueled by practical applications and by the issue of complexity. A graph with a spanning cycle is called a *Hamiltonian graph*. No easily testable characterization is known for Hamiltonian graphs. We will study necessary conditions and sufficient conditions. Loops and multiple edges are irrelevant to the existence of a spanning cycle. A graph is Hamiltonian if and only if the simple graph obtained by keeping one copy of each edge is Hamiltonian. *We focus on simple graphs, particularly when discussing degree conditions.*

9.2 NECESSARY CONDITIONS

Since deleting a vertex from a Hamiltonian graph leaves a subgraph with a spanning path, every Hamiltonian graph is 2-connected. Bipartite graphs suggest a way to strengthen this necessary condition.

Example 9.1

A spanning cycle in a bipartite graph visits the two partite sets alternately. Thus, there can be no such cycle unless the parts have the same size. Hence $K_{m,n}$ is Hamiltonian only if $m = n$. Alternatively, we can argue that the cycle returns to different vertices of one part after each visits the other part.

THEOREM 9.1 Let $G = (V, E)$ be a Hamiltonian graph and S be a subset of V. Then the graph $G - S$ has at most $|S|$ components.

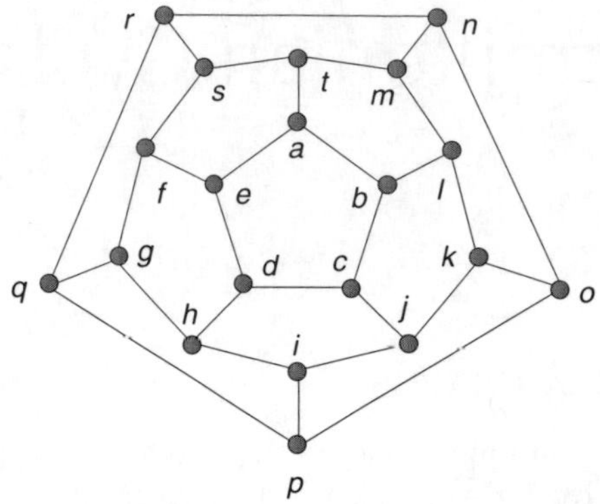

FIGURE 9.1 Traveler's Dodecahedron.

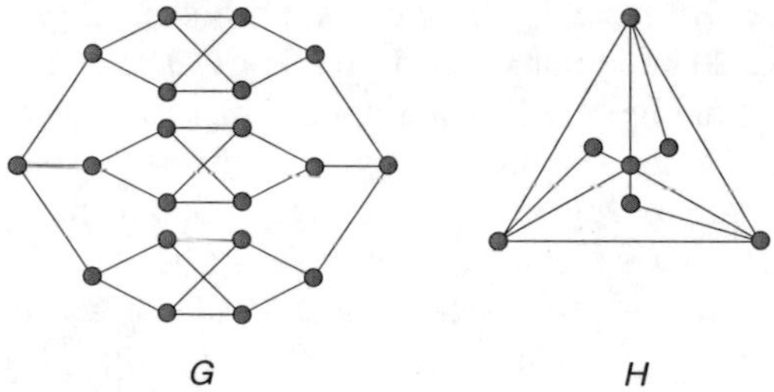

FIGURE 9.2 Graphs G and H for Example 11.2.

Proof: When leaving a component of $G - S$, a Hamiltonian cycle can go only to S, and the arrivals in S must visit different vertices of S. Hence S must have at least as many vertices as $G - S$ has components. □

We use $c(H)$ to denote the number of components of H. The necessary condition in Theorem 9.1 can be written as $c(G - S) \leq |S|$ for all $\emptyset \neq S \subseteq V$. This condition guarantees that G is 2-connected. However, this condition does not guarantee a Hamiltonian cycle.

Example 9.2

The graph G in Figure 9.2 fails this necessary condition, even though it is bipartite with partite sets of equal size. Hence it is not Hamiltonian. The graph G is 3-regular but not 3-connected. Tutte [312] conjectured that every 3-connected 3-regular bipartite graph is Hamiltonian. Horton [158] found a counterexample with 96 vertices, and the smallest known counterexample now has 50 vertices (Georges [122]).

The graph H in Figure 9.2 shows that the necessary condition is not sufficient. This graph satisfies the condition but no spanning cycle. All edges incident to 2-valent vertices must be used in a Hamiltonian cycle. In graph H, it uses three edges incident to the central vertex. Thus, H is not Hamiltonian.

The Petersen graph also satisfies the condition but is not Hamiltonian. A spanning cycle is a connected 2-factor. However, $2C_5$ is the only 2-factor of the Petersen graph.

9.3 SUFFICIENT CONDITIONS

The number of edges needed to guarantee an n-vertex graph to be Hamiltonian is quite large. Under conditions that *spread out* the edges, we can reduce the number of edges that guarantee Hamiltonian cycles. The simplest such condition is a lower bound on the minimum degree: $\delta(G) \geq n(G)/2$ suffices. We first note that this bound is tight.

EXAMPLE 9.3

Let G be a graph that consists of two cliques, one of order $\lceil (n+1)/2 \rceil$ and the other of order $\lfloor (n+1)/2 \rfloor$. Moreover, these two cliques share a vertex. Obviously, $\delta(G) = \lceil (n-1)/2 \rceil$. Since G is not 2-connected, it is not Hamiltonian.

THEOREM 9.2 **(Dirac [88])** Suppose that G is a graph with at least three vertices and $\delta(G) \geq n(G)/2$. Then G is Hamiltonian.

Proof: Since K_2 is not Hamiltonian but satisfies $\delta(K_2) = n(K_2)/2$, the condition $n(G) \geq 3$ is needed.

Suppose that there is a non-Hamiltonian graph satisfying the condition $\delta(G) \geq n(G)/2$ and $n(G) \geq 3$. Adding edges to G cannot reduce $\delta(G)$. Thus, we may restrict our attention to *maximal* non-Hamiltonian graphs with a minimum degree of at least $n/2$. Thus, no proper supergraph of G is also non-Hamiltonian. Hence, $G + (u, v)$ is Hamiltonian if u is not adjacent to v.

Since every spanning cycle in $G + (u, v)$ contains the new edge (u, v), G has a spanning path from $u = v_1$ to $v = v_n$. Suppose that some neighbor of u immediately follows a neighbor of v on the path; say, $u \leftrightarrow v_{i+1}$ and $v \leftrightarrow v_i$. Then G has the spanning cycle $\langle u = v_1, v_{i+1}, v_{i+2}, \ldots, v, v_i, v_{i-1}, \ldots, v_2, v_1 = u \rangle$ shown in Figure 9.3.

To prove that such a cycle exists, let $S = \{v_i \mid u \leftrightarrow v_{i+1}\}$ and $T = \{v_i \mid v \leftrightarrow v_i\}$. Summing the sizes of these sets yields

$$|S \cup T| + |S \cap T| = |S| + |T| = \deg(u) + \deg(v) \geq n$$

Obviously, $v_n \notin S \cup T$. Thus, $|S \cup T| < n$. Hence, $|S \cap T| \geq 1$. We have established a contradiction by finding a spanning cycle in G. Hence, there is no maximal non-Hamiltonian graph satisfying the hypotheses. □

Ore observed that the proof uses $\delta(G) \geq n(G)/2$ only to show $\deg(u) + \deg(v) \geq n$. Therefore, we can weaken the requirement of minimum degree $n/2$ to require only that $\deg(u) + \deg(v) \geq n$ whenever u is not adjacent to v. We also did not need the

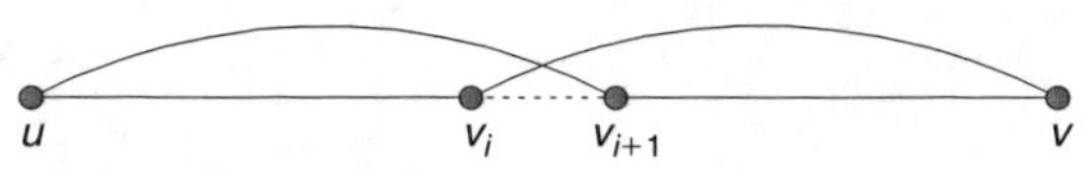

FIGURE 9.3 Illustration of Theorem 9.2.

fact that G was a maximal non-Hamiltonian graph. We only need that $G+(u, v)$ was Hamiltonian and thereby provided a spanning path from u to v.

THEOREM 9.3 **(Ore [262])** Suppose that G is a graph and u, v are distinct nonadjacent vertices of G with $\deg(u)+\deg(v) \geq n(G)$. Then G is Hamiltonian if and only if $G+(u, v)$ is Hamiltonian.

Proof: One direction is trivial. The proof of the other direction is the same as the previous theorem. □

With the previous theorem, we have the following theorem.

THEOREM 9.4 **(Ore [262])** Suppose that G is a graph with n vertices such that $\deg_G(x)+\deg_G(y) \geq n$ for every pair of nonadjacent vertices x, y in G. Then G is Hamiltonian.

Proof: Obviously, we can repeatedly join any two nonadjacent vertices u and v with $\deg(u)+\deg(v) \geq n(G)$ by an edge to obtain a complete graph. Applying Theorem 9.3, we prove the theorem. □

Bondy and Chvátal [30] observed the essence of Ore's argument in a much more general form that yields sufficient conditions for cycles of length l and other graphs. Here, we discuss only the case of spanning cycles. Using the previous theorem to add edges, we can test whether G is Hamiltonian by testing whether the larger graph is Hamiltonian.

The *(Hamiltonian) closure* of a graph G, denoted as $C(G)$, is the supergraph of G on $V(G)$ obtained by iteratively adding edges between pairs of nonadjacent vertices whose degree sum is at least n until no such pair remains. See Figure 9.4 for illustration.

Ore's lemma yields the following theorem.

THEOREM 9.5 An n-vertex graph is Hamiltonian if and only if its closure is Hamiltonian.

Mathematically, we need to check whether the closure does not depend on the order we choose to add edges when more than one is available. Otherwise, a graph may have two closures.

LEMMA 9.1 The closure of G is well-defined.

Proof: Suppose that adding $e_1, e_2, \ldots, e_r$ in sequence yields G_1 and that adding $f_1, f_2, \ldots, f_s$ yields G_2. Each addition joins vertices with degrees summing to at least

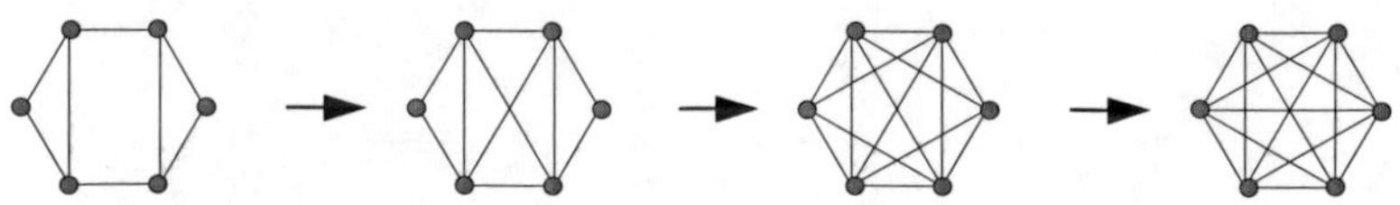

FIGURE 9.4 Illustration of Hamiltonian closure.

$n(G)$. Suppose that in either sequence a pair (u, v) become addable. Then it must eventually be added before the sequence ends. Hence, f_1, being initially addable to G, must belong to G_1. Similarly, if $f_1, f_2, \ldots, f_{i-1} \in E(G_1)$, then f_i become addable to G_1. Therefore it belongs to G_1. Hence, neither sequence contains a first edge omitted by the other sequence. We have $G_1 \subseteq G_2$ and $G_2 \subseteq G_1$. □

Now, we have a necessary and sufficient condition to test for Hamiltonian cycles in graphs. However, this necessary and sufficient condition does not help much. It requires us to test whether another graph is Hamiltonian. Yet, it does furnish a method for proving sufficient conditions. Any condition that forces $C(G)$ to be Hamiltonian implies a Hamiltonian cycle in G. For example, $C(G) = K_n$. Chvátal used this method to prove the following sufficient condition on vertex degrees.

THEOREM 9.6 **(Chvátal [70])** Suppose that G is a graph with vertex degrees $d_1 \leq d_2 \leq \ldots, d_n$. If $i < n/2$ implies that $d_i > i$ or $d_{n-i} \geq n - i$, then G is Hamiltonian.

Proof: Note that adding edges to form the closure reduces no entry in the degree sequence. It suffices to consider the special case where G is closed. Here, we prove that the condition implies $G = K_n$.

Suppose that $G = C(G) \neq K_n$. We need to find a value of i less than $n/2$ such that the Chvátal's condition is not satisfied; that is, at least i vertices have degree at most i and at least $n - i$ vertices have degree less than $n - i$.

Since $G \neq K_n$, among all pairs of nonadjacent vertices we can choose a pair u and v with a maximum degree sum. Since G is closed, $(u, v) \notin E(G)$ implies $\deg(u) + \deg(v) < n$. Without loss of generality, we may assume that $\deg(u) \leq \deg(v)$. As $\deg(u) + \deg(v) < n$, $\deg(u) < n/2$. We set $i = \deg(u)$ (see Figure 9.5).

Since we choose a nonadjacent pair with maximum degree sum, every vertex of $V - \{v\}$ that is not adjacent to v has degree at most $\deg(u) = i$. Moreover, there are at least $n - 1 - \deg(v) \geq \deg(u) = i$ of these vertices. Similarly, every vertex of $V - \{u\}$ that is not adjacent to u has degree at most $\deg(v) < n - \deg(u) = n - i$. Again, there are $n - 1 - \deg(u)$ of these vertices. Since $\deg(u) \leq \deg(v)$, we can also add u to the set of vertices with degree at most $\deg(v)$. Thus, we obtain $n - i$ vertices with degree less than $n - i$. Hence, we have proved $d_i \leq i$ and $d_{n-i} < n - i$ for this specially chosen i. We get a contradiction to the hypothesis. □

A sequence of real numbers $(p_1, p_2, \ldots, p_n)$ is said to be *majorized* by another such sequence $(q_1, q_2, \ldots, q_n)$ if $p_i \leq q_i$ for $1 \leq i \leq n$. A graph G is *degree-majorized* by a

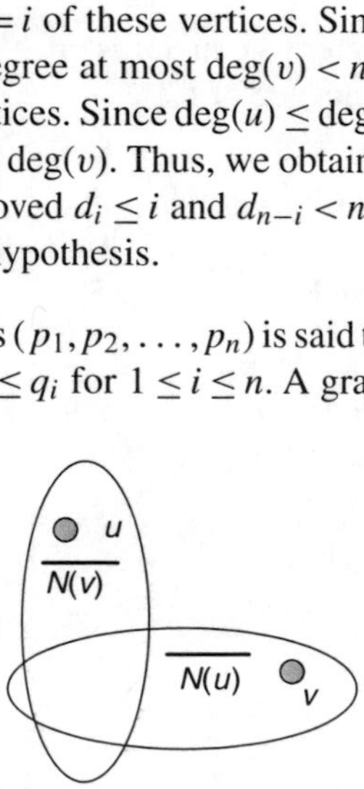

FIGURE 9.5 Illustration of Theorem 9.6.

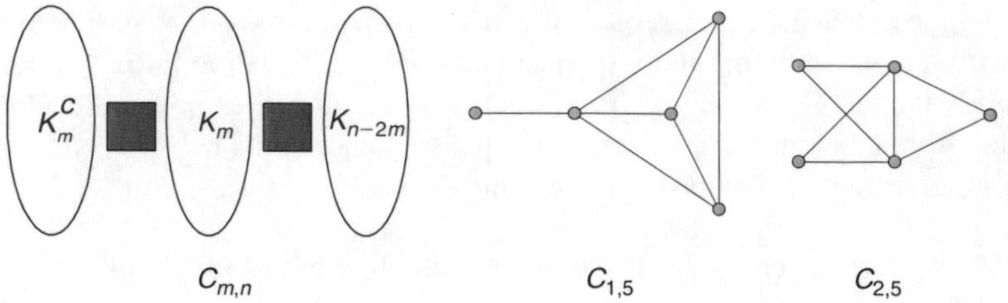

FIGURE 9.6 Illustration of $C_{m,n}$.

graph H if $n(G) = n(H)$ and the nondecreasing degree sequence of G is majorized by that of H. For instance, the 5-cycle is degree-majorized by $K_{2,3}$ because $(2, 2, 2, 2, 2)$ is majorized by $(2, 2, 2, 3, 3)$. For $1 \leq m < n/2$, let $C_{m,n}$ denote the graph $(\overline{K}_m + K_{n-2m}) \vee K_m$. See Figure 9.6 for illustration. We claim that $C_{m,n}$ is non-Hamiltonian. Let S be the set of m vertices of degree $n-1$ in $C_{m,n}$. We have $c(C_{m,n} - S) = m + 1 > |S|$. By Theorem 9.1, $C_{m,n}$ is not Hamiltonian. Chvátal [70] points out that the family of degree-maximal non-Hamiltonian graphs (those that are degree-majorized by no others) are exactly $C_{m,n}$.

THEOREM 9.7 **(Chvátal [70])** Suppose that G is a non-Hamiltonian graph with $n(G) = n \geq 3$. Then G is degree-majorized by some $C_{m,n}$.

Proof: Let G be a non-Hamiltonian graph with degree sequence $(d_1, d_2, \ldots, d_n)$, where $d_1 \leq d_2 \leq \cdots \leq d_n$ and $n \geq 3$. By Theorem 9.6, there exists $m < n/2$ such that $d_m \leq m$ and $d_{n-m} < n - m$. Therefore $(d_1, d_2, \ldots, d_n)$ is majorized by the sequence

$$(m, m, \ldots, m, n-m-1, n-m-1, \ldots, n-m-1, n-1, n-1, \ldots, n-1)$$

where m terms equal m, $n - 2m$ terms equal $n - m - 1$, and m terms equal $n - 1$. The latter sequence is the degree sequence of $C_{m,n}$. □

THEOREM 9.8 Suppose that G is a graph with $n \geq 3$ and $e > C(n-1, 2) + 1$. Then G is Hamiltonian. Moreover, the only non-Hamiltonian graphs with n vertices and $C(n-1, 2) + 1$ edges are $C_{1,n}$ and $C_{2,5}$ for $n = 5$.

Proof: Let G be a non-Hamiltonian graph with $n \geq 3$. By Theorem 9.7, G is degree-majorized by $C_{m,n}$ for some integer $m < n/2$. Therefore

$$\begin{aligned} e &\leq e(C_{m,n}) = \frac{1}{2}(m^2 + (n-2m)(n-m-1) + (m(n-1)) \\ &= C(n-1, 2) + 1 - \frac{1}{2}(m-1)(m-2) - (m-1)(n-2m-1) \\ &\leq C(n-1, 2) + 1 \end{aligned}$$

Furthermore, equality can hold if G has the same degree sequence as $C_{m,n}$, and if either $m = 2$ and $n = 5$, or $m = 1$. Hence, $e(G)$ can equal $C(n-1, 2) + 1$ only if G has

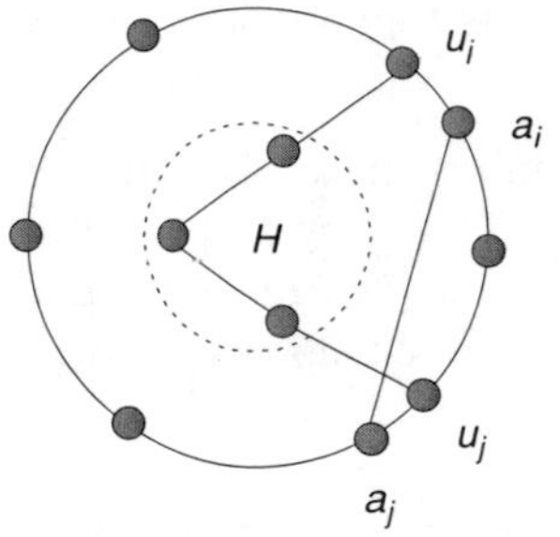

FIGURE 9.7 Illustration of Theorem 9.9.

the same degree sequence as $C_{1,n}$ or $C_{2,5}$, which is easily seen to imply that $G \cong C_{1,n}$ or $G \cong C_{2,5}$. □

There is another sufficient condition for Hamiltonian cycles that involves connectivity and independence, but not degrees. The proof gives a good algorithm to construct a Hamiltonian cycle or disprove the hypothesis.

THEOREM 9.9 **(Chvátal-Erdös [71])** Suppose that G is a graph with $\kappa(G) \geq \alpha(G)$. Then G has a Hamiltonian cycle or $G = K_2$.

Proof: Suppose that G is a graph with $\kappa(G) \geq \alpha(G)$ such that $G \neq K_2$. Suppose that $\kappa(G) = 1$. Then, $G = K_n$ and hence G is Hamiltonian. Thus, we assume $\kappa(G) > 1$. Let $k = \kappa(G) \geq \alpha(G)$ and let C be a longest cycle in G. Obviously, C forms a Hamiltonian cycle of G if we can prove that C spans G.

Suppose that C is not a Hamiltonian cycle of G. Let H be a component of $G - C$. Since $\delta(G) \geq \kappa(G)$ and every graph with $\delta(G) \geq 2$ has a cycle of length at least $\delta(G) + 1$, C has at least $k + 1$ vertices. Moreover, C has at least k vertices with edges to H. Otherwise, the vertices of C with edges to H contradicts $\kappa(G) = k$. Let $u_1, u_2, \ldots, u_k$ be k vertices of C with edges to H, in clockwise order. For $i = 1, 2, \ldots, k$, let a_i be the vertex immediately following u_i on C.

Suppose that any two of these vertices are adjacent; say, $a_i \leftrightarrow a_j$. Then we can construct a longer cycle by using the edge (a_i, a_j), the portions of C from a_i to u_j and a_j to u_i, and a path from u_i to u_j through H. See Figure 9.7 for illustration. This argument holds for the case $a_i = u_{i+1}$. Thus, we can conclude that no a_i has a neighbor in H. Hence $\{a_1, a_2, \ldots, a_k\}$ plus a vertex of H form an independent set of size $k + 1$. This contradiction implies that C is a Hamiltonian cycle. □

9.4 HAMILTONIAN-CONNECTED

A graph G is *Hamiltonian-connected* if there exists a Hamiltonian path joining any two different vertices of G. A Hamiltonian-connected graph G is *optimal* if G contains the least number of edges among all the Hamiltonian-connected graphs with the same number of vertices as that of G.

LEMMA 9.2 Assume that G is Hamiltonian-connected with at least four vertices. Then $\deg(v) \geq 3$ for any vertices v of G.

Proof: Suppose that the lemma is false. Then there exists a vertex v in G such that $\deg(v) \leq 2$. Suppose that $\deg(v) = 1$. Obviously, there is no Hamiltonian path joining x to y if $x \neq v$ and $y \neq v$. Suppose that $\deg(v) = 2$. Let x and y be the two neighbors of v. Obviously, there is no Hamiltonian path joining x to y. We get a contradiction. Hence, the lemma is proved. □

With this lemma, we have the following theorem.

THEOREM 9.10 **(Moon [255])** Every optimal Hamiltonian-connected graph has $\lceil 3n(G)/2 \rceil$ edges if $n(G) > 3$.

Proof: From the previous lemma, we know that any optimal Hamiltonian-connected graph with at least four vertices has at least $\lceil 3n(G)/2 \rceil$ edges. Assume that n is even. It can be checked whether the graph G in Figure 9.8 is Hamiltonian-connected. Assume that n is odd; it can be checked that the graph H in Figure 9.8 is Hamiltonian-connected. Hence, the theorem is proved. □

It is known that any bipartite graph with at least three vertices is not Hamiltonian-connected. The counterpart of Hamiltonian-connected graphs in bipartite graphs is known as Hamiltonian-laceable graphs. A bipartite graph with bipartition (X, Y) is *Hamiltonian-laceable* if there exists a Hamiltonian path joining any two vertices from different partite sets; that is, one in X and one in Y. This concept is proposed by Simmons [285]. It is easy to check whether every hypercube Q_n is Hamiltonian-laceable if $n \geq 1$. The following lemma can be obtained easily.

LEMMA 9.3 Suppose that m is an odd integer. Then $C_m \times K_2$ is optimal-Hamiltonian-connected. Suppose that m is an even integer. Then $(C_m \times K_2) - e$ Hamiltonian-laceable for any edge e in $C_m \times K_2$.

Comparing Lemma 9.3 with Theorem 9.10, we do not know how to prove that a Hamiltonian-laceable graph is optimal; that is, containing the least number of edges among all Hamiltonian-laceable graphs with the same number of vertices.

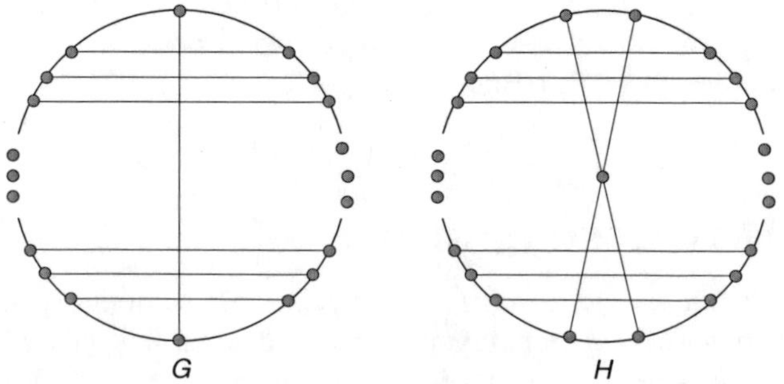

FIGURE 9.8 Examples of Hamiltonian-connected graphs.

THEOREM 9.11 Suppose that G is a graph and u, v are distinct nonadjacent vertices of G with $\deg(u)+\deg(v)\geq n(G)+1$. Then G is Hamiltonian-connected if and only if $G+(u,v)$ is Hamiltonian-connected.

Proof: Obviously, $G+(u,v)$ is Hamiltonian-connected if G is Hamiltonian-connected.

Suppose that $G+(u,v)$ is Hamiltonian-connected. Let x and y be any two distinct vertices of G. We need to find a Hamiltonian path of G between x and y.

Let P be a Hamiltonian path of $G+(u,v)$ between x and y. Obviously, P is a Hamiltonian path of G if (u,v) is not in P. Thus, we consider that (u,v) is in P. Without loss of generality, P can be written as $\langle x=z_1,\ldots,z_i,z_{i+1},\ldots,z_{n(G)}=y\rangle$ with $\{z_i,z_{i+1}\}=\{u,v\}$. Let H_1 be the subgraph of G induced by $\{z_1,z_2,\ldots,z_{i+1}\}$ and H_2 be the subgraph of G induced by $\{z_i,z_{i+1},\ldots,z_{n(G)}\}$. We have

$$\deg_{H_1}(u)+\deg_{H_2}(u)+\deg_{H_1}(v)+\deg_{H_2}(v)=\deg_G(u)+\deg_G(v)$$

We claim that there is an index $j\in\{1,2,\ldots,n\}-\{i,i+1\}$ such that (z_i,z_j) and (z_{i+1},z_{j+1}) are edges of G. Suppose that the claim is not true. Then $\deg_{H_j}(u)+\deg_{H_j}(v)\leq n(H_j)-1$ for $j\in\{1,2\}$. As $\deg_G(u)+\deg_G(v)=\deg_{H_1}(u)+\deg_{H_2}(u)+\deg_{H_1}(v)+\deg_{H_2}(v)$, $\deg_G(u)+\deg_G(v)\leq n(H_1)-1+n(H_2)-1=n(G)$. We obtain a contradiction. Hence, our claim holds.

Suppose that $j<i$. Then $\langle x=z_1,\ldots,z_j,z_i,z_{i-1},\ldots,z_{j+1},z_{j+1},\ldots,z_{n(G)}=y\rangle$ forms a Hamiltonian path of G joining x and y. Suppose that $j>i+1$. Then $\langle x=z_1,\ldots,z_i,z_j,z_{j-1},\ldots,z_{i+1},z_{j+1},\ldots,z_{n(G)}=y\rangle$ forms a Hamiltonian path of G joining x and y.

We find the required Hamiltonian path and the theorem is proved. □

THEOREM 9.12 **(Erdös [99])** Suppose that G is a graph such that any two nonadjacent vertices of G satisfying $\deg(u)+\deg(v)\geq n(G)+1$. Then G is Hamiltonian-connected.

Proof: Obviously, we can repeatedly join any two nonadjacent vertices u and v with $\deg(u)+\deg(v)\geq n(G)+1$ by an edge to obtain a complete graph. Applying Theorem 9.11, we prove the theorem. □

The following theorem is a consequence of the previous theorem.

THEOREM 9.13 Suppose that G is a graph with at least four vertices and $\delta(G)\geq n(G)/2+1$. Then G is Hamiltonian-connected.

Let $G=(V,E)$ be a graph. We use $\overline{E}$ to denote the edge set of the complement of G. Let $\overline{e}=|\overline{E}|$. The following theorem can be easily proved.

THEOREM 9.14 **(Ore [262])** Assume that $G=(V,E)$ is a graph with n vertices with $n\geq 4$. Then G is Hamiltonian if $\overline{e}\leq n-3$ and Hamiltonian-connected if $\overline{e}\leq n-4$.

Proof: With Theorem 9.8, G is Hamiltonian if $\overline{e} \leq n-3$. With Theorem 9.12, G is Hamiltonian-connected if $\overline{e} \leq n-4$. □

We can construct a Hamiltonian-connected graph using the operation $G_0 \oplus G_1$ discussed in Chapter 7. Here, we recall the definition of $G_0 \oplus G_1$. Suppose that $G_0 = (V_0, E_0)$ and $G_1 = (V_1, E_1)$ are two disjoint graphs with $|V_0| = |V_1|$. A *1-1 connection* between G_0 and G_1 is defined as an edge set $E_r = \{(v, \phi(v)) \mid v \in V_0, \phi(v) \in V_1$ and $\phi : V_0 \to V_1$ is a bijection$\}$. $G_0 \oplus G_1$ denotes the graph $G = (V_0 \cup V_1, E_0 \cup E_1 \cup E_r)$. Thus, ϕ induces a perfect matching in $G_0 \oplus G_1$. For convenience, G_0 and G_1 are called *components*. Let x be any vertex of $G_0 \oplus G_1$. We use $\overline{x}$ to denote the vertex matched under ϕ.

THEOREM 9.15 $G_0 \oplus G_1$ is Hamiltonian-connected if both G_0 and G_1 are Hamiltonian-connected and $|V(G_0)| = |V(G_1)| \geq 3$.

Proof: Assume that x and y are any two different vertices of $G_0 \oplus G_1$. Without loss of generality, we have the following two cases: (1) both x and y are in G_0 or (2) x is in G_0 and y is in G_1.

Assume that both x and y are in G_0. Since G_0 is Hamiltonian-connected, there exists a Hamiltonian path P of G_0 joining x and y. Obviously, P can be written as $\langle x, P_1, w, z, P_2, y \rangle$. Note that $x = w$ if $l(P_1) = 0$ and $z = y$ if $l(P_2) = 0$. Obviously, $\overline{w} \neq \overline{z}$ and there exists a Hamiltonian path Q of G_1 joining $\overline{w}$ and $\overline{z}$. Thus, $\langle x, P_1, w, \overline{w}, Q, \overline{z}, P_2, y \rangle$ forms a Hamiltonian path of $G_0 \oplus G_1$.

Assume that x is in G_0 and y is in G_1. Since $|V(G_0)| = |V(G_1)| \geq 3$, there exists a vertex z in G_0 such that $x \neq z$ and $\overline{z} \neq y$. Thus, there exists a Hamiltonian path P of G_0 joining x to z and there exists a Hamiltonian path Q of G_1 joining $\overline{z}$ to y. Obviously, $\langle x, P, z, \overline{z}, Q, y \rangle$ forms a Hamiltonian path of $G_0 \oplus G_1$.

Thus, the theorem is proved. □

Similarly, we can easily obtain the following theorem.

THEOREM 9.16 Assume that $r \geq 3$ and $k \geq 5$. Let $G_0, G_1, \ldots, G_{r-1}$ be r Hamiltonian-connected graphs with the same number of vertices. Then graph $G(G_0, G_1, \ldots, G_{r-1}; \mathcal{M})$ is Hamiltonian-connected.

Similarly, we can construct a Hamiltonian-laceable graph from two Hamiltonian-laceable graphs G_0 and G_1 with the same number of vertices and a 1-1 connection ϕ between G_0 and G_1 with the operation $G_0 \oplus G_1$. For $i = 0, 1$, let G_i be a graph in B'_n with bipartition V_0^i and V_1^i. To keep $G_0 \oplus G_1$ a bipartite graph, it is required that ϕ map vertices in V_i^0 to vertices in V_{1-i}^1 if $v \in V_i^0$. For the discussion of Hamiltonian properties, we assume that $|V_0^i| = |V_1^j|$ for $i, j \in \{0, 1\}$. *In the following discussion, we keep this assumption as we study the Hamiltonian-laceability of* $G_0 \oplus G_1$. Thus, $G_0 \oplus G_1$ is a bipartite graph with bipartition $V_0^0 \cup V_1^0$ and $V_0^1 \cup V_1^1$. With similar proof as Theorem 9.15, we have the following theorem.

THEOREM 9.17 Assume that G_0, G_1, and $G_0 \oplus G_1$ are bipartite graphs such that $|V(G_0)| = |V(G_1)| \geq 2$. Moreover, $G_0 \oplus G_1$ is bipartite. Then $G_0 \oplus G_1$ is Hamiltonian-laceable if both G_0 and G_1 are Hamiltonian-laceable.

Proof: By the symmetric property of $G_0 \oplus G_1$, without loss of generality we can assume the following two cases:

Case 1. $x \in V_0^0$ and $y \in V_0^1$. Since G_0 is Hamiltonian-laceble, there exists a Hamiltonian path P of G_0 joining x and y. Obviously, P can be written as $\langle x, P_1, w, z, P_2, y\rangle$ with $w \in V_0^0$ and $z \in V_0^1$. Note that it is possible that $l(P_1) = 0$ and $l(P_2) = 0$. Obviously, $\overline{w} \in V_1^1$ and $\overline{z} \in V_1^0$. Thus, there exists a Hamiltonian path Q of G_1 joining $\overline{w}$ and $\overline{z}$. Thus, $\langle x, P_1, w, \overline{w}, Q, \overline{z}, P_2, y\rangle$ forms a Hamiltonian path of $G_0 \oplus G_1$.

Case 2. $x \in V_0^0$ and y is in G_1^1. Since $|V(G_0)| = |V(G_1)| \geq 2$, there exists a vertex z in V_0^1. Obviously $\overline{z} \in V_1^0$. Thus, there exists a Hamiltonian path P of G_0 joining x to z and there exists a Hamiltonian path Q of G_1 joining $\overline{z}$ to y. Obviously, $\langle x, P, z, \overline{z}, Q, y\rangle$ forms a Hamiltonian path of $G_0 \oplus G_1$.

Thus, the theorem is proved. □

Similarly, we have the following theorem.

THEOREM 9.18 Assume that $r \geq 3$ and $k \geq 5$. Assume that $G_0, G_1, \ldots, G_{r-1}$ are bipartite with $|V(G_i)| = t$ for $1 \leq i < r$ with $t \geq 2$. Then $G(G_0, G_1, \ldots, G_{r-1}; \mathcal{M})$ is Hamiltonian-laceable if G_i is Hamiltonian laceable for $0 \leq i < r$.

THEOREM 9.19 $G_0 \oplus G_1$ is Hamiltonian-connected if G_0 is Hamiltonian-connected and G_1 is Hamiltonian-laceable and $|V(G_0)| = |V(G_1)| \geq 4$.

Proof: Let V_1^0 and V_1^1 be the bipartition of G_1. We have the following cases.

Case 1. $x \in V(G_0)$ and $y \in V_1^i$. Since $|V(G_0)| \geq 4$, there exists a vertex $z \in V(G_0) - \{x\}$ such that $\overline{z} \in V_1^{1-i}$. Obviously, there exists a Hamiltonian path P of G_0 joining x to z and there exists a Hamiltonian path Q of G_1 joining $\overline{z}$ to y. Obviously, $\langle x, P, z, \overline{z}, Q, y\rangle$ forms a Hamiltonian path of $G_0 \oplus G_1$.

Case 2. $\{x, y\} \subset V(G_0)$. Obviously, there exists a Hamiltonian path P of G_0 joining x to y. We can write P as $\langle x, P_1, w, z, P_2, y\rangle$ such that $\overline{w} \in V_1^i$ but $\overline{w} \in V_1^{1-i}$. Thus, there exists a Hamiltonian path Q of G_1 joining $\overline{w}$ to $\overline{z}$. Obviously, $\langle x, P_1, w, \overline{w}, Q, \overline{z}, z, P_2, y\rangle$ forms a Hamiltonian path of $G_0 \oplus G_1$.

Case 3. $x \in V_1^0$ and $y \in V_1^1$. Obviously, there exists a Hamiltonian path P of G_1 joining x to y. We can write P as $\langle x, P_1, w, z, P_2, y\rangle$. Again, there exists a Hamiltonian path Q of G_0 joining $\overline{w}$ to $\overline{z}$. Obviously, $\langle x, P_1, w, \overline{w}, Q, \overline{z}, z, P_2, y\rangle$ forms a Hamiltonian path of $G_0 \oplus G_1$.

Case 4. $x \in V_1^0$ and $y \in V_1^0$. Let z be any vertex in V_1^1. Obviously, there exists a Hamiltonian path P of G_1 joining x to z. We can write P as $\langle x, P_1, w, y, P_2, z\rangle$ for some

w in V_1^1. Again, there exists a Hamiltonian path Q of G_0 joining $\overline{w}$ to $\overline{z}$. Obviously, $\langle x, P_1, w, \overline{w}, Q, \overline{z}, z, (P_2)^{-1}, y\rangle$ forms a Hamiltonian path of $G_0 \oplus G_1$.

Thus, the theorem is proved. □

9.5 MUTUALLY INDEPENDENT HAMILTONIAN PATHS

Suppose that $P_1 = \langle v_1, v_2, v_3, \ldots, v_n\rangle$ and $P_2 = \langle u_1, u_2, u_3, \ldots, u_n\rangle$ are two Hamiltonian paths of G. We say that P_1 and P_2 are *independent* if $u_1 = v_1, u_n = v_n$, and $u_i \neq v_i$ for $1 < i < n$. We say a set of Hamiltonian paths $P_1, P_2, \ldots, P_s$ of G are *mutually independent* if any two distinct paths in the set are independent. The concept of mutually independent Hamiltonian paths arises from the following application. Suppose that there are k pieces of data that need to be sent from u to v. The data need to be processed at every node (and the process takes time). We need mutually independent Hamiltonian paths so that there will be no waiting time at a processor. The existence of mutually independent Hamiltonian paths is useful for communication algorithms.

LEMMA 9.4 Suppose that u and v are two distinct vertices of G. There are at most $\min\{\deg_G(u), \deg_G(v)\}$ mutually independent Hamiltonian paths between u and v if $(u, v) \notin E(G)$, and there are at most $\min\{\deg_G(u), \deg_G(v)\} - 1$ mutually independent Hamiltonian paths between u and v if $(u, v) \in E(G)$.

Let $[i]$ denote i $\mathbf{mod}(n-2)$.

THEOREM 9.20 Suppose that n is a positive integer with $n \geq 3$. There are $n-2$ mutually independent Hamiltonian paths between every two distinct vertices of K_n.

Proof: Let s and t be two distinct vertices of K_n. We can relabel the remaining $(n-2)$ vertices of K_n as $0, 1, 2, \ldots, n-3$. For $0 \leq i \leq n-3$, we set P_i as $\langle s, [i], [i+1], [i+2], \ldots, [i+(n-3)], t\rangle$. It is easy to see that $P_0, P_1, \ldots, P_{n-3}$ form $(n-2)$ mutually independent Hamiltonian paths joining s and t. □

Teng et al. [297] propose the following interesting result.

LEMMA 9.5 Suppose that G is a graph with $n \geq 4$ and $\overline{e} = n-4$. Then there are two independent Hamiltonian paths between any two distinct vertices of G except $n = 5$.

Proof: Assume that $n = 4$. Obviously, G is isomorphic to K_4. By Theorem 9.20, there are two independent Hamiltonian paths between any two distinct vertices of G. Assume that $n = 5$. Then G is isomorphic to $K_5 - \{f\}$ for some edge f. Without loss of generality, we assume that $V(G) = \{1, 2, 3, 4, 5\}$ and $f = (1, 2)$. It is easy to check whether $P_1 = \langle 3, 2, 5, 1, 4\rangle$ and $P_2 = \langle 3, 1, 5, 2, 4\rangle$ are the only two Hamiltonian paths between 3 and 4, but P_1 and P_2 are not independent.

Now, we assume that $n \geq 6$. Suppose that s and t are two distinct vertices of G. Let H be the subgraph of G induced by the remaining $(n-2)$ vertices of G. We have the following two cases.

Case 1. H is Hamiltonian. We can relabel the vertices of H with $\{0, 1, 2, \ldots, n-3\}$ so that $\langle 0, 1, 2, \ldots, n-3, 0\rangle$ forms a Hamiltonian cycle of H. We use Q to denote the set $\{i \mid (s, [i+1]) \in E(G) \text{ and } (i, t) \in E(G)\}$. Since $\overline{e} = n-4$, $|Q| \geq n-2-(n-4) = 2$. There are at least two elements in Q. Let q_1 and q_2 be the two elements in Q. For $j = 1, 2$, we set P_j as $\langle s, [q_j+1], [q_j+2], \ldots, [q_j], t\rangle$. Then P_1 and P_2 are two independent Hamiltonian paths between s and t.

Case 2. H is non-Hamiltonian. There are exactly $(n-2)$ vertices in H. By Theorem 9.8, there are exactly $(n-4)$ edges in the complement of H and H is isomorphic to $C_{1,n-2}$ or $C_{2,5}$. Since $\overline{e} = n-4$, $(s, v) \in E(G)$ and $(t, v) \in E(G)$ for every vertex v in H. We can construct two independent Hamiltonian paths between s and t as the following cases.

Case 2.1. H is isomorphic to $C_{2,5}$. We label the vertices of $C_{2,5}$ with $\{0, 1, 2, 3, 4\}$ as shown in Figure 9.9a. We set $P_1 = \langle s, 0, 1, 2, 3, 4, t\rangle$ and $P_2 = \langle s, 2, 3, 4, 1, 0, t\rangle$. Obviously, P_1 and P_2 form the required independent paths.

Case 2.2. H is isomorphic to $C_{1,n-2}$. We label the vertices of $C_{1,n-2}$ with $\{0, 1, \ldots, n-3\}$ as shown in Figure 9.9b. We set $P_1 = \langle s, 0, 1, 2, \ldots, n-3, t\rangle$ and $P_2 = \langle s, 2, 3, \ldots, n-3, 1, 0, t\rangle$. Obviously, P_1 and P_2 form the required independent paths. □

Teng et al. [297] further strengthen Lemma 9.5.

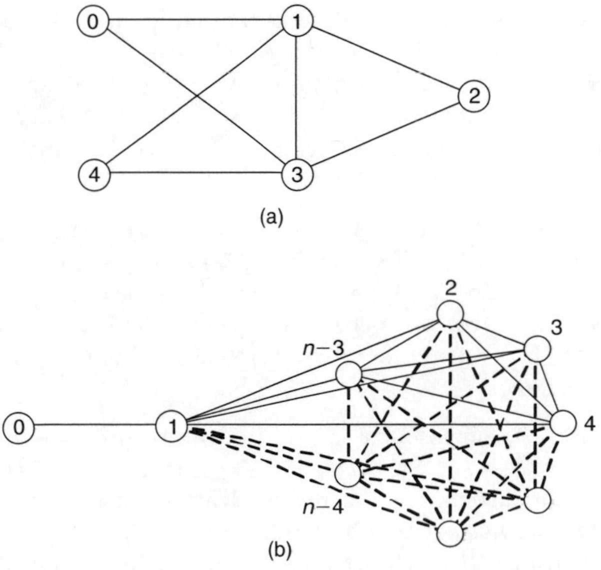

FIGURE 9.9 (a) $C_{2,5}$ and (b) $C_{1,n-2}$.

THEOREM 9.21 Suppose that G is a graph with $n \geq 4$ and $\overline{e} \leq n-4$. Then there are $n-2-\overline{e}$ mutually independent Hamiltonian paths between every two distinct vertices of G except $n=5$ and $\overline{e}=1$.

Proof: By Lemma 9.5, the theorem holds for $\overline{e}=n-4$. Now, we need to prove the theorem for $\overline{e}=n-4-r$ with $1 \leq r \leq n-4$. Suppose that s and t are two distinct vertices of G. Let H be the subgraph of G induced by the remaining $(n-2)$ vertices of G.

Then there are exactly $(n-2)$ vertices in H, and there are at most $n-4-r$ edges in the complement of H with $1 \leq r \leq n-4$. By Theorem 9.8, H is Hamiltonian. We can label the vertices of H with $\{0,1,2,\ldots,n-3\}$ so that $\langle 0,1,2,\ldots,n-3,0\rangle$ forms a Hamiltonian cycle of H. We use Q to denote the set $\{i \mid (s,[i+1]) \in E(G) \text{ and } (t,i) \in E(G)\}$. Since $\overline{e}=n-4-r$ with $1 \leq r \leq n-4$, $|Q| \geq n-2-(n-4-r)=n-2-\overline{e}$ for $1 \leq r \leq n-4$. Hence, there are at least $n-2-\overline{e}$ elements in Q. Let $q_1, q_2, \ldots, q_{n-2-\overline{e}}$ be the elements in Q. For $j=1,2,\ldots,n-2-\overline{e}$, we set $P_j=\langle s,[q_j+1],[q_j+2],\ldots,[q_j],t\rangle$. It is not difficult to see that $P_1, P_2, \ldots, P_{n-2-\overline{e}}$ are mutually independent paths between s and t. □

The following result, in a sense, generalizes that of Theorem 9.12.

THEOREM 9.22 Let G be a graph such that $\deg_G(x)+\deg_G(y) \geq n+2$ for any two vertices x and y with $(x,y) \notin E(G)$. Suppose that u and v are two distinct vertices of G. Then there are $\deg_G(u)+\deg_G(v)-n$ mutually independent Hamiltonian paths between u and v if $(u,v) \in E(G)$, and there are $\deg_G(u)+\deg_G(v)-n+2$ mutually independent Hamiltonian paths between u and v if $(u,v) \notin E(G)$.

Proof: Suppose that s and t are two distinct vertices of G, and H is the subgraph of G induced by the remaining $(n-2)$ vertices of G. Let u' and v' be any two distinct vertices in H. Obviously, $\deg_H(u')+\deg_H(v') \geq n+2-4=n-2=|V(H)|$. By Theorem 9.3, H is Hamiltonian. We can label the vertices of H with $\{0,1,\ldots,n-3\}$ so that $\langle 0,1,2,\ldots,n-3,0\rangle$ forms a Hamiltonian cycle of H. Let S denote the set $\{i \mid (s,[i+1]) \in E(G)\}$ and T denote the set $\{i \mid (i,t) \in E(G)\}$. Clearly, $|S \cup T| \leq n-2$. We have the following two cases:

Case 1. $(s,t) \in E(G)$. Assume that $|S \cap T| \leq \deg_G(s)+\deg_G(t)-n-1$. Then $\deg_G(s)+\deg_G(t)-2=|S|+|T|=|S \cup T|+|S \cap T| \leq \deg_G(s)+\deg_G(t)-n-1+n-2$. This is a contradiction. Thus, there are at least $w=\deg_G(s)+\deg_G(t)-n$ elements in $S \cap T$. Let $q_1, q_2, \ldots, q_w$ be the elements in $S \cap T$. For $j=1,2,\ldots,w$, we set $P_j=\langle s,[q_j+1],[q_j+2],\ldots,[q_j],t\rangle$. So $P_1, P_2, \ldots, P_w$ are mutually independent paths between s and t.

Case 2. $(s,t) \notin E(G)$. Suppose that $|S \cap T| \leq \deg_G(s)+\deg_G(t)-n+2-1$. Then $\deg_G(s)+\deg_G(t)=|S|+|T|=|S \cup T|+|S \cap T| \leq \deg_G(s)+\deg_G(t)-n+2-1+n-2$. This is a contradiction. Thus, there are at least $w=\deg_G(s)+\deg_G(t)-n+2$ elements in $S \cap T$. Let $q_1, q_2, \ldots, q_w$ be the elements in $S \cap T$. For $j=1,2,\ldots,w$, we set $P_j=\langle s,[q_j+1],[q_j+2],\ldots,[q_j],t\rangle$, and $P_1, P_2, \ldots, P_w$ are mutually independent paths between s and t. □

Example 9.4

Let G be the graph $(K_1 + K_{n-d-1}) \vee K_d$ where d is an integer with $4 \leq d < n-1$. Obviously, $\overline{e} = n-1-d \leq n-4$. Let x be the vertex corresponding to K_1, y be an arbitrary vertex in K_d, and z be a vertex in K_{n-d-1}. Then $\deg_G(x) = d$, $\deg_G(y) = n-1$, $\deg_G(z) = n-2$, $(x, y) \in E(G)$, $(y, z) \in E(G)$, and $(x, z) \notin E(G)$. By Theorem 9.21, there are $n-2-\overline{e} = n-2-(n-1-d) = d-1$ mutually independent Hamiltonian paths between any two distinct vertices of G. By Lemma 9.4, there are at most $(d-1)$ mutually independent Hamiltonian paths between x and y. Therefore, the result in Theorem 9.21 is optimal.

Let us consider the same example as previously. It is easy to check whether any two vertices u and v in G, $\deg_G(u) + \deg_G(v) \geq n+2$. Let x and y be the same vertices as described earlier. By Theorem 9.22, there are $\deg_G(x) + \deg_G(y) - n = d + (n-1) - n = d-1$ mutually independent Hamiltonian paths between x and y. By Lemma 9.4, there are at most $d-1$ mutually independent Hamiltonian paths between x and y. Hence, the result in Theorem 9.22 is also optimal.

Combining Theorems 9.12 and 9.22, we have the following corollary.

COROLLARY 9.1 *Suppose that r is a positive integer. Let G be a graph such that $\deg_G(x) + \deg_G(y) \geq n + r$ for any two distinct vertices x and y. Then there are at least r mutually independent Hamiltonian paths between any two distinct vertices of G.*

However, we would like to make the following conjecture. Assume that $r > 1$ and G is a graph such that $\deg_G(u) + \deg_G(v) \geq n + r$ for any two distinct vertices u and v in G. Then there are at least $r+1$ mutually independent Hamiltonian paths between any two distinct vertices of G.

9.6 DIAMETER FOR GENERALIZED SHUFFLE-CUBES

In Chapter 3, we introduced shuffle cubes, SQ_n. We only discuss the upper bound of the diameter of SQ_n. In Chapter 7, we discuss the connectivity of SQ_n. We only discuss the exact value of $D(SQ_n)$ after we introduce the concept of generalized shuffle-cubes.

In this section, the graph K_1 is also denoted by Q_0. For any positive integer l, we use $S(l)$ to denote the set of all binary strings of length l and we use $S^*(l)$ to denote $S(l) - \{\underbrace{00\cdots0}_{l}\}$. Assume that b and g are any positive integers satisfying $2^b \geq (2^g - 1)/g$. For each $i_1 i_2 \cdots i_b \in S(b)$, we associate it with a subset $A_{i_1 i_2 \cdots i_b}$ of $S^*(g)$ with the following properties: (1) $| A_{i_1 i_2 \cdots i_b} | = g$ and (2) $\bigcup_{i_1 i_2 \cdots i_b \in S(b)} A_{i_1 i_2 \cdots i_b} = S^*(g)$. A family $A = \{A_{i_1 i_2 \cdots i_b} \mid i_1 i_2 \cdots i_b \in S(b)\}$ is *a normal* (g, b) *family* if it satisfies these properties. For example, $\{A_{00}, A_{01}, A_{10}, A_{11}\}$ is the normal (4,2) family where A_{00}, A_{01}, A_{10}, and A_{11} are defined in SQ_n.

DEFINITION 9.1: Suppose that B is a b-regular graph with vertex set $S(b)$ and A is any normal (g, b) family. We can recursively define the n-dimensional generalized shuffle-cube $GSQ(n,A,B)$ for any $n=kg+b$ for $k \geq 0$ with its vertex set being $S(n)$ as follows:

1. Suppose that $n=b$. We set $GSQ(n,A,B)=B$.
2. Suppose that $n=kg+b$ for $k \geq 1$. Any two vertices u and v in $GSQ(n,A,B)$ are adjacent if and only if
   (a) $s_{n-g}(u)$ and $s_{n-g}(v)$ are adjacent in $GSQ(n-g,A,B)$, and $p_g(u)=p_g(v)$ or
   (b) $s_{n-g}(u)=s_{n-g}(v)$ and $p_g(u) \oplus p_g(v) \in A_{u_{b-1}u_{b-2}\cdots u_0}$.

For example, Q_n is $GSQ(n,A,B)$, where $A=\{A_0\}$ is a normal (1,0) family with $A_0=\{1\}$ and $B=Q_0$. Moreover, SQ_n is $GSQ(n,A,B)$, where $A=\{A_{00},A_{01},A_{10},A_{11}\}$ is a normal (4,2) family and B is Q_2.

Assume that $GSQ(n,A,B)$ is a generalized shuffle-cube. Obviously, $GSQ(n,A,B)$ is an n-regular graph with 2^n vertices. Suppose that u and v are two vertices in $GSQ(n,A,B)$. For $1 \leq j \leq k$, the *jth g-bit* of u, denoted by u_g^j, is $u_g^j = u_{gj+b-1}u_{gj+b-2}\cdots u_{gj+b-g}$. In particular, the 0th g-bit of u is $u_g^0=u_{b-1}u_{b-2}\cdots u_0$. The *g-bit Hamming distance* between u and v, denoted by $h_g(u,v)$, is the number of g-bits u_g^j with $0 \leq j \leq k$ such that $u_g^j \neq v_g^j$; that is, $h_g(u,v)=|\{j \mid u_g^j \neq v_g^j \text{ for } 0 \leq j \leq k\}|$. We also use $h_g^*(u,v)$ to denote the number of u_g^j for $1 \leq j \leq k$ such that $u_g^j \neq v_g^j$.

Applying a similar argument of Theorem 7.12, we have the following theorem.

THEOREM 9.23 $\kappa(GSQ(n,A,B))=n$ if $\kappa(B)=b$.

Now, we discuss the diameter of a generalized shuffle-cube $GSQ(n,A,B)$. We assume that B has some Hamiltonian properties.

Assume that B is Hamiltonian. Let $C=\langle x_0,x_1,\ldots,x_k=x_0\rangle$ be a Hamiltonian cycle of B. The cycle $\langle 00,01,11,10,00\rangle$, for example, is a Hamiltonian cycle of Q_2. We propose the routing algorithm **Route2**(u,v) for $GSQ(n,A,B)$ as follows:

Route2(u,v)

1. Suppose that $u=v$. Then accept the message.
2. Find a neighbor w of u such that $h_g^*(w,v)=h_g^*(u,v)-1$ if w exists. Then route into w.
3. Suppose that $h_g^*(u,v)>0$ and there is no neighbor w of u such that $h_g^*(w,v)=h_g^*(u,v)-1$. Then route into the neighbor w of u that changes u_g^0 in a cyclic manner with respect to C.
4. Suppose that $h_g^*(u,v)=0$. Find a neighbor z of $s_b(u)$ in B such that the distance between z and $s_b(v)$ is the distance between $s_b(u)$ and $s_b(v)$ minus 1. Then route into $p_{n-b}(u)z$.

So, we have the following theorem.

THEOREM 9.24 Suppose that B is Hamiltonian. Then $D(GSQ(n, A, B)) \leq [(n-b)/g] + 2^b - 1 + D(B)$.

The upper bound for $D(GSQ(n, A, B))$ can be further reduced if B is Hamiltonian-connected or Hamiltonian-laceable. Suppose that B is Hamiltonian-laceable. To route u to v, we first find a vertex sequence $Z(u, v)$ of $S(b)$ as follows: suppose that $s_b(u)$ and $s_b(v)$ are in different parts. We set $Z(u, v)$ to be any Hamiltonian path from $s_b(u)$ to $s_b(v)$. Suppose that $s_b(u)$ and $s_b(v)$ are in the same part. Find a neighborhood $s_b(z)$ of $s_b(v)$ in B. Let P be a Hamiltonian path from $s_b(u)$ to $s_b(z)$. We set $Z(u, v)$ to be the vertex sequence $\langle P, s_b(v)\rangle$. Then the path of $GSQ(n, A, B)$ from u to v can be determined by the following algorithm:

Route3(u, v)

1. Suppose that $u = v$. Then accept the message.
2. Find a neighbor w of u such that $h_g^*(w, v) = h_g^*(u, v) - 1$ if w exists. Then route into w.
3. Suppose that there is no neighbor w of u such that $h_g^*(w, v) = h_g^*(u, v) - 1$. Then route into the neighbor w of u that changes u_g^0 in the order of $Z(u, v)$.

EXAMPLE 9.5

As noted before, SQ_n is a generalized shuffle-cube $GSQ(n, A, B)$ with $B = Q_2$. It is known that Q_2 is Hamiltonian-laceable. Suppose that $u = 0001000101001000110000$ and $v = 0000000000000000000011$ are two vertices of SQ_{18}. Obviously, 00 and 11 are in the same part and 10 is a neighbor of 11. Hence, $\langle 00, 01, 11, 10\rangle$ is a Hamiltonian path from 00 to 10 in Q_2. We can set $Z(u, v)$ as $\langle 00, 01, 11, 10, 11\rangle$. The path obtained from **Route3**(u, v) is

0001000101001000110000, $\underline{0000}000101001000110000$, $0000\underline{0000}01001000110000$,

$00000000010010001100\underline{0}1$ → $000000000100100011000\underline{1}$, $00000000\underline{0000}1000110001$, $00000000000010001100\underline{1}1$,

$0000000000001000\underline{0000}11$, $000000000000100000001\underline{0}$, $000000000000\underline{0000}000010$,

$000000000000000000001\underline{1}$.

We note that this path is shorter than the path obtained in Example 3.2.

Note that we should apply Step 3 exactly $2^b - 1$ times to obtain a vertex w such that either $w = v$ or w is a neighbor of v. Thus, we have the following theorem.

THEOREM 9.25 Suppose that B is Hamiltonian-laceable. Then $(n-b)/g \leq D(GSQ(n, A, B)) \leq [(n-b)/g] + 2^b$.

Suppose that B is Hamiltonian-connected. To avoid a trivial case, we assume that $b \geq 1$. To route from u to v in $GSQ(n, A, B)$, we compute a vertex sequence $Z(u, v)$ of $S(b)$ as follows: suppose that $s_b(u) \neq s_b(v)$. Set $Z(u, v)$ to be any Hamiltonian path from $s_b(u)$ to $s_b(v)$. Suppose that $s_b(u) = s_b(v)$. Find a neighborhood $s_b(z)$ of $s_b(v)$ in

B. Let P be a Hamiltonian path from $s_b(u)$ to $s_b(z)$. We set $Z(u, v)$ to be the vertex sequence $\langle P, s_b(v)\rangle$. We can also apply **Route3**(u, v) to obtain a path from u to v. So we have the following theorem.

THEOREM 9.26 Suppose that B is Hamiltonian-connected. Then $D(GSQ(n, A, B)) \le n - b/g + 2^b$.

Now, we can determine the diameter of SQ_n for $n = 4k + 2$.

THEOREM 9.27 Assume that $n = 4k + 2$. The diameter of SQ_n is 2 if $n = 2$; 4 if $n = 6$; or $\lceil \frac{n}{4} \rceil + 3$ if $n \ge 10$.

Proof: Using breadth-first search (BFS), we can easily determine $D(SQ_2) = 2$, $D(SQ_6) = 4$, $D(SQ_{10}) = 6$, and $D(SQ_{14}) = 7$. Combining Lemma 3.1 and Theorem 9.25, we can conclude that $D(SQ_n) = \lceil \frac{n}{4} \rceil + 3$ if $n \ge 18$. The theorem is proved. □

9.7 CYCLES IN DIRECTED GRAPHS

The theory of cycles in digraphs is similar to that of cycles in graphs. For a digraph G, let $\delta^-(G) = \min\{\deg^-(v) \mid v \in G\}$ and $\delta^+(G) = \min\{\deg^+(v) \mid v \in G\}$.

Here, we consider spanning cycles in digraphs. Although cliques are trivially Hamiltonian, the question becomes interesting for tournaments. The orientation of a complete graph is called a *tournament*.

THEOREM 9.28 Every tournament has a Hamiltonian path.

Proof: Let D be a tournament. Obviously, the underlying graph of D is the complete graph with n vertices. Hence, $\chi(D) = n$. By Theorem 8.5, D has a Hamiltonian path. □

THEOREM 9.29 Assume that D is a strongly connected tournament with $n \ge 3$ vertices. D contains a directed k-cycle for any $3 \le k \le v$ that includes any vertex u. In particular, D is Hamiltonian.

Proof: Let D be a strongly connected tournament with $n \ge 3$. Let u be any vertex of D. Set $S = N^+(u)$ and $T = N^-(u)$. We first show that u is in a directed 3-cycle. Since D is strongly connected, neither S nor T can be empty. For the same reason, (S, T) must be nonempty; that is, there is a certain arc (v, w) in D with $v \in S$ and $w \in T$. See Figure 9.10 for illustration. Hence, u is in the directed 3-cycle $\langle u, v, w, u\rangle$.

The theorem is now proved by induction on k. Suppose that u is in a directed cycle of all lengths between 3 to k with $k \le n$. We shall show that u is in a directed $(k + 1)$-cycle.

Let $C = \langle u = v_0, v_1, \ldots, v_k = u\rangle$ be a directed k-cycle in D. Suppose that there is a vertex v in $V(D) - V(C)$ that is both the head of an arc with tail in C and the tail of an arc with head in C. Then there are adjacent vertices v_i and v_{i+1} on C such that both (v_i, v) and (v, v_{i+1}) are arcs of D. In this case, u is in the directed $(k + 1)$-cycle $\langle v_0, v_1, \ldots, v_i, v, v_{i+1}, \ldots, v_k\rangle$.

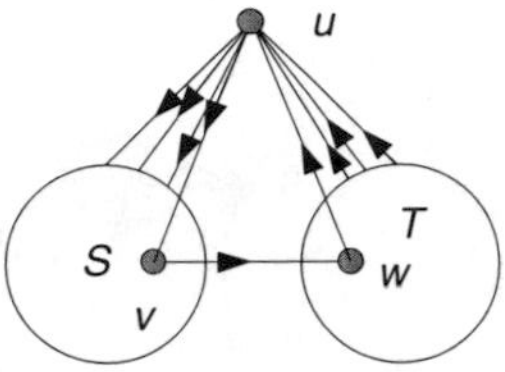

FIGURE 9.10 A 3-cycle in a strongly connected tournament.

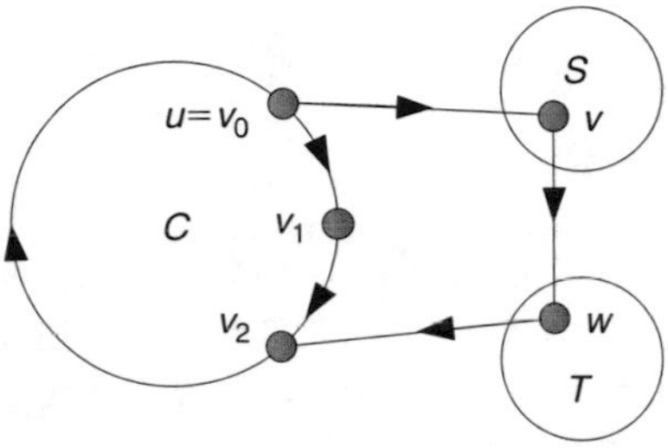

FIGURE 9.11 A directed $(k+1)$-cycle in a strongly connected tournament.

Otherwise, let S be the set of vertices in $V(D) - V(C)$ that are heads of arcs joined to C, and let T be the set of vertices in $V(D) - V(C)$ that are tails of arcs joined to C. See Figure 9.11 for illustration.

Since D is strongly connected, S, T, and (S, T) are all nonempty. Thus, there is some (v, w) in D with $v \in S$ and $w \in T$. Obviously, u is in the directed $(k+1)$-cycle $\langle v_0, v, w, v_2, \ldots, v_k \rangle$. The theorem is proved. □

For arbitrary digraphs, we consider degree conditions. The basic result is analogous to Dirac's Theorem. By applying it to the digraph obtained from a graph G by replacing every edge by a pair of opposed edges with the same endpoints, we obtained Dirac's Theorem as a special case. A digraph is *strict* if it has no loops and has at most one copy of ordered pair as an edge.

THEOREM 9.30 **(Ghouila-Houri [124])** Let D be a strict digraph with $\min\{\delta^-(G), \delta^+(G)\} \geq n(D)/2$. Then D is Hamiltonian.

Proof: Suppose this theorem is not true. Thus, there exists an n-vertex counterexample D. Let C be a longest cycle in D. Let the length of C be l. Obviously, $l < n$. Moreover, the length of l is at least $\max\{\delta^-, \delta^+\} + 1 \geq n/2$. Let P be a longest path in $D - V(C)$. In P, let u be the beginning vertex, w be the ending vertex, and m be the length. Obviously, $m \geq 0$. In summary, we have $l > n/2$, $n \geq |V(C)| + |V(P)| = l + m + 1$, and $m < n/2$.

Let S be the set of predecessors of u on C and T be the set of successors of w on C. (See Figure 9.12). By the maximality of P, all predecessors of u are a subset of $V(C) \cup V(P)$. Thus $|S| \geq \min\{\delta^+, \delta^-\} - m \geq n/2 - m > 0$. Similarly, all successors of w are a subset of $V(C) \cup V(P)$ and $|T| \geq n/2 - m > 0$.

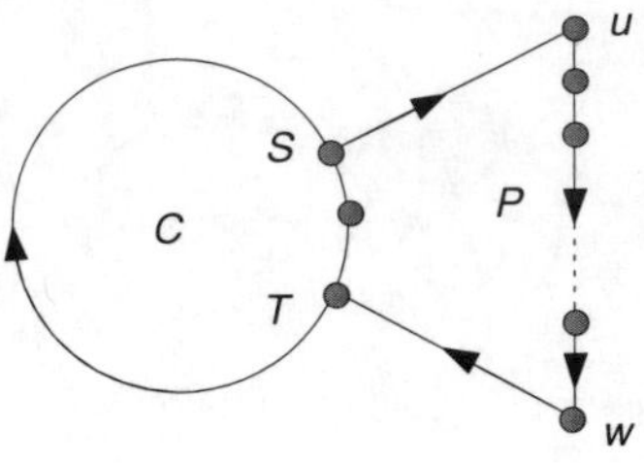

FIGURE 9.12 Illustration of Theorem 9.30.

Suppose that C contains a vertex x in S and a vertex y in T such that $d_C(x, y) \leq m$. Then we can form a longer cycle in G by using $E(C) \cup E(P)$ and deleting the edges in the section of C joining x to y. Thus, the distance along C from a vertex $u' \in S$ to a vertex $w' \in T$ must exceed $m + 1$. Hence, we may assume that every vertex of S is followed on C by more than m vertices not in T.

Since both S and T are nonempty, we may thus assume that there is a vertex of S followed on C by at least $m + 1$ vertices not in S. These vertices cannot be a vertex in T. Thus, $|V(C - T)| \geq |S| - 1 + m + 1 \geq n/2$. As $|T| \geq n/2 - m$, $l = |V(C)| = |V(T - C)| + |T| \geq (n/2) + (n/2) - m = n - m$. However, this contradicts $l \leq n - m + 1$. □

10 Planar Graphs

10.1 PLANAR EMBEDDINGS

The study of planar graphs is motivated by the famous Four Color Problem. Can the regions of any map on the globe be colored with four colors so that regions sharing a nontrivial boundary have different colors? Recently, the study of circuit layouts on silicon chips provides another motivation for such a study. Crossings cause problems in a layout. We want to know which circuits have layouts without crossings. Recently, the study of planar graphs is generalized into topological graph theory: layout graphs on different surfaces.

The following brain teaser appeared as early as Dudeney [90].

Example 10.1

In the woods, there live three sworn enemies A, B, C. We must find paths to install three utilities: gas, water, and electricity. In order to avoid confrontations, we want to avoid having any of these paths cross each other. It turns out that we cannot solve the problem. This question is equivalent to draw $K_{3,3}$ in the plane without edge crossings. (See Figure 10.1.) Later, we will give the reason why there is no solution.

A graph is said to be *embeddable in the plane*, or *planar*, if it can be drawn in the plane so that its edges intersect only at their ends. Such a drawing of a plane graph G is called a *planar embedding* of G. A planar embedding $\hat{G}$ of G can itself be regarded as a graph isomorphic to G. The vertex set of $\hat{G}$ is the set of points representing vertices of G. The edge set of $\hat{G}$ is the set of lines representing edges of G. A vertex of $\hat{G}$ is incident to all the edges of $\hat{G}$ that contain it. We therefore sometimes refer to a planar embedding of a planar graph as a *plane graph*. The graph in Figure 10.2b shows a planar embedding of the planar graph Figure 10.2a.

From this definition, the study of planar graphs necessarily involves the topology of the plane. However, we shall not attempt here to be strictly rigorous in topological matters. Instead, we use a naïve point of view toward them. This is done so as not to obscure the combinatorial aspect of the theory, which is our main interest.

The results of topology that are essentially relevant in the study of planar graphs are those that deal with Jordan curves. A *Jordan curve* is a continuous non-self-intersecting curve whose origin and terminus coincide. The union of the edges in a cycle of a plane graph constitutes a Jordan curve. This is the reason why properties of Jordan curves come into play in planar graph theory. We need a well-known theorem about Jordan curves and use it to demonstrate the nonplanarity of K_5 and $K_{3,3}$.

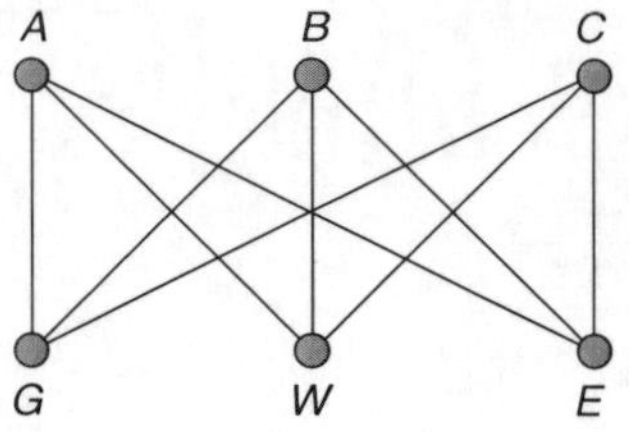

FIGURE 10.1 Gas–water–electricity graph.

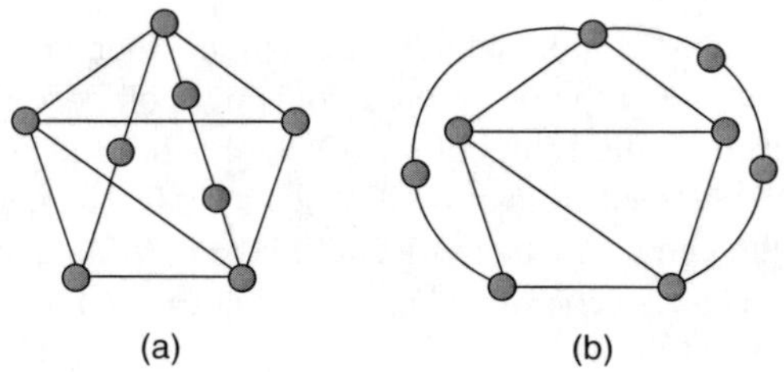

FIGURE 10.2 (a) A planar graph and (b) its planar embedding.

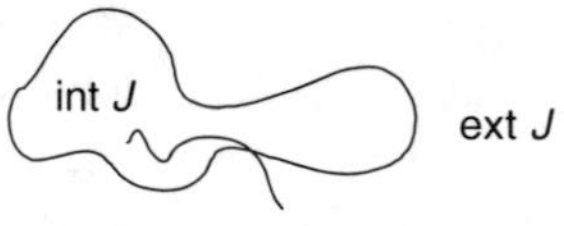

FIGURE 10.3 Illustration of the Jordan Curve Theorem.

Let J be a Jordan curve in the plane. The rest of the plane is partitioned into two disjoint open sets called the *interior* and *exterior* of J. We denote the interior and the exterior of J, respectively, by *int* J and *ext* J. The closure of *int* J is denoted by *Int* J and the closure of *ext* J is denoted by *Ext* J. Clearly, $Int\ J \cap Ext\ J = J$. The Jordan Curve Theorem states that any line joining a point in *int* J to a point in *ext* J must meet J in some point. See Figure 10.3 for an illustration. It seems that this theorem is intuitively obvious. However, a formal proof of it is quite difficult.

Let G be any planar graph. We draw G in the plane. Suppose that C is a spanning cycle of G. Then C is drawn as a closed curve. Suppose that the drawing is an embedding of G. Chords of C must be drawn inside or outside this curve. Two chords *conflict* if their endpoints on C occur in alternating order. By the Jordan Curve Theorem, conflicting chords must embed in opposite faces of C.

EXAMPLE 10.2

K_5 and $K_{3,3}$ are two important examples of nonplanar graphs. Let us draw $K_{3,3}$ in the plane. Let C be a 6-cycle from $K_{3,3}$. Obviously, three pairwise conflicting chords are left. We can put at most one inside and one outside. Hence $K_{3,3}$ is not planar.

Now, we draw K_5 in the plane. Let C be a 5-cycle from K_5. Obviously, five chords are left. However, at most two chords can go inside or outside. Hence K_5 is not planar.

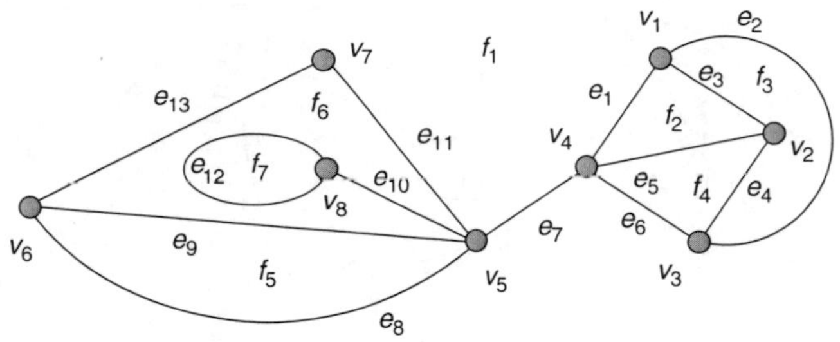

FIGURE 10.4 A plane graph with seven faces.

A plane graph G partitioned the rest of the plane into a number of connected regions. The closures of these regions are called the *faces* of G. A plane graph with seven faces, $f_1, f_2, f_3, f_4, f_5, f_6$, and f_7 is shown in Figure 10.4. We shall denote by $F(G)$ and $\phi(G)$, respectively, the set of faces and the number of faces of a plane graph. Each plane graph has exactly one unbounded face, called the *exterior face*. The face f_1 of the graph in Figure 10.4 is the exterior face.

The family of planar graphs and the family of graphs embeddable on the sphere are actually the same. Given an embedding of a graph G on the sphere, we can puncture the sphere within any face and project from there onto a plane tangent to the antipodal point to obtain a planar embedding of G. The punctured face on the sphere becomes the exterior face in the plane, and the process is reversible. Thus, any planar graph can be embedded in the plane so that any vertex v is on the exterior face of the embedding.

Let us view a geographic map on the plane or the sphere as a plane graph, in which the faces are the territories of the map, the vertices are places where several boundaries meet, and the edges are the portions of the boundaries that join two vertices. *Here, we allow graphs with loops and multiple edges.* From any plane graph G, we can build another plane graph called its *dual*.

Suppose that G is a plane graph. The *dual graph* G^* of G is a plane graph having a vertex for each region in G. The edges of G^* correspond to the edges of G as follows: suppose that e is an edge of G that has region X on one side and region Y on the other side. Then the corresponding dual edge $e^* \in E(G^*)$ is an edge joining the vertices x and y of G^* that correspond to the faces X and Y of G.

EXAMPLE 10.3

In Figure 10.5, we have a plane graph G with dashed edges and its dual G^* with solid edges. Since G has five vertices, seven edges, and four faces, G^* has five faces, seven edges, and four vertices. As in this example, a simple plane graph may have loops and multiple edges in its dual. A cut edge of G becomes a loop in G^*, because the faces on both sides of it are the same. Multiple edges arise in the dual when distinct regions of G have more than one common boundary edge. With this example, we know the reason why we allow graphs with loops and multiple edges as we discuss planar graphs.

A statement about a connected plane graph becomes a statement about the dual graph when we interchange the roles of vertices and faces. Edges incident to a vertex

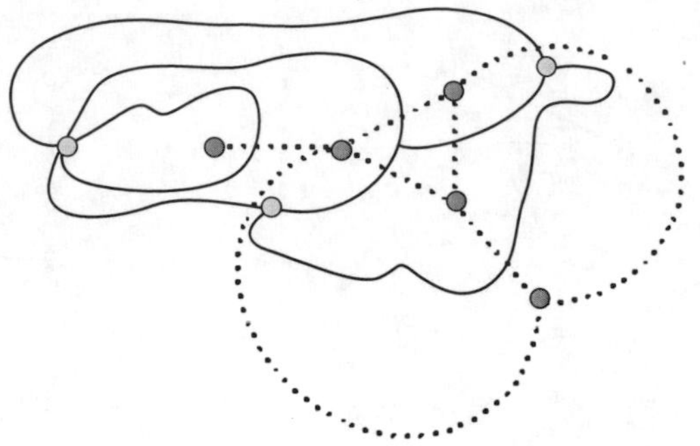

FIGURE 10.5 A plane graph G with dashed edges and its dual G^* with solid edges.

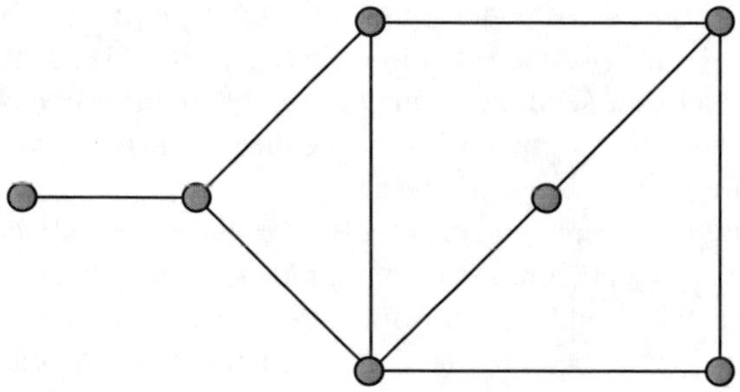

FIGURE 10.6 The plane graph with four faces of lengths 3,4,4,7.

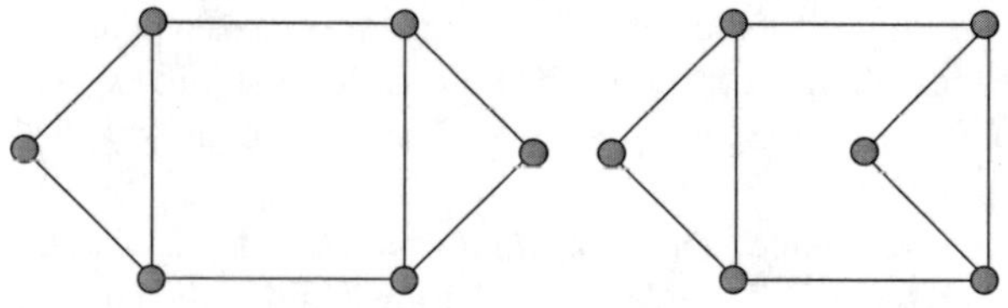

FIGURE 10.7 Different embeddings of a planar graph.

become edges bounding a face, and vice versa. For example, let us consider the degree-sum formula say about edges and faces instead of edges and vertices. The *length* of a face in a plane graph G is the length of the walk in G that bounds it. Note that a cut edge belongs to the boundary of only one face. Yet it contributes twice to the boundary of the face as we traverse the boundary. The plane graph in Figure 10.6 has four faces, with lengths 3,4,4,7. The sum of the lengths is 18, which is twice that of the number of edges.

Note that different embeddings of a planar graph may have nonisomorphic duals. For example, the plane graph in Figure 10.7a has faces of lengths 6,4,3,3. And the plane graph in Figure 10.7b has faces of lengths 5,5,3,3. It is easy to check whether these two graphs are isomorphic. However, their duals have degree sequences 6,4,3,3 and 5,5,3,3, respectively. Hence, the duals are not isomorphic. However, it can be proved that every 3-connected planar graph has essentially one embedding.

PROPOSITION 10.1 *Let $l(F_i)$ denote the length of face F_i in the plane graph G. Then $2e(G) = \sum l(F_i)$.*

Proof: It is observed that bounding edges for a face X correspond to dual edges incident to the dual vertex x. Since $e(G) = e(G^*)$, the statement $2e(G) = \sum l(F_i)$ is the same as the degree-sum formula $2e(G^*) = \sum \deg_{G^*}(x)$ for G^*. Alternatively, adding up the face lengths counts each edge twice. □

10.2 EULER'S FORMULA

Euler's formula is the basic computational tool for planar graphs.

THEOREM 10.1 **(Euler [99])** If a connected plane graph G has n vertices, e edges, and f faces, then $n - e + f = 2$.

Proof: We prove this theorem by induction on n. Suppose that $n = 1$. Then G is a *bouquet* of loops, each being a closed curve in the embedding. Assume that $e = 0$. Then we have one face. Obviously, the formula holds. By the Jordan Curve Theorem, each added loop passes through a region and partitions it into two regions. Thus, the formula holds for $n = 1$ and any $e \geq 0$.

Now suppose that the theorem holds for any connected plane graph G with k vertices for $1 \leq k < n(G)$. Since G is connected, we can find an edge that is not a loop. When we contract such an edge, we obtain a plane graph G' with n' vertices, e' edges, and f' faces. The contraction does not change the number of faces, but it reduces the number of edges and vertices by one. Applying the induction hypothesis, we find $n - e + f = n' + 1 - (e' + 1) + f = 2$. □

REMARK:

1. Euler's formula implies that all planar embeddings of a connected graph G have the same number of faces. Thus, although the dual depends on the emdedding chosen for G, the number of vertices in the dual does not.
2. Deleting an edge of G has the effect of contracting an edge in G^* because two faces of G merge into a single face in $G - e$. Similarly, contracting an edge of G has the effect of deleting an edge in G^*.
3. Euler's formula fails for disconnected graphs. The general formula for a plane graph with k components is $n - e + f = k + 1$.

THEOREM 10.2 Suppose that G is a simple planar graph with at least three vertices. Then $e(G) \leq 3n(G) - 6$. Moreover, $e(G) \leq 2n(G) - 4$ if G is triangle-free.

Proof: It suffices to consider connected graphs, as otherwise we could add all the edges in each connected component. We can use Euler's formula to relate $n(G)$ and $e(G)$ if we can dispose of f. Note that $n(G) \geq 3$ and loopless. Every face boundary in a simple graph contains at least three edges. Let $\{f_i\}$ be the sequence of face-lengths. By Proposition 10.1, $2e = \sum f_i \geq 3f$. Since $n - e + f = 2$,

$2 = n - e + f \leq n - e + (2e/3) = n - (e/3)$. Thus, $e \leq 3n - 6$. Suppose that G is triangle-free. The faces have a length of at least 4. Thus, $2e = \sum f_i \geq 4f$. Using a similar argument, we obtain $e \leq 2n - 4$. □

We can also use Euler's formula to show that K_5 and $K_{3,3}$ are nonplanar. For K_5, we have $e = 10 > 9 = 3n - 6$. Since $K_{3,3}$ is triangle-free, we have $e = 9 > 8 = 2n - 4$.

The following corollary is a direct consequence of Theorem 10.2.

COROLLARY 10.1 *Every simple planar graph has a vertex of degree at most 5.*

The proof of Theorem 10.2 shows that having $(3n - 6)$ edges in a simple n-vertex planar graph requires $2e = 3f$. Moreover, every face is a triangle. Suppose that G has some face that is not a triangle. We can add an edge between nonadjacent vertices on the boundary of this face to obtain a larger planar graph. Here, the family of simple plane graphs with $3n - 6$ edges, the family of triangularations, and the family of *maximal* plane graphs all belong to the same family.

Informally, we think of a regular polyhedron as a solid, whose boundary consists of regular polygons of same length, with same number of faces meeting at each vertex. When we lay the surface out in the plane, we obtain a regular planar graph with faces of the same length. Hence, the dual is also regular. We can prove that there are only five regular polyhedra by proving that there are only five regular planar graphs whose duals are also simple and regular.

Suppose that G is a plane graph with n vertices, e edges, and f faces. Suppose also that G is regular of degree k and that G^* is regular of degree l. Thus, G has faces of length l. By the degree-sum formula for G and G^*, we have $kn = 2e = lf$. Substituting for n and f into Euler's formula, we have $e[(2/k) - 1 + (2/l)] = 2$. Since e and 2 are positive, $(2/k) - 1 + (2/l) > 0$. Hence, $2l + 2k > kl$. This is equivalent to $(k - 2)(l - 2) < 4$. Because the dual of a 2-regular graph is not simple, we conclude that $k, l \geq 3$. By Corollary 10.1, $k, l \leq 5$. There are only five solution pairs for (k, l): (3,3), (3,4), (3,5), (4,3), (5,3). Once we specify k and l, there is only one way to lay out the plane graph when we start with any face. Hence, there are no more than five known Platonic solids.

k	l	$(k-2)(l-2)$	e	n	f	Name
3	3	1	6	4	4	Tetrahedron
3	4	2	12	8	6	Cube
4	3	2	12	6	8	Octahedron
3	5	3	30	20	12	Dodecahedron
5	3	3	30	12	20	Icosahedron

10.3 CHARACTERIZATION OF PLANAR GRAPHS

Before 1930, the most actively sought result in graph theory was the characterization of planar graphs. We have proved that K_5 and $K_{3,3}$ are not planar. In a natural sense, these are critical graphs and yield a characterization of planarity. It was Kuratowski who finally characterized the set of planar graphs using K_5 and $K_{3,3}$.

Note that *subdividing* an edge or performing an *elementary subdivision* means replacing the edge with a path of length 2. A *subdivision* of G is a graph obtained from G by a sequence of elementary subdivisions; that is, turning edges into paths through new vertices of degree 2.

It is observed that subdividing edges does not affect planarity. For this reason, we seek a characterization by finding a *topologically minimal* nonplanar graphs—those that are not subdivisions of other nonplanar graphs. We already know that a graph containing any subdivision of K_5 or $K_{3,3}$ is nonplanar. Kuratowski [213] proved the following theorem. We omit the proof because it is very complicated.

THEOREM 10.3 A graph G is planar if and only if G contains no subdivision of K_5 or $K_{3,3}$.

Wagner [329] proved another characterization. It is observed that deletion and contraction of edges preserve planarity. Again, we seek the minimal nonplanar graphs using these operations. Wagner proved the following theorem.

THEOREM 10.4 A graph G is planar if and only if it has no subgraph contractible to K_5 or $K_{3,3}$.

Mathematicians expand the study of planar embedding by putting additional requirements. For example, the *straight line embedding* requires that all edges embedded are straight line segments. Wagner [328], Fáry [105], and Stein [287] show that every finite planar graph has a straight line embedding. This is known as Fáry's Theorem. The *convex embedding* requiring each face boundary, including the unbounded face, is a convex polygon. Tutte [310,311] proved that every 3-connected planar graph has a convex embedding. This is the best possible result, in the sense that the 2-connected planar graph $K_{2,n}$ does not have a convex embedding if $n \geq 4$.

Computer scientists characterize planar graphs using algorithms. There are linear-time planarity-testing algorithms due to Hopcroft and Tarjan [157] and due to Booth and Lueker [31], but these are very complicated.

10.4 COLORING OF PLANAR GRAPHS

Sometimes, it is very difficult to study a property or a parameter on general graphs. We can restrict our attention to some families of graphs. Thus, every property and parameter we have studied for general graphs can be studied for planar graphs. The chromatic number of planar graphs is one of the greatest historical interests.

By Corollary 10.1, every simple planar graph has a vertex of degree at most 5. We can easily prove that every planar graph is 6-colorable by induction. Heawood improved this result.

THEOREM 10.5 **(Heawood [142])** Every planar graph is 5-colorable.

Proof: Obviously, the theorem holds for K_1. Suppose that the theorem is not true. Then there is a minimal counterexample G. Let v be the vertex of degree at most 5 in G. The choice of G implies that $G - v$ is 5-colorable. Let $f : V(G - v) \to [5]$

be a 5-coloring of $G-v$. Since G is not 5-colorable, each color appears at one of the neighbors of v. Hence, $\deg(v)=5$. We may label the colors to assume that the neighbors of v in a planar embedding of G are v_1, v_2, v_3, v_4, v_5 in clockwise order around v with $f(v_i)=i$.

Let G_{ij} denote the subgraph of G induced by the vertices of colors i and j. We can exchange the two colors on any component of G_{ij} to obtain another 5-coloring of $G-v$. Hence, the component of G_{ij} that contains v_i must also contain v_j. Otherwise, we could make the interchange on the component of G_{ij} containing v_i to remove color i from $N(v)$. Then we could assign color i to v to extend f to be a 5-coloring of G. Let P_{ij} be the path in G_{ij} from v_i to v_j.

Consider the cycle C completed with $P_{1,3}$ by v. Then C separates v_2 from v_4. By the Jordan Curve Theorem, the path $P_{2,4}$ must cross C. Since G is planar, such a crossing can happen only at a shared vertex. This is impossible, because the vertices of $P_{1,3}$ all have color 1 or 3, and the vertices of $P_{2,4}$ all have color 2 or 4. □

In October 23, 1852, Sir William Hamilton received a letter from Augustus de Morgan at University College in London. In this letter, the Four Color Problem first appeared. The problem was asked by de Morgan's student Frederick Guthrie, who later attributed it to his brother Francis Guthrie. The problem was phrased in terms of map coloring, in which the faces of a planar graph are to be colored.

In 1878, Cayley announced the problem to the London Mathematical Society. Within a year, Kempe [200] published a *solution*. Soon, Kempe was elected a Fellow of the Royal Society. In 1890, Heawood posed a refutation. Yet Kempe's idea was the alternating paths in the previous Five Color Theorem. Eventually, Appel and Haken [12–14] solved the Four Color Problem.

In 1878, Tait proved a theorem relating face-coloring of planar maps to proper edge-colorings of planar graphs. He used this in an approach to the Four Color Problem.

THEOREM 10.6 **(Tait [295])** A simple 2-edge-connected 3-regular planar graph is 3-edge-colorable if and only if it is 4-face-colorable.

Proof: Let G be a 2-edge-connected 3-regular planar simple graph. Suppose that G is 4-face-colorable. Let the four face colors be denoted by binary ordered pairs: $c_0=00, c_1=01, c_2=10$, and $c_3=11$. We obtain a proper 3-edge-coloring of G by assigning to the edge between faces with colors c_i and c_i the coloring obtained by adding c_i and c_j as vectors of length 2, using coordinate-wise addition modulo 2. We need to check whether such coloring is proper.

Since G is 2-edge-connected, each edge bounds two different faces. Hence the color 00 never occurs as a sum. It suffices to prove that the three edges at a vertex receive distinct colors. At vertex v the faces bordering the three incident edges are pairwise-adjacent. These three faces must have three distinct colors $\{c_i, c_j, c_k\}$. See Figure 10.8 for illustration. Suppose that color 00 is not in this set. Then the sum of any two of these is the third. Hence, $\{c_i, c_j, c_k\}$ is the set of colors on the three edges. Suppose that $c_k=00$. Then c_i and c_j appear on two of the edges, and the third receives color c_i+c_j, which is the color not in $\{c_i, c_j, c_k\}$.

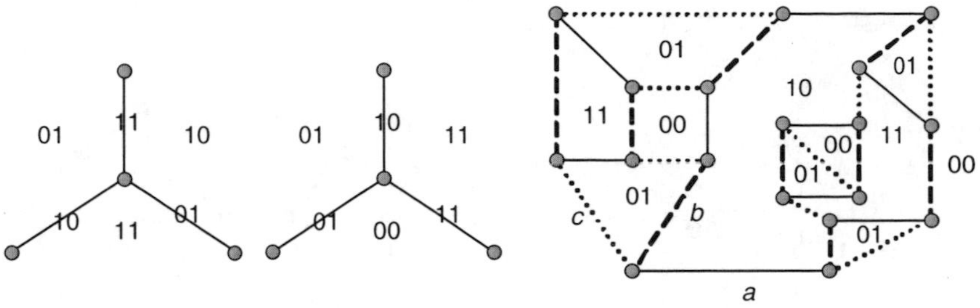

FIGURE 10.8 Illustration of Theorem 10.6.

Now, suppose that G has a proper 3-edge-coloring using colors a, b, c on the subgraphs E_a, E_b, and E_c. Since G is 3-regular, each color appears at every vertex. Thus, the union of any two of E_a, E_b, and E_c is 2-regular, which makes it a union of disjoint cycles. Each face of this subgraph is a union of faces of the original graph. Let $H_1 = E_a \cup E_b$ and $H_2 = E_b \cup E_c$. For $i \in \{1, 2\}$, we assign the color f_i to each face of G the parity of the number of cycles in H_i that contain it (0 for even, 1 for odd). Then, we set the function f as (f_1, f_2). We claim that f is a proper 4-face-coloring, as illustrated previously.

Suppose that two faces F and F' are separated by an edge e. Since G is 2-edge-connected, they are different faces. This edge belongs to a cycle C in at least one of H_1 or H_2. By the Jordan Curve Theorem, one of F and F' is inside C and the other is outside. In other words, $f_1(F) \neq f_1(F')$ if $e \in E_a \cup E_b$ and $f_2(F) \neq f_2(F')$ if $e \in E_b \cup E_c$. Hence, $f(F) \neq f(F')$. □

Due to this theorem, a proper 3-edge-coloring of a 3-regular graph is called a *Tait coloring*. Let G be a Hamiltonian 3-regular graph. Since G is 3-regular, $n(G)$ is even. We can alternatively color the edges on a Hamiltonian cycle C with two colors. Then all of the remaining edges are colored with the third color. We obtain a proper edge coloring of G. Hence, every Hamiltonian 3-regular graph has a Tait coloring. Tait believed that this gave a proof of the Four Color Theorem, because he assumed that every 3-connected 3-regular planar graph is Hamiltonian. Although the gap was noticed earlier, an explicit counterexample was found in 1946. Yet, the proof that a 3-connected 3-regular planar graph is not Hamiltonian is very tedious. Later, Grinberg [127] discovered a simple necessary condition that led to many 3-regular 3-connected non-Hamiltonian planar graphs.

THEOREM 10.7 **(Grinberg [127])** Suppose that G is a loopless plane graph having Hamiltonian cycle C, and G has f_i' faces of length i inside C and f_i'' faces of length i outside C. Then $\sum_i (i-2)(f_i' - f_i'') = 0$.

Proof: First, we prove that $\sum_i (i-2)f_i' = n-2$ by induction on the number of edges inside C.

Suppose that there are no edges inside C. Obviously, $f_i = 0$ if $i \neq n$ and $f_n = 1$. Hence, $\sum_i (i-2)f_i' = n-2$. Suppose that $\sum_i (i-2)f_i' = n-2$ for any graph with k

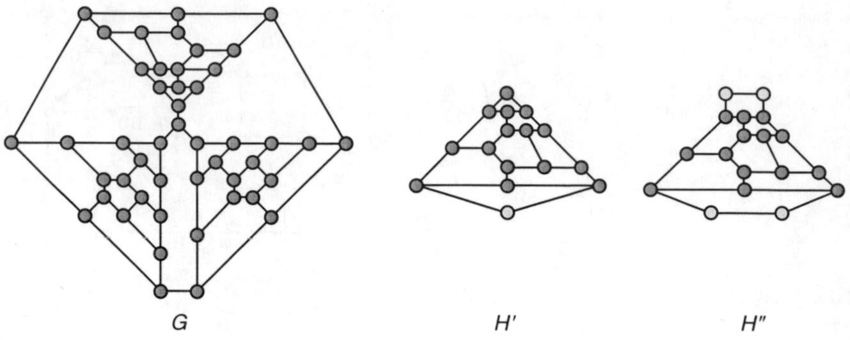

FIGURE 10.9 Graphs G, H', and H''.

edges inside C. Assume that G is a loopless plane graph having a Hamiltonian cycle C and having $(k+1)$ edges inside C. We can delete an edge e from the inside of C. Obviously, $G-e$ is a loopless plane graph having C as its Hamiltonian cycle and having k edges inside C. By induction, $\sum_i (i-2)g_i' = n-2$, where g_i' is the number of faces of length i for $G-e$ inside C. Now, we add e back to $G-e$ to obtain G. The edge addition cuts a face of some length r into two faces of lengths s and t. Then $s+t=r+2$, because the new edge contributes to each face and each of the edges on the old face contributes to one of the new faces. All other contributions to the sum remain unchanged. From this equality we obtain $(s-2)+(t-2)=(r-2)$, so that the total contribution from these faces is also the same as before. Thus, $\sum_i (i-2)f_i' = n-2$.

Since the inside and outside are symmetric with respect to C, $\sum_i (i-2)f_i'' = n-2$. Thus, $\sum_i (i-2)(f_i' - f_i'') = 0$. □

Grinberg's condition is used to show that graphs are not Hamiltonian. The arguments can often be simplified by using modular arithmetic. We apply such arguments to the first known non-Hamiltonian 3-connected 3-regular planar graph (Tutte [313]). Originally, Tutte used an *ad hoc* argument to prove that this graph is not Hamiltonian. For many years, this graph was the only known example.

The Tutte graph G is shown in Figure 10.9. Let H denote each component of the subgraph obtained by deleting the central vertex and the three long edges. Any Hamiltonian cycle must visit the central vertex of G. Thus, it must contain a Hamiltonian path in one copy of H between the two other entrances to that graphs. By adding a path of length 2 between the desired endpoints of the Hamiltonian path in H we obtain the graph H', shown in Figure 10.9. Obviously, H has a Hamiltonian path with the desired entrance if and only if H' is Hamiltonian. Furthermore, H' is Hamiltonian if and only if the graph H'' shown in Figure 10.9 is Hamiltonian.

Obviously, there are seven 5-faces, one 4-face, and one 11-face in H''. Grinberg's condition becomes $2a_4 + 3a_5 + 9a_9 = 0$, where $a_i = f_i' - f_i''$. Thus, $2a_4 \equiv 0$ **mod** 3. Since there is only one 4-face, $a_4 = \pm 1$. However, the equation $2 \times (\pm 1) \equiv 0$ **mod** 3 is impossible. Thus, G is not Hamiltonian.

11 Optimal k-Fault-Tolerant Hamiltonian Graphs

11.1 INTRODUCTION

An interconnection network connects the processors of the parallel computer. Its architecture can be represented as a graph in which the vertices correspond to the processors and the edges to the communication links. Hence, we use *graph* and *network* interchangeably. There are a lot of mutually conflicting requirements in designing the topology of computer networks. It is almost impossible to design a network that is optimal from all aspects. One has to design a suitable network depending on the requirements and their properties. The Hamiltonian properties are one of the major requirements in designing the topology of a network. For example, the *Token ring* approach is used in some distributed operation systems. An interconnection network requires the presence of Hamiltonian cycles in the structure to meet this approach. Fault-tolerance is also a desirable feature in massive parallel systems that have a relatively high probability of failure. A number of fault-tolerant designs for specific multiprocessor architectures have been proposed based on graph-theoretic models, in which the processor-to-processor interconnection structure is represented by a graph.

When faults occur in a network, it corresponds to removing edges and vertices from the graph. Let $G = (V, E)$ be a graph and let $V' \subseteq V$ and $E' \subseteq E$. We use $G - V'$ to denote the subgraph of G induced by $V - V'$, and $G - E'$ the subgraph obtained by removing E' from G. Faults can be in the combination of vertices and edges. Let $F \subseteq V \cup E$. We use $G - F$ to denote the subgraph induced by $V - F$ and deleting the edges in F from the induced subgraph.

Suppose that $G - V'$ is Hamiltonian for any $V' \subseteq V$ and $|V'| \leq k$. Then G is called a *k-vertex fault-tolerant Hamiltonian* graph. An n-vertex k-vertex fault-tolerant Hamiltonian graph is *optimal* if it contains the least number of edges among all n-vertex k-vertex fault-tolerant Hamiltonian graphs. Obviously, every k-vertex fault-tolerant Hamiltonian graph has at least $k + 3$ vertices. Moreover, the degree of every vertex in a k-vertex fault-tolerant Hamiltonian graph is at least $k + 2$. The *vertex fault-tolerant Hamiltonicity*, $\mathcal{H}_v(G)$, is defined as the maximum integer l such that $G - F$ remains Hamiltonian for every $F \subset V(G)$ with $|F| \leq l$ if G is Hamiltonian, and undefined if otherwise. Obviously, $\mathcal{H}_v(G) \leq \delta(G) - 2$. Moreover, a r-regular graph G is optimal vertex-fault-tolerant Hamiltonian if $\mathcal{H}_v(G) = r - 2$.

Suppose that $G - E'$ is Hamiltonian for any $E' \subseteq E$ and $|E'| \leq k$. Then G is called a *k-edge fault-tolerant Hamiltonian* graph. An n-vertex k-edge fault-tolerant Hamiltonian graph is *optimal* if it contains the least number of edges among all n-vertex k-edge fault-tolerant Hamiltonian graphs. Obviously, every k-edge fault-tolerant Hamiltonian graph has at least $k+3$ vertices. Moreover, the degree of every vertex in a k-edge fault-tolerant Hamiltonian graph is at least $k+2$. The *edge fault-tolerant Hamiltonicity*, $\mathcal{H}_e(G)$, is defined as the maximum integer l such that $G-F$ remains Hamiltonian for every $F \subset E(G)$ with $|F| \leq l$ if G is Hamiltonian, and undefined if otherwise. Obviously, $\mathcal{H}_e(G) \leq \delta(G)-2$. Moreover, a r-regular graph G is optimal edge-fault-tolerant Hamiltonian if $\mathcal{H}_e(G)=r-2$.

Suppose that $G-F$ is Hamiltonian for any $F \subseteq V \cup E$ and $|F| \leq k$; then G is called a *k-fault-tolerant Hamiltonian* graph. An n-vertex k-fault-tolerant Hamiltonian graph is *optimal* if it contains the least number of edges among all n-vertex k-fault-tolerant Hamiltonian graphs. Obviously, every k-fault-tolerant Hamiltonian graph has at least $k+3$ vertices. Moreover, the degree of every vertex in a k-fault-tolerant Hamiltonian graph is at least $k+2$. The *fault-tolerant Hamiltonicity*, $\mathcal{H}_f(G)$, is defined as the maximum integer l such that $G-F$ remains Hamiltonian for every $F \subset V(G) \cup E(G)$ with $|F| \leq l$ if G is Hamiltonian, and undefined if otherwise. Obviously, $\mathcal{H}_f(G) \leq \min\{\mathcal{H}_v(G), \mathcal{H}_e(G)\} \leq \delta(G)-2$. Moreover, a r-regular graph G is optimal fault-tolerant Hamiltonian if $\mathcal{H}_f(G)=r-2$.

The design of *k-fault-tolerant Hamiltonian* graph is equivalent to k-fault-tolerant design for token rings. Previous results have been focused mostly on the construction of either optimal k-vertex fault-tolerant Hamiltonian graph or k-edge fault-tolerant Hamiltonian graphs. Several families of optimal 1-fault-tolerant Hamiltonian graphs are proposed in Refs 139, 140, 234, and 256.

Let n and k be positive integers with $n \geq k+3$. Obviously, the complete graph K_n is k-fault-tolerant Hamiltonian. Thus, there exists an n-vertex k-fault-tolerant Hamiltonian graph for any integer n with $n \geq k+3$. Suppose that n and k are positive integers such that nk is even. Let $G=(V,E)$ be an n-vertex $(k+2)$-regular k-fault-tolerant Hamiltonian graph. From the previous discussion, G is an optimal k-fault-tolerant Hamiltonian graph. However, we have difficulty showing that any optimal k-fault-tolerant Hamiltonian graph is $(k+2)$-regular for $k \geq 3$. Similarly, suppose that n and k are positive integers such that nk is odd. Let $G=(V,E)$ be an n-vertex k-fault-tolerant Hamiltonian graph such that there is exactly one vertex $y \in V$ such that $\deg_G(y)=k+3$ and $\deg_G(x)=k+2$ for any $x \in V-\{y\}$. Since the number of vertices with odd degree in any graph is even, G is an optimal k-fault-tolerant Hamiltonian graph. Again, we cannot conclude that there is exactly one vertex of degree $k+3$ and all the remaining vertices are of degree $k+2$ in any optimal k-fault-tolerant Hamiltonian graph for $k \geq 3$. These two problems can be solved once we prove that every Harary graph $H_{k,n}$ is an optimal $(k-2)$-fault-tolerant Hamiltonian graph for $k \geq 3$.

Wong and Wong [345] and Paoli et al. [263] proved that the Harary graph $H_{k,n}$, introduced in Example 7.2, is optimal $(k-2)$-vertex fault-tolerant Hamiltonian and optimal $(k-2)$-edge fault-tolerant Hamiltonian for n being even and odd, respectively. Although the graph $H_{k,n}$ is both optimal $(k-2)$-vertex fault-tolerant Hamiltonian and optimal $(k-2)$-edge fault-tolerant Hamiltonian, it is not necessary for $H_{k,n}$ to be optimal $(k-2)$-fault-tolerant Hamiltonian. Sung et al. [291] showed that $H_{k,n}$ is optimally $(k-2)$-fault-tolerant Hamiltonian for $k=4,5$ and conjectured that the same is true for all $k \geq 6$.

Now, we concentrate on the construction of k-fault-tolerant Hamiltonian graphs. To make it simple, we concentrate only on the regular k-fault-tolerant Hamiltonian graphs. Thus nk is even, where n is the number of vertices.

11.2 NODE EXPANSION

The following theorem is proved by Ore [261].

THEOREM 11.1 **(Ore [261])** Assume that G is an n-vertex graph with $n \geq 4$. Then G is Hamiltonian if $\overline{e} \leq n-3$.

LEMMA 11.1 Assume that $n \geq 4$. Then K_n is $(n-3)$-fault-tolerant Hamiltonian.

Proof: Suppose that F is any subset of $V(K_n) \cup E(K_n)$. We use F_V to denote $F \cap V(K_n)$. Then $K_n - F$ is isomorphic to $K_{n-f} - F'$ where $f = |F_V|$ and F' is a subset of edges in the subgraph of K_n induced by $\langle n \rangle - F_V$. Obviously, $|F'| \leq |F| - f$. Thus, if $|F| \leq n-i$ then $\overline{E}(K_{n-f} - F') = |F'| \leq |F| - f \leq (n-f) - i$. Note that $n-f$ is the number of vertices of $K_{n-f} - F'$. The lemma follows from Theorem 11.1. □

COROLLARY 11.1 *The graph $K_n - F$ has a Hamiltonian path for $F \subset E(K_n)$ with $|F| \leq n-2$.*

Proof: Choose any $f \in F$. We set $F' = F - \{f\}$. Obviously, $|F'| \leq n-3$. Thus, $K_n - F'$ has a Hamiltonian cycle C. Suppose that $f \in C$. We delete f from C to obtain a path P. Obviously, P is a Hamiltonian path of $K_n - F$. Suppose $f \notin C$, we delete any edge of C to obtain a path P. Again, P is a Hamiltonian path of $K_n - F$. The corollary is proved. □

THEOREM 11.2 Let $K_n = (V, E)$ be the complete graph with n vertices and $F \subset V \cup E$ be a faulty set with $|F| \leq n-2$. There exists a set $V' \subseteq V(K_n - F)$ with $|V'| = n - |F|$ such that every pair of vertices in V' can be joined by a Hamiltonian path of $K_n - F$.

Proof: We prove this theorem by induction on n. This statement can be easily verified for $n = 3$ and 4. Assume that the statement holds for all K_j with $3 \leq j \leq n-1$ and $n \geq 5$.

First, we consider that $|F \cap V(K_n)| = i > 0$. Then the graph $K_n - F$ is isomorphic to $K_{n-i} - F'$ for some $|F'| \leq |F| - i$. By induction hypotheses, there exists a set $V' \subseteq V$ with $|V'| = n - i - |F'| \geq n - |F|$ such that every pair of vertices in V' can be joined by a Hamiltonian path of $K_{n-i} - F'$. Thus, the statement is also true for the graph $K_n - F$, since $K_{n-i} - F'$ is isomorphic to $K_n - F$.

Next, we consider that $F \subset E$. Suppose that $|F| = n-2$. It follows from Corollary 11.1 that the graph $K_n - F$ has a Hamiltonian path. Thus, there exists a set $V' \subset V$ with $|V'| = 2$ such that the pair of vertices in V' can be joined by a Hamiltonian path of $K_n - F$. Now consider that $F \subset E$ and $|F| \leq n-3$. Let H denote the subgraph of K_n given by (V, F). Since $\sum_{v \in V} \deg_H(v) \leq 2(n-3)$, there exists a vertex $v \in V$ with $\deg_H(v) \leq 1$. We distinguish the following two cases:

Case 1. There exists a vertex v with $\deg_H(v) = 0$. In other words, all of the edges in $K_n - F$ incident at v are not in F. Thus, the graph $K_n - v - F$ is isomorphic to $K_{n-1} - F$. By induction hypotheses, there exists a subset $V' \subseteq (V - \{v\})$ with $|V'| = n - 1 - |F|$ such that every two distinct vertices x and y in V' can be joined by a Hamiltonian path of $K_n - v - F$. Let P be a Hamiltonian path of $K_n - v - F$

joining x to y, which is written as $\langle x, x', P', y\rangle$, where x' is a vertex adjacent to x and P' is a path from x' to y. Then $\langle x, v, x', P', y\rangle$ and $\langle v, x, x', P', y\rangle$ form two Hamiltonian paths of $K_n - F$ from x to y and from v to y, respectively. Since x and y are arbitrary vertices in V', there always exists a Hamiltonian path of $K_n - F$ joining every pair of vertices in $V' \cup \{v\}$. Thus, this statement is true.

Case 2. There exists a vertex v with $\deg_H(v) = 1$. Since there is exactly one edge of $K_n - F$ incident at v which is also in F, it follows that the graph $K_n - v - F$ is isomorphic to the graph $K_{n-1} - F^*$ where $|F^*| = |F| - 1$. By induction hypotheses, there exists a subset $V' \subseteq (V - \{v\})$ with $|V'| = n - |F|$ such that every pair of vertices x and y in V' can be joined by a Hamiltonian path of $K_n - v - F$. Let P^* be a Hamiltonian path of $K_n - v - F$ joining x and y, which is written as $\langle x = u_0, u_1, \ldots, u_{n-2} = y\rangle$. Since $n \geq 5$, there exists u_j, $0 \leq j \leq n-3$, such that $(v, u_j) \notin F$ and $(v, u_{j+1}) \notin F$. Then $\langle x = u_0, u_1, \ldots, u_j, v, u_{j+1}, \ldots, u_{n-2} = y\rangle$ forms a Hamiltonian path of $K_n - F$ joining x to y. Hence, every pair of vertices in V' can be joined by a Hamiltonian path of $K_n - F$.

Thus, the theorem is proved. □

Let $G = (V, E)$ be any graph with x be a vertex in V with degree t and the set $\{x_1, x_2, \ldots, x_t\}$ consists of the neighborhood vertices of x. The *t-node expansion* $X(G, x)$ *of* G *on* x is the graph obtained from G by replacing x by the complete graph K_t, where $V(K_t) = \{k_1, k_2, \ldots, k_t\}$, with the edges (x, x_i), $i = 1, 2, \ldots, t$ deleted from G and the edges $(k_i, x_i), i = 1, 2, \ldots, t$ added to $X(G, x)$. More precisely, $V(X(G, x)) = (V - \{x\}) \cup \{k_1, k_2, \ldots, k_t\}$ and $E(X(G, x)) = (E - \{(x, x_i) \mid 1 \leq i \leq t\}) \cup \{(k_i, x_i) \mid 1 \leq i \leq t\} \cup \{(k_i, k_j) \mid 1 \leq i \neq j \leq t\}$.

Note that $\deg_{X(G,x)}(v) = \deg_G(x) = t$ for all $v \in V(K_t)$ and $\deg_{X(G,x)}(u) = \deg_G(u)$ for all $u \in (V - \{x\})$. In particular, $X(G, x)$ is also t-regular if G is t-regular. The graphs G and $X(G, x)$ are illustrated in Figure 11.1. Let $N^*_G(x)$ be the subgraph induced by $\{x\} \cup \{(x, x_i) \mid \text{for all } 1 \leq i \leq t\}$ of G and $M_G(x)$ be the set of $V(K_t) \cup E(K_t) \cup \{(k_i, x_i) \mid i = 1, 2, \ldots, t\}$ of G.

LEMMA 11.2 Let $G = (V, E)$ be any graph with x be a vertex in V with degree t and $F_1 \subset (V(G - x) \cup E(G - x))$. Suppose that we delete any f edges of $N^*_G(x)$ from

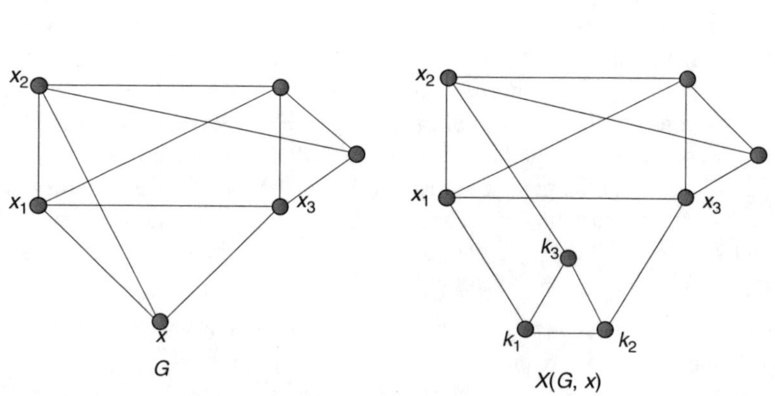

FIGURE 11.1 The graphs G and $X(G, x)$.

the graph $G - F_1$ such that the remaining graph is Hamiltonian for $f + |F_1| \le t - 2$. Then the graph $X(G,x) - (F_1 \cup F_3)$ is Hamiltonian where F_3 is a subset of $M_G(x)$ and $|F_3| = f$.

Proof: Since $F_3 \subset M_G(x)$ and $|F_3| \le t - 2$, there exists a set $V' \subset V(K_t)$ of size $t - f'$ such that every two distinct vertices in V' can be joined by a Hamiltonian path of the graph $K_t - F_3$ for $f' = |F_3 \cap (V(K_t) \cup E(K_t))|$. We define a faulty set F_2 of G as follows:

$$F_2 = \{(x, x_i) \mid k_i \notin V' \text{ or } (x_i, k_i) \in F_3, 1 \le i \le t\}$$

Thus, $|F_2| \le (|V(K_t)| - |V'|) + (|F_3| - f') = t - (t - f') + (f - f') = f \le t - 2$. The graph $(G - F_1) - F_2$ is Hamiltonian because we delete any f edges of $N_G^*(x)$ from $G - F_1$ such that the remaining graph is Hamiltonian. Thus, there is a Hamiltonian cycle $C = \langle x_i, x, x_j, P, x_i \rangle$ in the graph $G - (F_1 \cup F_2)$ where P is a path from x_j to x_i. By the definition of F_2, k_i and k_j are in V' and $(x_i, k_i), (x_j, k_j)$ are not in F_3. Thus, there exists a Hamiltonian path P' joining k_i and k_j in the graph $K_t - F_3$. Therefore, $\langle x_i, k_i, P', k_j, x_j, P, x_i \rangle$ forms a Hamiltonian cycle in the graph $X(G,x) - (F_1 \cup F_3)$. This lemma is proved. □

THEOREM 11.3 Let x be a vertex of $G = (V, E)$ with $\deg_G(x) = k + 2$. Then $X(G,x)$ is k-fault-tolerant Hamiltonian if G is k-fault-tolerant Hamiltonian.

Proof: Let F be any faulty set of the graph $X(G,x)$ where $|F| \le k$. We set $F_1 = F \cap (V(G - x) \cup E(G - x))$ and $F_3 = F - F_1$. Since G is k-fault-tolerant Hamiltonian, the graph $G - (F_1 \cup F_2)$ is Hamiltonian for every $F_2 \subset E(N_G^*(x))$ with $|F_2| = |F| - |F_1| = |F_3|$. Applying Lemma 11.2, the graph $X(G,x) - (F_1 \cup F_3)$ is Hamiltonian. Therefore, $X(G,x)$ is k-fault-tolerant Hamiltonian. The theorem is proved. □

COROLLARY 11.2 *Let $G = (V,E)$ be an n-vertex $(k + 2)$-regular k-fault-tolerant Hamiltonian graph. Then $X(G,x)$ is optimal k-fault-tolerant Hamiltonian for any vertex $x \in V$.*

Applying Theorem 11.3, we can obtain other optimal k-fault-tolerant Hamiltonian graphs from some known optimal k-fault-tolerant Hamiltonian graphs by $(k+2)$-node expansion.

The node expansion of $G = (V, E)$ on the set $U \subset V$, denoted by $X(G, U)$, is a graph that is obtained from G by a sequence of node-expansion operations on every vertex $u \in U$.

LEMMA 11.3 Let $G = (V, E)$ be a $(k + 2)$-regular and k-edge fault-tolerant Hamiltonian. Then $X(G, U) - F$ is Hamiltonian if $F \subset ((V(X(G, U)) \cup E(X(G, U)) - V)$ with $U \subseteq V$ and $|F| \le k$.

Proof: Let v be a vertex of U and $U' = U - \{v\}$. Assume that the graph $X(G, U') - F'$ is Hamiltonian for every $F' \subset ((V(X(G, U')) \cup E(X(G, U')) - V)$ for $|F'| \le k$. Let $F_3 = F \cap M_{X(G,U')}(v)$ and $F_1 = F - F_3$. Therefore, the graph which is deleted any $|F_3|$

edges, denoted by F_2, of $N^*_{X(G,U')}(v)$ from the graph $G - F_1$ is Hamiltonian because $F_1 \cup F_2$ is a subset of $(V(X(G,U')) \cup E(X(G,U')) - V$ and $|F_1 \cup F_2| = |F| \leq k$. By Lemma 11.2, the graph $X(G,U) - (F_1 \cup F_3)$ is Hamiltonian because $F = F_1 \cup F_3$ is an arbitrary subset of $(V(X(G,U)) \cup E(X(G,U)) - V$ and $|F| \leq k$. Thus, this lemma is proved. □

THEOREM 11.4 Suppose that the graph $G = (V, E)$ is $(k+2)$-regular and optimal k-edge fault-tolerant Hamiltonian. Then the graph $X(G, V)$ is $(k+2)$-regular and optimal k-fault-tolerant Hamiltonian.

Proof: By Lemma 11.3, $X(G,V) - F$ is Hamiltonian for every $F \subset ((V(X(G,V)) \cup E(X(G,V)) - V)$ for $|F| \leq k$. In fact, $(V(X(G,X)) \cup E(X(G,X))) \cap V = \emptyset$. Thus, $(V(X(G,X)) \cup E(X(G,X))) - V = V(X(G,X)) \cup E(X(G,X))$. Therefore, $X(G,V)$ is k-fault-tolerant Hamiltonian. Moreover, $X(G,V)$ is optimal k-fault-tolerant Hamiltonian, since it is $(k+2)$-regular. Hence, this theorem is proved. □

The following theorem is proved in Refs 51 and 281.

THEOREM 11.5 The n-dimensional hypercube Q_n is $(n-2)$-edge fault-tolerant Hamiltonian for $n \geq 2$.

By Theorem 11.4, we have the following corollary.

COROLLARY 11.3 *The graph $X(Q_n, V)$ is an optimal $(n-2)$-fault-tolerant Hamiltonian and vertex symmetric graph with $n \cdot 2^n$ vertices, degree n, and diameter $2n$ for $n \geq 2$.*

The following theorem is proved in Ref. 307.

THEOREM 11.6 The star graph S_n is $(n-3)$-edge fault-tolerant Hamiltonian for $n \geq 3$.

Again, we have the following corollary.

COROLLARY 11.4 *The graph $X(S_n, V)$ is an optimal $(n-3)$-fault-tolerant Hamiltonian and vertex-symmetric graph with $n \cdot n!$ vertices, degree $(n-1)$, and diameter $2\lfloor 3(n-1)/2 \rfloor$ for $n \geq 3$.*

With Corollary 11.2, we can easily obtain other optimal k-fault-tolerant Hamiltonian graphs from an known optimal k-fault-tolerant Hamiltonian graph by $(k+2)$-node expansions on a vertex of degree $k+2$. Note that the complete graph K_{k+3} of $k+3$ vertices is $(k+2)$-regular and is the smallest optimal k-fault-tolerant Hamiltonian graph. The graphs obtained by a sequence of $(k+2)$-node expansions from K_{k+3} are also $(k+2)$-regular, and thus optimal k-fault-tolerant Hamiltonian. One

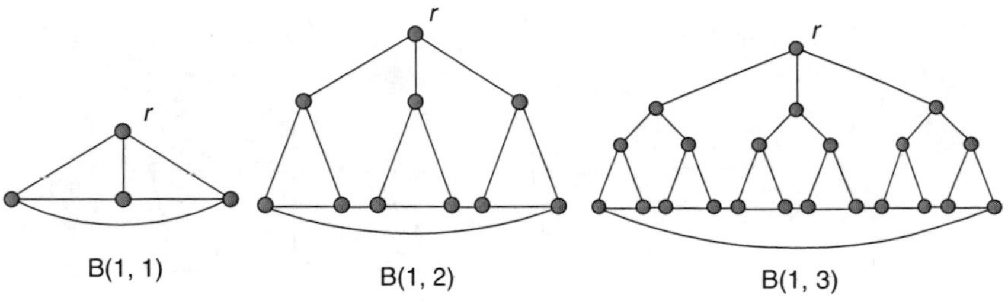

FIGURE 11.2 The graphs $B(1, 1), B(1, 2)$, and $B(1, 3)$.

possible sequence of $(k+2)$-node expansions to construct the optimal k-fault-tolerant Hamiltonian graphs $B(k, s)$ is as follows:

Procedure $B(k, s)$

1. $G = K_{k+3}$
2. pick any vertex r as the root of G
3. for $i = 1$ to $s - 1$ do
4. $B = \{v \mid d(v, r) = i\}$
5. For all $v \in B$
6. $G = X(G, v)$

The graphs $B(1, 1), B(1, 2)$, and $B(1, 3)$ are shown in Figure 11.2. The vertex labeled r indicates the root assigned by Procedure $B(k, s)$. It can be verified that the number of vertices in $B(k, s)$ is $[(k+2)(k+1)^s - 2]/k$. Moreover, the distance between a vertex v to the root r is at most s. Therefore, the diameter of $B(k, s)$ is at most $2s$. Thus, we have constructed a family of optimal k-fault-tolerant Hamiltonian graphs with diameter $2 \log_{k+1} n - c$.

Now, we concentrate on $k = 1$ and present some examples of cubic 1-Hamiltonian graphs. Let m be an even integer. Harary and Hayes [139,140] proposed a family of cubic 1-fault-tolerant Hamiltonian graphs, denoted by $H(m)$ for $m \geq 4$, where $V(H(m)) = \{0, 1, 2, \ldots, m-1\}$ and $E(H(m)) = \{(i, i+1) \mid 0 \leq i \leq m-1\} \cup \{(0, m/2), (0, m-1)\} \cup \{(i, m-i) \mid 1 \leq i \leq m/2 - 1\}$. Examples of $H(4)$ and $H(8)$ are shown in Figure 11.3. Note that $H(4)$ is indeed a complete graph K_4, which is the smallest 1-fault-tolerant Hamiltonian graph.

Mukhopadhyaya and Sinha [256] proposed another family of cubic 1-fault-tolerant Hamiltonian graphs, denoted by $M(m)$ with m even and $m \geq 4$. Let $m \geq 4$ be an integer and t be a nonnegative integer. To define $M(m)$, we first introduce $MS(i, t)$, as illustrated in Figure 11.4, which is defined as follows:

$$
\begin{aligned}
V(MS(i,t)) &= \left\{x^r_{i,j} \mid 1 \leq j \leq t\right\} \cup \left\{x^l_{i,j} \mid 1 \leq j \leq t\right\} \cup \{y_i, z_i\} \\
E(MS(i,t)) &= \left\{\left(x^l_{i,j}, x^l_{i,j-1}\right) \mid 1 < j \leq t\right\} \cup \left\{\left(x^r_{i,j}, x^r_{i,j+1}\right) \mid 1 \leq j \leq t-1\right\} \\
&\quad \cup \left\{\left(x^l_{i,j}, x^r_{i,j}\right) \mid 1 \leq j \leq t\right\} \cup \left\{\left(x^l_{i,1}, y_i\right), (y_i, z_i), \left(y_i, x^r_{i,1}\right)\right\}
\end{aligned}
$$

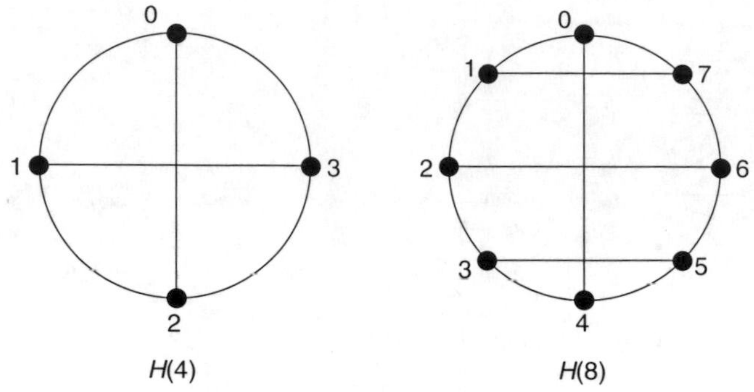

FIGURE 11.3 The graphs $H(m)$.

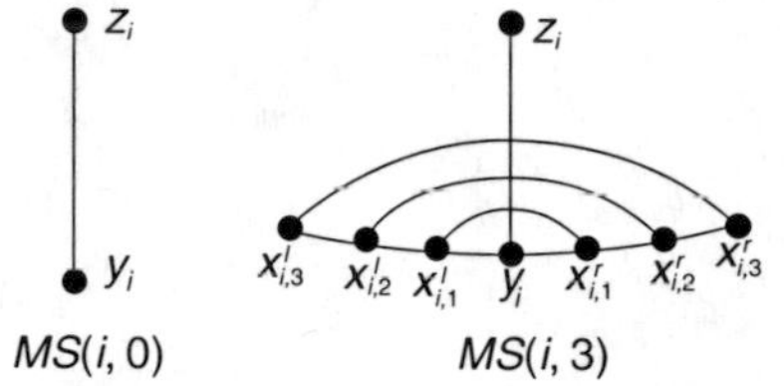

FIGURE 11.4 The graphs $MS(i,t)$.

It can be verified that $MS(i,t)$, for $t \geq 2$, is isomorphic to the graph obtained from performing a 3-node expansion on the vertex y_i of $MS(i,t-1)$. The graph $M(m)$ for m even is constructed as follows (for convenience, we write $y_i = x^l_{i,0} = x^r_{i,0}$ if $t = 0$):

1. Suppose that $m = 6t+4$. Then $M(m)$ is constructed from three $MS(i,t)$ for all $0 \leq i \leq 2$ by identifying z_1, z_2, and z_3 into a single vertex z and adding the edges $(x^l_{0,t}, x^r_{1,t})$, $(x^l_{1,t}, x^r_{2,t})$, and $(x^l_{2,t}, x^r_{0,t})$.
2. Suppose that $m = 6t+6$. Then $M(m)$ is constructed from $MS(0,t+1)$, $MS(1,t)$, and $MS(2,t)$ by identifying z_1, z_2, and z_3 into a single vertex z and adding the edges $(x^l_{0,t+1}, x^r_{1,t})$, $(x^l_{1,t}, x^r_{2,t})$, and $(x^l_{2,t}, x^r_{0,t+1})$.
3. Suppose that $m = 6t+8$. Then $M(m)$ is constructed from $MS(0,t+1)$, $MS(1,t+1)$, and $MS(2,t)$ by identifying z_1, z_2, and z_3 into a single vertex z and adding the edges $(x^l_{0,t+1}, x^r_{1,t+1})$, $(x^l_{1,t+1}, x^r_{2,t})$, and $(x^l_{2,t}, x^r_{0,t+1})$. See Figure 11.5.

Note that $M(4)$ is indeed a complete graph K_4.

Wang et al. [333] also presented a family of 1-fault-tolerant Hamiltonian graphs $W(m)$, as illustrated in Figure 11.6, which is constructed from $MS(i,m)$ for $0 \leq i \leq 2m$ by joining all z_i with a cycle $\langle z_0, z_1, \ldots, z_{2m}, z_0 \rangle$ and adding edges $(x^l_{2m,m}, x^r_{0,m})$ and $(x^l_{i,m}, x^r_{i+1,m})$ for all $0 \leq i \leq 2m-1$. Let $D(m)$ denote the graph obtained from

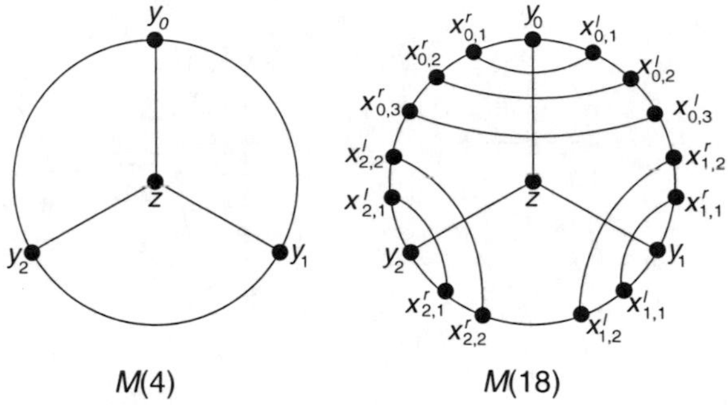

FIGURE 11.5 The graphs $M(m)$.

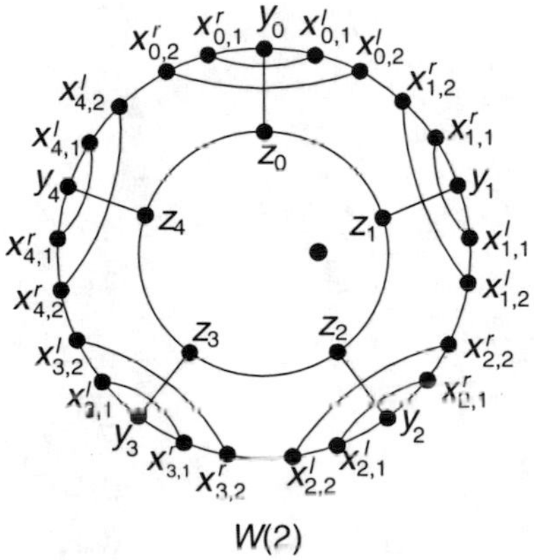

FIGURE 11.6 The graph $W(2)$.

$MS(i,0)$ for $0 \leq i \leq 2m$ by joining all z_i to y_i with cycles $\langle z_0, z_1, \ldots, z_{2m}, z_0\rangle$ and $\langle y_0, y_1, \ldots, y_{2m}, y_0\rangle$, respectively.

Note that $H(m)$ with m even and $m \geq 4$ can be constructed from K_4 by a sequence of node expansion as illustrated in Figure 11.7b; $M(m)$ with m even and $m \geq 4$ can be constructed from K_4 by a sequence of node expansions as illustrated in Figure 11.7d; and $W(m)$ with any integer m can be constructed from $D_m = C_{2m+1} \times K_2$ by a sequence of node expansions as illustrated in Figure 11.7f.

The family of 1-fault-tolerant Hamiltonian graphs $\{B(1,s) \mid s \text{ is a positive integer}\}$ is called *Christmas tree* [182]. Obviously, $B(1,1) = K_4$ and $B(1,s)$ can be constructed from K_4 by a sequence of node expansions.

Let n be the number of vertices in a graph. We note that the diameter of $H(m)$ is $\lfloor \frac{n}{4} \rfloor + 1$ if $n \geq 6$ and $H(4) = 1$, the diameter of $M(m)$ is $\lfloor \frac{n}{6} \rfloor + 2$ if $n \geq 8$, $M(6) = 2$, and $M(4) = 1$, the diameter of $W(m)$ is $O(\sqrt{n})$, and diameter for $B(1,s)$ is $2\log_2 n - c$. It is interesting to find other cubic 1-fault-tolerant Hamiltonian graphs with smaller diameters.

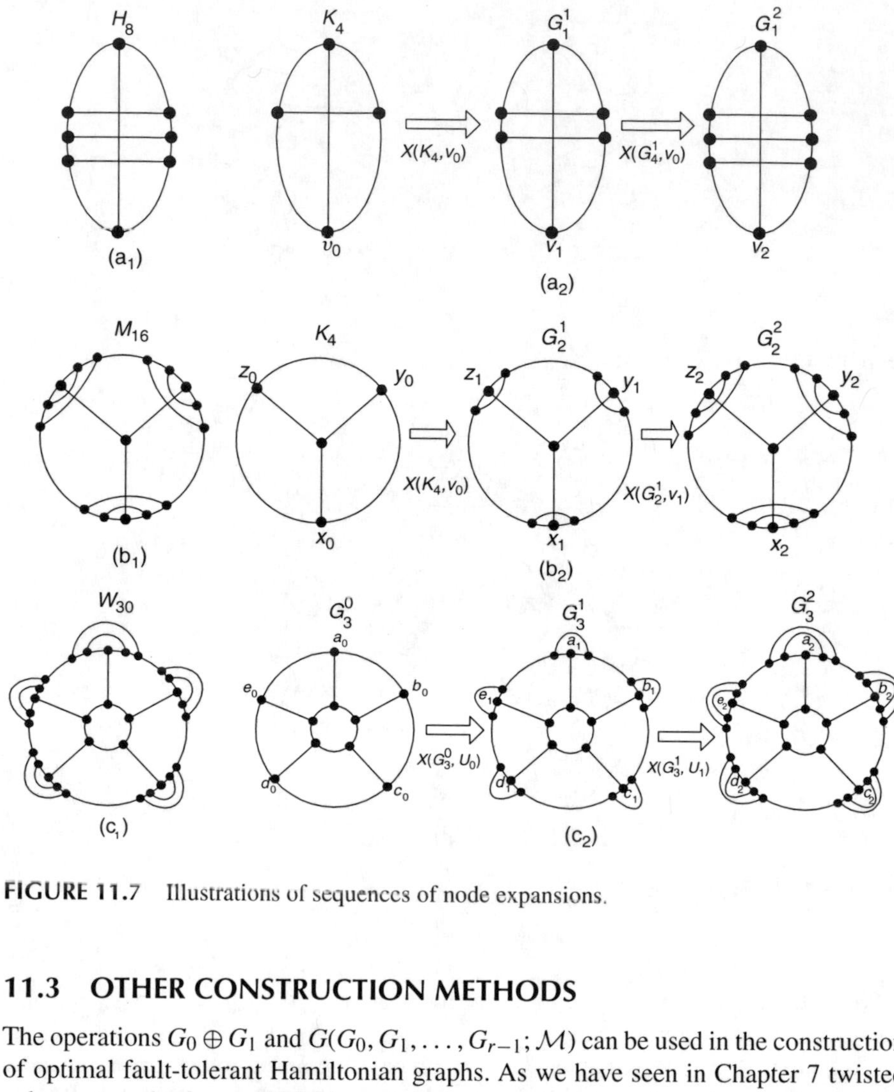

FIGURE 11.7 Illustrations of sequences of node expansions.

11.3 OTHER CONSTRUCTION METHODS

The operations $G_0 \oplus G_1$ and $G(G_0, G_1, \ldots, G_{r-1}; \mathcal{M})$ can be used in the construction of optimal fault-tolerant Hamiltonian graphs. As we have seen in Chapter 7 twisted cubes, crossed cubes, and Möbius cubes are recursively constructed with this scheme.

Yet, we need another concept called *fault-tolerant Hamiltonian-connected.* A graph G is called *l-fault-tolerant Hamiltonian-connected* if it remains Hamiltonian-connected after removing at most l vertices or edges. The fault-tolerant Hamiltonian connectivity, $\mathcal{H}_f^\kappa(G)$, is defined to be the maximum integer l such that $G - F$ remains Hamiltonian-connected for every $F \subset V(G) \cup E(G)$ with $|F| \leq l$ if G is Hamiltonian-connected, and undefined if otherwise. Obviously, $\mathcal{H}_f^\kappa(G) \leq \delta(G) - 3$. A r-regular graph G is *optimal fault-tolerant Hamiltonian-connected* if $\mathcal{H}_f^\kappa(G) = r - 3$.

The following theorem is proved by Ore [261].

THEOREM 11.7 **(Ore [261])** Assume that G is an n-vertex graph with $n \geq 4$. Then G is Hamiltonian-connected if $\overline{e} \leq n - 4$.

LEMMA 11.4 Assume that $n \geq 4$. Then K_n is $(n-4)$-fault-tolerant Hamiltonian-connected.

We say a k-regular graph is *super fault-tolerant Hamiltonian* if $\mathcal{H}_f(G) = k-2$ and $\mathcal{H}_f^{\kappa}(G) = k-3$.

Twisted cubes [180], crossed cubes [178], and Möbius cubes [179] are proved to be super fault-tolerant Hamiltonian. Chen et al. [54] observed the insight of the proofs of the aforementioned results and proposed the following result.

A vertex v is *healthy* if a vertex v is not faulty. An edge e (respectively, a matching edge e) is *healthy* if both edge e and its two endpoints are not faulty. We use F_i to denote the set of faults in G_i for $i = 0, 1$. Let $f_i = |F_i|$ for $i = 0, 1$.

Consider an interconnection network G, and suppose that there are faults in it. Let F_G be the set of faults in G, and $f_G = |F_G|$. Suppose that G is k-fault-tolerant Hamiltonian (k-fault-tolerant Hamiltonian-connected, respectively) and $f_G \leq k$. Suppose that u is a healthy vertex in G. It is clear that some of the edges incident to u are on a Hamiltonian cycle (Hamiltonian path, respectively) of $G - F_G$, but not every edge incident to u is on some Hamiltonian cycle (Hamiltonian path, respectively) of $G - F_G$.

LEMMA 11.5 Suppose that G is a k-fault-tolerant Hamiltonian graph, F_G is a set of faults in G with $|F_G| = f_G \leq k$, and u is a healthy vertex in G. Then there are at least $k - f_G + 2$ edges incident to vertex u, such that each one of them is on some Hamiltonian cycle in $G - F_G$.

Proof: Since G is k-fault-tolerant Hamiltonian and there are f_G faults in G, $G - F_G$ is still Hamiltonian even if we add $k - f_G$ more faults to $G - F_G$. Suppose $f_G < k$. Let C be a Hamiltonian cycle in $G - F_G$, and let e be an edge on C incident to vertex u. Deleting edge e, $G - F_G - \{e\}$ still contains a Hamiltonian cycle. Repeating this process $k - f_G$ times, we find $k - f_G + 2$ edges incident to vertex u, and each one of them is on some Hamiltonian cycle in $G - F_G$. □

LEMMA 11.6 Suppose that G is a k-fault-tolerant Hamiltonian-connected graph, F_G is a set of faults in G with $|F_G| = f_G \leq k$, and $\{x, y, u\}$ are three distinct healthy vertices in G. Then there are at least $k - f_G + 2$ edges incident to vertex u, such that each one of them is on some Hamiltonian path from x to y in $G - F_G$.

Proof: Since G is k-fault-tolerant Hamiltonian-connected and there are f_G faults in G, $G - F_G$ is still Hamiltonian-connected even if we add $k - f_G$ more faults to $G - F_G$. Suppose that $f_G < k$. Let P be a Hamiltonian path of $G - F_G$ joining x to y, and let e be an edge on P and incident to vertex u. Deleting edge e, $G - F_G - \{e\}$ still contains a Hamiltonian path joining x to y. Repeating this process $k - f_G$ times, we find $k - f_G + 2$ edges incident to vertex u, and each one of them is on some Hamiltonian path of $G - F_G$ joining x to y. □

LEMMA 11.7 Suppose that G_0 and G_1 are two k-regular graphs with the same number of vertices. Suppose that the total number of faults in $G_0 \oplus G_1$ is not greater than k. There exists at least one healthy matching edge $(z, \bar{z})$ between G_0 and G_1.

LEMMA 11.8 Suppose that G_0 and G_1 are two k-regular graphs with the same number of vertices. Let x and y be two healthy vertices in $G_0 \oplus G_1$. Suppose that the total number of faults in $G_0 \oplus G_1$ is not greater than $k-2$. There exists at least one healthy matching edge $(z, \bar{z})$ between G_0 and G_1 such that $\{x, y\} \cap \{z, \bar{z}\} = \emptyset$.

The Lemmas 11.7 and 11.8 result immediately from the fact that $|V(G_0)| = |V(G_1)| \geq k+1$.

OBSERVATION: To prove that a graph G is l-fault-tolerant Hamiltonian (respectively l-fault-tolerant Hamiltonian-connected), it suffices to show that $G - F_G$ is Hamiltonian (respectively Hamiltonian-connected) for any faulty set $F_G \subset V(G) \cup E(G)$ with $|F_G| = l$. Suppose that the total number of faults $|F_G|$ is strictly less than l. We may arbitrarily designate $l - |F_G|$ healthy edges as faulty to make exactly l faults.

THEOREM 11.8 Assume that $k \geq 4$. Suppose that G_0 and G_1 are two $(k-2)$-fault-tolerant Hamiltonian and $(k-3)$-fault-tolerant Hamiltonian-connected graphs with $|V(G_0)| = |V(G_1)|$. Then the graph $G = G_0 \oplus G_1$ is $(k-1)$-fault-tolerant Hamiltonian.

Proof: Obviously, $\delta(G) \geq (k+1)$. To prove that $G_0 \oplus G_1$ is $(k-1)$-fault-tolerant Hamiltonian, it suffices to show that $G - F$ is Hamiltonian for any faulty set $F \subset V(G) \cup E(G)$ with $|F| = k-1$.

Case 1. All $(k-1)$ faults are in the same component. We may assume without loss of generality that all faults are in G_0. Since G_0 is $(k-2)$-fault-tolerant Hamiltonian and $f_0 = k-1$, there exists a Hamiltonian path P_0 of $G_0 - F_0$ joining x to y. Since $f_1 = 0$ and G_1 is $(k-3)$-fault-tolerant Hamiltonian-connected, there exists a Hamiltonian path P_1 on G_1 joining $\bar{y}$ to $\bar{x}$. Therefore, $\langle x, P_0, y, \bar{y}, P_1, \bar{x}, x \rangle$ forms a Hamiltonian cycle of $G - F$. See Figure 11.8a.

Case 2. Not all $k-1$ faults are in the same component. Without loss of generality, we may assume that $f_1 \leq f_0 \leq k-2$. Since $k \geq 4$, $G_1 - F_1$ is Hamiltonian-connected.

By Lemma 11.7, there exists a healthy matching edge between G_0 and G_1; say, $(x, \bar{x})$. Now, we claim that there exists a vertex y incident to x such that (x, y) is on a Hamiltonian cycle in $G_0 - F_0$, and the edge $(y, \bar{y})$ is healthy. Obviously, such a Hamiltonian cycle can be written as $\langle x, P_0, y, x \rangle$.

By Lemma 11.5, among all the healthy vertices in $G_0 - F_0$ incident to x, there are at least $(k-2) - f_0 + 2 = k - f_0$ edges, which are on some Hamiltonian cycle in

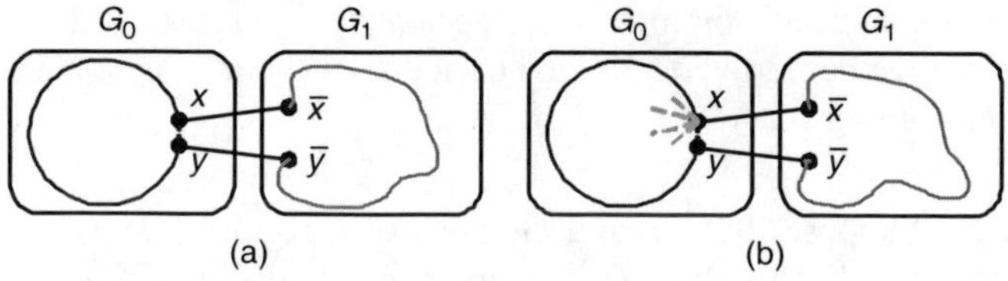

FIGURE 11.8 Illustrations of Theorem 11.8.

$G_0 - F_0$. Of all these $k - f_0$ edges, there is at least one edge—say, (x, y)—such that y, $\bar{y}$, and $(y, \bar{y})$ are healthy. Otherwise, G would contain at least $f_0 + (k - f_0) = k$ faults, which would contradict the fact that the total number of faults is $k - 1$. Thus, our claim holds.

Since $G_1 - F_1$ is Hamiltonian-connected, there exists a Hamiltonian path P_1 joining $\bar{y}$ to $\bar{x}$. Obviously, $\langle x, P_0, y, \bar{y}, P_1, \bar{x}, x\rangle$ forms a Hamiltonian cycle of $G - F$. See Figure 11.8b. This completes the proof of this theorem. □

THEOREM 11.9 Assume that $k \geq 5$. Suppose that G_0 and G_1 are two $(k-2)$-fault-tolerant Hamiltonian and $(k-3)$-fault-tolerant Hamiltonian-connected graphs with $|V(G_0)| = |V(G_1)|$. Then the graph $G = G_0 \oplus G_1$ is $(k-2)$-fault-tolerant Hamiltonian-connected.

Proof: Let F be a set of faults with $F \subset V(G) \cup E(G)$ and $|F| = k - 2$. Let x and y be two healthy vertices in G. To prove this theorem, we need to find a Hamiltonian path of $G - F$ joining x and y. The proof is classified into the following two cases:

Case 1. x and y are not in the same component. Without loss of generality, we may assume that x is in G_0, and y is in G_1. This case can be further divided into two subcases.

Case 1.1. All $k - 2$ faults are in the same component. Without loss of generality, we may assume that all $k - 2$ faults are in G_0. Thus, there is a Hamiltonian cycle in $G_0 - F_0$. On this cycle, there are two vertices incident to x. One of these two vertices is not $\bar{y}$; say, z. Obviously, this cycle can be written as $\langle x, P_0, z, x\rangle$. Since G_1 is Hamiltonian-connected, there is a Hamiltonian path P_1 of G_1 joining $\bar{z}$ to y. Thus, $\langle x, P_0, z, \bar{z}, P_1, y\rangle$ forms a Hamiltonian path of $G - F$ joining x to y. See Figure 11.9a.

Case 1.2. Not all $k - 2$ faults are in the same component. By Lemma 11.8, we can find a healthy matching edge $(z, \bar{z})$ between G_0 and G_1 where $z \in V(G_0) - \{x\}$ and $\bar{z} \in V(G_1) - \{y\}$. Since G_0 and G_1 are $(k-3)$-fault-tolerant Hamiltonian-connected, there is a Hamiltonian path P_0 of $G_0 - F_0$ joining x to z, and there is a Hamiltonian path P_1 of $G_1 - F_1$ joining $\bar{z}$ to y. Obviously, $\langle x, P_0, z, \bar{z}, P_1, y\rangle$ forms a Hamiltonian path of $G - F$ joining x to y. See Figure 11.9b.

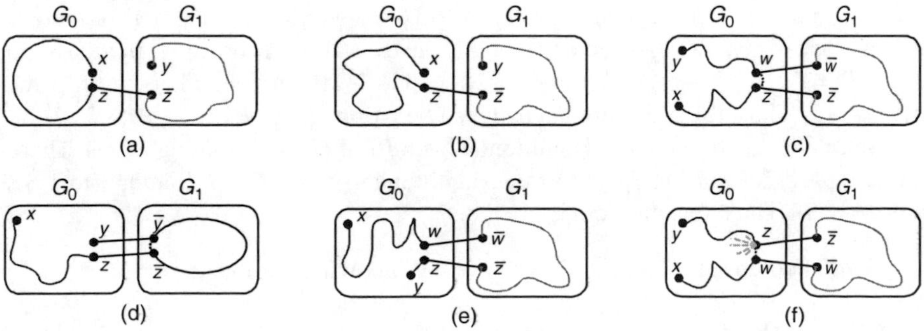

FIGURE 11.9 Illustrations of Theorem 11.9.

Case 2. x and y are in the same component. Without loss of generality, we may assume that x and y are in G_0. We divide this case into three subcases.

Case 2.1. All $k-2$ faults are in G_0. Thus, G_0 is $(k-3)$-Hamiltonian-connected and $f_0 = k-2$. Let g be a faulty edge or a faulty vertex. In $G_0 - (F_0 - \{g\})$, there is a Hamiltonian path P of $G_0 - (F_0 - \{g\})$ joining x and y. Removing the fault g, this Hamiltonian path is separated into two subpaths; say, $\langle x, P_0^1, z\rangle$ and $\langle w, P_0^2, y\rangle$, which cover all the vertices of $G_0 - F_0$. Obviously, there exists a Hamiltonian path P_1 of G_1 joining $\bar{z}$ to $\overline{w}$. Thus, $\langle x, P_0^1, z, \bar{z}, P_1, \overline{w}, w, P_0^2, y\rangle$ forms a Hamiltonian path of $G-F$ joining x to y. See Figure 11.9c.

Case 2.2. All $k-2$ faults are in G_1. This subcase can be further divided into two subcases.

Case 2.2.1. At least one of $\bar{x}$ and $\bar{y}$ is healthy. Without loss of generality, we may assume that $\bar{y}$ is healthy. Since $f_1 = k-2$, there exists a Hamiltonian cycle C of $G_1 - F_1$. On this cycle C, there are two vertices incident to vertex $\bar{y}$. At least one of these two vertices is not $\bar{x}$; say, $\bar{z}$. We can write C as $\langle \bar{z}, P_1, y, \bar{z}\rangle$. Since $k \geq 5$, there exists a Hamiltonian path P_0 of $G_0 - \{y\}$ joining x to z. Obviously, $\langle x, P_0, z, \bar{z}, P_1, \bar{y}, y\rangle$ forms a Hamiltonian path of $G-F$ joining x to y. See Figure 11.9d.

Case 2.2.2. Both $\bar{x}$ and $\bar{y}$ are faulty. In G_0, the number of healthy edges incident to y is k and $f_1 = k-2$. There exists a healthy vertex z incident to y such that $z \neq x$ and $\bar{z}$ is healthy. Since $f_1 = k-2$, there exists a Hamiltonian cycle of $G_1 - F_1$. Let $\overline{w}$ be a vertex on this cycle incident to $\bar{z}$. Obviously, C can be written as $\langle \overline{w}, P_1, \bar{z}, \overline{w}\rangle$. Since $k \geq 5$, there exists a Hamiltonian path P_0 of $G_0 - \{z, y\}$ joining x to w. Obviously, $\langle x, P_0, w, \overline{w}, P_1, \bar{z}, z, y\rangle$ forms a Hamiltonian path of $G-F$ joining x to y. See Figure 11.9e.

Case 2.3. Neither $F \subset V(G_0) \cup E(G_0)$ nor $F \subset V(G_1) \cup E(G_1)$. Since $|F| = k-2$ and not all faults are in one component, we have $f_0 \leq k-3$ and $f_1 \leq k-3$. Consequently, both $G_0 - F_0$ and $G_1 - F_1$ are Hamiltonian-connected. By Lemma 11.8, there is at least one healthy matching edge between G_0 and G_1—say, $(z, \bar{z})$—such that $z \in V(G_0) - \{x, y\}$.

By Lemma 11.6, there are at least $(k-3) - f_0 + 2 = k - 1 - f_0$ edges of $G_0 - F_0$ incident to vertex z such that each one of them is on some Hamiltonian path in $G_0 - F_0$ joining x to y. Among these $k - 1 - f_0$ edges, we claim that there is at least one—say, (z, w)—such that w, $\overline{w}$, and $(w, \overline{w})$ are healthy. If this is not true, $|F| = f_0 + (|F| - f_0) \geq f_0 + (k - 1 - f_0) = k-1$, which contradicts the fact that $|F| = k-2$. Thus, this Hamiltonian path can be written as $\langle x, P_0^1, z, w, P_0^2, y\rangle$.

Since $f_1 \leq k-3$, there is a Hamiltonian path P_1 of $G_1 - F_1$ joining $\bar{z}$ to $\overline{w}$. Therefore, $\langle x, P_0^1, z, \bar{z}, P_2, \overline{w}, w, P_0^2, y\rangle$ forms a Hamiltonian path of $G-F$ joining x to y. See Figure 11.9f. Thus, this theorem is proved. □

With Theorems 11.8 and 11.9, we have the following corollary.

COROLLARY 11.5 *Assume that G_0 and G_1 are k-regular and super fault-tolerant Hamiltonian where $k \geq 5$ and $|V(G_0)| = |V(G_1)|$. Then $G_0 \oplus G_1$ is $(k+1)$-regular super fault-tolerant Hamiltonian.*

Some recursive circulant graphs [302] and some k-ary n-cubes [349] are proved to be super fault-tolerant Hamiltonian. These graphs can be recursively constructed using the operation $G(G_0, G_1, \ldots, G_{r-1}; \mathcal{M})$ discussed in Chapter 7. Again, Chen et al. [53] observed that the insight of the proofs of the aforementioned results can be summarized as follows.

THEOREM 11.10 Assume that $G_0, G_1, \ldots, G_{r-1}$ are $(k-2)$-fault-tolerant Hamiltonian and $(k-3)$-fault-tolerant Hamiltonian graphs with the same number of vertices, where $r \geq 3$ and $k \geq 5$. Then graph $G(G_0, G_1, \ldots, G_{r-1}; \mathcal{M})$ is k-fault-tolerant Hamiltonian.

THEOREM 11.11 Assume that $G_0, G_1, \ldots, G_{r-1}$ are $(k-2)$-fault-tolerant Hamiltonian and $(k-3)$-fault-tolerant Hamiltonian graphs with the same number of vertices, where $r \geq 3$ and $k \geq 5$. Then $G(G_0, G_1, \ldots, G_{r-1}; \mathcal{M})$ is a $(k-1)$-fault-tolerant Hamiltonian-connected graph.

COROLLARY 11.6 *Assume that $G_0, G_1, \ldots, G_{r-1}$ are* k*-regular and super fault-tolerant Hamiltonian with the same number of vertices, where $r \geq 3$ and $k \geq 5$. Then $G(G_0, G_1, \ldots, G_{r-1}; \mathcal{M})$ is $(k+2)$-regular super fault-tolerant Hamiltonian.*

11.4 FAULT-TOLERANT HAMILTONICITY AND FAULT-TOLERANT HAMILTONIAN CONNECTIVITY OF THE FOLDED PETERSEN CUBE NETWORKS

The Petersen graph is a 3-regular graph with 10 vertices of diameter 2. Compared to this graph, the three-dimensional hypercube is a 3-regular graph with 8 vertices and of diameter 3. It has more vertices as compared to the three-dimensional hypercube and a smaller diameter. We call it the *simple Petersen graph*. As an extension, Öhring and Das [260] introduce the k-dimensional folded Petersen graph, FP_k, to be P^k. It is observed that FP_k possesses qualities of a good network topology for distributed systems with large number of sites, as it accommodates 10^k vertices and is a symmetric, $3k$-regular graph of diameter $2k$. Being an iterative Cartesian product on the Petersen graph, it is scalable. Moreover, Öhring and Das [260] define the folded Petersen cube networks $FPQ_{n,k}$ as $Q_n \times P^k$. In particular, $FPQ_{0,k} = P^k$ and $FPQ_{n,0} = Q_n$. The graph $FPQ_{0,2} = P^2$ is shown in Figure 11.10b.

In Ref. 260, it is proved that a number of standard topologies, such as linear arrays, rings, meshes, hypercubes, and so on, can be embedded into it. Recently, many studies on the folded Petersen cube networks have been published due to its favorite properties [79,260,280].

Lin et al. [228] prove that $\mathcal{H}_f(FPQ_{n,k}) = n + 3k - 2$ and $\mathcal{H}_f^\kappa(FPQ_{n,k}) = n + 3k - 3$ if $(n, k) \notin \{(0, 1)\} \cup \{(n, 0) \mid n \text{ is a positive integer}\}$. Moreover, $FPQ_{0,1}$ is neither Hamiltonian nor Hamiltonian-connected. Furthermore, $FPQ_{n,0}$ is Hamiltonian but not Hamiltonian-connected if $n > 1$; $FPQ_{1,0}$ is Hamiltonian-connected but not Hamiltonian. The basic idea is applying Theorems 11.8 through 11.11.

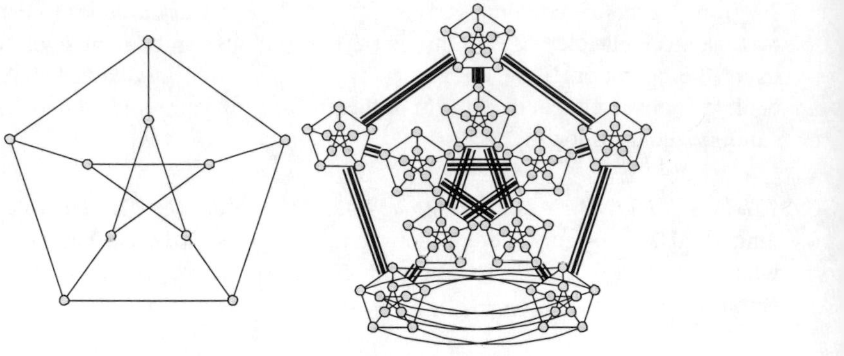

(a) (b)

FIGURE 11.10 (a) The Petersen graph P and (b) a schematic representation of P^2.

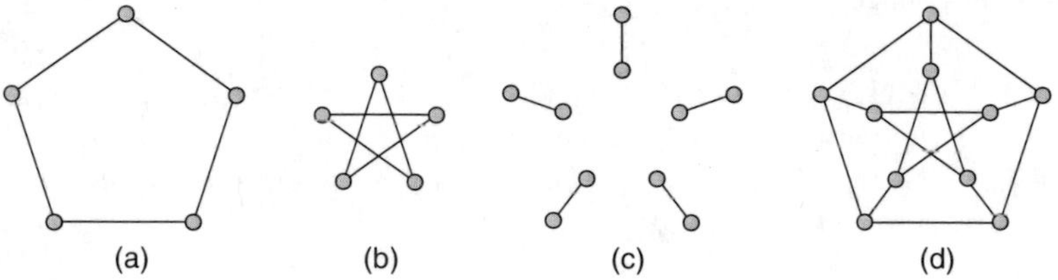

(a) (b) (c) (d)

FIGURE 11.11 (a) A copy of C_5, (b) another copy of C_5, (c) the matching M, and (d) the Petersen graph P.

We begin with the following observation. The Petersen graph P can be viewed as $C_5 \oplus C_5$ as shown in Figure 11.11. Moreover, $C_5 \times C_5$ can be viewed as $G(C_5, C_5, C_5, C_5, C_5; \mathcal{M})$. Furthermore, $P \times C_5$ can be viewed as $(C_5 \times C_5) \oplus (C_5 \times C_5)$. (See Figure 11.12 for illustration.) Thus, P^2 can be viewed as $(P \times C_5) \oplus (P \times C_5)$.

Theorems 11.8 through 11.11 are useful to construct super fault-tolerant Hamiltonian graphs. However, the drawback of these theorems are the restriction of k being rather high. Yet, we have some difficulty in improving upon Theorem 11.9 by including the case $k = 4$. For this reason, we introduce the concept of the extendable 4-regular super fault-tolerant Hamiltonian graph. A 4-regular super fault-tolerant Hamiltonian graph H is *extendable* if $H - \{x, y\}$ remains Hamiltonian-connected for any x and y such that $(x, y) \in E(H)$.

By brute force, we can prove the following lemma.

LEMMA 11.9 Both $P \times Q_1$ and $C_5 \times C_5$ are extendable 4-regular super fault-tolerant Hamiltonian graphs. Yet, $C_5 \times C_4$ is 4-regular super fault-tolerant Hamiltonian but not extendable.

THEOREM 11.12 Suppose that G_0 and G_1 are extendable 4-regular super fault-tolerant Hamiltonian graphs with $|V(G_0)| = |V(G_1)|$. Then $G_0 \oplus G_1$ is 5-regular super fault-tolerant Hamiltonian.

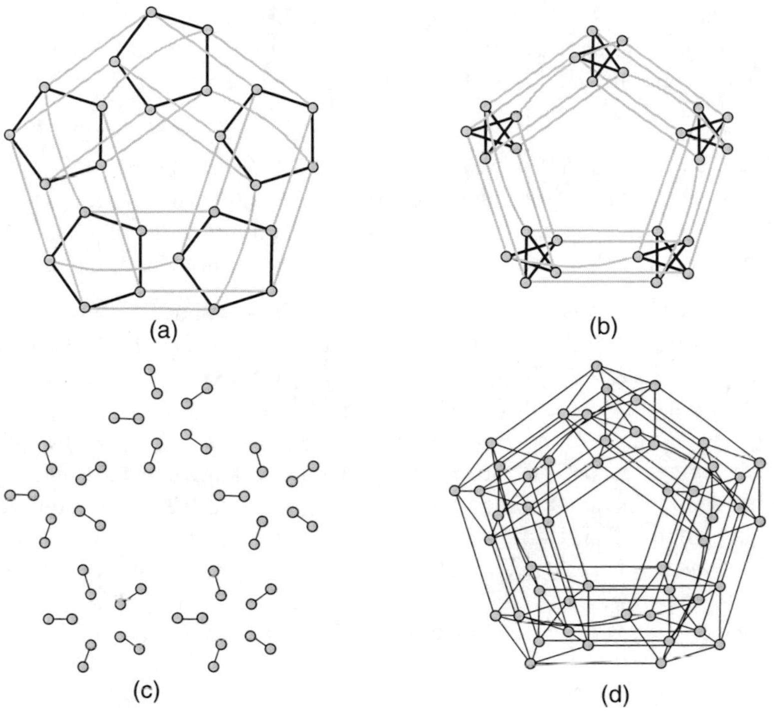

FIGURE 11.12 (a) A copy of $C_5 \times C_5$, (b) another copy of $C_5 \times C_5$, (c) the matching M, and (d) $P \times C_5$.

Proof: Since G_i is 4-regular, $|V(G_i)| \geq 5$. Thus $G_0 \oplus G_1$ is a 5-regular graph. By Theorem 11.8, $G_0 \oplus G_1$ is 3-fault-tolerant Hamiltonian.

Then, we need to prove $G_0 \oplus G_1$ is 2-fault-tolerant Hamiltonian-connected. Let F be any subset of $V(G_0 \oplus G_1) \cup E(G_0 \oplus G_1)$ with $|F| \leq 2$. We need to find a Hamiltonian path in $G_0 \oplus G_1 - F$ between any two different vertices u and v in $V(G_0 \oplus G_1) - F$.

We can easily follow the proof of Theorem 11.9 to find the corresponding Hamiltonian path except for Case 2.2.2. However, the Hamiltonian path in this case is guaranteed by the term of *extendable.* □

LEMMA 11.10 P^k is $(3k)$-regular super fault-tolerant Hamiltonian if and only if $k \geq 2$.

Proof: It is known that P is not Hamiltonian. Hence, it is not Hamiltonian-connected. We have observed that P^2 can be viewed as $(P \times C_5) \oplus (P \times C_5)$. Moreover, $P \times C_5$ can be viewed as $(C_5 \times C_5) \oplus (C_5 \times C_5)$. By Lemma 11.9, $C_5 \times C_5$ is extendable 4-regular super fault-tolerant Hamiltonian. Applying Theorem 11.12, $P \times C_5$ is 5-regular super fault-tolerant Hamiltonian. By Corollary 11.5, P^2 is 6-regular super fault-tolerant Hamiltonian. Let k be a positive integer with $k \geq 3$. Obviously, P^k can be viewed as $(P^{k-1} \times C_5) \oplus (P^{k-1} \times C_5)$. By Corollary 11.6, $P^{k-1} \times C_5$ is $(3k-1)$-regular super fault-tolerant Hamiltonian. By Corollary 11.5, P^k is $(3k)$-regular super fault-tolerant Hamiltonian. Thus, the lemma is proved. □

THEOREM 11.13 $\mathcal{H}_f(FPQ_{n,k}) = n + 3k - 2$ and $\mathcal{H}_f^{\kappa}(FPQ_{n,k}) = n + 3k - 3$ if $(n, k) \notin \{(0, 1)\} \cup \{(n, 0) \mid n$ is a positive integer$\}$. Moreover, $FPQ_{0,1}$ is neither Hamiltonian nor Hamiltonian-connected. Furthermore, $FPQ_{n,0}$ is Hamiltonian but not Hamiltonian-connected if $n > 1$; $FPQ_{1,0}$ is Hamiltonian-connected but not Hamiltonian.

Proof: Applying Lemma 11.10, $\mathcal{H}_f(FPQ_{0,k}) = 3k - 2$ and $\mathcal{H}_f^{\kappa}(FPQ_{0,k}) = 3k - 3$ if $k \geq 2$. By Lemma 11.9, $\mathcal{H}_f(FPQ_{1,1}) = 2$ and $\mathcal{H}_f^{\kappa}(FPQ_{1,1}) = 1$. By Lemma 11.9 and Theorem 11.12, $\mathcal{H}_f(FPQ_{2,1}) = 3$ and $\mathcal{H}_f^{\kappa}(FPQ_{2,1}) = 2$. By Corollary 11.5, $\mathcal{H}_f(FPQ_{n,k}) = n + 3k - 2$ and $\mathcal{H}_f^{\kappa}(FPQ_{n,k}) = n + 3k - 3$ if $n \geq 3$ and $k \geq 1$. Therefore, $\mathcal{H}_f(FPQ_{n,k}) = n + 3k - 2$ and $\mathcal{H}_f^{\kappa}(FPQ_{n,k}) = n + 3k - 3$ if $(n, k) \notin \{(0, 1)\} \cup \{(n, 0) \mid n$ is a positive integer$\}$. $FPQ_{0,1}$ is the Petersen graph. It is known that P is not Hamiltonian. Thus $FPQ_{0,1}$ is not Hamiltonian-connected. Note that $FPQ_{n,0}$ is isomorphic to Q_n. It is known that Q_1 is Hamiltonian-connected but not Hamiltonian. Moreover, Q_n is Hamiltonian and there is no Hamiltonian path of Q_n joining any two vertices in the same partite set if $n \geq 2$. Hence, $FPQ_{n,0}$ is Hamiltonian but not Hamiltonian-connected if $n > 1$ and $FPQ_{1,0}$ is Hamiltonian-connected but not Hamiltonian. □

We believe that our approach in this section can be applied to the same problem on other interconnection networks. Definitely, we can repeatedly apply Theorems 11.8 through 11.11 to obtain the result in this section. However, we need a lot of efforts to check the base cases for the requirement $k \geq 5$ in Theorems 11.9 through 11.11. By introducing the concept of the extendable 4-regular super fault-tolerant Hamiltonian graph, all difficulties are solved immediately. Thus, it would be a great improvement if Theorems 11.8 through 11.11 remain true for smaller k. For this reason, Kueng et al. prove that Theorems 11.10 and 11.11 also hold for $k = 4$ [210].

Let H be the graph shown in Figure 11.13a. Obviously, H is a 3-regular graph. We can confirm that H is 3-regular super fault-tolerant Hamiltonian by brute force. Let G_1 and G_2 be two copies of H. Let G be the graph shown in Figure 11.13b. Obviously,

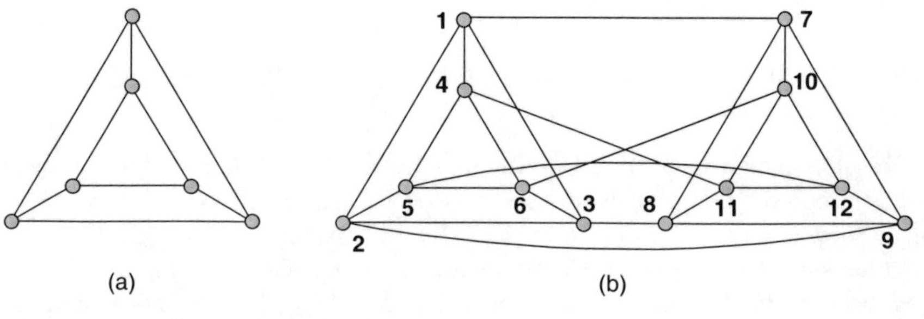

FIGURE 11.13 (a) The graph H and (b) the graph G.

$G = G_1 \oplus G_2$ for some 1-1 connection ϕ. Again, we can prove that $G - \{2, 12\}$ is not Hamiltonian by brute force. Thus, G is not 2-fault-tolerant Hamiltonian.

Therefore, the bound of k in Theorem 11.8 is optimal.

11.5 FAULT-TOLERANT HAMILTONICITY AND FAULT-TOLERANT HAMILTONIAN CONNECTIVITY OF THE PANCAKE GRAPHS

With the construction schemes of k-fault-tolerant Hamiltonian graphs, we can prove that several interconnection networks are optimal fault-tolerant Hamiltonian. However, we cannot use these construction schemes to construct all the known interconnection networks. However, the idea of combining fault-tolerant Hamiltonian and fault-tolerant Hamiltonian-connected together with the mathematical induction to discuss the fault-tolerant Hamiltonicity of some families of interconnection networks can be used to discuss the fault-tolerant Hamiltonicity of other families of interconnection networks. In this section, we discuss the fault-tolerant Hamiltonicity of the pancake graphs, presented by Hung et al. [181], as an example to illustrate this concept.

We will use boldface to denote a vertex of P_n. Hence, $\mathbf{u_1}, \mathbf{u_2}, \ldots, \mathbf{u_n}$ denote a sequence of vertices in P_n. In particular, $\mathbf{e}$ denotes the vertex $12\ldots n$. Let $\mathbf{u} = u_1 u_2 \ldots u_n$ be any vertex of P_n. We use $(\mathbf{u})_i$ to denote the ith component u_i of $\mathbf{u}$, and use $P_n^{\{i\}}$ to denote the ith subgraph of P_n induced by those vertices $\mathbf{u}$ with $(\mathbf{u})_n = i$. Obviously, P_n can be decomposed into n vertex disjoint subgraphs $P_n^{\{i\}}$ for every $i \in \langle n \rangle$ such that each $P_n^{\{i\}}$ is isomorphic to P_{n-1}. Thus, the pancake graph can be constructed recursively. Let $I \subseteq \langle n \rangle$; we use P_n^I to denote the subgraph of P_n induced by $\cup_{i \in I} V(P_n^{\{i\}})$. By definition, there is exactly one neighbor $\mathbf{v}$ of $\mathbf{u}$ such that $\mathbf{u}$ and $\mathbf{v}$ are adjacent through an i-dimensional edge with $2 \leq i \leq n$. For this reason, we use $(\mathbf{u})^i$ to denote the unique i-neighbor of $\mathbf{u}$. We have $((\mathbf{u})^i)^i = \mathbf{u}$ and $(\mathbf{u})^n \in P_n^{\{(\mathbf{u})_1\}}$. For $1 \leq i, j \leq n$ and $i \neq j$, we use $E^{i,j}$ to denote the set of edges between $P_n^{\{i\}}$ and $P_n^{\{j\}}$. Obviously, we have the following lemmas.

LEMMA 11.11 $|E^{i,j}| = (n-2)!$ for any $1 \leq i \neq j \leq n$.

LEMMA 11.12 Let $\mathbf{u}$ and $\mathbf{v}$ be two distinct vertices of P_n with $(\mathbf{u})_n = (\mathbf{v})_n$ such that $d(\mathbf{u}, \mathbf{v}) \leq 2$. Then $((\mathbf{u})^n)_n \neq ((\mathbf{v})^n)_n$. Moreover, $\{((\mathbf{u})^i)_1 \mid 2 \leq i \leq n-1\} = \langle n \rangle - \{(\mathbf{u})_1, (\mathbf{u})_n\}$ if $n \geq 3$.

Let $F \subset V(P_n) \cup E(P_n)$ be any faulty set of P_n. We use F^i to denote the set $F \cap (V(P_n^{\{i\}}) \cup E(P_n^{\{i\}}))$. Then we set F^0 to denote the set $F - \cup_{i=1}^n F^i$. An edge $(\mathbf{u}, \mathbf{v})$ is F-*fault* if $(\mathbf{u}, \mathbf{v}) \in F$, $\mathbf{u} \in F$, or $\mathbf{v} \in F$; and $(\mathbf{u}, \mathbf{v})$ is F-*fault free* if $(\mathbf{u}, \mathbf{v})$ is not F-fault. Let $H = (V', E')$ be a subgraph of P_n. We use $F(H)$ to denote the set $(V' \cup E') \cap F$.

LEMMA 11.13 Assume that $n \geq 5$ and $I = \{i_1, i_2, \ldots, i_m\}$ is a subset of $\langle n \rangle$ such that $|I| = m \geq 2$. Let $F \subset V(P_n) \cup E(P_n)$ be any faulty set such that $P_n^{\{i\}} - F^i$ is Hamiltonian-connected for any $i \in I$ and there are at least three F-fault free edges in $E^{i_j, i_{j+1}}$ for any $1 \leq j < m$. Then there exists a Hamiltonian path $P = \langle \mathbf{u} = \mathbf{x_1}, Q_1, \mathbf{y_1}, \mathbf{x_2}, Q_2, \mathbf{y_2}, \ldots, \mathbf{x_m}, Q_m, \mathbf{y_m} = \mathbf{v} \rangle$ of $P_n^I - F$ joining any two vertices **u** and **v** with $\mathbf{u} \in V(P_n^{\{i_1\}}) - F$ and $\mathbf{v} \in V(P_n^{\{i_m\}}) - F$ such that Q_i is a Hamiltonian path of $P_n^{\{a_i\}} - F^i$ joining $\mathbf{x_i}$ to $\mathbf{y_i}$ for every $1 \le i \le m$.

Proof: Let $\mathbf{u_1} = \mathbf{u}$ and $\mathbf{v} = \mathbf{v_m}$. Since there are at least three F-fault free edges in $E^{i_j, i_{j+1}}$ for any $1 \le j < m$, we can easily choose two different vertices $\mathbf{u_{i_j}}$ and $\mathbf{v_{i_{j+1}}}$ in $P_n^{\{i_j\}}$ such that $(\mathbf{v_{i_j}}, \mathbf{u_{i_{j+1}}})$ is F-fault free. Obviously, $\mathbf{u_{i_j}} \neq \mathbf{v_{i_j}}$. Since $P_n^{\{i_j\}} - F^{i_j}$ is Hamiltonian-connected for all $i_j \in I$, there is a Hamiltonian path Q_j of $P_n^{\{i_j\}} - F^{i_j}$ joining $\mathbf{u_{i_j}}$ and $\mathbf{v_{i_j}}$. Thus, $\langle \mathbf{u_{i_1}}, Q_1, \mathbf{v_{i_1}}, \mathbf{u_{i_2}}, Q_2, \ldots, \mathbf{v_{i_{m-1}}}, \mathbf{u_{i_m}}, Q_m, \mathbf{v_{i_m}} \rangle$ forms a Hamiltonian path of $P_n^I - F$ joining **u** and **v**. The lemma is proved. □

LEMMA 11.14 P_4 is 1-fault-tolerant Hamiltonian and Hamiltonian-connected.

Proof: To prove that P_4 is 1-fault-tolerant Hamiltonian we need to prove that $P_4 - F$ is Hamiltonian for any $F = \{f\}$ with $f \in V(P_4) \cup E(P_4)$. Without loss of generality, we may assume that $f = 1234$ if f is a vertex, or $f \in \{(1234, 2134), (1234, 3214), (1234, 4321)\}$ if f is an edge. The corresponding Hamiltonian cycles of $P_4 - F$ are listed as follows:

⟨3214, 2314, 4132, 1432, 2341, 4321, 3421, 2431, 1342, 3142, 2413, 4213, 1243, 2143, 3412, 4312, 2134, 3124, 1324, 4231, 3241, 1423, 4123, 3214⟩
⟨1234, 3214, 2314, 1324, 3124, 2134, 4312, 1342, 3142, 4132, 1432, 3412, 2143, 4123, 1423, 2413, 4213, 1243, 3421, 2431, 4231, 3241, 2341, 4321, 1234⟩
⟨1234, 4321, 3421, 2431, 4231, 3241, 2341, 1432, 4132, 3142, 1342, 4312, 3412, 2143, 1243, 4213, 2413, 1423, 4123, 3214, 2314, 1324, 3124, 2134, 1234⟩
⟨1234, 3214, 2314, 4132, 1432, 3412, 4312, 1342, 3142, 2413, 4213, 1243, 2143, 4123, 1423, 3241, 2341, 4321, 3421, 2431, 4231, 1324, 3124, 2134, 1234⟩

To prove that P_4 is Hamiltonian-connected, we have to find the Hamiltonian path joining any two vertices **u** and **v**. By the symmetric property of P_4, we may assume that $\mathbf{u} = 1234$ and **v** is any vertex in $V(P_4) - \{\mathbf{u}\}$. The corresponding Hamiltonian paths are listed as follows:

⟨1234, 3214, 2314, 1324, 4231, 3241, 2341, 4321, 3421, 2431, 1342, 3142, 4132, 1432, 3412, 4312, 2134, 3124, 4213, 2413, 1423, 4123, 2143, 1243⟩
⟨1234, 3214, 2314, 4132, 3142, 2413, 4213, 1243, 3421, 4321, 2341, 1432, 3412, 2143, 4123, 1423, 3241, 4231, 2431, 1342, 4312, 2134, 3124, 1324⟩
⟨1234, 3214, 2314, 4132, 3142, 2413, 4213, 1243, 2143, 4123, 1423, 3241, 4231, 1324, 3124, 2134, 4312, 3412, 1432, 2341, 4321, 3421, 2431, 1342⟩
⟨1234, 3214, 2314, 1324, 3124, 2134, 4312, 1342, 2431, 4231, 3241, 2341, 4321, 3421, 1243, 4213, 2413, 3142, 4231, 1432, 3412, 2143, 4123, 1423⟩
⟨1234, 3214, 2314, 1324, 3124, 2134, 4312, 3412, 2143, 4123, 1423, 2413, 4213, 1243, 3421, 4321, 2341, 2141, 4231, 2431, 1342, 3142, 4132, 1432⟩
⟨1234, 3214, 2314, 1324, 3124, 4213, 2413, 1423, 4123, 2143, 1243, 3421, 4321, 2341, 3241, 4231, 2431, 1342, 3142, 4132, 1432, 3412, 4312, 2134⟩
⟨1234, 3214, 2314, 1324, 3124, 2134, 4312, 3412, 4132, 4132, 3142, 1342, 2431, 4231, 3241, 2341, 4321, 3421, 1243, 4213, 2413, 1423, 4123, 2143⟩
⟨1234, 3214, 4123, 2143, 1243, 4213, 2413, 1423, 3241, 4231, 1324, 3124, 2134, 4312, 3412, 1432, 2341, 4321, 3421, 2431, 1342, 3142, 4132, 2314⟩
⟨1234, 3214, 2314, 1324, 3124, 2134, 4312, 1342, 2431, 4231, 3241, 1423, 4123, 2143, 3412, 1432, 4132, 3142, 2413, 4213, 1243, 3421, 4321, 2341⟩
⟨1234, 3214, 2314, 1324, 3124, 2134, 4312, 3412, 2143, 4123, 1423, 3241, 4231, 2431, 1342, 3142, 4132, 1432, 2341, 4321, 3421, 1243, 4213, 2413⟩
⟨1234, 3214, 2314, 1324, 3124, 2134, 4312, 1342, 3142, 4132, 1432, 3412, 2143, 4123, 1423, 2413, 4213, 1243, 3421, 4321, 2341, 3214, 4231, 2431⟩
⟨1234, 3214, 4123, 2143, 1243, 4213, 2413, 1423, 3241, 4231, 1324, 2314, 4132, 3142, 1342, 2431, 3421, 4321, 2341, 1432, 3412, 4312, 2134, 3124⟩
⟨1234, 3214, 2314, 1324, 3124, 2134, 4312, 1342, 2431, 4231, 3241, 2341, 4321, 3421, 1243, 4213, 2413, 1423, 4123, 2143, 3412, 1432, 4132, 3142⟩
⟨1234, 2134, 3124, 1324, 2314, 4132, 3142, 1342, 4312, 3412, 1432, 2341, 4321, 3421, 2431, 4231, 3241, 1423, 2413, 4213, 1243, 2143, 4123, 3214⟩
⟨1234, 3214, 2314, 1324, 3124, 2134, 4312, 3412, 2143, 4123, 1423, 2413, 4213, 1243, 3421, 4321, 2341, 1432, 4132, 3142, 1342, 2431, 4231, 3241⟩
⟨1234, 3214, 2314, 1324, 4231, 3241, 2341, 4321, 3421, 2431, 1342, 4312, 2134, 3124, 4213, 1243, 2143, 4123, 1423, 2413, 3142, 4132, 1432, 3412⟩
⟨1234, 3214, 2314, 4132, 1432, 3412, 4312, 2134, 3124, 1324, 4231, 2431, 1342, 3142, 2413, 4213, 1243, 2143, 4123, 1423, 3241, 2341, 4321, 3421⟩
⟨1234, 3214, 2314, 1324, 4231, 3241, 2341, 4321, 3421, 2431, 1342, 4312, 2134, 3124, 4213, 1243, 2143, 3412, 1432, 4132, 3142, 2413, 1423, 4123⟩
⟨1234, 3214, 2314, 1324, 3124, 2134, 4312, 1342, 2431, 4231, 3241, 1423, 4123, 2143, 3412, 1432, 2341, 4321, 3421, 1243, 4213, 2413, 3142, 4132⟩
⟨1234, 3214, 2314, 1324, 3124, 2134, 4312, 3412, 1432, 4132, 3142, 1342, 2431, 4231, 3241, 2341, 4321, 3421, 1243, 2143, 4123, 1423, 2413, 4213⟩
⟨1234, 4321, 3421, 2431, 1342, 3142, 4132, 2314, 3214, 4123, 2143, 1243, 4213, 2413, 1423, 3241, 2341, 1432, 3412, 4312, 2134, 3124, 1324, 4231⟩
⟨1234, 3214, 2314, 4132, 1432, 3412, 2143, 4123, 1423, 3241, 2341, 4321, 3421, 1243, 4213, 2413, 3142, 1342, 2431, 4231, 1324, 3124, 2134, 4312⟩
⟨1234, 3214, 2314, 1324, 3124, 2134, 4312, 1342, 3142, 4132, 1432, 3412, 2143, 4123, 1423, 2413, 4213, 1243, 3421, 2431, 4231, 3241, 2341, 4321⟩

Thus, the lemma is proved. □

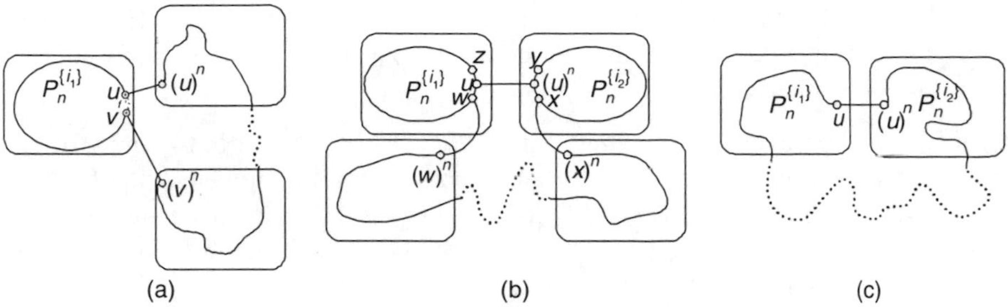

FIGURE 11.14 Illustrations for Lemma 11.15.

LEMMA 11.15 Suppose that $n \geq 5$. If P_{n-1} is $(n-4)$-fault-tolerant Hamiltonian and $(n-5)$-fault-tolerant Hamiltonian-connected, then P_n is $(n-3)$-fault-tolerant Hamiltonian.

Proof: Assume that F is any faulty set of P_n with $|F| \leq n-3$. Since $n \geq 5$, $|E^{i,j} - F| \geq (n-2)! - (n-3) \geq 4$ for any $1 \leq i, j \leq n$. Thus, there are at least four F-fault free edges between $P_n^{\{i\}}$ and $P_n^{\{j\}}$ for any $1 \leq i \neq j \leq n$. We may assume that $|F^{i_1}| \geq |F^{i_2}| \geq \ldots \geq |F^{i_n}|$.

Case 1. $|F^{i_1}| = n-3$. Thus, $F \subset P_n^{\{i_1\}}$. Choose any element f in $F(P_n^{\{i_1\}})$. By the assumption of this lemma, there exists a Hamiltonian cycle Q of $P_n^{\{i_1\}} - F + \{f\}$. We may write Q as $\langle \mathbf{u}, Q_1, \mathbf{v}, f', \mathbf{u} \rangle$ where $f' = f$ if f is incident with Q or f' is any edge of Q if otherwise. Obviously, $d(\mathbf{u}, \mathbf{v}) \leq 2$. By Lemma 11.2, $((\mathbf{u})^n)_n \neq ((\mathbf{v})^n)_n$. Then by Lemma 11.13, there exists a Hamiltonian path Q_2 of $P_n^{\langle n \rangle - \{i_1\}}$ joining $(\mathbf{u})^n$ and $(\mathbf{v})^n$. Then $\langle \mathbf{u}, (\mathbf{u})^n, Q_2, (\mathbf{v})^n, \mathbf{v}, Q_1, \mathbf{u} \rangle$ forms a Hamiltonian cycle of $P_n - F$. See Figure 11.14a for illustration.

Case 2. $|F^{i_1}| = n-4$. Thus, $|F - F^{i_1}| \leq 1$. Hence, there exists an index i_2 such that $|F(P_n^{\langle n \rangle - \{i_1, i_2\}})| = 0$. Since $|E^{i_1,i_2} - F| \geq (n-2)! - (n-3)$, there exists an F-fault free edge $(\mathbf{u}, (\mathbf{u})^n)$ in E^{i_1,i_2} such that $\mathbf{u} \in V(P_n^{\{i_1\}})$. By the assumption of this lemma, there exists a Hamiltonian cycle C_1 of $P_n^{\{i_1\}} - F$ and there exists a Hamiltonian cycle C_2 of $P_n^{\{i_2\}} - F$. We may write C_1 as $\langle \mathbf{u}, \mathbf{w}, Q_1, \mathbf{z}, \mathbf{u} \rangle$ and C_2 as $\langle (\mathbf{u})^n, \mathbf{y}, Q_2, \mathbf{x}, (\mathbf{u})^n \rangle$. Since $d(\mathbf{x}, \mathbf{y}) \leq 2$ and $d(\mathbf{w}, \mathbf{z}) \leq 2$, by Lemma 11.12 $((\mathbf{x})^n)_n \neq ((\mathbf{y})^n)_n$ and $((\mathbf{w})^n)_n \neq ((\mathbf{z})^n)_n$. Thus, we can choose a vertex from $\mathbf{x}$ and $\mathbf{y}$, say $\mathbf{x}$, and we can choose a vertex from $\mathbf{w}$ and $\mathbf{z}$, say $\mathbf{w}$, such that $((\mathbf{w})^n)_n \neq ((\mathbf{x})^n)_n$ and $(\mathbf{w}, (\mathbf{w})^n)$ and $(\mathbf{x}, (\mathbf{x})^n)$ are F-fault free. By Lemma 11.13, there exists a Hamiltonian path Q_3 of $P_n^{\langle n \rangle - \{i_1, i_2\}} - F$ joining $(\mathbf{w})^n$ and $(\mathbf{x})^n$. Hence, $\langle \mathbf{u}, (\mathbf{u})^n, \mathbf{y}, Q_2, \mathbf{x}, (\mathbf{x})^n, Q_3, (\mathbf{w})^n, \mathbf{w}, Q_1, \mathbf{u} \rangle$ forms a Hamiltonian cycle of $P_n - F$. See Figure 11.14b for illustration.

Case 3. $|F^{i_1}| \leq n-5$. We can choose any F-fault free edge $(\mathbf{u}, (\mathbf{u})^n)$ in E^{i_1,i_2} such that $\mathbf{u} \in V(P_n(i_1))$. By the assumption of this lemma, any $P_n^{\{i\}} - F$ is Hamiltonian-connected for $i \in \langle n \rangle$. Then by Lemma 11.13, there exists a Hamiltonian path Q_1 of $P_n - F$ joining $\mathbf{u}$ and $(\mathbf{u})^n$. Then $\langle \mathbf{u}, Q_1, (\mathbf{u})^n, \mathbf{u} \rangle$ forms a Hamiltonian cycle of $P_n - F$. See Figure 11.14c for illustration. □

LEMMA 11.16 Suppose that $n \geq 5$. If P_{n-1} is $(n-4)$-fault-tolerant Hamiltonian and $(n-5)$-fault-tolerant Hamiltonian-connected, then P_n is $(n-4)$-fault-tolerant Hamiltonian-connected.

Proof: Assume that F is any faulty set of P_n with $|F| \leq n-4$. Let $\mathbf{u}$ and $\mathbf{v}$ be any two arbitrary vertices of $P_n - F$. We want to construct a Hamiltonian path of $P_n - F$ joining $\mathbf{u}$ and $\mathbf{v}$. Obviously, $|E^{i,j} - F| \geq (n-2)! - (n-4) \geq 5$ for any $1 \leq i, j \leq n$ with $n \geq 5$. Thus, there are at least five F-fault free edges between $P_n^{\{i\}}$ and $P_n^{\{j\}}$ for any $1 \leq i \neq j \leq n$. We assume that $|F^{i_1}| \geq |F^{i_2}| \geq \ldots \geq |F^{i_n}|$.

Case 1. $|F^{i_1}| = n-4$. Hence, $F \subset P_n^{\{i_1\}}$.

Case 1.1. $(\mathbf{u})_n = (\mathbf{v})_n = i_1$. Choose any element f in $F(P_n^{\{i_1\}})$. By the assumption of this lemma, there exists a Hamiltonian path Q of $P_n^{\{i_1\}} - F + \{f\}$ joining $\mathbf{u}$ and $\mathbf{v}$. We may write Q as $\langle \mathbf{u}, Q_1, \mathbf{x}, f', \mathbf{y}, Q_2, \mathbf{v} \rangle$ where $f' = f$ if f is incident to Q or f' is any edge of Q if otherwise. Obviously, $d(\mathbf{x}, \mathbf{y}) \leq 2$. By Lemma 11.12, $((\mathbf{x})^n)_n \neq ((\mathbf{y})^n)_n$. By Lemma 11.13, there exists a Hamiltonian path Q_3 of $P_n^{\langle n \rangle - \{i_1\}}$ joining $(\mathbf{x})^n$ and $(\mathbf{y})^n$. Then $\langle \mathbf{u}, Q_1, \mathbf{x}, (\mathbf{x})^n, Q_3, (\mathbf{y})^n, \mathbf{y}, Q_2, \mathbf{v} \rangle$ forms a Hamiltonian path of $P_n - F$ joining $\mathbf{u}$ to $\mathbf{v}$. See Figure 11.15a for illustration.

Case 1.2. $(\mathbf{u})_n = i_1$ and $(\mathbf{v})_n = i_j$ with $j \neq 1$. By the assumption of this lemma, there exists a Hamiltonian cycle C_1 of $P_n^{\{i_1\}} - F$. We may write C_1 as $\langle \mathbf{u}, \mathbf{y}, Q_1, \mathbf{x}, \mathbf{u} \rangle$. Since $d(\mathbf{x}, \mathbf{y}) \leq 2$, by Lemma 11.12 $((\mathbf{x})^n)_n \neq ((\mathbf{y})^n)_n$. Thus, we can choose a vertex from $\mathbf{x}$ and $\mathbf{y}$, say $\mathbf{x}$, such that $((\mathbf{x})^n)_n \neq (\mathbf{v})_n$. By Lemma 11.13, there exists a Hamiltonian path Q_2 of $P_n^{\langle n \rangle - \{i_1\}}$ joining $(\mathbf{x})^n$ and $\mathbf{v}$. Then $\langle \mathbf{u}, \mathbf{y}, Q_1, \mathbf{x}, (\mathbf{x})^n, Q_2, \mathbf{v} \rangle$ forms a Hamiltonian path of $P_n - F$ joining $\mathbf{u}$ to $\mathbf{v}$. See Figure 11.15b for illustration.

Case 1.3. $(\mathbf{u})_n = (\mathbf{v})_n = i_j$ with $j \neq 1$. Since there are at least five F-fault free edges in E^{i_1,i_j}, there exists an F-fault free edge $(\mathbf{w}, (\mathbf{w})^n)$ in E^{i_1,i_j} such that $(\mathbf{w})_n = i_1$ and $(\mathbf{w})^n \neq \mathbf{v}$. By the assumption of this lemma, there exists a Hamiltonian cycle C_1 of

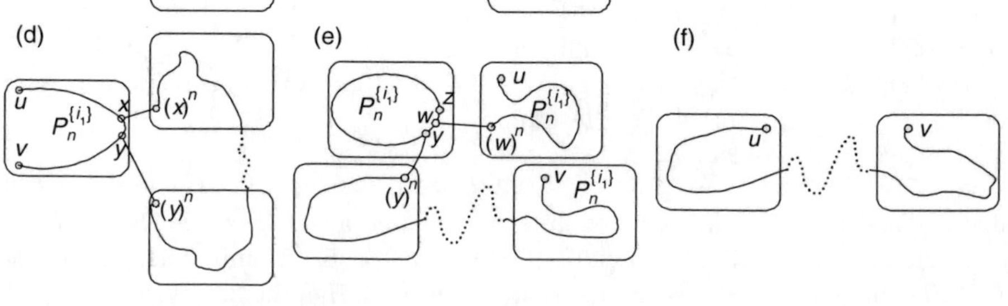

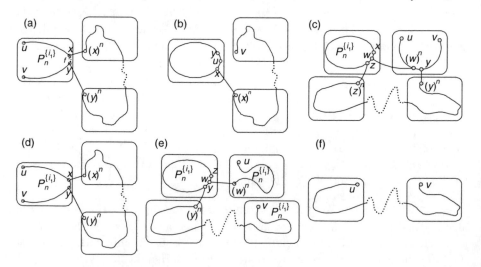

FIGURE 11.15 Illustrations for Lemma 11.16.

$P_n^{\{i_1\}} - F$ and a Hamiltonian path Q_1 of $P_n^{\{i_j\}}$ joining $\mathbf{u}$ and $\mathbf{v}$. We may write Q_1 as $\langle \mathbf{u}, Q_2, (\mathbf{w})^n, \mathbf{y}, Q_3, \mathbf{v}\rangle$ and C_1 as $\langle \mathbf{w}, \mathbf{x}, Q_4, \mathbf{z}, \mathbf{w}\rangle$. Since $d(\mathbf{z}', \mathbf{z}) \leq 2$, by Lemma 11.12 $((\mathbf{z})^n)_n \neq ((\mathbf{x})^n)_n$. Thus, we can choose a vertex from $\mathbf{z}$ and $\mathbf{x}$, say $\mathbf{z}$, such that $((\mathbf{z})^n)_n \neq ((\mathbf{y})^n)_n$. By Lemma 11.13, there exists a Hamiltonian path Q_5 of $P_n^{\langle n\rangle - \{i_1, i_j\}}$ joining $(\mathbf{y})^n$ and $(\mathbf{z})^n$. Then $\langle \mathbf{u}, Q_2, (\mathbf{w})^n, \mathbf{w}, \mathbf{x}, Q_4, \mathbf{z}, (\mathbf{z})^n, Q_5, (\mathbf{y})^n, \mathbf{y}, Q_3, \mathbf{v}\rangle$ forms a Hamiltonian path of $P_n - F$ joining $\mathbf{u}$ and $\mathbf{v}$. See Figure 11.15c for illustration.

Case 1.4. $(\mathbf{u})_n = i_j$ and $(\mathbf{v})_n = i_k$ with i_j, i_k and i_1 are all distinct. Since there are at least five F-fault free edges in E^{i_1, i_j}, there exists an F-fault free edge $(\mathbf{w}, (\mathbf{w})^n)$ in E^{i_1, i_j} such that $(\mathbf{w})_n = i_1$ and $(\mathbf{w})^n \neq \mathbf{u}$. By the assumption of this lemma, there exists a Hamiltonian cycle C_1 of $P_n^{\{i_1\}} - F$ and a Hamiltonian path Q_1 of $P_n^{\{i_j\}} - F$ joining $\mathbf{u}$ and $(\mathbf{w})^n$. We may write C_1 as $\langle \mathbf{w}, \mathbf{z}, Q_2, \mathbf{y}, \mathbf{w}\rangle$. Since $d(\mathbf{y}, \mathbf{z}) \leq 2$, by Lemma 11.12, $((\mathbf{y})^n)_n \neq ((\mathbf{z})^n)_n$. Thus, we can choose a vertex from $\mathbf{y}$ and $\mathbf{z}$, say $\mathbf{y}$, such that $((\mathbf{y})^n)_n \neq (\mathbf{v})_n$. By Lemma 11.13, there exists a Hamiltonian path Q_3 of $P_n^{\langle n\rangle - \{i_1, i_j\}}$ joining $(\mathbf{y})^n$ and $\mathbf{v}$. Thus, $\langle \mathbf{u}, Q_1, (\mathbf{w})^n, \mathbf{w}, \mathbf{z}, Q_2, \mathbf{y}, (\mathbf{y})^n, Q_3, \mathbf{v}\rangle$ forms a Hamiltonian path of $P_n - F$ joining $\mathbf{u}$ and $\mathbf{v}$. See Figure 11.15d for illustration.

Case 2. $|F^{i_1}| \leq n-5$. By the assumption of this lemma, $P_n^{\{i\}}$ is Hamiltonian-connected for every $1 \leq i \leq n$.

Case 2.1. $(\mathbf{u})_n = (\mathbf{v})_n = i_j$. By the assumption of this lemma, there exists a Hamiltonian path Q_1 of $P_n^{\{i_j\}} - F$ joining $\mathbf{u}$ to $\mathbf{v}$. We claim that there exists an F-fault free edge $(\mathbf{x}, \mathbf{y})$ in Q_1 such that $(\mathbf{x}, (\mathbf{x})^n)$ and $(\mathbf{y}, (\mathbf{y})^n)$ are F-fault free. Suppose there is no such edge, $|F| \geq |V(F(P_n^{\{i_j\}}))| + |(V(P_n^{\{i_j\}}) - V(F(P_n^{\{i_j\}}))|/2 \geq (n-1)!/2 > n-3$ for $n \geq 5$. However, $|F| \leq n-3$. We get a contradiction. Hence, such an edge exists.

Write Q_1 as $\langle \mathbf{u}, Q_2, \mathbf{x}, \mathbf{y}, Q_3, \mathbf{v}\rangle$. Since $d(\mathbf{x}, \mathbf{y}) = 1$, by Lemma 11.12 $((\mathbf{x})^n)_n \neq ((\mathbf{y})^n)_n$. By Lemma 11.13, there exists a Hamiltonian path Q_4 of $P_n^{\langle n\rangle - \{i_j\}}$ joining $(\mathbf{x})^n$ to $(\mathbf{y})^n$. Then $\langle \mathbf{u}, Q_2, \mathbf{x}, (\mathbf{x})^n, Q_4, (\mathbf{y})^n, \mathbf{y}, Q_3, \mathbf{v}\rangle$ forms a Hamiltonian path of $P_n - F$ joining $\mathbf{u}$ to $\mathbf{v}$. See Figure 11.15e for illustration.

Case 2.2. $(\mathbf{u})_n \neq (\mathbf{v})_n$. By Lemma 11.13, there exists a Hamiltonian path of $P_n - F$ joining $\mathbf{u}$ to $\mathbf{v}$. See Figure 11.15f for illustration. □

THEOREM 11.14 Let n be a positive integer with $n \geq 4$. Then P_n is $(n-3)$-fault-tolerant Hamiltonian and $(n-4)$-fault-tolerant Hamiltonian-connected.

Proof: We prove this theorem by induction. The induction base, $n = 4$, is proved in Lemma 11.14. With Lemmas 11.15 and 11.16, we prove the induction step. □

Since $\delta(P_n) = n-1$, we have the following theorem.

THEOREM 11.15 $\mathcal{H}_f(P_n) = n-3$ and $\mathcal{H}_f^\kappa(P_n) = n-4$ for any positive integer n with $n \geq 4$.

With a similar but much difficult argument, as for pancake graphs, we can discuss the fault-tolerant Hamiltonicity and the fault-tolerant Hamiltonian-connectivity of the (n, k)-star graph $S_{n,k}$.

The (n, k)-star graph is an attractive alternative to the n-star graph [2]. However, the growth of vertices is $n!$ for an n-star graph. In order to remedy this drawback, the (n, k)-star graph is proposed by Chiang and Chen [60]. The (n, k)-star graph is a generalization of the n-star graph. It has two parameters n and k. When $k = n - 1$, an $(n, n-1)$-star graph is isomorphic to an n-star graph, and when $k = 1$, an $(n, 1)$-star graph is isomorphic to a complete graph K_n.

Let n and k be positive integers with $n > k$. Let $\langle n \rangle$ denote the set $\{1, 2, \ldots, n\}$. The (n, k)-star graph, denoted by $S_{n,k}$, is a graph with the vertex set $V(S_{n,k}) = \{u_1 u_2 \ldots u_k \mid u_i \in \langle n \rangle \text{ and } u_i \neq u_j \text{ for } i \neq j\}$. Adjacency is defined as follows: a vertex $u_1 u_2 \ldots u_i \ldots u_k$ is adjacent to (1) the vertex $u_i u_2 u_3 \ldots u_1 \ldots u_k$, where $2 \leq i \leq k$ (that is, we swap u_i with u_1) and (2) the vertex $x u_2 u_3 \ldots u_k$, where $x \in \langle n \rangle - \{u_i \mid 1 \leq i \leq k\}$. The graph $S_{4,2}$ is shown in Figure 11.16. The edges of type 1 are referred to as *i-edges*, and the edges of type 2 are referred to as *1-edges*. By definition, $S_{n,k}$ is an $(n-1)$-regular graph with $n!/(n-k)!$ vertices. Moreover, it is vertex-transitive.

The following theorem is proved by Hsu et al. [172].

THEOREM 11.16 Suppose that n and k are two positive integers with $n > k \geq 1$. Then

1. $\mathcal{H}_f(S_{n,k}) = n - 3$ and $\mathcal{H}_f^\kappa(S_{n,k}) = n - 4$ if $n - k \geq 2$
2. $\mathcal{H}_f(S_{2,1})$ is undefined and $\mathcal{H}_f^\kappa(S_{2,1}) = 0$
3. $\mathcal{H}_f(S_{n,n-1}) = 0$ and $\mathcal{H}_f^\kappa(S_{n,n-1})$ is undefined if $n > 2$

Similarly, we can discuss the fault-tolerant Hamiltonicity and the fault-tolerant Hamiltonian-connectivity of the arrangement graphs.

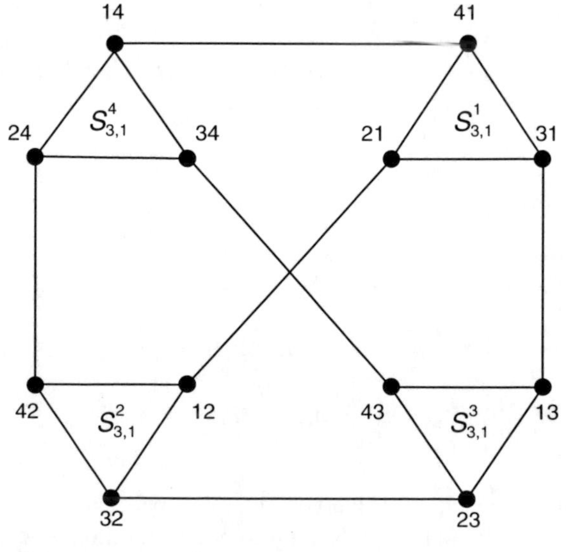

FIGURE 11.16 $S_{4,2}$.

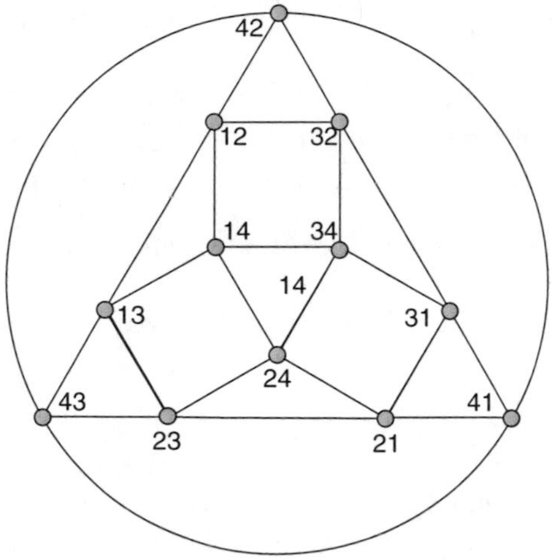

FIGURE 11.17 $A_{4,2}$.

The arrangement graph [83] was proposed by Day and Tripathi as a generalization of the star graph. It is more flexible in its size than the star graph. Let n and k be positive integers with $n > k$, the (n, k)-arrangement graph $A_{n,k}$ is the graph (V, E) where $V = \{p \mid p$ is an arrangement of k elements out of the n symbols: $1, 2 \ldots, n\}$ and $E = \{(p, q) \mid p, q \in V$ and p, q differ in exactly one position$\}$. By definition, $A_{n,k}$ is a regular graph of degree $k(n-k)$ with $n!/(n-k)!$ vertices. $A_{n,1}$ is isomorphic to the complete graph K_n, and $A_{n,n-1}$ is isomorphic to the n-dimensional star graph. Moreover, $A_{n,k}$ is vertex-symmetric and edge-symmetric [83]. Figure 11.17 illustrates $A_{4,2}$.

The following theorem is proved by Hsu et al. [173].

THEOREM 11.17 $\mathcal{H}_f(A_{n,k}) = k(n-k) - 2$ and $\mathcal{H}_f^{\kappa}(A_{n,k}) = k(n-k) - 3$ for any two positive integers with $n - k \geq 2$.

11.6 FAULT-TOLERANT HAMILTONICITY AND FAULT-TOLERANT HAMILTONIAN CONNECTIVITY OF AUGMENTED CUBES

We have discussed the fault-tolerant Hamiltonicity of several families of interconnection networks through the aid of fault-tolerant Hamiltonian connectivity. Yet we may need other properties to discuss the fault-tolerant Hamiltonicity for other interconnection networks. Hsu et al. [171] discussed the fault-tolerant Hamiltonicity of augmented cubes by adding other properties.

The augmented cube AQ_n, proposed by Choudum and Sunitha [66], is one of the variations of hypercubes. Assume that $n \geq 1$ is an integer. The graph of the *n-dimensional augmented cube*, denoted by AQ_n, has 2^n vertices, each labeled by an n-bit binary string $V(AQ_n) = \{u_1u_2 \ldots u_n \mid u_i \in \{0, 1\}\}$. AQ_1 is the graph K_2 with vertex set $\{0, 1\}$. For $n \geq 2$, AQ_n can be recursively constructed by two copies of AQ_{n-1}, denoted by AQ_{n-1}^0 and AQ_{n-1}^1, and by adding 2^n edges between AQ_{n-1}^0 and AQ_{n-1}^1 as follows:

Let $V(AQ_{n-1}^0) = \{0u_2u_3 \dots u_n \mid u_i = 0 \text{ or } 1 \text{ for } 2 \le i \le n\}$ and $V(AQ_{n-1}^1) = \{1v_2 v_3 \dots v_n \mid v_i = 0 \text{ or } 1 \text{ for } 2 \le i \le n\}$. A vertex $\mathbf{u} = 0u_2u_3 \dots u_n$ of AQ_{n-1}^0 is adjacent to a vertex $\mathbf{v} = 1v_2v_3 \dots v_n$ of AQ_{n-1}^1 if and only if one of the following cases holds.

1. $u_i = v_i$, for $2 \le i \le n$. In this case, $(\mathbf{u}, \mathbf{v})$ is called a *hypercube edge*. We set $\mathbf{v} = \mathbf{u}^h$.
2. $u_i = \overline{v}_i$, for $2 \le i \le n$. In this case, $(\mathbf{u}, \mathbf{v})$ is called a *complement edge*. We set $\mathbf{v} = \mathbf{u}^c$.

The augmented cubes AQ_1, AQ_2, AQ_3, and AQ_4 are illustrated in Figure 11.18. It is proved in Ref. 66 that AQ_n is a vertex-transitive, $(2n-1)$-regular, and $(2n-1)$-connected graph with 2^n vertices for any positive integer n. Let $E_n^h = \{(\mathbf{u}, \mathbf{u}^h) \mid \mathbf{u} \in V(AQ_{n-1}^0)\}$ and $E_n^c = \{(\mathbf{u}, \mathbf{u}^c) \mid \mathbf{u} \in V(AQ_{n-1}^0)\}$. Obviously, E_n^h and E_n^c are two perfect matchings between the vertices of AQ_{n-1}^0 and AQ_{n-1}^1. Let i be any index with $1 \le i \le n$ and $\mathbf{u} = u_1u_2u_3 \dots u_n$ be a vertex of AQ_n. We use $\mathbf{u}^i$ to denote the vertex $\mathbf{v} = v_1v_2v_3 \dots v_n$ such that $u_j = v_j$ with $1 \le j \ne i \le n$ and $u_i = \overline{v}_i$. Moreover, we use $\mathbf{u}^{i*}$ to denote the vertex $\mathbf{v} = v_1v_2v_3 \dots v_n$ such that $u_j = v_j$ for $j < i$ and $u_j = \overline{v}_j$ for $i \le j \le n$. Obviously, $\mathbf{u}^n = \mathbf{u}^{n*}$, $\mathbf{u}^1 = \mathbf{u}^h$, $\mathbf{u}^c = \mathbf{u}^{1*}$, and $Nbd_{AQ_n}(\mathbf{u}) = \{\mathbf{u}^i \mid 1 \le i \le n\} \cup \{\mathbf{u}^{i*} \mid 1 \le i < n\}$. Let $E_n^{2*} = \{(\mathbf{u}, \mathbf{u}^{2*}) \mid \mathbf{u} \in V(AQ_n)\}$.

The following lemmas are derived directly from the definition.

LEMMA 11.17 Let $\mathbf{x}$ and $\mathbf{y}$ be any two vertices in AQ_{n-1}^0 with $n \ge 4$. Assume that $(\mathbf{x}, \mathbf{y}) \notin E_n^{2*}$. Then $\mathbf{x}^h$, $\mathbf{x}^c$, $\mathbf{y}^h$, and $\mathbf{y}^c$ are all distinct.

LEMMA 11.18 Suppose that $(\mathbf{u}, \mathbf{v}) \in E(AQ_{n-1}^0)$. Then $(\mathbf{u}^h, \mathbf{v}^h) \in E(AQ_{n-1}^1)$ and $(\mathbf{u}^c, \mathbf{v}^c) \in E(AQ_{n-1}^1)$.

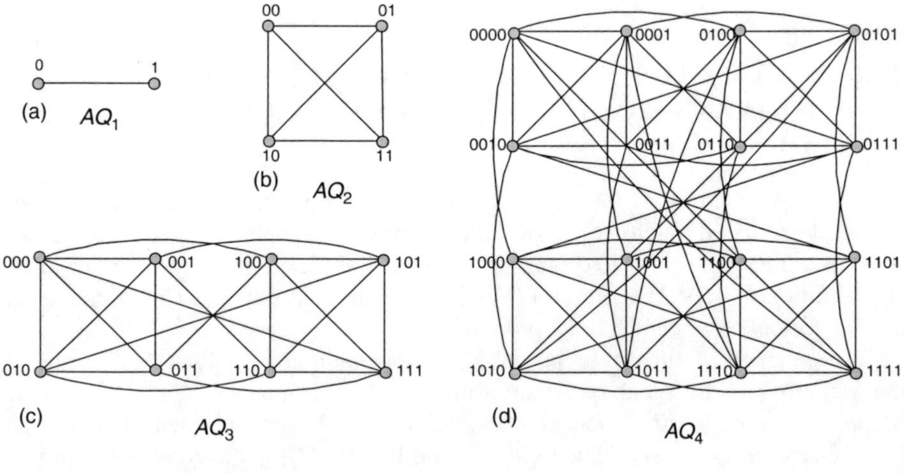

FIGURE 11.18 The augmented cubes AQ_1, AQ_2, AQ_3, and AQ_4.

LEMMA 11.19 Let $\mathbf{u}$ be a vertex in AQ_{n-1}^0. Let $A(\mathbf{u}) = \{\mathbf{v} \mid (\mathbf{u}, \mathbf{v}) \in E(AQ_{n-1}^0)\}$ and $B(\mathbf{u}) = \{\mathbf{x}^h \mid \mathbf{x} \in A(\mathbf{u})\} \cup \{\mathbf{x}^c \mid \mathbf{x} \in A(\mathbf{u})\}$. Then $|A(\mathbf{u})| = 2n-3$ and $|B(\mathbf{u})| = 4n-8$.

To discuss the fault-tolerant Hamiltonicity and the fault-tolerant Hamiltonian-connectivity of AQ_n, we need the following two terms for graphs.

A graph G satisfies *property 2H* if it processes the following condition:

Let $\{w, x\}$ and $\{y, z\}$ be any two pairs of four distinct vertices of G. There exist two disjoint paths P_1 and P_2 of G such that (1) P_1 joins w to x, (2) P_2 joins y to z, and (3) $P_1 \cup P_2$ span G.

A Hamiltonian-connected graph $G = (V, E)$ is said to be 1-*vertex fault-tolerant Hamiltonian-connected* if $G - \{u\}$ is still Hamiltonian-connected after removing any vertex $u \in V$.

LEMMA 11.20 Any 1-vertex fault-tolerant Hamiltonian-connected graph is Hamiltonian-connected.

Proof: Assume that G is a 1-vertex fault-tolerant Hamiltonian-connected graph. Obviously, the lemma is true if G has two or three vertices. Assume that G has at least three vertices. Then $\kappa(G) \geq \delta(G) \geq 3$. Let x and y be any two vertex of G. Let z be a neighbor of x such that $y \neq z$. Since G is 1-vertex fault-tolerant Hamiltonian-connected, there exists a Hamiltonian path P of $G - x$ joining z to y. Obviously, $\langle x, z, P, y \rangle$ forms a Hamiltonian path between x and y. Hence, G is Hamiltonian-connected. □

Furthermore, we say a graph G is *k-vertex fault-tolerant Hamiltonian-connected* if $G - F$ is Hamiltonian-connected for any $F \subset V$ with $|F| = k$.

COROLLARY 11.7 *Assume that G is a k-vertex fault-tolerant Hamiltonian-connected graph. Then G is r-vertex fault-tolerant Hamiltonian-connected for $0 \leq r \leq k$.*

Let $F \subset V(AQ_n) \cup E(AQ_n)$ be any faulty set of AQ_n. An edge $(\mathbf{u}, \mathbf{v})$ is *F-fault free* if $(\mathbf{u}, \mathbf{v}) \notin F$, $\mathbf{u} \notin F$, and $\mathbf{v} \notin F$; otherwise, it is *F-fault*. Let $H = (V', E')$ be a subgraph of AQ_n. We use $F(H)$ to denote the set $(V' \cup E') \cap F$.

By brute force, we have the following three lemmas.

LEMMA 11.21 $\mathcal{H}_f(AQ_3) = 2$.

LEMMA 11.22 $\mathcal{H}_f^k(AQ_3) = 1$.

LEMMA 11.23 $\mathcal{H}_f(AQ_4) = 5$ and $H_f^k(AQ_4) = 4$. Moreover, AQ_4 has property $2H$.

LEMMA 11.24 Assume that $n \geq 4$. Suppose that AQ_{n-1} is 1-vertex fault-tolerant Hamiltonian-connected and has property $2H$. Then AQ_n also has property $2H$.

Proof: Suppose that $\{\mathbf{w},\mathbf{x}\}$ and $\{\mathbf{y},\mathbf{z}\}$ are two pairs of four distinct vertices of AQ_n. We are going to construct two disjoint spanning paths R_1 and R_2 of AQ_n such that R_1 joins $\mathbf{w}$ and $\mathbf{x}$ and R_2 joins $\mathbf{y}$ and $\mathbf{z}$. We consider the following four cases:

Case 1. $\{\mathbf{w},\mathbf{x},\mathbf{y},\mathbf{z}\} \subset V(AQ^0_{n-1})$. With the assumption of this lemma, AQ_{n-1} has property $2H$. There exist two disjoint paths $\langle \mathbf{w}, P_1, \mathbf{x}\rangle$ and $\langle \mathbf{y}, P_2, \mathbf{z}\rangle$ spanning AQ^0_{n-1}. Without loss of generality, we can assume that $l(P_1) \le l(P_2)$. Since $n \ge 4$, $l(P_2) \ge 4$. Then P_2 can be written as $\langle \mathbf{y}, P_4, \mathbf{u}, \mathbf{v}, \mathbf{z}\rangle$. By Lemma 11.18, $(\mathbf{u}^h, \mathbf{v}^h) \in E(AQ^1_{n-1})$. With the assumption of this lemma, AQ^1_{n-1} is 1-vertex fault-tolerant Hamiltonian-connected. Then AQ^1_{n-1} is Hamiltonian-connected. Thus, there exists a Hamiltonian path $\langle \mathbf{u}^h, P_3, \mathbf{v}^h\rangle$ in AQ^1_{n-1}. Then $R_1 = \langle \mathbf{w}, P_1, \mathbf{x}\rangle$ and $R_2 = \langle \mathbf{y}, P_4, \mathbf{u}, \mathbf{u}^h, P_3, \mathbf{v}^h, \mathbf{v}, \mathbf{z}\rangle$ form the desired disjoint spanning paths of AQ_n. See Figure 11.19a for illustration.

Case 2. $\{\mathbf{w},\mathbf{x},\mathbf{y}\} \subset V(AQ^0_{n-1})$ and $\mathbf{z} \in V(AQ^1_{n-1})$. Let $\mathbf{y}^*$ be a vertex in $\{\mathbf{y}^h, \mathbf{y}^c\} - \{\mathbf{z}\}$. With the assumption of this lemma, there exists a Hamiltonian path $\langle \mathbf{y}^*, P_2, \mathbf{z}\rangle$ in AQ^1_{n-1} and there exists a Hamiltonian path $\langle \mathbf{w}, P_1, \mathbf{x}\rangle$ in $AQ^0_{n-1} - \{\mathbf{y}\}$. Obviously, $R_1 = \langle \mathbf{w}, P_1, \mathbf{x}\rangle$ and $R_2 = \langle \mathbf{y}, \mathbf{y}^*, P_2, \mathbf{z}\rangle$ form the desired disjoint spanning paths of AQ_n. See Figure 11.19b for illustration.

Case 3. $\{\mathbf{w},\mathbf{x}\} \subset V(AQ^0_{n-1})$ and $\{\mathbf{y},\mathbf{z}\} \subset V(AQ^1_{n-1})$. With the assumption of this lemma, there is Hamiltonian path $R_1 = \langle \mathbf{w}, P_1, \mathbf{x}\rangle$ in AQ^0_{n-1}, and there is a Hamiltonian path $R_2 = \langle \mathbf{y}, P_2, \mathbf{z}\rangle$ in AQ^1_{n-1}. Then R_1 and R_2 form the desired disjoint spanning paths of AQ_n. See Figure 11.19c for illustration.

Case 4. $\{\mathbf{w},\mathbf{y}\} \subset V(AQ^0_{n-1})$ and $\{\mathbf{x},\mathbf{z}\} \subset V(AQ^1_{n-1})$.

Suppose that $\{(\mathbf{w},\mathbf{x}),(\mathbf{y},\mathbf{z})\} \cap E(AQ_n) \ne \emptyset$. Without loss of generality, we can assume that $(\mathbf{w},\mathbf{x}) \in E(AQ_n)$. Obviously, there exists a vertex $\mathbf{u}$ in $V(AQ^0_{n-1}) - \{\mathbf{y}, \mathbf{z}^h\}$. With the assumption of this lemma, there exists a Hamiltonian path $\langle \mathbf{y}, P_1, \mathbf{u}\rangle$ of $AQ^0_{n-1} - \{\mathbf{w}\}$ and there exists a Hamiltonian path $\langle \mathbf{u}^h, P_2, \mathbf{z}\rangle$ of $AQ^1_{n-1} - \{\mathbf{x}\}$. Then $R_1 = \langle \mathbf{w}, \mathbf{x}\rangle$ and $R_2 = \langle \mathbf{y}, P_1, \mathbf{u}, \mathbf{u}^h, P_2, \mathbf{z}\rangle$ form the desired disjoint spanning paths of AQ_n. See Figure 11.19d for illustration.

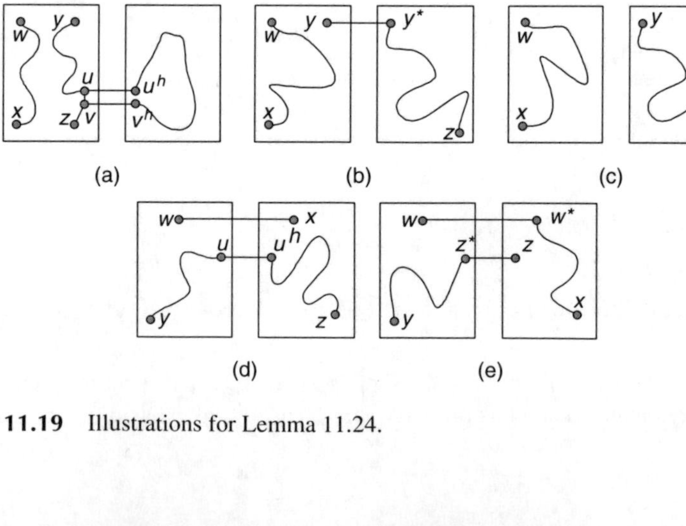

FIGURE 11.19 Illustrations for Lemma 11.24.

Suppose that $\{(\mathbf{w},\mathbf{x}),(\mathbf{y},\mathbf{z})\} \cap E(AQ_n) = \emptyset$. Let $\mathbf{w}^*$ be a vertex in $\{\mathbf{w}^h, \mathbf{w}^c\} - \{\mathbf{z}\}$ and $\mathbf{z}^*$ be a vertex in $\{\mathbf{z}^h, \mathbf{z}^c\} - \{\mathbf{z}^*\}$. Again, there exists a Hamiltonian path $\langle \mathbf{w}^*, P_1, \mathbf{x}\rangle$ in $AQ^1_{n-1} - \{\mathbf{z}\}$, and there exists a Hamiltonian path $\langle \mathbf{y}, P_2, \mathbf{z}^*\rangle$ in $AQ^0_{n-1} - \{\mathbf{w}\}$. Obviously, $R_1 = \langle \mathbf{w}, \mathbf{w}^*, P_1, \mathbf{x}\rangle$ and $R_2 = \langle \mathbf{y}, P_2, \mathbf{z}^*, \mathbf{z}\rangle$ form the desired disjoint spanning paths of AQ_n. See Figure 11.19e for illustration.

Hence, the lemma is proved. □

LEMMA 11.25 Let $n \geq 5$. Assume that AQ_{n-1} is $(2n-5)$-fault-tolerant Hamiltonian, $(2n-6)$-fault-tolerant Hamiltonian-connected, and has property $2H$. Then AQ_n is $(2n-3)$-fault-tolerant Hamiltonian.

Proof: Let F be any subset of $V(AQ_n) \cup E(AQ_n)$ with $|F| \leq 2n-3$. Without loss of generality, we assume that $|F(AQ^0_{n-1})| \geq |F(AQ^1_{n-1})|$. We are going to construct a Hamiltonian cycle of $AQ_n - F$ depending on the following cases:

Case 1. $|F(AQ^0_{n-1})| \leq 2n-6$. Then $AQ^0_{n-1} - F(AQ^0_{n-1})$ and $AQ^1_{n-1} - F(AQ^1_{n-1})$ are Hamiltonian-connected. Note that $|E^h_n \cup E^c_n| - |F| \geq 2^n - (2n-3) \geq 25$. There are two distinct F-fault free edges $(\mathbf{u},\mathbf{v})$ and $(\mathbf{x},\mathbf{y})$ such that $\mathbf{u}$ and $\mathbf{x}$ are in $AQ^0_{n-1} - F(AQ^0_{n-1})$, and $\mathbf{v}$ and $\mathbf{y}$ are in $AQ^1_{n-1} - F(AQ^1_{n-1})$. By assumption, there exist Hamiltonian paths $\langle \mathbf{u}, P_1, \mathbf{x}\rangle$ in $AQ^0_{n-1} - F(AQ^0_{n-1})$ and $\langle \mathbf{v}, P_2, \mathbf{y}\rangle$ in $AQ^1_{n-1} - F(AQ^1_{n-1})$. Obviously, $\langle \mathbf{u}, \mathbf{v}, P_2, \mathbf{y}, \mathbf{x}, P_1, \mathbf{u}\rangle$ forms a Hamiltonian cycle of $AQ_n - F$. See Figure 11.20a for illustration.

Case 2. $|F(AQ^0_{n-1})| = 2n-5$. Obviously, $|F - F(AQ^0_{n-1})| \leq 2$. Since AQ_{n-1} is $(2n-5)$-fault-tolerant Hamiltonian, there is a Hamiltonian cycle C in $AQ^0_{n-1} - F(AQ^0_{n-1})$. Note that $l(C)/2 \geq (2^{n-1} - 2n + 5)/2 > 5$. There exists an edge $(\mathbf{u},\mathbf{v})$ in C such that $(\mathbf{u},\mathbf{u}^h)$ and $(\mathbf{v},\mathbf{v}^h)$ are F-fault free. We can write C as $\langle \mathbf{u}, P_1, \mathbf{v}, \mathbf{u}\rangle$. Since AQ_{n-1} is $(2n-6)$-fault-tolerant Hamiltonian-connected, there is a Hamiltonian path $\langle \mathbf{u}^h, P_2, \mathbf{v}^h\rangle$ of $AQ^1_{n-1} - F(AQ^1_{n-1})$. Obviously, $\langle \mathbf{u}, \mathbf{u}^h, P_2, \mathbf{v}^h, \mathbf{v}, P_1, \mathbf{u}\rangle$ forms a Hamiltonian cycle of $AQ_n - F$. See Figure 11.20b for illustration.

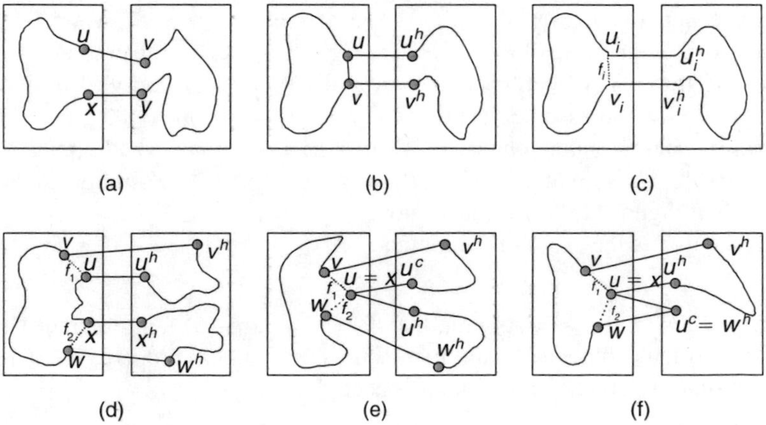

FIGURE 11.20 Illustrations for Lemma 11.25.

Case 3. $|F(AQ^0_{n-1})| = 2n-4$. Suppose that there exists a Hamiltonian cycle C of $AQ^0_{n-1} - F(AQ^0_{n-1})$. As in Case 2, there exists an edge $(\mathbf{u}, \mathbf{v})$ in C such that $(\mathbf{u}, \mathbf{u}^h)$ and $(\mathbf{v}, \mathbf{v}^h)$ are F-fault free. We can write C as $\langle \mathbf{u}, P_1, \mathbf{v}, \mathbf{u} \rangle$. Since AQ_{n-1} is $(2n-6)$-fault-tolerant Hamiltonian-connected, there is a Hamiltonian path P_2 of $AQ^1_{n-1} - F(AQ^1_{n-1})$ joining $\mathbf{u}^h$ and $\mathbf{v}^h$. Then $\langle \mathbf{u}, \mathbf{u}^h, P_2, \mathbf{v}^h, \mathbf{v}, P_1, \mathbf{u} \rangle$ forms a Hamiltonian cycle of $AQ_n - F$. Suppose that there is no Hamiltonian cycle of $AQ^0_{n-1} - F(AQ^0_{n-1})$. Since AQ_{n-1} is $(2n-5)$-fault-tolerant Hamiltonian, there exists a Hamiltonian cycle C_i of $AQ^0_{n-1} - (F(AQ^0_{n-1}) - \{f_i\})$ for every $f_i \in F(AQ^0_{n-1})$. We can write C_i as $\langle \mathbf{u}_i, P_i, \mathbf{v}_i, f_i, \mathbf{u}_i \rangle$. Since $|F(AQ^0_{n-1})| \geq 6$ and $|F - F(AQ^0_{n-1})| \leq 1$, there exists an index i such that $(\mathbf{u}_i, \mathbf{u}_i^h)$ and $(\mathbf{v}_i, \mathbf{v}_i^h)$ are F-fault free. Since AQ_{n-1} is $(2n-6)$-fault-tolerant Hamiltonian-connected, there is a Hamiltonian path Q_1 in $AQ^1_{n-1} - F(AQ^1_{n-1})$ joining $\mathbf{u}_i^h$ and $\mathbf{v}_i^h$. Obviously, $\langle \mathbf{u}_i, \mathbf{u}_i^h, Q_1, \mathbf{v}_i^h, \mathbf{v}_i, P_i, \mathbf{u}_i \rangle$ forms a Hamiltonian cycle in $AQ_n - F$. See Figure 11.20c for illustration.

Case 4. $|F(AQ^0_{n-1})| = 2n-3$. Thus, $|F - F(AQ^0_{n-1})| = 0$. Let f_1 and f_2 be any two elements in $F(AQ^0_{n-1})$. Since AQ_{n-1} is $(2n-5)$-fault-tolerant Hamiltonian, there exists a Hamiltonian cycle $C = \langle \mathbf{u}, f_1', \mathbf{v}, P_1, \mathbf{w}, f_2', \mathbf{x}, P_2, \mathbf{u} \rangle$ in $AQ^0_{n-1} - (F(AQ^0_{n-1}) - \{f_1, f_2\})$ where $f_1' = f_1$ if f_1 is on C or f_1' is an arbitrary edge of C otherwise and $f_2' = f_2$ if f_2 is on C or f_2' is an arbitrary edge of C otherwise. Without loss of generality, we assume that $l(P_2) \leq l(P_1)$.

Case 4.1. $l(P_2) \geq 1$. Then $\mathbf{u}^h$, $\mathbf{v}^h$, $\mathbf{w}^h$, and $\mathbf{x}^h$ are distinct. Since AQ_{n-1} has property $2H$, there exist two disjoint spanning paths $\langle \mathbf{u}^h, P_3, \mathbf{v}^h \rangle$ and $\langle \mathbf{w}^h, P_4, \mathbf{x}^h \rangle$ in AQ^1_{n-1}. Obviously, $\langle \mathbf{u}, \mathbf{u}^h, P_3, \mathbf{v}^h, \mathbf{v}, P_1, \mathbf{w}, \mathbf{w}^h, P_4, \mathbf{x}^h, \mathbf{x}, P_2, \mathbf{u} \rangle$ forms a Hamiltonian cycle of $AQ_n - F$. See Figure 11.20d for illustration.

Case 4.2. $l(P_2) = 0$. Then $\mathbf{u} = \mathbf{x}$. Moreover $\mathbf{w}$, $\mathbf{u}$, and $\mathbf{v}$ are distinct. Obviously, one of the following cases holds: (1) $\{\mathbf{w}^h, \mathbf{w}^c, \mathbf{u}^h, \mathbf{u}^c, \mathbf{v}^h, \mathbf{v}^c\}$ are all distinct, (2) $(\mathbf{w}^h = \mathbf{v}^c$ and $\mathbf{w}^c = \mathbf{v}^h)$, (3) $(\mathbf{w}^h = \mathbf{u}^c$ and $\mathbf{w}^c = \mathbf{u}^h)$, and (4) $(\mathbf{v}^h = \mathbf{u}^c$ and $\mathbf{v}^c = \mathbf{u}^h)$.

Suppose that (1) $\{\mathbf{w}^h, \mathbf{w}^c, \mathbf{u}^h, \mathbf{u}^c, \mathbf{v}^h, \mathbf{v}^c\}$ are all distinct or (2) $(\mathbf{w}^h = \mathbf{v}^c$ and $\mathbf{w}^c = \mathbf{v}^h)$. Since AQ_{n-1} has property $2H$, there exist two disjoint spanning paths $\langle \mathbf{w}^h, P_5, \mathbf{u}^h \rangle$ and $\langle \mathbf{u}^c, P_6, \mathbf{v}^h \rangle$ in AQ^1_{n-1}. Then $\langle \mathbf{u}, \mathbf{u}^c, P_6, \mathbf{v}^h, \mathbf{v}, P_1, \mathbf{w}, \mathbf{w}^h, P_5, \mathbf{u}^h, \mathbf{u} \rangle$ forms a Hamiltonian cycle of $AQ_n - F$. See Figure 11.20e for illustration.

Suppose that (3) $(\mathbf{w}^h = \mathbf{u}^c$ and $\mathbf{w}^c = \mathbf{u}^h)$ or (4) $(\mathbf{v}^h = \mathbf{u}^c$ and $\mathbf{v}^c = \mathbf{u}^h)$. Without loss of generality, we assume that $\mathbf{w}^h = \mathbf{u}^c$ and $\mathbf{w}^c = \mathbf{u}^h$. Since AQ_{n-1} is $(2n-6)$-fault-tolerant Hamiltonian-connected, there is a Hamiltonian path $\langle \mathbf{u}^h, P_7, \mathbf{v}^h \rangle$ in $AQ^1_{n-1} - \{\mathbf{w}^h\}$. Thus $\langle \mathbf{u}, \mathbf{u}^h, P_7, \mathbf{v}^h, \mathbf{v}, P_1, \mathbf{w}, \mathbf{w}^h, \mathbf{u} \rangle$ forms a Hamiltonian cycle of $AQ_n - F$. See Figure 11.20f for illustration.

This completes the proof of the lemma. □

LEMMA 11.26 Let $n \geq 5$. Assume that AQ_{n-1} is $(2n-5)$-fault-tolerant Hamiltonian, $(2n-6)$-fault-tolerant Hamiltonian-connected, and has property $2H$. Then AQ_n is $(2n-4)$-fault-tolerant Hamiltonian-connected.

Proof: Let F be any subset of $V(AQ_n) \cup E(AQ_n)$ with $|F| \leq 2n-4$. Without loss of generality, we can assume that $|F(AQ^0_{n-1})| \geq |F(AQ^1_{n-1})|$. We need to construct a

Hamiltonian path joining **u** and **v** in $AQ_n - F$ for any two different vertices **u** and $\mathbf{v} \in V(AQ_n) - F$.

Case 1. $|F(AQ^0_{n-1})| \leq 2n-6$. Since AQ_{n-1} is $(2n-6)$-fault-tolerant Hamiltonian-connected, both $AQ^0_{n-1} - F(AQ^0_{n-1})$ and $AQ^1_{n-1} - F(AQ^1_{n-1})$ are Hamiltonian-connected. By symmetric property of AQ_n, we have the following two sub cases:

Case 1.1. $\{\mathbf{u},\mathbf{v}\} \subset V(AQ^0_{n-1})$. Since $AQ^0_{n-1} - F(AQ^0_{n-1})$ is Hamiltonian-connected, there exists a Hamiltonian path P of $AQ^0_{n-1} - F(AQ^0_{n-1})$ joining **u** and **v**. We claim that there exists an edge $(\mathbf{w},\mathbf{x})$ on P such that both $(\mathbf{w},\mathbf{w}^h)$ and $(\mathbf{x},\mathbf{x}^h)$ are F-fault free. Assume that there is no such edge. Then at least $\lfloor l(P)/2 \rfloor$ edges in $E^h_n \cup E^c_n$ are F-fault. Since $l(P) \geq 2^{n-1} - |F(AQ^0_{n-1})|$, $|F| = |F(AQ^0_{n-1})| + \lfloor (2^{n-1} - |F(AQ^0_{n-1})|)/2 \rfloor \geq 2^{n-2} + |F(AQ^0_{n-1})|/2 > 2n-4$. We get a contradiction. Hence, such an edge exists as claimed. We can write P as $\langle \mathbf{u}, P_1, \mathbf{w}, \mathbf{x}, P_2, \mathbf{v} \rangle$. Since $AQ^1_{n-1} - F(AQ^1_{n-1})$ is also Hamiltonian-connected, there is a Hamiltonian path $\langle \mathbf{w}^h, P_3, \mathbf{x}^h \rangle$ in $AQ^1_{n-1} - F(AQ^1_{n-1})$. Obviously, $\langle \mathbf{u}, P_1, \mathbf{w}, \mathbf{w}^h, P_3, \mathbf{x}^h, \mathbf{x}, P_2, \mathbf{v} \rangle$ forms a Hamiltonian path of $AQ_n - F$ joining **u** and **v**. See Figure 11.21a for illustration.

Case 1.2. $\mathbf{u} \in AQ^0_{n-1}$ and $\mathbf{v} \in AQ^1_{n-1}$. Since $|F| \leq 2n-4 < 2^{n-1}$, there exists an F-fault free edge $(\mathbf{w},\mathbf{w}^h)$ such that $\mathbf{w} \in V(AQ^0_{n-1})$, $\mathbf{w} \neq \mathbf{u}$, and $\mathbf{w}^h \neq \mathbf{v}$. Since AQ_{n-1} is $(2n-6)$-fault-tolerant Hamiltonian-connected, there is a Hamiltonian path $\langle \mathbf{u}, P_1, \mathbf{w} \rangle$ in $AQ^0_{n-1} - F(AQ^0_{n-1})$ and there is a Hamiltonian path $\langle \mathbf{w}^h, P_2, \mathbf{v} \rangle$ in $AQ^1_{n-1} - F(AQ^1_{n-1})$. Obviously, $\langle \mathbf{u}, P_1, \mathbf{w}, \mathbf{w}^h, P_2, \mathbf{v} \rangle$ forms a Hamiltonian path of $AQ_n - F$ joining **u** and **v**. See Figure 11.21b for illustration.

Case 2. $|F(AQ^0_{n-1})| = 2n-5$. Thus, $|F - F(AQ^0_{n-1})| \leq 1$.

Case 2.1. $\mathbf{u},\mathbf{v} \in V(AQ^0_{n-1})$. Assume that there exists a Hamiltonian path P of $AQ^0_{n-1} - F(AQ^0_{n-1})$ joining **u** and **v**. Since $(l(P) - |F(AQ^0_{n-1})|)/2 \geq 2n-4$, there exists an edge $(\mathbf{w},\mathbf{x})$ in P such that $(\mathbf{w},\mathbf{w}^h)$ and $(\mathbf{x},\mathbf{x}^h)$ are F-fault free. We can write P as $\langle \mathbf{u}, P_1, \mathbf{w}, \mathbf{x}, P_2, \mathbf{v} \rangle$. Since AQ_{n-1} is $(2n-6)$-fault-tolerant Hamiltonian-connected, there exists a Hamiltonian path P_3 of $AQ^1_{n-1} - F(AQ^1_{n-1})$ joining $\mathbf{w}^h$ and $\mathbf{x}^h$. Therefore, $\langle \mathbf{u}, P_1, \mathbf{w}, \mathbf{w}^h, P_3, \mathbf{x}^h, \mathbf{x}, P_2, \mathbf{v} \rangle$ forms a Hamiltonian path of $AQ_n - F$ joining **u** and **v**. Suppose that there is no Hamiltonian path of $AQ^0_{n-1} - F(AQ^0_{n-1})$ joining **u** and **v**. Since AQ_{n-1} is $(2n-6)$-fault-tolerant Hamiltonian-connected,

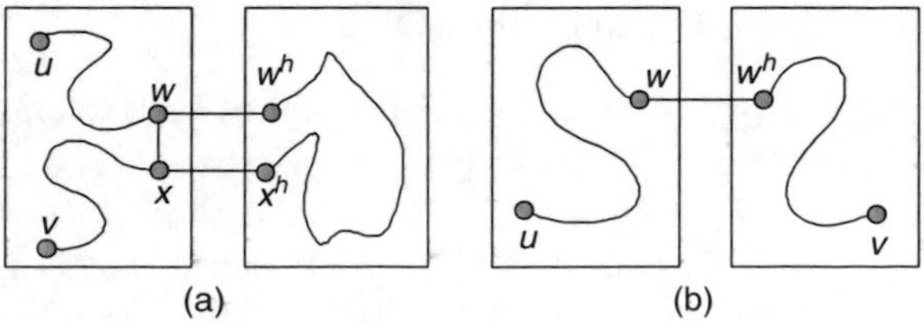

FIGURE 11.21 Illustrations for Case 1 of Lemma 11.26.

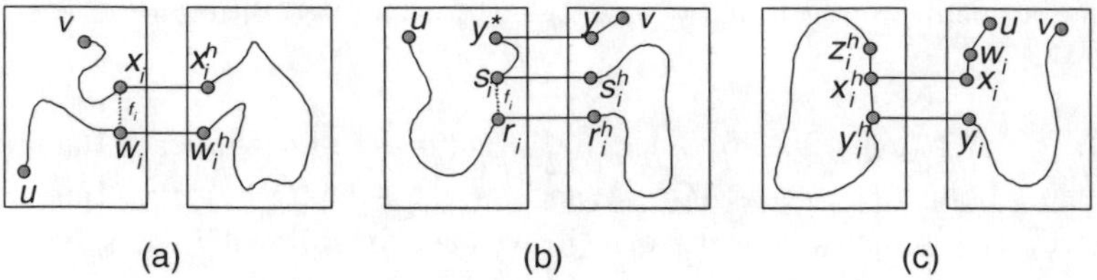

FIGURE 11.22 Illustrations for Case 2 of Lemma 11.26.

there exists a Hamiltonian path Q_i of $AQ^0_{n-1} - (F(AQ^0_{n-1}) - \{f_i\})$ joining **u** and **v** for every $f_i \in F(AQ^0_{n-1})$. We can write Q_i as $\langle \mathbf{u}, P^1_i, \mathbf{w}_i, f_i, \mathbf{x}_i, P^2_i, \mathbf{v}\rangle$. Since $|F(AQ^0_{n-1})| = 2n-5 \geq 5$ and $|F - F(AQ^0_{n-1})| \leq 1$, there exists an index i such that $(\mathbf{w}_i, \mathbf{w}^h_i)$ and $(\mathbf{x}_i, \mathbf{x}^h_i)$ are F-fault free. Since AQ^1_{n-1} is $(2n-6)$-fault-tolerant Hamiltonian-connected, there is a Hamiltonian path R_1 of $AQ^1_{n-1} - F(AQ^1_{n-1})$ joining $\mathbf{w}^h_i$ and $\mathbf{x}^h_i$. Thus, $\langle \mathbf{u}, P^1_i, \mathbf{w}_i, \mathbf{w}^h_i, R_1, \mathbf{x}^h_i, \mathbf{x}_i, P^2_i, \mathbf{v}\rangle$ forms a Hamiltonian path of $AQ_n - F$ joining **u** and **v**. See Figure 11.22a for illustration.

Case 2.2. $\mathbf{u} \in V(AQ^0_{n-1})$ and $\mathbf{v} \in V(AQ^1_{n-1})$. Let $A(\mathbf{v}) = \{\mathbf{x} \mid (\mathbf{x}, \mathbf{v}) \in E(AQ_n)$ and $\mathbf{x} \in V(AQ^1_{n-1})\}$. Then $|A(\mathbf{v})| = 2n-3$. Let $B(\mathbf{v}) = \{\mathbf{y}^h \mid \mathbf{y} \in A(\mathbf{v})\} \cup \{\mathbf{y}^c \mid \mathbf{y} \in A(\mathbf{v})\}$. By Lemma 11.19, $|B(\mathbf{v})| = 4n-8$. Since $|F| \leq 2n-4$, there exists a $\mathbf{y} \in A(\mathbf{v})$ such that either $(\mathbf{y}, \mathbf{y}^h)$ or $(\mathbf{y}, \mathbf{y}^c)$, say $(\mathbf{y}, \mathbf{y}^*)$, is F-fault free and $\mathbf{y}^* \neq \mathbf{u}$. Assume that there exists a Hamiltonian path P of $AQ^0_{n-1} - F(AQ^0_{n-1})$ joining **u** and $\mathbf{y}^*$. Since $(l(P) - |F(AQ^0_{n-1})|)/2 \geq 2n-4$, there exists an edge $(\mathbf{w}, \mathbf{x})$ in P such that $(\mathbf{w}, \mathbf{w}^h)$ and $(\mathbf{x}, \mathbf{x}^h)$ are F-fault free and $\mathbf{w}^h$, $\mathbf{x}^h$, and **v** are distinct. We can write P as $\langle \mathbf{u}, P_1, \mathbf{w}, \mathbf{x}, P_2, \mathbf{y}^*\rangle$. Since AQ_{n-1} is $(2n-6)$-fault-tolerant Hamiltonian-connected, there exists a Hamiltonian path P_3 of $AQ^1_{n-1} - F(AQ^1_{n-1}) - \{\mathbf{v}, \mathbf{y}\}$ joining $\mathbf{w}^h$ and $\mathbf{x}^h$. Then $\langle \mathbf{u}, P_1, \mathbf{w}, \mathbf{w}^h, P_3, \mathbf{x}^h, \mathbf{x}, P_2, \mathbf{y}^*, \mathbf{y}, \mathbf{v}\rangle$ forms a Hamiltonian path of $AQ_n - F$ joining **u** and **v**. Assume that there is no Hamiltonian path of $AQ^0_{n-1} - F(AQ^0_{n-1})$ joining **u** and $\mathbf{y}^*$. Since AQ_{n-1} is $(2n-6)$-fault-tolerant Hamiltonian-connected, there exists a Hamiltonian path Q_i of $AQ^0_{n-1} - (F(AQ^0_{n-1}) - \{f_i\})$ joining **u** and $\mathbf{y}^*$ for every $f_i \in F(AQ^0_{n-1})$. We can write Q_i as $\langle \mathbf{u}, P^1_i, \mathbf{r}_i, f_i, \mathbf{s}_i, P^2_i, \mathbf{y}^*\rangle$. Since $|F(AQ^0_{n-1})| = 2n-5 \geq 5$ and $|F - F(AQ^0_{n-1})| \leq 1$, there exists an index i such that $(\mathbf{r}_i, \mathbf{r}^h_i)$ and $(\mathbf{s}_i, \mathbf{s}^h_i)$ are F-fault free. Since AQ^1_{n-1} is $(2n-6)$-fault-tolerant Hamiltonian-connected, there is a Hamiltonian path R_1 of $AQ^1_{n-1} - (F(AQ^1_{n-1}) - \{\mathbf{y}, \mathbf{v}\})$ joining $\mathbf{r}^h_i$ and $\mathbf{s}^h_i$. Then $\langle \mathbf{u}, P^1_i, \mathbf{r}_i, \mathbf{r}^h_i, R_1, \mathbf{s}^h_i, \mathbf{s}_i, P^2_i, \mathbf{y}^*, \mathbf{y}, \mathbf{v}\rangle$ forms a Hamiltonian path of $AQ_n - F$ joining **u** and **v**. See Figure 11.22b for illustration.

Case 2.3. $\{\mathbf{u}, \mathbf{v}\} \subset V(AQ^1_{n-1})$. Since AQ_{n-1} is $(2n-5)$-fault-tolerant Hamiltonian, there is a Hamiltonian cycle C of $AQ^0_{n-1} - F$. Let $A = \{1x_2 \dots x_n \mid$ there exists some i and j in $\{2, \dots, n\}$ such that $x_i = \bar{u}_i, x_j = \bar{u}_j$, and $x_k = u_k$ if $k \neq i, j\}$ and let $B = \{1x_2 \dots x_n \mid$ there exists some $k \in \{2, \dots, n\}$ such that $x_k = u_k$ and $x_i = \bar{u}_i$ if $i \neq k\}$. Thus, $A \cup B$ is the subset of $\{\mathbf{x} \in V(AQ^1_{n-1}) \mid d_{AQ^1_{n-1}}(\mathbf{x}, \mathbf{u}) = 2\}$. Moreover, $|A \cup B| = \binom{n-1}{2} + \binom{n-1}{1} = \binom{n}{2}$. Furthermore, every vertex in $A \cup B$ has two common neighbors with $\mathbf{u}$ in AQ^1_{n-1}. Since $\binom{n}{2} - (2n-4) \geq 4$, there are at least four vertices in $A \cup B$, say $\{\mathbf{x}_i \mid 1 \leq i \leq 4\}$, such that $\mathbf{x}_i^h \in C$. Let $\mathbf{w}_i$ be a common neighbor of $\mathbf{u}$ and $\mathbf{x}_i$. Moreover, let $\mathbf{y}_i^h$ and $\mathbf{z}_i^h$ be the two neighbors of $\mathbf{x}_i^h$ in C. Thus, $|\{\mathbf{y}_i^h \mid 1 \leq i \leq 4\} \cup \{\mathbf{z}_i^h \mid 1 \leq i \leq 4\}| \geq 6$. There exists some i such that one of $\mathbf{y}_i^h$ and $\mathbf{z}_i^h$, say $\mathbf{y}_i^h$, satisfying $\mathbf{y}_i \notin \{\mathbf{u}, \mathbf{v}, \mathbf{w}_i\}$ and $\mathbf{y}_i$ is F-fault free. We can write C as $\langle \mathbf{x}_i^h, \mathbf{z}_i^h, P_1, \mathbf{y}_i^h, \mathbf{x}_i^h \rangle$. Since AQ_{n-1} is $(2n-6)$-fault-tolerant Hamiltonian, there exists a Hamiltonian path P_2 of $AQ^1_{n-1} - F(AQ^1_{n-1}) - \{\mathbf{u}, \mathbf{w}_i, \mathbf{x}_i\}$ joining $\mathbf{b}_i$ and $\mathbf{v}$. Then $\langle \mathbf{u}, \mathbf{w}_i, \mathbf{x}_i, \mathbf{x}_i^h, \mathbf{z}_i^h, P_1, \mathbf{y}_i^h, \mathbf{y}_i, P_2, \mathbf{v} \rangle$ forms a Hamiltonian path of $AQ_n - F$ joining $\mathbf{u}$ and $\mathbf{v}$. See Figure 11.22c for illustration.

Case 3. $|F(AQ^0_{n-1})| = 2n-4$. Thus, $|F - F(AQ^0_{n-1})| = 0$.

Case 3.1. $\{\mathbf{u}, \mathbf{v}\} \subset V(AQ^0_{n-1})$. Let f_1 and f_2 be any two elements in $F(AQ^0_{n-1})$. Since AQ_{n-1} is $(2n-6)$-fault-tolerant Hamiltonian-connected, there exists a Hamiltonian path P joining $\mathbf{u}$ and $\mathbf{v}$ in $AQ^0_{n-1} - (F(AQ^0_{n-1}) - \{f_1, f_2\})$, which may or may not include f_1 or f_2 on it. Removing f_1 and f_2, the path P is divided into three, two, or one pieces, depending on the cases in which f_1 and f_2 are on P or not. For the last two cases, we may arbitrarily delete one or two more edges from P to make it into three subpaths. Therefore, without loss of generality, we may write these three subpaths as $\langle \mathbf{u}, P_1, \mathbf{w}, f_1', \mathbf{x}, P_2, \mathbf{y}, f_2', \mathbf{z}, P_3, \mathbf{v} \rangle$.

Assume that $l(P_2) \geq 1$. Then $\mathbf{w}^h$, $\mathbf{x}^h$, $\mathbf{y}^h$, and $\mathbf{z}^h$ are distinct. Since AQ^1_{n-1} has property $2H$, there are two disjoint spanning paths $\langle \mathbf{w}^h, P_4, \mathbf{x}^h \rangle$ and $\langle \mathbf{y}^h, P_5, \mathbf{z}^h \rangle$ in AQ^1_{n-1}. Obviously, $\langle \mathbf{u}, P_1, \mathbf{w}, \mathbf{w}^h, P_4, \mathbf{x}^h, \mathbf{x}, P_2, \mathbf{y}, \mathbf{y}^h, P_5, \mathbf{z}^h, \mathbf{z}, P_3, \mathbf{v} \rangle$ forms a Hamiltonian path of $AQ_n - F$ joining $\mathbf{u}$ and $\mathbf{v}$. See Figure 11.23a for illustration.

Assume that $l(P_2) = 0$. Then $\mathbf{x} = \mathbf{y}$. Assume that $\{(\mathbf{x}, \mathbf{w}), (\mathbf{x}, \mathbf{z})\} \cap E_n^{2*} = \emptyset$. Then $\mathbf{x}^h$, $\mathbf{x}^c$, $\mathbf{w}^h$, and $\mathbf{z}^h$ are distinct. Since AQ^1_{n-1} has property $2H$, there are two disjoint spanning paths $\langle \mathbf{w}^h, P_6, \mathbf{x}^h \rangle$ and $\langle \mathbf{x}^c, P_7, \mathbf{z}^h \rangle$ in AQ^1_{n-1}. Obviously, $\langle \mathbf{u}, P_1, \mathbf{w}, \mathbf{w}^h, P_6, \mathbf{x}^h, \mathbf{x}, P_2, \mathbf{y}, \mathbf{y}^h, P_7, \mathbf{z}^h, \mathbf{z}, P_3, \mathbf{v} \rangle$ forms a Hamiltonian path of $AQ_n - F$ joining $\mathbf{u}$ and $\mathbf{v}$. See Figure 11.23b for illustration. Assume that $\{(\mathbf{x}, \mathbf{w}), (\mathbf{x}, \mathbf{z})\} \cap E_n^{2*} \neq \emptyset$. Without loss of generality, we can assume that $(\mathbf{x}, \mathbf{w})$ is in E_n^{2*}. Then $\mathbf{x}^h = \mathbf{w}^c$ and $\mathbf{x}^c = \mathbf{w}^h$. Since AQ_{n-1} is $(2n-6)$-fault-tolerant Hamiltonian-connected, there is a Hamiltonian path P_8 of $AQ^1_{n-1} - \{\mathbf{x}^c\}$ joining $\mathbf{x}^h$ and $\mathbf{z}^h$. Obviously, $\langle \mathbf{u}, P_1, \mathbf{w}, \mathbf{x}^c, \mathbf{x}, \mathbf{x}^h, P_8, \mathbf{z}^h, \mathbf{z}, P_3, \mathbf{v} \rangle$ forms a Hamiltonian path of $AQ_n - F$ joining $\mathbf{u}$ and $\mathbf{v}$. See Figure 11.23c for illustration.

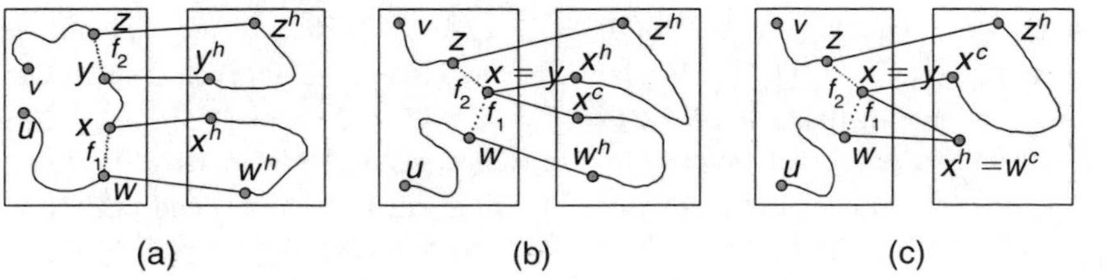

FIGURE 11.23 Illustrations for Case 3.1 of Lemma 11.26.

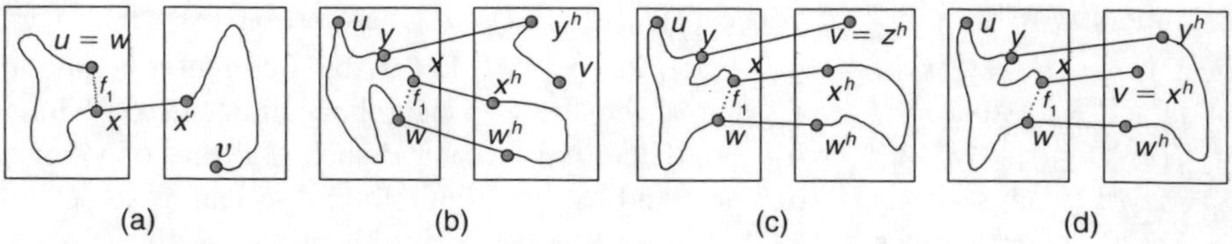

FIGURE 11.24 Illustrations for Case 3.2 of Lemma 11.26.

Case 3.2. $\mathbf{u} \in V(AQ^0_{n-1})$ and $\mathbf{v} \in V(AQ^1_{n-1})$. Let f_1 be any element in $F(AQ^0_{n-1})$. Since AQ_{n-1} is $(2n-5)$-fault-tolerant Hamiltonian, there exists a Hamiltonian cycle $C = \langle \mathbf{u}, P_1, \mathbf{w}, f_1', \mathbf{x}, P_2, \mathbf{u} \rangle$ in $AQ^0_{n-1} - (F(AQ^0_{n-1}) - f_1)$ where $f_1' = f_1$ if f_1 is on cycle C or f_1' is any edge of C otherwise. Without loss of generality, we can assume that $l(P_1) \leq l(P_2)$. Thus, we can rewrite C as $\langle \mathbf{u}, P_1, \mathbf{w}, f_1', \mathbf{x}, P_3, \mathbf{y}, \mathbf{u} \rangle$.

Assume that $l(P_1) = 0$. Then $\mathbf{u} = \mathbf{w}$. Thus, we can rewrite C as $\langle \mathbf{u}, f_1', \mathbf{x}, P_3', \mathbf{u} \rangle$. Then there exists a vertex $\mathbf{x}'$ in $\{\mathbf{x}^h, \mathbf{x}^c\} - \mathbf{v}$. Since AQ_{n-1} is $(2n-6)$-fault-tolerant Hamiltonian-connected, there exists a Hamiltonian path P_4 of AQ^1_{n-1} joining $\mathbf{x}'$ and $\mathbf{v}$. Obviously, $\langle \mathbf{u}, P_3', \mathbf{x}, \mathbf{x}', \mathbf{v} \rangle$ forms a Hamiltonian path of $AQ_n - F$ joining $\mathbf{u}$ and $\mathbf{v}$. See Figure 11.24a for illustration.

Assume that $l(P_1) \geq 1$. We have the following three cases:

1. $\mathbf{w}^h$, $\mathbf{x}^h$, $\mathbf{y}^h$, and $\mathbf{v}$ are distinct. Since AQ^1_{n-1} has property $2H$, there are two disjoint spanning paths $\langle \mathbf{w}^h, P_4, \mathbf{x}^h \rangle$ and $\langle \mathbf{y}^h, P_5, \mathbf{v} \rangle$ in AQ^1_{n-1}. Obviously, $\langle \mathbf{u}, P_1, \mathbf{w}, \mathbf{w}^h, P_4, \mathbf{x}^h, \mathbf{x}, P_3, \mathbf{y}, \mathbf{y}^h, P_5, \mathbf{v} \rangle$ forms a Hamiltonian path of $AQ_n - F$ joining $\mathbf{u}$ and $\mathbf{v}$. See Figure 11.24b for illustration.

2. $\mathbf{y}^h = \mathbf{v}$. Since AQ_{n-1} is $(2n-6)$ fault-tolerant Hamiltonian-connected, there is a Hamiltonian path $\langle \mathbf{w}^h, P_6, \mathbf{x}^h \rangle$ of $AQ^1_{n-1} - \{\mathbf{v}\}$. Obviously, $\langle \mathbf{u}, P_1, \mathbf{w}, \mathbf{w}^h, P_6, \mathbf{x}^h, \mathbf{x}, P_2, \mathbf{y}, \mathbf{v} \rangle$ forms a Hamiltonian path of $AQ_n - F$ joining $\mathbf{u}$ and $\mathbf{v}$. See Figure 11.24c for an illustration.

3. $\mathbf{x}^h = \mathbf{v}$ or $\mathbf{w}^h = \mathbf{v}$. Without loss of generality, we can assume that $\mathbf{x}^h = \mathbf{v}$. Since AQ_{n-1} is $(2n-6)$-fault-tolerant Hamiltonian-connected, there is a Hamiltonian path $\langle \mathbf{w}^h, P_7, \mathbf{y}^h \rangle$ of $AQ^1_{n-1} - \{\mathbf{v}\}$. Obviously, $\langle \mathbf{u}, P_1, \mathbf{w}, \mathbf{w}^h, P_7, \mathbf{y}^h, \mathbf{y}, P_2, \mathbf{x}, \mathbf{v} \rangle$ forms a Hamiltonian path of $AQ_n - F$ joining $\mathbf{u}$ and $\mathbf{v}$. See Figure 11.24d for illustration.

Case 3.3. $\mathbf{u}, \mathbf{v} \in V(AQ^1_{n-1})$. Assume that $F(AQ^0_{n-1}) \subset E^{2*}_n$. By definition, the hypercube Q_{n-1} is a spanning subgraph of $AQ^0_{n-1} - F(AQ^0_{n-1})$. Thus, there is a Hamiltonian cycle C in $AQ^0_{n-1} - F(AQ^0_{n-1})$. We claim that there is an edge $(\mathbf{w}, \mathbf{x})$ on C such that $\mathbf{w}^h$, $\mathbf{x}^h$, $\mathbf{u}$, and $\mathbf{v}$ are distinct. Since $l(C) = 2^{n-1}$ and $(2^{n-1}/2) - 3 = 2^{n-2} - 3 \geq 5$, there are at least five edges that we can choose. Hence, such an edge exists. Since AQ^1_{n-1} has property $2H$, there are two disjoint spanning paths $\langle \mathbf{u}, P_2, \mathbf{w}^h \rangle$ and $\langle \mathbf{x}^h, P_3, \mathbf{v} \rangle$ of AQ^1_{n-1}. Obviously, $\langle \mathbf{u}, P_2, \mathbf{w}^h, \mathbf{w}, P_1, \mathbf{x}, \mathbf{x}^h, P_3, \mathbf{v} \rangle$ is a Hamiltonian path of $AQ_n - F$ joining $\mathbf{u}$ and $\mathbf{v}$. See Figure 11.25a for illustration.

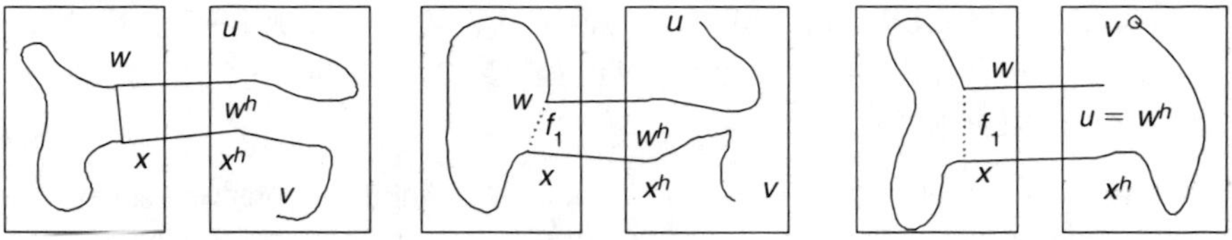

FIGURE 11.25 Illustrations for Case 3.3 of Lemma 11.26.

Assume that $F(AQ^0_{n-1}) \not\subset E^{2*}_n$. Let f_1 be any element in $F(AQ^0_{n-1}) - E^{2*}_n$. Since AQ_{n-1} is $(2n-5)$-fault-tolerant Hamiltonian, there exists a Hamiltonian cycle $C = \langle \mathbf{w}, P_1, \mathbf{x}, f'_1, \mathbf{w} \rangle$ in $AQ^0_{n-1} - (F(AQ^0_{n-1}) - f_1)$ where $f'_1 = f_1$ if f_1 is incident to C or f'_1 is any edge of C otherwise. Therefore, $\mathbf{w}^h$, $\mathbf{w}^c$, $\mathbf{x}^h$, and $\mathbf{x}^c$ are distinct. We have the following two cases:

1. $\mathbf{w}^h$, $\mathbf{x}^h$, $\mathbf{u}$, and $\mathbf{v}$ are distinct. Since AQ^1_{n-1} has property $2H$, there are two disjoint spanning paths $\langle \mathbf{u}, P_4, \mathbf{w}^h \rangle$ and $\langle \mathbf{x}^h, P_5, \mathbf{v} \rangle$ of AQ^1_{n-1}. Obviously, $\langle \mathbf{u}, P_4, \mathbf{w}^h, \mathbf{w}, P_1, \mathbf{x}, \mathbf{x}^h, P_5, \mathbf{v} \rangle$ forms a Hamiltonian path of $AQ_n - F$ joining $\mathbf{u}$ and $\mathbf{v}$. See Figure 11.25b for illustration.

2. $\mathbf{w}^h$, $\mathbf{x}^h$, $\mathbf{u}$, and $\mathbf{v}$ are not distinct. Without loss of generality, we assume that $\mathbf{w}^h = \mathbf{u}$. Let $\mathbf{x}'$ be a element in $\{\mathbf{x}^h, \mathbf{x}^c\}$ such that $\mathbf{x}' \neq \mathbf{v}$. Since AQ_{n-1} is $(2n-6)$-fault-tolerant Hamiltonian-connected, there is a Hamiltonian path P_6 of $AQ^1_{n-1} - \{\mathbf{u}\}$ joining $\mathbf{x}'$ and $\mathbf{v}$. Obviously, $\langle \mathbf{u}, \mathbf{w}, P_1, \mathbf{x}, \mathbf{x}', P_6, \mathbf{v} \rangle$ forms a Hamiltonian path of $AQ_n - F$ joining $\mathbf{u}$ and $\mathbf{v}$. See Figure 11.25c for illustration.

Hence, this lemma is proved. □

THEOREM 11.18 Assume that n is a positive integer with $n \geq 4$. Then AQ_n is $(2n-3)$-fault-tolerant Hamiltonian, $(2n-4)$-fault-tolerant Hamiltonian-connected, and has property $2H$.

Proof: We prove this theorem by induction on n. The induction base is $n=4$. By Lemmas 11.23 and 11.24, the theorem holds for $n=4$. With Lemmas 11.24 through 11.26, we prove the induction step. □

With Lemmas 11.21 and 11.22, we have the following theorem.

THEOREM 11.19 Assume that n is a positive integer:

1. $H_f(AQ_n) = 2n-3$ and $H^\kappa_f(AQ_n) = 2n-4$ if $n \notin \{1, 3\}$
2. $H_f(AQ_1)$ is undefined and $H^\kappa_f(AQ_1) = 0$
3. $H_f(AQ_3) = 2$ and $H^\kappa_f(AQ_3) = 1$

11.7 FAULT-TOLERANT HAMILTONICITY AND FAULT-TOLERANT HAMILTONIAN CONNECTIVITY OF THE WK-RECURSIVE NETWORKS

In the previous sections, we have discussed several methods to construct super fault-tolerant Hamiltonian graphs. It is very natural to ask whether the node expansion $X(G,x)$ is super fault-tolerant Hamiltonian once G is super fault-tolerant Hamiltonian. The statement is not true for $k=0$. For example, let G be the graph shown in Figure 11.26a, the graph $X(G,u)$ is shown in Figure 11.26b, and the graph $X(X(G,u),w)$ is shown in Figure 11.26c. By brute force, we can check that G is 1-fault-tolerant Hamiltonian and Hamiltonian-connected. Moreover, the graph $X(G,u)$ is 1-fault-tolerant Hamiltonian and Hamiltonian-connected. We will show that there is no Hamiltonian path of $X(X(G,u),w)$ between u_1 and w_1. Hence, $X(X(G,u),w)$ is not Hamiltonian-connected.

LEMMA 11.27 There exists no Hamiltonian path between u_1 and w_1 in $X(X(G,u),w)$.

Proof: Suppose that $X(X(G,u),w)$ has a Hamiltonian path, P, between u_1 and w_1.

Case 1. P contains (w_1,w_3). Then neither (w_1,w_2) nor (w_1,s) is in P. Hence, P contains the subpaths $\langle w_1,w_3,w_2,v\rangle$ and $\langle z,s,t,x,y\rangle$. Thus, (t,u_2) is not in P and $\langle u_3,u_2,u_1\rangle$ is a subpath of P. Therefore, $(y,z)\in P$. Thus, P contains a cycle $\langle z,s,t,x,y,z\rangle$, which is impossible.

Case 2. P contains (w_1,w_2). Then neither (w_1,w_3) nor (w_1,s) is in P. Hence, P contains the subpaths $\langle w_1,w_2,w_3,x\rangle$ and $\langle t,s,z,v,u_3\rangle$. Since $(z,y)\notin P$, $\langle x,y,u_1\rangle$ is also a subpath of P. Then (u_1,u_2) is not in P and $\langle u_3,u_2,t\rangle$ is a subpath of P. Thus, P contains a cycle $\langle t,s,z,v,u_3,u_2,t\rangle$, which is impossible.

Case 3. P contains (w_1,s). Then neither (w_1,w_2) nor (w_1,w_3) is in P. Hence, P contains the subpath $\langle v,w_2,w_3,x\rangle$.

Case 3.1. P contains (s,z). Then $\langle x,t,u_2\rangle$ is a subpath of P. Since (x,y) is not in P, $\langle z,y,u_1\rangle$ is a subpath of P. Therefore, neither (u_1,u_3) nor (u_1,u_2) is in P and $\langle v,u_3,u_2\rangle$ is a subpath of P. Thus, P has a cycle $\langle v,w_2,w_3,x,t,u_2,u_3,v\rangle$. This is a contradiction.

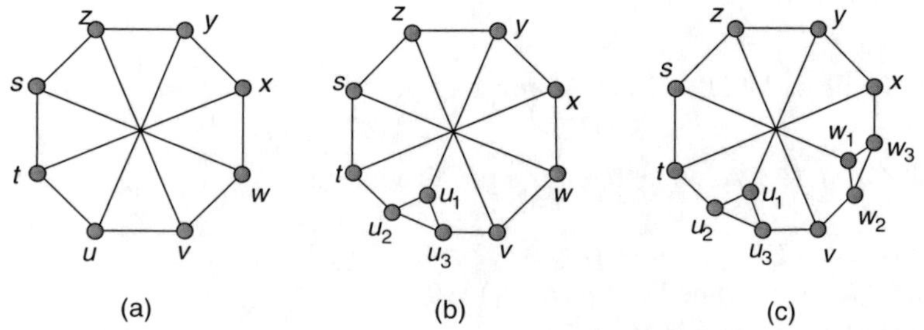

FIGURE 11.26 The graphs (a) G, (b) $X(G,u)$, and (c) $X(X(G,u),w)$.

Case 3.2. P contains (s,t). Then (s,z) is not in P. Hence, P contains the subpath $\langle y,z,v\rangle$. Since $(v,u_3)\notin P$, P contains the subpath $\langle u_1,u_3,u_2,t\rangle$ and the edge (y,x). Thus, P has a cycle $\langle v,w_2,w_3,x,y,z,v\rangle$. This is a contradiction.

The lemma is proved. □

Yet there are some interconnection networks that are recursively constructed using node expansion. We still want to know its fault-tolerant Hamiltonicity and its fault-tolerant Hamiltonian connectivity. In this section, we use the WK-recursive network as an example.

The WK-recursive network is proposed by Vecchia and Sanges [324]. We use $K(d,t)$ to denote the WK-recursive network of level t, each of whose basic module is a d-vertex complete graph, where $d>1$ and $t\geq 1$. It offers a high degree of scalability, which conforms very well to a modular design and the implementation of distributed systems involving a large number of computing elements. A transputer implementation of a 15-vertex WK-recursive network has been realized at the Hybrid Computing Center in Naples, Italy. In this implementation, each vertex is implemented with the IMS T414 Transputer [186]. Recently, the WK-recursive network has received much attention, due to its many favorable properties. In particular, it is proved that $\mathcal{H}_e(K(d,t))=d-3$ [108], $\mathcal{H}_v(K(d,t))=d-3$ [116], and $K(d,t)$ is Hamiltonian-connected [116]. Ho et al. [151] prove that $\mathcal{H}_f(K(d,t))=d-3$ and $\mathcal{H}_f^{\kappa}(K(d,t))=d-4$.

The WK-recursive network can be constructed hierarchically by grouping basic modules. A complete graph of any size d can serve as a basic module. We use $K(d,t)$ to denote a WK-recursive network of level t, each of whose basic module is a d-vertex complete graph, where $d>1$ and $t\geq 1$. The structures of $K(5,1)$, $K(5,2)$, and $K(5,3)$ are shown in Figure 11.27. $K(d,t)$ is defined in terms of a graph as follows.

Each vertex of $K(d,t)$ is labeled as a t-digit radix d number. Vertex $a_{t-1}a_{t-2}\ldots a_1a_0$ is adjacent to (1) $a_{t-1}a_{t-2}\ldots a_1b$, where $b\neq a_0$ and (2) $a_{t-1}a_{t-2}\ldots a_{j+1}a_{j-1}(a_j)^{j-1}$ if $a_j\neq a_{j-1}$ and $a_{j-1}=a_{j-2}=\ldots=a_0$, where $(a_j)^{j-1}$ denotes $j-1$ consecutive a_js. An *open edge* is incident to $a_{t-1}a_{t-2}\ldots a_0$ if $a_{t-1}=a_{t-2}=\ldots=a_0$. The open edge is reserved for further expansion. Hence, its other end vertex is unspecified. The *open vertex set* O_v of $K(d,t)$ is the set $\{a_{t-1}a_{t-2}\ldots a_0\mid a_i=a_{i+1}$ for $0\leq i\leq t-2\}$. In other words, O_v contains those vertices with open edges.

Obviously, $K(d,1)$ is a d-vertex complete graph augmented with d open edges. For $t\geq 1$, $K(d,t+1)$ consists d copies of $K(d,t)$, say $K_1(d,t)$, $K_2(d,t),\ldots,K_d(d,t)$. Thus, we consider $K_i(d,t)$ as the ith component of $K(d,t+1)$. Letting $I=\{w_1,\ldots,w_q\}$ to be any q subset of $\{1,2,\ldots,d\}$, we define graph $K_I(d,t)$ as the subgraph of $K(d,t+1)$ induced by $\bigcup_{i=1}^{q}V(K_{w_i}(d,t))$. For $t\geq 2$, the open vertices of $K_i(d,t)$ can be labeled as $o_{i,0}$ and $o_{i,j}$ for $1\leq i\neq j\leq d$, where $o_{i,0}$ is the only open vertex of $K(d,t+1)$ in $K_i(d,t)$ and $o_{i,j}$ is the vertex in $K_i(d,t)$ joining with the vertex $o_{j,i}$ in $K_j(d,t)$ with an open edge. Note that $(o_{i,j},o_{j,i})$ is the only edge joining $K_i(d,t)$ to $K_j(d,t)$.

Now, we define the *extended WK-recursive network* $\tilde{K}(d,t)$ as the graph obtained from $K(d,t)$ by replacing all open edges with the edges joining to x. For example, $\tilde{K}(5,1)$ and $\tilde{K}(5,2)$ are illustrated in Figure 11.28. Obviously, $\tilde{K}(d,1)$ is isomorphic to the complete graph K_{d+1}. It is observed that $\tilde{K}(d,t)$ is d-regular.

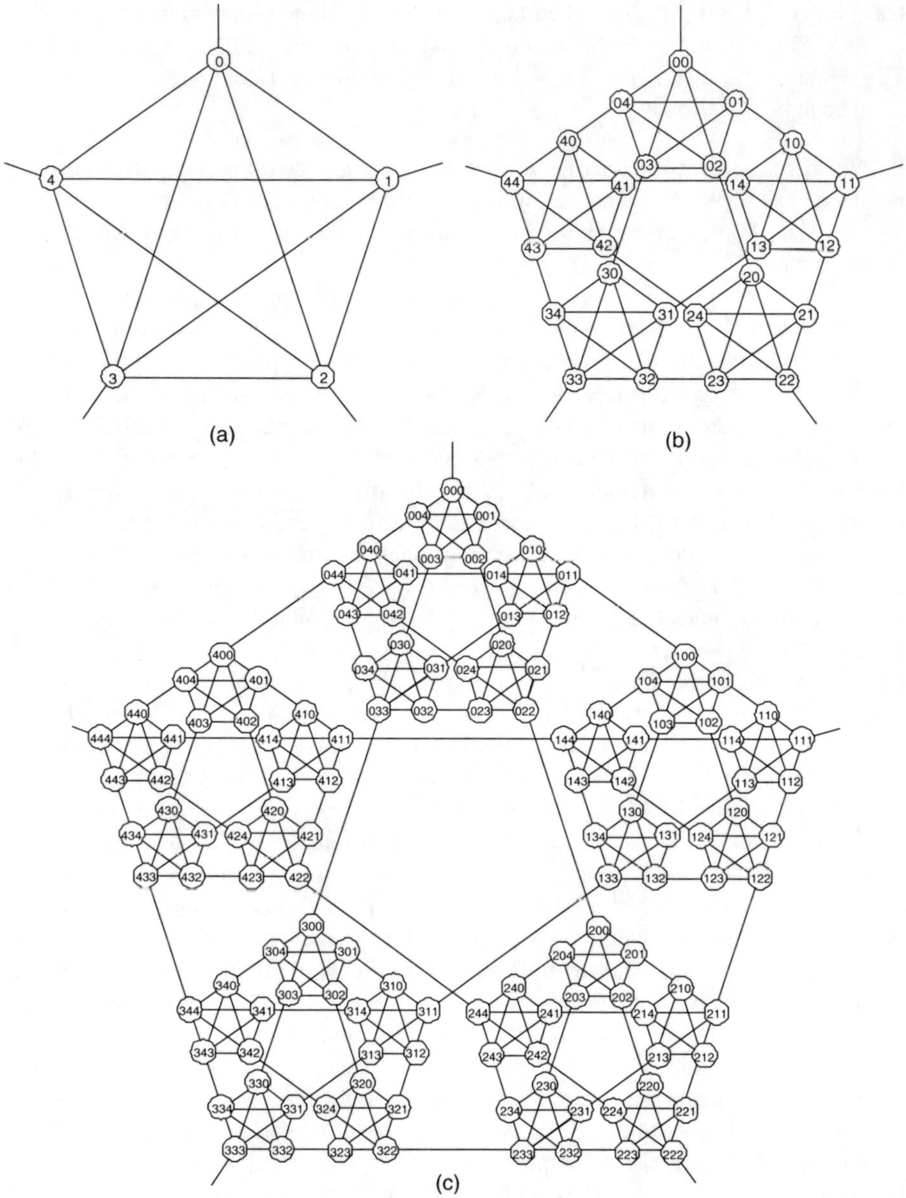

FIGURE 11.27 The graphs (a) $K(5, 1)$, (b) $K(5, 2)$, and (c) $K(5, 3)$.

From Figure 11.29, it is observed that $\tilde{K}(5, 2)$ is obtained from $\tilde{K}(5, 1)$ by taking node expansion at all vertices of $K(5, 1)$. With this observation, we have the following theorem.

THEOREM 11.20 $\tilde{K}(d, t+1)$ is obtained from the complete graph K_{d+1} by a sequence of node expansions.

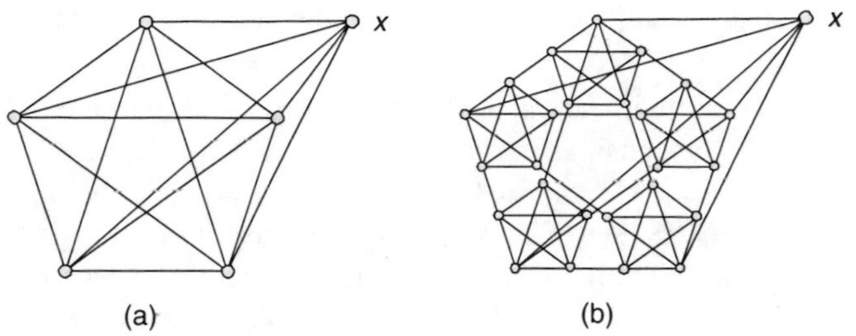

FIGURE 11.28 The graphs (a) $\tilde{K}(5,1)$ and (b) $\tilde{K}(5,2)$.

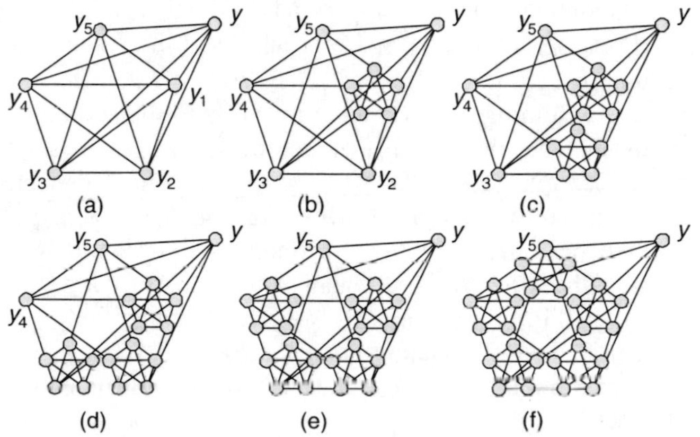

FIGURE 11.29 Illustrations of a sequence of node expansions.

With Theorems 11.3 and 11.20, and Lemma 11.1, we have the following theorem.

THEOREM 11.21 $\tilde{K}(d,t)$ is $(d-2)$-fault-tolerant Hamiltonian for $d \geq 3$ and $t \geq 1$.

THEOREM 11.22 $\mathcal{H}_f(K(d,t)) = d-3$ for $d \geq 3$ and $t \geq 1$.

Proof: Since $\delta(K(d,t)) = d-1$, $\mathcal{H}_f(K(d,t)) \leq d-3$. Let F be any faulty set of $K(d,t)$ with $|F| \leq d-3$. We set the faulty set $\overline{F} = F \cup \{x\}$ of $\tilde{K}(d,t)$, where x is the only vertex in $V(\tilde{K}(d,t)) - V(K(d,t))$. Obviously, $|\overline{F}| \leq d-2$. By Theorem 11.21, there exists a Hamiltonian cycle C in $\tilde{K}(d,t) - \overline{F}$. Obviously, cycle C is also a Hamiltonian cycle of $K(d,t) - F$. Thus, $\mathcal{H}_f(K(d,t)) \geq d-3$. Therefore, $\mathcal{H}_f(K(d,t)) = d-3$. □

Now, we prove that $\mathcal{H}_f^{\kappa}(K(d,t)) = d-4$ for $d \geq 4$ and $t \geq 1$ by induction. To get this goal, we need more observation about $\tilde{K}(d,t)$. We define the *weakly extended WK-recursive network* $\tilde{K}^i(d,t)$ as $V(\tilde{K}^i(d,t)) = V(K(d,t)) \cup \{x\}$ and $E(\tilde{K}^i(d,t)) = E(K(d,t)) \cup \{(o_{\alpha,0}, x) \mid \alpha \in \{1,2,\ldots,d\} - \{i\}\}$. Obviously, $\tilde{K}^i(d,t)$ is isomorphic to $\tilde{K}^j(d,t)$ for $1 \leq i \neq j \leq d$. Let $F \subset V(K(d,t+1)) \cup E(K(d,t+1))$ with $|F| \leq d-4$. For $1 \leq q \leq d$, we use F_q to denote $F \cap (V(K^q(d,t)) \cup E(K^q(d,t)))$. Note that it is possible that $F - \cup_{q=1}^{d} F_q \neq \emptyset$. For example, it is possible that $(o_{1,2}, o_{2,1}) \in F$ but $(o_{1,2}, o_{2,1}) \notin \cup_{q=1}^{d} F_q$.

Now, we construct another graph $H(F)$ from the complete graph K_d with vertex set $\{1, 2, \ldots, d\}$ by considering vertex i corresponding to the ith component of $K(d, t+1)$ for every i. Let $F' = \{(\alpha, \beta) \mid o_{\alpha,\beta} \in F, o_{\beta,\alpha} \in F, \text{ or } (o_{\alpha,\beta}, o_{\beta,\alpha}) \in F\}$. We set $H(F) = K_d - F'$. Since $|F| \leq d - 4$, by Lemma 11.4, $H(F)$ is Hamiltonian-connected. This result will help us find a Hamiltonian path between any two vertices in $K(d, t+1) - F$. However, there are several problems to be conquered. Let us consider the following example.

Assume that u is a vertex in $K^i(d, t)$ and v is a vertex in $K^j(d, t)$ with $1 \leq i \neq j \leq d$. Let $\langle i = w_1, w_2, \ldots, w_d = j\rangle$ be a Hamiltonian path of $H(F)$. Let P_i be a Hamiltonian path of $K^i(d, t) - F_i$ joining u to o_{i,w_2}, let P_q be a Hamiltonian path of $K^{w_q}(d, t) - F_{w_q}$ joining $o_{w_q,w_{q-1}}$ to $o_{w_q,w_{q+1}}$ for $2 \leq q \leq d - 1$; and let P_j be a Hamiltonian path of $K^j(d, t) - F_j$ joining $o_{j,w_{d-1}}$ to v. Obviously, $\langle P_i, P_2, \ldots, P_j\rangle$ forms a Hamiltonian path of $K(d, t+1) - F$ joining u to v. See Figure 11.30 for illustration.

Yet we need to guarantee the existence of required paths in each component. Later, we will prove that $K(d, t)$ is $(d - 4)$-fault-tolerant Hamiltonian-connected by induction. In the induction step, we assume $K(d, t)$ is $(d - 4)$-fault-tolerant Hamiltonian-connected and prove that $K(d, t+1)$ is $(d - 4)$-fault-tolerant Hamiltonian-connected. With the assumption, the required Hamiltonian path P_q exists for $2 \leq q \leq d - 1$. However, we cannot find P_i if $u = o_{i,w_2}$. Similarly, we cannot find P_j if $o_{j,w_{d-1}} = v$. To solve the problem, we can find another Hamiltonian path $\langle i = z_1, z_2, \ldots, z_d = j\rangle$ of $H(F)$ to meet the boundary conditions that $u \neq o_{i,z_2}$ and $v \neq o_{j,z_{d-1}}$. As a conclusion, we have the following lemma.

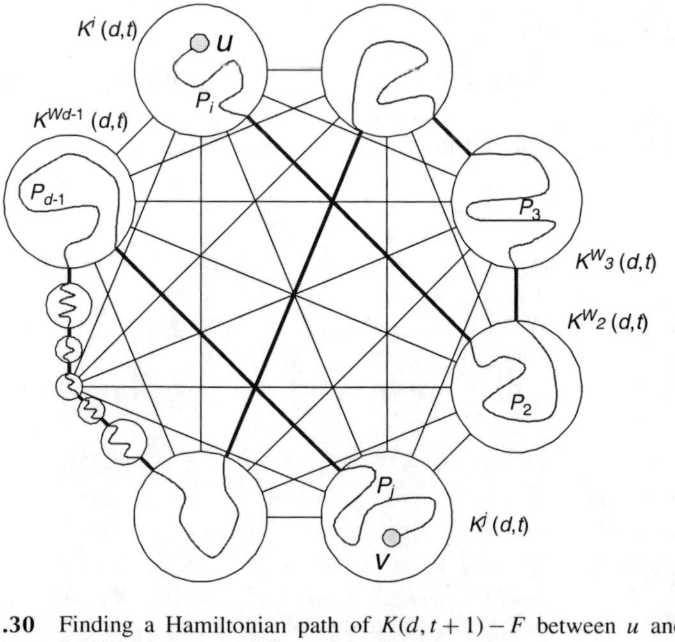

FIGURE 11.30 Finding a Hamiltonian path of $K(d, t+1) - F$ between u and v with a Hamiltonian path of $H(F)$ between i and j.

LEMMA 11.28 Assume that $K(d,t)$ is $(d-4)$-fault-tolerant Hamiltonian-connected. Let $F \subset V(K(d,t+1)) \cup E(K(d,t+1))$ with $|F| \leq d-4$. Let u be a vertex in $K^i(d,t)$ and let v be a vertex in $K^j(d,t)$ with $1 \leq i \neq j \leq d$. Suppose that $\langle i = w_1, w_2, \ldots, w_d = j \rangle$ be a Hamiltonian path of $H(F)$ that satisfies the boundary conditions: $u \neq o_{i,w_2}$ and $v \neq o_{j,w_{d-1}}$. Then there exists a Hamiltonian path of $K(d,t+1) - F$ joining u and v.

From the previous discussion, we have three problems to prove that $K(d,t+1) - F$ is Hamiltonian-connected; that is, there exists a Hamiltonian path of $K(d,t+1) - F$ between any two vertices u and v. First, assuming that u is a vertex in $K^i(d,t)$ and v is a vertex in $K^j(d,t)$ with $1 \leq i \neq j \leq d$, we need to find a Hamiltonian path in $H(F)$ that meets the boundary conditions. Second, find a Hamiltonian path of $K(d,t+1) - F$ joining u and v if we cannot find a Hamiltonian path in $H(F)$ that meets the boundary conditions. Finally, find a Hamiltonian path of $K(d,t+1) - F$ joining u and v if both u and v are in $K^i(d,t)$ for some i.

Now, we face the first problem. Let $P_1 = \langle u_1, u_2, \ldots, u_n \rangle$ and $P_2 = \langle v_1, v_2, \ldots, v_n \rangle$ be any two Hamiltonian paths of G. We say that P_1 and P_2 are *orthogonal* if $u_1 = v_1$, $u_n = v_n$, and $u_q \neq v_q$ for $q = 2$ and $q = n-1$. We say a set of Hamiltonian paths $\{P_1, P_2, \ldots, P_s\}$ of G are *mutually orthogonal* if any two distinct paths in the set are orthogonal. Suppose that there are three mutually orthogonal Hamiltonian paths between any two vertices of $H(F)$. By the pigeonhole principle, we can easily find a Hamiltonian path with the desired boundary conditions. For this reason, we would like to know all the cases in which there are at most two orthogonal Hamiltonian paths in $H(F)$. As mentioned earlier, $H(F)$ is isomorphic to a graph G with n vertices and $\bar{e} \leq n-4$.

The following theorem is proved by Chvátal [70].

THEOREM 11.23 **(Chvátal [70])** If G is an n-vertex graph where $n \geq 3$ and $|E(G)| > \{[(n-1)(n-2)]/2\} + 1$, then G is Hamiltonian. Moreover, the only non-Hamiltonian graphs with n vertices and $\{[(n-1)(n-2)]/2\} + 1$ edges are $C_{1,n}$ and, for $n = 5$, $C_{2,5}$.

Suppose that $G = (V, E)$ is an n-vertex graph with $\bar{e} \leq n-4$. Assume that $n = 4$. Obviously, G is isomorphic to K_4. It is easy to check whether there are exactly two orthogonal Hamiltonian paths between any two distinct vertices of G.

Assume that $n = 5$. Obviously, G is either isomorphic to K_5 or $K_5 - e$ where e is any edge of K_5. We can label the vertices of K_5 with $\{1,2,3,4,5\}$, and we set $e = (1,2)$. Suppose that G is isomorphic to K_5. It is easy to check whether there are exactly three mutually orthogonal Hamiltonian paths of G between any two vertices. Suppose that G is isomorphic to $K_5 - (1,2)$. By brute force, we can check whether there are exactly three mutually orthogonal Hamiltonian paths between vertices 1 and 2. However, there are exactly two orthogonal Hamiltonian paths between the remaining pairs.

Now, we assume that $n \geq 6$. Let s and t be any two distinct vertices of G. Let H be the subgraph of G induced by the remaining $(n-2)$ vertices of G. We have the following two cases:

Case 1. H is Hamiltonian. We can relabel the vertices of H with $\{0, 1, 2, \ldots, n-3\}$ so that $\langle 0, 1, 2, \ldots, n-3, 0 \rangle$ forms a Hamiltonian cycle of H. Let Q denote the set $\{i \mid (s, [i+1]) \in E(G) \text{ and } (i, t) \in E(G)\}$ where $[i]$ denotes i **mod** $(n-2)$. Since $\bar{e} \le n-4$, $|Q| \ge n-2-(n-4) = 2$. There are at least two elements in Q. Let q_1 and q_2 be the two elements in Q. For $j = 1, 2$, we set P_j as $\langle s, [q_j+1], [q_j+2], \ldots, [q_j], t\rangle$. Then P_1 and P_2 are two orthogonal Hamiltonian paths between s and t.

Suppose that $\bar{e} \le n-5$, $(s,t) \notin E$, or H is not isomorphic to the complete graph K_{n-2}. Then $|Q| \ge 3$. Let q_1, q_2, and q_3 be the three elements in Q. For $j = 1, 2$, and 3, we set P_j as $\langle s, [q_j+1], [q_j+2], \ldots, [q_j], t\rangle$. Then P_1, P_2, and P_3 are three mutually orthogonal Hamiltonian paths between s and t.

Thus, we consider $\bar{e} = n-4$, $(s,t) \in E$, and H is isomorphic to the complete graph K_{n-2}. Let ST be the set of vertices in H that are adjacent to s and t, let $S\overline{T}$ be the set of vertices in H that are adjacent to s but not adjacent to t, let $\overline{S}T$ be the set of vertices in H that are not adjacent to s but adjacent to t, and let $\overline{S}\,\overline{T}$ be the set of vertices in H that are not adjacent to s or adjacent to t.

Let $a = |ST|$, $b = |S\overline{T}|$, $c = |\overline{S}T|$, and $d = |\overline{S}\,\overline{T}|$. Without loss of generality, we assume that $\deg_G(s) \ge \deg_G(t)$. Then $b \ge c$, $b + c + 2d = n-4$, and $a + b + c + d = n-2$. Thus, $a - d = 2$. Hence, $a \ge 2$.

Suppose that $a \ge 3$. Let q_1, q_2, and q_3 be three vertices in ST and $q_4, q_5, \ldots, q_{n-2}$ are the remaining vertices of H. We set P_1 as $\langle s, q_1, q_2, X, q_3, t\rangle$, P_2 as $\langle s, q_2, q_3, Y, q_1, t\rangle$, and P_3 as $\langle s, q_3, Z, q_1, q_2, t\rangle$, where X, Y, and Z are any permutation of $q_4, q_5, \ldots, q_{n-2}$. Obviously, P_1, P_2, and P_3 are three mutually orthogonal Hamiltonian paths between s and t.

Suppose that $a = 2$. Then $d = 0$. Suppose that $c \ge 1$. Then $b \ge 1$. We rearrange the vertices of H so that 0 is a vertex in $S\overline{T}$, 1 and 2 are the vertices in ST, 3 is a vertex in $\overline{S}T$, and $4, 5, \ldots, n-3$ are the remaining vertices. Obviously, $\langle 0, 1, 2, \ldots, n-3, 0\rangle$ forms a Hamiltonian cycle of H. Let Q denote the set $\{i \mid (s, i) \in E(G) \text{ and } ([i+1], t) \in E(G)\}$. Obviously, $|Q| \ge 3$. Thus, there are three mutually orthogonal Hamiltonian paths between s and t.

Finally, we consider $a = 2$, $d = 0$, and $c = 0$. Thus, $b = n-4$. In this case, s is adjacent to t and adjacent to all the vertices in H; t is adjacent to s and adjacent to exactly two vertices in H; say, q_1 and q_2. Let $\langle s = v_1, v_2, \ldots, v_n = t\rangle$ be any Hamiltonian path of G between s and t. Obviously, v_{n-1} is either q_1 or q_2. Therefore, there are exactly two orthogonal Hamiltonian paths between s and t.

Case 2. H is non-Hamiltonian. There are exactly $(n-2)$ vertices in H. By Theorem 11.23, there are exactly $(n-4)$ edges in the complement of H, and H is isomorphic to $C_{1,n-2}$ or $C_{2,5}$. Hence, s is adjacent to $V(G) - \{s\}$ and t is adjacent to $V(G) - \{t\}$. We can construct two orthogonal Hamiltonian paths of G between s and t as the following cases.

Case 2.1. H is isomorphic to $C_{2,5}$. We label the vertices of $C_{2,5}$ with $\{1, 2, 3, 4, 5\}$ as shown in Figure 11.31a. Let $P_1 = \langle s, 1, 2, 3, 4, 5, t\rangle$ and $P_2 = \langle s, 3, 4, 5, 2, 1, t\rangle$. Then P_1 and P_2 form the required orthogonal paths. By brute force, we can check that there are exactly two orthogonal Hamiltonian paths between s and t.

Case 2.2. H is isomorphic to $C_{1,n-2}$. We label the vertices of $C_{1,n-2}$ with $\{1, 2, \ldots, n-2\}$, as shown in Figure 11.31b. Let $P_1 = \langle s, 1, 2, 3, \ldots, n-2, t\rangle$ and $P_2 = \langle s, 3, 4, \ldots, n-2, 2, 1, t\rangle$. Then P_1 and P_2 form the orthogonal Hamiltonian

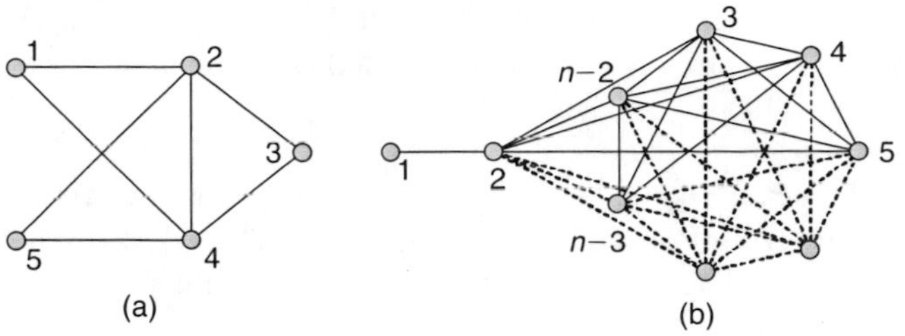

FIGURE 11.31 (a) $C_{2,5}$ and (b) $C_{1,n-2}$.

paths. Let $\langle s = v_1, v_2, \ldots, v_n = t\rangle$ be any Hamiltonian path of G between s and t. Obviously, 1 is either v_2 or v_{n-1}. Therefore, there are exactly two orthogonal Hamiltonian paths between s and t.

From the previous discussion, we have the following theorem.

THEOREM 11.24 Assume that $G = (V, E)$ is an n-vertex graph with $n \geq 4$ and $\overline{e} \leq n - 4$. Let s and t be any two vertices of G. Then there are at least two orthogonal Hamiltonian paths of G between s and t. Moreover, there are at least three mutually orthogonal Hamiltonian paths of G between s and t except the following cases:

Case 1. G is isomorphic to K_4, where s and t are any two vertices of G.

Case 2. G is isomorphic to $K_5 - (1, 2)$, where s and t are any two vertices except $\{s, t\} = \{1, 2\}$.

Case 3. The subgraph H induced by $V(G) - \{s, t\}$ is a complete graph with $n \geq 6$; s is adjacent to t and all the vertices in H; t is adjacent to s and exactly two vertices in H.

Case 4. The subgraph induced by $V(G) - \{s, t\}$ is isomorphic to $C_{2,5}$; s is adjacent to $V(G) - \{s\}$; t is adjacent to $V(G) - \{t\}$.

Case 5. The subgraph induced by $V(G) - \{s, t\}$ is isomorphic to $C_{1,n-2}$ with $n \geq 6$; s is adjacent to $V(G) - \{s\}$; t is adjacent to $V(G) - \{t\}$.

To solve the remaining problems, we need some path patterns.

LEMMA 11.29 Let $d \geq 4$, $t \geq 1$, and $I = \{w_1, \ldots, w_n\}$ be any subset with n of $\{1, 2, \ldots, d\}$. Let u be a vertex in $K_{w_1}(d, t)$ and v be a vertex in $K_{w_n}(d, t)$ such that $u \neq o_{w_1,w_2}$ and $v \neq o_{w_n,w_{n-1}}$. Let F be any subset of $V(K(d, t+1)) \cup E(K(d, t+1))$ such that (1) there exists an edge between $K_{w_q}(d, t) - F_q$ and $K_{w_{q+1}}(d, t) - F_{q+1}$ for $1 \leq q \leq n - 1$ and (2) $K_{w_q}(d, t) - F_q$ is Hamiltonian-connected for $1 \leq q \leq n$. Then there is a Hamiltonian path P of $K_I(d, t) - F$ joining u to v.

Proof: Since there exists an edge between $K_{w_q}(d, t) - F_q$ and $K_{w_{q+1}}(d, t) - F_{q+1}$ for $1 \leq q \leq n - 1$, the edge $(o_{w_q,w_{q+1}}, o_{w_{q+1},w_q})$ and the vertices $o_{w_q,w_{q+1}}$, o_{w_{q+1},w_q} are not in F for $1 \leq q \leq n - 1$. Since $K_{w_q}(d, t) - F_q$ is Hamiltonian-connected for $1 \leq q \leq n$, there exists a Hamiltonian path P_1 of $K_{w_1}(d, t) - F_1$ joining u to o_{w_1,w_2}, there exists a Hamiltonian path P_q of $K_{w_q}(d, t) - F_q$ joining $o_{w_q,w_{q-1}}$ to $o_{w_q,w_{q+1}}$ for

$2 \le q \le n-1$, and there exists a Hamiltonian path P_n of $K_{w_n}(d,t) - F_n$ joining $o_{w_n,w_{n-1}}$ to v. Therefore, $P = \langle u, P_1, P_2, \ldots, P_n, v\rangle$ is a Hamiltonian path of $K_I(d,t) - F$ joining u to v. □

LEMMA 11.30 Let $d \ge 4$ and $t \ge 1$. Assume that u and v are two vertices of $K(d,t)$. Let r and s be any two open vertices such that $|\{u,v\} \cap \{r,s\}| = 0$. Then there exist two disjoint paths R and S such that (1) R joins u to one of the vertices in $\{r,s\}$, say r, (2) S joins v to s and (3) $R \cup S$ spans $K(d,t)$.

Proof: We prove this lemma by induction on t. The lemma is obviously true for $t = 1$, because $K(d,1)$ is isomorphic to the complete graph K_d with $O_v = V(K(d,1))$. Thus, we assume that this lemma holds for $K(d,n)$ for every $1 \le n \le t$. We claim, that the statement holds for $K(d, t+1)$.

Let $u \in V(K_i(d,t))$, $v \in V(K_j(d,t))$, $r \in V(K_k(d,t))$, and $s \in V(K_l(d,t))$. Since there is only one open vertex in each component, we have $k \ne l$. Now, we consider the following cases:

Case 1. $i \ne j$.

Case 1.1. $|\{i,j\} \cap \{k,l\}| = 0$. Thus, there exists an index in $\{k,l\}$, say k, such that $u \ne o_{i,k}$. Let $I_1 = \{i,k\}$ and $I_2 = \{w_1, w_2, \ldots, w_{d-2}\} = \{1,2,\ldots,d\} - I_1$ such that $w_1 = j$, $w_{d-2} = l$, and $v \ne o_{j,w_2}$. By Lemma 11.29, there exists a Hamiltonian path R of $K_{I_1}(d,t)$ joining u to r; there exists a Hamiltonian path S of $K_{I_2}(d,t)$ joining v to s. Obviously, R and S are the required paths.

Case 1.2. $|\{i,j\} \cap \{k,l\}| = 1$. Without loss of generality, we assume that $i = k$. Let R be a Hamiltonian path of $K_i(d,t)$ joining u to r. Let $I = \{w_1, w_2, \ldots, w_{d-1}\} = \{1,2,\ldots,d\} - \{i\}$ such that $w_1 = j$ and $w_{d-1} = l$. By Lemma 11.29, there exists a Hamiltonian path S of $K_I(d,t)$ joining v to s. Obviously, R and S are the required paths.

Case 1.3. $|\{i,j\} \cap \{k,l\}| = 2$. Without loss of generality, we assume that $i = k$ and $j = l$. By assumption, $|\{u,v\} \cap \{r,s\}| = 0$. Let $I = \{w_2, \ldots, w_{d-1}\} = \{1,2,\ldots,d\} - \{i,j\}$. By induction, we can find two disjoint paths S_{j1} and S_{j2} such that (1) S_{j1} joins v to o_{j,w_2}, (2) S_{j2} joins $o_{j,w_{d-1}}$ to s, and (3) $S_{j1} \cup S_{j2}$ spans $K_j(d,t)$. Let R be a Hamiltonian path of $K_i(d,t)$ joining u and r. By Lemma 11.29, there exists a Hamiltonian path S' of $K_I(d,t)$ joining $o_{w_2 j}$ to $o_{w_{d-1},j}$. Obviously, R and $S = \langle v, S_{j1}, S', S_{j2}^{-1}, s\rangle$ are the required paths.

Case 2. $i = j$.

Case 2.1. $i \notin \{k,l\}$. Let $I = \{w_1, w_2, \ldots, w_{d-2}\} = \{1,2,\ldots,d\} - \{i,k\}$ such that $w_{d-2} = l$. By induction, we can find two disjoint paths S_{i1} and S_{i2} of $K_i(d,t)$ such that (1) S_{i1} joins u to $o_{i,k}$, (2) S_{i2} joins v to o_{i,w_1}, and (3) $S_{i1} \cup S_{i2}$ spans $K_i(d,t)$. Let R' be a Hamiltonian path of $K_k(d,t)$ joining $o_{k,i}$ and r. By Lemma 11.29, there exists a Hamiltonian path S' of $K_I(d,t)$ joining $o_{w_1,i}$ to s. Obviously, $R = \langle u, S_{i1}, R', r\rangle$ and $S = \langle v, S_{i2}, S', s\rangle$ are the required paths.

Case 2.2. $i \in \{k,l\}$. Without loss of generality, we assume that $i = k$. Let $I = \{w_1, w_2, \ldots, w_{d-1}\} = \{1,2,\ldots,d\} - \{i\}$ such that $w_{d-1} = l$. By induction, we can find two disjoint paths S_{i1} and S_{i2} of $K_i(d,t)$ such that (1) S_{i1} joins u to r, (2) S_{i2} joins v to o_{i,w_1}, and (3) $S_{i1} \cup S_{i2}$ spans $K_i(d,t)$. By Lemma 11.29, there exists a Hamiltonian path S' of $K_I(d,t)$ joining $o_{w_1,i}$ to s. Obviously, $R = S_{i1}$ and $S = \langle v, S_{i2}, S', s\rangle$ are the required paths. □

LEMMA 11.31 Both $K(d,t)$ and $\tilde{K}^i(d,t)$ are $(d-4)$-fault-tolerant Hamiltonian-connected for $d \geq 4$, $t \geq 1$, and $1 \leq i \leq d$.

Proof: Since $\tilde{K}^i(d,t)$ is isomorphic to $\tilde{K}^j(d,t)$ for $1 \leq i \neq j \leq d$, we consider $\tilde{K}^1(d,t)$ in the following cases.

Suppose $t = 1$. Note that $K(d,1)$ is isomorphic to K_d and $\tilde{K}^1(d,1)$ is isomorphic to $K_{d+1} - e$ where e is any edge in K_{d+1}. By Lemma 11.4, K_d and $K_{d+1} - e$ are $(d-4)$ fault-tolerant Hamiltonian-connected.

Assume that this lemma holds for $K(d,q)$ and $\tilde{K}^1(d,q)$ for every $1 \leq q \leq t$. We will claim that both $K(d,t+1)$ and $\tilde{K}^1(d,t+1)$ are also $(d-4)$ fault-tolerant Hamiltonian-connected.

First, we show that $K(d,t)$ is $(d-4)$-fault-tolerant Hamiltonian-connected. Since $K(d,t)$ is Hamiltonian-connected, $K(4,t)$ is 0-fault-tolerant Hamiltonian-connected. Thus, we assume that $d \geq 5$. Let u be a vertex in $K_i(d,t)$, v be a vertex in $K_j(d,t)$ with $u \neq v$ and F be the faulty set with $|F| \leq d-4$. We need to find a Hamiltonian path of $K(d,t+1) - F$ joining u to v.

Case A1. $i \neq j$. Assume that there exists a Hamiltonian path $\langle i = w_1, w_2, \ldots, w_d = j\rangle$ of $H(F)$ joining i and j that meet the boundary conditions: $u \neq o_{i,w_2}$ and $o_{j,w_{d-1}} \neq v$. By Lemma 11.28, there exists a Hamiltonian path of $K(d,t+1) - F$ joining u and v. By Theorem 11.24, such a Hamiltonian path in $H(F)$ that meets the boundary conditions except the Cases 2 through 5 of Theorem 11.24. We will show that this lemma holds when $H(F)$ is isomorphic to $K_5 - (1,2)$, the subgraph N of $H(F)$ induced by $V(H(F)) - \{i,j\}$ is a complete graph; isomorphic to $C_{2,5}$; isomorphic to $C_{1,n-2}$.

Case A1.1. $H(F)$ is isomorphic to the complete graph $K_5 - (1,2)$ and $\{i,j\} \neq \{1,2\}$. Obviously, $d = 5$ and $|F| = 1$. Thus, exactly one of $o_{1,2}$, $o_{2,1}$, or $(o_{1,2}, o_{2,1})$ is faulty. By the symmetric property of $H(F)$, we may assume that $(i,j) = (1,3)$ or $(i,j) = (5,3)$.

1. $(i,j) = (1,3)$. Obviously, $\langle i,5,4,2,j\rangle$ and $\langle i,4,2,5,j\rangle$ form two orthogonal Hamiltonian paths of $H(F)$. By Lemma 11.28, we can construct a Hamiltonian path between u and v of $K(d,t+1) - F$ unless (1) $(u = o_{i,4}$ and $v = o_{j,2})$ or (2) $(u = o_{i,5}$ and $v = o_{j,5})$.

Suppose that $u = o_{i,4}$ and $v = o_{j,2}$. Obviously, $\langle i,5,2,4,j\rangle$ is a Hamiltonian path in $H(F)$ satisfying the boundary conditions: $u \neq o_{i,5}$ and $v \neq o_{j,4}$. Suppose that $u = o_{i,5}$ and $v = o_{j,5}$. Since exactly one of $o_{1,2}$, $o_{2,1}$, or $(o_{1,2}, o_{2,1})$ is fault, $K_j(d,t) - \{o_{j,5}\}$ is Hamiltonian-connected. Let P_j be the Hamiltonian path of $K_j(d,t) - \{o_{j,5}\}$ joining $o_{j,4}$ to $o_{j,2}$. By induction, $K_q(d,t) - F$ is Hamiltonian-connected for $q \in \{i,5,4,2\}$. Let P_i be the Hamiltonian path of $K_i(d,t) - F$ joining u to $o_{i,4}$; let P_5 be the Hamiltonian path of $K_5(d,t)$ joining $o_{5,2}$ to $o_{5,j}$; let P_4 be the Hamiltonian path of $K_4(d,t)$ joining $o_{4,i}$ to $o_{4,j}$; let P_2 be the Hamiltonian path of $K_2(d,t)$ joining $o_{2,j}$ to $o_{2,5}$. Therefore, path $\langle u, P_i, P_4, P_j, P_2, P_5, v\rangle$ is the required path.

2. $(i,j)=(5,3)$. Obviously, $\langle i,1,4,2,j\rangle$ and $\langle i,2,4,1,j\rangle$ form two orthogonal Hamiltonian paths of $H(F)$. By Lemma 11.28, we can construct a Hamiltonian path between u and v of $K(d,t+1)-F$ unless (1) ($u=o_{i,1}$ and $v=o_{j,1}$) or (2) ($u=o_{i,2}$ and $v=o_{j,2}$).

Since the condition 1 is similar to the condition 2, we only consider condition 1. Let $u=o_{i,1}$ and $v=o_{j,1}$. Since exactly one of $o_{1,2}$, $o_{2,1}$, or $(o_{1,2},o_{2,1})$ is faulty, $K_j(d,t)-\{o_{j,1}\}$ is Hamiltonian-connected. Let P_j be the Hamiltonian path of $K_j(d,t)-\{o_{j,1}\}$ joining $o_{j,i}$ to $o_{j,2}$. By induction, $K_q(d,t)-F$ is Hamiltonian-connected for $q\in\{i,1,2,4\}$. Let P_i be the Hamiltonian path of $K_i(d,t)$ joining u to $o_{i,j}$; let P_1 be the Hamiltonian path of $K_1(d,t)$ joining $o_{1,4}$ to $o_{1,j}$; let P_4 be the Hamiltonian path of $K_4(d,t)$ joining $o_{4,2}$ to $o_{4,1}$; let P_2 be the Hamiltonian path of $K_2(d,t)$ joining $o_{2,j}$ to $o_{2,4}$. Therefore, path $\langle u,P_i,P_j,P_2,P_4,P_1,v\rangle$ is the required path.

Case A1.2. The subgraph N of $H(F)$ induced by $V(H(F))-\{i,j\}$ is a complete graph; vertex i is adjacent to j and all vertices in N; j is adjacent to i and exactly two vertices 1 and 2 in N. We label the remaining vertices in N as $3,\ldots,d-2$. See Figure 11.32a for illustration. It is easy to see that $\langle i,2,3,\ldots,d-2,1,j\rangle$ and $\langle i,3,\ldots,d-2,1,2,j\rangle$ form two orthogonal Hamiltonian paths of $H(F)$ between i and j. By Lemma 11.28, we can construct a Hamiltonian path of $K(d,t+1)-F$ joining u to v unless (1) ($u=o_{i,2}$ and $v=o_{j,2}$) or (2) ($u=o_{i,3}$ and $v=o_{j,1}$).

Suppose that $u=o_{i,2}$ and $v=o_{j,2}$. Obviously, $\langle i,3,2,4,\ldots,d-2,1,j\rangle$ is a Hamiltonian path in $H(F)$ satisfying the boundary conditions: $u\neq o_{i,3}$ and $v\neq o_{j,1}$. Suppose that $u=o_{i,3}$ and $v=o_{j,1}$. Thus, the Hamiltonian path $\langle i,1,d-2,\ldots,3,2,j\rangle$ in $H(F)$ satisfies the boundary conditions: $u\neq o_{i,1}$ and $v\neq o_{j,2}$. By Lemma 11.28, we can construct a Hamiltonian path of $K(d,t+1)-F$ joining u to v.

Case A1.3. The subgraph N of $H(F)$ induced by $V(H(F))-\{i,j\}$ is isomorphic to $C_{2,5}$; vertex i is adjacent to j and all vertices in N; j is adjacent to i and all vertices in N. We label the vertices in N as $1,2,\ldots,5$. See Figure 11.32b for illustration. Thus, $d=6$ and $|F|=3$. Obviously, $|F_i|=|F_j|=0$. Moreover, $\langle i,1,2,3,4,5,j\rangle$ and $\langle i,5,4,3,2,1,j\rangle$ form two orthogonal Hamiltonian paths of $H(F)$ between i and j. By Lemma 11.28, we can construct a Hamiltonian path of $K(d,t+1)$ joining u to v unless

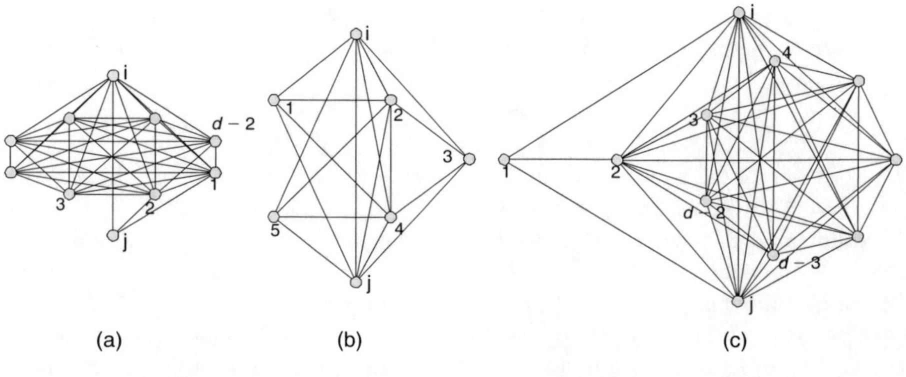

FIGURE 11.32 Illustrations for Case A1: (a) Case A1.2, (b) Case A1.3, and (c) Case A1.4.

(1) ($u = o_{i,1}$ and $v = o_{j,1}$) or (2) ($u = o_{i,5}$ and $v = o_{j,5}$). By the symmetric property of $H(F)$, we consider only the case $u = o_{i,1}$ and $v = o_{j,1}$.

Since $|F_i| = |F_j| = 0$, $K_j(d,t) - (F_j \cup \{o_{j,1}\})$ is Hamiltonian-connected. Let P_j be the Hamiltonian path of $K_j(d,t) - (F_j \cup \{o_{j,1}\})$ joining $o_{j,5}$ to $o_{j,3}$. By induction, $K_q(d,t) - F_q$ is Hamiltonian-connected for $q \in \{i, 1, \ldots, 5\}$. Let P_i be the Hamiltonian path of $K_i(d,t) - F_i$ joining u to $o_{i,2}$; let P_1 be the Hamiltonian path of $K_1(d,t) - F_1$ joining $o_{1,4}$ to $o_{1,j}$; let P_2 be the Hamiltonian path of $K_2(d,t) - F_2$ joining $o_{2,i}$ to $o_{2,5}$; let P_3 be the Hamiltonian path of $K_3(d,t) - F_3$ joining $o_{3,j}$ to $o_{3,4}$; let P_4 be the Hamiltonian path of $K_4(d,t) - F_4$ joining $o_{4,3}$ to $o_{4,1}$; let P_5 be the Hamiltonian path of $K_5(d,t) - F_5$ joining $o_{5,2}$ to $o_{5,j}$. Therefore, path $\langle u, P_i, P_2, P_5, P_j, P_3, P_4, P_1, v\rangle$ is the required path.

Case A1.4. The subgraph N of $H(F)$ induced by $V(H(F)) - \{i,j\}$ is isomorphic to $C_{1,n-2}$; vertex i is adjacent to j and all vertices in N; j is adjacent to i and all vertices in N. We label the vertices in N as $1, 2, \ldots, d-2$. See Figure 11.32c for illustration. Obviously, $\langle i, 1, 2, \ldots, d-2, j\rangle$ and $\langle i, d-2, d-3, \ldots, 1, j\rangle$ form two orthogonal Hamiltonian paths of $H(F)$ between i and j. By Lemma 11.28, we can construct a Hamiltonian path of $K(d, t+1)$ joining u to v unless (1) ($u = o_{i,d-2}$ and $v = o_{j,d-2}$) or (2) ($u = o_{i,1}$ and $v = o_{j,1}$).

Suppose that $u = o_{i,d-2}$ and $v = o_{j,d-2}$. Obviously, $\langle i, 3, d-2, \ldots, 4, 2, 1, j\rangle$ is a Hamiltonian path in $H(F)$ satisfying the boundary conditions: $u \neq o_{i,3}$ and $v \neq o_{j,1}$. By Lemma 11.28, we can construct a Hamiltonian path of $K(d, t+1)$ joining u to v.

Suppose that $u = o_{i,1}$ and $v = o_{j,1}$. Since $|F_i| = |F_j| = 0$, $K_j(d,t) - \{o_{j,1}\}$ is Hamiltonian-connected. Let P_j be the Hamiltonian path of $K_j(d,t) - \{o_{j,1}\}$ joining $o_{j,3}$ to $o_{j,4}$. By induction, $K_q(d,t) - F_q$ is Hamiltonian-connected for $q \in \{i, 2, \ldots, d-2\}$. Let P_i be the Hamiltonian path of $K_i(d,t)$ joining u to $o_{i,3}$; let P_3 be the Hamiltonian path of $K_3(d,t) - F_3$ joining $o_{3,i}$ to $o_{3,j}$; let P_4 be the Hamiltonian path of $K_4(d,t) - F_4$ joining $o_{4,j}$ to $o_{4,5}$ if $d \geq 7$; and let P_4 be the Hamiltonian path of $K_4(d,t) - F_4$ joining $o_{4,j}$ to $o_{4,2}$ if $d = 6$; let P_q be a Hamiltonian path of $K_q(d,t) - F_q$ joining $o_{q,q-1}$ to $o_{q,q+1}$ for $4 \leq q \leq d-3$; let P_{d-2} be the Hamiltonian path of $K_{d-2}(d,t) - F_{d-2}$ joining $o_{d-2,d-3}$ to $o_{d-2,2}$; let P_2 be a Hamiltonian path of $K_2(d,t) - F_2$ joining $o_{2,d-2}$ to $o_{2,1}$; let P_1 be a Hamiltonian path of $K_1(d,t) - F_1$ joining $o_{1,2}$ to $o_{1,j}$. Therefore, $\langle u, P_i, P_3, P_j, P_4, \ldots, P_{d-2}, P_2, P_1, v\rangle$ is the required path.

Case A2. $i = j$. Without loss of generality, we assume that $i = j = 1$. Let $A = K_1(d, t) \cup \{(o_{1,r}, o_{r,1}) \mid 2 \leq r \leq d\}$, $B = \{o_{r,1} \mid 2 \leq r \leq d\}$, and $C = K(d, t+1) - A - B$. We set $F_A = F \cap A$, $F_B = F \cap B$, and $F_C = F \cap C$.

Suppose that $|F_A| > 0$ or $|F_B| > 0$. We consider the graph $\tilde{K}_1^1(d,t)$. Let $F' = F_A \cup \{(o_{1,r}, x) \mid o_{r,1} \in F \text{ or } (o_{1,r}, o_{r,1}) \in F \text{ for } 2 \leq r \leq d\}$. Obviously, $|F'| \leq d-4$. By induction on $\tilde{K}_1^1(d,t)$, there exists a Hamiltonian path P_1 of $\tilde{K}_1^1(d,t) - F'$ joining u to v. Thus, path P_1 can be written as $\langle u, P_{11}, o_{1,a}, x, o_{1,b}, P_{12}, v\rangle$. Since $|F_A| > 0$ or $|F_B| > 0$, $|F_C| \leq d-5$. Therefore, $H(F) - \{1\}$ is Hamiltonian-connected. There exists a Hamiltonian path $\langle a = w_1, \ldots, w_{d-1} = b\rangle$ of $H(F) - \{1\}$ joining vertex a to vertex b. By Lemma 11.29, there is a Hamiltonian path Q of $K_I(d, t+1) - F$ joining $o_{a,1}$ to $o_{b,1}$ where $I = \{w_1, \ldots, w_{d-1}\}$. Hence, the Hamiltonian path $\langle u, P_{11}, o_{1,a}, o_{a,1}, Q, o_{b,1}, o_{1,b}, P_{12}, v\rangle$ is the required path.

Suppose that $|F_A| = 0$ and $|F_B| = 0$. Hence, $|F_C| \leq d - 4$. Therefore, $H(F) - \{1\}$ is Hamiltonian. Let $\langle w_1, \ldots, w_{d-1}, w_1 \rangle$ be the Hamiltonian cycle in $H(F) - \{1\}$ with $|\{o_{1,w_1}, o_{1,w_{d-1}}\} \cap \{u, v\}| = 0$. By Lemma 11.30, there exist two disjoint paths R and S such that (1) R joins u to one of the vertices in $\{o_{1,w_1}, o_{1,w_{d-1}}\}$, say o_{1,w_1}, (2) S joins v to $o_{1,w_{d-1}}$, and (3) $R \cup S$ spans $K_1(d,t)$. By Lemma 11.29, there is a Hamiltonian path P of $K_I(d,t) - F$ joining $o_{w_1,1}$ to $o_{w_{d-1},1}$, where $I = \{w_1, \ldots, w_{d-1}\}$. Hence, the Hamiltonian path $\langle u, R, o_{1,w_1}, P, o_{1,w_{d-1}}, S, v \rangle$ is the required path.

Second, we show that $\tilde{K}^1(d,t)$ is $(d-4)$ fault-tolerant Hamiltonian-connected for $d \geq 4$ and $t \geq 2$. Let u and v be any two distinct vertices in $\tilde{K}^1(d,t)$ and F be the faulty set with $|F| \leq d-4$. We have to show that there exists a Hamiltonian path of $\tilde{K}^1(d,t) - F$ joining u to v.

We construct graph $\tilde{H}^1(F)$ by setting $V(\tilde{H}^1(F)) = V(H(F)) \cup \{0\}$ and $E(\tilde{H}^1(F)) = E(H(F)) \cup \{(r,0) \mid o_{r,0} \notin F \text{ where } 2 \leq r \leq d\}$. Obviously, $\tilde{H}^1(F)$ is Hamiltonian-connected.

Case B1. $u \in K_i(d,t)$ and $v \in K_j(d,t)$ with $i \neq j$. Assume that there exists a Hamiltonian path $\langle i = w_1, w_2, \ldots, w_{d+1} = j \rangle$ of $\tilde{H}^1(F)$ joining i and j that meets the boundary conditions: $u \neq o_{i,w_2}$ and $o_{j,w_d} \neq v$. By Lemma 11.28, there exists a Hamiltonian path of $K(d,t+1) - F$ joining u and v. By Theorem 11.24, there exists such a Hamiltonian path in $\tilde{H}^1(F)$ that meets the boundary conditions except Cases 2 through 5. We will show that Lemma 11.28 holds when $\tilde{H}^1(F)$ is isomorphic to $K_5 - e$ where e is any edge in K_5, the subgraph N of $\tilde{H}^1(F)$ induced by $V(\tilde{H}^1(F)) - \{i,j\}$ is a complete graph; isomorphic to $C_{2,5}$; isomorphic to $C_{1,n-2}$.

Case B1.1. $\tilde{H}^1(F)$ is isomorphic to the complete graph $K_5 - e$ for any edge e in K_5. We label the vertices in K_5 as $0, 1, \ldots, 4$. See Figure 11.33a for illustration. By the definition of $\tilde{H}^1(F)$, $\{i,j\} \neq \{0,1\}$. Obviously, $d = 4$ and $|F| = 0$. By the symmetric property of $\tilde{H}^1(F)$, we may assume that $(i,j) = (1,2)$ or $(i,j) = (4,2)$.

1. $(i,j) = (1,2)$. Obviously, $\langle i, 4, 0, 3, j \rangle$ and $\langle i, 3, 0, 4, j \rangle$ form two orthogonal Hamiltonian paths of $\tilde{H}^1(F)$. By Lemma 11.28, we can construct a Hamiltonian path between u and v of $\tilde{K}^1(d,t+1) - F$ unless (1) ($u = o_{i,4}$ and $v = o_{j,4}$) or (2) ($u = o_{i,3}$ and $v = o_{j,3}$).

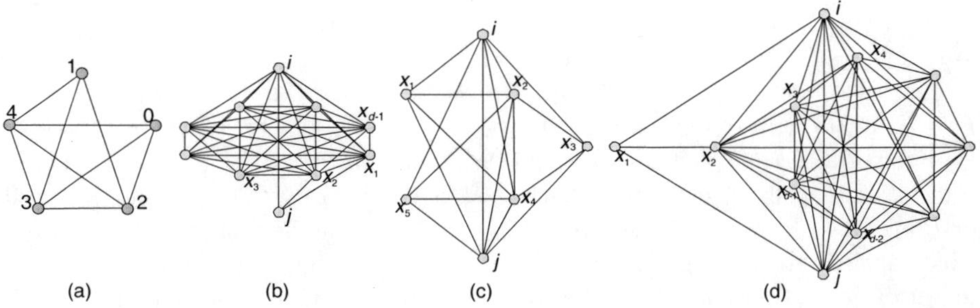

FIGURE 11.33 Illustrations for Case B1: (a) Case B1.1, (b) Case B1.2, (c) Case B1.3, and (d) Case B1.4.

Suppose that $u=o_{i,4}$ and $v=o_{j,4}$. Obviously, $\langle i,3,4,0,j\rangle$ is a Hamiltonian path in $\tilde{H}^1(F)$ satisfying the boundary conditions: $u\neq o_{i,3}$ and $v\neq o_{j,0}$. Suppose that $u=o_{i,3}$ and $v=o_{j,3}$. Obviously, $\langle i,4,3,0,j\rangle$ is a Hamiltonian path in $\tilde{H}^1(F)$ satisfying the boundary conditions: $u\neq o_{i,4}$ and $v\neq o_{j,0}$.

2. $(i,j)=(4,2)$. Obviously, $\langle i,0,3,1,j\rangle$ and $\langle i,1,3,0,j\rangle$ form two orthogonal Hamiltonian paths of $\tilde{H}^1(F)$. By Lemma 11.28, we can construct a Hamiltonian path between u and v of $\tilde{K}^1(d,t+1)-F$ unless (1) ($u=o_{i,0}$ and $v=o_{j,0}$) or (2) ($u=o_{i,1}$ and $v=o_{j,1}$).

Suppose that $u=o_{i,0}$ and $v=o_{j,0}$. Let P_i be the Hamiltonian path of $K_i(d,t)-\{u\}$ joining $o_{i,3}$ to $o_{i,1}$; let P_3 be the Hamiltonian path of $K_3(d,t)$ joining $o_{3,0}$ to $o_{3,2}$; let P_1 be the Hamiltonian path of $K_1(d,t)$ joining $o_{1,2}$ to $o_{1,4}$; P_j be the Hamiltonian path of $K_j(d,t)$ joining $o_{j,1}$ to v. Therefore, path $\langle u,x,P_3,P_i,P_1,P_j,v\rangle$ is the required path. Suppose that $u=o_{i,1}$ and $v=o_{j,1}$. Let P_i be the Hamiltonian path of $K_i(d,t)-\{u\}$ joining $o_{i,3}$ to $o_{i,0}$; let P_3 be the Hamiltonian path of $K_3(d,t)$ joining $o_{3,1}$ to $o_{3,2}$; let P_1 be the Hamiltonian path of $K_1(d,t)$ joining $o_{1,i}$ to $o_{1,3}$; let P_j be the Hamiltonian path of $K_j(d,t)$ joining $o_{j,0}$ to v. Therefore, path $\langle u,P_1,P_3,P_i,x,P_j,v\rangle$ is the required path.

Case B1.2. The subgraph N of $\tilde{H}^1(F)$ induced by $V(\tilde{H}^1(F))-\{i,j\}$ is a complete graph; vertex i is adjacent to j and all vertices in N; j is adjacent to i and exactly two vertices; say, x_1 and x_2, in N. Since $\deg_{\tilde{H}^1(F)}(0)<d$, x_1 or x_2 is not vertex 0. We label the remaining vertices in N as $x_3,\ldots,x_{d-1}$. See Figure 11.33b for illustration. It is easy to see that $\langle i,x_2,x_3,\ldots,x_{d-1},x_1,j\rangle$ and $\langle i,x_3,\ldots,x_{d-1},x_1,x_2,j\rangle$ form two orthogonal Hamiltonian paths of $\tilde{H}^1(F)$ between i and j. By Lemma 11.28, we can construct a Hamiltonian path between u and v of $\tilde{K}^1(d,t+1)-F$ unless (1) ($u=o_{i,x_2}$ and $v=o_{j,x_2}$) or (2) ($u=o_{i,x_3}$ and $v=o_{j,x_1}$).

Suppose that $u=o_{i,x_2}$ and $v=o_{j,x_2}$. Obviously, $\langle i,x_3,x_2,x_4,\ldots,x_{d-1},x_1,j\rangle$ is a Hamiltonian path in $\tilde{H}^1(F)$ satisfying the boundary conditions: $u\neq o_{i,x_3}$ and $v\neq o_{j,x_1}$. Suppose that $u=o_{i,x_3}$ and $v=o_{j,x_1}$. The Hamiltonian path $\langle i,x_1,x_{d-1},\ldots,x_3,x_2,j\rangle$ in $\tilde{H}^1(F)$ satisfying the boundary conditions: $u\neq o_{i,x_1}$ and $v\neq o_{j,x_2}$. By Lemma 11.28, we can construct a Hamiltonian paths between u and v of $\tilde{K}^1(d,t+1)-F$.

Case B1.3. The subgraph N of $\tilde{H}^1(F)$ induced by $V(\tilde{H}^1(F))-\{i,j\}$ is isomorphic to $C_{2,5}$; vertex i is adjacent to j and all vertices in N; j is adjacent to i and all vertices in N. We label the vertices of $C_{2,5}$ as indicated in Figure 11.33c. Obviously, $d=6$ and $|F|=2$. Thus, $|E(\overline{N})|=3$ and $\deg_{\overline{N}}(x_1)=\deg_{\overline{N}}(x_3)=\deg_{\overline{N}}(x_5)=2$. It is easy to see that $\langle i,x_1,x_2,x_3,x_4,x_5,j\rangle$ and $\langle i,x_5,x_4,x_3,x_2,x_1,j\rangle$ form two orthogonal Hamiltonian paths of $\tilde{H}^1(F)$ between i and j. By Lemma 11.28, we can construct a Hamiltonian path between u and v of $\tilde{K}^1(d,t+1)-F$ unless (1) ($u=o_{i,x_1}$ and $v=o_{j,x_1}$) or (2) ($u=o_{i,x_5}$ and $v=o_{j,x_5}$). By the symmetric property of $\tilde{H}^1(F)$, we consider only the case $u=o_{i,x_1}$ and $v=o_{j,x_1}$.

Since $|F_i|=|F_j|=0$, $K_j(d,t)-(F_j\cup\{o_{j,x_1}\})$ is Hamiltonian-connected. Let P_j be the Hamiltonian path of $K_j(d,t)-(F_j\cup\{o_{j,x_1}\})$ joining o_{j,x_5} to o_{j,x_3}. Let l be the index that x_l is vertex 0 in N. By induction, $K_q(d,t)-F_q$ is Hamiltonian-connected for $q\in\{i,x_1,\ldots,x_5\}-\{x_l\}$. Let P_i be the Hamiltonian path of $K_i(d,t)-F_i$ joining u to o_{i,x_2}; let P_1 be the Hamiltonian path of $K_1(d,t)-F_1$ joining o_{x_1,x_4} to $o_{x_1,j}$ if $l\neq 1$ and $P_1=\{x\}$ if otherwise; let P_2 be the Hamiltonian path of $K_2(d,t)-F_2$ joining $o_{x_2,i}$ to

o_{x_2,x_5}; let P_3 be the Hamiltonian path of $K_3(d,t)-F_3$ joining $o_{x_3,j}$ to o_{x_3,x_4} if $l \neq 3$ and $P_3 = \{x\}$ if otherwise; let P_4 be the Hamiltonian path of $K_4(d,t)-F_4$ joining o_{x_4,x_3} to o_{x_4,x_1}; let P_5 be the Hamiltonian path of $K_5(d,t)-F_5$ joining o_{x_5,x_2} to $o_{x_5,j}$ if $l \neq 5$ and $P_5 = \{x\}$ if otherwise. Therefore, path $\langle u, P_i, P_2, P_5, P_j, P_3, P_4, P_1, v\rangle$ is the required path.

Case B1.4. The subgraph N of $\tilde{H}^1(F)$ induced by $V(\tilde{H}^1(F))-\{i,j\}$ is isomorphic to $C_{1,n-2}$; vertex i is adjacent to j and all the vertices in N; j is adjacent to i and all the vertices in N. We label the vertices of $C_{1,n-2}$ as indicated in Figure 11.33d. Obviously, $E(\overline{N}) = d-3$ and $\deg_{\overline{N}}(1) = d-3$. It is easy to see that $\langle i, x_1, x_2, \ldots, x_{d-1}, j\rangle$ and $\langle i, x_{d-1}, x_{d-2}, \ldots, x_1, j\rangle$ form two orthogonal Hamiltonian paths of $\tilde{H}^1(F)$ between i and j. By Lemma 11.28, we can construct a Hamiltonian path between u and v in $\tilde{K}^1(d,t+1)-F$ unless (1) ($u = o_{i,x_{d-1}}$ and $v = o_{j,x_{d-1}}$) or (2) ($u = o_{i,x_1}$ and $v = o_{j,x_1}$).

Suppose that $u = o_{i,x_{d-1}}$ and $v = o_{j,x_{d-1}}$. Obviously, $\langle i, x_3, x_{d-1}, \ldots, x_4, x_2, x_1, j\rangle$ is a Hamiltonian path in $\tilde{H}^1(F)$ satisfying the boundary conditions: $u \neq o_{i,x_3}$ and $v \neq o_{j,x_1}$. By Lemma 11.28, we can construct a Hamiltonian path between u and v in $\tilde{K}^1(d,t+1)-F$.

Suppose that $u = o_{i,x_1}$ and $v = o_{j,x_1}$. Since $|F_i| = |F_j| = 0$, $K_j(d,t)-\{o_{j,x_1}\}$ is Hamiltonian-connected. Let P_j be the Hamiltonian path of $K_j(d,t)-\{o_{j,x_1}\}$ joining o_{j,x_3} to o_{j,x_4}. Let l be the index that x_l is vertex 0 in N. By induction, $K_q(d,t)-F_q$ is Hamiltonian-connected for $q \in \{i, x_2, \ldots, x_{d-1}\}-\{x_l\}$. Let P_i be the Hamiltonian path of $K_i(d,t)$ joining u to o_{i,x_3}; let P_3 be the Hamiltonian path of $K_3(d,t)-F_3$ joining $o_{x_3,i}$ to $o_{x_3,j}$ if $l \neq 3$ and $P_3 = \{x\}$ if otherwise. Suppose $l \neq 4$; let P_4 be the Hamiltonian path of $K_4(d,t)-F_4$ joining $o_{x_4,j}$ to o_{x_4,x_5} if $d \geq 7$ and let P_4 be the Hamiltonian path of $K_4(d,t)-F_4$ joining $o_{x_4,j}$ to o_{x_4,x_2} if $d=6$. Suppose $l=4$; let $P_4 = \{x\}$. Let P_q be a Hamiltonian path of $K_q(d,t)-F_q$ joining $o_{x_q,x_{q-1}}$ to $o_{x_q,x_{q+1}}$ for $4 \leq q \leq d-2$ if $l \neq q$ and $P_q = \{x\}$ if otherwise; let P_{d-1} be the Hamiltonian path of $K_{d-1}(d,t)-F_{d-1}$ joining $o_{x_{d-1},x_{d-2}}$ to o_{x_{d-1},x_2} if $l \neq d-1$ and $P_{d-1} = \{x\}$ if otherwise; let P_2 be a Hamiltonian path of $K_2(d,t)-F_2$ joining $o_{x_2,x_{d-1}}$ to o_{x_2,x_1}; let P_1 be a Hamiltonian path of $K_1(d,t)-F_1$ joining o_{x_1,x_2} to $o_{x_1,j}$ if $l \neq 1$ and $P_1 = \{x\}$ if otherwise. Therefore, path $\langle u, P_i, P_3, P_j, P_4, \ldots, P_{d-1}, P_2, P_1, v\rangle$ is the required path.

Case B2. $u \in K_i(d,t)$ and v is the vertex x. Since $|F| \leq d-4$, $|F \cap K(d,t+1)| \leq d-4$ and there exists at least one vertex $o_{r,x}$ with $\{o_{r,x}, (o_{r,x}, x)\} \cap F = \emptyset$. With the previous proof, $K(d,t+1)-F$ is Hamiltonian-connected. There exists a Hamiltonian path P_1 of $K(d,t+1)-F$ joining u to $o_{r,x}$. Hence, Hamiltonian path $\langle u, P_1, o_{r,x}, x = v\rangle$ is the required path.

Case B3. $u, v \in K_i(d,t)$. Let $A = K_i(d,t) \cup \{(o_{i,r}, o_{r,i}) \mid 1 \leq r \neq i \leq d\}$, $B = \{o_{r,i} \mid 1 \leq r \neq i \leq d\}$, and $C = K(d,t+1)-\{(o_{i,x}, x)\}-A-B$. We set $F_A = F \cap A$, $F_B = F \cap B$, and $F_C = F \cap C$.

Suppose that $|F_A| > 0$ or $|F_B| > 0$. Consider the graph $\tilde{K}_i^i(d,t)$. Let $F' = F_A \cup \{(o_{i,r}, x) \mid o_{r,i} \in F \text{ or } (o_{i,r}, o_{r,i}) \in F \text{ for } 1 \leq r \neq i \leq d\}$. Obviously, $|F'| \leq d-4$. By induction on $\tilde{K}_i^i(d,t)$, there exists a Hamiltonian path P_i of $\tilde{K}_i^i(d,t)-F'$ joining u to v. Path P_i can be written as $\langle u, P_{i1}, o_{i,a}, x, o_{i,b}, P_{i2}, v\rangle$. Since $|F_A| > 0$ or $|F_B| > 0$, $|F_C| \leq d-4$. Therefore, $\tilde{H}^1(F)-\{i\}$ is Hamiltonian-connected. There exists a Hamiltonian path $\langle a = w_1, \ldots, b = w_d\rangle$ of $\tilde{H}^1(F)-\{i\}$ joining vertex a to vertex b. By

Lemma 11.29, there is a Hamiltonian path Q of $K_I(d, t+1) - F$ joining $o_{a,i}$ to $o_{b,i}$ where $I = \{w_1, \ldots, w_d\}$. Hence, Hamiltonian path $\langle u, P_{i1}, o_{i,a}, o_{a,i}, Q, o_{b,i}, o_{i,b}, P_{i2}, v\rangle$ is the required path.

Suppose that $|F_A| = 0$ and $|F_B| = 0$. Hence, $|F_C| \leq d - 3$. Therefore, we know that $\tilde{H}^1(F) - \{i\}$ is Hamiltonian. Let $\langle w_1, \ldots, w_d, w_1\rangle$ be the Hamiltonian cycle in $\tilde{H}^1(F) - \{i\}$ with $|\{o_{i,w_1}, o_{i,w_d}\} \cap \{u, v\}| = 0$. By Lemma 11.30, there exist two disjoint paths R and S such that (1) R joins u to one of the vertices in $\{o_{i,w_1}, o_{i,w_d}\}$, say o_{i,w_1}, (2) S joins v to o_{i,w_d}, and (3) $R \cup S$ spans $K_i(d, t)$. By Lemma 11.29, there is a Hamiltonian path P of $K_I(d, t) - F$ joining $o_{w_1,i}$ to $o_{w_d,i}$, where $I = \{w_1, \ldots, w_d\}$. Hence, the Hamiltonian path $\langle u, R, o_{i,w_1}, P, o_{i,w_d}, S, v\rangle$ is the required path. □

THEOREM 11.25 $\mathcal{H}_f^\kappa(K(d,t)) = d - 4$ for $d \geq 4$ and $t \geq 1$.

Proof: Since $\delta(K(d,t)) = d - 1$, $\mathcal{H}_f^\kappa(K(d,t)) \leq d - 4$. By Lemma 11.31, $K(d,t)$ is $(d-4)$ fault-tolerant Hamiltonian-connected. Therefore, $\mathcal{H}_f^\kappa(K(d,t)) = d - 4$ for $d \geq 4$ and $t \geq 1$. □

11.8 FAULT-TOLERANT HAMILTONICITY OF THE FULLY CONNECTED CUBIC NETWORKS

Adding an extra vertex to the WK-recursive networks provides a technique to evaluate the corresponding fault-tolerant Hamiltonicity. Ho and Lin [148] use a similar technique to evaluate the fault-tolerant Hamiltonicity of the fully connected cubic networks.

Hierarchical interconnection networks (HINs) are appealing, mainly due to the following reasons:

1. They can provide good expandability. That is, with the growing size of the multicomputer systems, the alterations in both hardware configuration and communication software of each vertex can be minimized.
2. Compared with some nonhierarchical interconnection networks, such as the hypercube, they can integrate more vertices by using the same number of links.
3. They can integrate the positive features of two or more nonhierarchical networks.
4. They can be applied to new hybrid computer architectures utilizing both optical and electronic technologies. Specifically, processors are partitioned into two groups, where electronic interconnects are used to connect processors within the same group, and optical interconnects are used for intergroup communication.

In the past decade, a number of HINs were proposed. There are two different kinds of HINs: (1) HINs that consist of exactly two levels [82,125,177,264,265,336] and (2) HINs that can be defined recursively until a prescribed number of levels is reached [78,331,352,354]. The recursively defined HINs are desired because of their excellent expandability. The *fully connected cubic networks* (FCCNs), proposed in Ref. 331, are a class of HINs that are defined recursively by taking the three-dimensional cube as the basic graph. It indicates that FCCNs are a class of newly proposed hierachical networks for multisystems, which enjoy the strengths of constant vertex degree and good expandability. Some interesting properties about FCCNs are discussed in Ref. 331. In particular, Yang et al. [348] present a shortest-path routing algorithm.

Let $Z_8 = \{0, 1, 2, 3, 4, 5, 6, 7\}$. For $m \geq 1$ and $a \in Z_8$, let $a^m = \underbrace{aa \ldots a}_{m}$.

For $n \geq 1$, an n-level *FCCN*, $FCCN_n$, is a graph defined recursively as follows:

1. $FCCN_1$ is a graph with $V(FCCN_1) = Z_8$ and $E(FCCN_1) = \{(0,1), (0,2), (1,3), (2,3), (4,5), (4,6), (5,7), (6,7), (0,4), (1,5), (2,6), (3,7)\}$. Obviously, $FCCN_1$ is isomorphic to Q_3.
2. When $n \geq 2$, $FCCN_n$ is built from eight vertex-disjoint copies of $FCCN_{n-1}$ by adding 28 edges. For $0 \leq k \leq 7$, we let $kFCCN_{n-1}$ denote a copy of $FCCN_{n-1}$ with each vertex being prefixed with k, then $FCCN_n$ is defined by

$$V(FCCN_n) = \bigcup_{k=0}^{7} V(kFCCN_{n-1})$$

$$E(FCCN_n) = \left(\bigcup_{k=0}^{7} E(kFCCN_{n-1}) \right) \cup \{(pq^{n-1}, qp^{n-1}) \mid 0 \leq p < q \leq 7\}$$

Let $n \geq 2$. A vertex v in $FCCN_n$ is a *boundary vertex* if it is of the form p^n and v is an *intercubic vertex* if it is of the form pq^{n-1} with $p \neq q$. An *intercubic edge* is an edge joining two intercubic vertices.

Figures 11.34a through 11.34c show $FCCN_1$, $FCCN_2$, and $FCCN_3$. In essence, each vertex of an *FCCN* has four links, with each boundary vertex having one I/O channel link that is not counted in the vertex degree. Obviously, $iFCCN_{n-1}$ has seven intercubic vertices and one boundary vertex for $0 \leq i \leq 7$.

In order to discuss the fault-tolerant Hamiltonicity and fault-tolerant Hamiltonian connectivity of fully connected cubic networks, we need to introduce the extended fully connected cubic networks. For every $t \in Z_8$, the *extended fully connected cubic network* $FCCN_n^t$ is the graph obtained from $FCCN_n$ by joining the vertices in the set $\{p^n \mid p \in Z_8 - \{t\}\}$ to an extra vertex w. For example, $FCCN_2^0$ is illustrated in Figure 11.34d. Note that $FCCN_n^i$ is isomorphic to $FCCN_n^j$ for every i,j in Z_8. We discus only $FCCN_n^0$.

Since $FCCN_n$ has 64 vertices and $FCCN_n^0$ has 65 vertices, we can check the following two lemmas by brute force.

LEMMA 11.32 $FCCN_2$ and $FCCN_2^0$ are Hamiltonian-connected.

LEMMA 11.33 $FCCN_2 - f$ is Hamiltonian for any $f \in V(FCCN_2) \cup E(FCCN_2)$ with $|f| = 1$. Moreover, $FCCN_2^0 - f^*$ is Hamiltonian for any $f^* \in V(FCCN_2^0) \cup E(FCCN_2^0)$ with $|f^*| = 1$.

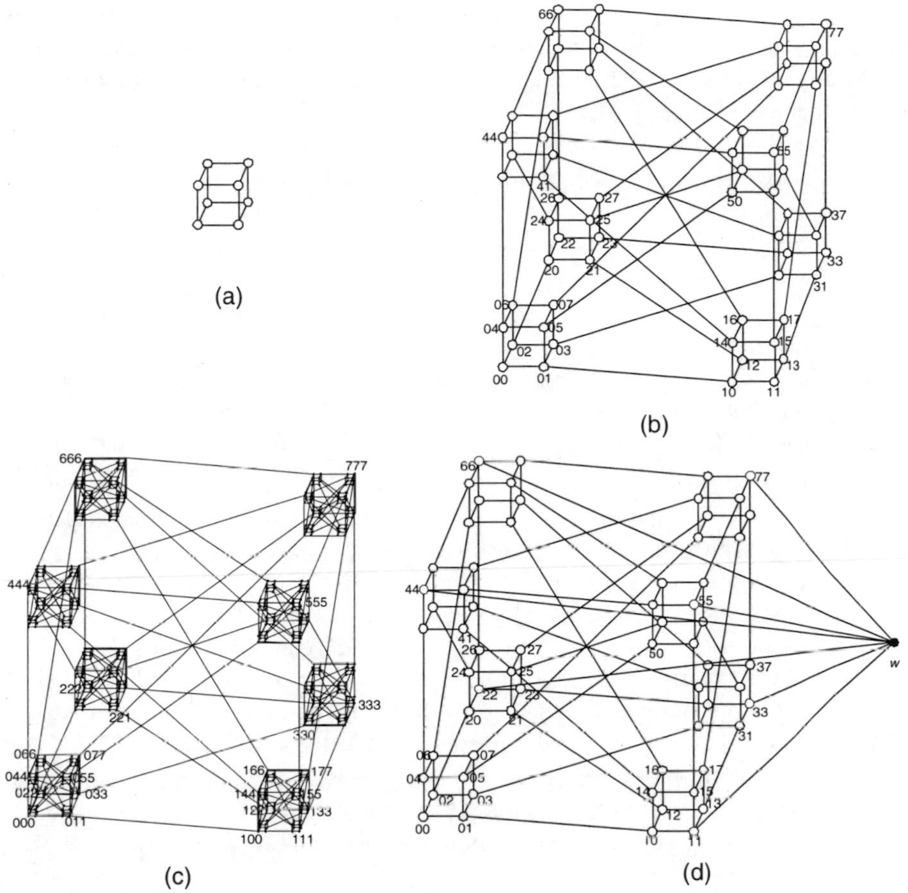

FIGURE 11.34 The graphs (a) $FCCN_1$, (b) $FCCN_2$, (c) $FCCN_3$, and (d) $FCCN_2^0$.

LEMMA 11.34 Both $FCCN_n$ and $FCCN_n^0$ are Hamiltonian-connected for $n \geq 2$.

Proof: We prove this lemma by induction. With Lemma 11.32, the statement holds for $n = 2$. We assume that the statement holds for l with $2 \leq l \leq n$.

First, we prove that $FCCN_{n+1}$ is Hamiltonian-connected. Let u and v be any two distinct vertices in $FCCN_{n+1}$. We need to find a Hamiltonian path of $FCCN_{n+1}$ joining u to v.

Suppose that $u \in bFCCN_n$ and $v \in eFCCN_n$ with $b \neq e$. Let $k_0, k_1, \ldots, k_7$ be any permutation of Z_8 such that $k_0 = b$, $k_7 = e$, $u \neq k_0 k_1^n$, and $v \neq k_7 k_6^n$. Let $x_i = k_i k_{i-1}^n$ for $1 \leq i \leq 7$, $y_j = k_j k_{j+1}^n$ for $0 \leq j \leq 6$, $x_0 = u$, and $y_7 = v$. By induction, there exists a Hamiltonian path P_i of $iFCCN_n$ joining x_i to y_i for $0 \leq i \leq 7$. Obviously, $P = \langle u, P_0, P_1, \ldots, P_7, v \rangle$ forms a Hamiltonian path of $FCCN_{n+1}$ joining u to v.

Suppose that $\{u, v\} \subset bFCCN_n$. By induction hypothesis, there is a Hamiltonian path P_b of $bFCCN_n^b$ joining u and v. Obviously, P_b can be written as $\langle u, P_{b1}, bk_1^n, w, bk_7^n, P_{b2}, v \rangle$. Clearly, $b \notin \{k_1, k_7\}$. Let $k_2, k_3, \ldots, k_6$ be any permutation of

$Z_8 - \{b, k_1, k_7\}$. We set $k_8 = k_0 = b$. Let $x_i = k_i k_{i-1}^n$ and $y_i = k_i k_{i+1}^n$ for $1 \le i \le 7$. By induction, there exists a Hamiltonian path P_i of $iFCCN_n$ joining x_i to y_i for $1 \le i \le 7$. Obviously, $P = \langle u, P_{b1}, P_1, \ldots, P_7, P_{b2}, v\rangle$ forms a Hamiltonian path of $FCCN_{n+1}$ joining u to v.

Second, we prove that $FCCN_{n+1}^0$ is Hamiltonian-connected. Let u and v be any two distinct vertices in $FCCN_{n+1}^0$. We need to find a Hamiltonian path of $FCCN_{n+1}^0$ joining u to v.

Suppose that $u \in bFCCN_n$ and v is the extra vertex w. Let $k_0, k_1, \ldots, k_7$ be any permutation of Z_8 such that $k_0 = b$, $u \ne k_0 k_1^n$, and $k_7 \ne 0$. Let $x_i = k_i k_{i-1}^n$ for $1 \le i \le 7$, $y_j = k_j k_{j+1}^n$ for $0 \le j \le 6$, $x_0 = u$, and $y_7 = k_7^{n+1}$. By induction, there exists a Hamiltonian path P_i of $iFCCN_n$ joining x_i to y_i for $0 \le i \le 7$. Obviously, $P = \langle u, P_0, P_1, \ldots, P_7, v = w\rangle$ forms a Hamiltonian path of $FCCN_{n+1}^0$ joining u to v.

Suppose that $u \in bFCCN_n$ and $v \in eFCCN_n$ with $b \ne e$. Let $k_0, k_1, \ldots, k_7$ be any permutation of Z_8 such that $k_0 = b$, $k_7 = e$, $u \ne k_0 k_1^n$, $v \ne k_7 k_6^n$, and $0 \notin \{k_1, k_2\}$. Let $x_i = k_i k_{i-1}^n$ for $3 \le i \le 7$, $y_j = k_j k_{j+1}^n$ for $2 \le j \le 6$, $x_0 = u$, $x_1 = k_1 k_0^n$, $x_2 = k_2^{n+1}$, $y_0 = k_0 k_1^n$, $y_1 = k_1^{n+1}$, and $y_7 = v$. By induction, there exists a Hamiltonian path P_i of $iFCCN_n$ joining x_i to y_i for $0 \le i \le 7$. Obviously, $P = \langle u, P_0, P_1, w, P_2, \ldots, P_7, v\rangle$ forms a Hamiltonian path of $FCCN_{n+1}^0$ joining u to v.

Suppose that $\{u, v\} \subset bFCCN_n$. By induction hypothesis, there exists a Hamiltonian path P_b of $bFCCN_n^b$ joining u and v. Obviously, P_b can be written as $\langle u, P_{b1}, bk_1^n, w, bk_7^n, P_{b2}, v\rangle$. Clearly, $b \notin \{k_1, k_7\}$. Obviously, at least one of k_1 and k_7 is not 0. Without loss of generality, we assume $k_1 \ne 0$. Let $k_2, k_3, \ldots, k_6$ be any permutation of $Z_8 - \{b, k_1, k_7\}$ such that $k_2 \ne 0$. We set $k_8 = b$. Let $x_i = k_i k_{i-1}^n$ for $3 \le i \le 7$, $y_i = k_i k_{i+1}^n$ for $2 \le i \le 7$, $x_1 = k_1 b^n$, $x_2 = k_2^{n+1}$, and $y_1 = k_1^{n+1}$. By induction, there exists a Hamiltonian path P_i of $iFCCN_n$ joining x_i to y_i for $1 \le i \le 7$. Obviously, $P = \langle u, P_{b1}, P_1, w, P_2, \ldots, P_7, P_{b2}, v\rangle$ forms a Hamiltonian path of $FCCN_{n+1}^0$ joining u to v. □

THEOREM 11.26 **(Ho [148])** $\mathcal{H}_f^\kappa(FCCN_n) = 0$ if $n \ge 2$ and is undefined if $n = 1$.

Proof: Assume $n \ge 2$. With Lemma 11.34, we have proved that $\mathcal{H}_f^\kappa(FCCN_n) \ge 0$. Obviously, 0^n is a boundary vertex of $FCCN_n$ with exactly three neighbors; say, x, y and z. It is easy to see that there is no Hamiltonian path of $FCCN_n - \{x\}$ joining y to z. Thus, $FCCN_n - \{x\}$ is not Hamiltonian-connected. Thus, $\mathcal{H}_f^\kappa(FCCN_n) = 0$ if $n \ge 2$. Note that $FCCN_1$ is isomorphic to Q_3. Since Q_3 is a bipartite graph with eight vertices, there is no Hamiltonian path joining two vertices in the same partite set. Thus, $FCCN_1$ is not Hamiltonian-connected. Therefore, $\mathcal{H}_f^\kappa(FCCN_1)$ is undefined. □

LEMMA 11.35 **(Ho [148])** Both $FCCN_n$ and $FCCN_n^0$ are 1 fault-tolerant Hamiltonian for $n \ge 2$.

Proof: We prove this lemma by induction. It is sufficient to prove that a graph G is 1 fault-tolerant Hamiltonian by proving that $G - f$ is Hamiltonian for any $f \in V(G) \cup E(G)$ with $|f| = 1$. By Lemma 11.33, the statement holds for $n = 2$. We assume the statement holds for l with $2 \le l \le n$.

First, we prove that $FCCN_{n+1} - f$ is Hamiltonian.

Suppose that $f \in V(bFCCN_n) \cup E(bFCCN_n)$. By induction, there is a Hamiltonian cycle $\langle w, bk_7^n, P_b, bk_1^n, w\rangle$ of $bFCCN_n^b - f$. Let $k_2, k_3, \dots, k_6$ be any permutation of $Z_8 - \{b, k_1, k_7\}$. We set $k_0 = k_8 = b$. Let $x_i = k_i k_{i-1}^n$ and $y_i = k_i k_{i+1}^n$ for $1 \le i \le 7$. By Theorem 11.25, there exists a Hamiltonian path P_i of $iFCCN_n$ joining x_i to y_i for $1 \le i \le 7$. Obviously, $\langle bk_7^n, P_b, P_1, \dots, P_7, bk_7^n\rangle$ forms a Hamiltonian cycle for $FCCN_{n+1} - f$.

Suppose that f is an intercubic edge between $bFCCN_n$ and $eFCCN_n$. Thus, $f = (be^n, eb^n)$. Let $k_0, k_1, \dots, k_7$ be any permutation of Z_8 such that $k_0 = b$ and $k_2 = e$. Let $x_i = k_i k_{i-1}^n$ for $1 \le i \le 7$, $y_j = k_j k_{j+1}^n$ for $0 \le j \le 6$, $x_0 = k_0 k_7^n$, and $y_7 = k_7 k_0^n$. By Theorem 11.25, there exists a Hamiltonian path P_i of $iFCCN_n$ joining x_i to y_i for $0 \le i \le 7$. Obviously, $\langle bk_7^n, P_0, P_1, \dots, P_7, bk_7^n\rangle$ forms a Hamiltonian cycle for $FCCN_{n+1} - f$.

Second, we prove that $FCCN_{n+1}^0 - f$ is Hamiltonian.

Suppose that f is the extra vertex w. Obviously, $FCCN_{n+1}^0 - \{w\} = FCCN_{n+1}$. By Theorem 11.25, $FCCN_{n+1}^0 - \{w\}$ is Hamiltonian-connected. Therefore, $FCCN_{n+1}^0 - f$ is Hamiltonian.

Suppose that $f \in V(bFCCN_n) \cup E(bFCCN_n)$. By the induction hypothesis, there is a Hamiltonian cycle C of $bFCCN_n^b - f$. Since C can be traversed forward and backward, we can assume that $C = \langle w, bk_7^n, P_b, bk_1^n, w\rangle$ with $k_1 \ne 0$. Let $k_2, k_3, \dots, k_6$ be any permutation of $Z_8 - \{b, k_1, k_7\}$ such that $k_2 \ne 0$. We set $k_8 = b$. Let $x_i = k_i k_{i-1}^n$ for $3 \le i \le 7$, $y_j = k_j k_{j+1}^n$ for $2 \le j \le 7$, $x_1 = k_1 b^n$, $x_2 = k_2^{n+1}$, and $y_1 = k_1^{n+1}$. By Theorem 11.25, there exists a Hamiltonian path P_i of $iFCCN_n$ joining x_i to y_i for $1 \le i \le 7$. Obviously, $\langle bk_7^n, P_b, P_1, w, P_2, \dots, P_7, bk_7^n\rangle$ forms a Hamiltonian cycle for $FCCN_{n+1}^0 - f$.

Suppose that f is an edge of the form (r^{n+1}, w). Let b and e be two indices with $\{b, e\} \cap \{0, r\} = \emptyset$. By Theorem 11.25, there is a Hamiltonian path P of $FCCN_{n+1}$ joining b^{n+1} to e^{n+1}. Obviously, $\langle w, b^{n+1}, P, e^{n+1}, w\rangle$ forms a Hamiltonian cycle for $FCCN_{n+1}^0 - f$.

Suppose that f is an intercubic edge between $bFCCN_n$ and $eFCCN_n$. Thus, $f = (be^n, eb^n)$. Let $k_0, k_1, \dots, k_7$ be any permutation of Z_8 such that $k_0 = b$, $k_1 \ne 0$, $k_2 \ne 0$, and $k_3 = e$. Let $x_i = k_i k_{i-1}^n$ for $3 \le i \le 7$, $y_j = k_j k_{j+1}^n$ for $2 \le j \le 6$, $x_0 = k_0 k_7^n$, $x_1 = k_1 k_0^n$, $x_2 = k_2^{n+1}$, $y_0 = k_0 k_1^n$, $y_1 = k_1^{n+1}$, and $y_7 = k_7 k_0^n$. By Theorem 11.25, there exists a Hamiltonian path P_i of $iFCCN_n$ joining x_i to y_i for $0 \le i \le 7$. Obviously, $\langle bk_7^n, P_0, P_1, w, P_2, \dots, P_7, bk_7^n\rangle$ forms a Hamiltonian cycle for $FCCN_{n+1}^0 - f$. □

THEOREM 11.27 $\mathcal{H}_f(FCCN_n) = 1$ if $n \ge 2$ and $\mathcal{H}_f(FCCN_1) = 0$.

Proof: Assume that $n \ge 2$. With Lemma 11.35, we have proved that $\mathcal{H}_f^\kappa(FCCN_n) \ge 1$. Obviously, 0^n is a boundary vertex of $FCCN_n$ with exactly three neighbors; say, x, y, and z. Since there is only one vertex z adjacent to 0^n in $FCCN_n - \{x, y\}$, $FCCN_n - \{x, y\}$ is not Hamiltonian. Thus, $\mathcal{H}_f(FCCN_n) = 1$ if $n \ge 2$. Note that $FCCN_1$ is isomorphic to Q_3. It is easy to check whether Q_3 is Hamiltonian but $Q_3 - \{0\}$ is not Hamiltonian. Therefore, $\mathcal{H}_f(FCCN_1) = 0$. □

12 Optimal 1-Fault-Tolerant Hamiltonian Graphs

12.1 INTRODUCTION

From previous discussions, we have several methods that recursively construct k-regular $(k-2)$-fault-tolerant Hamiltonian graphs for large k. Thus, we should put some effort on k-regular $(k-2)$-fault-tolerant Hamiltonian graphs for small k. A lot of optimal 1-Hamiltonian graphs have been proposed [140,182,256,333,335]. For any 1-Hamiltonian regular graph, it is proved that the graph is optimal if and only if the graph is 3-regular [183,334]. In this chapter, we are interested only in optimal 1-Hamiltonian regular graphs. Families of optimal 1-Hamiltonian regular graphs were proposed by Harary and Hayes [139,140], Mukhopadhyaya and Shani [256], Wang, Hung, and Hsu [333], and Hung, Hsu, and Sung [183]. From the mathematical point of view, it would be interesting to characterize all optimal 1-Hamiltonian regular graphs. One strategy to characterize all optimal 1-Hamiltonian regular graphs is to find some basic graphs and construction schemes with the hope that we can construct all optimal 1-Hamiltonian regular graphs. We also hope that these basic graphs and construction schemes are as simple as possible.

In this chapter, we present some construction schemes for 3-regular 1-fault-tolerant Hamiltonian graphs. Then, we discuss some families of 3-regular 1-fault-tolerant Hamiltonian graphs.

We have also observed that several interconnection networks are recursively constructed. Based on the recursive structure, we can use induction to prove that such networks are not only optimal fault-tolerant Hamiltonian but also optimal fault-tolerant Hamiltonian-connected. For a while, we were wondering whether any optimal k-fault-tolerant Hamiltonian graph is optimal $(k-1)$-fault-tolerant Hamiltonian-connected. Here, we will concentrate on the special case of $k = 1$.

Note that a 1-vertex fault-tolerant Hamiltonian graph is a Hamiltonian graph such that $G - f$ remains Hamiltonian for any $f \in V(G)$. There is another family of graph called the *hypohamiltonian graph*, which is very close to 1-vertex fault-tolerant hamiltonian graph. A hypohamiltonian graph is a non-Hamiltonian graph such that $G - f$ is Hamiltonian for any $f \in V(G)$. There are numerous studies on hypohamiltonian graphs. Readers can refer to Ref. 156 for a survey of hypohamiltonian graphs. A graph G is *1-edge fault-tolerant Hamiltonian* if $G - e$ is Hamiltonian for any $e \in E(G)$. Obviously, any 1-edge fault-tolerant Hamiltonian graph is Hamiltonian.

Kao et al. [192] show that the concepts of cubic 1-vertex fault-tolerant Hamiltonian graphs, cubic 1-edge fault-tolerant Hamiltonian graphs, and cubic Hamiltonian-connected graphs are independent of one another even if we restrict our attention to planar graphs. Let U be the set of cubic planar Hamiltonian graphs, A the set of

1-vertex fault-tolerant Hamiltonian graphs in U, B the set of 1-edge fault-tolerant Hamiltonian graphs in U, and C the set of Hamiltonian-connected graphs in U. With the inclusion or exclusion of the sets A, B, and C, the set U is divided into eight subsets; namely, $\overline{A} \cap \overline{B} \cap \overline{C}$, $\overline{A} \cap \overline{B} \cap C$, $\overline{A} \cap B \cap C$, $A \cap \overline{B} \cap C$, $A \cap \overline{B} \cap \overline{C}$, $\overline{A} \cap B \cap \overline{C}$, $A \cap B \cap \overline{C}$, and $A \cap B \cap C$. We will prove that there is an infinite number of elements in each of the eight subsets.

We have seen several variations of Hamiltonian undirected graphs, such as k-vertex fault-tolerant Hamiltonian graphs, k-edge fault-tolerant Hamiltonian graphs, and k-fault-tolerant Hamiltonian graphs. It is very natural to explore the corresponding study on directed graphs. However, not too many results are known. We study only the case where $k = 1$. It is easy to see that any 1-vertex fault-tolerant Hamiltonian 2-regular directed graph is optimal with respect to the property of 1-vertex fault-tolerant Hamiltonian. Similar statements hold for the 1-edge fault-tolerant Hamiltonian directed graph and the 1-fault-tolerant Hamiltonian directed graph. In this chapter, we discuss the corresponding properties with respect to the family of double loop networks.

12.2 3-JOIN

Wang et al. [334] proposed the operator, 3-join, to combine two cubic graphs. Let G_1 and G_2 be two graphs. We assume that $V(G_1) \cap V(G_2) = \emptyset$. Let x be a vertex of a graph. We use $N(x)$ to denote an *ordered set* that consists of all the neighbors of x. That is, $N(x)$ is an ordering of all of the neighbors of x. Hence, we use $N(x)$ as an ordered set. The *3-join*, as illustrated in Figure 12.1, is defined as follows.

Suppose that x is a vertex of degree 3 in G_1 and y a vertex of degree 3 in G_2. Moreover, assume that $N(x) = \{x_1, x_2, x_3\}$ and $N(y) = \{y_1, y_2, y_3\}$. The *3-join* of G_1 and G_2 at x and y is a graph K given by

$$\begin{aligned} V(K) &= (V(G_1) - \{x\}) \cup (V(G_2) - \{y\}) \\ E(K) &= (E(G_1) - \{(x, x_i) \mid 1 \leq i \leq 3\}) \cup (E(G_2) - \{(y, y_i) \mid 1 \leq i \leq 3\}) \\ &\quad \cup \{(x_i, y_i) \mid 1 \leq i \leq 3\} \end{aligned}$$

We note that different $N(x)$ and $N(y)$ generate different 3-joins of G_1 and G_2 at x and y. For example, as illustrated in Figure 12.1, given $N(y) = \{y_1, y_2, y_3\}$, the 3-join of G_1 and G_2 at x and y with $N(x) = \{x_1, x_2, x_3\}$ is different from the one with $N(x) = \{x_2, x_1, x_3\}$. Thus, each 3-join of G_1 and G_2 at x and y is uniquely determined by $N(x)$ and $N(y)$. In particular, $G = J(G_1, N(x); G_2, N(y))$ is the node expansion of G at x if $G_2 = K_4$.

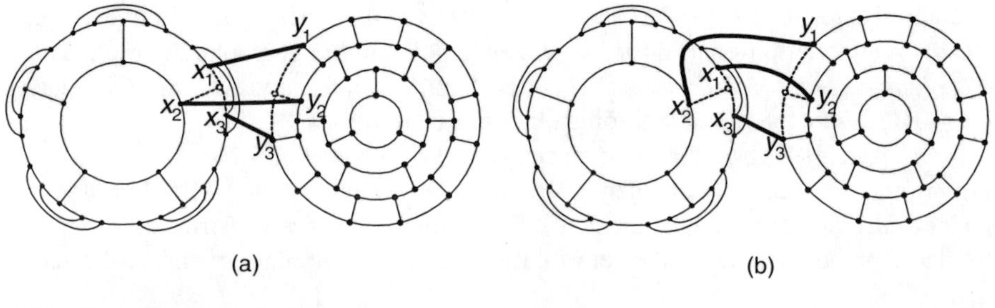

FIGURE 12.1 Examples of 3-joins of two cubic graphs.

Let x be a vertex of degree 3 in graph G_1, and y be a vertex of degree 3 in graph G_2. A graph K is said to be a *3-join of* G_1 *and* G_2 if K is a 3-join of G_1 and G_2 at x and y with some $N(x)$ and some $N(y)$. Clearly, 3-joins of two cubic graphs G_1 and G_2 are still cubic.

THEOREM 12.1 Suppose that G_1 and G_2 are two graphs and K is a 3-join of G_1 and G_2. Then K is a 1-fault-tolerant Hamiltonian graph if both G_1 and G_2 are 1-fault-tolerant Hamiltonian graphs.

Proof: Assume that K is a 3-join of G_1 and G_2 at x and y with $N(x) = \{x_1, x_2, x_3\}$ and $N(y) = \{y_1, y_2, y_3\}$. Let f be any fault, vertex, or edge, of K.

Suppose that $f \neq (x_i, y_i)$ for all $1 \leq i \leq 3$. Without loss of generality, we may assume that $f \in (V(G_1) - \{x\}) \cup (E(G_1) - \{(x, x_i) \mid 1 \leq i \leq 3\})$. Since G_1 is 1-fault-tolerant Hamiltonian, it follows that there is a Hamiltonian cycle H_1 in $G_1 - f$. Obviously, H_1 can be written as $\langle x, x_i, P, x_j, x\rangle$, where $i, j \in \{1, 2, 3\}$. Let k be the unique element in $\{1, 2, 3\} - \{i, j\}$. Since G_2 is 1-fault-tolerant Hamiltonian, there is a Hamiltonian cycle H_2 in $G_2 - (y, y_k)$, which can be written as $\langle y, y_i, Q, y_j, y\rangle$. Obviously, $\langle x_i, P, x_j, y_j, Q, y_i, x_i\rangle$ forms a Hamiltonian cycle of $K - f$.

Suppose that $f = (x_i, y_i)$ for some i. Without loss of generality, we may assume that $f = (x_1, y_1)$. Let H_1 be a Hamiltonian cycle of $G_1 - (x, x_1)$, which can be written as $\langle x, x_2, P, x_3, x\rangle$. Let H_2 be a Hamiltonian cycle of $G_2 - (y, y_1)$, which can be written as $\langle y, y_2, Q, y_3, y\rangle$. Obviously, $\langle x_2, P, x_3, y_3, Q, y_2, x_2\rangle$ forms a Hamiltonian cycle of $K - f$. Hence, the theorem is proved. □

12.3 CYCLE EXTENSION

Wang et al. [334] also proposed the following operation. Suppose that G is a graph and $C = \langle x_0, x_1, \ldots, x_{k-1}, x_0\rangle$ is a cycle of G, where $k \geq 3$ is an arbitrary integer. We introduce an operation called *cycle extension* that includes two aspects: first, augment G by replacing each edge in C with a path of odd length, and second, add a new cycle to the augmented graph in a specific manner. To be specific, we define cycle extension as follows.

The cycle extension of G around C is a graph denoted by $Ext_C(G)$ and given as follows:

$$V(Ext_C(G)) = \bigcup_{0 \leq i \leq k-1} \{p_{i,j}, q_{i,j} \mid \forall 1 \leq j \leq l_i\} \cup V(G)$$

$$\begin{aligned} E(Ext_C(G)) = {} & (E(G) - E(C)) \cup \bigcup_{0 \leq i \leq k-1} \{(p_{i,j}, q_{i,j}) \mid \forall 1 \leq j \leq l_i\} \\ & \cup \bigcup_{0 \leq i \leq k-1} (\{(x_i, p_{i,1}), (p_{i,l_i}, x_{i+1})\} \cup \{(p_{i,j}, p_{i,j+1}) \mid \forall 1 \leq j \leq l_i - 1\}) \\ & \cup \bigcup_{0 \leq i \leq k-1} (\{(q_{i,j}, q_{i,j+1}) \mid \forall 1 \leq j \leq l_i - 1\} \cup \{(q_{i,l_i}, q_{i+1,1})\}) \end{aligned}$$

where l_i is even for all i.

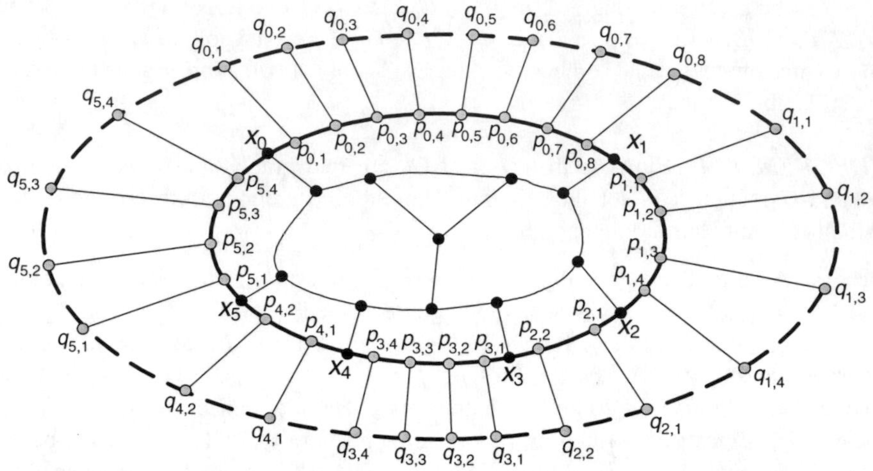

FIGURE 12.2 An example of $Ext_C(G)$.

The cycle induced by the vertices $\bigcup_{0 \le i \le k-1}\{q_{i,j} \mid \forall 1 \le j \le l_i\}$ is called the *extended cycle* of C, denoted by $\mathcal{O}_C$. An example of $Ext_C(G)$ is illustrated in Figure 12.2, where C is represented by darkened edges, and $\mathcal{O}_C$ by dashed edges. Here, we adopt the following notion:

- $C = \langle x_0, x_1, \ldots, x_{k-1}, x_0 \rangle$
- l_i: an even integer that is the number of vertices in $Ext_C(G)$ added between x_i and x_{i+1}
- $p_{i,j}$: a vertex in $Ext_C(G)$ added between x_i and x_{i+1}
- $q_{i,j}$: the vertex in $Ext_C(G)$ that is adjacent to $p_{i,j}$

Again, we focus our discussion of cycle extensions on cubic graphs only. Hence, we suppose that G is cubic in the following discussion of cycle extensions.

We use z_i to denote the unique neighbor of x_i that is not in C. The addition and subtraction involved in the subscript or index of a vertex in a cycle is taken modulo k, where k denotes the length of the cycle or k is clear from the context without ambiguity. For example, (x_i, x_{i+1}) for $i = k-1$ in C is simply (x_{k-1}, x_0), and $(q_{i-1,l_{i-1}}, q_{i,1})$ in $\mathcal{O}_C$ for $i = 0$ is simply $(q_{k-1,l_{k-1}}, q_{0,1})$.

Suppose that H is a cycle of G. Let H_C^* denote the cycle in $Ext_C(G)$ obtained from H by replacing all (x_j, x_{j+1}) in $E(H) \cap E(C)$ with $\langle x_j, p_{j,1}, p_{j,2}, \ldots, p_{j,l_j}, x_{j+1} \rangle$. Moreover, Ω_H denotes the cycle obtained from $\mathcal{O}_C$ by replacing every $\langle q_{i,1}, q_{i,2}, q_{i,3}, \ldots, q_{i,l_i} \rangle$ with $\langle q_{i,1}, p_{i,1}, p_{i,2}, q_{i,2}, q_{i,3}, p_{i,3}, \ldots, p_{i,l_i}, q_{i,l_i} \rangle$ if $\langle p_{i,1}, p_{i,2}, \ldots, p_{i,l_i} \rangle$ is not a subpath in H_C^*. Now, we introduce six operations $M_1(H,e,j)$, $M_2(H,e)$, and $M_i(H,x)$ for $3 \le i \le 6$ that augment the cycle H of G to a cycle of $Ext_C(G)$ with respect to some edge $e = (x_i, x_{i+1})$ or vertex $x = x_i$ or $x = x_{i+1}$ in C and a specific j. We use $M_1(H,e,j)$, $M_2(H,e)$, and $M_i(H,x)$ to mean the operation and the corresponding cycle interchangeably.

For ease of exposition, let us define

$$\Omega_{q_i} = \langle q_{i,1}, p_{i,1}, p_{i,2}, q_{i,2}, q_{i,3}, p_{i,3}, \ldots, p_{i,l_i}, q_{i,l_i} \rangle$$

$$\Omega_{p_i} = \langle p_{i,1}, q_{i,1}, q_{i,2}, p_{i,2}, p_{i,3}, q_{i,3}, \ldots, q_{i,l_i}, p_{i,l_i} \rangle$$

These notations will be used in the definitions of operations M_2, M_3, M_4, M_5, and M_6. To be specific, the six operations are defined as follows:

Operation M_1. Suppose that H is a cycle of G that contains the edge e. We define an operation $M_1(H, e, j)$ to construct a cycle in $Ext_C(G) - \{(p_{i,j}, p_{i,j+1}), (q_{i,j}, q_{i,j+1})\}$ for some $1 \leq j \leq l_i - 1$.

Let Q be the path $\Omega_H - (q_{i,j}, q_{i,j+1})$ and P be the path $H_C^* - (p_{i,j}, p_{i,j+1})$. We define

$$M_1(H, e, j) = \langle p_{i,j}, P, p_{i,j+1}, q_{i,j+1}, Q, q_{i,j}, p_{i,j} \rangle$$

as illustrated in Figure 12.3.

Operation M_2. Suppose that that z_i and z_{i+1} are adjacent in G, and H is a Hamiltonian cycle of G containing $\langle x_{i-1}, x_i, x_{i+1}, x_{i+2} \rangle$ as a subpath. We define an operation $M_2(H, e)$ to construct a Hamiltonian cycle of $Ext_C(G) - \{(q_{i-1,l_{i-1}}, q_{i,1}), (q_{i,l_i}, q_{i+1,1})\}$.

We use y_i to denote the unique neighbor of z_i different from x_i and z_{i+1}, and use y_{i+1} to denote the unique neighbor of z_{i+1} different from x_{i+1} and z_i. Since G is cubic and $\langle x_{i-1}, x_i, x_{i+1}, x_{i+2} \rangle$ is a subpath of H, $\langle y_i, z_i, z_{i+1}, y_{i+1} \rangle$ is a subpath of H in G. Moreover, $\langle q_{i-1,l_{i-1}}, q_{i,1}, \ldots, q_{i,l_i}, q_{i+1,1} \rangle$ is a subpath of Ω_H and $\langle p_{i-1,l_{i-1}}, x_i, p_{i,1}, p_{i,2}, \ldots, p_{i,l_i}, x_{i+1}, p_{i+1,1} \rangle$ is a subpath of H_C^*. Hence, we can obtain a path P from H_C^* by replacing $\langle p_{i-1,l_{i-1}}, x_i, p_{i,1}, p_{i,2}, \ldots, p_{i,l_i}, x_{i+1}, p_{i+1,1} \rangle$ with $\langle x_i, p_{i,1}, \Omega_{p_i}, p_{i,l_i}, x_{i+1} \rangle$ and replacing (z_i, z_{i+1}) with (z_i, x_i) and (z_{i+1}, x_{i+1}). Deleting $\langle q_{i-1,l_{i-1}}, q_{i,1}, \ldots, q_{i,l_i}, q_{i+1,1} \rangle$ from Ω_H yields a path Q. We define

$$M_2(H, e) = \langle p_{i-1,l_{i-1}}, q_{i-1,l_{i-1}}, Q, q_{i+1,1}, p_{i+1,1}, P, p_{i-1,l_{i-1}} \rangle$$

as illustrated in Figure 12.4.

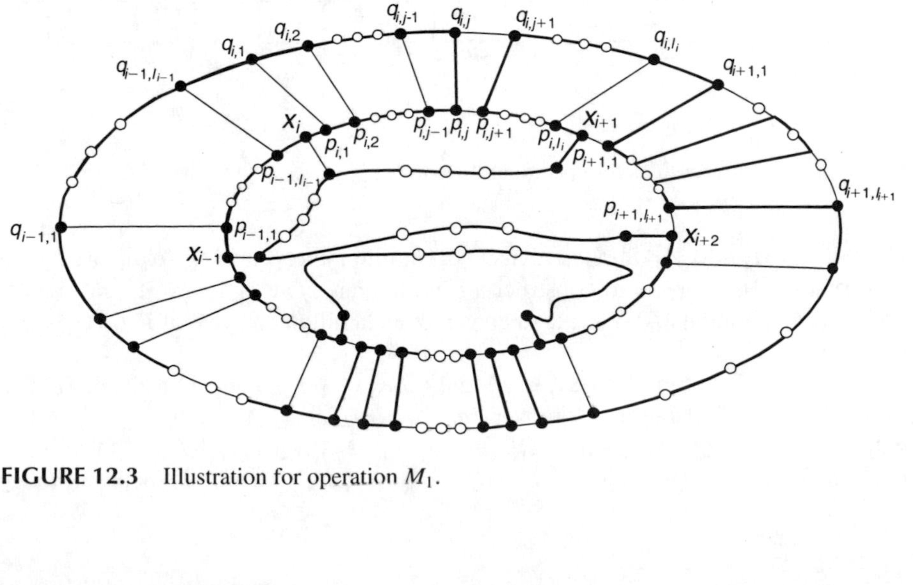

FIGURE 12.3 Illustration for operation M_1.

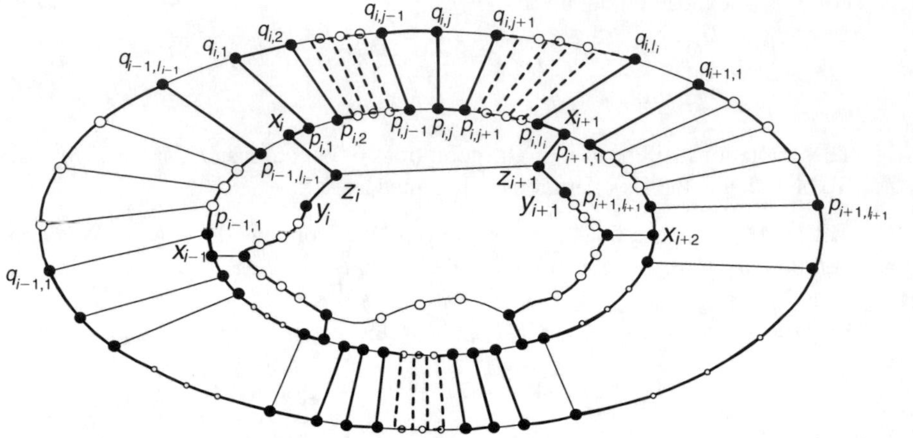

FIGURE 12.4 Illustration for operation M_2.

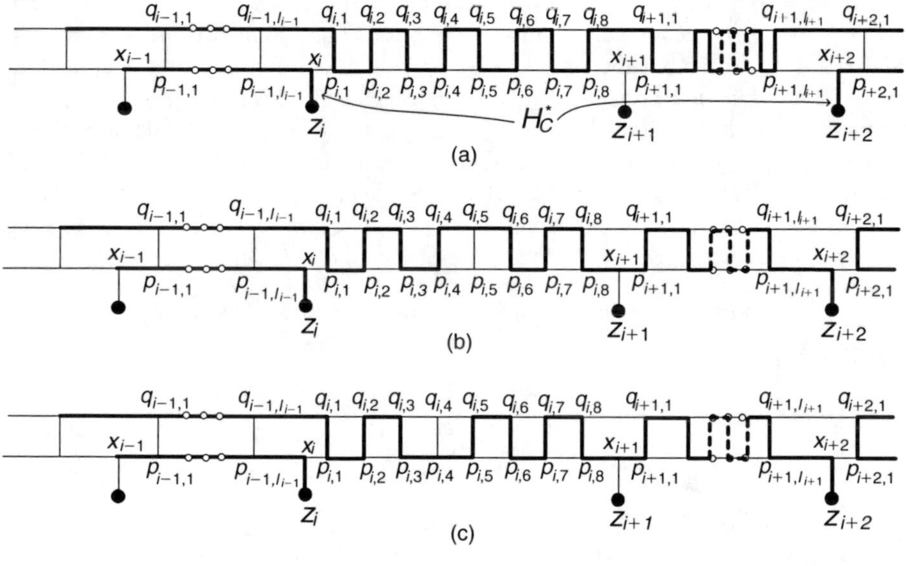

FIGURE 12.5 Illustrations for operations M_3 and M_4.

Operations M_3 and M_4. Suppose that H is a Hamiltonian cycle of $G - x_{i+1}$. We define an operation $M_3(H, x_{i+1})$ to construct a Hamiltonian cycle of $Ext_C(G) - p_{i,j}$ with j odd and an operation $M_4(H, x_{i+1})$ to construct a Hamiltonian cycle of $Ext_C(G) - q_{i,j}$ with j even.

Since x_{i+1} is not in H, $\langle x_{i-1}, x_i, z_i \rangle$ and $\langle x_{i+3}, x_{i+2}, z_{i+2} \rangle$ are subpaths of H. See Figure 12.5a for illustration. Moreover, $\langle q_{i,1}, \Omega_{q_i}, q_{i,l_i}, q_{i+1,1}, \Omega_{q_{i+1}}, q_{i+1,l_{i+1}} \rangle$ is a subpath of Ω_H. Let $Q = \Omega_H - \langle q_{i,1}, \Omega_{q_i}, q_{i,l_i}, q_{i+1,1}, \Omega_{q_{i+1}}, q_{i+1,l_{i+1}}, q_{i+2,1} \rangle$. Obviously, Q is a path from $q_{i+2,1}$ to $q_{i,1}$. Let $P = H_C^* - (x_{i+2}, p_{i+2,1})$. Then P is a path from x_{i+2}

to $p_{i+2,1}$. We define

$$M_3(H,e) = \langle p_{i+2,1}, q_{i+2,1}, Q, q_{i,1}, p_{i,1}, p_{i,2}, q_{i,2}, \ldots, p_{i,j-1}, q_{i,j-1}, q_{i,j}, q_{i,j+1}, p_{i,j+1}, p_{i,j+2}, q_{i,j+2}, \ldots, q_{i,l_i}, p_{i,l_i}, x_{i+1}, p_{i+1,1}, \Omega_{p_{i+1}}, p_{i+1,l_{i+1}}, x_{i+2}, P, p_{i+2,1}\rangle$$

$$M_4(H,e) = \langle p_{i+2,1}, q_{i+2,1}, Q, q_{i,1}, p_{i,1}, p_{i,2}, q_{i,2}, \ldots, q_{i,j-1}, p_{i,j-1}, p_{i,j}, p_{i,j+1}, q_{i,j+1}, q_{i,j+2}, p_{i,j+2}, \ldots, p_{i,l_i}, x_{i+1}, p_{i+1,1}, \Omega_{p_{i+1}}, p_{i+1,l_{i+1}}, x_{i+2}, P, p_{i+2,1}\rangle$$

as illustrated in Figures 12.5b and 12.5c.

Operations M_5 and M_6. Suppose that H is a Hamiltonian cycle of $G - x_i$ as illustrated in Figure 12.6a. We define an operation $M_5(H, x_i)$ to construct a Hamiltonian cycle of $Ext_C(G) - p_{i,j}$ with j even and an operation $M_6(H, x_i)$ to construct a Hamiltonian cycle of $Ext_C(G) - q_{i,j}$ with j odd.

Let Q denote $\Omega_H - \langle q_{i-2,l_{i-2}}, q_{i-1,1}, \Omega_{q_{i-1}}, q_{i-1,l_{i-1}}, q_{i,1}, \Omega_{q_i}, q_{i,l_i}\rangle$. Then Q is a path from q_{i,l_i} to $q_{i-2,l_{i-2}}$. Let $P = H_C^* - (x_{i-1}, p_{i-2,l_{i-2}})$. Then P is a path from $p_{i-2,l_{i-2}}$ to x_{i-1}. We define

$$M_5(H,x) = \langle x_{i-1}, p_{i-1,1}, \Omega_{p_{i-1}}, p_{i-1,l_{i-1}}, x_i, p_{i,1}, q_{i,1}, q_{i,2}, p_{i,2}, p_{i,3}, \ldots, p_{i,j-1}, q_{i,j-1}, q_{i,j}, q_{i,j+1}, p_{i,j+1}, p_{i,j+2}, q_{i,j+2}, \ldots, q_{i,l_i}, Q, q_{i-2,l_{i-2}}, p_{i-2,l_{i-2}}, P, x_{i-1}\rangle$$

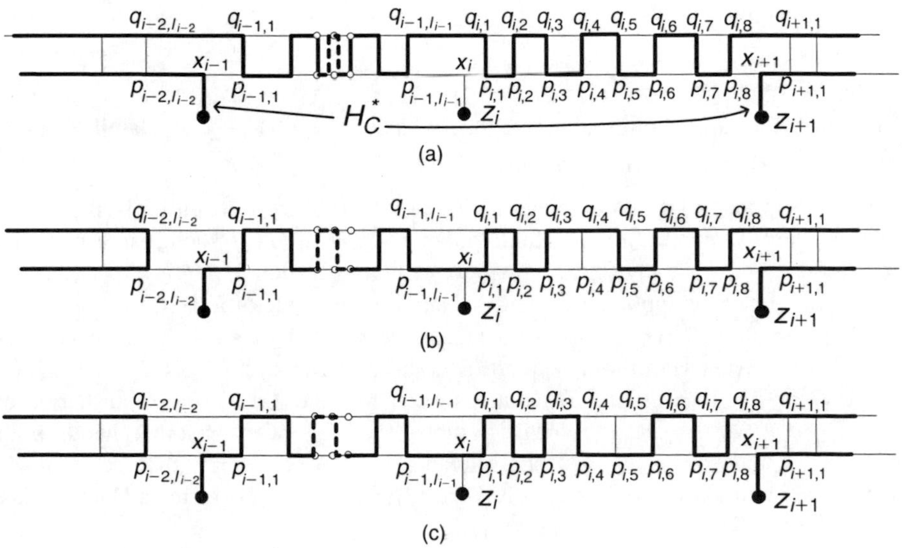

FIGURE 12.6 Illustrations for operations M_5 and M_6.

$$M_6(H,x) = \langle x_{i-1}, p_{i-1,1}, \Omega_{p_{i-1}}, p_{i-1,l_{i-1}}, x_i, p_{i,1}, q_{i,1}, q_{i,2}, p_{i,2}, p_{i,3}, \ldots, q_{i,j-1}, p_{i,j-1}, p_{i,j}, p_{i,j+1}, q_{i,j+1}, q_{i,j+2}, p_{i,j+2}, \ldots, q_{i,l_i}, Q, q_{i-2,l_{i-2}}, p_{i-2,l_{i-2}}, P, x_{i-1}\rangle$$

as illustrated in Figures 12.6b and 12.6c.

We will use these six operations in the proofs of the following lemmas and theorems.

LEMMA 12.1 Suppose that G is a cubic graph such that $G-f$ is Hamiltonian for any $f \in V(G)$ and $C = \langle x_0, x_1, \ldots, x_{k-1}, x_0\rangle$ is a cycle of G. Then $Ext_C(G) - f$ is Hamiltonian for any $f \in V(Ext_C(G))$.

Proof: Suppose that $f \in V(G) - V(C)$. By assumption of this lemma, there is a Hamiltonian cycle H of $G - f$ such that at least an edge in C, say (x_i, x_{i+1}), is in H. Obviously, $M_1(H, (x_i, x_{i+1}), j)$, $1 \le j \le l_i - 1$, forms a Hamiltonian cycle of $Ext_C(G) - f$.

Suppose that $f = x_i$ for $0 \le i \le k-1$. Since G is cubic and $k \ge 3$, two vertices x_{i-1} and x_{i+1} in C have degree 2 in $G - f$. Then we can always find a Hamiltonian cycle H of $G - f$ such that H contains an edge in C, say (x_j, x_{j+1}) for $j \ne i-1, i$. On the other hand, Ω_H contains the subpath $\langle q_{i-2,l_{i-2}}, q_{i-1,1}, \Omega_{q_{i-1}}, q_{i-1,l_{i-1}}, q_{i,1}, \Omega_{q_i}, q_{i,l_i}, q_{i+1,1}\rangle$, which does not contain the vertex x_i. Obviously, $M_1(H, (x_j, x_{j+1}), j')$, $1 \le j' \le l_j - 1$, is a Hamiltonian cycle of $Ext_C(G) - f$.

Suppose that $f \in \{p_{i,j}, p_{i,j'}, q_{i,j}, q_{i,j'}\}$ for some $1 \le j, j' \le l_i$ with j odd and j' even. Since G is 1-vertex fault-tolerant Hamiltonian, there are Hamiltonian cycles H_1 and H_2 of $G - x_{i+1}$ and $G - x_i$, respectively. Then the Operations $M_3(H_1, x_{i+1})$, $M_4(H_1, x_{i+1})$, $M_5(H_2, x_i)$, and $M_6(H_2, x_i)$ can be applied and they are indeed Hamiltonian cycles of $Ext_C(G) - p_{i,j}$, $Ext_C(G) - q_{i,j'}$, $Ext_C(G) - p_{i,j'}$, and $Ext_C(G) - q_{i,j}$, respectively. Hence, the lemma follows. □

LEMMA 12.2 Suppose that $C = \langle x_0, x_1, \ldots, x_{k-1}, x_0\rangle$ is a cycle of a cubic 1-edge fault-tolerant Hamiltonian graph G. Then $Ext_C(G) - f$ is Hamiltonian if $f \in E(Ext_C(G)) - \bigcup_{0 \le i \le k-1}\{(q_{i,l_i}, q_{i+1,1})\}$.

Proof: Suppose that f is an edge in $E(G) - E(C)$. Since G is cubic 1-edge fault-tolerant Hamiltonian, there is a Hamiltonian cycle H of $G - f$ such that H contains at least an edge in C, say (x_i, x_{i+1}) with $0 \le i \le k-1$. Then $M_1(H, (x_i, x_{i+1}), j)$ with $1 \le j \le l_i - 1$ can be applied and yields a Hamiltonian cycle of $Ext_C(G) - f$.

Suppose that $f = (x_i, p_{i,1})$ or (p_{i,l_i}, x_{i+1}) for some $0 \le i \le k-1$. Since G is 1-edge fault-tolerant Hamiltonian, there is a Hamiltonian cycle H of $G - (x_i, x_{i+1})$. Since G is a cubic graph, H contains (x_{i-1}, x_i) and (x_{i+1}, x_{i+2}). Furthermore, $\langle q_{i-1,l_{i-1}}, q_{i,1}, \Omega_{q_i}, q_{i,l_i}, q_{i+1,1}\rangle$ forms a subpath of Ω_H. On the other hand, both $\langle x_{i-1}, x_i, z_i\rangle$ and $\langle z_{i+1}, x_{i+1}, x_{i+2}\rangle$ are subpaths in H. We can apply the operation M_1 to (x_{i-1}, x_i) and j for $1 \le j \le l_{i-1} - 1$. Then $M_1(H, (x_{i-1}, x_i), j)$ forms a Hamiltonian cycle of $Ext_C(G) - \{(x_i, p_{i,1}), (p_{i,l_i}, x_{i+1})\}$.

Suppose that $f = (p_{i,j}, p_{i,j+1})$ or $(q_{i,j}, q_{i,j+1})$ for some $0 \le i \le k-1$ and some $1 \le j \le l_i - 1$. Let H be a Hamiltonian cycle of $G - (x_{i-1}, x_i)$. Then (x_i, x_{i+1}) is in H.

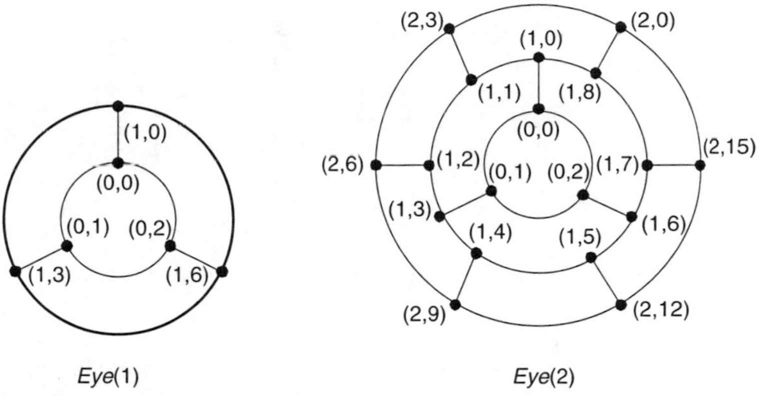

FIGURE 12.7 *Eye*(1) and *Eye*(2).

Therefore, $M_1(H,(x_i,x_{i+1}),j)$ with $1 \leq j \leq l_i - 1$ forms a Hamiltonian cycle of $Ext_C(G) - \{(p_{i,j},p_{i,j+1}),(q_{i,j},q_{i,j+1})\}$.

Suppose that $f=(p_{i,j},q_{i,j})$ for some $0 \leq i \leq k-1$ and some $1 \leq j \leq l_i$. Let H be a Hamiltonian cycle of $G - (x_{i-1},x_i)$. Obviously, both (x_i,x_{i+1}) and (x_{i-2},x_{i-1}) are in H. Moreover, H_C^* contains $\langle x_i,p_{i,1},p_{i,2},\ldots,p_{i,l_i},x_{i+1}\rangle$ as a subpath and Ω_H contains $\langle q_{i-1,l_{i-1}},q_{i,1},q_{i,2},\ldots,q_{i,l_i},q_{i+1,1}\rangle$ as a subpath. Hence, $M_1(H,(x_{i-2},x_{i-1}),j')$ for $1 \leq j' \leq l_{i-2} - 1$ forms a Hamiltonian cycle of $Ext_C(G) - (p_{i,j},q_{i,j})$. The lemma is proved. □

One may ask whether $Ext_C(G)$ is 1-fault-tolerant Hamiltonian if G is cubic 1-fault-tolerant Hamiltonian. The answer is no, and it can be verified by a counterexample shown in Figure 12.7. The graphs shown in Figure 12.7 were proposed by Wang et al. [335]; they are called *eye networks* and are denoted by $Eye(n)$ for $n \geq 1$. The graph $Eye(1)$ shown in Figure 12.7 is a cubic 1-fault-tolerant Hamiltonian graph. Let O_1 be the cycle indicated by darkened edges in $Eye(1)$ as shown in Figure 12.7. Obviously, $Eye(2) = Ext_{O_1}(Eye(1))$. Though $Eye(1)$ is cubic 1-fault-tolerant Hamiltonian, $Eye(2)$ is not 1-fault-tolerant Hamiltonian, because there is no Hamiltonian cycle in $Eye(2) - ((2,0),(2,3))$, $Eye(2) - ((2,6),(2,9))$, and $Eye(2) - ((2,12),(2,15))$.

Thus, that G is cubic 1-fault-tolerant Hamiltonian does not imply that $Ext_C(G)$ is 1-fault-tolerant Hamiltonian. However, we are interested in finding a sufficient condition on cycle C for $Ext_C(G)$ to be 1-fault-tolerant Hamiltonian. To get this goal, we define the *recoverable set* $R(C)$ of cycle C as follows:

$$R(C) = \{(x_i,x_{i+1}) \mid \langle x_{i-1},x_i,x_{i+1},x_{i+2}\rangle \text{is a subpath of some Hamiltonian cycle of } G \text{ and } (z_i,z_{i+1}) \text{ is an edge in } G \text{ for } 0 \leq i \leq k-1\}$$

The definition of $R(C)$ arises from the given condition for operation M_2 to be applicable. A cycle C of G is *recoverable* with respect to G if no two edges of $E(C) - R(C)$ are adjacent. For example, the cycle $C = \langle x_0,x_1,x_2,x_3,x_0\rangle$ shown in Figure 12.8 is recoverable with respect to G, since $\{(x_0,x_1),(x_2,x_3)\}$ is the recoverable set of C.

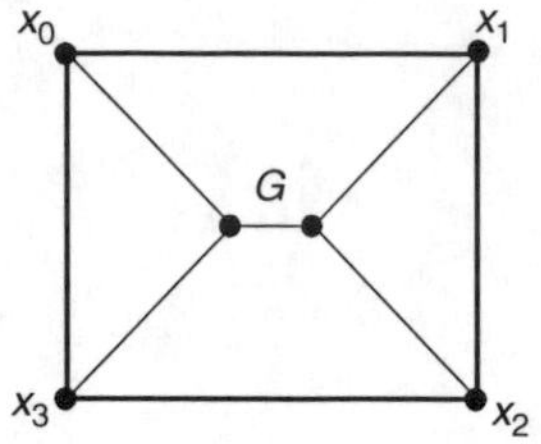

FIGURE 12.8 The cycle $\langle x_0, x_1, x_2, x_3, x_0\rangle$ is recoverable with respect to the graph G.

THEOREM 12.2 Suppose that $C = \langle x_0, x_1, \ldots, x_{k-1}, x_0\rangle$ is recoverable with respect to a cubic 1-fault-tolerant Hamiltonian graph G. Then $Ext_C(G)$ is cubic 1-fault-tolerant Hamiltonian.

Proof: By Lemma 12.1, $Ext_C(G) - f$ is Hamiltonian for any $f \in V(Ext_C(G))$. By Lemma 12.2, $Ext_C(G) - f$ is Hamiltonian for $f \in E(Ext_C(G)) - \bigcup_{0 \le i \le k-1} \{(q_{i,l_i}, q_{i+1,1})\}$. Now it suffices to show that $Ext_C(G) - f$ is Hamiltonian for $f \in \bigcup_{0 \le i \le k-1} \{(q_{i,l_i}, q_{i+1,1})\}$. Equivalently, we will construct a Hamiltonian cycle in $Ext_C(G)$ without using $(q_{i,l_i}, q_{i+1,1})$ for $0 \le i \le k-1$.

Since C is recoverable with respect to G, it follows that $R(C) \neq \emptyset$ and we have either $(x_i, x_{i+1}) \in R(C)$ or $(x_i, x_{i+1}) \notin R(C)$. For $(x_j, x_{j+1}) \in R(C)$, we use H_j to denote a Hamiltonian cycle of G such that H_j contains $\langle x_{j-1}, x_j, x_{j+1}, x_{j+2}\rangle$ as a subpath.

Suppose that $(x_i, x_{i+1}) \in R(C)$. Then $M_2(H_i, (x_i, x_{i+1}))$ is a Hamiltonian cycle of $Ext_C(G) - \{(q_{i-1,l_{i-1}}, q_{i,1}), (q_{i,l_i}, q_{i+1,1})\}$. Suppose that $(x_i, x_{i+1}) \notin R(C)$. Since any two edges in $E(C) - R(C)$ are not adjacent, (x_{i+1}, x_{i+2}) is in $R(C)$. Then $M_2(H_{i+1}, (x_{i+1}, x_{i+2}))$ forms a Hamiltonian cycle of $Ext_C(G) - \{(q_{i,l_i}, q_{i+1,1}), (q_{i+1,l_{i+1}}, q_{i+2,1})\}$ for $(x_{i+1}, x_{i+2}) \in R(C)$.

Therefore, $Ext_C(G)$ is 1-edge fault-tolerant Hamiltonian if G is 1-fault-tolerant Hamiltonian and C is recoverable with respect to G. In particular, $Ext_C(G)$ is Hamiltonian. Thus the theorem follows. □

In general, it is hard to verify whether a cycle C is recoverable with respect to G. Instead, we show that $\mathcal{O}_C$ is recoverable with respect to $Ext_C(G)$.

LEMMA 12.3 Suppose that G is a cubic graph and $C = \langle x_0, x_1, \ldots, x_{k-1}, x_0\rangle$ is a cycle of G. Then $\mathcal{O}_C$ is recoverable with respect to $Ext_C(G)$ if $G - (x_i, x_{i+1})$ is Hamiltonian for every $0 \le i \le k-1$.

Proof: It suffices to prove that $R(\mathcal{O}_C)$ is a collection of $(q_{i,j}, q_{i,j+1})$ for all $0 \le i \le k-1$ and $1 \le j \le l_i - 1$ because each $p_{i,j}$ is adjacent to $p_{i,j+1}$ in $Ext_C(G)$ but p_{i,l_i} is not adjacent to $p_{i+1,1}$.

Let H be a Hamiltonian cycle of $G - (x_{i-1}, x_i)$. Since G is cubic, both (x_{i-2}, x_{i-1}) and (x_i, x_{i+1}) are in H. Obviously, Ω_H contains $\langle q_{i-3,l_{i-3}}, q_{i-2,1}, q_{i-2,2}, \ldots, q_{i-2,l_{i-2}}, q_{i-1,1}\rangle$ and $\langle q_{i-1,l_{i-1}}, q_{i,1}, q_{i,2}, \ldots, q_{i,l_i}, q_{i+1,1}\rangle$ as subpaths. Therefore, the Hamiltonian cycle $M_1(H, (x_{i-2}, x_{i-1}), j')$, $1 \le j' \le l_{i-2} - 1$, of $Ext_C(G)$ contains $\langle q_{i-1,l_{i-1}}, q_{i,1}, q_{i,2}, \ldots, q_{i,l_i}, q_{i+1,1}\rangle$ as a subpath. Furthermore, $(q_{i,j}, q_{i,j+1})$ satisfies the definition of recoverable set $R(\mathcal{O}_C)$. Since (x_{i-1}, x_i) is an arbitrary edge and

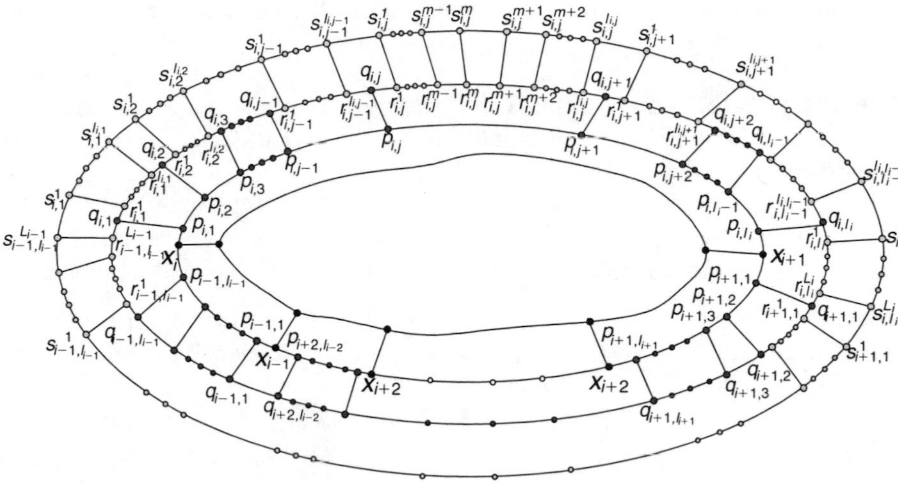

FIGURE 12.9 An example of $Ext_C^2(G)$.

$(q_{i,l_i}, q_{i+1,1})$ and $(q_{i',l_{i'}}, q_{i'+1,1})$ are not adjacent for $i \neq i'$, $R(\mathcal{O}_C)$ is a collection of $(q_{i,j}, q_{i,j+1})$ for all $0 \leq i \leq k-1$ and $1 \leq j \leq l_i - 1$ and moreover, $\mathcal{O}_C$ is recoverable with respect to $Ext_C(G)$. Hence, the theorem is proved. □

We can recursively apply the cycle extension operation. Let $Ext_C^0(G) = G$, $\mathcal{O}_C^0 = C$, $Ext_C^1(G) = Ext_C(G)$, and $\mathcal{O}_C^1 = \mathcal{O}_C$. We recursively define $Ext_C^n(G)$ for $n \geq 2$ as $Ext_C^n(G) = Ext_{\mathcal{O}_C^{n-1}}(Ext_C^{n-1}(G))$ and $\mathcal{O}_C^n$ being the extended cycle of $\mathcal{O}_C^{n-1}$ in $Ext_C^n(G)$.

Now, we consider in particular $Ext_C^2(G)$, which is the cycle extension of $Ext_C(G)$ around $\mathcal{O}_C$. Given $0 \leq i \leq k-1$ and $1 \leq j \leq l_i - 1$, let $l_{i,j}$ and L_i denote the number of vertices added between $q_{i,j}$ and $q_{i,j+1}$ and between q_{i,l_i} and $q_{i+1,1}$, respectively. These vertices are denoted by $r_{i,j}^m$ where $1 \leq m \leq l_{i,j}$ or $1 \leq m \leq L_i$. The vertex in $\mathcal{O}_C^2$, which is adjacent to $r_{i,j}^m$, is denoted by $s_{i,j}^m$. Note that $l_{i,j}$ and L_i are even. For ease of exposition, let $L^* = l_{i,j}$ for $1 \leq j \leq l_i - 1$ and $L^* = L_i$ for $j = l_i$. To be specific, the graph $Ext_C^2(G)$ is given as follows (see Figure 12.9):

$$V\left(Ext_C^2(G)\right) = V(Ext_C(G)) \cup \bigcup_{0 \leq i \leq k-1} \bigcup_{1 \leq j \leq l_i} \left\{ r_{i,j}^m, s_{i,j}^m \mid 1 \leq m \leq L^* \right\}$$

$$\begin{aligned} E\left(Ext_C^2(G)\right) = {} & (E(Ext_C(G)) - E(\mathcal{O}_C)) \\ & \cup \bigcup_{0 \leq i \leq k-1} \bigcup_{1 \leq j \leq l_i} \left(\left\{ \left(r_{i,j}^m, r_{i,j}^{m+1}\right), \left(s_{i,j}^m, s_{i,j}^{m+1}\right) \middle| 1 \leq m \leq L^* - 1 \right\} \right) \\ & \cup \bigcup_{0 \leq i \leq k-1} \bigcup_{1 \leq j \leq l_i} \left\{ \left(r_{i,j}^m, s_{i,j}^m\right) \middle| 1 \leq m \leq L^* \right\} \\ & \cup \bigcup_{0 \leq i \leq k-1} \left\{ \left(q_{i,l_i}, r_{i,l_i}^1\right), \left(r_{i,l_i}^{L_i}, q_{i+1,1}\right), \left(s_{i,l_i}^{L_i}, s_{i+1,1}^1\right) \right\} \\ & \cup \bigcup_{0 \leq i \leq k-1} \bigcup_{1 \leq j \leq l_i - 1} \left\{ \left(q_{i,j}, r_{i,j}^1\right), \left(r_{i,j}^{l_{i,j}}, q_{i,j+1}\right), \left(s_{i,j}^{l_{i,j}}, s_{i,j+1}^1\right) \right\} \end{aligned}$$

Using a proof technique similar to that in Lemma 12.3, we can prove the following corollary.

COROLLARY 12.1 *Suppose that G is a cubic graph and* $C = \langle x_0, x_1, \ldots, x_{k-1}, x_0\rangle$ *is a cycle of G. Then* $\mathcal{O}_C^2$ *is recoverable with respect to* $Ext_C^2(G)$ *if* $G-(x_i, x_{i+1})$ *is Hamiltonian for all* $0 \le i \le k-1$.

LEMMA 12.4 Suppose that $C = \langle x_0, x_1, x_2, \ldots, x_{k-1}, x_0\rangle$ is a cycle of a cubic 1-fault-tolerant Hamiltonian graph G. Then $Ext_C^2(G)$ is cubic 1-fault-tolerant Hamiltonian.

Proof: By Lemma 12.1, $Ext_C^2(G) - f$ is Hamiltonian for any $f \in V(Ext_C(G))$. It suffices to show that $Ext_C^2(G)$ is 1-edge fault-tolerant Hamiltonian. We divide the edge set of $Ext_C^2(G)$ into four sets as follows:

$$\begin{aligned}
\mathcal{E}_1 &= E(G) - E(C)\\
\mathcal{E}_2 &= \bigcup_{0\le i\le k-1} \left\{(q_{i,l_i}, r_{i,l_i}^1), \left(r_{i,l_i}^{L_i}, q_{i+1,1}\right), \left(s_{i,l_i}^{L_i}, s_{i+1,1}^1\right)\right\}\\
&\quad \cup \bigcup_{0\le i\le k-1}\ \bigcup_{1\le j\le l_i-1} \left\{\left(s_{i,j}^{l_{i,j}}, s_{i,j+1}^1\right)\right\}\\
\mathcal{E}_3 &= \bigcup_{0\le i\le k-1} \{(x_i, p_{i,1}), (p_{i,l_i}, x_{i+1})\}\\
&\quad \cup \bigcup_{0\le i\le k-1}\ \bigcup_{1\le j\le l_i-1} \left\{(p_{i,j}, p_{i,j+1}), \left(q_{i,j}, r_{i,j}^1\right), \left(r_{i,j}^{l_{i,j}}, q_{i,j+1}\right)\right\}\\
&\quad \cup \bigcup_{0\le i\le k-1}\ \bigcup_{1\le j\le l_i} \left\{\left(r_{i,j}^m, r_{i,j}^{m+1}\right), \left(s_{i,j}^m, s_{i,j}^{m+1}\right) \middle| 1 \le m \le L^* - 1\right\}\\
\mathcal{E}_4 &= \bigcup_{0\le i\le k-1}\ \bigcup_{1\le j\le l_i} \{(p_{i,j}, q_{i,j})\} \cup \bigcup_{0\le i\le k-1}\ \bigcup_{1\le j\le l_i} \left\{\left(r_{i,j}^m, s_{i,j}^m\right) \middle| 1 \le m \le L^*\right\}
\end{aligned}$$

Suppose that $f \in \mathcal{E}_1$. By Lemma 12.2, there is a Hamiltonian cycle H in $Ext_C(G) - f$. Since $Ext_C(G)$ is cubic, it follows that H contains at least an edge e in $\mathcal{O}_C$, say $e = (q_{i,j}, q_{i,j+1})$, for some $0 \le i \le k-1$ and $1 \le j \le l_i - 1$. Obviously, $M_1(H, e, m)$, $1 \le m \le l_{i,j} - 1$, is a Hamiltonian cycle of $Ext_C^2(G) - f$ for $f \in \mathcal{E}_1$.

Suppose that $f \in \mathcal{E}_2$. Since G is 1-fault-tolerant Hamiltonian, there is a Hamiltonian cycle in $G - (x_{i-1}, x_i)$. It follows from the proof of Lemma 12.3 that we have a Hamiltonian cycle, say H, in $Ext_C(G)$ containing $\langle q_{i-1,l_{i-1}}, q_{i,1}, q_{i,2}, \ldots, q_{i,l_i-1}, q_{i,l_i}, q_{i+1,1}\rangle$ as a subpath. Since $p_{i,j}$ is adjacent to $p_{i,j+1}$ for all $1 \le j \le l_i - 1$, we can apply operation M_2 to any $(q_{i,j}, q_{i,j+1})$. Obviously, $M_2(H, (q_{i,j}, q_{i,j+1}))$ forms a Hamiltonian cycle in $Ext_C^2(G) - \{(s_{i,j-1}^{l_{i,j-1}}, s_{i,j}^1), (s_{i,j}^{l_{i,j}}, s_{i,j+1}^1)\}$. Observing the Hamiltonian cycle $M_2(H, (q_{i,l_i-1}, q_{i,l_i}))$, which contains $\langle s_{i,l_i}^1, r_{i,l_i}^1, r_{i,l_i}^2, \ldots, r_{i,l_i}^{L_i}\rangle$ as a subpath, it is also a Hamiltonian cycle of $Ext_C^2(G) - (q_{i,l_i}, r_{i,l_i}^1)$. Similarly, we can use a Hamiltonian cycle of $G - (x_i, x_{i+1})$ to construct a Hamiltonian cycle H' of $Ext_C(G)$, which includes

$\langle q_{i,l_i}, q_{i+1,1}, q_{i+1,2}, \ldots, q_{i+1,l_{i+1}}, q_{i+2,1}\rangle$ as a subpath. Thus, $M_2(H', (q_{i+1,1}, q_{i+1,2}))$ forms a Hamiltonian cycle in $Ext_C^2(G) - \{(r_{i,l_i}^{L_i}, q_{i+1,1}), (s_{i,l_i}^{L_i}, s_{i+1,1}^{1})\}$.

Suppose that $f \in \mathcal{E}_3$. We first assume that $f = (x_i, p_{i,1})$ or (p_{i,l_i}, x_{i+1}). It follows from the proof of Lemma 12.2 that there is a Hamiltonian cycle H_1 in $Ext_C(G) - f$. Since $Ext_C(G)$ is cubic, H_1 contains at least an edge $(q_{i',j}, q_{i',j+1})$ in $\mathcal{O}_C$ for some $0 \le i' \le k-1, i \ne i'$, and $1 \le j \le l_{i'} - 1$. Thus, $M_1(H_1, (q_{i',j}, q_{i',j+1}), m)$ is a Hamiltonian cycle of $Ext_C^2(G) - f$, where $1 \le m \le l_{i',j} - 1$. Secondly, we assume that $f = (p_{i,j}, p_{i,j+1}), (q_{i,j}, r_{i,j}^{1})$, or $(r_{i,j}^{l_{i,j}}, q_{i,j+1})$. It follows from the proof of Lemma 12.2 that there is a Hamiltonian cycle H_2 in $Ext_C(G) - \{(p_{i,j}, p_{i,j+1}), (q_{i,j}, q_{i,j+1})\}$. Since $Ext_C(G)$ is cubic, H_2 contains $(q_{i,j+1}, q_{i,j+2})$. Hence, $M_1(H_2, (q_{i,j+1}, q_{i,j+2}), m)$ is a Hamiltonian cycle of $Ext_C^2(G) - \{(p_{i,j}, p_{i,j+1}), (q_{i,j}, r_{i,j}^{1}), (r_{i,j}^{l_{i,j}}, q_{i,j+1})\}$, where $1 \le m \le l_{i,j+1} - 1$. Finally, we assume that $f = (r_{i,j}^{m}, r_{i,j}^{m+1})$ or $(s_{i,j}^{m}, s_{i,j}^{m+1})$. Similarly, there is a Hamiltonian cycle H_3 in $Ext_C(G) - (q_{i,j-1}, q_{i,j})$. Hence, we can apply operation M_1 to $(q_{i,j}, q_{i,j+1})$ and m with $1 \le m \le l_{i,j} - 1$. Then $M_1(H_3, (q_{i,j}, q_{i,j+1}), m)$ is a Hamiltonian cycle of $Ext_C^2(G) - f$.

Suppose that $f \in \mathcal{E}_4$. Since G is 1-fault-tolerant Hamiltonian, there is a Hamiltonian cycle H of $G - (x_{i-1}, x_i)$. Then both (x_{i-2}, x_{i-1}) and (x_i, x_{i+1}) are in H. Therefore, $H_1 = M_1(H, (x_{i-2}, x_{i-1}), j)$ for some $1 \le j \le l_{i-2} - 1$ is a Hamiltonian cycle of $Ext_C(G)$, which contains $\langle q_{i-1,l_{i-1}}, q_{i,1}, q_{i,2}, \ldots, q_{i,l_i}, q_{i+1,1}\rangle$ as a subpath. Then we apply operation M_1 again on H_1 and $(q_{i-1,l_{i-1}}, q_{i,1})$. It follows that $M_1(H_1, (q_{i-1,l_{i-1}}, q_{i,1}), m)$ with $1 \le m \le L_{i-1} - 1$ is a Hamiltonian cycle of $Ext_C^2(G) - f$.

Therefore, $Ext_C^2(G)$ is 1-edge fault-tolerant Hamiltonian. The theorem follows. □

THEOREM 12.3 Suppose that G is a cubic 1-fault-tolerant Hamiltonian graph and $C = \langle x_0, x_1, x_2, \ldots, x_{k-1}, x_0\rangle$ is a cycle of G. Then $Ext_C^n(G)$ is cubic 1-fault-tolerant Hamiltonian for $n \ge 2$.

Proof: Since G is 1-fault-tolerant Hamiltonian, by Lemma 12.4 and Corollary 12.1 $Ext_C^2(G)$ is 1-fault-tolerant Hamiltonian and $\mathcal{O}_C^2$ is recoverable with respect to $Ext_C^2(G)$. Assume that $Ext_C^k(G)$ is 1-fault-tolerant Hamiltonian and $\mathcal{O}_C^k$ is recoverable with respect to $Ext_C^k(G)$ for $2 \le k \le l$. Consider $Ext_C^{l+1}(G) = Ext_{\mathcal{O}_C^l}(Ext_C^l(G))$. Then by Theorem 12.2, $Ext_{\mathcal{O}_C^l}(Ext_C^l(G)) = Ext_C^{l+1}(G)$ is 1-fault-tolerant Hamiltonian. By Lemma 12.3, $\mathcal{O}_C^{l+1}$ is recoverable with respect to $Ext_C^{l+1}(G)$. Thus, the induction is applicable. Hence, the theorem follows. □

We can recursively apply cycle extensions to construct a family of 1-fault-tolerant Hamiltonian graphs from a known cubic 1-fault-tolerant Hamiltonian graph. In the following discussion, we examine two such families of graphs; namely, eye networks and extended Petersen graphs. The eye networks proposed by Wang et al. [335] is indeed constructed by recursive cycle extensions from $Eye(1)$ as shown in Theorem 12.4. The 1-fault-tolerant Hamiltonicity of $Eye(n)$ is then a direct consequence of Theorem 12.3.

THEOREM 12.4

1. $Eye(n) = Ext_{O_1}^{n-1}(Eye(1))$ for $n \geq 2$, where O_1 is the outermost cycle of $Eye(1)$, as shown with the darkened edges in Figure 12.7.
2. $Eye(n)$ is 1-fault-tolerant Hamiltonian for $n \geq 3$.

Proof: Obviously, $Eye(2)$ is obtained by performing a cycle extension on $Eye(1)$ around O_1; that is, $Eye(2) = Ext_{O_1}(Eye(1))$. Let O_2 be the outermost cycle of $Eye(2)$. Note that O_2 is also the extended cycle of O_1 in $Eye(2)$. Let O_n denote the outermost cycle of $Eye(n)$ for $n \geq 1$. It can be easily verified that for $n \geq 3$, $Eye(n) = Ext_{O_{n-1}}(Eye(n-1))$ and O_n is the extended cycle of $Ext_{O_{n-1}}(Eye(n-1))$. Thus, $Eye(n) = Ext_{O_1}^{n-1}(Eye(1))$ for $n \geq 2$.

Note that $Eye(1)$ is also a 3-join of two K_4. Thus, $Eye(1)$ is also cubic 1-fault-tolerant Hamiltonian. By Theorem 12.3, $Ext_{O_1}^{n}(Eye(1))$; that is, $Eye(n+1)$, for $n \geq 2$, is 1-fault-tolerant Hamiltonian. Hence, the theorem follows. □

We recursively define *extended Petersen graphs*, denoted by $EP(n)$ for $n \geq 0$, as follows.

1. $EP(0)$ is the Petersen graph shown in Figure 12.10a with a specific cycle C indicated by the darkened edges. Define $\mathcal{O}_C^0 = C$.
2. Define $EP(1) = Ext_{\mathcal{O}_C^0}(EP(0))$ as shown in Figure 12.10b and $\mathcal{O}_C^1 = \mathcal{O}_C$ shown with the darkened edges in the figure.
3. For $n \geq 2$, define $EP(n) = Ext_{\mathcal{O}_C^{n-1}}(EP(n-1))$ and $\mathcal{O}_C^n$ as the extended cycle of $\mathcal{O}_C^{n-1}$ in $EP(n)$. Equivalently, we write $EP(n) = Ext_C^n(EP(0))$.

It is known that the Petersen graph is not Hamiltonian. Yet, it is Hamiltonian if any vertex is deleted. We can use the properties of cycle extensions to show the 1-fault-tolerant Hamiltonicity of $EP(n)$, as stated in the following theorem.

THEOREM 12.5 $EP(n)$ is 1-fault-tolerant Hamiltonian for $n \geq 1$.

Proof: Since the Petersen graph $EP(0)$ is hypohamiltonian, it follows from Lemma 12.1 that $EP(1)$ is also $EP(1) - f$ is Hamiltonian for any $f \in V(EP(1))$.

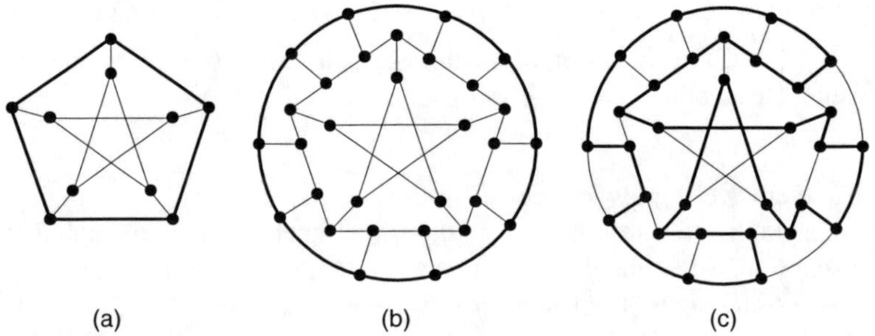

FIGURE 12.10 Extended Petersen graphs.

Furthermore, $EP(1)$ is Hamiltonian, where a Hamiltonian cycle of $EP(1)$ is illustrated in Figure 12.10c by the darkened edges. Using proper rotation of $EP(1)$, we can always obtain a Hamiltonian cycle without containing any specific edge. Therefore, $EP(1)$ is 1-edge fault-tolerant Hamiltonian, and, therefore, 1-fault-tolerant Hamiltonian. It can also be easily verified that $\mathcal{O}_C^1$ is recoverable with respect to $EP(1)$. It follows from Theorems 12.2 and 12.3 that $EP(n)$ is 1-fault-tolerant Hamiltonian for $n \geq 2$. □

12.4 CELLS FOR OPTIMAL 1-HAMILTONIAN REGULAR GRAPHS

For ease of description, we use *cells* to integrate the whole section. So we introduce cells first. Then we introduce the ideas and applications about flip-flops.

A *cell* is a 5-tuple (G, a, b, c, d), where G is a graph, and a, b, c, d are four distinct vertices in G. It is *cubic* if $\deg(x) = 3$ for every vertex x in $V(G) - \{a, b, c, d\}$ and $\deg(x) = 2$ for $x \in \{a, b, c, d\}$. The graph based on (G, a, b, c, d), denoted by $O_r((G, a, b, c, d))$, is the graph M obtained from G by adding two vertices u, v, and five edges $(u, v), (u, a), (u, d), (v, b)$, and (v, c). We also say that the cell (G, a, b, c, d) is *derived from M* by deleting two adjacent vertices u and v. See Figure 12.11.

Let $C_i = (G_i, a_i, b_i, c_i, d_i)$ be cells for $i = 1, 2, 3$. The cell $O_1(C_1, C_2)$ is defined to be the 5-tuple $(M_1, a_1, b_1, c_2, d_2)$, where M_1 is obtained from the disjoint union of G_1 and G_2 by adding edges (c_1, b_2) and (d_1, a_2). See Figure 12.12a. The cell $O_2(C_1, C_2, C_3)$ is defined as the 5-tuple $(M_2, a_1, b_1, c_3, d_3)$, where M_2 is obtained from the disjoint union of G_1, G_2, and G_3 by adding the vertices u_1, v_1, u_2, v_2, and the edges (u_1, v_1), (u_2, v_2), (c_1, u_1), (u_1, a_2), (d_1, v_1), (v_1, b_2), (c_2, u_2), (u_2, a_3), (d_2, v_2), (v_2, b_3). See Figure 12.12b. Obviously, both $O_1(C_1, C_2)$ and $O_2(C_1, C_2, C_3)$ are cubic if C_i is cubic for every i.

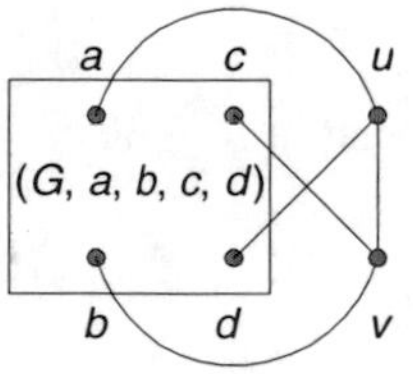

FIGURE 12.11 $O_r(G, a, b, c, d)$.

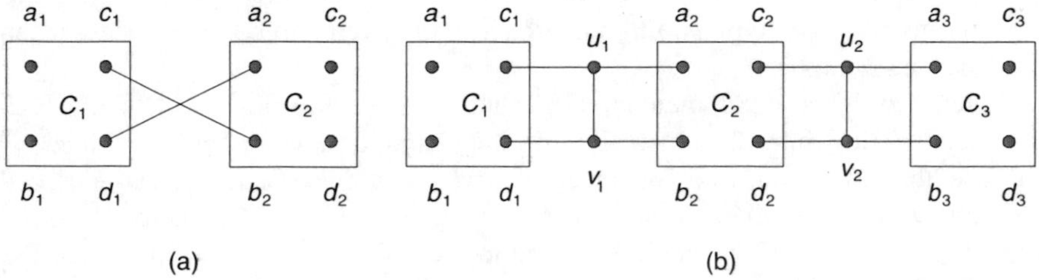

FIGURE 12.12 (a) $O_1(C_1, C_2)$ and (b) $O_2(C_1, C_2, C_3)$.

A pair of vertices (p, q) is *good* in a graph G if G contains a Hamiltonian path with end vertices p and q. The pair $((p, q), (r, s))$ is *good* in G if G has a spanning subgraph consisting of two disjoint paths whose end vertices are p, q, and r, s, respectively. Assume that $M = (V, E)$ is a hypohamiltonian graph with at least six distinct vertices u, v, a, b, c, d such that $\{v, a, d\}$ is the neighborhood of u and $\{u, b, c\}$ is the neighborhood of v. Let $G = M - \{u, v\}$. We call such a 5-tuple (G, a, b, c, d) a *J-cell*. The concept of flip-flops (explained shortly) originated from J-cells.

It is easy to see that $M - v$ and $M - u$ are Hamiltonian if and only if (a, d) and (b, c) are good in G. M is not Hamiltonian if and only if none of (a, b), (a, c), (b, d), (c, d), $((a, b), (c, d))$, and $((a, c), (b, d))$ are good in G. $M - x$ is Hamiltonian for each $x \in V(G)$ if and only if at least one of $(a, b), (a, c), (b, d), (c, d), ((a, b), (c, d))$, and $((a, c), (b, d))$ is good in $G - x$. Thus, (G, a, b, c, d) is a J-cell if and only if (G, a, b, c, d) satisfies the following properties:

J_1: (a, d) and (b, c) are good in G.
J_2: None of $(a, b), (a, c), (b, d), (c, d), ((a, b), (c, d))$ and $((a, c), (b, d))$ are good in G.
J_3: For each $x \in V(G)$, at least one of $(a, b), (a, c), (b, d), (c, d), ((a, b), (c, d))$, and $((a, c), (b, d))$ is good in $G - x$.

By adding extra conditions to J-cells, Chvátal [69] defined a cell (G, a, b, c, d) as a *flip-flop* if it has the following properties:

F_1: $(a, d), (b, c)$, and $((a, d), (b, c))$ are good in G.
F_2: None of $(a, b), (a, c), (b, d), (c, d), ((a, b), (c, d))$, and $((a, c), (b, d))$ are good in G.
F_3: For each $x \in V(G)$, at least one of $(a, c), (b, d), ((a, b), (c, d))$, and $((a, c), (b, d))$ is good in $G - x$.

The following theorem is proved in Ref. 69.

THEOREM 12.6 $O_1(C_1, C_2)$ and $O_2(C_1, C_2, C_3)$ are flip-flops if C_i is a flip-flop for every i.

Since flip-flops are J-cells, they can be used to construct back to a hypohamiltonian graph. By the theorem, Chvátal showed that there are infinite hypohamiltonian graphs. And by using some known hypohamiltonian graphs, he constructed the needed flip-flops to construct new hypohamiltonian graphs with given number of vertices, except for some small numbers.

Later, Collier and Schmeichel [75] studied those flip-flops with some extra conditions. They introduced *strong* flip-flops and a new operation. Since we will use the operation, we introduce it here. Let $C = (G, a, b, c, d)$ be a cell. The cell $O_3(C)$ is defined to be the 5-tuple (M_3, a', b', c', d'), where M_3 is the graph obtained from G by adding the vertices $a', b', c', d', p, q, r, s$, and the edges $(a, p), (b, q), (c, r), (d, s), (a', p), (p, c'), (c', r), (r, d')$, (d', s), (s, b'), (b', q) and (q, a'). Obviously, $O_3(C)$ is cubic if C is cubic. See Figure 12.13 for illustration.

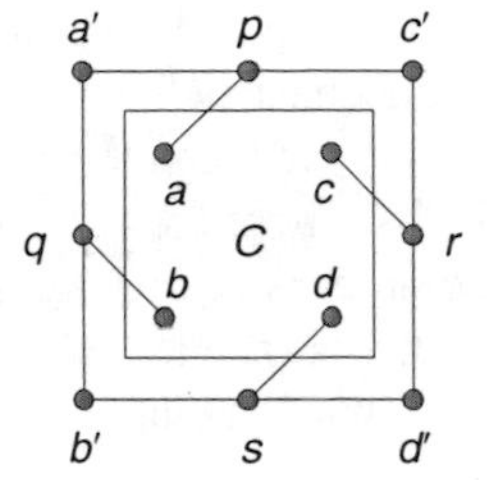

FIGURE 12.13 $O_3(C)$.

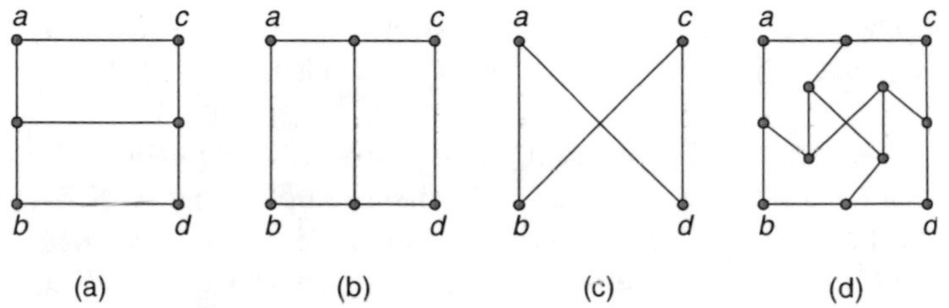

FIGURE 12.14 Examples of H-cells.

Li et al. [226] adopt Chvátal's method to construct optimal 1-fault-tolerant Hamiltonian regular graphs. Suppose that $M=(V,E)$ is an optimal 1-fault-tolerant Hamiltonian regular graph with at least six distinct vertices u, v, a, b, c, d such that $\{v, a, d\}$ is the neighborhood of u and $\{u, b, c\}$ is the neighborhood of v. Let $G=M-\{u,v\}$. We call the 5-tuple (G,a,b,c,d) an H-*cell*.

Obviously, $M-v$ and $M-u$ are Hamiltonian if and only if (a,d) and (b,c) are good in G. $M-(u,v)$ is Hamiltonian if and only if $((a,b),(c,d))$ or $((a,c),(b,d))$ is good in G. $M-e$ is Hamiltonian for $e \in \{(u,a),(u,d),(v,c),(v,b)\}$ if and only if (1) (a,b) and (c,d) are good in G or (2) (a,c) and (b,d) are good in G. $M-f$ is Hamiltonian for any $f \in V(G) \cup E(G)$ if and only if at least one of $(a,b),(a,c),(b,d),(c,d),((a,b),(c,d))$, and $((a,c),(b,d))$ is good in $G-f$. Thus, (G,a,b,c,d) is an H-cell if and only if it is cubic and satisfies all the following properties:

- H_1: (a,d) and (b,c) are good in G.
- H_2: (a,c) and (b,d) are good in G or (a,b) and (c,d) are good in G.
- H_3: $((a,b),(c,d))$ or $((a,c),(b,d))$ is good in G.
- H_4: At least one of $(a,b),(a,c),(b,d),(c,d),((a,b),(c,d))$, and $((a,c),(b,d))$ is good in $G-f$ for any $f \in V(G) \cup E(G)$.

Some examples are given in Figure 12.14. Obviously, (G,b,a,d,c), (G,c,d,a,b), (G,d,b,c,a), (G,a,c,b,d), and (G,d,c,b,a) are H-cells if (G,a,b,c,d) is an H-cell. In fact, the graphs $O_r((G,a,b,c,d))$, $O_r((G,b,a,d,c))$, $O_r((G,c,d,a,b))$, $O_r((G,d,b,c,a))$, $O_r((G,a,c,b,d))$, and $O_r((G,d,c,b,a))$ are all isomorphic.

To obtain more H-cells is our major goal. The more H-cells are found, the more 1-fault-tolerant Hamiltonian graphs are constructed. However, using the limited number of operations that we already know, we cannot generate an H-cell by combining arbitrary two or more known H-cells. There are two strategies to attack this problem: one is to find new operations; the other is to select some particular H-cells from the set of all H-cells. Here, we adopt the latter method; the way we do it is to put some more restrictions on the selection of the H-cells:

L_1: (a,d) and (b,c) are good in G. (The same as H_1.)
L_2: (a,b) and (c,d) are good in G.
L_3: $((a,c),(b,d))$ is good; or both $((a,b),(c,d))$ and $((a,d),(b,c))$ are good in G.
L_4: For any $f \in V(G) \cup E(G)$, at least one of $(a,b),(a,c),(b,d),(c,d),((a,b),(c,d))$, and $((a,c),(b,d))$ is good in $G-f$. (The same as H_4.)

We call the H-cells satisfying all of these properties *L-cells*. In Figure 12.14, the last three examples are L-cells; however, the first example is not.

Now we show that the set of L-cells is closed under O_1; that is, $O_1(C_1,C_2)$ is an L-cell if both C_1 and C_2 are L-cells. The proof is somewhat tedious. We support some figures to add readability. In fact, readers may directly check the proof by checking the figures, when familiar with the style of the proof.

THEOREM 12.7 $O_1(C_1,C_2)$ is an L-cell if both C_1 and C_2 are L-cells.

Proof: Let $C_1=(G_1,a_1,b_1,c_1,d_1)$, $C_2=(G_2,a_2,b_2,c_2,d_2)$, and $C=O_1(C_1,C_2)=(M_1,a_1,b_1,c_2,d_2)$. We show that $O_1(C_1,C_2)$ is an L-cell by the following steps:

1. ***C* satisfies L_1.** By property L_1, (a_1,d_1) and (b_1,c_1) are good in G_1, and (a_2,d_2) and (b_2,c_2) are good in G_2. So there are Hamiltonian paths $\langle a_1,P_1^{ad},d_1\rangle$ and $\langle b_1,P_1^{bc},c_1\rangle$ in G_1, and Hamiltonian paths $\langle a_2,P_2^{ad},d_2\rangle$ and $\langle b_2,P_2^{bc},c_2\rangle$ in G_2. Concatenating P_1^{ad} with P_2^{ad} and P_1^{bc} with P_2^{bc}, we obtain the two Hamiltonian paths of M_1, which show that C satisfies L_1. See Figure 12.15.
2. ***C* satisfies L_2.** By property L_2, there are Hamiltonian paths $\langle c_1,P_1^{cd},d_1\rangle$ in G_1 and $\langle a_2,P_2^{ab},b_2\rangle$ in G_2. By property L_3, at least one of $((a_1,d_1),(b_1,c_1))$ and $((a_1,c_1),(b_1,d_1))$ is good in G_1. Without loss of generality, assume that $((a_1,d_1),(b_1,c_1))$ is good in G_1. Similarly, assume that $((a_2,d_2),(b_2,c_2))$ is good in G_2. So there are two disjoint paths $\langle a_1,P_1^{ad},d_1\rangle$ and $\langle b_1,P_1^{bc},c_1\rangle$ covering all vertices in G_1, and two disjoint paths $\langle a_2,P_2^{ad},d_2\rangle$ and $\langle b_2,P_2^{bc},c_2\rangle$ covering all vertices in G_2. Then we have two Hamiltonian paths $\langle a_1,P_1^{ad},d_1,a_2,P_2^{ab},b_2,c_1,P_1^{bc},b_1\rangle$ and $\langle c_2,P_2^{bc},b_2,c_1,P_1^{cd},d_1,a_2,P_2^{ad},d_2\rangle$ in M_1, which show that C satisfies L_2. See Figure 12.15.
3. ***C* satisfies L_3.** By property L_2, there are Hamiltonian paths $\langle a_1,P_1^{ab},b_1\rangle$ in G_1 and $\langle c_2,P_2^{cd},d_2\rangle$ in G_2. The two paths are disjoint and cover all vertices in M_1. So $((a_1,b_1),(c_2,d_2))$ is good in M_1. As a result, it is sufficient to prove that at least one of $((a_1,d_2),(b_1,c_2))$ and $((a_1,c_2),(b_1,d_2))$ is good in M_1. Suppose that $((a_1,d_1),(b_1,c_1)$ is good in G_1, and $((a_2,c_2),(b_2,d_2))$ is good in G_2.

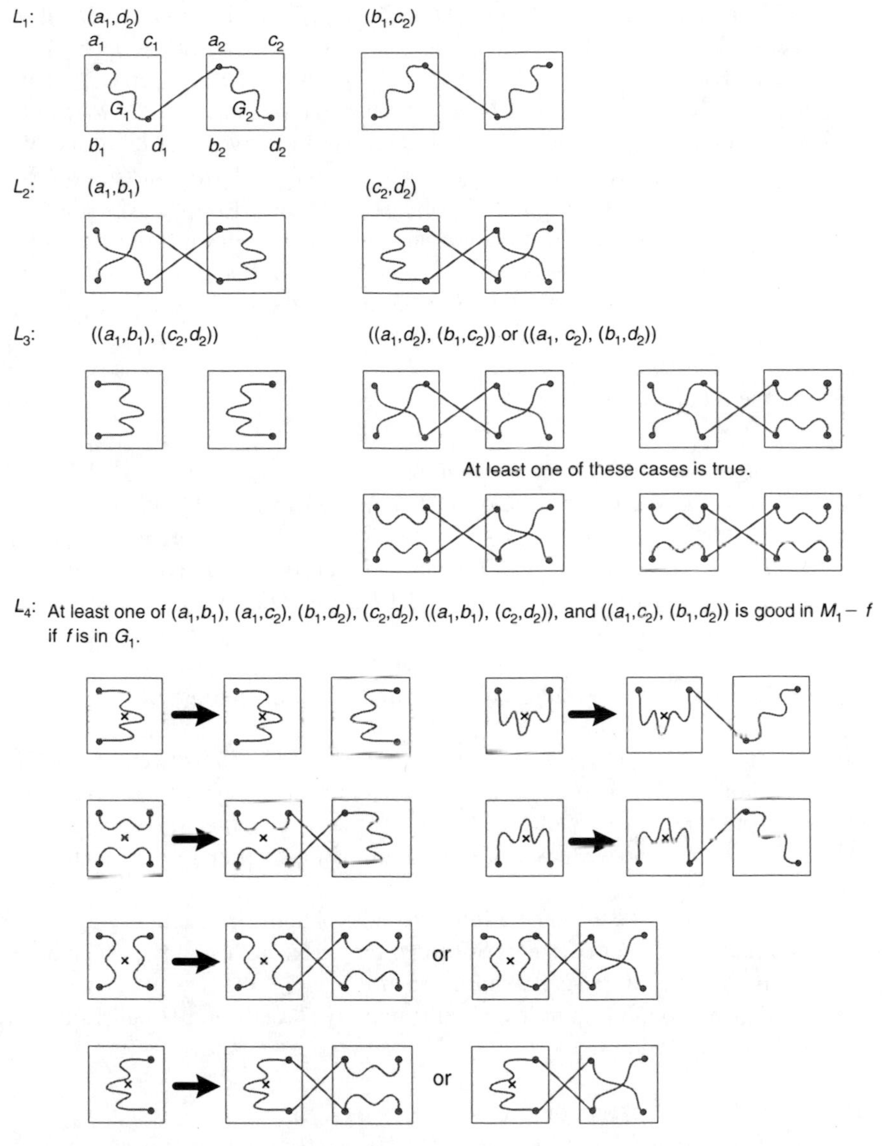

FIGURE 12.15 Proof of Theorem 12.7.

Then we have two disjoint paths $\langle a_1, P_1^{ad}, d_1\rangle$ and $\langle b_1, P_1^{bc}, c_1\rangle$ covering all vertices in G_1, and two disjoint paths $\langle a_2, P_2^{ac}, c_2\rangle$ and $\langle b_2, P_2^{bd}, d_2\rangle$ covering all vertices in G_2. Properly concatenating them, we have two disjoint paths $\langle a_1, P_1^{ad}, d_1, a_2, P_2^{ac}, c_2\rangle$ and $\langle b_1, P_1^{bc}, c_1, b_2, P_2^{bd}, d_2\rangle$ covering all vertices in M_1. And $((a_1, c_2), (b_1, d_2))$ is good in M_1. The proofs of other cases are similar. See Figure 12.15.

4. **C satisfies L_4.** Let $f \in V(M_1) \cup E(M_1)$. Clearly, f may be in G_1, in G_2, the edge (c_1, b_2), or the edge (d_1, a_2). If f is (c_1, b_2) or (d_1, a_2), by property L_2, there are Hamiltonian paths $\langle a_1, P_1^{ab}, b_1\rangle$ in G_1 and $\langle c_2, P_2^{cd}, d_2\rangle$ in G_2. Then $((a_1, b_1), (c_2, d_2))$ is good in $M_1 - f$. Suppose that f is in G_1 or G_2. Without loss of generality, assume that f is in G_1. By property L_4, at least one of $(a_1, b_1), (a_1, c_1), (b_1, d_1), (c_1, d_1), ((a_1, b_1), (c_1, d_1))$, and $((a_1, c_1), (b_1, d_1))$ is good in $G_1 - f$. So we should discuss the six situations. However, it is tedious. Here, we discuss the first situation and show the others in Figure 12.15. Assume that (a_1, b_1) is good in $G_1 - f$. By property L_2, there is a Hamiltonian path $\langle c_2, P_2^{cd}, d_2\rangle$ in G_2. Then $((a_1, b_2), (c_2, d_2))$ is good in $M_1 - f$. The proofs for other situations are similar. See Figure 12.15.

Hence, the theorem is proved. □

Furthermore, we define a new operation O_4. Let $C = (G, a, b, c, d)$ be a cell. $O_4(C)$ is a cell (M_4, a, b, u, v), where M_4 is obtained from G by adding two vertices u, v, and three edges (u, v), (u, d), and (v, c). See Figure 12.16. Consider the graph $I(\{u, v\}, \{(u, v)\})$, which is K_2. If we take (I, u, v, u, v) as a cell, we may easily check whether the cell satisfies all properties of L-cells. Thus, we have the following theorem.

THEOREM 12.8 The set of all L-cells is closed under operation O_4.

It is observed that $O_2(C_1, C_2, C_3)$ is isomorphic to $O_1(O_4(O_1(O_4(C_1), C_2)), C_3))$. Hence, the set of L-cells is also closed under O_2. As a result, using the two operations on arbitrary two or three known L-cells will generate a new L-cell; that is, we may generate more and more L-cells. So we may easily generate 1-fault-tolerant Hamiltonian graphs infinitely.

Although L-cells make the generation of 1-fault-tolerant Hamiltonian graphs easy, there is another requirement arising. Someone may ask: is there any particular cell operating with an arbitrary H-cell resulting in an H-cell?

We introduce two more families of cells; namely, T-cells and U-cells. First, we introduce T-cells as follows:

T_1: (a, d) and (b, c) are good in G. (The same as H_1.)
T_2: At least two of $((a, b), (c, d))$, $((a, c), (b, d))$, and $((a, d), (b, c))$ are good in G.

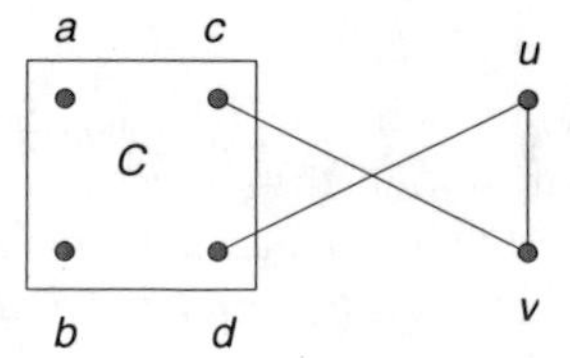

FIGURE 12.16 Operation O_4.

T_3: (a,c), (b,d), (a,b), and (c,d) are good in G.
T_4: For any $f \in V \cup E$, either

1. At least two of $((a,b),(c,d))$, $((a,c),(b,d))$, and $((a,d),(b,c))$ are good in $G-f$ or
2. $((a,c)$ or (b,d) is good in $G-f)$ and $((a,b)$ or (c,d) is good in $G-f)$.

A T-cell is a cubic cell satisfying these properties. Obviously, (G,b,a,c,d), (G,b,a,d,c), (G,c,d,a,b), (G,c,d,b,a), (G,d,c,a,b), and (G,d,c,b,a) are T-cells if (G,a,b,c,d) is a T-cell. It is clear to see that any T-cell is an L-cell. However, the converse is not true. In Figure 12.14, the last cell is a T-cell, but the others are not. Similarly, we have the following theorem. Since the proofs of the following theorems are similar to that of Theorem 12.7, we omit the proofs of the following theorems.

THEOREM 12.9 The set of all T-cells is closed under O_1, O_2, O_3, and O_4.

We have the following important theorem.

THEOREM 12.10 Assume that C_1 is a T-cell. Then $O_1(C_1,C_2)$ is an H-cell if C_2 is an H-cell.

By this theorem, once we have a T-cell, we may use operation O_1 to link the T-cell to all known H-cells, and then make the amount of H-cells double. Moreover, we can continue linking the T-cell to the newly generated H-cells to produce new H-cells. Thus, we can generate H-cells infinitely.

However, if a T-cell is difficult to find, such a method will be useless. Here, we propose a construction to construct a T-cell as follows. We define U-cells first. A U-cell (G,a,b,c,d) is an H-cell such that at least two of $((a,b),(c,d))$, $((a,c),(b,d))$, and $((a,d),(b,c))$ are good in G. Such a cell has a bit restriction on H-cells. We even do not find an H-cell that is not a U-cell, yet. Obviously, any L-cell is an U-cell. However, the converse is not true. For example, the cell in Figure 12.14a is an U-cell but not an L-cell. Moreover, (G,b,a,d,c), (G,c,d,a,b), (G,d,b,c,a), (G,a,c,b,d), and (G,d,c,b,a) are U-cells if (G,a,b,c,d) is an U-cell. The following theorem can be proved case by case.

THEOREM 12.11 $O_3(C)$ is a T-cell if C is an U-cell.

With the previous theorem, we can easily generate a T-cell. In the study of the T-cells, we find another interesting operation. Let $C_i = (G_i, a_i, b_i, c_i, d_i)$ be cells for $i = 1,2$. $O_5(C_1,C_2)$ is the graph M_5, which is obtained from the disjoint union of G_1 and G_2 by adding edges (a_1,c_2), (b_1,d_2), (d_1,a_2), and (c_1,b_2). See Figure 12.17. Obviously, $O_5(C_1,C_2)$ is a 3-regular graph if C_1 and C_2 are cubic. The following result can be proved by the same strategy in the proof of Theorem 12.7.

THEOREM 12.12 $O_5(C_1,C_2)$ is an optimal 1-fault-tolerant Hamiltonian regular graph if C_1 is an H-cell and C_2 is a T-cell.

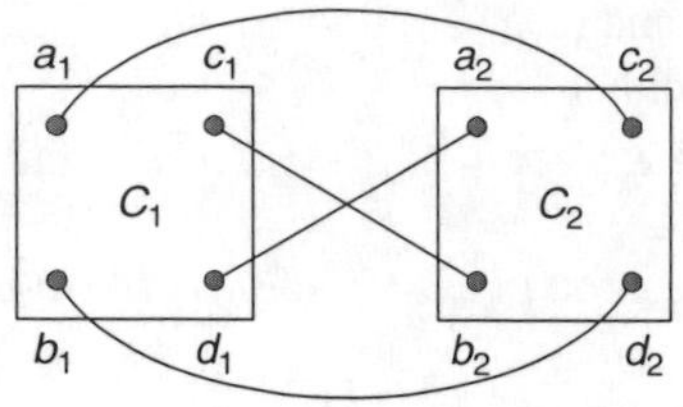

FIGURE 12.17 $O_5(C_1, C_2)$.

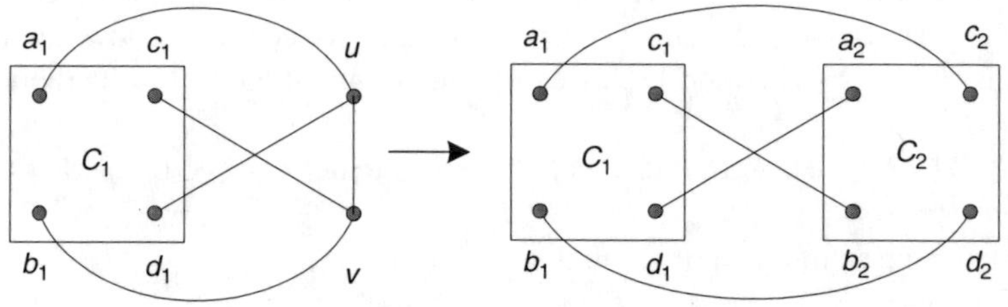

FIGURE 12.18 Substitution of an edge.

Why is Theorem 12.12 interesting? Consider Figure 12.18. The T-cell looks like a substitution of the edge (u, v) in the left graph. Remember that the left graph is just an optimal 1-fault-tolerant Hamiltonian graph. Since we do not add any restriction on the edge (u, v), the T-cell may be substituted for any edge in the left graph. This is the most intriguing property. For this reason, we loosen the restriction of T-cells and introduce another family of cells; namely, S-cells, which have the following properties:

S_1: (a, d) and (b, c) are good in G. (The same as H_1.)
S_2: $((a, c), (b, d))$ is good in G, or $((a, b), (c, d))$ and $((a, d), (b, c))$ are good in G.
S_3: (a, b) and (c, d) are good in G.
S_4: For any $f \in V \cup E$, at least one of the following statements is true:

1. At least one of (a, c), (b, d), and $((a, c), (b, d))$ is good in $G - f$.
2. (b, c) or (a, d) is good in $G - f$, and (a, b) or (c, d) is good in $G - f$.
3. $((a, b), (c, d))$ and $((a, d), (b, c))$ are good in $G - f$.

An S-cell is a cubic cell with these properties. It is easy to see that any S-cell is an L-cell and any T-cell is an S-cell. However, the converse is not true. The cell in Figure 12.19 is an S-cell but not a T-cell. The following theorem is about S-cells.

THEOREM 12.13 $O_5(C_1, C_2)$ is an optimal 1-fault-tolerant Hamiltonian regular graph if C_1 is an H-cell and C_2 is an S-cell.

It follows from the theorem that we can substitute any edge of an optimal 1-fault-tolerant Hamiltonian graph with an S-cell, and the graph obtained is also an optimal 1-fault-tolerant Hamiltonian graph.

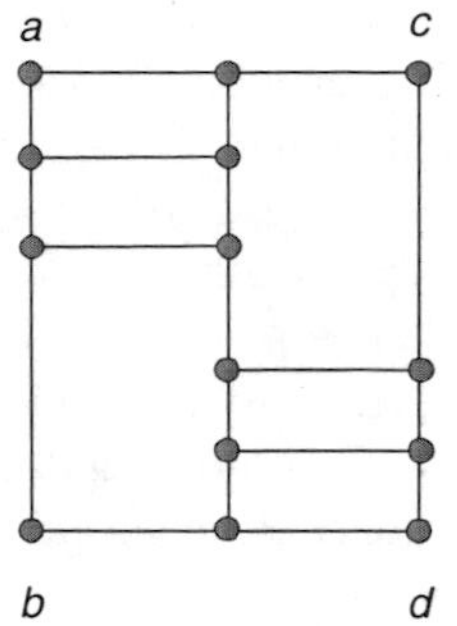

FIGURE 12.19 An example of an S-cell.

Here, we present five families of cells—H-cells, L-cells, T-cells, U-cells, and S-cells—and some operations for the construction of optimal 1-fault-tolerant Hamiltonian regular graphs. Apparently, we cannot generate all such graphs using our approaches. Usually, we say that an object is *primitive* if it cannot be constructed from smaller objects. Of course, the term *primitive* depends on the construction schemes proposed. Let X denote the set of graphs $\{C_n \times K_2 \mid n \text{ is odd and } n > 3\}$, where $\times$ is the cartesian product of graphs. In Ref. 334, any graph in X is viewed as primitive optimal 1-fault-tolerant Hamiltonian regular graphs because any graph in X cannot be constructed from K_4 using 3-join and cycle extension. However, it is proved in Ref. 334 that the graph $C_3 \times K_2$ is a 3-join of two K_4s. Let C denote the cell in Figure 12.14c. It can be verified that $O_r(C) = C_3 \times K_2$. Let us recursively define $O_4^n(C)$ as $O_4^0(C) = C$ and $O_4^n(C) = O_4(O_4^{n-1}(C))$. It can be verified that $C_{2k+1} \times K_2 = O_r(O_4^{2k-2}(C))$ for every $k \geq 1$. Hence, graphs in X cannot be viewed as primitive optimal 1-fault-tolerant Hamiltonian regular graphs.

12.5 GENERALIZED PETERSEN GRAPHS

In this section, we present some necessary conditions and some sufficient conditions about the 1-fault-tolerant Hamiltonian generalized Petersen graphs. We put this material in this book because its proof technique is very interesting.

Assume that n and k are positive integers with $n \geq 2k+1$. Let $\oplus$ denote addition in integer modular n, Z_n. The *generalized Petersen graph* $P(n,k)$ is the graph with vertex set $\{i \mid 0 \leq i \leq n-1\} \cup \{i' \mid 0 \leq i \leq n-1\}$ and edge set $\{(i, i \oplus 1) \mid 0 \leq i \leq n-1\} \cup \{(i, i') \mid 0 \leq i \leq n-1\} \cup \{(i', (i \oplus k)') \mid 0 \leq i \leq n-1\}$. In [9,18,28,277], it is proved that $P(n,k)$ is not Hamiltonian if and only if $P(n,k)$ is isomorphic to $P(n', k')$ $k' = 2$ and $n' \equiv 5 \pmod{6}$. For example, $P(11,5)$ is isomorphic to $P(11,2)$. Thus, $P(11,5)$ is not Hamiltonian. More precisely, $P(n,k)$ is not Hamiltonian if and only if (1) $n \equiv 5 \pmod{6}$ and (2) $k = 2$ or $k = (n-1)/2$. However, all non-Hamiltonian generalized Petersen graphs are proved to be hypohamiltonian [29].

We will prove that $P(n,k)$ is not 1-fault-tolerant Hamiltonian if n is even and k is odd. We also prove that $P(3k,k)$ is 1-fault-tolerant Hamiltonian if and only if k is odd. Moreover, $P(n,3)$ is 1-fault-tolerant Hamiltonian if and only if n is odd;

$P(n, 4)$ is 1-fault-tolerant Hamiltonian if and only if $n \neq 12$. Furthermore, $P(n, k)$ is 1-fault-tolerant Hamiltonian if k is even with $k \geq 6$ and $n \geq 2k + 2 + (4k - 1)(4k + 1)$, and $P(n, k)$ is 1-fault-tolerant hamiltonian if k is odd with $k \geq 5$ and n is odd with $n \geq 6k - 3 + 2k(6k - 2)$.

In Refs 9, 18, 29, and 277, it is proved that $P(n, k)$ is not Hamiltonian if and only if $k = 2$ and $n \equiv 5 \pmod{6}$. By the symmetric property of the generalized Petersen graph, $P(n, k)$ is not 1-edge fault-tolerant Hamiltonian if and only if $k = 2$ and $n \equiv 5 \pmod{6}$. Mai et al. [246] began the study of those 1-fault-tolerant Hamiltonian graphs by studying those Hamiltonian generalized Petersen graphs that are 1-vertex fault-tolerant Hamiltonian.

LEMMA 12.5 Any bipartite graph is not 1-fault-tolerant Hamiltonian.

Proof: Let G be a 1-fault-tolerant Hamiltonian bipartite graph with n vertices. Obviously, G has a cycle of length n and $n - 1$. This is impossible, because G has no cycle of odd length. Hence, the lemma is proved. □

The next lemma can easily be obtained from the definition of the generalized Petersen graphs.

LEMMA 12.6 $P(n, k)$ is bipartite if and only if n is even and k is odd. In particular, $P(n, k)$ is not 1-fault-tolerant Hamiltonian if n is even and k is odd.

With Lemma 12.6, we may ask whether $P(n, k)$ is not 1-fault-tolerant Hamiltonian if and only if n is even and k is odd. However, the statement is not true, because of the following theorem.

THEOREM 12.14 **[7]** $P(n, 1)$ is 1-fault-tolerant Hamiltonian if and only if n is odd and $P(n, 2)$ is 1-fault-tolerant Hamiltonian if and only if $n \equiv 1, 3 \pmod{6}$.

Therefore, $P(n, 2)$ with n even is neither bipartite nor 1-fault-tolerant Hamiltonian. Yet we desire to know whether there are other generalized Petersen graphs that are not 1-fault-tolerant Hamiltonian.

Let us consider the generalized Petersen graph $P(12, 4)$ shown in Figure 12.20a. Suppose that $P(12, 4)$ is 1-fault-tolerant Hamiltonian. Obviously, $P(12, 4) - 1$ is Hamiltonian. Note that the vertex set $\{0', 4', 8'\}$ induces a complete graph K_3. Any Hamiltonian cycle in $P(12, 4) - 1$ traverses the vertex set $\{0', 4', 8'\}$ consecutively, because $P(12, 4)$ is a cubic graph. Similarly, any Hamiltonian cycle traverses the vertex sets $\{1', 5', 9'\}$, $\{2', 6', 10'\}$, and $\{3', 7', 11'\}$ consecutively. Therefore, we can define a graph $P'(12, 4)$ obtained from $P(12, 4)$ by shrinking the vertices $0', 4'$, and $8'$ into a vertex $\langle 0 \rangle$, shrinking the vertices $1', 5'$, and $9'$ into a vertex $\langle 1 \rangle$, shrinking the vertices $2', 6'$, and $10'$ into a vertex $\langle 2 \rangle$, and shrinking the vertices $3', 7'$, and $11'$ into a vertex $\langle 3 \rangle$ as shown in Figure 12.20b. Obviously, $P(12, 4) - 1$ is Hamiltonian if and only if $P'(12, 4) - 1$ is Hamiltonian. However, $P'(12, 4)$ is a bipartite graph with eight vertices in each partite set. Therefore, $P'(12, 4) - 1$ is not Hamiltonian. We get a contradiction. Thus, $P(12, 4) - 1$ is not Hamiltonian.

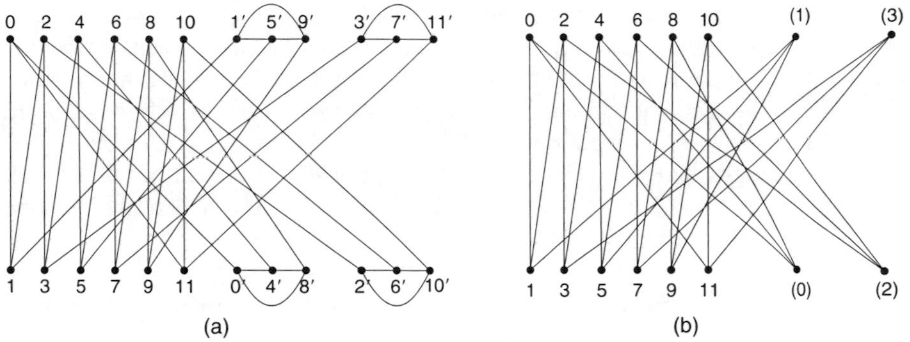

FIGURE 12.20 (a) $P(12,4)$ and (b) $P'(12,4)$.

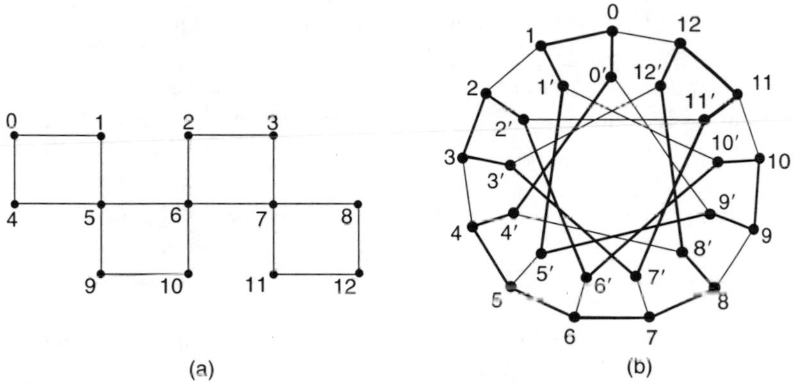

FIGURE 12.21 (a) A $D(13,4)$ and (b) its corresponding Hamiltonian cycle in $P(13,4)$.

Alspach et al. [11] proposed an interesting model, called the *lattice model*, to describe a Hamiltonian cycle of the generalized Petersen graphs. With the lattice model, a *lattice diagram* for a Hamiltonian generalized Petersen graph is a labeled graph in the (x, y)-plane that possesses a closed or an open Eulerian trail. By appropriately interpreting the edges in the diagram, the Eulerian trail corresponds to a Hamiltonian cycle of a generalized Petersen graph.

A lattice diagram $D(13,4)$ for a generalized Petersen graph $P(13,4)$, for example, is shown in Figure 12.21. In this diagram, there is an Eulerian trail $\langle 0, 1, 5, 9, 10, 6, 2, 3, 7, 11, 12, 8, 7, 6, 5, 4, 0\rangle$. By appropriately interpreting the edges in the diagram, it corresponds to a Hamiltonian cycle $\langle 0, 1, 1', 5', 9', 9, 10, 10', 6', 2', 2, 3, 3', 7', 11', 11, 12, 12', 8', 8, 7, 6, 5, 4, 4', 0', 0\rangle$ in $P(13,4)$.

Alspach et al. [11] introduced this scheme and proved that $P(n,k)$ is Hamiltonian for all $n \geq (4k-1)(4k+1)$ if k is even and for all $n \geq (2k-1)(3k+1)$ if k is an odd integer with $k \geq 3$.

The lattice model is described as follows. For more details, see Ref. 11. In a lattice model, a *lattice graph* L consists of lattice points in the (x, y)-plane. Two lattice points (a_1, b_1) and (a_2, b_2) in L are adjacent if and only if $|a_1 - a_2| + |b_1 - b_2| = 1$.

Hence, the degree of each vertex is less than or equal to 4. Suppose that n and k be positive integers with $n \geq 2k+1$. A *labeled lattice graph* $L(n, k)$ is obtained from the lattice graph L by labeling the lattice points following this rule: suppose that a lattice point (a, b) is labeled with an integer i with $0 \leq i \leq n-1$. Then $(a+1, b)$ is labeled with $i \oplus 1$ and $(a, b-1)$ is labeled with $i \oplus k$. A lattice diagram for $P(n, k)$, denoted as $D(n, k)$, is a subgraph of $L(n, k)$ induced by the vertices with labels $0, 1, \ldots, n-1$ such that it possesses either a closed or an open Eulerian trail. A traversal of the Eulerian trail in $D(n, k)$ obeys the following rules:

1. The trail does not change the direction when it passes through a vertex of degree 4.
2. Each label $0, 1, \ldots, n-1$ is encountered by the traversal once in a vertical direction and once in a horizontal direction.
3. Suppose that $D(n, k)$ has an open Eulerian trail. Then two vertices of odd degree must have the same label, and both are not of degree 3.

The correspondence between an Eulerian trail of a lattice diagram $D(n, k)$ and a Hamiltonian cycle of the generalized Petersen graph $P(n, k)$ is built by interpreting the edges in $L(n, k)$. The interpretations of edges in $L(n, k)$ are given as follows:

1. The vertical edge $(i, i \oplus k)$ in $L(n, k)$ corresponds to an edge $(i', (i \oplus k)')$ in $P(n, k)$.
2. The horizontal edge $(i, i \oplus 1)$ in $L(n, k)$ corresponds to an edge $(i, i \oplus 1)$ in $P(n, k)$.
3. Two edges in different directions incident to the vertex i of degree 2 in $L(n, k)$ correspond to an edge (i, i') in $P(n, k)$.

It is not difficult to see that an Eulerian trail in $L(n, k)$ corresponds to a Hamiltonian cycle of $P(n, k)$. Moreover, any Hamiltonian cycle of $P(n, k)$ can be converted into an Eulerian trail in $D(n, k)$. Hence, finding a Hamiltonian cycle of a generalized Petersen graph $P(n, k)$ is finding an appropriate lattice diagram for $P(n, k)$ [11].

Furthermore, Alspach et al. [11] proposed an amalgamating mechanism to generate lattice diagrams of various sizes. For example, we have a lattice diagram $D(4k, k)$ in Figure 12.22a and a lattice diagram $D(4k+2, k)$ in Figure 12.22b. By identifying the vertex with label $(4k-1)$ of $D(4k, k)$ and the vertex with label 0 of $D(4k+2, k)$, and relabeling all vertices of $D(4k+2, k)$ by adding $(4k-1)$ to all the labels, we obtain a lattice diagram $D(8k+1, k)$, as shown in Figure 12.22c. With this diagram, $P(8k+1, k)$ is proved to be Hamiltonian. With this mechanism, we can amalgamate r copies of $D(4k, k)$ and s copies of $D(4k+2, k)$, where $r \geq 1$ and $s \geq 0$, to make a lattice diagram $D(4k+(r-1)(4k-1)+s(4k+1), k)$. Hence, $P(4k+(r-1)(4k-1)+s(4k+1), k)$ is Hamiltonian for $r \geq 1$ and $s \geq 0$.

Here, we slightly modify the lattice diagram proposed in Ref. 11 to prove that a generalized Petersen graph is 1-fault-tolerant Hamiltonian.

First, we give examples to demonstrate the variation. In Figure 12.23a, an open Eulerian trail $\langle 0, 1, 7, 8, 2, 3, 9, 10, 4, 5, 11, 12, 6, 0 \rangle$ corresponds to the Hamiltonian cycle $\langle 0, 1, 1', 7', 7, 8, 8', 2', 2, 3, 3', 9', 9, 10, 10', 4', 4, 5, 5', 11', 11, 12, 12', 6', 0', 0 \rangle$

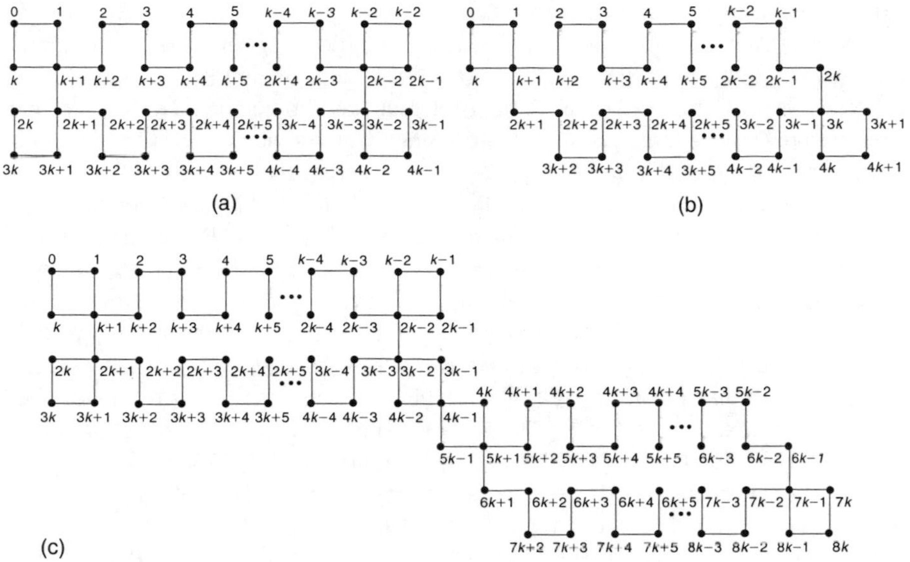

FIGURE 12.22 (a) A $D(4k, k)$, (b) a $D(4k+2, k)$, and (c) a $D(8k+1, k)$.

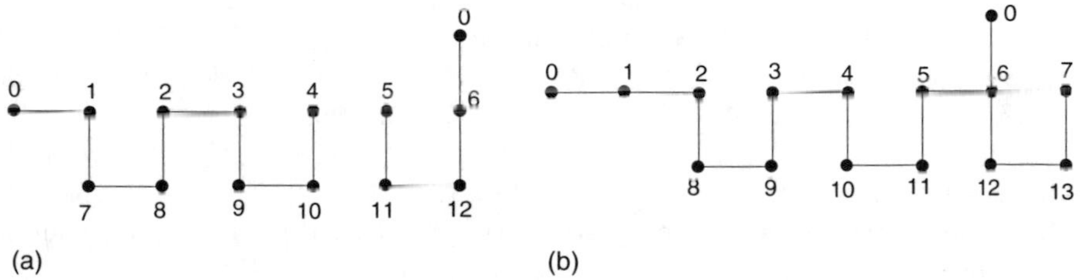

FIGURE 12.23 (a) A $D^v_6(13, 6)$ and (b) a $D^h_1(14, 6)$.

of $P(13, 6) - 6$ using the same interpretation described earlier. The reason why the open Eulerian trail corresponds to the Hamiltonian cycle of $P(13, 6) - 6$ is because the vertex with label 6 is traversed only in the vertical direction and all the other vertices are traversed in both directions. Again, the open Eulerian trail in Figure 12.23b, $\langle 0, 1, 2, 8, 9, 3, 4, 10, 11, 5, 6, 7, 13, 12, 6, 0\rangle$ corresponds to the Hamiltonian cycle $\langle 0, 1, 2, 2', 8', 8, 9, 9', 3', 3, 4, 4', 10', 10, 11, 11', 5', 5, 6, 7, 7', 13', 13, 12, 12', 6', 0', 0\rangle$ in $P(14, 6) - 1'$. In this diagram, the vertex with label 1 is traversed only in the horizontal direction, and all the other vertices are traversed in both directions.

In other words, finding a lattice diagram $D^v_i(n, k)$ with an Eulerian trail such that the vertex with label i is traversed only in the vertical direction and all the other vertices are traversed in both directions is finding a Hamiltonian cycle of $P(n, k) - i$. Similarly, finding a lattice diagram $D^h_i(n, k)$ with an Eulerian trail such that the vertex with label i is traversed only in a horizontal direction and all the other vertices are traversed in both directions is finding a Hamiltonian cycle of $P(n, k) - i'$. Note that the automorphism group of $P(n,k)$ has at most two orbits $0,1,2,\ldots,n-1$ and $0',1',2',\ldots,(n-1)'$. We need to construct a lattice diagram $D_i^v(n,k)$ and a lattice diagram $D_j^h(n,k)$ for some i and j with $0 \le i,j \le n-1$ to prove that $P(n,k)$ is 1-fault-tolerant Hamiltonian.

Now, we prove that $P(3k,k)$ is 1-fault-tolerant Hamiltonian if and only if k is odd. First, we prove that $P(3k,k)$ is not 1-fault-tolerant Hamiltonian if k is even. The proof is similar to the proof that $P(12,4)$ is not 1-fault-tolerant Hamiltonian.

Suppose that k is an even integer. Suppose that $P(3k,k)$ is 1-fault-tolerant Hamiltonian. Then, there exists a Hamiltonian cycle of $P(3k,k)-1$. By the definition of the generalized Petersen graph, the set $\{i',(i+k)',(i+2k)'\}$ induces a cycle of length 3 for $0 \le i \le k-1$. Therefore, any Hamiltonian cycle in $P(3k,k)-1$ traverses the vertex set $\{i',(i+k)',(i+2k)'\}$ consecutively for $0 \le i \le k-1$. Therefore, vertices $i',(i+k)'$, and $(i+2k)'$ can be regarded as a vertex for $0 \le i \le k-1$. Let $P'(3k,k)$ be the graph obtained from $P(3k,k)$ by shrinking the vertices $i',(i+k)'$, and $(i+2k)'$ into a new vertex, say $\langle i \rangle$, for $0 \le i \le k-1$. It is not difficult to see that $P(3k,k)-1$ is Hamiltonian if and only if $P'(3k,k)-1$ is Hamiltonian.

Now, we claim that $P'(3k,k)$ is a bipartite graph. Let $X=\{1,3,5,\ldots,(3k-1)\} \cup \{\langle i \rangle \mid 0 \le i \le k-1 \text{ and } i \text{ is even}\}$ and $Y=\{0,2,4,6,\ldots,(3k-2)\} \cup \{\langle i \rangle \mid 0 \le i \le k-1 \text{ and } i \text{ is odd}\}$. It is easy to check that (X,Y) forms a bipartition of $P(3k,k)$ with $|X|=|Y|=2k$. Hence, $P'(3k,k)$ is a bipartite graph with the same size of bipartition. This implies that $P'(3k,k)-1$ is not Hamiltonian. Hence, $P(3k,k)-1$ is not Hamiltonian. Therefore, $P(3k,k)$ is not 1-fault-tolerant Hamiltonian if k is even.

Thus, we consider k is odd. Suppose that $n=1$. Obviously, $\langle 1,1',0',2',2,1\rangle$ forms a Hamiltonian cycle for $P(3,1)-0$ and $\langle 0,1,1',2',2,0\rangle$ forms a Hamiltonian cycle for $P(3,1)-0'$. Thus, $P(3,1)$ is 1-fault-tolerant Hamiltonian.

Suppose that $k=3$. Obviously, $\langle 1,2,3,3',0',6',6,7,8,8',2',5',5,4,4',7',1',1\rangle$ forms a Hamiltonian cycle of $P(9,3)-0$ and $\langle 0,1,2,3,3',6',6,7,7',1',4',4,5,5',2',8',8,0\rangle$ forms a Hamiltonian cycle for $P(9,3)-0'$. Thus, $P(9,3)$ is 1-fault-tolerant Hamiltonian.

Suppose that $k \ge 5$. A lattice diagram $D_{2k-1}^v(3k,k)$ with $k \equiv 1 \pmod 4$ is shown in Figure 12.24a and a lattice diagram $D_{k-1}^v(3k,k)$ with $k \equiv 3 \pmod 4$ is shown in Figure 12.24b. Moreover, a lattice diagram $D_k^h(3k,k)$ is shown in Figure 12.24c. Therefore, $P(3k,k)$ is 1-fault-tolerant Hamiltonian.

Thus, we have the following theorem.

THEOREM 12.15 **[246]** $P(3k,k)$ is 1-fault-tolerant Hamiltonian if and only if k is odd.

THEOREM 12.16 **[246]** The generalized Petersen graph $P(n,3)$ is 1-fault-tolerant Hamiltonian if and only if n is odd.

Proof: By the definition of $P(n,k)$, we consider $n \ge 7$. By Lemma 12.6, $P(n,3)$ is not 1-fault-tolerant Hamiltonian if n is even. Thus, we consider the case n is odd.

Suppose that $n=7$. Obviously, $\langle 0,0',4',1',1,2,3,3',6',2',5',5,6,0\rangle$ forms a Hamiltonian cycle for $P(7,3)-4$ and $\langle 0,1,2,3,4,4',0',3',6',2',5',5,6,0\rangle$ forms a Hamiltonian cycle for $P(7,3)-1'$. Thus, $P(7,3)$ is 1-fault-tolerant Hamiltonian.

Suppose that $n=9$. By Theorem 12.15, $P(9,3)$ is 1-fault-tolerant Hamiltonian.

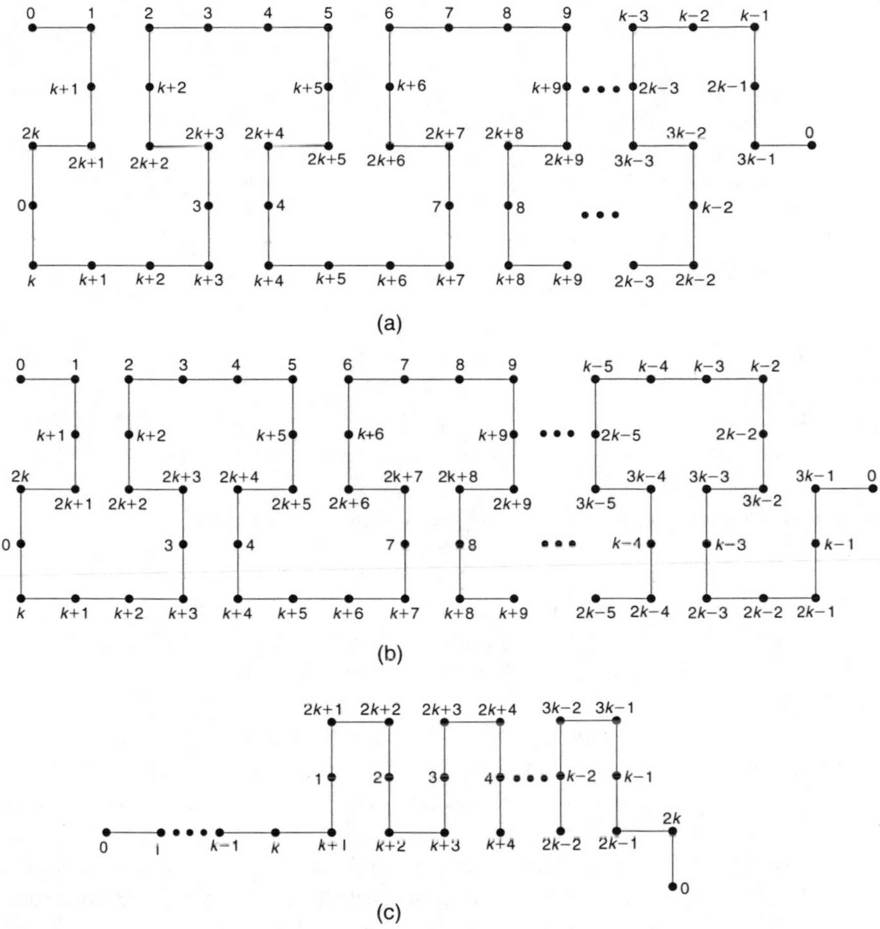

FIGURE 12.24 (a) A $D^{v}_{2k-1}(3k, k)$ with $k \equiv 1\,(\mathbf{mod}\,4)$, (b) a $D^{v}_{k-1}(3k, k)$ with $k \equiv 3\,(\mathbf{mod}\,4)$, and (c) a $D^{h}_{k}(3k, k)$ for odd k with $k \geq 5$.

Suppose that $n \geq 11$. A lattice diagram $D^{v}_{4}(n, 3)$ is shown in Figure 12.25a and a lattice diagram $D^{h}_{1}(n, 3)$ is shown in Figure 12.25b. Therefore, $P(n, 3)$ is 1-fault-tolerant Hamiltonian.

Thus, $P(n, 3)$ is 1-fault-tolerant Hamiltonian if and only if n is odd. □

THEOREM 12.17 **[246]** The generalized Petersen graph $P(n, 4)$ is 1-fault-tolerant Hamiltonian if and only if $n \neq 12$.

Proof: By the definition of $P(n, k)$, we consider $n \geq 9$. With Theorem 12.15, $P(12, 4)$ is not 1-fault-tolerant Hamiltonian. Thus, we consider the cases $n \geq 9$ and $n \neq 12$.

Suppose that $n = 9$. Obviously, $\langle 0, 1, 2, 2', 7', 7, 6, 6', 1', 5', 5, 4, 3, 3', 8', 4', 0', 0\rangle$ forms a Hamiltonian cycle for $P(9, 4) - 8$ and $\langle 0, 1, 1', 5', 0', 4', 8', 3', 7', 2', 2, 3, 4, 5, 6, 7, 8, 0\rangle$ forms a Hamiltonian cycle for $P(9, 4) - 6'$. Thus, $P(9, 4)$ is 1-fault-tolerant Hamiltonian.

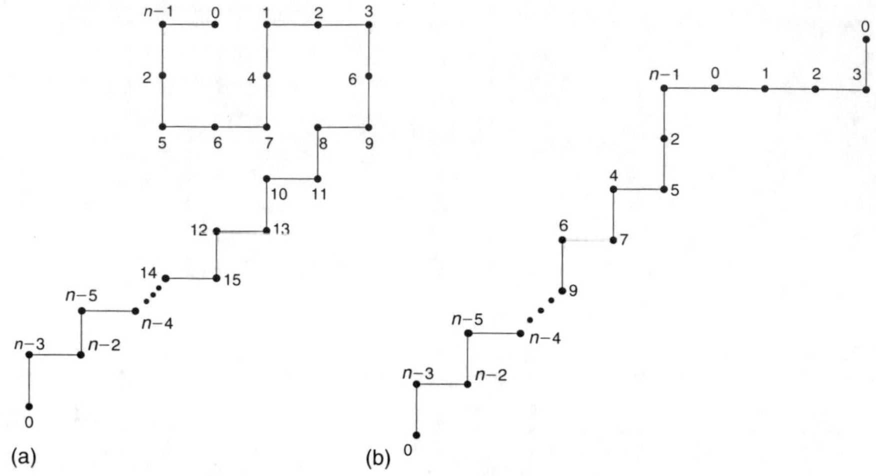

FIGURE 12.25 (a) A $D_4^v(n,3)$ and (b) a $D_1^h(n,3)$ for odd n with $n \geq 11$.

Suppose that $n = 10$. Obviously, $\langle 0, 1, 2, 3, 3', 9', 5', 1', 7', 7, 8, 8', 2', 6', 6, 5, 4, 4', 0', 0\rangle$ forms a Hamiltonian cycle for $P(10,4) - 9$ and $\langle 0, 1, 1', 5', 5, 4, 4', 8', 2', 2, 3, 3', 9', 9, 8, 7, 6, 6', 0', 0\rangle$ forms a Hamiltonian cycle for $P(10,4) - 7'$. Thus, $P(10,4)$ is 1-fault-tolerant Hamiltonian.

Suppose that $n = 11$. Obviously, $\langle 0, 1, 1', 8', 8, 9, 9', 5', 5, 6, 7, 7', 3', 10', 6', 2', 2, 3, 4, 4', 0', 0\rangle$ forms a Hamiltonian cycle for $P(11,4) - 10$ and $\langle 0, 1, 1', 5', 9', 2', 2, 3, 3', 10', 6', 6, 5, 4, 4', 0', 7', 7, 8, 9, 10, 0\rangle$ forms a Hamiltonian cycle for $P(11,4) - 8'$. Thus, $P(11,4)$ is 1-fault-tolerant Hamiltonian.

Suppose that $n \geq 13$. In Figures 12.26a through 12.26d, we have four lattice diagrams $D_{n-1}^v(n,4)$ depending on the value of $n\,(\mathbf{mod}\,4)$. In Figures 12.26e through 12.26h, we have four lattice diagrams $D_{n-3}^h(n,4)$ depending on the value of $n\,(\mathbf{mod}\,4)$. Thus, $P(n,4)$ is 1-fault-tolerant Hamiltonian.

Therefore, $P(n,4)$ is 1-fault-tolerant Hamiltonian if and only if $n \neq 12$. □

With Theorems 12.14, 12.16, and 12.17, we can recognize those 1-fault-tolerant Hamiltonian Petersen graphs $P(n,k)$ with $k \leq 4$. Now, we consider the case $k \geq 5$. By Lemma 12.6, $P(n,k)$ is not 1-fault-tolerant Hamiltonian if n is even and k is odd. We will prove that $P(n,k)$ is 1-fault-tolerant Hamiltonian if (1) $k \geq 5$ is odd and n is odd with n sufficiently large with respect to k and (2) k is even with $k \geq 6$ and n is sufficiently large with respect to k. Again, we will use the amalgamating mechanism proposed by Alspach et al. [11] and described earlier, to obtain suitable lattice diagrams.

We first consider the case that k is even with $k \geq 6$. We will use four basic lattice diagrams. We use $D(4k,k)$ and $D(4k+2,k)$ to denote the lattice diagrams shown in Figures 12.22a and 12.22b, respectively. Moreover, we use $D_k^v(2k+1,k)$ and $D_1^h(2k+2,k)$ to denote the lattice diagrams shown in Figures 12.27a and 12.27b, respectively. For illustration, we amalgamate a $D_6^v(13,6)$ and a $D(26,6)$ to obtain a $D_6^v(38,6)$ in Figure 12.27c. Note that $38 = 13 + 26 - 1$, because a vertex is duplicated

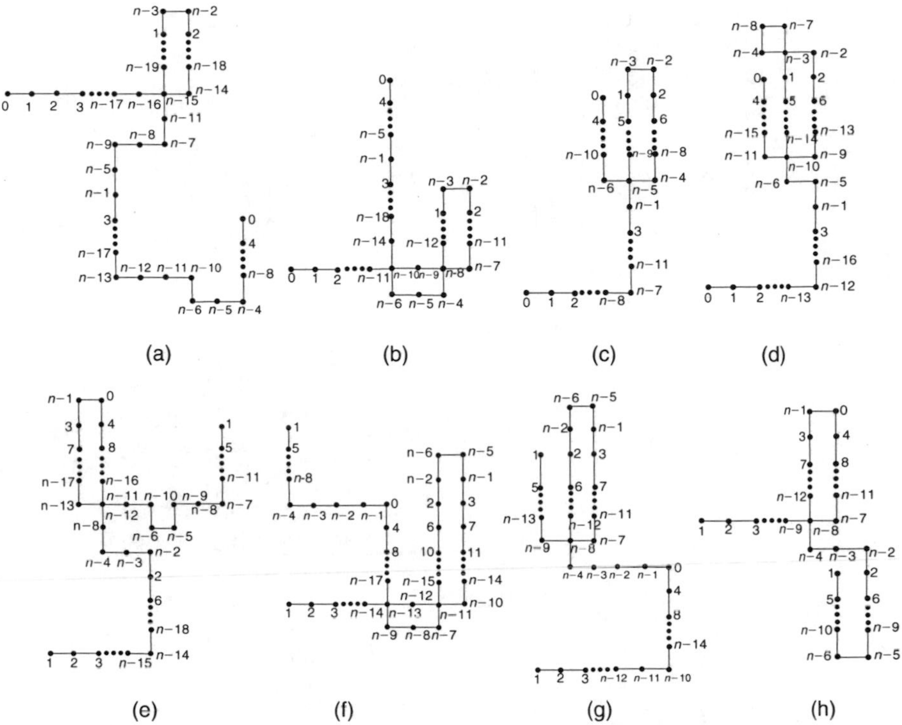

FIGURE 12.26 (a) A $D^{v}_{n-1}(n,4)$ with $n \equiv 0\,(\mathbf{mod}\,4)$, (b) a $D^{v}_{n-1}(n,4)$ with $n \equiv 1\,(\mathbf{mod}\,4)$, (c) a $D^{v}_{n-1}(n,4)$ with $n \equiv 2\,(\mathbf{mod}\,4)$, (d) a $D^{v}_{n-1}(n,4)$ with $n \equiv 3\,(\mathbf{mod}\,4)$, (e) a $D^{h}_{n-3}(n,4)$ with $n \equiv 0\,(\mathbf{mod}\,4)$, (f) a $D^{h}_{n-3}(n,4)$ with $n \equiv 1\,(\mathbf{mod}\,4)$, (g) a $D^{h}_{n-3}(n,4)$ with $n \equiv 2\,(\mathbf{mod}\,4)$, and (h) a $D^{h}_{n-3}(n,4)$ with $n \equiv 3\,(\mathbf{mod}\,4)$.

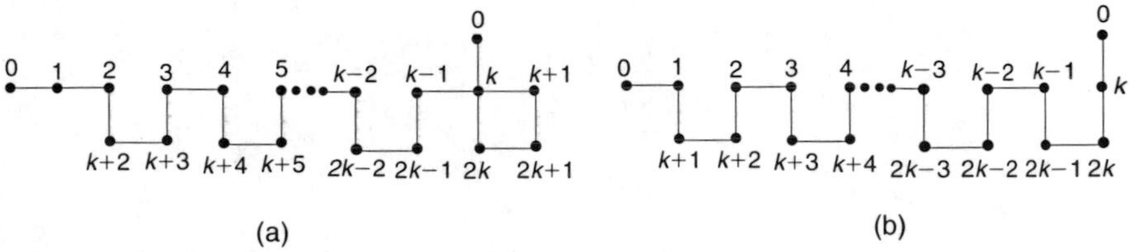

FIGURE 12.27 (a) A $D^{v}_{k}(2k+1,k)$ and (b) a $D^{h}_{1}(2k+2,k)$.

during the amalgamating mechanism. Similarly, we amalgamate a $D^{h}_{1}(14,6)$ and a $D(24,6)$ to obtain a $D^{h}_{1}(37,6)$ in Figure 12.27c.

Suppose that $n \geq 2k+2+(4k-1)(4k+1)$. Since gcd $(4k-1, 4k+1) = 1$, there exist nonnegative integers r and s such that $n = 2k+1+r(4k-1)+s(4k+1)$. Similarly, there exist nonnegative integers t and u such that $n = 2k+2+t(4k-1)+u(4k+1)$. We can amalgamate one copy of $D^{v}_{k}(2k+1,k)$, r copies of $D(4k,k)$, and s copies of $D(4k+2,k)$ to make a lattice diagram $D^{v}_{k}(n,k)$. Thus, $P(n,k)-k$ is Hamiltonian. Again, we can amalgamate one copy of $D^{h}_{1}(2k+2,k)$, t copies of $D(4k,k)$, and

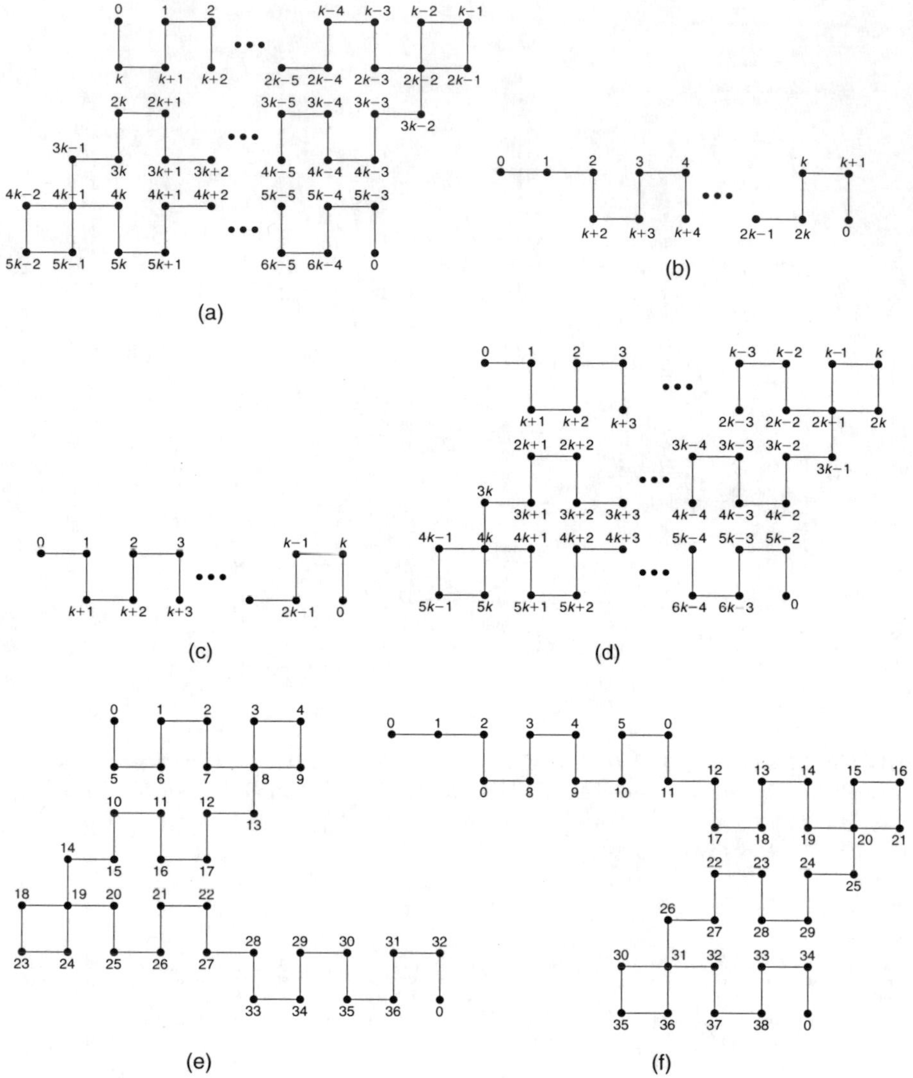

FIGURE 12.28 (a) A $D_0^v(6k-3,k)$, (b) a $D_1^h(2k+1,k)$, (c) a $D(2k,k)$, (d) a $D(6k-2,k)$, (e) a $D_0^h(27,5)$, and (f) a $D_1^h(39,5)$.

u copies of $D(4k+2,k)$ to make a lattice diagram $D_1^h(n,k)$. Therefore, $P(n,k)-1'$ is Hamiltonian.

Thus, we have the following theorem.

THEOREM 12.18 **[246]** Assume that k is even with $k \geq 6$. Then $P(n,k)$ is 1-fault-tolerant Hamiltonian if $n \geq 2k+2+(4k-1)(4k+1)$.

Now, we consider both k and n to be odd integers with $k \geq 5$. We will use four basic lattice diagrams. We use $D_0^v(6k-3,k)$ and $D_1^h(2k+1,k)$ to denote the lattice diagrams shown in Figures 12.28a and 12.28b respectively, and we use $D(2k,k)$ $D(6k-2,k)$ to denote the lattice diagrams in Figures 12.28c and 12.28d, respectively. For illustration, we amalgamate a $D_0^h(27,5)$ and a $D(10,5)$ to obtain a $D_0^h(37,5)$ in Figure 12.28e. We observe that $37 = 27 + 10$, which is different from the previous case. We note that the two vertices labeled with 0 in $D_0^h(27,5)$ are the terminal vertices of the open Eulerian trail. We identify one of the vertices labeled with 0 in $D_0^h(27,5)$ with one of the vertices labeled with 0 in $D(10,5)$ to obtain a $D_0^h(37,5)$. Similarly, we amalgamate a $D_1^h(11,5)$ with a $D(28,5)$ to obtain a $D_1^h(39,5)$ in Figure 12.28f.

Suppose that $n \geq 6k-3+(2k)(6k-2)$ and n is odd. Since gcd $(2k, 6k-2) = 2$, there exist nonnegative integers r and s such that $n = 6k-3+r(2k)+s(6k-2)$. Moreover, there exist nonnegative integers t and u such that $n = 6k-3+t(2k)+u(6k-2)$. We can amalgamate one copy of $D_0^v(6k-3,k)$, r copies of $D(2k,k)$, and s copies of $D(6k-2,k)$ to make a lattice diagram $D_0^v(n,k)$. Similarly, we can amalgamate one copy of $D_1^h(2k+1,k)$, t copies of $D(2k,k)$, and u copies of $D(6k-2,k)$ to make a lattice diagram $D_1^h(n,k)$. Therefore, $P(n,k)$ is 1-fault-tolerant Hamiltonian.

THEOREM 12.19 **[246]** Suppose that k and n are odd integers with $k \geq 5$. Then $P(n,k)$ is 1-fault-tolerant Hamiltonian if $n \geq 6k-3+(2k)(6k-2)$.

12.6 HONEYCOMB RECTANGULAR DISKS

In this section, we introduce another family of cubic 1-fault-tolerant Hamiltonian graphs. As discussed in Chapter 2, three different honeycomb meshes—the honeycomb rectangular mesh, the honeycomb rhombic mesh, and the honeycomb hexagonal mesh—are introduced by Stojmenovic [288]. Most of these meshes are not regular. Moreover, such meshes are not Hamiltonian unless they are small in size [251]. To remedy these drawbacks, the honeycomb rectangular torus, the honeycomb rhombic torus, and the honeycomb hexagonal torus are proposed [288]. Any such torus is 3-regular. Moreover, all honeycomb tori are not planar. Here, we propose a variation of honeycomb meshes, called the *honeycomb rectangular disk*. A honeycomb rectangular disk HReD(m,n) is obtained from the honeycomb rectangular mesh HReM(m,n) by adding a boundary cycle.

The *honeycomb rectangular mesh* HReM(m,n) is the graph with

$$
\begin{aligned}
V(\text{HReM}(m,n)) &= \{(i,j) \mid 0 \leq i < m, 0 \leq j < n\} \\
E(\text{HReM}(m,n)) &= \{((i,j),(k,l)) \mid i = k \text{ and } j = l \pm 1\} \\
&\quad \cup \{((i,j),(k,l)) \mid j = l \text{ and } k = i+1 \text{ with } i+j \text{ is odd}\}
\end{aligned}
$$

For example, the honeycomb rectangular mesh HReM$(8,6)$ is shown in Figure 12.29.

For easy presentation, we first assume that m and n are positive even integers with $m \geq 4$ and $n \geq 6$. A *honeycomb rectangular disk* HReD(m,n) is the graph obtained

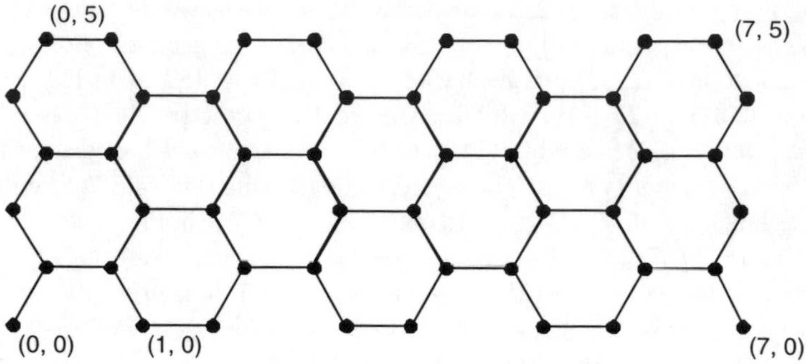

FIGURE 12.29 The honeycomb rectangular mesh HReM(8, 6).

from HReM(m, n) by adding a boundary cycle. More precisely,

$$\begin{aligned}
V(\mathrm{HReD}(m,n)) &= (\{(i,j) \mid 0 \le i < m, -1 \le j \le n\} - \{(0,-1),(m-1,-1)\}) \\
&\quad \cup \{(i,j) \mid i \in \{-1,m\}, 0 < j < n,\ j \text{ is even}\} \\
E(\mathrm{HReD}(m,n)) &= \{((i,j),(k,l)) \mid i = k \text{ and } j = l \pm 1\} \\
&\quad \cup \{((i,j),(k,l)) \mid j = l \text{ and } k = i+1 \text{ with } i+j \text{ is odd}\} \\
&\quad \cup \{(i,j),(k,l)) \mid i = k \in \{-1,m\} \text{ and } j = l \pm 2\} \\
&\quad \cup \{((0,0),(-1,2)),((-1,n-2),(0,n)),((m-1,n),(m,n-2))\} \\
&\quad \cup \{((m,2),(m-1,0)),((m-1,0),(m-2,-1)),((1,-1),(0,0))\}
\end{aligned}$$

For example, the honeycomb rectangular disk HReD(8, 6) is shown in Figure 12.30. Obviously, HReM(m, n) is a subgraph of HReD(m, n). Moreover, any honeycomb rectangular disk is a planar 3-regular graph.

We may also define HReD(m, n) for $m \geq 4$ and $n \geq 4$ by adding a boundary cycle to HReM(m, n). For example, the HReD(6, 4) is shown in Figure 12.31. By brute force, we can check whether such a honeycomb rectangular disk is 1-edge fault-tolerant Hamiltonian but not 1-vertex fault-tolerant Hamiltonian.

We may use a similar concept to define other cases of HReD(m, n). For example, the HReD(5, 6), HReD(5, 7), and HReD(6, 7) are shown in Figure 12.32. Obviously, HReM(m, n) is a subgraph of HReD(m, n). Moreover, any HReD(m, n) is a planar 3-regular Hamiltonian graph.

The following theorem is proved by Teng et al. [298]. The proof is omitted.

THEOREM 12.20 Any honeycomb rectangular disk HReD(m, n) is 1-edge fault-tolerant Hamiltonian if $m \geq 4$ and $n \geq 4$. Moreover, any honeycomb rectangular disk HReD(m, n) is 1-vertex fault-tolerant Hamiltonian if $m \geq 4$ and $n \geq 6$.

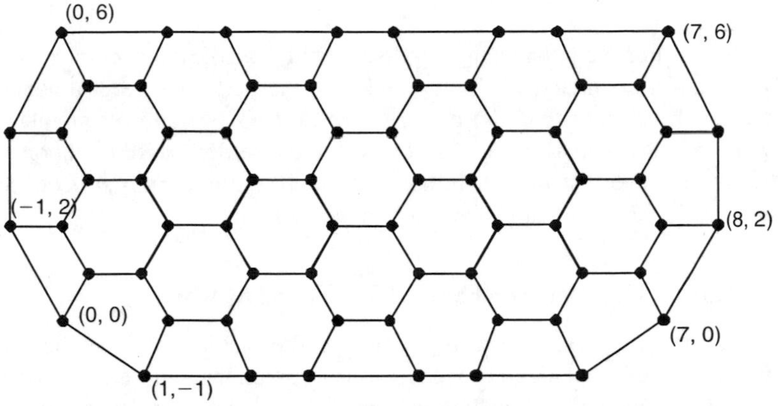

FIGURE 12.30 The honeycomb rectangular disk HReD(8, 6).

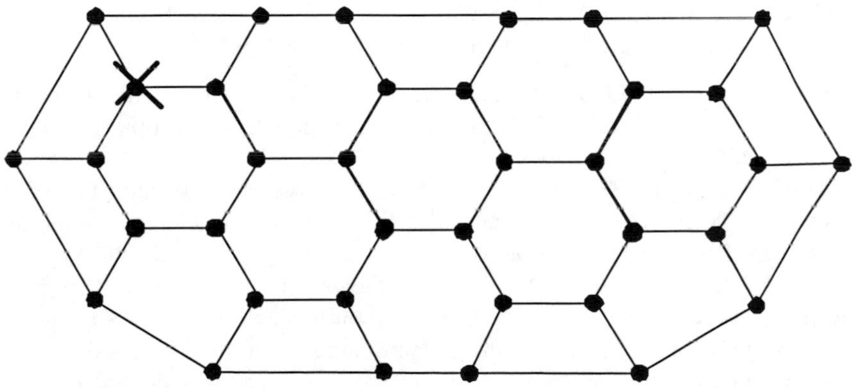

FIGURE 12.31 The honeycomb rectangular disk HReD(6, 4).

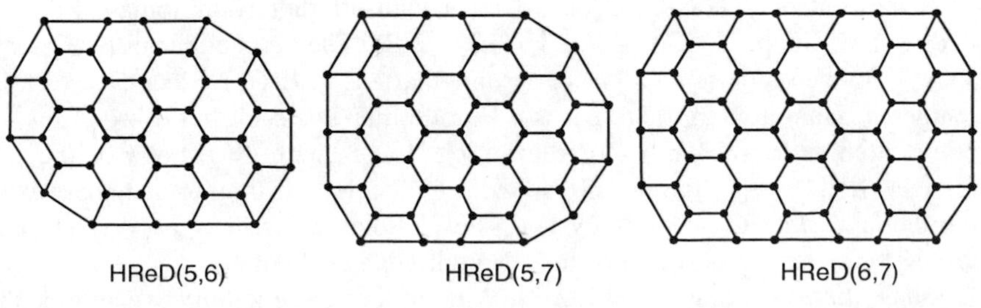

FIGURE 12.32 The honeycomb rectangular disks HReD(5, 6), HReD(5, 7), and HReD(6, 7).

Note that the honeycomb rectangular disk is built from the honeycomb rectangular mesh by adding a boundary. By Theorem 12.20, the performance of a honeycomb rectangular disk is better than the performance of a honeycomb rectangular mesh. Usually, the wireless network stations of a region are settled based on a honeycomb structure. We believe that the performance of such a honeycomb structure can be improved by adding a boundary, as we did in this section.

12.7 PROPERTIES WITH RESPECT TO THE 3-JOIN

In this section, we discuss more properties of the 3-join. Assume that G_1 is a graph with a vertex x of degree 3 and G_2 is a graph with a vertex y of degree 3. Let $G = J(G_1, N(x); G_2, N(y))$ where $N(x) = \{x_1, x_2, x_3\}$ and $N(y) = \{y_1, y_2, y_3\}$. Depending on the Hamiltonian properties of G_1 and G_2—that is, whether they are 1-vertex fault-tolerant Hamiltonian, 1-edge fault-tolerant Hamiltonian, or Hamiltonian-connected—and some local properties at x and y, we may have various Hamiltonian properties of G, as stated in the following lemmas proved by Kao et al. [192].

LEMMA 12.7 $G = J(G_1, N(x); G_2, N(y))$ is 1-edge fault-tolerant Hamiltonian if and only if both G_1 and G_2 are 1-edge fault-tolerant Hamiltonian.

Proof: Suppose that G is 1-edge fault-tolerant Hamiltonian. We claim that both G_1 and G_2 are 1-edge fault-tolerant Hamiltonian. By symmetry, it is sufficient to prove that G_1 is 1-edge fault-tolerant Hamiltonian.

Let e be any edge of G_1. Then e is either incident to x or not. Assume that e is not incident to x. Then e is an edge of G. Since G is 1-edge fault-tolerant Hamiltonian, there exists a Hamiltonian cycle C of $G - e$. Since there are exactly three edges of G joining $V(G_1) - \{x\}$ to $V(G_2) - \{y\}$, C can be written as $\langle x_i, P, x_j, y_j, Q, y_i, x_i \rangle$ for some $\{i, j\} \subset \{1, 2, 3\}$ with $i \neq j$, where P is a Hamiltonian path of $G_1 - x$ joining x_i to x_j. Obviously, $\langle x, x_i, P, x_j, x \rangle$ forms a Hamiltonian cycle of $G_1 - e$. Assume that e is an edge incident to x. Without loss of generality, we may assume that $e = (x, x_1)$. Since G is 1-edge fault-tolerant Hamiltonian, there exists a Hamiltonian cycle C of $G - (x_1, y_1)$. Since there are exactly three edges joining $V(G_1) - \{x\}$ to $V(G_2) - \{y\}$, C can be written as $\langle x_2, P, x_3, y_3, Q, y_2, x_2 \rangle$, where P is a Hamiltonian path of $G_1 - x$. Obviously, $\langle x, x_2, P, x_3, x \rangle$ forms a Hamiltonian cycle of $G_1 - (x, x_1)$. Then G_1 is 1-edge fault-tolerant Hamiltonian.

Assume that both G_1 and G_2 are 1-edge fault-tolerant Hamiltonian. Let e be any edge of G. Suppose that $e \notin \{(x_i, y_i) \mid i = 1, 2, 3\}$. Then e is either in $E(G_1)$ or in $E(G_2)$. Without loss of generality, we assume that e is in $E(G_1)$. Since G_1 is 1-edge fault-tolerant Hamiltonian, there exists a Hamiltonian cycle C_1 in $G_1 - e$. Thus, C_1 can be written as $\langle x, x_i, P, x_j, x \rangle$ for some $i, j \in \{1, 2, 3\}$ with $i \neq j$. Let k be the only element in $\{1, 2, 3\} - \{i, j\}$. Since G_2 is 1-edge fault-tolerant Hamiltonian, there exists a Hamiltonian cycle C_2 of $G_2 - (y, y_k)$. Thus, C_2 can be written as $\langle y, y_j, Q, y_i, y \rangle$. Obviously, $\langle x_i, P, x_j, y_j, Q, y_i, x_i \rangle$ forms a Hamiltonian cycle of $G - e$.

Assume that $e \in \{(x_i, y_i) \mid i = 1, 2, 3\}$. Without loss of generality, we assume that $e = (x_1, y_1)$. Since G_1 is 1-edge fault-tolerant Hamiltonian, there exists a Hamiltonian cycle C_1 in $G_1 - (x, x_1)$. Thus, C_1 can be written as $\langle x, x_2, P, x_3, x \rangle$. Since G_2 is

1-edge fault-tolerant Hamiltonian, there exists a Hamiltonian cycle C_2 in $G_2 - (y, y_1)$. Thus, C_2 can be written as $\langle y, y_3, Q, y_2, y\rangle$. Obviously, $\langle x_2, P, x_3, y_3, Q, y_2, x_2\rangle$ forms a Hamiltonian cycle of $G - e$. Hence, G is 1-edge fault-tolerant Hamiltonian. □

A vertex x in a graph G is *good* if $\deg_G(x) = 3$ and $G - e$ is Hamiltonian for any edge e incident to x. Let $Good(G)$ denote the set of good vertices in G. Obviously, $Good(G) = V(G)$ if G is a 3-regular 1-edge fault-tolerant Hamiltonian graph.

LEMMA 12.8 Suppose that both G_1 and G_2 are 1-vertex fault-tolerant Hamiltonian graphs. Then $G = J(G_1, N(x); G_2, N(y))$ is 1-vertex fault-tolerant Hamiltonian if and only if x is good in G_1 and y is good in G_2. Moreover, $Good(G) = (Good(G_1) \cup Good(G_2)) - \{x, y\}$ if x is good in G_1 and y is good in G_2.

Proof: We prove this lemma through the following steps:

1. Assume that G is 1-vertex fault-tolerant Hamiltonian. We claim that x is good in G_1 and y is good in G_2. By symmetry, it is sufficient to prove that x is good in G_1. Let e be any edge incident to x. Without loss of generality, we assume that $e = (x, x_1)$. Since G is 1-vertex fault-tolerant Hamiltonian, there exists a Hamiltonian cycle C in $G - y_1$. Thus, C can be written as $\langle x_2, P, x_3, y_3, Q, y_2, x_2\rangle$. Obviously, $\langle x, x_2, P, x_3, x\rangle$ forms a Hamiltonian cycle of $G_1 - (x, x_1)$. Hence, x is good in G_1.
2. Assume that x is good in G_1 and y is good in G_2. We claim that G is Hamiltonian. Since x is good in G_1, there exists a Hamiltonian cycle C_1 in $G_1 - (x, x_1)$. Thus, C_1 can be written as $\langle x, x_2, P, x_3, x\rangle$. Similarly, there exists a Hamiltonian cycle C_2 in $G_2 - (y, y_1)$. Again, C_2 can be written as $\langle y, y_3, Q, y_2, y\rangle$. Obviously, $\langle x_2, P, x_3, y_3, Q, y_2, x_2\rangle$ forms a Hamiltonian cycle of G.
3. We prove that G is 1-vertex fault-tolerant Hamiltonian. Let u be any vertex of G. Obviously, u is either in $(V(G_1) - \{x\})$ or in $(V(G_2) - \{y\})$. By symmetry, we assume that $u \in V(G_1) - \{x\}$.

   Suppose that $u \in N(x)$. Without loss of generality, we assume that $u = x_1$. Since G_1 is 1-vertex fault-tolerant Hamiltonian, there exists a Hamiltonian cycle C_1 in $G_1 - u$. Thus, C_1 can be written as $\langle x, x_2, P, x_3, x\rangle$. Since y is good in G_2, there exists a Hamiltonian cycle C_2 in $G_2 - (y, y_1)$. Thus, C_2 can be written as $\langle y, y_3, Q, y_2, y\rangle$. Obviously, $\langle x_2, P, x_3, y_3, Q, y_2, x_2\rangle$ forms a Hamiltonian cycle of $G - u$.

   Suppose that $u \notin N(x)$. Since G_1 is 1-vertex fault-tolerant Hamiltonian, there exists a Hamiltonian cycle C_1 in $G_1 - u$. Thus, C_1 can be written as $\langle x, x_i, P, x_j, x\rangle$ for some $i, j \in \{1, 2, 3\}$ with $i \neq j$. Let k be the only element in $\{1, 2, 3\} - \{i, j\}$. Since y is good in G_2, there exists a Hamiltonian cycle C_2 in $G_2 - (y, y_k)$. Thus, C_2 can be written as $\langle y, y_j, Q, y_i, y\rangle$. Obviously, $\langle x_i, P, x_j, y_j, Q, y_i, x_i\rangle$ forms a Hamiltonian cycle of $G - u$.

   Therefore, G is 1-vertex fault-tolerant Hamiltonian.
4. We claim that $Good(G) \subseteq (Good(G_1) \cup Good(G_2)) - \{x, y\}$. Let u be any good vertex in G. Note that $\deg_G(u) = 3$ and $u \in (V(G_1) - \{x\}) \cup (V(G_2) - \{y\})$. Without loss of generality, we assume that $u \in V(G_1) - \{x\}$. Thus, $\deg_{G_1}(u) = 3$. Let $e = (u, v)$ be any edge of G_1 incident to u.

Suppose that $u \in N_{G_1}(x)$. Without loss of generality, we assume that $u = x_1$. Suppose that $v = x$. Since u is good in G, there exists a Hamiltonian cycle C in $G - (x_1, y_1)$. Thus, C can be written as $\langle u, P, x_i, y_i, Q, y_j, x_j, R, u \rangle$ where $i, j \in \{2, 3\}$ with $i \neq j$. Obviously, $\langle u, P, x_i, x, x_j, R, u \rangle$ forms a Hamiltonian cycle of $G_1 - (u, v)$. Suppose that $v \neq x$. Since u is good in G, there exists a Hamiltonian cycle C in $G - (u, v)$. Thus, C can be written as $\langle u, y_1, Q, y_i, x_i, R, u \rangle$, where $i \in \{2, 3\}$. Obviously, $\langle u, x, x_i, R, u \rangle$ forms a Hamiltonian cycle of $G_1 - (u, v)$. Therefore, u is good in G_1.

Suppose that $u \notin N_{G_1}(x)$. Since u is good in G, there exists a Hamiltonian cycle C in $G - (u, v)$. Thus, C can be written as $\langle u, P, x_i, y_i, Q, y_j, x_j, R, u \rangle$, where $i, j \in \{1, 2, 3\}$ with $i \neq j$. Obviously, $\langle u, P, x_i, x, x_j, R, u \rangle$ forms a Hamiltonian cycle of $G_1 - (u, v)$. Hence, u is good in G_1.

Therefore, $Good(G) \subseteq (Good(G_1) \cup Good(G_2)) - \{x, y\}$.

5. We claim that $Good(G_1) - \{x\} \subseteq Good(G)$. Suppose that u is good in G_1 with $u \neq x$. Thus, $\deg_G(u) = 3$. Let $e = (u, v)$ be any edge of G incident to u.

   Suppose that $u = x_i$ and $e = (x_i, y_i)$ for some i with $i \in \{1, 2, 3\}$. Without loss of generality, we assume that $i = 1$. Since u is good in G_1, there exists a Hamiltonian cycle C_1 in $G_1 - (x_1, x)$. Thus, C_1 can be written as $\langle x, x_2, P, x_3, x \rangle$. Since y is good in G_2, there exists a Hamiltonian cycle C_2 in $G_2 - (y, y_1)$. Thus, C_2 can be written as $\langle y, y_3, Q, y_2, y \rangle$. Obviously, $\langle x_2, P, x_3, y_3, Q, y_2, x_2 \rangle$ forms a Hamiltonian cycle in $G - e$.

   Suppose that $u \neq x_i$ or $e \neq (x_i, y_i)$ for any i with $i \in \{1, 2, 3\}$. Since u is good in G_1, there exists a Hamiltonian cycle C_1 in $G_1 - e$. Thus, C_1 can be written as $\langle u, P_1, x_i, x, x_j, P_2, u \rangle$, where $i, j \in \{1, 2, 3\}$ with $i \neq j$. (Note that the length of P_t is 0 if $u = x_t$ for $t \in \{1, 2\}$.) Let k be the only element in $\{1, 2, 3\} - \{i, j\}$. Since y is good in G_2, there exists a Hamiltonian cycle C_2 in $G_2 - (y, y_k)$. Thus, C_2 can be written as $\langle y, y_i, Q, y_j, y \rangle$. Obviously, $\langle u, P_1, x_i, y_i, Q, y_j, x_j, P_2, u \rangle$ forms a Hamiltonian cycle in $G - e$.

   Therefore, $Good(G_1) - \{x\} \subseteq Good(G)$.
6. Similar to Step 5, we have $Good(G_2) - \{y\} \subseteq Good(G)$.
7. Combining Steps 4–6, $Good(G) = (Good(G_1) \cup Good(G_2)) - \{x, y\}$. □

LEMMA 12.9 Let $G = J(G_1, N(x); G_2, N(y))$ with $y \in Good(G_2)$. Suppose that a and b are two distinct vertices of G_1 such that $a \neq x$ and $b \neq x$. Then there exists a Hamiltonian path of G_1 joining a to b if and only if there exists a Hamiltonian path of G joining a to b.

Proof: Suppose that there exists a Hamiltonian path P_1 of G_1 joining a to b. We can write P_1 as $\langle a, P_1^1, x_i, x, x_j, P_1^2, b \rangle$ with $\{i, j\} \subset \{1, 2, 3\}$ and $i \neq j$. Note that the length of P_1^1 and the length of P_1^2 could be 0. Let k be the only element in $\{1, 2, 3\} - \{i, j\}$. Since y is good in G_2, there exists a Hamiltonian cycle C_2 in $G_2 - (y, y_k)$. We can write C_2 as $\langle y, y_i, P_2, y_j, y \rangle$. Obviously, $\langle a, P_1^1, x_i, y_i, P_2, y_j, x_j, P_1^2, b \rangle$ forms a Hamiltonian path of G joining a to b.

Suppose that there exists a Hamiltonian path P of G joining a to b. Since there are exactly three edges—namely, (x_1, y_1), (x_2, y_2), and (x_3, y_3)—between $V(G_1) - \{x\}$ and $V(G_2) - \{y\}$ in G, P can be written as $\langle a, P_1, x_i, y_i, P_2, y_j, x_j, P_3, b \rangle$ for some

$\{i, j\} \subset \{1, 2, 3\}$ with $i \neq j$. Note that the length of P_1 and the length of P_3 could be 0. Obviously, $\langle a, P_1, x_i, x, x_j, P_3, b\rangle$ forms a Hamiltonian path in G_1 joining a to b. □

A vertex x in a graph G is *nice* if it is good in G with the following property: *let* $N(x) = \{x_1, x_2, x_3\}$ *be the neighborhood of* x *in* G. *For any* $i \in \{1, 2, 3\}$, *there exists a Hamiltonian path of* $G - (x, x_i)$ *joining* u *to* x_i *for any vertex* u *of* G *with* $u \notin \{x, x_i\}$.

We use $Nice(G)$ to denote the set of nice vertices in G.

LEMMA 12.10 Suppose that both G_1 and G_2 are Hamiltonian-connected graphs, $x \in Nice(G_1)$, and $y \in Nice(G_2)$. Then $G = J(G_1, N(x); G_2, N(y))$ is Hamiltonian-connected.

Proof: To prove that G is Hamiltonian-connected, we want to show that there exists a Hamiltonian path joining a to b for any $a, b \in V(G)$ with $a \neq b$.

By symmetry, we need to consider only the following cases:

Case 1. $a, b \in V(G_1)$. Since G_1 is Hamiltonian-connected, there exists a Hamiltonian path $\langle a, P_1, x_i, x, x_j, P_2, b\rangle$ in G_1. Let k be the only element in $\{1, 2, 3\} - \{i, j\}$. Since y is good, there exists a Hamiltonian cycle $\langle y, y_i, Q, y_j, y\rangle$ in $G_2 - (y, y_k)$. Obviously, $\langle a, P_1, x_i, y_i, Q, y_j, x_j, P_2, b\rangle$ is a Hamiltonian path in G.

Case 2. $a \in V(G_1) - N(x)$ and $b \in V(G_2) - N(y)$. Since x is nice in G_1, $G_1 - (x, x_1)$ contains a Hamiltonian path P_1 joining a to x_1. Without loss of generality, P_1 can be written as $\langle a, P_1^1, x_2, x, x_3, P_1^2, x_1\rangle$. Since y is nice in G_2, $G_2 - (y, y_2)$ contains a Hamiltonian path P_2 joining b to y_2. P_2 can be written as $\langle y_2, P_2^1, y_j, y, y_k, P_2^2, b\rangle$ with $\{j, k\} = \{1, 3\}$. Suppose that $j = 1$. Then $k = 3$. Obviously, $\langle a, P_1^1, x_2, y_2, P_2^1, y_1, x_1, (P_1^2)^{-1}, x_3, y_3, P_2^2, b\rangle$ is a Hamiltonian path in G that joins a to b. Suppose that $j = 3$. Then $k = 1$. Obviously, $\langle a, P_1^1, x_2, y_2, P_2^1, y_3, x_3, P_1^2, x_1, y_1, P_2^2, b\rangle$ is a Hamiltonian path in G that joins a to b.

Case 3. $a \in N(x)$ and $b \in V(G_2)$. Without loss of generality, we assume that $a = x_1$.

Suppose that $b \neq y_1$. Since x is nice in G_1, there exists a Hamiltonian path P_1 of $G_1 - (x, x_2)$ joining a to x_2. We can write P_1 as $\langle a, x, x_3, Q_1, x_2\rangle$. Suppose that $b \neq y_2$. Since y is nice in G_2, there exists a Hamiltonian path P_2 of $G_2 - (y, y_1)$ joining y_1 to b. Thus, P_2 can be written as $\langle y_1, Q_2^1, y_2, y, y_3, Q_2^2, b\rangle$ or $\langle y_1, Q_2^1, y_3, y, y_2, Q_2^2, b\rangle$. Suppose that $P_2 = \langle y_1, Q_2^1, y_2, y, y_3, Q_2^2, b\rangle$. Then $\langle a, y_1, Q_2^1, y_2, x_2, Q_1^{-1}, x_3, y_3, Q_2^2, b\rangle$ is a Hamiltonian path joining a to b in G. Suppose that $P_2 = \langle y_1, Q_2^1, y_3, y, y_2, Q_2^2, b\rangle$. Then $\langle a, y_1, Q_2^1, y_3, x_3, Q_1, x_2, y_2, Q_2^2, b\rangle$ is a Hamiltonian path joining a to b in G.

Suppose that $b = y_1$. Since x is good in G_1, there exists a Hamiltonian cycle C_1 of $G - (x, x_2)$. Obviously, C_1 can be written as $\langle x, x_1 = a, Q_1, x_3, x\rangle$. Since y is good in G_2, there exists a Hamiltonian cycle $C_2 - (y, y_2)$. Obviously, C_2 can be written as $\langle y, y_3, Q_2, y_1 = b\rangle$. Then $\langle a, Q_1, x_3, y_3, Q_2, b\rangle$ is a Hamiltonian path joining a to b in G. □

LEMMA 12.11 Let G_1 be a Hamiltonian-connected graph with a nice vertex x and K_4 be the complete graph defined on $\{y, y_1, y_2, y_3\}$ and $G = J(G_1, N(x); K_4, N(y))$. Then G is Hamiltonian-connected. Moreover, $\{y_1, y_2, y_3\} \subseteq Nice(G)$.

Proof: Obviously, K_4 is Hamiltonian-connected. It is easy to check whether $Nice(K_4) = V(K_4)$. Using Lemma 12.10, G is Hamiltonian-connected. Now, we show that $\{y_1, y_2, y_3\} \subseteq Nice(G)$. Using symmetry, we need to verify only that y_1 is nice in G. Obviously, $\deg_G(y_1) = 3$.

We first claim that y_1 is good in G. Therefore, we show that $G - (y_1, z)$ is Hamiltonian for any $z \in \{x_1, y_2, y_3\}$. Since x is good in G_1, there exists a Hamiltonian cycle C_1^i of $G_1(x, x_i)$ for any $i \in \{1, 2, 3\}$. We may write C_1^i as $\langle x, x_j, P_i, x_k, x\rangle$ with $\{i, j, k\} = \{1, 2, 3\}$. We set $C^i = \langle x_j, P_i, x_k, y_k, y_i, y_j, x_j\rangle$. Obviously, C^1 is a Hamiltonian cycle in $G - (y_1, x_1)$; C^2 is a Hamiltonian cycle in $G - (y_1, y_3)$; and C^3 is a Hamiltonian cycle in $G - (y_1, y_2)$. Therefore, y_1 is good in G.

Let b be any element in $N(y_1) = \{x_1, y_2, y_3\}$. To show that y_1 is nice in G, we need to find a Hamiltonian path of $G - (y_1, b)$ that joins a to b for any vertex a of G with $a \notin \{y_1, b\}$.

Case 1. $a \in V(G_1) - \{x, x_1\}$ and $b = x_1$. Since x is nice in G_1, there exists a Hamiltonian path P_1 of $G_1(x, x_1)$ joining a to x_1. We can write P_1 as $\langle a, P_1^1, x_k, x, x_j, P_1^2, x_1\rangle$ where $\{k, j\} = \{2, 3\}$. Thus, $\langle a, P_1^1, x_k, y_k, y_1, y_j, x_j, P_1^2, x_1\rangle$ is a Hamiltonian path joining a to x_1 in $G - (x_1, y_1)$.

Case 2. $a \in V(G_1) - \{x, x_1\}$ and $b = y_k$ for some $k \in \{2, 3\}$. Since x is nice in G_1, there exists a Hamiltonian path P_1 joining a to x_k in $G_1(x, x_k)$. Write P_1 as $\langle a, P_1^1, x_i, x, x_j, P_1^2, x_k\rangle$ where $\{i, j, k\} = \{1, 2, 3\}$. Thus, $\langle a, P_1^1, x_i, y_i, y_j, x_j, P_1^2, x_k, y_k\rangle$ is a Hamiltonian path of $G - (y_1, y_k)$ joining a to y_k.

Case 3. $a \in \{y_2, y_3\}$ and $b = x_1$. Since x is good in G_1, there exists a Hamiltonian cycle C_1 in $G_1(x, x_2)$. We can write C_1 as $\langle x, x_3, P_1, x_1, x\rangle$. Thus, $\langle y_2, y_1, y_3, x_3, P_1, x_1\rangle$ forms a Hamiltonian path of $G - (y_1, x_1)$ joining y_2 to x_1. Since x is good in G_1, there exists a Hamiltonian cycle C_2 in $G_1(x, x_3)$. We can write C_2 as $\langle x, x_2, P_2, x_1, x\rangle$. Thus, $\langle y_3, y_1, y_2, x_2, P_2, x_1\rangle$ is a Hamiltonian path of $G - (y_1, x_1)$ joining y_3 to x_1.

Case 4. $a = x_1$ and $b \in \{y_2, y_3\}$. Let $b = y_k$ for some $k \in \{2, 3\}$ and j be the only index in $\{2, 3\} - \{k\}$. Since x is a nice vertex in G_1, there exists a Hamiltonian path P that joins x_1 to x_k in $G_1(x, x_k)$. We can write P as $\langle x_1, x, x_j, Q, x_k\rangle$. Thus, $\langle x_1, y_1, y_j, x_j, Q, x_k, y_k\rangle$ forms a Hamiltonian path of $G - (y_1, y_k)$ joining x_1 to y_k.

Case 5. $\{a, b\} = \{y_2, y_3\}$. Without loss of generality, we may assume that $a = y_2$ and $b = y_3$. Since x is good in G_1, there exists a Hamiltonian cycle C in $G_1(x, x_2)$. We can write C as $\langle x, x_1, R, x_3, x\rangle$. Obviously, $\langle y_2, y_1, x_1, R, x_3, y_3\rangle$ is a Hamiltonian path of $G - (y_1, y_3)$ joining y_2 to y_3. □

LEMMA 12.12 Let $G = J(G_1, N(x); G_2, N(y))$. Suppose that $y \in Good(G_2)$. Let $a \in V(G_1)$ such that $a \neq x$. Then $G_1 - a$ is Hamiltonian if and only if $G - a$ is Hamiltonian.

Proof: Suppose that there exists a Hamiltonian cycle C_1 in $G_1 - a$. We can write C_1 as $\langle x, x_i, P, x_j, x\rangle$ with $i, j \in \{1, 2, 3\}$ and $i \neq j$. Let k be the only element in $\{1, 2, 3\} - \{i, j\}$. Since y is good in G_2, there exists a Hamiltonian cycle C_2 in

$G_2 - (y, y_k)$. We can write C_2 as $\langle y, y_j, Q, y_i, y\rangle$. Obviously, $\langle x_i, P, x_j, y_j, Q, y_i, x_i\rangle$ forms a Hamiltonian cycle in $G - a$.

Suppose that there exists a Hamiltonian cycle C in $G - a$. Since there are exactly three edges between $V(G_1) - \{x\}$ and $V(G_2) - \{y\}$ in G, C can be written as $\langle x_i, P, x_j, y_j, Q, y_i, x_i\rangle$ for some $i, j \in \{1, 2, 3\}$ with $i \neq j$. Obviously, $\langle x, x_i, P, x_j, x\rangle$ forms a Hamiltonian cycle in $G_1 - a$. □

LEMMA 12.13 Let $a \in V(G_1)$ such that $a \neq x$, K_4 be the complete graph defined on $\{y, y_1, y_2, y_3\}$, and $G = J(G_1, N(x); K_4, N(y))$. Then $G - y_i$ is Hamiltonian if and only if $G_1 - (x, x_i)$ is Hamiltonian. Moreover, $G - a$ is Hamiltonian if and only if $G_1 - a$ is Hamiltonian.

Proof: Suppose that there exists a Hamiltonian cycle C in $G - y_i$. We can write C as $\langle x_j, y_j, y_k, x_k, P, x_j\rangle$, where $\{i, j, k\} = \{1, 2, 3\}$. Obviously, $\langle x_j, x, x_k, P, x_j\rangle$ forms a Hamiltonian cycle in $G_1 - (x, x_i)$.

Suppose that there exists a Hamiltonian cycle C_1 in $G_1 - (x, x_i)$. We can write C_1 as $\langle x_j, x, x_k, P, x_j\rangle$, where $\{i, j, k\} = \{1, 2, 3\}$. Obviously, $\langle x_j, y_j, y_k, x_k, P, x_j\rangle$ forms a Hamiltonian cycle in $G - y_i$.

Note that $Good(K_4) = V(K_4)$. Using Lemma 12.12, $G_1 - a$ is Hamiltonian if and only if $G - a$ is Hamiltonian. □

LEMMA 12.14 Suppose that G_1 is a Hamiltonian graph and G_2 is a 1-edge fault-tolerant Hamiltonian graph. Then $G = J(G_1, N(x); G_2, N(y))$ is Hamiltonian.

Proof: Since G_1 is Hamiltonian, G_1 has a Hamiltonian cycle C_1. We can write C_1 as $\langle x_i, P_1, x_j, x, x_i\rangle$, where $i, j \in \{1, 2, 3\}$ and $i \neq j$. Let k be the only element in $\{1, 2, 3\} - \{i, j\}$. Since G_2 is 1-edge fault-tolerant Hamiltonian, $G_2 - (y, y_k)$ has a Hamiltonian cycle C_2. Thus, we can write C_2 as $\langle y_i, y, y_j, P_2, y_i\rangle$. Obviously, G has a Hamiltonian cycle $\langle x_i, P_1, x_j, y_j, P_2, y_i, x_i\rangle$. □

Suppose that $G = (V, E)$ is a graph with a vertex x of degree 3. *The 3-node expansion of G*, $Ex(G, x)$, is the graph $J(G, N(x); K_4, N(y))$, where $y \in V(K_4)$. Any vertex in $Ex(G, x) - V(G)$ is called an *expanded vertex* of G at x. Obviously, $Ex(G, x)$ is cubic if G is cubic, $Ex(G, x)$ is planar if G is planar, and $Ex(G, x)$ is connected if G is connected.

12.8 EXAMPLES OF VARIOUS CUBIC PLANAR HAMILTONIAN GRAPHS

Let U be the set of cubic planar Hamiltonian graphs, A be the set of 1-vertex fault-tolerant Hamiltonian graphs in U, B be the set of 1-edge fault-tolerant Hamiltonian graphs in U, and C be the set of Hamiltonian-connected graphs in U. With the inclusion or exclusion of the sets A, B, and C, the set U is divided into eight subsets. Using the properties of a 3-join discussed in Section 12.7, Kao et al. [192] proved that there is an infinite number of elements in each of the eight subsets.

12.8.1 $A \cap B \cap C$

Obviously, K_4 is the smallest cubic planar Hamiltonian graph. It is easy to check that K_4 is a graph in $A \cap B \cap C$. Moreover, $Nice(K_4) = V(K_4)$. Let x_1 be any vertex of K_4. Using Lemmas 12.7, 12.8, and 12.11, $Ex(K_4, x_1)$ is a graph in $A \cap B \cap C$. With Lemma 12.11, any expanded vertex of K_4 at x_1 is nice. We can recursively define a sequence of graphs as follows: let $G_1 = K_4$ and $G_2 = Ex(K_4, x_1)$. Suppose that we have defined $G_1, G_2, \ldots, G_i$ with $i \geq 2$. Let x_i be any expanded vertex of G_{i-1} at x_{i-1}. We set G_{i+1} as $Ex(G_i, x_i)$. Recursively applying Lemmas 12.7, 12.8, and 12.11, $G_i \in A \cap B \cap C$ for every $i \geq 1$. We have the following theorem.

THEOREM 12.21 There is an infinite number of planar 1-vertex fault-tolerant Hamiltonian, 1-edge fault-tolerant Hamiltonian, and Hamiltonian-connected graphs.

12.8.2 $\overline{A} \cap B \cap \overline{C}$

Let Q_3 be the three-dimensional hypercube shown in Figure 12.33a. Obviously, Q_3 is planar. It is easy to check that Q_3 is 1-edge fault-tolerant Hamiltonian. Since Q_3 is a bipartite graph, there is no cycle of length 7. Therefore, $Q_3 - x$ is not Hamiltonian for any vertex x in Q_3. Thus, Q_3 is not 1-vertex fault-tolerant Hamiltonian. Since there are four vertices in each partite set, there is no Hamiltonian path joining any two vertices of the same partite set. Therefore, Q_3 is not Hamiltonian-connected. Thus, Q_3 is a graph in $\overline{A} \cap B \cap \overline{C}$. Let x_1 be any vertex in Q_3. With Lemma 12.7, $Ex(Q_3, x_1)$ is 1-edge fault-tolerant Hamiltonian. Applying Lemma 12.13, $Ex(Q_3, x_1)$ is not 1-vertex fault-tolerant Hamiltonian. By Lemma 12.9, $Ex(Q_3, x_1)$ is not Hamiltonian-connected. Therefore, $Ex(Q_3, x_1)$ is a graph in $\overline{A} \cap B \cap \overline{C}$. We can recursively define a sequence of graphs as follows: let $G_1 = Q_3$ and $G_2 = Ex(Q_3, x_1)$. Suppose that we have defined G_1, $G_2, \ldots, G_i$ with $i \geq 2$. Let x_i be any expanded vertex of G_{i-1} at x_{i-1}. We define G_{i+1} as $Ex(G_i, x_i)$. Recursively applying Lemmas 12.7, 12.9, and 12.13, $G_i \in \overline{A} \cap B \cap \overline{C}$ for every $i \geq 1$. We have the following theorem.

THEOREM 12.22 There is an infinite number of planar graphs that are 1-edge fault-tolerant Hamiltonian, but neither 1-vertex fault-tolerant Hamiltonian nor Hamiltonian-connected.

12.8.3 $\overline{A} \cap B \cap C$

Let Q be the graph in Figure 12.33b. Obviously, Q is obtained from the graph Q_3 by a sequence of 3-vertex expansions. From Section 12.8.2, $Q_3 \in \overline{A} \cap B \cap \overline{C}$. By Lemma 12.7, Q is 1-edge fault-tolerant Hamiltonian. Applying Lemma 12.13, $Q - g$ is not Hamiltonian. Hence, Q is not 1-vertex fault-tolerant Hamiltonian. By brute force, we can check whether Q is Hamiltonian-connected. Therefore, $Q \in \overline{A} \cap B \cap C$.

THEOREM 12.23 There is an infinite number of planar graphs that are 1-edge fault-tolerant Hamiltonian and Hamiltonian-connected but not 1-vertex fault-tolerant Hamiltonian.

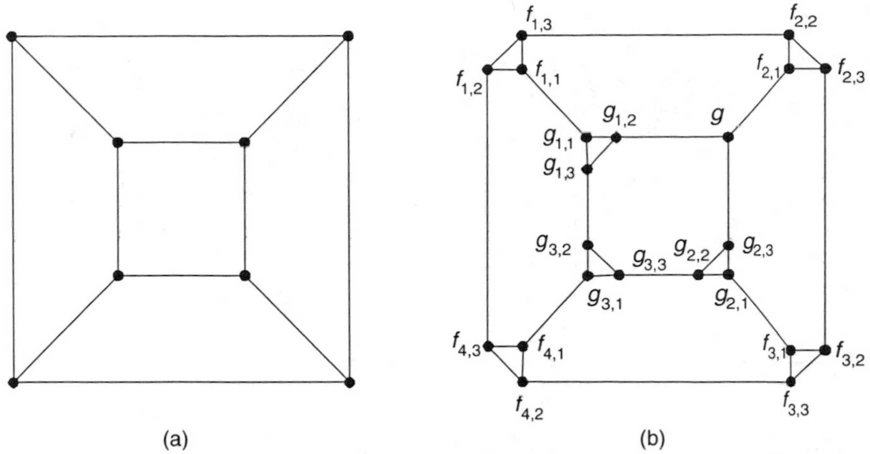

FIGURE 12.33 The graphs (a) Q_3 and (b) Q.

Proof: Let $g_{3,1}$ be the vertex of Q shown in Figure 12.33b. By brute force, we can check that $g_{3,1} \in Nice(Q)$. Let $Y = Ex(Q, g_{3,1})$. Applying Lemmas 12.7, 12.11, and 12.13, Y is a graph in $\overline{A} \cap B \cap C$. Moreover, by Lemma 12.11, any expanded vertex of Q at $g_{3,1}$ is nice. Now, we recursively define a sequence of graphs as follows: let $G_1 = Q$, $x_1 = g_{3,1}$, and $G_2 = Ex(Q, x_1)$. Suppose that we have defined G_1, $G_2, \ldots, G_i$ with $i \geq 2$. Let x_i be any expanded vertex of G_{i-1} at x_{i-1}. We define G_{i+1} as $Ex(G_i, x_i)$. Recursively applying Lemmas 12.7, 12.11, and 12.13, $G_i \in \overline{A} \cap B \cap C$ for every $i \geq 1$. □

12.8.4 $A \cap B \cap \overline{C}$

Let M denote the graph in Figure 12.34a and M_0 denote the graph in Figure 12.34b. Obviously, M is obtained from M_0 by a sequence of 3-node expansions.

LEMMA 12.15 The graph M is 1-edge fault-tolerant Hamiltonian and 1-vertex fault-tolerant Hamiltonian.

Proof: We first check that M_0 is 1-edge fault-tolerant Hamiltonian. By symmetry, we only need to prove that $M_0 - e$ is Hamiltonian where $e \in \{(s_1, s_2), (s_1, r_1), (r_1, q_2)\}$. The corresponding cycles are listed here:

$M_0 - (s_1, s_2)$	$\langle s_1, r_1, q_1, p_1, p_2, q_2, r_2, s_2, s_3, r_3, q_3, p_3, p_4, q_4, r_4, s_4, s_1 \rangle$
$M_0 - (s_1, r_1)$	$\langle s_1, s_2, r_2, q_3, p_3, p_4, p_1, p_2, q_2, r_1, q_1, r_4, q_4, r_3, s_3, s_4, s_1 \rangle$
$M_0 - (r_1, q_2)$	$\langle s_1, s_2, s_3, s_4, r_4, q_4, r_3, q_3, r_2, q_2, p_2, p_3, p_4, p_1, q_1, r_1, s_1 \rangle$

Note that M is obtained from M_0 by a sequence of 3-vertex expansions. Recursively applying Lemma 12.7, M is 1-edge fault-tolerant Hamiltonian. By Lemma 12.13, $M - v$ is Hamiltonian, where $v \in \{p_{i,j} \mid 1 \leq i \leq 4, 1 \leq j \leq 3\} \cup \{s_{i,j} \mid 1 \leq i \leq 4, 1 \leq j \leq 3\}$.

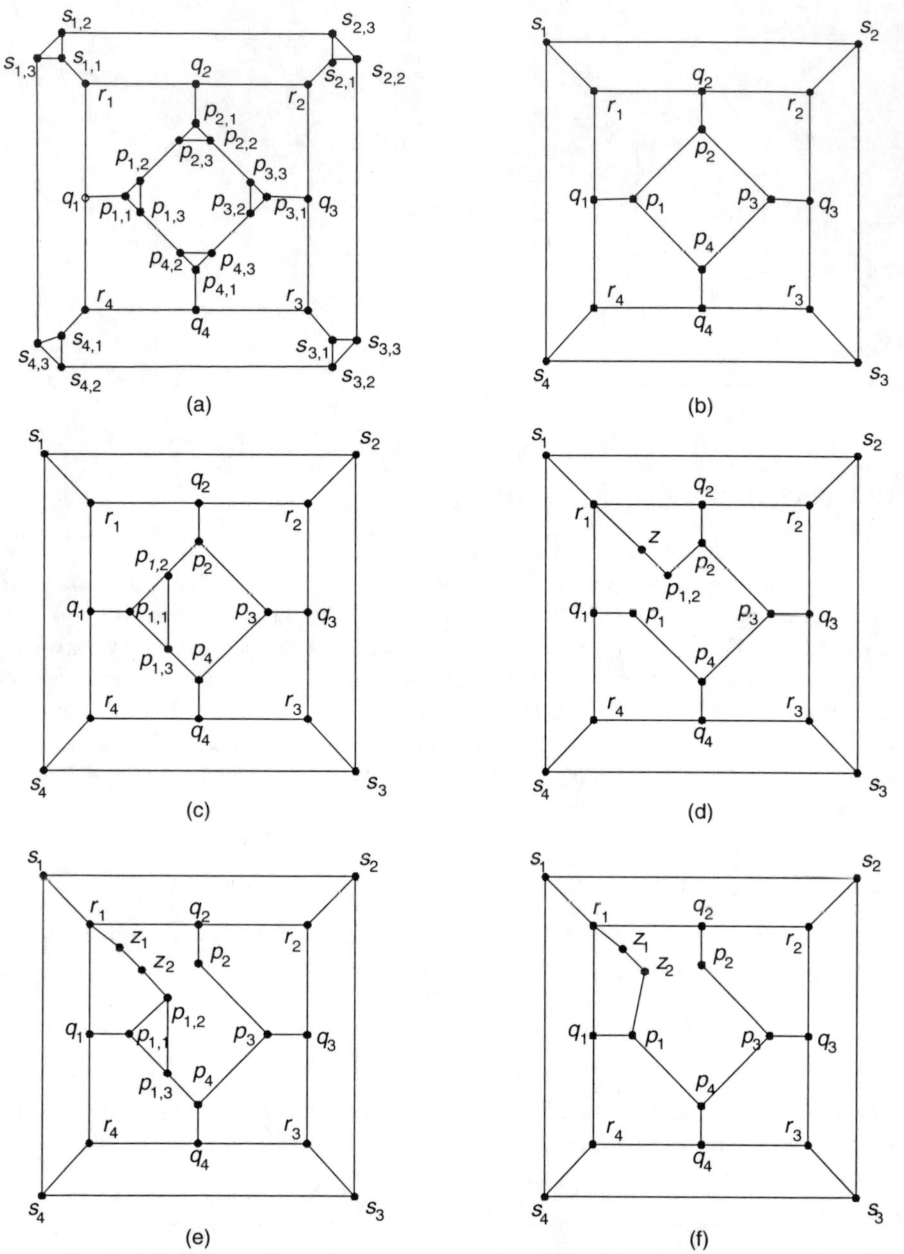

FIGURE 12.34 Graphs (a) M, (b) M_0, (c) M_1, (d) M_2, (e) M_3, and (f) M_4.

To prove that M is 1-vertex fault-tolerant Hamiltonian, we only need to check whether $M - r_1$ and $M - q_1$ are Hamiltonian by the symmetric property of M. By Lemma 12.13, it suffices to show that $M_0 - r_1$ and $M_0 - q_1$ are Hamiltonian. Obviously, $\langle q_1, p_1, p_2, q_2, r_2, s_2, s_1, s_4, s_3, r_3, q_3, p_3, p_4, q_4, r_4, q_1\rangle$ is a Hamiltonian cycle of $M_0 - r_1$, and $\langle r_1, q_2, r_2, s_2, s_3, r_3, q_3, p_3, p_2, p_1, p_4, q_4, r_4, s_4, s_1, r_1 \rangle$ is a Hamiltonian cycle of $M_0 - q_1$. Hence, M is 1-vertex fault-tolerant Hamiltonian. □

LEMMA 12.16 There is no Hamiltonian path of M joining $p_{1,2}$ to r_1. Therefore, M is not Hamiltonian-connected.

Proof: Let M_1 be the graph shown in Figure 12.34c. Obviously, M is obtained from M_1 by a sequence of 3-node expansions. By Lemma 12.9, it suffices to show that there is no Hamiltonian path in M_1 joining $p_{1,2}$ to r_1.

We prove it by contradiction. Suppose that there exists a Hamiltonian path P in M_1 joining $p_{1,2}$ to r_1. Let $P = \langle v_1 = p_{1,2}, v_2, \ldots, v_{18} \rangle$. Then v_2 is p_2, $p_{1,1}$, or $p_{1,3}$.

Case 1. $v_2 = p_2$. Then neither $(p_{1,1}, p_{1,2})$ nor $(p_{1,2}, p_{1,3})$ is in P. However, both $(p_{1,1}, p_{1,3})$ and $(p_{1,3}, p_4)$ are in P. Hence, the graph M_2 in Figure 12.34d is Hamiltonian. Let H be any Hamiltonian cycle of M_2. Since M_2 is a planar graph, M_2 and H satisfy the *Grinberg condition* [129] of Theorem 10.7.

Thus, $2(f'_4 - f''_4) + 3(f'_5 - f''_5) + 6(f'_8 - f''_8) = 0$, where f'_i is the number of faces of length i inside H and f''_i is the number of faces of length i outside H for $i = 4, 5, 8$. Thus, $2(f'_4 - f''_4) = 0 \pmod 3$. Since $|f'_4 - f''_4| = 1$, the equation cannot hold. We arrive at a contradiction.

Case 2. v_2 is either $p_{1,1}$ or $p_{1,3}$. Then the graph M_3 shown in Figure 12.34e is Hamiltonian. By Lemma 12.9, the graph M_4 in Figure 12.34f is Hamiltonian. It is easy to see that M_4 is isomorphic to M_2. Therefore, M_4 is not Hamiltonian. Again, we arrive at a contradiction. □

THEOREM 12.24 There is an infinite number of planar graphs that are 1-edge fault-tolerant Hamiltonian and 1-vertex fault-tolerant Hamiltonian but not Hamiltonian-connected.

Proof: Let $Y = Ex(M, q_1)$. Applying Lemmas 12.15 and 12.16, $M \in A \cap B \cap \overline{C}$. By Lemmas 12.7 through 12.9, $Y \in A \cap B \cap \overline{C}$. Let $G_1 = M$, $x_1 = q_1$, and $G_2 = Ex(M, q_1)$. Suppose that we have defined $G_1, G_2, \ldots, G_i$ with $i \geq 2$. Let x_i be any expanded vertex of G_{i-1} at x_{i-1}. We define G_{i+1} as $Ex(G_i, x_i)$. Recursively applying Lemmas 12.7 through 12.9, $G_i \in A \cap B \cap \overline{C}$ for every $i \geq 1$. □

12.8.5 $A \cap \overline{B} \cap C$

Let *Eye*(2) be the graph in Figure 12.35. By brute force, we can prove that *Eye*(2) is 1-vertex fault-tolerant Hamiltonian but not 1-edge fault-tolerant Hamiltonian. More precisely, $Eye(2) - e$ is not Hamiltonian for any edge in $\{(e_1, e_2), (e_3, e_4), (e_5, e_6)\}$. By brute force, we can also check that *Eye*(2) is Hamiltonian-connected. Therefore, *Eye*(2) is a graph in $A \cap \overline{B} \cap C$. Let e_{16} be the vertex of *Eye*(2) shown in Figure 12.35. By brute force, we can check that e_{16} is a nice vertex of *Eye*(2). Let $Y = Ex(Eye(2), e_{16})$. By Lemmas 12.7, 12.8, and 12.11, Y is a graph in $A \cap \overline{B} \cap C$. By Lemma 12.11, any expanded vertex of *Eye*(2) at e_{16} is nice in Y. We recursively define a sequence of

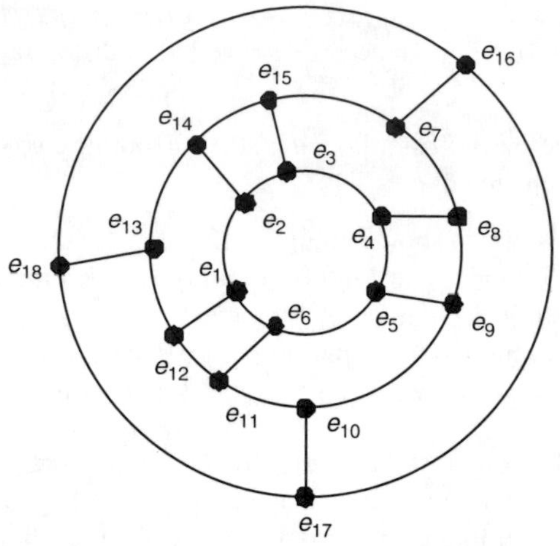

FIGURE 12.35 The graph *Eye*(2).

graphs as follows: let $G_1 = Eye(2)$ and $G_2 = Ex(Eye(2), e_{16})$. Suppose that we have defined $G_1, G_2, \ldots, G_i$ with $i \geq 2$. Let x_i be any expanded vertex of G_{i-1} at x_{i-1}. We define G_{i+1} as $Ex(G_i, x_i)$. Recursively applying Lemma 12.7, Lemma 12.8, and Lemma 12.11, $G_i \in A \cap \overline{B} \cap C$ for every $i \geq 1$. We have the following theorem.

THEOREM 12.25 There is an infinite number of planar graphs that are 1-vertex fault-tolerant Hamiltonian and Hamiltonian-connected but not 1-edge fault-tolerant Hamiltonian.

12.8.6 $A \cap \overline{B} \cap \overline{C}$

Let $N = J(Eye(2), N(e_{16}); M, N(s_{4,1}))$ be the graph in Figure 12.36. Obviously, N is a cubic planar graph. Since M is 1-edge fault-tolerant Hamiltonian, $s_{4,1} \in Good(M)$. From Section 12.8.5, we know that $e_{16} \in Good(Eye(2))$. By Lemma 12.8, N is 1-vertex fault-tolerant Hamiltonian. Again, we know that $Eye(2) \in A \cap \overline{B} \cap C$. By Lemmas 12.7, N is not 1-edge fault-tolerant Hamiltonian. Applying Lemma 12.16, there is no Hamiltonian path of M joining $p_{1,2}$ to r_1. By Lemma 12.9, there is no Hamiltonian path of N joining $p_{1,2}$ to r_1. Therefore, N is not Hamiltonian-connected. Thus, N is a graph in $A \cap \overline{B} \cap \overline{C}$.

THEOREM 12.26 There is an infinite number of planar graphs that are 1-vertex fault-tolerant Hamiltonian but neither 1-edge fault-tolerant Hamiltonian nor Hamiltonian-connected.

Proof: Let x be the vertex $p_{3,1}$ of N shown in Figure 12.36. Since M is 1-edge fault-tolerant Hamiltonian, $x \in Good(M)$. Using Lemma 12.8, $x \in Good(N)$. Let $Y = Ex(N, x)$. By Lemmas 12.7 through 12.9, Y is a graph in $A \cap \overline{B} \cap \overline{C}$. With

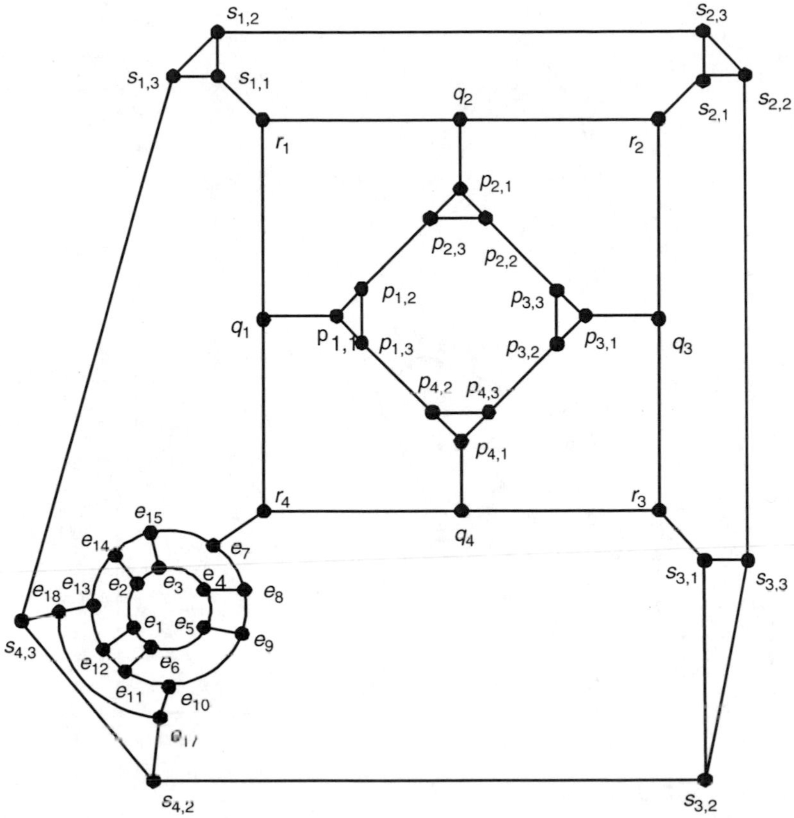

FIGURE 12.36 The graph N.

Lemma 12.8, any expanded vertex of N at x is good. We recursively define a sequence of graphs as follows: let $G_1 = N$ and $G_2 = Ex(N, x)$. Suppose that we have defined $G_1, G_2, \ldots, G_i$ with $i \geq 2$. Let x_i be any expanded vertex of G_{i-1} at x_{i-1}. We define G_{i+1} as $Ex(G_i, x_i)$. Recursively applying Lemmas 12.7 through 12.9, $G_i \in A \cap \overline{B} \cap \overline{C}$ for every $i \geq 1$. □

12.8.7 $\overline{A} \cap \overline{B} \cap C$

Let R be the graph $J(Eye(2), N(e_{16}); Q, N(g_{3,1}))$ shown in Figure 12.37. Obviously, R is a cubic planar graph. In Section 12.8.3, we know that $Q \in \overline{A} \cap B \cap C$, $Q - g$ is not Hamiltonian, and $g_{3,1}$ is nice in Q. From Section 12.8.5, we know that $Eye(2) \in A \cap \overline{B} \cap C$ and $e_{16} \in Nice(Eye(2))$. By Lemma 12.7, R is not 1-edge fault-tolerant Hamiltonian. With Lemma 12.12, $R - g$ is not Hamiltonian. Hence, R is not 1-vertex fault-tolerant Hamiltonian. By Lemma 12.10, R is Hamiltonian-connected. Thus, R is a graph in $\overline{A} \cap \overline{B} \cap C$.

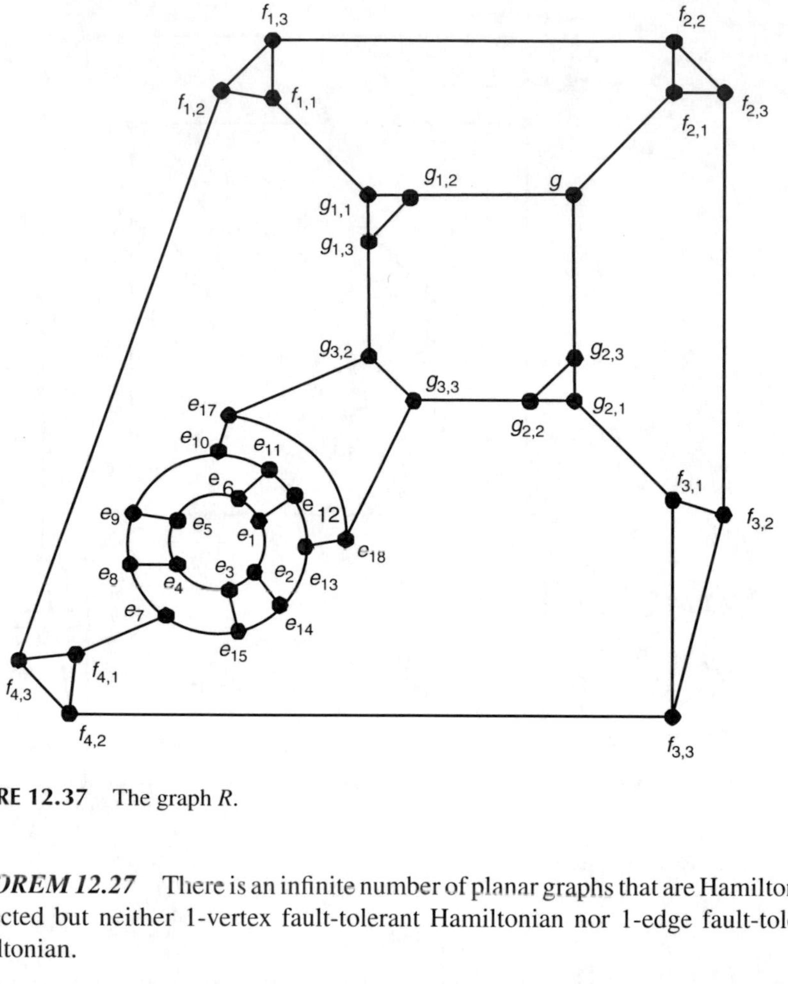

FIGURE 12.37 The graph R.

THEOREM 12.27 There is an infinite number of planar graphs that are Hamiltonian-connected but neither 1-vertex fault-tolerant Hamiltonian nor 1-edge fault-tolerant Hamiltonian.

Proof: Let x be the vertex e_{17} shown in Figure 12.37. By brute force, $x \in Nice(R)$. Let $Y = Ex(R, x)$. By Lemmas 12.7, 12.11, and 12.13, Y is a graph in $\overline{A} \cap \overline{B} \cap C$, and any expanded vertex of R at x is nice. We recursively define a sequence of graphs as follows: let $G_1 = R$ and $G_2 = Ex(R, x)$. Suppose that we have defined G_1, $G_2, \ldots, G_i$ with $i \geq 2$. Let x_i be any expanded vertex of G_{i-1} at x_{i-1}. We define G_{i+1} as $Ex(G_i, x_i)$. Recursively applying Lemmas 12.7, 12.11, and 12.13, $G_i \in \overline{A} \cap \overline{B} \cap C$ for every $i \geq 1$. □

12.8.8 $\overline{A} \cap \overline{B} \cap \overline{C}$

Let $Z = J(Eye(2), N(e_{16}); Q_3, N(f_4))$ shown in Figure 12.38. Obviously, Z is a connected planar cubic graph. From Sections 12.8.2 and 12.8.5, we know that $Q_3 \in \overline{A} \cap B \cap \overline{C}$ and $Eye(2) \in A \cap \overline{B} \cap C$. Moreover, $Q_3 - f_2$ is not Hamiltonian and there is no Hamiltonian path of Q_3 joining f_2 to g_2. By Lemma 12.7, Z is not 1-edge fault-tolerant Hamiltonian. From Section 12.8.5, we know that e_{16} is a good vertex

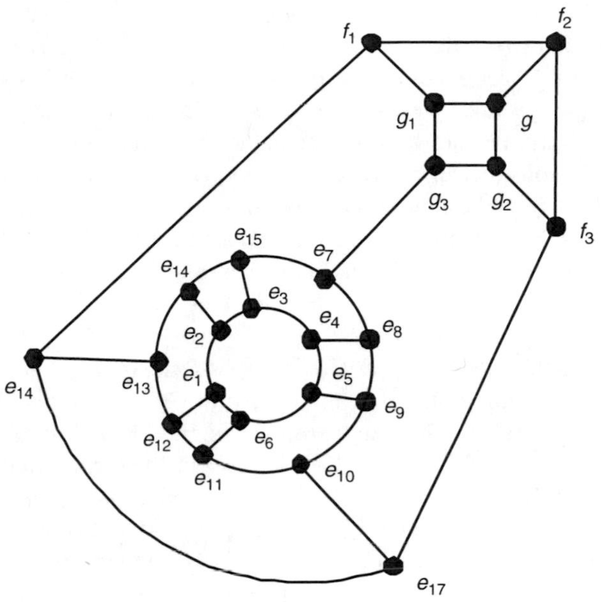

FIGURE 12.38 The graph Z.

of $Eye(2)$. By Lemma 12.12, $Z - f_2$ is not Hamiltonian. Hence, Z is not 1-vertex fault-tolerant Hamiltonian. Applying Lemma 12.9, there is no Hamiltonian path of Z joining f_2 to g_2. Therefore, Z is not Hamiltonian connected. Thus, Z is a graph in $\overline{A} \cap \overline{B} \cap \overline{C}$.

THEOREM 12.28 There is an infinite number of Hamiltonian planar graphs that are not Hamiltonian-connected, not 1-vertex fault-tolerant Hamiltonian, and not 1-edge fault-tolerant Hamiltonian.

Proof: Let x be the vertex g_1 in Z shown in Figure 12.38. Let $Y = Ex(Z, x)$. Using Lemma 12.14, Y is Hamiltonian. By Lemmas 12.7, 12.9, and 12.13, Y is a graph in $U - (A \cap B \cap C)$. We recursively define a sequence of graphs as follows: let $G_1 = Z$ and $G_2 = Ex(Z, x)$. Suppose that we have defined $G_1, G_2, \ldots, G_i$ with $i \geq 2$. Let x_i be any expanded vertex of G_{i-1} at x_{i-1}. We define G_{i+1} as $Ex(G_i, x_i)$. Recursively applying Lemmas 12.7, 12.9, and 12.13, $G_i \in U - (A \cap B \cap C)$ for every $i \geq 1$. □

12.9 HAMILTONIAN PROPERTIES OF DOUBLE LOOP NETWORKS

Since the complete digraph with n vertices is $(n-3)$-fault-tolerant Hamiltonian for $n \geq 3$, we can also study the optimal k-vertex fault-tolerant Hamiltonian digraphs, optimal k-edge fault-tolerant Hamiltonian digraphs, and k-fault-tolerant Hamiltonian digraphs. However, we do not get much of a result.

In this section, we discuss the only corresponding result we know in a digraph. A *double loop network*, denoted by $G(n; s_1, s_2)$, is a digraph with n vertices $\{0, 1, \ldots, n-1\}$ and $2n$ edges of the form $i \to i + s_1$ (**mod** n) and $i \to i + s_2$ (**mod** n), referred to as s_1-edges and s_2-edges, respectively. Double loop networks are extensions of ring networks and are widely used in the design and implementation of local area networks. To construct an n-vertex double loop network, the choice of s_1 and s_2 is a vital issue. In many studies, different values of s_1 and s_2 are chosen to achieve some desired properties. For example, Wong and Coppersmith [344] showed that choosing $s_1 = 1$ and s_2 around $\sqrt{n}$ yields a small diameter, approximate to $2\sqrt{n}$, and a small average distance, approximate to $\sqrt{n}$. Readers can refer to Refs 23 and 344 for a survey on double loop networks.

We say that a double loop network $G(n; s_1, s_2)$ is *1-edge fault-tolerant Hamiltonian* if $G(n; s_1, s_2) - e$ is Hamiltonian for any edge e in $G(n; s_1, s_2)$. Similarly, a double loop network $G(n; s_1, s_2)$ is *1-vertex fault-tolerant Hamiltonian* if $G(n; s_1, s_2) - v$ is Hamiltonian for any vertex v in $G(n; s_1, s_2)$. It can easily be verified that for $i = 1, 2$, all of the s_i edges form a Hamiltonian cycle if gcd $(n, s_i) = 1$. Thus, $G(n; s_1, s_2)$ is 1-edge fault-tolerant Hamiltonian if gcd $(s_1, n) = 1$ and gcd $(s_2, n) = 1$. However, the converse is not necessarily true. For example, $G(12; 5, 2)$ is 1-edge fault-tolerant Hamiltonian but gcd $(12, 2) \neq 1$. Furthermore, $G(n; 1, 2)$ is 1-vertex fault-tolerant Hamiltonian, but $G(n; 1, 3)$ is not 1-vertex fault-tolerant Hamiltonian if n is even.

Sung et al. [292] present necessary and sufficient conditions for 1-edge fault-tolerant Hamiltonian and 1-vertex fault-tolerant Hamiltonian double loop networks, respectively.

Let C be a cycle in $G(n; s_1, s_2)$. We write edge $e \in C$ to mean that e is an edge in C. In this section, we adopt the following notation:

$$s = s_1 - s_2 \,(\mathbf{mod}\, n)$$

$$d = \gcd(n, s)$$

$$T_i = \{j \mid 0 \le j < n \text{ and } j = i \,(\mathbf{mod}\, d)\} \text{ for } 0 \le i < d$$

$$[m] = m \,(\mathbf{mod}\, d) \quad \text{for } 0 \le m < n$$

It can be observed that vertex m is in $T_{[m]}$. Furthermore, all T_i with $0 \le i < d$ form a partition of $\{0, 1, \ldots, n-1\}$, and each T_i contains n/d elements.

Suppose that $s_1 = s_2$ (**mod** n). Obviously, $G(n; s_1, s_2)$ is 1-edge fault-tolerant Hamiltonian if and only if gcd $(n, s_1) = 1$. Thus, we consider only the case $s_1 \neq s_2$ (**mod** n).

Consider the double loop network $G(12; 3, 7)$ as an example. Thus, $s = 8$ (**mod** 12) and $d = 4$. Let C^* be a Hamiltonian cycle in $G(12; 3, 7)$ given by $\langle 0, 3, 10, 5, 8, 11, 6, 1, 4, 7, 2, 9, 0 \rangle$. It is observed that some 3-edge, say (3,6), and some 7-edge, say (0,7), are not included in C^*. Since $G(n; s_1, s_2)$ is vertex-symmetric, $G(12; 3, 7)$ is 1-edge fault-tolerant Hamiltonian. We can also observe 3-edge, 7-edge, 7-edge, 3-edge (this sequence is abbreviated as $3, 7, 7, 3$) in C*, and the previous sequence repeats in C^*. In other words, the type of edges in C^* can be represented by a periodic sequence of length 4 ($= d$); that is, $3, 7, 7, 3$. We will show in this section that any Hamiltonian cycle in $G(n; n_1, n_2)$ has such a periodic property.

Let C be a Hamiltonian cycle of $G(n; s_1, s_2)$. For any vertex i in $G(n; s_1, s_2)$, we define a function f_C as follows:

$$f_C(i) = \begin{cases} s_1 & \text{if } (i, i + s_1 \ (\mathbf{mod}\ n)) \in C \\ s_2 & \text{otherwise} \end{cases}$$

LEMMA 12.17 Let C be any Hamiltonian cycle in $G(n; s_1, s_2)$. For any two vertices j and k in T_i with $0 \le i < d$, we have $f_C(j) = f_C(k)$.

Proof: Without loss of generality, we assume that $f_C(i) = s_1$. Let $m = i + s\ (\mathbf{mod}\ n)$. Suppose that $f_C(m) = s_2$. It follows that $(i, i + s_1\ (\mathbf{mod}\ n)) \in C$ and $(m, m + s_2(\mathbf{mod}\ n)) \in C$. Equivalently, both $(i, i + s_1\ (\mathbf{mod}\ n))$ and $(m, i + s_1\ (\mathbf{mod}\ n))$ are in C. There are two edges in C entering the vertex $i + s_1\ (\mathbf{mod}\ n)$, which is contradictory to the fact that C is a Hamiltonian cycle. Thus, $f_C(m) = s_1$. Recursively, any vertex $k = i + ts\ (\mathbf{mod}\ n)$ for some integer t also satisfies $f_C(k) = s_1$. On the other hand, the set $\{i + ts\ (\mathbf{mod}\ n) \mid t \ge 0 \text{ and integer}\}$ constitutes T_i for every $0 \le i < d$. Therefore, the lemma is proved. □

For any Hamiltonian cycle C, we define a function g_C from $\{0, 1, \ldots, n-1\}$ into $\{T_i \mid 0 \le i < d\}$ as follows:

$$g_C(m) = T_{[m+f_C(m)]}$$

In other words, $g_C(m)$ denotes the set T_i to which the vertex next to m in C belongs. It follows from Lemma 12.17 that

$$g_C(T_i) = T_{[i+f_C(i)]} \qquad \text{for } 0 \le i < d$$

$$g_C(\{0, 1, \ldots, d-1\}) = \{T_i \mid \quad \text{for every } 0 \le i < d\}$$

Let H be a digraph with the vertex set $\{T_i \mid \text{for every } 0 \le i < d\}$ and edge set $\{(T_i, g_C(T_i)) \mid \text{for every } 0 \le i < d\}$. Digraph H is a directed cycle with d vertices, since otherwise it contradicts that C is a Hamiltonian cycle. For example, in $G(12; 3, 7)$ we have $g_{C^*}(T_0) = T_3$, $g_{C^*}(T_1) = T_0$, $g_{C^*}(T_2) = T_1$, $g_{C^*}(T_3) = T_2$, and H is given by $\langle (T_0, T_3), (T_3, T_2), (T_2, T_1), (T_1, T_0) \rangle$, which corresponds to the edge sequence 3, 7, 7, 3 in C^*.

LEMMA 12.18 Suppose that $G(n; s_1, s_2)$ is Hamiltonian. Then $\gcd(d, s_1) = \gcd(d, s_2) = 1$.

Proof: Let $r = \gcd(d, s_1)$. Obviously, $r = \gcd(d, s_2)$. Suppose that $G(n; s_1, s_2)$ has a Hamiltonian cycle C and $r > 1$. Let H be defined as earlier. It follows that H is a connected directed cycle. However, g_C maps the set $\{T_i \mid i \text{ is a multiple of } r\}$ onto itself. Thus, H is disconnected, which is a contradiction. Consequently, the lemma is proved. □

LEMMA 12.19 Suppose that $G(n; s_1, s_2)$ is a double loop network with $\gcd(d, s_1) = 1$. Let α, β, γ, and δ be nonnegative integers satisfying $\alpha \leq \gamma$, $\beta \leq \delta$, and $\alpha + \beta < \gamma + \delta < d$. Then $T_{[\alpha s_1 + \beta s_2]} \neq T_{[\gamma s_1 + \delta s_2]}$.

Proof: Let $G(n; s_1, s_2)$ be a double loop network with $\gcd(d, s_1) = 1$. Thus, $\gcd(d, s_2) = 1$. Let α, β, γ and δ be nonnegative integers satisfying $\alpha \leq \gamma$, $\beta \leq \delta$, $\alpha + \beta < \gamma + \delta < d$, and $T_{[\alpha s_1 + \beta s_2]} = T_{[\gamma s_1 + \delta s_2]}$. Since $T_{[\alpha s_1 + \beta s_2]} = T_{[\gamma s_1 + \delta s_2]}$, it follows that $\alpha s_1 + \beta s_2 \,(\mathbf{mod}\, d) = \gamma s_1 + \delta s_2 \,(\mathbf{mod}\, d)$. Therefore, $(\alpha - \gamma)s_1 + (\beta - \delta)s_2 = 0 \,(\mathbf{mod}\, d)$ and equivalently, $(\alpha - \gamma)(s_2 + s) + (\beta - \delta)s_2 = 0 \,(\mathbf{mod}\, d)$. Since $s = 0 \,(\mathbf{mod}\, d)$, it follows that $(\alpha + \beta - \gamma - \delta)s_2 = 0 \,(\mathbf{mod}\, d)$. Since $\gcd(d, s_2) = 1$, it follows that $\alpha + \beta - \gamma - \delta = 0 \,(\mathbf{mod}\, d)$, which contradicts $0 \leq \alpha + \beta < \gamma + \delta < d$. Hence, $T_{[\alpha s_1 + \beta s_2]} \neq T_{[\gamma s_1 + \delta s_2]}$, and the lemma is proved. □

THEOREM 12.29 $G(n; s_1, s_2)$ is Hamiltonian if and only if (1) $\gcd(d, s_1) = 1$, and (2) there exists an integer k, $0 \leq k \leq d$, satisfying $\gcd((ks_1 + (d-k)s_2)/d, n/d) = 1$.

Proof: To prove the necessity, assume that $G(n; s_1, s_2)$ is Hamiltonian. It follows from Lemma 12.18 that $\gcd(d, s_1) = 1$. Let $C = \langle x_0 = 0, x_1, \ldots, x_{n-1}, x_0 \rangle$ be a Hamiltonian cycle. We define a sequence of ordered pairs of nonnegative integers $\{(\alpha_i, \beta_i) \mid 0 \leq i \leq n-1\}$ as follows:

$$(\alpha_0, \beta_0) = (0, 0)$$
$$(\alpha_i, \beta_i) = \begin{cases} (\alpha_{i-1}, \beta_{i-1}) + (1, 0) & \text{if } x_i = x_{i-1} + s_1 \,(\mathbf{mod}\, n) \\ (\alpha_{i-1}, \beta_{i-1}) + (0, 1) & \text{otherwise} \end{cases}$$

It follows from the definition of (α_i, β_i) that $i = \alpha_i + \beta_i$ and $x_i = \alpha_i s_1 + \beta_i s_2 \,(\mathbf{mod}\, n)$. Since any two pairs (α_i, β_i), (α_j, β_j) with $0 \leq i < j < d$ satisfy $0 \leq \alpha_i + \beta_i < \alpha_j + \beta_j$, it follows from Lemma 12.19 that $T_{[x_i]} \neq T_{[x_j]}$. By the pigeonhole principle, $\{T_i \mid 0 \leq i < d\} = \{T_{[x_i]} \mid 0 \leq i < d\}$. Since $s = 0 \,(\mathbf{mod}\, d)$ and $x_d = ds_1 - \beta_d s \,(\mathbf{mod}\, d)$, we have $T_{[x_d]} = T_{[x_0]}$. Furthermore, it follows from Lemma 12.17 that $T_{[x_{d+1}]} = T_{[x_1]}$. Recursively, we have $T_{[x_j]} = T_{[x_{j\,(\mathbf{mod}\, d)}]}$. Let $k = \alpha_d = \mid \{i \mid fc(i) = s_1, 0 \leq i < d\} \mid$. It follows from Lemma 12.17 that $x_{td} = t(ks_1 + (d-k)s_2) \,(\mathbf{mod}\, n)$ for all $1 \leq t < n/d$. Suppose that $\gcd(n/d, x_d/d) = r > 1$. Then $n/d = ar$, $x_d/d = br$ for some integers a, b with $\gcd(a, b) = 1$. Obviously, $a < n/d$. We note that $ax_d/d = abr = bn/d$ is a multiple of n/d. That is, ax_d is a multiple of n which contradicts $x_{td} = tx_d \neq 0 \,(\mathbf{mod}\, n)$ for all $1 \leq t < n/d$. Thus, $\gcd(x_d/d, n/d) = \gcd((ks_1 + (d-k)s_2)/d, n/d) = 1$ for $0 \leq k \leq d$.

On the other hand, suppose that $\gcd(d, s_1) = 1$ and $\gcd((ks_1 + (d-k)s_2)/d, n/d) = 1$ for an integer k where $0 \leq k \leq d$. We construct a sequence $D = \langle y_0 = 0, y_1, \ldots, y_{n-1} \rangle$ as follows:

$$y_0 = 0$$
$$y_i = \begin{cases} y_{i-1} + s_1 \,(\mathbf{mod}\, n) & \text{if } 1 \leq i \,(\mathbf{mod}\, d) \leq k \\ y_{i-1} + s_2 \,(\mathbf{mod}\, n) & \text{otherwise} \end{cases}$$

In other words, $y_j = y_{j\,(\mathbf{mod}\, d)} + \lfloor \frac{j}{d} \rfloor (ks_1 + (d-k)s_2) \,(\mathbf{mod}\, n)$. Obviously, (y_{i-1}, y_i) is an edge of G. In order to prove that D forms a Hamiltonian cycle in

$G(n; s_1, s_2)$, we are required to show $y_i \neq y_j$ for all $0 \le i < j < n$. Since $\gcd(d, s_1) = 1$, it follows from Lemma 12.19 that $\{T_{[y_i]} \mid 0 \le i < d\} = \{T_i \mid 0 \le i < d\}$. Since $ks_1 + (d-k)s_2 = 0 (\bmod d)$, it follows that $y_j \in T_{[y_j (\bmod d)]}$. Hence, $y_i \neq y_j$ if $i \neq j\ (\mathbf{mod}\ d)$. Since $\gcd((ks_1 + (d-k)s_2)/d, n/d) = 1$, it follows that $y_{td} \neq 0\ (\mathbf{mod}\ n)$ for all $1 \le t < n/d$. Then we have $y_i \neq y_j$ for all $i = j\ (\mathbf{mod}\ d)$ and $0 \le i < j < n$. Hence, the theorem is proved. □

Using the proof of Lemma 12.19, we can easily obtain the following corollary.

COROLLARY 12.2 *$G(n; s_1, s_2)$ contains a Hamiltonian cycle with at least one s_1-edge and one s_2-edge if and only if (1) $\gcd(d, s_1) = 1$, and (2) there exists an integer k, $1 \le k < d$, satisfying $\gcd((ks_1 + (d-k)s_2)/d, n/d) = 1$.*

THEOREM 12.30 $G(n; s_1, s_2)$ is 1-edge fault-tolerant Hamiltonian if and only if at least one of the following statements holds:

1. $\gcd(d, s_1) = 1$, and there exists an integer k with $1 \le k < d$ such that $\gcd((ks_1 + (d-k)s_2)/d, n/d) = 1$
2. $\gcd(n, s_i) = 1$ for $i = 1$ and 2

Proof: To prove the necessity, suppose $G(n; s_1, s_2)$ is 1-edge fault-tolerant Hamiltonian. $G(n; s_1, s_2)$ contains a Hamiltonian cycle with at least one s_1-edge and one s_2-edge or two Hamiltonian cycles using only s_1-edges and s_2-edges, respectively. It follows from Corollary 12.2 that statement 1 holds in the former case. In the latter case, it is trivial that statement 2 holds.

On the other hand, statement 1 implies the existence of a Hamiltonian cycle with at one s_1-edge and one s_2-edge. Therefore, $G(n; s_1, s_2)$ is 1-edge fault-tolerant Hamiltonian since $G(n; s_1, s_2)$ is vertex-symmetric. When statement 2 holds, it is trivial that $G(n; s_1, s_2)$ is 1-edge fault-tolerant Hamiltonian. Hence, the theorem follows. □

Now, we discuss 1-vertex fault-tolerant Hamiltonian double loop networks. When $s_1 = s_2\ (\mathbf{mod}\ n)$, $G(n; s_1, s_1)$ cannot be 1-vertex fault-tolerant Hamiltonian. In the following discussion, we assume $s_1 \neq s_2\ (\mathbf{mod}\ n)$. Since $G(n; s_1, s_2)$ is vertex-symmetric, the existence of a Hamiltonian cycle in $G(n; s_1, s_2) - 0$ implies that $G(n; s_1, s_2)$ is 1-vertex fault-tolerant Hamiltonian. In order to obtain a Hamiltonian cycle in a 1-vertex fault-tolerant Hamiltonian double loop network $G(n; s_1, s_2) - 0$, we construct two sequences $N_1 = \{a_0^1, a_1^1, \ldots\}$ and $N_2 = \{a_0^2, a_1^2, \ldots\}$ as follows:

$$a_0^1 = n - s_2\ (\mathbf{mod}\ n)$$
$$a_i^1 = n - s_2 + is\ (\mathbf{mod}\ n) \text{ if } is \neq 0, s_2\ (\mathbf{mod}\ n)$$
$$a_0^2 = n - s_1\ (\mathbf{mod}\ n)$$
$$a_i^2 = n - s_1 - is\ (\mathbf{mod}\ n) \text{ if } is \neq 0, -s_1\ (\mathbf{mod}\ n)$$

The sequence N_1 terminates when $is = 0\,(\mathbf{mod}\, n)$ or $s_2\,(\mathbf{mod}\, n)$. It is obvious that $n + \frac{n}{d}s = 0\,(\mathbf{mod}\, n)$ or $is = s_2(\mathbf{mod}\, n)$ for some integer i. Thus the sequence N_1 is finite. Similarly, N_2 terminates when $is = 0\,(\mathbf{mod}\, n)$ or $is = -s_1(\mathbf{mod}\, n)$, and N_2 is also finite. Note that the vertex 0 is not in either N_1 or N_2; that is, $0 \notin N_1 \cup N_2$.

Consider the example of $G(12; 3, 8)$. It is observed that $G(12; 3, 8) - 0$ contains a Hamiltonian cycle $C^* = \langle 4, 7, 3, 6, 9, 5, 8, 11, 2, 10, 1, 4\rangle$. The construction of C^* first starts with vertex 4, which can only use the s_1-edge to connect to vertex 7 in C^*, since the s_2-edge of vertex 4, $(4, 0)$, is eliminated from $G(12; 3, 8)$. As a result, the s_2-edge $(11, 7)$ cannot be included in C^*; that is, vertex 11 must use s_1-edge to connect to vertex 2 in C^*. Subsequently, the s_2-edge $(6, 2)$ is excluded from C^* and equivalently, vertex 6 must use the s_1-edge in C^*. Iteratively, this motivates the definition of N_1, which is given by $N_1 = \{4, 11, 6, 1, 8, 3, 10, 5\}$ in this example. Furthermore, N_1 terminates at $i = 8$, since $8s = s_2\,(\mathbf{mod}\, n)$. Similarly, we define N_2 such that vertices in N_2 use s_2-edges to connect to other vertices in C^*, and N_2 is given by $N_2 = \{9, 2, 7\}$ in this example. We note that $N_1 \cup N_2 = \{1, 2, \ldots, 11\}$.

LEMMA 12.20 Let $G(n; s_1, s_2)$ be a double loop network with $\gcd(n, s) = 1$. Let $|N_1| = x$ and $|N_2| = y$. Then x and y are the smallest positive integers satisfying $xs = s_2\,(\mathbf{mod}\, n)$ and $-ys = s_1\,(\mathbf{mod}\, n)$, respectively. Moreover, $x + y = n - 1$, $N_1 \cap N_2 = \emptyset$, and $N_1 \cup N_2 = \{1, 2, \ldots, n - 1\}$.

Proof: Since $\gcd(n, s) = 1$, N_1 and N_2 terminate when $xs = s_2\,(\mathbf{mod}\, n)$ and $ys = -s_1\,(\mathbf{mod}\, n)$ are satisfied. Therefore, x is the smallest positive integer with $xs = s_2\,(\mathbf{mod}\, n)$ and y is the smallest positive integer with $-ys = s_1\,(\mathbf{mod}\, n)$. Obviously, $1 \le x < n$ and $1 \le y < n$. Since $(x + y)s = s_2 - s_1\,(\mathbf{mod}\, n)$ and $\gcd(n, s) = 1$, we have $x + y = -1\,(\mathbf{mod}\, n)$. Thus, $x + y = n - 1$ follows from $2 \le x + y \le 2n - 2$.

Suppose $N_1 \cap N_2 \neq \emptyset$. Let $a_i^1 = a_j^2$, where $0 \le i \le x - 1$ and $0 \le j \le y - 1$. It follows that $n - s_2 + is = n - s_1 - js\,(\mathbf{mod}\, n)$ and $(1 + i + j)s = 0\,(\mathbf{mod}\, n)$. Since $\gcd(s, n) = 1$, it follows that $1 + i + j = 0\,(\mathbf{mod}\, n)$. It implies that $i + j = n - 1$, which contradicts $i + j \le x + y - 2$ and $x + y = n - 1$. Thus, $N_1 \cap N_2 = \emptyset$. Since $x + y = n - 1$, $N_1 \cap N_2 = \emptyset$, and $0 \notin N_1 \cup N_2$, it follows that $N_1 \cup N_2 = \{1, 2, \ldots, n - 1\}$. The lemma follows. □

In the previous example, the Hamiltonian cycle C^* in $G(12; 3, 8) - \{0\}$ is obtained by using s_1-edges for vertices in N_1 and s_2-edges for vertices in N_2. This is true for all 1-vertex fault-tolerant Hamiltonian double loop networks. Let $G(n; s_1, s_2)$ be a 1-vertex fault-tolerant Hamiltonian double loop network, and let C be a Hamiltonian cycle in $G(n; s_1, s_2) - 0$. For any vertex $i \in \{1, 2, \ldots, n - 1\}$, we define a function h_C as follows:

$$h_C(i) = \begin{cases} s_1 & \text{if } (i, i + s_1\ (\mathbf{mod}\ n)) \in C \\ s_2 & \text{otherwise} \end{cases}$$

LEMMA 12.21 Suppose that $G(n; s_1, s_2)$ is a 1-vertex fault-tolerant Hamiltonian double loop network, and C is a Hamiltonian cycle in $G(n; s_1, s_2) - 0$. Then

1. $h_C(i) = s_1$ for all vertices $i \in N_1$
2. $h_C(i) = s_2$ for all vertices $i \in N_2$

Proof: Since $a_0^1 + s_2 = 0\,(\textbf{mod}\, n)$, it follows that $(a_0^1, a_0^1 + s_2\,(\textbf{mod}\, n)) \notin C$. Thus, $(a_0^1, a_0^1 + s_1\,(\textbf{mod}\, n)) \in C$ and $h_C(a_0^1) = s_1$. Assume that $h_C(a_{k-1}^1) = s_1$, where $k < |N_1|$; that is, $(a_{k-1}^1, a_{k-1}^1 + s_1\,(\textbf{mod}\, n)) \in C$. Note that $a_k^1 + s_2 = (a_{k-1}^1 + s) + s_2 = a_{k-1}^1 + s_1$ $(\textbf{mod}\, n)$. Thus, $(a_k^1, a_k^1 + s_2\,(\textbf{mod}\, n)) \notin C$. Consequently, $(a_k^1, a_k^1 + s_1\,(\textbf{mod}\, n)) \in C$ and $h_C(a_k^1) = s_1$. Therefore, $h_C(i) = s_1$ for all $i \in inN_1$. Similarly, we can prove $h_C(i) = s_2$ for all vertices $i \in N_2$. □

LEMMA 12.22 If $G(n; s_1, s_2)$ is a 1-vertex fault-tolerant Hamiltonian double loop network, then $\gcd(n, s) = 1$.

Proof: Suppose to the contrary that, $\gcd(n, s) = d > 1$. We first consider the case $d \mid s_1$. It follows that $d \mid s_2$ also holds. Then all vertices in $\{i \mid i = kd$ for some integer k and $0 \le i < n\}$ are adjacent to vertices in the same set. Thus, $G(n; s_1, s_2)$ is disconnected, and furthermore, $G(n; s_1, s_2)$ is not 1-vertex fault-tolerant Hamiltonian, which is a contradiction. Now, we consider the case $d \mid s_1$. It follows that $d \mid s_2$. Since s but not s_2 is a multiple of d, it follows that $is = s_2\,(\textbf{mod}\, n)$ has no solution. We also note that $is = 0\,(\textbf{mod}\, n)$ if and only if $i = 0\,(\textbf{mod}\, \frac{n}{d})$. Thus, $n - s_1 - (\frac{n}{d} - 1)s\,(\textbf{mod}\, n)$ is in N_2. However, $n - s_1 - (\frac{n}{d} - 1)s = n - s_2\,(\textbf{mod}\, n)$ is also an element in N_1. It follows from Lemma 12.22 that we have $s_1 = h_C(n - s_2) = s_2$, which is contradictory to $s_1 \neq s_2$. Hence, $\gcd(n, s) = 1$. □

It can be verified that the 1-vertex fault-tolerant Hamiltonian network $G(12; 3, 8)$ satisfies $\gcd(n, s) = 1$ and $\gcd(x, n-1) = 1$, where $s = 7$ and $x = 8$. On the other hand, we define a new sequence B by concatenating N_1 and N_2 with reversing the order of N_2; that is, $B = \{b_0, b_1, b_2, \ldots, b_{n-2}\} = \{4, 11, 6, 1, 8, 3, 10, 5, 7, 2, 9\}$. The constructed Hamiltonian cycle $C^* = \langle 4, 7, 3, 6, 9, 5, 8, 11, 2, 10, 1, 4\rangle$ in $G(12; 3, 8) - \{0\}$ is obtained by the sequence $\{b_{\langle 0\rangle}, b_{\langle x\rangle}, b_{\langle 2x\rangle}, \ldots, b_{\langle (n-2)x\rangle}, b_{\langle 0\rangle}\}$, where $\langle a\rangle = a$ $(\textbf{mod}\, n - 1)$. We use this observation in the proof of the following theorem.

THEOREM 12.31 $G(n; s_1, s_2)$ is 1-vertex fault-tolerant Hamiltonian if and only if the following conditions hold:

1. $s_1 \neq s_2$
2. $\gcd(n, s) = 1$
3. $\gcd(x, n - 1) = 1$, where x is the smallest positive integer satisfying $xs = s_2\,(\textbf{mod}\, n)$

Proof: To prove the necessity, let $G(n; s_1, s_2)$ be 1-vertex fault-tolerant Hamiltonian. It follows from Lemma 12.22 that $\gcd(n, s) = 1$. It follows from Lemma 12.20 that we can construct a sequence $B = \{b_0, b_1, \ldots, b_{n-2}\}$ as follows:

$$b_i = \begin{cases} a_i^1 = n - s_2 + is\,(\textbf{mod}\, n) & \text{if } 0 \le i \le x - 1 \\ a_{n-i-2}^2 = n - s_1 - (n - 2 - i)s\,(\textbf{mod}\, n) & \text{if } x \le i \le n - 2 \end{cases}$$

where x is the smallest integer satisfying $xs = s_2\,(\textbf{mod}\, n)$.

Let C be a Hamiltonian cycle in $G(n; s_1, s_2) - 0$. Let y be the smallest positive integer satisfying $-ys = s_1 (\mathbf{mod}\, n)$. We claim that for all $0 \le i < n-1$, $b_i + h_C(b_i) = b_j\, (\mathbf{mod}\, n)$, where $j = i + x\, (\mathbf{mod}\, n-1)$. To prove the claim, we assume without loss of generality that $x \le y$ for ease of exposition.

First consider $0 \le i \le x-1$. It follows from Lemma 12.21 that $h_C(b_i) = s_1$. Thus

$$\begin{aligned} b_i + h_C(b_i) &= n + (i+1)s = n - s_1 - ys + (i+1)s \\ &= a^2_{y-i-1} = b_j\, (\mathbf{mod}\, n) \end{aligned}$$

where $j = i + x\, (\mathbf{mod}\, n-1)$.

Consider $x \le i \le n-2-x$. It follows from Lemma 12.21 that $h_C(b_i) = s_2$. Thus, we have

$$\begin{aligned} b_i + h_C(b_i) &= n - (n-1-i)s = n - s_1 - (n-2-i-x)s \\ &= a^2_{n-2-i-x} = b_j\, (\mathbf{mod}\, n) \end{aligned}$$

where $j = i + x\, (\mathbf{mod}\, n-1)$.

For $n-1-x \le i < n-1$, we have $b_i = a^2_{n-2-i} = n - s_1 - (n-2-i)s\, (\mathbf{mod}\, n)$ and $h_C(b_i) = s_2$. Thus,

$$\begin{aligned} b_i + h_C(b_i) &= n - (n-1-i)s = n - s_2 + (i+x-n+1)s \\ &= a^1_{i+x-(n-1)} = b_j\, (\mathbf{mod}\, n) \end{aligned}$$

where $j = i + x\, (\mathbf{mod}\, n-1)$. The case $x > y$ can be similarly treated.

Thus, C is uniquely determined by the sequence $D = \{b_{jx\, (\mathbf{mod}\, n-1)} \mid 0 \le j < n-1\}$. In this case, $\{jx\, (\mathbf{mod}\, n-1) \mid 0 \le j < n-1\} = \{0, 1, \ldots, n-2\}$. Therefore, gcd $(x, n-1) = 1$.

On the other hand, assume that $\gcd(n, s) = 1$ and $\gcd(x, n-1) = 1$, where x is the smallest positive integer satisfying $xs = s_2\, (\mathbf{mod}\, n)$. To prove that $G(n; s_1, s_2)$ is 1-vertex fault-tolerant Hamiltonian, we need to construct a Hamiltonian cycle in $G(n; s_1, s_2) - 0$. Let $N_1 = \{a^1_0, a^1_1, \ldots, a^1_{x-1}\}$ and $N_2 = \{a^2_0, a^2_1, \ldots, a^2_{n-x-2}\}$. Since $\gcd(n, s) = 1$, it follows from Lemma 12.20 that $N_1 \cup N_2 = \{1, 2, \ldots, n-1\}$. We define a new sequence $B = \{b_0, b_1, \ldots, b_{n-2}\}$ by setting $b_i = a^1_i$ if $0 \le i \le x-1$ and $b_i = a^2_{n-i-2}$ if $x \le i \le n-2$. We claim that the sequence $\{b_{\langle 0\rangle}, b_{\langle x\rangle}, b_{\langle 2x\rangle}, \ldots, b_{\langle (n-2)x\rangle}, b_{\langle 0\rangle}\}$ forms a Hamiltonian cycle in $G(n; s_1, s_2) - 0$, where $\langle a\rangle = a\, (\mathbf{mod}\, n-1)$.

For ease of exposition, we assume without loss of generality that $x \le y$. Then for $0 \le i < x$,

$$\begin{aligned} b_{\langle i\rangle} &= a^1_i = n - s_2 + is\, (\mathbf{mod}\, n) \\ b_{\langle i+x\rangle} &= a^2_{y-i-1} = n - s_1 - (y-1-i)s = n + (i+1)s\, (\mathbf{mod}\, n) \end{aligned}$$

Thus, $b_{\langle i+x\rangle} - b_{\langle i\rangle} = s_1\, (\mathbf{mod}\, n)$ for $0 \le i < x$.

For $x \leq i < n-1-x$

$$\begin{aligned} b_{\langle i \rangle} &= a^2_{n-2-i} = n - s_1 - (n-2-i)s \,(\textbf{mod}\, n) \\ b_{\langle i+x \rangle} &= a^2_{n-2-i-x} = n - s_1 - (n-2-i-x)s = n - (n-1-i)s \,(\textbf{mod}\, n) \end{aligned}$$

Thus, $b_{\langle i+x \rangle} - b_{\langle i \rangle} = s_2 \,(\textbf{mod}\, n)$ for $x \leq i < n-1-x$.

For $n-1-x \leq i < n-1$

$$\begin{aligned} b_{\langle i \rangle} &= a^2_{n-2-i} = n - s_1 - (n-2-i)s \,(\textbf{mod}\, n) \\ b_{\langle i+x \rangle} &= a^1_{i+x-(n-1)} = n - s_2 + (i+x-n+1)s = n - (n-1-i)s \,(\textbf{mod}\, n) \end{aligned}$$

Thus, $b_{\langle i+x \rangle} - b_{\langle i \rangle} = s_2 \,(\textbf{mod}\, n)$ for $n-1-x \leq i < n-1$.

Therefore, $(b_{\langle jx \rangle}, b_{\langle (j+1)x \rangle})$ is an edge for all $0 \leq i < n-1$. Since $\gcd(x, n-1) = 1$, we have $\{\langle jx \rangle \mid 0 \leq j < n-1\} = \{0, 1, \ldots, n-2\}$. Thus $\{b_{\langle jx \rangle} \mid 0 \leq j < n-1\} = \{1, 2, \ldots, n-1\}$. Therefore, the sequence $\{b_{\langle 0 \rangle}, b_{\langle x \rangle}, b_{\langle 2x \rangle}, \ldots, b_{\langle (n-2)x \rangle}, b_{\langle 0 \rangle}\}$ forms a Hamiltonian cycle in $G(n; s_1, s_2) - 0$. Hence, the theorem follows. □

If a double loop network G is both 1-vertex fault-tolerant Hamiltonian and 1-edge fault-tolerant Hamiltonian, then G is said to be 1-fault-tolerant Hamiltonian. Let n be even, and let s_1 and s_2 have the same parity. Then $\gcd(n, s) \neq 1$. Therefore, when n is even and $s_1 - s_2 = 0 \,(\textbf{mod}\, 2)$, $G(n; s_1, s_2)$ is not 1-vertex fault-tolerant Hamiltonian following from Theorem 12.31. Let n be even, and let s_1 and s_2 have different parity. Without loss of generality, we assume that s_1 is even and s_2 is odd. Consider $G(n; s_1, s_2)$ is 1-vertex fault-tolerant Hamiltonian. It follows from Theorem 12.31 that $d = \gcd(n, s) = 1$. In this case, $\gcd(n, s_1) \neq 1$ and there is no integer k with $1 \leq k < d$, not to mention satisfying $\gcd((ks_1 + (d-k)s_2)/d, n/d) = 1$. Thus, when $G(n; s_1, s_2)$ is 1-vertex fault-tolerant Hamiltonian, $G(n; s_1, s_2)$ is not 1-edge fault-tolerant Hamiltonian. Therefore, when n is even, there is no 1-fault-tolerant Hamiltonian double loop network.

It is also observed that when n is odd, the number of double loop networks $G(n; s_1, s_2)$ that are 1-fault-tolerant Hamiltonian is not small. It is possible to choose a double loop network that is 1-fault-tolerant Hamiltonian. Moreover, when n is prime, every 1-vertex fault-tolerant Hamiltonian double loop network $G(n; s_1, s_2)$ is also 1-edge fault-tolerant Hamiltonian.

In addition to fault tolerance, the diameter is another performance measure of interconnection networks. Let $d(n; s_1, s_2)$ denote the diameter of $G(n; s_1, s_2)$. Let $d(n)$ denote the minimum diameter among all double loop networks having n vertices. Among those $G(n; s_1, s_2)$ achieving $d(n)$, is there always one 1-vertex fault-tolerant Hamiltonian or 1-edge fault-tolerant Hamiltonian or 1-fault-tolerant Hamiltonian? The answer is **no**, and it can be verified by counterexamples. But here we point out a class of 1-fault-tolerant Hamiltonian double loop networks with small diameters. Consider a double loop network $H = G(p^2; p+2, 1)$, where p is a prime number and $p \neq 2 \,(\textbf{mod}\, 3)$. It follows from Theorem 12.30 that H is 1-edge fault-tolerant Hamiltonian. Let x be the smallest positive integer satisfying $x(p+1) = 1 \,(\textbf{mod}\, p^2)$. Note that $(p+1)(p-1) = -1 \,(\textbf{mod}\, p^2)$ implies $x = p^2 - p + 1$. It is observed that

$\gcd(p^2-p+1,p^2-1)=\gcd(p^2-p+1,p-2)=\gcd(p+1,p-2)=\gcd(3,p-2)=1$. Therefore, we obtain $\gcd(p^2-p+1,p^2-1)=\gcd(p^2,p+1)=1$. It follows from Theorem 12.31 that H is 1-vertex fault-tolerant Hamiltonian. Note that $p+2$ is close to $\sqrt{p^2}$. Applying the results of Wong and Coppersmith [344], both the diameter and the average distance of H are $O(p)$. In other words, H is 1-fault-tolerant Hamiltonian and has a small diameter of $O(\sqrt{n})$, where $n=p^2$.

We also have the following two interesting results regarding double loop networks.

COROLLARY 12.3 *Suppose that $G(n;s_1,s_2)$ is Hamiltonian. Let $A(n;s_1,s_2)=\{k \mid 0 \le k \le d$ and $\gcd((ks_1+(d-k)s_2)/d, n/d)=1\}$. Then $G(n;s_1,s_2)$ has*

$$\sum_{k\in A(n;s_1,s_2)} \frac{d!}{k!(d-k)!}$$

different Hamiltonian cycles.

Proof: Any Hamiltonian cycle $C=\langle x_0=0,x_1,\ldots,x_{n-1},x_0\rangle$ uniquely determines a sequence $\{(\alpha_i,\beta_i) \mid 0 \le i \le d\}$. On the other hand, we can construct a Hamiltonian cycle by first setting a sequence $\{(\alpha_i,\beta_i) \mid 0 \le i \le d\}$ satisfying $\alpha_i+\beta_i=i$ and $\alpha_i,\beta_i \in \{0,1,\ldots,d\}$. The corollary follows from Theorem 12.29. □

It is observed that those double loop networks $G(n;s_1,s_2)$ with $\gcd(n,s_i)=1$ for $i=1,2$ have two disjoint Hamiltonian cycles. The classification of all such double loop networks is stated in the following corollary, which follows from the proof of Theorem 12.29.

COROLLARY 12.4 *Assume that $\gcd(d,s_1)=1$. Edges of $G(n;s_1,s_2)$ can be decomposed into two disjoint Hamiltonian cycles if and only if there exists an integer k with $0 \le k \le d$ such that $\gcd((ks_1+(d-k)s_2)/d, n/d)=1$ and $\gcd(((d-k)s_1+ks_2)/d, n/d)=1$.*

As an example, the double loop network $G(15;5,2)$ has two disjoint Hamiltonian cycles.

13 Optimal k-Fault-Tolerant Hamiltonian-Laceable Graphs

13.1 INTRODUCTION

In Chapter 11, we have discussed the fault-tolerant Hamiltonian of some families of interconnection networks. Yet we did not discuss the corresponding results for two important families of interconnection networks; namely, the hypercube family and the star graph family. Note that these two families of graphs are bipartite. Any cycle of a bipartite graph is of even length. Thus, any vertex fault in a Hamiltonian bipartite graph is not Hamiltonian. However, we can still discuss the edge fault-tolerant Hamiltonicity for bipartite graphs.

A Hamiltonian graph G is *k-edge fault-tolerant Hamiltonian* if $G - F$ remains Hamiltonian for every $F \subset E(G)$ with $|F| \leq k$. The *edge fault-tolerant Hamiltonicity*, $\mathcal{H}_f^e(G)$, is defined to be the maximum integer k such that G is k edge fault-tolerant Hamiltonian, and undefined if otherwise. Obviously, $\mathcal{H}_f^e(G) \leq \delta(G) - 2$. Moreover, a regular graph G is optimal edge fault-tolerant Hamiltonian if $\mathcal{H}_f^e(G) = \delta(G) - 2$.

As mentioned in Chapter 11, the counterpart of Hamiltonian-connected graphs in bipartite graphs is known as *Hamiltonian laceable*. Obviously, any Hamiltonian-laceable graphs, except for K_1 and K_2, is Hamiltonian and with at least $2k$ vertices with $k \geq 2$. Thus, $\delta(G) \geq 2$ if G is a Hamiltonian-laceable graph with at least four vertices. Again, we can discuss the edge fault-tolerant Hamiltonian laceability for bipartite graphs.

Similar to fault-tolerant Hamiltonian connectivity for nonbipartite Hamiltonian-connected graphs, Tsai et al. [305] proposed the concept of edge fault-tolerant Hamiltonian laceability. A Hamiltonian-laceable graph G is *k-edge fault-tolerant Hamiltonian* if $G - F$ remains Hamiltonian laceable for every $F \subset E(G)$ with $|F| \leq k$. The *edge fault-tolerant laceability*, $\mathcal{H}_e^L(G)$, is defined to be the maximum integer k such that G is k-edge fault-tolerant Hamiltonian laceable, and undefined if otherwise. Obviously, $\mathcal{H}_e^L(G) \leq \mathcal{H}_f^e(G) \leq \delta(G) - 2$ if G has at least four vertices. Moreover, a regular graph G with at least four vertices is optimal edge fault-tolerant Hamiltonian laceability if $\mathcal{H}_e^L(G) = \delta(G) - 2$.

Hsieh et al. [160] further extended the concept of *Hamiltonian laceable* into *strongly Hamiltonian laceable*. A Hamiltonian-laceable graph $G = (V_0 \cup V_1, E)$ is *strongly Hamiltonian laceable* if there is a simple path of length $|V_0 \cup V_1| - 2$ between any two vertices of the same partite set. Since a strongly Hamiltonian-laceable graph is Hamiltonian laceable, $\delta(G) \geq 2$ if G is a strongly Hamiltonian-laceable graph with at least four vertices. Lewinter et al. [221] also introduced the concept of *hyper Hamiltonian laceable*. A Hamiltonian-laceable graph $G=(V_0 \cup V_1, E)$ is *hyper Hamiltonian laceable* if for any vertex $v \in V_i$, $i=0,1$, there is a Hamiltonian path of $G-\{v\}$ between any two vertices of V_{1-i}. Assume that $G=(V_0 \cup V_1, E)$ is a bipartite graph with $|V_0|=|V_1|\geq 3$. Suppose that there exists a vertex x with exactly two neighbors y and z. Without loss of generality, we may assume that $x \in V_0$. Thus, y and z are in V_1. Let w be any vertex in $V_0-\{x\}$. Obviously, there is no Hamiltonian path of $G-\{w\}$ between y and z. Therefore, $\delta(G)\geq 3$ if G is a hyper Hamiltonian-laceable graph with at least six vertices.

Similarly, Tsai et al. [305] proposed the following concept. A strongly Hamiltonian-laceable graph G is *k-edge fault-tolerant strongly Hamiltonian laceable* if $G-F$ remains strongly Hamiltonian laceable for every $F \subset E(G)$ with $|F|\leq k$. The *edge fault-tolerant strongly Hamiltonian laceability*, $\mathcal{H}_e^{SL}(G)$, is defined to be the maximum integer k such that G is k-edge fault-tolerant strongly Hamiltonian laceable, and undefined if otherwise. Obviously, $\mathcal{H}_e^{SL}(G)\leq \mathcal{H}_e^{L}(G)\leq \delta(G)-2$ if G has at least four vertices. Moreover, a regular graph G with at least four vertices is optimal edge fault-tolerant strongly Hamiltonian laceability if $\mathcal{H}_e^{SL}(G)=\delta(G)-2$.

Again, the concept of edge fault-tolerant hyper Hamiltonian laceability is introduced [305]. A hyper Hamiltonian-laceable graph G is *k-edge fault-tolerant hyper Hamiltonian laceable* if $G-F$ remains hyper Hamiltonian laceable for every $F \subset E(G)$ with $|F|\leq k$. The *edge fault-tolerant hyper Hamiltonian laceability*, $\mathcal{H}_e^{HL}(G)$, is defined to be the maximum integer k such that G is k-edge fault-tolerant hyper Hamiltonian laceable, and undefined if otherwise. Obviously, $\mathcal{H}_e^{HL}(G)\leq \delta(G)-3$ if G has at least six vertices. Moreover, a regular graph G with at least six vertices has optimal edge fault-tolerant hyper Hamiltonian laceability if $\mathcal{H}_e^{HL}(G)=\delta(G)-3$.

For simplicity, a k-regular bipartite graph is *super fault-tolerant Hamiltonian laceable* if $\mathcal{H}_e^{SL}(G)=k-2$ and $\mathcal{H}_e^{HL}(G)=k-3$. Since $\mathcal{H}_e^{SL}(G)\leq \mathcal{H}_e^{L}(G)\leq \mathcal{H}_f^{e}(G)\leq \delta(G)-2$, $\mathcal{H}_e^{L}(G)=\mathcal{H}_f^{e}(G)=k-2$ if G is a k-regular super fault-tolerant Hamiltonian-laceable graph.

LEMMA 13.1 Let G be any bipartite graph with bipartition V_0 and V_1 such that $|V_0|=|V_1|$. Suppose that for any vertex $\mathbf{v}$ there exists a Hamiltonian path between any two different vertices in the partite set not containing $\mathbf{v}$. Then G is hyper Hamiltonian laceable.

Proof: Let $\mathbf{u}$ and $\mathbf{v}$ be any two vertices from different partite sets. Let $\mathbf{r}$ be any vertex adjacent to $\mathbf{v}$. Thus, $\mathbf{u}$ and $\mathbf{r}$ are in the same partite set. By assumption, there exists a Hamiltonian path P of $G-\{\mathbf{v}\}$ joining $\mathbf{u}$ to $\mathbf{r}$. Then $\langle \mathbf{u}, P, \mathbf{r}, \mathbf{v}\rangle$ forms the desired path. Thus, G is Hamiltonian laceable. Based on the assumption, G is hyper Hamiltonian laceable. □

LEMMA 13.2 Let G be a hyper Hamiltonian laceable bipartite graph and f be any edge of G. Then $G-\{f\}$ is strongly Hamiltonian laceable. Thus, any hyper Hamiltonian-laceable graph is 1-edge fault-tolerant Hamiltonian laceable.

Proof: By definition, G is Hamiltonian laceable. Let $(\mathbf{x}, \mathbf{y})$ be any edge of G, and $\mathbf{u}$ and $\mathbf{v}$ be any two vertices in the partite set not containing $\mathbf{x}$. By assumption, there exists a Hamiltonian path P of $G - \{\mathbf{x}\}$ joining $\mathbf{u}$ to $\mathbf{v}$. Then $\langle \mathbf{u}, P, \mathbf{v} \rangle$ forms the desired path. Thus, $G - \{(\mathbf{x}, \mathbf{y})\}$ is strongly Hamiltonian laceable. □

13.2 SUPER FAULT-TOLERANT HAMILTONIAN LACEABILITY OF HYPERCUBES

Tsai et al. [305] proved that Q_n is super fault-tolerant Hamiltonian laceable if $n \geq 3$. We use B and W to denote the bipartition of Q_n. Obviously, Q_n can be divided into two copies of Q_{n-1}, denoted by Q_n^0 and Q_n^1. Let E_c be the set of crossing edges; that is, $E_c = \{(\mathbf{u}, \mathbf{u}') \mid (\mathbf{u}, \mathbf{u}') \in E(Q_n), \mathbf{u} \in V(Q_n^0) \text{ and } \mathbf{u}' \in V(Q_n^1)\}$. Let F be the set of faulty edges in Q_n, $F_0 = F \cap E(Q_n^0)$, $F_1 = F \cap E(Q_n^1)$, and $F_c = F \cap E_c$. Also let $f_0 = |F_0|$, $f_1 = |F_1|$, and $f_c = |F_c|$. Since Q_n is edge-symmetric, it suffices to consider only the case that $f_c \geq 1$ if $F \neq \emptyset$. That means, Q_n can be split into Q_n^0 and Q_n^1 using any dimension d, where $1 \leq d \leq n$. Therefore, given a nonempty faulty edge set F, Q_n can be split into Q_n^0 and Q_n^1 such that $F_c \neq \emptyset$.

By brute force, we have the following lemma.

LEMMA 13.3 $\mathcal{H}_e^{SL}(Q_3) = \mathcal{H}_e^{L}(Q_3) = \mathcal{H}_f^{e}(Q_3) = 1$ and $\mathcal{H}_e^{HL}(Q_3) = 0$.

LEMMA 13.4 The hypercube Q_n is $(n-2)$-edge fault-tolerant Hamiltonian laceable for $n \geq 2$. Thus, $\mathcal{H}_e^{L}(Q_n) = n - 2$.

Proof: Since $\delta(Q_n) = n$, $\mathcal{H}_e^{L}(Q_n) \leq n - 2$. We prove $\mathcal{H}_e^{L}(Q_n) \geq n - 2$ by induction on n. Obviously, the statement holds for $n = 2$. By Lemma 13.3, the statement holds if $n = 3$. For $n \geq 4$, we assume that $\mathcal{H}_e^{L}(Q_n) \geq n - 2$ for every integer $k < n$. Let F be any faulty edge set with $|F| = n - 2$. Suppose that $\mathbf{x} \in W$ and $\mathbf{y} \in B$. To prove this lemma, we need to find a Hamiltonian path of $Q_n - F$ joining $\mathbf{x}$ and $\mathbf{y}$. Without loss of generality, we may assume that $f_c \geq 1, f_0 \leq n - 3$, and $f_1 \leq n - 3$.

Case 1. Either $\mathbf{x}, \mathbf{y} \in V(Q_n^0)$ or $\mathbf{x}, \mathbf{y} \in V(Q_n^1)$. Without loss of generality, we may assume that $\mathbf{x}$ and $\mathbf{y}$ are in Q_n^0. Since, $f_0 \leq n - 3$, by induction, there exists a Hamiltonian path P_0 of $Q_n^0 - F_0$ joining $\mathbf{x}$ to $\mathbf{y}$. Since $|E(P_0)| = 2^{n-1} - 1$ and $\left\lceil \frac{2^{n-1}-1}{2} \right\rceil > n - 2$ for $n \geq 4$, there exists an edge $(\mathbf{u}, \mathbf{v})$ in P_0 such that both crossing edges $(\mathbf{u}, \mathbf{u}') \notin F$ and $(\mathbf{v}, \mathbf{v}') \notin F$. Obviously, either ($\mathbf{u}' \in W$ and $\mathbf{v}' \in B$) or ($\mathbf{u}' \in B$ and $\mathbf{v}' \in W$). Since $f_1 \leq n - 3$, there exists a Hamiltonian path P_1 joining $\mathbf{u}'$ to $\mathbf{v}'$. Obviously, $(E(P_0) \cup E(P_1) \cup \{(\mathbf{u}, \mathbf{u}'), (\mathbf{v}, \mathbf{v}')\}) - \{(\mathbf{u}, \mathbf{v})\}$ forms a Hamiltonian path in $Q_n - F$ joining $\mathbf{x}$ to $\mathbf{y}$. See Figure 13.1a for illustration.

Case 2. Either $\mathbf{x} \in V(Q_n^0)$ and $\mathbf{y} \in V(Q_n^1)$, or $\mathbf{x} \in V(Q_n^1)$ and $\mathbf{y} \in V(Q_n^0)$. Without loss of generality, we assume that $\mathbf{x} \in V(Q_n^0)$ and $\mathbf{y} \in V(Q_n^1)$. Since $|V(Q_n^0) \cap B| = 2^{n-2}$ and $2^{n-2} > n - 2$ for $n \geq 4$, there exists a vertex $\mathbf{u}$ in $V(Q_n^0) \cap B$ such that the crossing edge $(\mathbf{u}, \mathbf{u}') \notin F$. Obviously, $\mathbf{u}' \in W$ and $y \in B$. Since $f_0 \leq n - 3$ and $f_1 \leq n - 3$, there exists a Hamiltonian path P_0 of $Q_n^0 - F_0$ joining $\mathbf{x}$ to $\mathbf{u}$ and there exists a Hamiltonian path P_1

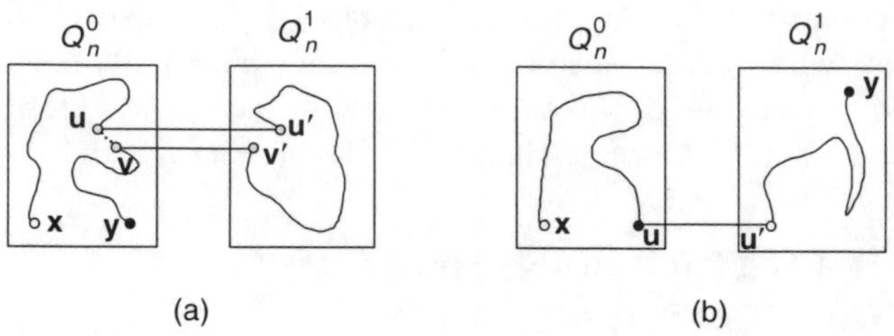

FIGURE 13.1 (a) Case 1: $\mathbf{x}, \mathbf{y} \in V(Q_n^0)$ and (b) Case 2: $\mathbf{x} \in V(Q_n^0)$ and $\mathbf{y} \in V(Q_n^1)$.

of $Q_n^1 - F_1$ joining $\mathbf{u}'$ to $\mathbf{y}$. Obviously, $E(P_0) \cup E(P_1)) \cup \{(\mathbf{u}, \mathbf{u}')\}$ forms a Hamiltonian path in $Q_n - F$ joining $\mathbf{x}$ to $\mathbf{y}$. See Figure 13.1b for illustration.

This completes the induction. □

COROLLARY 13.1 $\mathcal{H}_f^e(Q_n) = n - 2$ *if* $n \geq 2$.

LEMMA 13.5 The hypercube Q_n is $(n-2)$-edge fault-tolerant strongly Hamiltonian laceable for $n \geq 2$. Thus, $\mathcal{H}_e^{SL}(Q_n) = n - 2$.

Proof: Since $\delta(Q_n) = n$, $\mathcal{H}_e^{SL}(Q_n) \leq n - 2$. Let F be any faulty edge set with $|F| \leq n - 2$. By Lemma 13.4, $Q_n - F$ is Hamiltonian laceable. To finish the proof, we need to find a path of length $2^n - 2$ in $Q_n - F$ between any two vertices $\mathbf{x}$ and $\mathbf{y}$ in the same partite set. The proof is similar to the proof of Lemma 13.4. Hence, the detailed proof is omitted. □

LEMMA 13.6 The hypercube Q_n is $(n-3)$-edge fault-tolerant hyper Hamiltonian laceable for $n \geq 3$. Thus, $\mathcal{H}_e^{HL}(Q_n) = n - 3$.

Proof: Since $\delta(Q_n) = n$, $\mathcal{H}_e^{HL}(Q_n) \leq n - 3$. Let F be any faulty edge set with $|F| \neq n - 3$. We need to prove that $Q_n - F$ is hyper Hamiltonian laceable. By Lemma 13.4, $Q_n - F$ is Hamiltonian laceable. Let $\mathbf{w}$ be any vertex of Q_n. Without loss of generality, we may assume that $\mathbf{w} \in W$. Let $\mathbf{x}$ and $\mathbf{y}$ be two vertices in B. To finish the proof, we must find a Hamiltonian path of $(Q_n - F) - \{\mathbf{w}\}$ joining $\mathbf{x}$ to $\mathbf{y}$. We prove this statement by induction. By Lemma 13.3, the statement holds for $n = 3$. Now, we assume that the theorem is true for every integer $k < n$ with $n \geq 4$. Without loss of generality, we may assume that $f_c \geq 1, f_0 \leq n - 4$, and $f_1 \leq n - 4$.

Case 1. $\mathbf{x}, \mathbf{y} \in V(Q_n^0)$. Since $f_0 \leq n - 4$, by induction, there is a Hamiltonian path P_0 in $(Q_n^0 - F_0) - \{\mathbf{w}\}$ joining $\mathbf{x}$ to $\mathbf{y}$. Since $l(P_0) = 2^{n-1} - 2$ and $2^{n-2} > n - 2$ for $n \geq 4$, there exists an edge $(\mathbf{u}, \mathbf{v})$ in P_0 such that both crossing edges $(\mathbf{u}, \mathbf{u}') \notin F$ and $(\mathbf{v}, \mathbf{v}') \notin F$. Obviously, $\mathbf{u}'$ and $\mathbf{v}'$ are in different partite sets. Since $f_1 \leq n - 4$, by Lemma 13.4, there exists a Hamiltonian path P_1 of $Q_n^1 - F_1$ joining $\mathbf{u}'$ to $\mathbf{v}'$. Obviously, $(E(P_0) \cup E(P_1) \cup \{(\mathbf{u}, \mathbf{u}'), (\mathbf{v}, \mathbf{v}')\}) - \{(\mathbf{u}, \mathbf{v})\}$ forms a Hamiltonian path in $(Q_n - F) - \mathbf{w}$ joining $\mathbf{x}$ to $\mathbf{y}$. See Figure 13.2a for illustration.

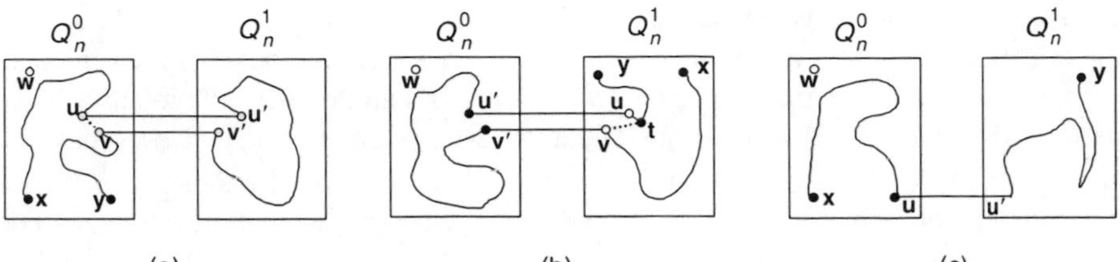

FIGURE 13.2 (a) Case 1: $\mathbf{x}, \mathbf{y} \in V(Q_n^0)$, (b) Case 2: $\mathbf{x}, \mathbf{y} \in V(Q_n^1)$, and (c) Case 3: $\mathbf{x} \in V(Q_n^0)$ and $\mathbf{y} \in V(Q_n^1)$.

Case 2. $\mathbf{x}, \mathbf{y} \in V(Q_n^1)$. Again, we can find a vertex $\mathbf{u}$ in $V(Q_n^1) \cap W$ such that the crossing edge $(\mathbf{u}, \mathbf{u}') \notin F$. Since $f_1 \leq n-4$, by induction, there exists a Hamiltonian path P in $(Q_n^1 - F_1) - \{\mathbf{u}\}$ joining $\mathbf{x}$ to $\mathbf{y}$. Since the number of neighboring vertices of $\mathbf{u}$ in Q_n^1 is $n-1$ and $|F| \leq n-3$, there exists a neighbor $\mathbf{t}$ of $\mathbf{u}$ in Q_n^1 such that $(\mathbf{u}, \mathbf{t}) \notin F$ and $(\mathbf{v}, \mathbf{v}')$ where $\mathbf{v}$ is a vertex adjacent to $\mathbf{t}$ in P.

Obviously, we can divide P into two sections P_0 and P_1, where either (P_0 is a path joining $\mathbf{x}$ to $\mathbf{v}$ and P_1 is a path joining $\mathbf{t}$ to $\mathbf{y}$) or (P_0 is a path joining $\mathbf{x}$ to $\mathbf{t}$ and P_1 is a path joining $\mathbf{v}$ to $\mathbf{y}$). Without loss of generality, we may assume that P_0 is a path joining $\mathbf{x}$ to $\mathbf{v}$ and P_1 is a path joining $\mathbf{t}$ to $\mathbf{y}$. Let $(\mathbf{u}, \mathbf{u}')$ and $(\mathbf{v}, \mathbf{v}')$ be two crossing edges incident to vertices $\mathbf{u}$ and $\mathbf{v}$, respectively. Obviously, $\{\mathbf{u}', \mathbf{v}'\} \subset B$. Since $f_0 \leq n-4$, by induction, there exists a Hamiltonian path R in $(Q_n^0 - F_0) - \{\mathbf{w}\}$ joining $\mathbf{u}'$ to $\mathbf{v}'$. Obviously, $E(P_0) \cup E(R) \cup E(P_1) \cup \{(\mathbf{u}, \mathbf{u}'), (\mathbf{v}, \mathbf{v}'), (\mathbf{t}, \mathbf{u})\}$ forms a Hamiltonian path in $(Q_n - F) - \{\mathbf{w}\}$ joining $\mathbf{x}$ to $\mathbf{y}$. See Figure 13.2b for illustration.

Case 3. $\mathbf{x} \in V(Q_n^0)$ and $\mathbf{y} \in V(Q_n^1)$, or $\mathbf{x} \in V(Q_n^1)$ and $\mathbf{y} \in V(Q_n^0)$. Without loss of generality, we assume that $\mathbf{x} \in V(Q_n^0)$ and $\mathbf{y} \in V(Q_n^1)$. Since $|V(Q_n^0) \cap B| = 2^{n-2}$ and $2^{n-2} - 1 > n-3 \geq |F|$ for $n \geq 3$, there exists a vertex $\mathbf{u}$ in $(V(Q_n^0) - \{\mathbf{x}\}) \cap B$ such that the crossing edge $(\mathbf{u}, \mathbf{u}') \notin F$. Obviously, $\mathbf{u}' \in W$. Since $f_0 \leq n-4$, there exists a Hamiltonian path P_0 of $(Q_n^0 - F_0) - \{\mathbf{w}\}$ joining $\mathbf{x}$ to $\mathbf{u}$. Also, since $f_1 \leq n-4$, by Lemma 13.4, there exists a Hamiltonian path P_1 of $Q_n^1 - F_1$ joining $\mathbf{u}'$ to $\mathbf{y}$. Obviously, $E(P_0) \cup E(P_1) \cup \{(\mathbf{u}, \mathbf{u}')\}$ forms a Hamiltonian path in $(Q_n - F) - \{\mathbf{w}\}$ joining $\mathbf{x}$ to $\mathbf{y}$. The proof is complete. See Figure 13.2c for illustration. □

Combining Lemmas 13.4 through 13.6, we have the following theorem.

THEOREM 13.1 The hypercube Q_n is super fault-tolerant Hamiltonian laceable for $n \geq 3$.

13.3 SUPER FAULT-TOLERANT HAMILTONIAN LACEABILITY OF STAR GRAPHS

Let S_n be the n-dimensional star graphs. In this section, we prove that S_n is super fault-tolerant Hamiltonian laceable if $n \geq 4$. Yet we need some basic background about the properties of star graphs.

It is known that S_n is a bipartite graph with one partite set containing all odd permutations and the other partite set containing all even permutations. For convenience, we refer to an even permutation as a *white vertex* and an odd permutation as a *black vertex*. Moreover, S_n is vertex-transitive and edge-transitive. Let $\mathbf{u} = u_1u_2 \ldots u_n$ be any vertex of S_n. We use $(\mathbf{u})_i$ to denote the ith component u_i of $\mathbf{u}$ and $S_n^{\{i\}}$ to denote the ith subgraph of S_n induced by those vertices $\mathbf{u}$ with $(\mathbf{u})_n = i$. Obviously, S_n can be decomposed into n vertex disjoint subgraphs $S_n^{\{i\}}$ for $1 \leq i \leq n$, such that each $S_n^{\{i\}}$ is isomorphic to S_{n-1}. Thus, the star graph can be constructed recursively. Let $H \subseteq \langle n \rangle$. We use S_n^H to denote the subgraph of S_n induced by $\cup_{i \in H} V(S_n^{\{i\}})$. By the definition of S_n, there is exactly one neighbor $\mathbf{v}$ of $\mathbf{u}$ such that $\mathbf{u}$ and $\mathbf{v}$ are adjacent through an i-dimensional edge with $2 \leq i \leq n$. For this reason, we use $(\mathbf{u})^i$ to denote the unique i-neighbor of $\mathbf{u}$. We have $((\mathbf{u})^i)^i = \mathbf{u}$ and $(\mathbf{u})^n \in S_n^{\{(\mathbf{u})_1\}}$. For $1 \leq i, j \leq n$ and $i \neq j$, we use $E^{i,j}$ to denote the set of edges between $S_n^{\{i\}}$ and $S_n^{\{j\}}$.

Let E_c be the edge set $\{(\mathbf{u}, (\mathbf{u})^n) \mid \mathbf{u} \in V(S_n)\}$. Let F be the set of faulty edges in S_n. We set $F_i = F \cap E(S_n^{\{i\}})$ for $1 \leq i \leq n$ and $F_c = F \cap E_c$. Also let $f_i = |F_i|$ for $1 \leq i \leq n$ and $f_c = |F_c|$. Since S_n is edge-symmetric, it suffices to consider only the case where $f_c \geq 1$ if $F \neq \emptyset$. That means that S_n can be split into $S_n^{\{i\}}$ for $1 \leq i \leq n$ using any dimension d for some d with $2 \leq d \leq n$. Therefore, given a nonempty faulty edge set F, S_n can be split into $S_n^{\{i\}}$ for $1 \leq i \leq n$ such that $F_c \neq \emptyset$.

The following lemmas can be easily obtained.

LEMMA 13.7 Assume that $n \geq 3$. $|E^{i,j}| = (n-2)!$ for any $1 \leq i \neq j \leq n$. Moreover, there are $(n-2)!/2$ edges joining black vertices of $S_n^{\{i\}}$ to white vertices of $S_n^{\{j\}}$.

LEMMA 13.8 Let $\mathbf{u}$ and $\mathbf{v}$ be any two distinct vertices of S_n with $d(\mathbf{u}, \mathbf{v}) \leq 2$. Then $(\mathbf{u})_1 \neq (\mathbf{v})_1$. Moreover, $\{((\mathbf{u})^i)_1 \mid 2 \leq i \leq n-1\} = \langle n \rangle - \{(\mathbf{u})_1, (\mathbf{u})_n\}$ if $n \geq 3$.

LEMMA 13.9 Let $n \geq 5$ and $I = \{k_1, k_2, \ldots, k_t\}$ be any t subset of $\langle n \rangle$ with $t \geq 2$. Let $\mathbf{u}$ be a white vertex of $S_n^{\{k_1\}}$ and $\mathbf{v}$ be a black vertex of $S_n^{\{k_t\}}$. Let F be any edge subset of S_n such that (1) $S_n^{\{k_i\}} - F_{k_i}$ is Hamiltonian laceable for $1 \leq i \leq t$ and (2) there exist's an edge in $E^{k_i,k_{i+1}} - F$ joining a black vertex of $S_n^{\{k_i\}}$ to a white vertex of $S_n^{\{k_{i+1}\}}$ for $1 \leq i < t$. Then there is a Hamiltonian path P of $S_n^I - F$ joining $\mathbf{u}$ to $\mathbf{v}$.

Proof: Obviously, $S_n^{\{k_i\}}$ is isomorphic to S_{n-1} for every $1 \leq i \leq t$. By assumption, there is a black vertex $\mathbf{y_i}$ in $S_n^{\{k_i\}}$ with $(\mathbf{y_i})^n \in S_n^{\{k_{i+1}\}}$ such that $(\mathbf{y_i}, (\mathbf{y_i})^n) \in E^{k_i,k_{i+1}} - F$ for every $1 \leq i < t$. We set $\mathbf{x_{i+1}} = (\mathbf{y_i})^n$ for every $i \in \langle t-1 \rangle$. Obviously, $\mathbf{x_i}$ is a white vertex for every $i \in \langle t \rangle$. By assumption, there is a Hamiltonian path H_i of $S_n^{\{k_i\}} - F^{k_i}$ joining $\mathbf{x_i}$ to $\mathbf{y_i}$ for every $1 \leq i \leq t$. Then $\langle \mathbf{u} = \mathbf{x_1}, H_1, \mathbf{y_1}, \mathbf{x_2}, H_2, \mathbf{y_2}, \ldots, \mathbf{x_t}, H_t, \mathbf{y_t} = \mathbf{v} \rangle$ forms the Hamiltonian path of $S_n^I - F$ joining $\mathbf{u}$ to $\mathbf{v}$. □

The following lemma can be proved using a computer program.

LEMMA 13.10 S_4 is 1-edge fault-tolerant Hamiltonian laceable, 1-edge fault-tolerant strongly Hamiltonian laceable, and hyper Hamiltonian laceable.

LEMMA 13.11 S_n is $(n-3)$-edge fault-tolerant Hamiltonian laceable for $n \geq 4$.

Proof: We prove this lemma by induction. By Lemma 13.10, S_4 is 1-edge fault-tolerant Hamiltonian laceable. Now, we assume that S_{n-1} is $(n-4)$-edge fault-tolerant Hamiltonian laceable with $n \geq 5$.

Let $F \subseteq E(S_n)$ be an arbitrary faulty edge set such that $|F| \leq n-3$. Since S_n is edge-symmetric, we may assume that $f_c \geq 1$ in $F \neq \emptyset$. Thus, $f_i \leq n-4$ for $1 \leq i \leq n$. So $S_n^{\{i\}} - F$ is still Hamiltonian laceable for each $1 \leq i \leq n$. Let $\mathbf{x}$ be a white vertex in $V(S_n^{\{r\}})$ and $\mathbf{y}$ be a black vertex in $V(S_n^{\{s\}})$. We need to find a Hamiltonian path of $S_n - F$ joining $\mathbf{x}$ to $\mathbf{y}$. Since $|F| \leq n-3 < (n-2)!/2$ for $n \geq 5$, $|E^{i,j} \cap F| < (n-2)!/2$ for any $i \neq j \in \langle n \rangle$. There exists an edge in $E^{i,j} - F$ joining a black vertex of $S_n^{\{i\}}$ to a white vertex of $S_n^{\{j\}}$ for $1 \leq i \neq j \leq n$.

Case 1. $r \neq s$. By Lemma 13.9, there is a Hamiltonian path of $S_n - F$ joining $\mathbf{x}$ to $\mathbf{y}$. See Figure 13.3a for illustration.

Case 2. $r = s$. Since $f_r \leq n-4$, by induction, there is a Hamiltonian path P of $S_n^{\{r\}}$ joining $\mathbf{x}$ to $\mathbf{y}$. Obviously, $l(P) = (n-1)! - 1$. Since $[(n-1)! - 1]/2 > n-3 \geq |F|$ for $n \geq 5$, there exists an edge $(\mathbf{u}, \mathbf{v})$ in P such that $\{(\mathbf{u}, (\mathbf{u})^n), (\mathbf{v}, (\mathbf{v})^n)\} \cap F = \emptyset$. Without loss of generality, we can write P as $\langle \mathbf{x}, P_1, \mathbf{u}, \mathbf{v}, P_2, \mathbf{y} \rangle$. Obviously, $\mathbf{u}$ is a black vertex or a white vertex. We consider only the case that $\mathbf{u}$ is a black vertex, because the other case can be similarly discussed. Then $(\mathbf{u})^n$ is a white vertex and $(\mathbf{v})^n$ is a black vertex. By Lemma 13.8, $(\mathbf{u})_1 \neq (\mathbf{v})_1$. By Lemma 13.9, there is a Hamiltonian path P_3 of $S_n^{\langle n \rangle - \{r\}} - F$ joining $(\mathbf{u})^n$ to $(\mathbf{v})^n$. Obviously, $\langle \mathbf{x}, P_1, \mathbf{u}, (\mathbf{u})^n, P_3, (\mathbf{v})^n, \mathbf{v}, P_2, \mathbf{y} \rangle$ forms a Hamiltonian path of S_n joining $\mathbf{x}$ to $\mathbf{y}$. See Figure 13.3b for illustration.

Hence, the lemma follows. □

LEMMA 13.12 S_n is $(n-3)$-edge fault-tolerant strongly Hamiltonian laceable for $n \geq 4$.

Proof: We also prove this lemma by induction. By Lemma 13.10, S_4 is 1-edge fault-tolerant strongly Hamiltonian laceable. Assume that S_{n-1} is $(n-4)$-edge fault-tolerant strongly Hamiltonian laceable with $n \geq 5$.

Let F be any faulty edge set in S_n with $|F| \leq n-3$. Let $\mathbf{x}$ and $\mathbf{y}$ be any two vertices of S_n in the same partite set. We need to find a path of length $n! - 2$ of $S_n - F$

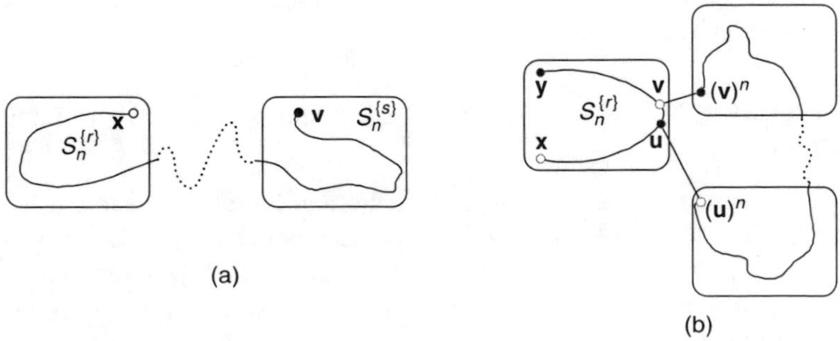

FIGURE 13.3 Illustration for Lemma 13.11.

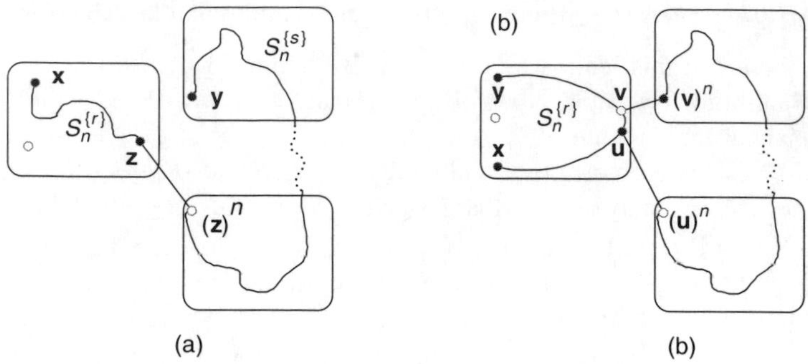

FIGURE 13.4 Illustration for Lemma 13.12.

joining **x** to **y**. Without loss of generality, we may assume that both **x** and **y** are black vertices. Assume that $(\mathbf{x})_n = r$ and $(\mathbf{y})_n = s$. Again, we can assume that $f_i \leq n-4$ for each $1 \leq i \leq n$. Since $|F| \leq n-3 < [(n-2)!/2] - 1$ for $n \geq 5$, there exists an edge in $E^{i,j} - F$ joining a black vertex of $S_n^{\{i\}}$ to a white vertex of $S_n^{\{j\}}$ for $1 \leq i \neq j \leq n$.

Case 1. $r \neq s$. Since $|F| \leq n-3 < [(n-2)!/2] - 1$ for $n \geq 5$, we can choose a black vertex **z** in $S_n^{\{r\}}$ such that $\mathbf{z} \neq \mathbf{x}$ and $(\mathbf{z})_1 \neq s$. By induction, there exists a path P_1 in $S_n^{\{r\}} - F$ of length $(n-1)! - 2$ joining **x** to **z**. Obviously, $(\mathbf{z})^n$ is a white vertex. By Lemma 13.9, there is a Hamiltonian path P_2 of $S_n^{\langle n\rangle - \{r\}} - F$ joining $(\mathbf{z})^n$ to **y**. Obviously, $\langle \mathbf{x}, P_1, \mathbf{z}, (\mathbf{z})^n, P_2, \mathbf{y}\rangle$ is a path of $S_n - F$ of length $n! - 2$ joining **x** to **y**. See Figure 13.4a for illustration.

Case 2. $r = s$. Since $f_r \leq n-4$, by induction, there is a path P of length $(n-1)! - 2$ in $S_n^{\{r\}} - F$ joining **x** to **y**. Since $[(n-1)! - 2]/2 > n-3 \geq |F|$ for $n \geq 5$, there exists an edge $(\mathbf{u}, \mathbf{v})$ in P such that $\{(\mathbf{u}, (\mathbf{u})^n), (\mathbf{v}, (\mathbf{v})^n)\} \cap F = \emptyset$. Without loss of generality, we can write P as $\langle \mathbf{x}, P_1, \mathbf{u}, \mathbf{v}, P_2, \mathbf{y}\rangle$. Obviously, **u** is a black vertex or a white vertex. We consider only the case that **u** is a black vertex, because the other case can be similarly discussed. Then $(\mathbf{u})^n$ is a white vertex and $(\mathbf{v})^n$ is a black vertex. By Lemma 13.8, $(\mathbf{u})_1 \neq (\mathbf{v})_1$. By Lemma 13.9, there is a Hamiltonian path P_3 of $S_n^{\langle n\rangle - \{r\}} - F$ joining $(\mathbf{u})^n$ to $(\mathbf{v})^n$. Obviously, $\langle \mathbf{x}, P_1, \mathbf{u}, (\mathbf{u})^n, P_3, (\mathbf{v})^n, \mathbf{v}, P_2, \mathbf{y}\rangle$ forms a path of length $n! - 2$ in $S_n - F$ joining **x** to **y**. See Figure 13.4b for illustration.

Hence, the lemma follows. □

LEMMA 13.13 **[222]** The S_n is $(n-4)$-edge fault-tolerant hyper Hamiltonian laceable for $n \geq 4$.

Proof: By Lemma 13.10, S_4 is hyper Hamiltonian laceable. Assume that S_{n-1} is $(n-5)$-edge fault-tolerant hyper Hamiltonian laceable with $n \geq 5$. Let F be a faulty edge set in S_n with $|F| \leq n-4$. Again, we may assume that $f_i \leq n-5$ for each $1 \leq i \leq n$. Thus, $S_n^{\{i\}} - F$ is hyper Hamiltonian laceable for $1 \leq i \leq n$. Let **w** be any vertex of S_n. Without loss of generality, we may assume that **w** is a white vertex. We need to construct a Hamiltonian path of $S_n - (F \cup \{\mathbf{w}\})$ between any two different white vertices **x** and **y**. Let $\mathbf{x} \in S_n^{\{r\}}$, $\mathbf{y} \in S_n^{\{s\}}$, and $\mathbf{w} \in S_n^{\{t\}}$. Since $|F| \leq n-4 < [(n-2)!/2] - 1$, there

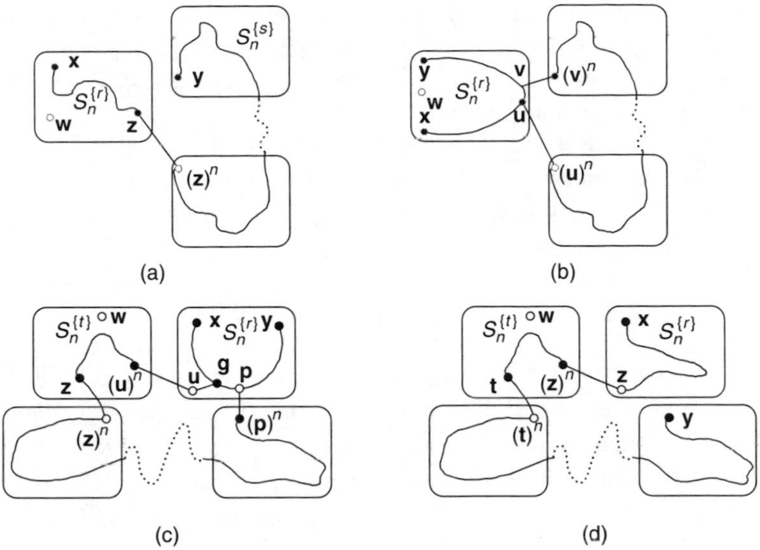

FIGURE 13.5 Illustration for Lemma 13.13.

exists an edge in $E^{i,j} - F$ joining a black vertex of $S_n^{\{i\}}$ to a white vertex of $S_n^{\{j\}}$ for $1 \leq i \neq j \leq n$.

Case 1. $r = t$ but $r \neq s$. Since $|F| \leq n-4 < [(n-2)!/2] - 1$, we can choose a black vertex $\mathbf{y_1}$ in $S_n^{\{r\}}$ such that $\mathbf{y_1} \neq \mathbf{x}$ and $((\mathbf{y_1})^n)_n \neq (\mathbf{y})_n$. By induction, there exists a Hamiltonian path P_1 of $S_n^{\{k_1\}} - F - \{\mathbf{w}\}$ joining $\mathbf{x}$ to $\mathbf{y_1}$. Obviously, $(\mathbf{y_1})^n$ is a white vertex. By Lemma 13.9, there is a Hamiltonian path P_2 of $S_n^{\langle n\rangle - \{r\}} - F$ joining $(\mathbf{y_1})^n$ to $\mathbf{y}$. Obviously, $\langle \mathbf{x}, P_1, \mathbf{y_1}, (\mathbf{y_1})^n, P_2, \mathbf{y}\rangle$ is a Hamiltonian path of $(S_n - F) - \{\mathbf{w}\}$ joining $\mathbf{x}$ to $\mathbf{y}$. See Figure 13.5a for illustration.

Case 2. $r = s = t$. Since $f_r \leq n-5$, by induction, there is a Hamiltonian path P of $(S_n^{\{r\}} - F) - \{\mathbf{w}\}$ joining $\mathbf{x}$ to $\mathbf{y}$. Obviously, $l(P) = (n-1)! - 1$. Since $[(n-1)! - 1]/2 > n-3 \geq |F|$ for $n \geq 5$, there exists an edge $(\mathbf{u}, \mathbf{v})$ in P such that $\{(\mathbf{u}, (\mathbf{u})^n), (\mathbf{v}, (\mathbf{v})^n)\} \cap F = \emptyset$. Without loss of generality, we can write P as $\langle \mathbf{x}, P_1, \mathbf{u}, \mathbf{v}, P_2, \mathbf{y}\rangle$. Obviously, $\mathbf{u}$ is a black vertex or a white vertex. We consider only the case that $\mathbf{u}$ is a black vertex, because the other case can be similarly discussed. Then $(\mathbf{u})^n$ is a white vertex and $(\mathbf{v})^n$ is a black vertex. By Lemma 13.8, $(\mathbf{u})_1 \neq (\mathbf{v})_1$. By Lemma 13.9, there is a Hamiltonian path P_3 of $S_n^{\langle n\rangle - \{r\}} - F$ joining $(\mathbf{u})^n$ to $(\mathbf{v})^n$. Obviously, $\langle \mathbf{x}, P_1, \mathbf{u}, (\mathbf{u})^n, P_3, (\mathbf{v})^n, \mathbf{v}, P_2, \mathbf{y}\rangle$ forms a Hamiltonian path of $(S_n - F) - \{\mathbf{w}\}$ joining $\mathbf{x}$ to $\mathbf{y}$. See Figure 13.5b for illustration.

Case 3. $r = s$ but $r \neq t$. Since there are $(n-2)!/2$ edges joining black vertices of $S_n^{\{t\}}$ to white vertices of $S_n^{\{r\}}$ and $|F| \leq n-4$, there exists a white vertex $\mathbf{u}$ in $S_n^{\{r\}}$ such that $(\mathbf{u}, (\mathbf{u}^n)) \in E^{r,t} - F$. Since $f_r \leq n-5$, by induction, there exists a fault-free Hamiltonian path P_1 in $(S_n^{\{r\}} - F) - \{\mathbf{u}\}$ joining $\mathbf{x}$ to $\mathbf{y}$. Since the number of neighboring vertices of $\mathbf{u}$ in $S_n^{\{r\}}$ is $n-2$ and $f_r \leq n-5$, there exists a neighbor $\mathbf{g}$ of $\mathbf{u}$ in $S_n^{\{r\}}$ such that $(\mathbf{u}, \mathbf{g}) \notin F$ and $(\mathbf{p}, (\mathbf{p})^n)$ where $\mathbf{p}$ is a vertex adjacent to $\mathbf{g}$ in P.

We can divide P into two sections P_0 and P_1 such that either (P_0 is a path joining $\mathbf{x}$ to $\mathbf{g}$ and P_1 is a path joining $\mathbf{p}$ to $\mathbf{y}$) or (P_0 is a path joining $\mathbf{x}$ to $\mathbf{p}$ and P_1 is a path joining $\mathbf{g}$ to $\mathbf{y}$). Without loss of generality, we may assume that P_0 is a path joining $\mathbf{x}$ to $\mathbf{g}$ and P_1 is a path joining $\mathbf{p}$ to $\mathbf{y}$.

Obviously, $\mathbf{u}$ and $\mathbf{p}$ are white vertices and $d(\mathbf{u},\mathbf{p})=2$. By Lemma 13.8, $(\mathbf{u})_1 \neq (\mathbf{q})_1$. We can choose a black vertex $\mathbf{z}$ in $S_n^{\{t\}}$ such that $(\mathbf{z})_1 \neq (\mathbf{p})_1$. By induction, there exists a Hamiltonian path P_2 of $(S_n^{\{t\}} - F) - \{\mathbf{w}\}$ joining $(\mathbf{u})^n$ to $\mathbf{z}$. By Lemma 13.9, there is a Hamiltonian path P_3 of $S_n^{\langle n\rangle - \{r,t\}} - F$ joining $(\mathbf{z})^n$ to $(\mathbf{p})^n$. Obviously, $\langle \mathbf{x}, P_0, \mathbf{g}, \mathbf{u}, (\mathbf{u})^n, P_2, \mathbf{z}, (\mathbf{z})^n, P_3, (\mathbf{p})^n, \mathbf{p}, P_1, \mathbf{y}\rangle$ forms a Hamiltonian path of $(S_n - F) - \{\mathbf{w}\}$ joining $\mathbf{x}$ to $\mathbf{y}$. See Figure 13.5c for illustration.

Case 4. r, s, and t are distinct. Again, there exists an edge $(\mathbf{z}, (\mathbf{z})^n)$ in $E^{r,t} - F$ such that $\mathbf{z}$ is a white vertex in $S_n^{\{r\}}$. Since $f_r \leq n-5$, there exists a Hamiltonian path of $S_n^{\{r\}} - F$ joining $\mathbf{x}$ to $\mathbf{z}$. Obviously, we cab choose a black vertex $\mathbf{t}$ such that $(\mathbf{t})_1 \notin \{r, s, t\}$. By induction, there exists a Hamiltonian path P_2 of $(S_n^{\{t\}} - F) - \{\mathbf{w}\}$ joining $(\mathbf{z})^n$ to $\mathbf{t}$. By Lemma 13.9, there is a Hamiltonian path P_3 of $S_n^{\langle n\rangle - \{r,t\}} - F$ joining $(\mathbf{t})^n$ to $\mathbf{y}$. Obviously, $\langle \mathbf{x}, P_1, \mathbf{z}, (\mathbf{z})^n, P_2, \mathbf{t}, (\mathbf{t})^n, P_3, \mathbf{y}\rangle$ forms a Hamiltonian path of $(S_n - F) - \{\mathbf{w}\}$ joining $\mathbf{x}$ to $\mathbf{y}$. See Figure 13.5d for illustration.

Hence, the lemma follows. □

Combining Lemmas 13.11 through 13.13, we have the following theorem.

THEOREM 13.2 The star graph S_n is super fault-tolerant Hamiltonian laceable for $n \geq 4$.

13.4 CONSTRUCTION SCHEMES

The operations $G_0 \oplus G_1$ and $G(G_0, G_1, \ldots, G_{r-1}; \mathcal{M})$ can also be used in the construction of super fault-tolerant Hamiltonian-laceable graphs.

LEMMA 13.14 The bipartite graph $G = G_0 \oplus G_1$ is hyper Hamiltonian laceable if both G_0 and G_1 are hyper Hamiltonian laceable.

Proof: Since both G_0 and G_1 are hyper Hamiltonian laceable by Theorem 9.17, G is Hamiltonian laceable.

Let B_i and W_i be the bipartition of G_i for $i \in \{0, 1\}$ and let $B_0 \cup B_1$ and $W_0 \cup W_1$ be the bipartition of G. Let u and v be two distinct vertices of $B_0 \cup B_1$ and let z be a vertex of $W_0 \cup W_1$. Without loss of generality, we may assume that $z \in W_0$. We need to construct a Hamiltonian path of $G - \{z\}$ between u and v.

Case 1. $u \in B_0$ and $v \in B_0$. Since G_0 is hyper Hamiltonian laceable, there is a Hamiltonian path H of $G_0 - \{z\}$ joining u to v. We write H as $\langle u, x, R, v\rangle$. Note that $\overline{u} \in W_1$ and $\overline{x} \in B_1$. By Lemma 13.1, there is a Hamiltonian path Q of G_1 joining $\overline{u}$ to $\overline{x}$. Then $\langle u, \overline{u}, Q, \overline{x}, x, R, v\rangle$ forms a Hamiltonian path of $G - \{z\}$ joining u to v.

Case 2. $u \in B_0$ and $v \in B_1$. Let x be any vertex in $B_0 - \{u\}$. Note that $\overline{x}$ is a vertex in W_1. Since G_0 is hyper Hamiltonian laceable, there is a Hamiltonian path R of $G_0 - \{z\}$ joining u to x. By Lemma 13.1, there is a Hamiltonian path H of G_1 joining $\overline{x}$ to v. Then $\langle u, R, x, \overline{x}, H, v\rangle$ forms a Hamiltonian path of $G - \{z\}$ joining u to v.

Case 3. $u \in B_1$ and $v \in B_1$. Let x be any vertex in G_1 such that $(x, v) \in E(G_1)$. Note that $\overline{x}$ is a vertex in B_0. Since G_1 is hyper Hamiltonian laceable, there is a Hamiltonian path H of $G_1 - \{x\}$ joining u to v. We write H as $\langle u, R, y, v\rangle$. Note that y is a vertex in W_1 and $\overline{y}$ is in B_0. Since G_0 is hyper Hamiltonian laceable, there is a Hamiltonian path Q of $G_0 - \{z\}$ joining $\overline{y}$ to $\overline{x}$. Then $\langle u, R, y, \overline{y}, Q, \overline{x}, x, v\rangle$ forms a Hamiltonian path of $G - \{z\}$ joining u to v. □

Yet we need the following term for bipartite graph. A bipartite graph $G = (V_0 \cup V_1, E)$ satisfies property $2H$ if for any two distinct vertices u and v in V_0 and for any two distinct x and y in V_1, there are two disjoint paths P_1 and P_2 of G such that (1) P_1 joins u to x, (2) P_2 joins v to y, and (3) $P_1 \cup P_2$ spans G.

LEMMA 13.15 A bipartite graph G is a Hamiltonian-laceable graph if G has property $2H$.

Proof: Let V_0 and V_1 be the bipartition of G. Let u be a vertex in V_0 and v be a vertex in V_1. Since G satisfies property $2H$, $|V_0| = |V_1| \geq 2$. We choose a vertex p in $V_0 - \{u\}$ and a vertex q in $V_1 - \{v\}$ such that $(p, q) \in E(G)$. Let P_1 and P_2 be two disjoint paths of G such that (1) P_1 joins u to q, (2) P_2 joins p to v, and $P_1 \cup P_2$ spans G. Then $\langle u, P_1, q, p, P_2, v\rangle$ is a Hamiltonian path of G joining u to v. □

By brute force, we have the following lemma.

LEMMA 13.16 The three-dimensional hypercube Q_3 satisfies the property $2H$.

LEMMA 13.17 Assume that G_0 and G_1 satisfy the property $2H$. Then the bipartite graph $G_0 \oplus G_1$ satisfies the property $2H$.

Proof: Let B_i and W_i be the bipartition of G_i for $i \in \{0, 1\}$. Without loss of generality, we assume that $B_0 \cup B_1$ and $W_0 \cup W_1$ are the bipartition of G. Let u and v be any two distinct vertices in $B_0 \cup B_1$ and let x and y be any two distinct vertices in $W_0 \cup W_1$. We have the following cases:

Case 1. $u \in B_0$, $v \in B_0$, $x \in W_0$, and $y \in W_0$. Since G_0 satisfies property $2H$, there are two disjoint paths H_1 and H_2 of G_0 such that (1) H_1 joins u to x, (2) H_2 joins v to y, and (3) $H_1 \cup H_2$ spans G_0. Without loss of generality, we write $H_1 = \langle u, z, Q, x\rangle$. By Lemma 13.15, there is a Hamiltonian path R of G_1 joining $\overline{u}$ to $\overline{z}$. We set $P_1 = \langle u, \overline{u}, Q, \overline{z}, z, R, x\rangle$ and $P_2 = R_2$. Then P_1 and P_2 form the desired paths.

Case 2. $u \in B_0$, $v \in B_0$, $x \in W_0$, and $y \in W_1$. By Lemma 13.15, there is a Hamiltonian path H of G_0 joining v to x. Without loss of generality, we write $R = \langle v, R_1, z, u, R_2, x\rangle$. By Lemma 13.15, there is a Hamiltonian path Q of G_1 joining $\overline{z}$ to y. We set $P_1 = R_2$ and $P_2 = \langle v, R_1, z, \overline{z}, Q, v\rangle$. Then P_1 and P_2 form the desired paths.

Case 3. $u \in B_0, v \in B_0, x \in W_1$, and $y \in W_1$. Let p and q be any two distinct vertices in W_0. Since G_0 satisfies property $2H$, there are two disjoint paths H_1 and H_2 of G_0 such that (1) H_1 joins u to p, (2) H_2 joins v to q, and (3) $H_1 \cup H_2$ spans G_0. Since G_1 satisfies property $2H$, there are two disjoint paths Q_1 and Q_2 of G_1 such that (1) Q_1 joins $\overline{p}$ to x, (2) Q_2 joins $\overline{q}$ to y, and (3) $Q_1 \cup Q_2$ spans G_1. We set $P_1 = \langle u, H_1, p, \overline{p}, Q_1, x\rangle$ and $P_2 = \langle v, H_2, q, \overline{q}, Q_2, y\rangle$. Then P_1 and P_2 form the desired paths.

Case 4. $u \in B_0, v \in B_1, x \in W_0$, and $y \in W_1$. By Lemma 13.15, there is a Hamiltonian path P_1 of G_0 joining u to x. Moreover, there is a Hamiltonian path P_2 of G_1 joining v to y. Then P_1 and P_2 form the desired paths.

Case 5. $u \in B_0$, $v \in B_1$, $x \in W_1$, and $y \in W_0$. Suppose that $|V(G_0)| = |V(G_1)| = 4$. Obviously, both G_0 and G_1 are isomorphic to a cycle with four vertices. Moreover, G is isomorphic to the three-dimensional hypercube Q_3. By Lemma 13.16, G satisfies property $2H$. We assume that $|V(G_0)| = |V(G_1)| \geq 6$. Let p be a vertex in $B_0 - \{u, \overline{x}\}$ and let q be a vertex in $W_0 - \{\overline{v}, y\}$. Since G_0 satisfies property $2H$, there are two disjoint paths H_1 and H_2 of G_0 such that (1) H_1 joins u to p, (2) H_2 joins q to y, and (3) $H_1 \cup H_2$ spans G_0. Since G_1 satisfies property $2H$, there are two disjoint paths Q_1 and Q_2 of G_2 such that (1) Q_1 joins $\overline{p}$ to x, (2) Q_2 joins v to $\overline{q}$, and (3) $Q_1 \cup Q_2$ spans G_1. We set $P_1 = \langle u, H_1, p, \overline{p}, Q_1, x\rangle$ and $P_2 = \langle v, Q_2, \overline{q}, q, H_1, y\rangle$. Then P_1 and P_2 form the desired paths. □

THEOREM 13.3 For $k \geq 3$, let $G = G_0 \oplus G_1$ be a bipartite graph such that both G_0 and G_1 are k-regular super fault-tolerant Hamiltonian laceable satisfies $2H$ property. Then G is $(k+1)$-regular $(k-1)$-edge fault-tolerant Hamiltonian laceable.

Proof: Let W_i and B_i be the bipartition of G_i for $i = 0, 1$. Without loss of generality, we assume that $B_0 \cup B_1$ and $W_0 \cup W_1$ are the bipartitions of G. Let F be any edge subset of G with $|F| = k - 1$. Let u be a vertex in $B_0 \cup B_1$ and let v be a vertex in $W_0 \cup W_1$. We need to show that there is a Hamiltonian path of $G - F$ joining u to v. We set $F_0 = F \cap E(G_0)$ and $F_1 = F \cap E(G_1)$. Let $2t = |V(G_0)|$. We have the following cases:

Case 1. $|F_0| \leq k - 2$ and $|F_1| \leq k - 2$.

Case 1.1. $u \in V_0^i$ and $v \in V_1^i$ for some i. Without loss of generality, we assume that $i = 0$. Since G_0 is $(k-2)$-edge fault-tolerant Hamiltonian laceable, there is a Hamiltonian path H of $G_0 - F_0$ joining u to v. Without loss of generality, we write $H = \langle u = x_0, x_1, \ldots, x_{2t} = v\rangle$. Since $k - 1 \leq t$, there is a positive integer i such that $(x_{2i-1}, \overline{x}_{2i-1}) \notin F$ and $(x_{2i}, \overline{x}_{2i}) \notin F$. Since G_1 is $(k-2)$-edge fault-tolerant Hamiltonian laceable, there is a Hamiltonian path R of $G_2 - F_1$ joining $\overline{x}_{2i-1}$ to $\overline{x}_{2i}$. Then $\langle u = x_1, x_2, \ldots, x_{2i-1}, \overline{x}_{2i-1}, R, \overline{x}_{2i}, x_{2i}, \ldots, x_{2t} = v\rangle$ forms the desired path.

Case 1.2. $u \in V_0^i$ and $v \in V_1^{1-i}$ for some i. Without loss of generality, we assume that $i = 0$. Since $|W_0| = t > k - 1$, there is a vertex $p \in W_0$ such that $(p, \overline{p}) \notin F$. Since G_0 is $(k-2)$-edge fault-tolerant Hamiltonian laceable, there is a Hamiltonian path H of $G_0 - F_0$ joining u to p. Since G_1 is $(k-2)$-edge fault-tolerant Hamiltonian laceable, there is a Hamiltonian path R of $G_1 - F_1$ joining $\overline{p}$ to v. Then $\langle u, H, p, \overline{p}, R, v\rangle$ forms the desired path.

Case 2. $|F_i| = k - 1$ for some i. Without loss of generality, we assume that $i = 0$.

Case 2.1. $u \in B_0$ and $v \in W_0$. Let f be any element in F_0. Since G_0 is $(k-2)$-edge fault-tolerant Hamiltonian laceable, there is a Hamiltonian path H of $G_0 - (F_0 - \{f\})$ joining u to v. We write $H = \langle u, H_1, p, q, H_2, v\rangle$ if $f = (p, q)$ in H or we write $H = \langle u, H_1, p, q, H_2, v\rangle$ by picking any edge (p, q) in H. Since G_1 is $(k-2)$-edge fault-tolerant Hamiltonian laceable, there is a Hamiltonian path R of G_1 joining $\overline{p}$ to $\overline{q}$. Then $\langle u, H_1, p, \overline{p}, R, \overline{q}, q, H_2, v\rangle$ forms the desired path.

Case 2.2. $u \in B_1$ and $v \in W_1$. Suppose that $k = 2$. Obviously, both G_0 and G_1 are isomorphic to a cycle with four vertices. Moreover, G is isomorphic to Q_3. It is easy to check that G is 1-edge fault-tolerant Hamiltonian laceable.

Suppose that $k \geq 3$. Thus, $k - 1 \geq 2$. Since $|F_0| = k - 1 \geq 2$, we can choose an edge $(x, y) \in F_0$ such that $|\{x, y\} \cap \{\overline{u}, \overline{v}\}| \leq 1$.

Case 2.2.1. $|\{x, y\} \cap \{\overline{u}, \overline{v}\}| = 0$. Without loss of generality, we assume that $x \in B_0$ and $y \in W_0$. Since G_0 is $(k-2)$-edge fault-tolerant Hamiltonian laceable, there is a Hamiltonian path H of $G_0 - (F_0 - \{(x, y)\})$ joining x to y. Since G_1 satisfies property $2H$, there are two disjoint paths Q_1 and Q_2 of G_1 such that (1) Q_1 joins u to $\overline{x}$, (2) Q_2 joins $\overline{q}$ to v, and (3) $Q_1 \cup Q_2$ spans G. Then $\langle u, Q_1, \overline{x}, x, H, y, \overline{y}, Q_2, v\rangle$ forms the desired path.

Case 2.2.2. $|\{x, y\} \cap \{\overline{u}, \overline{v}\}| = 1$. Without loss of generality, we assume that $x = \overline{u}$. Since G_0 is $(k-2)$-edge fault tolerant Hamiltonian laceable, there is a Hamiltonian path H of $G_0 - (F_0 - \{(x, y)\})$ joining x to y. Since G_1 is hyper Hamiltonian laceable, there is a Hamiltonian path R of $G_1 - \{u\}$ joining $\overline{y}$ to v. Then $\langle u, \overline{u} = x, H, y, \overline{y}, R, v\rangle$ forms the desired path.

Case 2.3. $u \in V_0^i$ and $v \in V_1^{1-i}$ for some i. Without loss of generality, we assume that $i = 0$. Let (x, y) be any element in F_0. Without loss of generality, we assume that $x \in B_0$ and $y \in W_0$.

Case 2.3.1. $x = u$. Since G_0 is $(k-2)$-edge fault-tolerant Hamiltonian laceable, there is a Hamiltonian path H of $G_0 - (F_0 - \{(x, y)\})$ joining u to y. Since G_1 is Hamiltonian laceable, there is a Hamiltonian path R of G_1 joining $\overline{y}$ to v. Then $\langle u, H, y, \overline{y}, R, v\rangle$ forms the desired path.

Case 2.3.2. $x \neq u$ and $x = \overline{v}$. Since G_0 is $(k-2)$-edge fault-tolerant Hamiltonian laceable, there is a Hamiltonian path H of $G_0 - (F_0 - \{(x, y)\})$ joining x to y. Without loss of generality, we write $H = \langle x, H_1, z, u, H_2, y\rangle$. Since G_1 is hyper Hamiltonian laceable, there is a Hamiltonian path R of $G_1 - \{v\}$ joining $\overline{y}$ to $\overline{z}$. Then $\langle u, H_2, y, \overline{y}, R, \overline{z}, z, H_1^{-1}, x = \overline{v}, v\rangle$ forms the desired path.

Case 2.3.3. $x \neq u$ and $x \neq \overline{v}$. Since G_0 is $(k-2)$-edge fault-tolerant Hamiltonian laceable, there is a Hamiltonian path H of $G_0 - (F_0 - \{(x, y)\})$ joining x to y. Without loss of generality, we write $H = \langle x, H_1, z, u, H_2, y\rangle$. Since G_1 satisfies property $2H$, there are two disjoint paths Q_1 and Q_2 of G_2 such that (1) Q_1 joins $\overline{y}$ to $\overline{x}$, (2) Q_2 joins $\overline{z}$ to v, and (3) $Q_1 \cup Q_2$ spans G. Then $\langle u, H_2, y, \overline{y}, Q_1, \overline{x}, x, H_1, z, \overline{z}, v\rangle$ forms the desired path. □

THEOREM 13.4 For $k \geq 3$, let $G = G_0 \oplus G_1$ be a bipartite graph such that both G_0 and G_1 are k-regular super fault-tolerant Hamiltonian laceable satisfies $2H$ property. Then G is $(k+1)$-regular $(k-1)$-edge fault-tolerant strongly Hamiltonian laceable.

Proof: Let W_i and B_i be the bipartition of G_i for $i = 0, 1$. Without loss of generality, we assume that $B_0 \cup B_1$ and $W_0 \cup W_1$ are the bipartition of G. Let F be any edge subset of G with $|F| = k-1$. Let u and v be two distinct vertices in $W_0 \cup W_1$. We need to show that there is a Hamiltonian path with length $|V(G)| - 2$ of $G - F$ joining u to v. We set $F_0 = F \cap E(G_0)$ and $F_1 = F \cap E(G_1)$. Without loss of generality, we assume that $|F_0| \geq |F_1|$. We have $|F_1| \leq k-2$ and $G_1 - F_1$ is Hamiltonian laceable and strongly Hamiltonian laceable. Let $2t = |V(G_0)|$. We have the following cases:

Case 1. $u \in V_0^1$ and $v \in V_0^1$.

Case 1.1. $|F_0| \leq k-2$. Since G_0 is $(k-2)$-edge fault-tolerant strongly Hamiltonian laceable, there is a path H with length $|V(G_0)| - 2$ of $G_0 - F_0$) joining u to v. Since G_0 is a k-regular bipartite graph, $|V(G_0)| \geq 2k$ and $l(H) \geq 2k-2$. Since $2k-2 > 2(k-2)$, there is an edge (p, q) of $V(H)$ such that $(p, \overline{p}) \notin F_0$ and $(q, \overline{q}) \notin F_0$. Without loss of generality, we write $H = \langle u, H_1, p, q, H_2, H_2, v \rangle$. Since G_1 is $(k-2)$-edge fault-tolerant Hamiltonian laceable, there is a Hamiltonian path R of $G_1 - F_1$ joining $\overline{p}$ to $\overline{q}$. Then $\langle u, H_1, p, \overline{p}, R, \overline{q}, H_2, v \rangle$ forms a path with length $|V(G)| - 2$ of $G - F$ joining u to v.

Case 1.2. $|F_0| = k-1$. Let (x, y) be any edge in F_0. Since G_0 is $(k-2)$-edge fault-tolerant strongly Hamiltonian laceable, there is a path H of $G_0 - (F_0 - \{(x, y)\})$ joining u to v. We write $H = \langle u, R_1, p, q, R_2, v \rangle$, where $\{x, y\} = \{p, q\}$ if $(x, y) \in E(H)$ or (p, q) is any edge in $E(H)$ if $(x, y) \notin E(H)$. Since G_1 is Hamiltonian laceable, there is a Hamiltonian path R of G_1 joining $\overline{p}$ to $\overline{q}$. Then $\langle u, H_1, p, \overline{p}, R, \overline{q}, H_2, v \rangle$ forms a path with length $|V(G)| - 2$ of $G - F$ joining u to v.

Case 2. $u \in G_0$ and $v \in G_1$.

Case 2.1. $|F_0| = 0$. Since G_0 is a k-regular bipartite graph, $|W_0| + |B_0| \geq 2k$. We can choose a vertex p in $V(G_0)$ such that $p \neq u$, $\overline{p} \neq v$, and $(p, \overline{p}) \notin F$.

Suppose that $p \in B_0$. Since G_0 is Hamiltonian laceable, there is a Hamiltonian path H of G_0 joining u to p. Since G_1 is strongly Hamiltonian laceable, there is a path R with length $|V(G_1)| - 2$ of G_1 joining $\overline{p}$ to v. Then $\langle u, H, p, \overline{p}, R, v \rangle$ forms a path with length $|V(G)| - 2$ of $G - F$ joining u to v.

Suppose that $p \in W_0$. Since G_0 is strongly Hamiltonian laceable, there is a path H of G_0 with length $V(G_0) - 2$ joining u to p. Since G_1 is Hamiltonian laceable, there is a Hamiltonian path R with length $|V(G_1)| - 2$ of G_1 joining $\overline{p}$ to v. Then $\langle u, H, p, \overline{p}, R, v \rangle$ forms a path with length $|V(G)| - 2$ of $G - F$ joining u to v.

Case 2.2. $1 \leq |F_0| \geq k-2$. Since $|B_0| \geq k$, we can choose a vertex z in G_0 such that $(z, \overline{z}) \notin F$ and $\overline{z} \neq v$. Since G_0 is $(k-2)$-edge fault-tolerant Hamiltonian laceable, there is a Hamiltonian path H of $G_0 - F_0$ joining u to z. Since G_1 is $(k-3)$-edge fault-tolerant strongly Hamiltonian laceable, there is a path R of $G_1 - F_1$ with length $|V(G_1)| - 2$ joining $\overline{z}$ to v. Then $\langle u, H, z, \overline{z}, R, v \rangle$ forms a path of $G - F$ with length $|V(G) - 2$ joining u to v.

Case 2.3. $|F_0| = k-1$. Let (p, q) be any edge in F_0 with $p \in W_0$ and $q \in B_0$. Since G_0 is $(k-2)$-edge fault-tolerant Hamiltonian laceable, there is a Hamiltonian laceable path H of $G_0 - (F_0 - \{(p, q)\})$ joining u to q. Without loss of generality, we write $H = \langle u, W, z, q\rangle$ where $z = p$ if $(p, q) \in E(H)$. Note that $z \in W_0$ and $\overline{z} \in B_1$. Since G_1 is Hamiltonian laceable, there is a Hamiltonian path R of G_1 joining $\overline{z}$ to v. Then $\langle u, W, z, \overline{z}, R, v\rangle$ forms a path of G with length $|V(G)| - 2$ joining u to v.

Case 3. $u \in G_1$ and $v \in G_1$.

Case 3.1. $|F_0| \leq k-2$. Since G_1 is $(k-2)$-edge fault-tolerant strongly Hamiltonian laceable, there is a path R with length $|V(G_1)| - 2$ of $G_1 - F_1$ joining u and v. Since G_1 is a k-regular bipartite graph, $|V(G_1)| \geq 2k$. Since $2k-1 > 2k-2$, there is a edge $(p, q) \in E(R)$ such that $(p, \overline{p}) \notin F$ and $(q, \overline{q}) \notin F$. Without loss of generality, we write $R = \langle u, R_1, p, q, R_2, v\rangle$. Since G_0 is $(k-2)$-edge fault-tolerant Hamiltonian laceable, there is a Hamiltonian path H of $G_0 - F_0$ joining $\overline{p}$ and $\overline{q}$. Then $\langle u, R_1, p, \overline{p}, H, \overline{q}, q, R_2, v\rangle$ forms a path with length $|V(G)| - 2$ of $G - F$ joining u to v.

Case 3.2. $|F_0| = k-1$. Let (p, q) be any edge in F_0. Without loss of generality, we assume that $p \in W_0$ and $q \in B_0$. Since G_0 is $(n-2)$-edge fault-tolerant Hamiltonian laceable, there is a Hamiltonian path H of $G_0 - F_0$ joining p to q. Without loss of generality, we write $H = \langle p, Q, z, q\rangle$. Since G_1 satisfies the $2H$ property, there are two disjoint paths R_1 and R_2 of G_1 such that (1) R_1 joins u to $\overline{p}$, (2) R_2 joins $\overline{z}$ to v, and (3) $R_1 \cup R_2$ spans G_1. Then $\langle u, R_1, \overline{p}, p, Q, z, \overline{z}, R_2, v\rangle$ forms the desired path of $G - F$. □

THEOREM 13.5 For $k \geq 3$, let $G = G_0 \oplus G_1$ be a bipartite graph such that both G_0 and G_1 are k-regular super fault-tolerant Hamiltonian laceable satisfies $2H$ property. Then G is $(k+1)$-regular $(k-2)$-edge fault-tolerant hyper Hamiltonian laceable.

Proof: Let W_i and B_i be the bipartition of G_i for $i = 0, 1$. Without loss of generality, we assume that $B_0 \cup B_1$ and $W_0 \cup W_1$ are the bipartition of G. Let F be any edge subset of G with $|F| = k-2$. Let z be a vertex in $B_0 \cup B_1$ and let u and v be two distinct vertices in $W_0 \cup W_1$. We need to show that there is a Hamiltonian path of $G - (F \cup \{z\})$ joining u to v. We set $F_0 = F \cap E(G_0)$ and $F_1 = F \cap E(G_1)$. Let $2t = |V(G_0)|$. Without loss of generality, we assume that $z \in B_0$. We have the following cases:

Case 1. $u \in G_0$ and $v \in G_0$.

Case 1.1. $|F_0| \leq k-3$. Since G_0 is $(k-3)$-edge fault-tolerant hyper Hamiltonian, there is a Hamiltonian path H of $G_0 - (F_0 \cup \{z\})$ joining u to v. Since G_0 is a k-regular bipartite graph, $|V(G_0)| \geq 2k$ and $l(H) \geq 2k-2$. Since $2k-2 > 2(k-2)$, there is an edge (p, q) of $V(H)$ such that $(p, \overline{p}) \notin F_0$ and $(q, \overline{q}) \notin F_0$. Without loss of generality, we write $H = \langle u, H_1, p, q, H_2, H_2, v\rangle$. Since G_1 is $(k-2)$-edge fault Hamiltonian laceable, there is a Hamiltonian path R of $G_1 - F_1$ joining $\overline{p}$ to $\overline{q}$. Then $\langle u, H_1, p, \overline{p}, R, \overline{q}, H_2, v\rangle$ forms a Hamiltonian path of $G - (F \cup \{z\})$ joining u to v.

Case 1.2. $|F_0| = k-2$. Since G_0 is $(k-2)$-edge fault-tolerant Hamiltonian laceable, there is a Hamiltonian path H of $G_0 - F_0$ joining u to z. Without loss of generality, we write $H = \langle u, H_1, p, v, H_2, q, z\rangle$. Note that $q = v$ if $l(H_2) = 0$. Since G_1 is Hamiltonian laceable, there is a Hamiltonian path R of G_1 joining $\overline{p}$ to $\overline{q}$. Then $\langle u, H_1, p, \overline{p}, R, \overline{q}, H_2^{-1}, v\rangle$ forms a Hamiltonian path of $G - (F \cup \{z\})$ joining u to v.

Case 2. $u \in G_0$ and $v \in G_1$.

Case 2.1. $|F_0| \leq k - 3$. Since G_0 is a k-regular bipartite graph, $|W_0| \geq k$. We can choose a vertex p in $W_0 - \{u\}$ such that $(p, \overline{p}) \notin F$. Since G_0 is $(k-3)$-edge fault-tolerant hyper Hamiltonian, there is a Hamiltonian path H of $G_0 - (F_0 \cup \{z\})$ joining u to p. Since G_1 is $(k-2)$-edge fault-tolerant Hamiltonian laceable, there is a Hamiltonian path R of $G_1 - F_1$ joining $\overline{p}$ to v. Then $\langle u, H, p, \overline{p}, R, v\rangle$ forms a Hamiltonian path of $G - (F \cup \{z\})$ joining u to v.

Case 2.2. $|F_0| = k - 2$. Since G_0 is $(k-2)$-edge fault-tolerant Hamiltonian laceable, there is a Hamiltonian path H of $G_0 - F_0$ joining u to z. Without loss of generality, we write $H = \langle u, Q, p, z\rangle$. Obviously, $p \in W_0$ and $\overline{p} \in B_1$. Since G_1 is Hamiltonian laceable, there is a Hamiltonian path R of G_1 joining $\overline{p}$ to v. Then $\langle u, Q, p, \overline{p}, R, v\rangle$ forms a Hamiltonian path of $G - (F \cup \{z\})$ joining u to v.

Case 3. $u \in G_1$ and $v \in G_1$.

Case 3.1. $|F_0| \leq k - 3$. Let p be a vertex in B_1 with $(p, \overline{p}) \notin F$. We set $F' = \{(v, z) \mid z \in B_1 \text{ and } (z, \overline{z}) \in F\}$. Obviously, $|F_1 \cup F'| \leq |F|$. Since G_1 is $(k-2)$-edge fault-tolerant Hamiltonian laceable, there is a Hamiltonian path R of $G_1 - F_1$ joining u to p. Without loss of generality, we write $R = \langle u, R_1, q, v, R_2, p\rangle$. Since G_0 is $(k-3)$-edge fault-tolerant hyper Hamiltonian laceable, there is a Hamiltonian path H of $G_0 - (F_0 \cup \{z\})$ joining $\overline{q}$ to $\overline{p}$. Then $\langle u, R_1, q, \overline{q}, H, \overline{p}, p, R_2^{-1}, v\rangle$ forms a Hamiltonian path of $G - (F \cup \{z\})$ joining u to v.

Case 3.2. $|F_0| = k - 2$. Let p be any vertex in W_1. Since G_0 is $(n-2)$-edge fault-tolerant Hamiltonian laceable, there is a Hamiltonian path H of $G_0 - F_0$ joining p to z. Without loss of generality, we write $H = \langle p, Q, q, z\rangle$. Since G_1 satisfies the $2H$ property, there are two disjoint paths H_1 and H_2 of G_1 such that (1) H_1 joins u to $\overline{u}$, (2) H_2 joins $\overline{q}$ to v, and (3) $H_1 \cup H_2$ spans G_1. Then $\langle u, H_1, \overline{p}, p, Q, q, \overline{q}, H_2, v\rangle$ forms the desired path of $G - F$. □

By Theorems 13.3 through 13.5, we have the following result.

THEOREM 13.6 For $k \geq 3$, let $G = G_0 \oplus G_1$ be a bipartite graph such that both G_0 and G_1 are k-regular super fault-tolerant Hamiltonian laceable satisfy the $2H$ property. Then G is super fault-tolerant Hamiltonian.

Similarly, we have the following result.

THEOREM 13.7 For $k \geq 3$, let $G = G(G_0, G_1, \ldots, G_{r-1}; \mathcal{M})$ be a bipartite graph such that G_i is k-regular super fault-tolerant Hamiltonian laceable satisfies the $2H$ property. Then G is super fault-tolerant Hamiltonian.

13.5 CUBIC HAMILTONIAN-LACEABLE GRAPHS

In Section 13.4, we mentioned that several k-regular graphs are $(k-2)$-edge fault-tolerant Hamiltonian, $(k-2)$-edge fault-tolerant Hamiltonian laceable, $(k-2)$-edge fault-tolerant strongly Hamiltonian laceable, and $(k-3)$-edge fault-tolerant hyper Hamiltonian laceable. It is reasonable to compare the differences among these concepts. However, not too many results are obtained. We begin with $k=3$. Let G be the cubic graph shown in Figure 13.6. By brute force, we can confirm that G is both 1-edge fault-tolerant Hamiltonian and Hamiltonian laceable. However, it is neither 1-edge fault-tolerant Hamiltonian laceable nor hyper Hamiltonian laceable. More precisely, there is no Hamiltonian path of $G-(1,2)$ joining vertex 3 to vertex 6, and there is no Hamiltonian path of $G-18$ joining vertex 15 to vertex 10.

We have observed that the operation 3-join is used for constructing 1-fault-tolerant Hamiltonian graphs. We would like to know the corresponding results for cubic bipartite graphs.

Let $G_1=(B_1\cup W_1,E_1)$ be a cubic bipartite graph and $u\in B_1$. Let $G_2=(B_2\cup W_2,E_2)$ be a cubic bipartite graph and $v\in W_2$. Let $N(u)=\{u_i\mid i=1,2,3\}$, $N(v)=\{v_i\mid i=1,2,3\}$, and $G=J(G_1,N(u);G_2,N(v))$. It is easy to confirm that G is a cubic bipartite graph with bipartition $B=(B_1\cup B_2)-\{u\}$ and $W=(W_1\cup W_2)-\{v\}$.

THEOREM 13.8 Let $G_1=(B_1\cup W_1,E_1)$ be a cubic 1-edge fault-tolerant Hamiltonian laceable and hyper Hamiltonian-laceable graph with $u\in B_1$ and $G_2=(B_2\cup W_2,E_2)$ be a cubic 1-edge fault-tolerant Hamiltonian laceable and hyper Hamiltonian-laceable graph with $v\in W_2$. Then $G=J(G_1,N(u);G_2,N(v))$ is Hamiltonian laceable.

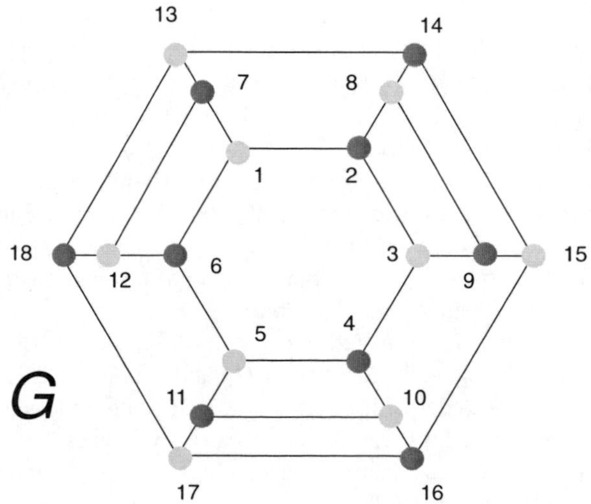

FIGURE 13.6 The graph G is both 1-edge fault-tolerant Hamiltonian and Hamiltonian laceable. However, it is neither 1-edge fault-tolerant Hamiltonian laceable nor hyper Hamiltonian laceable.

Moreover, G is neither 1-edge fault-tolerant Hamiltonian laceable nor hyper Hamiltonian laceable.

Proof: Suppose that both G_1 and G_2 are 1-edge fault-tolerant Hamiltonian laceable. We first prove that G is Hamiltonian laceable. Let x and y be two vertices from different partite sets of G. We want to find a Hamiltonian path of G joining x to y. We have the following cases:

Case 1. x and y are in $V(G_i)$ for $i \in \{1, 2\}$. Without loss of generality, we assume that x and y are in $V(G_1)$ with $x \in W_1$ and $y \in B_1$. Since G_1 is Hamiltonian laceable, there is a Hamiltonian path P_1 of G_1 joining x to y. Obviously, P_1 can be written as $\langle x, Q_1, u_i, u, u_j, Q_2, y\rangle$. Obviously, $\{u_i, u_j\} \subset W_1$, $\{v_i, v_j\} \subset B_2$, and $\{(u_i, v_i), (u_j, v_j)\} \in E(G)$. Since G_2 is hyper Hamiltonian laceable, there exists a Hamiltonian path Q_3 of $G_2 - v$ joining v_i and v_j. Obviously, $\langle x, Q_1, u_i, v_i, Q_3, v_j, Q_2, y\rangle$ forms a Hamiltonian path of G joining x to y.

Case 2. $x \in W_1$ and $y \in B_2$. Since G_1 is Hamiltonian laceable, there is a Hamiltonian path P_1 of G_1 joining x to u. Obviously, P_1 can be written as $\langle x, Q_1, u_i, u\rangle$. Obviously, $u_i \subset W_1$, $v_i \subset B_2$, and $(u_i, v_i) \in E(G)$. Since G_2 is hyper Hamiltonian laceable, there exists a Hamiltonian path Q_2 of $G_2 - v$ joining v_2 to y. Obviously, $\langle x, Q_1, u_i, v_i, Q_2, y\rangle$ forms a Hamiltonian path of G joining x to y.

Case 3. $x \in B_1$ and $y \in W_2$. Since G_1 is 1-edge fault-tolerant Hamiltonian laceable, there exists a Hamiltonian path P_1 of $G_1 - (u, u_1)$ joining x to u_1. Obviously, P_1 can be written as $\langle x, Q_1, u_i, u, u_j, Q_2, u_1\rangle$ with $\{i, j\} = \{2, 3\}$. Since G_2 is 1-edge fault-tolerant Hamiltonian laceable, there exists a Hamiltonian path P_2 of $G_2 - (v, v_i)$ joining y to v_i. Obviously, P_2 can either be written as $\langle y, Q_3, v_1, v, v_j, Q_4, v_i\rangle$ or $\langle y, Q_5, v_j, v, v_1, Q_6, v_i\rangle$.

Suppose that $P_2 = \langle y, Q_3, v_1, v, v_j, Q_4, v_i\rangle$. Then $\langle x, Q_1, u_i, v_i, Q_4^{-1}, v_j, u_j, Q_2, u_1, v_1, Q_3^{-1}, y\rangle$ forms a Hamiltonian path of G joining x to y. Suppose that $P_2 = \langle y, Q_5, v_j, v, v_1, Q_6, v_i\rangle$. Then $\langle x, Q_1, u_i, v_i, Q_6^{-1}, v_1, u_1, Q_2^{-1}, u_j, v_j, Q_5^{-1}, y\rangle$ forms a Hamiltonian path of G joining x to y.

Thus, G is Hamiltonian laceable.

Now, we prove that G is not 1-edge fault-tolerant Hamiltonian laceable. Let x be a vertex of G in B_1 and y is a vertex of G in W_2. We will claim that any Hamiltonian path P of G joining x and y contains the edge set $\{(u_i, v_i) \mid i = 1, 2, 3\}$. Obviously, there is no Hamiltonian path of $G - (u_1, v_1)$ joining x to y if our claim is true. Thus, G is not 1-edge fault-tolerant Hamiltonian laceable.

Note that P uses either exactly one or three edges of the edge set $\{(u_i, v_i) \mid i = 1, 2, 3\}$ because x is a vertex in G_1 and y is a vertex in G_2. Suppose that P includes exactly one edge of $\{(u_i, v_i) \mid i = 1, 2, 3\}$. Without loss of generality, we assume that P includes the edge (u_1, v_1). Then P can be written as $\langle x, Q_1, u_1, v_1, Q_2, y\rangle$. Obviously, $\langle x, Q_1, u_1, u\rangle$ forms a Hamiltonian path of G_1. This is impossible because both x and u are in the same partite set.

Finally, we prove that G is not hyper Hamiltonian laceable. Let x and y be two vertices in B_2. Suppose that G is hyper Hamiltonian laceable. Then there exists a Hamiltonian path P of $G - u_1$ joining x to y. This implies that there exists a

Hamiltonian path Q of G_2 joining x to y. This is impossible because x and y are of the same color in G_2. Thus, G is not hyper Hamiltonian laceable.

The theorem is proved. □

However, we would like to conjecture that $G = J(G_1, N(u); G_2, N(v))$ is Hamiltonian laceable if both G_1 and G_2 are cubic Hamiltonian-laceable graphs.

13.6 1_p-FAULT-TOLERANT HAMILTONIAN GRAPHS

Up to now, we have not discussed the Hamiltonian properties with vertex faults for bipartite graphs. Note that any cycle in a bipartite graph is of even length. Any vertex fault in a Hamiltonian bipartite graph is not Hamiltonian. However, some cubic bipartite Hamiltonian graphs remain Hamiltonian with a pair of vertices faults, one vertex fault from each partite set. A cubic bipartite Hamiltonian graph is said to be *1_p-fault-tolerant Hamiltonian* if it remains Hamiltonian with a pair of vertices faults. In this section, we propose two families of cubic bipartite Hamiltonian graphs as examples of 1_p-fault-tolerant Hamiltonian: spider web networks and brother trees.

13.6.1 SPIDER-WEB NETWORKS

To define the spider web network, we first introduce the honeycomb rectangular mesh. Assume that m and n are positive even integers with $m \geq 4$.

The *honeycomb rectangular mesh $HREM(m, n)$* is the graph with the vertex set $\{(i, j) \mid 0 \leq i < m, 0 \leq j < n\}$ such that (i, j) and (k, l) are adjacent if they satisfy one of the following conditions:

1. $i = k$ and $j = l \pm 1$
2. $j = l$ and $k = i + 1$ if $i + j$ is odd
3. $j = l$ and $k = i - 1$ if $i + j$ is even

For example, a honeycomb rectangular mesh $HREM(8, 6)$ is shown in Figure 13.7.

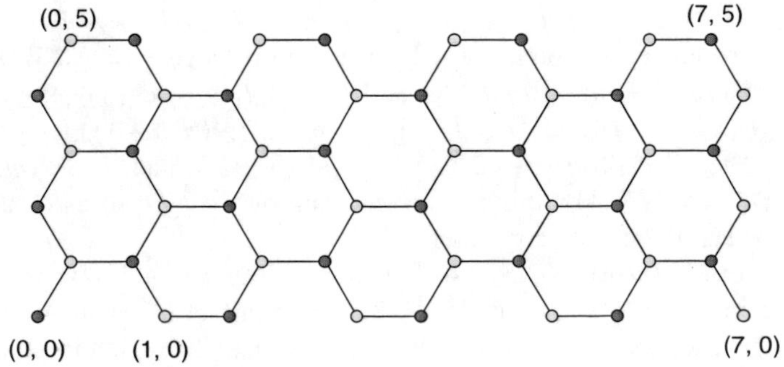

FIGURE 13.7 *HREM*(8, 6).

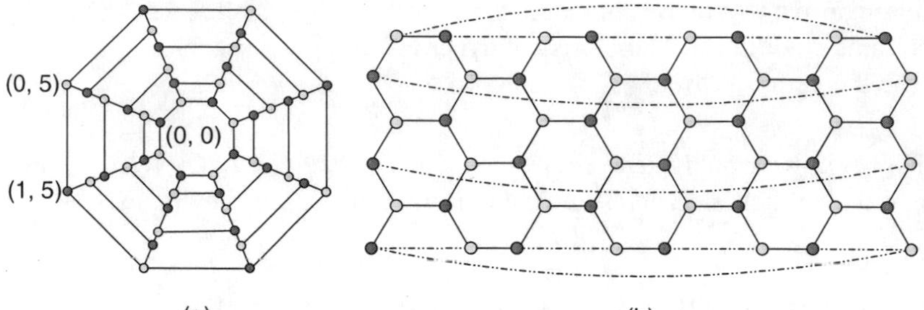

FIGURE 13.8 $SW(8,6)$.

A *spider web network* $SW(m,n)$, proposed by Kao and Hsu [190], is the graph with the vertex set $\{(i,j)\,|\,0 \le i < m, 0 \le j < n\}$ such that (i,j) and (k,l) are adjacent if they satisfy one of the following conditions:

1. $i=k$ and $j=l\pm 1$
2. $j=l$ and $k=[i+1]_m$ if $i+j$ is odd or $j=n-1$
3. $j=l$ and $k=[i-1]_m$ if $i+j$ is even or $j=0$

For example, a spider graph $SW(8,6)$ is shown in Figure 13.8a. Another layout of $SW(8,6)$ is shown in Figure 13.8b, with the dashed lines indicating the edges of $SW(m,n)$ that are not in $HREM(m,n)$. Obviously, $HREM(m,n)$ is a spanning subgraph of $SW(m,n)$. The *inner cycle* of $SW(m,n)$ is $\langle(0,0), (1,0),\ldots,(m-1,0),(0,0)\rangle$, whereas the *outer cycle* of $SW(m,n)$ is $\langle(0,n-1), (1,n-1),\ldots,(m-1,n-1),(0,n-1)\rangle$. It is obvious that any spider web network is a planar 3-regular bipartite graph. A vertex (i,j) is labeled black when $i+j$ is even and white if otherwise.

Since the honeycomb rectangular mesh $HREM(m,n)$ is a spanning subgraph of $SW(m,n)$, the spider web network can be viewed as a variation of the honeycomb mesh. Using the definition of a spider web network, $SW(m,n+2)$ can be constructed from $SW(m,n)$ as follows: let S denote the edge subset $\{((i,n-1),([i-1]_m,n-1))\,|\,i=0,2,4,\ldots,m-2\}$ of $SW(m,n)$. Let $SW^*(m,n)$ denote the spanning subgraph of $SW(m,n)$ with edge set $E(SW(m,n))-S$. Let $V^n=\{(i,k)\,|\,0\le i<m; k=n,n+1\}$ and $E^n=\{((i,k),(i,k+1))\,|\,0\le i<m;\ k=n-1,n\}\cup\{((i,n),([i-1]_m,n))\,|\,i=0,2,4,\ldots,m-2\}\cup\{((i,n+1),([i+1]_m,n+1))\,|\,0\le i<m\}$. Then $V(SW(m,n+2))=V(SW(m,n))\cup V^n$ and $E(SW(m,n+2))=(E(SW(m,n))-S)\cup E^n$. For this reason, we can view $SW(m,n)$ as a substructure of $SW(m,n+2)$ if there is no confusion.

Let $F'\subset V(SW^*(m,n))\cup E(SW^*(m,n))$ be a faulty set such that $|F'|=1$ if F' has an edge in $E(SW^*(m,n))$ and $|F'|=2$ if F' consists of a pair of vertices in $V(SW^*(m,n))$, one from each partite set. Suppose that C is a Hamiltonian cycle of $SW(m,n)-F'$, in which $(i,n-1)$ is fault-free for some $0\le i<m$. Now, we are going to construct a Hamiltonian cycle of $SW(m,n+2)$ as follows.

Case 1. There is some edge in $S \cap E(\mathcal{C})$. We can pick an edge $((r, n-1), ([r-1]_m, n-1)) \in \mathcal{C}$ for some even integer $0 \le r < m-1$. For $0 \le i \le m-2$, we define $e^* = (([r+i]_m, n-1), ([r+i+1]_m, n-1))$, and Q_i as

$$Q_i = \langle ([r+i]_m, n+1), ([r+i+1]_m, n+1) \rangle \quad \textit{if } [r+i]_2 = 0$$
$$Q_i = \langle ([r+i]_m, n+1), ([r+i+1]_m, n+1) \rangle \quad \textit{if } [r+i]_2 = 1 \textit{ and } e^* \in \mathcal{C}$$
$$Q_i = \langle ([r+i]_m, n+1), ([r+i]_m, n), ([r+i+1]_m, n), ([r+i+1]_m, n+1) \rangle \quad \textit{if otherwise.}$$

Then set the path Q as $\langle (r, n+1), Q_0, ([r+1]_m, n+1), Q_1, ([r+2]_m, n+1), \ldots, ([r-2]_m, n+1), Q_{m-2}, ([r-1]_m, n+1) \rangle$.

Now we perform the following algorithm on $\mathcal{C}$.

ALGORITHM 3 EXTEND$(\mathcal{C})$

1. Replace those edges $((i, n-1), ([i-1]_m, n-1)) \in \mathcal{C}$, where $i \neq r$ and i is even, with the path $\langle (i, n-1), (i, n), ([i-1]_m, n), ([i-1]_m, n-1) \rangle$.
2. Replace the edge $((r, n-1), (r-1, n-1))$ with the path $\langle (r, n-1), (r, n), (r, n+1), Q, (r-1, n+1), (r-1, n), (r-1, n-1) \rangle$.

Obviously, the resultant of Algorithm 3 is a Hamiltonian cycle of $SW(m, n+2) - F$.

Case 2. There is no edge in $S \cap E(\mathcal{C})$. Obviously, $((i, n-1), (i-1, n-1)) \in \mathcal{C}$ for every odd i with $1 \le i < m$. The Hamiltonian cycle of $SW(m, n+2) - F$ can easily be constructed by replacing every $((i, n-1), (i-1, n-1))$, where i is odd and $1 \le i < m$, with the path $\langle (i, n-1), (i, n), (i, n+1), (i-1, n+1), (i-1, n), (i-1, n-1) \rangle$.

Thus, we have the following theorem.

THEOREM 13.9 Assume that F' is a faulty subset of $V(SW^*(m, n)) \cup E(SW^*(m, n))$ such that some $(i, n-1)$ with $0 \le i < m$ is fault-free. Then $SW(m, n+2) - F'$ is Hamiltonian if $SW(m, n) - F'$ is Hamiltonian.

LEMMA 13.18 $SW(m, 2)$ is 1_p-fault-tolerant Hamiltonian for $m \ge 4$.

Proof: Let $F \in \mathcal{F}(SW(m, 2))$. By the symmetric property of $SW(m, 2)$, we may assume that $(0, 0) \in F$. So the other vertex in F is (x, y), where $x + y$ is odd. Define two paths:

$$p_i(k, k+1) = \langle (i-1, k), (i-1, k+1), (i, k+1), (i, k), (i+1, k) \rangle$$
$$q_i(k+1, k) = \langle (i-1, k+1), (i-1, k), (i, k), (i, k+1), (i+1, k+1) \rangle$$

To simplify the notation, $p_i = p_i(0, 1)$ and $q_i = q_i(1, 0)$.

Suppose that $y = 1$. Then we have a Hamiltonian cycle of $SW(m, 2) - F$:

$$\langle (1,0), (2,0), p_3, (4,0), p_5, (6,0), \ldots, (x,0), (x+1,0), p_{x+2},$$
$$(x+3,0), p_{x+4}, \ldots, (m-1,0), (m-1,1), (0,1), (1,1), (1,0) \rangle$$

Suppose that $y = 0$. There exists a Hamiltonian cycle of $SW(m, 2) - F$:

$$\langle (0,1), (1,1), q_2, (3,1), q_4, (5,1), \ldots, (x,1), (x+1,1), q_{x+2}, (x+3,1), \ldots,$$
$$q_{m-3}, (m-2,1), (m-2,0), (m-1,0), (m-1,1), (0,1) \rangle$$

Hence, $SW(m, 2)$ is 1_p-fault-tolerant Hamiltonian. □

LEMMA 13.19 There exist $(m/2) - 1$ disjoint paths, $P_1^n, P_2^n, \ldots, P_{(m/2)-1}^n$, that span $SW^*(m, n) - \{(0,0)\}$ such that P_l^n joins $(2l, n-1)$ to $(2l+1, n-1)$ for $1 \le l < (m/2) - 1$, and $P_{(m/2)-1}^n$ joins $(0, n-1)$ to $(m-2, n-1)$.

Proof: We prove this lemma by induction. For $n = 2$, we set P_l^2 as $\langle (2l, 1), (2l+1, 1) \rangle$ for $1 \le l < (m/2) - 1$ and $P_{(m/2)-1}^2$ as $\langle (0,1), (1,1), (1,0), I_0(1, m-1), (m-1, 0), (m-1, 1), (m-2, 1) \rangle$. Obviously, P_l^2's satisfy the requirement of the lemma for $0 \le l \le (m/2) - 1$. Now assume that the lemma holds for $n = k$, where k is even. Then there exist $(m/2) - 1$ disjoint paths, $P_1^k, P_2^k, \ldots, P_{(m/2)-1}^k$, that span $SW^*(m, k) - \{(0,0)\}$ such that P_l^k joins $(2l, k-1)$ to $(2l+1, k-1)$ for $1 \le l < (m/2) - 1$, and $P_{(m/2)-1}^k$ joins $(0, k-1)$ to $(m-2, k-1)$.

Now, we set P_l^{k+2} as $\langle (2l, k+1), (2l+1, k+1) \rangle$ for $1 \le l < (m/2) - 1$. Define $f_i = \langle (i, k-1), (i, k), (i+1, k), (i+1, k-1), P_{(i+1)/2}^k, (i+2, k-1) \rangle$ and $P_{(m/2)-1}^{k+2}$ as:

$$\Big\langle (0, k+1), (1, k+1), (1, k), (2, k), (2, k-1), P_1^k, (3, k-1), f_3, (5, k-1), f_5,$$
$$(7, k-1), \ldots, f_{m-5}, (m-3, k-1), (m-3, k), (m-2, k), (m-2, k-2),$$
$$\left(P_{\frac{m}{2}-1}^k\right)^{-1}, (0, k-1), (0, k), (m-1, k), (m-1, k+1), (m-2, k+1) \Big\rangle$$

P_l^{k+2}, $1 \le l \le (m/2) - 1$, satisfies the requirement of the lemma. Hence, the lemma is proved. See Figure 13.9a for illustration. □

LEMMA 13.20 Assume that r is an even integer, $0 < r \le m - 2$. There exist $r/2$ disjoint paths, $Q_1^n, Q_2^n, \ldots, Q_{r/2}^n$, that span $SW^*(m, n) - \{(r, 0)\}$ such that Q_l^n joins $(2l, n-1)$ to $(2l+1, n-1)$ for $1 \le l \le (r/2) - 1$ and $Q_{r/2}^n$ joins $(0, n-1)$ to $(r, n-1)$.

Proof: We prove this lemma by induction. For $n = 2$, we set Q_l^2 as $\langle (2l, 1), (2l, 0), (2l+1, 0), (2l+1, 1) \rangle$ for $1 \le l \le (r/2) - 1$ and $Q_{r/2}^2$ as $\langle (0,1), (1,1), (1,0),$

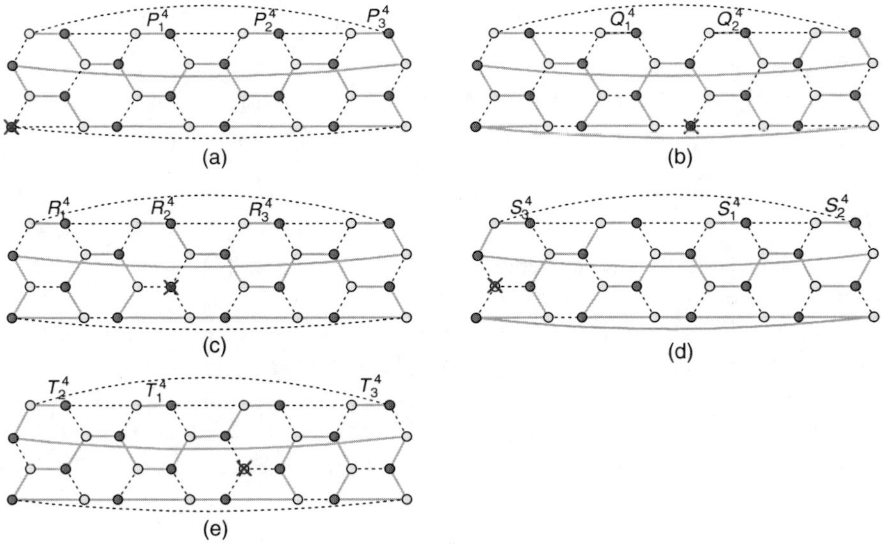

FIGURE 13.9 Illustration for Lemmas 13.19 through 13.23.

$(0,0),(m-1,0),(m-1,1), q^{-1}_{m-2}, (m-3,1), q^{-1}_{m-4}, (m-5,1), \ldots, q^{-1}_{r+2}, (r+1,1), (r,1)\rangle$. Obviously, Q_l^2's satisfy the requirement of the lemma for $1 \le l \le r/2$. We assume that the lemma holds for $n = k$ where k is even. Then there exist $r/2$ disjoint paths, Q_l^k, $1 \le l \le r/2$, that span $SW^*(m,k) - \{(r,0)\}$ such that Q_l^k joins $(2l, k-1)$ to $(2l+1, k-1)$ for $1 \le l < r/2$ and $Q_{r/2}^k$ joins $(0, k-1)$ to $(r, k-1)$.

Now, we set Q_l^{k+2} as $\langle(2l, k+1), (2l+1, k+1)\rangle$ for $1 \le l < r/2$. Define $g_i = \langle(i, k-1), Q_{i/2}^k, (i+1, k-1), (i+1, k), (i+2, k), (i+2, k-1)\rangle$ and $Q_{r/2}^{k+2}$ as:

$$\Big\langle(0,k+1),(1,k+1),(1,k),(2,k),(2,k-1),g_2,(4,k-1),g_4,(6,k-1),\ldots,g_{r-2},$$
$$(r,k-1),\left(Q^k_{\frac{r}{2}}\right)^{-1},(0,k-1),(0,k),(m-1,k),(m-1,k+1),q^{-1}_{m-2}(k+1,k),$$
$$(m-3,k+1),\ldots,q^{-1}_{r+2}(k+1,k),(r+1,k+1),(r,k+1)\Big\rangle$$

Obviously, Q_l^{k+2} for $1 \le l \le r/2$ satisfies the requirement of the lemma. Hence, the lemma is proved. See Figure 13.9b for illustration, where $r = 4$. □

LEMMA 13.21 Assume that s is a positive odd integer. There exist $(m/2) - 1$ disjoint paths, R_l^n, where $1 \le l < m/2$ that span $SW^*(m,n) - \{(s,1)\}$ such that R_l^n joins $(2(l-1), n-1)$ to $(2l-1, n-1)$ for $l \ne (s+1)/2$ and $R_{(s+1)/2}$ joins $(s-1, n-1)$ to $(m-2, n-1)$.

Proof: We prove this lemma by induction. For $n = 2$, we set

$$R_l = \langle(2(l-1), 1), (2(l-1), 0), (2l-1, 0), (2l-1, 1)\rangle \quad \text{for } 1 \leq l \leq \frac{s-1}{2}$$
$$R_l = \langle(2(l-1), 1), (2l-1, 1)\rangle \quad \text{for } \frac{s+3}{2} \leq l \leq \frac{m}{2} - 1$$

Besides, $R^2_{(s+1)/2}$ as $\langle(s-1, 1), (s-1, 0), I_0(s-1, m-1), (m-1, 0), (m-1, 1), (m-2, 1)\rangle$. Obviously, R^2_l satisfies the requirement of the lemma for $1 \leq l \leq (m/2) - 1$. Now assume that the lemma holds for $n = k$ where k is even. Then there exist $(m/2) - 1$ disjoint paths, R^k_l's, that span $SW^*(m, k) - \{(s, 1)\}$ such that R^k_l joins $(2(l-1), k-1)$ to $(2l-1, k-1)$ for $1 \leq l < m/2, l \neq (s+1)/2$ and $R^k_{(s+1)/2}$ joins $(s-1, k-1)$ to $(m-2, k-1)$.

Now, we set R^{k+2}_l as $\langle(2(l-1), k+1), (2l-1, k+1)\rangle$ for $1 \leq l < m/2, l \neq (s+1)/2$. Define $g_i = \langle(i, k-1), R^k_{i/2}, (i+1, k-1), (i+1, k), (i+2, k), (i+2, k-1)\rangle$ and $R^{k+2}_{(s+1)/2}$ as:

$$\Big\langle(s-1, k+1), (s, k+1), (s, k), (s+1, k), (s+1, k-1), g_{s+1}, (s+3, k-1), \ldots,$$
$$g_{m-4}, (m-2, k-1), \left(R^k_{\frac{s+1}{2}}\right)^{-1}, (s-1, k-1), g^{-1}_{s-3}, (s-3, k-1), \ldots,$$
$$g^{-1}_0, (0, k-1), (0, k), (m-1, k), (m-1, k+1), (m-2, k+1)\Big\rangle$$

Since R^{k+2}_l, for $1 \leq l \leq s/2$, satisfies the requirement of the lemma, the lemma is proved. See Figure 13.9c, where $s = 3$. □

LEMMA 13.22 There exist $(m/2) - 1$ disjoint paths, S^n_l, where $1 \leq l < m/2$ that span $SW^*(m, n) - \{(0, 1)\}$ such that S^n_l joins $(2l+2, n-1)$ to $(2l+3, n-1)$ for $1 \leq l \leq (m/2) - 2$ and $S^n_{(m/2)-1}$ joins $(1, n-1)$ to $(3, n-1)$.

Proof: We prove this lemma by induction. For $n = 2$, we set S^2_l as $\langle(2l+2, 1), (2l+3, 1)\rangle$ for $1 \leq l \leq (m/2) - 2$ and $S^2_{(m/2)-1}$ as $\langle(1, 1), (1, 0), (0, 0), (m-1, 0), I^{-1}_0(2, m-1), (2, 0), (2, 1), (3, 1)\rangle$. Obviously, S^2_l's satisfy the requirement of the lemma for $1 \leq l \leq (m/2) - 1$. Now assume that the lemma holds for $n = k$ where k is even. Then there exist $(m/2) - 1$ disjoint paths, S^k_l's, that span $SW^*(m, k) - \{(0, 1)\}$ such that S^k_l joins $(2l+2, k-1)$ to $(2l+3, k-1)$ for $1 \leq l \leq (m/2) - 2$ and $S^k_{(m/2)-1}$ joins $(1, k-1)$ to $(3, k-1)$.

Now, we set S^{k+2}_l as $\langle(2l+2, k+1), (2l+3, k+1)\rangle$ for $1 \leq l \leq (m/2) - 2$. Define $h_i = \langle(i, k-1), (i, k), (i+1, k), (i+1, k-1), S^k_{(i-1)/2}, (i+2, k-1)\rangle$ and $S^{k+2}_{(m/2)-1}$ as:

$$\Big\langle(1, k+1), (0, k+1), (0, k), (m-1, k), (m-1, k-1), h^{-1}_{m-3},$$
$$(m-3, k-1), h^{-1}_{m-5}, (m-5, k-1), \ldots, h^{-1}_3, (3, k-1), \left(S^k_{\frac{m}{2}-1}\right)^{-1},$$
$$(1, k-1), (1, k), (2, k), (2, k+1), (3, k+1)\Big\rangle$$

S_l^{k+2}, $1 \leq l \leq (m/2) - 1$, satisfies the requirement of the lemma, so the lemma is proved. See Figure 13.9d for illustration. □

LEMMA 13.23 Assume that t is an even integer, $0 < t \leq m - 2$. There exist $(m/2) - 1$ disjoint paths, T_l^n, where $1 \leq l < m/2$ that span $SW^*(m, n) - \{(t, 1)\}$ such that T_l^n joins $(2l, n-1)$ to $(2l+1, n-1)$ for $1 \leq l \leq (m/2) - 1$ and $l \neq t/2$, and $T_{t/2}^n$ joins $(1, n-1)$ to $(t+1, n-1)$.

Proof: We prove this lemma by induction. For $n = 2$, we set $T_{t/2}^2 = \langle (1, 1), (0, 1), (0, 0), I_0(0, t+1), (t+1, 0), (t+1, 1) \rangle$.

$$T_l = \langle (2l, 1), (2l+1, 1) \rangle \quad \text{for } 1 \leq l \leq \frac{t-2}{2}$$

$$T_l = \langle (2l, 1), (2l, 0), (2l+1, 0), (2l+1, 1) \rangle \quad \text{for } \frac{t+2}{2} \leq l \leq \frac{m-2}{2}$$

Obviously, T_l^2's satisfy the requirement of the lemma for $1 \leq l \leq (m/2) - 1$. Now assume that the lemma holds for $n = k$ where k is even. Then there exist $(m/2) - 1$ disjoint paths, T_l^k's, that span $SW^*(m, k) - \{(t, 1)\}$ such that T_l^k joins $(2l, k-1)$ to $(2l+1, k-1)$ for $1 \leq l \leq (m/2) - 1$ and $l \neq t/2$, and $T_{t/2}^k$ joins $(1, k-1)$ to $(t+1, k-1)$.

Now, we set T_l^{k+2} as $\langle (2l, k+1), (2l+1, k+1) \rangle$ for $1 \leq l \leq (m/2) - 1$ and $l \neq t/2$. Define $h_i = \langle (i, k-1), (i, k), (i+1, k), (i+1, k-1), T_{(i+1)/2}^k, (i+2, k-1) \rangle$ and set $T_{t/2}^{k+2}$ as:

$$\Big\langle (1, k+1), (0, k+1), (0, k), (m-1, k), (m-1, k-1), h_{m-3}^{-1}, (m-3, k-1), h_{m-5}^{-1},$$
$$(m-5, k-1), \ldots, h_{t+1}^{-1}, (t+1, k-1), \left(T_{\frac{t}{2}}^k\right)^{-1}, (1, k-1), h_1, (3, k-1), \ldots, h_{t-3},$$
$$(t-1, k-1), (t-1, k), (t, k), (t, k+1), (t+1, k+1) \Big\rangle$$

T_l^{k+2}, $1 \leq l \leq (m/2) - 1$, satisfies the requirement of the lemma, so the lemma is proved. See Figure 13.9e, where $t = 4$. □

THEOREM 13.10 $SW(m, n)$ is 1_p-fault-tolerant Hamiltonian for any even integer with $m \geq 4$ and $n \geq 2$.

Proof: This theorem is proved by induction. Using Lemma 13.18, $SW(m, 2)$ is 1_p-Hamiltonian. Assume that $SW(m, k)$ is 1_p-Hamiltonian for some positive integer k with $k \geq 2$. Now, we want to prove that $SW(m, n+2)$ is 1_p-fault-tolerant Hamiltonian. Let $F \in \mathcal{F}(SW(m, n+2))$.

Obviously, one of the following cases holds: (1) $\{(i, j) \mid 0 \leq i < m, j = n, n+1\} \cap F = \emptyset$, (2) $\{(i, j) \mid 0 \leq i < m, j = 0, 1\} \cap F = \emptyset$, and (3) $|\{(i, j) \mid 0 \leq i < m, j = n, n+1\} \cap F| = 1$ and $|\{(i, j) \mid 0 \leq i < m, j = 0, 1\} \cap F| = 1$.

Case 1. $\{(i,j) \mid 0 \le i < m, j = n, n+1\} \cap F = \emptyset$. Then $F \in \mathcal{F}(SW(m,n))$. By induction, $SW(m,n) - F$ is Hamiltonian. Applying Theorem 13.9, $SW(m,n+2) - F$ is Hamiltonian.

Case 2. $\{(i,j) \mid 0 \le i < m, j = 0, 1\} \cap F = \emptyset$. Since the inner cycle and the outer cycle are symmetrical in any spider web network, $SW(m,n+2) - F$ is Hamiltonian, as was Case 1.

Case 3. $|\{(i,j) \mid 0 \le i < m, j = n, n+1\} \cap F| = 1$ and $|\{(i,j) \mid 0 \le i < m, j = 0, 1\} \cap F| = 1$. By the symmetrical property of the spider web networks, we have the following five cases: (3.1) $F = \{(0,0), (0,n+1)\}$, (3.2) $F = \{(r,0), (0,n+1)\}$ with r an nonzero even integer, (3.3) $F = \{(s,1), (0,n+1)\}$ with s an odd integer, (3.4) $F = \{(0,1), (0,n)\}$, and (3.5) $F = \{(t,1), (0,n)\}$ with t an nonzero even integer.

Case 3.1. $F = \{(0,0), (0,n+1)\}$. Using Lemma 13.19, there exist $(m/2) - 1$ disjoint paths, $P_1^n, P_2^n, \ldots, P_{(m/2)-1}^n$, that span $SW^*(m,n) - \{(0,0)\}$ such that P_l^n joins $(2l, n-1)$ to $(2l+1, n-1)$ for $1 \le l < (m/2) - 1$ and $P_{(m/2)-1}^n$ joins $(0, n-1)$ to $(m-2, n-1)$.

Define $C_1(i) = \langle (i,n-1), (i,n), (i-1,n), (i-1,n-1), (P_{(i-2)/2}^n)^{-1}, (i-2,n-1) \rangle$. Obviously, $\langle (0,n-1), P_{(m/2)-1}^n, (m-2,n-1), C_1(m-2), (m-4,n-1), \ldots, C_1(4), (2,n-1), (2,n), (1,n), (1,n+1), I_{n+1}(1,m-1), (m-1,n+1), (m-1,n), (0,n), (0,n-1) \rangle$ forms a Hamiltonian cycle of $SW(m,n+2) - F$. See Figure 13.10a.

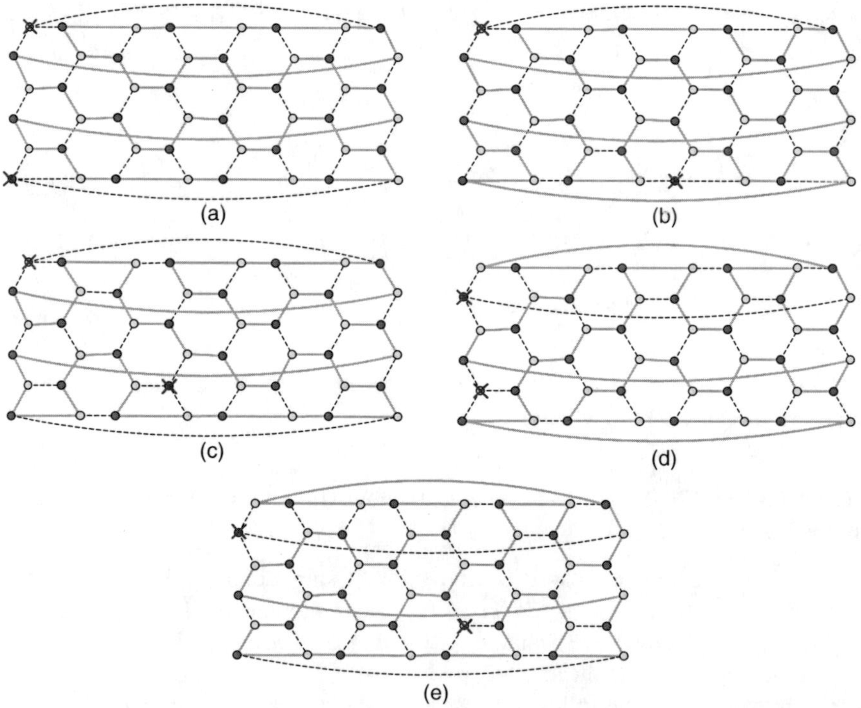

FIGURE 13.10 Illustration for Theorem 13.10, cases 3.1–3.5.

Case 3.2. $F = \{(r,0),(0,n+1)\}$. By Lemma 1, there exist $r/2$ disjoint paths, $Q_1^n, Q_2^n, \ldots, Q_{r/2}^n$, that span $SW^*(m,n) - \{(r,0)\}$ such that Q_l^n joins $(2l, n-1)$ to $(2l+1, n-1)$ for $1 \leq l < r/2$ and $Q_{r/2}^n$ joins $(0, n-1)$ to $(r, n-1)$.

Define $C_2(i) = \langle (i,n-1), (i,n), (i-1,n), (i-1,n-1), (Q_{(i-2)/2}^n)^{-1}, (i-2, n-1)\rangle$, $B(i) \equiv \langle (i,n+1),(i,n),(i+1,n),(i+1,n+1),(i+2,n+1)\rangle$. Obviously, $\langle (0,n-1), Q_{r/2}^n, (r,n-1), C_2(r), (r-2,n-1), \ldots, C_2(4), (2,n-1), (2,n), (1,n), (1, n+1), I_{n+1}(1,r+1), (r+1,n+1), B(r+1), (r+3,n+1), \ldots, B(m-3), (m-1,n+1), (m-1,n), (0,n), (0,n-1)\rangle$ forms a Hamiltonian cycle of $SW(m,n+2) - F$. See Figure 13.10b, where $r = 4$.

Case 3.3. $F = \{(s,1),(0,n+1)\}$. By Lemma 13.21, there exist $(m/2)-1$ disjoint paths, $R_1^n, R_2^n, \ldots, R_{(m/2)-1}^n$, that span $SW^*(m,n) - \{(s,1)\}$ such that R_l^n joins $(2l-2, n-1)$ to $(2l-1, n-1)$ for $1 \leq l < m/2$ and $l \neq (s+1)/2$, and $R_{(s+1)/2}^n$ joins $(s-1, n-1)$ to $(m-2, n-1)$.

Define $C_3(i) = \langle (i,n-1),(i,n),(i-1,n),(i-1,n-1),(R_{i/2}^n)^{-1},(i-2,n-1)\rangle$, $C_3'(i) \equiv \langle (i,n-1), R_{(i+2)/2}^n, (i+1,n-1),(i+1,n),(i+1,n+1),(i+2,n+1),(i+2, n), (i+2,n-1)\rangle$. Then $\langle (s-1,n-1), R_{(s+1)/2}^n, (m-2,n-1), C_3(m-2), (m-4, n-1), C_3(m-4), (m-6,n-1), \ldots, C_3(s+3), (s+1,n-1), (s+1,n), (s,n), (s,n+1), I_{n+1}(s, m-1), (m-1, n+1), (m-1, n), (0, n), (0, n-1), C_3'(0), (2, n-1), C_3'(2), (4,n-1), \ldots, C_3'(s-3), (s-1,n-1)\rangle$ forms a Hamiltonian cycle of $SW(m,n+2) - F$. See Figure 13.10c, where $s = 3$.

Case 3.4. $F = \{(0,1),(0,n)\}$. By Lemma 13.22, there exist $(m/2)-1$ disjoint paths, $S_1^n, S_2^n, \ldots, S_{(m/2)-1}^n$, that span $SW^*(m,n) - \{(0,1)\}$ such that S_l^n joins $(2l+2, n-1)$ to $(2l+3, n-1)$ for $1 \leq l \leq (m/2)-2$ and $S_{(m/2)-1}^n$ joins $(1, n-1)$ to $(3, n-1)$.

Define $C_4(i) = \langle (i,n-1),(i,n),(i,n+1),(i+1,n+1),(i+1,n),(i+1,n-1), S_{(i-1)/2}^n, (i+2,n-1)\rangle$. Then $\langle (1,n-1), S_{(m/2)-1}^n, (3,n-1), C_4(3), (5,n-1), C_4(5), (7,n-1), \ldots, C_4(m-3), (m-1,n-1), (m-1,n), (m-1,n+1), (0,n+1), (1,n+1), (2,n+1), (2,n), (1,n), (1,n-1)\rangle$ forms a Hamiltonian cycle of $SW(m,n+2) - F$. See Figure 13.10d.

Case 3.5. $F = \{(t,1),(0,n)\}$. By Lemma 13.23, there exist $(m/2)-1$ disjoint paths, $T_1^n, T_2^n, \ldots, T_{m/2-1}^n$, that span $SW^*(m,n) - \{(t,1)\}$ such that T_l^n joins $(2l, n-1)$ to $(2l+1, n-1)$ for $1 \leq l \leq (m/2)-1$ and $l \neq t/2$, and $T_{t/2}^n$ joins $(1, n-1)$ to $(t+1, n-1)$.

Define $C_5(i) = \langle (i,n-1),(i,n),(i,n+1),(i+1,n+1),(i+1,n),(i+1,n-1), T_{(i+1)/2}^n, (i+2,n-1)\rangle$ and $C_5'(i) = \langle (i,n-1), (T_{(i-1)/2}^n)^{-1}, (i-1,n-1), (i-1,n), (i-2,n), (i-2,n-1)\rangle$. Then $\langle (1,n-1), T_{t/2}, (t+1,n-1), C_5(t+1), (t+3,n-1), C_5(t+3), (t+5,n-1), \ldots, C_5(m-3), (m-1,n-1), (m-1,n), (m-1,n+1), (0,n+1), I_{n+1}(0,t), (t,n+1), (t,n), (t-1,n), (t-1,n-1), C_5'(t-1), (t-3,n-1), C_5'(t-3), (t-5,n-1), \ldots, C_5'(3), (1,n-1)\rangle$ forms a Hamiltonian cycle of $SW(m, n+2) - F$. See Figure 13.10e, where $t = 4$

The theorem is proved. □

Using a similar argument, we can prove the following theorem.

THEOREM 13.11 $SW(m, n)$ is 1-edge fault-tolerant strongly Hamiltonian laceable and hyper Hamiltonian laceable for any even integers m and n with $m \geq 4$ and $n \geq 2$.

Cho and Hsu [63] proved that the honeycomb rectangular torus HReT(m, n) is 1-edge fault-tolerant Hamiltonian and 1_p-fault-tolerant Hamiltonian for any even integers m and n with $n \geq 6$. It is very easy to see that the diameter of the spider web network $SW(m, n)$ and the diameter of the honeycomb rectangular torus HReT(m, n) are $O(m + n)$. By choosing $m = O(n)$, the diameter of these two families of graphs is $O(\sqrt{N})$, where $N = mn$ is the number of vertices.

13.6.2 BROTHER TREES

Kao and Hsu [189] present another family of 1-edge fault-tolerant Hamiltonian 1_p-fault-tolerant Hamiltonian graphs, called *brother tree's* $BT(n)$. To define brother trees, first we will define *brother cells*. Assume that k is an integer with $k \geq 2$. The *kth brother cell* $BC(k)$ is the 5-tuple $(G_k, w_k, x_k, y_k, z_k)$, where $G_k = (V, E)$ is a bipartite graph with bipartition W (white) and B (black) and contains four distinct vertices w_k, x_k, y_k, and z_k. w_k is the *white terminal*, x_k the *white root*, y_k the *black terminal*, and z_k the *black root*. We can recursively define $BC(k)$ as follows:

1. $BC(2)$ is the 5-tuple $(G_2, w_2, x_2, y_2, z_2)$ where $V(G_2) = \{w_2, x_2, y_2, z_2, s, t\}$ and $E(G_2) = \{(w_2, s), (s, x_2), (x_2, y_2), (y_2, t), (t, z_2), (w_2, z_2), (s, t)\}$.
2. The kth brother cell $BC(k)$ with $k \geq 3$ is composed of two disjoint copies of $(k-1)$th brother cells $BC^1(k-1) = (G^1_{k-1}, w^1_{k-1}, x^1_{k-1}, y^1_{k-1}, z^1_{k-1})$ and $BC^2(k-1) = (G^2_{k-1}, w^2_{k-1}, x^2_{k-1}, y^2_{k-1}, z^2_{k-1})$, a white root x_k and a black root z_k. To be specific, $V(G_k) = V(G^1_{k-1}) \cup V(G^2_{k-1}) \cup \{x_k, z_k\}$, $E(G_k) = E(G^1_{k-1}) \cup E(G^2_{k-1}) \cup \{(z_k, x^1_{k-1}), (z_k, x^2_{k-1}), (x_k, z^1_{k-1}), (x_k, z^2_{k-1}), (y^1_{k-1}, w^2_{k-1})\}$, $w_k = w^1_{k-1}$, and $y_k = y^2_{k-1}$.

$BC(2), BC(3)$, and $BC(4)$ are shown in Figure 13.11. We note that $BC^1(k-1)$ and $BC^2(k-1)$ are isomorphic for $k \geq 3$. This property is referred to as the *symmetrical property* of $BC(k)$. For this reason, we define the degenerated case, $BC(1)$, as the 5-tuple $(G_1, w_1, x_1, y_1, z_1)$ as $V(G_1) = \{w_1, y_1\}$, $E(G_1) = \{(w_1, y_1)\}$ such that $x_1 = w_1$ and $y_1 = z_1$.

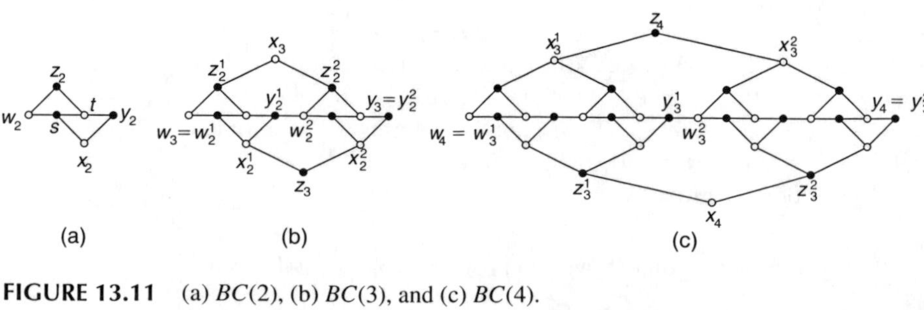

FIGURE 13.11 (a) $BC(2)$, (b) $BC(3)$, and (c) $BC(4)$.

We can also define the brother cell $BC(k)$ from the *complete binary tree* $B(k)$, where $V(B(k)) = \{1, 2, \ldots, 2^k - 1\}$ and $E(B(k)) = \{(i, j) \mid \lfloor j/2 \rfloor = i\}$. Assume that k is a positive integer with $k \geq 2$. The kth brother cell $BC(k) = (G_k, w_k, x_k, y_k, z_k)$ can be constructed by combining two $B(k)$'s, the *upper* tree $B(k)_u$ and the *lower* tree $B(k)_l$, and adding edges between their leaf vertices.

Let n be a positive integer with $n \geq 1$. The *brother tree*, $BT(n)$, is composed of an $(n+1)$th brother cell $BC(n+1) = (G^1_{n+1}, w^1_{n+1}, x^1_{n+1}, y^1_{n+1}, z^1_{n+1})$ and an nth brother cell $BC(n) = (G^2_n, w^2_n, x^2_n, y^2_n, z^2_n)$ with $V(G^1_{n+1}) \cap V(G^2_n) = \emptyset$. To be specific, $V(BT(n)) = V(G^1_{n+1}) \cup V(G^2_n)$ and $E(BT(n)) = E(G^1_{n+1}) \cup E(G^2_n) \cup \{(z^1_{n+1}, x^2_n), (y^1_{n+1}, w^2_n), (x^1_{n+1}, z^2_n), (w^1_{n+1}, y^2_n)\}$. $BT(1)$ and $BT(3)$ are shown in Figure 13.12. Obviously, $BT(n)$ is a 3-regular bipartite planar graph with $6 \times 2^n - 4$ vertices. Because the $(n+1)$th brother cell consists of two disjoint nth brother cells and two terminals, the nth brother tree $BT(n)$ is composed of three disjoint nth brother cells, $BC^1(n)$, $BC^2(n)$, $BC^3(n)$ and two terminals, $\{x^1_{n+1}, z^1_{n+1}\}$. Moreover, $BC^1(n)$, $BC^2(n)$, and $BC^3(n)$ are arranged in a cyclic order in $BT(n)$. Thus any two vertices of $BT(n)$ are in the union of two in the vertex set of $BC^1(n)$, $BC^2(n)$, $BC^3(n)$, and $\{x^1_{n+1}, z^1_{n+1}\}$. For this reason, we can assume without loss of generality that any two vertices of $BT(n)$ are in G^1_{n+1} and any edge of $BT(n)$ is in G^1_{n+1}. This property is referred to as the *symmetrical property* of $BT(n)$.

THEOREM 13.12 $D(BT(n)) = 2n + 1$ for any positive integer n with $n \geq 1$.

Proof: It is easy to prove by induction that $d_{BT(n)}(x^1_{n+1}, z^1_{n+1}) = 2n + 1$. Let u and v be any two vertices of $BT(n)$. We will prove that $d_{BT(n)}(u, v) \leq 2n + 1$. Using the symmetrical property of brother trees, we may assume that u and v are in the brother cell G^1_{n+1}. Note that the brother cell G^1_{n+1} is composed of two complete binary trees $B(n+1)_u$ and $B(n+1)_l$. Thus, (1) both u and v are in $V(B(n+1)_u)$, or both u and v are in $V(B(n+1)_l)$, or (2) $u \in V(B(n+1)_u)$ and $v \in V(B(n+1)_l)$. Now, we will introduce some notations before our proof. Let $V(B(n+1)_u) = \{1, 2, \ldots, 2^{n+1} - 1\}$

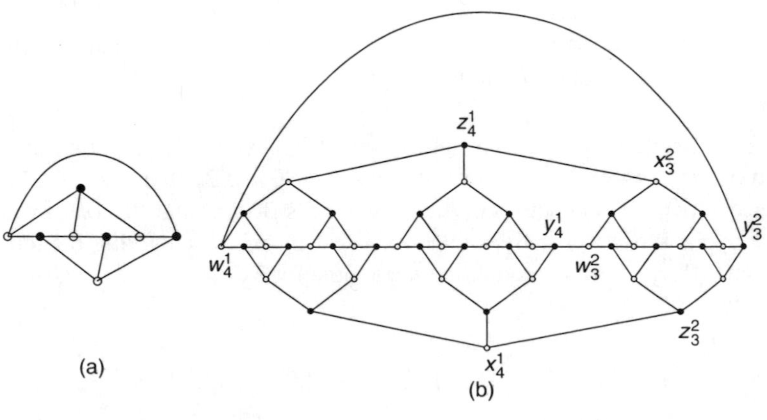

FIGURE 13.12 (a) $BT(1)$ and (b)$BT(3)$.

and $V(B(n+1)_l) = \{1', 2', \ldots, (2^{n+1}-1)'\}$. Now, join $B(n+1)_u$ and $B(n+1)_l$ with the edge set $\{(2^n+i, (2^n+i)'), ((2^n+i)', 2^n+i+1) \mid 0 \le i \le 2^n-2\} \cup \{(2^{n+1}-1, (2^{n+1}-1)')\}$ to obtain the brother cell $(G^1_{n+1}, w^1_{n+1}, x^1_{n+1}, y^1_{n+1}, z^1_{n+1})$, where $x_{n+1} = 1$ and $z_{n+1} = 1'$ if n is even and $x_{n+1} = 1'$ and $z_{n+1} = 1$ if otherwise. Moreover, $w_{n+1} = 2^n$ and $y_{n+1} = (2^{n+1}-1)'$.

Case 1. By symmetry, we may assume that both u and v are in $V(B(n+1)_u)$. Suppose that u is labeled i and v is labeled j. Obviously, $\max\{\log_2(i+1), \log_2(j+1)\} \le n+1$. Since $d_{B(n+1)}(i,1) = \lceil \log_2(i+1) \rceil - 1$, there exists a path P_1 of length $\lceil \log_2(i+1) \rceil - 1$ joining u to the root of $B(n+1)_u$. Similarly, there exists a path P_2 of length $\lceil \log_2(j+1) \rceil - 1$ joining v to 1. Thus, $\langle u, P_1, 1, P_2^{-1}, v \rangle$ forms a path joining u to v in $B(n+1)_u$. Thus, $d_{BT(n)}(u,v) \le \lceil \log_2(i+1) \rceil + \lceil \log_2(j+1) \rceil - 2 \le 2n+1$.

Case 2. We may assume that u is labeled i and v is labeled j'. Without loss of generality, we assume that $i \ge j$. There then exists a path P_1 from i to some leaf vertex h of $B(n+1)_u$ of length $n - (\lceil \log_2(i+1) \rceil - 1)$. Let h' be a neighborhood of h of $BT(n)$ in $B(n+1)_l$. Obviously, there exists a path of length n from h' to the root $1'$ of $B(n+1)_l$. Moreover, there exists a path P_3 of length $\lceil \log_2(j+1) \rceil - 1$ joining v to $1'$. Obviously, $\langle u, P_1, h, h', P_2, 1', P_3^{-1}, v \rangle$ forms a path joining u to v in G^1_{n+1}. Thus, $d_{BT(n)}(u,v) \le 2n+1-\lceil \log_2(i+1) \rceil + \lceil \log_2(j+1) \rceil$. Since $i \ge j$, $d_{BT(n)}(u,v) \le 2n+1$.

The theorem is proved. □

LEMMA 13.24 Assume that $BC(n) = (G_n, w_n, x_n, y_n, z_n)$ for some integer $n \ge 2$.

1. There exists a Hamiltonian path P^1_n of G_n joining w_n to y_n.
2. There exists a Hamiltonian path P^2_n of G_n joining w_n to z_n.
3. There exists a Hamiltonian path P^3_n of G_n joining x_n to y_n.
4. There exists a Hamiltonian path P^4_n of G_n joining x_n to z_n.
5. There exist two disjoint paths P^5_n and P^6_n such that (a) they span G_n, (b) P^5_n joins w_n to x_n, and (c) P^6_n joins y_n to z_n.
6. There exist two disjoint paths P^7_n and P^8_n such that (a) they span G_n, (b) P^7_n joins w_n to z_n, and (c) P^8_n joins x_n to y_n.

Proof: We prove this lemma by induction. It is easy to confirm that the lemma holds in $BC(2)$. Assume that the lemma holds for $BC(n)$. By definition, $BC(n+1)$ is composed of two disjoint copies of nth brother cells $BC^1(n)$ and $BC^2(n)$, a white root x_{n+1} and a black root z_{n+1}. By induction, $\{P^{i,1}_n\}_{i=1}^8$ exists, satisfying the lemma for $BC^1(n)$ and $\{P^{i,2}_n\}_{i=1}^8$ exists, satisfying the lemma for $BC^2(n)$.

We then set

$$P^1_{n+1} = \Big\langle w_{n+1} = w^1_n, P^{5,1}_n, x^1_n, z_{n+1}, x^2_n, \left(P^{5,2}_n\right)^{-1}, w^2_n, y^1_n, P^{6,1}_n, z^1_n, x_{n+1}, z^2_n, \left(P^{6,2}_n\right)^{-1}, y^2_n = y_{n+1} \Big\rangle$$

$$P_{n+1}^2 = \left\langle w_{n+1} = w_n^1, P_n^{2,1}, z_n^1, x_{n+1}, z_n^2, \left(P_n^{4,2}\right)^{-1}, x_n^2, z_{n+1}\right\rangle$$

$$P_{n+1}^3 = \left\langle x_{n+1}, z_n^2, \left(P_n^{7,2}\right)^{-1}, w_n^2, y_n^1, \left(P_n^{3,1}\right)^{-1}, x_n^1, z_{n+1}, x_n^2, P_n^{8,2}, y_n^2 = y_{n+1}\right\rangle$$

$$P_{n+1}^4 = \left\langle x_{n+1}, z_n^2, \left(P_n^{2,2}\right)^{-1}, w_n^2, y_n^1, \left(P_n^{3,1}\right)^{-1}, x_n^1, z_{n+1}\right\rangle$$

$$P_{n+1}^5 = \left\langle w_{n+1} = w_n^1, P_n^{1,1}, y_n^1, w_n^2, P_n^{7,2}, z_n^2, x_{n+1}\right\rangle$$

$$P_{n+1}^6 = \left\langle y_{n+1} = y_n^2, \left(P_n^{8,2}\right)^{-1}, x_n^2, z_{n+1}\right\rangle$$

$$P_{n+1}^7 = \left\langle w_{n+1} = w_n^1, P_n^{5,1}, x_n^1, z_{n+1}\right\rangle$$

$$P_{n+1}^8 = \left\langle x_{n+1}, z_n^1, \left(P_n^{6,1}\right)^{-1}, y_n^1, w_n^2, P_n^{1,2}, y_n^2 = y_{n+1}\right\rangle$$

The lemma is proved □

LEMMA 13.25 Assume that $BC(n) = (G_n, w_n, x_n, y_n, z_n)$ for some integer $n \geq 2$. Let e be any edge of $BC(n)$. Then at least one of the following properties holds:

1. There exists a Hamiltonian path $Q_n^1(e)$ of $G_n - e$ joining w_n to y_n.
2. There exists a Hamiltonian path $Q_n^2(e)$ of $G_n - e$ joining w_n to z_n.
3. There exists a Hamiltonian path $Q_n^3(e)$ of $G_n - e$ joining x_n to y_n.
4. There exists a Hamiltonian path $Q_n^4(e)$ of $G_n - e$ joining x_n to z_n.
5. There exist two disjoint paths $Q_n^5(e)$ and $Q_n^6(e)$ such that (a) they span $G_n - e$, (b) $Q_n^5(e)$ joins w_n to x_n, and (c) $Q_n^6(e)$ joins y_n to z_n.
6. There exist two disjoint paths $Q_n^7(e)$ and $Q_n^8(e)$ such that (a) they span $G_n - e$, (b) $Q_n^7(e)$ joins w_n to z_n, and (c) $Q_n^8(e)$ joins x_n to y_n.

Proof: We prove this lemma by induction. It is easy to confirm that the lemma holds for $BC(2)$. Assume that the lemma holds for $BC(n)$. By definition, $BC(n+1)$ is composed of two disjoint copies of nth brother cells $BC^1(n)$ and $BC^2(n)$, a white root x_{n+1} and a black root z_{n+1}. Let e be any edge of $BC(n+1)$. Using the symmetrical property of $BC(n+1)$, we may assume that e is (z_{n+1}, x_n^1), (z_n^1, x_{n+1}), (y_n^1, w_n^2), or an edge in $BC^1(n)$. By induction, there exists $\{P_n^{i,1}\}_{i=1}^8$, satisfying the lemma for $BC^1(n)$ if $e \notin E(BC^1(n))$ and there exists $\{P_n^{i,2}\}_{i=1}^8$, satisfying the lemma for $BC^2(n)$ if $e \notin E(BC^2(n))$.

Case 1. $e = (z_{n+1}, x_n^1)$. We set $Q_{n+1}^7(e)$ as $\langle w_{n+1} = w_n^1, P_n^{1,1}, y_n^1, w_n^2, P_n^{5,2}, x_n^2, z_{n+1}\rangle$ and $Q_{n+1}^8(e)$ as $\langle x_{n+1}, z_n^2, (P_n^{6,2})^{-1}, y_n^2 = y_{n+1}\rangle$.

Case 2. $e = (z_n^1, x_{n+1})$. We set $Q_{n+1}^5(e)$ as $\langle w_{n+1} = w_n^1, P_n^{1,1}, y_n^1, w_n^2, P_n^{7,2}, z_n^2, x_{n+1}\rangle$ and $Q_{n+1}^6(e)$ as $\langle y_{n+1} = y_n^2, (P_n^{8,2})^{-1}, x_n^2, z_{n+1}\rangle$.

Case 3. $e = (y_n^1, w_n^2)$. We set $Q_{n+1}^3(e)$ as $\langle x_{n+1}, z_n^1, (P_n^{4,1})^{-1}, x_n^1, z_{n+1}, x_n^2, P_n^{3,2}, y_n^2 = y_{n+1}\rangle$.

Case 4. e is in $BC^1(n)$. By induction hypothesis, one of the six properties of the lemma holds for $BC^1(n)$. In the following, we find the corresponding paths that satisfy the lemma for $BC(n+1)$:

1. $Q^7_{n+1}(e) = \langle w_{n+1} = w^1_n, Q^{1,1}_n(e), y^1_n, w^2_n, P^{5,2}_n, x^2_n, z_{n+1}\rangle$

   $Q^8_{n+1}(e) = \langle x_{n+1}, z^2_n, (P^{6,2}_n)^{-1}, y^2_n = y_{n+1}\rangle$
2. $Q^2_{n+1}(e) = \langle w_{n+1} = w^1_n, Q^{2,1}_n(e), z^1_n, x_{n+1}, z^2_n, (P^{4,2}_n)^{-1}, x^2_n, z_{n+1}\rangle$
3. $Q^4_{n+1}(e) = \langle x_{n+1}, z^2_n, (P^{2,2}_n)^{-1}, w^2_n, y^1_n, (Q^{3,1}_n(e))^{-1}, x^1_n, z_{n+1}\rangle$
4. $Q^3_{n+1}(e) = \langle x_{n+1}, z^1_n, (Q^{4,1}_n(e))^{-1}, x^1_n, z_{n+1}, x^2_n, P^{3,2}_n, y^2_n = y_{n+1}\rangle$
5. $Q^7_{n+1}(e) = \langle w_{n+1} = w^1_n, Q^{5,1}_n(e), x^1_n, z_{n+1}\rangle$

   $Q^8_{n+1}(e) = \langle x_{n+1}, z^1_n, (Q^{6,1}_n(e))^{-1}, y^1_n, w^2_n, P^{1,2}_n, y^2_n = y_{n+1}\rangle$
6. $Q^2_{n+1}(e) = \langle w_{n+1} = w^1_n, Q^{7,1}_n(e), z^1_n, x_{n+1}, z^2_n, (P^{2,2}_n)^{-1}, w^2_n, y^1_n, (Q^{8,1}_n(e))^{-1},$
   $x^1_n, z_{n+1}\rangle$

The lemma is proved. □

LEMMA 13.26 Assume that n is an integer with $n \geq 2$. Let $BC(n) = (G_n, w_n, x_n, y_n, z_n)$. Suppose that c is any vertex of G_n. There then exists a Hamiltonian path $R_n(c)$ of $G_n - c$ such that $R_n(c)$ joins y_n to z_n if c is a white vertex and $R_n(c)$ joins w_n to x_n if c is a black vertex.

Proof: We prove this lemma by induction. It is easy to confirm that the lemma holds for $BC(2)$. Assume that the lemma holds for $BC(n)$. By definition, $BC(n+1)$ is composed of two disjoint copies of nth brother cells $BC^1(n) = (G^1_n, w^1_n, x^1_n, y^1_n, z^1_n)$ and $BC^2(n) = (G^2_n, w^2_n, x^2_n, y^2_n, z^2_n)$, a white root x_{n+1} and a black root z_{n+1}. We prove only the case that c is a black vertex. Using the symmetrical property of $BC(n+1)$, we may assume that c is a vertex in $BC^1(n)$ or $c = z_{n+1}$. Using Lemma 13.24, there exists $\{P^{i,1}_n\}^8_{i=1}$ for $BC^1(n)$ if $c \notin V(BC^1(n))$ and $\{P^{i,2}_n\}^8_{i=1}$ for $BC^2(n)$ if $c \notin V(BC^2(n))$.

Suppose that c is in $BC^1(n)$. By induction, there exists a Hamiltonian path $R^1_n(c)$ of $G^1_n - c$ that joins w^1_n to x^1_n. Then we set $R_{n+1}(c)$ as $\langle w_{n+1} = w^1_n, R^1_n(c), x^1_n, z_{n+1}, x^2_n, P^{4,2}_n, z^2_n, x_{n+1}\rangle$.

Suppose that $c = z_{n+1}$. We set $R_{n+1}(c)$ as $\langle w_{n+1} = w^1_n, P^{1,1}_n, y^1_n, w^2_n, P^{2,2}_n, z^2_n, x_{n+1}\rangle$. The lemma is proved. □

LEMMA 13.27 Assume that n is an integer with $n \geq 2$. Let $BC(n) = (G_n, w_n, x_n, y_n, z_n)$. Let c be a white vertex of G_n and d be a black vertex of G_n. Then at least one of the following properties holds:

1. There exists a Hamiltonian path $S^1_n(c,d)$ of $G_n - \{c,d\}$ joining w_n to y_n.
2. There exists a Hamiltonian path $S^2_n(c,d)$ of $G_n - \{c,d\}$ joining w_n to z_n.
3. There exists a Hamiltonian path $S^3_n(c,d)$ of $G_n - \{c,d\}$ joining x_n to y_n.

4. There exists a Hamiltonian path $S_n^4(c,d)$ of $G_n - \{c,d\}$ joining x_n to z_n.
5. There exist two disjoint paths $S_n^5(c,d)$ and $S_n^6(c,d)$ such that (a) they span $G_n - \{c,d\}$, (b) $S_n^5(c,d)$ joins w_n to x_n, and (c) $S_n^6(c,d)$ joins y_n to z_n.
6. There exist two disjoint paths $S_n^7(c,d)$ and $S_n^8(c,d)$ such that (a) they span $G_n - \{c,d\}$, (b) $S_n^7(c,d)$ joins w_n to z_n, and (c) $S_n^8(c,d)$ joins x_n to y_n.

Proof: We prove this lemma by induction. It is easy to confirm that the lemma holds for $BC(2)$. Assume that the lemma holds for $BC(n)$. By definition, $BC(n+1)$ is composed of two disjoint copies of nth brother cells $BC^1(n)$ and $BC^2(n)$, a white root x_{n+1} and a black root z_{n+1}. Using the symmetrical property of $BC(n+1)$, we may assume that one of the following cases holds: (1) $c = x_{n+1}$ and $d = z_{n+1}$, (2) $c = x_{n+1}$ and $d \in V(BC^1(n))$, (3) $c \in V(BC^1(n))$ and $d = z_{n+1}$, (4) $c \in V(BC^1(n))$, $d \in V(BC^2(n))$, or (5) $\{c,d\} \subset V(BC^1(n))$. In the following discussion, we find the corresponding path(s) for each case.

Case 1. We set $S_{n+1}^1(c,d)$ as $\langle w_{n+1} = w_n^1, P_n^{1,1}, y_n^1, w_n^2, P_n^{1,2}, y_n^2\rangle$.

Case 2. With Lemma 13.26, we set $S_{n+1}^1(c,d)$ as $\langle w_{n+1} = w_n^1, R_n^1(d), x_n^1, z_{n+1}, x_n^2, P_n^{3,2}, y_n^2\rangle$.

Case 3. With Lemma 13.26, we set $S_{n+1}^3(c,d)$ as $\langle x_{n+1}, z_n^1, (R_n^1(c))^{-1}, y_n^1, w_n^2, P_n^{1,2}, y_n^2\rangle$.

Case 4. With Lemma 13.26, we set $S_{n+1}^4(c,d)$ as $\langle x_{n+1}, z_n^1, (R_n^1(c))^{-1}, y_n^1, w_n^2, R_n^2(d), x_n^2, z_{n+1}\rangle$.

Case 5. By induction hypothesis, one of the six properties of the lemma holds for $BC^1(n)$. Note that the endpoints of $S_n^i(c,d)$ are the same as the endpoints of $Q_n^i(e)$ stated in Lemma 13.25. With a similar argument for Case 4 in Lemma 13.25, we can prove that the lemma is true for this case.

Hence, the lemma is proved. □

THEOREM 13.13 The brother tree $BT(n)$ is 1_p-Hamiltonian for any positive integer n with $n \geq 1$.

Proof: Note that $BT(1)$ is isomorphic to the hypercube Q_3. By brute force, we can check whether $BT(1)$ is 1_p-fault-tolerant Hamiltonian. Now we consider $n \geq 2$. By definition, $BT(n)$ is composed of an $(n+1)$th brother cell, denoted by $BC^1(n+1) = (G_{n+1}^1, w_{n+1}^1, x_{n+1}^1, y_{n+1}^1, z_{n+1}^1)$ and an nth brother cell, denoted by $BC^2(n) = (G_n^2, w_n^2, x_n^2, y_n^2, z_n^2)$. Let $\{c,d\}$ be a pair of vertices in $BT(n)$ from different partite sets. By the symmetrical property of $BT(n)$, we assume that $\{c,d\} \subset V(G_{n+1}^1)$. Using Lemma 13.24, $\{P_n^{i,2}\}_{i=1}^8$ exists, satisfying the lemma for $BC^2(n)$. Obviously, one of the six properties of Lemma 13.27 holds for $BC^1(n+1)$. In the following discussion, we find the corresponding Hamiltonian cycle $H_{c,d}$ of $BT(n) - e$.

1. $H_{c,d} = \langle w_{n+1}^1, S_{n+1}^{1,1}(c,d), y_{n+1}^1, w_n^2, P_n^{1,2}, y_n^2, w_{n+1}^1\rangle$
2. $H_{c,d} = \langle w_{n+1}^1, S_{n+1}^{2,1}(c,d), z_{n+1}^1, x_n^2, P_n^{3,2}, y_n^2, w_{n+1}^1\rangle$

3. $H_{c,d} = \langle x^1_{n+1}, S^{3,1}_{n+1}(c,d), y^1_{n+1}, w^2_n, P^{2,2}_n, z^2_n, x^1_{n+1}\rangle$
4. $H_{c,d} = \langle x^1_{n+1}, S^{4,1}_{n+1}(c,d), z^1_{n+1}, x^2_n, P^{4,2}_n, z^2_n, x^1_{n+1}\rangle$
5. $H_{c,d} = \langle w^1_{n+1}, S^{5,1}_{n+1}(c,d), x^1_{n+1}, z^2_n, (P^{7,2}_n)^{-1}, w^2_n, y^1_{n+1}, S^{6,1}_{n+1}(c,d), z^1_{n+1}, x^2_n, P^{8,2}_n, y^2_n, w^1_{n+1}\rangle$
6. $H_{c,d} = \langle w^1_{n+1}, S^{7,1}_{n+1}(c,d), z^1_{n+1}, x^2_n, (P^{5,2}_n)^{-1}, w^2_n, y^1_{n+1}, (S^{8,1}_{n+1}(c,d))^{-1}, x^1_{n+1}, z^2_n, (P^{6,2}_n)^{-1}, y^2_n, w^1_{n+1}\rangle$

The theorem is proved. □

Using similar argument, we can prove the following theorem.

THEOREM 13.14 $BT(n)$ is 1-edge fault-tolerant strongly Hamiltonian laceable and hyper Hamiltonian laceable if $n \geq 2$.

It would be interesting to find other 3-regular 1-edge fault-tolerant Hamiltonian and 1_p-fault-tolerant Hamiltonian graphs with smaller diameters.

13.7 HAMILTONIAN LACEABILITY OF FAULTY HYPERCUBES

It seems that we can discuss the Hamiltonian problem on the bipartite Hamiltonian graph G with bipartition X and Y with a faulty vertex set. However, the faulty vertex set F_V must satisfy $|F_V \cap X| = |F_V \cap Y|$. However, this problem seems very difficult, even for hypercubes. We conjecture that $Q_n - F$ is Hamiltonian if F consists of vertex fault F_V and F_e with $|F_V \cap X| = |F_V \cap Y| = r$ and $r + |F_e| \leq n - 2$.

Sun et al. [289] restrict the faults on vertices occurring only on disjoint adjacent pairs. Let F be a subset of $V(Q_n) \cup E(Q_n)$ such that F can be decomposed into two parts F_{av} and F_e, where F_{av} is a union of f_{av} disjoint adjacent pairs of Q_n and F_e consists of f_e edges. More precisely, $F_{av} = \cup_{i=1}^{f_{av}} \{\mathbf{b_i}, \mathbf{w_i}\}$, where $(\mathbf{b_i}, \mathbf{w_i}) \in E(Q_n)$ and $\{\mathbf{b_i}, \mathbf{w_i}\} \cap \{\mathbf{b_j}, \mathbf{w_j}\} = \emptyset$ for $i \neq j$. Without loss of generality, we assume that $\{\mathbf{b_i} \mid 1 \leq i \leq f_{av}\}$ is a f_{av} set in one of the partite sets and $\{\mathbf{w_i} \mid 1 \leq i \leq f_{av}\}$ is a f_{av} set in the other partite set in Q_n. Then we define $\mathcal{H}_{av+e}(Q_n)$ to be the largest integer k such that $Q_n - F$ remains Hamiltonian for any $F = F_{av} \cup F_e$ with $f_{av} + f_e \leq k$. Again, we define $\mathcal{H}^L_{av+e}(Q_n)$ to be the largest integer k such that $Q_n - F$ remains Hamiltonian laceable for any $F = F_{av} \cup F_e$ with $f_{av} + f_e \leq k$. Similarly, we define $\mathcal{H}^{SL}_{av+e}(Q_n)$ to be the largest integer k such that $Q_n - F$ remains strongly Hamiltonian laceable for any $F = F_{av} \cup F_e$ with $f_{av} + f_e \leq k$. Then we define $\mathcal{H}^{HL}_{av+e}(Q_n)$ to be the largest integer k such that $Q_n - F$ remains hyper Hamiltonian laceable for any $F = F_{av} \cup F_e$ with $f_{av} + f_e \leq k$. Finally, we prove that $\mathcal{H}_{av+e}(Q_n) = n - 2$ if $n \geq 2$, and $\mathcal{H}^L_{av+e}(Q_n) = \mathcal{H}^{SL}_{av+e}(Q_n) = \mathcal{H}^{HL}_{av+e}(Q_n) = n - 3$ if $n \geq 3$.

We need the following properties for hypercubes.

LEMMA 13.28 $Q_n - \{\mathbf{x}, \mathbf{y}\}$ is Hamiltonian if $\mathbf{x}$ and $\mathbf{y}$ are any two vertices from different partite sets of Q_n with $n \geq 3$.

Proof: By the symmetric property of Q_n, we may assume that $\mathbf{x} \in Q^0_n$ and $\mathbf{y} \in Q^1_n$. Let $\mathbf{u}$ and $\mathbf{v}$ be any two different vertices in the partite set of Q^0_n not containing $\mathbf{x}$.

By Lemma 13.6, there exists a Hamiltonian path P of $Q_n^0 - \{\mathbf{x}\}$ joining $\mathbf{u}$ to $\mathbf{v}$. Again, there exists a Hamiltonian path S of $Q_n^1 - \{\mathbf{y}\}$ joining $\mathbf{u}^n$ to $\mathbf{v}^n$. Then $\langle \mathbf{u}, P, \mathbf{v}, \mathbf{v}^n, S^{-1}, \mathbf{u}^n, \mathbf{u}\rangle$ forms the desired cycle. □

The following lemma is proved in Ref. 290.

LEMMA 13.29 $Q_n - \{\mathbf{x}, \mathbf{y}\}$ is Hamiltonian laceable if $\mathbf{x}$ and $\mathbf{y}$ are any two vertices from different partite sets of Q_n with $n \geq 4$.

Proof: We prove this lemma by induction. The induction basis for $n=4$ can be proved by brute force. Assume that this lemma is true for Q_k with $4 \leq k < n$. By the symmetric property of Q_n, we can assume that $\mathbf{x}=\mathbf{e}$. Thus, $w(\mathbf{y})$ is odd. Let $\mathbf{u}$ and $\mathbf{v}$ be any two vertices from different partite sets of $Q_n - \{\mathbf{x}, \mathbf{y}\}$. Without loss of generality, we may assume that $\mathbf{u}$ is a white vertex. We need to find a Hamiltonian path of $Q_n - \{\mathbf{x}, \mathbf{y}\}$ joining $\mathbf{u}$ to $\mathbf{v}$.

Case 1. $w(\mathbf{y}) < n$. Without loss of generality, we assume that $(\mathbf{y})_n = 0$. Obviously, $\{\mathbf{x}, \mathbf{y}\} \subset Q_n^0$.

Case 1.1. $|\{\mathbf{u}, \mathbf{v}\} \cap Q_n^0| = 2$. By induction, there exists a Hamiltonian path P of $Q_n^0 - \{\mathbf{x}, \mathbf{y}\}$ joining $\mathbf{u}$ to $\mathbf{v}$. Write P as $\langle \mathbf{u}=\mathbf{y}_1, \mathbf{y}_2, \ldots, \mathbf{y}_{2^{n-1}-2}=\mathbf{v}\rangle$. Since Q_n^1 is Hamiltonian laceable, there exists a Hamiltonian path S of Q_n^1 joining $\mathbf{u}^n$ to $(\mathbf{y}_2)^n$. Then $\langle \mathbf{u}, \mathbf{u}^n, S, (\mathbf{y}_2)^n, \mathbf{y}_2, \mathbf{y}_3, \ldots, \mathbf{y}_{2^{n-1}-2}=\mathbf{v}\rangle$ forms the desired path.

Case 1.2. $|\{\mathbf{u}, \mathbf{v}\} \cap Q_n^0| = 1$. Without loss of generality, we assume that $\mathbf{u} \in Q_n^0$. Since there are 2^{n-2} black vertices in Q_n^0, we can choose a black vertex $\mathbf{r}$ of Q_n^0 such that $\mathbf{r} \neq \mathbf{y}$. By induction, there exists a Hamiltonian path P of $Q_n^0 - \{\mathbf{x}, \mathbf{y}\}$ joining $\mathbf{u}$ to $\mathbf{r}$. Since Q_n^1 is Hamiltonian laceable, there exists a Hamiltonian path S of Q_n^1 joining $\mathbf{r}^n$ to $\mathbf{v}$. Then $\langle \mathbf{u}, P, \mathbf{r}, \mathbf{r}^n, S, \mathbf{v}\rangle$ forms the desired path.

Case 1.3. $|\{\mathbf{u}, \mathbf{v}\} \cap Q_n^0| = 0$. Thus, $\mathbf{u}$ and $\mathbf{v}$ are in Q_n^1. Again, there exists a Hamiltonian path P of Q_n^1 between $\mathbf{u}$ and $\mathbf{v}$. We write P as $\langle \mathbf{u}=\mathbf{x}_1, \mathbf{x}_2, \ldots, \mathbf{x}_{2^n-1}=\mathbf{v}\rangle$. Obviously, there exists some index i with $1 \leq i < 2^n - 1$ such that $\{(\mathbf{x}_i)^n, (\mathbf{x}_{i+1})^n\} \cap \{\mathbf{x}, \mathbf{y}\} = \emptyset$. By induction, there exists a Hamiltonian path S of $Q_n^0 - \{\mathbf{x}, \mathbf{y}\}$ between $(\mathbf{x}_i)^n$ and $(\mathbf{x}_{i+1})^n$. Then $\langle \mathbf{u}=\mathbf{x}_1, \mathbf{x}_2, \ldots, \mathbf{x}_i, (\mathbf{x}_i)^n, S, (\mathbf{x}_{i+1})^n, \mathbf{x}_{i+1}, \mathbf{x}_{i+2}, \ldots, \mathbf{x}_{2^{n-1}-2}=\mathbf{v}\rangle$ forms the desired path.

Case 2. $w(\mathbf{y}) = n$. Since $w(\mathbf{y})$ is odd, n is odd. Moreover, $w(\mathbf{v}) < n$ because $\mathbf{y} \neq \mathbf{v}$. Without loss of generality, we assume that $(\mathbf{v})_n = 0$. Therefore, $\mathbf{v} \in Q_n^0$ and $\mathbf{y} \in Q_n^1$.

Case 2.1. $\mathbf{u} \in Q_n^0$. Since Q_n^0 is hyper Hamiltonian laceable, there exists a Hamiltonian path P of $Q_n^0 - \{\mathbf{v}\}$ joining $\mathbf{u}$ to $\mathbf{e}$. Write P as $\langle \mathbf{u}, S, \mathbf{r}, \mathbf{e}\rangle$. Similarly, there exists a Hamiltonian path R of $Q_n^1 - \{\mathbf{y}\}$ joining $\mathbf{r}^n$ to $\mathbf{v}^n$. Then $\langle \mathbf{u}, S, \mathbf{r}, \mathbf{r}^n, R, \mathbf{v}^n, \mathbf{v}\rangle$ forms the desired path.

Case 2.2. $\mathbf{u} \notin Q_n^0$. Since there are 2^{n-2} black vertices in Q_n^0, we can choose a black vertex $\mathbf{r}$ of Q_n^0 such that $\mathbf{r} \neq \mathbf{v}$. Since Q_n^0 is hyper Hamiltonian laceable, there exists a Hamiltonian path P of $Q_n^0 - \{\mathbf{e}\}$ joining $\mathbf{r}$ to $\mathbf{v}$. Similarly, there exists a Hamiltonian path S of $Q_n^0 - \{\mathbf{y}\}$ joining $\mathbf{u}$ to $\mathbf{r}^n$. Then $\langle \mathbf{u}, S, \mathbf{r}^n, \mathbf{r}, P, \mathbf{v}\rangle$ forms the desired path.

The lemma is proved. □

LEMMA 13.30 Let $(\mathbf{b}, \mathbf{w})$ be an edge in Q_4. $Q_4 - \{\mathbf{b}, \mathbf{w}\}$ is hyper Hamiltonian laceable.

Proof: Assume that $\mathbf{x}$ is any vertex of $Q_4 - \{\mathbf{b}, \mathbf{w}\}$. Let $\mathbf{u}$ and $\mathbf{v}$ be two vertices in the partite set of Q_4 not containing $\mathbf{x}$ such that $\{\mathbf{u}, \mathbf{v}\} \cap \{\mathbf{b}, \mathbf{w}\} = \emptyset$. We want to construct a Hamiltonian path of $Q_4 - \{\mathbf{b}, \mathbf{w}, \mathbf{x}\}$ joining $\mathbf{u}$ to $\mathbf{v}$. By the symmetric property of Q_4, we may assume that $\mathbf{u} \in Q_4^0$, $\mathbf{v} \in Q_4^1$, and $\{\mathbf{b}, \mathbf{w}\} \subset Q_4^0$. Without loss of generality, we assume that $\mathbf{u}$ and $\mathbf{b}$ are in the same partite set.

Case 1. $\mathbf{x} \in Q_4^0$. By Lemma 13.6, there exists a Hamiltonian path P of $Q_4^0 - \{\mathbf{x}\}$ joining $\mathbf{u}$ to $\mathbf{b}$. Write P as $\langle \mathbf{u} = \mathbf{y_1}, \mathbf{y_2}, \ldots, \mathbf{y_7} = \mathbf{b} \rangle$. Then $\mathbf{w} = \mathbf{y_i}$ for some $i \in \{2, 4, 6\}$.

Case 1.1. $i = 2$ or 4. Suppose that $(\mathbf{y_6})^4 = \mathbf{v}$. By Lemma 13.6, there exists a Hamiltonian path S of $Q_4^1 - \{\mathbf{v}\}$ joining $(\mathbf{y_{i-1}})^4$ to $(\mathbf{y_{i+1}})^4$. Then $\langle \mathbf{u}, \mathbf{y_2}, \ldots, \mathbf{y_{i-1}}, (\mathbf{y_{i-1}})^4, S, (\mathbf{y_{i+1}})^4, \mathbf{y_{i+1}}, \ldots, \mathbf{y_6}, \mathbf{v} \rangle$ forms the desired path. Suppose that $(\mathbf{y_6})^4 \neq \mathbf{v}$. By Lemma 13.17, there exist two disjoint paths P_1 and P_2 such that (1) P_1 joins $(\mathbf{y_{i-1}})^4$ to $(\mathbf{y_6})^4$, (2) P_2 joins $(\mathbf{y_{i+1}})^4$ to $\mathbf{v}$, and (3) $P_1 \cup P_2$ spans Q_4^1. Then $\langle \mathbf{u}, \mathbf{y_2}, \ldots, \mathbf{y_{i-1}}, (\mathbf{y_{i-1}})^4, P_1, (\mathbf{y_6})^4, \mathbf{y_6}, \mathbf{y_5}, \ldots, \mathbf{y_{i+1}}, (\mathbf{y_{i+1}})^4, P_2, \mathbf{v} \rangle$ forms the desired path.

Case 1.2. $i = 6$. By Lemma 13.4, there exists a Hamiltonian path S of Q_4^1 joining $(\mathbf{y_5})^4$ to $\mathbf{v}$. Then $\langle \mathbf{u}, \mathbf{y_2}, \ldots, \mathbf{y_5}, (\mathbf{y_5})^4, S, \mathbf{v} \rangle$ forms the desired path.

Case 2. $\mathbf{x} \in Q_4^1$. By Lemma 13.28, there exists a Hamiltonian cycle C of $Q_4^0 - \{\mathbf{b}, \mathbf{w}\}$. Write C as $\langle \mathbf{u} = \mathbf{y_1}, \mathbf{y_2}, \ldots, \mathbf{y_6}, \mathbf{u} \rangle$. Since C can be traversed backward and forward, we can assume that $(\mathbf{y_6})^4 \neq \mathbf{v}$. By Lemma 13.6, there exists a Hamiltonian path S of $Q_4^1 - \{\mathbf{x}\}$ joining $(\mathbf{y_6})^4$ to $\mathbf{v}$. Then $\langle \mathbf{u}, \mathbf{y_2}, \ldots, \mathbf{y_6}, (\mathbf{y_6})^4, S, \mathbf{v} \rangle$ forms the desired path. □

LEMMA 13.31 Let $\mathbf{x}$ and $\mathbf{y}$ be adjacent vertices of Q_4 and f be any edge of Q_4. Then $Q_4 - \{\mathbf{x}, \mathbf{y}, f\}$ is Hamiltonian laceable.

Proof: With Lemmas 13.2 and 13.30, the lemma is true. □

LEMMA 13.32 Let $F_{av} = \{\{\mathbf{b_i}, \mathbf{w_i}\} \mid (\mathbf{b_i}, \mathbf{w_i}) \in E(Q_4), 1 \leq i \leq 2\}$ be a set of adjacent vertices of Q_4 such that $\{\mathbf{b_1}, \mathbf{w_1}\} \cap \{\mathbf{b_2}, \mathbf{w_2}\} = \emptyset$. Then $Q_4 - F_{av}$ is Hamiltonian.

LEMMA 13.33 $\mathcal{H}^{L}_{av+e}(Q_n) \leq n - 3$ if $n \geq 3$.

Proof: Assume that $n \geq 3$. Let $F_{av} = \{\{(00\ldots0)^j, (010\ldots0)^j\} \mid 3 \leq j \leq n\}$. Obviously, $f_{av} = n - 2$. Since $d_{Q_n - F_{av}}((00\ldots0)) = d_{Q_n - F_{av}}(010\ldots0) = 2$, there is no Hamiltonian path from $(10\ldots0)$ to $(110\ldots0)$ in $Q_n - F_{av}$. Thus, $\mathcal{H}^{L}_{av+e}(Q_n) \leq n - 3$ if $n \geq 3$. See Figure 13.13 for illustration. □

LEMMA 13.34 $\mathcal{H}^{HL}_{av+e}(Q_n) \geq n - 3$ if $n \geq 3$.

Proof: We prove that $\mathcal{H}^{HL}_{av+e}(Q_n) \geq n - 3$ if $n \geq 3$ by induction. Suppose that $F_{av} = \emptyset$. By Lemma 13.6, $Q_n - F$ is hyper Hamiltonian laceable. Thus, we discuss the case $F_{av} \neq \emptyset$ only. With Lemmas 13.6 and 13.30, the theorem is true for $n = 3$ and 4. Assume that $\mathcal{H}^{HL}_{av+e}(Q_k) \geq k - 3$ for $4 \leq k < n$. Let $F = F_{av} \cup F_e$, where F_{av} is

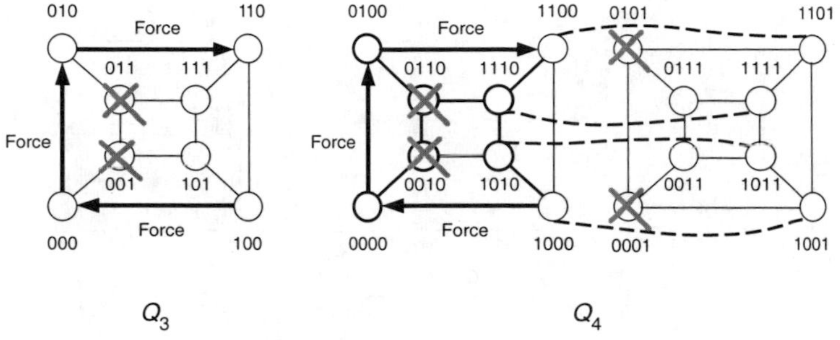

FIGURE 13.13 Illustration for Lemma 13.33.

a union of f_{av} disjoint adjacent vertices of Q_n and F_e consists of f_e edges of Q_n with $f_{av}+f_e \le n-3$. Let $F_{av} = \cup_{i=1}^{f_{av}} \{\mathbf{b_i}, \mathbf{w_i}\}$, where $(\mathbf{b_i}, \mathbf{w_i}) \in E(Q_n)$ and $\{\mathbf{b_i}, \mathbf{w_i}\} \cap \{\mathbf{b_j}, \mathbf{w_j}\} = \emptyset$ for $i \ne j$. Without loss of generality, we assume that $\{\mathbf{b_i} \mid 1 \le i \le f_{av}\}$ is a f_{av} set in one of the partite sets and $\{\mathbf{w_i} \mid 1 \le i \le f_{av}\}$ is a f_{av} set in the other partite set. Write F_e as $\{(\mathbf{b_i}, \mathbf{w_i}) \mid f_{av}+1 \le i \le n-3\}$. We want to prove that $Q_n - F$ is hyper Hamiltonian laceable.

Let $\mathbf{r}$ be any vertex of $Q_n - F_{av}$. Let $\mathbf{u}$ and $\mathbf{v}$ be any two different vertices of $Q_n - F_{av}$ in the partite set not containing $\mathbf{r}$. First, we must construct a Hamiltonian path of $Q_n - F - \{\mathbf{r}\}$ joining $\mathbf{u}$ to $\mathbf{v}$. Without loss of generality, we may assume that $\mathbf{r}$ and $\mathbf{b_i}$ are in the same partite set for $1 \le i \le f_{av}$. Let $t_j = |\{(\mathbf{b_i}, \mathbf{w_i}) \mid (\mathbf{b_i}, \mathbf{w_i})$ is a j-dimensional edge with $1 \le i \le n-3\}|$. Without loss of generality, we assume that $t_1 \ge t_2 \ge \ldots \ge t_n = 0$. Thus, both $\mathbf{b_i}$ and $\mathbf{w_i}$ are either in Q_n^0 or Q_n^1 for every i. Let $F_i = F \cap Q_n^i$, for $i \in \{0, 1\}$.

Let $r_j = |\{\mathbf{b_i} \mid \mathbf{b_i} \in Q_n^j, 1 \le i \le n-3\}|$ for $j = 0, 1$. Obviously, $r_0 + r_1 = n-3$. Since $F_{av} \ne \emptyset$, we may assume that $\{\mathbf{b_1}, \mathbf{w_1}\} \subset Q_n^0$. Let $F' = F'_{av} \cup F_e$, where $F'_{av} = F_{av} - \{\{\mathbf{b_1}, \mathbf{w_1}\}\}$. Hence, $f'_{av} = f_{av} - 1$ and $f'_{av} + f_e \le n-4$.

Case 1. $|\{\mathbf{u}, \mathbf{v}\} \cap Q_n^0| = 2$.

Case 1.1. $r_0 = n-3$.

Case 1.1.1. $\mathbf{r} \in Q_n^0$. By induction, there exists a Hamiltonian path P of $Q_n^0 - F' - \{\mathbf{r}\}$ joining $\mathbf{u}$ to $\mathbf{v}$. Suppose that $(\mathbf{b_1}, \mathbf{w_1}) \notin E(P)$. Since $\mathbf{u}$ and $\mathbf{v}$ are in the same partite set, we can write P as $\langle \mathbf{u}, P_1, \mathbf{w}, \mathbf{b_1}, \mathbf{x}, P_2, \mathbf{z}, \mathbf{w_1}, \mathbf{y}, P_3, \mathbf{v} \rangle$. Note that $\mathbf{w}^n$ and $\mathbf{x}^n$ are in a partite set, and $\mathbf{z}^n$ and $\mathbf{y}^n$ are in the other partite set. By Lemma 13.17, there exist two disjoint paths S_1 and S_2 such that (1) S_1 joins $\mathbf{w}^n$ to $\mathbf{z}^n$, (2) S_2 joins $\mathbf{x}^n$ to $\mathbf{y}^n$, and (3) $S_1 \cup S_2$ spans Q_n^1. Then $\langle \mathbf{u}, P_1, \mathbf{w}, \mathbf{w}^n, S_1, \mathbf{z}^n, \mathbf{z}, P_2^{-1}, \mathbf{x}, \mathbf{x}^n, S_2, \mathbf{y}^n, \mathbf{y}, P_3, \mathbf{v} \rangle$ forms the desired path. See Figure 13.14a for illustration. Suppose that $(\mathbf{b_1}, \mathbf{w_1}) \in E(P)$. Since $\mathbf{u}$ and $\mathbf{v}$ are in the same partite set, we can rewrite P as $\langle \mathbf{u}, P_1, \mathbf{w}, \mathbf{b_1}, \mathbf{w_1}, \mathbf{y}, P_2, \mathbf{v} \rangle$. Note that $\mathbf{w}^n$ and $\mathbf{y}^n$ are in different partite sets. Since every hypercube is Hamiltonian laceable, there exists a Hamiltonian path S of Q_n^1 joining $\mathbf{w}^n$ to $\mathbf{y}^n$. Then $\langle \mathbf{u}, P_1, \mathbf{w}, \mathbf{w}^n, S, \mathbf{y}^n, \mathbf{y}, P_2, \mathbf{v} \rangle$ forms the desired path. See Figure 13.14b for illustration.

Case 1.1.2. $\mathbf{r} \in Q_n^1$. Note that $\mathbf{u}$ and $\mathbf{v}$ are in the partite set not containing $\mathbf{b_1}$. By induction, there exists a Hamiltonian path P of $Q_n^0 - F' - \{\mathbf{b_1}\}$ joining $\mathbf{u}$ to $\mathbf{v}$. Write

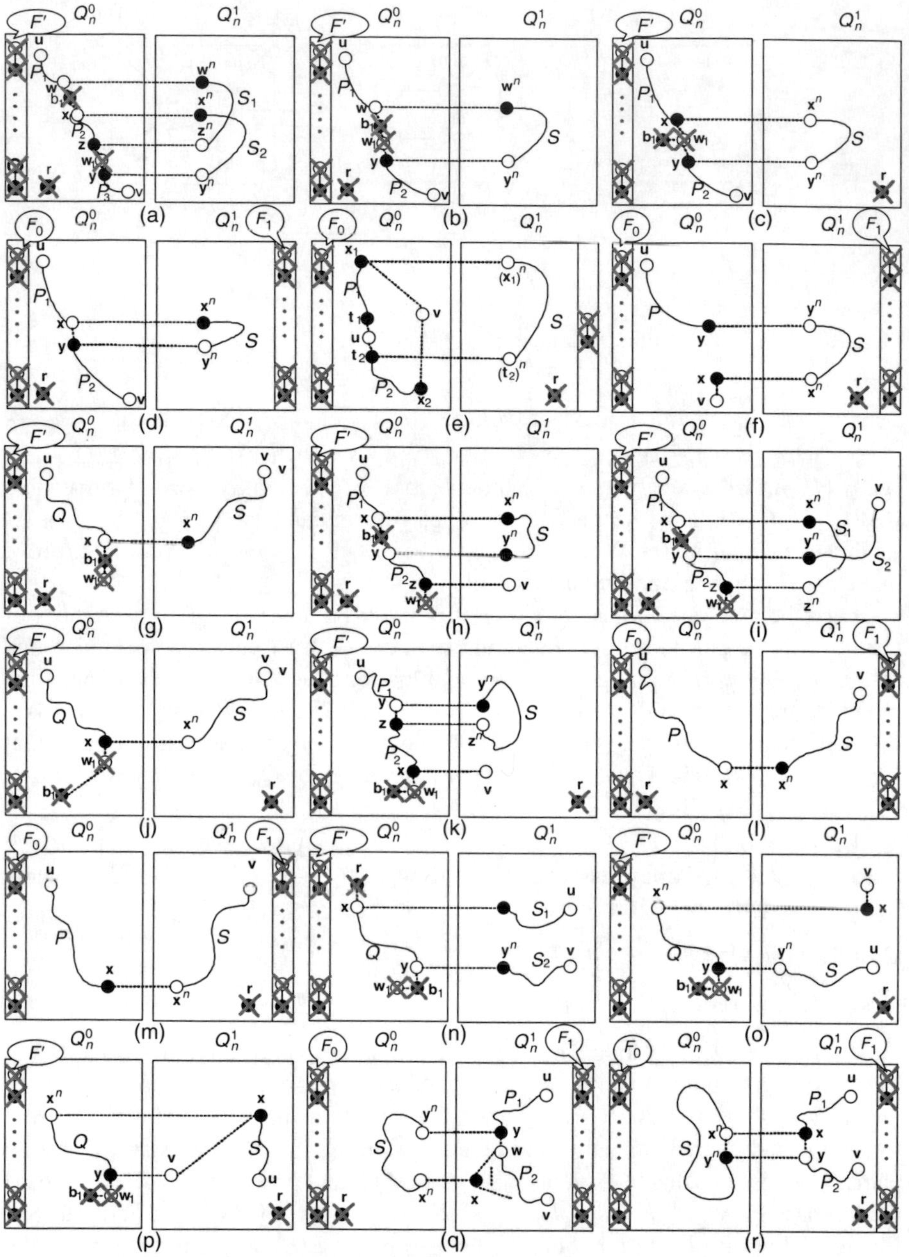

FIGURE 13.14 Illustration for Lemma 13.34.

P as $\langle \mathbf{u}, P_1, \mathbf{x}, \mathbf{w_1}, \mathbf{y}, P_2, \mathbf{v} \rangle$. Note that $\mathbf{x}^n$ and $\mathbf{y}^n$ are in the partite set not containing $\mathbf{r}$. By Lemma 13.6, there is a Hamiltonian path S of $Q_n^1 - \{\mathbf{r}\}$ joining $\mathbf{x}^n$ to $\mathbf{y}^n$. Then $\langle \mathbf{u}, P_1, \mathbf{x}, \mathbf{x}^n, S, \mathbf{y}^n, \mathbf{y}, P_2, \mathbf{v} \rangle$ forms the desired path. See Figure 13.14c for illustration.

Case 1.2. $r_0 \leq n-4$.

Case 1.2.1. $\mathbf{r} \in Q_n^0$. By induction, there exists a Hamiltonian path P of $Q_n^0 - F_0 - \{\mathbf{r}\}$ joining $\mathbf{u}$ to $\mathbf{v}$. Since there are at least $2^{n-1} - 2(n-4) - 1$ vertices in P, there exist adjacent vertices $\mathbf{x}$ and $\mathbf{y}$ of P such that $\{\mathbf{x}^n, \mathbf{y}^n\} \cap F_1 = \emptyset$. Write P as $\langle \mathbf{u}, P_1, \mathbf{x}, \mathbf{y}, P_2, \mathbf{v} \rangle$. Since $\{\mathbf{b_1}, \mathbf{w_1}\} \subset Q_n^0$, $r_1 \leq n-4$. By induction, there exists a Hamiltonian path S of $Q_n^1 - F_1$ joining $\mathbf{x}^n$ to $\mathbf{y}^n$. Then $\langle \mathbf{u}, P_1, \mathbf{x}, \mathbf{x}^n, S, \mathbf{y}^n, \mathbf{y}, P_2, \mathbf{v} \rangle$ forms the desired path. See Figure 13.14d for illustration.

Case 1.2.2. $\mathbf{r} \in Q_n^1$. Assume that $r_0 = n-4$. Since $d_{Q_n^0}(\mathbf{v}) = n-1$, there exist two neighbors $\mathbf{x_1}$ and $\mathbf{x_2}$ of $\mathbf{v}$ such that $\{\mathbf{x_1}, \mathbf{x_2}, (\mathbf{x_1})^n, (\mathbf{x_2})^n\} \cap F = \emptyset$. By induction, there exists a Hamiltonian path P of $Q_n^0 - F_0 - \{\mathbf{v}\}$ joining $\mathbf{x_1}$ to $\mathbf{x_2}$. Write P as $\langle \mathbf{x_1}, P_1, \mathbf{t_1}, \mathbf{u}, \mathbf{t_2}, P_2, \mathbf{x_2} \rangle$. Since $r_1 = 1$, $|\{(\mathbf{t_1})^n, (\mathbf{t_2})^n\} \cap F_1| \leq 1$. Without loss of generality, we may assume that $(\mathbf{t_2})^n \notin F_1$. By induction, there exists a Hamiltonian path S of $Q_n^1 - F_1 - \{\mathbf{r}\}$ joining $(\mathbf{x_1})^n$ to $(\mathbf{t_2})^n$. Then $\langle \mathbf{u}, \mathbf{t_1}, P_1^{-1}, \mathbf{x_1}, (\mathbf{x_1})^n, S, (\mathbf{t_2})^n, \mathbf{t_2}, P_2, \mathbf{x_2}, \mathbf{v} \rangle$ forms the desired path. See Figure 13.14e for illustration. Assume that $r_0 \leq n-5$. Since $d_{Q_n^0}(\mathbf{v}) = n-1$, there exists a neighbor $\mathbf{x}$ of $\mathbf{v}$ such that $\{\mathbf{x}, \mathbf{x}^n\} \cap F = \emptyset$. Since there are at least $2^{n-2} - (n-4)$ vertices in the partite set of $Q_n^0 - F_0 - \{\mathbf{x}\}$ not containing $\mathbf{u}$, we can choose a vertex $\mathbf{y}$ of $Q_n^0 - F_0 - \{\mathbf{x}\}$ such that $\{\mathbf{y}^n\} \cap F_1 = \emptyset$. By induction, there exists a Hamiltonian path P of $Q_n^0 - F_0 - \{\mathbf{v}, \mathbf{x}\}$ joining $\mathbf{u}$ to $\mathbf{y}$. Since $\{\mathbf{b_1}, \mathbf{w_1}\} \subset Q_n^0$, $r_1 \leq n-4$. By induction, there exists a Hamiltonian path S of $Q_n^1 - F_1 - \{\mathbf{r}\}$ joining $\mathbf{y}^n$ to $\mathbf{x}^n$. Then $\langle \mathbf{u}, P, \mathbf{y}, \mathbf{y}^n, S, \mathbf{x}^n, \mathbf{x}, \mathbf{v} \rangle$ forms the desired path. See Figure 13.14f for illustration.

Case 2. $|\{\mathbf{u}, \mathbf{v}\} \cap Q_n^0| = 1$. Without loss of generality, we assume that $\mathbf{u} \in Q_n^0$.

Case 2.1. $r_0 = n-3$.

Case 2.1.1. $\mathbf{r} \in Q_n^0$. Note that $\mathbf{u}$ and $\mathbf{w_1}$ are in the partite set not containing $\mathbf{r}$. By induction, there exists a Hamiltonian path P of $Q_n^0 - F' - \{\mathbf{r}\}$ joining $\mathbf{u}$ to $\mathbf{w_1}$. Suppose that $(\mathbf{b_1}, \mathbf{w_1}) \in E(P)$. Write P as $\langle \mathbf{u}, Q, \mathbf{x}, \mathbf{b_1}, \mathbf{w_1} \rangle$. Note that $\mathbf{x}^n$ and $\mathbf{v}$ are in different partite sets. Since every hypercube is Hamiltonian laceable, there exists a Hamiltonian path S of Q_n^1 joining $\mathbf{x}^n$ to $\mathbf{v}$. Then $\langle \mathbf{u}, Q, \mathbf{x}, \mathbf{x}^n, S, \mathbf{v} \rangle$ forms the desired path. See Figure 13.14g for illustration. Suppose that $(\mathbf{b_1}, \mathbf{w_1}) \notin E(P)$. We can write P as $\langle \mathbf{u}, P_1, \mathbf{x}, \mathbf{b_1}, \mathbf{y}, P_2, \mathbf{z}, \mathbf{w_1} \rangle$. Note that $\mathbf{x}^n$ and $\mathbf{y}^n$ are in the same partite set. Assume that $\mathbf{z}^n = \mathbf{v}$. By Lemma 13.6, there exists a Hamiltonian path S of $Q_n^1 - \{\mathbf{v}\}$ joining $\mathbf{x}^n$ to $\mathbf{y}^n$. Then $\langle \mathbf{u}, P_1, \mathbf{x}, \mathbf{x}^n, S, \mathbf{y}^n, \mathbf{y}, P_2, \mathbf{z}, \mathbf{v} \rangle$ forms the desired path. See Figure 13.14h for illustration. Assume that $\mathbf{z}^n \neq \mathbf{v}$. By Lemma 13.17, there exist two disjoint paths S_1 and S_2 such that (1) S_1 joins $\mathbf{x}^n$ to $\mathbf{z}^n$, (2) S_2 joins $\mathbf{y}^n$ to $\mathbf{v}$, and (3) $S_1 \cup S_2$ spans Q_n^1. Then $\langle \mathbf{u}, P_1, \mathbf{x}, \mathbf{x}^n, S_1, \mathbf{z}^n, \mathbf{z}, P_2^{-1}, \mathbf{y}, \mathbf{y}^n, S_2, \mathbf{v} \rangle$ forms the desired path. See Figure 13.14i for illustration.

Case 2.1.2. $\mathbf{r} \in Q_n^1$. Note that $\mathbf{u}$ and $\mathbf{w_1}$ are in the partite set not containing $\mathbf{b_1}$. By induction, there exists a Hamiltonian path P of $Q_n^0 - F' - \{\mathbf{b_1}\}$ joining $\mathbf{u}$ to $\mathbf{w_1}$. Write P as $\langle \mathbf{u}, Q, \mathbf{x}, \mathbf{w_1} \rangle$. Suppose that $\mathbf{x}^n \neq \mathbf{v}$. By Lemma 13.6, there exists a Hamiltonian path S of $Q_n^1 - \{\mathbf{r}\}$ joining $\mathbf{x}^n$ to $\mathbf{v}$. Then $\langle \mathbf{u}, Q, \mathbf{x}, \mathbf{x}^n, S, \mathbf{v} \rangle$ forms the desired path. See Figure 13.14j for illustration. Suppose that $\mathbf{x}^n = \mathbf{v}$. Since there are at least $2^{n-1} - 2(n-3)$ vertices in Q, we can choose adjacent vertices $\mathbf{y}$ and $\mathbf{z}$ of Q such that

$\{\mathbf{y}^n, \mathbf{z}^n\} \cap \{\mathbf{r}, \mathbf{v}\} = \emptyset$. Rewrite P as $\langle \mathbf{u}, P_1, \mathbf{y}, \mathbf{z}, P_2, \mathbf{x}, \mathbf{w_1} \rangle$. By Lemma 13.29, there exists a Hamiltonian path S of $Q_n^1 - \{\mathbf{r}, \mathbf{v}\}$ joining $\mathbf{y}^n$ to $\mathbf{z}^n$. Then $\langle \mathbf{u}, P_1, \mathbf{y}, \mathbf{y}^n, S, \mathbf{z}^n, \mathbf{z}, P_2, \mathbf{x}, \mathbf{v} \rangle$ forms the desired path. See Figure 13.14k for illustration.

Case 2.2. $r_0 \leq n-4$.

Case 2.2.1. $\mathbf{r} \in Q_n^0$. Let $\mathbf{x}$ be a vertex of $Q_n^0 - F_0$ such that (1) $\mathbf{x}$ and $\mathbf{u}$ are two different vertices in the same partite set of $Q_n^0 - F_0$ and (2) $\{\mathbf{x}^n\} \cap F_1 = \emptyset$. By induction, there exists a Hamiltonian path P of $Q_n^0 - F_0 - \{\mathbf{r}\}$ joining $\mathbf{u}$ to $\mathbf{x}$. Since $\{\mathbf{b_1}, \mathbf{w_1}\} \subset Q_n^0$, $r_1 \leq n-4$. By induction, there exists a Hamiltonian path S of $Q_n^1 - F_1$ joining $\mathbf{x}^n$ to $\mathbf{v}$. Then $\langle \mathbf{u}, P, \mathbf{x}, \mathbf{x}^n, S, \mathbf{v} \rangle$ forms the desired path. See Figure 13.14l for illustration.

Case 2.2.2. $\mathbf{r} \in Q_n^1$. Let $\mathbf{x}$ be a vertex of $Q_n^0 - F_0$ such that (1) $\mathbf{x}$ and $\mathbf{u}$ are two vertices from different partite sets of $Q_n^0 - F_0$ and (2) $\{\mathbf{x}^n\} \cap F_1 = \emptyset$. By induction, there exists a Hamiltonian path P of $Q_n^0 - F_0$ joining $\mathbf{u}$ to $\mathbf{x}$. Since $\{\mathbf{b_1}, \mathbf{w_1}\} \subset Q_n^0$, $r_1 \leq n-4$. Again, there exists a Hamiltonian path S of $Q_n^1 - F_1 - \{\mathbf{r}\}$ joining $\mathbf{x}^n$ to $\mathbf{v}$. Then $\langle \mathbf{u}, P, \mathbf{x}, \mathbf{x}^n, S, \mathbf{v} \rangle$ forms the desired path. See Figure 13.14m for illustration.

Case 3. $|\{\mathbf{u}, \mathbf{v}\} \cap Q_n^0| = 0$.

Case 3.1. $r_0 = n-3$.

Case 3.1.1. $\mathbf{r} \in Q_n^0$. Note that $\mathbf{r}$ and $\mathbf{b_1}$ are in the partite set not containing $\mathbf{w_1}$. By induction, there exists a Hamiltonian path P of $Q_n^0 - F' - \{\mathbf{w_1}\}$ joining $\mathbf{r}$ to $\mathbf{b_1}$. Write P as $\langle \mathbf{r}, \mathbf{x}, Q, \mathbf{y}, \mathbf{b_1} \rangle$. Note that $\mathbf{x}^n$ and $\mathbf{y}^n$ are in the partite set not containing $\mathbf{u}$. By Lemma 13.17, there exist two disjoint paths S_1 and S_2 such that (1) S_1 joins $\mathbf{u}$ to $\mathbf{x}^n$, (2) S_2 joins $\mathbf{y}^n$ to $\mathbf{v}$, and (3) $S_1 \cup S_2$ spans Q_n^1. Then $\langle \mathbf{u}, S_1, \mathbf{x}^n, \mathbf{x}, Q, \mathbf{y}, \mathbf{y}^n, S_2, \mathbf{v} \rangle$ forms the desired path. See Figure 13.14n for illustration.

Case 3.1.2. $\mathbf{r} \in Q_n^1$. Since $d_{Q_n^1}(\mathbf{v}) = n-1$, there is a neighbor $\mathbf{x}$ of $\mathbf{v}$ in $Q_n^1 - \{\mathbf{r}\}$ such that $\{\mathbf{x}^n\} \cap F_0 = \emptyset$. Note that $\mathbf{x}^n$ and $\mathbf{w_1}$ are in the partite set not containing $\mathbf{b_1}$. By induction, there exists a Hamiltonian path P of $Q_n^0 - F' - \{\mathbf{b_1}\}$ joining $\mathbf{x}^n$ to $\mathbf{w_1}$. Write P as $\langle \mathbf{x}^n, Q, \mathbf{y}, \mathbf{w_1} \rangle$. Suppose that $\{\mathbf{y}^n\} \cap \{\mathbf{u}, \mathbf{v}\} = \emptyset$. By induction, there exists a Hamiltonian path S of $Q_n^1 - \{\mathbf{r}, \mathbf{v}, \mathbf{x}\}$ joining $\mathbf{u}$ to $\mathbf{y}^n$. Then $\langle \mathbf{u}, S, \mathbf{y}^n, \mathbf{y}, Q^{-1}, \mathbf{x}^n, \mathbf{x}, \mathbf{v} \rangle$ forms the desired path. See Figure 13.14o for illustration. Suppose that $\{\mathbf{y}^n\} \cap \{\mathbf{u}, \mathbf{v}\} \neq \emptyset$. Assume that $\mathbf{y}^n = \mathbf{v}$. By Lemma 13.29, there exists a Hamiltonian path S of $Q_n^1 - \{\mathbf{r}, \mathbf{v}\}$ joining $\mathbf{u}$ to $\mathbf{x}$. Then $\langle \mathbf{u}, S, \mathbf{x}, \mathbf{x}^n, Q, \mathbf{y}, \mathbf{v} \rangle$ forms the desired path. See Figure 13.14p for illustration. Assume that $\mathbf{y}^n = \mathbf{u}$. By Lemma 13.29, there exists a Hamiltonian path S of $Q_n^1 - \{\mathbf{r}, \mathbf{u}\}$ joining $\mathbf{v}$ to $\mathbf{x}$. Then $\langle \mathbf{v}, S, \mathbf{x}, \mathbf{x}^n, Q, \mathbf{y}, \mathbf{u} \rangle$ forms the desired path.

Case 3.2. $r_0 \leq n-4$.

Case 3.2.1. $\mathbf{r} \in Q_n^0$. Since $\{\mathbf{b_1}, \mathbf{w_1}\} \subset Q_n^0$, $r_1 \leq n-4$. Since there are at least $2^{n-2} - (n-4)$ vertices in the partite set of $Q_n^1 - F_1$ not containing $\mathbf{u}$, we can choose a vertex $\mathbf{x}$ of $Q_n^1 - F_1$ such that $\{\mathbf{x}, \mathbf{x}^n\} \cap F = \emptyset$. By induction, there exists a Hamiltonian path P of $Q_n^1 - F_1 - \{\mathbf{x}\}$ joining $\mathbf{u}$ to $\mathbf{v}$. Since $d_{Q_n^1}(\mathbf{x}) = n-1$, we can choose an edge $(\mathbf{y}, \mathbf{w})$ in $E(P)$ such that (1) $\mathbf{w} \in N_{Q_n^1 - F_1}(\mathbf{x})$ and (2) $\{\mathbf{y}^n\} \cap F_0 = \emptyset$. Depending on the order of $\mathbf{y}$ and $\mathbf{w}$ on P, we write P as $\langle \mathbf{u}, P_1, \mathbf{y}, \mathbf{w}, P_2, \mathbf{v}\rangle$ or $\langle \mathbf{u}, P_1, \mathbf{w}, \mathbf{y}, P_2, \mathbf{v}\rangle$. Since the role of $\mathbf{u}$ and $\mathbf{v}$ are symmetric with respect to P, we can assume without loss of generality that $P = \langle \mathbf{u}, P_1, \mathbf{y}, \mathbf{w}, P_2, \mathbf{v}\rangle$. By induction, there exists a Hamiltonian path S of $Q_n^0 - F_0 - \{\mathbf{r}\}$ joining $\mathbf{y}^n$ to $\mathbf{x}^n$. Then $\langle \mathbf{u}, P_1, \mathbf{y}, \mathbf{y}^n, S, \mathbf{x}^n, \mathbf{x}, \mathbf{w}, P_2, \mathbf{v}\rangle$ forms the desired path. See Figure 13.14q for illustration.

Case 3.2.2. $\mathbf{r} \in Q_n^1$. Since $\{\mathbf{b_1}, \mathbf{w_1}\} \subset Q_n^0$, $r_1 \le n-4$. By induction, there exists a Hamiltonian path P of $Q_n^1 - F_1 - \{\mathbf{r}\}$ joining $\mathbf{u}$ to $\mathbf{v}$. Since there are at least $2^{n-1} - 2(n-4) - 1$ vertices in P, we can choose adjacent vertices $\mathbf{x}$ and $\mathbf{y}$ of P such that $\{\mathbf{x}^n, \mathbf{y}^n\} \cap F_{av} = \emptyset$. Write P as $\langle \mathbf{u}, P_1, \mathbf{x}, \mathbf{y}, P_2, \mathbf{v}\rangle$. By induction, there exists a Hamiltonian path S of $Q_n^0 - F_0$ joining $\mathbf{x}^n$ to $\mathbf{y}^n$. Then $\langle \mathbf{u}, P_1, \mathbf{x}, \mathbf{x}^n, S, \mathbf{y}^n, \mathbf{y}, P_2, \mathbf{v}\rangle$ forms the desired path. See Figure 13.14r for illustration.

Finally, by Lemma 13.1, $\mathcal{H}_{av+e}^{HL}(Q_n) \ge n-3$ if $n \ge 3$. □

THEOREM 13.15 $\mathcal{H}_{av+e}^{HL}(Q_n) = \mathcal{H}_{av+e}^{SL}(Q_n) = \mathcal{H}_{av+e}^{L}(Q_n) = n-3$ if $n \ge 3$.

Proof: By definition, $\mathcal{H}_{av+e}^{HL}(Q_n) \le \mathcal{H}_{av+e}^{SL}(Q_n) \le \mathcal{H}_{av+e}^{L}(Q_n) \le n-3$ if $n \ge 3$. With Lemmas 13.33 and 13.34, $\mathcal{H}_{av+e}^{HL}(Q_n) = \mathcal{H}_{av+e}^{SL}(Q_n) = \mathcal{H}_{av+e}^{L}(Q_n) = n-3$ if $n \ge 3$. □

THEOREM 13.16 $\mathcal{H}_{av+e}(Q_n) = n-2$ if $n \ge 3$.

Proof: We first prove that $\mathcal{H}_{av+e}(Q_n) \le n-2$ if $n \ge 3$. Let $n \ge 3$. Assume that $F = \{((00\ldots0), (00\ldots0)^j) \mid 1 \le j \le n-1\}$. Obviously, $f_e = n-1$. Since $d_{Q_n - F}((00\ldots0)) = 1$, there is no Hamiltonian cycle in $Q_n - F$. Thus $\mathcal{H}_{av+e}(Q_n) \le n-2$ if $n \ge 3$.

Now, we prove that $\mathcal{H}_{av+e}(Q_n) \ge n-2$ if $n \ge 3$ by induction. With Lemmas 13.6, 13.28, 13.31, and 13.32, the theorem is true for $n=3$ and 4. Assume that $\mathcal{H}_{av+e}(Q_k) \ge k-2$ for $4 \le k < n$. Let $F = F_{av} \cup F_e$, where F_{av} is a union of disjoint f_{av} adjacent vertices of Q_n and F_e consists of f_e edges of Q_n with $f_{av} + f_e \le n-2$. Let $F_{av} = \bigcup_{i=1}^{f_{av}} \{\mathbf{b_i}, \mathbf{w_i}\}$, where $(\mathbf{b_i}, \mathbf{w_i}) \in E(Q_n)$ and $\{\mathbf{b_i}, \mathbf{w_i}\} \cap \{\mathbf{b_j}, \mathbf{w_j}\} = \emptyset$ for $i \ne j$. Without loss of generality, we assume that $\{\mathbf{b_i} \mid 1 \le i \le f_{av}\}$ is a f_{av} set in one of the partite sets and $\{\mathbf{w_i} \mid 1 \le i \le f_{av}\}$ is a f_{av} set in the other partite set in Q_n. Write F_e as $\{(\mathbf{b_i}, \mathbf{w_i}) \mid f_{av}+1 \le i \le n-2\}$. We want to prove that $Q_n - F$ is Hamiltonian.

Suppose that $F_{av} = \emptyset$. By Lemma 13.6, $Q_n - F$ is Hamiltonian. Thus, we discuss the case $F_{av} \ne \emptyset$ only. We must construct a Hamiltonian cycle of $Q_n - F$. Let $t_j = |\{(\mathbf{b_i}, \mathbf{w_i}) \mid (\mathbf{b_i}, \mathbf{w_i})$ is a j-dimensional edge with $1 \le i \le n-2\}|$. Without loss of generality, we assume that $t_1 \ge t_2 \ge \cdots \ge t_n = 0$. Thus, both $\mathbf{b_i}$ and $\mathbf{w_i}$ are either in Q_n^0 or Q_n^1 for every i. Let $F_i = F \cap Q_n^i$, for $i \in \{0, 1\}$.

Let $r_j = |\{\mathbf{b_i} \mid \mathbf{b_i} \in Q_n^j, 1 \le i \le n-2\}|$ for $j = 0, 1$. Obviously, $r_0 + r_1 = n-2$. Since $F_{av} \ne \emptyset$, we may assume that $\{\mathbf{b_1}, \mathbf{w_1}\} \subset Q_n^0$. Let $F' = F'_{av} \cup F_e$, where $F'_{av} = F_{av} - \{\{\mathbf{b_1}, \mathbf{w_1}\}\}$. Hence, $f'_{av} = f_{av} - 1$ and $f'_{av} + f_e \le n-3$.

Case 1. $F \subset Q_n^0$. By induction, there exists a Hamiltonian cycle C of $Q_n^0 - F'$. Suppose that $(\mathbf{b_1}, \mathbf{w_1}) \in E(C)$. We can write C as $\langle \mathbf{b_1}, \mathbf{w_1}, \mathbf{x}, P, \mathbf{y}, \mathbf{b_1}\rangle$. Note that $\mathbf{x}$ and $\mathbf{y}$ are in different partite sets. Again, there exists a Hamiltonian path S of Q_n^1 joining $\mathbf{y}^n$ to

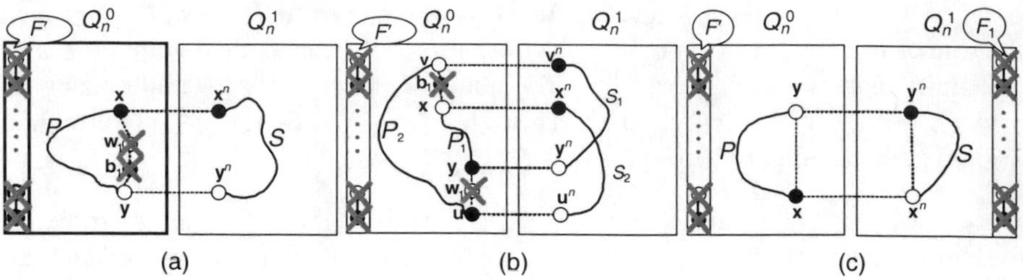

FIGURE 13.15 Illustration for Theorem 13.16.

$\mathbf{x}^n$. Then $\langle \mathbf{x}, P, \mathbf{y}, \mathbf{y}^n, S, \mathbf{x}^n, \mathbf{x}\rangle$ forms the desired cycle. See Figure 13.15a for illustration. Suppose that $(\mathbf{b_1}, \mathbf{w_1}) \notin E(C)$. We can write C as $\langle \mathbf{b_1}, \mathbf{x}, P_1, \mathbf{y}, \mathbf{w_1}, \mathbf{u}, P_2, \mathbf{v}, \mathbf{b_1}\rangle$. Note that $\mathbf{x}$ and $\mathbf{v}$ are in a partite set, and $\mathbf{y}$ and $\mathbf{u}$ are in the other partite set. By Lemma 13.17, there exist two disjoint paths S_1 and S_2 such that (1) S_1 joins $\mathbf{y}^n$ to $\mathbf{v}^n$, (2) S_2 joins $\mathbf{u}^n$ to $\mathbf{x}^n$, and (3) $S_1 \cup S_2$ spans Q_n^1. Then $\langle \mathbf{x}, P_1, \mathbf{y}, \mathbf{y}^n, S_1, \mathbf{v}^n, \mathbf{v}, P_2^{-1}, \mathbf{u}, \mathbf{u}^n, S_2, \mathbf{x}^n, \mathbf{x}\rangle$ forms the desired cycle. See Figure 13.15b for illustration.

Case 2. $F \not\subseteq Q_n^0$. Obviously, $r_0 \leq n-3$. Since $\{\mathbf{b_1}, \mathbf{w_1}\} \subset Q_n^0$, $r_1 \leq n-3$. Since $n \geq 5$ and $r_0 + r_1 \leq n-2$, $r_i \leq n-4$ for some $i \in \{0, 1\}$. Without loss of generality, we assume that $r_1 \leq n-4$. By induction, there exists a Hamiltonian cycle C of $Q_n^0 - F_0$. Since there are at least $2^{n-1} - 2(n-3)$ vertices in C, we can choose adjacent vertices $\mathbf{x}$ and $\mathbf{y}$ of C such that $\{\mathbf{x}^n, \mathbf{y}^n\} \cap F_1 = \emptyset$. Write C as $\langle \mathbf{x}, P, \mathbf{y}, \mathbf{x}\rangle$. By Theorem 13.15, there exists a Hamiltonian path S of $Q_n^1 - F_1$ joining $\mathbf{y}^n$ to $\mathbf{x}^n$. Then $\langle \mathbf{x}, P, \mathbf{y}, \mathbf{y}^n, S, \mathbf{x}^n, \mathbf{x}\rangle$ forms the desired cycle. See Figure 13.15c for illustration. □

COROLLARY 13.2 *If F is a vertex subset of Q_n with $|F| \leq n-2$, then $Q_n - F$ contains a path of length at least $2^n - 2|F| - 2$ between any two vertices in $Q_n - F$, $n \geq 2$.*

Proof: We prove this theorem by induction. Obviously, the theorem holds for $n = 2$ and 3. Assume that $Q_m - F$ contains a path of length at least $2^m - 2|F| - 2$ between any two vertices of $Q_m - F$ for $3 \leq m < n$ and $|F| \leq m-2$. Let $F = \{\mathbf{x_1}, \mathbf{x_2}, \ldots, \mathbf{x_k}\}$ be a subset of $V(Q_n)$ with $k \leq n-2$. Let $\mathbf{u}$ and $\mathbf{v}$ be any two vertices in $Q_n - F$. By the symmetric property of Q_n, we assume that $F \cap Q_n^j \neq \emptyset$ for $j = 0, 1$. Let $F \cap Q_n^j = F^j$ for $j = 0, 1$. Without loss of generality, we assume that $F^1 = \{\mathbf{x_1}, \ldots, \mathbf{x_t}\}$. Obviously, $t \leq n-3$.

Case 1. $|\{\mathbf{u}, \mathbf{v}\} \cap Q_n^j| = 2$ for $j = 0$ or 1. Without loss of generality, we may assume that $\{\mathbf{u}, \mathbf{v}\} \subset Q_n^0$. By induction, there exists a path P of length at least $2^{n-1} - 2(k-t) - 2$ joining $\mathbf{u}$ to $\mathbf{v}$ in $Q_n^0 - F^0$. Since there are $2^{n-1} - 2(k-t) - 1$ vertices in P, we can choose a path $\langle \mathbf{r}, \mathbf{z}, \mathbf{w}\rangle$ of P such that (1) $\mathbf{r}$, $\mathbf{w}$, and $\mathbf{x_1}$ are in the same partite set and (2) $\{\mathbf{r}^n, \mathbf{w}^n\} \cap F^1 = \emptyset$. Write P as $\langle \mathbf{u}, P_1, \mathbf{r}, \mathbf{z}, \mathbf{w}, P_2, \mathbf{v}\rangle$. Since $d_{Q_n^1}(\mathbf{x_i}) = n-1$, there exists a neighbor $\mathbf{y_i}$ of $\mathbf{x_i}$ such that (1) $\mathbf{y_i} \notin F^1$ and (2) $\mathbf{y_i} \neq \mathbf{y_j}$ for $i \neq j$. Let $F_{av} = \{\{\mathbf{x_i}, \mathbf{y_i}\} \mid 2 \leq i \leq t\}$. By Theorem 13.15, there exists a Hamiltonian path S of $Q_n^1 - F_{av} - \{\mathbf{x}_1\}$ joining $\mathbf{r}^n$ to $\mathbf{w}^n$. Obviously, the length of S is $2^{n-1} - 2t$. Then $\langle \mathbf{u}, P_1, \mathbf{r}, \mathbf{r}^n, S, \mathbf{w}^n, \mathbf{w}, P_2, \mathbf{v} \rangle$ forms the desired path.

Case 2. $|\{\mathbf{u}, \mathbf{v}\} \cap Q_n^0| = 1$. Without loss of generality, we may assume that $\mathbf{u} \in Q_n^0$ and $\mathbf{v} \in Q_n^1$. Since there are at least $2^{n-2} - (k - t)$ vertices in $Q_n^0 - F^0$, we can choose a vertex $\mathbf{r}$ in the partite set of $Q_n^0 - F^0$ not containing $\mathbf{u}$ such that $\{\mathbf{r}^n\} \cap \{F^1 \cup \{\mathbf{v}\}\} = \emptyset$. Note that $\mathbf{u}$ and $\mathbf{r}$ are in different partite sets. By induction, there exists a path P of length at least $2^{n-1} - 2(k - t) - 1$ joining $\mathbf{u}$ to $\mathbf{r}$ in $Q_n^0 - F^0$. Again, there exists a path S of length at least $2^{n-1} - 2t - 2$ joining $\mathbf{r}^n$ to $\mathbf{v}$ in $Q_n^1 - F^1$. Then $\langle \mathbf{u}, P, \mathbf{r}, \mathbf{r}^n, S, \mathbf{v} \rangle$ forms the desired path.

The theorem is proved. □

Recently, Fu [117] proved that $Q_n - F$ consists of a path of length $2^n - 2|F| - 2$ between any two vertices in $Q_n - F$ where $n \geq 2$ and F is a vertex subset of Q_n with $|F| \leq n - 2$. With Corollary 13.2, our result includes this as a special case.

13.8 CONDITIONAL FAULT HAMILTONICITY AND CONDITIONAL FAULT HAMILTONIAN LACEABILITY OF THE STAR GRAPHS

Obviously, $\delta(G) - 2$ is an immediate upper bound for $\mathcal{H}_f^e(G)$. For this reason, Chan and Lee [38] studied the existence of a Hamiltonian cycle in the n-dimensional hypercube when each vertex is incident to at least two nonfaulty edges. A graph G is called *k-edge-fault conditional Hamiltonian* if $G - F$ is Hamiltonian for every $F \subseteq E(G)$ with $|F| \leq k$ and $\delta(G - F) \geq 2$. Clearly, whenever a graph is k-edge-fault Hamiltonian, it is also k-edge-fault conditional Hamiltonian, but the conditional Hamiltonicity could be much larger. Chan and Lee [38] showed that the n-dimensional hypercube is $(2n - 5)$-edge-fault conditional Hamiltonian. Fu [115] showed that the n-dimensional star graph is $(2n - 7)$-edge-fault conditional Hamiltonian. Similarly, we can define a Hamiltonian-laceable graph G to be *k-edge-fault conditional Hamiltonian laceable* if $G - F$ is Hamiltonian laceable for every set of edges $F \subseteq E(G)$ with $|F| \leq k$ and $\delta(G - F) \geq 2$. Again, k-edge-fault Hamiltonian laceability implies k-edge-fault conditional Hamiltonian laceability, but the conditional Hamiltonian laceability could be much larger. Tsai [301] showed that the n-dimensional hypercube is $(2n - 5)$-edge-fault conditional Hamiltonian laceable. Lin et al. [237] proved that for $n \geq 4$, the n-dimensional star graph S_n is $(3n - 10)$-edge-fault conditional Hamiltonian and $(2n - 7)$-edge-fault conditional Hamiltonian laceable.

The following results are proved in Ref. 235.

LEMMA 13.35 **[235]** Assume that $\mathbf{r}$ and $\mathbf{s}$ are any two adjacent vertices of S_n with $n \geq 4$. Then, for any white vertex $\mathbf{u}$ in $S_n - \{\mathbf{r}, \mathbf{s}\}$ and for any $i \in \langle n \rangle$, there exists a Hamiltonian path P of $S_n - \{\mathbf{r}, \mathbf{s}\}$ joining $\mathbf{u}$ to some black vertex $\mathbf{v}$ of $S_n - \{\mathbf{r}, \mathbf{s}\}$ with $(\mathbf{v})_1 = i$.

Proof: Since S_n is vertex-transitive and edge-transitive, we assume that $\mathbf{r} = \mathbf{e}$ and $\mathbf{s} = (\mathbf{e})^2$. Obviously, both $\mathbf{e}$ and $(\mathbf{e})^2$ are in $S_n^{\{n\}}$. We prove this lemma by induction on n.

Suppose that $n = 4$. The required Hamiltonian paths of $S_4 - \{1234, 2134\}$ are listed here.

$\langle 1342, 2341, 4321, 1324, 3124, 4123, 2143, 3142, 4132, 1432, 3412, 4312, 2314, 3214, 4213, 2413, 1423, 3421, 2431, 4231, 3241, 1243 \rangle$
$\langle 1342, 2341, 4321, 1324, 3124, 4123, 2143, 3142, 4132, 1432, 3412, 4312, 2314, 3214, 4213, 1243, 3241, 4231, 2431, 3421, 1423, 2413 \rangle$
$\langle 1342, 2341, 4321, 1324, 3124, 4123, 2143, 3142, 4132, 1432, 2431, 4231, 3241, 1243, 4213, 3214, 2314, 4312, 3412, 2413, 1423, 3421 \rangle$
$\langle 1342, 2341, 4321, 1324, 3124, 4123, 2143, 3142, 4132, 1432, 2431, 3421, 1423, 2413, 3412, 4312, 2314, 3214, 4213, 1243, 3241, 4231 \rangle$
$\langle 1423, 3421, 4321, 2341, 1342, 4312, 3412, 2413, 4213, 3214, 2314, 1324, 3124, 4123, 2143, 3142, 4132, 1432, 2431, 4231, 3241, 1243 \rangle$
$\langle 1423, 3421, 2431, 4231, 3241, 1243, 4213, 3214, 2314, 4312, 1342, 2341, 4321, 1324, 3124, 4123, 2143, 3142, 4132, 1432, 3412, 2413 \rangle$
$\langle 1423, 3421, 2431, 4231, 3241, 1243, 4213, 2413, 3412, 1432, 4132, 3142, 2143, 4123, 3124, 1324, 4321, 2341, 1342, 4312, 2314, 3214 \rangle$
$\langle 1423, 3421, 2431, 4231, 3241, 1243, 2143, 3142, 4132, 1432, 3412, 2413, 4213, 3214, 2314, 4312, 1342, 2341, 4321, 1324, 3124, 4123 \rangle$
$\langle 2143, 3142, 4132, 1432, 3412, 2413, 1423, 4123, 3124, 1324, 4321, 3421, 2431, 4231, 3241, 2341, 1342, 4312, 2314, 3214, 4213, 1243 \rangle$
$\langle 2143, 3142, 4132, 1432, 2431, 4231, 3241, 1243, 4213, 3214, 2314, 1324, 3124, 4123, 1423, 3421, 4321, 2341, 1342, 4312, 3412, 2413 \rangle$
$\langle 2143, 3142, 4132, 1432, 2431, 4231, 3241, 1243, 4213, 3214, 2314, 1324, 3124, 4123, 1423, 2413, 3412, 4312, 1342, 2341, 4321, 3421 \rangle$
$\langle 2143, 3142, 4132, 1432, 2431, 3421, 4321, 2341, 1342, 4312, 3412, 2413, 1423, 4123, 3124, 1324, 2314, 3214, 4213, 1243, 3241, 4231 \rangle$
$\langle 2314, 3214, 4213, 1243, 3241, 4231, 2431, 3421, 1423, 2413, 3412, 4312, 1342, 2341, 4321, 1324, 3124, 4123, 2143, 3142, 4132, 1432 \rangle$
$\langle 2314, 3214, 4213, 1243, 3241, 4231, 2431, 1432, 4132, 3142, 2143, 4123, 3124, 1324, 4321, 3421, 1423, 2413, 3412, 4312, 1342, 2341 \rangle$
$\langle 2314, 4312, 1342, 2341, 4321, 1324, 3124, 4123, 2143, 3142, 4132, 1432, 3412, 2413, 1423, 3421, 2431, 4231, 3241, 1243, 4213, 3214 \rangle$
$\langle 2314, 3214, 4213, 1243, 3241, 4231, 2431, 3421, 1423, 2413, 3412, 1432, 4132, 3142, 2143, 4123, 3124, 1324, 4321, 2341, 1342, 4312 \rangle$
$\langle 2431, 4231, 3241, 1243, 4213, 3214, 2314, 4312, 1342, 2341, 4321, 3421, 1423, 2413, 3412, 1432, 4132, 3142, 2143, 4123, 3124, 1324 \rangle$
$\langle 2431, 4231, 3241, 1243, 4213, 3214, 2314, 1324, 3124, 4123, 2143, 3142, 4132, 1432, 3412, 4312, 1342, 2341, 4321, 3421, 1423, 2413 \rangle$
$\langle 2431, 4231, 3241, 1243, 4213, 3214, 2314, 4312, 1342, 2341, 4321, 1324, 3124, 4123, 2143, 3142, 4132, 1432, 3412, 2413, 1423, 3421 \rangle$
$\langle 2431, 3421, 4321, 1324, 3124, 4123, 1423, 2413, 3412, 1432, 4132, 3142, 2143, 1243, 4213, 3214, 2314, 4312, 1342, 2341, 3241, 4231 \rangle$
$\langle 3124, 4123, 2143, 3142, 4132, 1432, 2431, 4231, 3241, 2341, 1342, 4312, 3412, 2413, 1423, 3421, 4321, 1324, 2314, 3214, 4213, 1243 \rangle$
$\langle 3124, 4123, 2143, 3142, 4132, 1432, 2431, 4231, 3241, 1243, 4213, 3214, 2314, 1324, 4321, 3421, 1423, 2413, 3412, 4312, 1342, 2341 \rangle$
$\langle 3124, 4123, 2143, 3142, 4132, 1432, 2431, 4231, 3241, 1243, 4213, 3214, 2314, 1324, 4321, 2341, 1342, 4312, 3412, 2413, 1423, 3421 \rangle$
$\langle 3124, 4123, 2143, 3142, 4132, 1432, 2431, 3421, 1423, 2413, 3412, 4312, 1342, 2341, 4321, 1324, 2314, 3214, 4213, 1243, 3241, 4231 \rangle$
$\langle 3241, 4231, 2431, 1432, 4132, 3142, 1342, 2341, 4321, 3421, 1423, 2413, 3412, 4312, 2314, 3214, 4213, 1243, 2143, 4123, 3124, 1324 \rangle$
$\langle 3241, 4231, 2431, 1432, 4132, 3142, 2143, 1243, 4213, 3214, 2314, 1324, 3124, 4123, 1423, 3421, 4321, 2341, 1342, 4312, 3412, 2413 \rangle$
$\langle 3241, 4231, 2431, 1432, 4132, 3142, 2143, 1243, 4213, 3214, 2314, 1324, 3124, 4123, 1423, 2413, 3412, 4312, 1342, 2341, 4321, 3421 \rangle$
$\langle 3241, 1243, 2143, 4123, 3124, 1324, 4321, 2341, 1342, 3142, 4132, 1432, 3412, 4312, 2314, 3214, 4213, 2413, 1423, 3421, 2431, 4231 \rangle$
$\langle 3412, 2413, 1423, 3421, 2431, 4231, 3241, 1243, 4213, 3214, 2314, 4312, 1342, 2341, 4321, 1324, 3124, 4123, 2143, 3142, 4132, 1432 \rangle$
$\langle 3412, 2413, 1423, 3421, 4321, 1324, 3124, 4123, 2143, 3142, 4132, 1432, 2431, 4231, 3241, 1243, 4213, 3214, 2314, 4312, 1342, 2341 \rangle$
$\langle 3412, 2413, 1423, 3421, 4321, 1324, 3124, 4123, 2143, 1243, 4213, 3214, 2314, 4312, 1342, 2341, 3241, 4231, 2431, 1432, 4132, 3142 \rangle$
$\langle 3412, 2413, 1423, 3421, 2431, 1432, 4132, 3142, 1342, 4312, 2314, 3214, 4213, 1243, 2143, 4123, 3124, 1324, 4321, 2341, 3241, 4231 \rangle$
$\langle 4132, 3142, 2143, 4123, 3124, 1324, 4321, 2341, 1342, 4312, 2314, 3214, 4213, 1243, 3241, 4231, 2431, 3421, 1423, 2413, 3412, 1432 \rangle$
$\langle 4132, 3142, 2143, 4123, 3124, 1324, 4321, 3421, 1423, 2413, 3412, 1432, 2431, 4231, 3241, 1243, 4213, 3214, 2314, 4312, 1342, 2341 \rangle$
$\langle 4132, 3142, 2143, 4123, 3124, 1324, 4321, 2341, 1342, 4312, 2314, 3214, 4213, 1243, 3241, 4231, 2431, 1432, 3412, 2413, 1423, 3421 \rangle$
$\langle 4132, 3142, 2143, 4123, 3124, 1324, 2314, 3214, 4213, 1243, 3241, 4231, 2431, 1432, 3412, 2413, 1423, 3421, 4321, 2341, 1342, 4312 \rangle$
$\langle 4213, 3214, 2314, 4312, 1342, 2341, 4321, 1324, 3124, 4123, 2143, 3142, 4132, 1432, 3412, 2413, 1423, 3421, 2431, 4231, 3241, 1243 \rangle$
$\langle 4213, 3214, 2314, 4312, 1342, 2341, 4321, 1324, 3124, 4123, 1423, 3421, 2431, 4231, 3241, 1243, 2143, 3142, 4132, 1432, 3412, 2413 \rangle$
$\langle 4213, 3214, 2314, 4312, 1342, 2341, 4321, 1324, 3124, 4123, 2143, 1243, 3241, 4231, 2431, 3421, 1423, 2413, 3412, 1432, 4132, 3142 \rangle$
$\langle 4213, 3214, 2314, 4312, 1342, 2341, 3241, 1243, 2143, 3142, 4132, 1432, 3412, 2413, 1423, 4123, 3124, 1324, 4321, 3421, 2431, 4231 \rangle$
$\langle 4321, 1324, 3124, 4123, 2143, 3142, 4132, 1432, 3412, 2413, 1423, 3421, 2431, 4231, 3241, 2341, 1342, 4312, 2314, 3214, 4213, 1243 \rangle$
$\langle 4321, 1324, 3124, 4123, 2143, 3142, 4132, 1432, 3412, 2413, 1423, 3421, 2431, 4231, 3241, 1243, 4213, 3214, 2314, 4312, 1342, 2341 \rangle$
$\langle 4321, 1324, 3124, 4123, 2143, 1243, 4213, 3214, 2314, 4312, 1342, 2341, 3241, 4231, 2431, 3421, 1423, 2413, 3412, 1432, 4132, 3142 \rangle$
$\langle 4321, 1324, 3124, 4123, 2143, 1243, 3241, 2341, 1342, 3142, 4132, 1432, 3412, 4312, 2314, 3214, 4213, 2413, 1423, 3421, 2431, 4231 \rangle$

Assume that this result holds in S_k for every $4 \leq k \leq n-1$. We have the following cases:

Case 1. $\mathbf{u} \in S_n^{\{n\}}$. By induction, there is a Hamiltonian path P of $S_n^{\{n\}} - \{\mathbf{e}, (\mathbf{e})^2\}$ joining $\mathbf{u}$ to a black vertex $\mathbf{x}$ with $(\mathbf{x})_1 = n-1$. Note that $(\mathbf{x})^n$ is a white vertex of $S_n^{\{n-1\}}$. We choose a black vertex $\mathbf{v}$ in $S_n^{\langle n-2 \rangle}$ with $(\mathbf{v})_1 = i$. By Lemma 13.9, there is a Hamiltonian path Q of $S_n^{\langle n-1 \rangle}$ joining $(\mathbf{x})^n$ to $\mathbf{v}$. Then $\langle \mathbf{u}, P, \mathbf{x}, (\mathbf{x})^n, Q, \mathbf{v} \rangle$ forms the desired Hamiltonian path of $S_n - \{\mathbf{e}, (\mathbf{e})^2\}$ joining $\mathbf{u}$ to $\mathbf{v}$ with $(\mathbf{v})_1 = i$. See Figure 13.16a for illustration.

Case 2. $\mathbf{u} \in S_n^{\{k\}}$ for some $k \in \langle n-1 \rangle$. By Lemma 13.7, there are $(n-2)!/2 \geq 3$ edges joining black vertices of $S_n^{\{k\}}$ to white vertices of $S_n^{\{n\}}$. We can choose a white vertex $\mathbf{x} \in S_n^{\{n\}} - \{\mathbf{e}, (\mathbf{e})^2\}$ with $(\mathbf{x})_1 = k$. By Theorem 13.2, there is a Hamiltonian path P of $S_n^{\{k\}}$ joining $\mathbf{u}$ to the black vertex $(\mathbf{x})^n$. By induction, there is a Hamiltonian path Q of $S_n^{\{n\}} - \{\mathbf{e}, (\mathbf{e})^2\}$ joining $\mathbf{x}$ to a black vertex $\mathbf{y}$ with $(\mathbf{y})_1 \in \langle n-1 \rangle - \{k\}$. We choose a black vertex $\mathbf{v}$ in $S_n^{\langle n-1 \rangle - \{k, (\mathbf{y})_1\}}$ with $(\mathbf{v})_1 = i$. By Lemma 13.9, there exists a Hamiltonian path R of $S_n^{\langle n-1 \rangle - \{k\}}$ joining the white vertex $(\mathbf{y})^n$ to $\mathbf{v}$. Then $\langle \mathbf{u}, P, (\mathbf{x})^n, \mathbf{x}, Q, \mathbf{y}, (\mathbf{y})^n, R, \mathbf{v} \rangle$ forms the desired Hamiltonian path of $S_n - \{\mathbf{e}, (\mathbf{e})^2\}$ joining $\mathbf{u}$ to $\mathbf{v}$ with $(\mathbf{v})_1 = i$. See Figure 13.16b for illustration. □

THEOREM 13.17 Let $n \geq 5$ and $I = \{a_1, a_2, \ldots, a_r\}$ be a subset of $\langle n \rangle$ for some $r \in \langle n \rangle$. Then S_n^I is Hamiltonian laceable.

Proof: Let $\mathbf{u}$ be a white vertex and $\mathbf{v}$ be a black vertex of S_n^I. By Lemma 13.9, this theorem holds for either $r = 1$ or $r \geq 2$ and $(\mathbf{u})_n \neq (\mathbf{v})_n$. Thus, we assume that $r \geq 2$ and $(\mathbf{u})_n = (\mathbf{v})_n$. Without loss of generality, we assume that $(\mathbf{u})_n = (\mathbf{v})_n = a_1$.

Case 1. $(\mathbf{v})^n \in S_n^I$. Without loss of generality, we assume that $(\mathbf{v})^n \in S_n^{\{a_r\}}$. By Theorem 13.2, there is a Hamiltonian path P of $S_n^{\{a_1\}} - \{\mathbf{v}\}$ joining $\mathbf{u}$ to a white vertex $\mathbf{x}$ with $(\mathbf{x})_1 = a_2$. By Lemma 13.9, there is a Hamiltonian path Q of $S_n^{I - \{a_1\}}$ joining the black vertex $(\mathbf{x})^n$ to the white vertex $(\mathbf{v})^n$. Then $\langle \mathbf{u}, P, \mathbf{x}, (\mathbf{x})^n, Q, (\mathbf{v})^n, \mathbf{v} \rangle$ forms the desired Hamiltonian path of S_n^I joining $\mathbf{u}$ to $\mathbf{v}$. See Figure 13.17a for illustration.

Case 2. $(\mathbf{u})^n \notin S_n^I$ and $(\mathbf{v})^n \notin S_n^I$. We can choose a white vertex $\mathbf{y}$ with $\mathbf{y}$ being a neighbor of $\mathbf{v}$ in $S_n^{\{a_1\}}$ and $(\mathbf{y})_1 = a_r$. Obviously, $\mathbf{y} \neq \mathbf{u}$. By Lemma 13.35, there is a Hamiltonian path P of $S_n^{\{a_1\}} - \{\mathbf{v}, \mathbf{y}\}$ joining $\mathbf{u}$ to a black vertex $\mathbf{x}$ with $(\mathbf{x})_1 = a_2$. By Lemma 13.9, there is a Hamiltonian path Q of $S_n^{I - \{a_1\}}$ joining the white vertex $(\mathbf{x})^n$ to the black vertex $(\mathbf{y})^n$. Then $\langle \mathbf{u}, P, \mathbf{x}, (\mathbf{x})^n, Q, (\mathbf{y})^n, \mathbf{y}, \mathbf{v} \rangle$ is the desired Hamiltonian path of S_n^I joining $\mathbf{u}$ to $\mathbf{v}$. See Figure 13.17b for illustration. □

THEOREM 13.18 Let $I = \{i_1, i_2, \ldots, i_t\}$ be a nonempty subset of $\langle n \rangle$ with $n \geq 5$, and let F be a set of edges of S_n such that $|F \cap E(S_n^{\{i_k\}})| \leq n - 4$ for every $k \in \langle t \rangle$ and

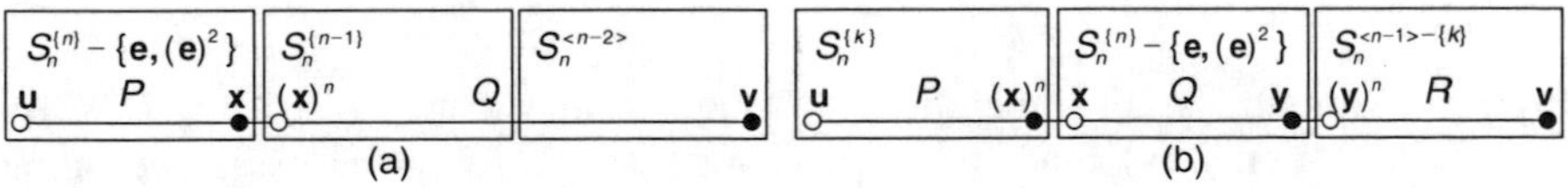

FIGURE 13.16 Illustration for Lemma 13.35.

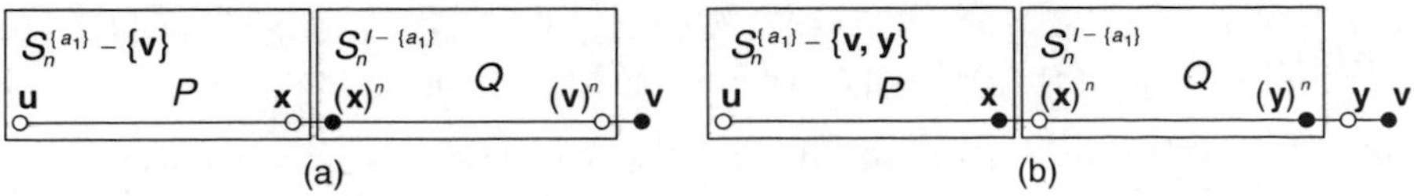

FIGURE 13.17 Illustration for Theorem 13.17.

$|F \cap E^{i_k,i_{k+1}}| \leq [(n-2)!/2] - 1$ for every $k \in \langle t-1 \rangle$. If $\mathbf{u}$ is a white vertex of $S_n^{\{i_1\}}$ and $\mathbf{v}$ is a black vertex of $S_n^{\{i_t\}}$, then there is a Hamiltonian path P of $S_n^I - F$ joining $\mathbf{u}$ to $\mathbf{v}$.

Proof: Since $S_n^{i_k}$ is isomorphic to S_{n-1} for every $k \in \langle t \rangle$ and S_{n-1} is hamiltonian laceable, this statement holds for $t = 1$. If $2 \leq t \leq n$, then we set $\mathbf{x_1} = \mathbf{u}$ and $\mathbf{y_t} = \mathbf{v}$. Since there are $(n-2)!/2$ edges joining black vertices of $S_n^{\{i\}}$ to white vertices of $S_n^{\{j\}}$ for every two distinct elements i and j in $\langle n \rangle$ and $(n-2)!/2 > [(n-2)!/2] - 1$ if $n \geq 5$, we can choose a black vertex $\mathbf{y_k}$ in $S_n^{\{i_k\}}$ such that $(\mathbf{y_k})n \in S_n^{\{i_{k+1}\}}$ and $(\mathbf{y_k}, (\mathbf{y_k})n) \notin F$ for every $k \langle t-1 \rangle$. Set $\mathbf{x_{k+1}} = (\mathbf{y_k})n$ for every $k \in \langle t-1 \rangle$. Obviously, $\mathbf{x_k}$ is a white vertex for every $k \in \langle t \rangle$. Since S_k is $(k-3)$-edge fault-tolerant Hamiltonian laceable for any $k \geq 4$, there is a Hamiltonian path H_k of $S_n^{\{i_k\}} - F$ joining $\mathbf{x_k}$ to $\mathbf{y_k}$ for every $k \in \langle t \rangle$. Hence, $\langle \mathbf{u} = \mathbf{x_1}, H_1, \mathbf{y_1}, \mathbf{x_2}, H_2, \mathbf{y_2}, \ldots, \mathbf{x_t}, H_t, \mathbf{y_t} = \mathbf{v} \rangle$ is a Hamiltonian path of $S_n^I - F$ joining $\mathbf{u}$ to $\mathbf{v}$. □

LEMMA 13.36 Let F be a subset of $E(S_n)$ with $|F| \leq 3n - 10$ such that $\delta(S_n - F) \geq 2$, for $n \geq 5$. If A is the set of vertices of degree 2 in $S_n - F$, then $|A| \leq 4$. Moreover, if $n \geq 6$, then $|A| \leq 3$.

Proof: Suppose first that $X = \{\mathbf{x_1}, \mathbf{x_2}, \mathbf{x_3}, \mathbf{x_4}, \mathbf{x_5}\} \subseteq A$, and let H be the subgraph with vertex set X consisting of those edges of F that join two vertices in X. Since S_n is bipartite, so is H. At least one partite set can be chosen such that it has at least three vertices. If we count the number of edges in F incident to these three vertices, then no edges of F is counted twice, so we get that $|F| \geq 3(n-3) \geq 3n - 9$, a contradiction. Thus, $|A| \leq 4$.

Next assume that $|A| = 4$. Again consider the subgraph H with vertex set A including those edges of F that join two vertices in A. Since H has no cycles, it is a forest and it has at most 3 edges. Counting the number of edges in F at each vertex of A, we count each edge of H twice, so we get that $|F| \geq 4(n-3) - 3 = 4n - 15$. Since $|F| \leq 3n - 10$, this implies that $n \leq 5$, and the lemma is proved. □

LEMMA 13.37 Let $\mathbf{x}$, $\mathbf{y}$, $\mathbf{p}$, and $\mathbf{q}$ be four distinct vertices of $S_n^{\{i\}}$ for some $i \in \langle n \rangle$ with $(\mathbf{x}, \mathbf{y}) \in E(S_n)$ and $(\mathbf{p}, \mathbf{q}) \in E(S_n)$, for $n \geq 5$. Then there are two disjoint paths P_1 and P_2 of $S_n^{\langle n \rangle - \{i\}}$ such that (1) P_1 joins $(\mathbf{x})^n$ to $(\mathbf{y})^n$, (2) P_2 joins $(\mathbf{p})^n$ to $(\mathbf{q})^n$, and (3) $P_1 \cup P_2$ spans $S_n^{\langle n \rangle - \{i\}}$.

Proof: Without loss of generality, we may assume that $\{\mathbf{x}, \mathbf{y}, \mathbf{p}, \mathbf{q}\} \subset S_n^{\{n\}}$. By Lemma 13.8, $(\mathbf{x})_1 \neq (\mathbf{y})_1$ and $(\mathbf{p})_1 \neq (\mathbf{q})_1$. We have the following cases depending on the location of the neighbors of $\mathbf{x}, \mathbf{y}, \mathbf{p}, \mathbf{q}$ outside $S_n^{\{n\}}$.

Case 1. $|\{(\mathbf{x})_1, (\mathbf{y})_1\} \cap \{(\mathbf{p})_1, (\mathbf{q})_1\}| = 0$. By Theorem 13.18, there is a Hamiltonian path P_1 of $S_n^{\{(\mathbf{x})_1, (\mathbf{y})_1\}}$ joining $(\mathbf{y})^n$ to $(\mathbf{x})^n$. Similarly, there is a Hamiltonian path P_2 of $S_n^{\langle n-1 \rangle - \{(\mathbf{x})_1, (\mathbf{y})_1\}}$ joining $(\mathbf{p})^n$ to $(\mathbf{q})^n$. Then P_1 and P_2 are the desired paths.

Case 2. $|\{(\mathbf{x})_1, (\mathbf{y})_1\} \cap \{(\mathbf{p})_1, (\mathbf{q})_1\}| = 1$. By symmetry, we may assume that $(\mathbf{x})_1 \notin \{(\mathbf{p})_1, (\mathbf{q})_1\}$ and $(\mathbf{y})_1 = (\mathbf{p})_1$. Since $((\mathbf{y})^n)_1 = ((\mathbf{p})^n)_1 = n$, by Lemma 13.8 there is a vertex $\mathbf{z}$ in $S_n^{\{(\mathbf{p})_1\}}$ such that $\mathbf{z}$ is a neighbor of $(\mathbf{y})^n$ and $(\mathbf{z})_1=(\mathbf{x})_1$. By Theorem 13.18, there is a Hamiltonian path H of $S_n^{\{(\mathbf{x})_1\}}$ joining $(\mathbf{x})^n$ to $(\mathbf{z})^n$. Let t be any integer in $\langle n-1\rangle-\{(\mathbf{x})_1,(\mathbf{p})_1,(\mathbf{q})_1\}$. By Lemma 13.35, there is a Hamiltonian path R of $S_n^{\{(\mathbf{p})_1\}}-\{(\mathbf{y})^n,\mathbf{z}\}$ joining $(\mathbf{p})^n$ to a vertex $\mathbf{w}$ such that $(\mathbf{w})_1=t$. By Theorem 13.18, there is a Hamiltonian path Q of $S_n^{\langle n-1\rangle-\{(\mathbf{x})_1,(\mathbf{p})_1\}}$ joining $(\mathbf{w})^n$ to $(\mathbf{q})^n$. Then $P_1=\langle(\mathbf{x})^n,H,(\mathbf{z})^n,\mathbf{z},(\mathbf{y})^n\rangle$ and $P_2=\langle(\mathbf{p})^n,R,\mathbf{w},(\mathbf{w})^n,Q,(\mathbf{q})^n\rangle$ are the desired paths.

Case 3. $|\{(\mathbf{x})_1,(\mathbf{y})_1\}\cap\{(\mathbf{p})_1,(\mathbf{q})_1\}|=2$. By symmetry, we may assume that $(\mathbf{x})_1=(\mathbf{p})_1$ and $(\mathbf{y})_1=(\mathbf{q})_1$. Let t and s be any two distinct integers in $\langle n-1\rangle-\{(\mathbf{p})_1,(\mathbf{q})_1\}$. Since $((\mathbf{x})_n)_1=n$ and $((\mathbf{p})^n)_1=n$, by Lemma 13.8 there is a vertex $\mathbf{w}$ in $S_n^{\{(\mathbf{p})_1\}}$ such that $\mathbf{w}$ is a neighbor of $(\mathbf{x})^n$ and $(\mathbf{w})_1=t$. Similarly, there is a vertex $\mathbf{z}$ in $S_n^{\{(\mathbf{q})_1\}}$ such that $\mathbf{z}$ is a neighbor of $(\mathbf{y})^n$ and $(\mathbf{z})_1=t$. By Theorem 13.18, there is a Hamiltonian path H of $S_n^{\{t\}}$ joining $(\mathbf{w})^n$ to $(\mathbf{z})^n$. By Lemma 13.35, there is a Hamiltonian path R_1 of $S_n^{\{(\mathbf{p})_1\}}-\{(\mathbf{x})^n,\mathbf{w}\}$ joining $(\mathbf{p})^n$ to a vertex $\mathbf{u}$ such that $(\mathbf{u})_1=s$. By Lemma 13.35, there is a Hamiltonian path R_2 of $S_n^{\{(\mathbf{q})_1\}}-\{(\mathbf{y})^n,\mathbf{z}\}$ joining a vertex $\mathbf{v}$ such that $(\mathbf{v})_1=s$ to $(\mathbf{q})^n$. By Theorem 15.18 there is a Hamiltonian path Q of $S_n^{\langle n-1\rangle-\{t,(\mathbf{p})_1,(\mathbf{q})_1\}}$ joining $(\mathbf{u})^n$ to $(\mathbf{v})^n$. Then $P_1=\langle(\mathbf{x})^n,\mathbf{w},(\mathbf{w})^n,H,(\mathbf{z})^n,\mathbf{z},(\mathbf{y})^n\rangle$ and $P_2=\langle(\mathbf{p})^n,R_1,\mathbf{u},(\mathbf{u})^n,Q,(\mathbf{v})^n,\mathbf{v},R_2,(\mathbf{q})^n\rangle$ are the desired paths. □

THEOREM 13.19 Let F be a subset of $E(S_n)$ with $|F|\le 3n-10$ and $\delta(S_n-F)\ge 2$, for $n\ge 4$. Then S_n-F is Hamiltonian.

Proof: We prove this statement by induction on n. We can prove that the statement holds for $n=4$ and 5 by brute force. The reader can read Ref. 237 for the detailed proof. Suppose now that the statement holds for S_k for every $5\le k\le n-1$, where $n\ge 6$. Let $A=\{\mathbf{u}\mid \deg_{S_n-F}(\mathbf{u})=2\}$. Since $n\ge 6$, by Lemma 13.36 we have $|A|\le 3$. We set $F_i=F\cap E(S_n^{\{i\}})$ for every $i\in\langle n\rangle$ and $F_{j,k}=F\cap E^{j,k}$ for every two distinct elements $j,k\in\langle n\rangle$. We consider cases depending on the size of A.

Case 1. $2\le|A|\le 3$. We consider cases depending on whether there is an edge of F joining two vertices in A.

Case 1.1. There are $\{\mathbf{u},\mathbf{v}\}\subset A$ with $(\mathbf{u},\mathbf{v})\in F$. Since S_n is vertex-transitive and edge-transitive, we may assume that $\mathbf{u}\in S_n^{\{n\}}$ and $\mathbf{v}\in S_n^{\{n-1\}}$. Clearly, $|F_n|\ge n-4$, $|F_{n-1}|\ge n-4$, and $|F_i|\le n-3$ for every $i\in\langle n-2\rangle$, and $|E^{i,j}|\le n-2$ for every distinct $i,j\in\langle n\rangle$.

Case 1.1.1. $\delta(S_n^{\{i\}}-F_i)\ge 2$ for every $i\in\langle n\rangle$. Since $|F|\le 3n-10$, $|F_i|\ge n-3$ can occur for at most two different $i\in\langle n\rangle$. Let a,b be integers in $\langle n\rangle$ such that $|F_a|\ge|F_i|$ for every $i\in\langle n\rangle$ and $|F_b|\ge|F_j|$ for every $j\in\langle n\rangle-\{a\}$. By induction, there is a Hamiltonian cycle C_1 of $S_n^{\{a\}}-F_a$ and there is a Hamiltonian cycle C_2 of $S_n^{\{b\}}-F_b$. For every vertex $\mathbf{p}\in S_n^{\{a\}}$ with $\mathbf{p}_1=b$, we set $A(\mathbf{p})=\{\mathbf{p}\}\cup N_{C_1}(\mathbf{p})\cup N_{C_2}((\mathbf{p})^n)$ and $B(\mathbf{p})=\{(\mathbf{q},\mathbf{q}^n)\mid\mathbf{q}\in A(\mathbf{p})\}$. Since $|E^{a,b}|=(n-2)!>n-2$ if $n\ge 5$, there is a vertex $\mathbf{z}\in S_n^{\{a\}}$ with $\mathbf{z}_1=b$ such that $B(\mathbf{z})\cap F=\emptyset$. Let $\mathbf{p}$ be a neighbor of $\mathbf{z}$ on C_1.

By Lemma 13.8, the two neighbors of $(\mathbf{z})^n$ on C_2 have different first coordinates, so at least one of them is different from $(\mathbf{p})1$. Let $\mathbf{q}$ be such a vertex. Then we have $C_1 = \langle \mathbf{z}, R_1, \mathbf{p}, \mathbf{z} \rangle$ and $C_2 = \langle \mathbf{q}, R_2, (\mathbf{z})^n, \mathbf{q} \rangle$ and $(\mathbf{p})^n \neq (\mathbf{q})^n$. Since $n \geq 6$, $(n-2)!/2 > n-2$. Hence, by Theorem 13.18, there is a Hamiltonian path H of $S_n^{\langle n \rangle - \{a,b\}} - F$ joining $(\mathbf{p})^n$ to $(\mathbf{q})^n$, and then $\langle \mathbf{z}, R_1, \mathbf{p}, (\mathbf{p})^n, H, (\mathbf{q})^n, \mathbf{q}, R_2, (\mathbf{z})^n, \mathbf{z} \rangle$ is a Hamiltonian cycle of $S_n - F$.

Case 1.1.2. $\delta(S_n^{\{t\}} - F_t) = 1$ for some $t \in \langle n \rangle$. Let $\mathbf{y}$ be the vertex in $S_n^{\{t\}}$ with $\deg_{S_n^{\{t\}} - F_t}(\mathbf{y}) = 1$. Obviously, $\mathbf{y}$ is neither $\mathbf{u}$ nor $\mathbf{v}$, and $(\mathbf{y}, (\mathbf{y})^n) \notin F$. Moreover, $|A| = 3$, and we have identified all but one edge in F.

Suppose first that $\mathbf{y}$ is adjacent to neither $\mathbf{u}$ nor $\mathbf{v}$. Then we have identified all edges of F; hence $|F_i| \leq n-4$ for every $i \in \langle n \rangle - \{t\}$. Since $\deg_{S_n^{\{t\}} - F_t}(\mathbf{y}) = 1$ and $n \geq 6$, we can choose an edge $(\mathbf{y}, \mathbf{z}) \in F_t$ such that $\mathbf{z}$ is neither $\mathbf{u}$ nor $\mathbf{v}$. By induction, there is a Hamiltonian cycle C of $S_n^{\{t\}} - (F_t - \{(\mathbf{y}, \mathbf{z})\})$. Clearly, edge $(\mathbf{y}, \mathbf{z})$ must be in C, so we can write $C = \langle \mathbf{y}, R, \mathbf{z}, \mathbf{y} \rangle$. By Theorem 13.18, there is a Hamiltonian path H of $S_n^{\langle n \rangle - \{t\}} - F$ joining $(\mathbf{z})^n$ to $(\mathbf{y})^n$. Then $\langle \mathbf{y}, R, \mathbf{z}, (\mathbf{z})^n, H, (\mathbf{y})^n, \mathbf{y} \rangle$ forms a Hamiltonian cycle of $S_n - F$.

Second, if $\mathbf{y}$ is adjacent to either $\mathbf{u}$ or $\mathbf{v}$, then without loss of generality we may assume that $(\mathbf{y}, \mathbf{u}) \in E(S_n)$, so $\mathbf{y} \in S_n^{\{n\}}$. Since all but one edge of F have been identified so far, out of the $n-3$ faulty edges incident to y in $S_n^{\{n\}}$, at most one can go to a vertex $\mathbf{z} \neq \mathbf{u}$ such that $(\mathbf{z}, (\mathbf{z})^n) \in F$. Since $n \geq 6$, we can choose an edge $(\mathbf{y}, \mathbf{z}) \in F_n$ such that $(\mathbf{z}, (\mathbf{z})^n) \notin F$. By induction, there is a Hamiltonian cycle C of $S_n^{\{n\}} - (F_n - \{(\mathbf{y}, \mathbf{z})\})$. Clearly, edge $(\mathbf{y}, \mathbf{z})$ must be in C, so we can write $C = \langle \mathbf{y}, R, \mathbf{z}, \mathbf{y} \rangle$. If the as-of-yet unidentified edge of F is not in F_{n-1}, then $|F_{n-1}| = n-4$, and by Theorem 13.18, there is a Hamiltonian path H of $S_n^{\langle n-1 \rangle} - F$ joining $(\mathbf{z})^n$ to $(\mathbf{y})^n$, hence, $\langle \mathbf{y}, R, \mathbf{z}, (\mathbf{z})^n, H, (\mathbf{y})^n, \mathbf{y} \rangle$ forms a Hamiltonian cycle of $S_n - F$. On the other hand, if the last unidentified edge of F is in F_{n-1}, then $|F_{n-1}| = n-3$, and we have $|F_i| = 0$ for every $i \in \langle n-1 \rangle$ and $|F_{i,j}| = 0$ for every two different $i, j \in \langle n-1 \rangle$. Since $(\mathbf{u})_1 = n-1$, and $\mathbf{y}$ is adjacent to both $\mathbf{u}$ and $\mathbf{z}$, by Lemma 13.8, it implies that $(\mathbf{y})_1 \neq (\mathbf{z})_1$, $(\mathbf{y})_1 \neq n-1$, and $(\mathbf{z})_1 \neq n-1$. Then we can choose a vertex $\mathbf{w}$ in $S_n^{\{n-1\}}$ such that its color is different from the color of $\mathbf{y}$, and $(\mathbf{w})_1 = (\mathbf{y})_1$. Since $n-3 \leq 3n-13$ for $n \geq 5$ and $\delta(S_n^{\{n-1\}} - F_{n-1}) = 2$, by induction, there is a Hamiltonian cycle C' in $S_n^{\{n-1\}} - F_{n-1}$. We can write $C' = \langle \mathbf{w}, \mathbf{p}, P, \mathbf{q}, \mathbf{w} \rangle$. By Lemma 13.8, $(\mathbf{p})_1$, $(\mathbf{w})_1$, and $(\mathbf{q})_1$ are all different, so without loss of generality, we may assume that $(\mathbf{p})_1 \neq n$. By Theorem 15.18, there is a Hamiltonian path H_1 of $S_n^{\{(\mathbf{y})_1\}}$ joining $(\mathbf{w})^n$ to $(\mathbf{y})^n$ and there is a Hamiltonian path H_2 of $S_n^{\langle n-2 \rangle - \{(\mathbf{y})_1\}}$ joining $(\mathbf{z})^n$ to $(\mathbf{p})^n$. Then $\langle \mathbf{y}, R, \mathbf{z}, (\mathbf{z})^n, H_2, (\mathbf{p})^n, \mathbf{p}, P, \mathbf{q}, \mathbf{w}, (\mathbf{w})^n, H_1, (\mathbf{y})^n, \mathbf{y} \rangle$ forms a Hamiltonian cycle of $S_n - F$.

Case 1.2. $(\mathbf{u}, \mathbf{v}) \notin F$ for every two distinct vertices $\mathbf{u}, \mathbf{v} \in A$. Counting the edges of F at each vertex of A, we get $|A|(n-3) \leq F \leq 3n-10$; hence $|A| \leq 2$. Thus $|A| = 2$, so let $A = \{\mathbf{u}, \mathbf{v}\}$. Since S_n is vertex-transitive and edge-transitive, we may assume that $(\mathbf{u})_n = n$ and $(\mathbf{u}, (\mathbf{u})^n) \in F$. Now we consider cases depending on the location of $\mathbf{v}$.

Case 1.2.1. $(\mathbf{v})_n = n$ and $(\mathbf{v}, (\mathbf{v})^n) \in F$. Clearly, $2n-8 \leq |F_n| \leq 3n-12$, so there can be at most $n-4$ edges of F outside $S_n^{\{n\}}$ different from $(\mathbf{u}, (\mathbf{u})^n)$ and $(\mathbf{v}, (\mathbf{v})^n)$. If $|F_n| \leq 3n-13$, then by induction, there is a Hamiltonian cycle C of $S_n^{\{n\}} - F_n$. Since $|V(S_n^{\{n\}})| = (n-1)! > 2(n-2)$, there is an edge $(\mathbf{p}, \mathbf{q}) \in E(C)$ such that $(\mathbf{p}, (\mathbf{p})^n) \notin F$ and $(\mathbf{q}, (\mathbf{q})^n) \notin F$. Then we can write $C = \langle \mathbf{p}, H, \mathbf{q}, \mathbf{p} \rangle$. Since $n-4 < (n-2)!/2$ for $n \geq 6$, Theorem 13.18 implies that there is a Hamiltonian path R of $S_n^{\langle n-1 \rangle}$ joining $(\mathbf{q})^n$ to $(\mathbf{p})^n$, and then $\langle \mathbf{p}, H, \mathbf{q}, (\mathbf{q})^n, R, (\mathbf{p})^n, \mathbf{p} \rangle$ is a Hamiltonian cycle of $S_n - F$.

On the other hand, if $|F_n| = 3n-12$, then every edge in F is located. Since $2(n-4) < 3n-12$, there is an edge $f \in F_n$ such that f is incident to neither $\mathbf{u}$ nor $\mathbf{v}$. By induction, there is a Hamiltonian cycle C of $S_n^{\{n\}} - (F_n - \{f\})$. However, f may be not in C. If $f \in C$, let $f = (\mathbf{p}, \mathbf{q})$; otherwise, pick an edge $(\mathbf{p}, \mathbf{q})$ in C such that $\{\mathbf{p}, \mathbf{q}\} \cap \{\mathbf{u}, \mathbf{v}\} = \emptyset$. Then we can write $C = \langle \mathbf{p}, H, \mathbf{q}, \mathbf{p} \rangle$. By Theorem 13.18, there is a Hamiltonian path R of $S_n^{\langle n-1 \rangle}$ joining $(\mathbf{q})^n$ to $(\mathbf{p})^n$; hence, $\langle \mathbf{p}, H, \mathbf{q}, (\mathbf{q})^n, R, (\mathbf{p})^n, \mathbf{p} \rangle$ is a Hamiltonian cycle of $S_n - F$.

Case 1.2.2. $(\mathbf{v})_n = n$ and $(\mathbf{v}, (\mathbf{v})^n) \notin F$. Within $S_n^{\{n\}}$, there are $n-4$ faulty edges incident to $\mathbf{u}$ and $n-3$ faulty edges incident to $\mathbf{v}$. Since $(n-3)+(n-4) > n-2$ for $n \geq 6$, there must be an integer $i \in \langle n \rangle - \{1\}$ such that $(\mathbf{u}, (\mathbf{u})^i) \in F$ and $(\mathbf{v}, (\mathbf{v})^i) \in F$. Using vertices of having the same ith coordinate instead of the same nth coordinate to define the sets $S_n^{\{j\}}$ for $j \in \langle n \rangle$ will change this case to Case 1.2.1.

Case 1.2.3. $(\mathbf{v})_n \neq n$ and $(\mathbf{v}, (\mathbf{v})^n) \in F$. Without loss of generality, we may assume that $(\mathbf{v})_n = n-1$. By induction, there is a Hamiltonian cycle C_1 of $S_n^{\{n\}} - F_n$ and there is a Hamiltonian cycle C_2 of $S_n^{\{n-1\}} - F_{n-1}$. For every vertex $\mathbf{p}$ in $S_n^{\{n\}}$ with $(\mathbf{p})_1 = n-1$, we set $A(\mathbf{p}) = \{\mathbf{p}\} \cup N_{C_1}(\mathbf{p}) \cup N_{C_2}((\mathbf{p})^n)$ and $B(\mathbf{p}) = \{(\mathbf{q}, (\mathbf{q})^n) \mid \mathbf{q} \in A(\mathbf{p})\}$. Since $|E^{n-1,n}| = (n-2)! > n-4$ if $n \geq 5$, there is a vertex $\mathbf{z} \in S_n^{\{n\}}$ with $(\mathbf{z})_1 = n-1$ such that $B(\mathbf{z}) \cap F = \emptyset$. Let $\mathbf{p}$ be a neighbor of $\mathbf{z}$ on C_1. Then by Lemma 13.8, the two neighbors of $(\mathbf{z})^n$ on C_2 have different first coordinates, so at least one of them is different from $(\mathbf{p})_1$. Let $\mathbf{q}$ be such a vertex. Then we can write $C_1 = \langle \mathbf{z}, R_1, \mathbf{p}, \mathbf{z} \rangle$ and $C_2 = \langle \mathbf{q}, R_2, (\mathbf{z})^n, \mathbf{q} \rangle$. By Theorem 13.18, there is a Hamiltonian path H of $S_n^{\langle n-2 \rangle} - F$ joining $(\mathbf{p})^n$ to $(\mathbf{q})^n$, and then $\langle \mathbf{z}, R_1, \mathbf{p}, (\mathbf{p})^n, H, (\mathbf{q})^n, \mathbf{q}, R_2, (\mathbf{z})^n, \mathbf{z} \rangle$ is a Hamiltonian cycle of $S_n - F$.

Case 1.2.4. $(\mathbf{v})_n \neq n$ and $(\mathbf{v}, (\mathbf{v})^n) \notin F$. There are $n-3$ faulty edges incident to both $\mathbf{u}$ and $\mathbf{v}$. Since $2(n-3) > n-1$ for $n \geq 6$, there must be an integer $i \in \langle n \rangle - \{1\}$ such that $(\mathbf{u}, (\mathbf{u})^i) \in F$ and $(\mathbf{v}, (\mathbf{v})^i) \in F$. Using vertices of having the same ith coordinate instead of the same nth coordinate to define the sets $S_n^{\{j\}}$ for $j \in \langle n \rangle$ will change this case to either Case 1.2.1 (if $\mathbf{u}$ and $\mathbf{v}$ have the same ith coordinate) or Case 1.2.3 (if $\mathbf{u}$ and $\mathbf{v}$ have different ith coordinates).

Case 2. $0 \leq |A| \leq 1$. Let $\mathbf{x}$ be a vertex in $S_n - F$ of minimum degree. Since S_n is edge-transitive, we may assume that $(\mathbf{x}, (\mathbf{x})^n) \in F$, thus $\delta(S_n^{\{i\}} - F_i) \geq 2$ and $|F_i| \leq 3n-11$ for every $i \in \langle n \rangle$. Without loss of generality, we may assume that $|F_n| \geq |F_{n-1}| \geq \cdots \geq |F_1|$. Thus, $|F_i| \leq n-4$ for every $i \in \langle n-2 \rangle$. Now we look at cases depending on the sizes of F_n and F_{n-1}.

Case 2.1. $|F_n| \leq 3n-13$ and $|F_{n-1}| \geq n-3$. By induction, there is a Hamiltonian cycle C_1 of $S_n^{\{n\}} - F_n$ and there is a Hamiltonian cycle C_2 of $S_n^{\{n-1\}} - F_{n-1}$. For every vertex $\mathbf{w} \in S_n^{\{n\}}$ with $(\mathbf{w})_1 = n-1$, let $\mathbf{w_1}$ and $\mathbf{w_2}$ be its two neighbors in $S_n^{\{n\}}$ on C_1, and let $\mathbf{w_3}$ and $\mathbf{w_4}$ be the two neighbors of $(\mathbf{w})^n$ in C_2. Since $(n-2)! > n-4$ for $n \geq 6$, there is a vertex $\mathbf{z}$ in $S_n^{\{n\}}$ such that $(\mathbf{z})_1 = n-1$, $(\mathbf{z},(\mathbf{z})^n) \notin F$, and $(\mathbf{z_i},(\mathbf{z_i})^n) \notin F$ for every $1 \leq i \leq 4$. By Lemma 13.8, $(\mathbf{z_1})_1 \neq (\mathbf{z_2})_1$ and $(\mathbf{z_3})_1 \neq (\mathbf{z_4})_1$. Without loss of generality, we may assume that $(\mathbf{z_2})_1 \neq (\mathbf{z_3})_1$. Then we can write $C_1 = \langle \mathbf{z}, \mathbf{z_1}, R_1, \mathbf{z_2}, \mathbf{z} \rangle$ and $C_2 = \langle (\mathbf{z})^n, \mathbf{z_3}, R_2, \mathbf{z_4}, (\mathbf{z})^n \rangle$. By Theorem 13.18, there is a Hamiltonian path H of $S_n^{\langle n-2 \rangle} - F$ joining $(\mathbf{z_2})^n$ to $(\mathbf{z_3})^n$, and then $\langle \mathbf{z}, \mathbf{z_1}, R_1, \mathbf{z_2}, (\mathbf{z_2})^n, H, (\mathbf{z_3})^n, \mathbf{z_3}, R_2, \mathbf{z_4}, (\mathbf{z})^n, \mathbf{z} \rangle$ is a Hamiltonian cycle of $S_n - F$.

Case 2.2. $|F_n| \leq 3n-13$ and $|F_{n-1}| \leq n-4$. If $|F_{i,j}| \leq [(n-2)!/2] - 1$ for every distinct $i,j \in \langle n-1 \rangle$, then by induction, there is a Hamiltonian cycle C of $S_n^{\{n\}} - F_n$. Since $|V(S_n^{\{n\}})| = (n-1)! > 2(3n-10)$ if $n \geq 6$, there is an edge $(\mathbf{u},\mathbf{v}) \in C$ such that $(\mathbf{u},(\mathbf{u})^n) \notin F$ and $(\mathbf{v},(\mathbf{v})^n) \notin F$. Then we can write $C = \langle \mathbf{u}, H, \mathbf{v}, \mathbf{u} \rangle$. By Theorem 13.18, there is a Hamiltonian path R of $S_n^{\langle n-1 \rangle} - F$ joining $(\mathbf{v})^n$ to $(\mathbf{u})^n$, and then $\langle \mathbf{u}, H, \mathbf{v}, (\mathbf{v})^n, R, (\mathbf{u})^n, \mathbf{u} \rangle$ is a Hamiltonian cycle of $S_n - F$. On the other hand, if $|F_{i,j}| > [(n-2)!/2] - 1$ for some $i,j \in \langle n-1 \rangle$, then since $|F| \leq 3n-10$ and $[(n-2)!/2] - 1 > 3n-10$ if $n \geq 6$, we get $n = 5$.

Case 2.3. $|F_n| = 3n-12$. Since $|F_n| = 3n-12 > 2(n-4)$ if $n \geq 6$, there is an edge $(\mathbf{p},\mathbf{q}) \in F_n$ such that edges $(\mathbf{p},(\mathbf{p})^n)$ and $(\mathbf{q},(\mathbf{q})^n)$ are not in F. By induction, there is a Hamiltonian cycle C of $S_n^{\{n\}} - (F_n - \{(\mathbf{p},\mathbf{q})\})$. We can write $C = \langle \mathbf{u}, H, \mathbf{v}, \mathbf{u} \rangle$, where $\{\mathbf{u},\mathbf{v}\} = \{\mathbf{p},\mathbf{q}\}$ if edge $(\mathbf{p},\mathbf{q})$ is part of C; otherwise, $(\mathbf{u},\mathbf{v})$ is an arbitrary edge of C such that edges $(\mathbf{u},(\mathbf{u})^n)$ and $(\mathbf{v},(\mathbf{v})^n)$ are not in F. By Theorem 13.18, there is a Hamiltonian path R of $S_n^{\langle n-1 \rangle} - F$ joining $(\mathbf{v})^n$ to $(\mathbf{u})^n$, and then $\langle \mathbf{u}, H, \mathbf{v}, (\mathbf{v})^n, R, (\mathbf{u})^n, \mathbf{u} \rangle$ is a Hamiltonian cycle of $S_n - F$.

Case 2.4. $|F_n| = 3n-11$ and $\deg_{S_n-F}(\mathbf{x}) = 2$. Clearly, $\mathbf{x} \in S_n^{\{n\}}$, and all edges of F are accounted for. Since $|F_n| = 3n-11 > 2(n-4)$ if $n \geq 5$, there are two edges $(\mathbf{u},\mathbf{v})$ and $(\mathbf{p},\mathbf{q})$ in F_n such that $(\mathbf{u},(\mathbf{u})^n) \notin F$, $(\mathbf{v},(\mathbf{v})^n) \notin F$, $(\mathbf{p},(\mathbf{p})^n) \notin F$, $(\mathbf{q},(\mathbf{q})^n) \notin F$, and vertices $\mathbf{u},\mathbf{v},\mathbf{p},\mathbf{q}$ are all different. By induction, there is a Hamiltonian cycle C in $S_n^{\{n\}} - (F_n - \{(\mathbf{u},\mathbf{v}),(\mathbf{p},\mathbf{q})\})$. If either $(\mathbf{u},\mathbf{v})$ or $(\mathbf{p},\mathbf{q})$ is not in C, then without loss of generality, we may assume that $(\mathbf{p},\mathbf{q}) \notin C$. We can write $C = \langle \mathbf{w}, R, \mathbf{z}, \mathbf{w} \rangle$, where $\{\mathbf{w},\mathbf{z}\} = \{\mathbf{u},\mathbf{v}\}$ if $(\mathbf{u},\mathbf{v}) \in E(C)$; otherwise, $(\mathbf{w},\mathbf{z})$ is chosen to be an arbitrary edge of C such that $(\mathbf{w},(\mathbf{w})^n) \notin F$ and $(\mathbf{z},(\mathbf{z})^n) \notin F$. By Theorem 13.18, there is a Hamiltonian path H of $S_n^{\langle n-1 \rangle}$ joining $(\mathbf{z})^n$ to $(\mathbf{w})^n$, and then $\langle \mathbf{w}, R, \mathbf{z}, (\mathbf{z})^n, H, (\mathbf{w})^n, \mathbf{w} \rangle$ is a Hamiltonian cycle of $S_n - F$.

On the other hand, if both $(\mathbf{u},\mathbf{v})$ and $(\mathbf{p},\mathbf{q})$ are in C, then without loss of generality, we can write $C = \langle \mathbf{u}, R_1, \mathbf{p}, \mathbf{q}, R_2, \mathbf{v}, \mathbf{u} \rangle$. By Lemma 13.37, there are two disjoint paths Q_1 and Q_2 of $S_n^{\langle n-1 \rangle}$ such that (1) Q_1 joins $(\mathbf{v})^n$ to $(\mathbf{u})^n$, (2) Q_2 joins $(\mathbf{p})^n$ to $(\mathbf{q})^n$, and (3) $Q_1 \cup Q_2$ spans $S_n^{\langle n-1 \rangle}$. Then $\langle \mathbf{u}, R_1, \mathbf{p}, (\mathbf{p})^n, Q_2, (\mathbf{q})^n, \mathbf{q}, R_2, \mathbf{v}, (\mathbf{v})^n, Q_1, (\mathbf{u})^n, \mathbf{u} \rangle$ is a Hamiltonian cycle of $S_n - F$.

Case 2.5. $|F_n| = 3n-11$ and $\deg_{S_n-F}(\mathbf{x}) \geq 3$. Since $\left\lceil \frac{3n-11}{n-2} \right\rceil \geq 2$ if $n \geq 6$, there is an integer $2 \leq i \leq n-1$ such that there are at least two faulty edges among the edges

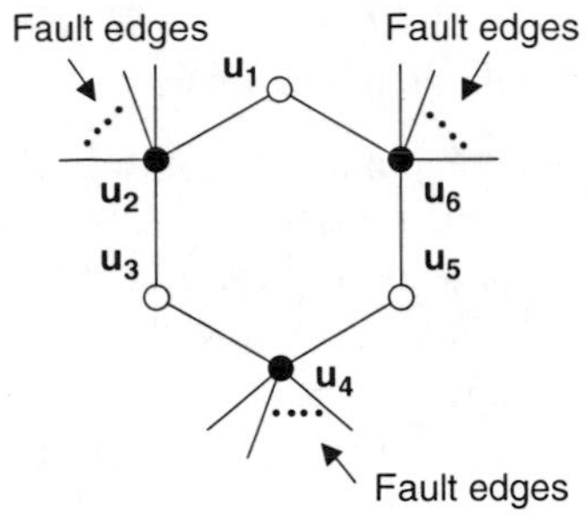

FIGURE 13.18 Illustration of S_n with $(3n-9)$ edge faults.

of dimension i in S_n. Define the sets $S_n^{\{j\}}$ for $j \in \langle n \rangle$ using vertices of having the same ith coordinate instead of the same nth coordinate. Then $\deg_{S_n - F}(\mathbf{x}) \geq 3$ will imply that the minimum degree in $S_n^{\{j\}} - F_j$ is at least 2 and $|F_j| \leq 3n - 12$ for every $j \in \langle n \rangle$. Hence, this case reduces to one of Cases 2.1 through 2.3.

This finishes the proof of the theorem. □

Let $n \geq 4$ and let $H = \langle \mathbf{u}_1, \mathbf{u}_2, \ldots, \mathbf{u}_6, \mathbf{u}_1 \rangle$ be a cycle with six vertices in S_n. We set the faulty set F to be all the edges incident to $\mathbf{u}_2$, $\mathbf{u}_4$, and $\mathbf{u}_6$ that are not edges of H shown in Figure 13.18. Obviously, $|F| = 3n - 9$. Since $\deg_{S_n - F}(\mathbf{v}) = 2$ for every vertex in $\{\mathbf{u}_2, \mathbf{u}_4, \mathbf{u}_6\}$, $S_n - F$ is not Hamiltonian. Thus, the bound $3n - 10$ in Theorem 13.19 is tight.

LEMMA 13.38 Let F be a subset of $E(S_n)$ with $|F| \leq 2n - 7$ and $\delta(S_n - F) \geq 2$ with $n \geq 4$. Let $A = \{\mathbf{u} \mid \deg_{S_n - F}(\mathbf{u}) = 2\}$. Then $|A| \leq 2$, moreover, if $A = \{\mathbf{u}, \mathbf{v}\}$, then $(\mathbf{u}, \mathbf{v}) \in F$.

Proof: Suppose that $|A| \geq 3$; say, $\{\mathbf{x}_1, \mathbf{x}_2, \mathbf{x}_3\} \subseteq A$. Let H be the subgraph of S_n with the vertex set $\{\mathbf{x}_1, \mathbf{x}_2, \mathbf{x}_3\}$ containing those edges of F that join two vertices in H. Since S_n is bipartite, so is H. At least one of the partite sets can be chosen to have at least two vertices. We count the number of edges in F at each of these vertices. Since no edge of F is counted twice, we get $|F| \geq 2(n-3) = 2n - 6$, which is a contradiction.

When $|A| = 2$, the same counting argument implies that the two vertices of A must be joined by an edge, which must belong to F. □

THEOREM 13.20 Let $I = \{i_1, i_2, \ldots, i_t\}$ be any nonempty subset of $\langle n \rangle$ with $n \geq 5$. Assume that S_{n-1} is $(2n-9)$-edge-fault conditional Hamiltonian laceable. Let F be a set of edges of S_n such that $|F \cap E(S_n^{\{i_k\}})| \leq 2n - 9$ with $\delta(S_n^{\{i_k\}} - F) \geq 2$ for every $k \in \langle t \rangle$, and $|F \cap E^{i_k, i_{k+1}}| \leq [(n-2)!/2] - 1$ for every $k \in \langle t - 1 \rangle$. If $\mathbf{u}$ is a white vertex of $S_n^{\{i_1\}}$ and $\mathbf{v}$ is a black vertex of $S_n^{\{i_t\}}$, then there is a Hamiltonian path P of $S_n^I - F$ joining $\mathbf{u}$ to $\mathbf{v}$.

Proof: The proof of this statement is similar to the proof of Theorem 13.18. □

THEOREM 13.21 Let F be a subset of $E(S_n)$ with $|F| \leq 2n-7$ and $\delta(S_n - F) \geq 2$ with $n \geq 4$. Then $S_n - F$ is Hamiltonian laceable.

Proof: We prove this statement by induction on n. Since S_4 is 1-edge fault-tolerant Hamitonian laceable, this statement holds for $n = 4$. Suppose that this statement holds for S_k for every $4 \leq k \leq n-1$ and show it for S_n. Let F be a subset of $E(S_n)$ with $|F| \leq 2n-7$ and $\delta(S_n - F) \geq 2$. We need to show that $S_n - F$ is Hamiltonian laceable. Without loss of generality, we may assume that $|F| = 2n-7$. We set $F_i = F \cap E(S_n^{\{i\}})$ for every $i \in \langle n \rangle$ and $F_{j,k} = F \cap E^{j,k}$ for every two distinct elements $j, k \in \langle n \rangle$. Let $A = \{\mathbf{x} \mid \deg_{S_n - F}(\mathbf{x}) = 2\}$. By Lemma 13.38, $|A| \leq 2$. We consider the following cases, depending on the size of A.

Case 1. $|A| = 2$. Let $A = \{\mathbf{x}, \mathbf{y}\}$. By Lemma 13.38, $(\mathbf{x}, \mathbf{y}) \in F$, and F is completely identified. Since S_n is edge-transitive, we may assume that $\mathbf{x} = (\mathbf{y})^n$. Hence, $\mathbf{x}_n \neq (\mathbf{y})_n$, $|F_i| = n-4$ for $i \in \{(\mathbf{x})_n, (\mathbf{y})_n\}$, $|F_j| = 0$ for $j \in \langle n \rangle - \{(\mathbf{x})_n, (\mathbf{y})_n\}$, and $|F_{s,t}| \leq 1$ for every two distinct $s, t \in \langle n \rangle$. Let $\mathbf{u}$ be any white vertex in S_n and let $\mathbf{v}$ be any black vertex in S_n. If $(\mathbf{u})_n \neq (\mathbf{v})_n$, then by Theorem 13.18, there is a Hamiltonian path H of $S_n - F$ joining $\mathbf{u}$ to $\mathbf{v}$. On the other hand, if $(\mathbf{u})_n = (\mathbf{v})_n$, then by induction, there is a Hamiltonian path Q of $S_n^{\{(\mathbf{u})_n\}} - F_{(\mathbf{u})_n}$ joining $\mathbf{u}$ to $\mathbf{v}$. Since $(n-1)! > 8$ if $n \geq 5$, there is an edge $(\mathbf{p}, \mathbf{q})$ in Q such that $|\{\mathbf{p}, \mathbf{q}\} \cap \{\mathbf{x}, \mathbf{y}, (\mathbf{x})^n, (\mathbf{y})^n\}| = 0$. Then we can write $Q = \langle \mathbf{u}, Q_1, \mathbf{p}, \mathbf{q}, Q_2, \mathbf{v} \rangle$. By Theorem 13.18, there is a Hamiltonian path H of $S_n^{\langle n \rangle - \{(\mathbf{u})_n\}} - F$ joining $(\mathbf{p})^n$ to $(\mathbf{q})^n$. Thus, $\langle \mathbf{u}, Q_1, \mathbf{p}, (\mathbf{p})^n, H, (\mathbf{q})^n, \mathbf{q}, Q_2, \mathbf{v} \rangle$ is a Hamiltonian path of $S_n - F$ joining $\mathbf{u}$ to $\mathbf{v}$.

Case 2. $|A| \leq 1$. Let $\mathbf{x}$ be a vertex of minimum degree in $S_n - F$. Since S_n is vertex-transitive and edge-transitive, we may assume that $(\mathbf{x}, (\mathbf{x})^n) \in F$. Thus $\delta(S_n^{\{i\}} - F_i) \geq 2$ for every $i \in \langle n \rangle$. Without loss of generality, we may assume that $|F_n| \geq |F_i|$ for every $i \in \langle n-1 \rangle$. Let $\mathbf{u}$ be any white vertex in S_n and let $\mathbf{v}$ be any black vertex in S_n. We now consider cases depending on the size of F_n.

Case 2.1. $|F_n| = 2n-8$. Then $F_i = 0$ for every $i \in \langle n-1 \rangle$ and $|F_{i,j}| = 0$ for every distinct $i, j \in \langle n \rangle$ except that $\{i, j\} = \{(\mathbf{x})_n, ((\mathbf{x})^n)n\}$. Since $3n - 13 \geq 2n - 8$ for $n \geq 5$, by Theorem 13.19 there is a Hamiltonian cycle C in $S_n^{\{n\}} - F_n$. Now we look at cases depending on the location of $\mathbf{u}$ and $\mathbf{v}$.

Case 2.1.1. $(\mathbf{u})_n = (\mathbf{v})_n = n$ and $(\mathbf{u}, \mathbf{v}) \notin E(C)$. Then we can write $C = \langle \mathbf{u}, \mathbf{p}, Q_1, \mathbf{q}, \mathbf{v}, \mathbf{w}, Q_2, \mathbf{z}, \mathbf{u} \rangle$ for some vertices $\mathbf{p}, \mathbf{q}, \mathbf{w}, \mathbf{z} \in S_n^{\{n\}}$. Obviously, either $\{\mathbf{x}, (\mathbf{x})^n\} \cap \{\mathbf{p}, \mathbf{w}\} = \emptyset$ or $\{\mathbf{x}, (\mathbf{x})^n\} \cap \{\mathbf{q}, \mathbf{z}\} = \emptyset$. By symmetry, we may assume that $\{\mathbf{x}, (\mathbf{x})^n\} \cap \{\mathbf{q}, \mathbf{z}\} = \emptyset$. By Theorem 13.18, there is a Hamiltonian path H of $S_n^{\langle n-1 \rangle}$ joining $(\mathbf{q})^n$ to $(\mathbf{z})^n$. Thus $\langle \mathbf{u}, \mathbf{p}, Q_1, \mathbf{q}, (\mathbf{q})^n, H, (\mathbf{z})^n, \mathbf{z}, Q_2^{-1}, \mathbf{w}, \mathbf{v} \rangle$ is a Hamiltonian path of $S_n - F$ joining $\mathbf{u}$ to $\mathbf{v}$.

Case 2.1.2. $(\mathbf{u})_n = (\mathbf{v})_n = n$ and $(\mathbf{u}, \mathbf{v}) \in E(C)$. Choose an edge $(\mathbf{p}, \mathbf{q}) \in E(C) - \{(\mathbf{u}, \mathbf{v})\}$ such that neither $\mathbf{p}$ nor $\mathbf{q}$ is $\mathbf{x}$ or $(\mathbf{x})^n$. Then we can write $C = \langle \mathbf{u}, Q_1, \mathbf{p}, \mathbf{q}, Q_2, \mathbf{v}, \mathbf{u} \rangle$. By Theorem 13.18, there is a Hamiltonian path H of $S_n^{\langle n-1 \rangle}$ joining $(\mathbf{p})^n$ to $(\mathbf{q})^n$; thus $\langle \mathbf{u}, Q_1, \mathbf{p}, (\mathbf{p})^n, H, (\mathbf{q})^n, \mathbf{q}, Q_2, \mathbf{v} \rangle$ is a Hamiltonian path of $S_n - F$ joining $\mathbf{u}$ to $\mathbf{v}$.

Case 2.1.3. $(\mathbf{u})_n = (\mathbf{v})_n = i$ for some $i \in \langle n-1 \rangle$. Since $|E^{n,(\mathbf{u})_n}| = (n-2)! > 2$ if $n \geq 5$, there is a black vertex $\mathbf{w}$ in $S_n^{\{(\mathbf{u})_n\}} - \{\mathbf{v}\}$ such that $(\mathbf{w}, (\mathbf{w})^n) \in E^{n,(\mathbf{u})_n} - \{(\mathbf{x}, (\mathbf{x})^n)\}$. Then we can write $C = \langle (\mathbf{w})^n, \mathbf{p}, Q, \mathbf{q}, (\mathbf{w})^n \rangle$. Clearly, either $\mathbf{p} \notin \{\mathbf{x}, (\mathbf{x})^n\}$ or $\mathbf{q} \notin \{\mathbf{x}, (\mathbf{x})^n\}$. By symmetry, we may assume that $\mathbf{p} \notin \{\mathbf{x}, (\mathbf{x})^n\}$. By Theorem 13.18, there is a Hamiltonian path R of $S_n^{\{(\mathbf{u})_n\}}$ joining $\mathbf{u}$ to $\mathbf{v}$. We can write $R = \langle \mathbf{u}, R_1, \mathbf{y}, \mathbf{w}, \mathbf{z}, R_2, \mathbf{v} \rangle$. By Lemma 13.8, $(\mathbf{y})_1 \neq (\mathbf{z})_1$, and by symmetry, we may assume that $(\mathbf{y})_1 \neq (\mathbf{p})1$. By Theorem 13.18, there is a Hamiltonian path H of $S_n^{\langle n-1 \rangle - \{(\mathbf{u})_n\}}$ joining $(\mathbf{y})^n$ to $(\mathbf{p})^n$, and then $\langle \mathbf{u}, R_1, \mathbf{y}, (\mathbf{y})^n, H, (\mathbf{p})^n, \mathbf{p}, Q, \mathbf{q}, (\mathbf{w})^n, \mathbf{w}, \mathbf{z}, R_2, \mathbf{v} \rangle$ is a Hamiltonian path of $S_n - F$ joining $\mathbf{u}$ to $\mathbf{v}$.

Case 2.1.4. $(\mathbf{u})_n = n$ and $(\mathbf{v})_n \in \langle n-1 \rangle$. Note that the cases where $(\mathbf{v})_n = n$ and $(\mathbf{u})_n \in \langle n-1 \rangle$ are identical, so it is enough to prove one of them. We can write $C = \langle \mathbf{u}, \mathbf{p}, Q, \mathbf{q}, \mathbf{u} \rangle$. Clearly, $\mathbf{p} \notin \{\mathbf{x}, (\mathbf{x})^n\}$ or $\mathbf{q} \notin \{\mathbf{x}, (\mathbf{x})^n\}$. By symmetry, we may assume that $\mathbf{q} \notin \{\mathbf{x}, (\mathbf{x})^n\}$. Since $\mathbf{u}$ is a white vertex and $\mathbf{q}$ is a black vertex.

If $(\mathbf{q})_1 \neq (\mathbf{v})_n$, then by Theorem 13.18, there is a Hamiltonian path H of $S_n^{\langle n-1 \rangle}$ joining $(\mathbf{q})^n$ to $\mathbf{v}$, and then $\langle \mathbf{u}, \mathbf{p}, Q, \mathbf{q}, (\mathbf{q})^n, \mathbf{v} \rangle$ is a Hamiltonian path of $S_n - F$ joining $\mathbf{u}$ to $\mathbf{v}$. On the other hand, if $(\mathbf{q})_1 = (\mathbf{v})_n$, then let $\mathbf{w}$ be any white vertex in $S_n^{\{(\mathbf{v})_n\}}$ such that $(\mathbf{v}, \mathbf{w}) \in E(S_n^{\{(\mathbf{v})_n\}})$ and $(\mathbf{w})_1 \in \langle n-1 \rangle - \{(\mathbf{v})_n\}$. Since there are no faulty edges in $S_n^{\{(\mathbf{v})_n\}}$, Lemma 13.35 implies that there is a Hamiltonian path R of $S_n^{\{(\mathbf{v})_n\}} - \{\mathbf{w}, \mathbf{v}\}$ joining $(\mathbf{q})^n$ to a vertex $\mathbf{z}$ such that $(\mathbf{z})_1 \in \langle n-1 \rangle - \{(\mathbf{v})_n, (\mathbf{w})_1\}$. By Theorem 13.18, there is a Hamiltonian path H of $S_n^{\langle n-1 \rangle - \{(\mathbf{v})_n\}}$ joining $(\mathbf{z})^n$ to $(\mathbf{w})^n$, and then $\langle \mathbf{u}, \mathbf{p}, Q, \mathbf{q}, (\mathbf{q})^n, R, \mathbf{z}, (\mathbf{z})^n, H, (\mathbf{w})^n, \mathbf{w}, \mathbf{v} \rangle$ is a Hamiltonian path of $S_n - F$ joining $\mathbf{u}$ to $\mathbf{v}$.

Case 2.1.5. $(\mathbf{u})_n \in \langle n-1 \rangle$ and $(\mathbf{v})_n \in \langle n-1 \rangle$ with $(\mathbf{u})_n \neq (\mathbf{v})_n$. Since there are $(n-2)!/2 > 2$ edges joining black vertices of $S_n^{\{(\mathbf{u})_n\}}$ to white vertices of $S_n^{\{n\}}$, we can choose a white vertex $\mathbf{w}$ in $S_n^{\{n\}}$ such that $(\mathbf{w})^n \in S_n^{\{(\mathbf{u})_n\}}$ and $\{\mathbf{x}, (\mathbf{x})^n\} \cap (N_C(\mathbf{w}) \cup \{\mathbf{w}\}) = \emptyset$. Then we can write $C = \langle \mathbf{w}, \mathbf{p}, Q, \mathbf{q}, \mathbf{w} \rangle$. By Lemma 13.8, $(\mathbf{p})_1 \neq (\mathbf{q})_1$. By symmetry, we may assume that $(\mathbf{q})_1 \neq (\mathbf{v})_n$. By Theorem 13.18, there is a Hamiltonian path H of $S_n^{\{(\mathbf{u})_n\}}$ joining $\mathbf{u}$ to $(\mathbf{w})^n$ and there is a Hamiltonian path R of $S_n^{\langle n-1 \rangle - \{(\mathbf{u})_n\}}$ joining $(\mathbf{q})^n$ to $\mathbf{v}$. Then $\langle \mathbf{u}, H, (\mathbf{w})^n, \mathbf{w}, \mathbf{p}, W, \mathbf{q}, (\mathbf{q})^n, R, \mathbf{v} \rangle$ is a Hamiltonian path of $S_n - F$ joining $\mathbf{u}$ to $\mathbf{v}$.

Case 2.2. $|F_n| \leq 2n-9$. Then $|F_i| \leq 2n-9$ for every $i \in \langle n \rangle$. If $(\mathbf{u})_n \neq (\mathbf{v})_n$ then by Theorem 13.20, there is a Hamiltonian path H of $S_n - F$ joining $\mathbf{u}$ to $\mathbf{v}$. On the other hand, if $(\mathbf{u})_n = (\mathbf{v})_n$, then by induction, there is a Hamiltonian path Q of $S_n^{\{(\mathbf{u})_n\}} - F_{(\mathbf{u})_n}$ joining $\mathbf{u}$ to $\mathbf{v}$. Since $(n-1)! - 1 > 2(2n-7)$ if $n \geq 5$, there is an edge $(\mathbf{p}, \mathbf{q}) \in E(Q)$

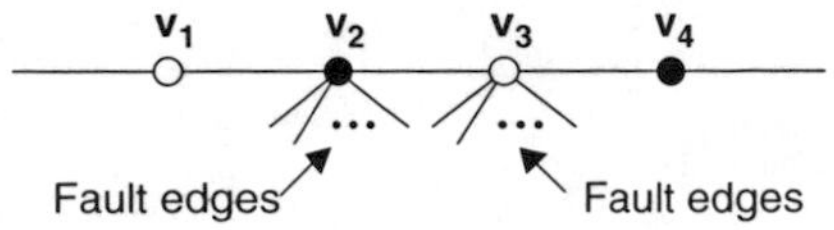

FIGURE 13.19 Illustration of S_n with $(2n-6)$ edge faults.

such that $(\mathbf{p}, (\mathbf{p})^n) \notin F$ and $(\mathbf{q}, (\mathbf{q})^n) \notin F$. Then we can write $Q = \langle \mathbf{u}, Q_1, \mathbf{p}, \mathbf{q}, Q_2, \mathbf{v} \rangle$. By Theorem 13.20, there is a Hamiltonian path H of $S_n^{\langle n \rangle - \{(\mathbf{u})_n\}}$ joining $(\mathbf{p})^n$ to $(\mathbf{q})^n$, and then $\langle \mathbf{u}, Q_1, \mathbf{p}, (\mathbf{p})^n, H, (\mathbf{q})^n, \mathbf{q}, \mathbf{v} \rangle$ is a Hamiltonian path of $S_n - F$ joining $\mathbf{u}$ to $\mathbf{v}$.

This finishes the proof. □

Let $n \geq 4$ and let $P = \langle \mathbf{v}_1, \mathbf{v}_2, \mathbf{v}_3, \mathbf{v}_4 \rangle$ be a path with four vertices in S_n. We set the faulty set F to be all the edges incident to $\mathbf{v}_2$ and $\mathbf{v}_3$ that are not edges of P shown in Figure 13.19. Obviously, $|F| = 2n - 6$. Since $\deg_{S_n - F}(\mathbf{v}_2) = 2$ and $\deg_{S_n - F}(\mathbf{v}_3) = 2$, there is no Hamiltonian path of $S_n - F$ joining $\mathbf{v}_1$ to $\mathbf{v}_4$. Thus, $S_n - F$ is not Hamiltonian laceable. Thus, the bound $2n - 7$ in Theorem 13.21 is tight.

14 Spanning Connectivity

14.1 INTRODUCTION

Assume that G is a k-connected graph. It follows from Menger's Theorem that there are k *internally vertex-disjoint* (abbreviated as *disjoint*) *paths* joining any two distinct vertices u and v. A k-container $C(u, v)$ of G is a set of k disjoint paths joining u to v. Here, we discuss another type of container, called *spanning container*. A *spanning k-container*, (abbreviated as *k^*-container*), $C(u, v)$ is a k-container that contains all vertices of G. A graph G is *k^*-connected* if there exists a k^*-container between any two distinct vertices. In particular, a graph G is 1^*-connected if and only if it is Hamiltonian-connected, and a graph G is 2^*-connected if and only if it is Hamiltonian. All 1^*-connected graphs except that K_1 and K_2 are 2^*-connected. Thus, we defined the *spanning connectivity* of a graph G, $\kappa^*(G)$, to be the largest integer k such that G is w^*-connected for all $1 \leq w \leq k$ if G is a 1^*-connected graph. Obviously, spanning connectivity is a hybrid concept of Hamiltonicity and connectivity. A graph G is *super spanning connected* if $\kappa^*(G) = \kappa(G)$. Obviously, the complete graph K_n is super spanning connected if $n \geq 2$.

A graph G is *bipartite* if its vertex set can be partitioned into two subsets V_1 and V_2 such that every edge joins vertices between V_1 and V_2. Let G be a k connected bipartite graph with bipartition V_1 and V_2 such that $|V_1| \geq |V_2|$. Suppose that there exists a k^*-container $C(u, v) = \{P_1, P_2, \ldots, P_k\}$ in a bipartite graph joining u to v with $u, v \in V_1$. Obviously, the number of vertices in P_i is $2k_i + 1$ for some integer k_i. There are $k_i - 1$ vertices of P_i in V_1 other than u and v, and k_i vertices of P_i in V_2. As a consequence, $|V_1| = \sum_{i=1}^{k} (k_i - 1) + 2$ and $|V_2| = \sum_{i=1}^{k} k_i$. Therefore, any bipartite graph G with $\kappa(G) \geq 3$ is not k^*-connected for any $3 \leq k \leq \kappa(G)$.

For this reason, a bipartite graph is *k^*-laceable* if there exists a k^*-container between any two vertices from different partite sets. Obviously, any bipartite k^*-laceable graph with $k \geq 2$ has the equal size of bipartition. A 1^*-laceable graph is also known as a *Hamiltonian-laceable graph*. Moreover, a graph G is 2^*-laceable if and only if it is Hamiltonian. All 1^*-laceable graphs except that K_1 and K_2 are 2^*-laceable. A bipartite graph G is *super spanning laceable* if G is i^*-laceable for all $1 \leq i \leq \kappa(G)$.

14.2 SPANNING CONNECTIVITY OF GENERAL GRAPHS

By Theorem 9.13, any graph G with at last four vertices and $\delta(G) \geq (n(G)/2) + 1$ is 1^*-connected. Obviously, every complete graph is super spanning connected. Lin et al. [230,232] begin the study of the spanning connectivity of a noncomplete graph with $\delta(G) \geq (n(G)/2) + 1$. We set $\overline{\kappa}(G) = 2\delta(G) - n(G) + 2$. Thus, $\overline{\kappa}(G) \geq 4$ for noncomplete graph with $\delta(G) \geq (n(G)/2) + 1$. Let H be a subgraph of G. The *neighborhood* of u with respect to H, denoted by $Nbd_H(u)$, is $\{v \in V(H) \mid (u, v) \in E(G)\}$. We use $\Gamma_H(u)$ to denote $|Nbd_H(u)|$. Obviously, $\Gamma_G(u) = \deg(u)$ for every $u \in V(G)$.

LEMMA 14.1 Let k be a positive integer. Suppose that there exist two nonadjacent vertices u and v with $\deg_G(u) + \deg_G(v) \geq n(G) + k$. Then for any two distinct vertices x and y, G has a $(k+2)^*$-container between x and y if and only if $G + (u, v)$ $(k+2)^*$-container between x and y.

Proof: Suppose that G has a $(k+2)^*$-container between x and y. Obviously, $G + (u, v)$ has a $(k+2)^*$-container between x and y.

For the other direction, suppose that $G + (u, v)$ has a $(k+2)^*$-container between x and y. Let x and y be any two different vertices of G. Let $C(x, y) = \{P_1, P_2, \ldots, P_{k+2}\}$ be a $(k+2)^*$-container of $G + (u, v)$.

Suppose that the edge $(u, v) \notin C(x, y)$. Then $C(x, y)$ forms a desired $(k+2)^*$-container of G. Thus, we suppose that $(u, v) \in C(x, y)$. Without loss of generality, we assume that $(u, v) \in P_1$. We write P_1 as $\langle x, H_1, u, v, H_2, y\rangle$. (Note that $l(H_1) = 0$ if $x = u$, and $l(H_2) = 0$ if $y = v$.) Let P_i' be the path obtained from P_i by deleting x and y_i. Thus, we can write P_i as $\langle x, P_i', y\rangle$ for $1 \leq i \leq k+2$. We set $C_i = \langle x, P_i', y, H_2^{-1}, v, u, H_1^{-1}, x\rangle$ for $2 \leq i \leq k+2$.

Case 1. $\Gamma_{C_i}(u) + \Gamma_{C_i}(v) \geq n(C_i)$ for some $2 \leq i \leq k+2$. Without loss of generality, we may assume that $\Gamma_{C_2}(u) + \Gamma_{C_2}(v) \geq n(C_2)$. By Theorem 9.4, there is a Hamiltonian cycle C of the subgraph of G induced by C_2. Let $C = \langle x, R_1, y, R_2, x\rangle$. We set $Q_1 = \langle x, R_1, y\rangle$, $Q_2 = \langle x, R_2^{-1}, y\rangle$, and $Q_i = P_i$ for $3 \leq i \leq k+2$. Then $\{Q_1, Q_2, \ldots, Q_{k+2}\}$ forms a $(k+2)^*$-container of G between x and y.

Case 2. $\Gamma_{C_i}(u) + \Gamma_{C_i}(v) \leq n(C_i) - 1$ for all $2 \leq i \leq k+2$. Since

$$\begin{aligned}
\sum_{i=2}^{k+2}(\Gamma_{C_i}(u) + \Gamma_{C_i}(v)) &= \sum_{i=2}^{k+2}(\Gamma_{P_i'}(u) + \Gamma_{P_1}(u) + \Gamma_{P_i'}(v) + \Gamma_{P_1}(v)) \\
&= \sum_{i=2}^{k+2}(\Gamma_{P_i'}(u) + \Gamma_{P_i'}(v)) + (k+1)(\Gamma_{P_1}(u) + \Gamma_{P_1}(v)) \\
&= \Gamma_G(u) + \Gamma_G(v) + k(\Gamma_{P_1}(u) + \Gamma_{P_1}(v)) \\
&\geq n(G) + k + k(\Gamma_{P_1}(u) + \Gamma_{P_1}(v))
\end{aligned}$$

$$\begin{aligned}
\sum_{i=2}^{k+2}(n(C_i) - 1) &= \sum_{i=2}^{k+2}(n(P_i') + n(P_1)) - (k+1) \\
&= \sum_{i=2}^{k+2} n(P_i') + (k+1)(n(P_1)) - (k+1) \\
&= n(G) + k(n(P_1)) - (k+1)
\end{aligned}$$

$n(G)+k+k(\Gamma_{P_1}(u)+\Gamma_{P_1}(v)) \le n(G)+k(n(P_1))-(k+1)$. Therefore, $\Gamma_{P_1}(u)+\Gamma_{P_1}(v) \le n(P_1)-1/k$. Since $k \ge 1$, $\Gamma_{P_1}(u)+\Gamma_{P_1}(v) \le n(P_1)$.

We claim that $\Gamma_{P'_i}(u)+\Gamma_{P'_i}(v) \ge n(P'_i)+2$ for some $2 \le i \le k+2$. Suppose that $\Gamma_{P'_i}(u)+\Gamma_{P'_i}(v) \le n(P'_i)+1$ for all $2 \le i \le k+2$. Then

$$\begin{aligned}\deg_G(u)+\deg_G(v) &= \sum_{i=2}^{k+2}(\Gamma_{P'_i}(u)+\Gamma_{P'_i}(v))+(\Gamma_{P_1}(u)+\Gamma_{P_1}(v))\\ &\le \sum_{i=2}^{k+2}(n(P'_i)+1)+n(P_1)\\ &= n(G)+k-1\end{aligned}$$

This contradicts with the fact that $\deg_G(u)+\deg_G(v) \ge n+k$.

Without loss of generality, we may assume that $\Gamma_{P'_2}(u)+\Gamma_{P'_2}(v) \ge n(P'_2)+2$. Obviously, $n(P'_2) \ge 2$. We write $P'_2 = \langle z_1, z_2, \ldots, z_r\rangle$. We claim that there exists an index j in $\{1,2,\ldots,r-1\}$ such that $(z_j, v) \in E(G)$ and $(z_{j+1}, u) \in E(G)$. Suppose there does not. Then $\Gamma_{P'_2}(u)+\Gamma_{P'_2}(v) \le r+r-(r-1) = r+1 = n(P'_2)+1$. We get a contradiction.

We set $Q_1 = \langle x, z_1, z_2, \ldots, z_j, v, H_2, y\rangle$, $Q_2 = \langle x, H_1, u, z_{j+1}, z_{j+2}, \ldots, z_r, y\rangle$, and $Q_i = P_i$ for $3 \le i \le k+2$. Then $\{Q_1, Q_2, \ldots, Q_{k+2}\}$ forms a $(k+2)^*$-container of G between x and y. See Figure 14.1 for illustration. □

Note that the statement of Theorem 14.1 is very similar to the statement of Theorems 9.2 and 9.3. Obviously, we have the following corollary.

COROLLARY 14.1 *Assume that k is any positive integer and there exist two nonadjacent vertices u and v with $\deg_G(u)+\deg_G(v) \ge n(G)+k$. Then G is $(k+2)^*$-connected if and only if $G+(u, v)$ is $(k+2)^*$-connected. Moreover, G is i^*-connected if and only if $G+(u, v)$ is i^*-connected for $1 \le i \le k+2$.*

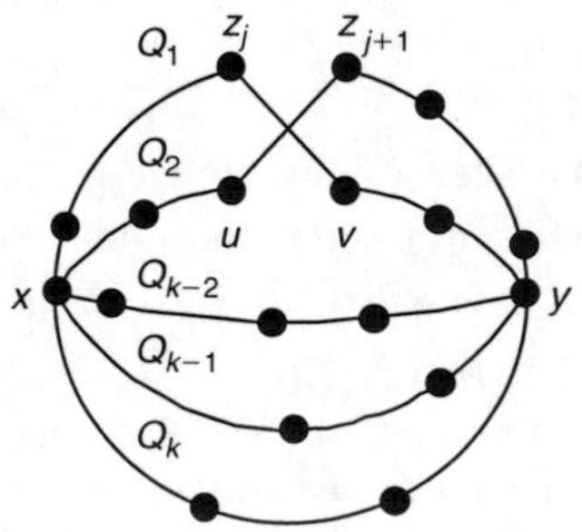

FIGURE 14.1 Illustration for Case 2 of Lemma 14.1.

THEOREM 14.1 Let k be a positive integer. Assume that $\deg_G(u) + \deg_G(v) \geq n(G) + k$ for all nonadjacent vertices u and v. Then G is r^*-connected for every $1 \leq r \leq k+2$.

Proof: By Theorem 9.4, G is 2^*-connected. By Theorem 9.12, G is 1^*-connected. Let x and y be two distinct vertices in G. Suppose there exists an r^*-container $\{P_1, P_2, \ldots, P_r\}$ of G between x and y for some $2 \leq r \leq k+1$. We need to construct only an $(r+1)^*$-container of G between x and y. We claim that $\deg_G(y) \geq k+2$. Suppose not. Let $w \notin N_G(y)$. Then $\deg_G(y) + \deg_G(w) \leq (k+1) + (n(G) - 2) = n(G) + k - 1$, given a contradiction.

We can choose a vertex u in $N_G(y) - \{x\}$ such that $(u, y) \notin E(P_i)$. Without loss of generality, we assume that $u \in P_r$. We write P_r as $\langle x, H_1, u, v, H_2, y \rangle$. We set $Q_i = P_i$ for $1 \leq i \leq r-1$, $Q_r = \langle x, H_1, u, y \rangle$, and $Q_{r+1} = \langle x, v, H_2, y \rangle$. Suppose that $(x, v) \in E(G)$. Then $\{Q_1, Q_2, \ldots, Q_{r+1}\}$ forms an $(r+1)^*$-container of G between x and y. Suppose that $(x, v) \notin E(G)$. Then, $\{Q_1, Q_2, \ldots, Q_{r+1}\}$ forms an $(r+1)^*$-container of $G + (x, u)$ between x and y. By Lemma 14.1, there exists an $(r+1)^*$-container of G between x and y. □

We give an example to show that the bound $(k+2)$ in the previous theorem is optimal. Let $G = (K_1 + K_b) \vee K_a$ where $b \geq 2$ and $a \geq 3$. We set $k = a - 2$. Obviously, $\delta(G) = a$ and $\deg_G(u) + \deg_G(v) \geq 2a + b - 1 = n(G) + a - 2 = n(G) + k$ for any two distinct vertices u and v. By the previous theorem, G is r^*-connected for any $r \leq a$. However, G is not r^*-connected for any $r > a$ because $\delta(G) = a$.

THEOREM 14.2 Assume that G is a noncomplete graph with $(n(G)/2) + 1 \leq \delta(G) \leq n(G) - 2$. Then G is r^*-connected for $1 \leq r \leq \overline{\kappa}(G)$.

Proof: Since $(n(G)/2) + 1 \leq \delta(G) \leq n(G) - 2$, $n(G) \geq 6$. Let k be a positive integer and $m \geq 3$. Suppose that $n(G) = 2m$ and $\delta(G) = m + k$ for some $m \geq 3$ and $1 \leq k \leq m - 2$. Then $\deg_G(u) + \deg_G(v) \geq 2\delta(G) = 2m + 2k$. By Theorem 14.1, G is r^*-connected for $1 \leq r \leq 2k + 2 = \overline{\kappa}(G)$. Suppose that $n(G) = 2m + 1$ and $\delta(G) = m + 1 + k$ for some $m \geq 3$ and $1 \leq k \leq m - 2$. We have $\deg_G(u) + \deg_G(v) \geq 2\delta(G) = 2m + 2 + 2k$. By Theorem 14.1, G is r^*-connected for $1 \leq r \leq 2k + 3 = \overline{\kappa}(G)$. □

COROLLARY 14.2 *Assume that G is a graph with $(n(G)/2) + 1 \leq \delta(G)$ with at least five vertices. Then $\kappa^*(G) \geq 4$.*

LEMMA 14.2 Suppose that G is a noncomplete graph with $\delta(G) \geq (n(G)/2) + 1$. Then $|N_G(u) \cap N_G(v)| \geq \overline{\kappa}(G) - 2$ if $(u, v) \in E(G)$, otherwise $|N_G(u) \cap N_G(v)| \geq \overline{\kappa}(G)$.

Proof: Obviously, $|N_G(u) \cap N_G(v)| = |N_G(u)| + |N_G(v)| - |N_G(u) \cup N_G(v)| \geq 2\delta(G) - |N_G(u) \cup N_G(v)|$. Suppose that $(u, v) \in E(G)$. Apparently, $|N_G(u) \cup N_G(v)| \leq n(G)$. Then $|N_G(u) \cap N_G(v)| \geq 2\delta(G) - n(G) = \overline{\kappa}(G) - 2$. Suppose that $(u, v) \notin E(G)$. Thus, $|N_G(u) \cup N_G(v)| \leq n(G) - 2$. Then $|N_G(u) \cap N_G(v)| \geq 2\delta(G) - (n(G) - 2) = \overline{\kappa}(G)$. □

With Theorem 14.2, we have the following corollary.

COROLLARY 14.3 *$\overline{\kappa}(G) \le \kappa^*(G) \le \kappa(G) \le \delta(G)$ if G is a graph with $(n(G)/2)+1 \le \delta(G) \le n(G)-2$. Moreover, G is super spanning connected if $\kappa(G)=\overline{\kappa}(G)$. In particular, G is super spanning connected if $\delta(G)=n(G)-2$.*

Among those graphs G with $\overline{\kappa}(G) \le \kappa(G) \le n(G)-3$, we give the following examples to illustrate the following cases: (1) $\overline{\kappa}(G)=\kappa^*(G)=\kappa(G)$, (2) $\overline{\kappa}(G)=\kappa^*(G)<\kappa(G)$, (3) $\overline{\kappa}(G)<\kappa^*(G)=\kappa(G)$, and (4) $\overline{\kappa}(G)<\kappa^*(G)<\kappa(G)$.

To verify the spanning connectivity of the following examples, we need the following lemmas.

LEMMA 14.3 Assume that G is a k^*-connected graph and S is a vertex subset of G. Then the graph $G-S$ has at most $|S|-k+2$ components.

LEMMA 14.4 Assume that a is any positive integer with $a \ge 3$. Let G be the complete 3-partite graph $K_{a,a,a}$ with the vertex partite sets V_1, V_2, and V_3. Then $\kappa^*(G)=a+2$.

Proof: It is easy to see that $\kappa(G)=2a$ and $\overline{\kappa}(G)=a+2$. Thus, $\kappa^*(G) \ge a+2$. To prove this lemma, we need to prove that there is no $(a+3)^*$-container between any two vertices u and v in V_1.

Suppose that there exists an $(a+3)^*$-container $C(u,v)=\{P_1,P_2,\ldots,P_{a+3}\}$ of G between u and v. Let $A=\{i \mid l(P_i)=2 \text{ for } 1 \le i \le a+3\}$ and $B=\{1,2,\ldots,a+3\}-A$. Obviously, $\sum_{i=1}^{a+3} |V(P_i) \cap (V_2 \cup V_3)| = 2a$.

Let P be any path of G joining u to v. Obviously, $|V(P) \cap (V_1-\{u,v\})| \le |V(P) \cap (V_2 \cup V_3)|-1$. Since $C(u,v)$ is an $(a+3)^*$-container of G between u and v, $\sum_{i=1}^{a+3} |V(P_i) \cap (V_1-\{u,v\})| = a-2$. Since $|V(P_i) \cap (V_2 \cup V_3)| \ge 1$ and $|V(P_i) \cap (V_1-\{u,v\})| \le |V(P_i) \cap (V_2 \cup V_3)|-1$ for every $i \in B$, $\sum_{i=1}^{a+3} |V(P_i) \cap (V_2 \cup V_3)| = \sum_{i \in B} |V(P_i) \cap (V_2 \cup V_3)| + |A| = \sum_{i \in B}(|V(P_i) \cap (V_2 \cup V_3)|-1)+|B|+|A| \ge \sum_{i \in B} |V(P_i) \cap (V_1-\{u,v\})| + a+3 = 2a+1$. We obtained a contradiction because $\sum_{i=1}^{a+3} |V(P_i) \cap (V_2 \cup V_3)| = 2a$. □

Recall that the Harary graph $H_{2r,n}$ can be described with $V(H_{2r,n})=\{0,1,\ldots,n-1\}$ and $E(H_{2r,n})=\{\{i,j\} \mid |i-j| \in \{1,2,\ldots,r\} \textbf{ mod } n\}$.

LEMMA 14.5 $H_{2r,n}$ is 1^*-spanning connected.

Proof: Let u and v be any two distinct vertices of $H_{2,n}$. Without loss of generality, we assume that $u=0$ and $1 \le v \le n/2$. Let t be any integer with $1 \le t \le n/4$. We set P as follows:

$$P = \langle 0, n-1, n-2, \ldots, 1 \rangle \quad \text{if } v=1$$

$$P = \langle 0, n-1, n-2, \ldots, 2t+1, 2t-1, 2t-3, \ldots, 1, 2, 4, \ldots, 2t \rangle \quad \text{if } v=2t$$

$$P = \langle 0, n-1, n-2, \ldots, 2t+2, 2t, 2t-2, \ldots, 2, 1, 3, \ldots, 2t+1 \rangle \quad \text{if } v=2t+1$$

Thus, P forms a Hamiltonian path of $H_{2,n}$ between u and v. Since $H_{2,n}$ is a spanning subgraph of $H_{2r,n}$, $H_{2r,n}$ is 1^*-spanning connected. □

LEMMA 14.6 Let v and k be any two distinct positive integers with $v > k$. Let G' be a graph with $V(G') = \{0, 1, \ldots, v\}$ and $E(G') = \{\{i, j\} \mid 1 \leq |i - j| \leq k\}$. Then there is a k^*-container of G' between $u = 0$ and v.

Proof: Let $v - 1 = qk + r$ for some two integers q and r with $0 \leq r \leq k - 1$. Since $v > k$, $p \geq 1$.

Case 1. $r = 0$. We set $P_i = \langle 0, i, i + k, \ldots, i + (q - 1)k, v\rangle$ for every $i \leq k$. Then $\{P_1, P_2, \ldots, P_k\}$ forms a k^*-container of G' between u and v.

Case 2. $r \geq 1$. We set $P_i = \langle 0, i, i + k, \ldots, i + qk, v\rangle$ for every $1 \leq i \leq r$ and $P_j = \langle 0, j, j + k, \ldots, j + (q - 1)k, v\rangle$ for every $r + 1 \leq j \leq k$. Then $\{P_1, P_2, \ldots, P_k\}$ forms a k^*-container of G' between u and v. □

LEMMA 14.7 Let $n \geq 4$ and $2 \leq r \leq (n/2) - 1$. Then $H_{2r,n}$ is w^*-connected with $w \in \{2r - 1, 2r\}$.

Proof: Let u and v be any two distinct vertices of $H_{2r,n}$. Without loss of generality, we assume that $u = 0$ and $1 \leq v \leq n/2$. We have the following cases:

Case 1. $v = 1$. We set $P_i = \langle 0, i + 1, v\rangle$ for every $1 \leq i \leq r - 1$, $P_j = \langle 0, n - 1 + r - j, v\rangle$ for every $r \leq j \leq 2r - 2$, $P_{2r-1} = \langle 0, n - r, n - r - 1, \ldots, r + 1, v\rangle$, and $P_{2r} = \langle 0, v\rangle$. Then $\{P_1, P_2, \ldots, P_{2r-1}\}$ forms a $(2r - 1)^*$-container of $H_{2r,n}$ between u and v and $\{P_1, P_2, \ldots, P_{2r}\}$ forms a $(2r)^*$-container of $H_{2r,n}$ between u and v.

Case 2. $1 < v < r$. We set $P_i = \langle 0, i, v\rangle$ for every $1 \leq i \leq v - 1$, $P_j = \langle 0, j + 1, v\rangle$ for every $v \leq j \leq r - 1$, and $P_t = \langle 0, n - 1 + r - t, v\rangle$ for every $r \leq t \leq 2r - v - 1$. Let H be a subgraph of $H_{2r,n}$ induced by $\{0, n - r + v - 1, n - r + v - 2, \ldots, r + 1, v\}$. By Lemma 14.6, there is a v^*-container $\{P_{2r-v}, P_{2r-v+1}, \ldots, P_{2r-1}\}$ of H between u and v. We set $P_{2r} = \langle 0, v\rangle$. Then $\{P_1, P_2, \ldots, P_{2r-1}\}$ forms a $(2r - 1)^*$-container of $H_{2r,n}$ between u and v and $\{P_1, P_2, \ldots, P_{2r}\}$ forms a $(2r)^*$-container of $H_{2r,n}$ between u and v.

Case 3. $v = r$. Let H be a subgraph of $H_{2r,n}$ induced by $\{0, n - 1, n - 2, \ldots, v\}$. By Lemma 14.6, there is a r^*-container $\{P_1, P_2, \ldots, P_r\}$ of H between u and v. We set $P_i = \langle 0, i - r, v\rangle$ for every $r + 1 \leq i \leq 2r - 1$ and $P_{2r} = \langle 0, v\rangle$. Then $\{P_1, P_2, \ldots, P_{2r-1}\}$ forms a $(2r - 1)^*$-container of $H_{2r,n}$ between 0 and v and $\{P_1, P_2, \ldots, P_{2r}\}$ forms a $(2r)^*$-container of $H_{2r,n}$ between u and v.

Case 4. $v > r$. Let H_1 be a subgraph of $H_{2r,n}$ induced by $\{0, 1, 2, \ldots, v\}$ and H_2 be a subgraph of $H_{2r,n}$ induced by $\{0, n - 1, n - 2, \ldots, v\}$. By Lemma 14.6, there is a r^*-container $\{P_1, P_2, \ldots, P_r\}$ of H_1 between u and v.

Case 4.1. By Lemma 14.6, there is a $(r - 1)^*$-container $\{P_{r+1}, P_{r+2}, \ldots, P_{2r-1}\}$ of H_2 between u and v. Then $\{P_1, P_2, \ldots, P_{2r-1}\}$ forms a $(2r - 1)^*$-container of $H_{2r,n}$ between u and v.

Case 4.2. By Lemma 14.6, there is a r^*-container $\{P_{r+1}, P_{r+2}, \ldots, P_{2r}\}$ of H_2 between u and v. Then $\{P_1, P_2, \ldots, P_{2r}\}$ forms a $2r^*$-container of $H_{2r,n}$ between u and v. □

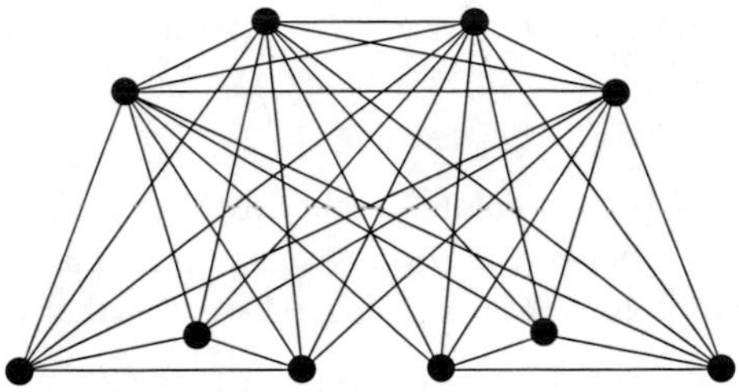

FIGURE 14.2 Illustration for $K_4 \vee (K_3 + K_3)$.

LEMMA 14.8 $H_{2r,n}$ is super spanning connected if $n \geq 4$ and $2 \leq r \leq n/2$.

Proof: Let $t = \lfloor \frac{n}{2} \rfloor$. Since $H_{2t,n} = K_n$, $H_{2t,n}$ is super spanning connected. By Lemmas 14.5 and 14.7, $H_{2,n}$ is super spanning connected. Thus, we assume that $3 \leq r \leq (n/2) - 1$. Since $3 \leq (n/2) - 1$, $n \geq 8$. Note that $H_{2i,n}$ is a spanning subgraph of $H_{2j,n}$ if $1 \leq i \leq j \leq (n/2) - 1$. By Lemma 14.7, $H_{2r,n}$ is w^*-connected for every $3 \leq w \leq 2r$. Hence, $H_{2r,n}$ is super spanning connected if $n \geq 4$ and $2 \leq r \leq (n/2)$. □

Example 14.1

$\overline{\kappa}(G) = \kappa^*(G) = \kappa(G)$. Let $G = K_a \vee (K_b + K_b)$ with $a \geq 4$ and $b \geq 2$. Obviously, $\kappa(G) = \overline{\kappa}(G) = a$. By Corollary 14.3, $\overline{\kappa}(G) = \kappa^*(G) = \kappa(G) = a$. See Figure 14.2. □

Example 14.2

$\overline{\kappa}(G) = \kappa^*(G) < \kappa(G)$. Let $G = K_{a,a,a}$ be the complete 3-partite graph with vertex partite set $V_1, V_2,$ and V_3 where $|V_1| = |V_2| = |V_3| = a \geq 3$. Obviously, $\kappa(G) = 2a$ and $\overline{\kappa}(G) = a + 2$. By Lemma 14.4, $\kappa^*(G) = a + 2$. □

Example 14.3

$\overline{\kappa}(G) < \kappa^*(G) = \kappa(G)$. Let G be the Harary graph $H_{2s,n}$ where s is a positive integer with $s > 3$. In Example 7.2, we know that $\kappa(G) = \delta(G) = 2s$. By Lemma 14.8, $\kappa^*(G) = 2s$. Thus, G is super spanning connected. Suppose that we choose $n = 4(s-1)$. Then $\overline{\kappa}(G) = 2\delta(G) - n(G) + 2 = 6$. Thus, $\overline{\kappa}(G) < \kappa^*(G) = \kappa(G)$. □

Example 14.4

$\overline{\kappa}(G) < \kappa^*(G) < \kappa(G)$. Let $G = K_a \vee (K_b^c + K_c)$ for $a \geq b + c$, $b \geq 2$, and $c \geq 2$. Obviously, $\kappa(G) = a$ and $\overline{\kappa}(G) = a - b - c + 2$. Since $G - K_a = K_b^c + K_c$, $G - K_a$ has $b + 1$ components. By Lemma 14.3, G is not w^*-connected for all $w > a - b + 1$. Thus, $\kappa(G) = a > a - b + 1 \geq \kappa^*(G)$. It is easy to construct a w^*-container joining any two vertices u, v in G. Hence, $\kappa^*(G) = a - b + 1$. □

We can also study the spanning connectivity of a faulty graph.

LEMMA 14.9 Assume that G is a graph with at least four vertices and $\delta(G) \geq (n(G)/2) + 1$. Let T be a vertex subset of G with $|T| \leq \overline{\kappa}(G) - 3$. Then the graph $G - T$ is 1*-connected.

Proof: Suppose that G is a complete graph. Obviously, $G - T$ is also a complete graph. Thus, $G - T$ is 1*-connected. Thus, we assume that $\delta(G) \leq n(G) - 2$. By Theorem 9.13, G is 1*-connected if $|T| = 0$. Thus, we assume that $1 \leq t = |T| \leq \overline{\kappa}(G) - 3$.

Case 1. $n(G) = 2s$ and $\delta(G) = s + r$ for some $1 \leq r \leq s - 2$. Since $t \leq \overline{\kappa}(G) - 3 = (2\delta(G) - n(G) + 2) - 3 = (2r + 2) - 3$, $r \geq (t + 1)/2$.

Case 1.1. $t = 2a$ for some $a \geq 1$. Obviously, $\delta(G - T) \geq \delta(G) - |T| = s + r - t \geq s + (t + 1)/2 - t = s - a + 1/2$. Since $\delta(G - T)$ is an integer, $\delta(G - T) \geq s - a + 1 = (n(G - T)/2) + 1$. By Theorem 9.13, $G - T$ is 1* connected.

Case 1.2. $t = 2a + 1$ for some $a \geq 0$. Choose an element z in T, and set $T' = T - \{z\}$. We claim that $G - T'$ is 1*-connected. Suppose that $T' = \emptyset$. Obviously, $G - T'$ is 1*-connected. Suppose that $T' \neq \emptyset$. Since $|T'| = 2a$, by Case 1.1, $G - T'$ is 1*-connected. Let $P = \langle x_1, \ldots, x_i, \ldots, x_{2s-2a} \rangle$ where $x_1 = u$, $x_{2s-2a} = v$, and $x_i = z$, be a Hamiltonian path of $G - T'$ joining u to v. Suppose that $\{x_{i-1}, x_{i+1}\} \in E(G)$. Then $\langle x_1, \ldots, x_{i-1}, x_{i+1}, \ldots, x_{2s-2a} \rangle$ forms a Hamiltonian path of $G - T$ joining u to v. Thus, we assume that $\{x_{i-1}, x_{i+1}\} \notin E(G)$. Since $r \geq (t + 1)/2 = a + 1$ and $n(G - T) = 2s - 2a - 1$, $\deg_{G-T}(x_{i-1}) + \deg_{G-T}(x_{i+1}) \geq 2\delta(G - T) \geq 2\delta(G) - 2|T| = 2s + 2r - 4a - 2 \geq n(G - T) + 1$. Therefore, there exists an index $j \in \{1, 2, \ldots, i - 2\} \cup \{i + 2, \ldots, 2s - 2a\}$ such that $\{x_j, x_{i-1}\} \in E(G)$ and $\{x_{j+1}, x_{i+1}\} \in E(G)$.

Suppose $j \in \{1, \ldots, i - 2\}$. Then $\langle x_1, \ldots, x_j, x_{i-1}, x_{i-2}, \ldots, x_{j+1}, x_{i+1}, x_{i+2}, \ldots, x_{2s-2a} \rangle$ forms a Hamiltonian path of $G - T$ joining u to v. Suppose $j \in \{i + 2, \ldots, 2s - 2a\}$. Then $\langle x_1, \ldots, x_{i-1}, x_j, x_{j-1}, x_{j-2}, \ldots, x_{i+1}, x_{j+1}, x_{j+2}, \ldots, x_{2s-2a} \rangle$ forms a Hamiltonian path of $G - T$ joining u to v.

Case 2. $n(G) = 2s + 1$ and $\delta(G) = s + r + 1$ for some $1 \leq r \leq s - 2$. Since $t \leq \overline{\kappa}(G) - 3 = (2\delta(G) - n(G) + 2) - 3 = 2r$, $\delta(G) \geq s + (t/2) + 1$.

Case 2.1. $t = 2a + 1$ for some $a \geq 0$. Since $\delta(G - T)$ is a positive integer and $\delta(G - T) \geq \delta(G) - t \geq s + (t/2) + 1 - t = s - ((2a + 1)/2) + 1$, $\delta(G - T) \geq ((2s + 1 - (2a + 1))/2) + 1 = (n(G - T)/2) + 1$. By Theorem 9.13, $G - T$ is 1*-connected.

Case 2.2. $t = 2a$ for some $a \geq 1$. Choose an element z in T, and set $T' = T - \{z\}$. Obviously, $|T'| = 2a - 1$. By Case 2.1, $G - T'$ is 1*-connected. Let $P = \langle x_1, x_2, \ldots, x_i, \ldots, x_{2s-2a+2} \rangle$ be a Hamiltonian path of $G - T'$ joining u to v with where $x_1 = u$, $x_{2s-2a+2} = v$, and $x_i = z$.

Suppose that $\{x_{i-1}, x_{i+1}\} \in E(G)$. Obviously, $\langle x_1, \ldots, x_{i-1}, x_{i+1}, \ldots, x_{2s-2a+2} \rangle$ forms a Hamiltonian path of $G - T$ joining u to v. Suppose that $\{x_{i-1}, x_{i+1}\} \notin E(G)$. Since $r \geq t/2 = a$ and $n(G - T) = 2s - 2a + 1$, $\deg_{G-T}(x_{i-1}) + \deg_{G-T}(x_{i+1}) \geq 2\delta(G - T) \geq 2\delta(G) - 2|T| = 2s + 2r + 2 - 4a \geq n(G - T) + 1$. There exists an index $j \in \{1, \ldots, i - 2\} \cup \{i + 2, \ldots, 2s - 2a + 2\}$ such that $\{x_j, x_{i-1}\} \in E(G)$ and $\{x_{j+1}, x_{i+1}\} \in E(G)$. We use the same reasoning as in Case 1.2 to find a Hamiltonian path of $G - T$ joining u to v. □

THEOREM 14.3 Assume that G is a graph with $(n(G)/2)+1 \le \delta(G) \le n(G)-2$ and $n(G) \ge 6$. Suppose that T is a vertex subset of G with $|T| \le \overline{\kappa}(G)-3$. Then $\kappa^*(G-T) \ge \overline{\kappa}(G)-|T|$.

Proof: Assume that $|T|=t$. By Theorem 14.2, this statement holds on $t=0$. Thus, we suppose that $t \ge 1$. By Lemma 14.9, $G-T$ is 1*-connected. Since every 1*-connected graph except K_1 and K_2 is also 2*-connected, $G-T$ is 2*-connected. To prove our theorem, we still need to construct a k^*-container of $G-T$ between any two distinct vertices u and v for any $3 \le k \le \overline{\kappa}(G)-t$. By Lemma 14.2, $|N_G(u) \cap N_G(v)| \ge \overline{\kappa}(G)-2$. Thus, $|N_{G-T}(u) \cap N_{G-T}(v)| \ge k-2$. Let $S=\{z_1, \ldots, z_{k-2}\}$ be a $(k-2)$ vertex subset of $N_{G-T}(u) \cap N_{G-T}(v)$. Since $k \le \overline{\kappa}(G)-t$, this implies that $\delta(G) \ge (n(G)+k+t-2)/2$ and $\delta(G-(S \cup T)) \ge \delta(G)-|S \cup T| \ge ((n(G)+k+t-2)/2)-(k-2+t) = (n(G)-k+t+2)/2 = n(G-(S \cup T))/2$. By Theorem 9.2, there is a 2*-container $\{R_1, R_2\}$ of $G-(S \cup T)$ between u and v. We set that $P_i = R_i$ for $1 \le i \le 2$, and $P_i = \langle u, z_{i-2}, v\rangle$ for $3 \le i \le k$. Then $\{P_1, P_2, \ldots, P_k\}$ forms a k^*-container of $G-T$ between u and v. □

The following example shows that the inequality in Theorem 14.3 is tight.

Example 14.5

Assume that a is any positive integer with $a \ge 3$. Let G be the complete 3-partite graph $K_{a,a,a}$ with the vertex partite sets V_1, V_2, and V_3. Obviously, $n(G)=3a$, $\delta(G)=2a$. Hence, $\overline{\kappa}(G)=2\delta(G)-n(G)+2=a+2$. Let T be a vertex subset of $V(G)$ with $|T|=t \le \overline{\kappa}(G)-3=a-1$. By Theorem 14.3, $\kappa^*(G-T) \ge \overline{\kappa}(G)-t=a+2-t$.

However, it is possible to find a vertex subset of $V(G)$ with $|T|=t \le a-1$ such that $\kappa^*(G-T)=\overline{\kappa}(G)-t=a+2-t$.

Let T be a subset of $V_2 \cup V_3$ with $|T|=t$. Let u and v be any two distinct vertices in V_1. Suppose that there exists an $(a-t+3)^*$-container $C(u,v)=\{P_1, P_2, \ldots, P_{a-t+3}\}$ of $G-T$ between u and v. Let $A=\{i \mid l(P_i)=2 \text{ for } 1 \le i \le a-t+3\}$ and $B=\{1,2,\ldots,a-t+3\}-A$. Obviously, $\sum_{i=1}^{a+3} |V(P_i) \cap ((V_2 \cup V_3)-T)| = 2a-t$.

Let P be any path of $G-T$ joining u to v. Obviously, $|V(P) \cap (V_1-\{u,v\})| \le |V(P) \cap ((V_2 \cup V_3)-T)|-1$. Since $C(u,v)$ is an $(a-t+3)^*$-container of G between u and v, $\sum_{i=1}^{a-t+3} |V(P_i) \cap (V_1-\{u,v\})| = a-2$. Since $|V(P_i) \cap ((V_2 \cup V_3)-T)| \ge 1$ and $|V(P_i) \cap (V_1-\{u,v\})| \le |V(P_i) \cap ((V_2 \cup V_3)-T)|-1$ for every $i \in B$, $\sum_{i=1}^{a-t+3} |V(P_i) \cap ((V_2 \cup V_3)-T)| = \sum_{i \in B} |V(P_i) \cap (V_2 \cup V_3)| + |A| = \sum_{i \in B} (|V(P_i) \cap (V_2 \cup V_3)|-1) + |B|+|A| \ge \sum_{i \in B} |V(P_i) \cap (V_1-\{u,v\})| + a-t+3 = 2a-t+1$. We obtained a contradiction because $\sum_{i=1}^{a-t+3} |V(P_i) \cap ((V_2 \cup V_3)-T)| = 2a-t$. □

14.3 SPANNING CONNECTIVITY AND SPANNING LACEABILITY OF THE HYPERCUBE-LIKE NETWORKS

Among all interconnection networks proposed in the literature, the hypercube Q_n is one of the most popular topologies [220]. Chang et al. [43] study the spanning laceability of Q_n. Lin et al. [239] discuss the spanning connectivity and the spanning laceability of hypercube-like networks.

As noted in Chapter 3, the hypercube does not have the smallest diameter for its resources. Various networks are proposed by twisting some pairs of links in hypercubes [1,76,93,94]. Because of the lack of the unified perspective on these variants, results of one topology are hard to extend to others. To make a unified study of these variants, Vaidya et al. introduced the class of hypercube-like graphs [322]. We denote these graphs as H'-graphs. The class of H'-graphs, consisting of simple, connected, and undirected graphs, contains most of the hypercube variants.

The hypercube-like graphs are constructed using the operation $G_0 \oplus G_1$ discussed in Chapter 7. Here, we recall the definition of $G_0 \oplus G_1$. Suppose that $G_0 = (V_0, E_0)$ and $G_1 = (V_1, E_1)$ are two disjoint graphs with $|V_0| = |V_1|$. A *1-1 connection* between G_0 and G_1 is defined as an edge set $E_r = \{(v, \phi(v)) \mid v \in V_0, \phi(v) \in V_1$ and $\phi : V_0 \to V_1$ is a bijection$\}$. $G_0 \oplus G_1$ denotes the graph $G = (V_0 \cup V_1, E_0 \cup E_1 \cup E_r)$. Thus, ϕ induces a perfect matching M in $G_0 \oplus G_1$. For convenience, G_0 and G_1 are called *components*. Let x be any vertex of $G_0 \oplus G_1$. We use $\overline{x}$ to denote the vertex matched under ϕ.

Now, we can define the set of n-dimensional H'-graph, H'_n, as follows:

1. $H'_1 = \{K_2\}$, where K_2 is the complete graph with two vertices.
2. Assume that $G_0, G_1 \in H'_n$. Then $G = G_0 \oplus G_1$ is a graph in H'_{n+1}.

Note that some n-dimensional H'-graphs are bipartite. We can define the set of bipartite n-dimensional H'-graph, B'_n, as follows:

1. $B'_1 = \{K_2\}$, where K_2 is the complete graph defined on $\{a, b\}$ with bipartition $V_0 = \{a\}$ and $V_1 = \{b\}$.
2. For $i = 0, 1$, let G_i be a graph in B'_n with bipartition V^i_0 and V^i_1. Then $G = G_0 \oplus G_1$ is a graph in B'_{n+1} where ϕ is a perfect matching between $V^0_0 \cup V^0_1$ and $V^1_0 \cup V^1_1$ such that M joining vertices in V^0_i to vertices in V^1_{1-i} if $v \in V^0_i$.

Every graph in H'_n is an n-regular graph with 2^n vertices, and every graph in B'_n contains 2^{n-1} vertices in each bipartition. We use N'_n to denote the set of nonbipartite graphs in H'_n. Clearly, we have $Q_n \in B'_n$.

By brute force, we can check whether Q_1 is the only graph in H'_1 and Q_2 is the only graph in H'_2. Thus, we have the following lemma.

LEMMA 14.10 Assume that G is graph in N'_n. Then $n \geq 3$.

The following lemma follows from Theorem 11.15.

LEMMA 14.11 Let $n \geq 3$. Every graph in N'_n is Hamiltonian connected and Hamiltonian.

The following lemma follows from Theorem 11.17.

LEMMA 14.12 Every graph in B'_n is Hamiltonian laceable and every graph in B'_n is Hamiltonian if $n \geq 2$.

THEOREM 14.4 **[268]** Let $n \geq 2$. Suppose that G is a graph in B'_n with bipartition V_0 and V_1. Suppose that u_1 and u_2 are two distinct vertices in V_i and that v_1 and v_2 are two distinct vertices in V_{1-i} with $i \in \{0,1\}$. Then there are two disjoint paths P_1 and P_2 of G such that (1) P_1 joins u_1 to v_1, (2) P_2 joins u_2 to v_2, and (3) $P_1 \cup P_2$ spans G.

Proof: We prove this theorem by induction on n. Clearly, Q_2 is the only graph in B'_2 and this theorem holds on Q_2. Moreover, Q_3 is the only graph in B'_3 and this theorem holds on Q_3. Suppose that this theorem holds for every graph in B'_{n-1} with $n \geq 4$. Let $G = G_0 \oplus G_1$ be a graph in B'_n. Let V_0^i and V_1^i be the bipartition of G_i for $i = 0, 1$. Without loss of generality, we assume that $V_0^0 \cup V_0^1$ and $V_1^0 \cup V_1^1$ form the bipartition of G. Suppose that u_1 and u_2 are two distinct vertices in $V_0^0 \cup V_0^1$ and v_1 and v_2 are two distinct vertices in $V_1^0 \cup V_1^1$. Without loss of generality, we assume that u_1 is a vertex in V_0^0. We have the following cases:

Case 1. u_2 in V_0^0, v_1 in V_1^0, and v_2 in V_1^0. By induction, there are two disjoint paths R_1 and R_2 of G_0 such that (1) R_1 joins u_1 to v_1, (2) R_2 joins u_2 to v_2, and (3) $R_1 \cup R_2$ spans G_0. Without loss of generality, we write $R_2 = \langle u_2, z, R, v_2 \rangle$. By Theorem 14.12, there is a Hamiltonian path Q of G_1 joining $\bar{u}_2$ to $\bar{z}$. We set $P_1 = R_1$ and $P_2 = \langle u_2, \bar{u}_2, Q, \bar{z}, z, R, v_2 \rangle$. Then P_1 and P_2 form the desired paths.

Case 2. u_2 in V_0^0, v_1 in V_1^0, and v_2 in V_1^1. Let z be any vertex in $V_0^1 - \{v_1\}$. By induction, there are two disjoint paths R_1 and R_2 of G_0 such that (1) R_1 joins u_1 to v_1, (2) R_2 joins u_2 to z, and (3) $R_1 \cup R_2$ spans G_0. By Theorem 14.12, there is a Hamiltonian path Q of G_1 joining $\bar{z}$ to v_2. We set $P_1 = R_1$ and $P_2 = \langle u_2, R_2, z, \bar{z}, Q, v_2 \rangle$. Then P_1 and P_2 form the desired paths.

Case 3. u_2 in V_0^0, v_1 in V_1^1, and v_2 in V_1^1. Let y and z be any two distinct vertices in V_0^1. By induction, there are two disjoint paths R_1 and R_2 of G_0 such that (1) R_1 joins u_1 to y, (2) R_2 joins u_2 to z, and (3) $R_1 \cup R_2$ spans G_0. Moreover, there are two disjoint paths Q_1 and Q_2 of G_1 such that (1) Q_1 joins $\bar{y}$ to v_1, (2) Q_2 joins $\bar{z}$ to v_2, and (3) $Q_1 \cup Q_2$ spans G_1. We set $P_1 = \langle u_1, R_1, y, \bar{y}, Q_1, v_1 \rangle$ and $P_2 = \langle u_2, R_2, z, \bar{z}, Q_2, v_2 \rangle$. Then P_1 and P_2 form the desired paths.

Case 4. u_2 in V_0^1, v_1 in V_1^0, and v_2 in V_1^1. By Theorem 14.12, there is a Hamiltonian path P_1 of G_0 joining u_1 to v_1. Moreover, there is a Hamiltonian path P_2 of G_1 joining u_2 to v_2. Then P_1 and P_2 form the desired paths.

Case 5. u_2 in V_0^1, v_1 in V_1^1, and v_2 in V_1^0. By Theorem 14.12, there is a Hamiltonian path P of G_0 joining u_1 to v_2. We write $P = \langle u_1 = z_1, z_2, z_3, \ldots, z_{2^{n-1}} = v_2 \rangle$. Since $n \geq 4$, $2^{n-1} \geq 8$. Thus, this is an index t in $\{1,2,3\}$ such that $\{z_{2t}, z_{2t+1}\} \cap \{\bar{u}_2, \bar{v}_1\} = \emptyset$. By induction, there are two disjoint paths Q_1 and Q_2 of G_1 such that (1) Q_1 joins $\bar{z}_{2t}$ to v_1, (2) Q_2 joins u_2 to $\bar{z}_{2t+1}$, and (3) $Q_1 \cup Q_2$ spans G_1. We set $P_1 = \langle u_1 = z_1, z_2, \ldots, z_{2t}, \bar{z}_{2t}, Q_1, v_1 \rangle$ and $P_2 = \langle u_2, Q_2, \bar{z}_{2t+1}, z_{2t+1}, z_{2t}, \ldots, z_{2^{n-1}} = v_2 \rangle$. Then P_1 and P_2 form the desired paths. □

THEOREM 14.5 Let G be a graph in B'_n with bipartition V_0 and V_1 for $n \geq 2$. Suppose that z is a vertex in V_i and that u and v are two distinct vertices in V_{1-i} with $i \in \{0,1\}$. Then there is a Hamiltonian path of $G - \{z\}$ joining u to v.

Proof: We prove this statement by induction on n. Since Q_2 is the only graph in B'_2, it is easy to check whether this statement holds for $n=2$. Thus, we assume that $G=G_0 \oplus G_1$ in B'_n with $n \geq 3$. We have $G_i \in B'_{n-1}$ for $i=0,1$. Let V_0^i and V_1^i be the bipartition of G_i for $i=0,1$. Without loss of generality, we assume that $V_0^0 \cup V_0^1$ and $V_1^0 \cup V_1^1$ form the bipartition of G. Let z be any vertex in $V_1^0 \cup V_1^1$, and let u and v be any two distinct vertices in $V_0^0 \cup V_0^1$. We need to show that there is a Hamiltonian path of $G-\{z\}$ joining u to v. Without loss of generality, we assume that $z \in V_1^0$. We have the following cases:

Case 1. $u \in V_0^0$ and $v \in V_0^0$. By induction, there is a Hamiltonian path Q in $G_0-\{z\}$ joining u to v. Without loss of generality, we write Q as $\langle u, x, R, v\rangle$. Since $u \in V_0^0$, $x \in V_1^0$. By Lemma 14.12, there is a Hamiltonian path W of G_1 joining the vertex $\bar{u} \in V_1^1$ to the vertex $\bar{x} \in V_0^1$. Then $\langle u, \bar{u}, W, \bar{x}, x, R, v\rangle$ is the Hamiltonian path of $G-\{z\}$ joining u to v. See Figure 14.3a for illustration.

Case 2. $u \in V_0^0$ and $v \in V_1^0$. Since $n \geq 3$, $|V_0^0| = 2^{n-1} \geq 2$. We can choose a vertex x in $V_0^0 - \{u\}$. By induction, there is a Hamiltonian path Q in $G_0-\{z\}$ joining u to x. Since $x \in V_0^0$, $\bar{x} \in V_1^1$. By Lemma 14.12, there is a Hamiltonian path W of G_1 joining $\bar{x}$ to v. Then $\langle u, Q, x, \bar{x}, W, v\rangle$ is the Hamiltonian path of $G-\{z\}$ joining u to v. See Figure 14.3b for illustration.

Case 3. $u \in V_0^1$ and $v \in V_0^1$. We can choose a vertex x in V_1^1. By Lemma 14.12, there is a Hamiltonian path W in G_1 joining u to x. Without loss of generality, we write W as $\langle u, W_1, y, v, W_2, x\rangle$. Since $v \in V_0^1$, $y \in V_1^1$. By induction, there is a Hamiltonian path Q in $G_0-\{z\}$ joining the vertex $\bar{y} \in V_0^0$ to the vertex $\bar{x} \in V_0^0$. Then $\langle u, W_1, y, \bar{y}, Q, \bar{x}, x, W_2^{-1}, v\rangle$ is the Hamiltonian path of $G-\{z\}$ joining u to v. See Figure 14.3c for illustration. □

Let n be any positive integer. To prove that every graph in B'_n is w^*-laceable for every w, $1 \leq w \leq n$, we need the concept of spanning fan. We note that there is another Menger-type theorem. Let u be a vertex of G and $S=\{v_1, v_2, \ldots, v_k\}$ be a subset of $V(G)$ not including u. An (u,S)-fan is a set of disjoint paths $\{P_1, P_2, \ldots, P_k\}$ of G such that P_i joins u and v_i. By Theorem 7.8, a graph G is k-connected if and only if there exists a (u,S)-fan between any vertex u and any k-subset of $V(G)$ such that $u \notin S$. Thus, we define a *spanning fan* as a fan that spans G. Naturally, we can study $\kappa^*_{fan}(G)$ as the largest integer k such that there exists a spanning (u,S)-fan between any vertex u and any k-vertex subset S with $u \notin S$. However, we defer such a study for the following reasons.

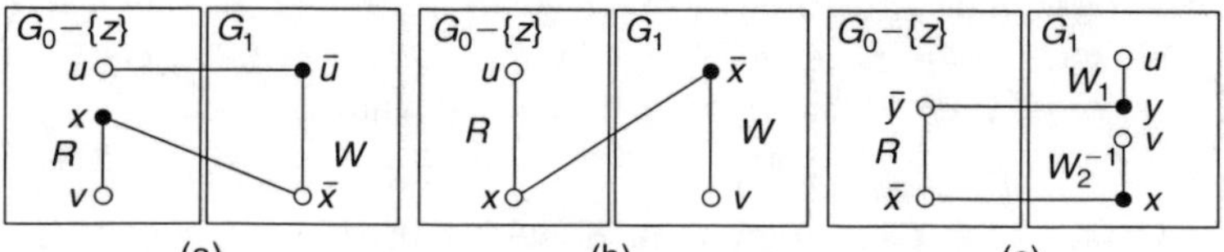

FIGURE 14.3 Illustrations for Theorem 14.5.

First, let S be a cut set of a graph G. Let u be any vertex of $V(G) - S$. It is easy to see that there is no spanning (u, S)-fan in G. Thus, $\kappa^*_{fan}(G) < \kappa(G)$ if G is not a complete graph.

Second, let G be a bipartite graph with bipartition V_0 and V_1 and $|V_0| = |V_1|$. Let u be a vertex in V_i, $S = \{v_1, v_2, \ldots, v_k\}$ be a subset of G not containing u, and $k \leq \kappa(G)$. Suppose that $|S \cap V_{1-i}| = r$. Without loss of generality, we assume that $\{v_1, v_2, \ldots, v_r\} \subset V_{1-i}$. Let $\{P_1, P_2, \ldots, P_k\}$ be any spanning (u, S)-fan of G. Then $l(P_i)$ is odd if $i \leq r$, and $l(P_i)$ is even if $r < i \leq k$. Let $l(P_i) = 2t_i + 1$ if $i \leq r$ and $l(P_i) = 2t_i$ if $i > r$. For $i \leq r$, there are $t_i - 1$ vertices of P_i in V_i other than u and there are t_i vertices of P_i in V_{1-i}. For $i > r$, there are r_i vertices of P_i in V_i other than u and there are t_i vertices of P_i in V_{1-i}. Thus, $|V_i| = 1 - r + \sum_{i=1}^{k} t_i$ and $|V_{1-i}| = \sum_{i=1}^{k} t_i$. Since $|V_i| = |V_{1-i}|$, $r = 1$. Thus, $r = 1$ is a requirement as we study the spanning fan of bipartite graphs with equal size of bipartition.

THEOREM 14.6 Let n and k be any two positive integers with $k \leq n$. Let G be a graph in B'_n with bipartition V_0 and V_1. There exists a spanning (u, S)-fan in G for any vertex u in V_i and any vertex subset S with $|S| \leq n$ such that $|S \cap V_{1-i}| = 1$ with $i \in \{0, 1\}$.

Proof: We prove this statement by induction on n. Let $G = G_0 \oplus G_1$ in B'_n such that V_0^i and V_1^i are the bipartition of G_i for every $i = 0, 1$. Without loss of generality, we assume that $V_0^0 \cup V_0^1$ and $V_1^0 \cup V_1^1$ form the bipartition of G. Let u be any vertex in $V_0^0 \cup V_1^0$ and $S = \{v_1, v_2, \ldots, v_k\}$ be any vertex subset in $G - \{u\}$ with v_1 being the unique vertex in $(V_1^0 \cup V_1^1) \cap S$. Without loss of generality, we assume that $u \in V_0^0$. By Lemma 14.12, this statement holds for $k = 1$. Thus, we assume that $k = 2$ and $n \geq 2$. By Lemma 14.12, there is a Hamiltonian path P of G joining v_1 to v_2. Without loss of generality, we write P as $\langle v_1, P_1, u, P_2, v_2 \rangle$. Then $\{P_1, P_2\}$ forms the spanning (u, S)-fan of G. Thus, this statement holds for $k = 2$. Moreover, this statement holds for $n = 2$. We assume that $3 \leq k \leq n$. Suppose that this statement holds for B'_{n-1}, and $G_i \in B'_{n-1}$ for $i = 0$ and 1. Without loss of generality, we assume that $u \in G_0$. Let $T = S - \{v_1\}$. We have the following cases:

Case 1. $|T \cap V_0^0| = |T|$. Then $v_i \in V_0^0$ for every i, $2 \leq i \leq k$.

Case 1.1. $v_1 \in V_0^1$. Let $H = S - \{v_k\}$. Obviously, $H \subset G_0$, $|H \cap V_1^0| = 1$, and $|H| = k - 1$. By induction, there is a spanning (u, H)-fan $\{P_1, P_2, \ldots, P_{k-1}\}$ of G_0. Without loss of generality, we assume that P_i is joining u to v_i for every i, $1 \leq i \leq k - 1$.

Suppose that $v_k \in V(P_1)$. Without loss of generality, we write P_1 as $\langle u, Q_1, v_k, x, Q_2, v_1 \rangle$. Since $v_k \in V_0^0$, $x \in V_1^0$. (Note that $x = v_1$ if $l(Q_2) = 0$.) By Lemma 14.12, there is a Hamiltonian path R of G_1 joining vertex $\overline{u} \in V_1^1$ to vertex $\overline{x} \in V_0^1$. We set $W_1 = \langle u, \overline{u}, R, \overline{x}, x, Q_2, v_1 \rangle$, $W_i = P_i$ for every i, $2 \leq i \leq k - 1$, and $W_k = \langle u, Q_1, v_k \rangle$. Then $\{W_1, W_2, \ldots, W_k\}$ is the spanning (u, S)-fan of G. See Figure 14.4a for illustration, where $k = 6$.

Suppose that $v_k \in V(P_i)$ for some $2 \leq i \leq k - 1$. Without loss of generality, we assume that $v_k \in V(P_{k-1})$ and we write P_{k-1} as $\langle u, Q_1, v_k, x, Q_2, v_{k-1} \rangle$. Since $v_k \in V_0^0$, $x \in V_1^0$. By Lemma 14.12, there is a Hamiltonian path R of G_1 joining vertex $\overline{u} \in V_1^1$ to vertex $\overline{x} \in V_0^1$. We set $W_i = P_i$ for every $i \in \langle k - 2 \rangle$, $W_{k-1} = \langle u, \overline{u}, R, \overline{x}, x, Q_2, v_{k-1} \rangle$,

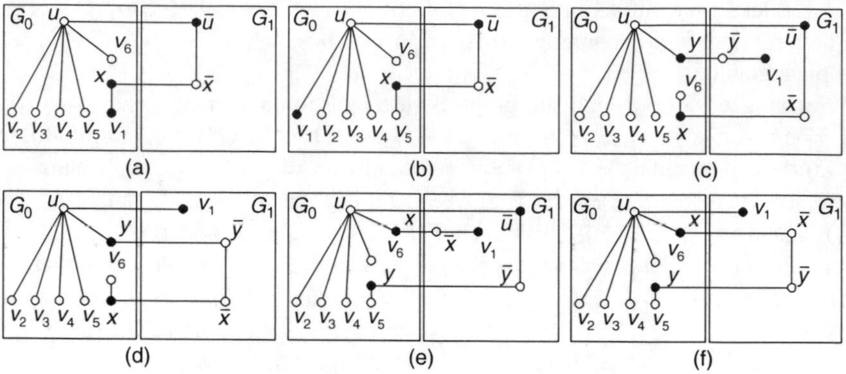

FIGURE 14.4 Illustrations for Case 1 of Theorem 14.6.

and $W_k = \langle u, Q_1, v_k\rangle$. Then $\{W_1, W_2, \ldots, W_k\}$ is the spanning (u, S)-fan of G. See Figure 14.4b for illustration, where $k = 6$.

Case 1.2. $v_1 \in V_1^1$. We choose a vertex x in V_1^0. Let $H = (T \cup \{x\}) - \{v_k\}$. So $H \subset G_0$, $|H \cap V_1^0| = 1$, and $|H| = k - 1$. By induction, there is a spanning (u, H)-fan $\{P_1, P_2, \ldots, P_{k-1}\}$ of G_0. Without loss of generality, we assume that P_1 is joining u to x and P_i is joining u to v_i for every $2 \le i \le k-1$. We have $\bar{u} \in V_1^1$ and $\bar{x} \in V_1^0$.

Case 1.2.1. $v_k \in V(P_1)$. Without loss of generality, we write P_1 as $\langle u, Q_1, y, v_k, Q_2, x\rangle$. Since $v_k \in V_0^0$, $y \in V_1^0$ and $\bar{y} \in V_0^1$.

Suppose that $v_1 \ne \bar{u}$. By Theorem 14.4, there are two disjoint paths R_1 and R_2 in G_1 such that (1) R_1 joins $\bar{y}$ to v_1, (2) R_2 joins $\bar{u}$ to $\bar{x}$, and (3) $R_1 \cup R_2$ spans G_1. We set $W_1 = \langle u, Q_1, y, \bar{y}, R_1, v_1\rangle$, $W_i = P_i$ for every $2 \le i \le k-1$, and $W_k = \langle u, \bar{u}, R_2, \bar{x}, x, Q_2^{-1}, v_k\rangle$. Then $\{W_1, W_2, \ldots, W_k\}$ is the spanning (u, S)-fan of G. See Figure 14.4c for illustration, where $k = 6$.

Suppose that $v_1 = \bar{u}$. By Theorem 14.5, there is a Hamiltonian path R of $G_1 - \{v_1\}$ joining $\bar{y}$ to $\bar{x}$. We set $W_1 = \langle u, \bar{u} = v_1\rangle$, $W_i = P_i$ for every $2 \le i \le k-1$, and $W_k = \langle u, Q_1, y, \bar{y}, R, \bar{x}, x, Q_2^{-1}, v_k\rangle$. Then $\{W_1, W_2, \ldots, W_k\}$ is the spanning (u, S)-fan of G. See Figure 14.4d for illustration, where $k = 6$.

Case 1.2.2. $v_k \in V(P_i)$ for some $2 \le i \le k-1$. Without loss of generality, we assume that $v_k \in V(P_{k-1})$ and we write P_{k-1} as $\langle u, Q_1, v_k, y, Q_2, v_{k-1}\rangle$. Since $v_k \in V_0^0$, $y \in V_1^0$ and $\bar{y} \in V_1^0$.

Suppose that $v_1 \ne \bar{u}$. By Theorem 14.4, there are two disjoint paths R_1 and R_2 in G_1 such that (1) R_1 joins $\bar{x}$ to v_1, (2) R_2 joins $\bar{u}$ to $\bar{y}$, and (3) $R_1 \cup R_2$ spans G_1. We set $W_1 = \langle u, P_1, x, \bar{x}, R_1, v_1\rangle$, $W_i = P_i$ for every $2 \le i \le k-2$, $W_{k-1} = \langle u, \bar{u}, R_2, \bar{y}, y, Q_2, v_{k-1}\rangle$, and $W_k = \langle u, Q_1, v_k\rangle$. Then $\{W_1, W_2, \ldots, W_k\}$ is the spanning (u, S)-fan of G. See Figure 14.4e for illustration, where $k = 6$.

Suppose that $v_1 = \bar{u}$. By Theorem 14.5, there is a Hamiltonian path R of $G_1 - \{v_1\}$ joining $\bar{x}$ to $\bar{y}$. We set $W_1 = \langle u, \bar{u} = v_1\rangle$, $W_i = P_i$ for every $2 \le i \le k-2$, $W_{k-1} = \langle u, P_1, x, \bar{x}, R, \bar{y}, y, Q_2, v_{k-1}\rangle$, and $W_k = \langle u, Q_1, v_k\rangle$. Then $\{W_1, W_2, \ldots, W_k\}$ is the spanning (u, S)-fan of G. See Figure 14.4f for illustration, where $k = 6$.

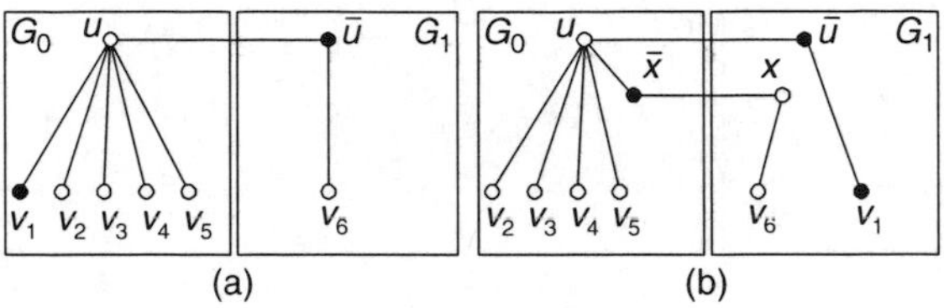

FIGURE 14.5 Illustrations for Case 2 of Theorem 14.6.

Case 2. $|T \cap V_0^1| = 1$. Without loss of generality, we assume that $v_k \in V_0^1$. We have $\overline{u} \in V_1^1$.

Case 2.1. $v_1 \in V_1^0$. Let $H = S - \{v_k\}$. Obviously, $H \subset G_0$, $|H \cap V_1^0| = 1$, and $|H| = k - 1$. By induction, there is a spanning (u, H)-fan $\{P_1, P_2, \ldots, P_{k-1}\}$ of G_0. Without loss of generality, we assume that P_i is joining u to v_i for every $1 \leq i \leq k - 1$. By Lemma 14.12, there is a Hamiltonian path R of G_1 joining $\overline{u}$ to v_k. We set $P_k = \langle u, \overline{u}, R, v_k \rangle$. Then $\{P_1, P_2, \ldots, P_k\}$ is the spanning (u, S)-fan of G. See Figure 14.5a for illustration, where $k = 6$.

Case 2.2. $v_1 \in V_1^1$. By Lemma 14.12, there is a Hamiltonian path R of G_1 joining v_1 to v_k. Without loss of generality, we write R as $\langle v_1, R_1, \overline{u}, x, R_2, v_k \rangle$. (Note that $v_1 = \overline{u}$ if $l(R_1) = 0$ and $x = v_k$ if $l(R_2) = 0$.) Since $\overline{u} \in V_1^1$, $x \in V_0^1$ and $\overline{x} \in V_1^0$. Let $H = (T \cup \{\overline{x}\}) - \{v_k\}$. Obviously, $H \subset G_0$, $|H \cap V_1^0| = 1$, and $|H| = k - 1$. By induction, there is a spanning (u, H)-fan $\{P_1, P_2, \ldots, P_{k-1}\}$ of G_0. Without loss of generality, we assume that P_1 is joining u to $\overline{x}$ and P_i is joining u to v_i for every $2 \leq i \leq k - 1$. We set $W_1 = \langle u, \overline{u}, R_1^{-1}, v_1 \rangle$, $W_i = P_i$ for every $2 \leq i \leq k - 1$, and $W_k = \langle u, P_1, \overline{x}, x, R_2, v_k \rangle$. Then $\{W_1, W_2, \ldots, W_k\}$ is the (u, S)-fan of G. See Figure 14.5b for illustration, where $k = 6$.

Case 3. $|T \cap V_0^1| = 2$. Without loss of generality, we assume that $\{v_{k-1}, v_k\} \subset V_0^1$. We have $|V_0^0| \geq n \geq k$. We can choose a vertex x in $V_0^0 - \{u, v_2, v_3, \ldots, v_{k-2}\}$. Obviously, $\{\overline{x}, \overline{u}\} \subset V_1^1$ with $\overline{x} \neq \overline{u}$. By Theorem 14.4, there are two disjoint paths R_1 and R_2 in G_1 such that (1) R_1 joins $\overline{x}$ to v_{k-1}, (2) R_2 joins $\overline{u}$ to v_k, and (3) $R_1 \cup R_2$ spans G_1.

Case 3.1. $v_1 \in V_1^0$. Let $H = (S \cup \{x\}) - \{v_{k-1}, v_k\}$. Obviously, $H \subset G_0$, $|H \cap V_1^0| = 1$, and $|H| = k - 1$. By induction, there is a spanning (u, H)-fan $\{P_1, P_2, \ldots, P_{k-1}\}$ of G_0. Without loss of generality, we assume that P_i is joining u to v_i for every $1 \leq i \leq k - 1$ and P_{k-1} is joining u to x. We set $W_i = P_i$ for every $1 \leq i \leq k - 2$, $W_{k-1} = \langle u, P_{k-1}, x, \overline{x}, R_1, v_{k-1} \rangle$, and $W_k = \langle u, \overline{u}, R_2, v_k \rangle$. Then $\{W_1, W_2, \ldots, W_k\}$ is the spanning (u, S)-fan of G. See Figure 14.6a for illustration, where $k = 6$.

Case 3.2. $v_1 \in V_1^1$ and $v_1 \in V(R_1)$. Without loss of generality, we write R_1 as $\langle \overline{x}, Q_1, v_1, y, Q_2, v_{k-1} \rangle$. Since $v_1 \in V_1^1$, $y \in V_0^1$ and $\overline{y} \in V_1^0$. Let $H = (T \cup \{x, \overline{y}\}) - \{v_{k-1}, v_k\}$. Obviously, $H \subset G_0$, $|H \cap V_1^0| = 1$, and $|H| = k - 1$. By induction, there is a spanning (u, H)-fan $\{P_1, P_2, \ldots, P_{k-1}\}$ of G_0. Without loss of generality, we assume that P_1 is joining u to x, P_i is joining u to v_i for every $i \in \langle k - 2 \rangle$, and P_{k-1} is joining u to $\overline{y}$. We set $W_1 = \langle u, P_1, x, \overline{x}, Q_1, v_1 \rangle$, $W_i = P_i$ for every $2 \leq i \leq k - 1$,

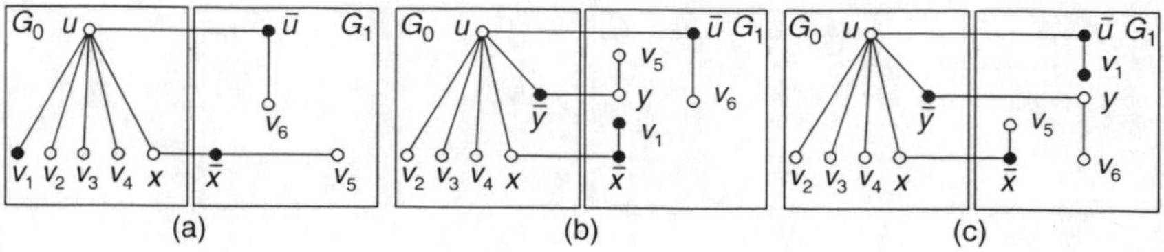

FIGURE 14.6 Illustrations for Case 3 of Theorem 14.6.

$W_{k-1}=\langle u,P_{k-1},\bar{y},y,Q_2,v_{k-1}\rangle$, and $W_k=\langle u,\bar{u},R_2,v_k\rangle$. Then $\{W_1,W_2,\ldots,W_k\}$ is the spanning (u,S)-fan of G. See Figure 14.6b for illustration, where $k=6$.

Case 3.3. $v_1\in V_1^1$ and $v_1\in V(R_2)$. Without loss of generality, we write R_2 as $\langle\bar{u},Q_1,v_1,y,Q_2,v_k\rangle$. Since $v_1\in V_1^1$, $y\in V_0^1$ and $\bar{y}\in V_1^0$. Let $H=(T\cup\{x,\bar{y}\})-\{v_{k-1},v_k\}$. Obviously, $H\subset G_0$, $|H\cap V_1^0|=1$, and $|H|=k-1$. By induction, there is a spanning (u,H)-fan $\{P_1,P_2,\ldots,P_{k-1}\}$ of G_0. Without loss of generality, we assume that P_1 is joining u to x, P_i is joining u to v_i for every $2\le i\le k-2$, and P_{k-1} is joining u to $\bar{y}$. We set $W_1=\langle u,\bar{u},Q_1,v_1\rangle$, $W_i=P_i$ for every $2\le i\le k-2$, $W_{k-1}=\langle u,P_1,x,\bar{x},R_1,v_{k-1}\rangle$, and $W_k=\langle u,P_{k-1},\bar{y},y,Q_2,v_k\rangle$. Then $\{W_1,W_2,\ldots,W_k\}$ is the spanning (u,S)-fan of G. See Figure 14.6c for illustration, where $k=6$.

Case 4. $|T\cap V_0^1|\ge 3$ and $|T\cap V_0^0|\ge 1$. We have $n\ge k=|S|\ge 5$. Without loss of generality, we assume that $A=T\cap V_0^0=\{v_2,v_3,\ldots,v_t\}$ and $B=T\cap V_0^1=\{v_{t+1},v_{t+2},\ldots,v_k\}$ for some $2\le t\le k-3$. Since $t\le k-3$ and $k\le n$, $|A|=t-1\le n-4$ and $|B|\le n-2$. Since $n\ge 5$, $(n-1)|A|+|B|\le(n-1)(n-4)+(n-2)<2^{n-2}=|V_1^1|$. Thus, we can choose a vertex x in V_1^0-B such that $\bar{v}_i\notin N_{G_1}(x)$ for every $2\le i\le t$. Since $2\le t\le k-3$ and $k\le n$, $k-t+1\le n-1$. Let $H=B\cup\{\bar{u}\}$. Obviously, $H\subset G_1$, $|H\cap V_1^1|=1$, and $|H|=k-t+1$. By induction, there is a spanning (x,H)-fan $\{P_1,P_2,\ldots,P_{k-t+1}\}$ of G_1. Without loss of generality, we assume that P_1 is joining x to $\bar{u}$ and P_i is joining x to v_{t+i-1} for every $2\le i\le k-t+1$. Moreover, we write $P_1=\langle x,x_1,R_1,\bar{u}\rangle$ and $P_i=\langle x,x_i,R_i,v_{t+i-1}\rangle$ for every $2\le i\le k-t+1$. Since $x\in V_1^0$, $x_i\in V_1^1$ and $\bar{x}_i\in V_0^0$ for every $1\le i\le k-t+1$. We set $C=\{\bar{x}_2,\bar{x}_3,\ldots,\bar{x}_{k-t}\}$.

Case 4.1. $v_1\in V_1^0$. Let $H'=A\cup C\cup\{v_1\}$. Obviously, $H'\subset G_0$, $|H'\cap V_1^0|=1$, and $|H'|=k-1$. By induction, there is a spanning (u,H')-fan $\{Q_1,Q_2,\ldots,Q_{k-1}\}$ of G_0. Without loss of generality, we assume that Q_i is joining u to v_i for every $1\le i\le t$ and Q_j is joining u to $\bar{x}_{j-t+2}$ for every $t+1\le j\le k-1$. We set $W_i=\langle u,Q_i,v_i\rangle$ for every $1\le i\le t$, $W_j=\langle u,Q_j,\bar{x}_{i-t+2},x_{i-t+2},R_{i-t+2},v_j\rangle$ for every $t+1\le j\le k-1$, and $W_k=\langle u,\bar{u},P_1^{-1},x,P_{k-t+1},v_k\rangle$. Then $\{W_1,W_2,\ldots,W_k\}$ is the spanning (u,S)-fan of G. See Figure 14.7a for illustration, where $k=6$ and $t=3$.

Case 4.2. $v_1\in V_1^1$ and $v_1\in V(P_1)$. Without loss of generality, we write P_1 as $\langle x,Z_1,y,v_1,Z_2,\bar{u}\rangle$. Since $v_1\in V_1^1$, $y\in V_0^1$ and $\bar{y}\in V_1^0$. Let $H'=A\cup C\cap\{\bar{y}\}$. Obviously, $H'\subset G_0$, $|H'\cap V_1^0|=1$, and $|H'|=k-1$. By induction, there is a spanning (u,H')-fan $\{Q_1,Q_2,\ldots,Q_{k-1}\}$ of G_0. Without loss of generality, we assume that Q_1 is joining u to $\bar{y}$, Q_i is joining u to v_i for every $2\le i\le t$, and Q_j is joining u to $\bar{x}_{j-t+2}$ for every $t+1\le j\le k-1$. We set $W_1=\langle u,\bar{u},Z_2^{-1},v_1\rangle$, $W_i=\langle u,Q_i,v_i\rangle$

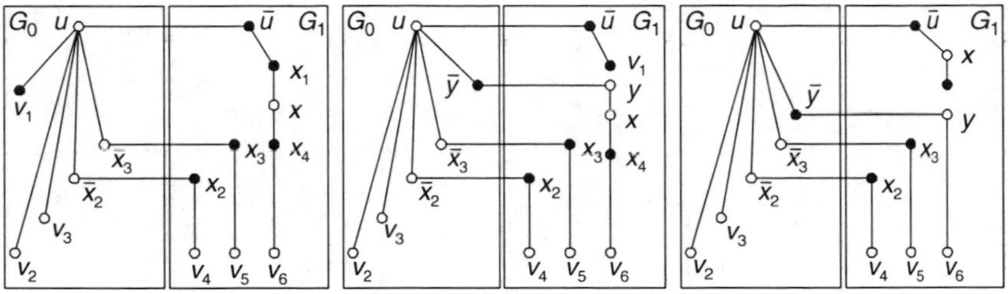

FIGURE 14.7 Illustrations for Case 4 of Theorem 14.6.

for every $2 \le i \le t$, $W_j = \langle u, Q_j, \overline{x}_{i-t+2}, x_{i-t+2}, R_{i-t+2}, v_j \rangle$ for every $t+1 \le j \le k-1$, and $W_k = \langle u, Q_1, \overline{y}, y, Z_1^{-1}, x, P_{k-t+1}, v_k \rangle$. Then $\{W_1, W_2, \ldots, W_k\}$ is the spanning (u, S)-fan of G. See Figure 14.7b for illustration, where $k = 6$ and $t = 3$.

Case 4.3. $v_1 \in V_1^1$ and $v_1 \in V(P_i)$ for some $2 \le i \le k-t+1$. Without loss of generality, we assume that $v_1 \in V(P_{k-t+1})$ and we write P_{k-t+1} as $\langle x, Z_1, v_1, y, Z_2, v_k \rangle$. Since $v_1 \in V_1^1$, $y \in V_0^1$ and $\overline{y} \in V_1^0$. Let $H' = A \cup C \cup \{\overline{y}\}$. Obviously, $H' \subset G_0$, $|H' \cap V_1^0| = 1$, and $|H'| = k-1$. By induction, there is a spanning (u, H')-fan $\{Q_1, Q_2, \ldots, Q_{k-1}\}$ of G_0. Without loss of generality, we assume that Q_1 is joining u to $\overline{y}$, Q_i is joining u to v_i for every $2 \le i \le t$, and Q_j is joining u to $\overline{x}_{j-t+2}$ for every $t+1 \le j \le k-1$. We set $W_1 = \langle u, \overline{u}, P_1^{-1}, x, Z_1, v_1 \rangle$, $W_i = \langle u, Q_i, v_i \rangle$ for every $2 \le i \le t$, $W_j = \langle u, Q_j, \overline{x}_{i-t+2}, x_{i-t+2}, R_{i-t+2}, v_j \rangle$ for every $t+1 \le j \le k-1$, and $W_k = \langle u, Q_1, y, y, Z_2, v_k \rangle$. Then $\{W_1, W_2, \ldots, W_k\}$ forms the spanning (u, S)-fan of G. See Figure 14.7c for illustration, where $k = 6$ and $t = 3$.

Case 5. $|T \cap V_0^1| = |T| \ge 3$. Let $H = (T \cup \{\overline{u}\}) - \{v_k\}$. Obviously, $H \subset G_1$, $|H \cap V_1^1| = 1$, and $|H| = k-1$. By induction, there is a spanning (v_k, H)-fan $\{P_1, P_2, \ldots, P_{k-1}\}$ of G_1. Without loss of generality, we assume that P_1 is joining v_k to $\overline{u}$ and P_i is joining v_k to v_i for every $2 \le i \le k-1$. Without loss of generality, we write $P_1 = \langle v_k, x_1, R_1, \overline{u} \rangle$ and write $P_i = \langle v_k, x_i, R_i, v_i \rangle$ for every $2 \le i \le k-1$. Since $v_k \in V_1^0$, $x_i \in V_1^1$ and $\overline{x}_i \in V_0^0$ for every $1 \le i \le k-1$. We set $C = \{\overline{x}_2, \overline{x}_3, \ldots, \overline{x}_{k-t}\}$.

Case 5.1. $v_1 \in V_1^0$. Let $H' = C \cup \{v_1\}$. Obviously, $H' \subset G_0$, $|H' \cap V_1^0| = 1$, and $|H'| = k-1$. By induction, there is a spanning (u, H')-fan $\{Q_1, Q_2, \ldots, Q_{k-1}\}$ of G_0. Without loss of generality, we assume that Q_1 is joining u to v_1 and Q_i is joining u to $\overline{x}_i$ for every $2 \le i \le k-1$. We set $W_1 = Q_1$, $W_i = \langle u, Q_i, \overline{x}_i, x_i, R_i, v_i \rangle$ for every $2 \le i \le k-1$, and $W_k = \langle u, \overline{u}, P_1^{-1}, v_k \rangle$. Then $\{W_1, W_2, \ldots, W_k\}$ is the spanning (u, S)-fan of G. See Figure 14.8a for illustration, where $k = 6$.

Case 5.2. $v_1 \in V_1^1$ and $v_1 \in V(P_1)$. Without loss of generality, we write $P_1 = \langle v_k, Z_1, y, v_1, Z_2, \overline{u} \rangle$. Since $v_1 \in V_1^1$, $y \in V_0^1$ and $\overline{y} \in V_1^0$. Let $H' = C \cup \{\overline{y}\}$. Obviously, $H' \subset G_0$, $|H' \cap V_0^1| = 1$, and $|H'| = k-1$. By induction, there is a spanning (u, H')-fan $\{Q_1, Q_2, \ldots, Q_{k-1}\}$ of G_0. Without loss of generality, we assume that Q_1 is joining u to $\overline{y}$ and Q_i is joining u to $\overline{x}_i$ for every $2 \le i \le k-1$. We set $W_1 = \langle u, \overline{u}, Z_2^{-1}, v_1 \rangle$, $W_i = \langle u, Q_i, \overline{x}_i, x_i, R_i, v_i \rangle$ for every $2 \le i \le k-1$, and

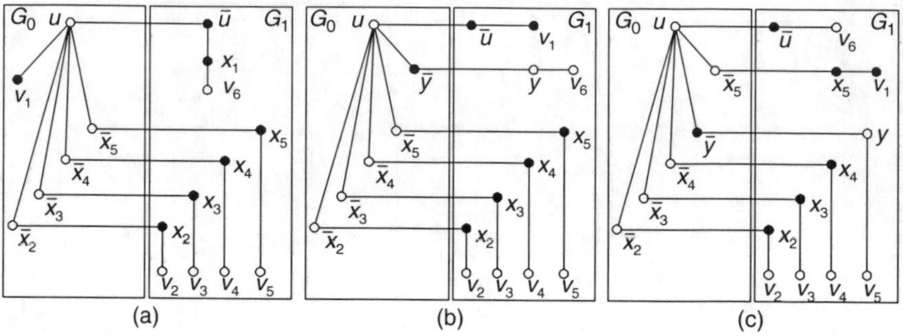

FIGURE 14.8 Illustrations for Case 5 of Theorem 14.6 .

$W_k = \langle u, Q_1, \bar{y}, y, Z_1^{-1}, v_k\rangle$. Then $\{W_1, W_2, \ldots, W_k\}$ is the spanning (u, S)-fan of G. See Figure 14.8b for illustration, where $k = 6$.

Case 5.3. $v_1 \in V_1^1$ and $v_1 \in V(P_i)$ for some $2 \le i \le k-1$. Without loss of generality, we assume that $v_1 \in V(P_{k-1})$ and write $P_{k-1} = \langle v_k, x_{k-1}, Z_1, v_1, y, Z_2, v_{k-1}\rangle$. Since $v_1 \in V_1^1$, $y \in V_0^1$ and $\bar{y} \in V_1^0$. Let $H' = C \cup \{\bar{y}\}$. Obviously, $H' \subset G_0$, $|H' \cap V_1^0| = 1$, and $|H'| = k-1$. By induction, there is a (u, H')-fan $\{Q_1, Q_2, \ldots, Q_{k-1}\}$ of G_0. Without loss of generality, we assume that Q_1 is joining u to $\bar{y}$ and Q_i is joining u to $\bar{x}_i$ for every $2 \le i \le k-1$. We set $W_1 = \langle u, Q_{k-1}, \bar{x}_{k-1}, x_{k-1}, Z_1, v_1\rangle$, $W_i = \langle u, Q_i, \bar{x}_i, x_i, R_i, v_i\rangle$ for every $2 \le i \le k-2$, $W_{k-1} = \langle u, Q_1, \bar{y}, y, Z_2, v_{k-1}\rangle$, and $W_k = \langle u, \bar{u}, P_1^{-1}, v_k\rangle$. Then $\{W_1, W_2, \ldots, W_k\}$ is the spanning (u, S)-fan of G. See Figure 14.8c for illustration, where $k = 6$. □

THEOREM 14.7 Every graph in B_n' is super spanning laceable for $n \ge 1$.

Proof: Suppose that $G = G_0 \oplus G_1$ in B_n' with bipartition V_0 and V_1. Let u be any vertex in V_0 and v be any vertex in V_1. We need to show that there is a k^*-container of G between u and v for every positive integer k with $k \le n$. By Lemma 14.12, there is a 1^*-container of G joining u to v. Thus, we assume that $k \ge 2$ and $n \ge 2$. Since $k \le n$ and $|N_G(v)| = n$, we can choose $(k-1)$ distinct vertices $x_1, x_2, \ldots, x_{k-1}$ in $N_G(v) - \{u\}$. Since v is in V_1, x_i is in $V_0 - \{u\}$ for $i = 1$ to $k-1$. We set $S = \{v, x_1, x_2, \ldots, x_{k-1}\}$. By Theorem 14.6, there is a spanning (u, S)-fan $\{R_1, R_2, \ldots, R_k\}$ of G. Without loss of generality, we assume that R_1 is joining u to v and R_i is joining u to x_{i-1} for every $2 \le i \le k$. We set $P_1 = R_1$ and $P_i = \langle u, R_i, x_{i-1}, v\rangle$ for every $2 \le i \le k$. Then $\{P_1, P_2, \ldots, P_k\}$ is the k^*-container of G between u and v. □

Let $n \ge 2$. Let $G = G_0 \oplus G_1 \in N_{n+1}'$ with $G_0 \in H_n'$ and $G_1 \in H_n'$. Depending on whether G_0 and G_1 is bipartite, we prove that $G = G_0 \oplus G_1$ is 3^*-connected with the following lemmas.

LEMMA 14.13 According to isomorphism, there is only one graph in N_3'. Moreover, this graph is 3^*-connected.

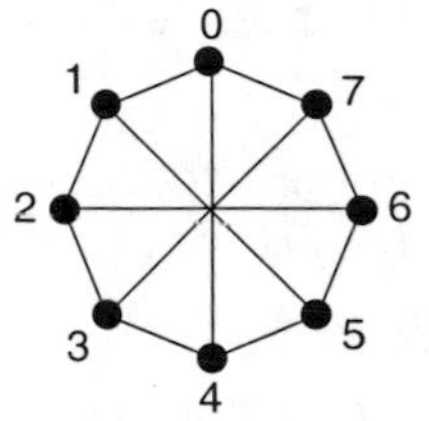

FIGURE 14.9 The only graph T in N'_3.

Proof: By brute force, we can check whether the graph T in Figure 14.9 is the only graph in N'_3.

Let x and y be two distinct vertices of T. By the symmetric of T, we can assume that $x = 0$ and $y \in \{1, 2, 3, 4\}$. The 3*-containers $\{P_1, P_2, P_3\}$ of T between x and y are listed here:

$y = 1$	$\{P_1 = \langle 0, 1\rangle, P_2 = \langle 0, 4, 3, 2, 1\rangle, P_3 = \langle 0, 7, 6, 5, 1\rangle\}$
$y = 2$	$\{P_1 = \langle 0, 1, 2\rangle, P_2 = \langle 0, 7, 3, 2\rangle, P_3 = \langle 0, 4, 5, 6, 2\rangle\}$
$y = 3$	$\{P_1 = \langle 0, 4, 3\rangle, P_2 = \langle 0, 7, 3\rangle, P_3 = \langle 0, 1, 5, 6, 2, 3\rangle\}$
$y = 4$	$\{P_1 = \langle 0, 4\rangle, P_2 = \langle 0, 1, 2, 3, 4\rangle, P_3 = \langle 0, 7, 6, 5, 4\rangle\}$

Thus, T is 3*-connected. □

LEMMA 14.14 Let $n \geq 3$. Assume that $G = G_0 \oplus G_1$ in N'_{n+1} with both G_0 and G_1 in N'_n. Then G is 3*-connected.

Proof: Let u and v be any two distinct vertices of G. We need to construct a 3*-container of G between u and v.

Case 1. $u, v \in G_0$. By Lemma 14.11, there is a 2*-container $\{P_1, P_2\}$ of G_0 between u and v. By Lemma 14.11 again, there is a Hamiltonian path P of G_1 joining $\overline{u}$ to $\overline{v}$. We set P_3 as $\langle u, \overline{u}, P, \overline{v}, v\rangle$. Then $\{P_1, P_2, P_3\}$ is the 3*-container of G between u and v.

Case 2. $u \in G_0$ and $v \in G_1$ with $\overline{u} = v$. Since there are 2^n vertices in G_0 and $2^n > 3$ for $n \geq 3$, we can choose two distinct vertices x and y in $G_0 - \{u\}$. By Lemma 14.11, there is a Hamiltonian path R of G_0 joining x to y. Again, there is a Hamiltonian path W of G_1 joining $\overline{x}$ to $\overline{y}$. We write $R = \langle x, R_1, u, R_2, y\rangle$ and $W = \langle \overline{x}, W_1, v, W_2, \overline{y}\rangle$. We set $P_1 = \langle u, R_1^{-1}, x, \overline{x}, W_1, v\rangle$, $P_2 = \langle u, R_2, y, \overline{y}, W_2^{-1}, v\rangle$, and $P_3 = \langle u, v\rangle$. Then $\{P_1, P_2, P_3\}$ is the 3*-container of G between u and v.

Case 3. $u \in G_0$ and $v \in G_1$ with $\overline{u} \neq v$. Since there are 2^n vertices in G_0, we choose a vertex x in $G_0 - \{u, \overline{v}\}$. By Lemma 14.11, there is a Hamiltonian path R of G_0 joining x to $\overline{v}$. Again, there is a Hamiltonian path W of G_1 joining $\overline{x}$ to $\overline{u}$. We write $R = \langle x, R_1, u, R_2, \overline{v}\rangle$ and $W = \langle \overline{x}, W_1, v, W_2, \overline{u}\rangle$. We set $P_1 = \langle u, \overline{u}, W_2^{-1}, v\rangle$, $P_2 = \langle u, R_1^{-1}, x, \overline{x}, W_1, v\rangle$, and $P_3 = \langle u, R_2, \overline{v}, v\rangle$. Then $\{P_1, P_2, P_3\}$ is the 3*-container of G between u and v. □

LEMMA 14.15 Let $n \geq 3$. Assume that $G = G_0 \oplus G_1$ in N'_{n+1} with G_0 in B'_n and G_1 in N'_n. Then G is 3^*-connected.

Proof: Let V_0 and V_1 be the bipartition of G_0. Let u and v be any two distinct vertices of G. We need to construct a 3^*-container of G between u and v.

Case 1. $u, v \in G_0$. By Lemma 14.12, there is a 2^*-container $\{P_1, P_2\}$ of G_0 between u and v. By Lemma 14.11, there is a Hamiltonian path P of G_1 joining $\overline{u}$ to $\overline{v}$. We set $P_3 = \langle u, \overline{u}, P, \overline{v}, v\rangle$. Then $\{P_1, P_2, P_3\}$ is the 3^*-container of G between u and v.

Case 2. $u, v \in G_1$. Without loss of generality, we assume that $\overline{u} \in V_0$.

Case 2.1. $\overline{v} \in V_0$. Since there are 2^{n-1} vertices in V_1 and $2^{n-1} \geq 4$ for $n \geq 3$, we can choose two distinct vertices x and y in V_1. By Lemma 14.11, there is a Hamiltonian path R of G_1 joining $\overline{x}$ to $\overline{y}$. Without loss of generality, we write $R = \langle \overline{x}, R_1, u, R_2, v, R_3, \overline{y}\rangle$. By Theorem 14.4, there are two disjoint paths T_1 and T_2 of G_0 such that (1) T_1 joins $\overline{u}$ to y, (2) T_2 joins x to $\overline{v}$, and (3) $T_1 \cup T_2$ spans G_1. We set $P_1 = \langle u, R_2, v\rangle$, $P_2 = \langle u, R_1^{-1}, \overline{x}, x, T_2, \overline{v}, v\rangle$, and $P_3 = \langle u, \overline{u}, T_1, y, \overline{y}, R_3^{-1}, v\rangle$. Then $\{P_1, P_2, P_3\}$ is the 3^*-container of G between u and v.

Case 2.2. $\overline{v} \in V_1$. By Lemma 14.11, there is a 2^*-container $\{P_1, P_2\}$ of G_1 between u and v. By Lemma 14.12, there is a Hamiltonian path P of G_0 joining $\overline{u}$ to $\overline{v}$. We set $P_3 = \langle u, \overline{u}, P, \overline{v}, v\rangle$. Then $\{P_1, P_2, P_3\}$ is the 3^*-container of G between u and v.

Case 3. $u \in G_0$ and $v \in G_1$ with $\overline{u} \neq v$. By Lemma 14.12, there is a Hamiltonian cycle C of G_0. Without loss of generality, we write $C = \langle u, R_1, \overline{v}, x, R_2, u\rangle$. By Lemma 14.11, there is a Hamiltonian path T of G_1 joining $\overline{u}$ to $\overline{x}$. Without loss of generality, we write $T = \langle \overline{u}, T_1, v, T_2, \overline{x}\rangle$. We set $P_1 = \langle u, R_1, \overline{v}, v\rangle$, $P_2 = \langle u, \overline{u}, T_1, v\rangle$, and $P_3 = \langle u, R_2^{-1}, x, \overline{x}, T_2^{-1}, v\rangle$. Then $\{P_1, P_2, P_3\}$ is the 3^*-container of G between u and v.

Case 4. $u \in G_0$ and $v \in G_1$ with $\overline{u} = v$. Without loss of generality, we assume that $u \in V_0$. We can choose a vertex x in $V_0 - \{u\}$ and a vertex y in V_1. By Lemma 14.12, there is a Hamiltonian path R of G_0 joining x to y. By Lemma 14.11, there is a Hamiltonian path T of G_1 joining $\overline{x}$ to $\overline{y}$. Without loss of generality, we write $R = \langle x, R_1, u, R_2, y\rangle$ and $T = \langle \overline{x}, T_1, v, T_2, \overline{y}\rangle$. We set $P_1 = \langle u, v\rangle$, $P_2 = \langle u, R_1^{-1}, x, \overline{x}, T_1, v\rangle$, and $P_3 = \langle u, R_2, y, \overline{y}, T_2^{-1}, v\rangle$. Then $\{P_1, P_2, P_3\}$ is the 3^*-container of G between u and v. □

LEMMA 14.16 Assume that $G = G_0 \oplus G_1$ in N'_{n+1} with both G_0 and G_1 in B'_n for $n \geq 2$. Then G is 3^*-connected.

Proof: Let V_0^i and V_1^i be the bipartition of G_i for $i = 0, 1$. Let u and v be two distinct vertices of G. Without loss of generality, we assume that $u \in V_0^0$ and $\overline{u} \in V_1^1$. We need to construct a 3^*-container of G between u and v.

Case 1. $v \in V_0^0 \cup V_1^0$ and $\overline{v} \in V_0^1$. By Lemma 14.12, there is a 2^*-container $\{P_1, P_2\}$ of G_0 between u and v. By Lemma 14.12, there is a Hamiltonian path P of G_1 joining $\overline{u}$ to $\overline{v}$. We set $P_3 = \langle u, \overline{u}, P, \overline{v}, v\rangle$. Then $\{P_1, P_2, P_3\}$ is the 3^*-container of G between u and v.

Case 2. $v \in V_0^0$ and $\overline{v} \in V_1^1$. Since $u \in V_0^0, \overline{u} \in V_1^1$, $v \in V_0^0$, and $\overline{v} \in V_1^1$, we can choose a vertex x in V_1^0 such that $\overline{x} \in V_0^1$ and choose a vertex y in V_0^0 such that $\overline{y} \in V_0^1$. By Lemma 14.12, there is a Hamiltonian path R of G_0 joining x to y. Without loss of generality, we write $R = \langle x, R_1, p, R_2, q, R_3, y \rangle$ where $\{p, q\} = \{u, v\}$. Without loss of generality, we assume that $p = u$ and $q = v$. By Theorem 14.4, there are two disjoint paths T_1 and T_2 of G_1 such that (1) T_1 joins $\overline{x}$ to $\overline{v}$, (2) T_2 joins $\overline{u}$ to $\overline{y}$, and (3) $T_1 \cup T_2$ spans G_1. We set $P_1 = \langle u, R_2, v \rangle$, $P_2 = \langle u, R_1^{-1}, x, \overline{x}, T_1, \overline{v}, v \rangle$, and $P_3 = \langle u, \overline{u}, T_2, \overline{y}, y, R_3^{-1}, v \rangle$. Then $\{P_1, P_2, P_3\}$ is the 3*-container of G between u and v.

Case 3. $v \in V_1^0$ and $\overline{v} \in V_1^1$. Since $u \in V_0^0$ and $\overline{u} \in V_1^1$, $v \in V_1^0$, and $\overline{v} \in V_1^1$, we can choose a vertex x in V_1^0 such that $\overline{x} \in V_0^1$ and choose a vertex y in V_0^0 such that $\overline{y} \in V_0^1$. By Lemma 14.12, there is a Hamiltonian path R of G_0 joining x to y. Without loss of generality, we write $R = \langle x, R_1, p, R_2, q, R_3, y \rangle$ where $\{p, q\} = \{u, v\}$. Without loss of generality, we assume that $p = u$ and $q = v$. By Theorem 14.4, there are two disjoint paths T_1 and T_2 of G_1 such that (1) T_1 joins $\overline{x}$ to $\overline{v}$, (2) T_2 joins $\overline{u}$ to $\overline{y}$, and (3) $T_1 \cup T_2$ spans G_1. We set $P_1 = \langle u, R_2, v \rangle$, $P_2 = \langle u, R_1^{-1}, x, \overline{x}, T_1, \overline{v}, v \rangle$, and $P_3 = \langle u, \overline{u}, T_2, \overline{y}, y, R_3^{-1}, v \rangle$. Then $\{P_1, P_2, P_3\}$ is the 3*-container of G between u and v.

Case 4. $v \in V_0^1 \cup V_1^1$ and $\overline{u} \neq v$.

Case 4.1. $\overline{v} \in V_0^0$. Since $u \in V_0^0, \overline{u} \in V_1^1$, and $\overline{v} \in V_0^0$, we can choose a vertex $x \in V_1^0$ such that $\overline{x} \in V_0^1$. By Lemma 14.12, there is a Hamiltonian path R of G_0 joining x to $\overline{v}$. Again, by Lemma 14.12, there is a Hamiltonian path T of G_1 joining $\overline{x}$ to $\overline{u}$. Write $R = \langle x, R_1, u, R_2, v \rangle$ and $T = \langle \overline{x}, T_1, v, T_2, \overline{u} \rangle$. We set $P_1 = \langle u, \overline{u}, T_2^{-1}, v \rangle$, $P_2 = \langle u, R_2, \overline{v}, v \rangle$, and $P_3 = \langle u, R_1^{-1}, x, \overline{x}, T_1, v \rangle$. Then $\{P_1, P_2, P_3\}$ is the 3*-container of G between u and v.

Case 4.2. $\overline{v} \in V_1^0$ and $v \in V_0^1$. Since $u \in V_0^0$, $\overline{u} \in V_1^1$, $v \in V_0^1$, and $\overline{v} \in V_1^0$, we can choose a vertex $x \in V_0^0$ such that $\overline{x} \in V_0^1$. By Lemma 14.12, there is a Hamiltonian path R of G_0 joining x to $\overline{v}$, and there is a Hamiltonian path T of G_1 joining $\overline{x}$ to $\overline{u}$. We write $R = \langle x, R_1, u, R_2, \overline{v} \rangle$ and $T = \langle \overline{x}, T_1, v, T_2, \overline{u} \rangle$. We set $P_1 = \langle u, \overline{u}, T_2^{-1}, v \rangle$, $P_2 = \langle u, R_2, \overline{v}, v \rangle$, and $P_3 = \langle u, R_1^{-1}, x, \overline{x}, T_1, v \rangle$. Then $\{P_1, P_2, P_3\}$ is the 3*-container of G between u and v.

Case 4.3. $\overline{v} \in V_1^0$ and $v \in V_1^1$. Since $u \in V_0^0$, $\overline{u} \in V_1^1$, and $v \in V_1^1$, we can choose a vertex $x \in V_0^0$ such that $\overline{x} \in V_0^1$. By Lemma 14.12, there is a Hamiltonian path R of G_0 joining x to $\overline{v}$, and there is a Hamiltonian path T of G_1 joining $\overline{x}$ to $\overline{u}$. We write $R = \langle x, R_1, u, R_2, \overline{v} \rangle$ and $T = \langle \overline{x}, T_1, v, T_2, \overline{u} \rangle$. We set $P_1 = \langle u, \overline{u}, T_2^{-1}, v \rangle$, $P_2 = \langle u, R_2, \overline{v}, v \rangle$, and $P_3 = \langle u, R_1^{-1}, x, \overline{x}, T_1, v \rangle$. Then $\{P_1, P_2, P_3\}$ is the 3*-container of G between u and v.

Case 5. $v = \overline{u}$. Since $u \in V_0^0$ and $\overline{u} \in V_1^1$, we can choose a vertex $x \in V_0^0$ such that $\overline{x} \in V_0^1$ and choose a vertex $y \in V_1^0$ such that $\overline{y} \in V_1^1$. By Lemma 14.12, there is a Hamiltonian path R of G_0 joining x to y, and there is a Hamiltonian path T of G_1 joining $\overline{x}$ to $\overline{y}$. Without loss of generality, we write that $R = \langle x, R_1, u, R_2, y \rangle$ and $T = \langle \overline{x}, T_1, v, T_2, \overline{y} \rangle$. We set $P_1 = \langle u, v \rangle$, $P_2 = \langle u, R_1^{-1}, x, \overline{x}, T_1, v \rangle$, and $P_is = \langle u, R_2, y, \overline{y}, T_2^{-1}, v \rangle$. Then $\{P_1, P_2, P_3\}$ forms the 3*-container of G between u and v. □

With Lemmas 14.13 through 14.16, we have the following theorem.

THEOREM 14.8 Every graph in N_n' is 3*-connected.

Now, we present an N_n'-graph H that is not 4*-connected. Note that Q_n is a bipartite graph with bipartition $\{u \mid w(u) \text{ is even}\}$ and $\{u \mid w(u) \text{ is odd}\}$. Let Q_n^i be the subgraph of Q_n induced by $\{u \in V(Q_n) \mid (u)_n = i\}$ for $i \in \{0, 1\}$. Then Q_n^i is isomorphic to Q_{n-1}. By the definition of Q_n, $Q_n \in B_n'$. Let $n \geq 4$ and let $e = \underbrace{00\ldots0}_{n}$ be a vertex in Q_n. We set $v = (e)^1$, $p = (e)^n$, and $q = ((e)^1)^n$.

Let H be the graph with $V(H) = V(Q_n)$ and $E(H) = (E(Q_n) - \{(e,p),(v,q)\}) \cup \{(e,q),(v,p)\}$. Obviously, $H - \{(e,q),(v,p)\}$ is a bipartite graph with bipartition $A = \{x \mid w(x) \text{ is even}\}$ and $B = \{x \mid w(x) \text{ is odd}\}$. Moreover, H is in N_n' and $H = G(Q_n^0; Q_n^1; M)$ for some perfect matching M. We will show that H is not k^*-connected for $k \geq 4$.

Suppose that there is a k^*-container $C = \{P_1, P_2, \ldots, P_k\}$ of H between e and q for some $k \geq 4$. We have the following cases:

Case 1. $(e,q) \in \bigcup_{i=1}^k P_i$ and $(v,p) \in \bigcup_{i=1}^k P_i$. Without loss of generality, we assume that $(e,q) \in P_1$. Thus, $P_1 = \langle e, q \rangle$. Again, we can assume without loss of generality that $(v,p) \in P_2$. Obviously, the number of vertices in P_2 is $2t_2$ for some integer t_2 and the number of vertices in P_i is $2t_i + 1$ for some integer t_i for every $3 \leq i \leq n$. Therefore, there are t_2 vertices of $V(P_1) \cap B$ and $(t_2 - 2)$ vertices of $V(P_1) \cap A$ other than e and q, and there are t_i vertices of $V(P_i) \cap B$ and $(t_i - 1)$ vertices of $V(P_i) \cap A$ other than e and q for every $3 \leq i \leq k$. As a consequence, $|A| = \sum_{i=2}^k t_i + 2 - k$ and $|B| = \sum_{i=2}^k t_i$. Thus, $|A| \neq |B|$.

Case 2. $(e,q) \in \bigcup_{i=1}^k P_i$ and $(v,p) \notin \bigcup_{i=1}^k P_i$. Without loss of generality, we assume that $(e,q) \in P_1$. Obviously, the number of vertices in P_i is $(2t_i + 1)$ for some integer t_i for every $2 \leq i \leq k$. Moreover, there are t_i vertices of $V(P_i) \cap B$, and $(t_i - 1)$ vertices of $V(P_i) \cap A$ other than e and q for every $2 \leq i \leq k$. As a consequence, $|A| = \sum_{i=2}^k t_i + 3 - k$ and $|B| = \sum_{i=2}^k t_i$. Thus, $|A| \neq |B|$.

Case 3. $(e,q) \notin \bigcup_{i=1}^k P_i$ and $(v,p) \in \bigcup_{i=1}^k P_i$. Without loss of generality, we assume that $(v,p) \in P_1$. Obviously, the number of vertices in P_1 is $2t_1$ for some integer t_1, and the number of vertices in P_i is $(2t_i + 1)$ for some integer t_i for every $2 \leq i \leq k$. Moreover, there are t_1 vertices of $V(P_1) \cap B$ and $(t_1 - 2)$ vertices of $V(P_1) \cap A$ other than e and q, and there are t_i vertices of $V(P_i) \cap B$ and $(t_i - 1)$ vertices of $V(P_i) \cap A$ other than e and q for every $2 \leq i \leq k$. As a consequence, $|A| = \sum_{i=1}^k t_i + 1 - k$ and $|B| = \sum_{i=1}^k t_i$. Thus, $|A| \neq |B|$.

Case 4. $(e,q) \notin \bigcup_{i=1}^k P_i$ and $(v,p) \notin \bigcup_{i=1}^k P_i$. Obviously, the number of vertices in P_i is $(2t_i + 1)$ for some integer t_i for every $1 \leq i \leq k$. Moreover, there are t_i vertices of $V(P_i) \cap B$, and $(t_i - 1)$ vertices of $V(P_i) \cap A$ other than e and q for every $1 \leq i \leq k$. As a consequence, $|A| = \sum_{i=1}^k t_i + 2 - k$ and $|B| = \sum_{i=1}^k t_i$. Thus, $|A| \neq |B|$.

With Cases 1 through 4, C is not a k^*-container of H between e and q. Thus, H is not k^*-connected for any k, where $4 \leq k \leq n$.

14.4 SPANNING CONNECTIVITY OF CROSSED CUBES

In Section 14.3, we have seen an N'_n-graph H that is not super spanning connected. However, some N'_n-graphs are super spanning connected. Yet, Lin et al. [227] proved that the n-dimensional cross cube CQ_n is super spanning connected if $n \geq 7$.

Recall that the *n-dimension crossed cube* CQ_n is recursively defined as follows. CQ_1 is the complete graph on two vertices which labeled by 0 and 1. Let $n \geq 2$. CQ_n consists of two identical $(n-1)$-dimension crossed cubes CQ^0_{n-1} and CQ^1_{n-1}. The vertex $0u_{n-2}\dots u_0 \in V(CQ^0_{n-1})$ and the vertex $1v_{n-2}\dots v_0 \in V(CQ^1_{n-1})$ are adjacent in CQ_n if and only if

1. $u_{n-2} = v_{n-2}$ if n is even
2. $(u_{2i+1}u_{2i}, v_{2i+1}v_{2i}) \in \{(00,00),(10,10),(01,11),(11,01)\}$ for every $0 \leq i < \lfloor \frac{n-1}{2} \rfloor$

According to the definition of the n-dimensional crossed cube, two distinct vertices $\mathbf{u} = u_{n-1}u_{n-2}\dots u_0$ and $\mathbf{v} = v_{n-1}v_{n-2}\dots v_0$ are adjacent in CQ_n if and only if there is an integer i such that

1. $u_j = v_j$ if $j > i$
2. $u_i = 1 - v_i$
3. $u_{i-1} = v_{i-1}$ if i is odd
4. $(u_{2j+1}u_{2j}, v_{2j+1}v_{2j}) \in \{(00,00),(10,10),(01,11),(11,01)\}$ for every $0 \leq j < \lfloor \frac{i-1}{2} \rfloor$

Now, we define another family of graphs G_n. Later, we will prove that G_n is isomorphic to CQ_n.

Let G_n be a graph with vertex set $V = V(CQ_n)$ and two distinct vertices $\mathbf{u} = u_{n-1}u_{n-2}\dots u_0$ and $\mathbf{v} = v_{n-1}v_{n-2}\dots v_0$ are adjacent in G_n if and only if there is an integer i such that

1. $u_j = v_j$ if $j > i$
2. $u_i = 1 - v_i$
3. $u_{i-1} = v_{i-1}$ if i is odd
4. $(u_{2j+1}u_{2j}, v_{2j+1}v_{2j}) \in \{(00,00),(01,01),(10,11),(11,10)\}$ for every $0 \leq j < \lfloor \frac{i-1}{2} \rfloor$

THEOREM 14.9 CQ_n is isomorphic to G_n.

Proof: Since G_1 is isomorphic to the complete graph with two vertices, G_1 is isomorphic to CQ_1. Since G_2 is isomorphic to a cycle with four vertices, G_2 is isomorphic to CQ_2. Thus, we assume that $n \geq 3$. We set $f : V(G_n) \to V(CQ_n)$ denoted $f(x_{n-1}x_{n-2}\dots x_0) = y_{n-1}y_{n-2}\dots y_0$ with (1) $y_{n-1} = x_{n-1}$ if n is odd and (2) $y_{2j+1} = x_{2j}$ and $y_{2j} = x_{2j+1}$ for every $0 \leq j < \lceil \frac{n-1}{2} \rceil$.

We claim that f is an isomorphism from G_n to CQ_n. Obviously, f is a bijection function. Let $R_0 = \{(00,00),(10,10),(01,11),(11,01)\}$ and $R_1 = \{(00,00),(01,01),(10,11),(11,10)\}$. Let **u** and **v** be any two adjacent vertices of G_n. Thus, there is an integer i such that (1) $u_j = v_j$ if $j > i$, (2) $u_i = 1 - v_i$, (3) $u_{i-1} = v_{i-1}$ if i is odd, and (4) $(u_{2j+1}u_{2j}, v_{2j+1}v_{2j}) \in R_1$ for every $0 \le j < \lfloor \frac{i-1}{2} \rfloor$. For fix i, let $\mathbf{p} = f(\mathbf{u})$ and $\mathbf{q} = f(\mathbf{v})$. We write $\mathbf{p} = p_{n-1}p_{n-2}\dots p_0$ and $\mathbf{q} = q_{n-1}q_{n-2}\dots q_0$. We have the following cases:

Case 1. Suppose that $i = 2t$ and $n - 1 = 2s$ for some positive integers t and s.

Case 1.1. Assume that $t = s$. Obviously, $p_{2s} = u_{2s} = 1 - v_{2s} = 1 - q_{2s}$. By condition 2 of the function f, we have $(u_{2j+1}u_{2j}, v_{2j+1}v_{2j}) \in R_1$ such that $(p_{2j+1}p_{2j}, q_{2j+1}q_{2j}) \in R_0$ for every $0 \le j \le s - 1$. Thus, **p** and **q** are adjacent in CQ_n.

Case 1.2. Assume that $t < s$. Obviously, $p_j = u_j = v_j = q_j$ for every $2t + 1 \le j \le 2s$. Since $u_{2t+1} = v_{2t+1}$ and $u_{2t} \neq v_{2t}$, $p_{2t+1} \neq q_{2t+1}$ and $p_{2t} = q_{2t}$. Also, we have $(u_{2j+1}u_{2j}, v_{2j+1}v_{2j}) \in R_1$ such that $(p_{2j+1}p_{2j}, q_{2j+1}q_{2j}) \in R_0$ for every $0 \le j \le t - 1$. Thus, **p** and **q** are adjacent in CQ_n.

Case 2. Suppose that $i = 2t + 1$ and $n - 1 = 2s$ for some positive integers t and s. Obviously, $p_j = u_j = v_j = q_j$ for every $2t + 2 \le j \le 2s$. Since $u_{2t+1} \neq v_{2t+1}$, we obtain $u_{2t} = v_{2t}$, $p_{2t+1} = q_{2t+1}$, and $p_{2t} \neq q_{2t}$. Similarly, we have $(u_{2j+1}u_{2j}, v_{2j+1}v_{2j}) \in R_1$ such that $(p_{2j+1}p_{2j}, q_{2j+1}q_{2j}) \in R_0$ for every $0 \le j \le t - 1$. Thus, **p** and **q** are adjacent in CQ_n.

Case 3. Suppose that $i = 2t$ and $n - 1 = 2s + 1$ for some positive integers t and s. Obviously, $p_j = u_j = v_j = q_j$ for every $2t + 1 \le j \le 2s + 1$. Since $u_{2t+1} = v_{2t+1}$ and $u_{2t} \neq v_{2t}$, we obtain $p_{2t+1} \neq q_{2t+1}$ and $p_{2t} = q_{2t}$. Also, we have $(u_{2j+1}u_{2j}, v_{2j+1}v_{2j}) \in R_1$ such that $(p_{2j+1}p_{2j}, q_{2j+1}q_{2j}) \in R_0$ for every $0 \le j \le t - 1$. Thus, **p** and **q** are adjacent in CQ_n.

Case 4. Suppose that $i = 2t + 1$ and $n - 1 = 2s + 1$ for some positive integers t and s.

Case 4.1. Assume that $t = s$. Since $u_{2t+1} \neq v_{2t+1}$, $u_{2t} = v_{2t}$. Thus, $p_{2t+1} = q_{2t+1}$ and $p_{2t} \neq q_{2t}$. Since $(u_{2j+1}u_{2j}, v_{2j+1}v_{2j}) \in R_1$ for every $0 \le j \le s - 1$, $(p_{2j+1}p_{2j}, q_{2j+1}q_{2j}) \in R_0$ for every $0 \le j \le s - 1$. Thus, **p** and **q** are adjacent in CQ_n.

Case 4.2. Assume that $t < s$. Obviously, $p_j = u_j = v_j = q_j$ for every $2t + 1 \le j \le 2s + 1$. Since $u_{2t+1} \neq v_{2t+1}$, we have $u_{2t} = v_{2t}$, $p_{2t+1} = q_{2t+1}$, and $p_{2t} \neq q_{2t}$. Since $(u_{2j+1}u_{2j}, v_{2j+1}v_{2j}) \in R_1$ every $0 \le j \le t - 1$, we obtain $(p_{2j+1}p_{2j}, q_{2j+1}q_{2j}) \in R_0$ for every $0 \le j \le t - 1$. Thus, $(\mathbf{p}, \mathbf{q}) \in E(CQ_n)$.

Thus, G_n is isomorphic to CQ_n. □

According to Theorem 14.9, we denote the n-dimensional crossed cube CQ_n as G_n. The crossed cubes of redefined CQ_3 and CQ_4 are illustrated in Figure 14.10. Let $\mathbf{u} = u_{n-1}u_{n-2}\dots u_0$. We say that u_i is the ith coordinate of **u**, denoted by $(\mathbf{u})_i$, for $0 \le i \le n - 1$. By the redefined of CQ_n, we say **u** and **v** are adjacent by an edge in dimension $(i + 1)$ and we use $(\mathbf{u})^i$ to denote **v** for every $0 \le i \le n - 1$. It follows immediately, $((\mathbf{u})^i)^i = \mathbf{u}$. Let $n \ge 3$ and i and j be two integers with $0 \le i \le 1$ and

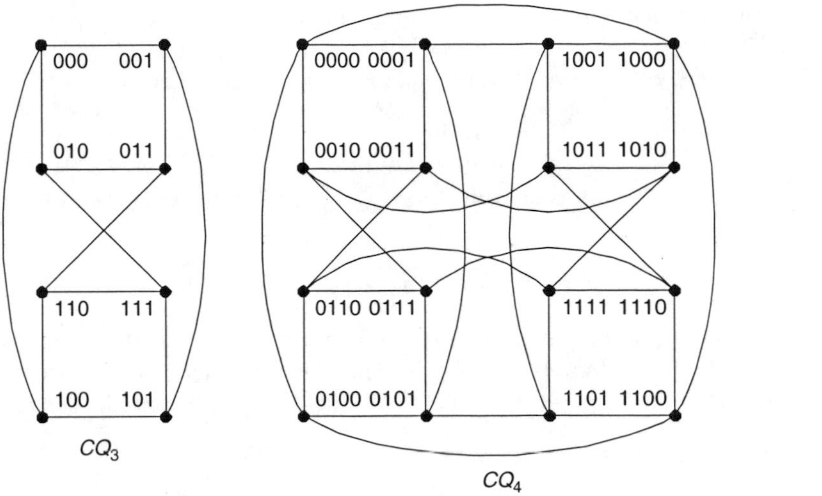

FIGURE 14.10 Illustrations for CQ_3 and CQ_4.

$0 \le j \le 1$. We set CQ^{ij}_{n-2} as a subgraph of CQ_n induced by $\{\mathbf{x} \mid \mathbf{x} \in V(CQ_n), (\mathbf{x})_n = i$ and $(\mathbf{x})_{n-1} = j\}$. Obviously, CQ^{ij}_{n-2} is isomorphic to CQ_{n-2}. Moreover, $CQ^{00}_{n-2} \cup CQ^{01}_{n-2}$, $CQ^{00}_{n-2} \cup CQ^{10}_{n-2}$, $CQ^{01}_{n-2} \cup CQ^{11}_{n-2}$, and $CQ^{10}_{n-2} \cup CQ^{11}_{n-2}$ are isomorphic to CQ_{n-1}.

A graph G is *t-fault-tolerant k^*-connected* if $G - F$ is k^*-connected for any F with $F \subset V(G) \cup E(G)$ and $|F| = t$.

The following theorem can be proved with Corollary 11.5.

THEOREM 14.10 **[178]** CQ_n is $(n-2)$-fault-tolerant 2^*-connected and $(n-3)$-fault-tolerant 1^*-connected if $n \ge 3$.

THEOREM 14.11 **[271]** Suppose that $\mathbf{u_1}$, $\mathbf{u_2}$, $\mathbf{v_1}$, and $\mathbf{v_2}$ are four distinct vertices of CQ_n for $n \ge 5$. Then there are two disjoint paths P_1 and P_2 of CQ_n such that (1) P_1 joins $\mathbf{u_1}$ to $\mathbf{v_1}$, (2) P_2 joins $\mathbf{u_2}$ to $\mathbf{v_2}$, and (3) $P_1 \cup P_2$ contains all the vertices in CQ_n.

Proof: We have the following cases:

Case 1. $\{\mathbf{u_1}, \mathbf{u_2}, \mathbf{v_1}, \mathbf{v_2}\} \subset V(CQ^t_{n-1})$ for some $t \in \{0, 1\}$. By Theorem 14.10, there is a Hamiltonian path R of CQ^t_{n-1} joining $\mathbf{u_1}$ to $\mathbf{v_1}$. Without loss of generality, we write $R = \langle \mathbf{u_1}, R_1, \mathbf{x}, \mathbf{p}, R_2, \mathbf{q}, \mathbf{y}, R_3, \mathbf{v_1}\rangle$ where $\{\mathbf{p}, \mathbf{q}\} = \{\mathbf{u_2}, \mathbf{v_2}\}$. By Theorem 14.10, there is a Hamiltonian path Q of CQ^{1-t}_{n-1} joining $(\mathbf{x})^n$ to $(\mathbf{y})^n$. We set $P_1 = \langle \mathbf{u_1}, R_1, \mathbf{x}, (\mathbf{x})^n, Q, (\mathbf{y})^n, \mathbf{y}, R_3, \mathbf{v_1}\rangle$ and $P_2 = R_2$. Then P_1 and P_2 form the desired paths.

Case 2. $\{\mathbf{u_1}, \mathbf{u_2}, \mathbf{v_1}\} \subset V(CQ^t_{n-1})$ and $\{\mathbf{v_2}\} \subset V(CQ^{1-t}_{n-1})$ for some $t \in \{0, 1\}$. By Theorem 14.10, there is a Hamiltonian path R of CQ^t_{n-1} joining $\mathbf{u_2}$ to $\mathbf{v_1}$. Without loss of generality, we write $R = \langle \mathbf{u_2}, R_1, \mathbf{x}, \mathbf{u_1}, \mathbf{y}, R_2, \mathbf{v_1}\rangle$.

Suppose that $(\mathbf{x})^n \ne \mathbf{v_2}$. By Theorem 14.10, there is a Hamiltonian path Q of CQ^{1-t}_{n-1} joining $(\mathbf{x})^n$ to $\mathbf{v_2}$. We set $P_1 = \langle \mathbf{u_1}, \mathbf{y}, R_2, \mathbf{v_1}\rangle$ and $P_2 = \langle \mathbf{u_2}, R_1, \mathbf{x}, (\mathbf{x})^n, Q, \mathbf{v_2}\rangle$. Then P_1 and P_2 form the desired paths.

Suppose that $(\mathbf{x})^n = \mathbf{v_2}$. By Theorem 14.10, there is a Hamiltonian path Q of $CQ_{n-1}^{1-t} - \{\mathbf{v_2}\}$ joining $(\mathbf{u_1})^n$ to $(\mathbf{y})^n$. We set $P_1 = \langle \mathbf{u_1}, (\mathbf{u_1})^n, Q, (\mathbf{y})^n, \mathbf{y}, R_2, \mathbf{v_1} \rangle$ and $P_2 = \langle \mathbf{u_2}, R_1, \mathbf{x}, (\mathbf{x})^n = \mathbf{v_2} \rangle$. Then P_1 and P_2 form the desired paths.

Case 3. $\{\mathbf{u_1}, \mathbf{v_1}\} \subset V(CQ_{n-1})^t$ and $\{\mathbf{u_2}, \mathbf{v_2}\} \subset V(CQ_{n-1}^{1-t})$ for some $t \in \{0,1\}$. By Theorem 14.10, there is a Hamiltonian path P_1 of CQ_{n-1}^t joining $\mathbf{u_1}$ to $\mathbf{v_1}$. Moreover, there is a Hamiltonian path P_2 of CQ_{n-1}^{1-t} joining $\mathbf{u_2}$ to $\mathbf{v_2}$. Then P_1 and P_2 form the desired paths.

Case 4. $\{\mathbf{u_1}, \mathbf{v_2}\} \subset V(CQ_{n-1}^t)$ and $\{\mathbf{u_2}, \mathbf{v_1}\} \subset V(CQ_{n-1}^{1-t})$ for some $t \in \{0,1\}$.

Case 4.1. $\mathbf{u_1} = (\mathbf{v_1})^n$ or $\mathbf{u_2} = (\mathbf{v_2})^n$. Without loss of generality, we assume that $\mathbf{u_1} = (\mathbf{v_1})^n$. By Theorem 14.10, there is a Hamiltonian path R of $CQ_n - \{\mathbf{u_1}, \mathbf{v_1}\}$ joining $\mathbf{u_2}$ to $\mathbf{v_2}$. We set $P_1 = \langle \mathbf{u_1}, \mathbf{v_1} \rangle$ and $P_2 = R$. Then P_1 and P_2 form the desired paths.

Case 4.2. $\mathbf{u_1} \neq (\mathbf{v_1})^n$ and $\mathbf{u_2} \neq (\mathbf{v_2})^n$.

Case 4.2.1. $\mathbf{u_1} \neq (\mathbf{u_2})^n$ or $\mathbf{v_2} \neq (\mathbf{v_1})^n$. Without loss of generality, we assume that $\mathbf{u_1} \neq (\mathbf{u_2})^n$. By Theorem 14.10, there is a Hamiltonian path R_1 of $CQ_{n-1}^{1-t} - \{\mathbf{u_2}\}$ joining $(\mathbf{u_1})^n$ to $\mathbf{v_1}$. Moreover, there is a Hamiltonian path R_2 of $CQ_{n-1}^t - \{\mathbf{u_1}\}$ joining $(\mathbf{u_2})^n$ to $\mathbf{v_2}$. We set $P_i = \langle \mathbf{u_i}, (\mathbf{u_i})^n, R_i, \mathbf{v_i} \rangle$ for every $i \in \{1,2\}$. Then P_1 and P_2 form the desired paths.

Case 4.2.2. $\mathbf{u_1} = (\mathbf{u_2})^n$ or $\mathbf{v_2} = (\mathbf{v_1})^n$.

Case 4.2. Either $\mathbf{u_1} = (\mathbf{v_1})^n$ and $\mathbf{v_2} \neq (\mathbf{u_2})^n$ or $\mathbf{u_1} \neq (\mathbf{v_1})^n$ and $\mathbf{v_2} = (\mathbf{u_2})^n$. Without loss of generality, we assume that $\mathbf{u_1} = (\mathbf{v_1})^n$ and $\mathbf{v_2} \neq (\mathbf{u_2})^n$. By Theorem 14.10, there is a Hamiltonian path R of $CQ_n - \{\mathbf{u_1}, \mathbf{v_1}\}$ joining $\mathbf{u_2}$ to $\mathbf{v_2}$. Moreover, there is a Hamiltonian path R_2 of $CQ_{n-1}^t - \{\mathbf{u_1}\}$ joining $(\mathbf{u_2})^n$ to $\mathbf{v_2}$. We set $P_1 = \langle \mathbf{u}, \mathbf{v_1} \rangle$ and $P_2 = R$. Then P_1 and P_2 form the desired paths.

Case 4.3. $\mathbf{u_1} = (\mathbf{v_1})^n$ and $\mathbf{v_2} = (\mathbf{u_2})^n$.

Suppose that $\{\mathbf{u_1}, \mathbf{v_2}\} \subset V(CQ_{n-2}^{ts})$ for some $s \in \{0,1\}$. We have $\{\mathbf{u_2}, \mathbf{v_1}\} \subset V(CQ_{n-2}^{(1-t)s})$. Since $CQ_{n-2}^{ts} \cup CQ_{n-2}^{(1-t)s}$ is isomorphic to CQ_{n-1}, by Theorem 14.10, there is a Hamiltonian path R of $CQ_{n-2}^{ts} \cup CQ_{n-2}^{(1-t)s}$ joining $\mathbf{u_1}$ to $\mathbf{v_1}$. Without loss of generality, we write $R = \langle \mathbf{u_1}, R_1, \mathbf{x}, \mathbf{p}, R_2, \mathbf{q}, \mathbf{y}, R_3, \mathbf{v_1} \rangle$ where $\{\mathbf{p}, \mathbf{q}\} = \{\mathbf{u_2}, \mathbf{v_2}\}$. Since $CQ_{n-2}^{t(1-s)} \cup CQ_{n-2}^{(1-t)(1-s)}$ is isomorphic to CQ_{n-1}, by Theorem 14.10, there is a Hamiltonian path Q of $CQ_{n-2}^{t(s-1)} \cup CQ_{n-2}^{(1-t)(1-s)}$ joining $(\mathbf{x})^{n-1}$ to $(\mathbf{y})^{n-1}$. We set $P_1 = \langle \mathbf{u_1}, R_1, \mathbf{x}, (\mathbf{x})^{n-1}, Q, (\mathbf{y})^{n-1}, \mathbf{y}, R_3, \mathbf{v_1} \rangle$ and $P_2 = R_2$. Then P_1 and P_2 form the desired paths.

Suppose that $\mathbf{u_1} \in V(CQ_{n-2}^{ts})$ and $\mathbf{v_2} \in V(CQ_{n-2}^{t(1-s)})$ for some $s \in \{0,1\}$. We have $\mathbf{u_2} \in V(CQ_{n-2}^{(1-t)s})$ and $\mathbf{v_1} \in V(CQ_{n-2}^{(1-t)(1-s)})$. Let $\mathbf{x}$ and $\mathbf{y}$ be two distinct vertices of $CQ_{n-2}^{ts} - \{\mathbf{u_1}, (\mathbf{v_2})^{n-1}\}$. By Theorem 14.10, there is a Hamiltonian path R_1 of CQ_{n-2}^{ts} joining $\mathbf{x}$ to $\mathbf{y}$. Moreover, there is a Hamiltonian path R_2 of $CQ_{n-2}^{t(1-s)}$ joining $(\mathbf{y})^{n-1}$ to $(\mathbf{x})^{n-1}$. Without loss of generality, we write $R_1 = \langle \mathbf{x}, Q_1, \mathbf{p}, \mathbf{u_1}, Q_2, \mathbf{y} \rangle$ and $R_2 = \langle (\mathbf{y})^{n-1}, Q_3, \mathbf{q}, \mathbf{v_2}, Q_2, (\mathbf{x})^{n-1} \rangle$. By Theorem 14.10, there is a Hamiltonian path W_1 of $CQ_{n-2}^{(1-t)(1-s)}$ joining $(\mathbf{q})^n$ to $\mathbf{v_1}$. Moreover, there is a Hamiltonian path

W_2 of $CQ_{n-2}^{(1-t)s}$ joining $\mathbf{u_2}$ to $(\mathbf{p})^n$. We set $P_1 = \langle \mathbf{u_1}, Q_2, \mathbf{y}, (\mathbf{y})^{n-1}, Q_3, \mathbf{q}, (\mathbf{q})^n, W_1, \mathbf{v_1} \rangle$ and $P_2 = \langle \mathbf{u_2}, W_2, (\mathbf{p})^n, \mathbf{p}, Q_1^{-1}, \mathbf{x}, (\mathbf{x})^{n-1}, Q_2^{-1}, \mathbf{v_1} \rangle$. Then P_1 and P_2 form the desired paths. □

THEOREM 14.12 Let $n \geq 3$ and $k \leq n-1$. Let F be any subset of $V(CQ_n) \cup E(CQ_n)$ with $|F| \leq n-3$ and $|F| + k \leq n-1$. Suppose that $\mathbf{u}$ is a vertex of $CQ_n - F$ and $S = \{\mathbf{v_1}, \mathbf{v_2}, \ldots, \mathbf{v_k}\}$ is a subset of vertices in $CQ_n - (F \cup \{\mathbf{u}\})$ with $\mathbf{v_i} \neq \mathbf{v_j}$ for every $i \neq j$. Then there is a set of k disjoint paths $\{P_1, P_2, \ldots, P_k\}$ of CQ_n such that P_i joins $\mathbf{u}$ to $\mathbf{v_i}$ for every $1 \leq i \leq k$ and $\bigcup_{i=1}^{k} P_i$ contains all the vertices in CQ_n.

Proof: We prove this theorem by induction on n. By Theorem 14.10, this theorem holds for CQ_3. Suppose that this theorem holds for CQ_{n-1} for $n \geq 4$. Let k be any positive integer with $k \leq n-1$ and let F be a subset of $V(CQ_n) \cup E(CQ_n)$ with $|F| \leq n-3$ and $|F| + k \leq n-1$. By Theorem 14.10, this theorem holds on $k = 1$ with $|F| \leq n-3$ and $k = 2$ with $|F| \leq n-3$. Thus, we assume that $k \geq 3$ and $|F| = n-k-1$. Suppose that $\mathbf{u}$ is a vertex of $CQ_n - F$ and $S = \{\mathbf{v_1}, \mathbf{v_2}, \ldots, \mathbf{v_k}\}$ is a subset of vertices in $CQ_n - (F \cup \mathbf{u})$ with $\mathbf{v_i} \neq \mathbf{v_j}$ for every $i \neq j$. By Theorem 14.10, this theorem holds for $k = 1$ and $k = 2$. Thus, we assume that $k \geq 3$. Without loss of generality, we assume that $\mathbf{u}$ is a vertex in CQ_{n-1}^0. We set $S_i = S \cap V(CQ_{n-1}^i)$ and $F_i = F \cap V(CQ_{n-1}^i) \cup E(CQ_{n-1}^i)$ for every $0 \leq i \leq 1$, and we set $F_2 = F - (F_0 \cup F_1)$.

We have the following cases:

Case 1. $|S_0| + |F_0| = |S| + |F|$. By induction, there is a set of $k-1$ disjoint paths $\{P_1, P_2, \ldots, P_{k-1}\}$ of CQ_{n-1}^0 such that P_i joins $\mathbf{u}$ to $\mathbf{v_i}$ for every $1 \leq i \leq k-1$ and $\bigcup_{i=1}^{k-1} P_i$ contains all the vertices in CQ_{n-1}^0. Without loss of generality, we assume that $\mathbf{v_k} \in P_{k-1}$ and we write $P_{k-1} = \langle \mathbf{u}, R_1, \mathbf{v_k}, \mathbf{z}, R_2, \mathbf{v_{k-1}} \rangle$. By Theorem 14.10, there is a Hamiltonian path Q of CQ_{n-1}^1 joining $(\mathbf{u})^n$ to $(\mathbf{z})^n$. We set $W_i = P_i$ for every $1 \leq i \leq k-2$, $W_{k-1} = \langle \mathbf{u}, (\mathbf{u})^n, Q, (\mathbf{z})^n, \mathbf{z}, R_2.\mathbf{v_{k-1}} \rangle$, and $P_k = \langle \mathbf{u}, R_1, \mathbf{v_k} \rangle$. Then $\{W_1, W_2, \ldots, W_k\}$ forms a desired set of paths of CQ_n.

Case 2. $|S_1| + |F_1| = |S| + |F|$. Without loss of generality, we assume that $\mathbf{v_k} \neq (\mathbf{u})^n$. By induction, there is a set of $k-1$ disjoint paths $\{P_1, P_2, \ldots, P_{k-1}\}$ of CQ_{n-1}^1 such that P_i joins $\mathbf{v_k}$ to $\mathbf{v_i}$ for every $1 \leq i \leq k-1$ and $\bigcup_{i=1}^{k-1} P_i$ contains all the vertices in CQ_{n-1}^1. Without loss of generality, we assume that $(\mathbf{u})^n \in V(P_{k-1})$, and we write $P_i = \langle \mathbf{v_k}, \mathbf{z_i}, P_i', \mathbf{v_i} \rangle$ for every $1 \leq i \leq k-2$.

Case 2.1. $(\mathbf{u})^n = \mathbf{v_{k-1}}$. We write $P_{k-1} = \langle \mathbf{v_k}, Q, \mathbf{x}, \mathbf{v_{k-1}} \rangle$. By induction, there is a set of $k-1$ disjoint paths $\{R_1, R_2, \ldots, R_{k-1}\}$ of CQ_{n-1}^0 such that R_i joins $\mathbf{u}$ to $(\mathbf{z_i})^n$ for every $1 \leq i \leq k-1$ and $\bigcup_{i=1}^{k-1} R_i$ contains all the vertices in CQ_{n-1}^0. We set $W_i = \langle \mathbf{u}, R_i, (\mathbf{z_i})^n, \mathbf{z_i}, P_i', \mathbf{v_i} \rangle$ for every $1 \leq i \leq k-2$, $W_{k-1} = \langle \mathbf{u}, \mathbf{v_{k-1}} \rangle$, and $W_k = \langle \mathbf{u}, R_{k-1}, (\mathbf{z_{k-1}})^n, \mathbf{z_{k-1}}, Q^{-1}, \mathbf{v_k} \rangle$. Then $\{W_1, W_2, \ldots, W_k\}$ forms a desired set of paths of CQ_n.

Case 2.2. $(\mathbf{u})^n \neq \mathbf{v_i}$ for every $1 \leq i \leq k-1$. We write $P_{k-1} = \langle \mathbf{v_k}, Q, \mathbf{x}, \mathbf{v_{k-1}} \rangle$. By induction, there is a set of $k-1$ disjoint paths $\{R_1, R_2, \ldots, R_{k-1}\}$ of CQ_{n-1}^0 such that R_i joins $\mathbf{u}$ to $(\mathbf{z_i})^n$ for every $1 \leq i \leq k-1$ and $\bigcup_{i=1}^{k-1} R_i$ contains all the vertices in CQ_{n-1}^0. We set $W_i = \langle \mathbf{u}, R_i, (\mathbf{z_i})^n, \mathbf{z_i}, P_i', \mathbf{v_i} \rangle$ for every $1 \leq i \leq k-2$, $W_{k-1} = \langle \mathbf{u}, \mathbf{v_{k-1}} \rangle$, and $W_k = \langle \mathbf{u}, R_{k-1}, (\mathbf{z_{k-1}})^n, \mathbf{z_{k-1}}, Q^{-1}, \mathbf{v_k} \rangle$. Then $\{W_1, W_2, \ldots, W_k\}$ forms a desired set of paths of CQ_n.

Case 3. $|S_0| + |F_0| < |S| + |F|$ and $|S_1| = 0$. By induction, there is a set of k disjoint paths $\{P_1, P_2, \ldots, P_k\}$ of CQ_{n-1}^0 such that P_i joins $\mathbf{u}$ to $\mathbf{v_i}$ for every $1 \le i \le k$ and $\bigcup_{i=1}^{k} P_i$ contains all the vertices in CQ_{n-1}^0. Since $(2^{n-1} - (n-2))/(n-2) > 2(n-3)$, there are two adjacent vertices $\mathbf{x}$ and $\mathbf{y}$ in some path P_i such that $\{(\mathbf{x})^n, (\mathbf{y})^n, (\mathbf{x}, (\mathbf{x})^n), (\mathbf{y}, (\mathbf{y})^n)\} \cap F = \emptyset$. Without loss of generality, we assume that both $\mathbf{x}$ and $\mathbf{y}$ are in P_k, and we write $P_k = \langle \mathbf{u}, R_1, \mathbf{x}, \mathbf{y}, R_2, \mathbf{v_k} \rangle$. By Theorem 14.10, there is a Hamiltonian path Q of CQ_{n-1}^1 joining $(\mathbf{x})^n$ to $(\mathbf{y})^n$. We set $W_i = P_i$ for every $1 \le i \le k-1$ and $W_k = \langle \mathbf{u}, R_1, \mathbf{x}, (\mathbf{x})^n, Q, (\mathbf{y})^n, \mathbf{y}, R_2, \mathbf{v_k} \rangle$. Then $\{W_1, W_2, \ldots, W_k\}$ forms a desired set of paths of CQ_n.

Case 4. $|S_1| + |F_1| < |S| + |F|$ and $|S_1| > 0$.

Case 4.1. $|F_0| < |F|$. Since $2^{n-1} - (n-3)(n-1) - (n-2) > 0$, there is a vertex $\mathbf{x}$ in $CQ_{n-1}^1 - (F_1 \cup \{(\mathbf{u})^n, \mathbf{v_1}, \mathbf{v_2}, \ldots, \mathbf{v_k}\})$ such that $\{\mathbf{y}, (\mathbf{y})^n, (\mathbf{y}, (\mathbf{y})^n)\} \cap F = \emptyset$ for every $\mathbf{y} \in N_{CQ_{n-1}^1}(\mathbf{x})$. By induction, there is a set of k disjoint paths $\{P_1, P_2, \ldots, P_k\}$ of CQ_{n-1}^1 such that P_i joins $\mathbf{x}$ to $\mathbf{v_i}$ for every $1 \le i \le k$ and $\bigcup_{i=1}^{k} P_i$ contains all the vertices in CQ_{n-1}^1. Without loss of generality, we write $P_i = \langle \mathbf{x}, \mathbf{y_i}, R_i, \mathbf{v_i} \rangle$ for every $1 \le i \le k-1$. By induction, there is a set of k disjoint paths $\{Q_1, Q_2, \ldots, Q_k\}$ of CQ_{n-1}^0 such that Q_i joins $\mathbf{u}$ to $(\mathbf{y_i})^n$ for every $1 \le i \le k-1$, Q_k joining $\mathbf{u}$ to $(\mathbf{x})^n$, and $\bigcup_{i=1}^{k} Q_i$ contains all the vertices in CQ_{n-1}^0. We set $W_i = \langle \mathbf{u}, Q_i, (\mathbf{y_i})^n, \mathbf{y_i}, R_i, \mathbf{v_i} \rangle$ for every $1 \le i \le k-1$ and $W_k = \langle \mathbf{u}, Q_k, (\mathbf{x})^n, \mathbf{x}, P_k, \mathbf{v_k} \rangle$. Then $\{W_1, W_2, \ldots, W_k\}$ forms a desired set of paths of CQ_n.

Case 4.2. $|F_0| = |F|$. Without loss of generality, we assume that $S_1 = \{\mathbf{v_t}, \mathbf{v_t} + 1, \ldots, \mathbf{v_k}\}$ for some $1 \le t \le k$.

Case 4.2.1. $t = k$. By induction, there is a set of k disjoint paths $\{P_1, P_2, \ldots, P_k\}$ of CQ_{n-1}^0 such that P_i joins $\mathbf{u}$ to $\mathbf{v_i}$ for every $1 \le i \le k-1$ and $\bigcup_{i=1}^{k-1} P_i$ contains all the vertices in CQ_{n-1}^0.

Suppose that $(\mathbf{u})^n = \mathbf{v_k}$. Since $2^{n-1} > n-1$, we assume that $l(P_{k-1}) > 1$ and we write $P_{k-1} = \langle \mathbf{u}, R, \mathbf{x}, \mathbf{v_{k-1}} \rangle$. By Theorem 14.10, there is a Hamiltonian path Q of $CQ_{n-1}^1 - \{\mathbf{v_k}\}$ joining $(\mathbf{x})^n$ to $(\mathbf{v_{k-1}})^n$. We set $W_i = P_i$ for every $1 \le i \le k-2$, $W_{k-1} = \langle \mathbf{u}, R, \mathbf{x}, (\mathbf{x})^n, Q, (\mathbf{v_{k-1}})^n, \mathbf{v_{k-1}} \rangle$, and $W_k = \langle \mathbf{u}, \mathbf{v_k} \rangle$. Then $\{W_1, W_2, \ldots, W_k\}$ forms a desired set of paths of CQ_n.

Suppose that $(\mathbf{u})^n \ne \mathbf{v_k}$. By Theorem 14.10, there is a Hamiltonian path Q of CQ_{n-1}^1 joining $(\mathbf{u})^n$ to $\mathbf{v_k}$. We set $W_i = P_i$ for every $1 \le i \le k-1$ and $W_k = \langle \mathbf{u}, (\mathbf{u})^n, Q, \mathbf{v_k} \rangle$. Then $\{W_1, W_2, \ldots, W_k\}$ forms a desired set of paths of CQ_n.

Case 4.2.2. $2 \le t \le k-1$. Suppose that $(\mathbf{u})^n \in S_1$. Without loss of generality, we assume that $(\mathbf{u})^n = \mathbf{v_k}$. Since $2^{n-1} - (n-2)(n-1) > 0$, there is a vertex $\mathbf{x}$ in $CQ_{n-1}^1 - S_1$ such that $(\mathbf{y})^n \notin F_0 \cup S_0$ for every $\mathbf{y} \in N_{CQ_{n-1}^1}(\mathbf{x})$. By induction, there is a set of $k-t$ disjoint paths $\{P_1, P_2, \ldots, P_{k-t}\}$ of $CQ_{n-1}^1 - \{\mathbf{v_k}\}$ such that P_i joins $\mathbf{x}$ to $\mathbf{v_{t+i-1}}$ for every $1 \le i \le k-t$ and $\bigcup_{i=1}^{k-t} P_i$ contains all the vertices in $CQ_{n-1}^1 - \{\mathbf{v_k}\}$. Without loss of generality, we write $P_i = \langle \mathbf{x}, \mathbf{y_i}, R_i, \mathbf{v_{t+i-1}} \rangle$ for every $1 \le i \le k-t-1$.

By induction, there is a set of $k-1$ disjoint paths $\{Q_1, Q_2, \ldots, Q_{k-1}\}$ of $CQ^0_{n-1} - F_0$ such that Q_i joins $\mathbf{u}$ to $\mathbf{v_i}$ for every $1 \le i \le t-1$, Q_j joining $\mathbf{u}$ to $(\mathbf{y_{t-j+1}})^n$ for every $t \le j \le k-2$, Q_{k-1} joining $\mathbf{u}$ to $(\mathbf{x})^n$, and $\bigcup_{i=1}^{k-1} Q_i$ contains all the vertices in CQ^0_{n-1}. We set $W_i = Q_i$ for every $1 \le i \le t-1$, $W_i = \langle \mathbf{u}, Q_i, (\mathbf{y_{t-j+1}})^n, \mathbf{y_{t-j+1}}, R_{t-j+1}, v_j \rangle$ for every $t \le j \le k-2$, $W_{k-2} = \langle \mathbf{u}, Q_{k-1}, (\mathbf{x})^n, P_{k-t-1}, \mathbf{v_{k-1}} \rangle$, and $W_k = \langle \mathbf{u}, \mathbf{v_k} \rangle$. Then $\{W_1, W_2, \ldots, W_k\}$ forms a desired set of paths of CQ_n.

Suppose that $(\mathbf{u})^n \notin S_1$. We set $A = \{((\mathbf{u})^n, (\mathbf{x})^n) \mid \mathbf{x} \in (F \cup S_0) - \{\mathbf{v_1}\}\}$. By induction, there is a set of $k-t$ disjoint paths $\{P_1, P_2, \ldots, P_{k-t+1}\}$ of $CQ^1_{n-1} - A$ such that P_i joins $(\mathbf{u})^n$ to $\mathbf{v_{t+i-1}}$ for every $1 \le i \le k-t+1$ and $\bigcup_{i=1}^{k-t+1} P_i$ contains all the vertices in CQ^1_{n-1}. Without loss of generality, we write $P_i = \langle (\mathbf{u})^n, \mathbf{y_i}, R_i, \mathbf{v_{t+i-1}} \rangle$ for every $1 \le i \le k-t+1$, and we assume that $\mathbf{y_j} \ne (\mathbf{v_1})^n$ for every $1 \le j \le k-t$. By induction, there is a set of $k-1$ disjoint paths $\{Q_1, Q_2, \ldots, Q_{k-1}\}$ of $CQ^0_{n-1} - F_0$ such that Q_i joins $\mathbf{u}$ to $\mathbf{v_i}$ for every $1 \le i \le t-1$, Q_j joining $\mathbf{u}$ to $(\mathbf{y_{t-j+1}})^n$ for every $t \le j \le k-1$, and $\bigcup_{i=1}^{k-1} Q_i$ contains all the vertices in CQ^0_{n-1}. We set $W_i = Q_i$ for every $1 \le i \le t-1$, $W_j = \langle \mathbf{u}, Q_j, (\mathbf{y_{t-j+1}})^n, \mathbf{y_{t-j+1}}, R_{t-j+1}, \mathbf{v_j} \rangle$ for every $t \le j \le k-1$, and $W_k = \langle \mathbf{u}, (\mathbf{u})^n, P_{k-t+1}, \mathbf{v_k} \rangle$. Then $\{W_1, W_2, \ldots, W_k\}$ forms a desired set of paths of CQ_n.

Case 4.2.3. $t = 1$. Suppose that $(\mathbf{u})^n \in S_1$. Without loss of generality, we assume that $(\mathbf{u})^n = \mathbf{v_k}$. By induction, there is a set of $(k-1)$ disjoint paths $\{P_1, P_2, \ldots, P_{k-1}\}$ of CQ^1_{n-1} such that P_i joins $\mathbf{v_k}$ to $\mathbf{v_i}$ for every $1 \le i \le k-1$ and $\bigcup_{i=1}^{k-1} P_i$ contains all the vertices in CQ^1_{n-1}. Without loss of generality, we write $P_i = \langle \mathbf{v_k}, \mathbf{y_i}, R_i, \mathbf{v_i} \rangle$ for every $1 \le i \le k-1$. By induction, there is a set of $k-1$ disjoint paths $\{Q_1, Q_2, \ldots, Q_{k-1}\}$ of $CQ^0_{n-1} - F_0$ such that Q_i joins $\mathbf{u}$ to $(\mathbf{y_i})^n$ for every $1 \le i \le k-1$ and $\bigcup_{i=1}^{k-1} Q_i$ contains all the vertices in $CQ^0_{n-1} - F_0$. We set $W_i = \langle \mathbf{u}, Q_i, (\mathbf{y_i})^n, \mathbf{y_i}, R_i, \mathbf{v_i} \rangle$ for every $1 \le i \le k-1$ and $W_k = \langle \mathbf{u}, \mathbf{v_k} \rangle$. Then $\{W_1, W_2, \ldots, W_k\}$ forms a desired set of paths of CQ_n.

Suppose that $(\mathbf{u})^n \notin S_1$. By induction, there is a set of $k-1$ disjoint paths $\{P_1, P_2, \ldots, P_{k-1}\}$ of CQ^1_{n-1} such that P_i joins $\mathbf{v_k}$ to $\mathbf{v_i}$ for every $1 \le i \le k-1$ and $\bigcup_{i=1}^{k-1} P_i$ contains all the vertices in CQ^1_{n-1}. Without loss of generality, we assume that $(\mathbf{u})^n \in P_{k-1}$, and we write $P_i = \langle \mathbf{v_k}, \mathbf{y_i}, R_i, \mathbf{v_i} \rangle$ for every $1 \le i \le k-2$ and $P_{k-1} = \langle \mathbf{v_k}, R_k, (\mathbf{u})^n, \mathbf{y_{k-1}}, R_{k-1}, \mathbf{v_{k-1}} \rangle$. By induction, there is a set of $k-1$ disjoint paths $\{Q_1, Q_2, \ldots, Q_{k-1}\}$ of $CQ^0_{n-1} - F_0$ such that Q_i joins $\mathbf{u}$ to $(\mathbf{y_i})^n$ for every $1 \le i \le k-1$ and $\bigcup_{i=1}^{k-1} Q_i$ contains all the vertices in $CQ^0_{n-1} - F_0$. We set $W_i = \langle \mathbf{u}, Q_i, (\mathbf{y_i})^n, \mathbf{y_i}, R_i, \mathbf{v_i} \rangle$ for every $1 \le i \le k-1$ and $W_k = \langle \mathbf{u}, (\mathbf{u})^n, R_k^{-1}, \mathbf{v_k} \rangle$. Then $\{W_1, W_2, \ldots, W_k\}$ forms a desired set of paths of CQ_n. □

LEMMA 14.17 CQ_n is k^*-connected for all $1 \le k \le n-1$ and $n \ge 3$.

Proof: Let $\mathbf{u}$ and $\mathbf{v}$ be two distinct vertices of CQ_n. By Theorem 14.10, this statement holds for $k = 1$. Thus, we consider that $2 \le k \le n-1$. Since $\deg_{CQ_n}(\mathbf{v}) = n$, we can choose a set of $(k-1)$ neighbors $\{\mathbf{x_1}, \mathbf{x_2}, \ldots, \mathbf{x_{k-1}}\}$ of $\mathbf{v}$ without $\mathbf{u}$. By Theorem 14.12, there is a set of k disjoint paths $\{R_1, R_2, \ldots, R_k\}$ of CQ_n such that (1) R_i joins $\mathbf{u}$ to $\mathbf{x_i}$ for every $1 \le i \le k-1$, (2) R_k joins $\mathbf{u}$ to $\mathbf{v}$, and (3) $\bigcup_{i=1}^{k} R_i$ contains all the vertices in CQ_n. We set $P_i = \langle \mathbf{u}, R_i, \mathbf{x_i}, \mathbf{v} \rangle$ for every $1 \le i \le k-1$ and $P_k = R_k$. Then $\{P_1, P_2, \ldots, P_k\}$ forms the k^*-container of CQ_n between $\mathbf{u}$ and $\mathbf{v}$. □

LEMMA 14.18 Suppose that **u** and **v** are two distinct vertices of CQ_n with $(\mathbf{u})_{n-1} = (\mathbf{v})_{n-1}$ for $n \geq 7$. Then there is an n^*-container of CQ_n between **u** and **v**.

Proof: Let $(\mathbf{u})_{n-1} = (\mathbf{v})_{n-1} = 0$. Since CQ^0_{n-1} is isomorphic to CQ_{n-1}, by Lemma 14.17, there is an $(n-2)^*$-container $\{R_1, R_2, \ldots, R_{n-2}\}$ of CQ^0_{n-1} between **u** and **v**. Since $\deg_{CQ^0_{n-1}}(\mathbf{u}) = n-1$, there is a vertex **x** in CQ^0_{n-1} such that $(\mathbf{x}, \mathbf{u}) \in E(CQ^0_{n-1})$ and $(\mathbf{x}, \mathbf{u}) \notin \bigcup_{i=1}^{n-2} E(R_i)$. Similarly, there is a vertex **y** in CQ^0_{n-1} such that $(\mathbf{y}, \mathbf{v}) \in E(CQ^0_{n-1})$ and $(\mathbf{y}, \mathbf{v}) \notin \bigcup_{i=1}^{n-2} E(R_i)$. We have the following cases:

Case 1. Suppose that $\mathbf{x} = \mathbf{v}$. Since $\mathbf{x} = \mathbf{v}$, $\mathbf{y} = \mathbf{u}$. By Theorem 14.10, there is a Hamiltonian path H of CQ^1_{n-1} joining $(\mathbf{u})^n$ to $(\mathbf{v})^n$. We set $P_i = R_i$ for every $1 \leq i \leq n-2$, $P_{n-1} = \langle \mathbf{u}, \mathbf{v} \rangle$, and $P_n = \langle \mathbf{u}, (\mathbf{u})^n, H, (\mathbf{v})^n, \mathbf{v} \rangle$. Then $\{P_1, P_2, \ldots, P_n\}$ is the n^*-container of CQ_n between **u** and **v**. See Figure 14.11a for illustration.

Case 2. Suppose that $\mathbf{x} \neq \mathbf{v}$. Since $\mathbf{x} \neq \mathbf{v}$, we have $\mathbf{y} \neq \mathbf{u}$.

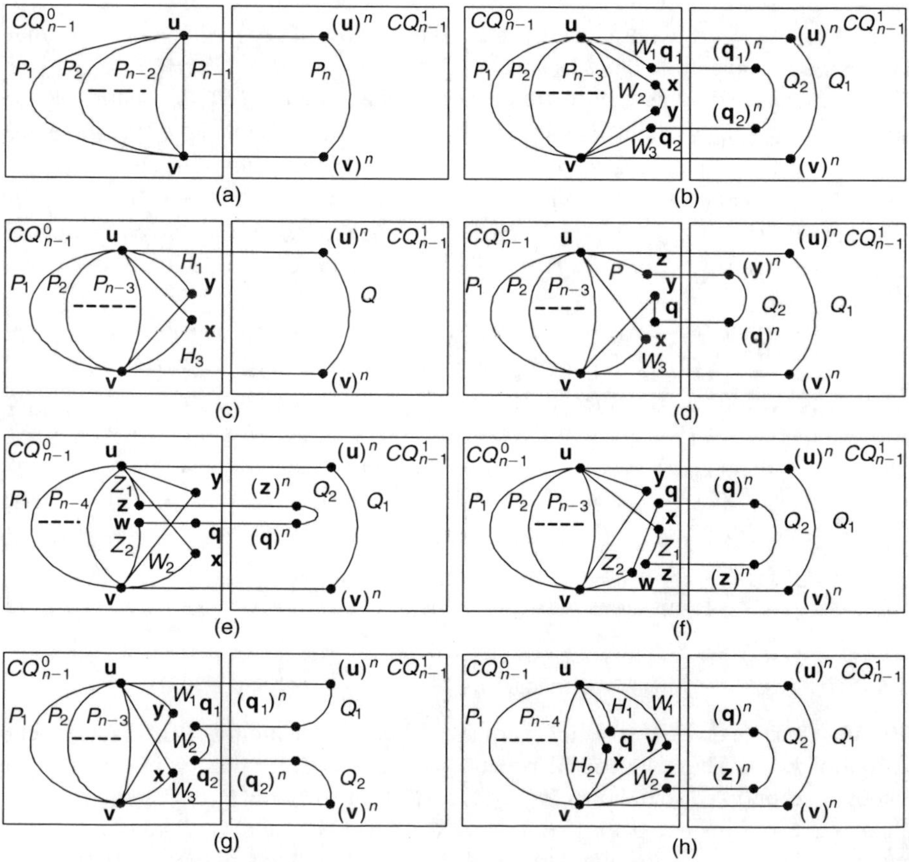

FIGURE 14.11 Illustrations for Lemma 14.18.

Case 2.1. Assume that $\mathbf{x} \in R_i$ and $\mathbf{y} \in R_i$ for $1 \le i \le n-2$. We assume that $\mathbf{x} \in R_{n-2}$ and $\mathbf{y} \in R_{n-2}$. We denote $R_{n-2} = \langle \mathbf{u}, H_1, \mathbf{p}_1, H_2, \mathbf{p}_2, H_3, \mathbf{v} \rangle$ with $\{\mathbf{p}_1, \mathbf{p}_2\} = \{\mathbf{x}, \mathbf{y}\}$. (Note that $\mathbf{p}_1 = \mathbf{p}_2$ if $\mathbf{x} = \mathbf{y}$.)

Case 2.1.1. Assume that $\mathbf{p}_1 = \mathbf{x}$ and $\mathbf{p}_2 = \mathbf{y}$. The R_{n-2} is written as $\langle \mathbf{u}, W_1, \mathbf{q}_1, \mathbf{x}, W_2, \mathbf{y}, \mathbf{q}_2, W_3, \mathbf{v} \rangle$. Since $(\mathbf{x}, \mathbf{u}) \notin E(R_{n-2})$ and $(\mathbf{y}, \mathbf{v}) \notin E(R_{n-2})$, $\mathbf{u} \neq \mathbf{q}_1$ and $\mathbf{v} \neq \mathbf{q}_2$. Since CQ_{n-1}^1 is isomorphic to CQ_{n-1}, by Theorem 14.11, there are two disjoint paths Q_1 and Q_2 of CQ_{n-1}^1 such that (1) Q_1 joins $(\mathbf{u})^n$ to $(\mathbf{v})^n$, (2) Q_2 joins $(\mathbf{q}_1)^n$ to $(\mathbf{q}_2)^n$, and (3) $Q_1 \cup Q_2$ contains all the vertices in CQ_{n-1}^1. We set

$$\begin{aligned} P_i &= R_i \quad \text{for every } 1 \le i \le n-3 \\ P_{n-2} &= \langle \mathbf{u}, \mathbf{x}, W_2, \mathbf{y}, \mathbf{v} \rangle \\ P_{n-1} &= \langle \mathbf{u}, W_1, \mathbf{q}_1, (\mathbf{q}_1)^n, Q_2, (\mathbf{q}_2)^n, \mathbf{q}_2, W_3, \mathbf{v} \rangle \\ P_n &= \langle \mathbf{u}, (\mathbf{u})^n, Q_1, (\mathbf{v})^n, \mathbf{v} \rangle \end{aligned}$$

Hence, $\{P_1, P_2, \dots, P_n\}$ is the n^*-container of CQ_n between $\mathbf{u}$ and $\mathbf{v}$ as illustrated in Figure 14.11b.

Case 2.1.2. Assume that $\mathbf{x} \neq \mathbf{y}$, $\mathbf{p}_1 = \mathbf{y}$, $\mathbf{p}_2 = \mathbf{x}$, and $l(H_2) = 1$. Since CQ_{n-1}^1 is isomorphic to CQ_{n-1}, by Theorem 14.10, there is a Hamiltonian path Q of CQ_{n-1}^1 joining $(\mathbf{u})^n$ to $(\mathbf{v})^n$. We set

$$\begin{aligned} P_i &= R_i \quad \text{for every } 1 \le i \le n-3 \\ P_{n-2} &= \langle \mathbf{u}, \mathbf{x}, H_3, \mathbf{v} \rangle \\ P_{n-1} &= \langle \mathbf{u}, H_1, \mathbf{y}, \mathbf{v} \rangle \\ P_n &= \langle \mathbf{u}, (\mathbf{u})^n, Q, (\mathbf{v})^n, \mathbf{v} \rangle \end{aligned}$$

Then $\{P_1, P_2, \dots, P_n\}$ forms the n^*-container of CQ_n between $\mathbf{u}$ and $\mathbf{v}$. See Figure 14.11c for illustration.

Case 2.1.3. Assume that $\mathbf{x} \neq \mathbf{y}$, $\mathbf{p}_1 = \mathbf{y}$, $\mathbf{p}_2 = \mathbf{x}$, and $l(H_2) = 2$. We rewrite $R_{n-2} = \langle \mathbf{u}, W_1, \mathbf{y}, \mathbf{q}, \mathbf{x}, W_2, \mathbf{v} \rangle$.

Case 2.1.3.1. Assume that $l(W_1) \ge 2$. Let $W_1 = \langle \mathbf{u}, P, \mathbf{z}, \mathbf{y} \rangle$. By Theorem 14.11, there are two disjoint paths Q_1 and Q_2 of CQ_{n-1}^1 such that (1) Q_1 joins $(\mathbf{u})^n$ to $(\mathbf{v})^n$, (2) Q_2 joins $(\mathbf{z})^n$ to $(\mathbf{q})^n$, and (3) $Q_1 \cup Q_2$ contains all the vertices in CQ_{n-1}^1. We set

$$\begin{aligned} P_i &= R_i \quad \text{for every } 1 \le i \le n-3 \\ P_{n-2} &= \langle \mathbf{u}, \mathbf{x}, W_2, \mathbf{v} \rangle \\ P_{n-1} &= \langle \mathbf{u}, (\mathbf{u})^n, Q_1, (\mathbf{v})^n, \mathbf{v} \rangle \\ P_n &= \langle \mathbf{u}, P, \mathbf{z}, (\mathbf{z})^n, Q_2, (\mathbf{q})^n, \mathbf{q}, \mathbf{y}, \mathbf{v} \rangle \end{aligned}$$

Then $\{P_1, P_2, \dots, P_n\}$ is the n^*-container of CQ_n between $\mathbf{u}$ and $\mathbf{v}$ as shown in Figure 14.11d.

Case 2.1.3.2. Assume that $l(W_1) = 1$. Thus, $W_1 = \langle \mathbf{u}, \mathbf{y} \rangle$. Since $N_{CQ^0_{n-1}}(\mathbf{u}) \neq N_{CQ^0_{n-1}}(\mathbf{q})$, there is a vertex $\mathbf{w}$ in $N_{CQ^0_{n-1}}(\mathbf{q}) - N_{CQ^0_{n-1}}(\mathbf{u})$. Note that $(\mathbf{u}, \mathbf{w}) \notin E(CQ^0_{n-1})$.

Case 2.1.3.2.1. Suppose that $\mathbf{w} \in R_{n-3}$. We write that $R_{n-3} = \langle \mathbf{u}, Z_1, \mathbf{z}, \mathbf{w}, Z_2, \mathbf{v} \rangle$. By Theorem 14.11, there are two disjoint paths Q_1 and Q_2 of CQ^1_{n-1} such that (1) Q_1 joins $(\mathbf{u})^n$ to $(\mathbf{v})^n$, (2) Q_2 joins $(\mathbf{z})^n$ to $(\mathbf{q})^n$, and (3) $Q_1 \cup Q_2$ contains all the vertices in CQ^1_{n-1}. We set

$$
\begin{aligned}
P_i &= R_i \quad \text{for every } 1 \leq i \leq n-4 \\
P_{n-3} &= \langle \mathbf{u}, \mathbf{x}, W_2, \mathbf{v} \rangle \\
P_{n-2} &= \langle \mathbf{u}, (\mathbf{u})^n, Q_1, (\mathbf{v})^n, \mathbf{v} \rangle \\
P_{n-1} &= \langle \mathbf{u}, Z_1, \mathbf{z}, (\mathbf{z})^n, Q_2, (\mathbf{q})^n, \mathbf{q}, \mathbf{w}, Z_2, \mathbf{v} \rangle \\
P_n &= \langle \mathbf{u}, \mathbf{y}, \mathbf{v} \rangle
\end{aligned}
$$

Then the n^*-container of CQ_n between $\mathbf{u}$ and $\mathbf{v}$ is composed of $\{P_1, P_2, \ldots, P_n\}$. See Figure 14.11e for illustration.

Case 2.1.3.2.2. Suppose that $\mathbf{w} \in W_2$. Since $\mathbf{w} \notin N_{CQ^0_{n-1}}(\mathbf{u})$, $\mathbf{w} \neq \mathbf{x}$. Let $W_2 = \langle \mathbf{x}, Z_1, \mathbf{z}, \mathbf{w}, Z_2, \mathbf{v} \rangle$. By Theorem 14.11, there are two disjoint paths Q_1 and Q_2 of CQ^1_{n-1} such that (1) Q_1 joins $(\mathbf{u})^n$ to $(\mathbf{v})^n$, (2) Q_2 joins $(\mathbf{z})^n$ to $(\mathbf{q})^n$, and (3) $Q_1 \cup Q_2$ contains all the vertices in CQ^1_{n-1}. We set

$$
\begin{aligned}
P_i &= R_i \quad \text{for every } 1 \leq i \leq n-3 \\
P_{n-2} &= \langle \mathbf{u}, \mathbf{x}, Z_1, \mathbf{z}, (\mathbf{z})^n, Q_2, (\mathbf{q})^n, \mathbf{q}, \mathbf{w}, Z_2, \mathbf{v} \rangle \\
P_{n-1} &= \langle \mathbf{u}, (\mathbf{u})^n, Q_1, (\mathbf{v})^n, \mathbf{v} \rangle \\
P_n &= \langle \mathbf{u}, \mathbf{y}, \mathbf{v} \rangle
\end{aligned}
$$

Then we have constructed the n^*-container $\{P_1, P_2, \ldots, P_n\}$ of CQ_n between $\mathbf{u}$ and $\mathbf{v}$ as illustration in Figure 14.11f.

Case 2.1.4. Assume that $\mathbf{x} \neq \mathbf{y}$, $\mathbf{p}_1 = \mathbf{y}$, $\mathbf{p}_2 = \mathbf{x}$, and $l(H_2) \geq 3$. We rewrite $R_{n-2} = \langle \mathbf{u}, W_1, \mathbf{y}, \mathbf{q}_1, W_2, \mathbf{q}_2, \mathbf{x}, W_3, \mathbf{v} \rangle$. By Theorem 14.11, there are two disjoint paths Q_1 and Q_2 of CQ^1_{n-1} such that (1) Q_1 joins $(\mathbf{u})^n$ to $(\mathbf{q}_1)^n$, (2) Q_2 joins $(\mathbf{q}_2)^n$ to $(\mathbf{v})^n$, and (3) $Q_1 \cup Q_2$ contains all the vertices in CQ^1_{n-1}. We set

$$
\begin{aligned}
P_i &= R_i \quad \text{for every } 1 \leq i \leq n-3 \\
P_{n-2} &= \langle \mathbf{u}, \mathbf{x}, W_3, \mathbf{v} \rangle \\
P_{n-1} &= \langle \mathbf{u}, W_1, \mathbf{y}, \mathbf{v} \rangle \\
P_n &= \langle \mathbf{u}, (\mathbf{u})^n, Q_1, (\mathbf{q}_1)^n, \mathbf{q}_1, W_2, \mathbf{q}_2, (\mathbf{q}_2)^n, Q_2, (\mathbf{v})^n, \mathbf{v} \rangle
\end{aligned}
$$

Then $\{P_1, P_2, \ldots, P_n\}$ forms the n^*-container of CQ_n between $\mathbf{u}$ and $\mathbf{v}$. See Figure 14.11g for illustration.

Case 2.2. Assume that $\mathbf{x} \in R_i$ and $\mathbf{y} \in R_j$ for some $1 \le i \le n-2$ and $1 \le j \le n-2$ with $i \ne j$. We assume that $\mathbf{x} \in R_{n-3}$ and $\mathbf{y} \in R_{n-2}$. We write $R_{n-3} = \langle \mathbf{u}, H_1, \mathbf{q}, \mathbf{x}, H_2, \mathbf{v} \rangle$ and $R_{n-2} = \langle \mathbf{u}, W_1, \mathbf{y}, \mathbf{z}, W_2, \mathbf{v} \rangle$. By Theorem 14.11, there are two disjoint paths Q_1 and Q_2 of CQ^1_{n-1} such that (1) Q_1 joins $(\mathbf{u})^n$ to $(\mathbf{v})^n$, (2) Q_2 joins $(\mathbf{q})^n$ to $(\mathbf{z})^n$, and (3) $Q_1 \cup Q_2$ contains all the vertices in CQ^1_{n-1}. We set

$$\begin{aligned} P_i &= R_i \quad \text{for every } 1 \le i \le n-4 \\ P_{n-3} &= \langle \mathbf{u}, \mathbf{x}, H_2, \mathbf{v} \rangle \\ P_{n-2} &= \langle \mathbf{u}, W_1, \mathbf{y}, \mathbf{v} \rangle \\ P_{n-1} &= \langle \mathbf{u}, H_1, \mathbf{q}, (\mathbf{q})^n, Q_2, (\mathbf{z})^n, \mathbf{z}, W_2, \mathbf{v} \rangle \\ P_n &= \langle \mathbf{u}, (\mathbf{u})^n, Q_1, (\mathbf{v})^n, \mathbf{v} \rangle \end{aligned}$$

Then $\{P_1, P_2, \ldots, P_n\}$ is the n^*-container of CQ_n between $\mathbf{u}$ and $\mathbf{v}$. See Figure 14.11h for illustration.

Thus, this lemma is proved. □

LEMMA 14.19 Suppose that $\mathbf{u}$ and $\mathbf{v}$ are two distinct vertices of CQ_n with $(\mathbf{u})_{n-2} = (\mathbf{v})_{n-2}$ and $(\mathbf{u})_{n-1} \ne (\mathbf{v})_{n-1}$ for $n \ge 7$. Then there is an n^*-container of CQ_n between $\mathbf{u}$ and $\mathbf{v}$.

Proof: We assume that $(\mathbf{u})_{n-2} = 0$ and $(\mathbf{u})_{n-1} = 0$. Thus, we have $\mathbf{u} \in CQ^{00}_{n-2}$ and $\mathbf{v} \in CQ^{10}_{n-2}$. Since $CQ^{00}_{n-2} \cup CQ^{10}_{n-2}$ is isomorphic to CQ_{n-1}, by Lemma 14.18, there is an n^*-container of CQ_n between $\mathbf{u}$ and $\mathbf{v}$. Hence, this lemma is proved. □

LEMMA 14.20 Let $n \ge 7$. Suppose that $\mathbf{u}$ and $\mathbf{v}$ are two distinct vertices of CQ_n with $(\mathbf{u})_{n-2} \ne (\mathbf{v})_{n-2}$ and $(\mathbf{u})_{n-1} \ne (\mathbf{v})_{n-1}$. Then there is an n^*-container of CQ_n between $\mathbf{u}$ and $\mathbf{v}$.

Proof: Assume that $(\mathbf{u})_{n-2} = 0$ and $(\mathbf{u})_{n-1} = 0$. Thus, $\mathbf{u} \in CQ^{00}_{n-2}$ and $\mathbf{v} \in CQ^{11}_{n-2}$. We have the following cases:

Case 1. $\mathbf{u} = ((\mathbf{v})^n)^{n-1}$. Since $|V(CQ^{01}_{n-2})| = 2^{n-2} \ge 2n-6$ if $n \ge 7$, there is a set of $(n-3)$ distinct vertices $\{\mathbf{x_1}, \mathbf{x_2}, \ldots, \mathbf{x_{n-3}}\}$ of $V(CQ^{01}_{n-2}) - \{(\mathbf{u})^{n-1}\}$ such that $(\mathbf{x_i})^{n-1}$ is not a neighbor of $\mathbf{u}$ for every $1 \le i \le n-3$. By Theorem 14.12, there are $(n-3)$ disjoint paths $R_1, R_2, \ldots, R_{n-3}$ of CQ^{01}_{n-2} such that R_i joins $\mathbf{x_i}$ to $(\mathbf{u})^{n-1}$ for every $1 \le i \le n-3$ and $\bigcup_{i=1}^{n-3} R_i$ contains all the vertices in CQ^{01}_{n-2}. We write $R_i = \langle \mathbf{x_i}, S_i, \mathbf{y_i}, (\mathbf{u})^{n-1} \rangle$ for every $1 \le i \le n-3$.

Also, there are $(n-3)$ disjoint paths $H_1, H_2, \ldots, H_{n-3}$ of CQ^{00}_{n-2} such that H_i joins $\mathbf{u}$ to $(\mathbf{x_i})^{n-1}$ for every $1 \le i \le n-3$ and $\bigcup_{i=1}^{n-3} H_i$ contains all the vertices in CQ^{00}_{n-2}.

Again, there are $(n-3)$ disjoint paths $W_1, W_2, \ldots, W_{n-3}$ of CQ^{11}_{n-2} such that W_i joins $(\mathbf{y_i})^n$ to $\mathbf{v}$ for every $1 \le i \le n-3$ and $\bigcup_{i=1}^{n-3} W_i$ contains all the vertices in CQ^{11}_{n-2}. Since $\deg_{CQ^{00}_{n-2}}(\mathbf{u}) = n-2$, there is a vertex $\mathbf{p}$ in CQ^{00}_{n-2} such that $(\mathbf{p}, \mathbf{u}) \in E(CQ^{00}_{n-2})$ and $(\mathbf{p}, \mathbf{u}) \notin \bigcup_{i=1}^{n-3} E(H_i)$.

Since $\deg_{CQ^{11}_{n-2}}(\mathbf{v}) = n-2$, there is a vertex $\mathbf{q}$ in CQ^{11}_{n-2} such that $(\mathbf{q},\mathbf{v}) \in E(CQ^{11}_{n-2})$ and $(\mathbf{q},\mathbf{v}) \notin \bigcup_{i=1}^{n-3} E(W_i)$. We assume that $\mathbf{p} \in H_{n-3}$ and $\mathbf{q} \in W_t$. Let $H_{n-3} = \langle \mathbf{u}, P_1, \mathbf{w}, \mathbf{p}, \mathbf{r}, P_2, (\mathbf{x_{n-3}})^{n-1} \rangle$ and $W_t = \langle (\mathbf{y_t})^n, Q_1, \mathbf{q}, \mathbf{z}, Q_2, \mathbf{v} \rangle$.

Case 1.1. Suppose that $t \in \{1, 2, \ldots, n-3\}$. Without loss of generality, we assume that $t = n-3$. We set

$$\begin{aligned} T_i &= \langle \mathbf{u}, H_i, (\mathbf{x_i})^{n-1}, \mathbf{x_i}, S_i, \mathbf{y_i}, (\mathbf{y_i})^n, W_i, \mathbf{v} \rangle \quad \text{for every } 1 \le i \le n-4 \\ T_{n-3} &= \langle \mathbf{u}, (\mathbf{u})^n = (\mathbf{v})^{n-1}, \mathbf{v} \rangle \\ T_{n-2} &= \langle \mathbf{u}, (\mathbf{u})^{n-1} = (\mathbf{v})^n, \mathbf{v} \rangle \end{aligned}$$

To construct two paths T_{n-1} and T_n, the following two sub cases are considered:

Case 1.1.1. Suppose that $(\mathbf{w})^n \neq (\mathbf{z})^{n-1}$. By Theorem 14.10, there is a Hamiltonian path Z of $CQ^{10}_{n-2} - \{(\mathbf{u})^n\}$ joining $(\mathbf{w})^n$ to $(\mathbf{z})^{n-1}$. We set

$$\begin{aligned} T_{n-1} &= \langle \mathbf{u}, P_1, \mathbf{w}, (\mathbf{w})^n, Z, (\mathbf{z})^{n-1}, \mathbf{z}, Q_2, \mathbf{v} \rangle \\ T_n &= \langle \mathbf{u}, \mathbf{p}, \mathbf{r}, P_2, (\mathbf{x_{n-3}})^{n-1}, \mathbf{x_{n-3}}, S_{n-3}, \mathbf{y_{n-3}}, (\mathbf{y_{n-3}})^n, Q_1, \mathbf{q}, \mathbf{v} \rangle \end{aligned}$$

Then $\{T_1, T_2, \ldots, T_n\}$ is the n^*-container of CQ_n between $\mathbf{u}$ and $\mathbf{v}$.

Case 1.1.2. Suppose that $(\mathbf{w})^n = (\mathbf{z})^{n-1}$. By Theorem 14.10, there is a Hamiltonian path Z of $CQ^{10}_{n-2} - \{(\mathbf{u})^n, (\mathbf{w})^n\}$ joining $(\mathbf{p})^n$ to $(\mathbf{r})^n$. We set

$$\begin{aligned} T_{n-1} &= \langle \mathbf{u}, P_1, \mathbf{w}, (\mathbf{w})^n = (\mathbf{z})^{n-1}, \mathbf{z}, Q_2, \mathbf{v} \rangle \\ T_n &= \langle \mathbf{u}, \mathbf{p}, (\mathbf{p})^n, Z, (\mathbf{r})^n, \mathbf{r}, P_2, (\mathbf{x_{n-3}})^{n-1}, \mathbf{x_{n-3}}, S_{n-3}, \mathbf{y_{n-3}}, (\mathbf{y_{n-3}})^n, Q_1, \mathbf{q}, \mathbf{v} \rangle \end{aligned}$$

Then $\{T_1, T_2, \ldots, T_n\}$ forms the n^*-container of CQ_n between $\mathbf{u}$ and $\mathbf{v}$.

Case 1.2. Suppose that $t = n-4$. The paths T_i are constructed as in Case 1.1 for every $1 \le i \le n-5$. Two paths T_{n-4} and T_{n-3} are the same as T_{n-3} and T_{n-2} in Case 1.1, respectively. The paths T_{n-2}, T_{n-1}, and T_n are constructed as follows.

Case 1.2.1. Suppose that $(\mathbf{w})^n \neq (\mathbf{z})^{n-1}$. By Theorem 14.10, there is a Hamiltonian path Z of $CQ^{10}_{n-2} - \{(\mathbf{u})^n\}$ joining $(\mathbf{w_3})^n$ to $(\mathbf{z})^{n-1}$. The path T_{n-2} is the same as T_{n-1} in Case 1.1.1. We set $T_{n-1} = \langle \mathbf{u}, H_{n-4}, (\mathbf{x_{n-4}})^{n-1}, \mathbf{x_{n-4}}, S_{n-4}, \mathbf{y_{n-4}}, (\mathbf{y_{n-4}})^n, Q_1, \mathbf{q}, \mathbf{v} \rangle$ and $T_n = \langle \mathbf{u}, \mathbf{p}, \mathbf{r}, P_2, (\mathbf{x_{n-3}})^{n-1}, \mathbf{x_{n-3}}, S_{n-3}, \mathbf{y_{n-3}}, (\mathbf{y_{n-3}})^n, W_{n-3}, \mathbf{v} \rangle$. Then the n^*-container of CQ_n between $\mathbf{u}$ and $\mathbf{v}$ is composed of $\{T_1, T_2, \ldots, T_n\}$.

Case 1.2.2. Suppose that $(\mathbf{w})^n = (\mathbf{z})^{n-1}$. By Theorem 14.10, there is a Hamiltonian path Z of $CQ^{10}_{n-2} - \{(\mathbf{u})^n, \mathbf{w}^n\}$ joining $(\mathbf{p})^n$ to $(\mathbf{r})^n$. Two paths T_{n-2} and T_{n-1} are the same as T_{n-2} and T_n in Case 1.2.1, respectively. We set $T_n = \langle \mathbf{u}, \mathbf{p}, (\mathbf{p})^n, Z, (\mathbf{r})^n, \mathbf{r}, P_2, (\mathbf{x_{n-3}})^{n-1}, \mathbf{x_{n-3}}, S_{n-3}, \mathbf{y_{n-3}}, (\mathbf{y_{n-3}})^n, W_{n-3}, \mathbf{v} \rangle$. Then we have constructed the n^*-container $\{T_1, T_2, \ldots, T_n\}$ of CQ_n between $\mathbf{u}$ and $\mathbf{v}$.

Case 2. Suppose that $\mathbf{u} \neq ((\mathbf{v})^n)^{n-1}$. Since $|V(CQ^{01}_{n-2})| = 2^{n-2} \ge 2n-6$ if $n \ge 7$, there is a set of $(n-3)$ distinct vertices $\{\mathbf{x_1}, \mathbf{x_2}, \ldots, \mathbf{x_{n-3}}\}$ of

$V(CQ^{01}_{n-2}) - \{(\mathbf{u})^{n-1}, (\mathbf{v})^n\}$ such that $(\mathbf{x_i})^{n-1}$ is not a neighbor of $\mathbf{u}$ for every $1 \leq i \leq n-3$.

By Theorem 14.12, there is a set of $(n-3)$ disjoint paths $\{R_1, R_2, \ldots, R_{n-3}\}$ of CQ^{01}_{n-2} such that (1) R_i joins $\mathbf{x_i}$ to $(\mathbf{u})^{n-1}$ for every $1 \leq i \leq n-3$ and (2) $\bigcup_{i=1}^{n-3} R_i$ contains all the vertices in CQ^{01}_{n-2}. Assume that $(\mathbf{v})^n$ in R_{n-3}. We write $R_i = \langle \mathbf{x_i}, S_i, \mathbf{y_i}, (\mathbf{u})^{n-1} \rangle$ for every $1 \leq i \leq n-4$ and $R_{n-3} = \langle \mathbf{x_{n-3}}, S_{n-3}, \mathbf{y_{n-3}}, (\mathbf{v})^n, S, (\mathbf{u})^{n-1} \rangle$. Note that S_i is a path joining $\mathbf{x_i}$ to $\mathbf{y_i}$ for every $1 \leq i \leq n-3$, S is a path joining $(\mathbf{v})^n$ to $(\mathbf{u})^{n-1}$, and $(\bigcup_{i=1}^{n-3} S_i) \cup S$ contains all the vertices in CQ^{01}_{n-2}.

Similarly, there is a set of $(n-3)$ disjoint paths $\{H_1, H_2, \ldots, H_{n-3}\}$ of CQ^{00}_{n-2} such that (1) H_i joins $\mathbf{u}$ to $(\mathbf{x_i})^{n-1}$ for every $1 \leq i \leq n-3$ and (2) $\bigcup_{i=1}^{n-3} H_i$ contains all the vertices in CQ^{00}_{n-2}.

Also, there is a set of $(n-3)$ disjoint paths $\{W_1, W_2, \ldots, W_{n-3}\}$ of CQ^{11}_{n-2} such that (1) W_i joins $(\mathbf{y_i})^n$ to $\mathbf{v}$ for every $1 \leq i \leq n-3$ and (2) $\bigcup_{i=1}^{n-3} W_i$ contains all the vertices in CQ^{11}_{n-2}.

Since $\deg_{CQ^{00}_{n-2}}(\mathbf{u}) = n-2$, there is a vertex $\mathbf{p}$ in CQ^{00}_{n-2} such that $(\mathbf{p}, \mathbf{u}) \in E(CQ^{00}_{n-2})$ and $(\mathbf{p}, \mathbf{u}) \notin \bigcup_{i=1}^{n-3} E(H_i)$. Since $\deg_{CQ^{11}_{n-2}}(\mathbf{v}) = n-2$, there is a vertex $\mathbf{q}$ in CQ^{00}_{n-2} such that $(\mathbf{q}, \mathbf{v}) \in E(CQ^{11}_{n-2})$ and $(\mathbf{q}, \mathbf{v}) \notin \bigcup_{i=1}^{n-3} E(H_i)$. Without loss of generality, we assume that $\mathbf{p} \in H_{n-3}$ and $\mathbf{q} \in W_t$. We write $H_{n-3} = \langle \mathbf{u}, P_1, \mathbf{w}, \mathbf{p}, \mathbf{r}, P_2, (\mathbf{x_{n-3}})^{n-1} \rangle$ and $W_t = \langle (\mathbf{y_t})^n, Q_1, \mathbf{q}, \mathbf{z}, Q_2, \mathbf{v} \rangle$.

Case 2.1. Suppose that $t = n-3$. We set $T_i = \langle \mathbf{u}, H_i, (\mathbf{x_i})^{n-1}, \mathbf{x_i}, S_i, \mathbf{y_i}, (\mathbf{y_i})^n, W_i, \mathbf{v} \rangle$ for every $1 \leq i \leq n-4$ and $T_{n-3} = \langle \mathbf{u}, (\mathbf{u})^{n-1}, S^{-1}, (\mathbf{v})^n, \mathbf{v} \rangle$.

Case 2.1.1. Suppose that $(\mathbf{u})^n \neq (\mathbf{z})^{n-1}$ and $(\mathbf{v})^{n-1} \neq (\mathbf{w})^n$. By Theorem 14.11, there are two disjoint paths Z_1 and Z_2 of CQ^{10}_{n-2} such that (1) Z_1 joins $(\mathbf{u})^n$ to $(\mathbf{v})^{n-1}$, (2) Z_2 joins $(\mathbf{w})^n$ to $(\mathbf{z})^{n-1}$, and (3) $Z_1 \cup Z_2$ contains all the vertices in CQ^{10}_{n-2}. We set

$$\begin{aligned} T_{n-2} &= \langle \mathbf{u}, P_1, \mathbf{w}, (\mathbf{w})^n, Z_2, (\mathbf{z})^{n-1}, \mathbf{z}, Q_2, \mathbf{v} \rangle \\ T_{n-1} &= \langle \mathbf{u}, (\mathbf{u})^n, Z_1, (\mathbf{v})^{n-1}, \mathbf{v} \rangle \\ T_n &= \langle \mathbf{u}, \mathbf{p}, \mathbf{r}, P_2, (\mathbf{x_{n-3}})^{n-1}, \mathbf{x_{n-3}}, S_{n-3}, \mathbf{y_{n-3}}, (\mathbf{y_{n-3}})^n, Q_1, \mathbf{q}, \mathbf{v} \rangle \end{aligned}$$

Then $\{T_1, T_2, \ldots, T_n\}$ forms the n^*-container of CQ_n between $\mathbf{u}$ and $\mathbf{v}$.

Case 2.1.2. Suppose that $(\mathbf{u})^n = (\mathbf{z})^{n-1}$ and $(\mathbf{v})^{n-1} \neq (\mathbf{w})^n$. By Theorem 14.10, there is a Hamiltonian path Z of $CQ^{10}_{n-2} - \{(\mathbf{u})^n\}$ joining $(\mathbf{w})^n$ to $(\mathbf{v})^{n-1}$. The path T_{n-2} is the same as T_n in Case 2.1.1. We set $T_{n-1} = \langle \mathbf{u}, P_1, \mathbf{w}, (\mathbf{w})^n, Z, (\mathbf{v})^{n-1}, \mathbf{v} \rangle$ and $T_n = \langle \mathbf{u}, (\mathbf{u})^n = (\mathbf{z})^{n-1}, \mathbf{z}, Q_2, \mathbf{v} \rangle$. Then we have constructed the n^*-container $\{T_1, T_2, \ldots, T_n\}$ of CQ_n between $\mathbf{u}$ and $\mathbf{v}$.

Case 2.1.3. Suppose that $(\mathbf{u})^n \neq (\mathbf{z})^{n-1}$ and $(\mathbf{v})^{n-1} = (\mathbf{w})^n$. The proof of this case is similar to that of Case 2.1.2.

Case 2.1.4. Suppose that $(\mathbf{u})^n = (\mathbf{z})^{n-1}$ and $(\mathbf{v})^{n-1} = (\mathbf{w})^n$. By Theorem 14.10, there is a Hamiltonian path Z of $CQ^{10}_{n-2} - \{(\mathbf{u})^n, (\mathbf{v})^{n-1}\}$ joining $(\mathbf{p})^n$ to $(\mathbf{r})^n$. The path T_{n-2} is the same as T_n in Case 2.1.2. We set $T_{n-1} = \langle \mathbf{u}, P_1, \mathbf{w}, (\mathbf{w})^n = (\mathbf{v})^{n-1}, \mathbf{v} \rangle$ and

$T_n = \langle \mathbf{u}, \mathbf{p}, (\mathbf{p})^n, Z, (\mathbf{r})^n, \mathbf{r}, P_2, (\mathbf{x_{n-3}})^{n-1}, \mathbf{x_{n-3}}, S_{n-3}, \mathbf{y_{n-3}}, (\mathbf{y_{n-3}})^n, Q_1, \mathbf{q}, \mathbf{v} \rangle$. Then the n^*-container of CQ_n between **u** and **v** is composed of $\{T_1, T_2, \ldots, T_n\}$.

Case 2.2. Suppose that $t = n-4$. We set $T_i = \langle \mathbf{u}, H_i, (\mathbf{x_i})^{n-1}, \mathbf{x_i}, S_i, \mathbf{y_i}, (\mathbf{y_i})^n, W_i, \mathbf{v} \rangle$ for every $1 \le i \le n-5$ and $T_{n-4} = \langle \mathbf{u}, (\mathbf{u})^{n-1}, S^{-1}, (\mathbf{v})^n, \mathbf{v} \rangle$.

Case 2.2.1. Suppose that $(\mathbf{u})^n \neq (\mathbf{z})^{n-1}$ and $(\mathbf{v})^{n-1} \neq (\mathbf{w})^n$. By Theorem 14.11, there are two disjoint paths Z_1 and Z_2 of CQ_{n-2}^{10} such that (1) Z_1 joins $(\mathbf{u})^n$ to $(\mathbf{v})^{n-1}$, (2) Z_2 joins $(\mathbf{w})^n$ to $(\mathbf{z})^{n-1}$, and (3) $Z_1 \cup Z_2$ contains all the vertices in CQ_{n-2}^{10}. Two paths T_{n-3} and T_{n-2} are the same as T_{n-2} and T_{n-1} in Case 2.1.1, respectively. Two paths T_{n-1} and T_n are the same as T_{n-1} and T_n in Case 1.2.1, respectively. Then $\{T_1, T_2, \ldots, T_n\}$ is the n^*-container of CQ_n between **u** and **v**.

Case 2.2.2. Suppose that $(\mathbf{u})^n = (\mathbf{z})^{n-1}$ and $(\mathbf{v})^{n-1} \neq (\mathbf{w})^n$. By Theorem 14.10, there is a Hamiltonian path Z of $CQ_{n-2}^{10} - \{(\mathbf{u})^n\}$ joining $(\mathbf{w})^n$ to $(\mathbf{v})^{n-1}$. The two paths T_{n-3} and T_{n-2} are the same as T_{n-1} and T_n in Case 2.1.2, respectively. Also, the two paths T_{n-1} and T_n are the same as T_{n-1} and T_n in Case 1.2.1, respectively. We say that $\{T_1, T_2, \ldots, T_n\}$ forms the n^*-container of CQ_n between **u** and **v**.

Case 2.2.3. Suppose that $(\mathbf{u})^n \neq (\mathbf{z})^{n-1}$ and $(\mathbf{v})^{n-1} = (\mathbf{w})^n$. The proof of this case is similar to that of Case 2.2.2.

Case 2.2.4. Suppose that $(\mathbf{u})^n = (\mathbf{z})^{n-1}$ and $(\mathbf{v})^{n-1} = (\mathbf{w})^n$. By Theorem 14.10, there is a Hamiltonian path Z of $CQ_{n-2}^{10} - \{(\mathbf{u})^n, (\mathbf{v})^{n-1}\}$ joining $(\mathbf{p})^n$ to $(\mathbf{r})^n$. The path T_{n-3} is the same as T_{n-1} in Case 1.2.1. Also, the path T_{n-2} is the same as T_n in Case 2.1.2. Two paths T_{n-2} and T_{n-1} are the same as T_{n-1} and T_n in Case 2.1.4, respectively. Then $\{T_1, T_2, \ldots, T_n\}$ is the n^*-container of CQ_n between **u** and **v**.

Thus, this lemma is proved. □

With Lemmas 14.17 through 14.20, we have the following result.

THEOREM 14.13 CQ_n is super spanning connected if $n \geq 7$.

14.5 SPANNING CONNECTIVITY AND SPANNING LACEABILITY OF THE ENHANCED HYPERCUBE NETWORKS

The *folded hypercube* FQ_n is an important variation of hypercube proposed by El-Amawy and Latifi [96]. The *enhanced hypercube* $Q_{n,m}$ $(2 \le m \le n)$ is another important variation of hypercube proposed by Tzeng and Wei [317]. The folded hypercube FQ_n is obtained from a hypercube Q_n with add-on edges defined by joining any vertex $\mathbf{u} = u_1 u_2 \ldots u_{n-1} u_n$ to $\bar{\mathbf{u}} = \bar{u}_1 \bar{u}_2 \ldots \bar{u}_{n-1} \bar{u}_n$, where $\bar{u}_i = 1 - u_i$ is the complement of u_i. The enhanced hypercube $Q_{n,m}$ is obtained from a hypercube Q_n with add-on edges defined by joining any vertex $\mathbf{u} = u_1 u_2 \ldots u_{n-1} u_n$ to $(\mathbf{u})^c = \bar{u}_1 \bar{u}_2 \ldots \bar{u}_m u_{m+1} u_{m+2} \ldots u_{n-1} u_n$. Obviously, $FQ_n = Q_{n,n}$ and FQ_n and $Q_{n,m}$ are $(n+1)$-regular. Moreover, FQ_n is a bipartite graph if and only if n is odd and $Q_{n,m}$ is a bipartite graph if and only if m is odd.

Chang et al. [44] prove that the folded hypercube FQ_n is super spanning laceable if n is an odd integer and super spanning connected if otherwise. Moreover, Chang et al. prove that the enhanced hypercube $Q_{n,m}$ is super spanning laceable if m is an odd integer and super spanning connected if otherwise.

Let $\mathbf{x}$ be a vertex of Q_n and $U = \{\mathbf{y}_1, \mathbf{y}_2, \ldots, \mathbf{y}_k\}$ be a subset of $V(Q_n) - \{\mathbf{x}\}$. If there exist k paths $\{P_1, P_2, \ldots, P_k\}$ of Q_n such that (1) P_i joining $\mathbf{x}$ to $\mathbf{y}_i$ for $1 \le i \le k$, (2) $\mathbf{x}$ is the only common vertex of P_i and P_j for any $1 \le i \ne j \le k$. We called $\{P_1, P_2, \ldots, P_k\}$ a k-container joining $\mathbf{x}$ to U, denoted by $C(\mathbf{x}, U)$. A k-container $C(\mathbf{x}, U)$ is a k^*-container of Q_n if it contains all vertices of Q_n.

THEOREM 14.14 Assume that $k \le n$ and $\mathbf{x}$ is a vertex of Q_n. Let $U = \{\mathbf{y}_1, \mathbf{y}_2, \ldots, \mathbf{y}_k\}$ be a subset of $V(Q_n) - \{\mathbf{x}\}$ with $\mathbf{y}_i \ne \mathbf{y}_j$ for every $i \ne j$ and $\mathbf{y}_k$ is the only vertex in $\{\mathbf{y}_1, \mathbf{y}_2, \ldots, \mathbf{y}_k\}$ such that $\mathbf{y}_k$ and $\mathbf{x}$ are in different partite sets. Then there are k disjoint paths $P_1, P_2, \ldots, P_k$ in Q_n joining $\mathbf{x}$ to U such that (1) P_i joins $\mathbf{x}$ to $\mathbf{y}_i$ for $1 \le i \le k$ and (2) $\bigcup_{i=1}^{k} P_i$ spans Q_n.

Proof: By Theorem 14.7, this statement holds for every Q_n if $k = 1$. Suppose that $k = 2$ and $n \ge 2$. By Theorem 14.7, there is a Hamiltonian path $P = \langle \mathbf{y}_1, R_1, \mathbf{x}, R_2, \mathbf{y}_2 \rangle$ of Q_n joining $\mathbf{y}_1$ to $\mathbf{y}_2$. We set $P_1 = \langle \mathbf{x}, R_1^{-1}, \mathbf{y}_1 \rangle$ and $P_2 = \langle \mathbf{x}, R_2, \mathbf{y}_2 \rangle$. Then P_1 and P_2 form the required paths. Thus, we assume that $3 \le k \le n$, and this theorem is true for Q_{n-1}. Since Q_n is vertex-transitive, we assume that $\mathbf{x} = 0^n$. Thus, $\mathbf{x}$ is an even vertex and $\mathbf{x} \in Q_{n-1}^0$. We have the following cases:

Case 1. $(\mathbf{y}_k)_i = 0$ for some $1 \le i \le n$. Since Q_n is edge-transitive, we assume that $(\mathbf{y}_k)_n = 0$. Thus, $\mathbf{y}_k \in Q_{n-1}^0$. For $0 \le j \le 1$, we set $U_j = \{\mathbf{y}_i \mid \mathbf{y}_i \in Q_{n-1}^j \text{ for } 1 \le i \le k\}$. Without loss of generality, we assume that $U_0 = \{\mathbf{y}_{m+1}, \mathbf{y}_{m+2}, \ldots, \mathbf{y}_k\} \subseteq Q_{n-1}^0$ and $U_1 = \{\mathbf{y}_1, \mathbf{y}_2, \ldots, \mathbf{y}_m\} \subseteq Q_{n-1}^1$ for some $0 \le m \le k-1$.

Case 1.1. $m = 0$. Let $\tilde{U} = U_0 - \{\mathbf{y}_{k-1}\}$. Obviously, $|\tilde{U}| = k - 1$. By induction, there are $(k-1)$ disjoint paths $\{R_1, R_2, \ldots, R_{k-1}\}$ of Q_{n-1}^0 joining $\mathbf{x}$ to $\tilde{U}$ such that (1) R_i joins $\mathbf{x}$ to $\mathbf{y}_i$ for every $1 \le i \le k-2$, (2) R_{k-1} joins $\mathbf{x}$ to $\mathbf{y}_k$, and (3) $\bigcup_{i=1}^{k-1} R_i$ spans Q_{n-1}^0. Obviously, $\mathbf{y}_{k-1} \in R_i$ for some $1 \le i \le k-1$.

Suppose that $\mathbf{y}_{k-1} \in R_i$ for some $1 \le i \le k-2$. Without loss of generality, we assume that $\mathbf{y}_{k-1} \in R_{k-2}$. We write R_{k-2} as $\langle \mathbf{x}, H_1, \mathbf{y}_{k-1}, \mathbf{z}, H_2, \mathbf{y}_{k-2} \rangle$. Since $\mathbf{y}_{k-1}$ is an even vertex, $\mathbf{z}$ is an odd vertex in Q_{n-1}^0. By Theorem 14.7, there is a Hamiltonian path W of Q_{n-1}^1 joining $(\mathbf{x})^n$ to $(\mathbf{z})^n$. We set $P_i = R_i$ for every $1 \le i \le k-3$, $P_{k-2} = \langle \mathbf{x}, (\mathbf{x})^n, W, (\mathbf{z})^n, \mathbf{z}, H_2, \mathbf{y}_{k-2} \rangle$, $P_{k-1} = \langle \mathbf{x}, H_1, \mathbf{y}_{k-1} \rangle$, and $P_k = R_{k-1}$. Then $\{P_1, P_2, \ldots, P_k\}$ forms a set of required paths of Q_n. See Figure 14.12a for illustration, for $k = 6$.

Suppose that $\mathbf{y}_{k-1} \in R_{k-1}$. We can write R_{k-1} as $\langle \mathbf{x}, H_1, \mathbf{y}_{k-1}, \mathbf{z}, H_2, \mathbf{y}_k \rangle$. (Note that $\mathbf{z} = \mathbf{y}_k$ if $l(H_2) = 0$.) By Theorem 14.7, there is a Hamiltonian path W of Q_{n-1}^1 joining $(\mathbf{x})^n$ to $(\mathbf{z})^n$. We set $P_i = R_i$ for every $1 \le i \le k-2$, $P_{k-1} = \langle \mathbf{x}, H_1, \mathbf{y}_{k-1} \rangle$, and $P_k = \langle \mathbf{x}, (\mathbf{x})^n, W, (\mathbf{z})^n, \mathbf{z}, H_2, \mathbf{y}_k \rangle$. Then $\{P_1, P_2, \ldots, P_k\}$ forms a set of required paths of Q_n. See Figure 14.12b for illustration, for $k = 6$.

Case 1.2. $m = 1$. Thus, $\mathbf{y}_1 \in Q_{n-1}^1$. By induction, there are $(k-1)$ disjoint paths $\{R_1, R_2, \ldots, R_{k-1}\}$ of Q_{n-1}^0 joining $\mathbf{x}$ to U_0 such that (1) R_i joins $\mathbf{x}$ to $\mathbf{y}_{i+1}$ for every

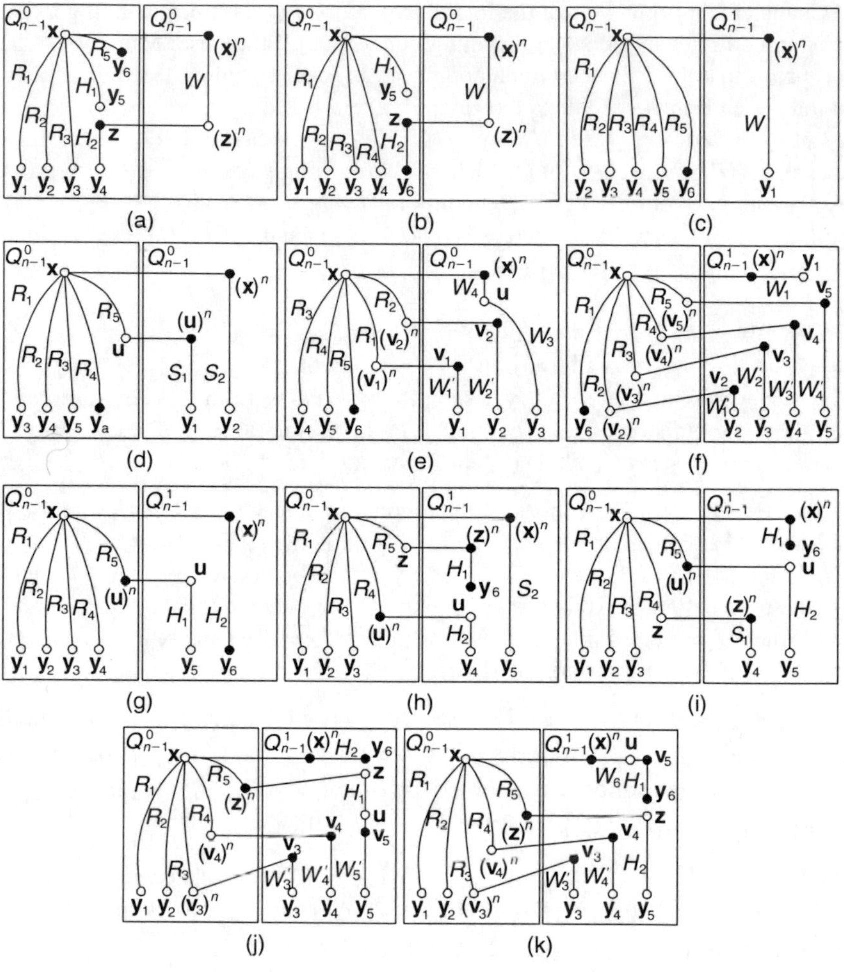

FIGURE 14.12 Illustrations for Theorem 14.14.

$1 \le i \le k-1$ and (2) $\bigcup_{i=1}^{k-1} R_i$ spans Q_{n-1}^0. By Theorem 14.7, there is a Hamiltonian path W of Q_{n-1}^1 joining $(\mathbf{x})^n$ to $\mathbf{y}_1$. We set $P_1 = \langle \mathbf{x}, (\mathbf{x})^n, W, \mathbf{y}_1 \rangle$ and $P_i = R_{i-1}$ for every $2 \le i \le k$. Then $\{P_1, P_2, \ldots, P_k\}$ forms a set of required paths of Q_n. See Figure 14.12c for illustration, for $k = 6$.

Case 1.3. $m = 2$. We have $\{\mathbf{y}_1, \mathbf{y}_2\} \subseteq Q_{n-1}^1$. Since there are 2^{n-2} even vertices in Q_{n-1}^0 and $2^{n-2} - |U_0 \cup \{\mathbf{x}\}| - 1 = 2^{n-2} - (k-2) \ge 2^{n-2} - n + 2 \ge 1$ if $n \ge 3$, we can choose an even vertex $\mathbf{u}$ in $Q_{n-1}^0 - (U_0 \cup \{\mathbf{x}\})$. By induction, there are $(k-1)$ disjoint paths $\{R_1, R_2, \ldots, R_{k-1}\}$ of Q_{n-1}^0 joining $\mathbf{x}$ to $U_0 \cup \{\mathbf{u}\}$ such that (1) R_i joins $\mathbf{x}$ to $\mathbf{y}_{i+2}$ for every $1 \le i \le k-2$, (2) R_{k-1} joins $\mathbf{x}$ to $\mathbf{u}$, and (3) $\bigcup_{i=1}^{k-1} R_i$ spans Q_{n-1}^0. By Theorem 14.4, there exist two disjoint paths S_1 and S_2 of Q_{n-1}^1 such that (1) S_1 joins $(\mathbf{u})^n$ to $\mathbf{y}_1$, (2) S_2 joins $(\mathbf{x})^n$ to $\mathbf{y}_2$, and (3) $S_1 \cup S_2$ spans Q_{n-1}^1. We set $P_1 = \langle \mathbf{x}, R_{k-1}, \mathbf{u}, (\mathbf{u})^n, S_1, \mathbf{y}_1 \rangle$, $P_2 = \langle \mathbf{x}, (\mathbf{x})^n, S_2, \mathbf{y}_2 \rangle$, and $P_i = R_{i-2}$ for every $3 \le i \le k$.

Then $\{P_1, P_2, \ldots, P_k\}$ forms a set of required paths of Q_n. See Figure 14.12d for illustrations, for $k = 6$.

Case 1.4. $3 \le m \le k-2$. We have $k \ge 5$. Hence, $n \ge 5$. Since $m \ge 3$ and $k \le n$, $|U_0 - \{\mathbf{y}_k\}| = k - m - 1 \le k - 4 \le n - 4$.

We claim that there exists an even vertex $\mathbf{u}$ in $Q^1_{n-1} - U_1$ such that $(\mathbf{y}_i)^n \notin N_{Q^1_{n-1}}(\mathbf{u})$ for every $m+1 \le i \le k-1$. Such a claim holds, because $(n-1)|U_0 - \{\mathbf{y}_k\}| + |U_1| \le (n-1)(n-4) + (n-2) \le (n-1)(n-3) - 1 < 2^{n-2}$ for all $n \ge 5$.

Since $m \le k-2$ and $k \le n$, $m+1 \le n-1$. By induction, there are $(m+1)$ disjoint paths $\{W_1, W_2, \ldots, W_{m+1}\}$ of Q^1_{n-1} joining $\mathbf{u}$ to $U_1 \cup \{(\mathbf{x})^n\}$ such that (1) W_i joins $\mathbf{u}$ to $\mathbf{y}_i$ for every $1 \le i \le m$, (2) W_{m+1} joins $\mathbf{u}$ to $(\mathbf{x})^n$, and (3) $\bigcup_{i=1}^{m+1} W_i$ spans Q^1_{n-1}. We write W_i as $\langle \mathbf{u}, \mathbf{v}_i, W'_i, \mathbf{y}_i \rangle$ for every $1 \le i \le m-1$. Since $\mathbf{u}$ is an even vertex in Q^1_{n-1}, $\mathbf{v}_i$ is an odd vertex in Q^1_{n-1} and $(\mathbf{v}_i)^n$ is an even vertex in Q^0_{n-1} for every $1 \le i \le m-1$. Let $\tilde{U}_0 = U_0 \cup \{(\mathbf{v}_i)^n \mid 1 \le i \le m-1\}$. Obviously, $|\tilde{U}_0| = (k-m) + (m-1) = k-1$. By induction, there are $(k-1)$ disjoint paths $\{R_1, R_2, \ldots, R_{k-1}\}$ of Q^0_{n-1} joining $\mathbf{x}$ to $\tilde{U}_0$ such that (1) R_i joins $\mathbf{x}$ to $(\mathbf{v}_i)^n$ for every $1 \le i \le m-1$, (2) R_i joins $\mathbf{x}$ to $\mathbf{y}_{i+1}$ for every $m \le i \le k-1$, and (3) $\bigcup_{i=1}^{k-1} R_i$ spans Q^0_{n-1}. We set $P_i = \langle \mathbf{x}, R_i, (\mathbf{v}_i)^n, \mathbf{v}_i, W'_i, \mathbf{y}_i \rangle$ for every $1 \le i \le m-1$, $P_m = \langle \mathbf{x}, (\mathbf{x})^n, W^{-1}_{m+1}, \mathbf{u}, W_m, \mathbf{y}_m \rangle$, and $P_i = R_{i-1}$ for every $m+1 \le i \le k$. Then $\{P_1, P_2, \ldots, P_k\}$ forms a set of required paths of Q_n. See Figure 14.12e for illustration, for $k = 6$ and $m = 3$.

Case 1.5. $m = k-1$ and $k - 1 \ge 3$. Let $\tilde{U}_1 = (U_1 - \{\mathbf{y}_1\}) \cup \{(\mathbf{x})^n\}$. Obviously, $|\tilde{U}_1| = k-1$. By induction, there are $(k-1)$ disjoint paths $\{W_1, W_2, \ldots, W_{k-1}\}$ of Q^1_{n-1} joining $\mathbf{y}_1$ to $\tilde{U}_1$ such that (1) W_1 joins $\mathbf{y}_1$ to $(\mathbf{x})^n$, (2) W_i joins $\mathbf{y}_1$ to $\mathbf{y}_i$ for every $2 \le i \le k-1$, and (3) $\bigcup_{i=1}^{k-1} W_i$ spans Q^1_{n-1}. We write W_i as $\langle \mathbf{y}_1, \mathbf{v}_i, W'_i, \mathbf{y}_i \rangle$ for every $2 \le i \le k-1$. Since $\mathbf{y}_1$ is an even vertex in Q^1_{n-1}, $\mathbf{v}_i$ is an odd vertex in Q^1_{n-1} and $(\mathbf{v}_i)^n$ is an even vertex in Q^0_{n-1} for every $2 \le i \le k-1$. Let $\tilde{U}_0 = \{\mathbf{y}_k\} \cup \{(\mathbf{v}_i)^n \mid 2 \le i \le k-1\}$. Obviously, $|\tilde{U}_0| = k-1$. By induction, there are $(k-1)$ disjoint paths $\{R_1, R_2, \ldots, R_{k-1}\}$ of Q^0_{n-1} joining $\mathbf{x}$ to $\tilde{U}_0$ such that (1) R_1 joins $\mathbf{x}$ to $\mathbf{y}_k$, (2) R_i joins $\mathbf{x}$ to $(\mathbf{v}_i)^n$ for every $2 \le i \le k-1$, and (3) $\bigcup_{i=1}^{k-1} R_i$ spans Q^0_{n-1}. We set $P_1 = \langle \mathbf{x}, (\mathbf{x})^n, W_1^{-1}, \mathbf{y}_1 \rangle$, $P_i = \langle \mathbf{x}, R_i, (\mathbf{v}_i)^n, \mathbf{v}_i, W'_i, \mathbf{y}_i \rangle$ for every $2 \le i \le k-1$, and $P_k = R_1$. Then $\{P_1, P_2, \ldots, P_k\}$ forms a set of required paths of Q_n. See Figure 14.12f for illustration, for $k = 6$.

Case 2. $(\mathbf{y}_k)_i = 1$ for every $1 \le i \le n$. Obviously, n is odd with $n \ge 3$ and $\mathbf{y}_k \in Q^1_{n-1}$. Since Q_n is edge-transitive, we assume that $U_0 = \{\mathbf{y}_1, \mathbf{y}_2, \ldots, \mathbf{y}_m\} \subseteq Q^0_{n-1}$ and $U_1 = \{\mathbf{y}_{m+1}, \mathbf{y}_{m+2}, \ldots, \mathbf{y}_k\} \subseteq Q^1_{n-1}$ for some $1 \le m \le k-2$.

Case 2.1. $m = k-2$. We have $\{\mathbf{y}_{k-1}, \mathbf{y}_k\} \subseteq Q^1_{n-1}$. Let H be a Hamiltonian path of Q^1_{n-1} joining $\mathbf{y}_{k-1}$ to $\mathbf{y}_k$. We write H as $\langle \mathbf{y}_{k-1}, H_1, \mathbf{u}, (\mathbf{x})^n, H_2, \mathbf{y}_k \rangle$. Since $(\mathbf{x})^n$ is an odd vertex, $\mathbf{u}$ is an even vertex and $(\mathbf{u})^n$ is an odd vertex in Q^0_{n-1}. (Note that $\mathbf{y}_{k-1} = \mathbf{u}$ if $l(H_1) = 0$ or $(\mathbf{x})^n = \mathbf{y}_k$ if $l(H_2) = 0$.) By induction, there are $(k-1)$ disjoint paths $\{R_1, R_2, \ldots, R_{k-1}\}$ of Q^0_{n-1} joining $\mathbf{x}$ to $U_0 \cup \{(\mathbf{u})^n\}$ such that (1) R_i joins $\mathbf{x}$ to $\mathbf{y}_i$ for $1 \le i \le k-2$, (2) R_{k-1} joins $\mathbf{x}$ to $(\mathbf{u})^n$, and (3) $\bigcup_{i=1}^{k-1} R_i$ spans Q^0_{n-1}. We set $P_i = R_i$ for

$1 \le i \le k-2$, $P_{k-1} = \langle \mathbf{x}, R_{k-1}, (\mathbf{u})^n, \mathbf{u}, H_1^{-1}, \mathbf{y}_{k-1}\rangle$, and $P_k = \langle \mathbf{x}, (\mathbf{x})^n, H_2, \mathbf{y}_k\rangle$. Then $\{P_1, P_2, \ldots, P_k\}$ forms a set of required paths of Q_n. See Figure 14.12g for illustration, for $k = 6$.

Case 2.2. $m = k-3$. We have $n \ge 5$ and $\{\mathbf{y}_{k-2}, \mathbf{y}_{k-1}, \mathbf{y}_k\} \subseteq Q^1_{n-1}$. Since $m+1 \le n-2 < 2^{n-2}$, we can pick an even vertex $\mathbf{z} \in Q^0_{n-1} - (\{\mathbf{y}_i \mid 1 \le i \le k-3\} \cup \{\mathbf{x}\})$. By Theorem 14.4, there exist two disjoint paths S_1 and S_2 of Q^1_{n-1} such that (1) S_1 joins $(\mathbf{z})^n$ to $\mathbf{y}_{k-2}$, (2) S_2 joins $(\mathbf{x})^n$ to $\mathbf{y}_{k-1}$, and (3) $S_1 \cup S_2$ spans Q^1_{n-1}. Obviously, $\mathbf{y}_k \in S_i$ for some $1 \le i \le 2$.

Suppose that $\mathbf{y}_k \in S_1$. We write S_1 as $\langle (\mathbf{z})^n, H_1, \mathbf{y}_k, \mathbf{u}, H_2, \mathbf{y}_{k-2}\rangle$. Obviously, $\mathbf{u}$ is an even vertex and $(\mathbf{u})^n$ is an odd vertex in Q^0_{n-1}. (Note that $(\mathbf{x})^n = \mathbf{y}_k$ if $l(H_1) = 0$ or $\mathbf{u} = \mathbf{y}_{k-2}$ if $l(H_2) = 0$.) Let $\tilde{U}_0 = U_0 \cup \{\mathbf{z}, (\mathbf{u})^n\}$. Obviously, $|\tilde{U}_0| = k-1$. By induction, there are $(k-1)$ disjoint paths $R_1, R_2, \ldots, R_{k-1}$ in Q^0_{n-1} joining $\mathbf{x}$ to $\tilde{U}_0$ such that (1) R_i joins $\mathbf{x}$ to $\mathbf{y}_i$ for $1 \le i \le k-3$, (2) R_{k-2} joins $\mathbf{x}$ to $\mathbf{z}$, (3) R_{k-1} joins $\mathbf{x}$ to $(\mathbf{u})^n$, and (4) $\bigcup_{i=1}^{k-1} R_i$ spans Q^0_{n-1}. We set $P_i = R_i$ for $1 \le i \le k-3$, $P_{k-2} = \langle \mathbf{x}, R_{k-1}, (\mathbf{u})^n, \mathbf{u}, H_2, \mathbf{y}_{k-2}\rangle$, $P_{k-1} = \langle \mathbf{x}, (\mathbf{x})^n, S_2, \mathbf{y}_{k-1}\rangle$, and $P_k = \langle \mathbf{x}, R_{k-2}, \mathbf{z}, (\mathbf{z})^n, H_1, \mathbf{y}_k\rangle$. Then $\{P_1, P_2, \ldots, P_k\}$ forms a set of required paths of Q_n. See Figure 14.12h for illustration, for $k = 6$.

Suppose that $\mathbf{y}_k \in S_2$. We write S_2 as $\langle (\mathbf{x})^n, H_1, \mathbf{y}_k, \mathbf{u}, H_2, \mathbf{y}_{k-1}\rangle$. Obviously, $\mathbf{u}$ is an even vertex in Q^1_{n-1} and $(\mathbf{u})^n$ is an odd vertex in Q^0_{n-1}. (Note that $(\mathbf{x})^n = \mathbf{y}_k$ if $l(H_1) = 0$ or $\mathbf{u} = \mathbf{y}_{k-1}$ if $l(H_2) = 0$.) Let $\tilde{U}_0 = U_0 \cup \{\mathbf{z}, (\mathbf{u})^n\}$. Obviously, $|\tilde{U}_0| = k-1$. By induction, there are $(k-1)$ disjoint paths $R_1, R_2, \ldots, R_{k-1}$ in Q^0_{n-1} joining $\mathbf{x}$ to $\tilde{U}_0$ such that (1) R_i joins $\mathbf{x}$ to $\mathbf{y}_i$ for $1 \le i \le k-3$, (2) R_{k-2} joins $\mathbf{x}$ to $\mathbf{z}$, (3) R_{k-1} joins $\mathbf{x}$ to $(\mathbf{u})^n$, and (4) $\bigcup_{i=1}^{k-1} R_i$ spans Q^0_{n-1}. We set $P_i = R_i$ for $1 \le i \le k-3$, $P_{k-2} = \langle \mathbf{x}, R_{k-2}, \mathbf{z}, (\mathbf{z})^n, S_1, \mathbf{y}_{k-2}\rangle$, $P_{k-1} = \langle \mathbf{x}, R_{k-1}, (\mathbf{u})^n, \mathbf{u}, H_2, \mathbf{y}_{k-1}\rangle$, and $P_k = \langle \mathbf{x}, (\mathbf{x})^n, H_1, \mathbf{y}_k\rangle$. Then $\{P_1, P_2, \ldots, P_k\}$ forms a set of required paths of Q_n. See Figure 14.12i for illustration, for $k = 6$.

Case 2.3. $1 \le m \le k-4$. We have $k \ge 5$. Moreover, $n \ge 5$. Since $m \le k-4$ and $k \le n$, $|U_0| = m \le k-4 \le n-4$.

We claim that there exists an even vertex $\mathbf{u}$ in $Q^1_{n-1} - U_1$ such that $(\mathbf{y}_i)^n \notin N_{Q^1_{n-1}}(\mathbf{u})$ for every $1 \le i \le m$. Such a claim holds, because $(n-1)|U_0| + |U_1 - \{\mathbf{y}_k\}| = (n-1)m + (k-m) - 1 = (n-2)m + k - 1 \le (n-2)(n-4) + n - 1 = (n-1)(n-4) + 3 < 2^{n-2}$ for all $n \ge 5$.

Let $\tilde{U}_1 = (U_1 - \{\mathbf{y}_k\}) \cup \{(\mathbf{x})^n\}$. Obviously, $|\tilde{U}_1| = k-m$. By induction, there are $(k-m)$ disjoint paths $\{W_{m+1}, W_{m+2}, \ldots, W_k\}$ of Q^1_{n-1} joining $\mathbf{u}$ to $\tilde{U}_1$ such that (1) W_i joins $\mathbf{u}$ to $\mathbf{y}_i$ for every $m+1 \le i \le k-1$, (2) W_k joins $\mathbf{u}$ to $(\mathbf{x})^n$, and (3) $\cup_{i=m+1}^{k} W_i$ spans Q^1_{n-1}. We write W_i as $\langle \mathbf{u}, \mathbf{v}_i, W'_i, \mathbf{y}_i\rangle$ for every $m+1 \le i \le k-1$. Since $\mathbf{u}$ is an even vertex in Q^1_{n-1}, $\mathbf{v}_i$ is an odd vertex in Q^1_{n-1} and $(\mathbf{v}_i)^n$ is an even vertex in Q^0_{n-1} for every $m+1 \le i \le k-2$.

Suppose that $\mathbf{y}_k \in W_k$. We write W_k as $\langle \mathbf{u}, H_1, \mathbf{z}, \mathbf{y}_k, H_2, (\mathbf{x})^n\rangle$. (Note that $\mathbf{u} = \mathbf{z}$ if $l(H_1) = 0$ or $\mathbf{y}_k = (\mathbf{x})^n$ if $l(H_2) = 0$.) Since $\mathbf{y}_k$ is an odd vertex in Q^1_{n-1}, $\mathbf{z}$ is an even vertex in Q^1_{n-1}, and $(\mathbf{z})^n$ is an odd vertex in Q^0_{n-1}. Let $\tilde{U}_0 = U_0 \cup \{(\mathbf{v}_i)^n \mid m+1 \le i \le k-2\} \cup \{(\mathbf{z})^n\}$. Obviously, $|\tilde{U}_0| = m + (k-m-2) + 1 = k-1$. By induction, there are $(k-1)$ disjoint paths

$\{R_1, R_2, \ldots, R_{k-1}\}$ of Q_{n-1}^0 joining $\mathbf{x}$ to $\tilde{U}_0$ such that (1) R_i joins $\mathbf{x}$ to $\mathbf{y}_i$ for $1 \le i \le m$, (2) R_i joins $\mathbf{x}$ to $(\mathbf{v}_i)^n$ for every $m+1 \le i \le k-2$, (3) R_{k-1} joins $\mathbf{x}$ to $(\mathbf{z})^n$, and (4) $\bigcup_{i=1}^{k-1} R_i$ spans Q_{n-1}^0. We set $P_i = R_i$ for every $1 \le i \le m$, $P_i = \langle \mathbf{x}, R_i, (\mathbf{v}_i)^n, \mathbf{v}_i, W_i', \mathbf{y}_i \rangle$ for every $m+1 \le i \le k-2$, $P_{k-1} = \langle \mathbf{x}, R_{k-1}, (\mathbf{z})^n, \mathbf{z}, H_1^{-1}, \mathbf{u}, \mathbf{v}_{k-1}, W_{k-1}', \mathbf{y}_{k-1} \rangle$, and $P_k = \langle \mathbf{x}, (\mathbf{x})^n, H_2^{-1}, \mathbf{y}_k \rangle$. Then $\{P_1, P_2, \ldots, P_k\}$ forms a set of required paths of Q_n. See Figure 14.12j for illustration, for $k=6$ and $m=2$.

Suppose that $\mathbf{y}_k \in W_i$ for some $1 \le i \le k-1$. Without loss of generality, we assume that $\mathbf{y}_k \in W_{k-1}$. We write W_{k-1} as $\langle \mathbf{u}, \mathbf{v}_{k-1}, H_1, \mathbf{y}_k, \mathbf{z}, H_2, \mathbf{y}_{k-1} \rangle$. (Note that $\mathbf{v}_{k-1} = \mathbf{y}_k$ if $l(H_1) = 0$ or $\mathbf{z} = \mathbf{y}_{k-1}$ if $l(H_2) = 0$.) Since $\mathbf{y}_k$ is an odd vertex in Q_{n-1}^1, $\mathbf{z}$ is an even vertex in Q_{n-1}^1 and $(\mathbf{z})^n$ is an odd vertex in Q_{n-1}^0. Let $\tilde{U}_0 = U_0 \cup \{(\mathbf{v}_i)^n \mid m+1 \le i \le k-2\} \cup \{(\mathbf{z})^n\}$. Obviously, $|\tilde{U}_0| = m + (k-m-2) + 1 = k-1$. By induction, there are $(k-1)$ disjoint paths $\{R_1, R_2, \ldots, R_{k-1}\}$ of Q_{n-1}^0 joining $\mathbf{x}$ to $\tilde{U}_0$ such that (1) R_i joins $\mathbf{x}$ to $\mathbf{y}_i$ for every $1 \le i \le m$, (2) R_i joins $\mathbf{x}$ to $(\mathbf{v}_i)^n$ for every $m+1 \le i \le k-2$, (3) R_{k-1} joins $\mathbf{x}$ to $(\mathbf{z})^n$, and (4) $\bigcup_{i=1}^{k-1} R_i$ spans Q_{n-1}^0. We set $P_i = R_i$ for every $1 \le i \le m$, $P_i = \langle \mathbf{x}, R_i, (\mathbf{v}_i)^n, \mathbf{v}_i, W_i', \mathbf{y}_i \rangle$ for every $m+1 \le i \le k-2$, $P_{k-1} = \langle \mathbf{x}, R_{k-1}, (\mathbf{z})^n, \mathbf{z}, H_2, \mathbf{y}_{k-1} \rangle$, and $P_k = \langle \mathbf{x}, (\mathbf{x})^n, W_k^{-1}, \mathbf{u}, \mathbf{v}_{k-1}, H_1, \mathbf{y}_k \rangle$. Then $\{P_1, P_2, \ldots, P_k\}$ forms a set of required paths of Q_n. See Figure 14.12k for illustration, for $k=6$ and $m=2$. □

Let $\mathbf{u} = u_1 u_2 \ldots u_{n-1} u_n$ be a vertex of FQ_n. The *c-neighbor* of $\mathbf{u}$ in FQ_n, $(\mathbf{u})^c$, is $\bar{u}_1 \bar{u}_2 \ldots \bar{u}_n$. Note that $(\mathbf{u})^c$ and $\mathbf{u}$ are of the same parity if and only if n is an even integer. Let $E^c = \{(u_1 u_2 \ldots u_n, \bar{u}_1 \bar{u}_2 \ldots \bar{u}_n) \mid u_1 u_2 \ldots u_n \in V(FQ_n)\}$. By definition, the n-dimensional folded hypercube FQ_n is obtained from Q_n by adding E^c. Let f be a function on $V(FQ_n)$ defined by $f(\mathbf{u}) = \mathbf{u}$ if $(\mathbf{u})_n = 0$ and $f(\mathbf{u}) = ((\mathbf{u})^c)^n$ if otherwise. The following theorem can be proved easily.

THEOREM 14.15 The function f is an isomorphism of FQ_n into itself.

Let FQ_{n-1}^j be the subgraph of FQ_n induced by $\{\mathbf{v} \in V(FQ_n) \mid (\mathbf{v})_n = j\}$ for $0 \le j \le 1$. Obviously, FQ_{n-1}^j is isomorphic to Q_{n-1} for $0 \le j \le 1$.

LEMMA 14.21 Let $\mathbf{x}$ be an even vertex and $\mathbf{y}$ be an odd vertex of FQ_n for any positive integer $n \ge 2$. Then there exists a k^*-container of FQ_n between $\mathbf{x}$ and $\mathbf{y}$ for every $1 \le k \le n+1$.

Proof: Since FQ_2 is isomorphic to the complete graph K_4, this statement holds for $n=2$. Suppose that $n \ge 3$. Since Q_n is a spanning subgraph of FQ_n, by Theorem 14.7, there exists a k^*-container between $\mathbf{x}$ and $\mathbf{y}$ for every $1 \le k \le n$. Thus, we need only to construct an $(n+1)^*$-container of FQ_n between $\mathbf{x}$ and $\mathbf{y}$. Since FQ_n is vertex-transitive, we assume that $\mathbf{x} = 0^n \in FQ_{n-1}^0$.

Case 1. $\mathbf{y} \in FQ_{n-1}^0$. We have the following sub cases:

Case 1.1. $n=3$. Without loss of generality, we assume that $\mathbf{y} = 100$. We set $P_1 = \langle 000, 001, 101, 100 \rangle$, $P_2 = \langle 000, 010, 110, 100 \rangle$, $P_3 = \langle 000, 100 \rangle$, and

$P_4 = \langle 000, 111, 011, 100\rangle$. Then $\{P_1, P_2, P_3, P_4\}$ forms a 4*-container of FQ_3 between **x** and **y**.

Case 1.2. $n \geq 4$ and $(\mathbf{x})^c \neq (\mathbf{y})^n$. Since FQ^0_{n-1} is isomorphic to Q_{n-1}, by Theorem 14.7, there is an $(n-1)^*$-container $\{P_1, P_2, \ldots, P_{n-1}\}$ of FQ^0_{n-1} between **x** and **y**. Obviously, $(\mathbf{x})^c$ and $(\mathbf{y})^c$ are of different parity. Since FQ^1_{n-1} is isomorphic to Q_{n-1}, by Theorem 14.4, there exist two disjoint paths S_1 and S_2 of FQ^1_{n-1} such that (1) S_1 joins $(\mathbf{x})^n$ to $(\mathbf{y})^n$, (2) S_2 joins $(\mathbf{x})^c$ to $(\mathbf{y})^c$, and (3) $S_1 \cup S_2$ spans FQ^1_{n-1}. We set $P_n = \langle \mathbf{x}, (\mathbf{x})^n, S_1, (\mathbf{y})^n, \mathbf{y}\rangle$ and $P_{n+1} = \langle \mathbf{x}, (\mathbf{x})^c, S_2, (\mathbf{y})^c, \mathbf{y}\rangle$. Then $\{P_1, P_2, \ldots, P_{n+1}\}$ forms an $(n+1)^*$-container of FQ_n between **x** and **y**. See Figure 14.13a for illustration, for $n = 5$.

Case 1.3. $n \geq 4$ and $(\mathbf{x})^c = (\mathbf{y})^n$. Then $(\mathbf{y})^c = (\mathbf{x})^n$ and n is even.

Suppose that $n = 4$. We have $\mathbf{x} = 0000$ and $\mathbf{y} = 1110$. We set $P_1 = \langle 0000, 0001, 1110\rangle$, $P_2 = \langle 0000, 0010, 0110, 1110\rangle$, $P_3 = \langle 0000, 0100, 0101, 0111, 0011, 1011, 1001, 1101, 1100, 1110\rangle$, $P_4 = \langle 0000, 1000, 1010, 1110\rangle$, and $P_5 = \langle 0000, 1111, 1110\rangle$. Then $\{P_1, P_2, P_3, P_4, P_5\}$ forms a 5*-container of FQ_4 between **x** and **y**.

Suppose that $n \geq 6$. By Theorem 14.7, there is an $(n-1)^*$-container $\{P_1, P_2, \ldots, P_{n-1}\}$ of FQ^0_{n-1} between **x** and **y**. Since $2^{n-1} - 2 \geq 3(n-1)$ for $n \geq 6$, there is one path P_i in $\{P_1, P_2, \ldots, P_{n-1}\}$ such that $I(P_i) \geq 3$. Without loss of generality, we may assume that $I(P_{n-1}) \geq 3$. We write P_{n-1} as $\langle \mathbf{x}, \mathbf{u}, \mathbf{v}, H, \mathbf{y}\rangle$ where **u** is an odd vertex and **v** is an even vertex. By Lemma 18.6, there is a Hamiltonian path W of $Q^1_{n-1} - \{(\mathbf{x})^n, (\mathbf{y})^n\}$ joining $(\mathbf{u})^n$ to $(\mathbf{v})^n$. We set $P'_{n-1} = \langle \mathbf{x}, \mathbf{u}, (\mathbf{u})^n, W, (\mathbf{v})^n, \mathbf{v}, H, \mathbf{y}\rangle$, $P_n = \langle \mathbf{x}, (\mathbf{x})^n = (\mathbf{y})^c, \mathbf{y}\rangle$, and $P_{n+1} = \langle \mathbf{x}, (\mathbf{x})^c = (\mathbf{y})^n, \mathbf{y}\rangle$. Then $\{P_1, P_2, \ldots, P_{n-2}, P'_{n-1}, P_n, P_{n+1}\}$ forms an $(n+1)^*$-container of FQ_n between **x** and **y**. See Figure 14.13b for illustration, for $n = 6$.

Case 2. $\mathbf{y} \in FQ^1_{n-1}$. We have the following sub cases:

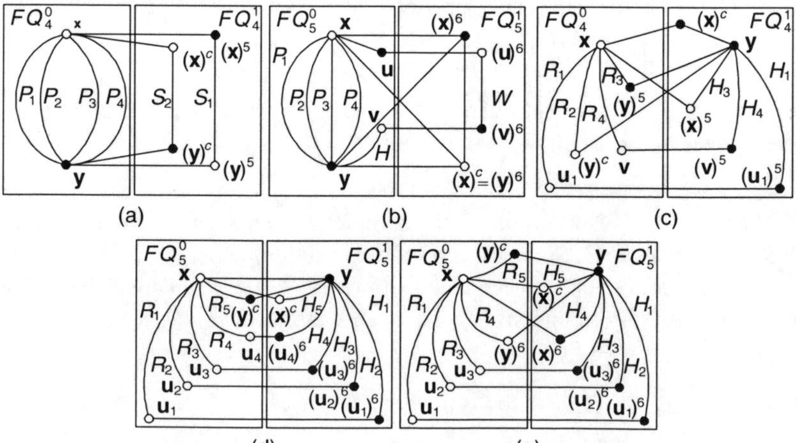

FIGURE 14.13 Illustrations for Lemma 14.21.

Case 2.1. n is odd and $\mathbf{y} \in \{(\mathbf{x})^n, (\mathbf{x})^c\}$. By Theorem 14.15, we consider only that $\mathbf{y} = (\mathbf{x})^c$. By Theorem 14.7, there is an n^*-container $\{P_1, P_2, \ldots, P_n\}$ of Q_n between $\mathbf{x}$ and $\mathbf{y}$. We set $P_{n+1} = \langle \mathbf{x}, \mathbf{y} \rangle$. Then $\{P_1, P_2, \ldots, P_{n+1}\}$ forms an $(n+1)^*$-container of FQ_n between $\mathbf{x}$ and $\mathbf{y}$.

Case 2.2. n is odd and $\mathbf{y} \notin \{(\mathbf{x})^c, (\mathbf{x})^n\}$. Since $\mathbf{y} \in FQ^1_{n-1}$ and $\mathbf{y}$ is an odd vertex, we have $\mathbf{y} = (\mathbf{x})^c$ or $\mathbf{y} = (\mathbf{x})^n$ if $n = 3$. Thus, $n \geq 5$. Since there are 2^{n-2} even vertices in FQ^0_{n-1} and $2^{n-2} \geq n - 1$ for $n \geq 5$, we can choose $(n-4)$ distinct even vertices $\{\mathbf{u}_1, \mathbf{u}_2, \ldots, \mathbf{u}_{n-4}\}$ of $FQ^0_{n-1} - \{\mathbf{x}, (\mathbf{y})^c, (\mathbf{y})^n\}$ such that $(\mathbf{u}_i)^n \neq (\mathbf{x})^c$ for $1 \leq i \leq n-4$. Let $\mathbf{v}$ be an odd vertex of FQ^0_{n-1} and let $U_0 = \{\mathbf{u}_i \mid 1 \leq i \leq n-4\} \cup \{(\mathbf{y})^c, (\mathbf{y})^n, \mathbf{v}\}$. Obviously, $|U_0| = n-1$. By Theorem 14.14, there are $(n-1)$ disjoint paths $\{R_1, R_2, \ldots, R_{n-1}\}$ of FQ^0_{n-1} joining $\mathbf{x}$ to U_0 such that (1) R_i joins $\mathbf{x}$ to $\mathbf{u}_i$ for $1 \leq i \leq n-4$, (2) R_{n-3} joins $\mathbf{x}$ to $(\mathbf{y})^c$, (3) R_{n-2} joins $\mathbf{x}$ to $(\mathbf{y})^n$, (4) R_{n-1} joins $\mathbf{x}$ to $\mathbf{v}$, and (5) $\bigcup_{i=1}^{n-1} R_i$ spans FQ^0_{n-1}. Let $U_1 = \{(\mathbf{u}_i)^n \mid 1 \leq i \leq n-4\} \cup \{(\mathbf{x})^c, (\mathbf{x})^n, (\mathbf{v})^n\}$. Obviously, $|U_1| = n-1$. By Theorem 14.14, there are $(n-1)$ disjoint paths $\{H_1, H_2, \ldots, H_{n-1}\}$ of FQ^1_{n-1} joining U_1 to $\mathbf{y}$ such that (1) H_i joins $(\mathbf{u}_i)^n$ to $\mathbf{y}$ for $1 \leq i \leq n-4$, (2) H_{n-3} joins $(\mathbf{x})^c$ to $\mathbf{y}$, (3) H_{n-2} joins $(\mathbf{x})^n$ to $\mathbf{y}$, (4) H_{n-1} joins $(\mathbf{v})^n$ to $\mathbf{y}$, and (5) $\bigcup_{i=1}^{n-1} H_i$ spans FQ^1_{n-1}. We set $P_i = \langle \mathbf{x}, R_i, \mathbf{u}_i, (\mathbf{u}_i)^n, H_i, \mathbf{y} \rangle$ for $1 \leq i \leq n-4$, $P_{n-3} = \langle \mathbf{x}, R_{n-3}, (\mathbf{y})^c, \mathbf{y} \rangle$, $P_{n-2} = \langle \mathbf{x}, R_{n-2}, (\mathbf{y})^n, \mathbf{y} \rangle$, $P_{n-1} = \langle \mathbf{x}, R_{n-1}, \mathbf{v}, (\mathbf{v})^n, H_{n-1}, \mathbf{y} \rangle$, $P_n = \langle \mathbf{x}, (\mathbf{x})^c, H_{n-3}, \mathbf{y} \rangle$, and $P_{n+1} = \langle \mathbf{x}, (\mathbf{x})^n, H_{n-2}, \mathbf{y} \rangle$. Then $\{P_1, P_2, \ldots, P_{n+1}\}$ forms an $(n+1)^*$-container of FQ_n between $\mathbf{x}$ and $\mathbf{y}$. See Figure 14.13c for illustration, for $n = 5$.

Case 2.3. n is even with $n \geq 4$ and $\mathbf{y} = (\mathbf{x})^n$. Since there are 2^{n-2} even vertices in FQ^0_{n-1} and $2^{n-2} \geq n-1$ for $n \geq 4$, we can choose $(n-2)$ distinct even vertices $\{\mathbf{u}_1, \mathbf{u}_2, \ldots, \mathbf{u}_{n-2}\}$ of $FQ^0_{n-1} - \{\mathbf{x}\}$. Let $U_0 = \{\mathbf{u}_i \mid 1 \leq i \leq n-2\} \cup \{(\mathbf{y})^c\}$. Obviously, $|U_0| = n-1$. By Theorem 14.14, there are $(n-1)$ disjoint paths $\{R_1, R_2, \ldots, R_{n-1}\}$ of FQ^0_{n-1} joining $\mathbf{x}$ to U_0 such that (1) R_i joins $\mathbf{x}$ to $\mathbf{u}_i$ for $1 \leq i \leq n-2$, (2) R_{n-1} joins $\mathbf{x}$ to $(\mathbf{y})^c$, and (3) $\bigcup_{i=1}^{n-1} R_i$ spans FQ^0_{n-1}. Let $U_1 = \{(\mathbf{u}_i)^n \mid 1 \leq i \leq n-2\} \cup \{(\mathbf{x})^c, \}$. Obviously, $|U_1| = n-1$. By Theorem 14.14, there are $(n-1)$ disjoint paths $\{H_1, H_2, \ldots, H_{n-1}\}$ of FQ^1_{n-1} joining U_1 to $\mathbf{y}$ such that (1) H_i joins $(\mathbf{u}_i)^n$ to $\mathbf{y}$ for $1 \leq i \leq n-2$, (2) H_{n-1} joins $(\mathbf{x})^c$ to $\mathbf{y}$, and (3) $\bigcup_{i=1}^{n-1} H_i$ spans FQ^1_{n-1}. We set $P_i = \langle \mathbf{x}, R_i, \mathbf{u}_i, (\mathbf{u}_i)^n, H_i, \mathbf{y} \rangle$ for $1 \leq i \leq n-2$, $P_{n-1} = \langle \mathbf{x}, R_{n-1}, (\mathbf{y})^c, \mathbf{y} \rangle$, $P_n = \langle \mathbf{x}, (\mathbf{x})^c, H_{n-1}, \mathbf{y} \rangle$, and $P_{n+1} = \langle \mathbf{x}, \mathbf{y} = (\mathbf{x})^n \rangle$. Then $\{P_1, P_2, \ldots, P_{n+1}\}$ forms an $(n+1)^*$-container of FQ_n between $\mathbf{x}$ and $\mathbf{y}$. See Figure 14.13d for illustration, for $n = 6$.

Case 2.4. n is even with $n \geq 4$ and $\mathbf{y} \neq (\mathbf{x})^n$. Since there are 2^{n-2} even vertices in FQ^0_{n-1} and $2^{n-2} \geq n-1$ for $n \geq 4$, we can choose $(n-3)$ distinct even vertices $\{\mathbf{u}_1, \mathbf{u}_2, \ldots, \mathbf{u}_{n-3}\}$ of $FQ^0_{n-1} - \{\mathbf{x}, (\mathbf{y})^n\}$ such that $(\mathbf{u}_i)^n \neq (\mathbf{x})^n$ for $1 \leq i \leq n-3$. Let $U_0 = \{\mathbf{u}_i \mid 1 \leq i \leq n-3\} \cup \{(\mathbf{y})^n, (\mathbf{y})^c\}$. Obviously, $|U_0| = n-1$. By Theorem 14.14, there are $(n-1)$ disjoint paths $\{R_1, R_2, \ldots, R_{n-1}\}$ of FQ^0_{n-1} such that (1) R_i joins $\mathbf{x}$ to $\mathbf{u}_i$ for $1 \leq i \leq n-3$, (2) R_{n-2} joins $\mathbf{x}$ to $(\mathbf{y})^n$, (3) R_{n-1} joins $\mathbf{x}$ to $(\mathbf{y})^c$, and (4) $\bigcup_{i=1}^{n-1} R_i$ spans FQ^0_{n-1}. Let $U_1 = \{(\mathbf{u}_i)^n \mid 1 \leq i \leq n-3\} \cup \{(\mathbf{x})^n, (\mathbf{x})^c\}$. Obviously, $|U_1| = n-1$. By Theorem 14.14, there are $(n-1)$ disjoint paths $\{H_1, H_2, \ldots, H_{n-1}\}$ of FQ^1_{n-1} joining U_1 to $\mathbf{y}$ such that (1) H_i joins $(\mathbf{u}_i)^n$ to $\mathbf{y}$ for $1 \leq i \leq n-3$, (2) H_{n-2} joins $(\mathbf{x})^n$ to $\mathbf{y}$, (3) H_{n-1} joins $(\mathbf{x})^c$ to $\mathbf{y}$, and (4) $\bigcup_{i=1}^{n-1} H_i$ spans

FQ^1_{n-1}. We set $P_i = \langle \mathbf{x}, R_i, \mathbf{u}_i, (\mathbf{u}_i)^n, H_i, \mathbf{y}\rangle$ for $1 \le i \le n-3$, $P_{n-2} = \langle \mathbf{x}, R_{n-2}, (\mathbf{y})^n, \mathbf{y}\rangle$, $P_{n-1} = \langle \mathbf{x}, R_{n-1}, (\mathbf{y})^c, \mathbf{y}\rangle$, $P_n = \langle \mathbf{x}, (\mathbf{x})^n, H_{n-2}, \mathbf{y}\rangle$, and $P_{n+1} = \langle \mathbf{x}, (\mathbf{x})^c, H_{n-1}, \mathbf{y}\rangle$. Then $\{P_1, P_2, \ldots, P_{n+1}\}$ forms an $(n+1)^*$-container of FQ_n between $\mathbf{x}$ and $\mathbf{y}$. See Figure 14.13e for illustration, for $n = 6$. □

THEOREM 14.16 FQ_n is super spanning laceable if n is an odd integer and FQ_n is super spanning connected if n is an even integer.

Proof: Since FQ_1 is isomorphic to Q_1, this statement holds for $n = 1$. By Lemma 14.21, this statement holds if n is odd and $n \ge 3$. Thus, we assume that n is even. Let $\mathbf{x}$ and $\mathbf{y}$ be any two different vertices of FQ_n. We need to find a k^*-container of FQ_n between $\mathbf{x}$ and $\mathbf{y}$ for $1 \le k \le n+1$. Without loss of generality, we assume that $\mathbf{x}$ is an even vertex. By Lemma 14.21, this statement holds if $\mathbf{y}$ is an odd vertex. Thus, we assume that $\mathbf{y}$ is an even vertex. Without loss of generality, we assume that $(\mathbf{x})_n = 0$ and $(\mathbf{y})_n = 1$. Let f be the function on $V(FQ_n)$ defined by $f(\mathbf{u}) = \mathbf{u}$ if $(\mathbf{u})_n = 0$ and $f(\mathbf{u}) = ((\mathbf{u})^c)^n$ if otherwise. By Theorem 14.15, f is an isomorphism from FQ_n into itself. In other words, we still get FQ_n if we relabel all the vertices $\mathbf{u}$ with $f(\mathbf{u})$. However, $f(\mathbf{x}) = \mathbf{x}$ is an even vertex and $f(\mathbf{y}) = ((\mathbf{y})^c)^n$ is an odd vertex. By Lemma 14.21, there exists a k^*-container of FQ_n between $f(\mathbf{x})$ and $f(\mathbf{y})$ for every $1 \le k \le n+1$. Thus, there exists a k^*-container of FQ_n between $\mathbf{x}$ and $\mathbf{y}$ for every $1 \le k \le n+1$. This theorem is proved. □

Let $\mathbf{u} = u_1u_2\ldots u_{n-1}u_n$ be a vertex of $Q_{n,m}$. Similar to before, the *c-neighbor* of $\mathbf{u}$ in $Q_{n,m}$, $(\mathbf{u})^c$, is $\bar{u}_1\bar{u}_2\ldots\bar{u}_m u_{m+1}u_{m+2}\ldots u_{n-1}u_n$. Note that $(\mathbf{u})^c$ and $\mathbf{u}$ are of the same parity if and only if m is even. Let $E^c = \{(u_1u_2\ldots u_n, \bar{u}_1\bar{u}_2\ldots \bar{u}_m u_{m+1}u_{m+2}\ldots u_{n-1}u_n) \mid u_1u_2\ldots u_n \in V(Q_{n,m})\}$. By definition, the n-dimensional enhanced hypercube $Q_{n,m}$ is obtained from Q_n by adding E^c. Obviously, $Q_{n,m}$ is FQ_n if $m = n$. We use $Q^j_{n,m}$ to denote the subgraph of $Q_{n,m}$ induced by $\{\mathbf{v} \in V(Q_{n,m}) \mid (\mathbf{v})_n = j\}$ for $0 \le j \le 1$. Moreover, we use $Q^{ij}_{n,m}$ to denote the subgraph of $Q_{n,m}$ induced by $\{\mathbf{v} \in V(Q_{n,m}) \mid (\mathbf{v})_{n-1} = i \text{ and } (\mathbf{v})_n = j\}$ for $0 \le i, j \le 1$.

LEMMA 14.22 Let $\mathbf{x}$ and $\mathbf{y}$ be any two distinct vertices of $Q^j_{n,m}$ with $n - m \ge 1$ for some j. Suppose that there is a k^*-container of $Q^j_{n,m}$ between $\mathbf{x}$ and $\mathbf{y}$ and there is 1^*-container of $Q^{1-j}_{n,m}$ between $(\mathbf{x})^n$ and $(\mathbf{y})^n$. Then there is a $(k+1)^*$-container of $Q_{n,m}$ between $\mathbf{x}$ and $\mathbf{y}$.

Proof: Let $\{P_1, P_2, \ldots, P_k\}$ be a k^*-container of $Q^j_{n,m}$ between $\mathbf{x}$ and $\mathbf{y}$ and W be a Hamiltonian path of $Q^{1-j}_{n,m}$ joining $(\mathbf{x})^n$ to $(\mathbf{y})^n$. Set $P_{k+1} = \langle \mathbf{x}, (\mathbf{x})^n, W, (\mathbf{y})^n, \mathbf{y}\rangle$. Then $\{P_1, P_2, \ldots, P_{k+1}\}$ forms a $(k+1)^*$-container of $Q_{n,m}$ between $\mathbf{x}$ and $\mathbf{y}$. □

LEMMA 14.23 Let $\mathbf{x}$ be an even vertex and $\mathbf{y}$ be an odd vertex of $Q_{n,n-1}$ for any positive integer $n \ge 3$. Then there exists a k^*-container of $Q_{n,n-1}$ between $\mathbf{x}$ and $\mathbf{y}$ for every $1 \le k \le n+1$.

Proof: Since Q_n is a spanning subgraph of $Q_{n,n-1}$, by Theorem 14.7, there exists a k^*-container of $Q_{n,n-1}$ between $\mathbf{x}$ and $\mathbf{y}$ for every $1 \le k \le n$. Thus, we need only to construct an $(n+1)^*$-container of $Q_{n,n-1}$ between $\mathbf{x}$ and $\mathbf{y}$. Without loss of generality, we assume that $\mathbf{x} \in Q^{00}_{n,n-1}$. We have the following cases:

Case 1. $\mathbf{y} \in Q^{00}_{n,n-1} \cup Q^{10}_{n,n-1}$. Since $Q^{00}_{n,n-1} \cup Q^{10}_{n,n-1} = Q^{0}_{n,n-1}$ is isomorphic to FQ_{n-1}, by Lemma 14.21, there exists an n^*-container of $Q^{0}_{n,n-1}$ between $\mathbf{x}$ and $\mathbf{y}$. Since $Q^{01}_{n,n-1} \cup Q^{11}_{n,n-1} = Q^{1}_{n,n-1}$ is isomorphic to FQ_{n-1}, by Lemma 14.21, there exists a Hamiltonian path of $Q^{1}_{n,n-1}$ joining $(\mathbf{x})^n$ to $(\mathbf{y})^n$. By Lemma 14.22, there exists an $(n+1)^*$-container of $Q_{n,n-1}$ between $\mathbf{x}$ and $\mathbf{y}$.

Case 2. $\mathbf{y} \in Q^{01}_{n,n-1}$. Suppose that $n=3$. We have $\mathbf{x}=000$ and $\mathbf{y}=001$. We set $P_1 = \langle 000, 001\rangle$, $P_2 = \langle 000, 010, 011, 001\rangle$, $P_3 = \langle 000, 100, 101, 001\rangle$, and $P_4 = \langle 000, 110, 111, 001\rangle$. Then $\{P_1, P_2, P_3, P_4\}$ forms a 4*-container of $Q_{3,2}$ between $\mathbf{x}$ and $\mathbf{y}$.

Now, we consider $n \geq 4$. Since $Q^{00}_{n,n-1} \cup Q^{01}_{n,n-1}$ is isomorphic to Q_{n-1}, by Lemma 14.7, there exists an $(n-1)^*$-container $\{P_1, P_2, \ldots, P_{n-1}\}$ of $Q^{00}_{n,n-1} \cup Q^{01}_{n,n-1}$ joining $\mathbf{x}$ to $\mathbf{y}$. Obviously, $(\mathbf{x})^c$ and $(\mathbf{y})^c$ are of different parity. Note that $(\mathbf{x})^{n-1}$ is an odd vertex and $(\mathbf{y})^{n-1}$ is an even vertex. By Theorem 14.4, there exist two disjoint paths S_1 and S_2 of $Q^{10}_{n,n-1} \cup Q^{11}_{n,n-1}$ such that (1) S_1 joins $(\mathbf{x})^{n-1}$ to $(\mathbf{y})^{n-1}$, (2) S_2 joins $(\mathbf{x})^c$ to $(\mathbf{y})^c$, and (3) $S_1 \cup S_2$ spans $Q^{10}_{n,n-1} \cup Q^{11}_{n,n-1}$. We set $P_n = \langle \mathbf{x}, (\mathbf{x})^{n-1}, S_1, (\mathbf{y})^{n-1}, \mathbf{y}\rangle$ and $P_{n+1} = \langle \mathbf{x}, (\mathbf{x})^c, S_2, (\mathbf{y})^c, \mathbf{y}\rangle$. Then $\{P_1, P_2, \ldots, P_{n+1}\}$ forms an $(n+1)^*$-container of $Q_{n,m}$ between $\mathbf{x}$ and $\mathbf{y}$. See Figure 14.14a for illustration.

Case 3. $\mathbf{y} \in Q^{11}_{n,n-1}$. Suppose that $n=3$. We have $\mathbf{x}=000$ and $\mathbf{y}=111$. We set $P_1 = \langle 000, 100, 101, 111\rangle$, $P_2 = \langle 000, 010, 011, 111\rangle$, $P_3 = \langle 000, 001, 111\rangle$, and $P_4 = \langle 000, 110, 111\rangle$. Then $\{P_1, P_2, P_3, P_4\}$ forms a 4*-container of $Q_{3,2}$ between $\mathbf{x}$ and $\mathbf{y}$.

Now, we consider $n \geq 4$. Since $\mathbf{y}$ is adjacent to $(n-2)$ even vertices in $Q^{11}_{n,n-1}$, we can choose an even vertex $\mathbf{z} \in Q^{11}_{n,n-1}$ that is a neighbor of $\mathbf{y}$ such that $(\mathbf{z})^n \neq (\mathbf{x})^c$ and $(\mathbf{z})^n \neq (\mathbf{x})^{n-1}$. Let $\mathbf{v} = (\mathbf{z})^n$. Obviously, $\mathbf{v}$ is an odd vertex. Since $Q^{00}_{n,n-1} \cup Q^{10}_{n,n-1} = Q^{0}_{n,n-1}$ is isomorphic to FQ_{n-1}, by Lemma 14.21, there exists an n^*-container $\{R_1, R_2, \ldots, R_n\}$ of $Q^{00}_{n,n-1} \cup Q^{10}_{n,n-1}$ between $\mathbf{x}$ and $\mathbf{v}$. Since $\mathbf{v}$ is adjacent to n vertices in $Q^{00}_{n,n-1} \cup Q^{10}_{n,n-1}$, by relabeling, we can write R_i as $\langle \mathbf{x}, R'_i, \mathbf{u}_i, \mathbf{v}\rangle$ for $1 \leq i \leq n-3$, write R_{n-2} as $\langle \mathbf{x}, R'_{n-2}, (\mathbf{y})^n, \mathbf{v}\rangle$, write R_{n-1} as $\langle \mathbf{x}, R'_{n-1}, (\mathbf{v})^c, \mathbf{v}\rangle$, and write R_n as $\langle \mathbf{x}, R'_n, (\mathbf{v})^{n-1}, \mathbf{v}\rangle$. Let $A = \{(\mathbf{u}_i)^n \mid 1 \leq i \leq n-3\}$. Obviously, A is a set of $(n-3)$ odd vertices of $Q^{11}_{n,n-1}$. Since $Q^{11}_{n,n-1}$ is isomorphic to Q_{n-2}, by Theorem 14.14, there are $(n-2)$ disjoint paths $\{H_1, H_2, \ldots, H_{n-2}\}$ of $Q^{11}_{n,n-1}$ joining $\mathbf{y}$ to $A \cup \{\mathbf{z}\}$ such that (1) H_i joins $(\mathbf{u}_i)^n$ to $\mathbf{y}$ for $1 \leq i \leq n-3$, (2) H_{n-2} joins $\mathbf{z}$ to $\mathbf{y}$, and (3) $\bigcup_{i=1}^{n-2} H_i$ spans $Q^{11}_{n,n-1}$. We set $P_i = \langle \mathbf{x}, R'_i, \mathbf{u}_i, (\mathbf{u}_i)^n, H_i, \mathbf{y}\rangle$ for $1 \leq i \leq n-3$ and $P_{n-2} = \langle \mathbf{x}, R'_{n-2}, (\mathbf{y})^n, \mathbf{y}\rangle$.

Suppose that $(n-1)$ is an odd integer. We set $P_{n-1} = \langle \mathbf{x}, R'_{n-1}, (\mathbf{v})^c, \mathbf{v}, \mathbf{z}, H_{n-2}, \mathbf{y}\rangle$. Since $Q^{01}_{n,n-1}$ is isomorphic to Q_{n-2}, by Theorem 14.4, there exist two disjoint paths S_1 and S_2 of $Q^{01}_{n,n-1}$ such that (1) S_1 joins $((\mathbf{v})^{n-1})^n$ to $(\mathbf{y})^c$, (2) S_2 joins $(\mathbf{x})^n$ to $(\mathbf{y})^{n-1}$, and (3) $S_1 \cup S_2$ spans $Q^{01}_{n,n-1}$. Let $P_n = \langle \mathbf{x}, R'_n, (\mathbf{v})^{n-1}, ((\mathbf{v})^{n-1})^n, S_1, (\mathbf{y})^c, \mathbf{y}\rangle$, and $P_{n+1} = \langle \mathbf{x}, (\mathbf{x})^n, S_2, (\mathbf{y})^{n-1}, \mathbf{y}\rangle$. Then $\{P_1, P_2, \ldots, P_{n+1}\}$ forms an $(n+1)^*$-container of $Q_{n,n-1}$ between $\mathbf{x}$ and $\mathbf{y}$. See Figure 14.14b for illustration.

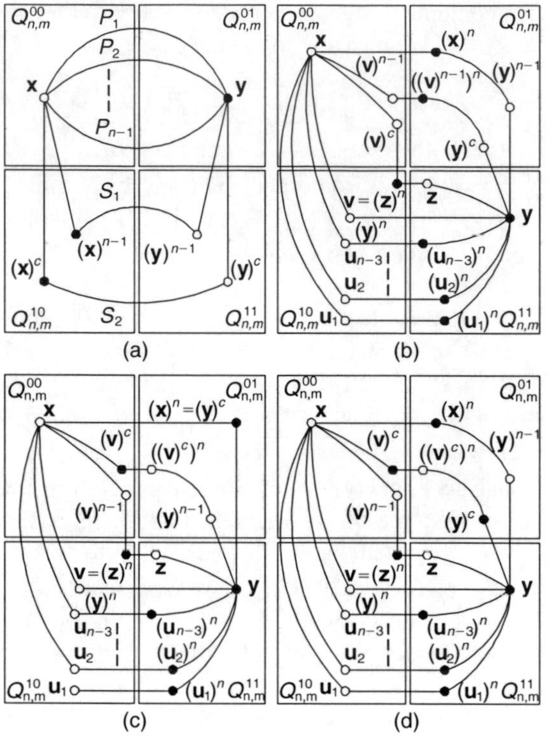

FIGURE 14.14 Illustrations for Lemma 14.23.

Suppose that $(n-1)$ is an even integer. We set $P_{n-1} = \langle \mathbf{x}, R'_n, (\mathbf{v})^{n-1}, \mathbf{v}, \mathbf{z}, H_{n-2}, \mathbf{y} \rangle$. Suppose that $(\mathbf{y})^c = (\mathbf{x})^n$. Note that Q_{n-2} is a spanning subgraph of $Q^{01}_{n,n-1}$. Since Q_{n-2} is hyper Hamiltonian laceable, there exists a Hamiltonian path S of $Q^{01}_{n,n-1} - \{(\mathbf{x})^n\}$ joining $((\mathbf{v})^c)^n$ to $(\mathbf{y})^{n-1}$. Set $P_n = \langle \mathbf{x}, R'_{n-1}, (\mathbf{v})^c, ((\mathbf{v})^c)^n, S, (\mathbf{y})^{n-1}, \mathbf{y} \rangle$ and $P_{n+1} = \langle \mathbf{x}, (\mathbf{x})^n = (\mathbf{y})^c, \mathbf{y} \rangle$. Then $\{P_1, P_2, \ldots, P_{n+1}\}$ forms an $(n+1)^*$-container of $Q_{n,n-1}$ between $\mathbf{x}$ and $\mathbf{y}$. See Figure 14.14c for illustration. Thus, we assume that $(\mathbf{y})^c \neq (\mathbf{x})^n$. By Theorem 14.4, there exist two disjoint paths S_1 and S_2 of $Q^{01}_{n,n-1}$ such that (1) S_1 joins $((\mathbf{v})^c)^n$ to $(\mathbf{y})^c$, (2) S_2 joins $(\mathbf{x})^n$ to $(\mathbf{y})^{n-1}$, and (3) $S_1 \cup S_2$ spans $Q^{01}_{n,n-1}$. Let $P_n = \langle \mathbf{x}, R'_{n-1}, (\mathbf{v})^c, ((\mathbf{v})^c)^n, S_1, (\mathbf{y})^c, \mathbf{y} \rangle$ and $P_{n+1} = \langle \mathbf{x}, (\mathbf{x})^n, S_2, (\mathbf{y})^{n-1}, \mathbf{y} \rangle$. Then $\{P_1, P_2, \ldots, P_{n+1}\}$ forms an $(n+1)^*$-container of $Q_{n,n-1}$ between $\mathbf{x}$ and $\mathbf{y}$. See Figure 14.14d for illustration. □

LEMMA 14.24 Let $\mathbf{x}$ be an even vertex and $\mathbf{y}$ be an odd vertex of $Q_{n,m}$ for any two positive integers $n \geq m \geq 2$. Then there exists a k^*-container of $Q_{n,m}$ between $\mathbf{x}$ and $\mathbf{y}$ for every $1 \leq k \leq n+1$.

Proof: Since $Q_{2,2}$ is isomorphic to complete graph K_4, this statement holds for $n = 2$. Suppose that $n \geq 3$.

Since Q_n is a spanning subgraph of $Q_{n,m}$, by Theorem 14.7, there exists a k^*-container of $Q_{n,m}$ between $\mathbf{x}$ and $\mathbf{y}$ for every $1 \leq k \leq n$. Thus, we need only to construct an $(n+1)^*$-container of $Q_{n,m}$ between $\mathbf{x}$ and $\mathbf{y}$. Without loss of generality, we assume that $\mathbf{x} \in Q^{00}_{n,m}$. We prove our claim by induction on $t = n - m$. The induction bases are $t = 0$ and 1. By Lemma 14.21, our claim holds for $t = 0$. With Lemma 14.23, our claim holds for $t = 1$. Consider $t \geq 2$ and assume that our claim holds for $(t-1)$. We have the following cases:

Case 1. $\mathbf{y} \in Q^{00}_{n,m} \cup Q^{10}_{n,m}$. Since $Q^{00}_{n,m} \cup Q^{10}_{n,m}$ is isomorphic to $Q_{n-1,m}$, by induction, there exists an n^*-container of $Q^{00}_{n,m} \cup Q^{10}_{n,m}$ between $\mathbf{x}$ and $\mathbf{y}$. Since $Q^{01}_{n,m} \cup Q^{11}_{n,m}$ is isomorphic to $Q_{n-1,m}$, by induction, there is a Hamiltonian path of $Q^{01}_{n,m} \cup Q^{11}_{n,m}$ joining $(\mathbf{x})^n$ to $(\mathbf{y})^n$. Thus, by Lemma 14.22, there exists an $(n+1)^*$-container of $Q_{n,m}$ between $\mathbf{x}$ and $\mathbf{y}$.

Case 2. $\mathbf{y} \in Q^{01}_{n,m}$. Note that $Q^{01}_{n,m}$ and $Q^{10}_{n,m}$ are symmetric with respect to $Q_{n,m}$ and $Q^{00}_{n,m} \cup Q^{01}_{n,m}$ is isomorphic to $Q_{n-1,m}$. Similar to Case 1, there is an $(n+1)^*$-container of $Q_{n,m}$ between $\mathbf{x}$ and $\mathbf{y}$.

Case 3. $\mathbf{y} \in Q^{11}_{n,m}$. Since $\mathbf{y}$ is adjacent to $(n-1)$ vertices in $Q^{11}_{n,m}$, we can choose a neighbor $\mathbf{z}$ of $\mathbf{y}$ in $Q^{11}_{n,m}$ such that $\mathbf{z} \neq (\mathbf{y})^c$ and $(\mathbf{z})^n \neq (\mathbf{x})^{n-1}$. Let $\mathbf{v} = (\mathbf{z})^n$. Obviously, $\mathbf{v}$ is an odd vertex. Since $Q^{00}_{n,m} \cup Q^{10}_{n,m}$ is isomorphic to $Q_{n-1,m}$, by induction, there exists an n^*-container $\{R_1, R_2, \ldots, R_n\}$ of $Q^{00}_{n,m} \cup Q^{10}_{n,m}$ joining $\mathbf{x}$ to $\mathbf{v}$. Since $\mathbf{v}$ is adjacent to n vertices in $Q^{00}_{n,m} \cup Q^{10}_{n,m}$, by relabeling, we can write R_i as $\langle \mathbf{x}, R'_i, \mathbf{u}_i, \mathbf{v} \rangle$ for $1 \leq i \leq n-2$, write R_{n-1} as $\langle \mathbf{x}, R'_{n-1}, (\mathbf{y})^n, \mathbf{v} \rangle$, and write R_n as $\langle \mathbf{x}, R'_n, (\mathbf{v})^{n-1}, \mathbf{v} \rangle$. Since $Q^{11}_{n,m}$ is isomorphic to $Q_{n-2,m}$, by induction, there exists an $(n-1)^*$-container $\{H_1, H_2, \ldots, H_{n-1}\}$ of $Q^{11}_{n,m}$ joining $\mathbf{z}$ to $\mathbf{y}$. Since $\mathbf{y}$ is adjacent to $(n-1)$ vertices in $Q^{11}_{n,m}$ and $(\mathbf{z}, \mathbf{y}) \in E(Q^{11}_{n,m})$, one of these paths is $\langle \mathbf{z}, \mathbf{y} \rangle$. Without loss of generality, we assume that $H_i = \langle \mathbf{z}, (\mathbf{u}_i)^n, H'_i, \mathbf{y} \rangle$ for $1 \leq i \leq n-2$ and $H_{n-1} = \langle \mathbf{z}, \mathbf{y} \rangle$. We set $P_i = \langle \mathbf{x}, R'_i, \mathbf{u}_i, (\mathbf{u}_i)^n, H'_i, \mathbf{y} \rangle$ for $1 \leq i \leq n-2$, $P_{n-1} = \langle \mathbf{x}, R'_{n-1}, (\mathbf{y})^n, \mathbf{y} \rangle$, and $P_n = \langle \mathbf{x}, R'_n, (\mathbf{v})^{n-1}, \mathbf{v}, \mathbf{z}, \mathbf{y} \rangle$. Since $Q^{01}_{n,m}$ is isomorphic to $Q_{n-2,m}$, by induction, there exists a Hamiltonian path W in $Q^{01}_{n,m}$ joining $(\mathbf{x})^n$ to $(\mathbf{y})^{n-1}$. We set $P_{n+1} = \langle \mathbf{x}, (\mathbf{x})^n, W, (\mathbf{y})^{n-1}, \mathbf{y} \rangle$. Then $\{P_1, P_2, \ldots, P_{n+1}\}$ forms an $(n+1)^*$-container of $Q_{n,m}$ between $\mathbf{x}$ and $\mathbf{y}$. See Figure 14.15 for illustration. □

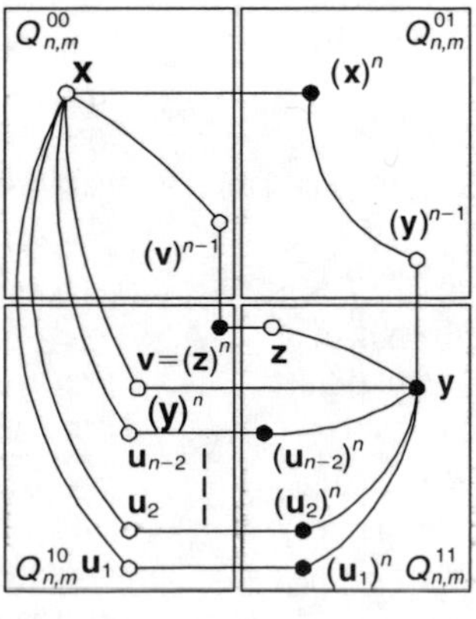

FIGURE 14.15 Illustration for Lemma 14.24.

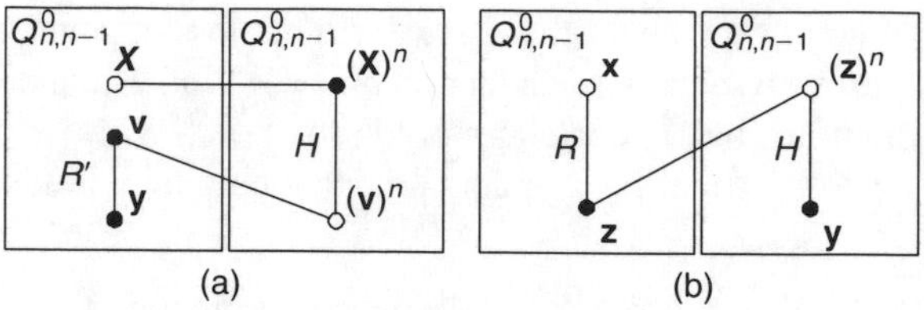

FIGURE 14.16 Illustration for Lemma 14.25.

LEMMA 14.25 $Q_{n,n-1}$ is 1*-connected and 2*-connected if n is an odd integer with $n \geq 3$.

Proof: Since any 1*-connected graph with more than three vertices is 2*-connected, we need to show only that $Q_{n,n-1}$ is 1*-connected. Suppose that $\mathbf{x}$ and $\mathbf{y}$ are two distinct vertices of $Q_{n,n-1}$. Without loss of generality, we assume that $\mathbf{x} \in Q^0_{n,n-1}$.

Suppose that $\mathbf{y} \in Q^0_{n,n-1}$. By Theorem 14.16, there exists a Hamiltonian path $R = \langle \mathbf{x}, \mathbf{v}, R', \mathbf{y} \rangle$ in $Q^0_{n,n-1}$ joining $\mathbf{x}$ to $\mathbf{y}$ and there exists a Hamiltonian path H in $Q^1_{n,n-1}$ joining $(\mathbf{x})^n$ to $(\mathbf{v})^n$. Set $P = \langle \mathbf{x}, (\mathbf{x})^n, H, (\mathbf{v})^n, \mathbf{v}, R', \mathbf{y} \rangle$. Thus, P forms a Hamiltonian path in $Q_{n,n-1}$ joining $\mathbf{x}$ to $\mathbf{y}$. See Figure 14.16a for illustration.

Suppose that $\mathbf{y} \in Q^1_{n,n-1}$. Note that there are $(2^{n-1} - 1)$ vertices in $Q^0_{n,n-1} - \{\mathbf{x}\}$ and $2^{n-1} - 1 \geq 3$ for $n \geq 3$. We can pick a vertex $\mathbf{z}$ in $Q^0_{n,n-1}$ such that $(\mathbf{z})^n \neq \mathbf{y}$. By Theorem 14.16, there exists a Hamiltonian path R in $Q^0_{n,n-1}$ joining $\mathbf{x}$ to $\mathbf{z}$ and there exists a Hamiltonian path H in $Q^1_{n,n-1}$ joining $(\mathbf{z})^n$ to $\mathbf{y}$. Set $P = \langle \mathbf{x}, R, \mathbf{z}, (\mathbf{z})^n, H, \mathbf{y} \rangle$. Thus, P forms a Hamiltonian path in $Q_{n,n-1}$ joining $\mathbf{x}$ to $\mathbf{y}$. See Figure 14.16b for illustration. □

LEMMA 14.26 $Q_{3,2}$ is super spanning connected.

Proof: Let $\mathbf{x}$ and $\mathbf{y}$ be any two different vertices of $Q_{3,2}$. By Lemma 14.25, $Q_{3,2}$ is 1*-connected and 2*-connected. Hence, we need to construct a 3*-container and a 4*-container between $\mathbf{x}$ and $\mathbf{y}$. Without loss of generality, we assume that $\mathbf{x} = 000$. By Lemma 14.23, this statement holds if $\mathbf{y}$ is an odd vertex. Thus, we assume that $\mathbf{y}$ is an even vertex. We list all possible cases, as follows:

y	3*-container	4*-container
110	⟨000, 010, 110⟩ ⟨000, 100, 110⟩ ⟨000, 001, 011, 101, 111, 110⟩	⟨000, 010, 110⟩ ⟨000, 100, 110⟩ ⟨000, 001, 011, 101, 111, 110⟩ ⟨000, 110⟩
011	⟨000, 010, 011⟩ ⟨000, 100, 101, 001, 011⟩ ⟨000, 110, 111, 011⟩	⟨000, 001, 011⟩ ⟨000, 010, 011⟩ ⟨000, 100, 101, 011⟩ ⟨000, 110, 111, 011⟩
101	⟨000, 001, 011, 101⟩ ⟨000, 010, 110, 111, 101⟩ ⟨000, 100, 101⟩	⟨000, 001, 101⟩ ⟨000, 010, 011, 101⟩ ⟨000, 100, 101⟩ ⟨000, 110, 111, 101⟩

□

LEMMA 14.27 Suppose that $n \geq 3$ is an odd integer. Let $\mathbf{x}$ and $\mathbf{y}$ be any two different even vertices of $Q_{n,n-1}$. Then there exists a k^*-container of $Q_{n,n-1}$ between $\mathbf{x}$ and $\mathbf{y}$ for every $1 \leq k \leq n+1$.

Proof: By Lemma 14.26, this statement holds for $Q_{3,2}$. Thus, we assume that $n \geq 5$. By Lemma 14.25, $Q_{n,n-1}$ is 1*-connected and 2*-connected. Thus, we need to construct a k^*-container between $\mathbf{x}$ and $\mathbf{y}$ for every $3 \leq k \leq n+1$. Without loss of generality, we assume that $\mathbf{x} \in Q^{00}_{n,n-1}$.

Case 1. $(\mathbf{y})_n = 0$. Since $Q^{00}_{n,n-1} \cup Q^{10}_{n,n-1}$ is isomorphic to FQ_{n-1}, by Theorem 14.16, there exists a $(k-1)^*$-container $\{P_1, P_2, \ldots, P_{k-1}\}$ of $Q^{00}_{n,n-1} \cup Q^{10}_{n,n-1}$ between $\mathbf{x}$ and $\mathbf{y}$ for every $2 \leq k-1 \leq n$. By Lemma 14.22, there is a k^*-container of $Q_{n,n-1}$ between $\mathbf{x}$ and $\mathbf{y}$ for every $3 \leq k \leq n+1$.

Case 2. $(\mathbf{y})_n = 1$. Since $\mathbf{y}$ is an even vertex, $|\{i \mid i \neq n \text{ and } (\mathbf{y})_i = 1\}|$ is odd. Without loss of generality, we assume that $(\mathbf{y})_{n-1} = 1$. Thus, $\mathbf{y} \in Q^{11}_{n,n-1}$. We have the following cases.

Case 2.1. $n \leq k \leq n+1$. Since $\mathbf{y}$ is adjacent to $(n-2)$ vertices in $Q^{11}_{n,n-1}$, we can choose a neighbor $\mathbf{z}$ of $\mathbf{y}$ in $Q^{11}_{n,n-1}$ such that $(\mathbf{z})^n \neq (\mathbf{x})^{n-1}$. Let $\mathbf{v} = (\mathbf{z})^n$. Obviously, $\mathbf{v}$ is an even vertex. Since $Q^{00}_{n,n-1} \cup Q^{10}_{n,n-1}$ is isomorphic to FQ_{n-1}, by Theorem 14.16, there exists an n^*-container $\{R_1, R_2, \ldots, R_n\}$ of $Q^{00}_{n,n-1} \cup Q^{10}_{n,n-1}$ between $\mathbf{x}$ and $\mathbf{v}$. Since $\mathbf{v}$ is adjacent to n vertices in $Q^{00}_{n,m} \cup Q^{10}_{n,m}$, by relabeling, we can write R_i as $\langle \mathbf{x}, R'_i, \mathbf{u}_i, \mathbf{v} \rangle$ for $1 \leq i \leq n-3$, write R_{n-2} as $\langle \mathbf{x}, R'_{n-2}, (\mathbf{v})^c, \mathbf{v} \rangle$, write R_{n-1} as $\langle \mathbf{x}, R'_{n-1}, (\mathbf{y})^n, \mathbf{v} \rangle$, and write R_n as $\langle \mathbf{x}, R'_n, (\mathbf{v})^{n-1}, \mathbf{v} \rangle$. Let $A = \{(\mathbf{u}_i)^n \mid 1 \leq i \leq n-3\}$. Obviously, A is a set of $(n-3)$ even vertices of $Q^{11}_{n,n-1}$.

Case 2.1.1. $k = n+1$. Since Q_{n-2} is a spanning subgraph of $Q^{11}_{n,n-1}$, by Theorem 14.14, there are $(n-2)$ disjoint paths $H_1, H_2, \ldots, H_{n-2}$ in $Q^{11}_{n,n-1}$ joining $\mathbf{y}$ to $A \cup \{\mathbf{z}\}$ such that (1) H_i joins $(\mathbf{u}_i)^n$ to $\mathbf{y}$ for $1 \leq i \leq n-3$, (2) H_{n-2} joins $\mathbf{z}$ to $\mathbf{y}$, and (3) $\bigcup_{i=1}^{n-2} H_i$ spans $Q^{11}_{n,n-1}$. We set $P_i = \langle \mathbf{x}, R'_i, \mathbf{u}_i, (\mathbf{u}_i)^n, H_i, \mathbf{y} \rangle$ for $1 \leq i \leq n-3$, $P_{n-2} = \langle \mathbf{x}, R'_{n-2}, (\mathbf{v})^c, \mathbf{v}, \mathbf{z}, H_{n-2}, \mathbf{y} \rangle$, and $P_{n-1} = \langle \mathbf{x}, R'_{n-1}, (\mathbf{y})^n, \mathbf{y} \rangle$.

Suppose that $(\mathbf{y})^{n-1} = (\mathbf{x})^n$. Note that Q_{n-2} is a spanning subgraph of $Q^{01}_{n,n-1}$. Since Q_{n-2} is hyper Hamiltonian laceable, there exists a Hamiltonian path S of $Q^{01}_{n,n-1} - \{(\mathbf{x})^n\}$ joining $((\mathbf{v})^{n-1})^n$ to $(\mathbf{y})^c$. Set $P_n = \langle \mathbf{x}, R'_n, (\mathbf{v})^{n-1}, ((\mathbf{v})^{n-1})^n, S, (\mathbf{y})^c, \mathbf{y} \rangle$ and $P_{n+1} = \langle \mathbf{x}, (\mathbf{x})^n = (\mathbf{y})^{n-1}, \mathbf{y} \rangle$. Then $\{P_1, P_2, \ldots, P_{n+1}\}$ forms an $(n+1)^*$-container of $Q_{n,n-1}$ between $\mathbf{x}$ and $\mathbf{y}$. See Figure 14.17a for illustration.

Now, we consider $(\mathbf{y})^{n-1} \neq (\mathbf{x})^n$. Since Q_{n-2} is a spanning subgraph of $Q^{01}_{n,n-1}$, by Lemmas 13.16 and 13.17, there exist two disjoint paths S_1 and S_2 of $Q^{01}_{n,n-1}$ such that (1) S_1 joins $((\mathbf{v})^{n-1})^n$ to $(\mathbf{y})^{n-1}$, (2) S_2 joins $(\mathbf{x})^n$ to $(\mathbf{y})^c$, and (3) $S_1 \cup S_2$ spans $Q^{01}_{n,n-1}$. Let $P_n = \langle \mathbf{x}, R'_n, (\mathbf{v})^{n-1}, ((\mathbf{v})^{n-1})^n, S_1, (\mathbf{y})^{n-1}, \mathbf{y} \rangle$ and $P_{n+1} = \langle \mathbf{x}, (\mathbf{x})^n, S_2, (\mathbf{y})^c, \mathbf{y} \rangle$. Then $\{P_1, P_2, \ldots, P_{n+1}\}$ forms an $(n+1)^*$-container of $Q_{n,n-1}$ between $\mathbf{x}$ and $\mathbf{y}$. See Figure 14.17b for illustration.

Case 2.1.2. $k = n$. Obviously, $A \cup \{((\mathbf{v})^{n-1})^n\}$ is a set of $(n-2)$ even vertices of $Q^{01}_{n,n-1} \cup Q^{11}_{n,n-1}$. Since Q_{n-1} is a spanning subgraph of $Q^{01}_{n,n-1} \cup Q^{11}_{n,n-1}$, by Theorem 14.14, there are $(n-1)$ disjoint paths $\{H_1, H_2, \ldots, H_{n-1}\}$ of

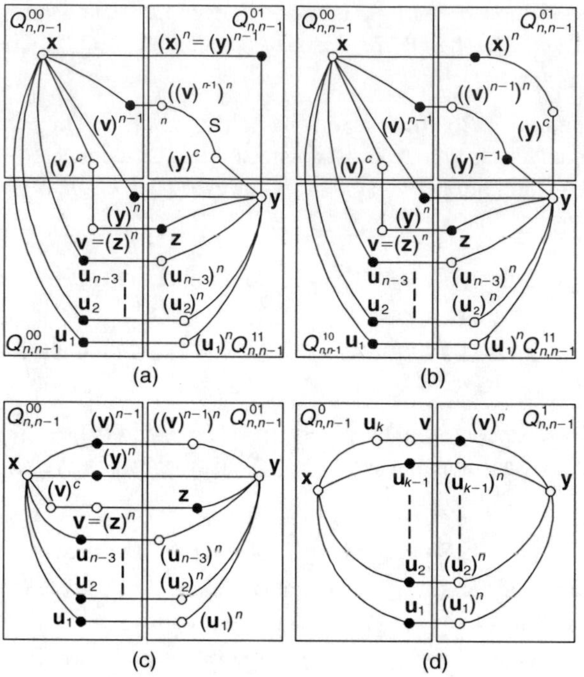

FIGURE 14.17 Illustrations for Lemma 14.27.

$Q^{01}_{n,n-1} \cup Q^{11}_{n,n-1}$ joining $\mathbf{y}$ to $A \cup \{\mathbf{z}, ((\mathbf{v})^{n-1})^n\}$ such that (1) H_i joins $(\mathbf{u}_i)^n$ to $\mathbf{y}$ for $1 \le i \le n-3$, (2) H_{n-2} joins $\mathbf{z}$ to $\mathbf{y}$, (3) H_{n-1} joins $((\mathbf{v})^{n-1})^n$ to $\mathbf{y}$, and (4) $\bigcup_{i=1}^{n-1} H_i$ spans $Q^{01}_{n,n-1} \cup Q^{11}_{n,n-1}$. We set $P_i = \langle \mathbf{x}, R'_i, \mathbf{u}_i, (\mathbf{u}_i)^n, H_i, \mathbf{y}\rangle$ for $1 \le i \le n-3$, $P_{n-2} = \langle \mathbf{x}, R'_{n-2}, (\mathbf{v})^c, \mathbf{v}, \mathbf{z}, H_{n-2}, \mathbf{y}\rangle$, $P_{n-1} = \langle \mathbf{x}, R'_{n-1}, (\mathbf{y})^n, \mathbf{y}\rangle$, and $P_n = \langle \mathbf{x}, R'_n, (\mathbf{v})^{n-1}, ((\mathbf{v})^{n-1})^n, H_{n-1}, \mathbf{y}\rangle$. Then $\{P_1, P_2, \ldots, P_n\}$ forms an n^*-container of $Q_{n,n-1}$ between $\mathbf{x}$ and $\mathbf{y}$. See Figure 14.17c for illustration.

Case 2.2. $3 \le k \le n-1$. Let $\mathbf{v}$ be an even vertex of $Q^{00}_{n,n-1} \cup Q^{10}_{n,n-1}$ such that $\mathbf{v} \ne \mathbf{x}$ and $(\mathbf{y})^n$ is not a neighbor of $\mathbf{v}$. Since $Q^{00}_{n,n-1} \cup Q^{10}_{n,n-1}$ is isomorphic to FQ_{n-1}, by Theorem 14.16, there exists a k^*-container $\{R_1, R_2, \ldots, R_k\}$ of $Q^{00}_{n,n-1} \cup Q^{10}_{n,n-1}$ between $\mathbf{x}$ and $\mathbf{v}$. We write $R_i = \langle \mathbf{x}, R'_i, \mathbf{u}_i, \mathbf{v}\rangle$ for $1 \le i \le k$. Let $A = \{\mathbf{u}_1, \mathbf{u}_2, \ldots, \mathbf{u}_k\}$. Since $k \ge 3$, at most one vertex of A is an even vertex. Without loss of generality, we assume that $\{\mathbf{u}_1, \mathbf{u}_2, \ldots, \mathbf{u}_{k-1}\}$ is a set of $(k-1)$ odd vertices. Obviously, $\{(\mathbf{u}_i)^n \mid 1 \le i \le k-1\}$ is a set of $(k-1)$ even vertices of $Q^{01}_{n,n-1} \cup Q^{11}_{n,n-1}$. Since Q_{n-1} is a spanning subgraph of $Q^{01}_{n,n-1} \cup Q^{11}_{n,n-1}$, by Theorem 14.14, there are k disjoint paths $\{H_1, H_2, \ldots, H_k\}$ of $Q^{01}_{n,n-1} \cup Q^{11}_{n,n-1}$ joining $\mathbf{y}$ to $\{(\mathbf{u}_i)^n \mid 1 \le i \le k-1\} \cup \{(\mathbf{v})^n\}$ such that (1) H_i joins $(\mathbf{u}_i)^n$ to $\mathbf{y}$ for $1 \le i \le k-1$, (2) H_k joins $(\mathbf{v})^n$ to $\mathbf{y}$, and (3) $\bigcup_{i=1}^{k} H_i$ spans $Q^{01}_{n,n-1} \cup Q^{11}_{n,n-1}$. We set $P_i = \langle \mathbf{x}, R'_i, \mathbf{u}_i, (\mathbf{u}_i)^n, H_i, \mathbf{y}\rangle$ for $1 \le i \le k-1$ and $P_k = \langle \mathbf{x}, R'_k, \mathbf{u}_k, \mathbf{v}, (\mathbf{v})^n, H_k, \mathbf{y}\rangle$. Then $\{P_1, P_2, \ldots, P_k\}$ forms a k^*-container of $Q_{n,n-1}$ between $\mathbf{x}$ and $\mathbf{y}$. See Figure 14.17d for illustration. □

LEMMA 14.28 Suppose that $n \geq 3$ and m is even. Let $\mathbf{x}$ and $\mathbf{y}$ be any two different even vertices of $Q_{n,m}$. Then there exists a Hamiltonian path P of $Q_{n,m}$ between $\mathbf{x}$ and $\mathbf{y}$.

Proof: For the fixed number m, we prove this statement by induction on $t = n - m$. Suppose that $\mathbf{x}$ and $\mathbf{y}$ are any two different even vertices of $Q_{n,m}$. By Lemma 14.26, this statement holds for $t = 1$. Consider $t \geq 2$ and assume that our claim holds for $(t-1)$. Without loss of generality, we assume that $\mathbf{x} \in Q_{n,m}^0$.

Suppose that $\mathbf{y} \in Q_{n,m}^0$. Since $Q_{n,m}^0$ is isomorphic to $Q_{n-1,m}$, by induction, there exists a Hamiltonian path $R = \langle \mathbf{x}, \mathbf{v}, R', \mathbf{y} \rangle$ in $Q_{n,m}^0$ joining $\mathbf{x}$ to $\mathbf{y}$ and there exists a Hamiltonian path H in $Q_{n,m}^1$ joining $(\mathbf{x})^n$ to $(\mathbf{v})^n$. Set $P = \langle \mathbf{x}, (\mathbf{x})^n, H, (\mathbf{v})^n, \mathbf{v}, R', \mathbf{y} \rangle$. Thus, P forms a Hamiltonian path in $Q_{n,m}$ joining $\mathbf{x}$ to $\mathbf{y}$.

Suppose that $\mathbf{y} \in Q_{n,m}^1$. We pick an even vertex $\mathbf{z}$ in $Q_{n,m}^0$ such that $\mathbf{z} \neq \mathbf{x}$. By induction, there exists a Hamiltonian path R in $Q_{n,m}^0$ joining $\mathbf{x}$ to $\mathbf{z}$. Obviously, $(\mathbf{z})^n$ is an odd vertex of $Q_{n,m}^1$. Since Q_{n-1} is a spanning subgraph of $Q_{n,m}^1$, by Theorem 14.7, there exists a Hamiltonian path H in $Q_{n,m}^1$ joining $(\mathbf{z})^n$ to $\mathbf{y}$. Set $P = \langle \mathbf{x}, R, \mathbf{z}, (\mathbf{z})^n, H, \mathbf{y} \rangle$. Thus, P forms a Hamiltonian path in $Q_{n,m}$ joining $\mathbf{x}$ to $\mathbf{y}$. □

LEMMA 14.29 Suppose that $n \geq 3$ and m is even. Let $\mathbf{x}$ and $\mathbf{y}$ be any two different even vertices of $Q_{n,m}$. Then there exists a k^*-container of $Q_{n,m}$ between $\mathbf{x}$ and $\mathbf{y}$ for every $1 \leq k \leq n+1$.

Proof: By Lemma 14.26, this statement holds for $n = 3$. Suppose that $n \geq 4$. We claim that there exists a k^*-container between $\mathbf{x}$ and $\mathbf{y}$ for every $1 \leq k \leq n+1$. By Lemma 14.28, this statement holds for $k = 1$. Note that Q_n is a spanning subgraph of $Q_{n,m}$ and Q_n is Hamiltonian. So, this statement holds for $k = 2$. We claim that there exists a k^*-container between $\mathbf{x}$ and $\mathbf{y}$ for every $3 \leq k \leq n+1$. Without loss of generality, we assume that $\mathbf{x} \in Q_{n,m}^{00}$. We prove our claim by induction on $t = n - m$. The induction bases are $t = 0$ and 1. By Theorem 14.16, our claim holds for $t = 0$. With Lemma 14.27, this statement holds for $t = 1$. Consider $t \geq 2$ and assume that this statement holds for $(t-1)$. We have the following cases:

Case 1. $\mathbf{y} \in Q_{n,m}^{00} \cup Q_{n,m}^{10}$. Since $Q_{n,m}^{00} \cup Q_{n,m}^{10}$ is isomorphic to $Q_{n-1,m}$, by induction, there exists a $(k-1)^*$-container $\{P_1, P_2, \ldots, P_{k-1}\}$ of $Q_{n,m}^{00} \cup Q_{n,m}^{10}$ between $\mathbf{x}$ and $\mathbf{y}$ for every $2 \leq k-1 \leq n$. By Lemma 14.22, there exists a k^*-container of $Q_{n,m}$ between $\mathbf{x}$ and $\mathbf{y}$.

Case 2. $\mathbf{y} \in Q_{n,m}^{01}$. Note that $Q_{n,m}^{01}$ and $Q_{n,m}^{10}$ are symmetric with respect to $Q_{n,m}$ and $Q_{n,m}^{00} \cup Q_{n,m}^{01}$ is isomorphic to $Q_{n-1,m}$. Similar to Case 1, there is a k^*-container of $Q_{n,m}$ between $\mathbf{x}$ and $\mathbf{y}$ for every $3 \leq k \leq n+1$.

Case 3. $\mathbf{y} \in Q_{n,m}^{11}$.

Case 3.1. $n \leq k \leq n+1$. Since $\mathbf{y}$ is adjacent to $(n-1)$ vertices in $Q_{n,m}^{11}$, we can choose a neighbor $\mathbf{z}$ of $\mathbf{y}$ in $Q_{n,m}^{11}$ such that $\mathbf{z} \neq (\mathbf{y})^c$ and $(\mathbf{z})^n \neq (\mathbf{x})^{n-1}$. Let $\mathbf{v} = (\mathbf{z})^n$. Obviously, both $\mathbf{v}$ and $((\mathbf{v})^{n-1})^n$ are even vertices. Since $Q_{n,m}^{00} \cup Q_{n,m}^{10}$ is isomorphic to $Q_{n-1,m}$, by induction, there exists an n^*-container $\{R_1, R_2, \ldots, R_n\}$ of $Q_{n,m}^{00} \cup Q_{n,m}^{10}$ between $\mathbf{x}$ and $\mathbf{v}$. Since $\mathbf{v}$ is adjacent to n vertices in $Q_{n,m}^{00} \cup Q_{n,m}^{10}$, by relabeling, we can write R_i as $\langle \mathbf{x}, R'_i, \mathbf{u}_i, \mathbf{v}\rangle$ for $1 \le i \le n-3$, write R_{n-2} as $\langle \mathbf{x}, R'_{n-2}, (\mathbf{v})^c, \mathbf{v}\rangle$, write R_{n-1} as $\langle \mathbf{x}, R'_{n-1}, (\mathbf{y})^n, \mathbf{v}\rangle$, and write R_n as $\langle \mathbf{x}, R'_n, (\mathbf{v})^{n-1}, \mathbf{v}\rangle$. Let $A = \{(\mathbf{u}_i)^n \mid 1 \le i \le n-3\}$. Obviously, A is a set of $(n-3)$ even vertices of $Q^{11}_{n,m}$.

Case 3.1.1. $k = n+1$. Since Q_{n-2} is a spanning subgraph of $Q^{11}_{n,m}$, by Theorem 14.7, there exists an $(n-2)^*$-container $\{H_1, H_2, \ldots, H_{n-2}\}$ of $Q^{11}_{n,m}$ between $\mathbf{z}$ and $\mathbf{y}$. Since $\mathbf{y}$ is adjacent to $(n-1)$ vertices in $Q^{11}_{n,m}$ and $(\mathbf{z}, \mathbf{y}) \in E(Q^{11}_{n,m})$, one of these paths is $\langle \mathbf{z}, \mathbf{y}\rangle$ and $(\mathbf{y})^c \in H_i$ for some $1 \le i \le n-2$. Without loss of generality, we assume that $(\mathbf{y})^c \in H_{n-3}$. We can write H_i as $\langle \mathbf{z}, (\mathbf{u}_i)^n, H'_i, \mathbf{y}\rangle$ for $1 \le i \le n-4$, write H_{n-3} as $\langle \mathbf{z}, (\mathbf{u}_{n-3})^n, H'_{n-3}, (\mathbf{y})^c, \mathbf{w}, H''_{n-3}, \mathbf{y}\rangle$, and write H_{n-2} as $\langle \mathbf{z}, \mathbf{y}\rangle$. Obviously, $\mathbf{w}$ is an odd vertex. We set $P_i = \langle \mathbf{x}, R'_i, \mathbf{u}_i, (\mathbf{u}_i)^n, H'_i, \mathbf{y}\rangle$ for $1 \le i \le n-4$, $P_{n-3} = \langle \mathbf{x}, R'_{n-3}, \mathbf{u}_{n-3}, (\mathbf{u}_{n-3})^n, H'_{n-3}, (\mathbf{y})^c, \mathbf{y}\rangle$, $P_{n-2} = \langle \mathbf{x}, R'_{n-2}, (\mathbf{v})^c, \mathbf{v}, \mathbf{z}, H_{n-2}, \mathbf{y}\rangle$, and $P_{n-1} = \langle \mathbf{x}, R'_{n-1}, (\mathbf{y})^n, \mathbf{y}\rangle$.

Suppose that $(\mathbf{y})^{n-1} = (\mathbf{x})^n$. Note that Q_{n-2} is a spanning subgraph of $Q^{01}_{n,n-1}$. Since Q_{n-2} is hyper Hamiltonian laceable, there exists a Hamiltonian path S of $Q^{01}_{n,m} - \{(\mathbf{x})^n\}$ joining $((\mathbf{v})^{n-1})^n$ to $(\mathbf{w})^{n-1}$. Set $P_n = \langle \mathbf{x}, R'_n, (\mathbf{v})^{n-1}, ((\mathbf{v})^{n-1})^n, S, (\mathbf{w})^{n-1}, \mathbf{w}, H''_{n-3}, \mathbf{y}\rangle$ and $P_{n+1} = \langle \mathbf{x}, (\mathbf{x})^n = (\mathbf{y})^{n-1}, \mathbf{y}\rangle$. Then $\{P_1, P_2, \ldots, P_{n+1}\}$ forms an $(n+1)^*$-container of $Q_{n,m}$ between $\mathbf{x}$ and $\mathbf{y}$. See Figure 14.18a for illustration.

Now, we consider $(\mathbf{y})^{n-1} \ne (\mathbf{x})^n$. Since Q_{n-2} is a spanning subgraph of $Q^{11}_{n,m}$, by Lemmas 13.16 and 13.17, there exist two disjoint paths S_1 and S_2 of $Q^{11}_{m,n}$ such that (1) S_1 joins $((\mathbf{v})^{n-1})^n$ to $(\mathbf{y})^{n-1}$, (2) S_2 joins $(\mathbf{x})^n$ to $(\mathbf{w})^{n-1}$, and (3) $S_1 \cup S_2$ spans $Q^{01}_{n,n-1}$. Set $P_n = \langle \mathbf{x}, R'_n, (\mathbf{v})^{n-1}, ((\mathbf{v})^{n-1})^n, S_1, (\mathbf{y})^{n-1}, \mathbf{y}\rangle$ and $P_{n+1} = \langle \mathbf{x}, (\mathbf{x})^n, S_2, (\mathbf{w})^{n-1}, \mathbf{w}, H''_{n-3}, \mathbf{y}\rangle$. Then $\{P_1, P_2, \ldots, P_{n+1}\}$ forms an $(n+1)^*$-container of $Q_{n,m}$ between $\mathbf{x}$ and $\mathbf{y}$. See Figure 14.18b for illustration.

Case 3.1.2. $k = n$. Obviously, $A \cup \{((\mathbf{v})^{n-1})^n\}$ is a set of $(n-2)$ even vertices of $Q^{01}_{n,m} \cup Q^{11}_{n,m}$. Since Q_{n-1} is a spanning subgraph of $Q^{01}_{n,m} \cup Q^{11}_{n,m}$, by Theorem 14.14, there are $(n-1)$ disjoint paths $\{H_1, H_2, \ldots, H_{n-1}\}$ of $Q^{01}_{n,m} \cup Q^{11}_{n,m}$ joining $\mathbf{y}$ to $A \cup \{\mathbf{z}, ((\mathbf{v})^{n-1})^n\}$ such that (1) H_i joins $(\mathbf{u}_i)^n$ to

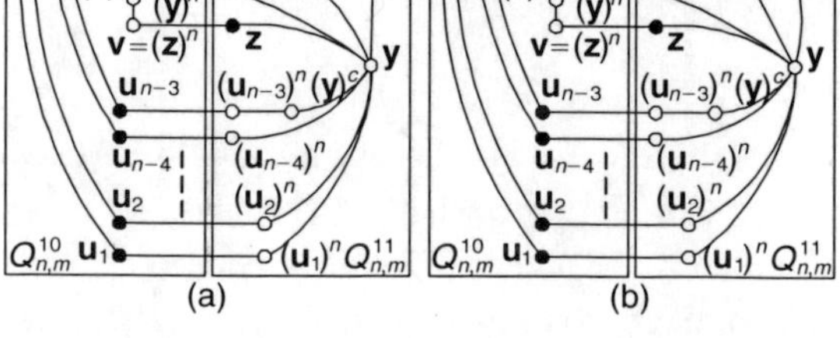

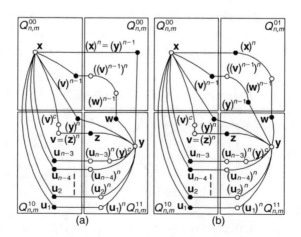

FIGURE 14.18 Illustration for Lemma 14.28.

$\mathbf{y}$ for $1 \le i \le n-3$, (2) H_{n-2} joins $\mathbf{z}$ to $\mathbf{y}$, (3) H_{n-1} joins $((\mathbf{v})^{n-1})^n$ to $\mathbf{y}$, and (4) $\bigcup_{i=1}^{n-1} H_i$ spans $Q_{n,m}^{01} \cup Q_{n,m}^{11}$. We set $P_i = \langle \mathbf{x}, R_i', \mathbf{u}_i, (\mathbf{u}_i)^n, H_i, \mathbf{y}\rangle$ for $1 \le i \le n-3$, $P_{n-2} = \langle \mathbf{x}, R_{n-2}', (\mathbf{v})^c, \mathbf{v}, \mathbf{z}, H_{n-2}, \mathbf{y}\rangle$, $P_{n-1} = \langle \mathbf{x}, R_{n-1}', (\mathbf{y})^n, \mathbf{y}\rangle$, and $P_n = \langle \mathbf{x}, R_n', (\mathbf{v})^{n-1}, ((\mathbf{v})^{n-1})^n, H_{n-1}, \mathbf{y}\rangle$. Then $\{P_1, P_2, \ldots, P_n\}$ forms an n^*-container of $Q_{n,m}$ between $\mathbf{x}$ and $\mathbf{y}$.

Case 3.2. $3 \le k \le n-1$. Let $\mathbf{v}$ be an even vertex of $Q_{n,m}^{00} \cup Q_{n,m}^{10}$ such that $\mathbf{v} \neq \mathbf{x}$ and $(\mathbf{y})^n$ is not a neighbor of $\mathbf{v}$. Since $Q_{n,m}^{00} \cup Q_{n,m}^{10}$ is isomorphic to $Q_{n-1,m}$, by induction, there exists a k^*-container $\{R_1, R_2, \ldots, R_k\}$ of $Q_{n,m}^{00} \cup Q_{n,m}^{10}$ between $\mathbf{x}$ and $\mathbf{v}$. We write $R_i = \langle \mathbf{x}, R_i', \mathbf{u}_i, \mathbf{v}\rangle$ for $1 \le i \le k$. Let $A = \{\mathbf{u}_1, \mathbf{u}_2, \ldots, \mathbf{u}_k\}$. Since $k \ge 3$, at most one vertex of A is an even vertex. Without loss of generality, we assume that $\{\mathbf{u}_1, \mathbf{u}_2, \ldots, \mathbf{u}_{k-1}\}$ is a set of $(k-1)$ odd vertices. Obviously, $\{(\mathbf{u}_i)^n \mid 1 \le i \le k-1\}$ is a set of $(k-1)$ even vertices of $Q_{n,m}^{01} \cup Q_{n,m}^{11}$. Since Q_{n-1} is a spanning subgraph of $Q_{n,m}^{01} \cup Q_{n,m}^{11}$, by Theorem 14.14, there are k disjoint paths $\{H_1, H_2, \ldots, H_k\}$ of $Q_{n,m}^{01} \cup Q_{n,m}^{11}$ joining $\mathbf{y}$ to $\{(\mathbf{u}_i)^n \mid 1 \le i \le k-1\} \cup \{(\mathbf{v})^n\}$ such that (1) H_i joins $(\mathbf{u}_i)^n$ to $\mathbf{y}$ for $1 \le i \le k-1$, (2) H_k joins $(\mathbf{v})^n$ to $\mathbf{y}$, and (3) $\bigcup_{i=1}^{k} H_i$ spans $Q_{n,m}^{01} \cup Q_{n,m}^{11}$. We set $P_i = \langle \mathbf{x}, R_i', \mathbf{u}_i, (\mathbf{u}_i)^n, H_i, \mathbf{y}\rangle$ for $1 \le i \le k-1$ and $P_k = \langle \mathbf{x}, R_k', \mathbf{u}_k, \mathbf{v}, (\mathbf{v})^n, H_k, \mathbf{y}\rangle$. Then $\{P_1, P_2, \ldots, P_k\}$ forms a k^*-container of $Q_{n,m}$ between $\mathbf{x}$ and $\mathbf{y}$. □

With Lemmas 14.24 and 14.29, we have the following theorem.

THEOREM 14.17 The enhanced hypercube $Q_{n,m}$ is super spanning laceable if m is an odd integer and $Q_{n,m}$ is super spanning connected if m is an even integer.

Proof: Since $Q_{2,2}$ is isomorphic to the complete graph K_4. Obviously, this theorem holds for $n = 2$. By Lemma 14.24, this theorem holds if $n \ge 3$ and m is an odd integer. Thus, we suppose that $n \ge 3$ and m is an even integer. Let $\mathbf{x}$ and $\mathbf{y}$ be any two different vertices of $Q_{n,m}$. We need to find a k^*-container of $Q_{n,m}$ between $\mathbf{x}$ and $\mathbf{y}$ for every $1 \le k \le n+1$. Without loss of generality, we assume that $\mathbf{x}$ is an even vertex. By Lemma 14.24, this theorem holds if $\mathbf{y}$ is an odd vertex. By Lemma 14.29, this theorem holds if $\mathbf{y}$ is an even vertex. Thus, this theorem is proved. □

14.6 SPANNING CONNECTIVITY OF THE PANCAKE GRAPHS

Lin et al. [233] discuss the spanning connectivity of the pancake graphs. Now, we review some basic background information on the properties of pancake graphs.

We will use boldface to denote a vertex of P_n. Hence, $\mathbf{u_1}, \mathbf{u_2}, \ldots, \mathbf{u_n}$ denote a sequence of vertices in P_n. In particular, $\mathbf{e}$ denotes the vertex $12\ldots n$. Let $\mathbf{u} = u_1 u_2 \ldots u_n$ be any vertex of P_n. We use $(\mathbf{u})_i$ to denote the ith component u_i of $\mathbf{u}$, and use $P_n^{\{i\}}$ to denote the ith subgraph of P_n induced by those vertices $\mathbf{u}$ with $(\mathbf{u})_n = i$. Obviously, P_n can be decomposed into n vertex-disjoint subgraphs $P_n^{\{i\}}$ for every $i \in \langle n\rangle$ such that each $P_n^{\{i\}}$ is isomorphic to P_{n-1}. Thus, the pancake graph can be constructed recursively. Let $H \subseteq \langle n\rangle$, we use P_n^H to denote the subgraph of P_n induced by $\cup_{i \in H} V(P_n^{\{i\}})$. By definition, there is exactly one neighbor $\mathbf{v}$ of $\mathbf{u}$ such that $\mathbf{u}$ and $\mathbf{v}$ are adjacent through an i-dimensional edge with $2 \le i \le n$. For this reason, we use

$(\mathbf{u})^i$ to denote the unique i-neighbor of $\mathbf{u}$. We have $((\mathbf{u})^i)^i = \mathbf{u}$ and $(\mathbf{u})^n \in P_n^{\{(\mathbf{u})_1\}}$. For $1 \le i, j \le n$ and $i \ne j$, we use $E^{i,j}$ to denote the set of edges between $P_n^{\{i\}}$ and $P_n^{\{j\}}$.

The following lemma follows from Theorem 11.15.

LEMMA 14.30 P_n is 1^*-connected if $n \ne 3$, and P_n is 2^*-connected if $n \ge 3$.

LEMMA 14.31 Let $n \ge 5$. Let $\mathbf{u}$ and $\mathbf{v}$ be any two distinct vertices in $P_n^{\{t\}}$ for some $t \in \langle n \rangle$. If P_{n-1} is k^*-connected, then there is a $(k+1)^*$-container of P_n between $\mathbf{u}$ and $\mathbf{v}$.

Proof: Since $P_n^{\{t\}}$ is isomorphic to P_{n-1}, there is a k^*-container $\{Q_1, Q_2, \ldots, Q_k\}$ of $P_n^{\{t\}}$ joining $\mathbf{u}$ to $\mathbf{v}$. We need to find a $(k+1)^*$-container of P_n joining $\mathbf{u}$ to $\mathbf{v}$. We set $p = (\mathbf{u})_1$ and $q = (\mathbf{v})_1$.

Case 1. $p = q$. Thus, $(\mathbf{u})^n$ and $(\mathbf{v})^n$ are in $P_n^{\{p\}}$. By Lemma 11.13, there is a Hamiltonian path Q of $P_n^{\{p\}}$ joining $(\mathbf{u})^n$ to $(\mathbf{v})^n$. We write Q as $\langle (\mathbf{u})^n, Q', \mathbf{y}, \mathbf{z}, (\mathbf{v})^n \rangle$. By Lemma 11.12, $(\mathbf{y})_1 \ne (\mathbf{z})_1$, $(\mathbf{y})_1 \ne t$, and $(\mathbf{z})_1 \ne t$. By Lemma 11.13, there is a Hamiltonian path R of $P_n^{\langle n \rangle - \{t,p\}}$ joining $(\mathbf{y})^n$ to $(\mathbf{z})^n$. We set Q_{k+1} as $\langle \mathbf{u}, (\mathbf{u})^n, Q', \mathbf{y}, (\mathbf{y})^n, R, (\mathbf{z})^n, \mathbf{z}, (\mathbf{v})^n, \mathbf{v} \rangle$. Then $\{Q_1, Q_2, \ldots, Q_{k+1}\}$ forms a $(k+1)^*$-container of P_n joining $\mathbf{u}$ to $\mathbf{v}$. See Figure 14.19a for illustration.

Case 2. $p \ne q$. Thus, $(\mathbf{u})^n$ and $(\mathbf{v})^n$ are in different subgraphs $P_n^{\{p\}}$ and $P_n^{\{q\}}$. By Lemma 11.13, there is a Hamiltonian path Q of $P_n^{\langle n \rangle - \{t\}}$ joining $(\mathbf{u})^n$ to $(\mathbf{v})^n$. We set Q_{k+1} as $\langle \mathbf{u}, (\mathbf{u})^n, Q, (\mathbf{v})^n, \mathbf{v} \rangle$. Then $\{Q_1, Q_2, \ldots, Q_{k+1}\}$ forms a $(k+1)^*$-container of P_n joining $\mathbf{u}$ to $\mathbf{v}$. See Figure 14.19b for illustration.

Thus, the theorem is proved. □

LEMMA 14.32 Let $n \ge 5$ and k be any positive integer with $3 \le k \le n-1$. Let $\mathbf{u}$ be any vertex in $P_n^{\{s\}}$ and $\mathbf{v}$ be any vertex in $P_n^{\{t\}}$ such that $s \ne t$. Suppose that P_{n-1} is k^*-connected. Then there is a k^*-container of P_n between $\mathbf{u}$ and $\mathbf{v}$ not using the edge $(\mathbf{u}, \mathbf{v})$ if $(\mathbf{u}, \mathbf{v}) \in E(P_n)$.

Proof: Since $|E^{s,t}| = (n-2)! \ge 6$, we can choose a vertex $\mathbf{y}$ in $P_n^{\{s\}} - \{\mathbf{u}\}$ and a vertex $\mathbf{z}$ in $P_n^{\{t\}} - \{\mathbf{v}\}$ with $(\mathbf{y}, \mathbf{z}) \in E^{s,t}$. Note that $P_n^{\{s\}}$ and $P_n^{\{t\}}$ are both isomorphic to P_{n-1}. Let $\{R_1, R_2, \ldots, R_k\}$ be a k^*-container of $P_n^{\{s\}}$ joining $\mathbf{u}$ to $\mathbf{y}$, and $\{H_1, H_2, \ldots, H_k\}$ be a k^*-container of $P_n^{\{t\}}$ joining $\mathbf{z}$ to $\mathbf{v}$. We write $R_i = \langle \mathbf{u}, R_i', \mathbf{y_i}, \mathbf{y} \rangle$ and $H_i = \langle \mathbf{z}, \mathbf{z_i}, H_i', \mathbf{v} \rangle$. (Note that $\mathbf{y_i} = \mathbf{u}$ if the length of R_i' is zero and $\mathbf{z_i} = \mathbf{v}$ if the length of H_i' is zero.)

Let $I = \{\mathbf{y_i} \mid 1 \le i \le k\}$ and $J = \{\mathbf{z_i} \mid 1 \le i \le k\}$. Note that $(\mathbf{y_i})_1 = (\mathbf{y})_j$ for some $j \in \{2, 3, \ldots, n-1\}$, and $(\mathbf{y})_l \ne (\mathbf{y})_m$ if $l \ne m$. By Lemma 11.12, $\{(\mathbf{y_i})_1 \mid 1 \le i \le k\} \cap$

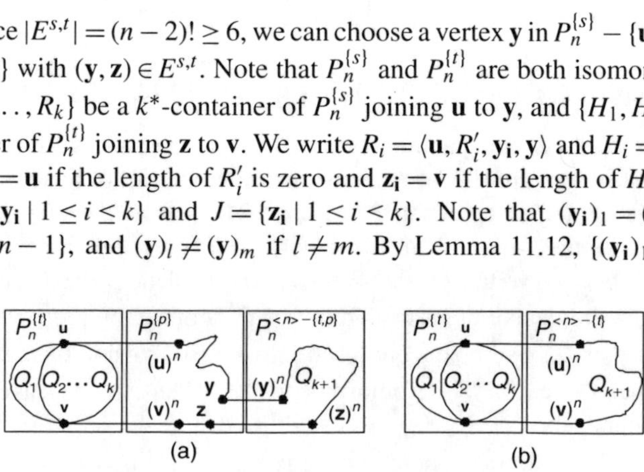

FIGURE 14.19 Illustration for Lemma 14.31.

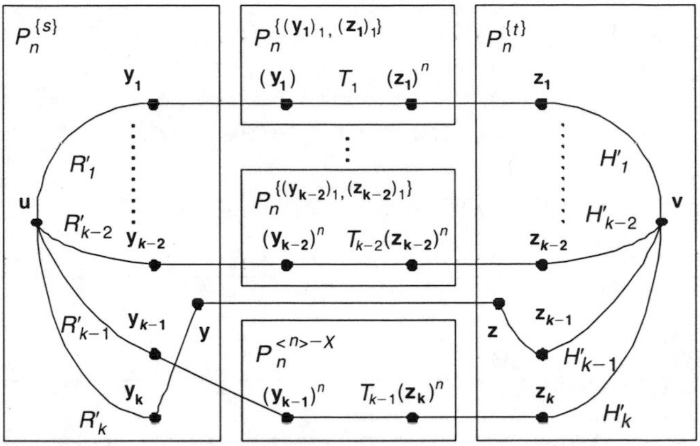

FIGURE 14.20 Illustration for Lemma 14.32.

$\{s,t\}=\emptyset$. Similarly, $\{(\mathbf{z_i})_1 \mid 1 \leq i \leq k\} \cap \{s,t\} = \emptyset$. Let $A = \{\mathbf{y_i} \mid \mathbf{y_i} \in I$ and there exists an element $\mathbf{z_j} \in J$ such that $(\mathbf{y_i})_1 = (\mathbf{z_j})_1\}$. Then we relabel the indices of I and J such that $(\mathbf{y_i})_1 = (\mathbf{z_i})_1$ for $1 \leq i \leq |A|$. We set X as $\{(\mathbf{y_i})_1 \mid 1 \leq i \leq k-2\} \cup \{(\mathbf{z_i})_1 \mid 1 \leq i \leq k-2\} \cup \{s,t\}$. By Lemma 11.13, there is a Hamiltonian path T_i of $P_n^{\{(\mathbf{y_i})_1,(\mathbf{z_i})_1\}}$ joining $(\mathbf{y_i})^n$ to $(\mathbf{z_i})^n$ for every $1 \leq i \leq k-2$, and there is a Hamiltonian path T_{k-1} of $P_n^{\langle n \rangle - X}$ joining $(\mathbf{y_{k-1}})^n$ to $(\mathbf{z_k})^n$. (Note that $\{(\mathbf{y_i})_1, (\mathbf{z_i})_1\} = \{(\mathbf{y_i})_1\}$ if $(\mathbf{y_i})_1 = (\mathbf{z_i})_1$.) We set

$$\begin{aligned} Q_i &= \langle \mathbf{u}, R'_i, \mathbf{y_i}, (\mathbf{y_i})^n, T_i, (\mathbf{z_i})^n, \mathbf{z_i}, H'_i, \mathbf{v} \rangle \quad \text{for } 1 \leq i \leq k-2 \\ Q_{k-1} &= \langle \mathbf{u}, R'_{k-1}, \mathbf{y_{k-1}}, (\mathbf{y_{k-1}})^n, T_{k-1}, (\mathbf{z_k})^n, \mathbf{z_k}, H'_k, \mathbf{v} \rangle \\ Q_k &= \langle \mathbf{u}, R'_k, \mathbf{y_k}, \mathbf{y}, \mathbf{z}, \mathbf{z_{k-1}}, H'_{k-1}, \mathbf{v} \rangle \end{aligned}$$

It is easy to check whether $\{Q_1, Q_2, \ldots, Q_k\}$ forms a k^*-container of P_n joining $\mathbf{u}$ to $\mathbf{v}$ not using the edge $(\mathbf{u},\mathbf{v})$ if $(\mathbf{u},\mathbf{v}) \in E(P_n)$. See Figure 14.20 for illustration. □

THEOREM 14.18 P_n is $(n-1)^*$-connected if $n \geq 2$.

Proof: It is easy to see that P_2 is 1^*-connected and P_3 is 2^*-connected. Since P_4 is vertex-transitive, we claim that P_4 is 3^*-connected by listing all 3^*-containers from 1234 to any vertex, as follows:

⟨(1234), (2134), (4312)⟩ ⟨(1234), (3214), (4123), (2143), (3412), (4312)⟩ ⟨(1234), (4321), (2341), (1432), (4132), (2314), (1324), (3124), (4213), (1243), (3421), (2431), (4231), (3241), (1423), (2413), (3142), (1342), (4312)⟩
⟨(1234), (2134), (3124), (1324), (4231), (2431), (1342)⟩ ⟨(1234), (3214), (2314), (4132), (1432), (2341), (3241), (1423), (4123), (2143), (3412), (4312), (1342)⟩ ⟨(1234), (4321), (3421), (1243), (4213), (2413), (3142), (1342)⟩
⟨(1234),(2134),(3124),(4213),(2413),(1423),(3241),(2341),(1432),(3412),(4312),(1342),(3142),(4132),(2314),(1324), (4231), (2431), (3421)⟩ ⟨(1234), (3214), (4123), (2143), (1243), (3421)⟩ ⟨(1234), (4321), (3421)⟩

⟨(1234), (2134), (3124), (1324), (4231), (2431), (3421), (4321)⟩ ⟨(1234), (3214), (2314), (4132), (1432), (3412), (4312), (1342), (3142), (2413), (4213), (1243), (2143), (4123), (1423), (3241), (2341), (4321)⟩ ⟨(1234), (4321)⟩
⟨(1234), (2134), (3124), (1324), (4231), (2431)⟩ ⟨(1234), (3214), (2314), (4132), (3142), (2413), (4213), (1243), (2143), (4123), (1423), (3241), (2341), (1432), (3412), (4312), (1342), (2431)⟩ ⟨(1234), (4321), (3421), (2431)⟩
⟨(1234), (2134), (3124), (4213), (1243), (2143), (3412), (4312), (1342), (3142), (2413), (1423)⟩ ⟨(1234), (3214), (4123), (1423)⟩ ⟨(1234), (4321), (3421), (2431), (4231), (1324), (2314), (4132), (1432), (2341), (3241), (1423)⟩
⟨(1234), (2134), (3124), (4213), (2413), (3142), (1342), (4312), (3412), (1432), (4132), (2314), (1324), (4231), (2431), (3421), (1243), (2143), (4123)⟩ ⟨(1234), (3214), (4123)⟩ ⟨(1234), (4321), (2341), (3241), (1423), (4123)⟩
⟨(1234), (2134), (3124), (4213), (2413), (1423), (3241), (4231)⟩ ⟨(1234), (3214), (4123), (2143), (1243), (3421), (2431), (4231)⟩ ⟨(1234), (4321), (2341), (1432), (3412), (4312), (1342), (3142), (4132), (2314), (1324), (4231)⟩
⟨(1234), (2134), (3124), (1324), (2314), (4132), (3142), (1342), (4312), (3412), (1432), (2341), (3241)⟩ ⟨(1234), (3214), (4123), (2143), (1243), (4213), (2413), (1423), (3241)⟩ ⟨(1234), (4321), (3421), (2431), (4231), (3241)⟩
⟨(1234), (2134), (4312), (1342), (2431), (3421), (1243), (4213), (3124), (1324), (4231), (3241), (2341)⟩ ⟨(1234), (3214), (2314), (4132), (3142), (2413), (1423), (4123), (2143), (3412), (1432), (2341)⟩ ⟨(1234), (4321), (2341)⟩
⟨(1234), (2134), (3124), (4213), (2413)⟩ ⟨(1234), (3214), (4123), (2143), (1243), (3421), (2431), (4231), (1324), (2314), (4132), (1432), (3412), (4312), (1342), (3142), (2413)⟩ ⟨(1234), (4321), (2341), (3241), (1423), (2413)⟩
⟨(1234), (2134), (3124), (1324), (4231), (2431), (3421), (1243)⟩ ⟨(1234), (3214), (2314), (4132), (1432), (3412), (4312), (1342), (3142), (2413), (4213), (1243)⟩ ⟨(1234), (4321), (2341), (3241), (1423), (4123), (2143), (1243)⟩
⟨(1234), (2134), (3124), (1324), (2314), (3214)⟩ ⟨(1234), (3214)⟩ ⟨(1234), (4321), (3421), (2431), (4231), (3241), (2341), (1432), (4132), (3142), (1342), (4312), (3412), (2143), (1243), (4213), (2413), (1423), (4123), (3214)⟩
⟨(1234), (2134), (3124), (4213), (2413), (3142), (4132), (2314)⟩ ⟨(1234), (3214), (2314)⟩ ⟨(1234), (4321), (3421), (1243), (2143), (4123), (1423), (3241), (2341), (1432), (3412), (4312), (1342), (2431), (4231), (1324), (2314)⟩
⟨(1234), (2134), (4312), (3412), (1432), (4132), (3142), (1342), (2431), (3421), (1243), (2143), (4123), (1423), (2413), (4213), (3124), (1324)⟩ ⟨(1234), (3214), (2314), (1324)⟩ ⟨(1234), (4321), (2341), (3241), (4231), (1324)⟩
⟨(1234), (2134), (3124)⟩ ⟨(1234), (3214), (4123), (1423), (3241), (4231), (2431), (3421), (1243), (2143), (3412), (4312), (1342), (3142), (2413), (4213), (3124)⟩ ⟨(1234), (4321), (2341), (1432), (4132), (2314), (1324), (3124)⟩
⟨(1234), (2134)⟩ ⟨(1234), (3214), (2314), (1324), (3124), (2134)⟩ ⟨(1234), (4321), (2341), (3241), (4231), (2431), (3421), (1243), (4213), (2413), (1423), (4123), (2143), (3412), (1432), (4132), (3142), (1342), (4312), (2134)⟩
⟨(1234), (2134), (4312), (1342), (3142)⟩ ⟨(1234), (3214), (4123), (1423), (2413), (3142)⟩ ⟨(1234), (4321), (3421), (2431), (4231), (3241), (2341), (1432), (3412), (2143), (1243), (4213), (3124), (1324), (2314), (4132), (3142)⟩
⟨(1234), (2134), (3124), (1324), (2314), (4132)⟩ ⟨(1234), (3214), (4123), (1423), (2413), (4213), (1243), (2143), (3412), (4312), (1342), (3142), (4132)⟩ ⟨(1234), (4321), (3421), (2431), (4231), (3241), (2341), (1432), (4132)⟩
⟨(1234), (2134), (4312), (1342), (3142), (2413), (1423), (3241), (4231), (2431), (3421), (1243), (4213), (3124), (1324), (2314), (4132), (1432)⟩ ⟨(1234), (3214), (4123), (2143), (3412), (1432)⟩ ⟨(1234), (4321), (2341), (1432)⟩

⟨(1234), (2134), (3124), (1324), (2314), (4132), (3142), (1342), (4312), (3412)⟩
⟨(1234), (3214), (4123), (1423), (2413), (4213), (1243), (2143), (3412)⟩
⟨(1234), (4321), (3412), (2431), (4231), (3241), (2341), (1432), (3412)⟩
⟨(1234), (2134), (4312), (1342), (3142), (4132), (2314), (1324), (3124), (4213)⟩
⟨(1234), (3214), (4123), (1423), (2413), (4213)⟩
⟨(1234), (4321), (3421), (2431), (4231), (3241), (2341), (1432), (3412), (2143), (1243), (4213)⟩
⟨(1234), (2134), (4312), (1342), (3142), (4132), (1432), (3412), (2143)⟩
⟨(1234), (3214), (2314), (1324), (3124), (4213), (2413), (1423), (4123), (2143)⟩
⟨(1234), (4321), (2341), (3241), (4231), (2431), (3421), (1243), (2143)⟩

Assume that P_k is $(k-1)^*$-connected for every $4 \leq k \leq n-1$. Let $\mathbf{u}$ and $\mathbf{v}$ be any two distinct vertices of P_n with $\mathbf{u} \in P_n^{\{s\}}$ and $\mathbf{v} \in P_n^{\{t\}}$. We need to find an $(n-1)^*$-container between $\mathbf{u}$ and $\mathbf{v}$ of P_n. Suppose that $s = t$. By Lemma 14.31, there is an $(n-1)^*$-container of P_n joining $\mathbf{u}$ to $\mathbf{v}$. Thus, we assume that $s \neq t$. We set $p = (\mathbf{u})_1$ and $q = (\mathbf{v})_1$.

Case 1. $p = t$ and $q = s$. Thus, $(\mathbf{u})^n \in P_n^{\{t\}}$ and $(\mathbf{v})^n \in P_n^{\{s\}}$.

Case 1.1. $\mathbf{u} = (\mathbf{v})^n$. Thus, $(\mathbf{u}, \mathbf{v}) \in E(P_n)$. By Lemma 14.32, there is an $(n-2)^*$-container $\{Q_1, Q_2, \ldots, Q_{n-2}\}$ of P_n joining $\mathbf{u}$ to $\mathbf{v}$ not using the edge $(\mathbf{u}, \mathbf{v})$. We set Q_{n-1} as $\langle \mathbf{u}, \mathbf{v} \rangle$. Then $\{Q_1, Q_2, \ldots, Q_{n-1}\}$ forms an $(n-1)^*$-container of P_n joining $\mathbf{u}$ to $\mathbf{v}$.

Case 1.2. $\mathbf{u} \neq (\mathbf{v})^n$. We set $\mathbf{y} = (\mathbf{v})^n$ and $\mathbf{z} = (\mathbf{u})^n$. Let $\{R_1, R_2, \ldots, R_{n-2}\}$ be an $(n-2)^*$-container of $P_n^{\{s\}}$ joining $\mathbf{u}$ to $\mathbf{y}$, and let $\{H_1, H_2, \ldots, H_{n-2}\}$ be an $(n-2)^*$-container of $P_n^{\{t\}}$ joining $\mathbf{z}$ to $\mathbf{v}$. We write $R_i = \langle \mathbf{u}, R_i', \mathbf{y_i}, \mathbf{y} \rangle$ and $H_i = \langle \mathbf{z}, \mathbf{z_i}, H_i', \mathbf{v} \rangle$. We set $I = \{(\mathbf{y_i})_1 \mid 1 \leq i \leq n-2\}$ and $J = \{(\mathbf{z_i})_1 \mid 1 \leq i \leq n-2\}$. Note that $(\mathbf{y_i})_1 = (\mathbf{y})_j$ for some $j \in \{2, 3, \ldots, n-1\}$, and $(\mathbf{y})_k \neq (\mathbf{y})_l$ if $k \neq l$. By Lemma 11.12, $I = \{(\mathbf{y})_i \mid 2 \leq i \leq n-1\} = \langle n \rangle - \{s, t\}$. Similarly, $J = \langle n \rangle - \{s, t\}$. We have $I = J$. Without loss of generality, we assume that $(\mathbf{y_i})_1 = (\mathbf{z_i})_1$ for every $1 \leq i \leq n-2$. By Lemma 11.13, there is a Hamiltonian path T_i of $P_n^{\{(\mathbf{y_i})_1\}}$ joining $(\mathbf{y_i})^n$ to $(\mathbf{z_i})^n$ for every $1 \leq i \leq n-4$, and there is a Hamiltonian path T_{n-3} of $P_n^{\{(\mathbf{y_{n-3}})_1, (\mathbf{y_{n-2}})_1\}}$ joining $(\mathbf{y_{n-3}})^n$ to $(\mathbf{z_{n-2}})^n$. We set

$$
\begin{aligned}
Q_i &= \langle \mathbf{u}, R_i', \mathbf{y_i}, (\mathbf{y_i})^n, T_i, (\mathbf{z_i})^n, \mathbf{z_i}, H_i', \mathbf{v} \rangle \quad \text{for } 1 \leq i \leq n-4 \\
Q_{n-3} &= \langle \mathbf{u}, R_{n-3}', \mathbf{y_{n-3}}, (\mathbf{y_{n-3}})^n, T_{n-3}, (\mathbf{z_{n-2}})^n, \mathbf{z_{n-2}}, H_{n-2}', \mathbf{v} \rangle \\
Q_{n-2} &= \langle \mathbf{u}, \mathbf{z}, \mathbf{z_{n-3}}, H_{n-3}', \mathbf{v} \rangle \\
Q_{n-1} &= \langle \mathbf{u}, R_{n-2}', \mathbf{y_{n-2}}, \mathbf{y}, \mathbf{v} \rangle
\end{aligned}
$$

Then $\{Q_1, Q_2, \ldots, Q_{n-1}\}$ forms an $(n-1)^*$-container of P_n joining $\mathbf{u}$ to $\mathbf{v}$. See Figure 14.21a for illustration.

Case 2. $p = t$ and $q \in \langle n \rangle - \{s, t\}$. Since $|E^{s,q}| = (n-2)! \geq 6$, we can choose a vertex $\mathbf{y}$ in $P_n^{\{s\}} - \{\mathbf{u}\}$ with $(\mathbf{y})^n \in P_n^{\{q\}}$. We set $\mathbf{z} = (\mathbf{u})^n \in P_n^{\{t\}}$. Let $\{R_1, R_2, \ldots, R_{n-2}\}$ be an $(n-2)^*$-container of $P_n^{\{s\}}$ joining $\mathbf{u}$ to $\mathbf{y}$, and $\{H_1, H_2, \ldots, H_{n-2}\}$ be an $(n-2)^*$-container of $P_n^{\{t\}}$ joining $\mathbf{z}$ to $\mathbf{v}$. We write $R_i = \langle \mathbf{u}, R_i', \mathbf{y_i}, \mathbf{y} \rangle$ and $H_i = \langle \mathbf{z}, \mathbf{z_i}, H_i', \mathbf{v} \rangle$. We have $\{(\mathbf{y_i})_1 \mid 1 \leq i \leq n-2\} = \{(\mathbf{y})_i \mid 2 \leq i \leq n-1\}$. By Lemma 11.12, $\{(\mathbf{y_i})_1 \mid 1 \leq i \leq n-2\} = \langle n \rangle - \{s, q\}$. Similarly, $\{(\mathbf{z_i})_1 \mid 1 \leq i \leq n-2\} = \langle n \rangle - \{s, t\}$. Without loss of generality, we assume that $(\mathbf{y_i})_1 = (\mathbf{z_i})_1$ for every

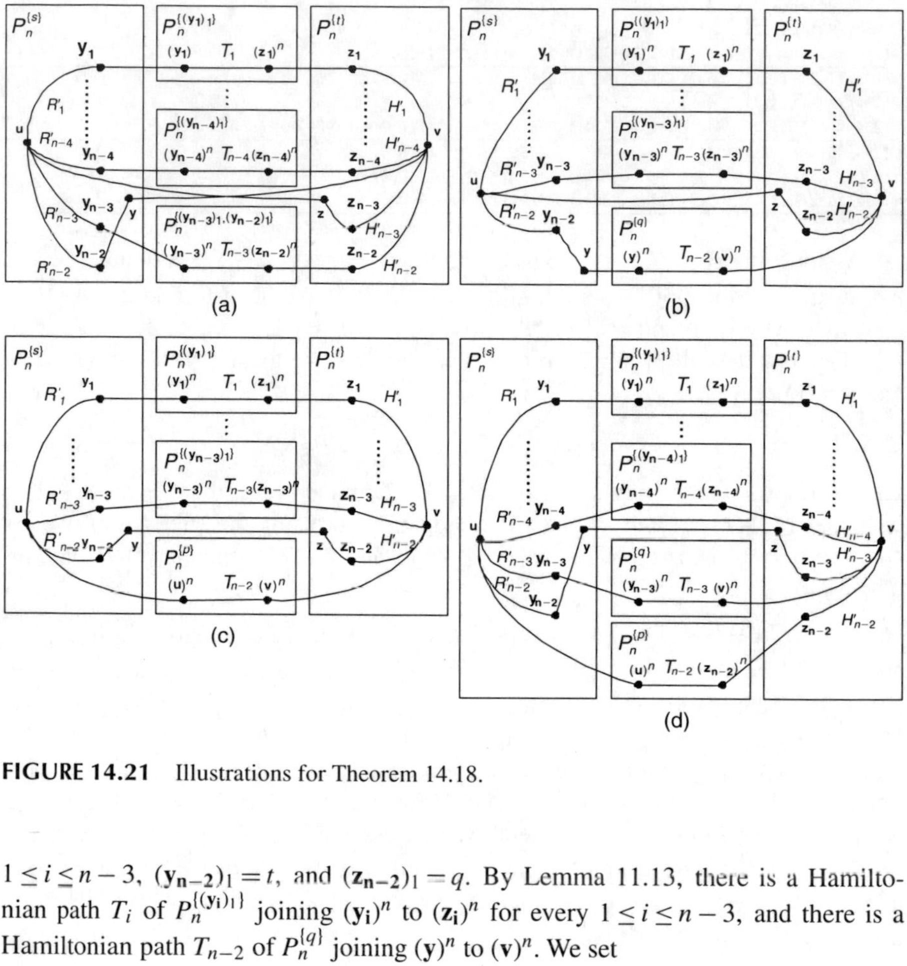

FIGURE 14.21 Illustrations for Theorem 14.18.

$1 \le i \le n-3$, $(\mathbf{y_{n-2}})_1 = t$, and $(\mathbf{z_{n-2}})_1 = q$. By Lemma 11.13, there is a Hamiltonian path T_i of $P_n^{\{(\mathbf{y_i})_1\}}$ joining $(\mathbf{y_i})^n$ to $(\mathbf{z_i})^n$ for every $1 \le i \le n-3$, and there is a Hamiltonian path T_{n-2} of $P_n^{\{q\}}$ joining $(\mathbf{y})^n$ to $(\mathbf{v})^n$. We set

$$\begin{aligned} Q_i &= \langle \mathbf{u}, R_i', \mathbf{y_i}, (\mathbf{y_i})^n, T_i, (\mathbf{z_i})^n, \mathbf{z_i}, H_i', \mathbf{v}\rangle \quad \text{for } 1 \le i \le n-3 \\ Q_{n-2} &= \langle \mathbf{u}, R_{n-2}', \mathbf{y_{n-2}}, \mathbf{y}, (\mathbf{y})^n, T_{n-2}, (\mathbf{v})^n, \mathbf{v}\rangle \\ Q_{n-1} &= \langle \mathbf{u}, \mathbf{z}, \mathbf{z_{n-2}}, H_{n-2}', \mathbf{v}\rangle \end{aligned}$$

Then $\{Q_1, Q_2, \ldots, Q_{n-1}\}$ forms an $(n-1)^*$-container of P_n joining $\mathbf{u}$ to $\mathbf{v}$. See Figure 14.21b for illustration.

Case 3. $p, q \in \langle n\rangle - \{s,t\}$. Since $|E^{s,t}| = (n-2)! \ge 6$, there exists an edge $(\mathbf{y}, \mathbf{z})$ in $E^{s,t}$ with $\mathbf{y} \in P_n^{\{s\}} - \{\mathbf{u}\}$ and $\mathbf{z} \in P_n^{\{t\}} - \{\mathbf{v}\}$. Let $\{R_1, R_2, \ldots, R_{n-2}\}$ be an $(n-2)^*$-container of $P_n^{\{s\}}$ joining $\mathbf{u}$ to $\mathbf{y}$, and let $\{H_1, H_2, \ldots, H_{n-2}\}$ be an $(n-2)^*$-container of $P_n^{\{t\}}$ joining $\mathbf{z}$ to $\mathbf{v}$. We write $R_i = \langle \mathbf{u}, R_i', \mathbf{y_i}, \mathbf{y}\rangle$ and $H_i = \langle \mathbf{z}, \mathbf{z_i}, H_i', \mathbf{v}\rangle$. We set $I = \{(\mathbf{y_i})_1 \mid 1 \le i \le n-2\}$ and $J = \{(\mathbf{z_i})_1 \mid 1 \le i \le n-2\}$. We have $I = \{(\mathbf{y})_i \mid 2 \le i \le n-1\}$. By Lemma 11.12, $I = \langle n\rangle - \{s,t\}$. Similarly,

$J = \langle n\rangle - \{s, t\}$. We have $I = J$. Without loss of generality, we assume that $(\mathbf{y_i})_1 = (\mathbf{z_i})_1$ for every $1 \le i \le n-2$ with $(\mathbf{y_{n-2}})_1 = p$.

Case 3.1. $p = q$. By Lemma 11.13, there is a Hamiltonian path T_i of $P_n^{\{(\mathbf{y_i})_1\}}$ joining $(\mathbf{y_i})^n$ to $(\mathbf{z_i})^n$ for every $i \in \langle n-3\rangle$, and there is a Hamiltonian path T_{n-2} in $P_n^{\{p\}}$ joining $(\mathbf{u})^n$ to $(\mathbf{v})^n$. We set

$$\begin{aligned} Q_i &= \langle \mathbf{u}, R_i', \mathbf{y_i}, (\mathbf{y_i})^n, T_i, (\mathbf{z_i})^n, \mathbf{z_i}, H_i', \mathbf{v}\rangle \text{ for } 1 \le i \le n-3 \\ Q_{n-2} &= \langle \mathbf{u}, R_{n-2}', \mathbf{y_{n-2}}, \mathbf{y}, \mathbf{z}, \mathbf{z_{n-2}}, H_{n-2}', \mathbf{v}\rangle \\ Q_{n-1} &= \langle \mathbf{u}, (\mathbf{u})^n, T_{n-2}, (\mathbf{v})^n, \mathbf{v}\rangle \end{aligned}$$

Then $\{Q_1, Q_2, \ldots, Q_{n-1}\}$ forms an $(n-1)^*$-container of P_n joining $\mathbf{u}$ and $\mathbf{v}$. See Figure 14.21c for illustration.

Case 3.2. $p \ne q$. Without loss of generality, we assume that $(\mathbf{y_{n-3}})_1 = q$. By Lemma 14.30, there is a Hamiltonian path T_i of $P_n^{\{(\mathbf{y_i})_1\}}$ joining $(\mathbf{y_i})^n$ to $(\mathbf{z_i})^n$ for every $1 \le i \le n-4$, there is a Hamiltonian path T_{n-3} of $P_n^{\{q\}}$ joining $(\mathbf{y_{n-3}})^n$ to $(\mathbf{v})^n$, and there is a Hamiltonian path T_{n-2} of $P_n^{\{p\}}$ joining $(\mathbf{u})^n$ to $(\mathbf{z_{n-2}})^n$. We set

$$\begin{aligned} Q_i &= \langle \mathbf{u}, R_i', \mathbf{y_i}, (\mathbf{y_i})^n, T_i, (\mathbf{z_i})^n, H_i', \mathbf{v}\rangle \quad \text{for } 1 \le i \le n-4 \\ Q_{n-3} &= \langle \mathbf{u}, R_{n-3}', \mathbf{y_{n-3}}, (\mathbf{y_{n-3}})^n, T_{n-3}, (\mathbf{v})^n, \mathbf{v}\rangle \\ Q_{n-2} &= \langle \mathbf{u}, (\mathbf{u})^n, T_{n-2}, (\mathbf{z_{n-2}})^n, \mathbf{z_{n-2}}, H_{n-2}', \mathbf{v}\rangle \\ Q_{n-1} &= \langle \mathbf{u}, R_{n-2}', \mathbf{y_{n-2}}, \mathbf{y}, \mathbf{z}, \mathbf{z_{n-3}}, H_{n-3}', \mathbf{v}\rangle \end{aligned}$$

It is easy to check whether $\{Q_1, Q_2, \ldots, Q_{n-1}\}$ is an $(n-1)^*$-container of P_n from $\mathbf{u}$ to $\mathbf{v}$. See Figure 14.21d for illustration.

Thus, the theorem is proved. □

THEOREM 14.19 **[233]** The P_n is super connected if and only if $n \ne 3$.

Proof: We prove this theorem by induction. Obviously, this theorem is true for P_1 and P_2. Since P_3 is isomorphic to a cycle with six vertices, P_3 is not 1^*-connected. Thus, P_3 is not super connected. By Lemma 14.30 and Theorem 14.18, this theorem holds for P_4. Assume that P_k is super connected for every $4 \le k \le n-1$. By Lemma 14.30 and Theorem 14.18, P_n is k^*-connected for any $k \in \{1, 2, n-1\}$. Thus, we still need to construct a k^*-container of P_n between any two distinct vertices $\mathbf{u} \in P_n^{\{s\}}$ and $\mathbf{v} \in P_n^{\{t\}}$ for every $3 \le k \le n-2$.

Suppose that $s = t$. By induction, P_{n-1} is $(k-1)^*$-connected. By Lemma 14.31, there is a k^*-container of P_n joining $\mathbf{u}$ to $\mathbf{v}$. Suppose that $s \ne t$. By induction, P_{n-1} is k^*-connected. By Lemma 14.32, there is a k^*-container of P_n joining $\mathbf{u}$ to $\mathbf{v}$.

Hence, the theorem is proved. □

Lin et al. further discuss the spanning connectivity of the faulty pancake graphs [229]. Let F be any subset of $V(P_n) \cup E(P_n)$. We use F^i to denote the set $F \cap (V(P_n^{\{i\}}) \cup E(P_n^{\{i\}}))$. Then we set F^0 to denote the set $F - \bigcup_{i=1}^n F^i$.

THEOREM 14.20 Suppose that F is a subset of $V(P_n) \cup E(P_n)$ with $|F| = f \le n-4$ and $n \ge 4$. Then $P_n - F$ is i^*-connected for every $1 \le i \le n-1-f$.

Proof: We prove this statement by induction on n. Suppose that $n=4$. By Theorem 14.19, this statement is true for P_4. Suppose that this statement is true for P_m for every $4 \le m \le n-1$. By Theorems 14.16 and 14.19, $P_n - F$ is k^*-connected if $f=0$ or $f \ge 1$ with $k \in \{1,2\}$. Thus, we assume that $f \ge 1$ and $k \ge 3$. Let $\mathbf{u}$ and $\mathbf{v}$ be any two distinct vertices of $P_n - F$. We need to show that there is a k^*-container of $P_n - F$ between $\mathbf{u}$ and $\mathbf{v}$ for every $3 \le k \le n-f-1$. We have the following cases:

Case 1. $|F^i| \le f-1$ for every $i \in \langle n \rangle$.

Case 1.1. $\mathbf{u}, \mathbf{v} \in P_n^{\{r\}}$ for some $r \in \langle n \rangle$. By induction hypothesis, there is a k^*-container $\{R_1, R_2, \ldots, R_k\}$ of $P_n^{\{r\}} - F^r$ between $\mathbf{u}$ and $\mathbf{v}$. Without loss of generality, we write R_i as $\langle \mathbf{x_{i_1}}, \mathbf{x_{i_2}}, \ldots, \mathbf{x_{i_m}} \rangle$ with $\mathbf{x_{i_1}} = \mathbf{u}$ and $\mathbf{x_{i_m}} = \mathbf{v}$ for every $1 \le i \le k$. Note that $|\cup_{i=1}^{k}(V(R_i) - \{\mathbf{x_{i_1}}, \mathbf{x_{i_m}}\})| = |V(P_n^{\{r\}}) - (\{\mathbf{u},\mathbf{v}\} \cup V(F^r))| \ge (n-1)! - n + 3$. There are at least $(n-1)! - n + 3 - k$ edges in $\cup_{i=1}^{k}(E(R_i) - \{(\mathbf{x_{i_1}}, \mathbf{x_{i_2}}), (\mathbf{x_{i_{m-1}}}, \mathbf{x_{i_m}})\})$. (Note that $\{(\mathbf{x_{i_1}}, \mathbf{x_{i_2}}), (\mathbf{x_{i_{m-1}}}, \mathbf{x_{i_m}})\} = \{(\mathbf{x_{i_1}}, \mathbf{x_{i_2}})\}$ if $m=2$.) Since $|F| \le n-4$, there are at most $(n-4)$ vertices $\mathbf{z}$ in $P_n^{\{r\}}$ such that either the nth neighbors of them are in F or the edges between them and their nth neighbors are in F. Since $k \le n-2$ and $n \ge 5$, $(n-1)! - n + 3 - k \ge (n-1)! - 2n + 5 > 2(n-4)$. Thus, there is an edge $(\mathbf{x_{s_t}}, \mathbf{x_{s_{t+1}}}) \in E(R_s)$ for some $s \in \langle k \rangle$ and for some $t \in \langle s_m - 1 \rangle - \{1\}$ such that $\{(\mathbf{x_{s_t}})^n, (\mathbf{x_{s_{t+1}}})^n, (\mathbf{x_{s_t}}, (\mathbf{x_{s_t}})^n), (\mathbf{x_{s_{t+1}}}, (\mathbf{x_{s_{t+1}}})^n)\} \cap F = \emptyset$. Without loss of generality, we assume that $s=k$. Since $d(\mathbf{x_{k_t}}, \mathbf{x_{k_{t+1}}}) = 1$, by Lemma 11.12, $(\mathbf{x_{k_t}})_1 \ne (\mathbf{x_{k_{t+1}}})_1$. By Theorem 11.13, there is a Hamiltonian path W of $P_n^{\langle n \rangle - \{r\}} - F$ joining the vertex $(\mathbf{x_{k_t}})^n$ to the vertex $(\mathbf{x_{k_{t+1}}})^n$. We set $H = \langle \mathbf{x_{k_1}}, \mathbf{x_{k_2}}, \ldots, \mathbf{x_{k_t}}, (\mathbf{x_{k_t}})^n, W, (\mathbf{x_{k_{t+1}}})^n, \mathbf{x_{k_{t+1}}}, \mathbf{x_{k_{t+2}}}, \ldots, \mathbf{x_{k_m}} \rangle$. Then $\{R_1, R_2, \ldots, R_{k-1}, H\}$ forms the k^*-container of $P_n - F$ between $\mathbf{u}$ and $\mathbf{v}$. See Figure 14.22a for illustration.

Case 1.2. $\mathbf{u} \in P_n^{\{r\}}$ and $\mathbf{v} \in P_n^{\{s\}}$ for some $r, s \in \langle n \rangle$ with $r \ne s$. By Lemma 11.12, $|E^{r,s}| = (n-2)! > |F \cup \{\mathbf{u},\mathbf{v}\}|$ if $n \ge 5$. We can choose two adjacent vertices $\mathbf{x}$ and $\mathbf{y}$ with (1) $\mathbf{x} \in P_n^{\{r\}} - \{\mathbf{u}\}$, (2) $\mathbf{y} \in P_n^{\{s\}} - \{\mathbf{v}\}$, (3) $(N_{P_n}(\mathbf{x}) \cup N_{P_n}(\mathbf{y})) \cap F = \emptyset$, and (4) $\{(\mathbf{z}, (\mathbf{z})^n) \mid \mathbf{z} \in N_{P_n}(\mathbf{x}) \cup N_{P_n}(\mathbf{y})\} \cap F = \emptyset$. By induction, there is a k^*-container

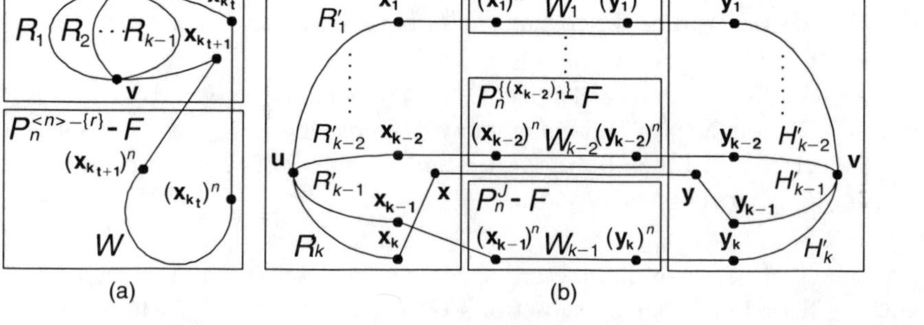

FIGURE 14.22 Illustrations for Case 1 of Theorem 14.20.

$\{R_1, R_2, \ldots, R_k\}$ of $P_n^{\{r\}} - F^r$ joining $\mathbf{u}$ to $\mathbf{x}$, and $\{H_1, H_2, \ldots, H_k\}$ is a k^*-container of $P_n^{\{s\}} - F^s$ joining $\mathbf{y}$ to $\mathbf{v}$.

Without loss of generality, we write $R_i = \langle \mathbf{u}, R_i', \mathbf{x_i}, \mathbf{x}\rangle$ and $H_i = \langle \mathbf{y}, \mathbf{y_i}, H_i', \mathbf{v}\rangle$ for every $1 \le i \le k$. (Note that $\mathbf{u} = \mathbf{x_i}$ is $l(R_i') = 0$ and $\mathbf{v} = \mathbf{y_i}$ is $l(H\prime_i) = 0$.) We set $A = \{(\mathbf{x_i})_1 \mid (\mathbf{x}, \mathbf{x_i}) \in E(R_i) \text{ and } i \in \langle k\rangle\}$ and $B = \{(\mathbf{y_i})_1 \mid (\mathbf{y}, \mathbf{y_i}) \in E(H_i) \text{ and } i \in \langle k\rangle\}$. By Lemma 11.12, $(\mathbf{x_i})_1 \neq (\mathbf{x_j})_1$ for every $i, j \in \langle k\rangle$ with $i \neq j$, and $A \subset \langle n\rangle - \{r, s\}$. Similarly, $(\mathbf{y_i})_1 \neq (\mathbf{y_j})_1$ for every $i, j \in \langle k\rangle$ with $i \neq j$, and $B \subset \langle n\rangle - \{r, s\}$.

Let $I = \{\mathbf{x_i} \mid \mathbf{x_i} \in A$, and there exists an element $\mathbf{y_j} \in B$ such that $(\mathbf{x_i})_1 = (\mathbf{y_j})_1\}$. Let $|I| = m$. (Note that $|I| = 0$ if $I = \emptyset$.) Then we relabel the indices of A and B such that $(\mathbf{x_i})_1 = (\mathbf{y_i})_1$ for all $i \le m$. By Lemma 11.13, there is a Hamiltonian path W_i of $P_n^{\{(\mathbf{x_i})_1, (\mathbf{y_i})_1\}} - F$ joining the vertex $(\mathbf{x_i})^n$ to the vertex $(\mathbf{y_i})^n$ for every $1 \le i \le k-2$. (Note that $\{(\mathbf{x_i})_1, (\mathbf{y_i})_1\} = \{(\mathbf{x_i})_1\}$ if $(\mathbf{x_i})_1 = (\mathbf{y_i})_1$.) We set $J = \langle n\rangle - (A \cup B \cup \{r, s\} - \{\mathbf{y_{k-1}}, \mathbf{y_k}\})$. By Lemma 11.13, there is a Hamiltonian path W_{k-1} of $P_n^J - F$ joining the vertex $(\mathbf{x_{k-1}})^n$ to the vertex $(\mathbf{y_k})^n$. We set

$$\begin{aligned} T_i &= \langle \mathbf{u}, R_i', \mathbf{x_i}, (\mathbf{x_i})^n, T_i, (\mathbf{y_i})^n, \mathbf{y_i}, H_i', \mathbf{v}\rangle \quad \text{for every } 1 \le i \le k-2 \\ T_{k-1} &= \langle \mathbf{u}, R_{k-1}', \mathbf{x_{k-1}}, (\mathbf{x_{k-1}})^n, W_{k-1}, (\mathbf{y_k})^n, \mathbf{y_k}, H_k', \mathbf{v}\rangle \\ T_k &= \langle \mathbf{u}, R_k', \mathbf{x_k}, \mathbf{x}, \mathbf{y}, \mathbf{y_{k-1}}, H_{k-1}', \mathbf{v}\rangle \end{aligned}$$

Then $\{T_1, T_2, \ldots, T_k\}$ forms the k^*-container of $P_n - F$ between $\mathbf{u}$ and $\mathbf{v}$. See Figure 14.22b for illustration.

Case 2. $|F^t| = f$ for some $t \in \langle n\rangle$. Without loss of generality, we assume that $t = n$.

Case 2.1. $\mathbf{u}, \mathbf{v} \in P_n^{\{n\}}$. By induction and Theorem 11.16, there is a $(k-1)^*$-container $\{R_1, R_2, \ldots, R_{k-1}\}$ of $P_n^{\{n\}} - F^n$ between $\mathbf{u}$ and $\mathbf{v}$.

Suppose that $(\mathbf{u})_1 \neq (\mathbf{v})_1$. By Lemma 11.13, there is a Hamiltonian path Q of $P_n^{\langle n-1\rangle}$ joining the vertex $(\mathbf{u})^n$ to the vertex $(\mathbf{v})^n$. We set $R_k = \langle \mathbf{u}, (\mathbf{u})^n, Q, (\mathbf{v})^n, \mathbf{v}\rangle$. Then $\{R_1, R_2, \ldots, R_k\}$ forms the k^*-container of $P_n - F$ between $\mathbf{u}$ and $\mathbf{v}$. See Figure 14.23a for illustration.

Suppose that $(\mathbf{u})_1 = (\mathbf{v})_1$. By Lemma 11.13, there is a Hamiltonian path Q of $P_n^{\{(\mathbf{u})_1\}}$ joining the vertex $(\mathbf{u})^n$ to the vertex $(\mathbf{v})^n$. We can write Q as $\langle (\mathbf{u})^n, \mathbf{x}, \mathbf{y}, Q', (\mathbf{v})^n\rangle$. We have $d(\mathbf{x}, \mathbf{y}) = 1$, $d(\mathbf{x}, (\mathbf{u})^n) = 1$, and $d(\mathbf{y}, (\mathbf{u})^n) = 2$. By Lemma 11.12, $(\mathbf{y})_1 \neq n$, $(\mathbf{x})_1 \neq n$, and $(\mathbf{x})_1 \neq (\mathbf{y})_1$. By Lemma 11.13, there is a Hamiltonian path W of $P_n^{\langle n-1\rangle - \{(\mathbf{u})_1\}}$ joining the vertex $(\mathbf{x})^n$ to the vertex $(\mathbf{y})^n$. We set $R_k = \langle \mathbf{u}, (\mathbf{u})^n, \mathbf{x}, (\mathbf{x})^n, W, (\mathbf{y})^n, \mathbf{y}, Q', (\mathbf{v})^n, \mathbf{v}\rangle$. Then $\{R_1, R_2, \ldots, R_k\}$ forms the k^*-container of $P_n - F$ between $\mathbf{u}$ and $\mathbf{v}$. See Figure 14.23b for illustration.

Case 2.2. $\mathbf{u}, \mathbf{v} \in P_n^{\{r\}}$ for some $r \in \langle n-1\rangle$. By Theorem 14.19, there is a k^*-container $\{R_1, R_2, \ldots, R_k\}$ of $P_n^{\{r\}}$ joining $\mathbf{u}$ to $\mathbf{v}$. By Lemma 11.11, $|E^{r,n}| = (n-2)! > n-3$ if $n \ge 5$. We can choose a vertex $\mathbf{x} \in P_n^{\{r\}} - \{\mathbf{v}\}$ with $(\mathbf{x})^n \in P_n^{\{n\}} - F^n$. Without loss of generality, we assume that $\mathbf{x} \in R_k$ and we write R_k as $\langle \mathbf{u}, Q_1, \mathbf{x}, \mathbf{y}, Q_2, \mathbf{v}\rangle$. (Note that $\mathbf{x} = \mathbf{u}$ if $l(Q_1) = 0$ and $\mathbf{y} = \mathbf{v}$ if $l(Q_2) = 0$.) We have $d(\mathbf{x}, \mathbf{y}) = 1$. By Lemma 13.12, $(\mathbf{x})_1 \neq (\mathbf{y})_1$. By Theorem 11.16, there is a Hamiltonian cycle C of $P_n^{\{n\}} - F^n$.

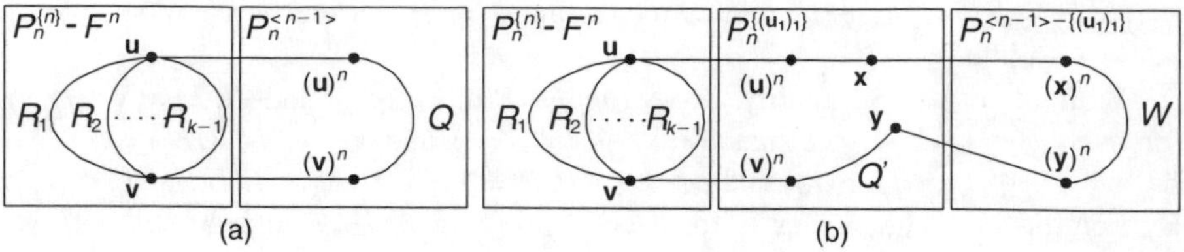

FIGURE 14.23 Illustrations for Case 2.1 of Theorem 14.20.

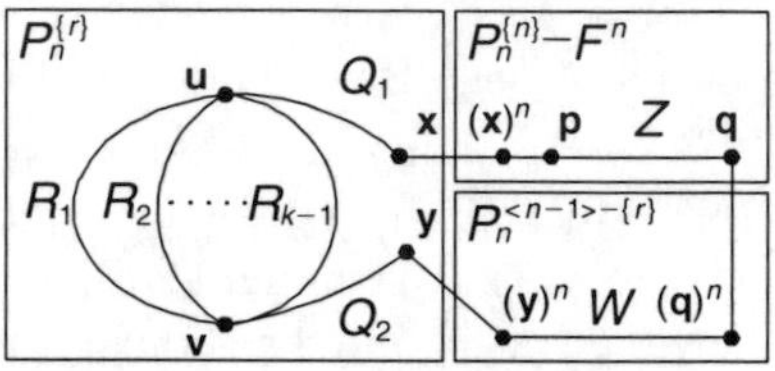

FIGURE 14.24 Illustration for Case 2.2 of Theorem 14.20.

Without loss of generality, we write C as $\langle(\mathbf{x})^n, \mathbf{p}, Z, \mathbf{q}, (\mathbf{x})^n\rangle$. Since $d((\mathbf{x})^n, \mathbf{p}) = 1$ and $d((\mathbf{x})^n, \mathbf{q}) = 1$, $d(\mathbf{p}, \mathbf{q}) = 2$. By Lemma 11.12, $(\mathbf{p})_1 \neq (\mathbf{q})_1$, $(\mathbf{p})_1 \neq ((\mathbf{x})^n)_1$, and $(\mathbf{q})_1 \neq ((\mathbf{x})^n)_1$. Since $(\mathbf{p})_1 \neq (\mathbf{q})_1$, either $(\mathbf{p})_1 \neq (\mathbf{y})_1$ or $(\mathbf{q})_1 \neq (\mathbf{y})_1$. Without loss of generality, we assume that $(\mathbf{q})_1 \neq (\mathbf{y})_1$. By Lemma 11.13, there is a Hamiltonian path W of $P_n^{\langle n-1\rangle - \{r\}}$ joining the vertex $(\mathbf{q})^n$ to the vertex $(\mathbf{y})^n$. We set $H = \langle \mathbf{u}, Q_1, \mathbf{x}, (\mathbf{x})^n, \mathbf{p}, Z, \mathbf{q}, (\mathbf{q})^n, W, (\mathbf{y})^n, \mathbf{y}, Q_2, \mathbf{v}\rangle$. Then $\{R_1, R_2, \ldots, R_{k-1}, H\}$ forms the k^*-container of $P_n - F$ between $\mathbf{u}$ and $\mathbf{v}$. See Figure 14.24 for illustration.

Case 2.3. $\mathbf{u} \in P_n^{\{n\}}$ and $\mathbf{v} \in P_n^{\{r\}}$ for some $r \in \langle n-1\rangle$.

Case 2.3.1. Suppose that $(\mathbf{u})^n = \mathbf{v}$. By Lemma 14.12, $|E^{r,n}| = (n-2)! > n-3$. We can choose a vertex $\mathbf{x}$ in $P_n^{\{n\}} - (F_n^n \cup \{\mathbf{u}\})$ such that $(\mathbf{x})_1 = r$. We set $\mathbf{y} = (\mathbf{x})^n$. By induction and Theorem 11.16, there is a $(k-1)^*$-container $\{R_1, R_2, \ldots, R_{k-1}\}$ of $P_n^{\{n\}} - F^n$ joining $\mathbf{u}$ to $\mathbf{x}$.

Without loss of generality, we write R_i as $\langle \mathbf{u}, R_i', \mathbf{x_i}, \mathbf{x}\rangle$ for every $1 \leq i \leq k-1$. Let $\{\mathbf{y_1}, \mathbf{y_2}, \ldots, \mathbf{y_{k-1}}\}$ be the set of $(k-1)$ distinct neighbors of $\mathbf{y}$ in $P_n^{\{r\}}$ with $(\mathbf{y_i})_1 = (\mathbf{x_i})_1$ for every $1 \leq i \leq k-1$. We set $F' = \{(\mathbf{y}, \mathbf{z}) \mid \mathbf{z} \in N_{P_n^{\{r\}}}(\mathbf{y}) - \{\mathbf{y_i} \mid 1 \leq i \leq k-1\}\}$. Obviously, $|F'| = (n-1) - (k-1)$. By induction and Theorem 11.16, there is a $(k-1)^*$-container $\{H_1, H_2, \ldots, H_{k-1}\}$ of $P_n^{\{r\}} - F'$ joining $\mathbf{y}$ to $\mathbf{v}$. Without loss of generality, we write H_i as $\langle \mathbf{y}, \mathbf{y_i}, H_i', \mathbf{v}\rangle$ for every $1 \leq i \leq k-1$.

Suppose that $k = 3$. We set $J = \langle n-1\rangle - \{r\}$. By Lemma 11.13, there is a Hamiltonian path W of P_n^J joining the vertex $(\mathbf{x_1})^n$ to the vertex $(\mathbf{y_2})^n$. We set

$$
\begin{aligned}
T_1 &= \langle \mathbf{u}, R_1', \mathbf{x_1}, (\mathbf{x_1})^n, W, (\mathbf{y_2})^n, \mathbf{y_2}, H_2', \mathbf{v}\rangle \\
T_2 &= \langle \mathbf{u}, R_2', \mathbf{x_2}, \mathbf{x}, \mathbf{y}, \mathbf{y_1}, H_1', \mathbf{v}\rangle \\
T_3 &= \langle \mathbf{u}, \mathbf{v}\rangle
\end{aligned}
$$

Then $\{T_1, T_2, T_3\}$ forms the 3*-container of $P_n - F$ between $\mathbf{u}$ and $\mathbf{v}$. See Figure 14.25a for illustration.

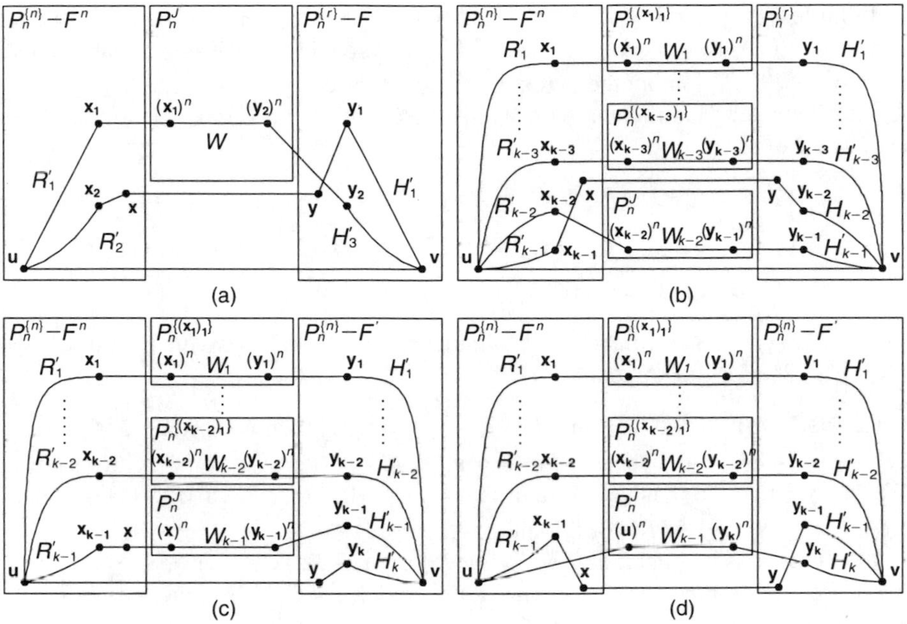

FIGURE 14.25 Illustrations for Case 2.3 of Theorem 14.20.

Suppose that $k > 3$. We set $J = \langle n-1\rangle - (\{(\mathbf{x_i})_1 \mid 1 \le i \le k-3\} \cup \{r\})$. By Lemma 11.13, there is a Hamiltonian path W_i of $P_n^{\{(\mathbf{x_i})_1\}}$ joining the vertex $(\mathbf{x_i})^n$ to the vertex $(\mathbf{y_i})^n$ for every $1 \le i \le k-3$. Again, there is a Hamiltonian path W_{k-2} of P_n^J joining the vertex $(\mathbf{x_{k-2}})^n$ to the vertex $(\mathbf{y_{k-1}})^n$. We set

$$\begin{aligned} T_i &= \langle \mathbf{u}, R_i', \mathbf{x_i}, (\mathbf{x_i})^n, W_i, (\mathbf{y_i})^n, \mathbf{y_i}, H_i', \mathbf{v}\rangle \quad \text{for every } 1 \le i \le k-3 \\ T_{k-2} &= \langle \mathbf{u}, R_{k-2}', \mathbf{x_{k-2}}, (\mathbf{x_{k-2}})^n, W_{k-2}, (\mathbf{y_{k-1}})^n, \mathbf{y_{k-1}}, H_{k-1}', \mathbf{v}\rangle \\ T_{k-1} &= \langle \mathbf{u}, R_{k-1}', \mathbf{x_{k-1}}, \mathbf{x}, \mathbf{y}, \mathbf{y_{k-2}}, H_{k-2}', \mathbf{v}\rangle \\ T_k &= \langle \mathbf{u}, \mathbf{v}\rangle \end{aligned}$$

Then $\{T_1, T_2, \ldots, T_k\}$ forms the k^*-container of $P_n - F$ between $\mathbf{u}$ and $\mathbf{v}$. See Figure 14.25b for illustration.

Case 2.3.2. Suppose that $(\mathbf{u})^n \in P_n^{\{r\}}$ and $(\mathbf{u})^n \ne \mathbf{v}$. Let $t \in \langle n-1\rangle = \{r\}$. By Lemma 11.12, $|E^{r,n}| = (n-2)! > n-3$. We can choose a vertex $\mathbf{x}$ in $P_n^{\{n\}} - (F_n^n \cup \{\mathbf{u}\})$ such that $(\mathbf{x})_1 = t$. By induction and Theorem 11.16, there is a $(k-1)^*$-container $\{R_1, R_2, \ldots, R_{k-1}\}$ of $P_n^{\{n\}} - F^n$ joining $\mathbf{u}$ to $\mathbf{x}$.

Without loss of generality, we write R_i as $\langle \mathbf{u}, R_i', \mathbf{x_i}, \mathbf{x}\rangle$ for every $1 \le i \le k-1$. Moreover, we assume that $(\mathbf{x_i})_1 \ne r$ for every $1 \le i \le k-2$. We set $\mathbf{y} = (\mathbf{u})^n$. Let $\{\mathbf{y_1}, \mathbf{y_2}, \ldots, \mathbf{y_k}\}$ be the set of k distinct neighbors of $\mathbf{y}$ in $P_n^{\{r\}}$ with $(\mathbf{y_i})_1 = (\mathbf{x_i})_1$ for every $1 \le i \le k-2$ and $(\mathbf{y_{k-1}})_1 = (\mathbf{x})_1$. We set $F' = \{(\mathbf{y}, \mathbf{z}) \mid \mathbf{z} \in N_{P_n^{\{r\}}}(\mathbf{y}) - \{\mathbf{y_i} \mid 1 \le i \le k\}\}$. By induction, there is a k^*-container $\{H_1, H_2, \ldots, H_k\}$ of $P_n^{\{r\}} - F'$ joining $\mathbf{y}$ to $\mathbf{v}$.

Without loss of generality, we write H_i as $\langle \mathbf{y}, \mathbf{y_i}, H_i', \mathbf{v}\rangle$ for every $1 \leq i \leq k-1$. We set $J = \langle n\rangle - (\{(\mathbf{x_i})_1 \mid 1 \leq i \leq k-2\} \cup \{r, n\})$. By Lemma 11.13, there is a Hamiltonian path W_i of $P_n^{\{(\mathbf{x_i})_1\}}$ joining the vertex $(\mathbf{x_i})^n$ to the vertex $(\mathbf{y_i})^n$ for every $1 \leq i \leq k-2$. Again, there is a Hamiltonian path W_{k-1} of P_n^J joining the vertex $(\mathbf{x})^n$ to the vertex $(\mathbf{y_{k-1}})^n$. We set

$$\begin{aligned} T_i &= \langle \mathbf{u}, R_i', \mathbf{x_i}, (\mathbf{x_i})^n, W_i, (\mathbf{y_i})^n, \mathbf{y_i}, H_i', \mathbf{v}\rangle \quad \text{for every } 1 \leq i \leq k-2 \\ T_{k-1} &= \langle \mathbf{u}, R_{k-1}', \mathbf{x_{k-1}}, \mathbf{x}, (\mathbf{x})^n, W_{k-1}, (\mathbf{y_{k-1}})^n, \mathbf{y_{k-1}}, H_{k-1}', \mathbf{v}\rangle \\ T_k &= \langle \mathbf{u}, (\mathbf{u})^n = \mathbf{y}, \mathbf{y_k}, H_k', \mathbf{v}\rangle \end{aligned}$$

Then $\{T_1, T_2, \ldots, T_k\}$ forms the k^*-container of $P_n - F$ between $\mathbf{u}$ and $\mathbf{v}$. See Figure 14.25c for illustration.

Case 2.3.3. $(\mathbf{u})^n \in P_n^{\{s\}}$ for some $s \in \langle n-1\rangle - \{r\}$. By Lemma 11.12, $|E^{r,n}| = (n-2)! > n-2$. We can choose a vertex $\mathbf{x}$ in $P_n^{\{n\}} - (F^n \cup \{\mathbf{u}, (\mathbf{v})^n\})$ with $(\mathbf{x})_1 = r$. We set $\mathbf{y} = (\mathbf{x})^n$. By induction and Theorem 11.16, there is a $(k-1)^*$-container $\{R_1, R_2, \ldots, R_{k-1}\}$ of $P_n^{\{n\}} - F^n$ joining $\mathbf{u}$ to $\mathbf{x}$.

Without loss of generality, we write R_i as $\langle \mathbf{u}, R_i', \mathbf{x_i}, \mathbf{x}\rangle$ for every $1 \leq i \leq k-1$. Moreover, we assume that $(\mathbf{x_i})_1 \neq (\mathbf{u})_1$ for every $1 \leq i \leq k-2$. Let $\{\mathbf{y_1}, \mathbf{y_2}, \ldots, \mathbf{y_k}\}$ be the set of k distinct neighbors of $\mathbf{y}$ in $P_n^{\{r\}}$ with $(\mathbf{y_i})_1 = (\mathbf{x_i})_1$ for every $1 \leq i \leq k-2$, and $(\mathbf{y_{k-1}})_1 = (\mathbf{u})_1$. We set $F' = \{(\mathbf{y}, \mathbf{z}) \mid \mathbf{z} \in N_{P_n^{\{r\}}}(\mathbf{y}) - \{\mathbf{y_i} \mid 1 \leq i \leq k\}\}$. By Theorem 11.16, there is a k^*-container $\{H_1, H_2, \ldots, H_k\}$ of $P_n^{\{r\}} - F'$ joining $\mathbf{y}$ to $\mathbf{v}$. Without loss of generality, we write H_i as $\langle \mathbf{y}, \mathbf{y_i}, H_i', \mathbf{v}\rangle$ for every $1 \leq i \leq k$. We set $J = \langle n-1\rangle - (\{(\mathbf{x_i})_1 \mid 1 \leq i \leq k-2\} \cup \{r\})$. By Lemma 11.13, there is a Hamiltonian path W_i of $P_n^{\{(\mathbf{x_i})_1\}}$ joining the vertex $(\mathbf{x_i})^n$ to the vertex $(\mathbf{y_i})^n$ for every $i \leq k-2$. Again, there is a Hamiltonian path W_{k-2} of P_n^J joining the vertex $(\mathbf{u})^n$ to the vertex $(\mathbf{y_k})^n$. We set

$$\begin{aligned} T_i &= \langle \mathbf{u}, R_i', \mathbf{x_i}, (\mathbf{x_i})^n, W_i, (\mathbf{y_i})^n, \mathbf{y_i}, H_i', \mathbf{v}\rangle \quad \text{for every } 1 \leq i \leq k-2 \\ T_{k-1} &= \langle \mathbf{u}, (\mathbf{u})^n, W_{k-1}, (\mathbf{y_k})^n, \mathbf{y_k}, H_k', \mathbf{v}\rangle \\ T_k &= \langle \mathbf{u}, R_{k-1}', \mathbf{x_{k-1}}, \mathbf{x}, \mathbf{y}, \mathbf{y_{k-1}}, H_{k-1}', \mathbf{v}\rangle \end{aligned}$$

Then $\{T_1, T_2, \ldots, T_k\}$ forms the k^*-container of $P_n - F$ between $\mathbf{u}$ and $\mathbf{v}$. See Figure 14.25d for illustration.

Case 2.4. $\mathbf{u} \in P_n^{\{r\}}$ and $\mathbf{v} \in P_n^{\{s\}}$ for some $r, s \in \langle n-1\rangle$ with $r \neq s$. By Theorem 11.16, there is a Hamiltonian cycle C of $P_n^{\{n\}} - F^n$. By Lemma 11.12, $E^{r,n} = (n-2)! > n-2$. We can choose two adjacent vertices $\mathbf{w}$ and $\mathbf{x}$ in $P_n^{\{r\}} - \{\mathbf{u}\}$ with $(\mathbf{w})^n \in P_n^{\{r\}} - \{\mathbf{v}\}$ and $(\mathbf{x})^n \in P_n^{\{n\}} - F^n$. Without loss of generality, we write C as $\langle (\mathbf{x})^n, \mathbf{y}, Q, \mathbf{z}, (\mathbf{x})^n\rangle$. By Lemma 11.12, $(\mathbf{y})_1 \neq (\mathbf{x})_1$, $(\mathbf{z})_1 \neq (\mathbf{x})_1$, and $(\mathbf{y})_1 \neq (\mathbf{z})_1$. Without loss of generality, we assume that $(\mathbf{z})_1 \in \langle n-1\rangle - \{r, s\}$. Let $\{\mathbf{q_1}, \mathbf{q_2}, \ldots, \mathbf{q_{k-1}}\}$ be a $(k-1)$ subset of $N_{P_n^{\{r\}}}(\mathbf{w})$ with $(\mathbf{q_{k-1}})_1 = (\mathbf{z})_1$. We set $F' = \{(\mathbf{w}, \mathbf{z}) \mid \mathbf{z} \in N_{P_n^{\{r\}}}(\mathbf{w}) - \{\mathbf{x}, \mathbf{q_1}, \mathbf{q_2}, \ldots, \mathbf{q_{k-1}}\}$. By induction, there is a k^*-container $\{R_1, R_2, \ldots, R_k\}$ of $P_n^{\{r\}} - F'$ joining $\mathbf{u}$ to $\mathbf{x}$. Without loss of generality, we write R_i as $\langle \mathbf{u}, R_i', \mathbf{q_i}, \mathbf{w}\rangle$ for every $1 \leq i \leq k-1$ and R_k as $\langle \mathbf{u}, R_k', \mathbf{x}, \mathbf{w}\rangle$.

Suppose that $k = n-2$. Let $\{\mathbf{p_1}, \mathbf{p_2}, \ldots, \mathbf{p_{n-2}}\}$ be the set of neighbors of the vertex $(\mathbf{w})^n$ in $P_n^{\{s\}}$. Without loss of generality, we assume that $(\mathbf{p_i})_1 = (\mathbf{q_i})_1$ for every $1 \le i \le n-3$ and $(\mathbf{p_{n-2}})_1 = (\mathbf{x})_1$. By induction, there is an $(n-2)^*$-container $\{H_1, H_2, \ldots, H_{n-2}\}$ of $P_n^{\{s\}}$ joining $(\mathbf{w})^n$ to $\mathbf{v}$. Without loss of generality, we write R_i as $\langle (\mathbf{w})^n, \mathbf{p_i}, H_i', \mathbf{v}\rangle$ for every $1 \le i \le n-2$. By Lemma 11.13, there is a Hamiltonian path W_i of $P_n^{\{(\mathbf{q_i})_1\}}$ joining the vertex $(\mathbf{q_i})^n$ to the vertex $(\mathbf{p_i})^n$ for every $i \le n-4$. Again, there is a Hamiltonian path W_{n-3} of $P_n^{\{(\mathbf{z})_1\}}$ joining the vertex $(\mathbf{z})^n$ to the vertex $(\mathbf{p_{n-3}})^n$. We set

$$\begin{aligned} T_i &= \langle \mathbf{u}, R_i', \mathbf{q_i}, (\mathbf{q_i})^n, W_i, (\mathbf{p_i})^n, \mathbf{p_i}, H_i', \mathbf{v}\rangle \quad \text{for every } 1 \le i \le n-4 \\ T_{n-3} &= \langle \mathbf{u}, R_{n-2}', \mathbf{x}, \mathbf{y}, Q, \mathbf{z}, (\mathbf{z})^n, W_{n-3}, (\mathbf{p_{n-3}})^n, \mathbf{p_{n-3}}, H_{n-3}', \mathbf{v}\rangle \\ T_{n-2} &= \langle \mathbf{u}, R_{n-3}', \mathbf{q_{n-3}}, \mathbf{w}, (\mathbf{w})^n, \mathbf{p_{n-2}}, H_{n-2}', \mathbf{v}\rangle \end{aligned}$$

Then $\{T_1, T_2, \ldots, T_{n-2}\}$ forms the $(n-2)^*$-container of $P_n - F$ between $\mathbf{u}$ and $\mathbf{v}$. See Figure 14.26a for illustration.

Suppose that $3 \le k \le n-3$. Obviously, $n \ge 6$. Let $t \in \langle n\rangle - \{(\mathbf{q_i})_1 \mid 1 \le i \le k-1\} \cup \{r, s\}$. Let $\{\mathbf{p_1}, \mathbf{p_2}, \ldots, \mathbf{p_k}\}$ be the set of k distinct neighbors of the

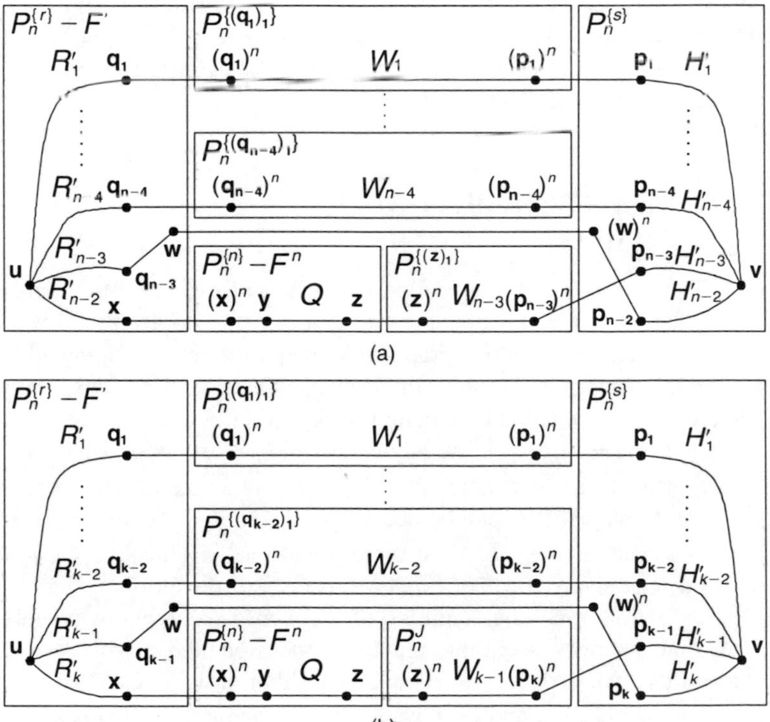

FIGURE 14.26 Illustration for Case 2.4 of Theorem 14.20.

vertex $(\mathbf{w})^n$ in $P_n^{\{s\}}$ with $(\mathbf{p_i})_1 = (\mathbf{q_i})_1$ for every $1 \le i \le k-1$ and $(\mathbf{p_k})_1 = t$. We set $F' = \{((\mathbf{w})^n, \mathbf{z}) \mid \mathbf{z} \in N_{P_n^{\{s\}}}((\mathbf{w})^n) - \{\mathbf{p_i} \mid 1 \le i \le k\}$. By induction, there is a k^*-container $\{H_1, H_2, \ldots, H_k\}$ of $P_n^{\{s\}} - F'$ joining $(\mathbf{w})^n$ to $\mathbf{v}$. Without loss of generality, we write R_i as $\langle (\mathbf{w})^n, \mathbf{p_i}, H_i', \mathbf{v} \rangle$ for every $1 \le i \le k$. We set $J = \langle n-1 \rangle - (\{(\mathbf{q_i})_1 \mid 1 \le i \le k-2\} \cup \{r, s\})$. By Lemma 11.13, there is a Hamiltonian path W_i of $P_n^{\{(\mathbf{q_i})_1\}}$ joining the vertex $(\mathbf{q_i})^n$ to the vertex $(\mathbf{p_i})^n$ for every $i \le k-2$. Again, there is a Hamiltonian path W_{k-1} of $P_n^{\{J\}}$ joining the vertex $(\mathbf{z})^n$ to the vertex $(\mathbf{p_k})^n$. We set

$$\begin{aligned} T_i &= \langle \mathbf{u}, R_i', \mathbf{q_i}, (\mathbf{q_i})^n, W_i, (\mathbf{p_i})^n, \mathbf{p_i}, H_i', \mathbf{v} \rangle \quad \text{for every } 1 \le i \le k-2 \\ T_{k-1} &= \langle \mathbf{u}, R_k', \mathbf{x}, \mathbf{y}, Q, \mathbf{z}, (\mathbf{z})^n, W_{k-1}, (\mathbf{p_k})^n, \mathbf{p_k}, H_k', \mathbf{v} \rangle \\ T_k &= \langle \mathbf{u}, R_{k-1}', \mathbf{q_{k-1}}, \mathbf{w}, (\mathbf{w})^n, \mathbf{p_{k-1}}, H_{k-1}', \mathbf{v} \rangle \end{aligned}$$

Then $\{T_1, T_2, \ldots, T_k\}$ forms the k^*-container of $P_n - F$ between $\mathbf{u}$ and $\mathbf{v}$. See Figure 14.26b for illustration. □

THEOREM 14.21 $\kappa_f^*(P_n) = n-4$ if $n \ge 4$.

Proof: By Theorem 14.20, $\kappa_f^*(P_n) \ge n-4$ if $n \ge 4$. Let $\mathbf{u}$ be any vertex of P_n and F be any subset of the neighborhood of u with $|F| = n-3$. Obviously, $\deg_{P_n - F}(\mathbf{u}) = 2$. Let $\mathbf{x}$ and $\mathbf{y}$ be two distinct vertices in $N_{P_n - F}(\mathbf{u})$. Since $\deg_{P_n - F}(\mathbf{u}) = 2$ and $n(P_n - F) \ge n! - (n-f) \ge 4$, there is not an 1^*-container of $P_n - F$ between $\mathbf{x}$ and $\mathbf{y}$. Hence, $P_n - F$ is not 1^*-connected. Thus, $\kappa_f^*(P_n) = n-4$ if $n \ge 4$. □

14.7 SPANNING LACEABILITY OF THE STAR GRAPHS

Lin et al. [233] also discuss the spanning laceability of the star graphs. Again, we review some basic background about the properties of star graphs.

It is known that S_n is a bipartite graph with one partite set containing all odd permutations and the other partite set containing all even permutations. For convenience, we refer to an even permutation as a white vertex, and refer to an odd permutation as a black vertex. Let $\mathbf{u} = u_1 u_2 \ldots u_n$ be any vertex of S_n. We use $(\mathbf{u})_i$ to denote the ith component u_i of $\mathbf{u}$ and $S_n^{\{i\}}$ to denote the ith subgraph of S_n induced by those vertices $\mathbf{u}$ with $(\mathbf{u})_n = i$. Obviously, S_n can be decomposed into n vertex disjoint subgraphs $S_n^{\{i\}}$ for $1 \le i \le n$, such that each $S_n^{\{i\}}$ is isomorphic to S_{n-1}. Thus, the star graph can be constructed recursively. Let $H \subseteq \langle n \rangle$. We use S_n^H to denote the subgraph of S_n induced by $\cup_{i \in H} V(S_n^{\{i\}})$. By the definition of S_n, there is exactly one neighbor $\mathbf{v}$ of $\mathbf{u}$ such that $\mathbf{u}$ and $\mathbf{v}$ are adjacent through an i-dimensional edge with $2 \le i \le n$. For this reason, we use $(\mathbf{u})^i$ to denote the unique i-neighbor of $\mathbf{u}$. We have $((\mathbf{u})^i)^i = \mathbf{u}$ and $(\mathbf{u})^n \in S_n^{\{(\mathbf{u})_1\}}$. For $1 \le i, j \le n$ and $i \ne j$, we use $E^{i,j}$ to denote the set of edges between $S_n^{\{i\}}$ and $S_n^{\{j\}}$. By Theorem 13.2, S_n is 1^*-laceable if $n \ne 3$, and S_n is 2^*-connected if $n \ge 3$.

LEMMA 14.33 Let $\{a_1, a_2, \ldots, a_r\}$ be a subset of $\langle n\rangle$ for some $r \in \langle n\rangle$ with $n \geq 5$. Assume that $\mathbf{u}$ is a white vertex in $S_n^{\{a_1\}}$ and $\mathbf{v}$ is a black vertex in $S_n^{\{a_r\}}$. Then there is a Hamiltonian path $\langle \mathbf{u} = \mathbf{x_1}, H_1, \mathbf{y_1}, \mathbf{x_2}, H_2, \mathbf{y_2}, \ldots, \mathbf{x_r}, H_r, \mathbf{y_r} = \mathbf{v}\rangle$ of $\bigcup_{i=1}^{r} S_n^{\{a_i\}}$ joining $\mathbf{u}$ to $\mathbf{v}$ such that $\mathbf{x_1} = \mathbf{u}$, $\mathbf{y_r} = \mathbf{v}$, and H_i is a Hamiltonian path of $S_n^{\{a_i\}}$ joining $\mathbf{x_i}$ to $\mathbf{y_i}$ for every i, $1 \leq i \leq r$.

Proof: Note that $S_n^{\{i_j\}}$ is isomorphic to S_{n-1} for every $1 \leq j \leq m$. We set $\mathbf{x_1} = \mathbf{u}$ and $\mathbf{y_m} = \mathbf{v}$. By Theorem 13.2, this theorem holds for $m = 1$. Assume that $m \geq 2$. By Lemma 13.7, we choose $(\mathbf{y_j}, \mathbf{x_{j+1}}) \in E^{i_j, i_{j+1}}$ with $\mathbf{y_j}$ is a black vertex of $S_n^{\{j\}}$ and $\mathbf{x_{j+1}}$ is a white vertex of $S_n^{\{j+1\}}$ for every $1 \leq j \leq m-1$. By Theorem 13.2, there is a Hamiltonian path Q_j of $S_n^{\{i_j\}}$ joining $\mathbf{x_j}$ to $\mathbf{y_j}$. The path $\langle \mathbf{x_1}, Q_1, \mathbf{y_1}, \mathbf{x_2}, Q_2, \mathbf{y_2}, \ldots, \mathbf{x_m}, Q_m, \mathbf{y_m}\rangle$ forms a desired path. □

LEMMA 14.34 Let $n \geq 5$ and k be any positive integer with $3 \leq k \leq n-1$. Let $\mathbf{u}$ be any white vertex and $\mathbf{v}$ be any black vertex of S_n. Suppose that S_{n-1} is k^*-laceable. Then there is a k^*-container of S_n between $\mathbf{u}$ and $\mathbf{v}$ not using the edge $(\mathbf{u}, \mathbf{v})$ if $(\mathbf{u}, \mathbf{v}) \in E(S_n)$.

Proof: Since S_n is edge-transitive, we may assume that $\mathbf{u} \in S_n^{\{n\}}$ and $\mathbf{v} \in S_n^{\{n-1\}}$. By Lemma 13.7, there are $((n-2)!/2) \geq 3$ edges joining black vertices of $S_n^{\{n\}}$ to white vertices of $S_n^{\{n-1\}}$. We can choose an edge $(\mathbf{y}, \mathbf{z}) \in E^{n-1,n}$ where $\mathbf{y}$ is a black vertex in $S_n^{\{n\}}$ and $\mathbf{z}$ is a white vertex in $S_n^{\{n-1\}}$. By induction, there is a k^*-container $\{R_1, R_2, \ldots, R_k\}$ of $S_n^{\{n\}}$ joining $\mathbf{u}$ to $\mathbf{y}$, and there is a k^*-container $\{H_1, H_2, \ldots, H_k\}$ of $S_n^{\{n-1\}}$ joining $\mathbf{z}$ to $\mathbf{v}$. We write $R_i = \langle \mathbf{u}, R_i', \mathbf{y_i}, \mathbf{y}\rangle$ and $H_i = \langle \mathbf{z}, \mathbf{z_i}, H_i', \mathbf{v}\rangle$. Note that $\mathbf{y_i}$ is a white vertex and $\mathbf{z_i}$ is a black vertex for every $1 \leq i \leq k$. Let $I = \{\mathbf{y_i} \mid (\mathbf{y_i}, \mathbf{y}) \in E(R_i)$ and $1 \leq i \leq k\}$, and $J = \{\mathbf{z_i} \mid (\mathbf{z_i}, \mathbf{x}) \in E(H_i)$ and $1 \leq i \leq k\}$. Note that $(\mathbf{y_i})_1 = (\mathbf{y})_j$ for some $j \in \{2, 3, \ldots, n-1\}$, and $(\mathbf{y})_l \neq (\mathbf{y})_m$ if $l \neq m$. By Lemma 13.8, $\{(\mathbf{y_i})_1 \mid 1 \leq i \leq k\} \cap \{n-1, n\} = \emptyset$. Similarly, $\{(\mathbf{z_i})_1 \mid 1 \leq i \leq k\} \cap \{n-1, n\} = \emptyset$. Let $A = \{\mathbf{y_i} \mid \mathbf{y_i} \in I$ and there exists an element $\mathbf{z_j} \in J$ such that $(\mathbf{y_i})_1 = (\mathbf{z_j})_1\}$. Then we relabel the indices of I and J such that $(\mathbf{y_i})_1 = (\mathbf{z_i})_1$ for $1 \leq i \leq |A|$. We set X as $\{(\mathbf{y_i})_1 \mid 1 \leq i \leq k-2\} \cup \{(\mathbf{z_i})_1 \mid 1 \leq i \leq k-2\} \cup \{n-1, n\}$. By Lemma 14.33, there is a Hamiltonian path T_i of $S_n^{\{(\mathbf{y_i})_1, (\mathbf{z_i})_1\}}$ joining the black vertex $(\mathbf{y_i})^n$ to the white vertex $(\mathbf{z_i})^n$ for every $1 \leq i \leq k-2$, and there is a Hamiltonian path T_{k-1} of $S_n^{\langle n\rangle - X}$ joining the black vertex $(\mathbf{y_{k-1}})^n$ to the white vertex $(\mathbf{z_k})^n$. (Note that $\{(\mathbf{y_i})_1, (\mathbf{z_i})_1\} = \{(\mathbf{y_i})_1\}$ if $(\mathbf{y_i})_1 = (\mathbf{z_i})_1$.) We set

$$
\begin{aligned}
Q_i &= \langle \mathbf{u}, R_i', \mathbf{y_i}, (\mathbf{y_i})^n, T_i, (\mathbf{z_i})^n, \mathbf{z_i}, H_i', \mathbf{v}\rangle \quad \text{for } 1 \leq i \leq k-2 \\
Q_{k-1} &= \langle \mathbf{u}, R_{k-1}', \mathbf{y_{k-1}}, (\mathbf{y_{k-1}})^n, T_{k-1}, (\mathbf{z_k})^n, \mathbf{z_k}, H_k', \mathbf{v}\rangle \\
Q_k &= \langle \mathbf{u}, R_k', \mathbf{y_k}, \mathbf{y}, \mathbf{z}, \mathbf{z_{k-1}}, H_{k-1}', \mathbf{v}\rangle
\end{aligned}
$$

It is easy to check whether $\{Q_1, Q_2, \ldots, Q_k\}$ forms a k^*-container of S_n joining $\mathbf{u}$ to $\mathbf{v}$ not using the edge $(\mathbf{u}, \mathbf{v})$ if $(\mathbf{u}, \mathbf{v}) \in E(S_n)$. □

THEOREM 14.22 S_n is $(n-1)^*$-laceable if $n \geq 2$.

Proof: It is easy to see that S_2 is 1^*-laceable and S_3 is 2^*-laceable. Since the S_4 is vertex-transitive, we claim that S_4 is 3^*-laceable by listing all 3^*-containers from the white vertex 1234 to any black vertex as follows:

⟨(1234), (2134)⟩
⟨(1234), (3214), (2314), (4312), (1342), (2341), (4321), (1324), (3124), (2134)⟩
⟨(1234), (4231), (3241), (1243), (4213), (2413), (3412), (1432), (2431), (3421), (1423), (4123), (2143), (3142), (4132), (2134)⟩
⟨(1234), (3214)⟩
⟨(1234), (4231), (3241), (2341), (1342), (3142), (2143), (1243), (4213), (3214)⟩
⟨(1234), (2134), (4132), (1432), (2431), (3421), (4321), (1324), (3124), (4123), (1423), (2413), (3412), (4312), (2314), (3214)⟩
⟨(1234), (4231)⟩
⟨(1234), (2134), (4132), (3142), (1342), (4312), (3412), (1432), (2431), (4231)⟩
⟨(1234), (3214), (2314), (1324), (3124), (4123), (2143), (1243), (4213), (2413), (1423), (3421), (4321), (2341), (3241), (4231)⟩
⟨(1234), (2134), (3124), (1324), (2314), (4312), (1342), (3142), (4132), (1432), (3412), (2413), (1423), (4123), (2143), (1243)⟩
⟨(1234), (3214), (4213), (1243)⟩
⟨(1234), (4231), (2431), (3421), (4321), (2341), (3241), (1243)⟩
⟨(1234), (2134), (4132), (1432)⟩
⟨(1234), (3214), (2314), (1324), (3124), (4123), (1423), (2413), (4213), (1243), (2143), (3142), (1342), (4312), (3412), (1432)⟩
⟨(1234), (4231), (3241), (2341), (4321), (3421), (2431), (1432)⟩
⟨(1234), (2134), (4132), (3142), (1342), (4312), (3412), (1432), (2431), (3421), (1423), (2413), (4213), (1243), (2143), (4123), (3124), (1324)⟩
⟨(1234), (3214), (2314), (1324)⟩
⟨(1234), (4231), (3241), (2341), (4321), (1324)⟩
⟨(1234), (2134), (3124), (1324), (2314), (4312), (3412), (1432), (4132), (3142), (1342), (2341)⟩
⟨(1234), (3214), (4213), (1423), (2413), (4123), (2143), (1243), (3241), (2341)⟩
⟨(1234), (4231), (2431), (3421), (4321), (2341)⟩
⟨(1234), (2134), (4132), (3142), (1342), (4312), (3412), (1432), (2431), (3421)⟩
⟨(1234), (3214), (2314), (1324), (3124), (4123), (2143), (1243), (4213), (2413), (1423), (3421)⟩
⟨(1234), (4231), (3241), (2341), (4321), (3421)⟩
⟨(1234), (2134), (3124), (1324), (2314), (4312)⟩
⟨(1234), (3214), (4213), (1243), (2143), (4123), (1423), (2413), (3412), (4312)⟩
⟨(1234), (4231), (3241), (2341), (4321), (3421), (2431), (1432), (4132), (3142), (1342), (4312)⟩
⟨(1234), (2134), (4132), (1432), (3412), (4312), (1342), (3142), (2143), (4123)⟩
⟨(1234), (3214), (2314), (1324), (3124), (4123)⟩
⟨(1234), (4231), (2431), (3421), (4321), (2341), (3241), (1243), (4213), (2413), (1423), (4123)⟩
⟨(1234), (2134), (4132), (3142)⟩
⟨(1234), (3214), (2314), (1324), (3124), (4123), (1423), (2413), (4213), (1243), (2143), (3142)⟩
⟨(1234), (4231), (3241), (2341), (4321), (3421), (2431), (1432), (3412), (4312), (1342), (3142)⟩
⟨(1234), (2134), (3124), (1324), (2314), (4312), (1342), (3142), (4132), (1432), (3412), (2413)⟩
⟨(1234), (3214), (4213), (2413)⟩
⟨(1234), (4231), (2431), (3421), (4321), (2341), (3241), (1243), (2143), (4123), (1423), (2413)⟩

Assume that S_k is $(k-1)^*$-laceable for every $4 \leq k \leq n-1$. We need to construct an $(n-1)^*$-container of S_n between any white vertex $\mathbf{u}$ to any black vertex $\mathbf{v}$.

Case 1. $d(\mathbf{u}, \mathbf{v}) = 1$. We have $(\mathbf{u}, \mathbf{v}) \in E(S_n)$. By induction, S_{n-1} is $(n-2)^*$-laceable. By Lemma 14.34, there exists an $(n-2)^*$-container $\{Q_1, Q_2, \ldots, Q_{n-2}\}$ of S_n joining $\mathbf{u}$ to $\mathbf{v}$ not using the edge $(\mathbf{u}, \mathbf{v})$. We set Q_{n-1} as $\langle \mathbf{u}, \mathbf{v} \rangle$. Then $\{Q_1, Q_2, \ldots, Q_{n-1}\}$ forms an $(n-1)^*$-container of S_n joining $\mathbf{u}$ to $\mathbf{v}$.

Case 2. $d(\mathbf{u}, \mathbf{v}) \geq 3$. We have a star graph that is edge-transitive. Without loss of generality, we may assume that $\mathbf{u} \in S_n^{\{n\}}$ and $\mathbf{v} \in S_n^{\{n-1\}}$ with $(\mathbf{u})_1 \neq n-1$ and $(\mathbf{v})_1 \neq n$. By Lemma 13.7, there are $((n-2)!/2) \geq 3$ edges joining black vertices of $S_n^{\{n\}}$ to white vertices of $S_n^{\{n-1\}}$. We can choose an edge $(\mathbf{y},\mathbf{z})\in E^{n-1,n}$ where $\mathbf{y}$ is a black vertex in $S_n^{\{n\}}$ and $\mathbf{z}$ is a white vertex in $S_n^{\{n-1\}}$. Let $\{R_1,R_2,\ldots,R_{n-2}\}$ be an $(n-2)^*$-container of $S_n^{\{n\}}$ joining $\mathbf{u}$ to $\mathbf{y}$, and let $\{H_1,H_2,\ldots,H_{n-2}\}$ be an $(n-2)^*$-container of $S_n^{\{n-1\}}$ joining $\mathbf{z}$ to $\mathbf{v}$. We write $R_i=\langle \mathbf{u},R'_i,\mathbf{y_i},\mathbf{y}\rangle$ and $H_i=\langle \mathbf{z},\mathbf{z_i},H'_i,\mathbf{v}\rangle$. Note that $\mathbf{y_i}$ is a white vertex and $\mathbf{z_i}$ is a black vertex for every $1\le i\le n-2$. We have $\{(\mathbf{y_i})_1 \mid 1\le i\le n-2\}=\{(\mathbf{z_i})_1 \mid 1\le i\le n-2\}=\langle n-2\rangle$. Without loss of generality, we assume that $(\mathbf{y_i})_1=(\mathbf{z_i})_1$ for every $1\le i\le n-2$ with $(\mathbf{y_{n-2}})_1=(\mathbf{u})_1$.

Case 2.1. $(\mathbf{u})_1=(\mathbf{v})_1$. By Theorem 13.2, there is a Hamiltonian path T_i of $S_n^{\{(\mathbf{y_i})_1\}}$ joining the black vertex $(\mathbf{y_i})^n$ to the white vertex $(\mathbf{z_i})^n$ for every $i\in\langle n-3\rangle$, and there is a Hamiltonian path H of $S_n^{\{(\mathbf{y_{n-2}})_1\}}$ joining the black vertex $(\mathbf{u})^n$ to the white vertex $(\mathbf{v})^n$. We set

$$\begin{aligned}
Q_i &= \langle \mathbf{u},R'_i,\mathbf{y_i},(\mathbf{y_i})^n,T_i,(\mathbf{z_i})^n,\mathbf{z_i},H'_i,\mathbf{v}\rangle \quad \text{for } 1\le i\le n-3\\
Q_{n-1} &= \langle \mathbf{u},R'_{n-2},\mathbf{y_{n-2}},\mathbf{y},\mathbf{z},\mathbf{z_{n-2}},H'_{n-2},\mathbf{v}\rangle\\
Q_{n-2} &= \langle \mathbf{u},(\mathbf{u})^n,H,(\mathbf{v})^n,\mathbf{v},\rangle
\end{aligned}$$

Then $\{Q_1,Q_2,\ldots,Q_{n-1}\}$ forms an $(n-1)^*$-container of S_n joining $\mathbf{u}$ and $\mathbf{v}$.

Case 2.2. $(\mathbf{u})_1\neq(\mathbf{v})_1$. Without loss of generality, we assume that $(\mathbf{y_{n-3}})_1=(\mathbf{v})_1$. By Theorem 13.2, there is a Hamiltonian path T_i of $S_n^{\{(\mathbf{y_i})_1\}}$ joining $(\mathbf{y_i})^n$ to $(\mathbf{z_i})^n$ for every $i\in\langle n-4\rangle$, there is a Hamiltonian path H of $S_n^{\{(\mathbf{y_{n-3}})_1\}}$ joining the black vertex $(\mathbf{y_{n-3}})^n$ to the white vertex $(\mathbf{v})^n$, and there is a Hamiltonian path P of $S_n^{\{(\mathbf{y_{n-2}})_1\}}$ joining the black vertex $(\mathbf{u})^n$ to the white vertex $(\mathbf{z_{n-2}})^n$. We set

$$\begin{aligned}
Q_i &= \langle \mathbf{u},R'_i,\mathbf{y_i},(\mathbf{y_i})^n,T_i,(\mathbf{z_i})^n,\mathbf{z_i},H'_i,\mathbf{v}\rangle \quad \text{for } 1\le i\le n-4\\
Q_{n-3} &= \langle \mathbf{u},R'_{n-3},\mathbf{y_{n-3}},(\mathbf{y_{n-3}})^n,H,(\mathbf{v})^n,\mathbf{v},\rangle\\
Q_{n-2} &= \langle \mathbf{u},(\mathbf{u})^n,P,(\mathbf{z_{n-2}})^n,\mathbf{z_{n-2}},H'_{n-2},\mathbf{v},\rangle\\
Q_{n-1} &= \langle \mathbf{u},R'_{n-2},\mathbf{y_{n-2}},\mathbf{y},\mathbf{z},\mathbf{z_{n-3}},H'_{n-3},\mathbf{v}\rangle
\end{aligned}$$

It is easy to check whether $\{Q_1,Q_2,\ldots,Q_{n-1}\}$ is an $(n-1)^*$-container of S_n joining $\mathbf{u}$ to $\mathbf{v}$.

Thus, this theorem is proved. □

THEOREM 14.23 **[233]** S_n is super spanning laceable if and only if $n\neq 3$.

Proof: It is easy to see that this theorem is true for S_1 and S_2. Since S_3 is isomorphic to a cycle with six vertices, S_3 is not 1^*-laceable. Thus, S_3 is not super laceable. By Theorems 13.2 and 14.22, this theorem holds for S_4. Assume that S_k is super spanning laceable for every $4\le k\le n-1$. By Theorems 13.2 and 14.22, S_n is k^*-laceable for any $k\in\{1,2,n-1\}$. Thus, we still need to construct a k^*-container of S_n between any white vertex $\mathbf{u}$ and any black vertex $\mathbf{v}$ for every $3\le k\le n-2$. By induction, S_{n-1} is k^*-laceable. By Lemma 14.34, there is a k^*-container of S_n joining $\mathbf{u}$ to $\mathbf{v}$. □

Again, Chang et al. [43] discuss the spanning connectivity of a bipartite graph with faults. However, we should only discuss the edge fault. Suppose that a bipartite graph G is a super spanning laceable graph with $\kappa(G)=k$. Let F be any subset of $E(G)$ with $|F|=f\leq k-1$. Obviously, $\kappa(G-F)\geq k-f$. We say that G is *f-fault-tolerant spanning laceable* if $G-F$ is i^*-connected for any $1\leq i\leq k-f$ and $F\subset E(S_n)$ with $|F|=f$. The *fault-tolerance spanning laceability* of a bipartite graph G, $\kappa^*_{f,L}(G)$, is defined as the maximum integer f such that G is f-fault-tolerance spanning laceable.

LEMMA 14.35 Let $n\geq 5$ and $I=\{k_1,k_2,\ldots,k_t\}$ be any nonempty subset of $\langle n\rangle$. Suppose that F^i is a subset of $E(S_n^{\{i\}})$ with $|F^i|\leq n-4$ for every $1\leq i\leq n$, and F^0 is a subset of $\{(\mathbf{x},(\mathbf{x})^n)\mid \mathbf{x}\in V(S_n)\}$ with $|F^0|\leq n-3$. Suppose that $\mathbf{u}$ is a white vertex of $S_n^{\{k_1\}}$ and $\mathbf{v}$ is a black vertex of $S_n^{\{k_t\}}$. Then there is a Hamiltonian path P of $S_n^I-(\bigcup_{i=0}^n F^i)$ joining $\mathbf{u}$ to $\mathbf{v}$.

Proof: Obviously, $S_n^{\{k_i\}}$ is isomorphic to S_{n-1} for every $1\leq i\leq t$. By Theorem 13.2, this statement is true for $t=1$. Thus, we suppose that $2\leq t\leq n$. We set $\mathbf{x_1}=\mathbf{u}$ and $\mathbf{y_t}=\mathbf{v}$. By Lemma 13.7, $(n-2)!/2$ edges joining black vertices of $S_n^{\{i\}}$ to white vertices of $S_n^{\{j\}}$ for any $i,j\in\langle n\rangle$ with $i\neq j$. Since $(n-2)!/2>n-3$ if $n\geq 5$, we choose a black vertex $\mathbf{y_i}$ in $S_n^{\{k_i\}}$ with $(\mathbf{y_i})^n\in S_n^{\{k_{i+1}\}}$ and $(\mathbf{y_i},(\mathbf{y_i})^n)\notin F^0$ for every $i\in\langle t-1\rangle$. We set $\mathbf{x_{i+1}}=(\mathbf{y_i})^n$ for every $i\in\langle t-1\rangle$. Obviously, $\mathbf{x_i}$ is a white vertex for every $i\in\langle t\rangle$. By Theorem 13.2, there is a Hamiltonian path H_i of $S_n^{\{k_i\}}-F^{k_i}$ joining $\mathbf{x_i}$ to $\mathbf{y_i}$ for every $1\leq i\leq t$. Then $\langle \mathbf{u}=\mathbf{x_1},H_1,\mathbf{y_1},\mathbf{x_2},H_2,\mathbf{y_2},\ldots,\mathbf{x_t},H_t,\mathbf{y_t}=\mathbf{v}\rangle$ forms the Hamiltonian path of $S_n^I-(\bigcup_{i=0}^n F^i)$ joining $\mathbf{u}$ to $\mathbf{v}$. □

THEOREM 14.24 Suppose that F is an edge subset of $E(S_n)$ with $|F|=f\leq n-3$ and $n\geq 4$. Then S_n-F is i^*-laceable for every $1\leq i\leq n-1-f$.

Proof: We prove this theorem by induction on n. Suppose that $n=4$. By Theorems 13.2 and 14.23, this statement is true for S_4. Suppose that this statement is true for S_k for every $4\leq k\leq n-1$. By Theorems 13.2 and 14.23, this theorem is true for S_n-F if $f\in\{0,n-3\}$. Thus, we assume that $1\leq f\leq n-4$. Let $\mathbf{u}$ be a white vertex and $\mathbf{v}$ be a black vertex of S_n. By Theorem 13.2, there is a k^*-container of S_n-F between $\mathbf{u}$ and $\mathbf{v}$ if $k=1$ or $k=2$. Thus, we need to show that there is a k^*-container of S_n-F between $\mathbf{u}$ and $\mathbf{v}$ for every $3\leq k\leq n-f-1$. We set $F^i=F\cap E(S_n^{\{i\}})$ for every $1\leq i\leq n$, and $F^0=F-(\bigcup_{i=1}^n F_i)$. Since S_n is edge-transitive, we assume that $|F^0|\geq 1$. Thus, $|F^i|\leq f-1\leq n-4$ for every $i\in\langle n\rangle$.

Case 1. $\mathbf{u},\mathbf{v}\in S_n^{\{r\}}$ for some $r\in\langle n\rangle$. By induction hyphothesis, there is a k^*-container $\{R_1,\ldots,R_k\}$ of $S_n^{\{r\}}-F^r$ between $\mathbf{u}$ and $\mathbf{v}$. Without loss of generality, we can write R_i as $\langle \mathbf{x_{i_1}},\mathbf{x_{i_2}},\ldots,\mathbf{x_{i_m}}\rangle$ with $\mathbf{x_{i_1}}=\mathbf{u}$ and $\mathbf{x_{i_m}}=\mathbf{v}$ for every $1\leq i\leq k$. Note that $|\bigcup_{i=1}^k(V(R_i)-\{\mathbf{x_{i_1}},\mathbf{x_{i_m}}\})|=(n-1)!-2$. There are at least $(n-1)!-k-2$ edges in $\bigcup_{i=1}^k(E(R_i)-\{(\mathbf{x_{i_1}},\mathbf{x_{i_2}}),(\mathbf{x_{i_{m-1}}},\mathbf{x_{i_m}})\})$. (Note that $\{(\mathbf{x_{i_1}},\mathbf{x_{i_2}}),(\mathbf{x_{i_{m-1}}},\mathbf{x_{i_m}})\}=\{(\mathbf{x_{i_1}},\mathbf{x_{i_2}})\}$ if $m=2$.) Since $|F|\leq n-3$, there are at most $(n-3)$ vertices in $S_n^{\{r\}}$ such that the edges between them and their nth neighbors are in F. Since $k\leq n-2$

and $n \geq 5$, $(n-1)! - k - 2 \geq (n-1)! - n + 1 > 2(n-3)$. Thus, there is an edge $(\mathbf{x_{s_t}}, \mathbf{x_{s_{t+1}}}) \in E(R_s)$ for some $s \in \langle k \rangle$ and for some $t \in \langle s_m - 1 \rangle - \{1\}$ such that $\{(\mathbf{x_{s_t}}, (\mathbf{x_{s_t}})^n), (\mathbf{x_{s_{t+1}}}, (\mathbf{x_{s_{t+1}}})^n)\} \cap F = \emptyset$. Without loss of generality, we assume that $s = k$. Since $d(\mathbf{x_{k_t}}, \mathbf{x_{k_{t+1}}}) = 1$, by Lemma 13.8, $(\mathbf{x_{k_t}})_1 \neq (\mathbf{x_{k_{t+1}}})_1$. Obviously, the color of $\mathbf{x_{k_t}}$ and the color of $\mathbf{x_{k_{t+1}}}$ are distinct. Moreover, the color of $(\mathbf{x_{k_t}})^n$ and the color of $(\mathbf{x_{k_{t+1}}})^n$ are distinct. By Theorem 14.35, there is a Hamiltonian path H of $S_n^{\langle n \rangle - \{r\}} - F$ joining the vertex $(\mathbf{x_{k_t}})^n$ to the vertex $(\mathbf{x_{k_{t+1}}})^n$. We set $T_i = R_i$ for every $1 \leq i \leq k-1$, and $T_k = \langle \mathbf{x_{k_1}}, \mathbf{x_{k_2}}, \ldots, \mathbf{x_{k_t}}, (\mathbf{x_{k_t}})^n, H, (\mathbf{x_{k_{t+1}}})^n, \mathbf{x_{k_{t+1}}}, \mathbf{x_{k_{t+2}}}, \ldots, \mathbf{x_{k_m}} \rangle$. Then $\{T_1, T_2, \ldots, T_k\}$ forms the k^*-container of $S_n - F$ between $\mathbf{u}$ and $\mathbf{v}$.

Case 2. $\mathbf{u} \in S_n^{\{r\}}$ and $\mathbf{v} \in S_n^{\{s\}}$ for some $r, s \in \langle n \rangle$ with $r \neq s$. By Lemma 13.8, there are $(n-2)!/2$ edges between the black vertices of $S_n^{\{r\}}$ and the white vertices of $S_n^{\{s\}}$. Since $(n-2)!/2 > n-3$ if $n \geq 5$, we can choose a black vertex $\mathbf{x}$ in $S_n^{\{r\}}$ and a white vertex $\mathbf{y}$ in $S_n^{\{s\}}$ with (1) $(\mathbf{x}, \mathbf{y}) \in E^{r,s}$ and (2) $\{(\mathbf{z}, (\mathbf{z})^n) \mid \mathbf{z} \in N_{S_n}(\mathbf{x}) \cup N_{S_n}(\mathbf{y})\} \cap F = \emptyset$. By induction, there is a k^*-container $\{R_1, R_2, \ldots, R_k\}$ of $S_n^{\{r\}} - F^r$ joining $\mathbf{u}$ to $\mathbf{x}$. Again, there is a k^*-container $\{H_1, H_2, \ldots, H_k\}$ of $S_n^{\{s\}} - F^s$ joining $\mathbf{y}$ to $\mathbf{v}$. Without loss of generality, we write R_i as $\langle \mathbf{u}, R_i', \mathbf{x_i}, \mathbf{x} \rangle$ and write H_i as $\langle \mathbf{y}, \mathbf{y_i}, H_i', \mathbf{v} \rangle$ for every $1 \leq i \leq k$. We set $A = \{(\mathbf{x_i})_1 \mid (\mathbf{x}, \mathbf{x_i}) \in E(R_i) \text{ and } i \in \langle k \rangle\}$ and $B = \{(\mathbf{y_i})_1 \mid (\mathbf{y}, \mathbf{y_i}) \in E(H_i) \text{ and } i \in \langle k \rangle\}$. By Lemma 13.8, $(\mathbf{x_i})_1 \neq (\mathbf{x_j})_1$ for every $i, j \in \langle k \rangle$ with $i \neq k$, and $A \subset \langle n \rangle - \{r, s\}$. Similarly, $(\mathbf{y_i})_1 \neq (\mathbf{y_j})_1$ for every $i, j \in \langle k \rangle$ with $i \neq k$, and $B \subset \langle n \rangle - \{r, s\}$. Let $I = \{\mathbf{x_i} \mid \mathbf{x_i} \in A$ and there exists an element $\mathbf{y_j} \in B$ such that $(\mathbf{x_i})_1 = (\mathbf{y_j})_1\}$. Let $|I| = m$. (Note that $|I| = 0$ if $I = \emptyset$.) Then we relabel the indices of A and B such that $(\mathbf{x_i})_1 = (\mathbf{y_i})_1$ for all $i \leq m$. By Lemma 14.35, there is a Hamiltonian path W_i of $S_n^{\{(\mathbf{x_i})_1, (\mathbf{y_i})_1\}} - F$ joining the black vertex $(\mathbf{x_i})^n$ to the white vertex $(\mathbf{y_i})^n$ for every $1 \leq i \leq k-2$. (Note that $\{(\mathbf{x_i})_1, (\mathbf{y_i})_1\} = \{(\mathbf{x_i})_1\}$ if $(\mathbf{x_i})_1 = (\mathbf{y_i})_1$.) We set $J = \langle n \rangle - (\{(\mathbf{x_1})_1 \mid 1 \leq i \leq k-2\} \cup \{(\mathbf{y_1})_1 \mid 1 \leq i \leq k-2\} \cup \{r, s\})$. By Lemma 14.35, there is a Hamiltonian path W_{k-1} of $S_n^J - F$ joining the black vertex $(\mathbf{x_{k-1}})^n$ to the white vertex $(\mathbf{y_k})^n$. We set

$$\begin{aligned} T_i &= \langle \mathbf{u}, R_i', \mathbf{x_i}, (\mathbf{x_i})^n, T_i, (\mathbf{y_i})^n, \mathbf{y_i}, H_i', \mathbf{v} \rangle \quad \text{for every } 1 \leq i \leq k-2 \\ T_{k-1} &= \langle \mathbf{u}, R_{k-1}', \mathbf{x_{k-1}}, (\mathbf{x_{k-1}})^n, W_{k-1}, (\mathbf{y_k})^n, \mathbf{y_k}, H_k', \mathbf{v} \rangle \\ T_k &= \langle \mathbf{u}, R_k', \mathbf{x_k}, \mathbf{x}, \mathbf{y}, \mathbf{y_{k-1}}, H_{k-1}', \mathbf{v} \rangle \end{aligned}$$

Then $\{T_1, T_2, \ldots, T_k\}$ forms the k^*-container of $S_n - F$ between $\mathbf{u}$ and $\mathbf{v}$. □

THEOREM 14.25 $\kappa_{f,L}^*(S_n) = n - 3$ if $n \geq 4$.

Proof: By Theorem 14.24, $\kappa_{f,L}^*(S_n) \geq n-3$ if $n \geq 4$. Let $\mathbf{u}$ be any vertex of S_n and F be any subset of $\{(\mathbf{u}, \mathbf{v}) \mid \mathbf{v} \in N_{S_n}(\mathbf{u})\}$ with $|F| = n-2$. Since $\deg_{S_n - F}(\mathbf{u}) = 1$ and $n(S_n - F) = n! > 3$, $S_n - F$ is not 1^*-laceable. Thus, $\kappa_{f,L}^*(S_n) = n-3$ if $n \geq 4$. □

There are some studies of other families of interconnection networks. For example, Tsai et al. [304] study the spanning connectivity of the recursive circulant graphs $RCG(2^m, 4)$. Ho et al. [152] study the spanning connectivity of the augmented cubes.

14.8 SPANNING FAN-CONNECTIVITY AND SPANNING PIPE-CONNECTIVITY OF GRAPHS

There is a Menger-type theorem similar to the spanning connectivity of a graph. Let x be a vertex in a graph G and let $U = \{y_1, y_2, \ldots, y_t\}$ be a subset of $V(G)$ where x is not in U. A *t-(x, U)-fan*, $F_t(x, U)$, is a set of internally disjoint paths $\{P_1, P_2, \ldots, P_t\}$ such that P_i is a path connecting x and y_i for $1 \leq i \leq t$. It follows from Theorem 7.8 that a graph G is k-connected if and only if it has at least $(k+1)$ vertices and there exists a t-(x, U)-fan for every choice of x and U with $|U| \leq k$ and $x \notin U$. Similarly, we can introduce the concept of a spanning fan. A *spanning k-(x, U)-fan* is a k-(x, U)-fan $\{P_1, P_2, \ldots, P_k\}$ such that $\bigcup_{i=1}^{k} V(P_i) = V(G)$. A graph G is *k^*-fan-connected* (also written as *k_f^*-connected*) if there exists a spanning k-(x, U)-fan for every choice of x and U with $|U| = k$ and $x \notin U$. The *spanning fan-connectivity* of a graph G, $\kappa^*_{fan}(G)$, is defined as the largest integer k such that G is w^*_{fan}-connected for $1 \leq w \leq k$ if G is a 1^*_{fan}-connected graph.

There is another Menger-type theorem similar to the spanning connectivity and spanning fan-connectivity of a graph. Let $U = \{x_1, x_2, \ldots x_t\}$ and $W = \{y_1, y_2, \ldots, y_t\}$ be two t-subsets of $V(G)$. An *(U, W)-pipeline* is a set of internally disjoint paths $\{P_1, P_2, \ldots, P_t\}$ such that P_i is a path connecting x_i to $y_{\pi(i)}$ where π is a permutation of $\{1, 2, \ldots, t\}$. It is known that a graph G is k-connected if and only if it has at least $k+1$ vertices and there exists an (U, W)-pipeline for every choice of U and W with $|U| = |W| \leq k$ and $U \neq W$. Similarly, we can introduce the concept of spanning pipeline. A *spanning (U, W)-pipeline* is an (U, W)-pipeline $\{P_1, P_2, \ldots, P_k\}$ such that $\bigcup_{i=1}^{k} V(P_i) = V(G)$. A graph G is *k^*-pipeline-connected* (or *k^*_{pipe}-connected*) if there exists a spanning (U, W)-pipeline for every choice of U and W with $|U| = |W| \leq k$ and $U \neq W$. The *spanning pipeline-connectivity* of a graph G, $\kappa^*_{pipe}(G)$, is defined as the largest integer k such that G is w^*_{pipe}-connected for $1 \leq w \leq k$ if G is a 1^*_{pipe}-connected graph.

Lin et al. [240] begin the study on the relationships between $\kappa(G)$, $\kappa^*(G)$, $\kappa^*_{fan}(G)$, and $\kappa^*_{pipe}(G)$.

LEMMA 14.36 Every 1^*-connected graph is 1^*_{fan}-connected. Moreover, every 1^*_{fan}-connected graph that is not K_2 is 2^*_{fan}-connected. Thus, $\kappa^*_{fan}(G) \geq 2$ if G is a Hamiltonian-connected graph with at least three vertices.

Proof: Let G be a 1^*-connected graph with at least three vertices and let x be any vertex of G. Assume that $U = \{y\}$ with $x \neq y$. Obviously, there exists a Hamiltonian path P_1 joining x and y. Apparently, $\{P_1\}$ forms a spanning 1-(x, U)-fan. Thus, G is 1^*_{fan}-connected. Assume that $U = \{y_1, y_2\}$ with $x \notin U$. Let Q be a Hamiltonian path of G connecting y_1 and y_2. We write Q as $\langle y_1, Q_1, x, Q_2, y_2 \rangle$. We set P_1 as $\langle x, Q_1^{-1}, y_1 \rangle$ and P_2 as $\langle x, Q_2, y_2 \rangle$. Then $\{P_1, P_2\}$ forms a spanning 2-(x, U)-fan. Thus, G is 2^*_{fan}-connected and $\kappa^*_{fan}(G) \geq 2$. □

Similarly, we have the following lemma.

LEMMA 14.37 Every 1^*-connected graph is 1^*_{pipe}-connected.

THEOREM 14.26 $\kappa^*_{fan}(G) \le \kappa^*(G) \le \kappa(G)$ for any 1^*_{fan}-connected graph. Moreover, $\kappa^*_{fan}(G) = \kappa^*(G) = \kappa(G) = n(G) - 1$ if and only if G is a complete graph.

Proof: Obviously, $\kappa^*(G) \le \kappa(G)$. Now, we prove that $\kappa^*_{fan}(G) \le \kappa^*(G)$. Assume that $\kappa^*_{fan}(G) = k$. Let x and y be any two vertices of G. We need to show that there is a k^*-container of G between x and y.

Suppose that $k = 1$. Since G is 1^*_{fan}-connected, there is a spanning $1-(x, \{y\})$-fan, $\{P_1\}$, of G. Then $\{P_1\}$ forms a spanning container of G between x and y.

Suppose that $k \ge 2$. Let $U' = \{y_1, y_2, \dots, y_{k-1}\}$ be a set of $(k-1)$ neighbors of y not containing x. We set $U = U' \cup \{y\}$. By assumption, there exists a spanning k-(x, U)-fan. Obviously, we can extend the spanning k-(x, U)-fan by adding the edges $\{(y_i, y) \mid y_i \in U'\}$ to obtain a k^*-container between x to y. Hence, G is k^*-connected. Therefore, $\kappa^*_{fan}(G) \le \kappa^*(G)$ for every 1^*_{fan}-connected graph.

Suppose that G is not a complete graph. There exists a vertex cut S of size $\kappa(G)$. Let x and y be any two vertices in different connected components of $G - S$. Obviously, y is not in any (x, S)-fan of G. Thus, $\kappa^*_{fan}(G) < \kappa(G)$ □

Similarly, we have the following theorem.

THEOREM 14.27 $\kappa^*_{pipe}(G) \le \kappa^*_{fan}(G) \le \kappa^*(G) \le \kappa(G)$ for any 1^*_{pipe}-connected graph. Moreover, $\kappa^*_{pipe}(G) = \kappa^*_{fan}(G) = \kappa^*(G) = \kappa(G)$ if and only if G is a complete graph.

LEMMA 14.38 Let u and v be two non-adjacent vertices of G with $d_G(u) + d_G(v) \ge n(G) + 1$, and let x and y be any two distinct vertices of G. Then G has a Hamiltonian path joining x to y if and only if $G + (u, v)$ has a Hamiltonian path joining x to y.

Proof: Since every path in G is a path in $G + (u, v)$, there is a Hamiltonian path of $G + (u, v)$ joining x to y if G has a Hamiltonian path joining x to y [4].

Suppose that there is a Hamiltonian path P of $G + (u, v)$ joining x to y. We need to show there is a Hamiltonian path of G between x and y. If $(u, v) \notin E(P)$, then P is a Hamiltonian path of G between x and y . Thus, we consider that $(u, v) \in E(P)$. Without loss of generality, we write P as $\langle z_1, z_2, \dots, z_i, z_{i+1}, \dots, z_{n(G)}\rangle$ where $z_1 = x$, $z_i = u$, $z_{i+1} = v$, and $z_{n(G)} = y$. Since $d_G(u) + d_G(v) \ge n(G) + 1$, there is an index k in $\{1, 2, \dots, n(G)\} - \{i - 1.i, i + 1\}$ such that $(z_i, z_k) \in E(G)$ and $(z_{i+1}, z_{k+1}) \in E(G)$. We set $R = \langle z_1, z_2, \dots, z_k, z_i, z_{i-1}, \dots, z_{k+1}, z_{i+1}, z_{i+2}, \dots, z_{n(G)}\rangle$ if $1 \le k \le i - 2$ and $R = \langle z_1, z_2, \dots, z_i, z_k, z_{k-1}, \dots, z_{i+1}, z_{k+1}, z_{k+2}, \dots, z_{n(G)}\rangle$ if $i + 2 \le k \le n(G)$. Then R is a Hamiltonian path of G between x and y. □

Lin et al. [240] also discuss some sufficient conditions for a graph to be κ^*_{fan}-connected and κ^*_{pipe}-connected. Let H be a subgraph of G. The *neighborhood* of u with respect to H, denoted by $Nbd_H(u)$, is $\{v \in V(H) \mid (u, v) \in E(G)\}$. We use $\Gamma_H(u)$ to denote $|Nbd_H(u)|$. Obviously, $\Gamma_G(u) = \deg(u)$ for every $u \in V(G)$. The following theorem on spanning fan-connectivity is analogous to that on spanning connectivity in Lemma 14.1.

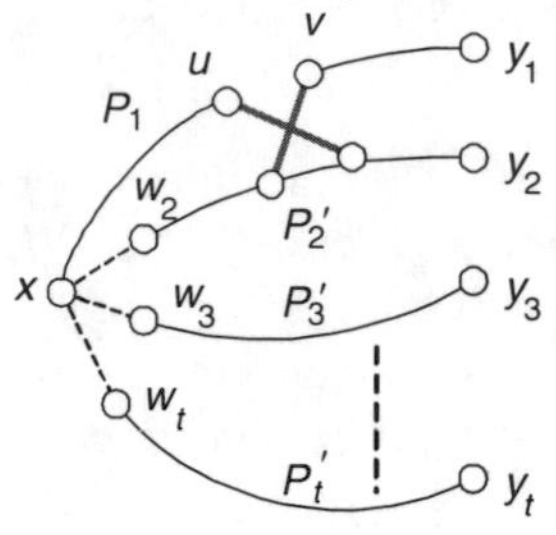

FIGURE 14.27 Illustration for Case 1 of Theorem 14.28

THEOREM 14.28 Assume that k is a positive integer. Let u and v be two nonadjacent vertices of G with $\deg_G(u) + \deg_G(v) \geq n(G) + k$. Then $\kappa^*_{fan}(G) \geq k+1$ if and only if $\kappa^*_{fan}(G+(u,v)) \geq k+1$.

Proof: Obviously, $\kappa^*_{fan}(G+(u,v)) \geq k+1$ if $\kappa^*_{fan}(G) \geq k+1$. Suppose that $\kappa^*_{fan}(G+(u,v)) \geq k+1$. Let x be any vertex of G and $U = \{y_1, y_2, \ldots, y_t\}$ be any subset of $V(G)$ such that $x \notin U$ and $t \leq k+1$. We need to find a spanning t-(x, U)-fan of G.

Since $G+(u,v)$ is $(k+1)^*_{fan}$-connected, there exists a spanning t-(x, U)-fan $\{P_1, P_2, \ldots, P_t\}$ of $G+(u,v)$ with P_i joining x to y_i for $1 \leq i \leq t$. Obviously, $\{P_1, P_2, \ldots, P_t\}$ is a spanning t-(x, U)-fan of G if (u, v) is not in $\bigcup_{i=1}^t E(P_i)$. Thus, we consider (u, v) is in $\bigcup_{i=1}^t E(P_i)$. By Theorems 19.3 and 19.11, we can find a spanning (x, U)-fan of G if $t = 1, 2$. Thus, we consider the case $t \geq 3$. Without loss of generality, we may assume that $(u, v) \in P_1$. Thus, we can write P_1 as $\langle x, H_1, u, v, H_2, y_1 \rangle$. Let P_i' be the path obtained from P_i by deleting x. Thus, we can write P_i as $\langle x, w_i, P_i', y_i \rangle$ for $1 \leq i \leq t$. Note that $x \neq w_i$ and $P_i = \langle y_i \rangle$ if $w_i = y_i$ for every $2 \leq i \leq t$.

Case 1. $\Gamma_{P_i'}(u) + \Gamma_{P_i'}(v) \geq n(P_i') + 2$ for some $2 \leq i \leq t$. Without loss of generality, we may assume that $\Gamma_{P_2'}(u) + \Gamma_{P_2'}(v) \geq n(P_2') + 2$. Obviously, $n(P_2') \geq 2$. We write $P_2' = \langle w_2 = z_1, z_2, \ldots, z_r = y_2 \rangle$. We claim that there exists an index j in $\{1, 2, \ldots, r-1\}$ such that $(z_j, v) \in E(G)$ and $(z_{j+1}, u) \in E(G)$. Suppose that this is not the case. Then $\Gamma_{P_2'}(u) + \Gamma_{P_2'}(v) \leq r + r - (r-1) = r + 1 = n(P_2') + 1$. We get a contradiction.

We set $Q_1 = \langle x, w_2 = z_1, z_2, \ldots, z_j, v, H_2, y_1 \rangle$, $Q_2 = \langle x, H_1, u, z_{j+1}, z_{j+2}, \ldots, z_r = y_2 \rangle$, and $Q_i = P_i$ for $3 \leq i \leq t$. Then $\{Q_1, Q_2, \ldots, Q_t\}$ forms a spanning t-(x, U)-fan of G. See Figure 14.27 for illustration.

Case 2. $\Gamma_{P_i'}(u) + \Gamma_{P_i'}(v) \leq n(P_i') + 1$ for every $2 \leq i \leq t$.

Case 2.1. $\Gamma_{P_i'}(u) + \Gamma_{P_i'}(v) < n(P_i') + 1$ for some $2 \leq i \leq t$. Without loss of generality, we may assume that $\Gamma_{P_2'}(u) + \Gamma_{P_2'}(v) \leq n(P_2')$. Thus,

$$\begin{aligned}\Gamma_{P_1}(u) + \Gamma_{P_1}(v) &= \deg_G(u) + \deg_G(v) - \sum_{i=2}^{t} (\Gamma_{P_i'}(u) + \Gamma_{P_i'}(v)) \\ &= \deg_G(u) + \deg_G(v) - (\Gamma_{P_2'}(u) + \Gamma_{P_2'}(v)) - \sum_{i=3}^{t} (\Gamma_{P_i'}(u) + \Gamma_{P_i'}(v))\end{aligned}$$

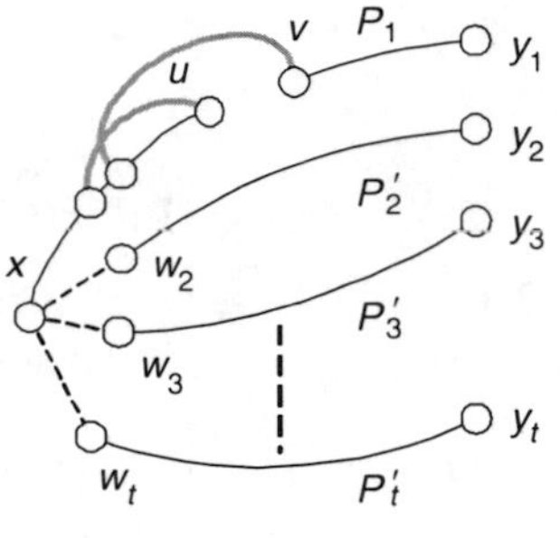

FIGURE 14.28 Illustration for Case 2.1 of Theorem 14.28.

$$\begin{aligned}&\geq n(G)+k-n(P_2')-\sum_{i=3}^{t}(n(P_i')+1)\\&=n(P_1)+k-(t-2)\\&\geq n(P_1)+1\end{aligned}$$

By Theorem X, there is a Hamiltonian path Q_1 of $G[P_1]$ joining x to y_1. We set $Q_i=P_i$ for $2\leq i\leq t$. Then $\{Q_1,Q_2,\ldots,Q_t\}$ forms a spanning t-(x,U)-fan of G. See Figure 14.28 for illustration.

Case 2.2. $\Gamma_{P_i'}(u)+\Gamma_{P_i'}(v)=n(P_i')+1$ for every $2\leq i\leq t$. We have

$$\begin{aligned}\Gamma_{P_1}(u)+\Gamma_{P_1}(v)&=\deg_G(u)+\deg_G(v)-\sum_{i=2}^{t}(\Gamma_{P_i'}(u)+\Gamma_{P_i'}(v))\\&=n(G)+k-\sum_{i=2}^{t}(n(P_i')+1)\\&=n(P_1)+k-(t-1)\\&\geq n(P_1)\end{aligned}$$

We set $R=\langle y_1,H_2^{-1},v,u,H_1^{-1},x,w_2,P_2',y_2\rangle$. Then

$$\begin{aligned}\Gamma_R(u)+\Gamma_R(v)&=\Gamma_{P_1}(u)+\Gamma_{P_1}(v)+\Gamma_{P_2'}(u)+\Gamma_{P_2'}(v)\\&\geq n(P_1)+n(P_2')+1\\&=n(R)+1\end{aligned}$$

By Theorem X, there is a Hamiltonian path W of $G[R]$ joining y_1 to y_2. Thus, W can be written as $\langle y_1,W_1,x,W_2,y_2\rangle$. We set $Q_1=\langle x,W_1^{-1},y_1\rangle$, $Q_2=\langle x,W_2,y_2\rangle$, and $Q_i=P_i$ for $3\leq i\leq t$. Then $\{Q_1,Q_2,\ldots,Q_t\}$ forms a spanning (x,U)-fan of G. □

Using the closure technique, the following theorem is obtained.

THEOREM 14.29 Let k be a positive integer. Then $\kappa_{fan}^*(G)\geq k+1$ if $\deg_G(u)+\deg_G(v)\geq n(G)+k$ for all nonadjacent vertices u and v.

Again, Theorem 14.29 is an analog result of Theorem 14.1 in spanning fan-connectivity.

THEOREM 14.30 $\kappa^*_{fan}(G) \geq 2\delta(G) - n(G) + 1$ if $(n(G)/2) + 1 \leq \delta(G)$.

Proof: Suppose that $n(G)/2 + 1 \leq \delta(G)$. Obviously, $\delta(G) \leq n(G) - 1$ and $n(G) \geq 4$. Suppose that $n(G) = 2m$ for some integer $m \geq 2$. Then $\delta(G) = m + k$ for some integer k with $1 \leq k \leq m - 1$. Obviously, $\deg_G(u) + \deg_G(v) \geq 2\delta(G) = 2m + 2k$. By Theorem 14.29, $\kappa^*_{fan}(G) \geq 2k + 1 = 2\delta(G) - n(G) + 1$. Suppose that $n(G) = 2m + 1$ for some integer $m \geq 2$. Then $\delta(G) = m + 1 + k$ for some integer k with $1 \leq k \leq m - 1$. By Theorem 14.29, $\kappa^*_{fan}(G) \geq 2k + 2 = 2\delta(G) - n(G) + 1$. The theorem is proved. □

COROLLARY 14.4 $\kappa^*_{\mathrm{fan}}(\mathrm{G}) = \mathrm{n}(\mathrm{G}) - 3$ *if* $\delta(\mathrm{G}) = \mathrm{n}(\mathrm{G}) - 2$ *and* $\mathrm{n}(\mathrm{G}) \geq 5$.

Proof: By Theorem 14.2, $\kappa^*(G) \geq n(G) - 2$. Since $n(G) - 2 \leq \kappa^*(G) \leq \kappa(G) \leq \delta(G) = n(G) - 2$, $\kappa(G) = n(G) - 2$. By Theorem 14.30, $\kappa^*_{fan}(G) \geq n(G) - 3$. By Theorem 14.26, $\kappa^*_{fan}(G) < \kappa(G)$. Thus, $\kappa^*_{fan}(G) = n(G) - 3$. □

We have similar result for spanning pipeline connectivity.

THEOREM 14.31 Assume that k is a positive integer. Let u and v be two nonadjacent vertices of G. Suppose that $\deg_G(u) + \deg_G(v) \geq n(G) + k$. Then $\kappa^*_{pipe}(G) \geq k$ if and only if $\kappa^*_{pipe}(G + (u, v)) \geq k$.

Proof: Obviously, $\kappa^*_{pipe}(G + (u, v)) \geq k$ if $\kappa^*_{pipe}(G) \geq k$. Suppose that $\kappa^*_{pipe}(G + (u, v)) \geq k$. Let $U = \{x_1, x_2, \ldots, x_t\}$ and $W = \{y_1, y_2, \ldots, y_t\}$ be any two subsets of G such that $U \neq W$ and $t \leq k$. We need to find a spanning (U, W)-pipeline of G.

Since $G + (u, v)$ is k^*_{pipe}-connected, there exists a spanning (U, W)-pipeline of $G + (u, v)$. Let $\{P_1, P_2, \ldots, P_t\}$ be a spanning (U, W)-pipeline with P_i joining x_i to $y_{\pi(i)}$ for $1 \leq i \leq t$. Without loss of generality, we assume that $\pi(i) = i$. Obviously, $\{P_1, P_2, \ldots, P_t\}$ is a spanning (U, W)-pipeline of G if (u, v) is not in P. Thus, we consider the case that (u, v) is in P. By Theorems 19.3 and 19.11, we can find a spanning (U, W)-pipeline of G if $t = 1$. Thus, we consider the case $t \geq 2$.

Case 1. $U \cap W = \emptyset$. Without loss of generality, we may assume that $(u, v) \in P_1$. Thus, we can write P_1 as $\langle x_1, H_1, u, v, H_2, y_1 \rangle$. (Note that $H_1 = \langle x \rangle$ if $x = u$, and $H_2 = \langle y \rangle$ if $y = v$.) Let P'_i be the path obtained from P_i by deleting x and y_i. Thus, we can write P_i as $\langle x_i, P'_i, y_i \rangle$ for $1 \leq i \leq t$.

Case 1.1. $\Gamma_{P_1}(u) + \Gamma_{P_1}(v) \geq n(P_1) + 1$. By Theorem X, there is a Hamiltonian path Q_1 of $G[P_1]$ joining x_1 to y_1. We set $Q_i = P_i$ for $2 \leq i \leq t$. Then $\{Q_1, Q_2, \ldots, Q_t\}$ forms a spanning (U, W)-pipeline of G. See Figure 14.29 for illustration.

Case 1.2. $\Gamma_{P_1}(u) + \Gamma_{P_1}(v) \leq n(P_1)$. We claim that $\Gamma_{P_i}(u) + \Gamma_{P_i}(v) \geq n(P_i) + 2$ for some $2 \leq i \leq t$.

Suppose that $\Gamma_{P_i}(u)+\Gamma_{P_i}(v) \le n(P_i)+1$ for every $2 \le i \le t$. Then

$$\begin{aligned}
\deg_G(u)+\deg_G(v) &= \Gamma_{P_1}(u)+\Gamma_{P_1}(v)+\sum_{i=2}^{t}(\Gamma_{P_i}(u)+\Gamma_{P_i}(v)) \\
&\le n(P_1)+\sum_{i=2}^{t}(n(P_i)+1) \\
&= n(G)+t-1 \\
&\le n(G)+k-1
\end{aligned}$$

We obtain a contradiction. Thus, $\Gamma_{P_i}(u)+\Gamma_{P_i}(v) \ge n(P_i)+2$ for some $2 \le i \le t$. Without loss of generality, we assume that $\Gamma_{P_2}(u)+\Gamma_{P_2}(v) \ge n(P_2)+2$. Obviously, $n(P'_2) \ge 2$. We write $P'_2 = \langle x_2 = z_1, z_2, \ldots, z_r = y_2\rangle$. We claim that there exists an index j in $\{1, 2, \ldots, r-1\}$ such that $(z_j, v) \in E(G)$ and $(z_{j+1}, u) \in E(G)$. Suppose this is not the case. Then $\Gamma_{P'_2}(u)+\Gamma_{P'_2}(v) \le r+r-(r-1) = r+1 = n(P'_2)+1$. We get a contradiction.

We set $Q_1 = \langle x_2 = z_1, z_2, \ldots, z_j, v, H_2, y_1\rangle$, $Q_2 = \langle x_1, H_1, u, z_{j+1}, z_{j+2}, \ldots, z_r = y_2\rangle$, and $Q_i = P_i$ for $3 \le i \le t$. Then $\{Q_1, Q_2, \ldots, Q_t\}$ forms a spanning (U, W)-pipeline of G. See Figure 14.30 for illustration.

Case 2. $U \cap W \neq \emptyset$. Let $|U \cap W| = r$. Without loss of generality, we assume that $x_i = y_i$ for $t-r+1 \le i \le t$. Let $G' = G[V(G) - (U \cap W)]$, $U' = U - W$, and $W' = W - U$. Obviously, $d_{G'}(u)+d_{G'}(v) \ge \deg_G(u)+\deg_G(v)-2r \ge n(G)+k-2r = n(G')+k-r$, $|U'| = |W'| = t-r \le k-r$, and $U' \cap W' = \emptyset$. By Case 1, there exists a spanning (U', W')-pipeline $\{Q_1, Q_2, \ldots, Q_{t-r}\}$ of G'. We set $Q_i = \langle x_i\rangle$ for $t-r+1 \le i \le t$. Then $\{Q_1, Q_2, \ldots, Q_t\}$ forms a spanning (U, W)-pipeline of G. □

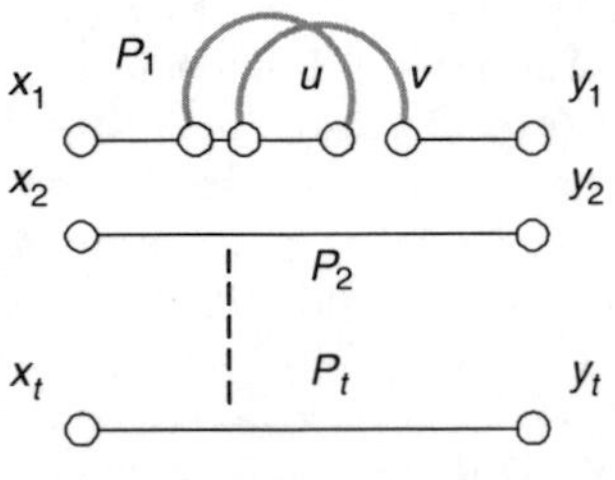

FIGURE 14.29 Illustration for Case 1.1 of Theorem 14.31.

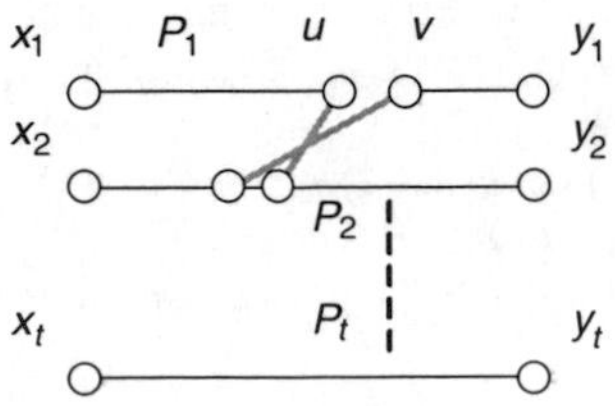

FIGURE 14.30 Illustration for Case 1.2 of Theorem 14.31.

THEOREM 14.32 Let k be a positive integer. Then $\kappa^*_{pipe}(G) \geq k$ if $\deg_G(u) + \deg_G(v) \geq n(G) + k$ for all nonadjacent vertices u and v.

THEOREM 14.33 $\kappa^*_{pipe}(G) \geq 2\delta(G) - n(G)$ if $(n(G)/2) + 1 \leq \delta(G)$.

COROLLARY 14.5 $\kappa^*_{\text{pipe}}(G) = n(G) - 4$ *if* $\delta(G) = n(G) - 2$ *and* $n(G) \geq 5$.

Proof: By Theorem 14.33, $\kappa^*_{pipe}(G) \geq n(G) - 4$. Let $V(G) = \{x_1, x_2, \ldots, x_{n(G)}\}$. Without loss of generality, we assume that $(x_1, x_2) \notin E(G)$. We set $U = \{x_3, x_5, x_6, \ldots, x_{n(G)}\}$ and $W = \{x_4, x_5, x_6 \ldots, x_{n(G)}\}$. Obviously, $U \neq W$ and $|U| = |W| = n(G) - 3$. Since there is not a Hamiltonian path of $G[V(G) - \{x_5, x_6, \ldots, x_{n(G)}\}]$ joining x_3 to x_4, there is not a spanning (U, W)-pipeline of G. Thus, $\kappa^*_{pipe}(G) = n(G) - 4$. □

We use the following example to illustrate that $\kappa(G)$, $\kappa^*(G)$, $\kappa^*_{fan}(G)$, and $\kappa^*_{pipe}(G)$ are really different concepts and, in general, have different values.

EXAMPLE 14.6

Suppose that n is a positive integer with $n \geq 2$. Let $H(n)$ be the complete 3-partite graph $K_{2n,2n,n-1}$ with vertex partite sets $V_1 = \{x_1, x_2, \ldots x_{2n}\}$, $V_2 = \{y_1, y_2, \ldots, y_{2n}\}$, and $V_3 = \{z_1, z_2, \ldots, z_{n-1}\}$. Let $G(n)$ be the graph obtained from $H(n)$ by adding the edge set $\{(z_i, z_j) \mid 1 \leq i \neq j < n\}$. Thus, $G[V_3]$ is the complete graph K_{n-1}. Obviously, $n(G(n)) = 5n - 1$, $\delta(G(n)) = 2n + (n - 1) = 3n - 1$, and $\kappa(G(n)) = \delta(G(n))$. In the following discussion, we show that $\kappa^*(G(n)) = n + 1$, $\kappa^*_{fan}(G(n)) = n$, and $\kappa^*_{pipe}(G(n)) = n - 1$.

By Theorem 14.2, $\kappa^*(G(n)) \geq 2\delta(G(n)) - n(G(n)) + 2 = n + 1$. To show $\kappa^*(G(n)) = n + 1$, we claim that there is no $(n + 2)^*$-container of $G(n)$ between x_1 and x_2. Suppose this is not the case. Let $\{P_1, P_2, \ldots, P_{n+2}\}$ be an $(n + 2)^*$-container of $G(n)$ between x_1 and x_2. Obviously, $|V(P_i) \cap (V_1 - \{x_1, x_2\})| \leq |V(P_i) \cap (V_2 \cup V_3)| - 1$ for $1 \leq i \leq n + 2$. Thus $\sum_{i=1}^{n+2} |V(P_i) \cap (V_1 - \{x_1, y_1\})| = (\sum_{i=1}^{n+2} |(V_2 \cup V_3) \cap V(P_i)|) - (n + 2)$. Therefore, $|V_1 - \{x_1, y_1\}| \leq |V_2 \cup V_3| - (n + 2)$. However, $|V_1 - \{x_1, y_1\}| = 2n - 2$ but $|V_2 \cup V_3| - (n + 2) = 2n - 3$. This leads to a contradiction.

By Theorem 14.30, $\kappa^*_{fan}(G(n)) \geq 2\delta(G(n)) - n(G(n)) + 1 = n$. To show $\kappa^*_{fan}(G(n)) = n$, we claim that there is no spanning (x_1, U)-fan of $G(n)$, where $U = \{y_1, y_2, \ldots, y_{n+1}\}$. Suppose not. Let $\{P_1, P_2, \ldots, P_{n+1}\}$ be a spanning (x_1, U)-fan of $G(n)$. Without loss of generality, we assume that P_i is a path joining x_1 to y_i for $1 \leq i \leq n + 1$. Obviously, $|(V_1 - \{x_1\}) \cap V(P_i)| \leq |((V_2 \cup V_3) - \{y_i\}) \cap V(P_i)|$. Thus, $\sum_{i=1}^{n+1} |(V_1 - \{x_1\}) \cap V(P_i)| \leq \sum_{i=1}^{n+1} |((V_2 \cup V_3) - \{y_i\}) \cap V(P_i)|$. Therefore, $|V_1 - \{x_1\}| \leq |(V_2 \cup V_3) - U|$. However, $|V_1 - \{x_1\}| = 2n - 1$ but $|(V_2 \cup V_3) - U| = 2n - 2$. We get a contradiction.

By Theorem 14.33, $\kappa^*_{pipe}(G(n)) \geq 2\delta(G(n)) - n(G(n)) = n - 1$. To prove that $\kappa^*_{pipe}(G(n)) = n - 1$, we claim that there is no spanning (U, W)-pipeline, where $U = \{x_1, x_2, \ldots, x_n\}$ and $W = \{x_{n+1}, x_{n+2}, \ldots, x_{2n}\}$. Suppose that there exists a spanning (U, W)-pipeline $\{P_1, P_2, \ldots, P_n\}$. Obviously, $|V_2 \cap V(P_i)| - 1 \leq |V_3 \cap V(P_i)|$ for $1 \leq i \leq n$. Then $\sum_{i=1}^{n} (|V_2 \cap V(P_i)| - 1) \leq \sum_{i=1}^{n} |V_3 \cap V(P_i)|$. Therefore, $(\sum_{i=1}^{n} |V_2 \cap V(P_i)|) - n \leq \sum_{i=1}^{n} |V_3 \cap V(P_i)|$. However, $(\sum_{i=1}^{n} |V_2 \cap V(P_i)|) - n = |V_2| - n = n$ but $\sum_{i=1}^{n} |V_3 \cap V(P_i)| = |V_3| = n - 1$. We get a contradiction.

15 Cubic 3*-Connected Graphs and Cubic 3*-Laceable Graphs

15.1 PROPERTIES OF CUBIC 3*-CONNECTED GRAPHS

From our observation, super k-spanning connected graphs are closely related to $(k-2)$-fault-tolerant Hamiltonian graphs. There are numerous interesting problems we can investigate regarding super k-spanning connected graphs. For example, we wonder whether every super k-spanning connected graph is $(k-2)$-fault-tolerant Hamiltonian, because all examples we have indicate that the statement is true. Yet we would like to point out that we have proved in Theorem 14.8 that the Harary graph $H_{2r,n}$ is proved to be super spanning connected. However, we have difficulty in proving that every $H_{2r,n}$ is super fault-tolerant Hamiltonian. Similarly, we have proved in Theorem 14.16 that the folded hypercube FQ_n is super spanning connected if n is odd. Again, we have difficulty in proving that FQ_n is super fault-tolerant Hamiltonian. Another interesting question is the existence of any graph that is k^*-connected but not $(k-t)^*$-connected for some $1 \leq t \leq k-1$.

To investigate these questions, we begin with $k = 3$. With the following theorem, we can easily combine two 3^*-connected graphs to get another 3^*-connected graph.

THEOREM 15.1 [7] Assume that both G_1 and G_2 are cubic graphs, x is a vertex in G_1, and y is a vertex in G_2. Then $J(G_1, N(x); G_2, N(y))$ is 3^*-connected if and only if both G_1 and G_2 are 3^*-connected.

Proof: Assume that both G_1 and G_2 are 3^*-connected. We claim that $J(G_1, N(x); G_2, N(y))$ is 3^*-connected. Let u and v be any two different vertices of $J(G_1, N(x); G_2, N(y))$. Without loss of generality, we have the following two cases:

Case 1. Both u and v are in G_1. Since G_1 is 3^*-connected, there are three disjoint paths P_1, P_2, and P_3 of G_1 joining u to v such that $P_1 \cup P_2 \cup P_3$ spans G_1. Obviously, x is in exactly one of the paths of P_1, P_2, and P_3. Without loss of generality, we may assume that x is in P_3. Thus, P_3 can be written as $\langle u, Q_1, x_1, x, x_2, Q_2, v\rangle$ where $\{x_1, x_2, x_3\}$ are neighborhoods of x and $\{y_1, y_2, y_3\}$ are neighborhoods of y with x_t joining with y_t in $J(G_1, N(x); G_2, N(y))$ for $t \in \{1, 2, 3\}$. By Theorem 15.2, there exists a Hamiltonian cycle of $G_2 - (y, y_3)$. Obviously, R can be written as $\langle y, y_1, Q_3, y_2, y\rangle$. We set $P_3' = \langle u, Q_1, x_1, y_1, Q_3, y_2, x_2, Q_2, v\rangle$. Obviously, P_1, P_2, and P_3' form three disjoint paths joining u and v such that $P_1 \cup P_2 \cup P_3$ spans $J(G_1, N(x); G_2, N(y))$.

Case 2. u is in G_1 and v is in G_2. Since G_1 is 3*-connected, there exist three disjoint paths P_1, P_2, and P_3 of G_1 joining u to x such that $P_1 \cup P_2 \cup P_3$ spans G_1. Similarly, there exist three disjoint paths Q_1, Q_2, and Q_3 of G_2 joining y to v such that $Q_1 \cup Q_2 \cup Q_3$ spans G_2. Without loss of generality, we can write $P_i = \langle u, R_i, x_i, x \rangle$ and $Q_i = \langle y, y_i, S_i, v \rangle$ where $\{x_1, x_2, x_3\}$ are neighborhoods of x and $\{y_1, y_2, y_3\}$ are neighborhoods of y with x_t joining with y_t in $J(G_1, N(x); G_2, N(y))$ for $t \in \{1, 2, 3\}$. We set T_i as $\langle u, R_i, x_i, y_i, S_i, v \rangle$ for $i = 1, 2$, and 3. Obviously, T_1, T_2, and T_3 form three disjoint paths joining u and v such that $P_1 \cup P_2 \cup P_3$ spans $J(G_1, N(x); G_2, N(y))$.

Thus, $J(G_1, N(x); G_2, N(y))$ is 3*-connected.

On the other hand, suppose that $J(G_1, N(x); G_2, N(y))$ is 3*-connected. We need to prove that both G_1 and G_2 are 3*-connected. By symmetry, it is sufficient to prove that G_1 is 3*-connected. Let u and v be any two different vertices of G_1. Without loss of generality, we have the following two cases:

Case 1. $x \notin \{u, v\}$. Thus, u and v are vertices of $J(G_1, N(x); G_2, N(y))$. Note that $J(G_1, N(x); G_2, N(y))$ is 3*-connected. There are three disjoint paths P_1, P_2, and P_3 of $J(G_1, N(x); G_2, N(y))$ joining u to v such that $P_1 \cup P_2 \cup P_3$ spans $J(G_1, N(x); G_2, N(y))$. Let $N(x) = \{x_1, x_2, x_3\}$ and $N(y) = \{y_1, y_2, y_3\}$. Obviously, $\{(x_i, y_i) \mid i \in \{1, 2, 3\}\}$ are all the edges joining vertices of G_1 to vertices of G_2. Thus, exactly one of P_1, P_2, and P_3, say P_1, is of the form $\langle u, Q_1, x_i, y_i, R, y_j, Q_2, v \rangle$. Moreover, P_2 and P_3 are paths in G_1. We set P_1' as $\langle u, Q_1, x_i, x, x_j, Q_2, v \rangle$. Obviously, P_1', P_2, and P_3 form three disjoint paths joining u and v such that $P_1' \cup P_2 \cup P_3$ spans G_1.

Case 2. $u \in V(G_1) - \{x\}$ and $v = x$. Let w be any vertex in $V(G_2) - \{y\}$. Thus, u and v are vertices of $J(G_1, N(x); G_2, N(y))$. Since $J(G_1, N(x); G_2, N(y))$ is 3*-connected, there are three disjoint paths P_1, P_2, and P_3 of $J(G_1, N(x); G_2, N(y))$ joining u to v such that $P_1 \cup P_2 \cup P_3$ spans $J(G_1, N(x); G_2, N(y))$. Obviously, $\{(x_i, y_i) \mid i \in \{1, 2, 3\}\}$ are all the edges joining vertices of G_1 to vertices of G_2. Without loss of generality, we can assume that $P_i = \langle u, Q_i, x_i, y_i, R_i, v \rangle$ for $i \in \{1, 2, 3\}$. We set P_i' as $\langle u, Q_i, x_i, x \rangle$. Obviously, P_1', P_2', and P_3' form three disjoint paths joining u and v such that $P_1' \cup P_2' \cup P_3'$ spans G_1.

Thus, G_1 is 3*-connected. The theorem is proved. □

With the following theorem, every cubic 3*-connected graph is 1-fault-tolerant Hamiltonian.

THEOREM 15.2 Every cubic 3*-connected graph is 1-fault-tolerant Hamiltonian.

Proof: Suppose that G is a cubic 3*-connected graph. Let F be a subset of $V \cup E$ with $|F| = 1$. Suppose that $e = (x, y)$ is an edge in F. Since G is cubic 3*-connected, there are three disjoint paths P_1, P_2, and P_3 of G_1 joining x to y such that $P_1 \cup P_2 \cup P_3$ spans G. Note that G is cubic. One of P_1, P_2, and P_3, say P_3, is $\langle x, y \rangle$. Obviously, $P_1 \cup P_2$ forms a Hamiltonian cycle of $G - e$. Suppose that v is a vertex in F. Let x and y be two distinct neighbors of v. Since G is cubic 3*-connected, there are three disjoint paths P_1, P_2, and P_3 of G_2 joining x to y such that $P_1 \cup P_2 \cup P_3$ spans G.

Note that G is cubic. One of P_1, P_2, and P_3, say P_3, is $\langle x, v, y\rangle$. Obviously, $P_1 \cup P_2$ forms a Hamiltonian cycle of $G - v$. Thus, every cubic 3*-connected graph is 1-fault-tolerant Hamiltonian. □

15.2 EXAMPLES OF CUBIC SUPER 3*-CONNECTED GRAPHS

The following lemma is needed for further discussion.

LEMMA 15.1 Assume that G is a cubic 1*-connected graph. Let u and v be two distinct vertices in G with $d(u, v) = 2$. Then either $G - u$ or $G - v$ is Hamiltonian.

Proof: Let w be the common neighbor of u and v. Since G is 1*-connected, there exists a Hamiltonian path P between u and v. Then P can be written as either $\langle u, w, P_1, v\rangle$ or $\langle u, P_2, w, v\rangle$. If $P = \langle u, w, P_1, v\rangle$, then $\langle w, P_1, v, w\rangle$ forms a Hamiltonian cycle of $G - u$. If $P = \langle u, P_2, w, v\rangle$, then $\langle u, P_2, w, u\rangle$ forms a Hamiltonian cycle of $G - v$. □

Throughout this section, we use $\oplus$ and $\ominus$ to denote addition and subtraction in integer modular n, Z_n.

Assume that n is a positive even integer with $n \geq 4$. The graph $H(n)$ is the graph with vertex set $\{0, 1, \ldots, n-1\}$ and edge set $\{(i, n-i) | 1 \leq i < \frac{n}{2}\} \cup \{(i, i \oplus 1) \mid 0 \leq i \leq n-1\} \cup \{(0, \frac{n}{2})\}$. Examples of $H(4)$ and $H(8)$ are shown in Figure 11.3 of Chapter 11. We can easily prove the following lemma by brute force.

LEMMA 15.2 Every $H(n)$ is Hamiltonian-connected for every even integer with $n \geq 4$.

THEOREM 15.3 Every $H(n)$ is super 3*-connected for every even integer with $n \geq 4$.

Proof: In Section 11.2, we saw that $H(n)$ can be obtained form K_4 by a sequence of node expansion. By Theorem 15.1, every $H(n)$ is 3*-connected. Hence, it is 2*-connected. Combined with Lemma 15.2, $H(n)$ is super 3*-connected. □

Assume that n and k are two positive integers with $n = 2k$ and $k \geq 2$. The *projective cycle graph* $PJ(k)$ is the graph with vertex set $\{0, 1, \ldots, 2k-1\}$ and edge set $\{(i, k+i) \mid 0 \leq i < k\} \cup \{(i, i \oplus 1) | 0 \leq i < 2k\}$. The projective cycle graphs $PJ(3)$ and $PJ(4)$ are shown in Figure 15.1. The following theorem is proved by Kao et al. [194].

THEOREM 15.4 $PJ(k)$ is super 3*-connected if and only if k is even.

Kao et al. [194] also prove that the generalized Petersen graph $P(n, 1)$ is super 3*-connected if and only if n is odd. Now, we try to classify the super 3*-connected generalized Petersen graph $P(n, 2)$.

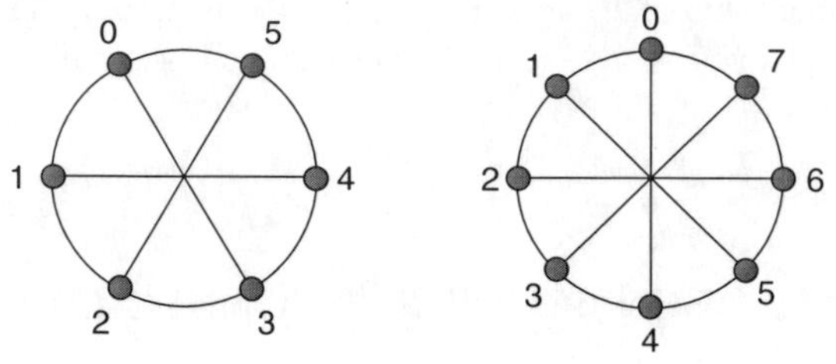

FIGURE 15.1 The projective cycle graphs (a) $PJ(3)$ and (b) $PJ(4)$.

Kao et al. [194] also prove that the generalized Petersen graph $P(n,2)$ is 1*-connected if and only if $n=1,3$ (**mod** 6) and $n>6$. We present this fact as depending on the value of n (**mod** 6). However, we first observe the symmetric property of $P(n,2)$. For any $i \in Z_n$, let σ_i be the function from $V(P(n,2))$ to $V(P(n,2))$ defined by $\sigma_i(x)=x+i$, and $\sigma_i(x')=(x+i)'$. It is easy to see that σ_i is an automorphism of $P(n,2)$ for any $i \in Z_n$. Assume that n is an odd integer. Let ϕ be the function from $V(P(n,2))$ to $V(P(n,2))$ defined by $\phi(0)=0$, $\phi(0')=0'$, $\phi(i)=n-i+1$, and $\phi(i')=(n-i+1)'$ if $i \neq 0$. It is easy to see that ϕ is also an automorphism of $P(n,2)$. With functions σ_0 and ϕ, we have the following observation: for any $i \in Z_n-\{0\}$, there exists an automorphism ψ such that $\psi(0)=0$, $\psi(i)=k$, and $\psi(i')=k'$ for some even $k \in Z_n$.

THEOREM 15.5 $P(n,2)$ is 1*-connected if and only if $n=1,3$ (**mod** 6) and $n>6$.

Proof: Suppose that n is even. It is easy to see that $P(n,2)$ is planar. We will prove that $P(n,2)$ is not Hamiltonian-connected using the Grinberg condition.

Case 1. $n=0$ (**mod** 6). Then $n=6k$ with $k>1$. Suppose that $P(n,2)$ is 1*-connected. Note that $d_{P(n,2)}(0,n-2)=2$. By Lemma 15.1, either $P(n,2)-\{0\}$ or $P(n,2)-\{n-2\}$ is Hamiltonian. With the symmetric property of $P(n,2)$, $P(n,2)-\{0\}$ is Hamiltonian. Let C be a Hamiltonian cycle of $P(n,2)-\{0\}$. Obviously, $P(n,2)-\{0\}$ and C satisfy the Grinberg condition. Thus, $3(f_5-f_5')+7(f_9-f_9')+(3k-2)(f_{3k}-f_{3k}')=0$ where f_i is the number of the faces of length i inside C and f_i' is the number of the faces of length i outside C for $i=5,9,3k$. Obviously, $(f_9-f_9')+(f_{3k}-f_{3k}')=0$ (**mod** 3). We suppose that $f_9=1$ and $f_9'=0$. The equation holds only when $f_{3k}-f_{3k}'=2$. This is impossible, since the face of length 9 and the face of length $3k$ are separated by the path $\langle 2',0',(n-2)'\rangle$. Hence, $P(n,2)-\{0\}$ is not Hamiltonian. We get a contradiction. See Figure 15.2a for illustration.

Case 2. $n=2$ (**mod** 6). Then $n=6k+2$ with $k\geq 1$. Suppose that $P(n,2)$ is Hamiltonian-connected. Note that $d_{P(n,2)}(0',(n-4)')=2$. By Lemma 15.1, either $P(n,2)-0'$ or $P(n,2)-(n-4)'$ is Hamiltonian. With the symmetric property of $P(n,2)$, $P(n,2)-0'$ is Hamiltonian. Let C be a Hamiltonian cycle of $P(n,2)-0'$. Obviously, $P(n,2)-0'$ and C satisfy the Grinberg condition. Thus, $3(f_5-f_5')+(3k-1)(f_{3k+1}-f_{3k+1}')+3(k+1)(f_{3k+5}-f_{3k+5}')=0$ where f_i is the number of the faces of length i inside C and f_i' is the number of the faces of length i outside C for

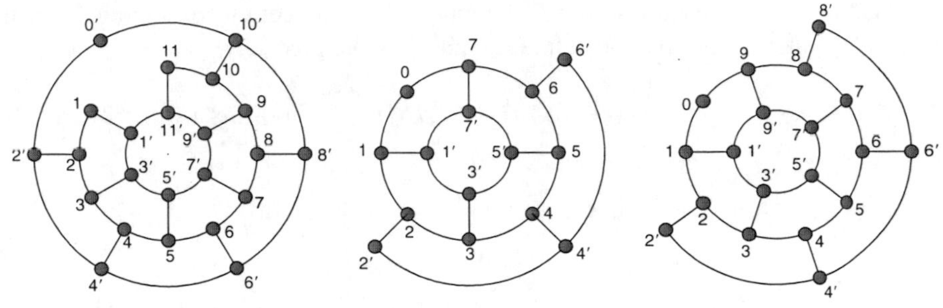

(a) (b) (c)

FIGURE 15.2 The graphs (a) $P(12,2)-0$, (b) $P(8,2)-0'$, and (c) $P(10,2)-0'$.

$i=5, 3k+1, 3k+5$. Obviously, $f_{3k+1}-f'_{3k+1}=0\ (\mathbf{mod}\ 3)$. Since $f_{3k+1}-f'_{3k+1}=\pm 1$, the equation cannot hold. Hence, $P(n,2)-0'$ is not Hamiltonian. We get a contradiction. See Figure 15.2b for illustration.

Case 3. $n=4\ (\mathbf{mod}\ 6)$. Then $n=6k+4$ with $k\geq 1$. Suppose that $P(n,2)$ is Hamiltonian-connected. Note that $d_{P(n,2)}(0',(n-4)')=2$. By Lemma 15.1, either $P(n,2)-0'$ or $P(n,2)-(n-4)'$ is Hamiltonian. With the symmetric property of $P(n,2)$, $P(n,2)-0'$ is Hamiltonian. Let C be a Hamiltonian cycle of $P(n,2)-0'$. Obviously, $P(n,2)-0'$ and C satisfy the Grinberg condition. Thus, $3(f_5-f'_5)+3k(f_{3k+2}-f'_{3k+2})+(3k+4)(f_{3k+6}-f'_{3k+6})=0$ where f_i is the number of the faces of length i inside C and f'_i is the number of the faces of length i outside C for $i=5, 3k+2, 3k+6$. Obviously, $f_{3k+6}-f'_{3k+6}=0\ (\mathbf{mod}\ 3)$. Since $f_{3k+6}-f'_{3k+6}=\pm 1$, the equation cannot hold. Hence, $P(n,2)-0'$ is not Hamiltonian. We get a contradiction. See Figure 15.2c for illustration.

Now, we consider that n is odd.

Case 4. $n=5\ (\mathbf{mod}\ 6)$. It is proved by Bondy [29] that $P(n,2)$ is Hamiltonian if and only if $n\neq 5\ (\mathbf{mod}\ 6)$. Thus, $P(n,2)$ is not 1*-connected if $n=5\ (\mathbf{mod}\ 6)$. Finally, we consider $n=1$ and $3\ (\mathbf{mod}\ 6)$. To prove that $P(n,2)$ is 1*-connected, we need to find a Hamiltonian path of $P(n,2)$ between any two different vertices a and b. With the symmetric property of $P(n,2)$, we have the following three cases: (1) $a=0$ and $b=i$ for all odd $i\in Z_n$, (2) $a=0'$ and $b=i$ for all odd $i\in Z_n$, and (3) $a=0'$ and $b=i'$ for all odd $i\in Z_n$. To describe the required Hamiltonian paths, we define five basic path patterns. Each path pattern is associated with its beginning vertex and end vertex. The path patterns are as follows:

$$
\begin{aligned}
P(i,i\oplus t) &= \langle i, i\oplus 1, i\oplus 2, \ldots, i\oplus (t-1), i\oplus t\rangle \\
Q(i',(i\oplus 2t)') &= \langle i', (i\oplus 2)', (i\oplus 4)', \ldots, (i\oplus 2(t-1))', (i\oplus 2t)'\rangle \\
D(i',(i\oplus 6)') &= \langle i', i, i\oplus 1, i\oplus 2, (i\oplus 2)', (i\oplus 4)', (i\oplus 6)'\rangle \\
E(i,i\oplus 6) &= \langle i, i\oplus 1, (i\oplus 1)', (i\oplus 3)', (i\oplus 5)', i\oplus 5, i\oplus 6\rangle \\
F(i',(i\oplus 6)') &= \langle i', (i\oplus 2)', i\oplus 2, i\oplus 3, i\oplus 4, (i\oplus 4)', (i\oplus 6)'\rangle
\end{aligned}
$$

Then we define the path pattern D^t for any positive integer t by executing the path pattern D t times. Similarly, we define E^t and F^t. More precisely,

$$\begin{aligned}
D^t(i', (i \oplus 6t)') &= \langle i', D(i', (i \oplus 6)'), (i \oplus 6)', D((i \oplus 6)', (i \oplus 12)'), (i \oplus 12)', \ldots, \\
&\quad (i \oplus 6(t-1))', D((i \oplus 6(t-1))', (i \oplus 6t)'), (i \oplus 6t)' \rangle \\
E^t(i, i \oplus 6t) &= \langle i, E(i, i \oplus 6), i \oplus 6, E(i \oplus 6, i \oplus 12), i \oplus 12, \ldots, \\
&\quad i \oplus 6(t-1), E(i \oplus 6(t-1), i \oplus 6t), i \oplus 6t \rangle \\
F^t(i', (i \oplus 6t)') &= \langle i', F(i', (i \oplus 6)'), (i \oplus 6)', F((i \oplus 6)', (i \oplus 12)'), (i \oplus 12)', \ldots, \\
&\quad (i \oplus 6(t-1))', F((i \oplus 6(t-1))', (i \oplus 6t)'), (i \oplus 6t)' \rangle
\end{aligned}$$

We also define P^{-1}, Q^{-1}, D^{-1}, E^{-1}, and F^{-1} by executing the corresponding path pattern backward. More precisely,

$$\begin{aligned}
P^{-1}(i, i \ominus t) &= \langle i, i \ominus 1, i \ominus 2, \ldots, i \ominus (t-1), i \ominus t \rangle \\
Q^{-1}(i', (i \ominus 2t)') &= \langle i', (i \ominus 2)', (i \ominus 4)', \ldots, (i \ominus 2(t \ominus 1))', (i \ominus 2t)' \rangle \\
D^{-1}(i', (i \ominus 6)') &= \langle i', i, i \ominus 1, i \ominus 2, (i \ominus 2)', (i \ominus 4)', (i \ominus 6)' \rangle \\
E^{-1}(i, i \ominus 6) &= \langle i, i \ominus 1, (i \ominus 1)', (i \ominus 3)', (i \ominus 5)', i \ominus 5, i \ominus 6 \rangle \\
F^{-1}(i', (i \ominus 6)') &= \langle i', (i \ominus 2)', i \ominus 2, i \ominus 3, i \ominus 4, (i \ominus 4)', (i \ominus 6)' \rangle
\end{aligned}$$

Similarly, we can define the path patterns $D^{-t}(i, i \ominus 6t)$, $E^{-t}(i, i \ominus 6t)$, and $F^{-t}(i', (i \ominus 6t)')$. See Figure 15.3 for illustrations of these path patterns.

Case 5. $n = 1$ (**mod** 6).

The following three sub cases consider the situation $a = 0$ and $b = i$ with i being odd.

Case 5.1. $i = 1$ (**mod** 6). Obviously, $\langle 0, 0', Q(0', (i \ominus 1)'), (i \ominus 1)', F^{\frac{n-i}{6}}((i \ominus 1)', (n \ominus 1)'), (n \ominus 1)', n \ominus 1, n \ominus 2, (n \ominus 2)', (n \ominus 4)', F^{-\frac{n-i-6}{6}}((n \ominus 4)', i \oplus 2)'), (i \oplus 2)', Q^{-1}((i \oplus 2)', 1'), 1', 1, P(1, i), i \rangle$ forms a Hamiltonian path of $P(n, 2)$ between a and b.

Case 5.2. $i = 3$ (**mod** 6). Obviously, $\langle 0, 0', Q^{-1}(0', (i \ominus 2)'), (i \ominus 2)', F^{-\frac{i-3}{6}}((i \ominus 2)', 1'), 1', 1, 2, 2', 4', F^{\frac{i-3}{6}}(4', (i \oplus 1)'), (i \oplus 1)', Q((i \oplus 1)', (n \ominus 1)'), (n \ominus 1)', n \ominus 1, P^{-1}(n \ominus 1, i), i \rangle$ forms a Hamiltonian path of $P(n, 2)$ between a and b.

Case 5.3. $i = 5$ (**mod** 6). Obviously, $\langle 0, 1, 1', 3', F^{\frac{i-5}{6}}(3', (i \ominus 2)'), (i \ominus 2)', Q((i \ominus 2)', 0'), 0', F^{\frac{i+1}{6}}(0', (i \oplus 1)'), (i \oplus 1)', Q((i \oplus 1)', (n \ominus 1)'), (n \ominus 1)', n \ominus 1, P^{-1}(n \ominus 1, i), i \rangle$ forms a Hamiltonian path of $P(n, 2)$ between a and b.

The following three sub cases consider the situation $a = 0'$ and $b = i$ with i being odd.

Case 5.4. $i = 1$ (**mod** 6). Obviously, $\langle 0', Q(0', (i \ominus 1)'), (i \ominus 1)', F^{\frac{n-i}{6}}((i \ominus 1)', (n \ominus 1)'), (n \ominus 1)', Q((n \ominus 1)', (i \oplus 2)'), (i \oplus 2)', F^{\frac{n-i-6}{6}}((i \oplus 2)', (n \ominus 4)'), (n \ominus 4)', (n \ominus 2)', n \ominus 2, P(n \ominus 2, i), i \rangle$ forms a Hamiltonian path of $P(n, 2)$ between a and b.

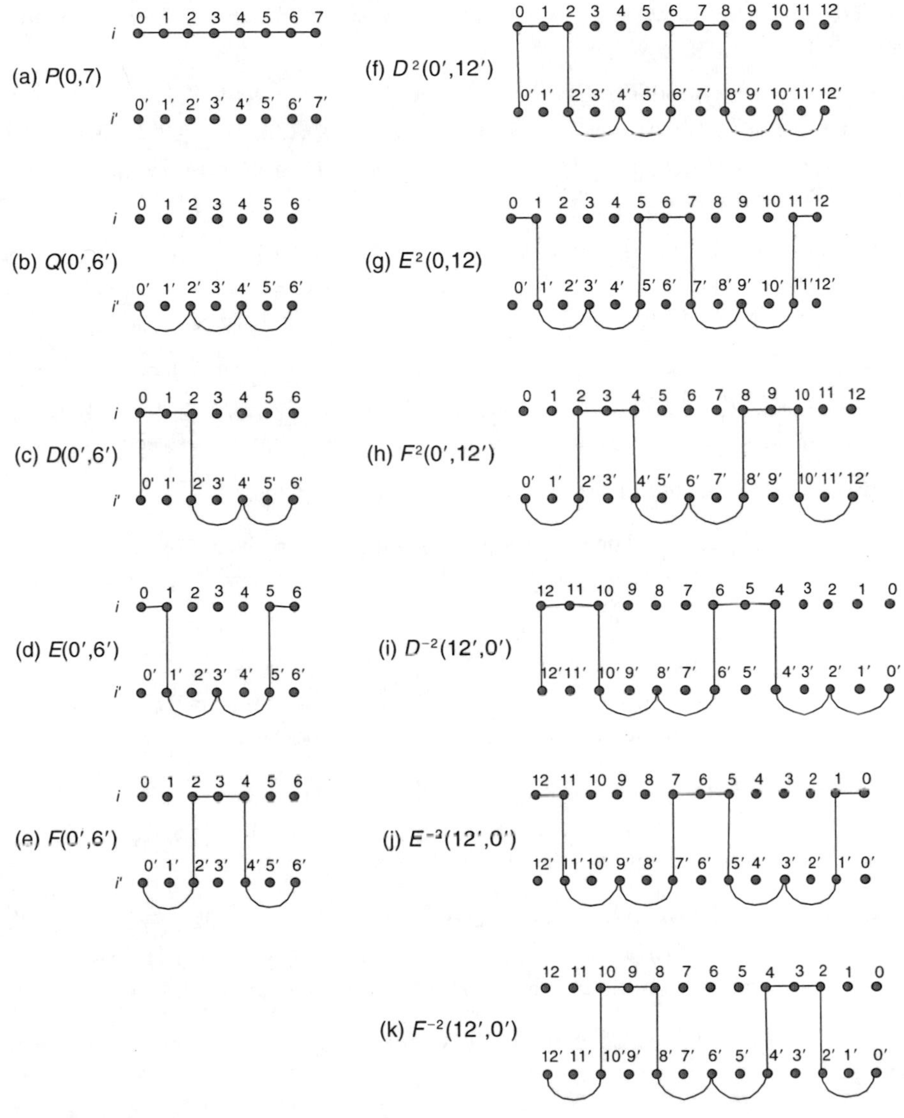

FIGURE 15.3 Illustrations for path patterns.

Case 5.5. $i = 3$ (**mod** 6). Obviously, $\langle 0', Q(0', (i \ominus 1)'), (i \ominus 1)', F^{\frac{n-i-4}{6}}((i \ominus 1)', (n \ominus 5)'), (n \ominus 5)', (n \ominus 3)', n \ominus 3, n \ominus 2, (n \ominus 2)', F^{-\frac{n-i-4}{6}}((n \ominus 2)', (i \oplus 2)'), (i \oplus 2)', Q^{-1}((i \oplus 2)', (n \ominus 1)'), (n \ominus 1)', n \ominus 1, P(n \ominus 1, i), i \rangle$ forms a Hamiltonian path of $P(n, 2)$ between a and b.

Case 5.6. $i = 5$ (**mod** 6). Obviously, $\langle 0', F^{-\frac{n-i-2}{6}}(0', (i \oplus 2)'), (i \oplus 2)', Q^{-1}((i \oplus 2)', 3'), 3', F^{-\frac{4 \ominus i}{6}}(3', (i \ominus 1)'), (i \ominus 1)', Q^{-1}((i \ominus 1)', 2'), 2', 2, P(2, i), i \rangle$ forms a Hamiltonian path of $P(n, 2)$ between a and b.

The following three sub cases consider the situation $a = 0'$ and $b = i'$ with i being odd.

Case 5.7. $i = 1$ (**mod** 6). Obviously, $\langle 0', Q(0', (i \oplus 1)'), (i \oplus 1)', i \oplus 1, P^{-1}(i \oplus 1, n \ominus 2), n \ominus 2, (n \ominus 2)', (n \ominus 4)', F^{-\frac{n-i-6}{6}}((n \ominus 4)', (i \oplus 2)'), (i \oplus 2)', i \oplus 2, i \oplus 3, (i \oplus 3)', (i \oplus 5)', F^{\frac{n-i-6}{6}}((i \oplus 5)', (n \ominus 1)'), (n \ominus 1)', Q((n \ominus 1)', i'), i' \rangle$ forms a Hamiltonian path of $P(n, 2)$ between a and b.

Case 5.8. $i = 3$ (**mod** 6). Obviously, $\langle 0', Q(0', (i \ominus 3)'), F^{\frac{n-i+2}{6}}((i \ominus 3)', (n \ominus 1)'), (n \ominus 1)', Q((n \ominus 1)', (i \ominus 2)'), (i \ominus 2)', i \ominus 2, P^{-1}(i \ominus 2, n \ominus 2), n \ominus 2, (n \ominus 2)', (n \ominus 4)', F^{-\frac{n-i-4}{6}}((n \ominus 4)', i'), i' \rangle$ forms a Hamiltonian path of $P(n, 2)$ between a and b.

Case 5.9. $i = 5$ (**mod** 6). Obviously, $\langle 0', 0, n \ominus 1, (n \ominus 1)', Q^{-1}((n \ominus 1)', (i \ominus 1)'), F^{-\frac{i-5}{6}}((i \ominus 1)', 4'), 4', 2', 2, 1, 1', 3', D^{\frac{i-5}{6}}(3', (i \ominus 2)'), i \ominus 2, P(i \ominus 2, n \ominus 2), n \ominus 2, (n \ominus 2)', Q^{-1}((n \ominus 2)', i'), i' \rangle$ forms a Hamiltonian path of $P(n, 2)$ between a and b.

See Figure 15.4 for illustrations of the Hamiltonian paths.

Case 6. $n = 3$ (**mod** 6). The corresponding Hamiltonian paths are listed here.

The following three sub cases consider the situation $a = 0$ and $b = i$ with i being odd.

Case 6.1. $i = 1$ (**mod** 6). Obviously, $\langle 0, n \ominus 1, (n \ominus 1)', (n \ominus 3)', F^{-\frac{n-i-2}{6}}((n \ominus 3)', (i \ominus 1)'), (i \ominus 1)', Q^{-1}((i \ominus 1)', 0'), 0', F^{-\frac{n-i-2}{6}}(0', (i \oplus 2)'), (i \oplus 2)', Q^{-1}((i \oplus 2)', 1'), 1', 1, P(1, i), i \rangle$ forms a Hamiltonian path of $P(n, 2)$ between a and b.

Case 6.2. $i = 3$ (**mod** 6). Obviously, $\langle 0, 0', Q^{-1}(0', (i \ominus 2)'), (i \ominus 2)', F^{-\frac{i-3}{6}}((i \ominus 2)', 1'), 1', 1, 2, 2', 4', F^{\frac{i-3}{6}}(4', (i \oplus 1)'), (i \oplus 1)', Q((i \oplus 1)', (n \ominus 1)'), (n \ominus 1)', n \ominus 1, P^{-1}(n \ominus 1, i), i \rangle$ forms a Hamiltonian path of $P(n, 2)$ between a and b.

Case 6.3. $i = 5$ (**mod** 6). Obviously, $\langle 0, 1, 1', 3', F^{\frac{i-5}{6}}(3', (i \ominus 2)'), (i \ominus 2)', Q((i \ominus 2)', 0'), 0', F^{\frac{i+1}{6}}(0', (i \oplus 1)'), (i \oplus 1)', Q((i \oplus 1)', (n \ominus 1)'), (n \ominus 1)', n \ominus 1, P^{-1}(n \ominus 1, i), i \rangle$ forms a Hamiltonian path of $P(n, 2)$ between a and b.

The following three sub cases consider the situation $a = 0'$ and $b = i$ with i being odd.

Case 6.4. $i = 1$ (**mod** 6). Obviously, $\langle 0', Q^{-1}(0', i'), i', F^{-\frac{i-1}{6}}(i', 1'), 1', Q^{-1}(1', (i \oplus 1)'), (i \oplus 1)', i \oplus 1, P(i \oplus 1, 1), 1, E^{\frac{i-1}{6}}(1, i), i \rangle$ forms a Hamiltonian path of $P(n, 2)$ between a and b.

Case 6.5. $i = 3$ (**mod** 6) Obviously, $\langle 0', Q(0', (i \ominus 1)'), (i \ominus 1)', F^{\frac{n-i}{6}}((i \ominus 1)', (n \ominus 1)'), (n \ominus 1)', Q((n \ominus 1)', (i \oplus 2)'), (i \oplus 2)', F^{\frac{n-i-6}{6}}((i \oplus 2)', (n \ominus 4)'), (n \ominus 4)', (n \ominus 2)', n \ominus 2, P(n \ominus 2, i), i \rangle$ forms a Hamiltonian path of $P(n, 2)$ between a and b.

Case 6.6. $i = 5$ (**mod** 6). Obviously, $\langle 0', Q(0', (i \ominus 1)'), (i \ominus 1)', i \ominus 1, P^{-1}(i \ominus 1, n \ominus 2), n \ominus 2, (n \ominus 2)', (n \ominus 4)', F^{-\frac{n-i-4}{6}}((n \ominus 4)', i'), i', Q^{-1}(i', (n \ominus 1)'), (n \ominus 1)', F^{-\frac{n-i-4}{6}}((n \ominus 1)', (i \oplus 3)'), (i \oplus 3)', (i \oplus 1)', i \oplus 1, i \rangle$ forms a hamiltonian path of $P(n, 2)$ between a and b.

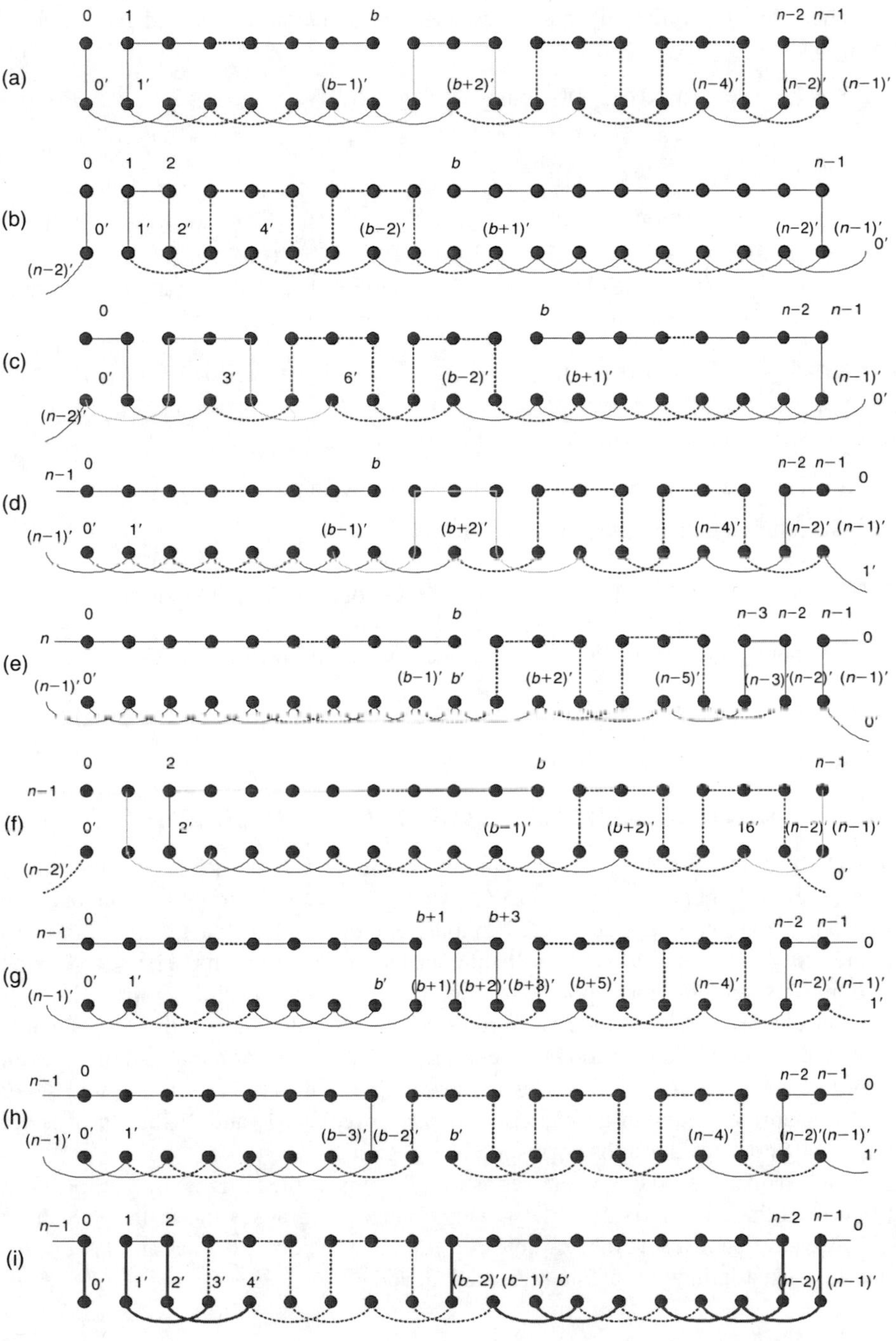

FIGURE 15.4 Illustrations of the Hamiltonian paths of $P(n, 2)$, where $n = 1$ (**mod** 6) in Case 5 of Theorem 15.5.

The following three sub cases consider the situation $a=0'$ and $b=i'$ with i being odd.

Case 6.7. $i=1$ (**mod** 6). Obviously, $\langle 0', Q^{-1}(0', (i\oplus 2)'), (i\oplus 2)', i\oplus 2, P(i\oplus 2, 2), 2, 2', 4', F^{\frac{i-1}{6}}(4', (i\oplus 3)'), (i\oplus 3)', Q((i\oplus 3)', 1'), 1', F^{\frac{i-1}{6}}(1', i'), i'\rangle$ forms a Hamiltonian path of $P(n,2)$ between a and b.

Case 6.8. $i=3$ (**mod** 6). Obviously, $\langle 0', Q(0', (i\oplus 1)'), (i\oplus 1)', F^{\frac{n-i-6}{6}}((i\oplus 1)', (n\ominus 5)'), (n\ominus 5)', (n\ominus 3)', n\ominus 3, n\ominus 2, (n\ominus 2)', F^{-\frac{n-i-6}{6}}((n\ominus 2), (i\oplus 4)'), (i\oplus 4)', i\oplus 2)', i\oplus 2, P^{-1}(i\oplus 2, n\ominus 1), n\ominus 1, (n\ominus 1)', Q((n\ominus 1)', i'), i'\rangle$ forms a Hamiltonian path of $P(n,2)$ between a and b.

Case 6.9. $i=5$ (**mod** 6). Obviously, $\langle 0', F^{-\frac{n-i-4}{6}}(0', (i\oplus 4)'), (i\oplus 4)', (i\oplus 2)', i\oplus 2, P^{-1}(i\oplus 2, 2), 2, 2', Q(2', (i\oplus 1)'), (i\oplus 1)', F^{\frac{n-i+2}{6}}((i\oplus 1)', 3'), 3', Q(3', i'), i'\rangle$ forms a Hamiltonian path of $P(n,2)$ between a and b.

See Figure 15.5 for illustrations of the Hamiltonian paths. □

The following theorem is proved in Ref. 7.

THEOREM 15.6 $P(n,2)$ is 3*-connected if and only if $n=1,3$ (**mod** 6) and $n>6$.

Combining Theorems 15.5 and 15.6, we have the following theorem.

THEOREM 15.7 $P(n,2)$ is super 3*-connected if and only if $n=1,3$ (**mod** 6) and $n>6$.

15.3 COUNTEREXAMPLES OF 3*-CONNECTED GRAPHS

Obviously, a 1*-connected 1-fault-tolerant Hamiltonian cubic graph is super fault-tolerant Hamiltonian. In Figure 15.6, we use a Venn diagram to illustrate the relations among cubic 1*-connected graphs, cubic 2*-connected graphs, cubic 3*-connected graphs, and cubic 1-fault-tolerant Hamiltonian graphs. The set of cubic 1*-connected graphs corresponds to regions 1, 3, and 6; the set of cubic 2*-connected graphs corresponds to regions 1, 2, 3, 4, 5, and 6; the set of cubic 3*-connected graphs corresponds to regions 1 and 2; the set of cubic 1-fault-tolerant Hamiltonian graphs corresponds to regions 1, 2, 3, and 4; the set of cubic super fault-tolerant Hamiltonian graphs corresponds to regions 1, and 3; the set of super 3-spanning connected graphs corresponds to region 1.

We consider the existence of graphs for all the possible regions. In Section 12.8, it is proved that there exist infinitely many graphs in regions 5 and 6. In the Section 15.2, we have shown several examples of graphs in region 1. Kao et al. [194] prove that there exist infinite graphs in regions 2, 3, and 4.

Example 15.1

(Cubic 1-fault-tolerant hamiltonian graphs that are 3*-connected but not 1*-connected.)

Let T be the graph in Figure 15.7. Obviously, T is obtained from $PJ(4)$ by a sequence

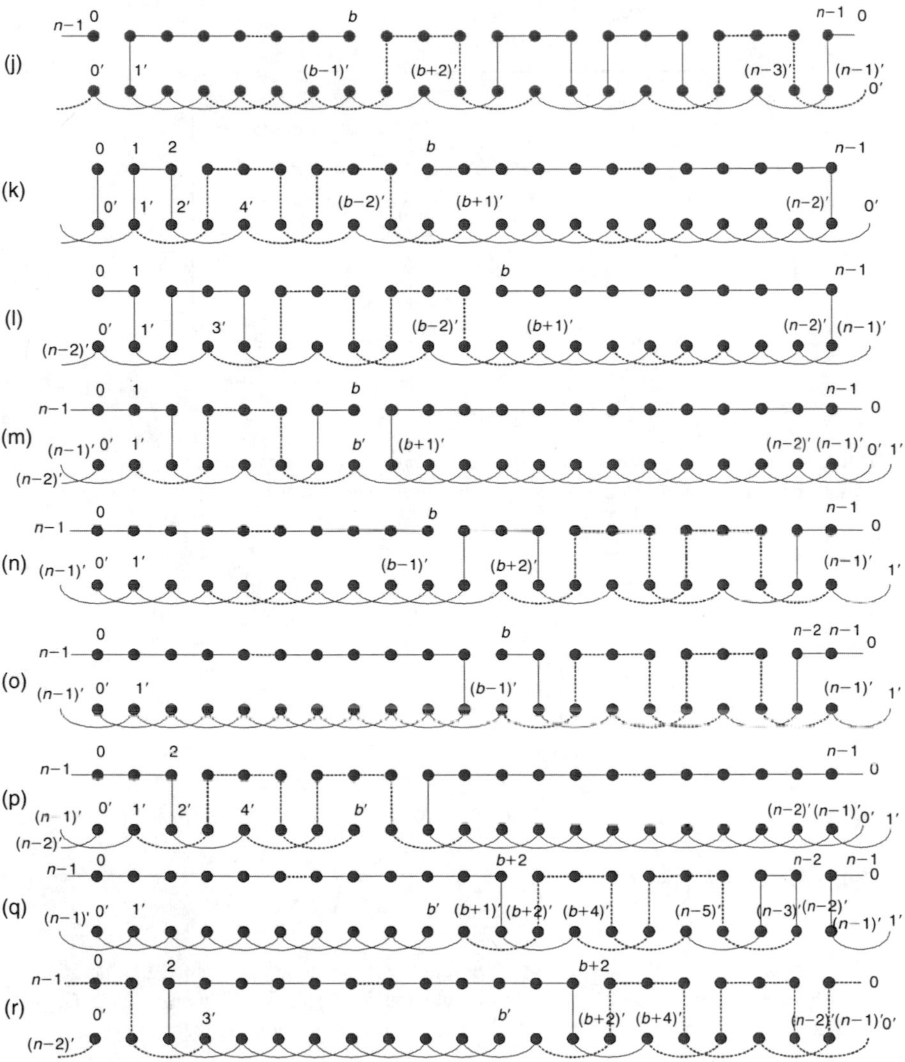

FIGURE 15.5 Illustrations of the Hamiltonian paths of $P(n, 2)$, where $n = 3(\mathbf{mod}\ 6)$ in Case 6 of Theorem 15.5.

of 3-vertex expansions of $PJ(4)$. By Theorem 15.4, $PJ(4)$ is 3*-connected. Since K_4 is isomorphic to $PJ(2)$, K_4 is 3*-connected. With Theorem 15.1, T is 3*-connected. With Lemma 11.27, T is not 1*-connected. Hence, T is a graph that is 3*-connected, but not super 3*-spanning connected.

Let w be the vertex of T shown in Figure 15.7. With Lemma 11.27 and Theorem 12.9, there exists no Hamiltonian path between vertices u and v in $J(T, N(w); K_4, N(y))$. Thus, $J(T, N(w); K_4, N(y))$ is not 1*-connected. Now, we recursively define a sequence of graphs as follows: let $G_1 = T$, $x_1 = w$, and $G_2 = J(G_1, N(x_1); K_4, N(y))$. Suppose that we have defined $G_1, G_2, \ldots, G_i$ with $i \geq 2$. Let x_i be any expanded vertex of G_{i-1} at x_{i-1}.

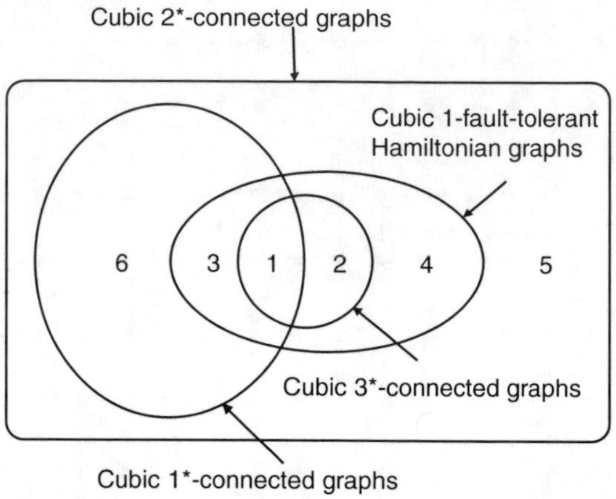

FIGURE 15.6 Venn diagram of relations among cubic 1*-connected graphs, cubic 2*-connected graphs, cubic 3*-connected graphs, and cubic 1-fault-tolerant Hamiltonian graphs.

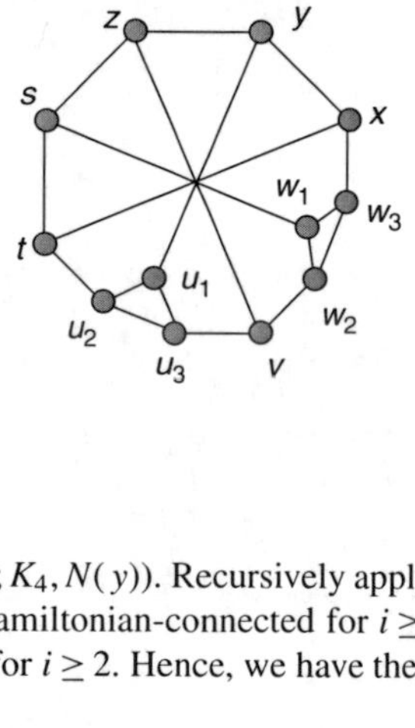

FIGURE 15.7 The graph T.

We define G_{i+1} as $J(G_i, N(x_i); K_4, N(y))$. Recursively applying Theorems 12.9 and 15.1, G_i is 3*-connected but not Hamiltonian-connected for $i \geq 2$. With Theorem 15.2, G_i is 1-fault-tolerant Hamiltonian for $i \geq 2$. Hence, we have the following theorem.

THEOREM 15.8 There are infinite cubic 1-fault-tolerant Hamiltonian graphs that are 3*-connected but not 1*-connected.

EXAMPLE 15.2

(Cubic 1-fault-tolerant Hamiltonian graphs that are 1*-connected but not 3*-connected.)
Let Q be the graph in Figure 15.8a. Let Q_3 be the three-dimensional hypercube. Obviously, Q is obtained from Q_3 by a sequence of node-expansions at every vertex of Q_3. Note that Q_3 is 1-edge fault-tolerant Hamiltonian. By Theorem 11.4, Q is 1-fault-tolerant Hamiltonian. By brute force, we can check whether Q is 1*-connected. Note that Q_3 is bipartite. By Lemma 1.2, Q_3 is not 1-fault-tolerant Hamiltonian. By Theorem 15.2, Q_3 is not 3*-connected. With Theorem 15.1, Q is not 3*-connected.

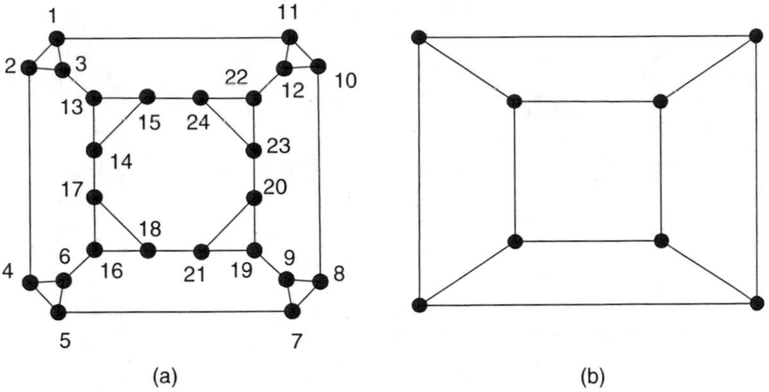

FIGURE 15.8 The graphs (a) Q and (b) Q_3.

Let x be the vertex 1 of Q shown in Figure 15.8a. By brute force, we can check whether x is nice in Q. Moreover, by Theorem 12.11, any expanded vertex of Q at 1 is nice. Now, we recursively define a sequence of graphs as follows: let $G_1 = Q$, $x_1 = x$, and $G_2 = J(G_1, N(x_1); K_4, N(y))$. Suppose that we have defined $G_1, G_2, \ldots, G_i$ with $i \geq 2$. Let x_i be any expanded vertex of G_{i-1} at x_{i-1}. We define G_{i+1} as $J(G_i, N(x_i); K_4, N(y))$. Recursively applying Theorems 12.1, 12.11, and 15.1, G_i is 1-fault-tolerant Hamiltonian and 1*-connected but not 3*-connected for $i \geq 2$. Hence, we have the following theorem.

THEOREM 15.9 There are infinite cubic 1-fault-tolerant Hamiltonian graphs that are 1*-connected but not 3*-connected.

EXAMPLE 15.3

(Examples of cubic 1-fault-tolerant Hamiltonian graphs that are not 1*-connected or 3*-connected.)
Let M be the graph in Figure 15.9a. Obviously, M can be obtained from $P(8, 2)$, shown in Figure 15.9b, by a sequence of 3-vertex expansions of $P(8, 2)$. With Theorem 15.6, $P(8, 2)$ is not 3*-connected. Using Theorem 15.1, M is not 3*-connected. Yet, it is proved in Lemmas 12.15 and 12.16 that M is 1-fault-tolerant Hamiltonian but not 1*-connected.

Let x, a, and b be three distinct vertices of M shown in Figure 15.9a. By Lemma 14.9, there is no Hamiltonian path between a and b in $J(M, N(x); K_4, N(y))$. Thus, $J(M, N(x); K_4, N(y))$ is not 1*-connected. Let $G_1 = M$, $x_1 = x$, and $G_2 = J(G_1, N(x_1); K_4, N(y))$. Suppose that we have defined G_1, G_2,..., G_i with $i \geq 2$. Let x_i be any expanded vertex of G_{i-1} at x_{i-1}. We define G_{i+1} as $J(G_i, N(x_i); K_4, N(y))$. Recursively applying Theorems 12.1 and 15.1, and Lemma 12.9, G_i is a 1-fault-tolerant Hamiltonian graph that is not 1*-connected or 3*-connected for every $i \geq 2$. Hence, we have the following theorem.

THEOREM 15.10 There are infinite cubic 1-fault-tolerant Hamiltonian graphs that are not 1*-connected or 3*-connected.

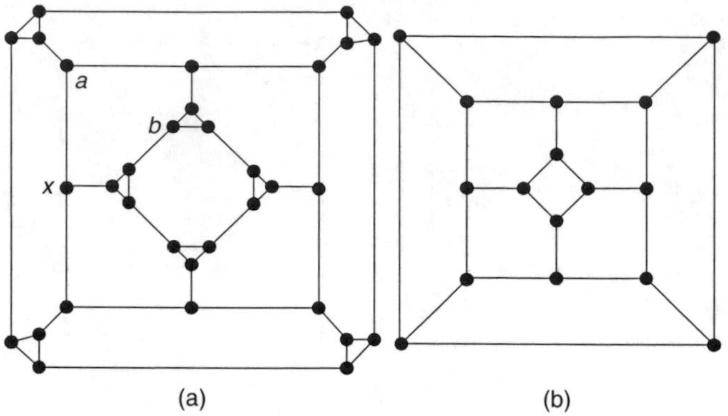

FIGURE 15.9 The graphs (a) M and (b) $P(8, 2)$.

15.4 PROPERTIES OF CUBIC 3*-LACEABLE GRAPHS

We can also study similar problems for cubic 3*-laceable graph as discussed previously. Kao et al. [193] observed the following theorem.

THEOREM 15.11 Every cubic 3*-laceable graph is 1-edge fault-tolerant Hamiltonian.

Proof: Let $G = (B \cup W, E)$ be a cubic 3*-laceable graph, and (x, y) be any edge in G. Apparently, there exists a 3*-container $C(x, y) = \{P_1, P_2, P_3\}$ in G. Obviously, one of these paths is $\langle x, y\rangle$; say, P_1. Then $P_2 \cup P_3$ forms a Hamiltonian cycle of $G - (x, y)$. The theorem is proved. □

Similarly to hyper Hamiltonian laceable and strong Hamiltonian laceable, we can define hyper 3*-laceable and strong 3*-laceable. A 3*-laceable graph is *hyper* if there exists a 3*-container $C(x, y)$ in $G - \{z\}$ for any three vertices x,y, and z of the same partite set of G. A 3*-laceable graph is *strong* if for any x and y in the same partite set of G, there exists a vertex z of the same partite set as the one that contains x and y such that $G - \{z\}$ has a 3*-container $C(x, y)$. Obviously, any 3*-laceable is strong if it is hyper. The concepts of 3*-laceable, hyper 3*-laceable, and strong 3*-laceable were proposed by Kao et al. [193]. Kao et al. [193] also observed the following theorem.

THEOREM 15.12 Every cubic hyper 3*-laceable graph is 1_p-fault-tolerant Hamiltonian.

Proof: Let $G = (B \cup W, E)$ be a cubic hyper 3*-laceable graph. We want to show that $G - \{u, v\}$ is Hamiltonian for any pair of vertices $\{u, v\}$ with $u \in B$ and $v \in W$. Let x and y be two neighbors of v such that $u \notin \{x, y\}$. Obviously, $x, y \in B$. Since G is hyper 3*-laceable, there is a 3*-container $C(x, y) = \{P_1, P_2, P_3\}$ in $G - u$. Obviously, one of these paths is $\langle x, v, y\rangle$; say, P_1. Then $P_2 \cup P_3$ forms a Hamiltonian cycle of $G - \{u, v\}$. This proves the theorem. □

15.5 EXAMPLES OF CUBIC HYPER 3*-LACEABLE GRAPHS

We assume that m and n are positive even integers with $n \geq 4$. For any two positive integers r and s, we use $[r]_s$ to denote r (**mod** s). In Chapter 2, we introduced the honeycomb rectangular torus. The honeycomb rectangular torus HReT(m, n) is the graph with the vertex set $\{(i,j) \mid 0 \leq i < m, 0 \leq j < n\}$ such that (i,j) and (k,l) are adjacent if they satisfy one of the following conditions:

1. $i = k$ and $j = [l \pm 1]_n$
2. $j = l$ and $k = [i + 1]_m$ if $i + j$ is odd
3. $j = l$ and $k = [i - 1]_m$ if $i + j$ is even

Teng et al. [299] prove that any honeycomb rectangular torus HReT(m, n) is strongly 3*-laceable. Moreover, HReT(m, n) is hyper 3*-laceable if and only if $n \geq 6$ or $m = 2$.

We first present an algorithm. The purpose of this algorithm is to extend a 3-container $C(x, y) = \{P_1, P_2, P_3\}$ of HReT(m, n) to a 3-container of HReT$(m + 2, n)$.

ALGORITHM 1: For $0 \leq i \leq m - 1$, let $f_i : V(\text{HReT}(m, n)) \to V(\text{HReT}(m + 2, n))$ be a function so assigned

$$f_i(k, l) = \begin{cases} (k, l) & \text{if } i \geq k \geq 0 \\ (k + 2, l) & \text{otherwise} \end{cases}$$

For $0 \leq i \leq m - 1$ and $0 \leq j, k \leq n - 1$, let $Q_i(j, [j + k]_n)$ denote the path $\langle (i, [j]_n), (i, [j + 1]_n), (i, [j + 2]_n), \ldots, (i, [j + k]_n) \rangle$ in HReT(m, n). Suppose that $C(x, y)$ is a 3-container of HReT(m, n) containing at least one edge joining vertices of column i to vertices of column $[i + 1]_m$; that is, $((i, j), ([i + 1]_m, j))$ in $E(C(x, y))$ for some $0 \leq j \leq n - 1$. Let $0 \leq k_0 < k_1 < \cdots < k_t \leq n - 1$ be the indices such that $((i, k_j), (i + 1, k_j)) \in E(C(x, y))$. We construct $C_i(x, y)$ as follows.

Let $\overline{C_i(x, y)}$ be the image of $C(x, y) - \{((i, k_j), (i + 1, k_j)) \mid 0 \leq k_j \leq n - 1\}$ under f_i. We set $j' = [j]_{(t+1)}$ and define A_j as

$$\langle (i, [k_j]_n), ([i + 1]_{m+2}, [k_j]_n), Q_{[i+1]_{m+2}}([k_j]_n, [k_{j'} - 1]_n), ([i + 1]_{m+2}, [k_{j'} - 1]_n),$$
$$([i + 2]_{m+2}, [k_{j'} - 1]_n), Q^{-1}_{[i+2]_{m+2}}([k_j]_n, [k_{j'} - 1]_n), ([i + 2]_{m+2},$$
$$[k_j]_n), ([i + 3]_{m+2}, [k_j]_n) \rangle$$

Obviously, A_j is a path joining $(i, [k_j]_n)$ and $(i + 3, [k_j]_n)$ for $0 \leq j < t$. It is easy to see that edges of $\overline{C_i(x, y)}$ together with edges of A_j, with $0 \leq j \leq t$ form a 3-container $Ci(x, y)$ of HReT$(m + 2, n)$. For example, a 3*-container $C((0, 0), (2, 2))$ of HReT$(4, 12) - \{(1, 7)\}$ is shown in Figure 15.10a. The corresponding $C_1((0, 0), (2, 2))$ is shown in Figure 15.10b. We have the following lemma.

LEMMA 15.3 Suppose that $C(x, y)$ is a 3-container of HReT(m, n) containing at least one edge joining vertices of column i to vertices of column $[i + 1]_m$. Then

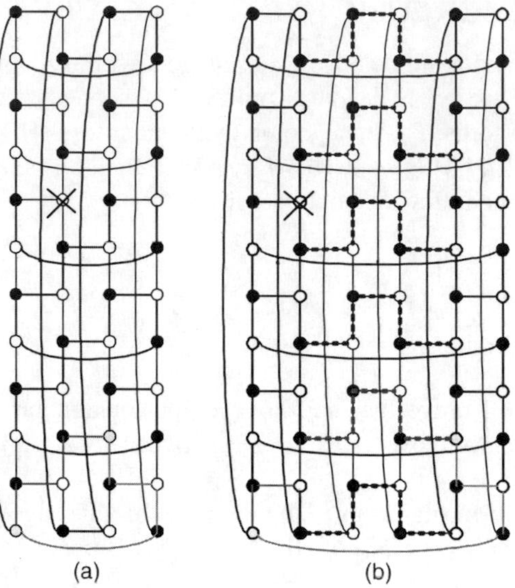

FIGURE 15.10 Illustrations for Algorithm 1.

$C_i(x, y)$ forms a 3-container of HReT$(m+2, n)$ containing at least one edge joining the vertices of column l to the vertices of column $[l+1]_m$ for any $l \in \{i, [i+1]_m, [i+2]_m\}$. Moreover, $C_i(x, y)$ is a 3*-container of HReT$(m+2, n)$ if $C(x, y)$ is a 3*-container of HReT(m, n). Furthermore, $C_i(x, y)$ is a 3*-container of HReT$(m+2, n) - \{f_i(z)\}$ if $C(x, y)$ is a 3*-container of HReT$(m, n) - \{z\}$.

LEMMA 15.4 Suppose that $C(x, y)$ is a 3-container of HReT$(2, n)$ containing at least one edge in $\{((0, j), (1, j)) \mid j \text{ is odd}\}$ and at least one edge in $\{((0, j), (1, j)) \mid j \text{ is even}\}$. Then $C_i(x, y)$ with $i \in \{0, 1\}$ forms a 3-container of HReT$(4, n)$ containing at least one edge joining the vertices of column l to the vertices of column $l+1$ for any $l \in \{0, 1, 2, 3\}$. Moreover, $C_i(x, y)$ is a 3*-container of HReT$(m+2, n)$ if $C(x, y)$ is a 3*-container of HReT(m, n). Furthermore, $C_i(x, y)$ is a 3*-container of HReT$(m+2, n) - \{f_i(z)\}$ if $C(x, y)$ is a 3*-container of HReT$(m, n) - \{z\}$.

With Lemmas 15.3 and 15.4, we say that a 3-container $C(x, y)$ of HReT$(2, n)$ is *regular* if $C(x, y)$ contains at least one edge in $\{((0, j), (1, j)) \mid j \text{ is odd}\}$ and at least one edge in $\{((0, j), (1, j)) \mid j \text{ is even}\}$. Assume that $m \geq 4$. We say a 3-container $C(x, y)$ of HReT(m, n) is *regular* if $C(x, y)$ contains at least one edge joining vertices in column i to vertices in column $[i+1]_m$ for $0 \leq i \leq m-1$. We have the following lemma.

LEMMA 15.5 Suppose that $C(x, y)$ is a regular 3*-container for HReT(m, n). Then $C_i(x, y)$ is a regular 3*-container for HReT$(m+2, n)$ for every $0 \leq i < m$. Moreover, suppose that $C(x, y)$ is a regular 3*-container for HReT$(m, n) - \{z\}$. Then $C_i(x, y)$ is a regular 3*-container for HReT$(m+2, n) - \{f_i(z)\}$ for every $0 \leq i < m$.

Now, we prove that HReT$(4,n)$ is hyper 3*-laceable. For $h=\{0,1\}$ and $0\leq j, k\leq n-1$, let $R_h(j,[j+k]_n)$ denote the path $\langle (h,[j]_n),(h,[j+1]_n),([h+1]_m,[j+1]_n),([h+1]_m,[j+2]_n),(h,[j+2]_n),\ldots,([h+1]_m,[j+k-1]_n),(h,[j+k-1]_n),(h,[j+k]_n)\rangle$ in HReT$(2,n)$.

LEMMA 15.6 Let x and y be any two vertices of HReT$(2,n)=(V_0\cup V_1,E)$ with $x\in V_0$ and $y\in V_1$. Then there exists a regular 3*-container $C(x,y)$ of HReT$(2,n)$. Hence, HReT$(2,n)$ is 3*-laceable.

Proof: Without loss of generality, we may assume that $x=(0,0)$ and $y=(i,j)$. In order to prove this lemma, we will construct a regular 3*-container $C(x,y)=\{P_1,P_2,P_3\}$ in HReT$(2,n)$. We have the following cases:

Case 1. $i=0$ and j is odd. The corresponding paths are

$$P_1 = \langle (0,0),Q_0(0,j),(0,j)\rangle$$
$$P_2 = \langle (0,j),R_0(j,0),(0,0)\rangle$$
$$P_3 = \langle (0,0),(1,0),Q_1(0,j),(1,j),(0,j)\rangle$$

Case 2. $i=1$ and j is even.

Case 2.1. $j=0$. The corresponding paths are

$$P_1 = \langle (0,0),Q_0(0,n-2),(0,n-2),(1,n-2),Q_1^{-1}(0,n-2),(1,0)\rangle$$
$$P_2 = \langle (0,0),(1,0)\rangle$$
$$P_3 = \langle (0,0),(0,n-1),(1,n-1),(1,0)\rangle$$

Case 2.2. $j>0$. The corresponding paths are

$$P_1 = \langle (0,0),Q_0(0,j),(0,j),(1,j)\rangle$$
$$P_2 = \langle (1,j),(1,j+1),(0,j+1),R_0(j+1,0),(0,0)\rangle$$
$$P_3 = \langle (0,0),(1,0),Q_1(0,j),(1,j)\rangle$$

Hence, HReT$(2,n)$ is 3*-laceable. See Figure 15.11 for illustrations. □

LEMMA 15.7 Let x,y, and z be any three different vertices of HReT$(2,n)=(V_0\cup V_1,E)$ in V_0. Then there exists a regular 3*-container $C(x,y)$ of $HReT(2,n)-\{z\}$. Hence, $HReT(2,n)$ is hyper 3*-laceable.

Proof: Without loss of generality, we may assume that $x=(0,0)$, $y=(i,j)$, and $z=(k,l)$. In order to prove this lemma, we will construct a regular 3*-container $C(x,y)=\{P_1,P_2,P_3\}$ in HReT$(2,n)-\{z\}$. We have the following cases:

Case 1. $i=0$. Then j is even.

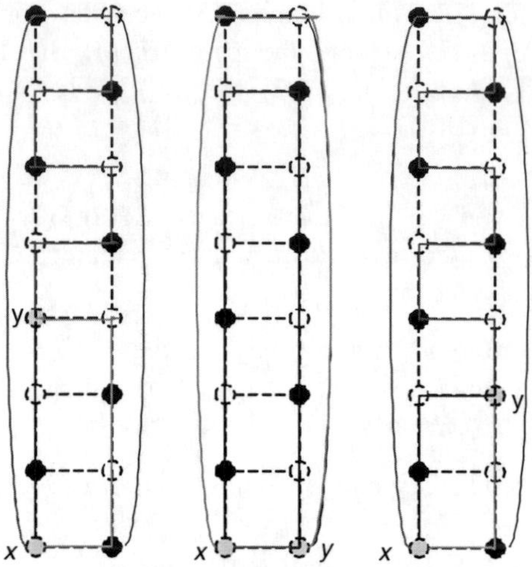

FIGURE 15.11 Illustrations for Lemma 15.6: (a) Case 1, (b) Case 2.1, and (c) Case 2.2.

Case 1.1. $k=0$. Then l is even. By the symmetric property of HReT$(2,n)$, we may assume that $l<j$. The corresponding paths are

$$\begin{aligned}
P_1 &= \langle (0,j), Q_0(j,0), (0,0)\rangle \\
P_2 &= \langle (0,0), R_0(0,l-1), (0,l-1), (1,l-1), (1,l), (1,l+1), (0,l+1), \\
&\quad R_0(l+1,j), (0,j)\rangle \\
P_3 &= \langle (0,j), (1,j), Q_1(j,0), (1,0), (0,0)\rangle
\end{aligned}$$

Case 1.2. $k=1$. Then l is odd. By the symmetric property of HReT$(2,n)$, we may assume that $l<j$. The corresponding paths are

$$\begin{aligned}
P_1 &= \langle (0,j), Q_0(j,0), (0,0)\rangle \\
P_2 &= \langle (0,0), R_0(0,l), (0,l), R_0(l,j), (0,j)\rangle \\
P_3 &= \langle (0,j), (1,j), Q_1(j,0), (1,0), (0,0)\rangle
\end{aligned}$$

Case 2. $i=1$. Then j is odd. $k=0$. Then l is even. By the symmetric property of HReT$(2,n)$, we may assume that $l<j$. The corresponding paths are

$$\begin{aligned}
P_1 &= \langle (1,j), (0,j), Q_0(j,0), (0,0)\rangle \\
P_2 &= \langle (0,0), R_0(0,l-1), (0,l-1), (1,l-1), (1,l), (1,l+1), (0,l+1), \\
&\quad R_0(l+1,j-1), (0,j-1), (1,j-1), (1,j)\rangle \\
P_3 &= \langle (1,j), Q_1(j,0), (1,0), (0,0)\rangle
\end{aligned}$$

Hence, HReT$(2,n)$ is hyper 3*-laceable. See Figure 15.12 for illustrations. □

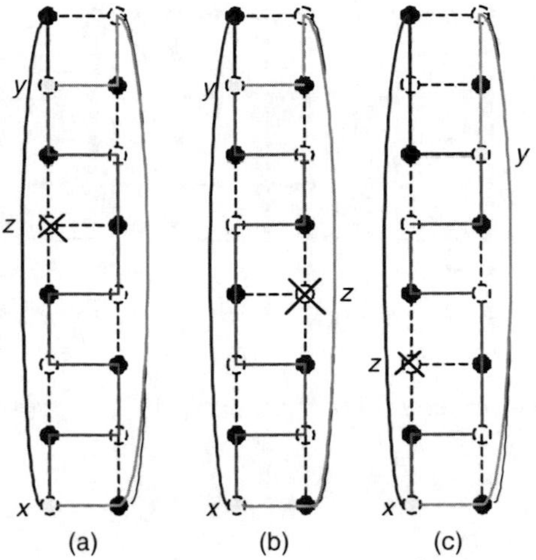

FIGURE 15.12 Illustrations for Lemma 15.7: (a) Case 1.1, (b) Case 1.2, and (c) Case 2.

Now, we discuss the 3*-laceability for HReT$(4,n)$. We need the following path patterns. For $0 \le i \le m-1$ and $0 \le j,k \le n-1$, we set

$$\begin{aligned} S_i^L(j) = \langle & ([i]_m,[j]_n),([i-1]_m,[j]_n),([i-1]_m,[j+1]_n),([i-2]_m,[j+1]_n),\\ & ([i-2]_m,[j+2]_n),([i-3]_m,[j+2]_n),([i-3]_m,[j+3]_n),\\ & ([i-4]_m,[j+3]_n),([i-4]_m,[j+2]_n)\rangle \end{aligned}$$

$$\begin{aligned} S_i^R(j) = \langle & ([i]_m,[j]_n),([i+1]_m,[j]_n),([i+1]_m,[j+1]_n),([i+2]_m,[j+1]_n),\\ & ([i+2]_m,[j+2]_n),([i+3]_m,[j+2]_n),([i+3]_m,[j+3]_n),\\ & ([i+4]_m,[j+3]_n),([i+4]_m,[j+2]_n)\rangle \end{aligned}$$

$$\begin{aligned} S_i^L(j,k) = \langle & ([i]_m,[j]_n),S_{[i]_m}^L(j),([i-4]_m,[j+2]_n),S_{[i-4]_m}^L([j+2]_n),\\ & ([i-8]_m,[j+4]_n),\ldots,([i-2(k-j-2)]_m,[k-2]_n),\\ & S_{[i-2(k-j-2)]_m}^L([k-2]_n),([i-2(k-j)]_m,[k]_n)\rangle \end{aligned}$$

$$\begin{aligned} S_i^R(j,k) = \langle & ([i]_m,[j]_n),S_{[i]_m}^R(j),([i+4]_m,[j+2]_n),S_{[i+4]_m}^R([j+2]_n),\\ & ([i+8]_m,[j+4]_n),\ldots,([i+2(k-j-2)]_m,[k-2]_n),\\ & S_{[i+2(k-j-2)]_m}^R([k-2]_n),([i+2(k-j)]_m,[k]_n)\rangle \end{aligned}$$

See Figure 15.13 for illustrations.

LEMMA 15.8 Let x and y be any two vertices of HReT$(4,n) = (V_0 \cup V_1, E)$ with $x \in V_0$ and $y \in V_1$. Then there exists a regular 3*-container $C(x,y)$ of HReT$(4,n)$. Hence, HReT$(4,n)$ is 3*-laceable.

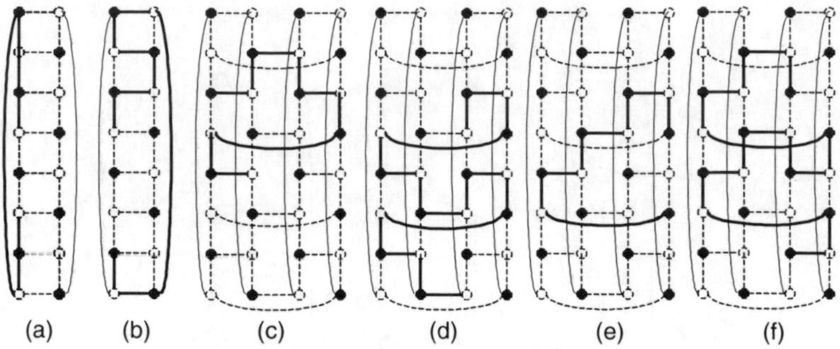

FIGURE 15.13 The path patterns (a) $Q_0(4,2)$, (b) $R_0(4,1)$, (c) $S_1^L(3)$, (d) $S_2^L(0,4)$, (e) $S_3^R(2)$, and (f) $S_2^R(1,5)$.

Proof: Without loss of generality, we may assume that $x=(0,0)$ and $y=(i,j)$. In order to prove this lemma, we will construct a regular 3*-container $C(x,y)=\{P_1,P_2,P_3\}$ in HReT(4, n). By the symmetric property of HReT(4, n), we may assume that $i \in \{0,1,2\}$. Hence, we have the following cases:

Case 1. Suppose that $i \in \{0,1\}$. By Lemma 15.6, there exists a regular 3*-container $C((0,0),(i,j))$ of HReT(2, n). By Lemma 15.5, $C1((0,0),(i,j))$ forms a 3*-container of HReT(4, n).

Case 2. $i=2$. Then j is odd.

Case 2.1. Suppose that $j=1$. The corresponding paths are

$$P_1 = \langle (0,0),(0,n-1),(1,n-1),Q_1^{-1}(0,n-1),(1,0),(2,0),(2,1)\rangle$$

$$P_2 = \langle (0,0),Q_0(0,n-2),(0,n-2),(3,n-2),Q_3^{-1}(1,n-2),(3,1),(2,1)\rangle$$

$$P_3 = \langle (0,0),(3,0),(3,n-1),(2,n-1),Q_2^{-1}(1,n-1),(2,1)\rangle$$

Case 2.2. Suppose that $j \neq 1$. The corresponding paths are

$$P_1 = \langle (0,0),Q_0(0,j-1),(0,j-1),(3,j-1),(3,j),(2,j)\rangle$$

$$P_2 = \langle (0,0),(3,0),Q_3(0,j-2),(3,j-2),(2,j-2),Q_2^{-1}(0,j-2),(2,0),(1,0),$$
$$Q_1(0,j-1),(1,j-1),(2,j-1),(2,j)\rangle$$

$$P_3 = \langle (0,0),(0,n-1),S_L^{-1}(j+3,n-1),(0,j+3),(0,j+2),(1,j+2),(1,j+1),$$
$$(1,j),(0,j),(0,j+1),(3,j+1),(3,j+2),(2,j+2),(2,j+1),(2,j)\rangle$$

Hence, HReT(4, n) is 3*-laceable. See Figure 15.14 for illustrations. □

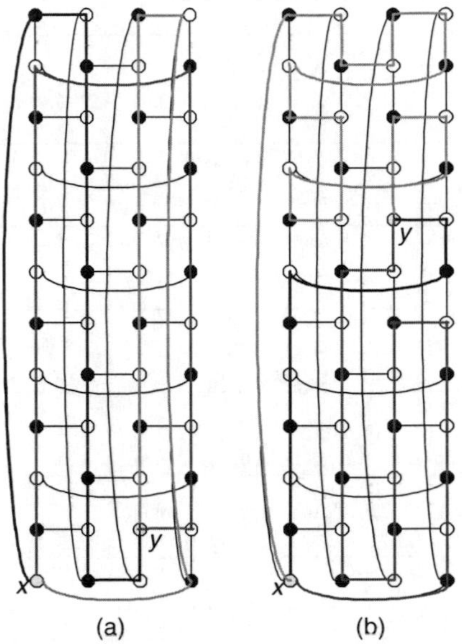

FIGURE 15.14 Illustrations for Lemma 15.8: (a) Case 2.1 and (b) Case 2.2.

LEMMA 15.9 Let x, y, and z be any three different vertices of $\mathrm{HReT}(4,6) = (V_0 \cup V_1, E)$ in V_0. Then there exists a regular 3^*-container $C(x,y)$ of $\mathrm{HReT}(4,6) - \{z\}$. Hence, HReT(4, 6) is hyper 3*-laceable.

Proof: Without loss of generality, we may assume that $x=(0,0)$, $y=(i,j)$, and $z=(k,l)$. The corresponding regular 3^*-container $C(x,y)=\{P_1,P_2,P_3\}$ in $\mathrm{HReT}(4,6)-\{z\}$ are listed as follows:

y	*z*	*C(x,y)*
(0, 2)	(2, 2)	$\langle (0,0),(0,1),(0,2)\rangle$ $\langle (0,0),(0,5),(1,5),(1,0),Q_1(0,4),(1,4),(2,4),(2,3),(3,3),(3,2),(0,2)\rangle$ $\langle (0,0),(3,0),(3,1),(2,1),(2,0),(2,5),(3,5),(3,4),(0,4),(0,3),(0,2)\rangle$
(0, 2)	(2, 4)	$\langle (0,0),(0,1),(0,2)\rangle$ $\langle (0,2),(0,3),(0,4),(3,4),(3,5),(2,5),(2,0),(1,0),Q_1(0,5),(1,5),(0,5),(0,0)\rangle$ $\langle (0,0),(3,0),(3,1),(2,1),(2,2),(2,3),(3,3),(3,2),(0,2)\rangle$
(0, 4)	(0, 2)	$\langle (0,0),(0,5),(0,4)\rangle$ $\langle (0,0),(3,0),Q_3(0,3),(3,3),(2,3),(2,4),(2,5),(3,5),(3,4),(0,4)\rangle$ $\langle (0,0),(0,1),(1,1),(1,2),(2,2),(2,1),(2,0),(1,0),(1,5),(1,4),(1,3),(0,3),(0,4)\rangle$
(0, 4)	(1, 1)	$\langle (0,0),(0,5),(0,4)\rangle$ $\langle (0,0),(0,1),(0,2),(3,2),(3,3),(2,3),(2,2),(1,2),(1,3),(0,3),(0,4)\rangle$ $\langle (0,0),(3,0),(3,1),(2,1),(2,0),(1,0),(1,5),(1,4),(2,4),(2,5),(3,5),(3,4),(0,4)\rangle$
(1, 3)	(0, 2)	$\langle (0,0),(0,5),(0,4),(0,3),(1,3)\rangle$ $\langle (0,0),(0,1),(1,1),(1,2),(1,3)\rangle$ $\langle (0,0),(3,0),Q_3(0,5),(3,5),(2,5),Q_2^{-1}(0,5),(2,0),(1,0),(1,5),(1,4),(1,3)\rangle$
(1, 5)	(0, 2)	$\langle (0,0),(0,5),(1,5)\rangle$ $\langle (0,0),(0,1),(1,1),(1,2),(1,3),(0,3),(0,4),(3,4),(3,5),(2,5),(2,4),(1,4),(1,5)\rangle$ $\langle (0,0),(3,0),Q_3(0,3),(3,3),(2,3),Q_2^{-1}(0,3),(2,0),(1,0),(1,5)\rangle$

(*Continued*)

y	*z*	*C(x,y)*
(1, 1)	(2, 0)	$\langle(0,0),(0,1),(1,1)\rangle$
		$\langle(0,0),(3,0),(3,1),(2,1),(2,2),(1,2),(1,1)\rangle$
		$\langle(0,0),(0,5),(0,4),(3,4),(3,5),(2,5),(2,4),(2,3),(3,3),(3,2),(0,2),(0,3),(1,3),(1,4),(1,5),$ $(1,0),(1,1)\rangle$
(1, 1)	(2, 2)	$\langle(1,1),Q_1(1,4),(1,4),(2,4),(2,3),(3,3),(3,2),(0,2),(0,3),(0,4),(3,4),(3,5),(2,5),(2,0),(2,1),(3,1),$ $(3,0),(0,0)\rangle\langle(0,0),(0,1),(1,1)\rangle$
		$\langle(0,0),(0,5),(1,5),(1,0),(1,1)\rangle$
(1, 1)	(2, 4)	$\langle(1,1),(1,0),(2,0),(2,5),(3,5),(3,4),(0,4),(0,3),(0,2),(3,2),(3,3),(2,3),(2,2),(2,1),(3,1),(3,0),(0,0)\rangle$
		$\langle(0,0),(0,1),(1,1)\rangle$
		$\langle(0,0),(0,5),(1,5),Q_1^{-1}(1,5),(1,1)\rangle$
(1, 3)	(2, 0)	$\langle(0,0),(0,1),(1,1),(1,0),(1,5),(1,4),(1,3)\rangle$
		$\langle(0,0),(3,0),(3,1),(2,1),(2,2),(1,2),(1,3)\rangle$
		$\langle(0,0),(0,5),(0,4),(3,4),(3,5),(2,5),(2,4),(2,3),(3,3),(3,2),(0,2),(0,3),(1,3)\rangle$
(1, 3)	(2, 2)	$\langle(0,0),(0,5),(1,5),(1,4),(1,3)\rangle$
		$\langle(0,0),(3,0),(3,5),(2,5),(2,4),(2,3),(3,3),(3,4),(0,4),(0,3),(1,3)\rangle$
		$\langle(0,0),(0,1),(0,2),(3,2),(3,1),(2,1),(2,0),(1,0),Q_1(0,3),(1,3)\rangle$
(1, 3)	(2, 4)	$\langle(0,0),(0,5),(1,5),(1,4),(1,3)\rangle$
		$\langle(0,0),(3,0),(3,5),(2,5),(2,0),(1,0),Q_1(0,3),(1,3)\rangle$
		$\langle(0,0),(0,1),(0,2),(3,2),(3,1),(2,1),(2,2),(2,3),(3,3),(3,4),(0,4),(0,3),(1,3)\rangle$
(2, 0)	(0, 2)	$\langle(0,0),(3,0),Q_3(0,3),(3,3),(2,3),(2,4),(1,4),(1,3),(0,3),(0,4),(3,4),(3,5),(2,5),(2,0)\rangle$
		$\langle(0,0),(0,5),(1,5),(1,0),(2,0)\rangle$
		$\langle(0,0),(0,1),(1,1),(1,2),(2,2),(2,1),(2,0)\rangle$
(2, 2)	(0, 2)	$\langle(0,0),(0,1),(1,1),(1,0),(2,0),(2,1),(2,2)\rangle$
		$\langle(0,0),(3,0),Q_3(0,3),(3,3),(2,3),(2,2)\rangle$
		$\langle(0,0),(0,5),(1,5),(1,4),(2,4),(2,5),(3,5),(3,4),(0,4),(0,3),(1,3),(1,2),(2,2)\rangle$
(2, 2)	(0, 4)	$\langle(0,0),(0,5),(1,5),(1,0),(2,0),(2,5),(3,5),Q_3^{-1}(2,5),(3,2),(0,2),(0,3),(1,3),(1,4),(2,4),(2,3),(2,2)\rangle$
		$\langle(0,0),(0,1),(1,1),(1,2),(2,2)\rangle$
		$\langle(0,0),(3,0),(3,1),(2,1),(2,2)\rangle$
(2, 2)	(1, 1)	$\langle(0,0),(0,5),(1,5),(1,0),(2,0),(2,1),(2,2)\rangle$
		$\langle(0,0),(3,0),Q_3(0,3),(3,3),(2,3),(2,2)\rangle$
		$\langle(0,0),Q_0(0,4),(0,4),(3,4),(3,5),(2,5),(2,4),(1,4),(1,3),(1,2),(2,2)\rangle$

Hence, HReT(4, 6) is hyper 3*-laceable. □

LEMMA 15.10 Assume that $n \geq 8$. Let x, y, and z be any three different vertices of HReT$(4,n) = (V_0 \cup V_1, E)$ in V_0. Then there exists a regular 3*-container $C(x,y)$ of HReT$(4,n) - \{z\}$. Hence, HReT$(4,n)$ is hyper 3*-laceable.

Proof: Without loss of generality, we may assume that $x=(0,0)$, $y=(i,j)$, and $z=(k,l)$. In order to prove this lemma, we will construct a regular 3*-container $C(x,y)=\{P_1,P_2,P_3\}$ in HReT$(4,n)-\{z\}$. By the symmetric property of HReT$(4,n)$, we may assume that $i \in \{0,1,2\}$. We have the following cases:

Case 1. Suppose that $i \in \{0,1\}$ and $z \in \{0,1\}$. By Lemma 15.7, there exists a regular 3*-container $C((0,0),(i,j))$ of HReT$(2,n) - \{(k,l)\}$. By Lemma 15.5, $C_1((0,0),(i,j))$ forms a 3*-container of HReT$(4,n) - \{(k,l)\}$.

Case 2. $i=0$ and $k=2$. Then j and l are even. By the symmetric property, we have the following subcases:

Case 2.1. Suppose that $j=4$ and $l=2$. The corresponding paths are

$$P_1 = \langle(0,0),Q_0(0,4),(0,4)\rangle$$
$$P_2 = \langle(0,0),(0,n-1),(0,n-2),(3,n-2),Q_3^{-1}(4,n-2),(3,4),(0,4)\rangle$$

$$P_3 = \langle (0,4), Q_0(4,n-3), (0,n-3), (1,n-3), Q_1^{-1}(0,n-3), (1,0), (1,n-1), (1,n-2), (2,n-2), Q_2^{-1}(3,n-2), (2,3), (3,3), (3,2), (3,1), (2,1), (2,0), (2,n-1), (3,n-1), (3,0), (0,0) \rangle$$

Case 2.2. Suppose that $n-4 > j \geq 2$ and $l = j+2$. The corresponding paths are

$$P_1 = \langle (0,0), Q_0(0,j), (0,j) \rangle$$
$$P_2 = \langle (0,0), (3,0), Q_3(0,j), (3,j), (0,j) \rangle$$
$$P_3 = \langle (0,j), Q_0(j,j+4), (0,j+4), (3,j+4), (3,j+5), (2,j+5), (2,j+4), (2,j+3), (3,j+3), (3,j+2), (3,j+1), (2,j+1), Q_2^{-1}(0,j+1), (2,0), (1,0), Q_1(0,j+5), (1,j+5), (0,j+5), (0,j+6), (3,j+6), (3,j+7), (2,j+7), (2,j+6), S_2^L(j+6,n-2), (2,n-2), (1,n-2), (1,n-1), (0,n-1), (0,0) \rangle$$

Case 2.3. Suppose that $n-6 > j \geq 2$ and $n-4 > l > j+2$. The corresponding paths are

$$P_1 = \langle (0,0), Q_0(0,j), (0,j) \rangle$$
$$P_2 = \langle (0,0), (3,0), Q_3(0,j), (3,j), (0,j) \rangle$$
$$P_3 = \langle (1,j), (1,j+1), (1,j+2), (3,j+2), (3,j+1), (2,j+1), Q_2^{-1}(0,j+1), (2,0), (1,0), Q_1(0,j+2), (1,j+2), (2,j+2), (2,j+3), (3,j+3), (3,j+4), (0,j+4), (0,j+3), S_0^R(j+3,l-3), (0,l-3), (1,l-3), (1,l-2), (2,l-2), (2,l-1), (3,l-1), (3,l), (3,l+1), (2,l+1), (2,l+2), (2,l+3), (3,l+3), (3,l+2), (0,l+2), Q_0^{-1}(l-1,l+2), (0,l-1), (1,l-1), Q_1(l-1,l+3), (1,l+3), (0,l+3), (0,l+4), (3,l+4), S_2^L(l+4,n-2), (2,n-2), (1,n-2), (1,n-1), (0,n-1), (0,0) \rangle$$

Case 2.4. Suppose that $n > 8$ and $j = l \geq 2$. The corresponding paths are

$$P_1 = \langle (0,0), Q_0(0,j), (0,j) \rangle$$
$$P_2 = \langle (0,0), (3,0), Q_3(0,j-1), (3,j-1), (2,j-1), Q_2^{-1}(0,j-1), (2,0), (1,0), Q_1(0,j+1), (1,j+1), (0,j+1), (0,j) \rangle$$
$$P_3 = \langle (0,j), (3,j), (3,j+1), (2,j+1), (2,j+2), (1,j+2), (1,j+3), (1,j+4), (2,j+4), (2,j+3), (3,j+3), (3,j+2), (0,j+2), (0,j+3), (0,j+4), (3,j+4), (3,j+5), (2,j+5), (2,j+6), (1,j+6), (1,j+5), S_1^L(j+5,n-5), (1,n-5), (0,n-5), (0,n-4), (3,n-4), (3,n-3), (2,n-3), (2,n-2), (2,n-1), (3,n-1), (3,n-2), (0,n-2), (0,n-3), (1,n-3), (1,n-2), (1,n-1), (0,n-1), (0,0) \rangle$$

Case 2.5. Suppose that $n=8, j=2$, and $l=2$. The corresponding paths are

$$P_1 = \langle(0,0),(0,1),(0,2)\rangle$$
$$P_2 = \langle(0,2),(0,3),(0,4),(3,4),Q_3(4,7),(3,7),(2,7),(2,0),(2,1),(3,1),(3,0),(0,0)\rangle$$
$$P_3 = \langle(0,2),(3,2),(3,3),(2,3),Q_2(3,6),(2,6),(1,6),(1,7),(1,0),Q_1(0,5),(1,5),(0,5),(0,6),(0,7),(0,0)\rangle$$

Case 2.6. Suppose that $n=8, j=4$, and $l=4$. The corresponding paths are

$$P_1 = \langle(0,0),Q_0(0,4),(0,4)\rangle$$
$$P_2 = \langle(0,0),(0,7),(1,7),(1,0),Q_1(0,6),(1,6),(2,6),(2,5),(3,5),(3,4),(0,4)\rangle$$
$$P_3 = \langle(0,0),(3,0),Q_3(0,3),(3,3),(2,3),Q_2^{-1}(0,3),(2,0),(2,7),(3,7),(3,6),(0,6),(0,5),(0,4)\rangle$$

Case 3. $i=1$ and $k=2$. Then j is odd and l is even. By the symmetric property, we have the following subcases:

Case 3.1. Suppose that $n-5 > j\geq 1$ and $n-4>l>j+2$. The corresponding paths are

$$P_1 = \langle(0,0),Q_0(0,j),(0,j),(1,j)\rangle$$
$$P_2 = \langle(0,0),(3,0),Q_3(0,j),(3,j),(2,j),Q_2^{-1}(0,j),(2,0),(1,0),Q_1(0,j),(1,j)\rangle$$
$$P_3 = \langle(1,j),(1,j+1),(2,j+1),(2,j+2),(3,j+2),(3,j+1),S_3^L(j+1,l-2),(3,l-2),(0,l-2),(0,l-1),(1,l-1),(1,l),(1,l+1),(1,l+2),(2,l+2),(2,l+1),(3,l+1),(3,l),(0,l),(0,l+1),(0,l+2),(3,l+2),(3,l+3),(2,l+3),(2,l+4),(1,l+4),(1,l+3),S_1^L(l+3,n-5),(1,n-5),(0,n-5),(0,n-4),(3,n-4),(3,n-3),(2,n-3),(2,n-2),(2,n-1),(3,n-1),(3,n-2),(0,n-2),(0,n-3),(1,n-3),(1,n-2),(1,n-1),(0,n-1),(0,0)\rangle$$

Case 3.2. Suppose that $n-5 > j\geq 1$ and $l=j+1$. The corresponding paths are

$$P_1 = \langle(0,0),Q_0(0,j),(0,j),(1,j)\rangle$$
$$P_2 = \langle(0,0),(3,0),Q_3(0,j),(3,j),(2,j),Q_2^{-1}(0,j),(2,0),(1,0),Q_1(0,j),(1,j)\rangle$$

$$\begin{aligned}P_3 = \langle &(1,j), Q_1(j,j+3), (1,j+3), (2,j+3), (2,j+2), (3,j+2), (3,j+1),\\ &(0,j+1), (0,j+2), (0,j+3), (3,j+3), (3,j+4), (2,j+4), (2,j+5),\\ &(1,j+5), (1,j+4), S_1^L(j+4,n-5), (1,n-5), (0,n-5), (0,n-4),\\ &(3,n-4), (3,n-3), (2,n-3), (2,n-2), (2,n-1), (3,n-1), (3,n-2),\\ &(0,n-2), (0,n-3), (1,n-3), (1,n-2), (1,n-1), (0,n-1), (0,0)\rangle\end{aligned}$$

Case 3.3. Suppose that $n-5 > j \geq 1$ and $l = n-4$. The corresponding paths are

$$\begin{aligned}P_1 &= \langle (0,0), Q_0(0,j), (0,j), (1,j)\rangle\\ P_2 &= \langle (0,0), (0,n-1), (1,n-1), (1,0), Q_1(0,j), (1,j)\rangle\\ P_3 &= \langle (1,j), (1,j+1), (2,j+1), (2,j+2), (3,j+2), (3,j+1), S_3^L(j+1,n-6),\\ &\quad (0,n-6), (0,n-5), (1,n-5), Q_1(n-5,n-2), (1,n-2), (2,n-2),\\ &\quad (2,n-3), (3,n-3), (3,n-4), (0,n-4), (0,n-3), (0,n-2), (3,n-2),\\ &\quad (3,n-1), (2,n-1), (2,0), (2,1), (3,1), (3,0), (0,0)\rangle\end{aligned}$$

Case 3.4. Suppose that $j = n-5$ and $l = n-4$. The corresponding paths are

$$\begin{aligned}P_1 &= \langle (0,0), Q_0(0,n-5), (0,n-5), (1,n-5)\rangle\\ P_2 &= \langle (0,0), (0,n-1), (1,n-1), (1,0), Q_1(0,n-5), (1,n-5)\rangle\\ P_3 &= \langle (1,n-5), Q_1(n-5,n-2), (1,n-2), (2,n-2), (2,n-3), (3,n-3),\\ &\quad (3,n-4), (0,n-4), (0,n-3), (0,n-2), (3,n-2), (3,n-1), (2,n-1),\\ &\quad (2,0), Q_2(0,n-5), (2,n-5), (3,n-5), Q_3^{-1}(0,n-5), (3,0), (0,0)\rangle\end{aligned}$$

Case 3.5. Suppose that $n-5 > j \geq 1$ and $l = n-2$. The corresponding paths are

$$\begin{aligned}P_1 &= \langle (0,0), Q_0(0,j), (0,j), (1,j)\rangle\\ P_2 &= \langle (0,0), (3,0), (3,n-1), (2,n-1), (2,0), (1,0), Q_1(0,j), (1,j)\rangle\\ P_3 &= \langle (1,j), (1,j+1), (1,j+2), (0,j+2), (0,j+1), (3,j+1), Q_3^{-1}(1,j+1), (3,1),\\ &\quad (2,1), Q_2(1,j+2), (2,j+2), (3,j+2), (3,j+3), (0,j+3), (0,j+4),\\ &\quad (1,j+4), (1,j+3), S_1^R(j+3,n-6), (1,n-6), (2,n-6), (2,n-5),\\ &\quad (3,n-5), (3,n-4), (0,n-4), (0,n-3), (0,n-2), (3,n-2), (3,n-3),\\ &\quad (2,n-3), (2,n-4), (1,n-4), Q_1(n-4,n-1), (1,n-1), (0,n-1), (0,0)\rangle\end{aligned}$$

Case 4. $i = 2$ and $k = 0$. Then j and l are even. By the symmetric property, we have the following subcases:

Case 4.1. Suppose that $j = 0$ and $l > 0$. The corresponding paths are

$$\begin{aligned}
P_1 &= \langle(0,0),(0,n-1),(1,n-1),(1,0),(2,0)\rangle \\
P_2 &= \langle(0,0),(0,1),(1,1),(1,2),(2,2),(2,1),(2,0)\rangle \\
P_3 &= \langle(0,0),(3,0),(3,1),(3,2),(0,2),(0,3),(1,3),(1,4),(2,4),(2,3), \\
&\quad S_2^R(3,j-1),(2,j-1),(3,j-1),(3,j),(3,j+1),(2,j+1),(2,j+2),(1,j+2), \\
&\quad (1,j+1),S_1^L(j+1,n-3),(1,n-3),(0,n-3),(0,n-2),(3,n-2), \\
&\quad (3,n-1),(2,n-1),(2,0)\rangle
\end{aligned}$$

Case 4.2. Suppose that $l > j > 0$. The corresponding paths are

$$\begin{aligned}
P_1 &= \langle(0,0),(0,1),(1,1),Q_1(1,j),(1,j),(2,j)\rangle \\
P_2 &= \langle(0,0),(3,0),(3,1),(2,1),Q_2(1,j),(2,j)\rangle \\
P_3 &= \langle(2,j),(2,j+1),(2,j+2),(1,j+2),(1,j+1),(0,j+1),Q_0^{-1}(2,j+1), \\
&\quad (0,2),(3,2),Q_3(2,j+2),(3,j+2),(0,j+2),(0,j+3),(1,j+3),(1,j+4), \\
&\quad (2,j+4),(2,j+3),S_2^R(j+3,l-1),(2,l-1),(3,l-1),(3,l),(3,l+1), \\
&\quad (2,l+1),(2,l+2),(1,l+2),(1,l+1),S_1^L(l+1,n-1),(1,n-1),(0,n-1), \\
&\quad (0,0)\rangle
\end{aligned}$$

Case 4.3. Suppose that $j = l > 0$. The corresponding paths are

$$\begin{aligned}
P_1 &= \langle(0,0),Q_0(0,j-1),(0,j-1),(1,j-1),Q_1^{-1}(0,j-1),(1,0),(2,0),Q_2(0,j), \\
&\quad (2,j)\rangle \\
P_2 &= \langle(0,0),(3,0),Q_3(0,j+1),(3,j+1),(2,j+1),(2,j)\rangle \\
P_3 &= \langle(2,j),S_2^L(j,n-1),(2,n-2),(1,n-2),(1,n-1),(0,n-1),(0,0)\rangle
\end{aligned}$$

Case 5. $i = 2$ and $k = 1$. Then j is even and l is odd. By the symmetric property, we have the following subcases:

Case 5.1. Suppose that $j = 0$ and $l = 1$. The corresponding paths are

$$\begin{aligned}
P_1 &= \langle(0,0),(0,n-1),(1,n-1),(1,0),(2,0)\rangle \\
P_2 &= \langle(0,0),(0,1),(0,2),(3,2),(3,1),(2,1),(2,0)\rangle \\
P_3 &= \langle(0,0),(3,0),(3,n-1),(3,n-2),(0,n-2),Q_0^{-1}(3,n-2),(0,3),(1,3), \\
&\quad (1,2),(2,2),(2,3),(3,3),Q_3(3,n-3),(3,n-3),(2,n-3),Q_2^{-1}(4,n-3), \\
&\quad (2,4),(1,4),Q_1(4,n-2),(1,n-2),(2,n-2),(2,n-1),(2,0)\rangle
\end{aligned}$$

Case 5.2. Suppose that $j = 0$ and $n-1 > l > 1$. The corresponding paths are

$$P_1 = \langle (0,0),(0,n-1),(1,n-1),(1,0),(2,0)\rangle$$
$$P_2 = \langle (0,0),(3,0),(3,1),(2,1),(2,0)\rangle$$
$$P_3 = \langle (0,0),(0,1),(1,1),(1,2),(2,2),(2,3),(3,3),(3,2),S_3^L(2,j-3),(3,j-3),(0,j-3),(0,j-2),(1,j-2),(1,j-1),(2,j-1),(2,j),(2,j+1),(1,j+1),(1,j+2),(1,j+3),(2,j+3),(2,j+2),(3,j+2),Q_3^{-1}(j-1,j+2),(3,j-1),(0,j-1),Q_0(j-1,j+3),(0,j+3),(3,j+3),(3,j+4),(2,j+4),(2,j+5),(1,j+5),(1,j+4),S_1^L(j+4,n-3),(1,n-3),(0,n-3),(0,n-2),(3,n-2),(3,n-1),(2,n-1),(2,0)\rangle$$

Case 5.3. Suppose that $n-1 > l > j+2$ and $j > 0$. The corresponding paths are

$$P_1 = \langle (0,0),(0,1),(1,1),Q_1(1,j),(1,j),(2,j)\rangle$$
$$P_2 = \langle (0,0),(3,0),(3,1),(2,1),Q_2(1,j),(2,j)\rangle$$
$$P_3 = \langle (2,j),(2,j+1),(3,j+1),(3,j),S_3^L(j,l-3),(3,l-3),(0,l-3),(0,l-2),(1,l-2),(1,l-1),(2,l-1),(2,l),(2,l+1),(1,l+1),(1,l+2),(1,l+3),(2,l+3),(2,l+2),(3,l+2),(3,l+1),(3,l),(3,l-1),(0,l-1),Q_0(l-1,l+3),(0,l+3),(3,l+3),(3,l+4),(2,l+4),(2,l+5),(1,l+5),(1,l+4),S_1^L(l+4,n-1),(1,n-1),(0,n-1),(0,0)\rangle$$

Case 5.4. Suppose that $n-2 > j$ and $l = n-1$. The corresponding paths are

$$P_1 = \langle (0,0),Q_0(0,j+1),(0,j+1),(1,j+1),(1,j+2),(2,j+2),(2,j+1),(2,j)\rangle$$
$$P_2 = \langle (0,0),(3,0),(3,n-1),(2,n-1),(2,0),(1,0),Q_1(0,j),(1,j),(2,j)\rangle$$
$$P_3 = \langle (2,j),Q_2^{-1}(1,j),(2,1),(3,1),Q_3(1,j+2),(3,j+2),(0,j+2),(0,j+3),(1,j+3),(1,j+4),(2,j+4),(2,j+3),S_2^R(j+3,n-3),(2,n-3),(3,n-3),(3,n-2),(0,n-2),(0,n-1),(0,0)\rangle$$

Case 5.5. Suppose that $j = n-2$ and $l = n-1$. The corresponding paths are

$$P_1 = \langle (0,0),(0,n-1),(0,n-2),(0,n-3),(1,n-3),(1,n-2),(2,n-2)\rangle$$
$$P_2 = \langle (0,0),Q_0(0,n-4),(3,n-4),Q_3(n-4,n-1),(2,n-1),(2,n-2)\rangle$$
$$P_3 = \langle (0,0),(3,0),Q_3(0,n-5),(3,n-5),(2,n-5),Q_2^{-1}(0,n-5),(2,0),Q_1(0,n-4),(1,n-4),(2,n-4),(2,n-3),(2,n-2)\rangle$$

Hence, HReT$(4,n)$ is hyper 3*-laceable for $n \geq 8$. See Figure 15.15 for illustrations. □

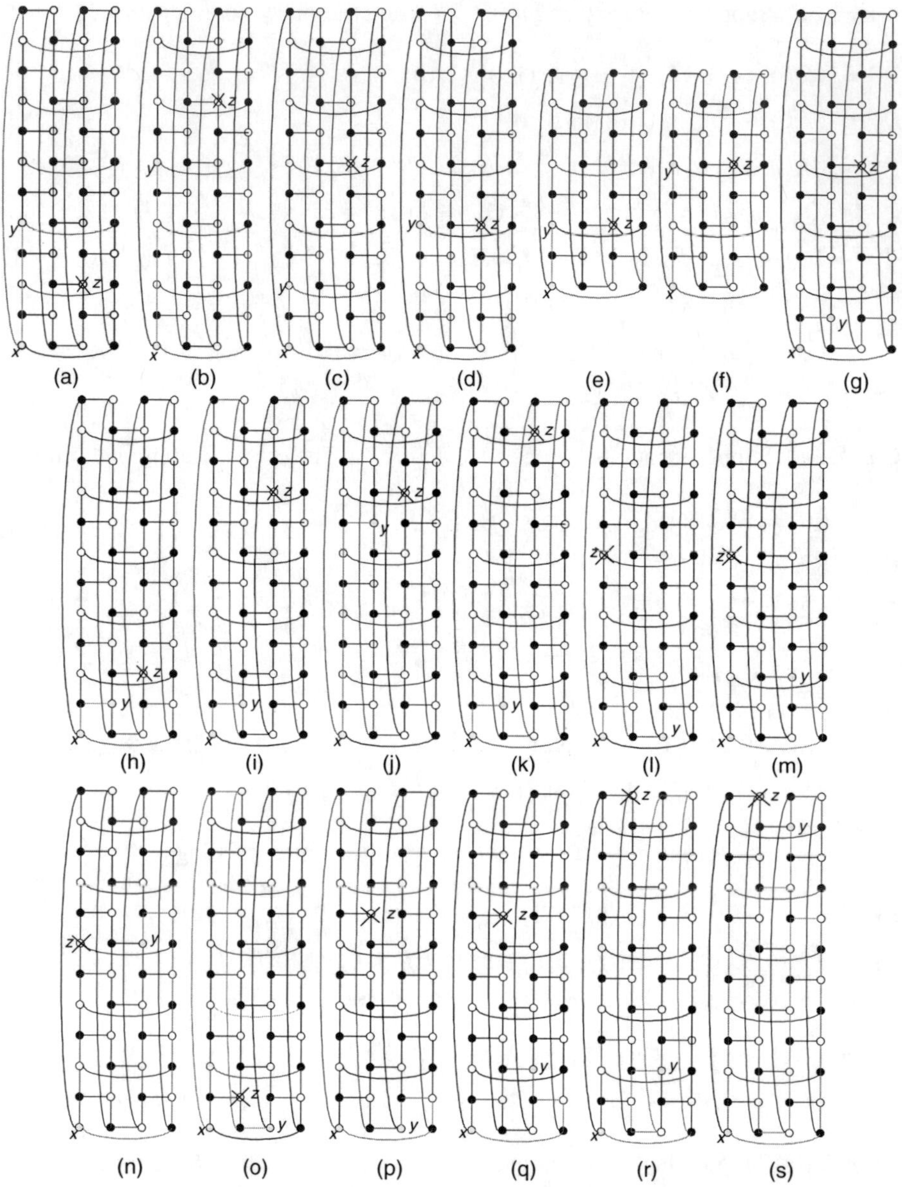

FIGURE 15.15 Illustrations for Lemma 15.10: (a) Case 2.1, (b) Case 2.2, (c) Case 2.3, (d) Case 2.4, (e) Case 2.5, (f) Case 2.6, (g) Case 3.1, (h) Case 3.2, (i) Case 3.3, (j) Case 3.4, (k) Case 3.5, (l) Case 4.1, (m) Case 4.2, (n) Case 4.3, (o) Case 5.1, (p) Case 5.2, (q) Case 5.3, (r) Case 5.4, and (s) Case 5.5.

Now, we discuss the 3*-laceability for general HReT(m, n).

LEMMA 15.11 Assume that m and n are positive even integers with $m, n \geq 4$. Let x and y be any two vertices of HReT$(m, n) = (V_0 \cup V_1, E)$ with $x \in V_0$ and $y \in V_1$. Then there exists a regular 3*-container $C(x, y)$ of HReT(m, n).

Proof: Without loss of generality, we may assume that $x=(0,0)$ and $y=(i,j)$. In order to prove this lemma, we will construct a regular 3*-container $C(x,y)=\{P_1,P_2,P_3\}$ in HReT(m,n). We prove the lemma by induction on m. With Lemma 15.8, our theorem holds for $m=4$. Now, we consider the case where $m\geq 6$.

Suppose that $i<m-2$. By induction, there exists a regular 3*-container $C(x,y)=\{P_1,P_2,P_3\}$ in HReT$(m-2,n)$. By Lemma 15.5, $Cm-3((0,0),(i,j))$ forms a 3*-container of HReT(m,n). Suppose that $i\geq m-2$. By induction, there exists a regular $C(x,(i-2,j))=\{P_1,P_2,P_3\}$ in HReT$(m-2,n)$. By Lemma 15.5, $C_1((0,0),(i,j))$ forms a 3*-container of HReT(m,n). □

LEMMA 15.12 Assume that m and n are positive even integers with $m\geq 4$ and $n\geq 6$. Let x, y, and z be any three different vertices of HReT$(m,n)=(V_0\cup V_1,E)$ in V_0. Then there exists a regular 3*-container $C(x,y)$ of HReT$(m,n)-\{z\}$.

Proof: Without loss of generality, we may assume that $x=(0,0)$, $y=(i,j)$, and $z=(k,l)$. In order to prove this lemma, we will construct a regular 3*-container $C(x,y)=\{P_1,P_2,P_3\}$ in HReT$(m,n)-\{z\}$. We prove the lemma by induction on m. With Lemmas 15.9 and 15.10, our theorem holds for $m=4$. Now, we consider the case that $m\geq 6$.

Suppose that $i<m-2$ and $k<m-2$. By induction, there exists a regular 3*-container $C(x,y)=\{P_1,P_2,P_3\}$ in HReT$(m-2,n)-\{z\}$. By Lemma 15.5, $C_{m-3}((0,0),(i,j))$ forms a 3*-container of HReT$(m,n)-\{z\}$. Suppose that $i<m-2$ and $k\geq m-2$. By induction, there exists a regular 3*-container $C(x,y)=\{P_1,P_2,P_3\}$ in HReT$(m-2,n)-(k-2,l)$. By Lemma 15.5, $C_i((0,0),(i,j))$ forms a 3*-container of HReT$(m,n)-\{z\}$. Suppose that $i\geq m-2$ and $k<m-2$. By induction, there exists a regular 3*-container $C(x,(i-2,j))=\{P_1,P_2,P_3\}$ in HReT$(m-2,n)-\{z\}$. By Lemma 15.5, $C_k((0,0),(i,j))$ forms a 3*-container of HReT$(m,n)-\{z\}$. Suppose that $i\geq m-2$ and $k\geq m-2$. By induction, there exists a regular 3*-container $C(x,(i-2,j))=\{P_1,P_2,P_3\}$ in HReT$(m-2,n)-(k-2,l)$. By Lemma 15.5, $C1((0,0),(i,j))$ forms a 3*-container of HReT$(m,n)-\{z\}$. □

THEOREM 15.13 Assume that m and n are positive even integers with $n\geq 4$. Then HReT(m,n) is strongly 3*-laceable. Moreover, HReT(m,n) is hyper 3*-laceable if and only if $n\geq 6$ or $m=2$.

Proof: With Lemmas 15.6 and 15.11, HReT(m,n) is 3*-laceable if m, n are even integers with $n\geq 4$.

By Lemmas 15.7 and 15.12, HReT(m,n) is hyper 3*-laceable if m, n are even integers with $n\geq 6$ or $m=2$.

Now we consider the case HReT$(m,4)$ with m being an even integer and $m\geq 4$. We first prove that HReT$(m,4)$ is not hyper 3*-laceable.

To prove this fact, let $x=(1,1)$, $y=(1,3)$, and $z=(0,2)$. Suppose that there exists a 3*-container $C(x,y)=\{P_1,P_2,P_3\}$ of HReT$(m,4)-\{z\}$. Since $\deg_{\text{HReT}(m,4)-z}(v)=2$ for $v\in\{(0,1),(0,3),(3,2)\}$, $\langle(1,1),(1,2),(1,3)\rangle$ and $\langle(1,1),(0,1),(0,0),(0,3),(1,3)\rangle$ are two paths in $C_{3^*}(x,y)$. Without loss of generality, we assume that $P_1=\langle(1,1),(1,2),(1,3)\rangle$ and $P_2=\langle(1,1),(0,1),(0,0),(0,3),(1,3)\rangle$. Since $\deg_{\text{HReT}(m,4)-z}((1,1))=\deg_{\text{HReT}(m,4)-z}((1,3))=3$, $((1,3),(1,0))$ and $((1,0),$

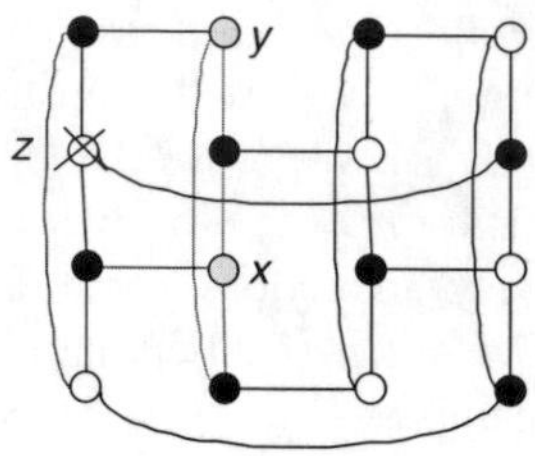

FIGURE 15.16 Illustration for Theorem 15.13.

(1, 1)) are edges in P_3. Thus $P_3 = \langle(1,1),(1,0),(1,3)\rangle$. Obviously, $\{P_1 \cup P_2 \cup P_3\}$ does not span HReT$(m,4) - \{z\}$. See Figure 15.16 for illustration. Hence, HReT$(m,4)$ is not hyper 3*-laceable.

Although any HReT$(m,4)$ with m is an even integer and $m \geq 4$ is not hyper 3*-laceable, we will prove that such HReT$(m,4)$ is strongly 3*-laceable by induction.

We first prove that HReT(4, 4) is strongly bi-3*-connected. Let x and y be any two different vertices in the same partite set of HReT(4, 4). Without loss of generality, we may assume that x and y are vertices in V_0 and $x = (0,0)$. We need to find a vertex z in $V_0 - \{x,y\}$ such that there exists a 3*-container $C(x,y) = \{P_1, P_2, P_3\}$ of HReT(4,4) $- \{z\}$. The corresponding vertex z and 3*-container $C(x,y)$ are listed as follows:

y	*z*	*C(x,y)*
(0, 2)	(1, 3)	$\langle(0,0),(0,1),(0,2)\rangle$
		$\langle(0,0),(0,3),(0,2)\rangle$
		$\langle(0,0),(3,0),(3,1),(2,1),(2,0),(1,0),(1,1),(1,2),(2,2),(2,3),(3,3),(3,2),(0,2)\rangle$
(1, 1)	(1, 3)	$\langle(0,0),(0,1),(1,1)\rangle$
		$\langle(0,0),(3,0),(3,1),(2,1),(2,0),(1,0),(1,1),\rangle$
		$\langle(0,0),(0,3),(0,2),(3,2),(3,3),(2,3),(2,2),(1,2),(1,1)\rangle$
(1, 3)	(0, 2)	$\langle(0,0),(0,3),(1,3)\rangle$
		$\langle(0,0),(0,1),(1,1),(1,2),(1,3)\rangle$
		$\langle(0,0),(3,0),Q_3(0,3),(3,3),(2,3),Q_2^{-1}(0,3),(2,0),(1,0),(1,3)\rangle$
(2, 0)	(0, 2)	$\langle(0,0),(0,3),(1,3),(1,0),(2,0)\rangle$
		$\langle(0,0),(3,0),(3,3),(3,2),(3,1),(2,1),(2,0)\rangle$
		$\langle(0,0),(0,1),(1,1),(1,2),(2,2),(2,3),(2,0)\rangle$
(2, 2)	(0, 2)	$\langle(0,0),(3,0),Q_3(0,3),(3,3),(2,3),(2,2)\rangle$
		$\langle(0,0),(0,3),(1,3),(1,0),(2,0),(2,1),(2,2)\rangle$
		$\langle(0,0),(0,1),(1,1),(1,2),(2,2)\rangle$
(3, 1)	(0, 2)	$\langle(0,0),(3,0),(3,1)\rangle$
		$\langle(0,0),(0,3),(1,3),(1,0),(2,0),(2,1),(3,1)\rangle$
		$\langle(0,0),(0,1),(1,1),(1,2),(2,2),(2,3),(3,3),(3,2),(3,1)\rangle$
(3, 3)	(0, 2)	$\langle(0,0),(3,0),(3,3)\rangle$
		$\langle(0,0),(0,3),(1,3),(1,0),(2,0),(2,1),(3,1),(3,2),(3,3)\rangle$
		$\langle(0,0),(0,1),(1,1),(1,2),(2,2),(2,3),(3,3)\rangle$

Obviously, all these 3*-containers of HReT$(4,4) - \{z\}$ are regular.

Now we consider the case HReT$(m,4)$ with $m > 4$. Without loss of generality, we may assume that $x = (0,0)$, $y = (i,j)$, and $z = (k,l)$. Suppose that $i < m-2$ and

$k < m-2$. By induction, there exists a regular 3*-container $C(x,y) = \{P_1, P_2, P_3\}$ in HReT$(m-2,4) - \{z\}$. By Lemma 15.5, $C_{m-3}((0,0),(i,j))$ forms a 3*-container of HReT$(m,4) - \{z\}$. Suppose that $i < m-2$ and $k \geq m-2$. By induction, there exists a regular 3*-container $C(x,y) = \{P_1, P_2, P_3\}$ in HReT$(m-2,4) - (k-2,l)$. By Lemma 15.5, $C_i((0,0),(i,j))$ forms a 3*-container of HReT$(m,4) - \{z\}$. Suppose that $i \geq m-2$ and $k < m-2$. By induction, there exists a regular $C(x,(i-2,j)) = \{P_1, P_2, P_3\}$ in HReT$(m-2,4) - \{z\}$. By Lemma 15.5, $C_k((0,0),(i,j))$ forms a 3*-container of HReT$(m,4) - \{z\}$. Suppose that $i \geq m-2$ and $k \geq m-2$. By induction, there exists a regular 3*-container $C(x,(i-2,j)) = \{P_1, P_2, P_3\}$ in HReT$(m-2,4) - (k-2,l)$. By Lemma 15.5, $C_1((0,0),(i,j))$ forms a 3*-container of HReT$(m,4) - \{z\}$.

Thus, the theorem is proved. □

Using similar approach as shown previously, we have the following theorem.

THEOREM 15.14 Assume that m and n are positive even integers with $n \geq 4$. Then HReT(m,n) is strongly Hamiltonian laceable. Moreover, HReT(m,n) is hyper Hamiltonian laceable if and only if $n \geq 6$ or $m = 2$.

Combining Theorems 15.11, 15.13, and 15.14, we have the following theorem.

THEOREM 15.15 Assume that m and n are positive even integers with $n \geq 4$. Then HReT(m,n) is super 3*-laceable if and only if $n \geq 6$ or $m = 2$.

In Ref. 193, Kao et al. proved that $PJ(k)$ is hyper 3*-laceable if and only if k is an odd integer with $k \geq 3$. Moreover, $P(n,1)$ is hyper 3*-laceable if and only if n is even and $n \geq 4$. Furthermore, Kao and Hsu [191] prove that every spider web network $SW(m,n)$ is hyper bi-3*-connected for $m \geq 4$ and $n \geq 2$.

15.6 COUNTEREXAMPLES OF 3*-LACEABLE GRAPHS

With Theorem 15.11, every cubic 3*-laceable graph is 1-edge fault-tolerant Hamiltonian. We would like to find examples of cubic 1-edge fault-tolerant Hamiltonian graphs that are not 3*-laceable.

Let $G_1 = (B_1 \cup W_1, E_1)$ be a bipartite graph and $u \in B_1$. Let $G_2 = (B_2 \cup W_2, E_2)$ be a bipartite graph and $v \in W_2$. Let $N(u) = \{u_i \mid i = 1,2,3\}$, $N(v) = \{v_i \mid i = 1,2,3\}$, and $G = J(G_1, N(u); G_2, N(v))$. Obviously, G is a bipartite graph with bipartition $B = (B_1 \cup B_2) - \{u\}$ and $W = (W_1 \cup W_2) - \{v\}$.

Suppose that both G_1 and G_2 are two 3*-laceable graphs with $u \in B_1$ and $v \in W_2$. By Theorem 15.11, G_1 and G_2 are 1-edge fault-tolerant Hamiltonian. By Theorem 12.7, G is also 1-edge fault-tolerant Hamiltonian.

Suppose that G is a 3*-laceable graph. Let $x \in (B_1 - \{u\})$ and $y \in (W_2 - \{v\})$. Then there exists a 3*-container $C(x,y) = \{P_1, P_2, P_3\}$ in G. Without loss of generality, we can write P_i as $\langle x, P_i^1, u_i, v_i, P_i^2, y\rangle$ for $i = 1,2,3$. Let $P_i' = \langle x, P_i^1, u_i, u\rangle$ for $i = 1,2,3$. Obviously, $\{P_i' \mid i = 1,2,3\}$ is a 3*-container between x and u in G_1. This is impossible, because x and u belong to the same partite set B_1. Thus G cannot be 3*-laceable. We have the following theorem.

THEOREM 15.16 Let G_1 and G_2 be two cubic 3*-laceable graphs. Let $u \in V(G_1)$ and $v \in V(G_2)$. Then $J(G_1, N(u); G_2, N(v))$ is 1-edge Hamiltonian, but not 3*-laceable.

With Theorem 15.16, we can easily construct cubic 1-edge Hamiltonian graphs that are not 3*-laceable.

With Theorem 15.12, every cubic hyper 3*-laceable graph is 1_p-fault-tolerant Hamiltonian. However, we have difficulty in finding a cubic 1_p-fault-tolerant Hamiltonian graph that is not hyper 3*-laceable.

16 Spanning Diameter

16.1 INTRODUCTION

Graph containers do exist in engineering designed information and telecommunication networks and in biological and neural systems. See [3,164] and their references. The study of w-container, w-wide distance, and their w^*-versions plays a pivotal role in design and implementation of parallel routing and efficient information transmission in large-scale networking systems. In biological informatics and neuroinformatics, the existence and structure of a w^*-container signifies the cascade effect in the signal transduction system and the reaction in a metabolic pathway.

Let G be a w^*-connected graph. We can define the *w^*-connected distance* between any two vertices u and v, $d_w(u, v)$, as $\min\{l(C(u, v)) \mid C(u, v)$ is a w^*-container$\}$. The *w^*-diameter* of G, denoted by $D^s_w(G)$, is defined as $\max\{d^s_w(u, v) \mid u$ and v are two different vertices of $G\}$. In particular, the *spanning wide diameter* of G is $D^s_{\kappa(G)}(G)$.

Similarly, let G be a w^*-laceable bipartite graph with bipartite vertex sets V_1 and V_2 and $|V_1 \cup V_2| \geq 2$. From the previous discussion, $|V_1| = |V_2|$. Similarly, we can define the *w^*-laceable distance* between any two vertices u and v from different partite sets, $d^{sL}_w(u, v)$, as $\min\{l(C(u, v)) \mid C(u, v)$ is a w^*-container$\}$. The *w^{*L}-diameter* of G, denoted by $D^{sL}_w(G)$, is defined as $\max\{d^{sL}_w(u, v) \mid u$ and v are vertices from different partite sets$\}$. In particular, the *spanning wide diameter* of G is $D^{sL}_{\kappa(G)}(G)$.

In this chapter, we evaluate the spanning wide diameter and the 2^{*L}-diameter of the star graph S_n. We also discuss the *w^{*L}-diameter* for hypercube Q_n with $1 \leq w \leq n-4$. Then, we evaluate the spanning wide diameter of some (n, k)-star graph $S_{n,k}$.

16.2 SPANNING DIAMETER FOR THE STAR GRAPHS

Lin et al. [235] started the study of the spanning diameter for the star graphs. We need some properties concerning the star graphs.

THEOREM 16.1 Assume that $\mathbf{r}$ and $\mathbf{s}$ are two adjacent vertices of S_n with $n \geq 5$. Then $S_n - \{\mathbf{r}, \mathbf{s}\}$ is Hamiltonian laceable.

Proof: Since S_n is vertex-transitive and edge-transitive, we assume that $\mathbf{r} = \mathbf{e}$ and $\mathbf{s} = (\mathbf{e})^2$. Obviously, both $\mathbf{e}$ and $(\mathbf{e})^2$ are in $S_n^{\{n\}}$. Let $\mathbf{u}$ be a white vertex and $\mathbf{v}$ be a black vertex of $S_n - \{\mathbf{e}, (\mathbf{e})^2\}$. We want to find a Hamiltonian path of $S_n - \{\mathbf{e}, (\mathbf{e})^2\}$ joining $\mathbf{u}$ to $\mathbf{v}$.

Case 1. $\mathbf{u}, \mathbf{v} \in S_n^{\{n\}}$. By Lemma 13.35, there is a Hamiltonian path P of $S_n^{\{n\}} - \{\mathbf{e}, (\mathbf{e})^2\}$ joining $\mathbf{u}$ to a black vertex $\mathbf{y}$ with $(\mathbf{y})_1 = 1$. We write P as $\langle \mathbf{u}, Q_1, \mathbf{x}, \mathbf{v}, Q_2, \mathbf{y}\rangle$. (Note that $l(Q_1) = 0$ if $\mathbf{u} = \mathbf{x}$ and $l(Q_2) = 0$ if $\mathbf{v} = \mathbf{y}$.) By Theorem 13.17, there is

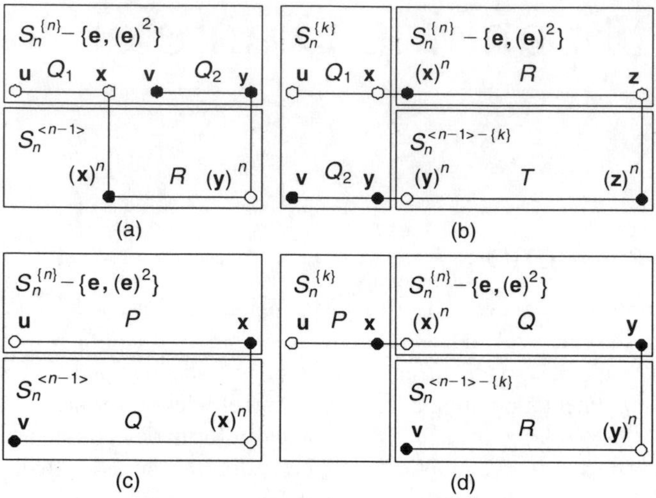

FIGURE 16.1 Illustration for Theorem 16.1.

a Hamiltonian path R of $S_n^{\langle n-1\rangle}$ joining the black vertex $(\mathbf{x})^n$ to the white vertex $(\mathbf{y})^n$. Then $\langle \mathbf{u}, Q_1, \mathbf{x}, (\mathbf{x})^n, R, (\mathbf{y})^n, \mathbf{y}, (Q_2)^{-1}, \mathbf{v}\rangle$ is the desired Hamiltonian path of $S_n - \{\mathbf{e}, (\mathbf{e})^2\}$ joining $\mathbf{u}$ to $\mathbf{v}$. See Figure 16.1a for illustration.

Case 2. $\mathbf{u}, \mathbf{v} \in S_n^{\{k\}}$ for some $k \in \langle n-1\rangle$. By Theorem 13.2, there is a Hamiltonian path P of $S_n^{\{k\}}$ joining $\mathbf{u}$ to $\mathbf{v}$. By Lemma 13.7, there are $(n-2)!/2 \geq 3$ edges joining white vertices of $S_n^{\{k\}}$ to black vertices of $S_n^{\{n\}}$. We can choose a white vertex $\mathbf{x}$ of $S_n^{\{k\}}$ with $(\mathbf{x})^n$ being a black vertex of $S_n^{\{n\}} - \{\mathbf{e}, (\mathbf{e})^2\}$. We write P as $\langle \mathbf{u}, Q_1, \mathbf{x}, \mathbf{y}, Q_2, \mathbf{v}\rangle$. (Note that $l(Q_1) = 0$ if $\mathbf{u} = \mathbf{x}$ and $l(Q_2) = 0$ if $\mathbf{v} = \mathbf{y}$.) Since $d(\mathbf{x}, \mathbf{y}) = 1$, by Lemma 13.8, $(\mathbf{y})_1 \in \langle n-1\rangle - \{k\}$. By Lemma 13.35, there is a Hamiltonian path R of $S_n^{\{n\}} - \{\mathbf{e}, (\mathbf{e})^2\}$ joining $(\mathbf{x})^n$ to a white vertex $\mathbf{z}$ with $(\mathbf{z})_1 \in \langle n-1\rangle - \{k\}$. By Theorem 13.17, there is a Hamiltonian path T of $S_n^{\langle n-1\rangle - \{k\}}$ joining the black vertex $(\mathbf{z})^n$ to the white vertex $(\mathbf{y})^n$. Then $\langle \mathbf{u}, Q_1, \mathbf{x}, (\mathbf{x})^n, R, \mathbf{z}, (\mathbf{z})^n, T, (\mathbf{y})^n, \mathbf{y}, Q_2, \mathbf{v}\rangle$ is the desired Hamiltonian path of $S_n - \{\mathbf{e}, (\mathbf{e})^2\}$ joining $\mathbf{u}$ to $\mathbf{v}$. See Figure 16.1b for illustration.

Case 3. $\mathbf{u} \in S_n^{\{n\}}$ and $\mathbf{v} \in S_n^{\{k\}}$ for some $k \in \langle n-1\rangle$. By Lemma 13.35, there is a Hamiltonian path P of $S_n^{\{n\}} - \{\mathbf{e}, (\mathbf{e})^2\}$ joining $\mathbf{u}$ to a black vertex $\mathbf{x}$ with $(\mathbf{x})_1 \in \langle n-1\rangle$. By Theorem 13.17, there is a Hamiltonian path Q of $S_n^{\langle n-1\rangle}$ joining the white vertex $(\mathbf{x})^n$ to $\mathbf{v}$. Then $\langle \mathbf{u}, P, \mathbf{x}, (\mathbf{x})^n, Q, \mathbf{v}\rangle$ is the desired Hamiltonian path of $S_n - \{\mathbf{e}, (\mathbf{e})^2\}$ joining $\mathbf{u}$ to $\mathbf{v}$. See Figure 16.1c for illustration.

Case 4. $\mathbf{u} \in S_n^{\{k\}}$ and $\mathbf{v} \in S_n^{\{l\}}$ with k, l, and n being distinct. By Lemma 13.7, there are $(n-2)!/2 \geq 3$ edges joining black vertices of $S_n^{\{k\}}$ to white vertices of $S_n^{\{n\}}$. We choose a black vertex $\mathbf{x}$ of $S_n^{\{k\}}$ with $(\mathbf{x})^n$ being a white vertex of $S_n^{\{n\}} - \{\mathbf{e}, (\mathbf{e})^2\}$. By Theorem 13.2, there is a Hamiltonian path P of $S_n^{\{k\}}$ joining $\mathbf{u}$ to $\mathbf{x}$. By Lemma 13.35, there is a Hamiltonian path Q of $S_n^{\{n\}} - \{\mathbf{e}, (\mathbf{e})^2\}$ joining $(\mathbf{x})^n$ to a black vertex $\mathbf{y}$ with $(\mathbf{y})_1 \in \langle n-1\rangle - \{k\}$. By Theorem 14.33, there is a Hamiltonian path R of

$S_n^{\langle n-1\rangle-\{k\}}$ joining the white vertex $(\mathbf{y})^n$ to $\mathbf{v}$. Then $\langle \mathbf{u}, P, \mathbf{x}, (\mathbf{x})^n, Q, \mathbf{y}, (\mathbf{y})^n, R, \mathbf{v}\rangle$ is the desired Hamiltonian path of $S_n - \{\mathbf{e}, (\mathbf{e})^2\}$ joining $\mathbf{u}$ to $\mathbf{v}$. See Figure 16.1d for illustration. □

LEMMA 16.1 Assume that $n \geq 5$. Suppose that $\mathbf{p}$ and $\mathbf{q}$ are two different white vertices of S_n, and $\mathbf{r}$ and $\mathbf{s}$ are two different black vertices of S_n. Then there exist two disjoint paths P_1 and P_2 such that (1) P_1 joins $\mathbf{p}$ to $\mathbf{r}$, (2) P_2 joins $\mathbf{q}$ to $\mathbf{s}$, and (3) $P_1 \cup P_2$ spans S_n.

Proof: Since S_n is edge-transitive, we assume that $\mathbf{p} \in S_n^{\{n\}}$ and $\mathbf{q} \in S_n^{\{n-1\}}$. Suppose that $\mathbf{r} \in S_n^{\{i\}}$ and $\mathbf{s} \in S_n^{\{j\}}$.

Case 1. $i, j \in \langle n-2\rangle$ with $i \neq j$. By Theorem 13.17, there is a Hamiltonian path P_1 of $S_n^{\{i,n\}}$ joining $\mathbf{p}$ to $\mathbf{r}$. Again, there is a Hamiltonian path P_2 of $S_n^{\langle n-1\rangle-\{i\}}$ joining $\mathbf{q}$ to $\mathbf{s}$. Then P_1 and P_2 are the desired paths.

Case 2. $i, j \in \langle n-2\rangle$ with $i = j$. We can choose a white vertex $\mathbf{x}$ with $\mathbf{x}$ being a neighbor of $\mathbf{s}$ in $S_n^{\{i\}}$ and $(\mathbf{x})_1 \in \langle n-1\rangle - \{i\}$. By Lemma 13.35, there is a Hamiltonian path Q of $S_n^{\{i\}} - \{\mathbf{s}, \mathbf{x}\}$ joining $\mathbf{r}$ to a white vertex $\mathbf{y}$ with $(\mathbf{y})_1 = n$. By Theorem 13.17, there is a Hamiltonian path P of $S_n^{\{n\}}$ joining $\mathbf{p}$ to the black vertex $(\mathbf{y})^n$. Moreover, there is a Hamiltonian path R of $S_n^{\langle n-1\rangle-\{i\}}$ joining $\mathbf{q}$ to the black vertex $(\mathbf{x})^n$. Then $P_1 = \langle \mathbf{p}, P, (\mathbf{y})^n, \mathbf{y}, Q^{-1}, \mathbf{r}\rangle$ and $P_2 = \langle \mathbf{q}, R, (\mathbf{x})^n, \mathbf{x}, \mathbf{s}\rangle$ are the desired paths.

Case 3. Either ($i = n$ and $j \in \langle n-1\rangle$) or ($i \in \langle n\rangle - \{n-1\}$ and $j = n-1$). By symmetry, we assume that $i = n$ and $j \in \langle n-1\rangle$. By Theorem 13.17, there is a Hamiltonian path P_1 of $S_n^{\{n\}}$ joining $\mathbf{p}$ to $\mathbf{r}$. Moreover, there is a Hamiltonian path P_2 of $S_n^{\langle n-1\rangle}$ joining $\mathbf{q}$ to $\mathbf{s}$. Then P_1 and P_2 are the desired paths.

Case 4. Either ($i = n-1$ and $j \in \langle n-2\rangle$) or ($i \in \langle n-2\rangle$ and $j = n$). By symmetry, we assume that $i = n-1$ and $j \in \langle n-2\rangle$. By Lemma 13.7, there exist $(n-2)!/2 \geq 3$ edges joining white vertices of $S_n^{\{n-1\}}$ to black vertices of $S_n^{\{n\}}$. We can choose a white vertex $\mathbf{x}$ in $S_n^{\{n-1\}} - \{\mathbf{q}\}$ with $(\mathbf{x})_1 = n$. By Theorem 13.17, there is a Hamiltonian path R of $S_n^{\{n-1\}}$ joining $\mathbf{q}$ to $\mathbf{r}$. We write R as $\langle \mathbf{q}, R_1, \mathbf{y}, \mathbf{x}, R_2, \mathbf{r}\rangle$. By Theorem 13.17, there is a Hamiltonian path P of $S_n^{\{n\}}$ joining $\mathbf{p}$ to the black vertex $(\mathbf{x})^n$. Since $d(\mathbf{x}, \mathbf{y}) = 1$, by Lemma 13.8, $(\mathbf{y})^n \in S_n^{\langle n-2\rangle}$. By Theorem 13.17, there exists a Hamiltonian path Q of $S_n^{\langle n-2\rangle}$ joining the white vertex $(\mathbf{y})^n$ to $\mathbf{s}$. Then $P_1 = \langle \mathbf{p}, P, (\mathbf{x})^n, \mathbf{x}, R_2, \mathbf{r}\rangle$ and $P_2 = \langle q, R_1, \mathbf{y}, (\mathbf{y})^n, Q, \mathbf{s}\rangle$ are the desired paths.

Case 5. $i = n-1$ and $j = n$. By Theorem 13.17, there is a Hamiltonian path Q of $S_n^{\{n\}}$ joining $\mathbf{p}$ to $\mathbf{s}$. Again, there is a Hamiltonian path R of $S_n^{\{n-1\}}$ joining $\mathbf{q}$ to $\mathbf{r}$. We choose a white vertex $\mathbf{x} \in S_n^{\{n\}}$ with $(\mathbf{x})_1 = n-1$. We write Q as $\langle \mathbf{p}, Q_1, \mathbf{x}, \mathbf{y}, Q_2, \mathbf{s}\rangle$ and write R as $\langle \mathbf{q}, R_1, \mathbf{w}, (\mathbf{x})^n, R_2, \mathbf{r}\rangle$. Obviously, $\mathbf{y}$ is a black vertex and $\mathbf{w}$ is a white vertex. Since $d(\mathbf{x}, \mathbf{y}) = 1$, by Lemma 13.8, $(\mathbf{y})_1 \in \langle n-2\rangle$. Since $d((\mathbf{x})^n, \mathbf{w}) = 1$, by Lemma 13.8, $(\mathbf{w})_1 \in \langle n-2\rangle$. By Theorem 13.17, there exists a Hamiltonian path W of $S_n^{\langle n-2\rangle}$ joining the black vertex $(\mathbf{w})^n$ to the white vertex $(\mathbf{y})^n$. Then $P_1 = \langle \mathbf{p}, Q_1, \mathbf{x}, (\mathbf{x})^n, R_2, \mathbf{r}\rangle$ and $P_2 = \langle \mathbf{q}, R_1, \mathbf{w}, (\mathbf{w})^n, W, (\mathbf{y})^n, \mathbf{y}, Q_2, \mathbf{s}\rangle$ are the desired paths.

Case 6. Either $i=j=n$ or $i=j=n-1$. By symmetry, we assume that $i=j=n$. By Theorem 13.17, there is a Hamiltonian path P of $S_n^{\{n\}}$ joining $\mathbf{p}$ to $\mathbf{s}$. We can write P as $\langle \mathbf{p}, R_1, \mathbf{r}, \mathbf{x}, R_2, \mathbf{s} \rangle$. By Theorem 13.17, there is a Hamiltonian path Q of $S_n^{\langle n-1 \rangle}$ joining $\mathbf{q}$ to the black vertex $(\mathbf{x})^n$. Then $P_1 = \langle \mathbf{p}, R_1, \mathbf{r} \rangle$ and $P_2 = \langle \mathbf{q}, Q, (\mathbf{x})^n, \mathbf{x}, R_2, \mathbf{s} \rangle$ are the desired paths. □

Let $\mathbf{u}$ be a vertex of S_n with $n \geq 4$ and let m be any integer with $3 \leq m \leq n$. We set $F_m(\mathbf{u}) = \{(\mathbf{u})^i \mid 3 \leq i \leq m\} \cup \{((\mathbf{u})^i)^{i-1} \mid 3 \leq i \leq m\}$.

LEMMA 16.2 Assume that $\mathbf{u}$ is a white vertex of S_n and $j \in \langle n \rangle$ with $n \geq 4$. Then there is a Hamiltonian path P of $S_n - F_n(\mathbf{u})$ joining $\mathbf{u}$ to some black vertex $\mathbf{v}$ with $(\mathbf{v})_1 = j$.

Proof: We prove this lemma by induction on n. Since S_n is vertex-transitive, we assume that $\mathbf{u} = \mathbf{e}$. Suppose that $n = 4$. The required Hamiltonian paths of $S_4 - F_4(\mathbf{e})$ are listed here:

$j=1$	⟨1234, 2134, 3124, 4123, 1423, 3421, 2431, 1432, 4132, 3142, 2143, 1243, 4213, 2413, 3412, 4312, 1342, 2341, 4321, 1324⟩
$j=2$	⟨1234, 2134, 4132, 3142, 1342, 4312, 3412, 1432, 2431, 3421, 1423, 2413, 4213, 1243, 2143, 4123, 3124, 1324, 4321, 2341⟩
$j=3$	⟨1234, 2134, 4132, 1432, 2431, 3421, 1423, 4123, 3124, 1324, 4321, 2341, 1342, 4312, 3412, 2413, 4213, 1243, 2143, 3142⟩
$j=4$	⟨1234, 2134, 3124, 1324, 4321, 2341, 1342, 3142, 4132, 1432, 2431, 3421, 1423, 4123, 2143, 1243, 4213, 2413, 3412, 4312⟩

Assume that this statement holds for any S_k for every $4 \leq k \leq n-1$. We have $F_n(\mathbf{e}) = F_{n-1}(\mathbf{e}) \cup \{(\mathbf{e})^n, ((\mathbf{e})^n)^{n-1}\}$. By induction, there is a Hamiltonian path P of $S_n^{\{n\}} - F_{n-1}(\mathbf{e})$ joining $\mathbf{e}$ to a black vertex $\mathbf{x}$ with $(\mathbf{x})_1 = 1$. By Lemma 13.35, there is a Hamiltonian path Q of $S_n^{\{1\}} - \{(\mathbf{e})^n, ((\mathbf{e})^n)^{n-1}\}$ joining the white vertex $(\mathbf{x})^n$ to a black vertex $\mathbf{y}$ with $(\mathbf{y})_1 = 2$. We can choose a black vertex $\mathbf{z}$ of $S_n^{\langle n-1 \rangle - \{1\}}$ with $(\mathbf{z})_1 = j$. By Theorem 13.17, there exists a Hamiltonian path R of $S_n^{\langle n-1 \rangle - \{1\}}$ joining the white vertex $(\mathbf{y})^n$ to $\mathbf{z}$. Then $\langle \mathbf{e}, P, \mathbf{x}, (\mathbf{x})^n, Q, \mathbf{y}, (\mathbf{y})^n, R, \mathbf{z} \rangle$ is a desired Hamiltonian path. □

LEMMA 16.3 Let $\mathbf{u} = u_1u_2u_3u_4$ be any white vertex of S_4. There exist three paths P_1, P_2, and P_3 such that (1) P_1 joins $\mathbf{u}$ to the black vertex $u_2u_4u_1u_3$ with $l(P_1) = 7$, (2) P_2 joins $\mathbf{u}$ to the white vertex $u_3u_4u_1u_2$ with $l(P_2) = 8$, (3) P_3 joins $\mathbf{u}$ to the white vertex $u_4u_1u_3u_2$ with $l(P_3) = 8$, and (4) $P_1 \cup P_2 \cup P_3$ spans S_4.

Proof: Since S_4 is vertex-transitive, we assume that $\mathbf{u} = 1234$. Then we set

$$P_1 = \langle 1234, 3214, 4213, 1243, 2143, 4123, 1423, 2413 \rangle$$

$$P_2 = \langle 1234, 4231, 3241, 2341, 4321, 3421, 2431, 1432, 3412 \rangle$$

$$P_3 = \langle 1234, 2134, 3124, 1324, 2314, 4312, 1342, 3142, 4132 \rangle$$

Obviously, P_1, P_2, and P_3 are the desired paths. □

LEMMA 16.4 Let $\mathbf{u} = u_1u_2u_3u_4$ be any white vertex of S_4. Let $i_1i_2i_3$ be a permutation of u_2, u_3, and u_4. There exist four paths P_1, P_2, P_3, and P_4 of S_4 such that (1) P_1 joins $\mathbf{u}$ to a white vertex $\mathbf{w}$ with $(\mathbf{w})_1 = i_1$ and $l(P_1) = 2$, (2) P_2 joins $\mathbf{u}$ to a white vertex $\mathbf{x}$ with $(\mathbf{x})_1 = i_2$ and $l(P_2) = 2$, (3) P_3 joins $\mathbf{u}$ to a black vertex $\mathbf{y}$ with $(\mathbf{y})_1 = i_3$ and $l(P_3) = 19$, (4) P_4 joins $\mathbf{u}$ to a black vertex $\mathbf{z}$ with $\mathbf{z} \neq \mathbf{y}$,

$(\mathbf{z})_1 = i_3$, and $l(P_4) = 19$, (5) $P_1 \cup P_2 \cup P_3$ spans S_4, (6) $P_1 \cup P_2 \cup P_4$ spans S_4, (7) $V(P_1) \cap V(P_2) \cap V(P_3) = \{\mathbf{u}\}$, and (8) $V(P_1) \cap V(P_2) \cap V(P_4) = \{\mathbf{u}\}$.

Proof: Since S_4 is vertex-transitive, we assume that $\mathbf{u} = 1234$. Since $\mathbf{u} = 1234$, we have $\{i_1, i_2\} \subset \{2, 3, 4\}$ and $i_3 \in \{2, 3, 4\} - \{i_1, i_2\}$. Without loss of generality, we suppose that $i_1 < i_2$. The required four paths are listed here:

$i_1 = 2$	$P_1 = \langle 1234, 4231, 2431 \rangle$
$i_2 = 3$	$P_2 = \langle 1234, 2134, 3124 \rangle$
$i_3 = 4$	$P_3 = \langle 1234, 3214, 2314, 1324, 4321, 3421, 1423, 4123, 2143, 3142, 4132, 1432, 3412, 2413, 4213, 1243, 3241, 2341, 1342, 4312 \rangle$
	$P_4 = \langle 1234, 3214, 2314, 1324, 4321, 3421, 1423, 2413, 4213, 1243, 3241, 2341, 1342, 4312, 3412, 1432, 4132, 3142, 2143, 4123 \rangle$
$i_1 = 2$	$P_1 = \langle 1234, 4231, 2431 \rangle$
$i_2 = 4$	$P_2 = \langle 1234, 3214, 4213 \rangle$
$i_3 = 3$	$P_3 = \langle 1234, 2134, 3124, 4123, 2143, 1243, 3214, 2314, 1342, 4312, 2314, 1324, 4321, 3421, 1423, 2413, 3412, 1432, 4132, 3142 \rangle$
	$P_4 = \langle 1234, 2134, 3142, 1324, 2314, 4312, 1342, 3142, 4132, 1432, 3412, 2413, 1423, 4123, 2143, 1243, 3241, 2341, 4321, 3421 \rangle$
$i_1 = 3$	$P_1 = \langle 1234, 2134, 3124 \rangle$
$i_2 = 4$	$P_2 = \langle 1234, 3214, 4213 \rangle$
$i_3 = 2$	$P_3 = \langle 1234, 4231, 3241, 1243, 2143, 4123, 1423, 2413, 3412, 4312, 2314, 1324, 4321, 3421, 2431, 1432, 4132, 3142, 1342, 2341 \rangle$
	$P_4 = \langle 1234, 4231, 3241, 1243, 2143, 4123, 1423, 3421, 2431, 1432, 4132, 3142, 1342, 2341, 4321, 1324, 2314, 4312, 3412, 2413 \rangle$

Thus, this statement is proved. □

LEMMA 16.5 Assume that $n \geq 5$ and $i_1 i_2 \ldots i_{n-1}$ is an $(n-1)$-permutation on $\langle n \rangle$. Let $\mathbf{u}$ be any white vertex of S_n. Then there exist $(n-1)$ paths $P_1, P_2, \ldots, P_{n-1}$ of S_n such that (1) P_1 joins $\mathbf{u}$ to a black vertex $\mathbf{y_1}$ with $(\mathbf{y_1})_1 = i_1$ and $l(P_1) = n(n-2)! - 1$, (2) P_j joins $\mathbf{u}$ to a white vertex $\mathbf{y_j}$ with $(\mathbf{y_j})_1 = i_j$ and $l(P_j) = n(n-2)!$ for every $2 \leq j \leq n-1$, (3) $\bigcup_{j=1}^{n-1} P_j$ spans S_n, and (4) $\bigcap_{j=1}^{n-1} V(P_j) = \{\mathbf{u}\}$.

Proof: The proof of this lemma is rather tedious. The authors strongly suggest that the reader skim over the proof first and examine the details later.

Since S_n is vertex-transitive, we assume that $\mathbf{u} = \mathbf{e}$. Without loss of generality, we suppose that $i_2 < i_3 < \cdots < i_{n-1}$.

Case 1. $n = 5$. Hence, $n(n-2)! = 30$. We have $i_2 \neq 4$, $i_3 \geq 2$, and $i_4 \geq 3$. We set $\mathbf{x_1} = (\mathbf{e})^5$ and $\mathbf{x_i} = ((\mathbf{x_{i-1}})^i)^5$ for every $2 \leq i \leq 4$, and $\mathbf{x_5} = ((\mathbf{x_4})^3)^5$. Note that $\mathbf{x_i}$ is a black vertex in $S_n^{\{i\}}$ for every $i \in \langle 4 \rangle$ and $\mathbf{x_5}$ is a black vertex in $S_n^{\{1\}}$. Obviously, $\mathbf{x_1} \neq \mathbf{x_5}$. We set $H = \langle \mathbf{e}, \mathbf{x_1}, (\mathbf{x_1})^2, \mathbf{x_2}, (\mathbf{x_2})^3, \mathbf{x_3}, (\mathbf{x_3})^4, \mathbf{x_4}, (\mathbf{x_4})^3, \mathbf{x_5} \rangle$.

Case 1.1. $i_1 = 3$. We have $i_2 \neq 4$, $i_3 \neq 3$, and $i_4 \neq 1$. Let $\mathbf{u_1} = 24135$, $\mathbf{u_2} = 41325$, and $\mathbf{u_3} = 34125$. We set

$$W_1 = \langle \mathbf{e} = 12345, 32145, 42135, 12435, 21435, 41235, 14235, 24135 = \mathbf{u_1} \rangle$$

$$W_2 = \langle \mathbf{e} = 12345, 21345, 31245, 13245, 23145, 43125, 13425, 31425, 41325 = \mathbf{u_2} \rangle$$

$$W_3 = \langle \mathbf{e} = 12345, 42315, 32415, 23415, 43215, 34215, 24315, 14325, 34125 = \mathbf{u_3} \rangle$$

Obviously, $W_1 \cup W_2 \cup W_3$ spans $S_5^{\{5\}}$ and $V(W_i) \cap V(W_j) = \{\mathbf{e}\}$ for every $i, j \in \langle 3 \rangle$ with $i \neq j$. By Lemma 13.35, there exists a Hamiltonian path Q_1 of $S_5^{\{2\}} - \{\mathbf{x_2}, (\mathbf{x_2})^3\}$ joining the white vertex $(\mathbf{u_1})^5$ to a black vertex $\mathbf{y_1}$ with $(\mathbf{y_1})_1 = i_1$. Again, there exists a Hamiltonian path Q_2 of $S_5^{\{4\}} - \{\mathbf{x_4}, (\mathbf{x_4})^3\}$ joining the black vertex $(\mathbf{u_2})^5$ to a white vertex $\mathbf{y_2}$ with $(\mathbf{y_2})_1 = i_2$. Moreover, there exists a Hamiltonian path Q_3 of $S_5^{\{3\}} - \{\mathbf{x_3}, (\mathbf{x_3})^4\}$ joining the black vertex $(\mathbf{u_3})^5$ to a white vertex $\mathbf{y_3}$ with $(\mathbf{y_3})_1 = i_3$. Similarly, there exists a Hamiltonian path Q_4 of $S_5^{\{1\}} - \{\mathbf{x_1}, (\mathbf{x_1})^2\}$ joining the black vertex $\mathbf{x_5}$ to a white vertex $\mathbf{y_4}$ with $(\mathbf{y_4})_1 = i_4$. We set

$$\begin{aligned}
P_1 &= \langle \mathbf{e}, W_1, \mathbf{u_1}, (\mathbf{u_1})^5, Q_1, \mathbf{y_1} \rangle \\
P_2 &= \langle \mathbf{e}, W_2, \mathbf{u_2}, (\mathbf{u_2})^5, Q_2, \mathbf{y_2} \rangle \\
P_3 &= \langle \mathbf{e}, W_3, \mathbf{u_3}, (\mathbf{u_3})^5, Q_3, \mathbf{y_3} \rangle \\
P_4 &= \langle \mathbf{e}, H, \mathbf{x_5}, Q_4, \mathbf{y_4} \rangle
\end{aligned}$$

Obviously, $l(P_1) = 29$ and $l(P_i) = 30$ for every $2 \le i \le 4$. Apparently, P_1, P_2, P_3, and P_4 are the desired paths. See Figure 16.2a for illustration.

Case 1.2. $i_1 \neq 3$. We have $i_2 \neq 4$, $i_3 \neq 1$, and $i_4 \neq 2$. Let $\mathbf{u_1} = 31425$, $\mathbf{u_2} = 42135$, and $\mathbf{u_3} = 21435$. We set

$$\begin{aligned}
W_1 &= \langle \mathbf{e} = 12345, 21345, 41325, 14325, 34125, 43125, 13425, 31425 = \mathbf{u_1} \rangle \\
W_2 &= \langle \mathbf{e} = 12345, 32145, 23145, 13245, 31245, 41235, 14235, 24135, 42135 = \mathbf{u_2} \rangle \\
W_3 &= \langle \mathbf{e} = 12345, 42315, 24315, 34215, 43215, 23415, 32415, 12435, 21435 = \mathbf{u_3} \rangle
\end{aligned}$$

Obviously, $W_1 \cup W_2 \cup W_3$ spans $S_5^{\{5\}}$ and $V(W_i) \cap V(W_j) = \{\mathbf{e}\}$ for every $i, j \in \langle 3 \rangle$ with $i \neq j$. By Lemma 13.35, there exists a Hamiltonian path Q_1 of $S_5^{\{3\}} - \{\mathbf{x_3}, (\mathbf{x_3})^4\}$ joining

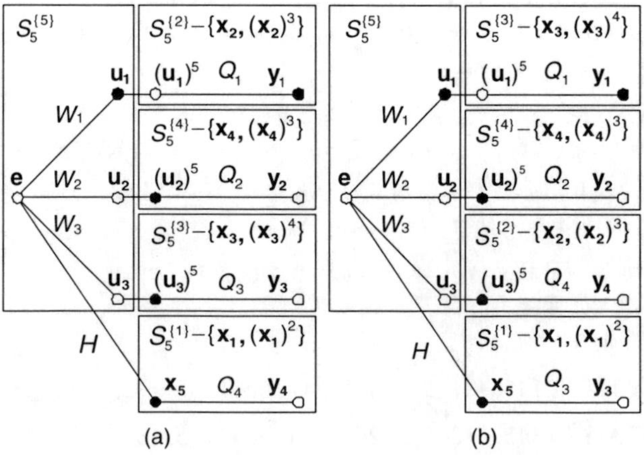

FIGURE 16.2 Illustration of Case 1.

the white vertex $(\mathbf{u_1})^5$ to a black vertex $\mathbf{y_1}$ with $(\mathbf{y_1})_1 = i_1$. Again, there exists a Hamiltonian path Q_2 of $S_5^{\{4\}} - \{\mathbf{x_4}, (\mathbf{x_4})^3\}$ joining the black vertex $(\mathbf{u_2})^5$ to a white vertex $\mathbf{y_2}$ with $(\mathbf{y_2})_1 = i_2$. Moreover, there exists a Hamiltonian path Q_3 of $S_5^{\{1\}} - \{\mathbf{x_1}, (\mathbf{x_1})^2\}$ joining the black vertex $\mathbf{x_5}$ to a white vertex $\mathbf{y_3}$ with $(\mathbf{y_3})_1 = i_3$. Similarly, there exists a Hamiltonian path Q_4 of $S_5^{\{2\}} - \{\mathbf{x_2}, (\mathbf{x_2})^3\}$ joining the black vertex $(\mathbf{u_3})^5$ to a white vertex $\mathbf{y_4}$ with $(\mathbf{y_4})_1 = i_4$. We set

$$\begin{aligned} P_1 &= \langle \mathbf{e}, W_1, \mathbf{u_1}, (\mathbf{u_1})^5, Q_1, \mathbf{y_1} \rangle \\ P_2 &= \langle \mathbf{e}, W_2, \mathbf{u_2}, (\mathbf{u_2})^5, Q_2, \mathbf{y_2} \rangle \\ P_3 &= \langle \mathbf{e}, H, \mathbf{x_5}, Q_3, \mathbf{y_3} \rangle \\ P_4 &= \langle \mathbf{e}, W_3, \mathbf{u_3}, (\mathbf{u_3})^5, Q_4, \mathbf{y_4} \rangle \end{aligned}$$

Obviously, $l(P_1) = 29$ and $l(P_i) = 30$ for every $2 \le i \le 4$. Apparently, P_1, P_2, P_3, and P_4 are the desired paths. See Figure 16.2b for illustration.

Case 2. $n \ge 6$. Since $n - 1 \ge 5$, we have $i_k \ne k + 2$ for every $2 \le k \le n - 4$, $i_{n-3} \ne 1$, $i_{n-2} \ne 2$, and $i_{n-1} \ne 3$. We set $\mathbf{u_j} = (\mathbf{e})^{j+2}$ and $\mathbf{v_j} = ((\mathbf{e})^{j+2})^{j+1}$ for every $j \in \langle n-4 \rangle$. Thus, $\mathbf{u_j}$ is a black vertex in $S_n^{\{(n-1,n)\}}$ and $\mathbf{v_j}$ is a white vertex in $S_n^{\{(n-1,n)\}}$ for every $j \in \langle n-4 \rangle$. Note that $F_{n-2}(\mathbf{e}) = \{\mathbf{u_j} \mid j \in \langle n-4 \rangle\} \cup \{\mathbf{v_j} \mid j \in \langle n-4 \rangle\}$.

By Lemma 16.2, there is a Hamiltonian path P of $S_n^{\{(n-1,n)\}} - F_{n-2}(\mathbf{e})$ joining $\mathbf{e}$ to a black vertex $\mathbf{x_1}$ with $(\mathbf{x_1})_1 = 2$. We recursively set $\mathbf{x_j}$ as the unique neighbor of $(\mathbf{x_{j-1}})^{n-1}$ in $S_n^{\{(j,n)\}}$ with $(\mathbf{x_j})_1 = j + 1$ for every $2 \le j \le n - 4$, and we set $\mathbf{x_{n-3}}$ as the unique neighbor of $(\mathbf{x_{n-4}})^{n-1}$ in $S_n^{\{(n-3,n)\}}$ with $(\mathbf{x_{n-3}})_1 = n - 1$. It is easy to see that $\mathbf{x_j}$ is a black vertex for $1 \le j \le n - 3$ and $\{(\mathbf{x_j})^{n-1}, \mathbf{x_{j+1}}\} \subset S_n^{\{(j+1,n)\}}$ for $1 \le j \le n-4$. We construct P_j for every $1 \le j \le n - 1$ as follows:

1. $j \in \langle n-4 \rangle - \{1\}$. By Lemma 13.35, there is a Hamiltonian path T_j of $S_n^{\{(j+1,n)\}} - \{(\mathbf{x_j})^{n-1}, \mathbf{x_{j+1}}\}$ joining the black vertex $(\mathbf{v_j})^{n-1}$ to a white vertex $\mathbf{z_j}$ with $(\mathbf{z_j})_1 = j+2$. Again, there is a Hamiltonian path T_j' of $S_n^{\{j+2\}}$ joining the black vertex $(\mathbf{z_j})^n$ to a white vertex $\mathbf{y_j}$ with $(\mathbf{y_j})_1 = i_j$. Then we set P_j as $\langle \mathbf{e}, \mathbf{u_j}, \mathbf{v_j}, (\mathbf{v_j})^{n-1}, T_j, \mathbf{z_j}, (\mathbf{z_j})^n, T_j', \mathbf{y_j} \rangle$. Obviously, $l(P_j) = n(n-2)!$.
2. $j = n - 3$. We choose a white vertex $\mathbf{y_{n-3}}$ in $S_n^{\{1\}}$ with $(\mathbf{y_{n-3}})_1 = i_{n-3}$. Note that there are $((n-3)!/2)$ edges joining some black vertices of $S_n^{\{(n-2,n)\}}$ to some white vertices of $S_n^{\{1\}}$ and there are $((n-3)!/2)$ edges joining some white vertices of $S_n^{\{(n-2,n)\}}$ to some black vertices of $S_n^{\{1\}}$. We choose a white vertex $\mathbf{r}$ in $S_n^{\{1\}}$ with $(\mathbf{r})^n$ being a black vertex in $S_n^{\{(n-2,n)\}}$ and choose a black vertex $\mathbf{s}$ in $S_n^{\{1\}}$ with $(\mathbf{s})^n$ being a white vertex in $S_n^{\{(n-2,n)\}}$. By Lemma 16.1, there exist two disjoint paths H_1 and H_2 of $S_n^{\{1\}}$ such that (1) H_1 joins $(\mathbf{e})^n$ to $\mathbf{r}$, (2) H_2 joins $\mathbf{s}$ to $\mathbf{y_{n-3}}$, and (3) $H_1 \cup H_2$ spans $S_n^{\{1\}}$. By Theorem 13.17, there is a Hamiltonian path H of $S_n^{\{(n-2,n)\}}$ joining the black vertex $(\mathbf{r})^n$ to the white vertex $(\mathbf{s})^n$. We set P_{n-3} as $\langle \mathbf{e}, (\mathbf{e})^n, H_1, \mathbf{r}, (\mathbf{r})^n, H, (\mathbf{s})^n, \mathbf{s}, H_2, \mathbf{y_{n-3}} \rangle$. Obviously, $l(P_{n-3}) = n(n-2)!$.

3. $j = n-1$. By Lemma 13.35, there is a Hamiltonian path Q_1 of $S_n^{\{(2,n)\}} - \{(\mathbf{x_1})^{n-1}, \mathbf{x_2}\}$ joining the black vertex $(\mathbf{v_1})^{n-1}$ to a white vertex $\mathbf{q}$ with $(\mathbf{q})_1 = 3$. Again, there is a Hamiltonian path Q_2 of $S_n^{\{3\}}$ joining the black vertex $(\mathbf{q})^n$ to a white vertex $\mathbf{y_{n-1}}$ with $(\mathbf{y_{n-1}})_1 = i_{n-1}$. We set P_{n-1} as $\langle \mathbf{e}, \mathbf{u_1}, \mathbf{v_1}, (\mathbf{v_1})^{n-1}, Q_1, \mathbf{q}, (\mathbf{q})^n, Q_2, \mathbf{y_{n-1}} \rangle$. Obviously, $l(P_{n-1}) = n(n-2)!$.
4. We construct P_1 and P_{n-2} dependent on whether $i_1 = n-1$. We set L as $\langle \mathbf{x_1}, (\mathbf{x_1})^{n-1}, \mathbf{x_2}, (\mathbf{x_2})^{n-1}, \ldots, \mathbf{x_{n-4}}, (\mathbf{x_{n-4}})^{n-1}, \mathbf{x_{n-3}}, (\mathbf{x_{n-3}})^n \rangle$. By Theorem 13.2, there is a Hamiltonian path W of $S_n^{\{(1,n)\}}$ joining the black vertex $(\mathbf{e})^{n-1}$ to a white vertex $\mathbf{p}$ with $(\mathbf{p})_1 = 2$.

Suppose that $i_1 \neq n-1$. By Theorem 13.2, there is a Hamiltonian path R of $S_n^{\{n-1\}}$ joining the white vertex $(\mathbf{x_{n-3}})^n$ to a black vertex $\mathbf{y_1}$ with $(\mathbf{y_1})_1 = i_1$. Again, there exists a Hamiltonian path Z of $S_{n-1}^{\{2\}}$ joining the black vertex $(\mathbf{p})^n$ to a white vertex $\mathbf{y_{n-2}}$ with $(\mathbf{y_{n-2}})_1 = i_{n-2}$. We set P_1 as $\langle \mathbf{e}, P, \mathbf{x_1}, L, (\mathbf{x_{n-3}})^n, R, \mathbf{y_1} \rangle$ and P_{n-2} as $\langle \mathbf{e}, (\mathbf{e})^{n-1}, W, \mathbf{p}, (\mathbf{p})^n, Z, \mathbf{y_{n-2}} \rangle$. Obviously, $l(P_1) = n(n-2)! - 1$ and $l(P_{n-2}) = n(n-2)!$. Apparently, $P_1, P_2, \ldots, P_{n-1}$ are the desired paths. See Figure 16.3a for illustration for the case $n = 7$.

Suppose that $i_1 = n-1$. Note that $i_{n-2} \neq n-1$. Since $(\mathbf{x_{n-3}})^n$ is a white vertex in $S_n^{\{n-1\}}$ with $((\mathbf{x_{n-3}})^n)_1 = n$ and $((\mathbf{x_{n-3}})^n)_n = n-1$, there is a black vertex $\mathbf{z}$ in $S_n^{\{n-1\}}$ such that $\mathbf{z}$ is the unique neighbor of $(\mathbf{x_{n-3}})^n$ with $(\mathbf{z})_1 = 2$. Since $((\mathbf{x_{n-3}})^n)_{n-1} = n-3$, we have $(\mathbf{z})_{n-1} = n-3$. Note that $(\mathbf{z})^n$ is a white vertex in $S_n^{\{2\}}$ with $((\mathbf{z})^n)_{n-1} = n-3$. Since $(\mathbf{p})^n$ is a black vertex in $S_n^{\{2\}}$ with $((\mathbf{p})^n)_1 = n$ and $((\mathbf{p})^n)_n = 2$, there is a white vertex $\mathbf{t}$ in $S_n^{\{2\}}$ such that $\mathbf{t}$ is the unique neighbor

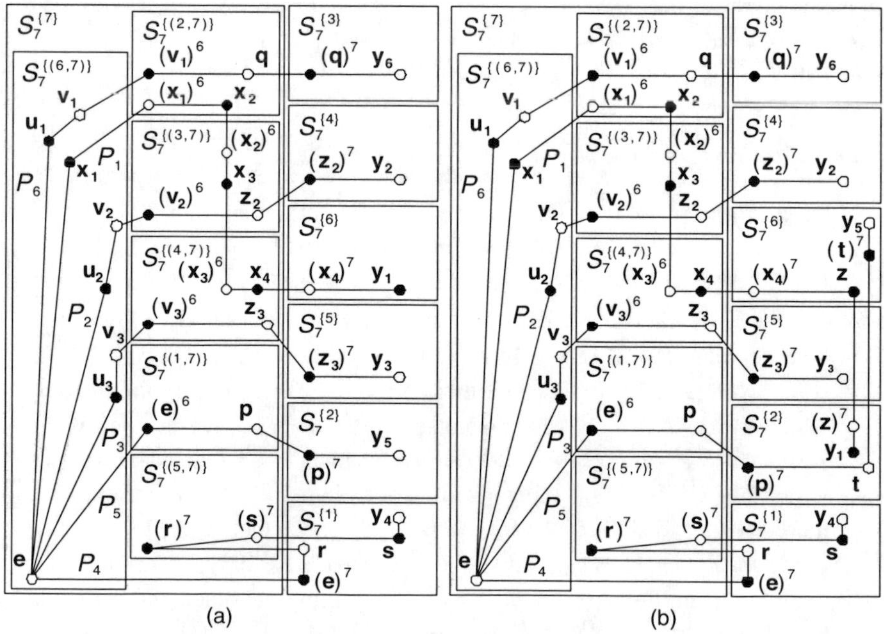

FIGURE 16.3 Illustration of Case 3 with $n = 7$.

TABLE 16.1
All Hamiltonian Cycles in S_4

⟨1234, 4231, 3241, 1243, 4213, 3214, 2314, 1324, 4321, 2341, 1342, 4312, 3412, 2413, 1423, 3421, 2431, 1432, 4132, 3142, 2143, 4123, 3124, 2134, 1234⟩
⟨1234, 4231, 3241, 1243, 4213, 2413, 3412, 1432, 2431, 3421, 1423, 4123, 2143, 3142, 4132, 2134, 3124, 1324, 4321, 2341, 1342, 4312, 2314, 3214, 1234⟩
⟨1234, 4231, 3241, 1243, 2143, 4123, 3124, 2134, 4132, 3142, 1342, 2341, 4321, 1324, 2314, 4312, 3412, 1432, 2431, 3421, 1423, 2413, 4213, 3214, 1234⟩
⟨1234, 4231, 3241, 2341, 1342, 4312, 2314, 1324, 4321, 3421, 2431, 1432, 3412, 2413, 1423, 4123, 3124, 2134, 4132, 3142, 2143, 1243, 4213, 3214, 1234⟩
⟨1234, 4231, 3241, 2341, 1342, 3142, 2143, 1243, 4213, 3214, 2314, 4312, 3412, 2413, 1423, 4123, 3124, 1324, 4321, 3421, 2431, 1432, 4132, 2134, 1234⟩
⟨1234, 4231, 3241, 2341, 4321, 3421, 2431, 1432, 4132, 3142, 1342, 4312, 3412, 2413, 1423, 4123, 2143, 1243, 4213, 3214, 2314, 1324, 3124, 2134, 1234⟩
⟨1234, 4231, 2431, 1432, 3412, 4312, 2314, 3214, 4213, 2413, 1423, 3421, 4321, 1324, 3124, 4123, 2143, 1243, 3241, 2341, 1342, 3142, 4132, 2134, 1234⟩
⟨1234, 4231, 2431, 1432, 4132, 2134, 3124, 1324, 4321, 3421, 1423, 4123, 2143, 3142, 1342, 2341, 3241, 1243, 4213, 2413, 3412, 4312, 2314, 3214, 1234⟩
⟨1234, 4231, 2431, 1432, 4132, 3142, 2143, 1243, 3241, 2341, 1342, 4312, 3412, 2413, 4213, 3214, 2314, 1324, 4321, 3421, 1423, 4123, 3124, 2134, 1234⟩
⟨1234, 4231, 2431, 3421, 1423, 2413, 3412, 1432, 4132, 2134, 3124, 4123, 2143, 3142, 1342, 4312, 2314, 1324, 4321, 2341, 3241, 1243, 4213, 3214, 1234⟩
⟨1234, 4231, 2431, 3421, 1423, 4123, 3124, 1324, 4321, 2341, 3241, 1243, 2143, 3142, 1342, 4312, 2314, 3214, 4213, 2413, 3412, 1432, 4132, 2134, 1234⟩
⟨1234, 4231, 2431, 3421, 4321, 2341, 3241, 1243, 4213, 2413, 1423, 4123, 2143, 3142, 1342, 4312, 3412, 1432, 4132, 2134, 3124, 1324, 2314, 3214, 1234⟩
⟨1234, 3214, 4213, 1243, 3241, 4231, 2431, 1432, 3412, 2413, 1423, 3421, 4321, 2341, 1342, 4312, 2314, 1324, 3124, 4123, 2143, 3142, 4132, 2134, 1234⟩
⟨1234, 3214, 4213, 2413, 3412, 4312, 2314, 1324, 3124, 4123, 1423, 3421, 4321, 2341, 1342, 3142, 2143, 1243, 3241, 4231, 2431, 1432, 4132, 2134, 1234⟩
⟨1234, 3214, 4213, 2413, 1423, 4123, 2143, 1243, 3241, 4231, 2431, 3421, 4321, 2341, 1342, 3142, 4132, 1432, 3412, 4312, 2314, 1324, 3124, 2134, 1234⟩
⟨1234, 3214, 2314, 4312, 1342, 3142, 4132, 1432, 3412, 2413, 4213, 1243, 2143, 4123, 1423, 3421, 2431, 4231, 3241, 2341, 4321, 1324, 3124, 2134, 1234⟩
⟨1234, 3214, 2314, 1324, 4321, 3421, 2431, 4231, 3241, 2341, 1342, 4312, 3412, 1432, 4132, 3142, 2143, 1243, 4213, 2413, 1423, 4123, 3124, 2134, 1234⟩
⟨1234, 3214, 2314, 1324, 3124, 4123, 2143, 1243, 4213, 2413, 1423, 3421, 4321, 2341, 3241, 4231, 2431, 1432, 3412, 4312, 1342, 3142, 4132, 2134, 1234⟩

of $(\mathbf{p})^n$ with $(\mathbf{t})_1 = n-1$. Since $(\mathbf{p})_{n-1} = 1$ and $(\mathbf{p})_n = n$, we have $((\mathbf{p})^n)_{n-1} = 1$ and $((\mathbf{p})^n)_1 = n$. Since $((\mathbf{p})^n)_{n-1} = 1$, we have $(\mathbf{t})_{n-1} = 1$. Since $((\mathbf{z})^n)_{n-1} = n-3$ and $(\mathbf{t})_{n-1} = 1$, we have $(\mathbf{z})^n \neq \mathbf{t}$. By Theorem 16.1, there is a Hamiltonian path W_1 of $S_n^{\{2\}} - \{(\mathbf{p})^n, \mathbf{t}\}$ joining $(\mathbf{z})^n$ to a black vertex $\mathbf{y_1}$ with $(\mathbf{y_1})_1 = i_1$. Again, there is a Hamiltonian path W_2 of $S_n^{\{n-1\}} - \{(\mathbf{x_{n-3}})^n, \mathbf{z}\}$ joining $(\mathbf{t})^n$ to a white vertex $\mathbf{y_{n-2}}$ with $(\mathbf{y_{n-2}})_1 = i_{n-2}$. We set P_1 as $\langle \mathbf{e}, P, \mathbf{x_1}, L, (\mathbf{x_{n-3}})^n, \mathbf{z}, (\mathbf{z})^n, W_1, \mathbf{y_1} \rangle$ and P_{n-2} as $\langle \mathbf{e}, (\mathbf{e})^{n-1}, W, \mathbf{p}, (\mathbf{p})^n, \mathbf{t}, (\mathbf{t})^n, W_2, \mathbf{y_{n-2}} \rangle$. Obviously, $l(P_1) = n(n-2)! - 1$ and $l(P_{n-2}) = n(n-2)!$. Apparently, $P_1, P_2, \ldots, P_{n-1}$ are the desired paths. See Figure 16.3b for illustration for the case where $n = 7$. □

Using depth-first search, we list all Hamiltonian cycles in S_4 in Table 16.1.

LEMMA 16.6 $D_3^{SL}(S_4) = 15$.

Proof: Let $\mathbf{u}$ be any white vertex in S_4 and let $\mathbf{v}$ be any black vertex in S_4. Since S_4 is vertex-transitive, we assume that $\mathbf{u} = 1234$. Suppose that $d(\mathbf{u}, \mathbf{v}) = 1$. Since S_4 is edge-transitive, we assume that $\mathbf{v} = 2134$. Let $\{P_1, P_2, P_3\}$ be a 3*-container joining $\mathbf{u}$ to $\mathbf{v}$. Since S_4 is 3-regular, one of three paths, say P_3, is $\langle \mathbf{u}, \mathbf{v} \rangle$. Thus, $P_1 \cup P_2^{-1}$ forms a Hamiltonian cycle of S_4 not using the edge $(\mathbf{u}, \mathbf{v})$. From Table 16.1, we obtain $d_3^{SL}(\mathbf{u}, \mathbf{v}) = 15$. Thus, $D_3^{SL}(S_4) \geq 15$. Suppose that $d(\mathbf{u}, \mathbf{v}) \neq 1$. Then $\mathbf{v} \in \{1324, 1243, 1432, 2413, 2341, 3142, 4123, 4312, 3421\}$. We find the following set of 3*-containers of S_4 between $\mathbf{u} = 1234$ and $\mathbf{v}$:

	$P_1 = \langle 1234, 4231, 3241, 1243, 2143, 4123, 3124, 1324 \rangle$
$C(1234, 1324)$	$P_2 = \langle 1234, 2134, 4132, 3142, 1342, 2341, 4321, 1324 \rangle$
	$P_3 = \langle 1234, 3214, 4213, 2413, 1423, 3421, 2431, 1432, 3412, 4312, 2314, 1324 \rangle$
	$P_1 = \langle 1234, 2134, 4132, 1432, 3412, 2413, 4213, 1243 \rangle$
$C(1234, 1243)$	$P_2 = \langle 1234, 3214, 2314, 4312, 1342, 3142, 2143, 1243 \rangle$
	$P_3 = \langle 1234, 4231, 2431, 3421, 1423, 4123, 3124, 1324, 4321, 2341, 3241, 1243 \rangle$

	$P_1 = \langle 1234, 3214, 2314, 1324, 4321, 3421, 2431, 1432\rangle$
$C(1234, 1432)$	$P_2 = \langle 1234, 4231, 3241, 2341, 1342, 4312, 3412, 1432\rangle$
	$P_3 = \langle 1234, 2134, 3124, 4123, 1423, 2413, 4213, 1243, 2143, 3142, 4132, 1432\rangle$
	$P_1 = \langle 1234, 3214, 4213, 2413\rangle$
$C(1234, 2413)$	$P_2 = \langle 1234, 4231, 2431, 3421, 4321, 2341, 3241, 1243, 2143, 4123, 1423, 2413\rangle$
	$P_3 = \langle 1234, 2134, 3124, 1324, 2314, 4312, 1342, 3142, 4132, 1432, 3412, 2413\rangle$
	$P_1 = \langle 1234, 4231, 3241, 2341\rangle$
$C(1234, 2341)$	$P_2 = \langle 1234, 3214, 2314, 4312, 3412, 2413, 4213, 1243, 2143, 3142, 1342, 2341\rangle$
	$P_3 = \langle 1234, 2134, 4132, 1432, 2431, 3421, 1423, 4123, 3124, 1324, 4321, 2341\rangle$
	$P_1 = \langle 1234, 2134, 4132, 3142\rangle$
$C(1234, 3142)$	$P_2 = \langle 1234, 4231, 3241, 2341, 4321, 3421, 2431, 1432, 3412, 4312, 1342, 3142\rangle$
	$P_3 = \langle 1234, 3241, 2341, 4321, 3421, 1423, 4123, 2143, 1243, 4213, 2413, 3412\rangle$
	$P_1 = \langle 1234, 2134, 3124, 4123\rangle$
$C(1234, 4123)$	$P_2 = \langle 1234, 3214, 4213, 2413, 3412, 4312, 2314, 1324, 4321, 3421, 1423, 4123\rangle$
	$P_3 = \langle 1234, 4231, 2431, 1432, 4132, 3142, 1342, 2341, 3241, 1243, 2143, 4213\rangle$
	$P_1 = \langle 1234, 3214, 2314, 4312\rangle$
$C(1234, 4312)$	$P_2 = \langle 1234, 2134, 4132, 3142, 2143, 4123, 3124, 1324, 4321, 2341, 1342, 4312\rangle$
	$P_3 = \langle 1234, 4231, 3241, 1243, 4213, 2413, 1423, 3421, 2431, 1432, 3412, 4312\rangle$
	$P_1 = \langle 1234, 4231, 2431, 3421\rangle$
$C(1234, 3421)$	$P_2 = \langle 1234, 2134, 3124, 4123, 2143, 3142, 4132, 1432, 3412, 2413, 1423, 3421\rangle$
	$P_3 = \langle 1234, 3214, 4213, 1243, 3241, 2341, 1342, 4312, 2314, 1324, 4321, 3421\rangle$

From this table, $d_3^{SL}(\mathbf{u}, \mathbf{v}) \leq 15$ if $d(\mathbf{u}, \mathbf{v}) \neq 1$. Hence, $D_3^{SL}(S_4) = 15$. □

LEMMA 16.7 $D_{n-1}^{SL}(S_n) \geq n!/(n-2) + 1 = (n-1)! + 2(n-2)! + 2(n-3)! + 1$ if $n \geq 5$.

Proof: Let $\mathbf{u}$ and $\mathbf{v}$ be two adjacent vertices of S_n. Obviously, $\mathbf{u}$ and $\mathbf{v}$ are in different partite sets. Let $\{P_1, P_2, \ldots, P_{n-1}\}$ be any $(n-1)^*$-container of S_n joining $\mathbf{u}$ to $\mathbf{v}$. Obviously, one of these paths is $\langle \mathbf{u}, \mathbf{v}\rangle$. Thus, $\max\{l(P_i) \mid 1 \leq i \leq n-1\} \geq \left\lceil \frac{n!-2}{n-2} \right\rceil + 1 = \left\lceil \frac{n!}{n-2} - \frac{2}{n-2} \right\rceil + 1 = n!/(n-2) + 1$. Hence, $d_{n-1}^{SL}(\mathbf{u}, \mathbf{v}) \geq n!/(n-2) + 1$ and $D_{n-1}^{SL}(S_n) \geq n!/(n-2) + 1$. □

LEMMA 16.8 $D_4^{SL}(S_5) \leq 41$.

Proof: Let $\mathbf{u}$ be any white vertex and $\mathbf{v}$ be any black vertex of S_5. Obviously, $d(\mathbf{u}, \mathbf{v})$ is odd.

Case 1. $d(\mathbf{u}, \mathbf{v}) = 1$. Since S_5 is vertex-transitive and edge-transitive, we may assume that $\mathbf{u} = \mathbf{e} = 12345$ and $\mathbf{v} = (\mathbf{e})^5 = 52341$. By Lemma 16.3, there exist three paths P_1, P_2, and P_3 of $S_5^{\{5\}}$ such that (1) P_1 joins 12345 to the black vertex 24135 with $l(P_1) = 7$, (2) P_2 joins 12345 to the white vertex 34125 with $l(P_2) = 8$, (3) P_3 joins 12345 to the white vertex 41325 with $l(P_3) = 8$, and (4) $P_1 \cup P_2 \cup P_3$ spans $S_5^{\{5\}}$. Similarly, there exist three paths Q_1, Q_2, and Q_3 of $S_5^{\{1\}}$ such that (1) Q_1 joins 52341 to the white vertex 24531 with $l(Q_1) = 7$, (2) Q_2 joins 52341 to the black vertex 34521 with $l(Q_2) = 8$, (3) Q_3 joins 52341 to the black vertex 45321 with $l(Q_3) = 8$, and (4) $Q_1 \cup Q_2 \cup Q_3$ spans $S_5^{\{1\}}$. By Theorem 13.2, there is a Hamiltonian path R_1 of $S_5^{\{2\}}$

joining the white vertex 54132 to the black vertex 14532, there is a Hamiltonian path R_2 of $S_5^{\{3\}}$ joining the black vertex 54123 to the white vertex 14523, and there is a Hamiltonian path R_3 of $S_5^{\{4\}}$ joining the black vertex 51324 to the white vertex 15324. Then we set

$$\begin{aligned}T_1 &= \langle \mathbf{e} = 12345, P_1, 24135, 54132, R_1, 14532, 24531, (Q_1)^{-1}, 52341 = (\mathbf{e})^5\rangle\\ T_2 &= \langle \mathbf{e} = 12345, P_2, 34125, 54123, R_2, 14523, 34521, (Q_2)^{-1}, 52341 = (\mathbf{e})^5\rangle\\ T_3 &= \langle \mathbf{e} = 12345, P_3, 41325, 51324, R_3, 15324, 45321, (Q_3)^{-1}, 52341 = (\mathbf{e})^5\rangle\\ T_4 &= \langle \mathbf{e} = 12345, 52341 = (\mathbf{e})^5\rangle\end{aligned}$$

Obviously, $\{T_1, T_2, T_3, T_4\}$ is a 4*-container of S_5 between $\mathbf{e}$ and $(\mathbf{e})^5$. Moreover, $l(T_1) = 39$, $l(T_2) = l(T_3) = 41$, and $l(T_4) = 1$. Thus, $d_4^{SL}(\mathbf{e}, (\mathbf{e})^5) \leq 41$.

Case 2. $d(\mathbf{u}, \mathbf{v}) \geq 3$. Since $d(\mathbf{u}, \mathbf{v}) \geq 3$, there is $i \in \{2, 3, 4, 5\}$ such that $(\mathbf{u})_i \neq (\mathbf{v})_i$ and $\{(\mathbf{u})_i, (\mathbf{v})_i\} \cap \{(\mathbf{u})_1, (\mathbf{v})_1\} = \emptyset$. Without loss of generality, we assume that $(\mathbf{u})_5 \neq (\mathbf{v})_5$ and $\{(\mathbf{u})_5, (\mathbf{v})_5\} \cap \{(\mathbf{u})_1, (\mathbf{v})_1\} = \emptyset$. Moreover, we assume that $(\mathbf{u})_5 = 5$, $(\mathbf{v})_5 = 4$, $(\mathbf{u})_1 = 1$, and $(\mathbf{v})_1 \neq 5$. Since $(\mathbf{u})_1 = 1$ and $(\mathbf{u})_5 = 5$, we have $\{(\mathbf{u})_2, (\mathbf{u})_3, (\mathbf{u})_4\} = \{2, 3, 4\}$.

Case 2.1. $(\mathbf{v})_1 = 1$. We have $\{(\mathbf{v})_2, (\mathbf{v})_3, (\mathbf{v})_4\} = \{2, 3, 5\}$. By Lemma 16.4, there exist four paths P_1, P_2, P_3, and P_4 of $S_5^{\{5\}}$ such that (1) P_1 joins $\mathbf{u}$ to a white vertex $\mathbf{w}$ with $(\mathbf{w})_1 = 2$ and $l(P_1) = 2$, (2) P_2 joins $\mathbf{u}$ to a white vertex $\mathbf{x}$ with $(\mathbf{x})_1 = 3$ and $l(P_1) = 2$, (3) P_3 joins $\mathbf{u}$ to a black vertex $\mathbf{y}$ with $(\mathbf{y})_1 = 4$ and $l(P_3) = 19$, (4) P_4 joins $\mathbf{u}$ to a black vertex $\mathbf{z} \neq \mathbf{y}$ with $(\mathbf{z})_1 = 4$ and $l(P_4) = 19$, (5) $P_1 \cup P_2 \cup P_3$ spans $S_5^{\{5\}}$, (6) $P_1 \cup P_2 \cup P_4$ spans $S_5^{\{5\}}$, (7) $V(P_1) \cap V(P_2) \cap V(P_3) = \{\mathbf{u}\}$, and (8) $V(P_1) \cap V(P_2) \cap V(P_4) = \{\mathbf{u}\}$.

Similarly, there exist four paths Q_1, Q_2, Q_3, and Q_4 of $S_5^{\{4\}}$ such that (1) Q_1 joins $\mathbf{v}$ to a black vertex $\mathbf{p}$ with $(\mathbf{p})_1 = 2$ and $l(Q_1) = 2$, (2) Q_2 joins $\mathbf{v}$ to a black vertex $\mathbf{q}$ with $(\mathbf{q})_1 = 3$ and $l(Q_2) = 2$, (3) Q_3 joins $\mathbf{v}$ to a white vertex $\mathbf{r}$ with $(\mathbf{r})_1 = 5$ and $l(Q_3) = 19$, (4) Q_4 joins $\mathbf{v}$ to a white vertex $\mathbf{s} \neq \mathbf{r}$ with $(\mathbf{s})_1 = 5$ and $l(Q_4) = 19$, (5) $Q_1 \cup Q_2 \cup Q_3$ spans $S_5^{\{4\}}$, (6) $Q_1 \cup Q_2 \cup Q_4$ spans $S_5^{\{4\}}$, (7) $V(Q_1) \cap V(Q_2) \cap V(Q_3) = \{\mathbf{v}\}$, and (8) $V(Q_1) \cap V(Q_2) \cap V(Q_4) = \{\mathbf{v}\}$.

By Lemma 13.7, there are exactly three edges joining some black vertices of $S_5^{\{5\}}$ to some white vertices of $S_5^{\{4\}}$. By the pigeonhole principle, at least one vertex in $\{\mathbf{y}, \mathbf{z}\}$ is adjacent to a vertex in $\{\mathbf{r}, \mathbf{s}\}$. Without loss of generality, we assume that $\mathbf{y}$ is adjacent to $\mathbf{r}$. Let T_1 be the Hamiltonian path of $S_5^{\{1\}}$ joining the black vertex $(\mathbf{u})^5$ to the white vertex $(\mathbf{v})^5$, T_2 be the Hamiltonian path of $S_5^{\{2\}}$ joining the black vertex $(\mathbf{w})^5$ to the white vertex $(\mathbf{p})^5$, and T_3 be the Hamiltonian path of $S_5^{\{3\}}$ joining the black vertex $(\mathbf{x})^5$ to the white vertex $(\mathbf{q})^5$. We set

$$\begin{aligned}H_1 &= \langle \mathbf{u}, (\mathbf{u})^5, T_1, (\mathbf{v})^5, \mathbf{v}\rangle\\ H_2 &= \langle \mathbf{u}, P_1, \mathbf{w}, (\mathbf{w})^5, T_2, (\mathbf{p})^5, \mathbf{p}, Q_1^{-1}, \mathbf{v}\rangle\\ H_3 &= \langle \mathbf{u}, P_2, \mathbf{x}, (\mathbf{x})^5, T_3, (\mathbf{q})^5, \mathbf{q}, Q_2^{-1}, \mathbf{v}\rangle\\ H_4 &= \langle \mathbf{u}, P_3, \mathbf{y}, \mathbf{r}, Q_3^{-1}, \mathbf{v}\rangle\end{aligned}$$

Obviously, $\{H_1, H_2, H_3, H_4\}$ is a 4*-container of S_5 between $\mathbf{u}$ and $\mathbf{v}$. Moreover, $l(H_1)=25$, $l(H_2)=l(H_3)=29$, and $l(H_4)=39$. Thus, $d_4^{SL}(\mathbf{u},\mathbf{v}) \leq 41$.

Case 2.2. $(\mathbf{v})_1 = a \in \{2,3\}$. We have $\{(\mathbf{v})_2, (\mathbf{v})_3, (\mathbf{v})_4\} = \{1,2,3,5\} - \{a\}$. Let b be the only element in $\{2,3\} - \{a\}$. By Lemma 16.4, there exist four paths P_1, P_2, P_3, and P_4 of $S_5^{\{5\}}$ such that (1) P_1 joins $\mathbf{u}$ to a white vertex $\mathbf{w}$ with $(\mathbf{w})_1 = a$ and $l(P_1)=2$, (2) P_2 joins $\mathbf{u}$ to a white vertex $\mathbf{x}$ with $(\mathbf{x})_1 = b$ and $l(P_2)=2$, (3) P_3 joins $\mathbf{u}$ to a black vertex $\mathbf{y}$ with $(\mathbf{y})_1 = 4$ and $l(P_3)=19$, (4) P_4 joins $\mathbf{u}$ to a black vertex $\mathbf{z} \neq \mathbf{y}$ with $(\mathbf{z})_1 = 4$ and $l(P_4)=19$, (5) $P_1 \cup P_2 \cup P_3$ spans $S_5^{\{5\}}$, (6) $P_1 \cup P_2 \cup P_4$ spans $S_5^{\{5\}}$, (7) $V(P_1) \cap V(P_2) \cap V(P_3) = \{\mathbf{u}\}$, and (8) $V(P_1) \cap V(P_2) \cap V(P_4) = \{\mathbf{u}\}$.

Again, there exist four paths Q_1, Q_2, Q_3, and Q_4 of $S_5^{\{4\}}$ such that (1) Q_1 joins $\mathbf{v}$ to a black vertex $\mathbf{p}$ with $(\mathbf{p})_1 = 1$ and $l(Q_1)=2$, (2) Q_2 joins $\mathbf{v}$ to a black vertex $\mathbf{q}$ with $(\mathbf{q})_1 = b$ and $l(Q_2)=2$, (3) Q_3 joins $\mathbf{v}$ to a white vertex $\mathbf{r}$ with $(\mathbf{r})_1 = 5$ and $l(Q_3)=19$, (4) Q_4 joins $\mathbf{v}$ to a white vertex $\mathbf{s} \neq \mathbf{r}$ with $(\mathbf{s})_1 = 5$ and $l(Q_4)=19$, (5) $Q_1 \cup Q_2 \cup Q_3$ spans $S_5^{\{4\}}$, (6) $Q_1 \cup Q_2 \cup Q_4$ spans $S_5^{\{4\}}$, (7) $V(Q_1) \cap V(Q_2) \cap V(Q_3) = \{\mathbf{v}\}$, and (8) $V(Q_1) \cap V(Q_2) \cap V(Q_4) = \{\mathbf{v}\}$.

By Lemma 13.7, there are exactly three edges joining some black vertices of $S_5^{\{5\}}$ to some white vertices of $S_5^{\{4\}}$. By the pigeonhole principle, at least one vertex in $\{\mathbf{y}, \mathbf{z}\}$ is adjacent to a vertex in $\{\mathbf{r}, \mathbf{s}\}$. Without loss of generality, we assume that $\mathbf{y}$ is adjacent to $\mathbf{r}$. Let T_1 be the Hamiltonian path of $S_5^{\{1\}}$ joining the black vertex $(\mathbf{u})^5$ to the white vertex $(\mathbf{p})^5$, T_2 be the Hamiltonian path of $S_5^{\{a\}}$ joining the black vertex $(\mathbf{w})^5$ to the white vertex $(\mathbf{v})^5$, and T_3 be the Hamiltonian path of $S_5^{\{b\}}$ joining the black vertex $(\mathbf{x})^5$ to the white vertex $(\mathbf{q})^5$. We set

$$\begin{aligned} H_1 &= \langle \mathbf{u}, (\mathbf{u})^5, T_1, (\mathbf{p})^5, \mathbf{p}, Q_1^{-1}, \mathbf{v} \rangle \\ H_2 &= \langle \mathbf{u}, P_1, \mathbf{w}, (\mathbf{w})^5, T_2, (\mathbf{v})^5, \mathbf{v} \rangle \\ H_3 &= \langle \mathbf{u}, P_2, \mathbf{x}, (\mathbf{x})^5, T_3, (\mathbf{q})^5, \mathbf{q}, Q_2^{-1}, \mathbf{v} \rangle \\ H_4 &= \langle \mathbf{u}, P_3, \mathbf{y}, \mathbf{r}, Q_3^{-1}, \mathbf{v} \rangle \end{aligned}$$

Obviously, $\{H_1, H_2, H_3, H_4\}$ is a 4*-container of S_5 between $\mathbf{u}$ and $\mathbf{v}$. Moreover, $l(H_1)=l(H_2)=27$, $l(H_3)=29$, and $l(H_4)=39$. Thus, $d_4^{SL}(\mathbf{u},\mathbf{v}) \leq 41$. □

LEMMA 16.9 $d_{n-1}^{SL}(\mathbf{u},\mathbf{v}) \leq (n-1)! + 2(n-2)! + 2(n-3)! + 1 = n!/(n-2) + 1$ for every $n \geq 6$.

Proof: Let $\mathbf{u}$ be any white vertex and $\mathbf{v}$ be any black vertex of S_n. Obviously, $d(\mathbf{u},\mathbf{v})$ is odd.

Case 1. $d(\mathbf{u},\mathbf{v}) = 1$. Since the star graph is vertex-transitive and edge-transitive, we may assume that $\mathbf{u} = \mathbf{e}$ and $\mathbf{v} = (\mathbf{e})^n$.

By Lemma 16.5, there exist $(n-2)$ paths $P_1, P_2, \ldots, P_{n-2}$ of $S_n^{\{n\}}$ such that (1) P_1 joins $\mathbf{e}$ to a black vertex $\mathbf{x_1}$ with $(\mathbf{x_1})_1 = 2$ and $l(P_1) = (n-1)(n-3)! - 1$, (2) P_i joins $\mathbf{e}$ to a white vertex $\mathbf{x_i}$ with $(\mathbf{x_i})_1 = i+1$ and $l(P_i) = (n-1)(n-3)!$ for $2 \leq i \leq n-2$, (3) $\bigcup_{i=1}^{n-2} P_i$ spans $S_n^{\{n\}}$, and (4) $\bigcap_{i=1}^{n-2} V(P_i) = \{\mathbf{e}\}$. Again, there exist

$n-2$ paths $Q_1, Q_2, \ldots, Q_{n-2}$ of $S_n^{\{1\}}$ such that (1) Q_1 joins $(\mathbf{e})^n$ to a white vertex $\mathbf{y_1}$ with $(\mathbf{y_1})_1 = 2$ and $l(Q_1) = (n-1)(n-3)! - 1$, (2) Q_i joins $(\mathbf{e})^n$ to a black vertex $\mathbf{y_i}$ with $(\mathbf{y_i})_1 = i+1$ and $l(Q_i) = (n-1)(n-3)!$ for $2 \le i \le n-2$, (3) $\bigcup_{i=1}^{n-2} Q_i$ spans $S_n^{\{1\}}$, and (4) $\bigcap_{i=1}^{n-2} V(Q_i) = \{\mathbf{v}\}$.

By Theorem 13.2, there is a Hamiltonian path R_1 of $S_n^{\{2\}}$ joining the white vertex $(\mathbf{x_1})^n$ to the black vertex $(\mathbf{y_1})^n$. Again, there is a Hamiltonian path R_i of $S_n^{\{i+1\}}$ joining the black vertex $(\mathbf{x_i})^n$ to the black vertex $(\mathbf{y_i})^n$ for every $2 \le i \le n-2$.

We set $H_i = \langle \mathbf{e}, P_i, \mathbf{x_i}, (\mathbf{x_i})^n, R_i, (\mathbf{y_i})^n, \mathbf{y_i}, Q_i^{-1}, (\mathbf{e})^n \rangle$ for every $1 \le i \le n-2$ and $H_{n-1} = \langle \mathbf{e}, (\mathbf{e})^n \rangle$. Then $\{H_1, H_2, \ldots, H_{n-1}\}$ is an $(n-1)^*$-container between $\mathbf{e}$ and $(\mathbf{e})^n$. Obviously, $l(H_1) = (n-1)! + 2(n-2)! + 2(n-3)! - 1$, $l(H_i) = (n-1)! + 2(n-2)! + 2(n-3)! + 1$ for $2 \le i \le n-2$, and $l(H_{n-1}) = 1$. Hence, $d_{n-1}^{S_L}(\mathbf{e}, (\mathbf{e})^n) \le (n-1)! + 2(n-2)! + 2(n-3)! + 1$.

Case 2. $d(\mathbf{u}, \mathbf{v}) \ge 3$. Since $d(\mathbf{u}, \mathbf{v}) \ge 3$, there is $i \in \langle n \rangle - \{1\}$ such that $(\mathbf{u})_i \ne (\mathbf{v})_i$ and $\{(\mathbf{u})_i, (\mathbf{v})_i\} \cap \{(\mathbf{u})_1, (\mathbf{v})_1\} = \emptyset$. Without loss of generality, we assume that $(\mathbf{u})_n \ne (\mathbf{v})_n$ and $\{(\mathbf{u})_n, (\mathbf{v})_n\} \cap \{(\mathbf{u})_1, (\mathbf{v})_1\} = \emptyset$. Moreover, we assume that $(\mathbf{u})_n = n$, $(\mathbf{v})_n = n-1$, $(\mathbf{u})_1 = 1$, and $(\mathbf{v})_1 \ne 5$.

Case 2.1. $(\mathbf{v})_1 = 1$. By Lemma 16.5, there are $(n-2)$ paths $P_1, P_2, \ldots, P_{n-2}$ of $S_n^{\{n\}}$ such that (1) P_1 joins $\mathbf{u}$ to a black vertex $\mathbf{x_1}$ with $(\mathbf{x_1})_1 = 1$ and $l(P_1) = (n-1)(n-3)! - 1$, (2) P_i joins $\mathbf{u}$ to a white vertex $\mathbf{x_i}$ with $(\mathbf{x_i})_1 = i$ and $l(P_i) = (n-1)(n-3)!$ for $2 \le i \le n-2$, (3) $\bigcup_{i=1}^{n-2} P_i$ spans $S_n^{\{n\}}$, and (4) $\bigcap_{i=1}^{n-2} V(P_i) = \{\mathbf{u}\}$. Again, there are $(n-2)$ paths $Q_1, Q_2, \ldots, Q_{n-2}$ of $S_n^{\{n-1\}}$ such that (1) Q_1 joins $\mathbf{v}$ to a white vertex $\mathbf{y_1}$ with $(\mathbf{y_1})_1 = 1$ and $l(Q_1) = (n-1)(n-3)! - 1$, (2) Q_i joins $\mathbf{v}$ to a black vertex $\mathbf{y_i}$ with $(\mathbf{y_i})_1 = i$ and $l(Q_i) = (n-1)(n-3)!$ for $2 \le i \le n-2$, (3) $\bigcup_{i=1}^{n-2} Q_i$ spans $S_n^{\{n-1\}}$, and (4) $\bigcap_{i=1}^{n-2} V(Q_i) = \{\mathbf{v}\}$.

By Lemma 16.1, there are two disjoint paths H_1 and H_2 of $S_n^{\{1\}}$ such that (1) H_1 joins the white vertex $(\mathbf{x_1})^n$ to the black vertex $(\mathbf{y_1})^n$, (2) H_2 joins the black vertex $(\mathbf{u})^n$ to the white vertex $(\mathbf{v})^n$, and (3) $H_1 \cup H_2$ spans $S_n^{\{1\}}$. By Theorem 13.2, there is a Hamiltonian path R_i of $S_n^{\{i\}}$ joining the black vertex $(\mathbf{x_i})^n$ to the white vertex $(\mathbf{y_i})^n$ for every $2 \le i \le n-2$. We set

$$T_1 = \langle \mathbf{u}, P_1, \mathbf{x_1}, (\mathbf{x_1})^n, H_1, (\mathbf{y_1})^n, \mathbf{y_1}, Q_1^{-1}, \mathbf{v} \rangle$$

$$T_i = \langle \mathbf{u}, P_i, \mathbf{x_i}, (\mathbf{x_i})^n, R_i, (\mathbf{y_i})^n, \mathbf{y_i}, Q_i^{-1}, \mathbf{v} \rangle \text{ for } 2 \le i \le n-2$$

$$T_{n-1} = \langle \mathbf{u}, (\mathbf{u})^n, H_2, (\mathbf{v})^n, \mathbf{v} \rangle$$

Obviously, $\{T_1, T_2, \ldots, T_{n-1}\}$ is an $(n-1)^*$-container of S_n between $\mathbf{u}$ and $\mathbf{v}$. Moreover, $l(T_i) \le (n-1)! + 2(n-2)! + 2(n-3)! + 1$. Thus, $d_{n-1}^{S_L}(\mathbf{u}, \mathbf{v}) \le (n-1)! + 2(n-2)! + 2(n-3)! + 1$.

Case 2.2. $(\mathbf{v})_1 = t \in \langle n-2 \rangle - \{1\}$. By Lemma 16.5, there are $(n-2)$ paths $P_1, P_2, \ldots, P_{n-2}$ of $S_n^{\{n\}}$ such that (1) P_1 joins $\mathbf{u}$ to a black vertex $\mathbf{x_1}$ with $(\mathbf{x_1})_1 = 1$ and $l(P_1) = (n-1)(n-3)! - 1$, (2) P_i joins $\mathbf{u}$ to a white vertex $\mathbf{x_i}$ with $(\mathbf{x_i})_1 = i$ and $l(P_i) = (n-1)(n-3)!$ for $2 \le i \le n-2$, (3) $\bigcup_{i=1}^{n-2} P_i$ spans $S_n^{\{n\}}$, and (4) $\bigcap_{i=1}^{n-2} V(P_i) = \{\mathbf{u}\}$. Again, there are $(n-2)$ paths $Q_1, Q_2, \ldots, Q_{n-2}$ of $S_n^{\{n-1\}}$ such that (1) Q_1 joins $\mathbf{v}$ to a white vertex $\mathbf{y_1}$ with $(\mathbf{y_1})_1 = 1$ and $l(Q_1) = (n-1)(n-3)! - 1$, (2) Q_i joins $\mathbf{v}$ to a black vertex $\mathbf{y_i}$ with $(\mathbf{y_i})_1 = i$ and $l(Q_i) = (n-1)(n-3)!$ for $2 \le i \le n-2$, (3) $\cup_{i=1}^{n-2} Q_i$ spans $S_n^{\{n-1\}}$, and (4) $\cap_{i=1}^{n-2} V(Q_i) = \{\mathbf{v}\}$.

Since $(\mathbf{v})^n$ is a white vertex in $S_n^{\{t\}}$ with $((\mathbf{v})^n)_1 = (\mathbf{v})_n = n-1$ and $((\mathbf{v})^n)_n = (\mathbf{v})_1 = t \neq 1$, we can choose a black vertex $\mathbf{w}$ in $N_{S_n^{\{t\}}}((\mathbf{v})^n)$ with $(\mathbf{w})_1 = 1$. By Lemma 16.1, there exist two disjoint paths H_1 and H_2 of $S_n^{\{1\}}$ such that (1) H_1 joins the white vertex $(\mathbf{x_1})^n$ to the black vertex $(\mathbf{y_1})^n$, (2) H_2 joins the black vertex $(\mathbf{u})^n$ to the white vertex $(\mathbf{w})^n$, and (3) $H_1 \cup H_2$ spans $S_n^{\{1\}}$.

By Theorem 16.1, there exists a Hamiltonian path R_t of $S_n^{\{t\}} - \{(\mathbf{v})^n, \mathbf{w}\}$ joining the black vertex $(\mathbf{x_t})^n$ to the white vertex $(\mathbf{y_t})^n$. By Theorem 13.2, there exists a Hamiltonian path R_i of $S_n^{\{i\}}$ joining the black vertex $(\mathbf{x_i})^n$ to the white vertex $(\mathbf{y_i})^n$ for every $2 \le i \le n-2$ with $i \neq t$. We set

$$T_1 = \langle \mathbf{u}, P_1, \mathbf{x_1}, (\mathbf{x_1})^n, H_1, (\mathbf{y_1})^n, \mathbf{y_1}, Q_1^{-1}, \mathbf{v} \rangle$$

$$T_i = \langle \mathbf{u}, P_i, \mathbf{x_i}, (\mathbf{x_i})^n, R_i, (\mathbf{y_i})^n, \mathbf{y_i}, Q_i^{-1}, \mathbf{v} \rangle \text{ for } 2 \le i \le n-2$$

$$T_{n-1} = \langle \mathbf{u}, (\mathbf{u})^n, H_2, (\mathbf{w})^n, \mathbf{w}, (\mathbf{v})^n, \mathbf{v} \rangle$$

Obviously, $\{T_1, T_2, \ldots, T_{n-1}\}$ is an $(n-1)^*$-container of S_n between $\mathbf{u}$ and $\mathbf{v}$. Moreover, $l(T_i) \le (n-1)! + 2(n-2)! + 2(n-3)! + 1$. Thus, $d_{n-1}^{sL}(\mathbf{u}, \mathbf{v}) \le (n-1)! + 2(n-2)! + 2(n-3)! + 1$. □

THEOREM 16.2

$$D_{n-1}^{sL}(S_n) = \begin{cases} 1 & \text{if } n = 2 \\ 5 & \text{if } n = 3 \\ 15 & \text{if } n = 4 \\ (n-1)! + 2(n-2)! + 2(n-3)! + 1 & \text{if } n \ge 5 \end{cases}$$

Proof: It is easy to check whether $D_1^{sL}(S_2) = 1$ and $D_2^{sL}(S_3) = 5$. By Lemma 16.6, $D_3^{sL}(S_4) = 15$. By Lemmas 16.7 through 16.9, we have $D_{n-1}^{sL}(S_n) = (n-1)! + 2(n-2)! + 2(n-3)! + 1$ if $n \ge 5$. Hence, this statement is proved. □

LEMMA 16.10 $D_2^{sL}(S_4) = 15$.

Proof: Let $\mathbf{u}$ and $\mathbf{v}$ be any two distinct vertices of S_4. Since S_4 is vertex-transitive, we assume that $\mathbf{u} = 1234$. Let $\{P_1, P_2\}$ be a 2^*-container joining $\mathbf{u}$ to $\mathbf{v}$. Thus, $P_1 \cup P_2^{-1}$ is a Hamiltonian cycle of S_4. In Table 16.1, we list all the possible Hamiltonian cycles of S_4. From Table 16.1, we know that $d_2^{sL}(1234, \mathbf{v}) = 13$ if $\mathbf{v} \in \{1423, 3142, 2413, 1243, 3421, 1432, 2341, 4123, 4312\}$ and $d_2^{sL}(1234, \mathbf{v}) = 15$ if $\mathbf{v} \in \{2134, 3214, 4231\}$. Hence, $D_2^{sL}(S_4) = 15$. □

LEMMA 16.11 Assume that a and b are any two distinct elements of $\langle 4 \rangle$ and $\mathbf{u}$ is any white vertex of S_4. There exist two paths P_1 and P_2 of S_4 such that (1) P_1 joins $\mathbf{u}$

to a black vertex $\mathbf{x}$ with $(\mathbf{x})_1 = a$ and $l(P_1) = 5$, (2) P_2 joins $\mathbf{u}$ to a white vertex $\mathbf{y}$ with $(\mathbf{y})_1 = b$ and $l(P_2) = 18$, and (3) $P_1 \cup P_2$ spans S_4.

Proof: Since S_4 is vertex-transitive, we may assume that $\mathbf{u} = 1234$. The required two paths are listed here:

$P_1 = \langle 1234, 3214, 4213, 2413, 3412, 1432 \rangle$ $P_2 = \langle 1234, 4231, 2431, 3421, 1423, 4123, 3124, 2134, 4132, 3142, 1342, 4312, 2314, 1324, 4321, 2341, 3241, 1243, 2143 \rangle$
$P_1 = \langle 1234, 3214, 4213, 2413, 3412, 1432 \rangle$ $P_2 = \langle 1234, 4231, 2431, 3421, 1423, 4123, 2143, 1243, 3241, 2341, 4321, 1324, 2314, 4312, 1342, 3142, 4132, 2134, 3124 \rangle$
$P_1 = \langle 1234, 3214, 4213, 2413, 3412, 1432 \rangle$ $P_2 = \langle 1234, 4231, 2431, 3421, 1423, 4123, 3124, 2134, 4132, 3142, 2143, 1243, 3241, 2341, 1342, 4312, 2314, 1324, 4321 \rangle$
$P_1 = \langle 1234, 3214, 2314, 4312, 1342, 2341 \rangle$ $P_2 = \langle 1234, 2134, 3124, 1324, 4321, 3421, 2431, 4231, 3241, 1243, 4213, 2413, 3412, 1432, 4132, 3142, 2143, 4123, 1423 \rangle$
$P_1 = \langle 1234, 2134, 4132, 3142, 1342, 2341 \rangle$ $P_2 = \langle 1234, 3214, 4213, 2413, 1423, 3421, 4321, 1324, 2314, 4312, 3412, 1432, 2431, 4231, 3241, 1243, 2143, 4123, 3124 \rangle$
$P_1 = \langle 1234, 3214, 2314, 4312, 1342, 2341 \rangle$ $P_2 = \langle 1234, 2134, 4132, 3142, 2143, 4123, 3124, 1324, 4321, 3421, 1423, 2413, 3412, 1432, 2431, 4231, 3241, 1243, 4213 \rangle$
$P_1 = \langle 1234, 2134, 3124, 1324, 2314, 3214 \rangle$ $P_2 = \langle 1234, 4231, 3241, 2341, 4321, 3421, 2431, 1432, 4132, 3142, 1342, 4312, 3412, 2413, 4213, 1243, 2143, 4123, 1423 \rangle$
$P_1 = \langle 1234, 3214, 4213, 2413, 1423, 3421 \rangle$ $P_2 = \langle 1234, 4231, 3241, 1243, 2143, 4123, 3124, 2134, 4132, 3142, 1342, 2341, 4321, 1324, 2314, 4312, 3412, 1432, 2431 \rangle$
$P_1 = \langle 1234, 3214, 4213, 2413, 1423, 3421 \rangle$ $P_2 = \langle 1234, 2134, 3124, 4123, 2143, 1243, 3241, 4231, 2431, 1432, 3412, 4312, 2314, 1324, 4321, 2341, 1342, 3142, 4132 \rangle$
$P_1 = \langle 1234, 3214, 4213, 1243, 3241, 4231 \rangle$ $P_2 = \langle 1234, 2134, 4132, 3142, 2143, 4123, 3124, 1324, 2314, 4312, 1342, 2341, 4321, 3421, 2431, 1432, 3412, 2413, 1423 \rangle$
$P_1 = \langle 1234, 3214, 4213, 1243, 3241, 4231 \rangle$ $P_2 = \langle 1234, 2134, 4132, 3142, 2143, 4123, 3124, 1324, 2314, 4312, 1342, 2341, 4321, 3421, 1423, 2413, 3412, 1432, 2431 \rangle$
$P_1 = \langle 1234, 3214, 4213, 1243, 3241, 4231 \rangle$ $P_2 = \langle 1234, 2134, 4132, 3142, 2143, 4123, 1423, 2413, 3412, 1432, 2431, 3421, 4321, 2341, 1342, 4312, 2314, 1324, 3124 \rangle$

Hence, this statement is proved. □

THEOREM 16.3

$$D_2^{SL}(S_n) = \begin{cases} 5 & \text{if } n = 3 \\ 15 & \text{if } n = 4 \\ \frac{n!}{2} + 1 & \text{if } n \geq 5 \end{cases}$$

Proof: It is easy to check whether $D_2^{SL}(S_3) = 5$. By Lemma 16.10, we have that $D_2^{SL}(S_4) = 15$. Thus, we assume that $n \geq 5$. Let $\mathbf{u}$ be a white vertex and $\mathbf{v}$ be a black vertex of S_n. Let P_1 and P_2 be any 2*-container of S_n joining $\mathbf{u}$ to $\mathbf{v}$. Obviously, $\max\{l(P_1), l(P_2)\} \geq (n!/2) + 1$. Hence, $d_2^{SL}(\mathbf{u}, \mathbf{v}) \geq (n!/2) + 1$ and $D_2^{SL}(S_n) \geq (n!/2) + 1$. Hence, we only need to show that $d_2^{SL}(\mathbf{u}, \mathbf{v}) \leq (n!/2) + 1$. Since S_n is edge-transitive, we assume that $\mathbf{u} \in S_n^{\{n\}}$ and $\mathbf{v} \in S_n^{\{n-1\}}$.

Case 1. $n = 5$. By Lemma 16.11, there exist two paths H_1 and H_2 of $S_5^{\{5\}}$ such that (1) H_1 joins $\mathbf{u}$ to a black vertex $\mathbf{x}$ with $(\mathbf{x})_1 = 1$ and $l(H_1) = 5$, (2) H_2 joins $\mathbf{u}$ to a white vertex $\mathbf{y}$ with $(\mathbf{y})_1 = 3$ and $l(H_2) = 18$, and (3) $H_1 \cup H_2$ spans $S_5^{\{5\}}$. Again, there exist two paths T_1 and T_2 of $S_5^{\{4\}}$ such that (1) T_1 joins $\mathbf{v}$ to a white vertex $\mathbf{p}$ with $(\mathbf{p})_1 = 2$ and $l(T_1) = 5$, (2) T_2 joins $\mathbf{v}$ to a black vertex $\mathbf{q}$ with $(\mathbf{q})_1 = 3$ and $l(T_2) = 18$, and

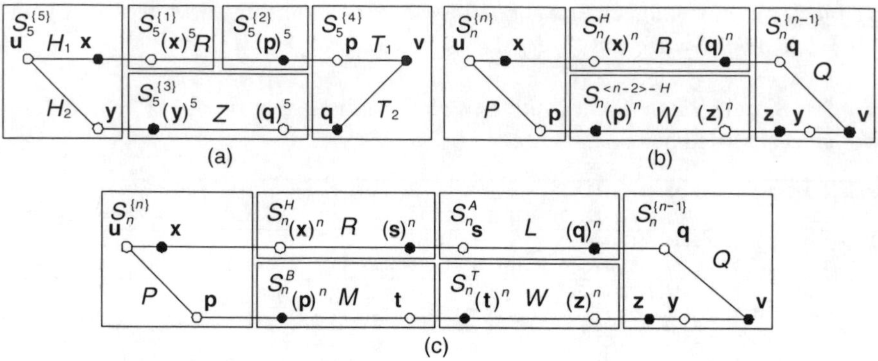

FIGURE 16.4 Illustration for Theorem 16.3.

(3) $T_1 \cup T_2$ spans $S_5^{\{4\}}$. By Theorem 13.17, there is a Hamiltonian path R of $S_5^{\{1,2\}}$ joining the white vertex $(\mathbf{x})^5$ to the black vertex $(\mathbf{p})^5$. Again, there is a Hamiltonian path Z of $S_5^{\{3\}}$ joining the black vertex $(\mathbf{y})^5$ to the white vertex $(\mathbf{q})^5$. We set

$$L_1 = \langle \mathbf{u}, H_1, \mathbf{x}, (\mathbf{x})^5, R, (\mathbf{p})^5, \mathbf{p}, T_1^{-1}, \mathbf{v} \rangle$$
$$L_2 = \langle \mathbf{u}, H_2, \mathbf{y}, (\mathbf{y})^5, Z, (\mathbf{q})^5, \mathbf{q}, T_2^{-1}, \mathbf{v} \rangle$$

Obviously, $\{L_1, L_2\}$ is a 2*-container. Moreover, $l(L_1) = 59$ and $l(L_2) = 61$. Hence, $d_2^{SL}(\mathbf{u}, \mathbf{v}) \le (n!/2) + 1$. See Figure 16.4a for illustration.

Case 2. $n \ge 6$ is even. Let $\mathbf{x}$ be a neighbor of $\mathbf{u}$ in $S_n^{\{n\}}$ with $(\mathbf{x})_1 \in \langle n-2 \rangle$. Let $\mathbf{y}$ be a neighbor of $\mathbf{v}$ in $S_n^{\{n-1\}}$. Let $\mathbf{z}$ be a neighbor of $\mathbf{y}$ in $S_n^{\{n-1\}}$ with $(\mathbf{z})_1 \in \langle n-2 \rangle - \{(\mathbf{v})_1, (\mathbf{y})_1, (\mathbf{x})_1\}$. Let $a_1 a_2 \ldots a_{n-2}$ be a permutation of $\langle n-2 \rangle$ such that $a_1 = (\mathbf{x})_1$ and $a_{n-2} = (\mathbf{z})_1$. Let $H = \{a_1, a_2, \ldots, a_{(n-2)/2}\}$. By Theorem 13.2, there is a Hamiltonian path P of $S_n^{\{n\}} - \{\mathbf{x}\}$ joining $\mathbf{u}$ to a white vertex $\mathbf{p}$ with $(\mathbf{p})_1 = a_{n-2}$. By Theorem 16.1, there is a Hamiltonian path Q of $S_n^{\{n-1\}} - \{\mathbf{y}, \mathbf{z}\}$ joining a white vertex $\mathbf{q}$ with $(\mathbf{q})_1 = a_1$ to $\mathbf{v}$. By Theorem 13.17, there is a Hamiltonian path R of S_n^H joining the white vertex $(\mathbf{x})^n$ to the black vertex $(\mathbf{q})^n$. Again, there is a Hamiltonian path W of $S_n^{\langle n-2 \rangle - H}$ joining the black vertex $(\mathbf{p})^n$ to the white vertex $(\mathbf{z})^n$. We set

$$L_1 = \langle \mathbf{u}, \mathbf{x}, (\mathbf{x})^n, R, (\mathbf{q})^n, \mathbf{q}, Q, \mathbf{v} \rangle$$
$$L_2 = \langle \mathbf{u}, P, \mathbf{p}, (\mathbf{p})^n, W, (\mathbf{z})^n, \mathbf{z}, \mathbf{y}, \mathbf{v} \rangle$$

Obviously, $\{L_1, L_2\}$ is a 2*-container of S_n between $\mathbf{u}$ and $\mathbf{v}$. Since $l(L_1) = (n!/2) - 1$ and $l(L_2) = (n!/2) + 1$, we have $d_2^{SL}(\mathbf{u}, \mathbf{v}) \le (n!/2) + 1$. See Figure 16.4b for illustration.

Case 3. $n \ge 7$ is odd. Let $\mathbf{x}$ be a neighbor of $\mathbf{u}$ in $S_n^{\{n\}}$ with $(\mathbf{x})_1 \in \langle n-2 \rangle$. Let $\mathbf{y}$ be a neighbor of $\mathbf{v}$ in $S_n^{\{n-1\}}$. Let $\mathbf{z}$ be a neighbor of $\mathbf{y}$ in $S_n^{\{n-1\}}$ with $(\mathbf{z})_1 \in \langle n-2 \rangle - \{(\mathbf{v})_1, (\mathbf{y})_1, (\mathbf{x})_1\}$. Let $a_1 a_2 \ldots a_{n-2}$ be a permutation of $\langle n-2 \rangle$ such that $a_1 = (\mathbf{x})_1$ and $a_{n-3} = (\mathbf{z})_1$. Let $H = \{a_1, a_2, \ldots, a_{(n-3)/2}\}$ and

$T=\{a_{\frac{n-3}{2}+1}, a_{\frac{n-3}{2}+2}, \ldots, a_{n-3}\}$. We set $A=\{(i, a_{n-2}) \mid i \in H \cup \{n-1\}\}$ and $B=\{(i, a_{n-2}) \mid i \in T \cup \{n\}\}$. Let S_n^A denote the subgraph of S_n induced by $\cup_{i \in H \cup \{n-1\}} S_n^{\{(i,a_{n-2})\}}$, and let S_n^B denote the subgraph of S_n induced by $\cup_{i \in T \cup \{n\}} S_n^{\{(i,a_{n-2})\}}$. By Theorem 13.2, there is a Hamiltonian path P of $S_n^{\{n\}} - \{\mathbf{x}\}$ joining $\mathbf{u}$ to a white vertex $\mathbf{p}$ with $(\mathbf{p})_1 = a_{n-2}$ and $(\mathbf{p})_{n-2} = a_{n-3}$. By Theorem 16.1, there is a Hamiltonian path Q of $S_n^{\{n-1\}} - \{\mathbf{y}, \mathbf{z}\}$ joining a white vertex $\mathbf{q}$ with $(\mathbf{q})_1 = a_{n-2}$ and $(\mathbf{q})_{n-1} = a_1$ to $\mathbf{v}$. By Theorem 13.17, there is a Hamiltonian path L of S_n^A joining a white vertex $\mathbf{s}$ with $(\mathbf{s})_1 = a_1$ to the black vertex $(\mathbf{q})^n$. Again, there is a Hamiltonian path M of S_n^B joining the black vertex $(\mathbf{p})^n$ to a white vertex $\mathbf{t}$ with $(\mathbf{t})_1 = a_{n-3}$. By Theorem 13.17, there is a Hamiltonian path R of S_n^H joining the white vertex $(\mathbf{x})^n$ to the black vertex $(\mathbf{s})^n$. Again, there is a Hamiltonian path W of S_n^T joining the black vertex $(\mathbf{t})^n$ to the white vertex $(\mathbf{z})^n$. We set

$$L_1 = \langle \mathbf{u}, \mathbf{x}, (\mathbf{x})^n, R, (\mathbf{s})^n, \mathbf{s}, L, (\mathbf{q})^n, \mathbf{q}, Q, \mathbf{v} \rangle$$
$$L_2 = \langle \mathbf{u}, P, \mathbf{p}, (\mathbf{p})^n, M, \mathbf{t}, (\mathbf{t})^n, W, (\mathbf{z})^n, \mathbf{z}, \mathbf{y}, \mathbf{v} \rangle$$

Obviously, $\{L_1, L_2\}$ is a 2*-container of S_n between $\mathbf{u}$ and $\mathbf{v}$. Since $l(L_1) = (n!/2) - 1$ and $l(L_2) = (n!/2) + 1$, we have $d_2^{SL}(\mathbf{u}, \mathbf{v}) \leq (n!/2) + 1$. See Figure 16.4c for illustration. □

Actually, we prove that $d_2^{SL}(\mathbf{u}, \mathbf{v}) = (n!/2) + 1$ for any two vertices $\mathbf{u}$ and $\mathbf{v}$ from different bipartite sets of S_n.

16.3 SPANNING DIAMETER OF HYPERCUBES

Because of the important role of hypercubes in interconnection networks, we should discuss the spanning diameter of hypercubes. However, this is a difficult job. The reason we can compute the spanning diameter for pancake graphs and star graphs is because we can easily set the bound for spanning diameter. Then, we construct the desired spanning containers. To overcome this problem, Chang et al. [45] introduce the concept of the equitable container. A k^*-container $C_k^*(\mathbf{u}, \mathbf{v}) = \{P_1, \ldots, P_k\}$ is *equitable* if $||V(P_i)| - |V(P_j)|| \leq 2$ for all $1 \leq i, j \leq k$. A graph is equitably k^*-laceable if there is an equitable k^*-container joining any two vertices in different partite sets. In this section, we will prove that the hypercube Q_n is equitably k^*-laceable for all $k \leq n-4$ and $n \geq 5$. With this result, we can conclude that $D_k^{SL}(Q_n) = 2^n/k$ for all $k \leq n-4$ and $n \geq 5$.

For convenience, let $\{\mathbf{u}, \mathbf{v}\}$ be an edge of Q_n. We call this edge $\{\mathbf{u}, \mathbf{v}\}$ *parallel to* $\mathbf{e}_i$ if $\mathbf{u} - \mathbf{v} = \mathbf{e}_i$. The following result referred as *parallel spanning property*, proved by Kobeissi and Mollard [209], is an important lemma. We shall use this property to construct equitable k^*-containers in Q_n. Let $\mathbf{u}_1, \mathbf{u}_2, \ldots, \mathbf{u}_k$ be k distinct white vertices of Q_n with $k \leq n-2$. We set $\mathbf{v}_i = \mathbf{u}_i + \mathbf{e}_1$ for $1 \leq i \leq k$. We call $\{(\mathbf{u}_1, \ldots, \mathbf{u}_k), (\mathbf{v}_1, \ldots, \mathbf{v}_k)\}$ a *parallel configuration of order* k.

LEMMA 16.12 **[202]** Let $\{(\mathbf{u}_1, \ldots, \mathbf{u}_k), (\mathbf{v}_1, \ldots, \mathbf{v}_k)\}$ be a parallel configuration of order k with $k \leq n-2$. Suppose that $a_1, a_2, \ldots, a_k$ are any positive even integers with $a_1 + a_2 + \cdots + a_k = 2^n$. Then there exists a k^*-container

$C_k^*((\mathbf{u}_1, \mathbf{u}_2, \ldots, \mathbf{u}_k), (\mathbf{v}_1, \mathbf{v}_2, \ldots, \mathbf{v}_k)) = \{P_1, \ldots, P_k\}$ such that $|V(P_i)| = a_i$ for all $i = 1, 2, \ldots, k$.

LEMMA 16.13 [202] Let $a_1, a_2, \ldots, a_k$ be any positive even integer with $a_1 + a_2 + \cdots + a_k = 2^n$ and $k \leq n-1$. Then there exists a k^*-container $C_k^*(\mathbf{u}, \mathbf{v}) = \{P_1, \ldots, P_k\}$ between any adjacent pair of vertices $\mathbf{u}$ and $\mathbf{v}$ of Q_n such that $|V(P_i)| = a_i$, $i = 1, \ldots, k$.

COROLLARY 16.1 *For $k \leq n-1$ with $n \geq 3$, there are equitable k^*-containers $C_k^*(\mathbf{u}, \mathbf{v})$ for any adjacent pair* **u** *and* **v**.

Let $\{(\mathbf{u}_1, \ldots, \mathbf{u}_k), (\mathbf{v}_1, \ldots, \mathbf{v}_k)\}$ be a parallel configuration of order k. $\{(\mathbf{u}_0, \mathbf{u}_1, \ldots, \mathbf{u}_k), (\mathbf{v}_0, \mathbf{v}_1, \ldots, \mathbf{v}_k)\}$ is referred as an *extended parallel configuration* of order k in Q_n where $k \leq n-4$ if the following conditions are satisfied: (1) $\mathbf{v}_i = \mathbf{u}_1 + \mathbf{e}_i$, for all $i \in \{1, 2, \ldots, k\}$, (2) $\mathbf{v}_0 = \mathbf{u}_1 + \mathbf{e}_n$, and (3) $\mathbf{u}_0$ is an arbitrary white vertex other than $\mathbf{u}_1, \ldots, \mathbf{u}_k$.

Recall that Q_n is vertex-symmetric. In the following discussion, we assume that $\mathbf{u}_1 = \mathbf{0}$, $\mathbf{v}_i = \mathbf{e}_i \; \forall\, i = 1, \ldots, k$, $\mathbf{u}_i = \mathbf{e}_i + \mathbf{e}_1 \; \forall\, i = 2, \ldots, k$, $\mathbf{v}_0 = \mathbf{e}_n$, and $\mathbf{u}_0$ is an arbitrary white vertex other than $\mathbf{u}_1, \ldots, \mathbf{u}_k$. Let $2 \leq j \leq k$ be a fixed positive integer. Suppose we decompose Q_n into 2 subcubes Q_n^0, Q_n^1 where $Q_n^0 = \{(x_1, \ldots, x_n) \mid x_j = 0\}$, $Q_n^1 = \{(x_1, \ldots, x_n) \mid x_j = 1\}$, then $\mathbf{u}_j, \mathbf{v}_j \in Q_n^1$ and $\mathbf{u}_i, \mathbf{v}_i \in Q_n^0$ for all $0 \neq i \neq j$. For convenience, we use $\hat{\mathbf{x}}$ to denote $\mathbf{x} + \mathbf{e}_1$ for all $\mathbf{x} \in Q_n$.

LEMMA 16.14 Let $\{(\mathbf{u}_0, \mathbf{u}_1), (\mathbf{v}_0, \mathbf{v}_1)\}$ be an extended parallel configuration of order 1 in Q_n, where $n \geq 4$. Suppose that a_0, a_1 are positive even integers with $a_0 \geq 2n$ and $a_0 + a_1 = 2^n$; then there exists a 2^*-container $C_2^*((\mathbf{u}_0, \mathbf{u}_1), (\mathbf{v}_0, \mathbf{v}_1)) = \{P_0, P_1\}$ in Q_n such that (1) $|V(P_i)| = a_i$, for all $i = 0, 1$, (2) P_0 and P_1 that contain an edge parallel to $\mathbf{e}_1$.

Proof: We prove this statement by induction on n. By brute force, we can check that this lemma is true for $n = 4$. Now assume that $n \geq 5$ and statement is true for $n - 14$. Without loss of generality, we assume that $\{\mathbf{u}_1, \mathbf{v}_1\} \in Q_n^0$, $\mathbf{v}_0 \in Q_n^1$, where $Q_n^0 = \{(x_1, \ldots, x_n) \mid x_n = 0\}$, $Q_n^1 = \{(x_1, \ldots, x_n) \mid x_n = 1\}$.

Case 1. $\mathbf{u}_0 \in Q_n^1$.

Case 1.1. $a_1 < 2^{n-1}$. We choose a black vertex $\mathbf{x}$ in Q_n^0 such that (1) $\mathbf{x} \neq \mathbf{v}_1$ and $\hat{\mathbf{x}} \neq \mathbf{u}_1$, (2) $\mathbf{x} + \mathbf{e}_n \neq \mathbf{u}_0$ and $\hat{\mathbf{x}} + \mathbf{e}_n \neq \mathbf{v}_0$.

By Lemma 16.12, there exist two vertex disjoint paths P_1 and R in Q_n^0 such that (1) P_1 joins $\mathbf{u}_1$ to $\mathbf{v}_1$ with $|V(P_1)| = a_1$ and (2) R joins $\hat{\mathbf{x}}$ to $\mathbf{x}$ with $|V(R)| = 2^{n-1} - a_1$.

By Lemma 13.17, there exist two vertex disjoint paths H_1 and H_2 in Q_n^1 such that (1) H_1 joins $\mathbf{u}_0$ to $\hat{\mathbf{x}} + \mathbf{e}_n$, (2) H_2 joins $\mathbf{x} + \mathbf{e}_n$ to $\mathbf{v}_0$, and (3) $H_1 \cup H_2$ spans Q_n^1.

Set $P_0 = \langle \mathbf{u}_0, H_1, \hat{\mathbf{x}} + \mathbf{e}_n, \hat{\mathbf{x}}, R, \mathbf{x}, \mathbf{x} + \mathbf{e}_n, H_2, \mathbf{v}_0 \rangle$. Then $\{P_0, P_1\}$ satisfies the requirement of Lemma 16.14. See Figure 16.5 for illustration.

Case 1.2. $a_1 > 2^{n-1}$. We choose a black vertex $\mathbf{x}$ in Q_n^1 such that (1) $\mathbf{x} \neq \mathbf{v}_0$ and $\hat{\mathbf{x}} \neq \mathbf{u}_0$, and (2) $\mathbf{x} + \mathbf{e}_n \neq \mathbf{u}_1$ and $\hat{\mathbf{x}} + \mathbf{e}_n \neq \mathbf{v}_1$.

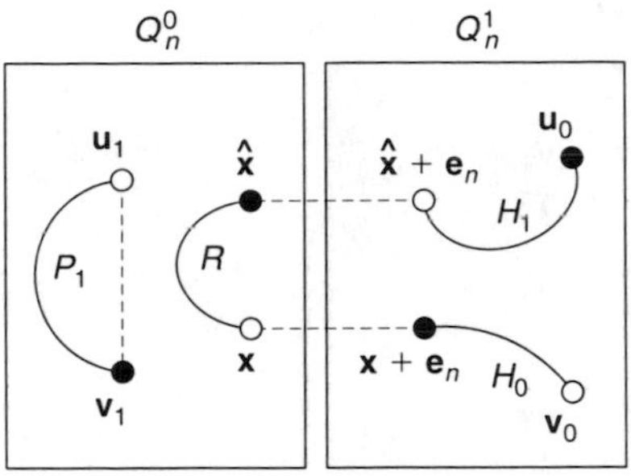

FIGURE 16.5 Illustration for Case 1.1 of Lemma 16.14.

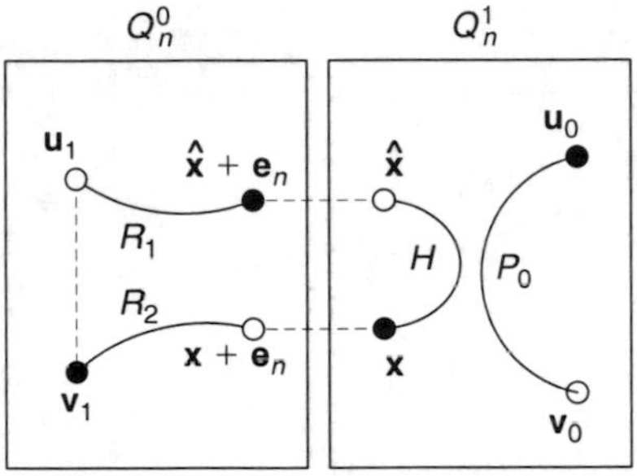

FIGURE 16.6 Illustration for Case 1.2 of Lemma 16.14.

By induction hypothesis, there exist two vertex disjoint paths H and P_0 in Q_n^1 such that (1) H joins $\hat{\mathbf{x}}$ to $\mathbf{x}$ with $|V(H)| = a_1 - 2^{n-1}$ and (2) P_0 joins $\mathbf{u}_0$ to $\mathbf{v}_0$ with $|V(P_0)| = a_0$.

By Lemma 13.17, there exist two vertex disjoint paths R_1 and R_2 in Q_n^0 such that (1) R_1 joins $\mathbf{u}_1$ to $\hat{\mathbf{x}} + \mathbf{e}_n$, (2) R_2 joins $\mathbf{x} + \mathbf{e}_n$ to $\mathbf{v}_1$, and (3) $R_1 \cup R_2$ spans Q_n^0.

Set $P_1 = \langle \mathbf{u}_1, R_1, \hat{\mathbf{x}} + \mathbf{e}_n, \hat{\mathbf{x}}, H, \mathbf{x}, \mathbf{x} + \mathbf{e}_n, R_2, \mathbf{v}_1 \rangle$. Then $\{P_0, P_1\}$ satisfies the requirement of Lemma 16.14. See Figure 16.6 for illustration.

Case 1.3. $a_1 = 2^{n-1}$. Since a hypercube is hyper Hamiltonian laceable, we set P_0, P_1 to be Hamiltonian paths joining $\mathbf{u}_0$ to $\mathbf{v}_0$ and $\mathbf{u}_1$ to $\mathbf{v}_1$ in Q_n^1, Q_n^0, respectively.

Case 2. $\mathbf{u}_0 \in Q_n^0$.

Case 2.1. $a_1 < 2^{n-1}$. By Lemma 16.12, there exist two vertex disjoint paths P_1 and R in Q_n^0 such that (1) P_1 is joining $\mathbf{u}_1$ to $\mathbf{v}_1$ with $|V(P_1)| = a_1$, (2) R is joining $\mathbf{u}_0$ to $\hat{\mathbf{u}}_0$ with $|V(R)| = 2^{n-1} - a_1$, and (3) R contains an edge parallel to $\mathbf{e}_1$.

Since Q_n^1 is Hamiltonian laceable, there exists a Hamiltonian path H in Q_n^1 joining $\hat{\mathbf{u}}_0 + \mathbf{e}_n$ to $\mathbf{v}_0$. Set $P_0 = \langle \mathbf{u}_0, R, \hat{\mathbf{u}}_0, \hat{\mathbf{u}}_0 + e_n, H, \mathbf{v}_0 \rangle$. Thus, $\{P_0, P_1\}$ satisfies the requirement of Lemma 16.14. See Figure 16.7 for illustration.

Case 2.2. $a_1 \geq 2^{n-1}$. By Lemma 16.12, there exist two vertex disjoint paths $R_0 = \langle \mathbf{u}_0, \hat{\mathbf{u}}_0 \rangle$ and R_1 in Q_n^0 such that (1) R_0 is joining $\mathbf{u}_0$ to $\hat{\mathbf{u}}_0$ with $|V(R_0)| = 2$ and (2) R_1 is joining $\mathbf{u}_1$ to $\mathbf{v}_1$ with $|V(R_1)| = 2^{n-1} - 2$.

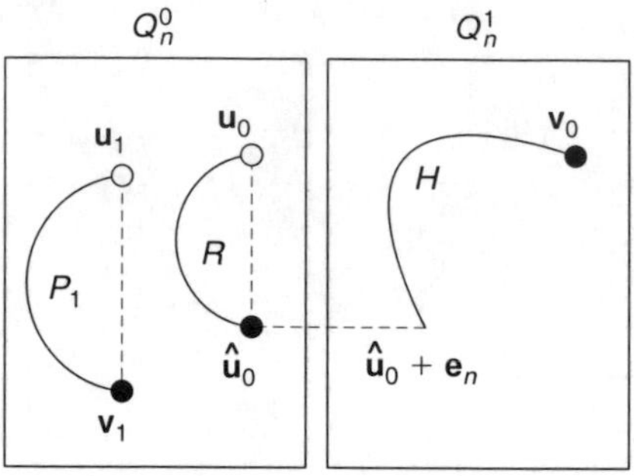

FIGURE 16.7 Illustration for Case 2.1 of Lemma 16.14.

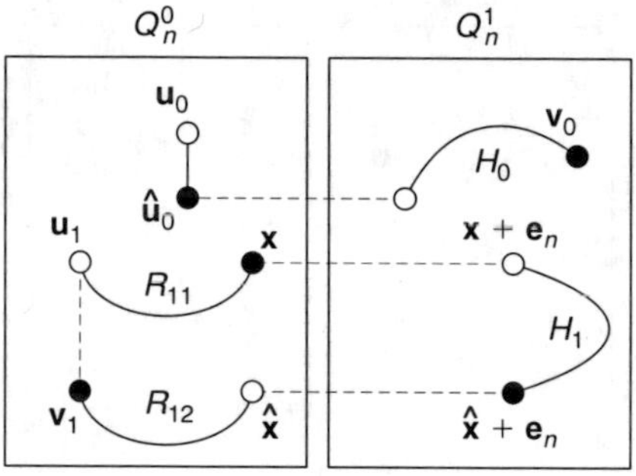

FIGURE 16.8 Illustration for Case 2.2 of Lemma 16.14.

Since $R_1 \cup \{\mathbf{u}_1, \mathbf{v}_1\}$ is a cycle, R_1 contains an edge $\{\mathbf{x}, \hat{\mathbf{x}}\}$ parallel to $\mathbf{e}_1$. Let $R_1 = \langle \mathbf{u}_1, R_{11}, \mathbf{x}, \hat{\mathbf{x}}, R_{12}, \mathbf{v}_1 \rangle$. By induction hypothesis, there exist two vertex disjoint paths H_0 and H_1 in Q_n^1 such that (1) H_0 is joining $\hat{\mathbf{u}}_0 + \mathbf{e}_n$ to $\mathbf{v}_0$ with $|V(H_0)| = a_0 - 2$ and (2) H_1 is joining $\mathbf{x} + \mathbf{e}_n$ to $\hat{\mathbf{x}} + \mathbf{e}_n$ with $|V(H_1)| = a_1 - 2^{n-1} - 2$.

Set $P_0 = \langle \mathbf{u}_0, \hat{\mathbf{u}}_0, \hat{\mathbf{u}}_0 + \mathbf{e}_n, H_0, \mathbf{v}_0 \rangle$ and $P_1 = \langle \mathbf{u}_1, R_{11}, \mathbf{x}, \mathbf{x} + \mathbf{e}_n, H_1, \hat{\mathbf{x}} + \mathbf{e}_n, \hat{\mathbf{x}}, R_{12}, \mathbf{v}_1 \rangle$. Thus, $\{P_0, P_1\}$ satisfies the requirement of Lemma 16.14. See Figure 16.8 for illustration. □

LEMMA 16.15 Let $\{(\mathbf{u}_0, \mathbf{u}_1, \mathbf{u}_2), (\mathbf{v}_0, \mathbf{v}_1, \mathbf{v}_2)\}$ be an extended parallel configuration of order 2 in Q_n where $n \geq 6$. Suppose that a_0, a_1, a_2 are positive even integers with (1) $a_0 \geq 2n$, (2) $a_i < 2^{n-1}, i = 1, 2$, and (3) $a_0 + a_1 + a_2 = 2^n$; then there exists a 3*-container $C_3^*((\mathbf{u}_0, \mathbf{u}_1, \mathbf{u}_2), (\mathbf{v}_0, \mathbf{v}_1, \mathbf{v}_2)) = \{P_0, P_1, P_2\}$ in Q_n such that (1) $|V(P_i)| = a_i$, for all $i = 0, 1, 2$ (2) P_0 contains an edge parallel to $\mathbf{e}_1$.

Proof: Since $a_2 < 2^{n-1}$, $a_0 + a_1 = 2^n - a_2 > 2^{n-1}$. There exist positive even integers $\tilde{a_0}, \tilde{a_1}$ such that $2(n-1) \leq \tilde{a_0} < a_0$, $\tilde{a_1} \leq a_1$, and $\tilde{a_0} + \tilde{a_1} = 2^{n-1}$. Without loss of generality, we assume that $\{\mathbf{u}_1, \mathbf{v}_1\} \in Q_n^0$, $\mathbf{v}_0 \in Q_n^0$, and $\{\mathbf{u}_2, \mathbf{v}_2\} \in Q_n^1$, where $Q_n^0 = \{(x_1, \ldots, x_n) \mid x_2 = 0\}$, $Q_n^1 = \{(x_1, \ldots, x_n) \mid x_2 = 1\}$.

Case 1. $\mathbf{u}_0 \in Q_n^0$. By Lemma 16.14, there exist two vertex disjoint paths R_0 and R_1 in Q_n^0 joining $\mathbf{u}_0$ to $\mathbf{v}_0$ and $\mathbf{u}_1$ to $\mathbf{v}_1$ of order $\tilde{a_0}, \tilde{a_1}$, respectively, such that R_0, R_1 contain edges $\{\mathbf{x}_0, \hat{\mathbf{x}}_0\}$, $\{\mathbf{x}_1, \hat{\mathbf{x}}_1\}$ parallel to $\mathbf{e}_1$, respectively. Let $R_i = \langle \mathbf{u}_i, R_{i1}, \mathbf{x}_i, \hat{\mathbf{x}}_i, R_{i2}, \mathbf{v}_i \rangle$ for

$i=0,1$. By Lemma 16.12, there exist three vertex disjoint paths H_0, H_1, and P_2 in Q_n^1 such that (1) H_i is joining $\mathbf{x}_i+\mathbf{e}_2$ to $\hat{\mathbf{x}}_i+\mathbf{e}_2$ of order $a_i-\tilde{a}_i$ for $i=0,1$ and (2) P_2 is joining $\mathbf{u}_2$ to $\mathbf{v}_2$ of order a_2.

Set $P_i=\langle \mathbf{u}_i, R_{i1}, \mathbf{x}_i, \mathbf{x}_i+\mathbf{e}_2, H_i, \hat{\mathbf{x}}_i+\mathbf{e}_2, \hat{\mathbf{x}}_i, R_{i2}, \mathbf{v}_i\rangle$ for $i=0,1$. Thus, $C_3^*((\mathbf{u}_0,\mathbf{u}_1,\mathbf{u}_2),(\mathbf{v}_0,\mathbf{v}_1,\mathbf{v}_2))=\{P_0,P_1,P_2\}$ is the required 3*-container. See Figure 16.9 for illustration.

REMARK: $P_1=\langle \mathbf{u}_1, R_{11}, \mathbf{x}_1, \hat{\mathbf{x}}_1, R_{12}, \mathbf{v}_1\rangle$ in case $a_1-\tilde{a}_1=0$.

Case 2. $\mathbf{u}_0\in Q_n^1$. Obviously, $\hat{\mathbf{u}}_0+\mathbf{e}_2$ is a white vertex in Q_n^0. By Lemma 16.14, there exist two vertex disjoint paths R_0 and R_1 in Q_n^0 joining $\hat{\mathbf{u}}_0+\mathbf{e}_2$ to $\mathbf{v}_0$ and $\mathbf{u}_1$ to $\mathbf{v}_1$ of order $\tilde{a}_0,\tilde{a}_1$, respectively, such that R_1 contains an edge $\{\mathbf{x}_1,\hat{\mathbf{x}}_1\}$ parallel to $\mathbf{e}_1$. Let $R_1=\langle \mathbf{u}_1, R_{11}, \mathbf{x}_1, \hat{\mathbf{x}}_1, R_{12}, \mathbf{v}_1\rangle$. By Lemma 16.12, there exist three vertex disjoint paths H_0, H_1, and P_2 in Q_n^1 such that (1) H_0 is joining $\mathbf{u}_0$ to $\hat{\mathbf{u}}_0$ of order $a_0-\tilde{a}_0$, (2) H_1 is joining $\mathbf{x}_1+\mathbf{e}_2$ to $\hat{\mathbf{x}}_1+\mathbf{e}_2$ of order $a_1-\tilde{a}_1$, and (3) P_2 is joining $\mathbf{u}_2$ to $\mathbf{v}_2$ of order a_2.

Set $P_0=\langle \mathbf{u}_0, H_0, \hat{\mathbf{u}}_0, \hat{\mathbf{u}}_0+\mathbf{e}_2, R_0, \mathbf{v}_0\rangle$ and $P_1=\langle \mathbf{u}_1, R_{11}, \mathbf{x}_1, \mathbf{x}_1+\mathbf{e}_2, H_1, \hat{\mathbf{x}}_1+\mathbf{e}_2, \hat{\mathbf{x}}_1, R_{12}, \mathbf{v}_1\rangle$. Thus, $C_3^*((\mathbf{u}_0,\mathbf{u}_1,\mathbf{u}_2),(\mathbf{v}_0,\mathbf{v}_1,\mathbf{v}_2))=\{P_0,P_1,P_2\}$ is the required 3*-container. See Figure 16.10 for illustration. □

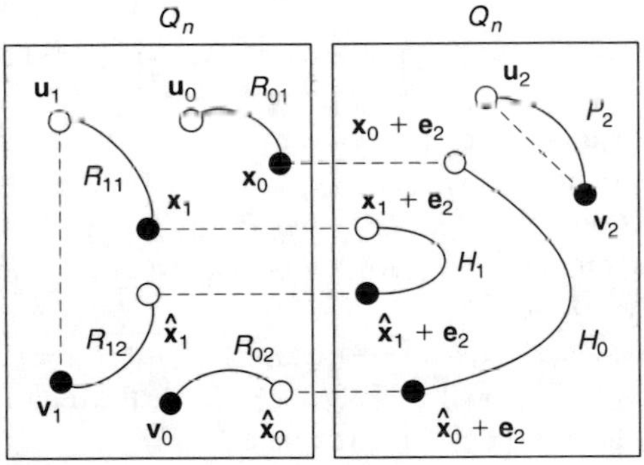

FIGURE 16.9 Illustration for Case 1 of Lemma 16.15.

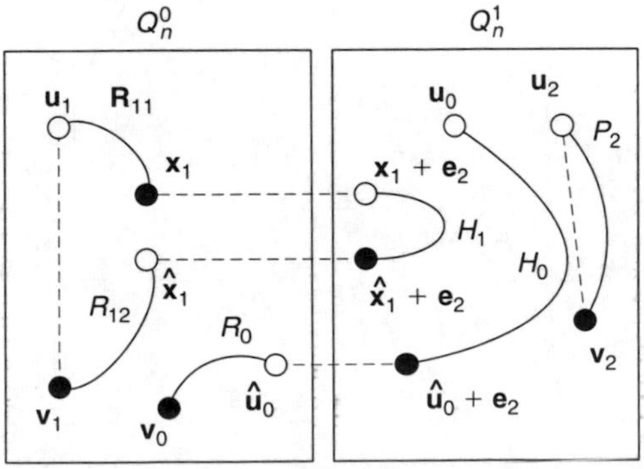

FIGURE 16.10 Illustration for Case 2 of Lemma 16.15.

Let $\Gamma = \{a_0, a_1, \ldots, a_k\}$ be a set of $k+1$ positive even integers where $3 \le k \le n-4$. We say that Γ satisfies the (n,k) condition if the following conditions are satisfied: (1) $a_i < a_0$ for all $1 \le i \le k$, (2) $a_0 \ge 2^{n+1}/(k+1)$, and (3) $a_0 + a_1 + \cdots + a_k = 2^n$.

THEOREM 16.4 Let $\{(\mathbf{u}_0, \mathbf{u}_1, \mathbf{u}_2, \ldots, \mathbf{u}_k), (\mathbf{v}_0, \mathbf{v}_1, \mathbf{v}_2, \ldots, \mathbf{v}_k)\}$ be an extended parallel configuration of order k in Q_n where $n-4 \ge k \ge 3$. If $\{a_0, a_1, \ldots a_k\}$ satisfies the (n,k) condition, then there exists a $(k+1)^*$-container $C^*_{k+1}((\mathbf{u}_0, \mathbf{u}_1, \ldots, \mathbf{u}_k), (\mathbf{v}_0, \mathbf{v}_1, \ldots, \mathbf{v}_k)) = \{P_0, P_1, \ldots, P_k\}$ such that (1) $|V(P_i)| = a_i$ for all $i \in \{1, 2, \ldots, k\}$ (2) P_0 contains an edge parallel to $\mathbf{e}_1$ with $|V(P_0)| = a_0$.

Proof: First, we shall prove that this is true for $k=3$ and then use induction to show that it is true for all $k \ge 4$. Assume $k=3$, $n \ge k+4 \ge 7$. Let $Q_n = Q_n^0 \cup Q_n^1$ where $Q_n^0 = \{(x_1, \ldots, x_n) \mid x_3 = 0\}$, $Q_n^1 = \{(x_1, \ldots, x_n) \mid x_3 = 1\}$. Clearly, edges $\{\mathbf{u}_1, \mathbf{v}_1\}, \{\mathbf{u}_2, \mathbf{v}_2\} \in Q_n^0$, $\mathbf{v}_0 \in Q_n^0$, and $\{\mathbf{u}_3, \mathbf{v}_3\} \in Q_n^1$.

If $\{a_0, a_1, a_2, a_3\}$ satisfies the $(n,3)$ condition then we have $a_1 + a_2 + a_3 = 2^n - a_0 \le 2^n - (2^{n+1})/4 = 2^{n-1}$. This implies that $a_1 + a_2 < 2^{n-1}$ and $\min\{a_1, a_2\} < 2^{n-2}$. Without loss of generality, we assume that $a_2 < 2^{n-2}$. Since $2^{n-2} > 2(n-1)$ for $n \ge 7$ and $a_0 + a_1 + a_2 \ge 2^{n-1}$, there exist even positive integers $\tilde{a}_0, \tilde{a}_1, \tilde{a}_2$ that satisfy (1) $a_0 > \tilde{a}_0 \ge 2(n-1)$, (2) $\tilde{a}_1 \le 2^{n-2}$ and $\tilde{a}_1 \le a_1$, (3) $\tilde{a}_2 = a_2$, and (4) $\tilde{a}_0 + \tilde{a}_1 + \tilde{a}_2 = 2^{n-1}$.

Case 1. $\mathbf{u}_0 \in Q_n^0$. By induction hypothesis, there exist three vertex disjoint paths R_i in Q_n^0 joining $\mathbf{u}_i$ to $\mathbf{v}_i$ of order $\tilde{a}_i$ for $0 \le i \le 2$ such that R_0 contains an edge $\{\mathbf{x}_0, \hat{\mathbf{x}}_0\}$ parallel to $\mathbf{e}_1$. Let $\{\mathbf{x}_1, \hat{\mathbf{x}}_1\}$ be an edge on the paths R_1 parallel to $\mathbf{e}_1$. We can write $R_i = \langle \mathbf{u}_i, R_{i1}, \mathbf{x}_i, \hat{\mathbf{x}}_i, R_{i2}, \mathbf{v}_i \rangle$. for $i = 0, 1$. By Lemma 16.12, there exist three vertex disjoint paths H_0, H_1, and P_3 in Q_n^1 such that (1) H_i is joining $\mathbf{x}_i + \mathbf{e}_3$ to $\hat{\mathbf{x}}_i + \mathbf{e}_3$ of order $a_i - \tilde{a}_i$ for $i = 0, 1$ and (2) P_3 is joining $\mathbf{u}_3$ to $\mathbf{v}_3$ of order a_3.

Set $P_i = \langle \mathbf{u}_i, R_{i1}, \mathbf{x}_i, \mathbf{x}_i + \mathbf{e}_3, H_i, \hat{\mathbf{x}}_i + \mathbf{e}_3, \hat{\mathbf{x}}_i, R_{i2}, \mathbf{v}_i \rangle$ for $i = 0, 1$, $P_2 = R_2$. Thus, $C_4^*((\mathbf{u}_0, \mathbf{u}_1, \mathbf{u}_2, \mathbf{u}_3), (\mathbf{v}_0, \mathbf{v}_1, \mathbf{v}_2, \mathbf{v}_3)) = \{P_0, P_1, P_2, P_3\}$ is the required 4^*-container. See Figure 16.11 for an illustration of Case 1, where $k=3$.

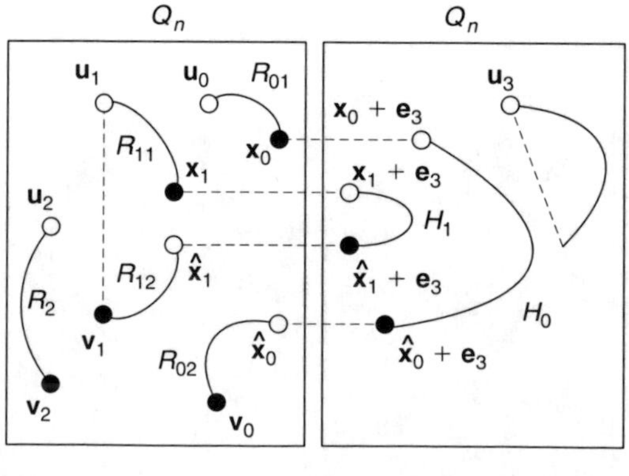

FIGURE 16.11 Illustration for Case 1, where $k=3$, of Theorem 16.4.

Case 2. $\mathbf{u}_0 \in Q_n^1$.

Case 2.1. $\hat{\mathbf{u}}_0 + \mathbf{e}_3 \neq \mathbf{u}_2$. Obviously, $\hat{\mathbf{u}}_0 + \mathbf{e}_3$ is a white vertex in Q_n^0. Note that $\mathbf{v}_3 + \mathbf{e}_3 = \mathbf{u}_1$ and $\hat{\mathbf{u}}_0 + \mathbf{e}_3 \neq \mathbf{u}_1$.

By induction hypothesis, there exist three vertex disjoint paths R_0, R_1, R_2 in Q_n^0 such that (1) R_i is joining $\mathbf{u}_i$ to $\mathbf{v}_i$ of order $\tilde{a}_i$ for $i = 1, 2$ and (2) R_0 is joining $\hat{\mathbf{u}}_0 + \mathbf{e}_3$ to $\mathbf{v}_0$ of order $\tilde{a}_0$, which contains an edge parallel to $\mathbf{e}_1$. Let $\{\mathbf{x}_1, \hat{\mathbf{x}}_1\}$ be an edge on the paths R_1 parallel to $\mathbf{e}_1$. We can write $R_1 = \langle \mathbf{u}_1, R_{11}, \mathbf{x}_1, \hat{\mathbf{x}}_1, R_{12}, \mathbf{v}_1 \rangle$. By Lemma 16.12, there exist three vertex disjoint paths H_0, H_1, and P_3 in Q_n^1 such that (1) H_0 is joining $\mathbf{u}_0$ to $\hat{\mathbf{u}}_0$ of order $a_0 - \tilde{a}_0$, (2) H_1 is joining $\mathbf{x}_1 + \mathbf{e}_3$ to $\hat{\mathbf{x}}_1 + \mathbf{e}_3$ of order $a_1 - \tilde{a}_1$, and (3) P_3 is joining $\mathbf{u}_3$ to $\mathbf{v}_3$ of order a_3.

We set $\{P_0, P_1, P_2\}$ as

$$\begin{aligned} P_0 &= \langle \mathbf{u}_0, H_0, \hat{\mathbf{u}}_0, \hat{\mathbf{u}}_0 + \mathbf{e}_3, R_0, \mathbf{v}_0 \rangle \\ P_1 &= \langle \mathbf{u}_1, R_{11}, \mathbf{x}_1, \mathbf{x}_1 + \mathbf{e}_3, H_1, \hat{\mathbf{x}}_1 + \mathbf{e}_3, \hat{\mathbf{x}}_1, R_{12}, \mathbf{v}_1 \rangle \\ P_2 &= R_2 \end{aligned}$$

Thus, $C_4^*((\mathbf{u}_0, \mathbf{u}_1, \mathbf{u}_2, \mathbf{u}_3), (\mathbf{v}_0, \mathbf{v}_1, \mathbf{v}_2, \mathbf{v}_3)) = \{P_0, P_1, P_2, P_3\}$ is the required 4^*-container. See Figure 16.12 for an illustration of Case 2.1, where $k = 3$.

Case 2.2. $\hat{\mathbf{u}}_0 + \mathbf{e}_3 = \mathbf{u}_2$. Set $\mathbf{z} = \mathbf{u}_0 + \mathbf{e}_4$. Obviously, $\hat{\mathbf{z}} = \hat{\mathbf{u}}_0 + \mathbf{e}_4$ and $\mathbf{z} + \mathbf{e}_3$ is a white vertex in Q_n^0.

By induction hypothesis, there exist three vertex disjoint paths R_0, R_1, R_2 in Q_n^0 such that (1) R_i is joining $\mathbf{u}_i$ to $\mathbf{v}_i$ of order $\tilde{a}_i$, for $i = 1, 2$ and (2) R_0 is joining $\mathbf{z} + \mathbf{e}_3$ to $\mathbf{v}_0$ of order $\tilde{a}_0$, which contains an edge parallel to $\mathbf{e}_1$.

Let $\{\mathbf{x}_1, \hat{\mathbf{x}}_1\}$ be an edge on the path R_1 parallel to $\mathbf{e}_1$. We can write $R_1 = \langle \mathbf{u}_1, R_{11}, \mathbf{x}_1, \hat{\mathbf{x}}_1, R_{12}, \mathbf{v}_1 \rangle$. By Lemma 16.12, there exists a $C_4^*((\mathbf{u}_0, \mathbf{x}_1 + \mathbf{e}_3, \hat{\mathbf{z}}, \mathbf{u}_3), (\hat{\mathbf{u}}_0, \hat{\mathbf{x}}_1 + \mathbf{e}_3, \mathbf{z}, \mathbf{v}_3)) = \{\langle \mathbf{u}_0, \hat{\mathbf{u}}_0 \rangle, H_1, H_2, P_3\}$ of order 2, $a_1 - \tilde{a}_1$, $a_0 - 2 - \tilde{a}_0$, and a_3, respectively. We set $\{P_0, P_1, P_2\}$ as

$$\begin{aligned} P_0 &= \langle \mathbf{u}_0, \hat{\mathbf{u}}_0, \hat{\mathbf{z}}, H_2, \mathbf{z}, \mathbf{z} + \mathbf{e}_3, R_0, \mathbf{v}_0 \rangle \\ P_1 &= \langle \mathbf{u}_1, R_{11}, \mathbf{x}_1, \mathbf{x}_1 + \mathbf{e}_3, H_1, \hat{\mathbf{x}}_1 + \mathbf{e}_3, \hat{\mathbf{x}}_1, R_{12}, \mathbf{v}_1 \rangle \\ P_2 &= R_2 \end{aligned}$$

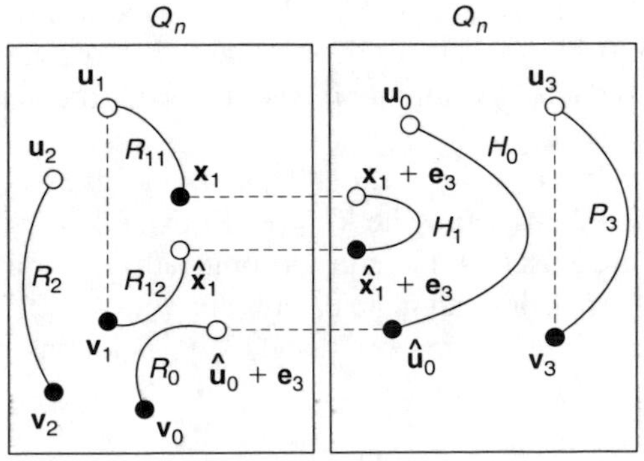

FIGURE 16.12 Illustration for Case 2.1, where $k = 3$, of Theorem 16.4.

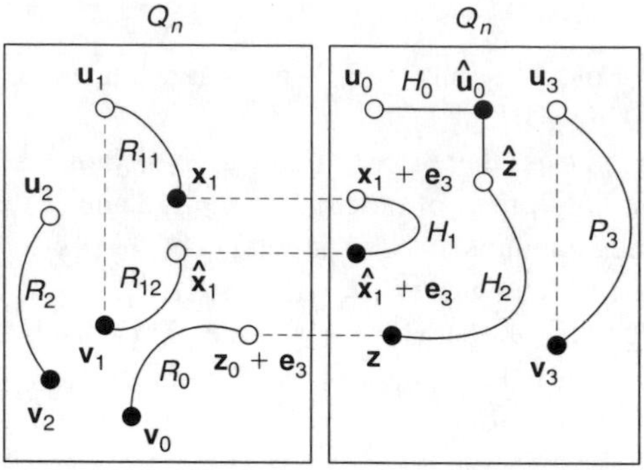

FIGURE 16.13 Illustration for Case 2.2, where $k = 3$, of Theorem 16.4.

Thus, $C_4^*((\mathbf{u}_0, \mathbf{u}_1, \mathbf{u}_2, \mathbf{u}_3), (\mathbf{v}_0, \mathbf{v}_1, \mathbf{v}_2, \mathbf{v}_3)) = \{P_0, P_1, P_2, P_3\}$ is the required 4*-container. See Figure 16.13 for an illustration of Case 2.2, where $k = 3$.

For $k \geq 4$, we prove this theorem is true by induction as follows.

Since $a_0 + a_1 + \cdots + a_{k-1} > 2^{n-1}$, there exist $\Gamma = \{\tilde{a}_0, \tilde{a}_1, \ldots, \tilde{a}_{k-1}\}$, which satisfies the $(n-1, k-1)$ condition with $\tilde{a}_i \leq a_i$, for all $0 \leq i \leq k-1$. Let $Q_n = Q_n^0 \cup Q_n^1$, where $Q_n^0 = \{(x_1, \ldots, x_n) \mid x_k = 0\}$ and $Q_n^1 = \{(x_1, \ldots, x_n) \mid x_k = 1\}$. Clearly, edge $\{\mathbf{u}_i, \mathbf{v}_i\} \in Q_n^0$ for $1 \leq i \leq k-1$, $\mathbf{v}_0 \in Q_n^0$, and $\{\mathbf{u}_k, \mathbf{v}_k\} \in Q_n^1$.

Case 1. $\mathbf{u}_0 \in Q_n^0$. By induction hypothesis, there exist k-vertex disjoint paths R_i in Q_n^0 joining $\mathbf{u}_i$ to $\mathbf{v}_i$ of order $\tilde{a}_i$, $0 \leq i \leq k-1$ such that R_0 contains an edge $\{\mathbf{x}_0, \hat{\mathbf{x}}_0\}$ parallel to $\mathbf{e}_1$. Obviously, $\tilde{a}_i \leq a_i$ for $0 \leq i \leq k-1$. Let $\{\mathbf{x}_i, \hat{\mathbf{x}}_i\}$ be an edge on the paths R_i parallel to $\mathbf{e}_1$ for $1 \leq i \leq k-1$. We can write $R_i = \langle \mathbf{u}_i, R_{i1}, \mathbf{x}_i, \hat{\mathbf{x}}_i, R_{i2}, \mathbf{v}_i \rangle$ for all $0 \leq i \leq k-1$. By Lemma 16.12, there exist $(k+1)$ vertex disjoint paths H_i for $0 \leq i \leq k-1$ and P_k in Q_n^1 such that H_i is joining $\mathbf{x}_i + \mathbf{e}_k$ to $\hat{\mathbf{x}}_i + \mathbf{e}_k$ of order $a_i - \tilde{a}_i$ for $i = 0 \leq i \leq k-1$ and P_k is joining $\mathbf{u}_k$ to $\mathbf{v}_k$ of order a_k. Set $P_i = \langle \mathbf{u}_i, R_{i1}, \mathbf{x}_i, \mathbf{x}_i + \mathbf{e}_k, H_i, \hat{\mathbf{x}}_i + \mathbf{e}_k, \hat{\mathbf{x}}_i, R_{i2}, \mathbf{v}_i \rangle$ for $0 \leq i \leq k-1$. Thus, $C_{k+1}^*((\mathbf{u}_0, \mathbf{u}_1, \ldots, \mathbf{u}_k), (\mathbf{v}_0, \mathbf{v}_1, \ldots, \mathbf{v}_k)) = \{P_i \mid 0 \leq i \leq k\}$ is the required $(k+1)^*$-container.

Case 2. $\mathbf{u}_0 \in Q_n^1$.

Case 2.1. $\mathbf{u}_0 \neq \mathbf{e}_j + \mathbf{e}_k$ for all $j \in \{2, 3, \ldots, k-1\}$. Obviously, $\hat{\mathbf{u}}_0 + \mathbf{e}_k$ is a white vertex in Q_n^0 other than $\mathbf{u}_1, \ldots, \mathbf{u}_{k-1}$. By induction hypothesis, there exist k-vertex disjoint paths R_i, for $0 \leq i \leq k-1$ in Q_n^0 such that (1) R_i is joining $\mathbf{u}_i$ to $\mathbf{v}_i$ of order $\tilde{a}_i$, for $1 \leq i \leq k-1$ and (2) R_0 is joining $\hat{\mathbf{u}}_0 + \mathbf{e}_k$ to $\mathbf{v}_0$ of order $\tilde{a}_0$, which contains an edge parallel to $\mathbf{e}_1$.

Obviously, $\tilde{a}_i \leq a_i$ for $0 \leq i \leq k-1$. Let $\{\mathbf{x}_i, \hat{\mathbf{x}}_i\}$ be an edge on the paths R_i of parallel to $\mathbf{e}_1$, for $1 \leq i \leq k-1$. We can write $R_i = \langle \mathbf{u}_i, R_{i1}, \mathbf{x}_i, \hat{\mathbf{x}}_i, R_{i2}, \mathbf{v}_i \rangle$. For $1 \leq i \leq k-1$. By Lemma 16.12, there exist $(k+1)$-vertex disjoint paths H_i for $0 \leq i \neq k \leq k+1$ and P_k in Q_n^1 such that (1) H_0 is joining $\mathbf{u}_0$ to $\hat{\mathbf{u}}_0$ of order $a_0 - \tilde{a}_0$, (2) H_i is joining $\mathbf{x}_i + \mathbf{e}_k$ to $\hat{\mathbf{x}}_i + \mathbf{e}_k$ of order $a_i - \tilde{a}_i$ for $1 \leq i \leq k-1$, and (3) P_k is joining $\mathbf{u}_k$ to $\mathbf{v}_k$ of order a_k.

We set $\{P_0, P_1, \ldots, P_k\}$ as

$$P_0 = \langle \mathbf{u}_0, H_0, \hat{\mathbf{u}}_0, \hat{\mathbf{u}}_0 + \mathbf{e}_k, R_0, \mathbf{v}_0 \rangle$$

$$P_i = \langle \mathbf{u}_i, R_{i1}, \mathbf{x}_i, \mathbf{x}_i + \mathbf{e}_k, H_i, \hat{\mathbf{x}}_i + \mathbf{e}_k, \hat{\mathbf{x}}_i, R_{i2}, \mathbf{v}_i \rangle \text{ for } 1 \leq i \leq k-1$$

Thus, $C^*_{k+1}((\mathbf{u}_0, \mathbf{u}_1, \ldots, \mathbf{u}_k), (\mathbf{v}_0, \mathbf{v}_1, \ldots, \mathbf{v}_k)) = \{P_i \mid 0 \le i \le k\}$ is the required $(k+1)^*$-container.

Case 2.2. $\mathbf{u}_0 = \mathbf{e}_j + \mathbf{e}_k$ for some $j \in \{2, 3, \ldots, k-1\}$. We decomposed Q_n into Q_n^0 and Q_n^1 by dimension j; that is, $Q_n = Q_n^0 \bigcup Q_n^1$, where $Q_n^0 = \{(x_1, \ldots, x_n) \mid x_j = 0\}$, and $Q_n^1 = \{(x_1, \ldots, x_n) \mid x_j = 1\}$. Clearly, edge $\{\mathbf{u}_i, \mathbf{v}_i\} \in Q_n^0$, for $0 \le i \ne j \le k$ and $\{\mathbf{u}_j, \mathbf{v}_j\} \in Q_n^1$. The proof is similar to that of Case 1. □

A set of even positive integers $\{a_1, \ldots, a_k\}$ is an *equitable* set if $|a_i - a_j| \le 2$ for all $1 \le i, j \le k$. A container $\{P_1, \ldots, P_k\}$ is an *equitable container* if $||V(P_i)| - |V(P_j)|| \le 2$ for all $1 \le i, j \le k$.

LEMMA 16.16 Q_n is equitably 2^*-laceable for all $n \ge 3$.

Proof: Let $Q_n = Q_n^0 \cup Q_n^1$ where $Q_n^0 = \{(x_1, \ldots, x_n) \mid x_n = 0\}$, $Q_n^1 = \{(x_1, \ldots, x_n) \mid x_n = 1\}$. Since Q_n is vertex-transitive, we need only to find an equitable 2^*-container joining $\mathbf{0} = (0, 0, \ldots, 0)$ to any black vertex $\mathbf{v}$ of Q_n.

Case 1. $\mathbf{v} \in Q_n^0$. Since Q_n^0 is Hamiltonian laceable, there exists a Hamiltonian path P_1 joining $\mathbf{0}$ to $\mathbf{v}$ in Q_n^0. Since Q_n^1 is Hamiltonian laceable, there exists a Hamiltonian path H joining $\mathbf{e}_n$ to $\mathbf{v} + \mathbf{e}_n$ in Q_n^1. Set $P_2 = \langle \mathbf{0}, \mathbf{e}_n, H, \mathbf{v} + \mathbf{e}_n, \mathbf{v} \rangle$. Obviously, $\{P_1, P_2\}$ is an equitable 2^*-container of Q_n joining $\mathbf{0}$ to $\mathbf{v}$.

Case 2. $\mathbf{v} \in Q_n^1$. Let $\mathbf{z}$ be a neighbor of $\mathbf{v}$ in Q_n^1. There exists a Hamiltonian path R joining $\mathbf{0}$ to $\mathbf{z} + \mathbf{e}_n$ in Q_n^0. Since every hypercube is hyper Hamiltonian laceable, there exists a Hamiltonian path H joining $\mathbf{e}_n$ to $\mathbf{v}$ in $Q_n^1 - \{\mathbf{z}\}$. Set $P_1 = \langle \mathbf{0}, R, \mathbf{z} + \mathbf{e}_n, \mathbf{z}, \mathbf{v} \rangle$ and $P_2 = \langle \mathbf{0}, \mathbf{e}_n, H, \mathbf{v} \rangle$. Obviously, $\{P_1, P_2\}$ is an equitable 2^*-container of Q_n joining $\mathbf{0}$ to $\mathbf{v}$. □

Let $\mathbf{x}$ to be the any real number. We define $\lfloor\lfloor \mathbf{x} \rfloor\rfloor$ to be the largest even integer smaller or equal to $\mathbf{x}$.

THEOREM 16.5 Q_n is equitable k^*-laceable for all $k \le n-4$.

Proof: By Lemma 16.16, the theorem is true if $k = 2$. We shall use induction on k to prove that the theorem is true for all $k \le n-4$. Assume that Q_n is equitable $(k-1)^*$-laceable. Since Q_n is vertex-transitive, we need only to find an equitable k^*-container of Q_n between the white vertex $\mathbf{0}$ and any black vertex $\mathbf{v}$.

By Corollary 16.1, there are equitable k^*-containers $C_k^*(\mathbf{0}, \mathbf{v})$ for any neighbor $\mathbf{v}$ of $\mathbf{0}$. Thus we assume that $\mathbf{v}$ is not adjacent to $\mathbf{0}$. By the symmetric property of Q_n, we assume that $\mathbf{0} \in Q_n^0$ and $\mathbf{v} \in Q_n^1$ where $Q_n^0 = \{(x_1, \ldots, x_n) \mid x_n = 0\}$, $Q_n^1 = \{(x_1, \ldots, x_n) \mid x_n = 1\}$. Let $\mathbf{u}_1 = \mathbf{v} + \mathbf{e}_1$ and $\mathbf{z} = \mathbf{u}_1 + \mathbf{e}_n$. Obviously, $\mathbf{z}$ is a black vertex of Q_n^0. By induction hypothesis, there exists an equitable $(k-1)^*$-container $C_{k-1}^*(\mathbf{0}, \mathbf{z}) = \{R_1, R_2, \ldots, R_{k-1}\}$ of Q_n^0 joining $\mathbf{0}$ to $\mathbf{z}$. Without loss of generality, we assume that $|V(R_i)| = b_i$ for all $1 \le i \le k-1$ and $b_1 \ge b_2 \ge \cdots \ge b_{k-1}$. Let $R_i = \langle \mathbf{0}, T_i, \mathbf{w}_i, \mathbf{z} \rangle$ for all $1 \le i \le k-1$. Without loss of generality, we assume that $\mathbf{w}_i = \mathbf{z} + \mathbf{e}_i$ for all $1 \le i \le k-1$. Let $\mathbf{v}_i = \mathbf{w}_i + \mathbf{e}_n$ for all $1 \le i \le k-1$ and let $\mathbf{u}_i = \mathbf{v} + \mathbf{e}_i$ for all $2 \le i \le k-1$. It is easy checked that $\mathbf{v}_1 = \mathbf{v}$ and $\mathbf{u}_i = \mathbf{v}_i + \mathbf{e}_1$ for all $1 \le i \le k-1$.

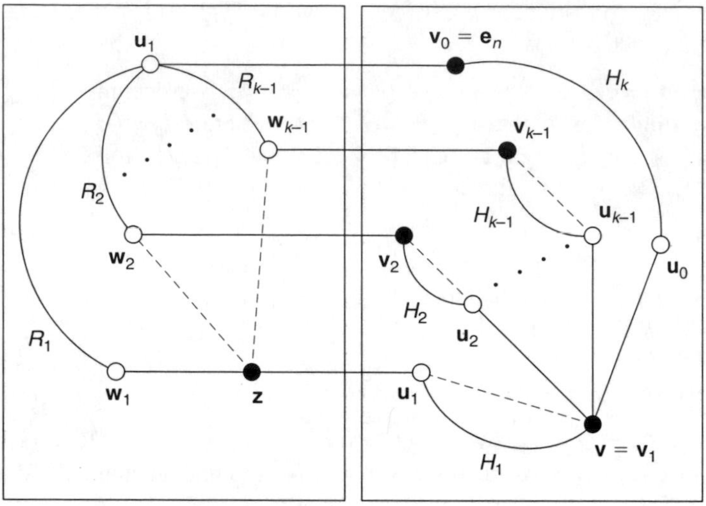

FIGURE 16.14 Illustration for Theorem 16.5.

Let $\mathbf{u}_0 = \mathbf{v} + \mathbf{e}_k$ and $\mathbf{v}_0 = \mathbf{e}_n$. Then $\{(\mathbf{u}_0, \mathbf{u}_1, \ldots, \mathbf{u}_{k-1}), (\mathbf{v}_0, \mathbf{v}_1, \ldots, \mathbf{v}_{k-1})\}$ is an extended parallel configuration of order $k-1$ in Q_n^1. We choose positive even integers $\{a_0, a_1, \ldots, a_{k-1}\}$ satisfying $(n-1, k-1)$ condition with (1) $a_0 = \left\lfloor\left\lfloor \frac{2^n-2}{k} \right\rfloor\right\rfloor$, (2) $a_1 \leq a_2 \leq \cdots \leq a_{k-1}$, and (3) $a_{k-1} - a_1 \leq 2$.

By Theorem 16.4, there exist k-vertex disjoint paths $H_1, H_2, \ldots, H_k$ in Q_n^1 such that (1) H_1 joins $\mathbf{u}_1$ and $\mathbf{v}$ with $|V(H_1)| = a_1$, (2) H_i joins $\mathbf{v}_i$ and $\mathbf{u}_i$ with $|V(H_i)| = a_i$, $i = 2, \ldots, k-1$, and (3) H_k joins $\mathbf{v}_0$ and $\mathbf{u}_0$ with $|V(H_k)| = a_0$.

We set $\{P_1, P_2, \ldots, P_k\}$ as

$$\begin{aligned} P_1 &= \langle \mathbf{0}, T_1, \mathbf{w}_1, \mathbf{z}, \mathbf{u}_1, H_1, \mathbf{v} \rangle \\ P_i &= \langle \mathbf{0}, T_i, \mathbf{w}_i, \mathbf{v}_i, H_i, \mathbf{u}_i, \mathbf{v} \rangle \quad \text{for} \quad 2 \leq i \leq k-1 \\ P_k &= \langle \mathbf{0}, \mathbf{v}_0, H_k, \mathbf{u}_0, \mathbf{v} \rangle \end{aligned}$$

Clearly, $|V(P_i)| = a_i + b_i$ $i = 1, \ldots, k-1$, $|V(P_k)| = a_0 + 2$. It is easy to observe that $\{|V(P_i)| \mid i = 1, \ldots, k-1\}$ is an equitable set. Moreover, it is easy to verify that $\min_{1 \leq i \leq k-1} (a_i + b_i - 2) \geq a_0$ and $\max_{1 \leq i \leq k-1} (a_i + b_i - 2) \leq a_0 + 2$. Therefore, $\{P_1, P_2, \ldots, P_k\}$ is an equitable k^*-container of Q_n joining $\mathbf{0}$ to $\mathbf{v}$. See Figure 16.14 for illustration. □

16.4 SPANNING DIAMETER FOR SOME (*n, k*)-STAR GRAPHS

In Section 16.3, we computed the spanning diameter of the star graphs and the hypercubes. These two families of graphs are bipartite. Here, we discuss the spanning diameter of a family of nonbipartite graphs.

The (n, k)-star graph is an attractive alternative to the n-star graph [3]. However, the growth of vertices is $n!$ for an n-star graph. To remedy this drawback, the (n, k)-star graph is proposed by Chiang and Chen [60]. The (n, k)-star graph is a generalization of the n-star graph. It has two parameters n and k. When $k = n - 1$, an $(n, n - 1)$-star graph is isomorphic to an n-star graph, and when $k = 1$, an $(n, 1)$-star graph is isomorphic to a complete graph K_n.

Assume that n and k are two positive integers with $n > k$. We use $\langle n \rangle$ to denote the set $\{1, 2, \ldots, n\}$. The (n, k)-star graph, $S_{n,k}$, is a graph with the node set $V(S_{n,k}) = \{u_1 u_2 \ldots u_k \mid u_i \in \langle n \rangle$ and $u_i \neq u_j$ for $i \neq j\}$. Adjacency is defined as follows: a node $u_1 u_2 \ldots u_i \ldots u_k$ is adjacent to (1) the node $u_i u_2 u_3 \ldots u_{i-1} u_1 u_{i+1} \ldots u_k$, where $2 \leq i \leq k$ (i.e., swap u_i with u_1) and (2) the node $x u_2 u_3 \ldots u_k$ where $x \in \langle n \rangle - \{u_i \mid 1 \leq i \leq k\}$. The edges of type (1) are referred to as *i-edges* and the edges of type (2) are referred to as *1-edges*.

By definition, $S_{n,k}$ is an $(n - 1)$-regular graph with $n!/(n - k)!$ nodes. Moreover, it is node-transitive [60]. We use boldface to denote nodes in $S_{n,k}$. Hence, $\mathbf{u}_1, \mathbf{u}_2, \ldots, \mathbf{u}_m$ denotes a sequence of nodes in $S_{n,k}$. Let $\mathbf{u} = u_1 u_2 \ldots u_k$ be any node of $S_{n,k}$. We say that u_i is the ith coordinate of $\mathbf{u}$, denoted by $(\mathbf{u})_i$, for $1 \leq i \leq k$. By the definition of $S_{n,k}$, there is exactly one neighbor $\mathbf{v}$ of $\mathbf{u}$ such that $\mathbf{u}$ and $\mathbf{v}$ are adjacent through an i-edge with $2 \leq i \leq k$. For this reason, we use $(\mathbf{u})^i$ to denote the unique i-neighbor of $\mathbf{u}$. Obviously, $((\mathbf{u})^i)^i = \mathbf{u}$. For $1 \leq i \leq n$, let $S_{n,k}^{\{i\}}$ denote the subgraph of $S_{n,k}$ induced by those nodes $\mathbf{u}$ with $(\mathbf{u})_k = i$. In Ref. 60, it is showed that $S_{n,k}$ can be decomposed into n subgraphs $S_{n,k}^{\{i\}}$, $1 \leq i \leq n$, such that each subgraph $S_{n,k}^{\{i\}}$ is isomorphic to $S_{n-1,k-1}$. Thus, the (n, k)-star graph can be constructed recursively. Obviously, $\mathbf{u} \in S_{n,k}^{\{(\mathbf{u})_k\}}$. Let $I \subseteq \langle n \rangle$. We use $S_{n,k}^I$ to denote the subgraph of $S_{n,k}$ induced by those nodes $\mathbf{u}$ with $(\mathbf{u})_k \in I$. For $1 \leq i \leq n$ and $1 \leq j \leq n$ with $i \neq j$, we use $E_{n,k}^{i,j}$ to denote the set of edges between $S_{n,k}^{\{i\}}$ and $S_{n,k}^{\{j\}}$. $S_{3,1}$, $S_{4,1}$, $S_{4,2}$, and $S_{5,2}$ are shown in Figure 16.15.

Owing to the nice structure of (n, k)-star graphs, there are a lot of studies on its topological properties. In particular, Chang and Kim [41] studied the cycles embedding in faulty (n, k)-star graphs. Hsu et al. [172] are studied the fault Hamiltonicity and fault Hamiltonian connectivity of the (n, k)-star graphs. Chang et al. study the embedding of mutually independent hamiltonian paths between any two vertices [40]. In this section, we discuss only the spanning diameter of some (n, k)-star graphs. Yet we need some background about (n, k)-star graphs. The following lemma can easily be obtained from the definition of (n, k)-star graphs.

LEMMA 16.17 $|E_{n,k}^{i,j}| = (n - 2)!/(n - k)!$ if $k \geq 2$.

Since, $S_{n,1}$ is isomorphic to the complete graph K_n with n vertices, we have the following result.

LEMMA 16.18 Let n be any positive integer with $n \geq 3$. Then $S_{n,1} - F$ is Hamiltonian-connected if $F \subseteq V(S_{n,1})$ with $|F| \leq n - 2$.

The following theorem is a direct consequence of Theorem 11.16.

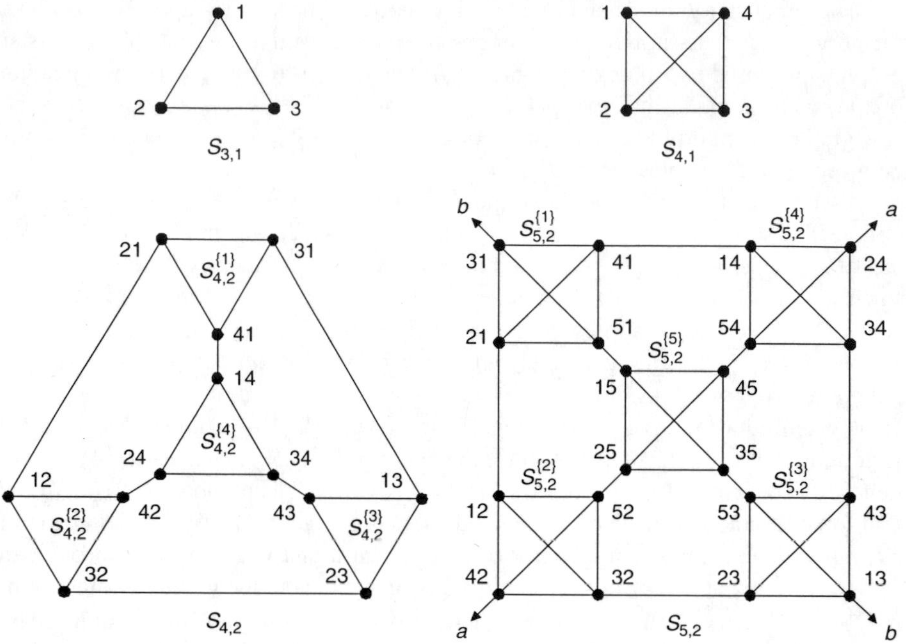

FIGURE 16.15 Some examples of the (n, k)-star graphs: $S_{3,1}$, $S_{4,1}$, $S_{4,2}$, and $S_{5,2}$.

THEOREM 16.6 Let n and k be any two positive integers with $n-k \geq 2$, and let $F \subseteq V(S_{n,k}) \cup E(S_{n,k})$. Then $S_{n,k} - F$ is Hamiltonian-connected if $|F| \leq n-4$ and $S_{n,k} - F$ is Hamiltonian if $|F| \leq n-3$.

THEOREM 16.7 Let n and k be any two integers with $n-k \geq 2$ and $k \geq 3$. Let $I = \{i_1, i_2, \ldots, i_r\}$ be any subset of $\langle n \rangle$ for $1 \leq r \leq n$. Assume that $\mathbf{u}$ and $\mathbf{v}$ are two distinct vertices of $S_{n,k}$ with $\mathbf{u} \in S_{n,k}^{\{i_1\}}$ and $\mathbf{v} \in S_{n,k}^{\{i_r\}}$. Then there is a Hamiltonian path $H = \langle \mathbf{u} = \mathbf{x}_1, H_1, \mathbf{y}_1, \mathbf{x}_2, H_2, \mathbf{y}_2, \ldots, \mathbf{x}_r, H_r, \mathbf{y}_r = \mathbf{v} \rangle$ of $S_{n,k}^I$ joining $\mathbf{u}$ to $\mathbf{v}$ such that H_j is a Hamiltonian path of $S_{n,k}^{\{i_j\}}$ joining $\mathbf{x}_j$ to $\mathbf{y}_j$ for every $1 \leq j \leq r$.

Proof: Let $\mathbf{x}_1 = \mathbf{u}$ and $\mathbf{y}_r = \mathbf{v}$. Since $S_{n,k}^{\{i_j\}}$ is isomorphic to $S_{n-1,k-1}$ for every $j \in \langle r \rangle$, by Theorem 16.6, this statement holds for $r = 1$. Thus, we assume that $r \geq 2$.

Since $k \geq 3$ and $n-k \geq 2$, by Lemma 16.17, $|E_{n,k}^{i_j,i_{j+1}}| = (n-2)!/(n-k)! \geq 3$ for every $j \in \langle r-1 \rangle$. We can choose $(\mathbf{y}_j, \mathbf{x}_{j+1}) \in E_{n,k}^{i_j,i_{j+1}}$ for every $j \in \langle r-1 \rangle$ with $\mathbf{y}_j \in S_{n,k}^{\{i_j\}}$, $\mathbf{x}_{j+1} \in S_{n,k}^{\{i_{j+1}\}}$, $\mathbf{y}_1 \neq \mathbf{u}$, and $\mathbf{x}_r \neq \mathbf{v}$. By Theorem 16.6, there is a Hamiltonian path H_j of $S_{n,k}^{\{i_j\}}$ joining $\mathbf{x}_j$ to $\mathbf{y}_j$ for every $j \in \langle r \rangle$. Then $H = \langle \mathbf{u} = \mathbf{x}_1, H_1, \mathbf{y}_1, \mathbf{x}_2, H_2, \mathbf{y}_2, \ldots, \mathbf{x}_r, H_r, \mathbf{y}_r = \mathbf{v} \rangle$ forms a desired path. See Figure 16.16 for illustration. □

Since $S_{n,1}$ is isomorphic to K_n, it is easy to see that $w^*(S_{n,1}) = 2$ for $n \geq 2$. Thus, we study $w^*(S_{n,2})$.

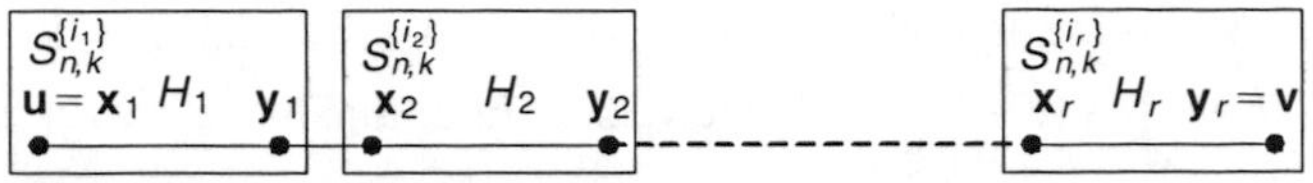

FIGURE 16.16 Illustration for Theorem 16.7.

THEOREM 16.8 For $n \geq 4$, $w^*(S_{n,2}) \geq n^2 - 2n$.

THEOREM 16.9 For $n \geq 4$, $w^*(S_{n,2}) = n^2 - 2n$.

Proof: Since $S_{n,2}$ is vertex-transitive, we assume that $\mathbf{u} = 1n$. We have the following cases:

Case 1. Suppose that $(\mathbf{v})_2 = n$. By Theorem 16.7, there is a Hamiltonian path R of $S_{n,2}^{\langle n-1 \rangle}$ joining $(\mathbf{u})^k$ to $(\mathbf{v})^k$. We set $\{\mathbf{z_1}, \mathbf{z_2}, \ldots, \mathbf{z_{n-3}}\} = V(S_{n,2}^{\{n\}}) - \{\mathbf{u}, \mathbf{v}\}$. We set $P_1 = \langle \mathbf{u}, \mathbf{v} \rangle$, $P_2 = \langle \mathbf{u}, (\mathbf{u})^2, R, (\mathbf{v})^2, \mathbf{v} \rangle$, and $P_i = \langle \mathbf{u}, \mathbf{z_{i-2}}, \mathbf{v} \rangle$ for every $3 \leq i \leq n-1$. Then $\{P_1, P_2, \ldots, P_{n-1}\}$ forms a $(n-1)^*$-container of $S_{n,2}$ between **u** and **v**. Moreover, $l(P_1) = 1$, $l(P_2) = n^2 - 2n$, and $l(P_i) = 2$ for every $3 \leq i \leq n-1$.

Case 2. Suppose that $(\mathbf{v})_2 = 1$ and $(\mathbf{v})_1 = n$. We set $\mathbf{x_i} = (i+1)n$ and $\mathbf{y_i} = (i+1)1$ for every $1 \leq i \leq n-2$. By Theorem 16.7, there is a Hamiltonian path R_i of $S_{n,2}^{\{i+1\}}$ joining $(\mathbf{x_i})^k$ to $(\mathbf{y_i})^k$ for every $1 \leq i \leq n-2$. We set $P_1 = \langle \mathbf{u}, \mathbf{v} \rangle$ and $P_i = \langle \mathbf{u}, \mathbf{x_{i-1}}, (\mathbf{x_{i-1}})^2, R_{i-1}, (\mathbf{y_{i-1}})^2, \mathbf{y_{i-1}}, \mathbf{v} \rangle$ for every $2 \leq i \leq n-1$. Then $\{P_1, P_2, \ldots, P_{n-1}\}$ forms a $(n-1)^*$-container of $S_{n,2}$ between **u** and **v**. Moreover, $l(P_1) = 1$ and $l(P_i) = n+3$ for every $2 \leq i \leq n-1$.

Case 3. Suppose that either $(\mathbf{v})_2 = 1$ and $(\mathbf{v})_1 = t$ for some $t \neq n$ or $(\mathbf{v})_2 = t$ for some $t \neq 1$ and $(\mathbf{v})_1 = n$. By symmetric rule, we assume that $(\mathbf{v})_2 = 1$ and $(\mathbf{v})_1 = t$ for some $t \neq n$. We set $\mathbf{x_i} = (i+1)n$ for every $1 \leq i \leq n-2$ and $\mathbf{y_j} = (j+1)1$ for every $j \in \langle n-2 \rangle - \{t-1\}$. By Theorem 16.7, there is a Hamiltonian path R_i of $S_{n,2}^{\{i+1\}}$ joining $(\mathbf{x_i})^2$ to $(\mathbf{y_i})^2$ for every $i \in \langle n-2 \rangle - \{t-1\}$ and there is a Hamiltonian path R_{t-1} of $S_{n,2}^{\{t\}}$ joining $(\mathbf{x_{t-1}})^2$ to $(\mathbf{v})^2$. We set $P_i = \langle \mathbf{u}, \mathbf{x_i}, (\mathbf{x_i})^2, R_i, (\mathbf{y_i})^2, \mathbf{y_i}, \mathbf{v} \rangle$ for every $i \in \langle n-2 \rangle - \{t-1\}$, $P_{t-1} = \langle \mathbf{u}, \mathbf{x_{t-1}}, (\mathbf{x_{t-1}})^2, R_{t-1}, (\mathbf{v})^2, \mathbf{v} \rangle$, and $P_{n-1} = \langle \mathbf{u}, (\mathbf{u})^2, \mathbf{v} \rangle$. Then $\{P_1, P_2, \ldots, P_{n-1}\}$ forms a $(n-1)^*$-container of $S_{n,2}$ between **u** and **v**. Moreover, $l(P_i) = n+3$ for every for every $i \in \langle n-2 \rangle - \{t-1\}$, $l(P_{t-1}) = n+1$, and $l(P_{n-1}) = 2$.

Case 4. Suppose that $(\mathbf{v})_2 = t$ and $(\mathbf{v})_1 = 1$ for some $t \neq n$. We set $\mathbf{x_i} = (i+1)n$ for every $1 \leq i \leq n-2$ and $\mathbf{y_j} = (j+1)1$ for every $j \in \langle n-2 \rangle - \{t-1\}$. By Theorem 16.7, there is a Hamiltonian path R_i of $S_{n,2}^{\{i+1\}}$ joining $(\mathbf{x_i})^2$ to $(\mathbf{y_i})^2$ for every $i \in \langle n-2 \rangle - \{t-1\}$ and there is a Hamiltonian path R_{t-1} of $S_{n,2}^{\{1\}}$ joining $(\mathbf{u})^2$ to $(\mathbf{v})^2$. We set $P_i = \langle \mathbf{u}, \mathbf{x_i}, (\mathbf{x_i})^2, R_i, (\mathbf{y_i})^2, \mathbf{y_i}, \mathbf{v} \rangle$ for every $i \in \langle n-2 \rangle - \{t-1\}$, $P_{t-1} = \langle \mathbf{u}, \mathbf{x_{t-1}}, (\mathbf{x_{t-1}})^2 = \mathbf{y_{n-1}}, \mathbf{v} \rangle$, and $P_{n-1} = \langle \mathbf{u}, (\mathbf{u})^k, R_{t-1}, (\mathbf{v})^2, \mathbf{v} \rangle$. Then $\{P_1, P_2, \ldots, P_{n-1}\}$ forms a $(n-1)^*$-container of $S_{n,2}$ between **u** and **v**. Moreover, $l(P_i) = n+3$ for every for every $i \in \langle n-2 \rangle - \{t-1\}$, $l(P_{t-1}) = 3$, and $l(P_{n-1}) = n$.

Case 5. Suppose that $(\mathbf{v})_2 = s$ and $(\mathbf{v})_1 = t$ for some $s \neq n$ and for some $t \neq 1$. We set $\mathbf{x_i} = (i+1)n$ for every $1 \leq i \leq n-2$ and $\mathbf{y_j} = js$ for every $j \in \langle n \rangle - \{s, t\}$.

By Theorem 16.7, there is a Hamiltonian path R_i of $S_{n,2}^{\{i+1\}}$ joining $(\mathbf{x_i})^2$ to $(\mathbf{y_{i+1}})^2$ for every $i \in \langle n-2 \rangle - \{s-1, t-1\}$, there is a Hamiltonian path R_{t-1} of $S_{n,2}^{\{t\}}$ joining $(\mathbf{x_{t-1}})^2$ to $(\mathbf{v})^2$, and there is a Hamiltonian path Q of $S_{n,2}^{\{1\}}$ joining $(\mathbf{u})^2$ to $(\mathbf{y_1})^2$. We set $P_i = \langle \mathbf{u}, \mathbf{x_i}, (\mathbf{x_i})^2, R_i, (\mathbf{y_i})^2, \mathbf{y_i}, \mathbf{v} \rangle$ for every $i \in \langle n-2 \rangle - \{s-1, t-1\}$, $P_{t-1} = \langle \mathbf{u}, \mathbf{x_{t-1}}, (\mathbf{x_{t-1}})^2, R_{t-1}, (\mathbf{v})^2, \mathbf{v} \rangle$, $P_{s-1} = \langle \mathbf{u}, \mathbf{x_{s-1}}, (\mathbf{x_{s-1}})^k = \mathbf{y_s}, \mathbf{v} \rangle$, and $P_{n-1} = \langle \mathbf{u}, (\mathbf{u})^2, Q, (\mathbf{y_1})^2, \mathbf{y_1}, \mathbf{v} \rangle$. Then $\{P_1, P_2, \ldots, P_{n-1}\}$ forms a $(n-1)^*$-container of $S_{n,2}$ between $\mathbf{u}$ and $\mathbf{v}$. Moreover, $l(P_i) = 4$ for every for every $i \in \langle n-2 \rangle - \{s-1, t-1\}$, $l(P_{t-1}) = 4$, $l(P_{s-1}) = 3$, and $l(P_{n-1}) = 4$.

Hence, we have $w^*(S_{n,2}) \leq n^2 - 2n$. By Theorem 16.8, $w^*(S_{n,2}) = n^2 - 2n$. □

17 Pancyclic and Panconnected Property

17.1 INTRODUCTION

The graph embedding problem is a central issue in evaluating a network. The graph embedding problem asks whether the quest graph is a subgraph of a host graph, and an important benefit of the graph embeddings is that we can apply an existing algorithm for guest graphs to host graphs. This problem has attracted a burst of studies in recent years. Cycle networks and path networks are suitable for designing simple algorithms with low communication costs. The cycle embedding problem, which deals with all possible length of the cycles in a given graph, is investigated in a lot of interconnection networks [84,123,176,185,187,225,243]. The path embedding problem, which deals with all possible lengths of the paths between given two vertices in a given graph, is also investigated in a lot of interconnection networks [48,104–106,159,161,162,225,243,244,247]. In graph theory, we use the term *pancyclic property* to discuss the cycle embedding problem and the term *panconnected property* to discuss the path embedding problem.

A graph is *pancyclic* if it contains a cycle of every length from 3 to $|V(G)|$ inclusive. The concept of pancyclic graphs is proposed by Bondy [28] and is used in interconnection network for embedding all the possible lengths of the cycles in a given graph [84,123,184]. The pancyclic property has been extended to vertex-pancyclic [155] and edge-pancyclic [10]. It is known that there is no odd cycle in any bipartite graph. Hence, any bipartite graph is not pancyclic. For this reason, the concept of bipancyclicity is proposed [254]. A bipartite graph is *bipancyclic* if it contains a cycle of every even length from 4 to $|V(G)|$ inclusive. It is proved that the hypercube is bipancyclic [225,279]. A bipartite graph is *vertex-bipancyclic* [254] if every vertex lies on a cycle of every even length from 4 to $|V(G)|$ inclusive. Similarly, a bipartite graph is *edge-bipancyclic* if every edge lies on a cycle of every even length from 4 to $|V(G)|$ inclusive. Obviously, every edge-bipancyclic graph is vertex-bipancyclic.

A graph G is *panconnected* if there exists a path of length l joining any two different vertices x and y with $d_G(x,y) \leq l \leq |V(G)| - 1$. Obviously, every panconnected graph is pancyclic. The concept of panconnected graphs is proposed by Alavi and Williamson [5]. It is obvious that any bipartite graph with at least three vertices is not panconnected. For this reason, we say a bipartite graph is *bipanconnected* if there exists a path of length l joining any two different vertices x and y with $d_G(x,y) \leq l \leq |V(G)| - 1$ and $(l - d_G(x,y))$ is even.

17.2 BIPANCONNECTED AND BIPANCYCLIC PROPERTIES OF HYPERCUBES

To discuss the pancyclic property and the panconnected property for an interconnection network, we may need more information about its topological property. For this reason, we discuss only the bipanconnected property and edge-fault-tolerant bipancyclic property of Q_n.

Assume that n is any positive integer with $n \geq 2$. Let $\mathbf{u}$ and $\mathbf{x}$ be two distinct white vertices of Q_n and $\mathbf{v}$ and $\mathbf{y}$ be two distinct black vertices of Q_n. With Lemma 13.17, there are two disjoint paths P_1 and P_2 such that (1) P_1 is a path joining $\mathbf{u}$ to $\mathbf{v}$, (2) P_2 is a path joining $\mathbf{x}$ to $\mathbf{y}$, and (3) $P_1 \cup P_2$ spans Q_n. We call such property the $2H$ property. The $2H$ property has been used in many applications of hypercubes [174,242]. Obviously, the lengths of P_1 and P_2 satisfy $l(P_1) + l(P_2) = 2^n - 2$. Yet we can further require that the length of P_1, and hence the length of P_2 can be any odd integer such that $l(P_1) \geq h(\mathbf{u}, \mathbf{v})$ and $l(P_2) \geq h(\mathbf{x}, \mathbf{y})$. We call such a property the $2RH$ property. More precisely, let $\mathbf{u}$ and $\mathbf{x}$ be two distinct white vertices of Q_n and $\mathbf{v}$ and $\mathbf{y}$ be two distinct black vertices of Q_n. Let l_1 and l_2 are odd integers with $l_1 \geq h(\mathbf{u}, \mathbf{v})$, $l_2 \geq h(\mathbf{x}, \mathbf{y})$, and $l_1 + l_2 = 2^n - 2$. Then there are two disjoint paths P_1 and P_2 such that (1) P_1 is a path joining $\mathbf{u}$ to $\mathbf{v}$ with $l(P_1) = l_1$, (2) P_2 is a path joining $\mathbf{x}$ to $\mathbf{y}$ with $l(P_2) = l_2$, and (3) $P_1 \cup P_2$ spans Q_n.

For $i = 0, 1$, let Q_n^i denote the subgraph of Q_n induced by $\{\mathbf{u} = u_n u_{n-1} \ldots u_2 u_1 \mid u_n = i\}$. Obviously, Q_n^i is isomorphic to Q_{n-1}. For any vertex $\mathbf{u} = u_n u_{n-1} \ldots u_2 u_1$, we use $\mathbf{u}^i$ to denote the vertex $\mathbf{v} = v_n v_{n-1} \ldots v_2 v_1$ with $u_j = v_j$ for $1 \leq i \neq j \leq n-1$ and $u_i = 1 - v_i$. An edge joining vertex $\mathbf{u}$ to vertex $\mathbf{u}^i$ is an *i-dimensional edge*.

THEOREM 17.1 **[218]** The hypercube Q_n satisfies the $2RH$ property if and only if $n \neq 3$.

Proof: Let $\mathbf{u} = 000$, $\mathbf{v} = 001$, $\mathbf{x} = 010$, and $\mathbf{y} = 101$. There are just two paths $P_1 = \langle \mathbf{u} = 000, 010, 011, 001 = \mathbf{v} \rangle$ and $P_2 = \langle \mathbf{u} = 000, 100, 101, 001 = \mathbf{v} \rangle$ between $\mathbf{u}$ and $\mathbf{v}$ with length 3. However, there is not another path of length 3 between $\mathbf{u}$ and $\mathbf{v}$ such that the path does not contain $\mathbf{x}$ or $\mathbf{y}$. Thus, Q_3 is not satisfies the $2RH$. It is easy to see that Q_2 satisfies the $2RH$. For $n \geq 4$, we show that the Q_n satisfies the $2RH$ by induction. By brute force, we can check whether the theorem holds for $n = 4$. Assume that the theorem holds for any Q_k with $4 \leq k < n$. Without loss of generality, we can assume that $l_1 \geq l_2$. Thus, $l_2 \leq 2^{n-1} - 1$. Since Q_n is edge-symmetric, we can assume that $\mathbf{u} \in V(Q_n^0)$ and $\mathbf{x} \in V(Q_n^1)$. We have the following cases:

Case 1. $\mathbf{v} \in V(Q_n^0)$ and $\mathbf{y} \in V(Q_n^1)$.

Suppose that $l_2 < 2^{n-1} - 1$. Since Q_n is Hamiltonian laceable, there exists a Hamiltonian path R of Q_n^0 joining $\mathbf{u}$ and $\mathbf{v}$. Since the length of R is $2^{n-1} - 1$, we can write R as $\langle \mathbf{u}, R_1, \mathbf{p}, \mathbf{q}, R_2, \mathbf{v} \rangle$ for some black vertex $\mathbf{p}$ with $\mathbf{p}^n \neq \mathbf{x}$ and some white vertex $\mathbf{q}$ with $\mathbf{q}^n \neq \mathbf{y}$. Obviously, $h(\mathbf{p}^n, \mathbf{q}^n) = 1$. By induction, there exist two disjoint paths S_1 and S_2 such that (1) S_1 is a path joining $\mathbf{p}^n$ to $\mathbf{q}^n$ with $l(S_1) = l_1 - 2^{n-1}$, (2) S_2 is a path joining $\mathbf{x}$ to $\mathbf{y}$ with $l(S_2) = l_2$, and (3) $S_1 \cup S_2$ spans Q_n^1. We set P_1 as $\langle \mathbf{u}, R_1, \mathbf{p}, \mathbf{p}^n, S_1, \mathbf{q}^n, \mathbf{q}, R_2, \mathbf{v} \rangle$ and P_2 as S_2. Obviously, P_1 and P_2 are the required paths. See Figure 17.1a for illustration.

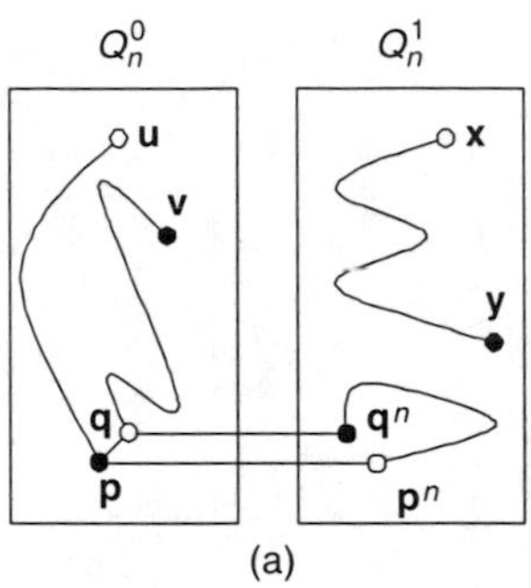

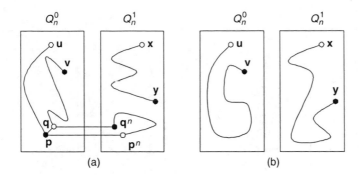

FIGURE 17.1 Illustration for Case 1 of Theorem 17.1.

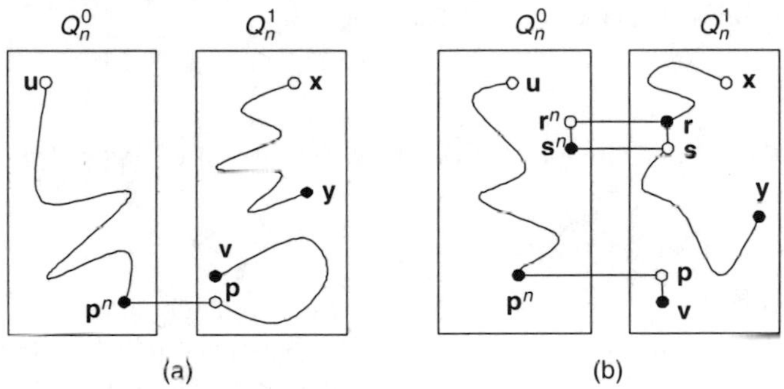

FIGURE 17.2 Illustration for Case 2 of Theorem 17.1.

Suppose that $l_2 = 2^{n-1} - 1$. Since Q_n is Hamiltonian laceable, there exists a Hamiltonian path P_1 of Q_n^0 joining **u** and **v** and there exists a Hamiltonian path P_2 of Q_n^1 joining **x** and **y**. Obviously, P_1 and P_2 are the required paths. See Figure 17.1b for illustration.

Case 2. $\{\mathbf{v}, \mathbf{y}\} \subset V(Q_n^1)$.

Suppose that $l_2 < 2^{n-1} - 1$ there exists a neighbor **p** of **v** such that $\mathbf{p} \neq \mathbf{x}$. Obviously, **p** is a white vertex. By induction, there exist two disjoint paths S_1 and S_2 such that (1) S_1 is a path joining **p** to **v** with $l(S_1) = l_1 - 2^{n-1}$, (2) S_2 is a path joining **x** to **y** with $l(S_2) = l_2$, and (3) $S_1 \cup S_2$ spans Q_n^1. Since Q_n is Hamiltonian laceable, there exists a Hamiltonian path R of Q_n^0 joining **u** and $\mathbf{p}^n$. We set P_1 as $\langle \mathbf{u}, R, \mathbf{p}^n, \mathbf{p}, S_1, \mathbf{v} \rangle$ and P_2 as S_2. Obviously, P_1 and P_2 are the required paths. See Figure 17.2a for illustration.

Suppose that $l_2 = 2^{n-1} - 1$. Again, there exists a neighbor **p** of **v** such that $\mathbf{p} \neq \mathbf{x}$. By induction, there exist two disjoint paths S_1 and S_2 such that (1) S_1 is a path joining **p** to **v** with $l(S_1) = 1$, (2) S_2 is a path joining **x** to **y** with $l(S_2) = 2^{n-1} - 3$, and (3) $S_1 \cup S_2$ spans Q_n^1. Obviously, we can write S_2 as $\langle \mathbf{x}, S_2^1, \mathbf{r}, \mathbf{s}, S_2^2, \mathbf{y} \rangle$ for some black vertex **r** with $\mathbf{r}^n \neq \mathbf{u}$. Again, by induction, there exist two disjoint paths R_1 and R_2 such that (1) R_1 is a path joining **u** to $\mathbf{p}^n$ with $l(R_1) = 2^{n-1} - 3$, (2) R_2 is a path joining $\mathbf{r}^n$ to $\mathbf{s}^n$ with $l(R_2) = 1$, and (3) $R_1 \cup R_2$ spans Q_n^0. We set P_1 as $\langle \mathbf{u}, R_1, \mathbf{p}^n, \mathbf{p}, \mathbf{v} \rangle$ and P_2 as

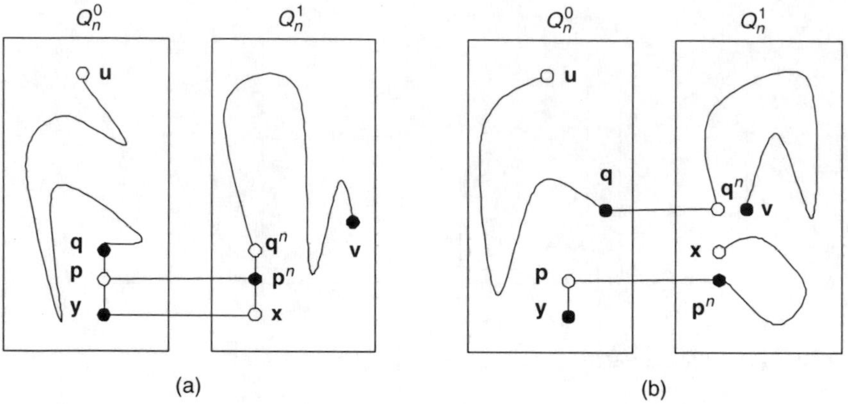

FIGURE 17.3 Illustration for Case 3 of Theorem 17.1.

$\langle \mathbf{x}, S_2^1, \mathbf{r}, \mathbf{r}^n, \mathbf{s}^n, \mathbf{s}, S_2^2, \mathbf{y} \rangle$. Obviously, P_1 and P_2 are the required paths. See Figure 17.2b for illustration.

Case 3. $\mathbf{y} \in V(Q_n^0)$ and $\mathbf{v} \in V(Q_n^1)$.

Suppose that $l_2 = 1$. Obviously, $\mathbf{x} = \mathbf{y}^n$. Let $\mathbf{p}$ be a neighbor of $\mathbf{y}$ in Q_n^0 such that $\mathbf{y}^n \neq \mathbf{v}$ and let $\mathbf{q}$ be a neighbor of $\mathbf{p}$ in Q_n^0 such that $\mathbf{p} \neq \mathbf{y}$. By induction, there exist two disjoint paths R_1 and R_2 such that (1) R_1 is a path joining $\mathbf{u}$ to $\mathbf{q}$ and $l(R_1) = 2^{n-1} - 3$, (2) R_2 is a path joining $\mathbf{p}$ to $\mathbf{y}$ and $l(R_2) = 1$, and (3) $R_1 \cup R_2$ spans Q_n^0. Obviously, $\mathbf{p}^n$ is a black vertex and $\mathbf{q}^n$ is a white vertex. Again, by induction, there exist two disjoint paths S_1 and S_2 such that (1) S_1 is a path joining $\mathbf{q}^n$ to $\mathbf{v}$ with $l(S_1) = 2^{n-1} - 3$, (2) S_2 is a path joining $\mathbf{x}$ to $\mathbf{p}^n$ with $l(S_2) = 1$, and (3) $S_1 \cup S_2$ spans Q_n^1. We set P_1 as $\langle \mathbf{u}, R_1, \mathbf{q}, \mathbf{p}, \mathbf{p}^n, \mathbf{q}^n, S_1, \mathbf{v} \rangle$ and P_2 as $\langle \mathbf{x}, \mathbf{y} \rangle$. Obviously, P_1 and P_2 are the required paths. See Figure 17.3a for illustration.

Suppose that $l_2 \geq 3$. We set $\mathbf{p}$ be a neighbor in Q_n^0 of $\mathbf{y}$ with $\mathbf{p} \neq \mathbf{u}$ if $h(\mathbf{x}, \mathbf{y}) = 1$ and $\mathbf{p}$ be a neighbor of $\mathbf{y}$ in Q_n^0 with $\mathbf{p} \neq \mathbf{u}$ and $h(\mathbf{p}, \mathbf{y}) = h(\mathbf{x}, \mathbf{y}) - 1$ if $h(\mathbf{x}, \mathbf{y}) \geq 3$. Let $\mathbf{q}$ be a neighbor of $\mathbf{v}^n$ in Q_n^0 such that $\mathbf{q} \neq \mathbf{y}$ and $\mathbf{q}^n \neq \mathbf{x}$. Thus, $h(\mathbf{q}^n, \mathbf{v}) = 1$. By induction, there exist two disjoint paths R_1 and R_2 such that (1) R_1 is a path joining $\mathbf{u}$ to $\mathbf{p}$ with $l(R_1) = 2^{n-1} - 3$, (2) R_2 is a path joining $\mathbf{q}$ to $\mathbf{y}$ with $l(R_2) = 1$, and (3) $R_1 \cup R_2$ spans Q_n^0. Again, by induction, there exist two disjoint paths S_1 and S_2 such that (1) S_1 is a path joining $\mathbf{q}^n$ to $\mathbf{v}$ with $l(S_1) = l_1 - 2^{n-1} + 2$, (2) S_2 is a path joining $\mathbf{x}$ to $\mathbf{p}^n$ with $l(S_2) = l_2 - 2$, and (3) $S_1 \cup S_2$ spans Q_n^1. We set P_1 as $\langle \mathbf{u}, R_1, \mathbf{q}, \mathbf{q}^n, S_1, \mathbf{v} \rangle$ and P_2 as $\langle \mathbf{x}, S_2, \mathbf{p}^n, \mathbf{p}, \mathbf{y} \rangle$. Obviously, P_1 and P_2 are the required paths. See Figure 17.3b for illustration.

Case 4. $\{\mathbf{v}, \mathbf{y}\} \subset V(Q_n^0)$.

Suppose that $l_2 = 1$. Obviously, $\mathbf{y} = \mathbf{x}^n$. Since Q_n is Hamiltonian laceable, there exists a Hamiltonian path R of Q_n^0 joining $\mathbf{u}$ to $\mathbf{v}$. Obviously, R can be written as $\langle \mathbf{u}, R_1, \mathbf{p}, \mathbf{y}, \mathbf{q}, R_2, \mathbf{v} \rangle$. Note that $\mathbf{u} = \mathbf{p}$ if $l(R_1) = 0$. Obviously, $\mathbf{p}$ and $\mathbf{q}$ are white vertices. Thus, $\mathbf{p}^n$ and $\mathbf{q}^n$ are black vertices. Let $\mathbf{r}$ be a neighbor of $\mathbf{q}^n$ in Q_n^1 such that

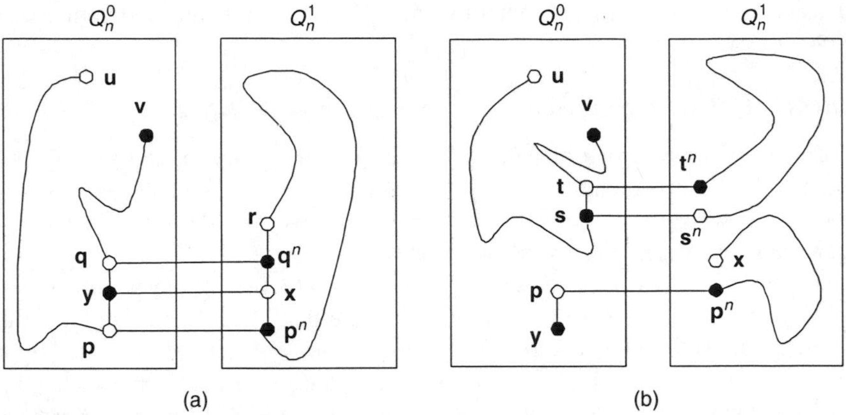

FIGURE 17.4 Illustration for Case 4 of Theorem 17.1.

$\mathbf{r} \neq \mathbf{x}$. By induction, there exist two disjoint paths S_1 and S_2 such that (1) S_1 is a path joining $\mathbf{p}^n$ to $\mathbf{r}$ with $l(S_1) = 2^{n-1} - 3$, (2) S_2 is a path joining $\mathbf{q}^n$ to $\mathbf{x}$ with $l(S_2) = 1$, and (3) $S_1 \cup S_2$ spans Q_n^1. We set P_1 as $\langle \mathbf{u}, R_1, \mathbf{p}, \mathbf{p}^n, S_1, \mathbf{r}, \mathbf{q}^n, \mathbf{q}, R_2, \mathbf{v} \rangle$ and P_2 as $\langle \mathbf{x}, \mathbf{y} \rangle$. Obviously, P_1 and P_2 are the required paths. See Figure 17.4a for illustration.

Suppose that $l_2 \geq 3$. We set $\mathbf{p}$ be a neighbor of $\mathbf{y}$ in Q_n^0 with $\mathbf{p} \neq \mathbf{u}$ if $h(\mathbf{x}, \mathbf{y}) = 1$ and $\mathbf{p}$ be a neighbor of $\mathbf{y}$ in Q_n^0 with $\mathbf{p} \neq \mathbf{u}$ and $h(\mathbf{p}, \mathbf{y}) = h(\mathbf{x}, \mathbf{y}) - 1$ if $h(\mathbf{x}, \mathbf{y}) \geq 3$. By induction, there exist two disjoint paths R_1 and R_2 such that (1) R_1 is a path joining $\mathbf{u}$ to $\mathbf{v}$ with $l(R_1) = 2^{n-1} - 3$, (2) R_2 is a path joining $\mathbf{p}$ to $\mathbf{y}$ with $l(R_2) = 1$, and (3) $R_1 \cup R_2$ spans Q_n^0. Obviously, we can write R_1 as $\langle \mathbf{u}, R_1^1, \mathbf{s}, \mathbf{t}, R_1^2, \mathbf{v} \rangle$ for some black vertex $\mathbf{s}$ such that $\mathbf{s}^n \neq \mathbf{x}$. By induction, there exist two disjoint paths S_1 and S_2 such that (1) S_1 is a path joining $\mathbf{s}^n$ to $\mathbf{t}^n$ with $l(S_1) = l_1 - 2^{n-1} - 2$, (2) S_2 is a path joining $\mathbf{x}$ to $\mathbf{p}^n$ with $l(S_2) = l_2 - 2$, and (3) $S_1 \cup S_2$ spans Q_n^1. We set P_1 as $\langle \mathbf{u}, R_1^1, \mathbf{s}, \mathbf{s}^n, S_1, \mathbf{t}^n, \mathbf{t}, R_1^2, \mathbf{v} \rangle$ and P_2 as $\langle \mathbf{x}, S_2, \mathbf{p}^n, \mathbf{p}, \mathbf{y} \rangle$. Obviously, P_1 and P_2 are the required paths. See Figure 17.4b for illustration.

The theorem is proved. □

By changing the roles of the vertices in bipartite sets in Theorem 17.1, we have the following theorem. The proof is similar to the proof of Theorem 17.1.

THEOREM 17.2 Assume that n is a positive integer with $n \geq 4$. Let $\mathbf{u}$ and $\mathbf{x}$ be two distinct white vertices of Q_n and $\mathbf{v}$ and $\mathbf{y}$ be two distinct black vertices of Q_n. Let l_1 and l_2 be even integers with $l_1 \geq h(\mathbf{u}, \mathbf{x})$, $l_2 \geq h(\mathbf{v}, \mathbf{y})$, and $l_1 + l_2 = 2^n - 2$. There exist two disjoint paths P_1 and P_2 such that (1) P_1 is a path joining $\mathbf{u}$ to $\mathbf{x}$ and $l(P_1) = l_1$, (2) P_2 is a path joining $\mathbf{v}$ to $\mathbf{y}$ and $l(P_2) = l_2$, and (3) $P_1 \cup P_2$ spans Q_n.

It is easy to confirm that the theorem does not hold for $n = 2$. Suppose that $n = 3$. Let $\mathbf{u} = 000$, $\mathbf{v} = 001$, $\mathbf{x} = 011$, and $\mathbf{y} = 010$. There are just two paths $P_1 = \langle \mathbf{u} = 000, 001, 011 = \mathbf{x} \rangle$ and $P_2 = \langle \mathbf{u} = 000, 010, 011 = \mathbf{x} \rangle$ of Q_3 with length 2 between $\mathbf{u}$ and $\mathbf{v}$. Moreover, there is not another path of Q_3 with length 2 between $\mathbf{u}$

and $\mathbf{x}$ where the path does not contain $\mathbf{v}$ and $\mathbf{y}$. Thus, the previous theorem does not hold for $n=3$.

THEOREM 17.3 The hypercube Q_n is bipanconnected if $n \geq 2$.

Proof: We need to prove that for any two different vertices $\mathbf{x}$ and $\mathbf{y}$ of Q_n there exists a path $P_l(\mathbf{x},\mathbf{y})$ of length l for any l with $h(\mathbf{x},\mathbf{y}) \leq l \leq 2^n - 1$ and $2 \mid (l - h(\mathbf{x},\mathbf{y}))$. Obviously, this statement holds for $n = 1, 2, 3$. Now, we consider $n \geq 4$. Without loss of generality, we assume that $\mathbf{x}$ is a white vertex.

Suppose that $\mathbf{y}$ is a black vertex. Thus, $h(\mathbf{x},\mathbf{y})$ is odd. Let l be any odd integer with $h(\mathbf{x},\mathbf{y}) \leq l \leq 2^n - 1$. Suppose that $l = 2^n - 1$. Note that Q_n is Hamiltonian laceable. Obviously, the Hamiltonian path of Q_n joining $\mathbf{x}$ and $\mathbf{y}$ is of length $2^n - 1$. Suppose that $l < 2^n - 1$. Since $n \geq 4$, there exists a pair of adjacent vertices $\mathbf{u}$ and $\mathbf{v}$ such that $\mathbf{u}$ is a white vertex with $\mathbf{u} \neq \mathbf{x}$ and $\mathbf{v}$ is a black vertex with $\mathbf{v} \neq \mathbf{y}$. Obviously, $h(\mathbf{u},\mathbf{v}) = 1$. By Theorem 17.1, there exist two disjoint paths P_1 and P_2 such that (1) P_1 is a path joining $\mathbf{u}$ to $\mathbf{v}$ with $l(P_1) = 2^n - 2 - l$, (2) P_2 is a path joining $\mathbf{x}$ to $\mathbf{y}$ with $l(P_2) = l$, and (3) $P_1 \cup P_2$ spans Q_n. Obviously, P_2 is a path of length l joining $\mathbf{x}$ to $\mathbf{y}$.

Suppose that $\mathbf{y}$ is a white vertex. Thus, $h(\mathbf{x},\mathbf{y})$ is even. Let l be any even integer with $h(\mathbf{x},\mathbf{y}) \leq l < 2^n - 1$. Since $n \geq 4$, there exist two different neighbors $\mathbf{u}$ and $\mathbf{v}$ of $\mathbf{y}$ such that $h(\mathbf{x},\mathbf{u}) = h(\mathbf{x},\mathbf{y}) - 1$. By Theorem 17.1, there exist two disjoint paths P_1 and P_2 such that (1) P_1 is a path joining $\mathbf{x}$ to $\mathbf{u}$ with $l(P_1) = l - 1$, (2) P_2 is a path joining $\mathbf{y}$ to $\mathbf{v}$ with $l(P_2) = 2^n - l - 1$, and (3) $P_1 \cup P_2$ spans Q_n. Obviously, $\langle \mathbf{x}, P_1, \mathbf{u}, \mathbf{y} \rangle$ is a path of length l joining $\mathbf{x}$ to $\mathbf{y}$.

Thus, the theorem is proved. □

With the bipanconnected property of Q_n, for any two different vertices $\mathbf{x}$ and $\mathbf{y}$ of Q_n there exists a path $P_l(\mathbf{x},\mathbf{y})$ of length l for any l with $h(\mathbf{x},\mathbf{y}) \leq l \leq 2^n - 1$ and $2 \mid (l - h(\mathbf{x},\mathbf{y}))$. We expect such a path $P_l(\mathbf{x},\mathbf{y})$ can be further extended by including the vertices not in $P_l(\mathbf{x},\mathbf{y})$ into a Hamiltonian path from $\mathbf{x}$ to a fixed vertex $\mathbf{z}$ or a Hamiltonian cycle.

THEOREM 17.4 Assume that n be any positive integer with $n \geq 2$. Let $\mathbf{x}$ and $\mathbf{z}$ be two vertices from different partite set of Q_n and $\mathbf{y}$ be a vertex of Q_n that is not in $\{\mathbf{x},\mathbf{z}\}$. For any integer l with $h(\mathbf{x},\mathbf{y}) \leq l \leq 2^n - 1 - h(\mathbf{y},\mathbf{z})$ and $2 \mid (l - h(\mathbf{x},\mathbf{y}))$, there exists a Hamiltonian path $R(\mathbf{x},\mathbf{y},\mathbf{z};l)$ from $\mathbf{x}$ to $\mathbf{z}$ such that $d_{R(\mathbf{x},\mathbf{y},\mathbf{z};l)}(\mathbf{x},\mathbf{y}) = l$.

Proof: By brute force, we can confirm that the theorem holds for $n = 2, 3$. Now, we consider $n \geq 4$. Without loss of generality, we assume that $\mathbf{x}$ is a white vertex and $\mathbf{z}$ is a black vertex.

Suppose that $\mathbf{y}$ is a black vertex. Obviously, $h(\mathbf{y},\mathbf{z}) \geq 2$. There exists a neighbor $\mathbf{w}$ of $\mathbf{y}$ such that $\mathbf{w} \neq \mathbf{x}$ and $h(\mathbf{w},\mathbf{z}) = h(\mathbf{y},\mathbf{z}) - 1$. Obviously, $\mathbf{w}$ is a white vertex. By Theorem 17.1, there exist two disjoint paths R_1 and R_2 such that (1) R_1 is a path joining $\mathbf{x}$ to $\mathbf{y}$ with $l(R_1) = l$, (2) R_2 is a path joining $\mathbf{w}$ to $\mathbf{z}$ with $l(R_2) = 2^n - l - 2$, and (3) $R_1 \cup R_2$ spans Q_n. We set R as $\langle \mathbf{x}, R_1, \mathbf{y}, \mathbf{w}, R_2, \mathbf{z} \rangle$. Obviously, R is the required Hamiltonian path.

Suppose that $\mathbf{y}$ is a white vertex. Obviously, $h(\mathbf{x},\mathbf{y}) \geq 2$. There exists a neighbor $\mathbf{w}$ of $\mathbf{y}$ such that $\mathbf{w} \neq \mathbf{z}$ and $h(\mathbf{x},\mathbf{w}) = h(\mathbf{x},\mathbf{y}) - 1$. Obviously, $\mathbf{w}$ is a black vertex.

By Theorem 17.1, there exist two disjoint paths R_1 and R_2 such that (1) R_1 is a path joining $\mathbf{x}$ to $\mathbf{w}$ with $l(R_1) = l - 1$, (2) R_2 is a path joining $\mathbf{y}$ to $\mathbf{z}$ with $l(R_2) = 2^n - l - 1$, and (3) $R_1 \cup R_2$ spans Q_n. We set R as $\langle \mathbf{x}, R_1, \mathbf{w}, \mathbf{y}, R_2, \mathbf{z} \rangle$. Obviously, R is the required Hamiltonian path. □

COROLLARY 17.1 *Assume that n is a positive integer with $n \geq 2$. Let $\mathbf{x}$ and $\mathbf{y}$ be any two different vertices of Q_n. For any integer l with $h(\mathbf{x}, \mathbf{y}) \leq l \leq 2^{n-1}$, there exists a Hamiltonian cycle $S(\mathbf{x}, \mathbf{y}; l)$ such that $d_{S(\mathbf{x},\mathbf{y};l)}(\mathbf{x}, \mathbf{y}) = l$ and $2 \mid (l - h(\mathbf{x}, \mathbf{y}))$.*

Proof: Let $\mathbf{z}$ be a neighbor of $\mathbf{x}$ such that $\mathbf{z} \neq \mathbf{y}$. By Theorem 17.4, there exists a Hamiltonian path R joining $\mathbf{x}$ to $\mathbf{z}$ such that $d_{R(\mathbf{x},\mathbf{y},\mathbf{z};l)}(\mathbf{x}, \mathbf{y}) = l$. We set S as $\langle \mathbf{x}, R, \mathbf{z}, \mathbf{x} \rangle$. Obviously, S forms the required Hamiltonian cycle. □

17.3 EDGE FAULT-TOLERANT BIPANCYCLIC PROPERTIES OF HYPERCUBES

With Theorem 17.1, we can easily prove that Q_n is edge-bipancyclic. Li et al. [225] study the edge-fault-tolerant edge-bipancyclic of hypercubes. A bipartite graph G is *k-edge fault-tolerant edge-bipancyclic* if $G - F$ remains edge-bipancyclic for any $F \subset E(G)$ with $|F| \leq k$.

THEOREM 17.5 Q_n is $(n-2)$-edge fault-tolerant edge-bipancyclic.

Proof: Obviously, the theorem is true for $n = 2$. We can prove that the theorem is true for $n = 3$ by brute force. Assume that the theorem is true for all $3 \leq k < n$. Let F be any subset of $E(Q_n)$ with $|F| \leq n - 2$. For $1 \leq i \leq n$, let F_i denote the set of i-dimensional edges in F. Thus, $\sum_{i=1}^{n} |F_i| = |F|$. Without loss of generality, we assume that $|F_1| \leq |F_2| \leq \cdots \leq |F_n|$. Moreover, we use F^0 to denote the set $E(Q_n^0) \cap F$ and F^1 to denote the set $E(Q_n^1) \cap F$. Thus, $F = F^0 \cup F_0 \cup F^1$ and $|F^0| + |F^1| \leq n - 3$.

Let e be any edge of $E(Q_n) - F$ and l be any even integer with $4 \leq l \leq 2^n$. To prove this theorem, we need to construct a cycle of length l containing e.

Case 1. e is not of dimension n. Without loss of generality, we may assume that $e \in E(Q_n^0)$.

Suppose that $4 \leq l \leq 2^{n-1}$. Since $|F^0| \leq n - 3$, by induction hypothesis there exists a cycle of length l in $Q_n^0 - F$ containing e. In particular, we use C_0 to denote such a cycle of length 2^{n-1}.

Suppose that $2^{n-1} + 2 \leq l \leq 2^n$. Let $l_1 = l - 2^{n-1}$. Then $2 \leq l_1 \leq 2^{n-1}$. Since $|E(C_0) - \{e\}| = 2^{n-1} - 1 > 2(n-2) = 2|F|$ for $n \geq 3$, there exists an edge $(\mathbf{u}, \mathbf{v})$ on C_0 such that $(\mathbf{u}, \mathbf{v}) \neq e$ and $\{(\mathbf{u}, \mathbf{u}^n), (\mathbf{v}, \mathbf{v}^n), (\mathbf{u}^n, \mathbf{v}^n)\} \cap F = \emptyset$. We may write C_0 as $\langle \mathbf{u}, P_0, \mathbf{v}, \mathbf{u} \rangle$. Obviously, e lies on P_0, $h(\mathbf{u}^n, \mathbf{v}^n) = 1$, and $\{\mathbf{u}^n, \mathbf{v}^n\} \subseteq V(Q_n^1)$. Suppose that $l_1 = 2$. Then $\langle \mathbf{u}, P_0, \mathbf{v}, \mathbf{v}^n, \mathbf{u}^n, \mathbf{u} \rangle$ is a cycle of length l in $Q_n - F$. Suppose that $l_1 \geq 4$. Since $|F^1| \leq n - 3$, by induction hypothesis there exists a cycle C_1 of length l_1 in $Q_n^1 - F$ containing $(\mathbf{u}^n, \mathbf{v}^n)$. We can write C_1 as $\langle \mathbf{u}^n, \mathbf{v}^n, P_1, \mathbf{u}^n \rangle$. Then $\langle \mathbf{u}, P_0, \mathbf{v}, \mathbf{v}^n, P_1, \mathbf{u}^n, \mathbf{u} \rangle$ is a cycle of length l in $Q_n - F$ containing e. See Figure 17.5a for illustration.

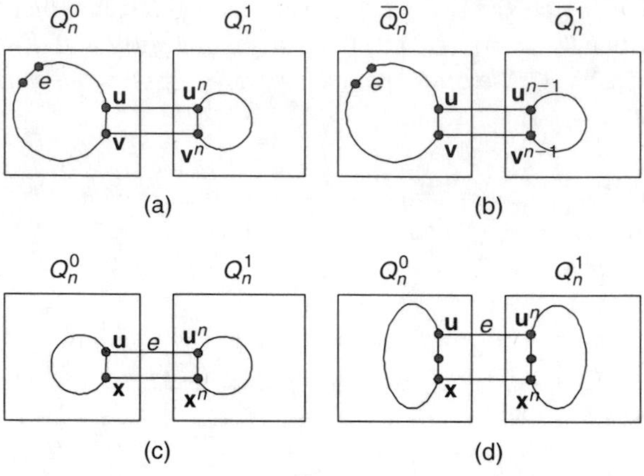

FIGURE 17.5 Illustration for Theorem 17.5.

Case 2. e is of dimension n.

Case 2.1. $|F_n| < n-2$. Let $\overline{Q}_n^0$ denote the subgraph of Q_n induced by $\{\mathbf{x} \in V(Q_n) \mid x_{n-1} = 0\}$ and $\overline{Q}_n^1$ denote the subgraph of Q_n induced by $\{\mathbf{x} \in V(Q_n) \mid x_{n-1} = 1\}$. Thus, $\overline{Q}_n^0$ and $\overline{Q}_n^1$ are isomorphic to Q_n. Without loss of generality, we may assume that $e \in E(\overline{Q}_n^0)$. We claim that $|E(\overline{Q}_n^0) \cap F| + |E(\overline{Q}_n^1) \cap F| \leq n-3$.

Suppose that $|F| \leq n-3$. Obviously, $|E(\overline{Q}_n^0) \cap F| + |E(\overline{Q}_n^1) \cap F| \leq n-3$. Suppose that $|F| = n-2$. Then, $|F_{n-1}| \geq 1$. Again, $|E(\overline{Q}_n^0) \cap F| + |E(\overline{Q}_n^1) \cap F| \leq n-3$. Accordingly, $|E(\overline{Q}_n^0) \cap F| + |E(\overline{Q}_n^1) \cap F| \leq n-3$.

Suppose that $4 \leq l \leq 2^{n-1}$. Since $|E(\overline{Q}_n^0) \cap F| \leq n-3$, by induction hypothesis there exists a cycle of length l in $\overline{Q}_n^0 - F$ containing e. In particular, we use C_0 to denote such a cycle of length 2^{n-1}.

Suppose that $2^{n-1} + 2 \leq l \leq 2^n$. Let $l_1 = l - 2^{n-1}$. Then $2 \leq l_1 \leq 2^{n-1}$. Since $|E(C_0) - \{e\}| = 2^{n-1} - 1 > 2(n-2) = 2|F|$ for $n \geq 3$, there exists an edge $(\mathbf{u}, \mathbf{v})$ on C_0 such that $(\mathbf{u}, \mathbf{v}) \neq e$ and $\{(\mathbf{u}, \mathbf{u}^{n-1}), (\mathbf{v}, \mathbf{v}^{n-1}), (\mathbf{u}^{n-1}, \mathbf{v}^{n-1})\} \cap F = \emptyset$. We may write C_0 as $\langle \mathbf{u}, P_0, \mathbf{v}, \mathbf{u} \rangle$. Obviously, e is on P_0, $h(\mathbf{u}^{n-1}, \mathbf{v}^{n-1}) = 1$ and $\{\mathbf{u}^{n-1}, \mathbf{v}^{n-1}\} \subseteq V(\overline{Q}_n^1)$. Suppose that $l_1 = 2$. Then $\langle \mathbf{u}, P_0, \mathbf{v}, \mathbf{v}^{n-1}, \mathbf{u}^{n-1}, \mathbf{u} \rangle$ is a cycle of length l in $Q_n - F$. Suppose that $l_1 \geq 4$. Since $|E(\overline{Q}_n^1) \cap F| \leq n-3$, by induction hypothesis there exists a cycle C_1 of length l_1 in $\overline{Q}_n^1 - F$ containing $(\mathbf{u}^{n-1}, \mathbf{v}^{n-1})$. We can write C_1 as $\langle \mathbf{u}^{n-1}, \mathbf{v}^{n-1}, P_1, \mathbf{u}^{n-1} \rangle$. Then $\langle \mathbf{u}, P_0, \mathbf{v}, \mathbf{v}^{n-1}, P_1, \mathbf{u}^{n-1}, \mathbf{u} \rangle$ is a cycle of length l in $Q_n - F$ containing e. See Figure 17.5b for illustration.

Case 2.2. $|F_n| = n-2$. Then $E(Q_n^0) \cap F = \emptyset$ and $E(Q_n^1) \cap F = \emptyset$. Assume that $e = (\mathbf{u}, \mathbf{u}^n)$ with $\mathbf{u} \in V(Q_n^0)$.

Suppose that $l = 4l'$ for $1 \leq l' \leq 2^{n-2}$. Since there are $(n-1)$ neighbors of $\mathbf{u}$ in Q_n^0, there exists a neighbor $\mathbf{x}$ of $\mathbf{u}$ in Q_n^0 such that $(\mathbf{x}, \mathbf{x}^n) \notin F$. Obviously,

$h(\mathbf{u},\mathbf{x})=h(\mathbf{u}^n,\mathbf{v}^n)=1$. By Theorem 17.3, there exists a path P_0 of length $2l'-1$ in Q_n^0 joining $\mathbf{u}$ and $\mathbf{x}$ and there exists a path P_1 in Q_n^1 of length $2l'-1$ joining $\mathbf{x}^n$ and $\mathbf{u}^n$. Then $\langle \mathbf{u},P_0,\mathbf{x},\mathbf{x}^n,P_1,\mathbf{u}^n,\mathbf{u}\rangle$ is a cycle in Q_n-F containing e of length l. See Figure 17.5c for illustration.

Suppose that $l=4l'+2$ for $1\le l'\le 2^{n-2}-1$. Let $A=\{\mathbf{w}\mid \mathbf{w}\in V(Q_n^0)$ and $h(\mathbf{u},\mathbf{w})=2\}$. Obviously, $|A|=C(^{n-1}_{\ 2})\ge n-2$. There exists an element $\mathbf{x}$ in A such that $(\mathbf{x},\mathbf{x}^n)\notin F$. Obviously, $h(\mathbf{u},\mathbf{x})=h(\mathbf{u}^n,\mathbf{x}^n)=2$. By Theorem 17.3, there exists a path P_0 in Q_n^0 of length $2l'$ joining $\mathbf{u}$ and $\mathbf{x}$ and there exists a path P_1 in Q_n^1 of length $2l'$ joining $\mathbf{x}^n$ and $\mathbf{u}^n$. Then $\langle \mathbf{u},P_0,\mathbf{x},\mathbf{x}^n,P_1,\mathbf{u}^n,\mathbf{u}\rangle$ is a cycle in Q_n-F containing e of length l. See Figure 17.5d for illustration.

The theorem is proved. □

Let $\mathbf{u}$ be any vertex of Q_n and F be $\{(\mathbf{u},\mathbf{u}^i)\mid 1\le i<n\}$. Obviously, $|F|=n-1$ and $\deg_{Q_n-F}(\mathbf{u})=1$. Thus, $(\mathbf{u},\mathbf{u}^n)$ does not lie on any cycle of Q_n-F. Hence, the previous result is optimal.

Yang et al. [350] observed that Q_n-F may still bipancyclic for $F\subset E(Q_n)$ with $|F|\le 2n-5$ if we assume that every vertex is incident with at least two healthy edges. Formally, we say that a bipartite graph G is *k conditional edge-fault-tolerant bipancyclic* (abbreviated as *k conditional fault-bipancyclic*) if $G-F$ is bipancyclic for every $F\subset E(G)$ with $|F|\le k$ under the condition that every vertex is incident with at least two nonfaulty edges; that is, at least two edges not in F.

THEOREM 17.6 The hypercube Q_n is $(2n-5)$ conditional fault-bipancyclic for $n\ge 3$.

Proof: We prove this by induction on n. By Theorem 17.5, Q_3 is 1-edge fault-tolerant bipancyclic. Since $2\times 3-5=1$, the theorem holds for $n=3$. Assume that Q_{n-1} is $2(n-1)-5=2n-7$ conditional fault-bipancyclic for some $n\ge 4$. We shall prove that Q_n is $(2n-5)$ conditional fault-bipancyclic. Let $F\subset E(Q_n)$ that satisfies the condition that every vertex is incident with at least two nonfaulty edges and $|F|\le 2n-5$. We note that there are three possible fault distributions:

1. There is only one vertex incident with $n-2$ faulty edges. Without loss of generality, we may assume that one of these $n-2$ faulty edges is an n-dimensional edge.
2. There are two vertices that share a faulty edge and are both incident with $n-2$ faulty edges. Without loss of generality, we may assume that the faulty edge they share is an n-dimensional edge.
3. Every vertex is incident with less than $n-2$ faulty edges. We may assume without loss of generality that one of them is an n-dimensional edge.

Note that there cannot be more than two vertices that are incident with $n-2$ faulty edges for $n\ge 3$. Then, we can divide Q_n into Q_n^0 and Q_n^1 along dimension n. So both of Q_n^0 and Q_n^1 satisfy the condition that every vertex is incident with at least two nonfaulty edges. Let $F^0=F\cap E(Q_n^0)$ and $F^1=F\cap E(Q_n^1)$. Without loss of generality, we may assume that $|F^0| \geq |F^1|$. Since $[(2n-5)-1]/2 = n-3$, $|F^1| \leq n-3$. We discuss the existence of cycles of all even lengths from 4 to 2^n in the following two cases:

Case 1. Cycles of even lengths from 4 to 2^{n-1}. Note that $|F^1| \leq n-3 \leq 2n-7$ for $n \geq 4$. By induction hypothesis, Q_n^1 is $(2n-7)$ conditional fault-bipancyclic, so Q_n^1 contains cycles of all even lengths from 4 to 2^{n-1}.

Case 2. Cycles of even lengths from $2^{n-1}+2$ to 2^n. We divide this case further into two subcases.

Case 2.1. $|F^0| = 2n-6$. Hence, there is only one n-dimensional faulty edge; say, e. Let $\mathbf{x} \in V(Q_n^0)$ be the vertex incident with e. Notice that there are at most $n-3$ faulty edges incident with $\mathbf{x}$ in Q_n^0. Since $2n-6 > n-3$ for $n \geq 4$, there must be a faulty edge in Q_n^0, say e', such that it is not incident to $\mathbf{x}$. Let $F' = F^0 - \{e'\}$. Clearly, $|F^0| = 2n-7$. By induction hypothesis, Q_n^0 is $(2n-7)$ conditional fault-bipancyclic, so $Q_n^0 - F'$ contains a Hamiltonian cycle, say C. Then, $Q_n^0 - F^0$ contains a Hamiltonian path on C, say $P = \langle \mathbf{u_1}, \mathbf{u_2}, \ldots, \mathbf{u_{2^{n-1}}} \rangle$, such that $\mathbf{u_1} \neq \mathbf{x}$ and $\mathbf{u_{2^{n-1}}} \neq \mathbf{x}$. Let $2 \leq l \leq 2^{n-1}$ be an even integer. We construct a cycle of length $2^{n-1}+l$ as follows. Since the edge e is the only faulty edge in n-dimension, there must exist two vertices $\mathbf{u_i}$ and $\mathbf{u_j}$ such that the two n-dimensional edges incident to $\mathbf{u_i}$ and $\mathbf{u_j}$, respectively, are nonfaulty and $j - i = l - 1$. Since $j - i$ is odd, $\mathbf{u_i}$ and $\mathbf{u_j}$ are in different partite sets, and then $(\mathbf{u_i})^n$ and $(\mathbf{u_j})^n$ are also in different partite sets. By Theorem 13.1, Q_n^1 is $(n-3)$-fault Hamiltonian laceable. Since $|F^1| \leq n-3$, $Q_n^1 - F^1$ contains a Hamiltonian path, say Q joining $(\mathbf{u_j})^n$ to $(\mathbf{u_i})^n$. Obviously, $\langle \mathbf{u_i}, \mathbf{u_{i+1}}, \ldots, \mathbf{u_j}, (\mathbf{u_j})^n, Q, (\mathbf{u_i})^n, \mathbf{u_i} \rangle$ is a cycle of length $2^{n-1}+l$ in $Q_n - F$. See Figure 17.6a for illustration.

Case 2.2. $|F^0| \leq 2n-7$. By induction hypothesis, Q_n^0 is $(2n-7)$ conditional fault-bipancyclic. Therefore, $Q_n^0 - F^0$ contains a Hamiltonian cycle, say $C = \langle \mathbf{u_1}, \mathbf{u_2}, \ldots, \mathbf{u_{2^{n-1}}}, \mathbf{u_1} \rangle$. Let l be an even integer for $2 \leq l \leq 2^{n-1}$. We construct a cycle of length $2^{n-1}+l$ as follows. First, we claim that there exist two vertices $\mathbf{u_i}$ and $\mathbf{u_j}$ on C such that the two n-dimensional edges incident to $\mathbf{u_i}$ and $\mathbf{u_j}$, respectively, are nonfaulty, and $j - i \ (\mathbf{mod}\ 2^{n-1}) = l - 1$. Suppose on the contrary that there do

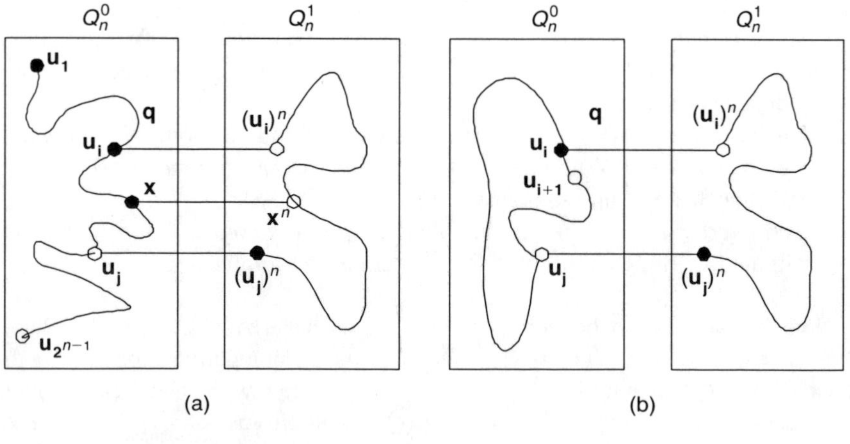

FIGURE 17.6 Illustration for Theorem 17.6.

not exist such $\mathbf{u_i}$ and $\mathbf{u_j}$. Then there are at least $2^{n-1}/2 = 2^{n-2}$ n-dimensional faulty edges. However, $2^{n-2} > 2n-5$ for $n \geq 4$. We obtain a contradiction. Thus, there exist such $\mathbf{u_i}$ and $\mathbf{u_j}$. By Theorem 13.1, Q_n^1 is $(n-3)$ fault-Hamiltonian laceable. Since $|F^1| \leq n-3$, $Q_n^1 - F^1$ is still Hamiltonian laceable. Since $\mathbf{u_i}$ and $\mathbf{u_j}$ are in different partite sets, $(\mathbf{u_i})^n$ and $(\mathbf{u_j})^n$ are also in different partite sets. There is a Hamiltonian path P in $Q_n^1 - F^1$ between $(\mathbf{u_j})^n$ and $(\mathbf{u_i})^n$. Then $\langle \mathbf{u_i}, \mathbf{u_{i+1}}, \ldots, \mathbf{u_j}, (\mathbf{u_j})^n, P, (\mathbf{u_i})^n, \mathbf{u_i} \rangle$ forms a cycle of length $2^{n-1} + l$. See Figure 17.6b for illustration.

The theorem is proved. □

Let $\mathbf{a}$ be any vertex of Q_n and $\mathbf{b} = (\mathbf{a}^1)^2$. Let $F = \{(\mathbf{a}, \mathbf{a}^i) \mid 3 \leq i \leq n\} \cup \{(\mathbf{b}, \mathbf{b}^i) \mid 3 \leq i \leq n\}$. Thus, $|F| = 2n-4$. It is easy to see that there is no Hamiltonian cycle in $Q_n - F$. Thus, the previous result is optimal.

17.4 PANCONNECTED AND PANCYCLIC PROPERTIES OF AUGMENTED CUBES

As before, we need a family of nonbipartite interconnection connection networks to discuss the corresponding panconnected and pancyclic properties. Here, we use augmented cubes. Now, we cover some background information on augmented cubes.

Assume that $n \geq 1$ is an integer. The graph of the *n-dimensional augmented cube*, denoted by AQ_n, has 2^n vertices, each labeled by an n-bit binary string $V(AQ_n) = \{u_1 u_2 \ldots u_n \mid u_i \in \{0, 1\}\}$. For $n = 1$, AQ_1 is the graph K_2 with vertex set $\{0, 1\}$. For $n \geq 2$, AQ_n can be recursively constructed by two copies of AQ_{n-1}, denoted by AQ_{n-1}^0 and AQ_{n-1}^1, and by adding 2^n edges between AQ_{n-1}^0 and AQ_{n-1}^1 as follows.

Let $V(AQ_{n-1}^0) = \{0u_2u_3 \ldots u_n \mid u_i = 0 \text{ or } 1 \text{ for } 2 \leq i \leq n\}$ and $V(AQ_{n-1}^1) = \{1v_2v_3 \ldots v_n \mid v_i = 0 \text{ or } 1 \text{ for } 2 \leq i \leq n\}$. A vertex $\mathbf{u} = 0u_2u_3 \ldots u_n$ of AQ_{n-1}^0 is adjacent to a vertex $\mathbf{v} = 1v_2v_3 \ldots v_n$ of AQ_{n-1}^1 if and only if one of the following cases holds:

1. $u_i = v_i$, for $2 \leq i \leq n$. In this case, $(\mathbf{u}, \mathbf{v})$ is called a *hypercube edge*. We set $\mathbf{v} = \mathbf{u}^h$.
2. $u_i = \overline{v}_i$, for $2 \leq i \leq n$. In this case, $(\mathbf{u}, \mathbf{v})$ is called a *complement edge*. We set $\mathbf{v} = \mathbf{u}^c$.

The augmented cubes AQ_1, AQ_2, AQ_3, and AQ_4 are illustrated in Figure 17.7. It is proved in Ref. 66 that AQ_n is a vertex-transitive, $(2n-1)$-regular, and $(2n-1)$-connected graph with 2^n vertices for any positive integer n. Let i be any index with $1 \leq i \leq n$ and $\mathbf{u} = u_1u_2u_3 \ldots u_n$ be a vertex of AQ_n. We use $\mathbf{u}^i$ to denote the vertex $\mathbf{v} = v_1v_2v_3 \ldots v_n$ such that $u_j = v_j$ with $1 \leq j \neq i \leq n$ and $u_i = \overline{v}_i$. Moreover, we use $\mathbf{u}^{i*}$ to denote the vertex $\mathbf{v} = v_1v_2v_3 \ldots v_n$ such that $u_j = v_j$ for $j < i$ and $u_j = \overline{v}_j$ for $i \leq j \leq n$. Obviously, $\mathbf{u}^n = \mathbf{u}^{n*}$, $\mathbf{u}^1 = \mathbf{u}^h$, $\mathbf{u}^c = \mathbf{u}^{1*}$, and $Nbd_{AQ_n}(\mathbf{u}) = \{\mathbf{u}^i \mid 1 \leq i \leq n\} \cup \{\mathbf{u}^{i*} \mid 1 \leq i < n\}$.

LEMMA 17.1 Assume that $n \geq 2$. Then $|Nbd_{AQ_n}(\mathbf{u}) \cap Nbd_{AQ_n}(\mathbf{v})| \geq 2$ if $(\mathbf{u}, \mathbf{v}) \in E(G)$.

Proof: We prove this lemma by induction. Since AQ_2 is isomorphic to the complete graph K_4, the lemma holds for $n = 2$. Assume that the lemma

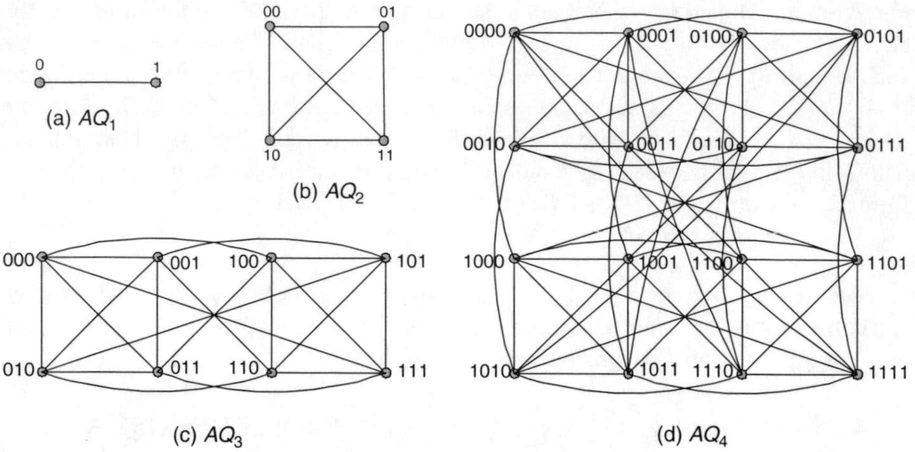

FIGURE 17.7 The augmented cubes AQ_1, AQ_2, AQ_3, and AQ_4.

holds for $2 \leq k < n$. Suppose that $\{\mathbf{u}, \mathbf{v}\} \subset V(AQ^i_{n-1})$ for some $i \in \{0, 1\}$. By induction, $|Nbd_{AQ_n}(\mathbf{u}) \cap Nbd_{AQ_n}(\mathbf{v})| \geq 2$. Thus, consider the case that either $\mathbf{v} = \mathbf{u}^h$ or $\mathbf{v} = \mathbf{u}^c$. Obviously, $\{\mathbf{u}^{2*}, \mathbf{u}^c\} \subset Nbd_{AQ_n}(\mathbf{u}) \cap Nbd_{AQ_n}(\mathbf{v})$ if $\mathbf{v} = \mathbf{u}^h$; and $\{\mathbf{u}^{2*}, \mathbf{u}^h\} \subset Nbd_{AQ_n}(\mathbf{u}) \cap Nbd_{AQ_n}(\mathbf{v})$ if $\mathbf{v} = \mathbf{u}^c$. Then the statement holds. □

The following lemma can easily obtained from the definition of AQ_n.

LEMMA 17.2 Assume that $n \geq 3$. For any two different vertices $\mathbf{u}$ and $\mathbf{v}$ of AQ_n, there exist two other vertices $\mathbf{x}$ and $\mathbf{y}$ of AQ_n such that the subgraph of $\{\mathbf{u}, \mathbf{v}, \mathbf{x}, \mathbf{y}\}$ contains a 4-cycle.

The following lemma can easily obtained from Theorem 11.19.

LEMMA 17.3 Let F be a subset of $V(AQ_n)$. Then there exists a Hamiltonian path between any two vertices of $V(AQ_n) - F$ if $|F| \leq 2n - 4$ for $n \geq 4$ and $|F| \leq 1$ for $n = 3$.

LEMMA 17.4 **[66]** Let $\mathbf{u}$ and $\mathbf{v}$ be any two vertices in AQ_n with $n \geq 2$. Suppose that both $\mathbf{u}$ and $\mathbf{v}$ are in AQ^i_{n-1} for $i = 0, 1$. Then $d_{AQ_n}(\mathbf{u}, \mathbf{v}) = d_{AQ^i_{n-1}}(\mathbf{u}, \mathbf{v})$. Suppose that $\mathbf{u}$ is a vertex in AQ^i_{n-1} and $\mathbf{v}$ is a vertex in AQ^{1-i}_{n-1}. Then there exist two shortest paths P_1 and P_2 of AQ_n joining $\mathbf{u}$ to $\mathbf{v}$ such that $(V(P_1) - \{\mathbf{u}\}) \subset V(AQ^{1-i}_{n-1})$ and $(V(P_2) - \{\mathbf{v}\}) \subset V(AQ^i_{n-1})$.

Proof: We prove this theorem through the following steps.

Case 1. We first prove that $d_{AQ_n}(\mathbf{u}, \mathbf{v}) = d_{AQ^i_{n-1}}(\mathbf{u}, \mathbf{v})$ if both $\mathbf{u}$ and $\mathbf{v}$ are in AQ^i_{n-1}. Without loss of generality, we assume that $i = 0$. Let P be a shortest path of AQ_n joining $\mathbf{u}$ to $\mathbf{v}$ that containing the most number of vertices in AQ^0_n. Obviously, $d_{AQ_n}(\mathbf{u}, \mathbf{v}) = d_{AQ^0_{n-1}}(\mathbf{u}, \mathbf{v})$ if $V(P) \subset V(AQ^0_n)$. Thus, we assume that $V(P) \cap V(AQ_n^1) \neq \emptyset$. We can write P as $\langle \mathbf{u}, P_1, \mathbf{a}, \mathbf{b}, P_2, \mathbf{c}, \mathbf{d}, P_3, \mathbf{v} \rangle$ where $(\mathbf{a}, \mathbf{b})$ and $(\mathbf{c}, \mathbf{d})$ are the first and the second edges on P such that $\{\mathbf{a}, \mathbf{d}\} \subset V(AQ_n^0)$ and $\{\mathbf{b}, \mathbf{c}\} \subset V(AQ_n^1)$. Note that $\mathbf{b} = \mathbf{c}$ if $l(P_2) = 0$. However, $\mathbf{a} = \mathbf{d}^{2*}$ if $\mathbf{b} = \mathbf{c}$. Then $\langle \mathbf{u}, P_1, \mathbf{a}, \mathbf{d}, P_3, \mathbf{v} \rangle$ is a path joining $\mathbf{u}$ to $\mathbf{v}$ shorter that P. Thus, $\mathbf{b} \neq \mathbf{c}$. Then $\mathbf{b}$ is either $\mathbf{a}^h$ or $\mathbf{a}^c$. Similarly, $\mathbf{c}$ is either $\mathbf{d}^h$ or $\mathbf{d}^c$. We consider all four combinations and get a contradiction to the length on P or of the same length but with more vertices in AQ_n^0. Let P^0 be the copy of $\langle \mathbf{b}, P_2, \mathbf{c} \rangle$ in AQ_n^0 under the isomorphism $f(\mathbf{z}) = \mathbf{z}^h$.

Case 1.1. $\mathbf{b} = \mathbf{a}^h$ and $\mathbf{c} = \mathbf{d}^h$. Obviously, P^0 is a path joining $\mathbf{a}$ to $\mathbf{d}$. Then $\langle \mathbf{u}, P_1, \mathbf{a}, P^0, \mathbf{d}, P_3, \mathbf{v} \rangle$ forms a path joining $\mathbf{u}$ to $\mathbf{v}$ shorter that P.

Case 1.2. $\mathbf{b} = \mathbf{a}^h$ and $\mathbf{c} = \mathbf{d}^c$. Obviously, P^0 is a path joining $\mathbf{a}$ to $\mathbf{d}^{2*}$. Then $\langle \mathbf{u}, P_1, \mathbf{a}, P^0, \mathbf{d}^{2*}, \mathbf{d}, P_3, \mathbf{v} \rangle$ forms a path joining $\mathbf{u}$ to $\mathbf{v}$ shorter that P.

Case 1.3. $\mathbf{b} = \mathbf{a}^c$ and $\mathbf{c} = \mathbf{d}^h$. Obviously, P^0 is a path joining $\mathbf{a}^{2*}$ to $\mathbf{d}$. Then $\langle \mathbf{u}, P_1, \mathbf{a}, \mathbf{a}^{2*}, P^0, \mathbf{d}, \mathbf{d}, P_3, \mathbf{v} \rangle$ forms a path joining $\mathbf{u}$ to $\mathbf{v}$ shorter that P.

Case 1.4. $\mathbf{b} = \mathbf{a}^c$ and $\mathbf{c} = \mathbf{d}^c$. Obviously, P^0 is a path joining $\mathbf{a}^{2*}$ to $\mathbf{d}^{2*}$. Then $\langle \mathbf{u}, P_1, \mathbf{a}, \mathbf{a}^{2*}, P^0, \mathbf{d}^{2*}, \mathbf{d}, \mathbf{d}, P_3, \mathbf{v} \rangle$ forms a path joining $\mathbf{u}$ to $\mathbf{v}$ of the same length as P but with more vertices in AQ_n^0.

Case 2. Suppose that $\mathbf{u}$ is a vertex in AQ_{n-1}^0 and $\mathbf{v}$ is a vertex in AQ_{n-1}^1. We prove that there exists a shortest path P_1 of AQ_n joining $\mathbf{u}$ to $\mathbf{v}$ such that $(V(P_1) - \{\mathbf{u}\}) \subset V(AQ_{n-1}^1)$. Let P be a shortest path of AQ_n joining $\mathbf{u}$ to $\mathbf{v}$. We can write P as $\langle \mathbf{u}, R_1, \mathbf{a}, \mathbf{b}, R_2, \mathbf{v} \rangle$ where $(\mathbf{a}, \mathbf{b})$ is the last edge on P such that $\mathbf{a} \in V(AQ_n^0)$ and $\mathbf{b} \in V(AQ_n^1)$. Note that $l(R_1)$ if $\mathbf{u} = \mathbf{a}$ and $l(R_2)$ if $\mathbf{b} = \mathbf{v}$. Obviously, the assertion holds if $\mathbf{u} = \mathbf{a}$. Thus, we consider $\mathbf{u} \neq \mathbf{a}$. With the previous proof, we may assume that $V(R_1) \subset V(AQ_n^0)$ and $V(R_2) \subset V(AQ_n^1)$. Obviously, $\mathbf{b}$ is either $\mathbf{a}^h$ or $\mathbf{a}^c$. $\mathbf{a}^h$. Let R be the copy of $\langle \mathbf{u}, P_1, \mathbf{a} \rangle$ in AQ_n^1 under the isomorphism $f(\mathbf{z}) = \mathbf{z}^h$.

Case 2.1. $\mathbf{b} = \mathbf{a}^h$. Obviously, R is a path joining $\mathbf{u}^c$ to $\mathbf{b}$. Then $\langle \mathbf{u}, \mathbf{u}^c, R, \mathbf{b}, R_2, \mathbf{v} \rangle$ forms a shortest path P_1 joining $\mathbf{u}$ to $\mathbf{v}$ such that $(V(P_1) - \{\mathbf{u}\}) \subset V(AQ_{n-1}^1)$.

Case 2.2. $\mathbf{b} = \mathbf{a}^c$. Suppose that $\mathbf{a}^{2*}$ is a vertex in R_1. Let S be the section of R_1 joining $\mathbf{u}$ to $\mathbf{a}^{2*}$. Then $\langle \mathbf{u}, S, \mathbf{a}^{2*}, \mathbf{b} = (\mathbf{a}^{2*})^h, R_2\mathbf{v} \rangle$ forms a shorter path joining $\mathbf{u}$ to $\mathbf{v}$. Thus, $\mathbf{a}^{2*}$ is not a vertex in R_1. Similarly, $\mathbf{a}^h$ is not a vertex in R_2. Obviously, R is a path joining $\mathbf{x}^h$ to $\mathbf{a}^h$. Let R^* be the copy of R in AQ_n^1 under the isomorphism $f(\mathbf{z}) = \mathbf{z}^{2*}$. Then $\langle \mathbf{u}, \mathbf{u}^c, R^*, \mathbf{b}, R_2, \mathbf{v} \rangle$ forms a shortest path P_1 joining $\mathbf{u}$ to $\mathbf{v}$ such that $(V(P_1) - \{\mathbf{u}\}) \subset V(AQ_{n-1}^1)$.

Case 3. Suppose that $\mathbf{u}$ is a vertex in AQ_{n-1}^0 and $\mathbf{v}$ is a vertex in AQ_{n-1}^1. We prove that there exist a shortest path P_2 of AQ_n joining $\mathbf{u}$ to $\mathbf{v}$ such that $(V(P_2) - \{\mathbf{v}\}) \subset V(AQ_{n-1}^0)$. The proof in this case is exactly as earlier.

The theorem is proved. □

With Lemma 17.4, we have the following corollary.

COROLLARY 17.2 *Assume that* $n \geq 3$. *Let* $\mathbf{x}$ *and* $\mathbf{y}$ *be two vertices of* AQ_n *with* $d_{AQ_n}(\mathbf{x}, \mathbf{y}) \geq 2$. *Then, there are two vertices* $\mathbf{p}$ *and* $\mathbf{q}$ *in* $Nbd_{AQ_n}(\mathbf{x})$ *with* $d_{AQ_n}(\mathbf{p}, \mathbf{y}) = d_{AQ_n}(\mathbf{q}, \mathbf{y}) = d_{AQ_n}(\mathbf{x}, \mathbf{y}) - 1$.

LEMMA 17.5 **[171]** Let $\{\mathbf{u}, \mathbf{v}, \mathbf{x}, \mathbf{y}\}$ be any four distinct vertices of AQ_n with $n \geq 2$. Then there exist two disjoint paths P_1 and P_2 such that (1) P_1 is a path joining $\mathbf{u}$ and $\mathbf{v}$, (2) P_2 is a path joining $\mathbf{x}$ and $\mathbf{y}$, and (3) $P_1 \cup P_2$ spans AQ_n.

We refer to Lemma 17.5 as $2H$-property of the augmented cube. This property is used for many applications of the augmented cubes [152,171]. Obviously, $l(P_1) \geq d_{AQ_n}(\mathbf{u}, \mathbf{v})$ and $l(P_2) \geq d_{AQ_n}(\mathbf{x}, \mathbf{y})$, and $l(P_1) + l(P_2) = 2^n - 2$. We expect that $l(P_1)$, and hence $l(P_2)$, can be an arbitrarily integer with the previous constraint. However, such an expectation is almost true. Let us consider AQ_3. Suppose that $\mathbf{u} = 001$, $\mathbf{v} = 110$, $\mathbf{x} = 101$, and $\mathbf{y} = 010$. Thus, $d_{AQ_3}(\mathbf{u}, \mathbf{v}) = 1$ and $d_{AQ_3}(\mathbf{x}, \mathbf{y}) = 1$. We can find P_1 and P_2 with $l(P_1) \in \{1, 3, 5\}$. Note that $\{\mathbf{x}, \mathbf{y}\} = Nbd_{AQ_3}(\mathbf{u}) \cap Nbd_{AQ_3}(\mathbf{v})$. We cannot find P_1 with $l(P_1) = 2$. Again, $\{\mathbf{u}, \mathbf{v}\} = Nbd_{AQ_3}(\mathbf{x}) \cap Nbd_{AQ_3}(\mathbf{y})$. We cannot find P_2 with $l(P_2) = 2$. Hence, we cannot find P_1 with $l(P_1) = 4$. Similarly, we consider AQ_4. Suppose that $\mathbf{u} = 0000$, $\mathbf{v} = 1001$, $\mathbf{x} = 0001$, and $\mathbf{y} = 1000$. Thus, $d_{AQ_4}(\mathbf{u}, \mathbf{v}) = 2$ and $d_{AQ_4}(\mathbf{x}, \mathbf{y}) = 2$. We can find P_1 and P_2 with $l(P_1) \in \{3, 4, \ldots, 11\}$. Note that $\{\mathbf{x}, \mathbf{y}\} = Nbd_{AQ_4}(\mathbf{u}) \cap Nbd_{AQ_4}(\mathbf{v})$. We cannot find P_1 with $l(P_1) = 2$. Again, $\{\mathbf{u}, \mathbf{v}\} = Nbd_{AQ_4}(\mathbf{x}) \cap Nbd_{AQ_4}(\mathbf{y})$. We cannot find P_2 with $l(P_2) = 2$.

Now, we propose the $2RH$ property of AQ_n with $n \geq 2$. Let $\{\mathbf{u}, \mathbf{v}, \mathbf{x}, \mathbf{y}\}$ be any four distinct vertices of AQ_n. Let l_1 and l_2 be two integers with $l_1 \geq d_{AQ_n}(\mathbf{u}, \mathbf{v})$, $l_2 \geq d_{AQ_n}(\mathbf{x}, \mathbf{y})$, and $l_1 + l_2 = 2^n - 2$. Then there exist two disjoint paths P_1 and P_2 such that (1) P_1 is a path joining $\mathbf{u}$ and $\mathbf{v}$ with $l(P_1) = l_1$, (2) P_2 is a path joining $\mathbf{x}$ and $\mathbf{y}$ with $l(P_2) = l_2$, and (3) $P_1 \cup P_2$ spans AQ_n, except for the following cases: (a) $l_1 = 2$ with $d_{AQ_n}(\mathbf{u}, \mathbf{v}) = 1$ such that $\{\mathbf{x}, \mathbf{y}\} = Nbd_{AQ_n}(\mathbf{u}) \cap Nbd_{AQ_n}(\mathbf{v})$, (b) $l_2 = 2$ with $d_{AQ_n}(\mathbf{x}, \mathbf{y}) = 1$ such that $\{\mathbf{u}, \mathbf{v}\} = Nbd_{AQ_n}(\mathbf{x}) \cap Nbd_{AQ_n}(\mathbf{y})$, (c) $l_1 = 2$ with $d_{AQ_n}(\mathbf{u}, \mathbf{v}) = 2$ such that $\{\mathbf{x}, \mathbf{y}\} = Nbd_{AQ_n}(\mathbf{u}) \cap Nbd_{AQ_n}(\mathbf{v})$, and (d) $l_2 = 2$ with $d_{AQ_n}(\mathbf{x}, \mathbf{y}) = 2$ such that $\{\mathbf{u}, \mathbf{v}\} = Nbd_{AQ_n}(\mathbf{x}) \cap Nbd_{AQ_n}(\mathbf{y})$.

THEOREM 17.7 **[219]** Assume that n is a positive integer with $n \geq 2$. Then AQ_n satisfies the $2RH$ property.

Proof: We prove this theorem by induction. By brute force, we check whether the theorem holds for $n = 2, 3, 4$. Assume that the theorem holds for any AQ_k with $4 \leq k < n$. Without loss of generality, we can assume that $l_1 \geq l_2$. Thus, $l_2 \leq 2^{n-1} - 1$. By the symmetric property of AQ_n, we can assume that at least one of $\mathbf{u}$ and $\mathbf{v}$, say $\mathbf{u}$, is in $V(AQ_{n-1}^0)$. Thus we have the following cases:

Case 1. $\mathbf{v} \in V(AQ_{n-1}^0)$ and $\{\mathbf{x}, \mathbf{y}\} \subset V(AQ_{n-1}^1)$.

Case 1.1. $d_{AQ_n}(\mathbf{x}, \mathbf{y}) \leq l_2 \leq 2^{n-1} - 3$ except that (1) $l_2 = 2^{n-1} - 4$ and (2) $l_2 = 2$ if $d_{AQ_n}(\mathbf{x}, \mathbf{y}) = 1$ or 2 with $\{\mathbf{u}, \mathbf{v}\} \neq Nbd_{AQ_n}(\mathbf{x}) \cap Nbd_{AQ_n}(\mathbf{y})$. By Lemma 17.3, there exists a Hamiltonian path R of AQ_{n-1}^0 joining $\mathbf{u}$ and $\mathbf{v}$. Since $l(R) = 2^{n-1} - 1$, we can write R as $\langle \mathbf{u}, R_1, \mathbf{p}, \mathbf{q}, R_2, \mathbf{v} \rangle$ for some vertices $\mathbf{p}$ and $\mathbf{q}$ such that $\{\mathbf{p}^h, \mathbf{q}^h\} \cap \{\mathbf{x}, \mathbf{y}\} = \emptyset$. By induction, there exist two disjoint paths S_1 and S_2 such that (1) S_1 is a path joining $\mathbf{p}^h$ to $\mathbf{q}^h$ with $l(S_1) = 2^{n-1} - l_2 - 2$, (2) S_2 is a path joining $\mathbf{x}$ to $\mathbf{y}$ with $l(S_2) = l_2$, and (3) $S_1 \cup S_2$ spans AQ_{n-1}^1. We set P_1 as $\langle \mathbf{u}, R_1, \mathbf{p}, \mathbf{p}^h, S_1, \mathbf{q}^h, \mathbf{q}, R_2, \mathbf{v} \rangle$ and P_2 as S_2. Obviously, P_1 and P_2 are the required paths.

Case 1.2. $l_2 = 2$ if $d_{AQ_n}(\mathbf{x}, \mathbf{y}) = 1$ or 2 with $\{\mathbf{u}, \mathbf{v}\} \neq Nbd_{AQ_n}(\mathbf{x}) \cap Nbd_{AQ_n}(\mathbf{y})$. Obviously, there exists a path P_2 of length 2 in $AQ_n - \{\mathbf{u}, \mathbf{v}\}$ joining $\mathbf{x}$ to $\mathbf{y}$. By Lemma 17.3, there exists a Hamiltonian path P_1 of $AQ_n - V(P_2)$ joining $\mathbf{u}$ to $\mathbf{v}$. Obviously, P_1 and P_2 are the required paths.

Case 1.3. $l_2 = 2^{n-1} - 4$. Obviously, there exists a vertex $\mathbf{p}$ in $V(AQ^1_{n-1}) - \{\mathbf{x}, \mathbf{y}, \mathbf{u}^h, \mathbf{v}^h\}$, a vertex $\mathbf{q}$ in $Nbd_{AQ^1_{n-1}}(\mathbf{p}) - \{\mathbf{x}, \mathbf{y}\}$, and a vertex $\mathbf{r}$ in $Nbd_{AQ^1_{n-1}}(\mathbf{q}) - \{\mathbf{x}, \mathbf{y}, \mathbf{p}\}$. Suppose that $\mathbf{r}^h \notin \{\mathbf{u}, \mathbf{v}\}$. By induction, there exist two disjoint paths Q_1 and Q_2 such that (1) Q_1 is a path joining $\mathbf{u}$ to $\mathbf{p}^h$, (2) Q_2 is a path joining $\mathbf{r}^h$ to $\mathbf{v}$, and (3) $Q_1 \cup Q_2$ spans AQ^0_{n-1}. By Lemma 17.3, there exists a Hamiltonian path P_2 of $AQ^1_{n-1} - \{\mathbf{p}, \mathbf{q}, \mathbf{r}\}$ joining $\mathbf{x}$ to $\mathbf{y}$. We set P_1 as $\langle \mathbf{u}, Q_1, \mathbf{p}^h, \mathbf{p}, \mathbf{q}, \mathbf{r}, \mathbf{r}^h, Q_2, \mathbf{v} \rangle$. Suppose that $\mathbf{r}^h \in \{\mathbf{u}, \mathbf{v}\}$. Without loss of generality, we assume that $\mathbf{r}^h = \mathbf{v}$. By Lemma 17.3, there exists a Hamiltonian path R of $AQ^0_{n-1} - \{\mathbf{v}\}$ joining $\mathbf{u}$ to $\mathbf{p}^h$. We set P_1 as $\langle \mathbf{u}, R, \mathbf{p}^h, \mathbf{p}, \mathbf{q}, \mathbf{r}, \mathbf{r}^h = \mathbf{v} \rangle$. Obviously, P_1 and P_2 are the required paths.

Case 1.4. $l_2 = 2^{n-1} - 2$. Obviously, there exists a vertex $\mathbf{p} \in V(AQ^1_{n-1}) - \{\mathbf{x}, \mathbf{y}, \mathbf{u}^h, \mathbf{u}^c, \mathbf{v}^h, \mathbf{v}^c\}$. By Lemma 17.5, there exist two disjoint paths Q_1 and Q_2 such that (1) Q_1 is a path joining $\mathbf{u}$ and $\mathbf{p}^h$, (2) Q_2 is a path joining $\mathbf{p}^c$ and $\mathbf{v}$, and (3) $Q_1 \cup Q_2$ spans AQ^0_{n-1}. By Lemma 17.3, there exists a Hamiltonian path P_2 of $AQ^0_{n-1} - \{\mathbf{p}\}$ joining $\mathbf{x}$ to $\mathbf{y}$. We set P_1 as $\langle \mathbf{u}, Q_1, \mathbf{p}^h, \mathbf{p}, \mathbf{p}^c, Q_2, \mathbf{v} \rangle$. Obviously, P_1 and P_2 are the required paths.

Case 1.5. $l_2 = 2^{n-1} - 1$. By Lemma 17.3, there exists a Hamiltonian path P_1 of AQ^0_{n-1} joining $\mathbf{u}$ and $\mathbf{v}$ and there exists a Hamiltonian path P_2 of AQ^1_{n-1} joining $\mathbf{x}$ and $\mathbf{y}$. Obviously, P_1 and P_2 are the required paths.

Case 2. $\mathbf{v} \in V(AQ^0_{n-1})$ and exactly one of $\mathbf{x}$ and $\mathbf{y}$ is in $V(AQ^0_{n-1})$. Without loss of generality, we assume that $\mathbf{x} \in V(AQ^0_{n-1})$.

Case 2.1. $l_2 = 1$. Obviously, $d_{AQ_n}(\mathbf{x}, \mathbf{y}) = 1$. We set P_2 as $\langle \mathbf{x}, \mathbf{y} \rangle$. By Lemma 17.3, there exists a Hamiltonian path P_1 of $AQ_n - \{\mathbf{x}, \mathbf{y}\}$ joining $\mathbf{u}$ to $\mathbf{v}$. Obviously, P_1 and P_2 are the required paths.

Case 2.2. $l_2 = 2$ if $d_{AQ_n}(\mathbf{x}, \mathbf{y}) = 1$ or 2 with $\{\mathbf{u}, \mathbf{v}\} \neq Nbd_{AQ_n}(\mathbf{x}) \cap Nbd_{AQ_n}(\mathbf{y})$. The proof is the same to Case 1.2.

Case 2.3. $l_2 = 3$.

Suppose that $d_{AQ_n}(\mathbf{x}, \mathbf{y}) = 1$. There exists a vertex $\mathbf{p}$ in $Nbd_{AQ^0_{n-1}}(\mathbf{x}) - \{\mathbf{u}, \mathbf{v}\}$. By Lemma 17.3, there exists a Hamiltonian path P_1 of $AQ_n - \{\mathbf{x}, \mathbf{y}, \mathbf{p}, \mathbf{p}^h\}$ joining $\mathbf{u}$ to $\mathbf{v}$. We set P_2 as $\langle \mathbf{x}, \mathbf{p}, \mathbf{p}^h, \mathbf{y} \rangle$. Obviously, P_1 and P_2 are the required paths.

Suppose that $d_{AQ_n}(\mathbf{x}, \mathbf{y}) = 2$. By Lemma 17.4, there exists a path $\langle \mathbf{x}, \mathbf{p}, \mathbf{y} \rangle$ from $\mathbf{x}$ to $\mathbf{y}$ such that $\mathbf{p} \in V(AQ^1_{n-1})$. By Lemma 17.1, there exists a vertex $\mathbf{q} \in Nbd_{AQ^1_{n-1}}(\mathbf{p}) \cap Nbd_{AQ^1_{n-1}}(\mathbf{y})$. By Lemma 17.3, there exists a Hamiltonian path P_1 of $AQ_n - \{\mathbf{x}, \mathbf{y}, \mathbf{p}, \mathbf{q}\}$ joining $\mathbf{u}$ to $\mathbf{v}$. We set P_2 as $\langle \mathbf{x}, \mathbf{p}, \mathbf{q}, \mathbf{y} \rangle$. Obviously, P_1 and P_2 are the required paths.

Suppose that $d_{AQ_n}(\mathbf{x}, \mathbf{y}) = 3$. By Lemma 17.4, there exists a path P_2 from $\mathbf{x}$ to $\mathbf{y}$ such that $(V(P_2) - \{\mathbf{x}\}) \subset V(AQ^1_{n-1})$. By Lemma 17.3, there exists a Hamiltonian path P_1 of $AQ_n - V(P_2)$ joining $\mathbf{u}$ to $\mathbf{v}$. Obviously, P_1 and P_2 are the required paths.

Case 2.4. $4 \le l_2 \le 2^{n-1} - 2$ except $l_2 = 2^{n-1} - 3$.

Suppose that $d_{AQ_n}(\mathbf{x}, \mathbf{y}) = 1$ or 2. We first claim that there exists a vertex $\mathbf{p}$ in $Nbd_{AQ_n}(\mathbf{x}) \cap Nbd_{AQ_n}(\mathbf{y})$. Assume that $d_{AQ_n}(\mathbf{x}, \mathbf{y}) = 1$. Obviously, either $\mathbf{y} = \mathbf{x}^h$ or $\mathbf{y} = \mathbf{x}^c$. We set $\mathbf{p} = \mathbf{x}^c$ if $\mathbf{y} = \mathbf{x}^h$ and $\mathbf{p} = \mathbf{x}^h$ if $\mathbf{y} = \mathbf{x}^c$. Assume that $d_{AQ_n}(\mathbf{x}, \mathbf{y}) = 2$. By Lemma 17.4, there exists a path $\langle \mathbf{x}, \mathbf{p}, \mathbf{y} \rangle$ from $\mathbf{x}$ to $\mathbf{y}$ such that $\mathbf{p} \in V(AQ_{n-1}^1)$. Obviously, $\mathbf{p}$ satisfies our claim. By Lemma 17.3, there exists a Hamiltonian path R of $AQ_{n-1}^0 - \{\mathbf{x}\}$ joining $\mathbf{u}$ to $\mathbf{v}$. Since $l(R) = 2^{n-1} - 3$, we can write R as $\langle \mathbf{u}, R_1, \mathbf{s}, \mathbf{t}, R_2, \mathbf{v} \rangle$ such that $\{\mathbf{s}^h, \mathbf{t}^h\} \cap \{\mathbf{p}, \mathbf{y}\} = \emptyset$. By induction, there exist two disjoint paths S_1 and S_2 such that (1) S_1 is a path joining $\mathbf{s}^h$ to $\mathbf{t}^h$ with $l(S_1) = 2^{n-1} - 1 - l_2$, (2) S_2 is a path joining $\mathbf{p}$ to $\mathbf{y}$ with $l(S_2) = l_2 - 1$, and (3) $S_1 \cup S_2$ spans AQ_{n-1}^1. We set P_1 as $\langle \mathbf{u}, R_1, \mathbf{s}, \mathbf{s}^h, S_1, \mathbf{t}^h, \mathbf{t}, R_2, \mathbf{v} \rangle$ and P_2 as $\langle \mathbf{x}, \mathbf{p}, S_2, \mathbf{y} \rangle$. Obviously, P_1 and P_2 are the required paths.

Suppose that $d_{AQ_n}(\mathbf{x}, \mathbf{y}) \ge 3$. By Lemma 17.4, there exists a vertex $\mathbf{p}$ in $V(AQ_{n-1}^1)$ such that $d_{AQ_n}(\mathbf{p}, \mathbf{y}) = d_{AQ_n}(\mathbf{x}, \mathbf{y}) - 1$. By Lemma 17.3, there exists a Hamiltonian path R of $AQ_{n-1}^0 - \{\mathbf{x}\}$ joining $\mathbf{u}$ to $\mathbf{v}$. We can write R as $\langle \mathbf{u}, R_1, \mathbf{s}, \mathbf{t}, R_2, \mathbf{v} \rangle$ such that $\{\mathbf{s}^h, \mathbf{t}^h\} \cap \{\mathbf{p}, \mathbf{y}\} = \emptyset$. By induction, there exist two disjoint paths S_1 and S_2 such that (1) S_1 is a path joining $\mathbf{s}^h$ to $\mathbf{t}^h$ with $l(S_1) = 2^{n-1} - 1 - l_2$, (2) S_2 is a path joining $\mathbf{p}$ to $\mathbf{y}$ with $l(S_2) = l_2 - 1$, and (3) $S_1 \cup S_2$ spans AQ_{n-1}^1. We set P_1 as $\langle \mathbf{u}, R_1, \mathbf{s}, \mathbf{s}^h, S_1, \mathbf{t}^h, \mathbf{t}, R_2, \mathbf{v} \rangle$ and P_2 as $\langle \mathbf{x}, \mathbf{p}, S_2, \mathbf{y} \rangle$. Obviously, P_1 and P_2 are the required paths.

Case 2.5. $l_2 = 2^{n-1} - 3$ or $l_2 = 2^{n-1} - 1$. Let $k = 3$ if $l_2 = 2^{n-1} - 3$ and $k = 1$ if $l_2 = 2^{n-1} - 1$. There exists a vertex $\mathbf{p}$ in $Nbd_{AQ_{n-1}^0}(\mathbf{x}) - \{\mathbf{u}, \mathbf{v}, \mathbf{y}^n\}$. By Lemma 17.3, there exists a Hamiltonian path R of $AQ_{n-1}^0 - \{\mathbf{x}, \mathbf{p}\}$ joining $\mathbf{u}$ to $\mathbf{v}$. We can write R as $\langle \mathbf{u}, R_1, \mathbf{s}, \mathbf{t}, R_2, \mathbf{v} \rangle$ such that $\{\mathbf{s}, \mathbf{t}\} \cap \{\mathbf{p}, \mathbf{y}^n\} = \emptyset$. By induction, there exist two disjoint paths S_1 and S_2 such that (1) S_1 is a path joining $\mathbf{s}^n$ to $\mathbf{t}^n$ with $l(S_1) = k$, (2) S_2 is a path joining $\mathbf{p}^n$ to $\mathbf{y}$ with $l(S_2) = 2^{n-1} - k - 2$, and (3) $S_1 \cup S_2$ spans AQ_{n-1}^1. We set P_1 as $\langle \mathbf{u}, R_1, \mathbf{s}, \mathbf{s}^n, S_1, \mathbf{t}^n, \mathbf{t}, R_2, \mathbf{v} \rangle$ and P_2 as $\langle \mathbf{x}, \mathbf{p}, \mathbf{p}^n, S_2, \mathbf{y} \rangle$. Obviously, P_1 and P_2 are the required paths.

Case 3. $\{\mathbf{v}, \mathbf{x}, \mathbf{y}\} \subset V(AQ_{n-1}^0)$.

Case 3.1. $l_2 = 1$. The proof is the same as that of Case 2.1.

Case 3.2. $l_2 = 2$ if $d_{AQ_n}(\mathbf{x}, \mathbf{y}) = 1$ or 2 with $\{\mathbf{u}, \mathbf{v}\} \ne Nbd_{AQ_n}(\mathbf{x}) \cap Nbd_{AQ_n}(\mathbf{y})$. The proof is the same as that of Case 1.2.

Case 3.3. $d_{AQ_n}(\mathbf{x}, \mathbf{y}) \le l_2 \le 2^{n-2} - 1$. By induction, there exist two disjoint paths R_1 and R_2 such that (1) R_1 is a path joining $\mathbf{u}$ to $\mathbf{v}$ with $l(R_1) = 2^{n-1} - l_2 - 2$, (2) R_2 is a path joining $\mathbf{x}$ to $\mathbf{y}$ with $l(R_2) = l_2$ and (3) $R_1 \cup R_2$ spans AQ_{n-1}^0. We can write R_1 as $\langle \mathbf{u}, R_3, \mathbf{p}, \mathbf{q}, R_4, \mathbf{v} \rangle$. By Lemma 17.3, there exists a Hamiltonian path S of AQ_{n-1}^1 joining $\mathbf{p}^h$ to $\mathbf{q}^h$. We set P_1 as $\langle \mathbf{u}, R_3, \mathbf{p}, \mathbf{p}^h, S, \mathbf{q}^h, \mathbf{q}, R_4, \mathbf{v} \rangle$ and P_2 as R_2. Obviously, P_1 and P_2 are the required paths.

Case 3.4. $2^{n-2} + 1 \le l_2 \le 2^{n-1} - 1$ except $l_2 = 2^{n-2} + 2$. By induction, there exist two disjoint paths R_1 and R_2 such that (1) R_1 is a path joining $\mathbf{u}$ to $\mathbf{v}$ with $l(R_1) = 2^{n-2} - 1$, (2) R_2 is a path joining $\mathbf{x}$ to $\mathbf{y}$ with $l(R_2) = 2^{n-2} - 1$, and

(3) $R_1 \cup R_2$ spans AQ^0_{n-1}. We can write R_1 as $\langle \mathbf{u}, R_3, \mathbf{p}, \mathbf{q}, R_4, \mathbf{v}\rangle$ and write R_2 as $\langle \mathbf{x}, R_5, \mathbf{s}, \mathbf{t}, R_6, \mathbf{y}\rangle$. By induction, there exist two disjoint paths S_1 and S_2 such that (1) S_1 is a path joining $\mathbf{p}^h$ to $\mathbf{q}^h$ with $l(S_1) = 2^{n-1} - l_2 + 2^{n-2} - 2$, (2) S_2 is a path joining $\mathbf{s}^h$ to $\mathbf{t}^h$ with $l(S_2) = l_2 - 2^{n-2}$, and (3) $S_1 \cup S_2$ spans AQ^1_{n-1}. We set P_1 as $\langle \mathbf{u}, R_3, \mathbf{p}, \mathbf{p}^h, S_1, \mathbf{q}^h, \mathbf{q}, R_4, \mathbf{v}\rangle$ and P_2 as $\langle \mathbf{x}, R_5, \mathbf{s}, \mathbf{s}^h, S_2, \mathbf{t}^h, \mathbf{t}, R_6, \mathbf{y}\rangle$. Obviously, P_1 and P_2 are the required paths.

Case 3.5. $l_2 = 2^{n-2}$ or $2^{n-2}+2$. Let $k=0$ if $l_2 = 2^{n-2}$ and $k=2$ if $l_2 = 2^{n-2}+2$. By induction, there exist two disjoint paths R_1 and R_2 such that (1) R_1 is a path joining $\mathbf{u}$ to $\mathbf{v}$ with $l(R_1) = 2^{n-2} - k$, (2) R_2 is a path joining $\mathbf{x}$ to $\mathbf{y}$ with $l(R_2) = 2^{n-2} + k - 2$, and (3) $R_1 \cup R_2$ spans AQ^0_{n-1}. We can write R_1 as $\langle \mathbf{u}, R_3, \mathbf{p}, \mathbf{q}, R_4, \mathbf{v}\rangle$ and write R_2 as $\langle \mathbf{x}, R_5, \mathbf{s}, \mathbf{t}, R_6, \mathbf{y}\rangle$. By Lemma 17.3, there exists a Hamiltonian path S of $AQ^1_{n-1} - \{\mathbf{s}^n, \mathbf{t}^n\}$ joining $\mathbf{p}^n$ to $\mathbf{q}^n$. We set P_1 as $\langle \mathbf{u}, R_3, \mathbf{p}, \mathbf{p}^n, S, \mathbf{q}^n, \mathbf{q}, R_4, \mathbf{v}\rangle$ and P_2 as $\langle \mathbf{x}, R_5, \mathbf{s}, \mathbf{s}^n, \mathbf{t}^n, \mathbf{t}, R_6, \mathbf{y}\rangle$. Obviously, P_1 and P_2 are the required paths.

Case 4. $\{\mathbf{x}, \mathbf{v}, \mathbf{y}\} \subset V(AQ^1_{n-1})$.

Case 4.1. $d_{AQ_n}(\mathbf{x}, \mathbf{y}) \le l_2 \le 2^{n-1} - 3$ except that (1) $l_2 = 2^{n-1} - 4$ and (2) $l_2 = 2$ if $d_{AQ_n}(\mathbf{x}, \mathbf{y}) = 1$ or 2 with $\{\mathbf{u}, \mathbf{v}\} \ne Nbd_{AQ_n}(\mathbf{x}) \cap Nbd_{AQ_n}(\mathbf{y})$. Obviously, there exists a vertex $\mathbf{p}$ in $Nbd_{AQ^1_{n-1}}(\mathbf{v}) - \{\mathbf{x}, \mathbf{y}, \mathbf{u}^h\}$. By induction, there exist two disjoint paths S_1 and S_2 such that (1) S_1 is a path joining $\mathbf{p}$ to $\mathbf{v}$ with $l(S_1) = l_1 - 2^{n-1}$, (2) S_2 is a path joining $\mathbf{x}$ to $\mathbf{y}$ with $l(S_2) = l_2$, and (3) $S_1 \cup S_2$ spans AQ^1_{n-1}. By Lemma 17.3, there exists a Hamiltonian path R of AQ^0_{n-1} joining $\mathbf{u}$ and $\mathbf{p}^h$. We set P_1 as $\langle \mathbf{u}, R, \mathbf{p}^h, \mathbf{p}, S_1, \mathbf{v}\rangle$ and P_2 as S_2. Obviously, P_1 and P_2 are the required paths.

Case 4.2. $l_2 = 2$ if $d_{AQ_n}(\mathbf{x}, \mathbf{y}) = 1$ or 2 with $\{\mathbf{u}, \mathbf{v}\} \ne Nbd_{AQ_n}(\mathbf{x}) \cap Nbd_{AQ_n}(\mathbf{y})$. The proof is the same as that of Case 1.2.

Case 4.3. $l_2 = 2^{n-1} - 4$. Obviously, there exists a vertex $\mathbf{p}$ in $Nbd_{AQ^1_{n-1}}(\mathbf{v}) - \{\mathbf{x}, \mathbf{y}\}$, and there exists a vertex $\mathbf{q}$ in $Nbd_{AQ^1_{n-1}}(\mathbf{p}) - \{\mathbf{x}, \mathbf{y}, \mathbf{v}, \mathbf{u}^h\}$. By Lemma 17.3, there exists a Hamiltonian path R of AQ^0_{n-1} joining $\mathbf{u}$ to $\mathbf{q}^h$, and there exists a Hamiltonian path P_2 of $AQ^1_{n-1} - \{\mathbf{v}, \mathbf{p}, \mathbf{q}\}$ joining $\mathbf{x}$ to $\mathbf{y}$. We set P_1 as $\langle \mathbf{u}, R, \mathbf{q}^h, \mathbf{q}, \mathbf{p}, \mathbf{v}\rangle$. Obviously, P_1 and P_2 are the required paths.

Case 4.4. $l_2 = 2^{n-1} - 2$. Let $\mathbf{v}'$ be an element in $\{\mathbf{v}^h, \mathbf{v}^c\} - \{\mathbf{u}\}$. By Lemma 17.3, there exists a Hamiltonian path R of AQ^0_{n-1} joining $\mathbf{u}$ and $\mathbf{v}'$, and there exists a Hamiltonian path P_2 of $AQ^1_{n-1} - \{\mathbf{v}\}$ joining $\mathbf{x}$ to $\mathbf{y}$. We set P_1 as $\langle \mathbf{u}, R, \mathbf{v}', \mathbf{v}\rangle$. Obviously, P_1 and P_2 are the required paths.

Case 4.5. $l_2 = 2^{n-1} - 1$. Obviously, there exists a vertex $\mathbf{p}$ in $Nbd_{AQ^1_{n-1}}(\mathbf{v}) - \{\mathbf{x}, \mathbf{y}\}$. By induction, there exist two disjoint paths S_1 and S_2 such that (1) S_1 is a path joining $\mathbf{p}$ to $\mathbf{v}$ with $l(S_1) = 1$, (2) S_2 is a path joining $\mathbf{x}$ to $\mathbf{y}$ with $l(S_2) = 2^{n-1} - 3$, and (3) $S_1 \cup S_2$ spans AQ^1_{n-1}. Obviously, we can write S_2 as $\langle \mathbf{x}, S_2^1, \mathbf{r}, \mathbf{s}, S_2^2, \mathbf{y}\rangle$ for some vertex $\mathbf{r}$ and $\mathbf{s}$ such that $\mathbf{u} \notin \{\mathbf{r}^h, \mathbf{s}^h\}$. Again, by induction, there exist two disjoint paths R_1 and R_2 such that (1) R_1 is a path joining $\mathbf{u}$ to $\mathbf{p}^h$ with $l(R_1) = 2^{n-1} - 3$, (2) R_2 is a path joining $\mathbf{r}^h$ to $\mathbf{s}^h$ with $l(R_2) = 1$, and (3) $R_1 \cup R_2$ spans AQ^0_{n-1}. We set P_1 as $\langle \mathbf{u}, R_1, \mathbf{p}^h, \mathbf{p}, \mathbf{v}\rangle$ and P_2 as $\langle \mathbf{x}, S_2^1, \mathbf{r}, \mathbf{r}^h, \mathbf{s}^h, \mathbf{s}, S_2^2, \mathbf{y}\rangle$. Obviously, P_1 and P_2 are the required paths.

Case 5. $\mathbf{v} \in V(AQ_{n-1}^1)$ and exactly one of $\mathbf{x}$ and $\mathbf{y}$ is in $V(AQ_{n-1}^0)$. Without loss of generality, we assume that $\mathbf{x} \in V(AQ_{n-1}^0)$.

Case 5.1. $l_2 = 1$. The proof is the same to Case 2.1.

Case 5.2. $l_2 = 2$ if $d_{AQ_n}(\mathbf{x}, \mathbf{y}) = 1$ or 2 with $\{\mathbf{u}, \mathbf{v}\} \neq Nbd_{AQ_n}(\mathbf{x}) \cap Nbd_{AQ_n}(\mathbf{y})$. The proof is the same of that for Case 1.2.

Case 5.3. $l_2 = 3$.

Suppose that $d_{AQ_n}(\mathbf{x}, \mathbf{y}) = 1$. Obviously, there exists a vertex $\mathbf{p}$ in $Nbd_{AQ_{n-1}^0}(\mathbf{x}) - \{\mathbf{u}, \mathbf{v}^h\}$. We set P_2 as $\langle \mathbf{x}, \mathbf{p}, \mathbf{p}^h, \mathbf{y} \rangle$. By Lemma 17.3, there exists a Hamiltonian path P_1 of $AQ_n - V(P_2)$ joining $\mathbf{u}$ to $\mathbf{v}$. Obviously, P_1 and P_2 are the required paths.

Suppose that $d_{AQ_n}(\mathbf{x}, \mathbf{y}) = 2$. Assume that $\{\mathbf{u}, \mathbf{v}\} = Nbd_{AQ_n}(\mathbf{x}) \cap Nbd_{AQ_n}(\mathbf{y})$. Thus, we have either $\mathbf{v} = \mathbf{x}^h$ or $\mathbf{v} = \mathbf{x}^c$. Moreover, $\mathbf{u} = \mathbf{x}^\alpha$ and $\mathbf{y} = \mathbf{v}^\alpha$ for some $\alpha \in \{i \mid 2 \le i \le n\} \cup \{i* \mid 2 \le i \le n-1\}$. We set P_2 as $\langle \mathbf{x}, \mathbf{x}^{h*}, (\mathbf{x}^{h*})^\alpha, ((\mathbf{x}^h)^\alpha) = \mathbf{y} \rangle$ in the case of $\mathbf{v} = \mathbf{x}^h$. Otherwise, we set P_2 as $\langle \mathbf{x}, \mathbf{x}^h, (\mathbf{x}^h)^\alpha, ((\mathbf{x}^{h*})^\alpha) = \mathbf{y} \rangle$. By Lemma 17.3, there exists a Hamiltonian path P_1 of $AQ_n - V(P_2)$ joining $\mathbf{u}$ to $\mathbf{v}$. Obviously, P_1 and P_2 are the required paths. Now, assume that $\{\mathbf{u}, \mathbf{v}\} \neq Nbd_{AQ_n}(\mathbf{x}) \cap Nbd_{AQ_n}(\mathbf{y})$. By Lemma 17.1, there exists a vertex $\mathbf{p}$ in $(Nbd_{AQ_n}(\mathbf{x}) \cap Nbd_{AQ_n}(\mathbf{y})) - \{\mathbf{u}, \mathbf{v}\}$. Without loss of generality, we may assume that $\mathbf{p}$ is in AQ_{n-1}^0. By Lemma 17.1, there exists a vertex $\mathbf{q}$ in $(Nbd_{AQ_{n-1}^0}(\mathbf{p}) \cap Nbd_{AQ_{n-1}^0}(\mathbf{x})) - \{\mathbf{u}\}$. By Lemma 17.3, there exists a Hamiltonian path P_1 of $AQ_n - \{\mathbf{x}, \mathbf{q}, \mathbf{p}, \mathbf{y}\}$ joining $\mathbf{u}$ to $\mathbf{v}$. We set P_2 as $\langle \mathbf{x}, \mathbf{q}, \mathbf{p}, \mathbf{y} \rangle$. Obviously, P_1 and P_2 are the required paths.

Suppose that $d_{AQ_n}(\mathbf{x}, \mathbf{y}) = 3$. By Lemma 17.4, there are two shortest paths R_1 and R_2 of AQ_n joining $\mathbf{x}$ and $\mathbf{y}$ such that R_1 can be written as $\langle \mathbf{x}, \mathbf{r}_1, \mathbf{r}_2, \mathbf{y} \rangle$ with $\{\mathbf{r}_1, \mathbf{r}_2\} \subset V(AQ_{n-1}^0)$ and R_2 can be written as $\langle \mathbf{x}, \mathbf{s}_1, \mathbf{s}_2, \mathbf{y} \rangle$ with $\{\mathbf{s}_1, \mathbf{s}_2\} \subset V(AQ_{n-1}^1)$. Suppose that $\mathbf{u} \neq \mathbf{r}_2$ or $\mathbf{v} \neq \mathbf{s}_1$. Without loss of generality, we assume that $\mathbf{u} \neq \mathbf{r}_2$. By Corollary 17.2, there exist a vertex $\mathbf{t} \in Nbd_{AQ_{n-1}^0}(\mathbf{x}) \cap Nbd_{AQ_{n-1}^0}(\mathbf{r}_2) - \{\mathbf{u}\}$. We set P_2 as $\langle \mathbf{x}, \mathbf{t}, \mathbf{r}_2, \mathbf{y} \rangle$. By Lemma 17.3, there exists a Hamiltonian path P_1 of $AQ_n - V(P_2)$ joining $\mathbf{u}$ to $\mathbf{v}$. Obviously, P_1 and P_2 are the required paths. Thus, we consider $\mathbf{u} = \mathbf{r}_2$ and $\mathbf{v} = \mathbf{s}_1$. By Corollary 17.2, there exists a vertex $\mathbf{p}$ in $Nbd_{AQ_{n-1}^0}(\mathbf{x}) \cap Nbd_{AQ_{n-1}^0}(\mathbf{u})$. Obviously, $d_{AQ_n}(\mathbf{p}, \mathbf{y}) = 2$. By Lemma 17.4, there exists a vertex $\mathbf{q}$ in $V(AQ_{n-1}^1) \cap Nbd_{AQ_n}(\mathbf{p}) \cap Nbd_{AQ_n}(\mathbf{y})$. Since $d_{AQ_n}(\mathbf{q}, \mathbf{y}) = 1$ and $d_{AQ_n}(\mathbf{v}, \mathbf{y}) = 2$, $\mathbf{q} \neq \mathbf{v}$. We set P_2 as $\langle \mathbf{x}, \mathbf{p}, \mathbf{q}, \mathbf{y} \rangle$. By Lemma 17.3, there exists a Hamiltonian path P_1 of $AQ_n - V(P_2)$ joining $\mathbf{u}$ to $\mathbf{v}$. Obviously, P_1 and P_2 are the required paths.

Case 5.4. $4 \le l_2 \le 2^{n-1} - 1$ with $d_{AQ_n}(\mathbf{x}, \mathbf{y}) = 1$.

Suppose that $l_2 = 4$. Obviously, there exists a vertex $\mathbf{p}$ in $Nbd_{AQ_{n-1}^0}(\mathbf{x}) - \{\mathbf{u}, \mathbf{v}^h\}$. By Lemma 17.1, there exists a vertex $\mathbf{q}$ in $(Nbd_{AQ_{n-1}^0}(\mathbf{x}) \cap Nbd_{AQ_{n-1}^0}(\mathbf{p})) - \{\mathbf{u}\}$. By Lemma 17.3, there exists a Hamiltonian path P_1 of $AQ_n - \{\mathbf{x}, \mathbf{y}, \mathbf{p}, \mathbf{p}^h, \mathbf{q}\}$ joining $\mathbf{u}$ to $\mathbf{v}$. We set P_2 as $\langle \mathbf{x}, \mathbf{q}, \mathbf{p}, \mathbf{p}^h, \mathbf{y} \rangle$. Obviously, P_1 and P_2 are the required paths.

Suppose that $5 \le l_2 \le 2^{n-1} - 1$ except $l_2 = 2^{n-1} - 2$. Obviously, there exists a vertex $\mathbf{p}$ in $Nbd_{AQ_{n-1}^0}(\mathbf{x}) - \{\mathbf{u}, \mathbf{v}^h, \mathbf{y}^h\}$ and a vertex $\mathbf{s}$ in $Nbd_{AQ_{n-1}^0}(\mathbf{u}) - \{\mathbf{x}, \mathbf{p}, \mathbf{v}^h, \mathbf{y}^h\}$. By induction, there exist two disjoint paths R_1 and R_2 such that (1) R_1 is a path joining

$\mathbf{u}$ to $\mathbf{s}$ with $l(R_1)=2^{n-1}-2-l_2$, (2) R_2 is a path joining $\mathbf{p}$ to $\mathbf{x}$ with $l(R_2)=l_2-2$, and (3) $R_1 \cup R_2$ spans AQ_{n-1}^0. By Lemma 17.3, there exists a Hamiltonian path S of $AQ_{n-1}^1 - \{\mathbf{y}, \mathbf{p}^h\}$ joining $\mathbf{s}^h$ to $\mathbf{v}$. We set P_1 as $\langle \mathbf{u}, R_1, \mathbf{s}, \mathbf{s}^h, S, \mathbf{v}\rangle$ and P_2 as $\langle \mathbf{x}, R_2, \mathbf{p}, \mathbf{p}^h, \mathbf{y}\rangle$. Obviously, P_1 and P_2 are the required paths.

Suppose that $l_2 = 2^{n-1}-2$. Let $\mathbf{s}$ and $\mathbf{p}$ be two vertices in $V(AQ_{n-1}^0) - \{\mathbf{u}, \mathbf{x}, \mathbf{v}^h, \mathbf{y}^h\}$. By induction, there exist two disjoint paths R_1 and R_2 such that (1) R_1 is a path joining $\mathbf{u}$ to $\mathbf{s}$ with $l(R_1)=2^{n-2}$, (2) R_2 is a path joining $\mathbf{p}$ to $\mathbf{x}$ with $l(R_2)=2^{n-2}-2$, (3) $R_1 \cup R_2$ spans AQ_{n-1}^0. Similarly, there exist two disjoint paths S_1 and S_2 such that (1) S_1 is a path joining $\mathbf{s}^h$ to $\mathbf{v}$ with $l(S_1)=2^{n-2}-1$, (2) S_2 is a path joining $\mathbf{p}^h$ to $\mathbf{y}$ with $l(S_2)=2^{n-2}-1$, and (3) $S_1 \cup S_2$ spans AQ_{n-1}^1. We set P_1 as $\langle \mathbf{u}, R_1, \mathbf{s}, \mathbf{s}^h, S_1, \mathbf{v}\rangle$ and P_2 as $\langle \mathbf{x}, R_2, \mathbf{p}, \mathbf{p}^h, S_2, \mathbf{y}\rangle$. Obviously, P_1 and P_2 are the required paths.

Case 5.5. $4 \le l_2 \le 2^{n-1}-1$ except $l_2 = 2^{n-1}-3$ with $d_{AQ_n}(\mathbf{x}, \mathbf{y}) \ge 2$.

Suppose that $d_{AQ_n}(\mathbf{x}, \mathbf{y})=2$ with $\{\mathbf{u}, \mathbf{v}\} = Nbd_{AQ_n}(\mathbf{x}) \cap Nbd_{AQ_n}(\mathbf{y})$. Thus, we have either $\mathbf{v}=\mathbf{x}^h$ or $\mathbf{v}=\mathbf{x}^c$. Moreover, $\mathbf{u}=\mathbf{x}^\alpha$ and $\mathbf{y}=(\mathbf{x}^h)^\alpha$ for some $\alpha \in \{i \mid 2 \le i \le n\} \cup \{i* \mid 2 \le i \le n-1\}$. Obviously, there exists a vertex $\mathbf{t}$ in $Nbd_{AQ_{n-1}^1}(\mathbf{v}) - \{\mathbf{x}^h, \mathbf{y}, \mathbf{x}^c, \mathbf{u}^h\}$. By induction, there exist two disjoint paths R_1 and R_2 such that (1) R_1 is a path joining $\mathbf{t}$ to $\mathbf{v}$ with $l(R_1)=2^{n-1}-1-l_2$, (2) R_2 is a path joining $\mathbf{x}^c$ to $\mathbf{y}$ with $l(R_2)=l_2-1$ in the case of $\mathbf{v}=\mathbf{x}^h$; otherwise R_2 is a path joining $\mathbf{x}^h$ to $\mathbf{y}$ with $l(R_2)=l_2-1$, and (3) $R_1 \cup R_2$ spans AQ_{n-1}^1. By Lemma 17.3, there exists a Hamiltonian path S of $AQ_{n-1}^0 - \{\mathbf{x}\}$ joining $\mathbf{t}^h$ to $\mathbf{u}$. We set P_1 as $\langle \mathbf{u}, S, \mathbf{t}^h, \mathbf{t}, R_1, \mathbf{v}\rangle$ and P_2 as $\langle \mathbf{x}, \mathbf{x}^c, R_2, \mathbf{y}\rangle$ in the case of $\mathbf{v}=\mathbf{x}^h$; otherwise, we set P_2 as $\langle \mathbf{x}, \mathbf{x}^h, R_2, \mathbf{y}\rangle$. Obviously, P_1 and P_2 are the required paths.

Suppose that $d_{AQ_n}(\mathbf{x}, \mathbf{y})=2$ with $\{\mathbf{u}, \mathbf{v}\} \ne Nbd_{AQ_n}(\mathbf{x}) \cap Nbd_{AQ_n}(\mathbf{y})$. Then, there exists a vertex $\mathbf{p}$ in $(Nbd_{AQ_n}(\mathbf{x}) \cap Nbd_{AQ_n}(\mathbf{y})) - \{\mathbf{u}, \mathbf{v}\}$. Without loss of generality, we may assume that $\mathbf{p} \in V(AQ_{n-1}^1)$. Obviously, there exists a vertex $\mathbf{t}$ in $Nbd_{AQ_{n-1}^1}(\mathbf{v}) - \{\mathbf{y}, \mathbf{p}, \mathbf{u}^h, \mathbf{x}^h\}$. By induction, there exist two disjoint paths R_1 and R_2 such that (1) R_1 is a path joining $\mathbf{t}$ to $\mathbf{v}$ with $l(R_1)=2^{n-1}-1-l_2$, (2) R_2 is a path joining $\mathbf{p}$ to $\mathbf{y}$ with $l(R_2)=l_2-1$, and (3) $R_1 \cup R_2$ spans AQ_{n-1}^1. By Lemma 17.3, there exists a Hamiltonian path S of $AQ_{n-1}^0 - \{\mathbf{x}\}$ joining $\mathbf{t}^h$ to $\mathbf{u}$. We set P_1 as $\langle \mathbf{u}, S, \mathbf{t}^h, \mathbf{t}, R_1, \mathbf{v}\rangle$ and P_2 as $\langle \mathbf{x}, \mathbf{p}, R_2, \mathbf{y}\rangle$. Obviously, P_1 and P_2 are the required paths.

Suppose that $d_{AQ_n}(\mathbf{x}, \mathbf{y})=k \ge 3$. By Lemma 17.4, there are two shortest paths S_1 and S_2 of AQ_n joining $\mathbf{x}$ and $\mathbf{y}$ such that S_1 can be written as $\langle \mathbf{x}=\mathbf{r_0}, \mathbf{r_1}, \mathbf{r_2}, \ldots, \mathbf{r_{k-1}}, \mathbf{y}\rangle$ with $(V(S_1)-\{\mathbf{y}\}) \subset V(AQ_{n-1}^0)$ and S_2 can be written as $\langle \mathbf{x}, \mathbf{s_1}, \mathbf{s_2}, \ldots, \mathbf{s_{k-1}}, \mathbf{y}\rangle$ with $(V(S_2)-\{\mathbf{x}\}) \subset V(AQ_{n-1}^1)$. Suppose that $\mathbf{u} \ne \mathbf{r_{k-1}}$. We set $\mathbf{p}=\mathbf{r_{k-1}}$. Again, there exists a vertex $\mathbf{s}$ in $Nbd_{AQ_{n-1}^0}(\mathbf{u}) - \{\mathbf{x}, \mathbf{p}, \mathbf{y}^h, \mathbf{v}^h\}$. By induction, there exist two disjoint paths R_1 and R_2 such that (1) R_1 is a path joining $\mathbf{u}$ to $\mathbf{s}$ with $l(R_1)=2^{n-1}-1-l_2$, (2) R_2 is a path joining $\mathbf{p}$ to $\mathbf{x}$ with $l(R_2)=l_2-1$, and (3) $R_1 \cup R_2$ spans AQ_{n-1}^0. By Lemma 17.3, there exists a Hamiltonian path S of $AQ_{n-1}^1 - \{\mathbf{y}\}$ joining $\mathbf{s}^h$ to $\mathbf{v}$. We set P_1 as $\langle \mathbf{u}, R_1, \mathbf{s}, \mathbf{s}^h, S, \mathbf{v}\rangle$ and P_2 as $\langle \mathbf{x}, R_2, \mathbf{p}, \mathbf{y}\rangle$. Obviously, P_1 and P_2 are the required paths.

Now we assume that $\mathbf{r_{k-1}}=\mathbf{u}$ and $\mathbf{s_1}=\mathbf{v}$. Since $d_{AQ_n}(\mathbf{r_{k-2}}, \mathbf{y})=2$, by Lemma 17.4, there exists a vertex $\mathbf{p} \in Nbd_{AQ_n}(\mathbf{r_{k-2}})$ in $V(AQ_{n-1}^1)$ such that $d_{AQ_n}(\mathbf{p}, \mathbf{y})=1$. Suppose that $l_2=4$ with $d_{AQ_n}(\mathbf{x}, \mathbf{y})=3$. Thus, $\langle \mathbf{x}, \mathbf{r_1}, \mathbf{p}, \mathbf{y}\rangle$ is a shortest path joining $\mathbf{x}$ and $\mathbf{y}$. By Lemma 17.1, there exists a vertex $\mathbf{q} \in Nbd_{AQ^1_{n-1}}(\mathbf{p}) \cap Nbd_{AQ^1_{n-1}}(\mathbf{y}) - \{\mathbf{v}\}$. By Lemma 17.3, there exists a Hamiltonian path P_1 of $AQ_n - \{\mathbf{x}, \mathbf{r_1}, \mathbf{p}, \mathbf{q}, \mathbf{y}\}$ joining $\mathbf{u}$ to $\mathbf{v}$. We set P_2 as $\langle \mathbf{x}, \mathbf{r_1}, \mathbf{p}, \mathbf{q}, \mathbf{y} \rangle$. Obviously, P_1 and P_2 are the required paths. Suppose that $l_2 = 4$ with $d_{AQ_n}(\mathbf{x}, \mathbf{y}) = 4$. Thus, $P_2 = \langle \mathbf{x}, \mathbf{r_1}, \mathbf{r_2}, \mathbf{p}, \mathbf{y} \rangle$ is a shortest path joining $\mathbf{x}$ and $\mathbf{y}$. By Lemma 17.3, there exists a Hamiltonian path P_1 of $AQ_n - \{\mathbf{x}, \mathbf{r_1}, \mathbf{r_2}, \mathbf{p}, \mathbf{y}\}$ joining $\mathbf{u}$ to $\mathbf{v}$. Obviously, P_1 and P_2 are the required paths. Suppose that $5 \le l_2 \le 2^{n-2}$ with $d_{AQ_n}(\mathbf{x}, \mathbf{y}) \ge 3$. Obviously, there exists a vertex $\mathbf{s}$ in $Nbd_{AQ^0_{n-1}}(\mathbf{u}) - \{\mathbf{x}, \mathbf{r_{k-2}}, \mathbf{y}^h, \mathbf{v}^h\}$. By induction, there exist two disjoint paths R_1 and R_2 such that (1) R_1 is a path joining $\mathbf{u}$ to $\mathbf{s}$ with $l(R_1) = 2^{n-1} - l_2$, (2) R_2 is a path joining $\mathbf{r_{k-2}}$ to $\mathbf{x}$ with $l(R_2) = l_2 - 2$, and (3) $R_1 \cup R_2$ spans AQ^0_{n-1}. By Lemma 17.3, there exists a Hamiltonian path S of $AQ^1_{n-1} - \{\mathbf{p}, \mathbf{y}\}$ joining $\mathbf{s}^h$ to $\mathbf{v}$. We set P_1 as $\langle \mathbf{u}, R_1, \mathbf{s}, \mathbf{s}^h, S, \mathbf{v} \rangle$ and P_2 as $\langle \mathbf{x}, R_2, \mathbf{r_{k-2}}, \mathbf{p}, \mathbf{y} \rangle$. Obviously, P_1 and P_2 are the required paths. Suppose that $2^{n-2} + 1 \le l_2 < 2^{n-1} - 1$ except $2^{n-1} - 3$ with $d_{AQ_n}(\mathbf{x}, \mathbf{y}) \ge 3$. Obviously, there exists a vertex $\mathbf{s}$ in $Nbd_{AQ^0_{n-1}}(\mathbf{u}) - \{\mathbf{x}, \mathbf{r_{k-2}}, \mathbf{y}^h, \mathbf{v}^h\}$. By induction, there exist two disjoint paths R_1 and R_2 such that (1) R_1 is a path joining $\mathbf{u}$ to $\mathbf{s}$ with $l(R_1) = 2^{n-2} + 1$, (2) R_2 is a path joining $\mathbf{r_{k-2}}$ to $\mathbf{x}$ with $l(R_2) = 2^{n-2} - 3$, and (3) $R_1 \cup R_2$ spans AQ^0_{n-1}. Again, by induction, there exist two disjoint paths S_1 and S_2 such that (1) S_1 is a path joining $\mathbf{s}^h$ to $\mathbf{v}$ with $l(S_1) = 2^{n-1} - l_2 + 2^{n-2} - 4$, (2) S_2 is a path joining $\mathbf{p}$ to $\mathbf{y}$ with $l(S_2) = l_2 - 2^{n-2} + 2$, and (3) $S_1 \cup S_2$ spans AQ^1_{n-1}. We set P_1 as $\langle \mathbf{u}, R_1, \mathbf{s}, \mathbf{s}^h, S_1, \mathbf{v} \rangle$ and P_2 as $\langle \mathbf{x}, R_2, \mathbf{r_{k-2}}, \mathbf{p}, S_2, \mathbf{y} \rangle$. Obviously, P_1 and P_2 are the required paths.

Case 5.6. $l_2 = 2^{n-1} - 3$ or $l_2 = 2^{n-1} - 1$ with $d_{AQ_n}(\mathbf{x}, \mathbf{y}) \ge 2$. Let $t = 0$ if $l_2 = 2^{n-1} - 3$ and $t = 1$ if $l_2 = 2^{n-1} - 1$. Obviously, there exist two vertices $\mathbf{s}$ and $\mathbf{p}$ in $AQ^0_{n-1} - \{\mathbf{u}, \mathbf{x}, \mathbf{v}^n, \mathbf{y}^n\}$. By induction, there exist two disjoint paths R_1 and R_2 such that (1) R_1 is a path joining $\mathbf{u}$ to $\mathbf{s}$ with $l(R_1) = 2^{n-2} - t$, (2) R_2 is a path joining $\mathbf{p}$ to $\mathbf{x}$ with $l(R_2) = 2^{n-2} + t - 2$, and (3) $R_1 \cup R_2$ spans AQ^0_{n-1}. Similarly, there exist two disjoint paths S_1 and S_2 such that (1) S_1 is a path joining $\mathbf{s}^n$ to $\mathbf{v}$ with $l(S_1) = 2^{n-2} - t$, (2) S_2 is a path joining $\mathbf{p}^n$ to $\mathbf{y}$ with $l(S_2) = 2^{n-2} + t - 2$, and (3) $S_1 \cup S_2$ spans AQ^1_{n-1}. We set P_1 as $\langle \mathbf{u}, R_1, \mathbf{s}, \mathbf{s}^n, S_1, \mathbf{v} \rangle$ and P_2 as $\langle \mathbf{x}, R_2, \mathbf{p}, \mathbf{p}^n, S_2, \mathbf{y} \rangle$. Obviously, P_1 and P_2 are the required paths.

Thus, Theorem 17.7 is proved. □

COROLLARY 17.3 *AQ_n is panconnected for every positive integer n.*

Proof: We prove that AQ_n is panconnected by Theorem 17.7. Obviously, AQ_n is panconnected for $n = 1, 2$. Now, we consider that $n \ge 3$.

Suppose that $l = 2^n - 1$. By Remark 1, AQ_n is Hamiltonian connected. Obviously, the Hamiltonian path of AQ_n joining $\mathbf{x}$ and $\mathbf{y}$ is of length $2^n - 1$. Suppose that $l = 2^n - 2$. Let $\mathbf{u}$ be a vertex in $Nbd_{AQ_n}(\mathbf{y}) - \{\mathbf{x}\}$. By Lemma 17.1, there exists a vertex $\mathbf{v}$ in $(Nbd_{AQ_n}(\mathbf{u}) \cap Nbd_{AQ_n}(\mathbf{y})) - \{\mathbf{x}\}$. By Theorem 17.7, there exist two disjoint paths P_1 and P_2 such that (1) P_1 is a path joining $\mathbf{x}$ to $\mathbf{u}$ with $l(P_1) = 2^n - 3$, (2) P_2 is a path joining $\mathbf{y}$ to $\mathbf{v}$ with $l(P_2) = 1$, and (3) $P_1 \cup P_2$ spans AQ_n. Obviously, $\langle \mathbf{x}, P_1, \mathbf{u}, \mathbf{y} \rangle$ is a path of length $2^n - 2$ joining $\mathbf{x}$ to $\mathbf{y}$. Suppose that $l = 2^n - 3$. We can find two adjacent vertices $\mathbf{u}$ and $\mathbf{v}$ such that $\{\mathbf{u}, \mathbf{v}\} \cap \{\mathbf{x}, \mathbf{y}\} = \emptyset$. By Theorem 17.7, there exist two disjoint paths P_1 and P_2 such that (1) P_1 is a path joining $\mathbf{x}$ to $\mathbf{y}$ with $l(P_1) = 2^n - 3$, (2) P_2 is a path joining $\mathbf{u}$ to $\mathbf{v}$ with $l(P_2) = 1$, and (3) $P_1 \cup P_2$ spans AQ_n. Obviously, P_1 is a path of length $2^n - 3$ joining $\mathbf{x}$ to $\mathbf{y}$. Suppose that $l \leq 2^n - 4$. By Lemma 17.2, there exist two vertices $\mathbf{u}$ and $\mathbf{v}$ such that $d_{AQ_n}(\mathbf{u}, \mathbf{v}) = 2$, $\{\mathbf{x}, \mathbf{y}\} \neq Nbd_{AQ_n}(\mathbf{u}) \cap Nbd_{AQ_n}(\mathbf{v})$, and $\{\mathbf{u}, \mathbf{v}\} \neq Nbd_{AQ_n}(\mathbf{x}) \cap Nbd_{AQ_n}(\mathbf{y})$. By Theorem 17.7, there exist two disjoint paths P_1 and P_2 such that (1) P_1 is a path joining $\mathbf{x}$ to $\mathbf{y}$ with $l(P_1) = l$, (2) P_2 is a path joining $\mathbf{u}$ to $\mathbf{v}$ with $l(P_2) = 2^n - 2 - l$, and (3) $P_1 \cup P_2$ spans AQ_n. Obviously, P_1 is a path of length l joining $\mathbf{x}$ to $\mathbf{y}$. Thus, AQ_n is panconnected. □

COROLLARY 17.4 *AQ_n is edge-pancyclic for every positive integer n.*

Proof: Obviously, AQ_n is edge-pancyclic for $n = 2$. Thus, we consider that $n \geq 3$. Suppose that $l = 3$. By Lemma 17.1, there exists $\mathbf{u} \in Nbd_{AQ_n}(\mathbf{x}) \cap Nbd_{AQ_n}(\mathbf{y})$. Obviously, $\langle \mathbf{x}, \mathbf{y}, \mathbf{u}, \mathbf{x} \rangle$ forms a cycle of length three containing e. Now, we consider that $l = 2^n$ and $l = 2^n - 1$. By Lemma 17.1, there exists $\mathbf{v} \in (Nbd_{AQ_n}(\mathbf{u}) \cap Nbd_{AQ_n}(\mathbf{y})) - \{\mathbf{x}\}$. By Theorem 17.7, there exist two disjoint paths P_1 and P_2 such that (1) P_1 is a path joining $\mathbf{x}$ to $\mathbf{u}$ with $l(P_1) = 2^n - 3$, (2) P_2 is a path joining $\mathbf{v}$ to $\mathbf{y}$ with $l(P_2) = 1$, and (3) $P_1 \cup P_2$ spans AQ_n. Obviously, $\langle \mathbf{x}, P_1, \mathbf{u}, \mathbf{v}, \mathbf{y}, \mathbf{x} \rangle$ forms a cycle of length 2^n containing e and $\langle \mathbf{x}, P_1, \mathbf{u}, \mathbf{y}, \mathbf{x} \rangle$ forms a cycle of length $2^n - 1$ containing e. Suppose $l = 2^n - 2$. By Theorem 17.7, there exist two disjoint paths Q_1 and Q_2 such that (1) Q_1 is a path joining $\mathbf{x}$ to $\mathbf{y}$ with $l(Q_1) = 2^n - 3$, (2) Q_2 is a path joining $\mathbf{u}$ to $\mathbf{v}$ with $l(Q_2) = 1$, and (3) $Q_1 \cup Q_2$ spans AQ_n. Obviously, $\langle \mathbf{x}, Q_1, \mathbf{y}, \mathbf{x} \rangle$ forms a cycle of length $2^n - 2$ containing e. Suppose that $4 \leq l \leq 2^n - 3$. By Lemma 17.2, there exist two vertices $\mathbf{p}$ and $\mathbf{q}$ of AQ_n such that $d_{AQ_n}(\mathbf{p}, \mathbf{q}) = 2$, $\{\mathbf{x}, \mathbf{y}\} \neq Nbd_{AQ_n}(\mathbf{p}) \cap Nbd_{AQ_n}(\mathbf{q})$, and $\{\mathbf{p}, \mathbf{q}\} \neq Nbd_{AQ_n}(\mathbf{x}) \cap Nbd_{AQ_n}(\mathbf{y})$. By Theorem 17.7, there exist two disjoint paths R_1 and R_2 such that (1) R_1 is a path joining $\mathbf{x}$ to $\mathbf{y}$ with $l(R_1) = l - 1$, (2) R_2 is a path joining $\mathbf{u}$ to $\mathbf{v}$ with $l(R_2) = 2^n - l - 1$, and (3) $R_1 \cup R_2$ spans AQ_n. Obviously, $\langle \mathbf{x}, R_1, \mathbf{y}, \mathbf{x} \rangle$ forms a cycle of length l containing e. □

With the panconnected property of AQ_n, there exists a path $P_l(\mathbf{x}, \mathbf{y})$ of length l if $d_{AQ_n}(\mathbf{x}, \mathbf{y}) \leq l \leq 2^n - 1$ between any two distinct vertices $\mathbf{x}$ and $\mathbf{y}$ of AQ_n. Again, we expect such path $P_l(\mathbf{x}, \mathbf{y})$ can be further extended by including the vertices not in $P_l(\mathbf{x}, \mathbf{y})$ into a Hamiltonian path from $\mathbf{x}$ to a fixed vertex $\mathbf{z}$ or a Hamiltonian cycle.

THEOREM 17.8 Assume that n is a positive integer with $n \geq 2$. For any three distinct vertices $\mathbf{x}$, $\mathbf{y}$ and $\mathbf{z}$ of AQ_n and for any $d_{AQ_n}(\mathbf{x}, \mathbf{y}) \leq l \leq 2^n - 1 - d_{AQ_n}(\mathbf{y}, \mathbf{z})$, there exists a Hamiltonian path $R(\mathbf{x}, \mathbf{y}, \mathbf{z}; l)$ from $\mathbf{x}$ to $\mathbf{z}$ such that $d_{R(\mathbf{x},\mathbf{y},\mathbf{z};l)}(\mathbf{x}, \mathbf{y}) = l$.

Proof: Obviously, the theorem holds for $n = 2$. Thus, we consider that $n \geq 3$. We have the following cases:

Case 1. $d_{AQ_n}(\mathbf{x}, \mathbf{y}) = 1$ and $d_{AQ_n}(\mathbf{y}, \mathbf{z}) = 1$. By Lemma 17.1, there exists a vertex $\mathbf{w}$ in $(Nbd_{AQ_n}(\mathbf{y}) \cap Nbd_{AQ_n}(\mathbf{z})) - \{\mathbf{x}\}$. Similarly, there exists a vertex $\mathbf{p}$ in $(Nbd_{AQ_n}(\mathbf{x}) \cap Nbd_{AQ_n}(\mathbf{y})) - \{\mathbf{z}\}$. Suppose that $l = 2$. By Theorem 17.7, there exist two disjoint paths S_1 and S_2 such that (1) S_1 is a path joining $\mathbf{x}$ to $\mathbf{p}$ with $l(S_1) = 1$, (2) S_2 is a path joining $\mathbf{y}$ to $\mathbf{z}$ with $l(S_2) = 2^n - 3$, and (3) $S_1 \cup S_2$ spans AQ_n. We set R as $\langle \mathbf{x}, \mathbf{p}, \mathbf{y}, S_2, \mathbf{z} \rangle$. Obviously, R forms a Hamiltonian path from $\mathbf{x}$ to $\mathbf{z}$ such that $d_R(\mathbf{x}, \mathbf{y}) = l$. Suppose that $l = 2^n - 3$. By Theorem 17.7, there exist two disjoint paths Q_1 and Q_2 such that (1) Q_1 is a path joining $\mathbf{x}$ to $\mathbf{y}$ with $l(Q_1) = 2^n - 3$, (2) Q_2 is a path joining $\mathbf{w}$ to $\mathbf{z}$ with $l(Q_2) = 1$, and (3) $Q_1 \cup Q_2$ spans AQ_n. We set R as $\langle \mathbf{x}, Q_1, \mathbf{y}, \mathbf{w}, \mathbf{z} \rangle$. Obviously, R forms a Hamiltonian path from $\mathbf{x}$ to $\mathbf{z}$ such that $d_R(\mathbf{x}, \mathbf{y}) = l$. Suppose that $1 \le l \le 2^n - 2$ with $l \notin \{2, 2^n - 3\}$. By Theorem 17.7, there exist two disjoint paths P_1 and P_2 such that (1) P_1 is a path joining $\mathbf{x}$ to $\mathbf{y}$ with $l(P_1) = l$, (2) P_2 is a path joining $\mathbf{w}$ to $\mathbf{z}$ with $l(P_2) = 2^n - 2 - l$, and (3) $P_1 \cup P_2$ spans AQ_n. We set R as $\langle \mathbf{x}, P_1, \mathbf{y}, \mathbf{w}, P_2, \mathbf{z} \rangle$. Obviously, R forms a Hamiltonian path from $\mathbf{x}$ to $\mathbf{z}$ such that $d_R(\mathbf{x}, \mathbf{y}) = l$.

Case 2. $d_{AQ_n}(\mathbf{x}, \mathbf{y}) = 1$ and $d_{AQ_n}(\mathbf{y}, \mathbf{z}) \neq 1$. By Lemma 17.1, there exists a vertex $\mathbf{p}$ in $Nbd_{AQ_n}(\mathbf{x}) \cap Nbd_{AQ_n}(\mathbf{y})$. Suppose that $l = 2$. By Theorem 17.7, there exist two disjoint paths S_1 and S_2 such that (1) S_1 is a path joining $\mathbf{x}$ to $\mathbf{p}$ with $l(S_1) = 1$, (2) S_2 is a path joining $\mathbf{y}$ to $\mathbf{z}$ with $l(S_2) = 2^n - 3$, and (3) $S_1 \cup S_2$ spans AQ_n. We set R as $\langle \mathbf{x}, \mathbf{p}, \mathbf{y}, S_2, \mathbf{z} \rangle$. Obviously, R forms a Hamiltonian path from $\mathbf{x}$ to $\mathbf{z}$ such that $d_R(\mathbf{x}, \mathbf{y}) = l$. Suppose that $1 \le l \le 2^n - 1 - d_{AQ_n}(\mathbf{y}, \mathbf{z})$ with $l \neq 2$. By Corollary 17.2, there exists a vertex $\mathbf{w}$ in $Nbd_{AQ_n}(\mathbf{y}) - \{\mathbf{x}\}$ such that $d_{AQ_n}(\mathbf{w}, \mathbf{z}) = d_{AQ_n}(\mathbf{y}, \mathbf{z}) - 1$. By Theorem 17.7, there exist two disjoint paths P_1 and P_2 such that (1) P_1 is a path joining $\mathbf{x}$ to $\mathbf{y}$ with $l(P_1) = l$, (2) S_2 is a path joining $\mathbf{w}$ to $\mathbf{z}$ with $l(P_2) = 2^n - 2 - l$, and (3) $P_1 \cup P_2$ spans AQ_n. We set R as $\langle \mathbf{x}, P_1, \mathbf{y}, \mathbf{w}, P_2, \mathbf{z} \rangle$. Obviously, R forms a Hamiltonian path from $\mathbf{x}$ to $\mathbf{z}$ such that $d_R(\mathbf{x}, \mathbf{y}) = l$.

Case 3. $d_{AQ_n}(\mathbf{x}, \mathbf{y}) \neq 1$ and $d_{AQ_n}(\mathbf{y}, \mathbf{z}) = 1$. This case is similar to Case 2, but interchanging the roles of $\mathbf{x}$ and $\mathbf{z}$.

Case 4. $d_{AQ_n}(\mathbf{x}, \mathbf{y}) \neq 1$ and $d_{AQ_n}(\mathbf{y}, \mathbf{z}) \neq 1$. Let l be any integer with $d_{AQ_n}(\mathbf{x}, \mathbf{y}) \le l \le 2^n - 1 - d_{AQ_n}(\mathbf{y}, \mathbf{z})$. Let $\mathbf{w}$ be a vertex in $Nbd_{AQ_n}(\mathbf{y})$. By Theorem 17.7, there exist two disjoint paths S_1 and S_2 such that (1) S_1 is a path joining $\mathbf{x}$ to $\mathbf{y}$ with $l(S_1) = l$, (2) S_2 is a path joining $\mathbf{w}$ to $\mathbf{z}$ with $l(S_2) = 2^n - 2 - l$, and (3) $S_1 \cup S_2$ spans AQ_n. We set R as $\langle \mathbf{x}, S_1, \mathbf{y}, \mathbf{w}, S_2, \mathbf{z} \rangle$. Obviously, R forms a Hamiltonian path from $\mathbf{x}$ to $\mathbf{z}$ such that $d_R(\mathbf{x}, \mathbf{y}) = l$.

The theorem is proved. □

COROLLARY 17.5 *Assume that n is a positive integer with $n \ge 2$. For any two distinct vertices $\mathbf{x}$ and $\mathbf{y}$ and for any $d_{AQ_n}(\mathbf{x}, \mathbf{y}) \le l \le 2^{n-1}$, there exists a Hamiltonian cycle $S(\mathbf{x}, \mathbf{y}; l)$ such that $d_{S(\mathbf{x},\mathbf{y};l)}(\mathbf{x}, \mathbf{y}) = l$.*

Proof: Let $\mathbf{z}$ be a vertex in $Nbd_{AQ_n}(\mathbf{x}) - \{\mathbf{y}\}$. By Theorem 17.8, there exists a Hamiltonian path R joining $\mathbf{x}$ to $\mathbf{z}$ such that $d_{R(\mathbf{x},\mathbf{y},\mathbf{z};l)}(\mathbf{x}, \mathbf{y}) = l$. We set S as $\langle \mathbf{x}, R, \mathbf{z}, \mathbf{x} \rangle$. Obviously, S forms the required Hamiltonian cycle. □

17.5 COMPARISON BETWEEN PANCONNECTED AND PANPOSITIONABLE HAMILTONIAN

With Corollary 17.5, we can define the following term. A Hamiltonian graph G is *panpositionable* if for any two different vertices x and y of G and for any integer l satisfying $d(x, y) \le l \le |V(G)| - d(x, y)$, there exists a Hamiltonian cycle C of G such that the relative distance between x and y on C is l; more precisely, $d_C(x,y)=l$ if $l \leq \left\lfloor \frac{|V(G)|}{2} \right\rfloor$ or $D_C(x,y)=l$ if $l>|V(G)|/2$. Given a Hamiltonian cycle C, if $d_C(x,y)=l$, we have $D_C(x,y)=|V(G)|-d_C(x,y)$. Therefore, a graph is panpositionable Hamiltonian if for any integer l with $d(x,y)\leq l \leq |V(G)|/2$, there exists a Hamiltonian cycle C of G with $d_C(x,y)=l$. With Corollary 17.5, we have seen every AQ_n with $n\geq 2$ is panpositionable Hamiltonian. It is easy to see that the length of the shortest cycle for any panpositionable Hamiltonian graph is 3. Moreover, every panpositionable Hamiltonian graph is pancyclic. A Hamiltonian bipartite graph G is *bipanpositionable* if for any two different vertices x and y of G and for any integer k with $d_G(x,y)\leq k\leq |V(G)|/2$ and $(k-d_G(x,y))$ is even, there exists a Hamiltonian cycle C of G such that $d_C(x,y)=k$. With Corollary 17.1, we have seen every Q_n with $n\geq 2$ is bipanpositionable Hamiltonian.

Assume that G is a panpositionable Hamiltonian graph with n vertices. Obviously, $d_2^*(u,v)=\left\lceil \frac{n}{2} \right\rceil$ if u and v are two different vertices in G. Hence $D_2^*(G)=\left\lceil \frac{n}{2} \right\rceil$. Similarly, let G be a bipanpositionable Hamiltonian graph with n vertices. Obviously, $d_2^*(u,v)$ is either $\left\lceil \frac{n}{2} \right\rceil+1$ or $\left\lceil \frac{n}{2} \right\rceil$ depending on the parity of $d(u,v)$.

In the following, we would like to compare the difference among pancyclic, panconnected, and panpositionable Hamiltonian. Assume that G is a panconnected graph with at least three vertices. Let u and v be any two vertices with $d_G(u,v)=1$. Obviously, there exists a path P_k of length k for any $1\leq k\leq n(G)-1$. Obviously, $P_1\cup P_k$ constitute a cycle of length $k+1$ for any $2\leq k\leq n(G)$. Thus, G contains a cycle of length l for any $3\leq l\leq n(G)$. Therefore, any panconnected graph is pancyclic.

With the following theorem, we have examples of a pancyclic graph that is neither panconnected nor panpositionable Hamiltonian.

THEOREM 17.9 Let m be an integer with $m\geq 3$. $K_m\times K_2$ is a pancyclic graph that is neither panconnected nor panpositionable Hamiltonian.

Proof: We use G to denote the graph $K_m\times K_2$. We can describe G with $V(G)=\{0,1,2,\dots,m-1\}\cup\{0',1',2',\dots,(m-1)'\}$ and $E(G)=\{(i,j)\mid i<j, 0\leq i, j\leq m-1\}\cup\{(i',j')\mid i<j, 0\leq i,j\leq m-1\}\cup\{(i,i')\mid 0\leq i\leq m-1\}$.

Let $P=\langle m-2,m-3,m-4,\dots,1\rangle$. Since K_m is a subgraph of G, G contains a cycle of every length from 3 to m. Obviously, $\langle 0,m-2,P,1,1',0',0\rangle$ is a cycle of length $m+1$, $\langle 0,m-2,P,1,1',2',0',0\rangle$ is a cycle of length $m+2$, $\langle 0,m-2,P,1,1',2',3',\dots,k',0',0\rangle$ is a cycle of length $m+k$ with $3\leq k<m$, and $\langle 0,m-1,m-2,P,1,1',2',3',\dots,(m-1)',0',0\rangle$ is a cycle of length $2m$. Thus, G is pancyclic.

Now, we want to prove that G is not panposionable Hamiltonian. Let $u=0$ and $v=0'$ of G. Obviously, $d_G(u,v)=1$. Since $N(u)\cap N(v)=\emptyset$, it is impossible to have a path with length 2 between u and v. Thus, G is not panconnected. For the same reason, G cannot be panpositionable Hamiltonian. The theorem is proved. □

In the following, we will prove that every Harary graph $H_{2,n}$ is panconnected. Moreover, $H_{2,n}$ is panpositionable Hamiltonian if and only if $n\in\{5,6,7,8,9,11\}$.

THEOREM 17.10 Let n be an integer with $n \geq 5$. Then the Harary graph $H_{2,n}$ is a panconnected graph.

Proof: Let $G = H_{2,n}$. To show that G is panconnected, we prove that there exists a path of length l for $d_G(x,y) \leq l \leq n-1$ between any two distinct vertices x, y of G. Without loss of generality, let $x = 0$ and $y \leq \lfloor \frac{n}{2} \rfloor$. We define some paths as follows:

$$
\begin{aligned}
I(i,j) &= \langle i, i+1, i+2, \ldots, j-1, j \rangle \\
S(i) &= \langle i, i+2 \rangle \\
S^t(i) &= \langle i, i+2, i+4, \ldots, i+2(t-1), i+2t \rangle \\
S^{-t}(i) &= \langle i, i-2, i-4, \ldots, i-2(t-1), i-2t \rangle
\end{aligned}
$$

Suppose that $l \leq y$. Let $k = 2(l - d_G(0,y))$. Then $\langle 0, I(0,k), S^{\lfloor \frac{y-k}{2} \rfloor}(k), y \rangle$ is the required path. Suppose that $l > y$. Let $k = \lceil \frac{l-y}{2} \rceil$. Then the required path is $\langle 0, I(0, y-1), S^k(y-1), y+2k-1, y+2k, S^{-k}(y+2k), y \rangle$ if $l-y$ is even and $\langle 0, I(0,y-1), S^k(y-1), y+2k-1, y+2k-2, S^{-(k-1)}(y+2k-2), y \rangle$ if otherwise. The theorem is proved. □

THEOREM 17.11 $H_{2,n}$ is panpositionable Hamiltonian if and only if $n \in \{5, 6, 7, 8, 9, 11\}$.

Proof: We first show that $H_{2,n}$ is panpositionable Hamiltonian if $n \in \{5,6,7,8,9,11\}$. With the symmetric property of $H_{2,n}$, it suffices to show that for any vertex u with $1 \leq u \leq n/2$ and for any integer k with $d_{H_{2,n}}(0,u) \leq k \leq n/2$, there exists a Hamiltonian cycle C of $G(n; 1, 2)$ such that $d_C(0,u) = k$. It is easy to see that $d_{H_{2,n}}(0,u) = \lceil \frac{u}{2} \rceil$. We set $r = \lceil \frac{u}{2} \rceil$. To describe the required Hamiltonian cycles, we define some path patterns:

$$
\begin{aligned}
p(i,j) &= \langle i, i+1, i+2, \ldots, j-1, j \rangle \\
q(i,j) &= \langle i, i+2, i+4, \ldots, j-2, j \rangle \\
q^{-1}(j,i) &= \langle j, j-2, j-4, \ldots, i+2, i \rangle
\end{aligned}
$$

Case 1. $n \in \{5, 7, 9, 11\}$. Suppose that $u = 1$. Obviously, $\langle 0, 1, 3, 2, 4, p(4, n-1), n-1, 0 \rangle$ forms a Hamiltonian cycle of $H_{2,n}$ such that $d_C(0,u) = 1$ and $\langle 0, 2, 1, 3, p(3, n-1), n-1, 0 \rangle$ forms a Hamiltonian cycle of $H_{2,n}$ such that $d_C(0,u) = 2$. Suppose we want to find a Hamiltonian cycle of $H_{2,n}$ such that $d_C(0,u) = 3$. Then $n \in \{7, 9, 11\}$. Then $\langle 0, 2, 3, 1, n-1, q^{-1}(n-1, 4), 4, 5, q(5, n-2), n-2, 0 \rangle$ forms the required Hamiltonian cycle. Similarly, $\langle 0, 2, 4, 3, 1, n-1, q^{-1}(n-1, 6), 6, 5, q(5, n-2), n-2, 0 \rangle$ forms a Hamiltonian cycle of $H_{2,n}$ such that $d_C(0,u) = 4$ with $n \in \{9, 11\}$. Moreover, $\langle 0, 9, q^{-1}(9, 1), 1, 2, q(2, 10), 10, 0 \rangle$ forms a Hamiltonian cycle of $H_{2,n}$ such that $d_C(0,u) = 4$ with $n = 11$.

Suppose that $u = 2$. Then $\langle 0, 2, 1, 3, p(3, n-1), n-1, 0 \rangle$ forms a Hamiltonian cycle of $H_{2,n}$ such that $d_C(0,u) = 1$; $\langle 0, p(0, n-1), n-1, 0 \rangle$ forms a Hamiltonian cycle of $H_{2,n}$ such that $d_C(0,u) = 2$; $\langle 0, n-1, 1, p(1, n-2), n-2, 0 \rangle$ forms a Hamiltonian cycle of $H_{2,n}$ such that $d_C(0,u) = 3$ with $n \in \{7, 9, 11\}$;

$\langle 0, n-1, 1, 3, 2, 4, p(4, n-2), n-2, 0\rangle$ forms a Hamiltonian cycle of $H_{2,n}$ such that $d_C(0, u) = 4$ with $n \in \{9, 11\}$; and $\langle 0, 9, q^{-1}(9, 3), 3, 2, q(2, 10), 10, 1, 0\rangle$ forms a Hamiltonian cycle of $H_{2,n}$ such that $d_C(0, u) = 5$ with $n = 11$.

Suppose $u = 3$. Then $d_{H_{2,n}}(0, u) = 2$ and $n \in \{7, 9, 11\}$. Then $\langle 0, 2, p(2, n-1), n-1, 1, 0\rangle$ forms a Hamiltonian cycle of $H_{2,n}$ such that $d_C(0, u) = 2$; $\langle 0, p(0, n-1), n-1, 0\rangle$ forms a Hamiltonian cycle of $H_{2,n}$ such that $d_C(0, u) = 3$; $\langle 0, 1, 2, 4, 3, 5, p(5, n-1), n-1, 0\rangle$ forms a Hamiltonian cycle of $H_{2,n}$ such that $d_C(0, u) = 4$ with $n \in \{9, 11\}$; and $\langle 0, 10, 1, 2, 4, 3, 5, p(5, 9), 9, 0\rangle$ forms a Hamiltonian cycle of $H_{2,n}$ such that $d_C(0, u) = 5$ with $n = 11$.

Suppose that $u = 4$. Then $\langle 0, q(0, n-1), n-1, 1, q(1, n-2), n-2, 0\rangle$ forms a Hamiltonian cycle of $H_{2,n}$ such that $d_C(0, u) = 2$; $\langle 0, 2, p(2, n-1), n-1, 1, 0\rangle$ forms a Hamiltonian cycle of $H_{2,n}$ such that $d_C(0, u) = 3$; $\langle 0, p(0, n-1), n-1, 0\rangle$ forms a Hamiltonian cycle of $H_{2,n}$ such that $d_C(0, u) = 4$; and $\langle 0, p(0, 3), 3, 5, 4, 6, p(6, 10), 10, 0\rangle$ forms a Hamiltonian cycle of $H_{2,n}$ such that $d_C(0, u) = 5$ with $n = 11$.

Suppose $u = 5$. Obviously, $n = 11$. Then $\langle 0, 1, q(1, 9), 9, 10, q^{-1}(10, 0), 0\rangle$ forms a Hamiltonian cycle of $H_{2,n}$ such that $d_C(0, u) = 3$; $\langle 0, 10, 1, q(1, 9), 9, 8, q^{-1}(8, 0), 0\rangle$ forms a Hamiltonian cycle of $H_{2,n}$ such that $d_C(0, u) = 4$; and $\langle 0, p(0, 10), 10, 0\rangle$ forms a Hamiltonian cycle of $H_{2,n}$ such that $d_C(0, u) = 5$.

Case 2. $n \in \{6, 8\}$. Suppose that $u = 1$. Then $\langle 0, p(0, n-1), n-1, 0\rangle$ forms a Hamiltonian cycle of $H_{2,n}$ such that $d_C(0, u) = 1$; $\langle 0, 2, 1, 3, p(3, n-1), n-1, 0\rangle$ forms a Hamiltonian cycle of $H_{2,n}$ such that $d_C(0, u) = 2$; $\langle 0, 2, 3, 1, n-1, q^{-1}(n-1, 5), 5, 4, q(4, n-2), n-2, 0\rangle$ forms a Hamiltonian cycle of $H_{2,n}$ such that $d_C(0, u) = 3$; and $\langle 0, 2, 4, 3, 1, 7, 5, 6, 0\rangle$ forms a Hamiltonian cycle of $H_{2,n}$ such that $d_C(0, u) = 4$ with $n = 8$.

Suppose that $u = 2$. Then $\langle 0, 2, 1, 3, p(3, n-1), n-1, 0\rangle$ forms a Hamiltonian cycle of $H_{2,n}$ such that $d_C(0, u) = 1$; $\langle 0, p(0, n-1), n-1, 0\rangle$ forms a Hamiltonian cycle of $H_{2,n}$ such that $d_C(0, u) = 2$; $\langle 0, 1, 3, 2, 4, p(4, n-1), n-1, 0\rangle$ forms a Hamiltonian cycle of $H_{2,n}$ such that $d_C(0, u) = 3$; and $\langle 0, 7, 1, 3, 2, 4, 5, 6, 0\rangle$ forms a Hamiltonian cycle of $H_{2,n}$ such that $d_C(0, u) = 4$ with $n = 8$.

Suppose that $u = 3$. Then $\langle 0, 2, p(2, n-1), n-1, 1, 0\rangle$ forms a Hamiltonian cycle of $H_{2,n}$ such that $d_C(0, u) = 2$; $\langle 0, p(0, n-1), n-1, 0\rangle$ forms a Hamiltonian cycle of $H_{2,n}$ such that $d_C(0, u) = 3$; and $\langle 0, 1, 2, 4, 3, 5, 6, 7, 0\rangle$ forms a Hamiltonian cycle of $H_{2,n}$ such that $d_C(0, u) = 4$ with $n = 8$.

Suppose that $u = 4$. Then $\langle 0, q(0, 6), 7, q^{-1}(7, 1), 1, 0\rangle$ forms a Hamiltonian cycle of $H_{2,n}$ such that $d_C(0, u) = 2$; $\langle 0, 2, p(2, 7), 7, 1, 0\rangle$ forms a Hamiltonian cycle of $H_{2,n}$ such that $d_C(0, u) = 3$; and $\langle 0, p(0, 7), 7, 0\rangle$ forms a Hamiltonian cycle of $H_{2,n}$ such that $d_C(0, u) = 4$.

To show that H is not panpositionable Hamiltonian if $n = 10$ or $n \geq 12$, we prove that there exists no Hamiltonian cycle in H such that the distance between 0 and 2 is 5. Suppose that C is a Hamiltonian cycle of H with $d_C(0, 2) = 5$. Obviously, $P_1 = \langle 0, n-2, n-1, 1, 3, 2\rangle$, $P_2 = \langle 0, n-1, 1, 3, 4, 2\rangle$, and $P_3 = \langle 0, 1, 3, 5, 4, 2\rangle$ are all the possible paths of length 5 joining 0 and 2. Then C contains exactly one of P_1, P_2, and P_3.

Suppose that C contains P_1. Then $(0, 1)$ and $(0, n-1)$ are in C. Thus, C contains $\langle n-2, 0, 2\rangle$. This means that C contains a cycle $\langle 0, P_1, 2, 0\rangle$, which is impossible.

Suppose that C contains P_2 or P_3. Then (2, 1) and (2, 3) are not in C. Thus, C contains $\langle 0, 2, 4 \rangle$. This means that C contains a cycle $\langle 0, P_2, 2, 0 \rangle$ or $\langle 0, P_3, 2, 0 \rangle$, which is impossible. The theorem is proved. □

Thus, we have some examples of pancyclic graphs that are neither panconnected nor panpositionable Hamiltonian, some examples of panconnected graphs that are not panpositionable Hamiltonian, and some examples of graphs that are panconnected and panpositionable Hamiltonian. The existence of panpositionable Hamiltonian graphs that are not panconnected is still an open problem.

17.6 BIPANPOSITIONABLE BIPANCYCLIC PROPERTY OF HYPERCUBE

Again, we would like to compare the difference among bipancyclic, bipanconnected, and bipanpositionable Hamiltonian. Obviously, any bipanconnected graph is bipancyclic. However, we introduce another interesting concept. A k-cycle is a cycle of length k. A bipartite graph G is *k-cycle bipanpositionable* if for every different vertices x and y of G and for any integer l with $d_G(\mathbf{x}, \mathbf{y}) \leq l \leq k/2$ and $(l - d_G(x, y))$ being even, there exists a k-cycle C of G such that $d_C(x, y) = l$. (Note that $d_C(x, y) \leq k/2$ for every cycle C of length k.) A bipartite graph G is *bipanpositionable bipancyclic* if G is k-cycle bipanpositionable for every even integer k with $4 \leq k \leq |V(G)|$.

Now, we prove that the hypercube Q_n is bipanpositionable bipancyclic if and only if $n \geq 2$, by induction.

LEMMA 17.6 Q_3 is bipanpositionable bipancyclic.

Proof: Let $\mathbf{x}$ and $\mathbf{y}$ be two different vertices in Q_3. It is known that the diameter of Q_n is n, so $d_{Q_3}(\mathbf{x}, \mathbf{y}) = 1, 2$, or 3. Since the hypercube is vertex-symmetric, without loss of generality, we may assume that $\mathbf{x} = 000$.

Case 1. Suppose that $d_{Q_3}(\mathbf{x}, \mathbf{y}) = 1$. Since Q_3 is edge-symmetric, we may assume that $\mathbf{y} = 001$.

$\mathbf{y} = 001$	4-cycle	$d_C(\mathbf{x}, \mathbf{y}) = 1$	$\langle 000, 001, 011, 010, 000 \rangle$
	6-cycle	$d_C(\mathbf{x}, \mathbf{y}) = 1$	$\langle 000, 001, 101, 111, 110, 100, 000 \rangle$
		$d_C(\mathbf{x}, \mathbf{y}) = 3$	$\langle 000, 100, 101, 001, 011, 010, 000 \rangle$
	8-cycle	$d_C(\mathbf{x}, \mathbf{y}) = 1$	$\langle 000, 001, 101, 111, 011, 010, 110, 100, 000 \rangle$
		$d_C(\mathbf{x}, \mathbf{y}) = 3$	$\langle 000, 100, 101, 001, 011, 111, 110, 010, 000 \rangle$

Case 2. Suppose that $d_{Q_3}(\mathbf{x}, \mathbf{y}) = 2$. We have $\mathbf{y} \in \{011, 101, 110\}$.

$\mathbf{y} = 011$	4-cycle	$d_C(\mathbf{x}, \mathbf{y}) = 2$	$\langle 000, 001, 011, 010, 000 \rangle$
	6-cycle	$d_C(\mathbf{x}, \mathbf{y}) = 2$	$\langle 000, 001, 011, 010, 110, 100, 000 \rangle$
	8-cycle	$d_C(\mathbf{x}, \mathbf{y}) = 2$	$\langle 000, 001, 011, 010, 110, 111, 101, 100, 000 \rangle$
		$d_C(\mathbf{x}, \mathbf{y}) = 4$	$\langle 000, 001, 101, 111, 011, 010, 110, 100, 000 \rangle$

$\mathbf{y} = 101$	4-cycle	$d_C(\mathbf{x},\mathbf{y}) = 2$	$\langle 000, 001, 101, 100, 000\rangle$
	6-cycle	$d_C(\mathbf{x},\mathbf{y}) = 2$	$\langle 000, 001, 101, 111, 110, 100, 000\rangle$
	8-cycle	$d_C(\mathbf{x},\mathbf{y}) = 2$	$\langle 000, 001, 101, 111, 011, 010, 110, 100, 000\rangle$
		$d_C(\mathbf{x},\mathbf{y}) = 4$	$\langle 000, 001, 011, 111, 101, 100, 110, 010, 000\rangle$
$\mathbf{y} = 110$	4-cycle	$d_C(\mathbf{x},\mathbf{y}) = 2$	$\langle 000, 010, 110, 100, 000\rangle$
	6-cycle	$d_C(\mathbf{x},\mathbf{y}) = 2$	$\langle 000, 100, 110, 111, 101, 001, 000\rangle$
	8-cycle	$d_C(\mathbf{x},\mathbf{y}) = 2$	$\langle 000, 100, 110, 010, 011, 111, 101, 001, 000\rangle$
		$d_C(\mathbf{x},\mathbf{y}) = 4$	$\langle 000, 100, 101, 111, 110, 010, 011, 001, 000\rangle$

Case 3. Suppose that $d_{Q_3}(\mathbf{x},\mathbf{y}) = 3$. We have $\mathbf{y} = 111$.

$\mathbf{y} = 111$	6-cycle	$d_C(\mathbf{x},\mathbf{y}) = 3$	$\langle 000, 001, 011, 111, 110, 100, 000\rangle$
	8-cycle	$d_C(\mathbf{x},\mathbf{y}) = 3$	$\langle 000, 001, 011, 111, 101, 100, 110, 010, 000\rangle$

Thus, Q_3 is bipanpositionable bipancyclic. □

THEOREM 17.12 Q_n is bipanpositionable bipancyclic if and only if $n \geq 2$.

Proof: We observe that Q_1 is not bipanpositionable bipancyclic. So we start with $n \geq 2$. We prove that Q_n is bipanpositionable bipancyclic by induction on n. It is easy to see that Q_2 is bipanpositionable bipancyclic. By Lemma 17.7, this statement holds for $n = 3$. Suppose that Q_{n-1} is bipanpositionable bipancyclic for some $n \geq 4$. Let $\mathbf{x}$ and $\mathbf{y}$ be two distinct vertices in Q_n, and let k be an even integer with $k \geq \max\{4, 2d_{Q_n}(\mathbf{x},\mathbf{y})\}$ and $k \leq 2^n$. For every integer l with $d_{Q_n}(\mathbf{x},\mathbf{y}) \leq l \leq k/2$ and $(l - d_{Q_n}(\mathbf{x},\mathbf{y}))$ being even, we need to construct a k-cycle C of Q_n with $d_C(\mathbf{x},\mathbf{y}) = l$.

Case 1. $d_{Q_n}(\mathbf{x},\mathbf{y}) = 1$. Without loss of generality, we may assume that both $\mathbf{x}$ and $\mathbf{y}$ are in Q_n^0. $(l - d_{Q_n}(\mathbf{x},\mathbf{y}))$ is even, so l is an odd number.

Case 1.1. $l = 1$. Suppose that $k \leq 2^{n-1}$. By induction, there is a k-cycle C of Q_n^0 with $d_C(\mathbf{x},\mathbf{y}) = 1$. Suppose that $k \geq 2^{n-1} + 2$. By induction, there is a 2^{n-1}-cycle C' of Q_n^0 with $d_C(\mathbf{x},\mathbf{y}) = 1$. Without loss of generality, we write $C' = \langle \mathbf{x}, P, \mathbf{z}, \mathbf{y}, \mathbf{x}\rangle$ such that $d_P(\mathbf{x},\mathbf{z}) = k - 2$. Suppose that $k - 2^{n-1} = 2$. Then $C = \langle \mathbf{x}, P, \mathbf{z}, \overline{\mathbf{z}}, \overline{\mathbf{y}}, \mathbf{y}, \mathbf{x}\rangle$ forms a $(2^{n-1} + 2)$-cycle with $d_C(\mathbf{x},\mathbf{y}) = 1$. Suppose that $k - 2^{n-1} \geq 4$. By induction, there is a $(k - 2^{n-1})$-cycle C'' of Q_n^1 such that $d_{C''}(\overline{\mathbf{z}},\overline{\mathbf{y}}) = 1$. We write $C'' = \langle \overline{\mathbf{z}}, R, \overline{\mathbf{y}}, \overline{\mathbf{z}}\rangle$ with $d_R(\overline{\mathbf{z}},\overline{\mathbf{y}}) = k - 2^{n-1} - 1$. Then $C = \langle \mathbf{x}, P, \mathbf{z}, \overline{\mathbf{z}}, R, \overline{\mathbf{y}}, \mathbf{y}, \mathbf{x}\rangle$ forms a k-cycle of Q_n with $d_C(\mathbf{x},\mathbf{y}) = l$.

Case 1.2. $l \geq 3$. Suppose that $k - l - 1 \leq 2^{n-1}$. By induction, there is a $(l+2)$-cycle C' of Q_n^0 with $d_{C'}(\mathbf{x},\mathbf{y}) = 1$. We write $C' = \langle \mathbf{x}, P, \mathbf{y}, \mathbf{x}\rangle$ where $d_P(\mathbf{x},\mathbf{y}) = l$. By induction, there is a $(k - l - 1)$-cycle C'' of Q_n^1 with $d_{C''}(\overline{\mathbf{x}},\overline{\mathbf{y}}) = 1$. We then write $C'' = \langle \overline{\mathbf{y}}, R, \overline{\mathbf{x}}, \overline{\mathbf{y}}\rangle$ such that $d_R(\overline{\mathbf{y}},\overline{\mathbf{x}}) = k - l - 1$. Then $C = \langle \mathbf{x}, P, \mathbf{y}, \overline{\mathbf{y}}, R, \overline{\mathbf{x}}, \mathbf{x}\rangle$ forms a k-cycle of Q_n with $d_C(\mathbf{x},\mathbf{y}) = l$. Suppose that $k - l - 2 \geq 2^{n-1} + 1$. By induction, there is a $(k - 2^{n-1})$-cycle C' of Q_n^0 with $d_{C'}(\mathbf{x},\mathbf{y}) = l$. We write $C' = \langle \mathbf{x}, P, \mathbf{y}, \mathbf{u}, R, \mathbf{x}\rangle$ with $d_P(\mathbf{x},\mathbf{y}) = l$ and $d_R(\mathbf{u},\mathbf{x}) = k - (2^{n-1} - 1) - l - 2$. By induction, there is a (2^{n-1})-cycle C'' of Q_n^1 with $d_{C''}(\overline{\mathbf{x}},\overline{\mathbf{u}}) = 1$. We write $C'' = \langle \overline{\mathbf{x}}, \overline{\mathbf{u}}, S, \overline{\mathbf{x}}\rangle$ with

$d_S(\bar{\mathbf{u}}, \bar{\mathbf{x}}) = 2^{n-1} - 1$. Then $C = \langle \mathbf{x}, P, \mathbf{y}, R, \mathbf{u}, \bar{\mathbf{u}}, S, \bar{\mathbf{x}}, \mathbf{x} \rangle$ forms a k-cycle of Q_n with $d_C(\mathbf{x}, \mathbf{y}) = l$.

Case 2. $d_{Q_n}(\mathbf{x}, \mathbf{y}) \geq 2$ and $l = 2$. Since $d_{Q_n}(\mathbf{x}, \mathbf{y}) \leq l$ and $l = 2$, so $d_{Q_n}(\mathbf{x}, \mathbf{y}) = 2$. Without loss of generality, we may assume that $\mathbf{x}$ is in Q_n^0 and $\mathbf{y}$ is in Q_n^1. Then $d_{Q_n}(\bar{\mathbf{x}}, \mathbf{y}) = 1$ and $d_{Q_n}(\bar{\mathbf{y}}, \mathbf{x}) = 1$.

Suppose that $k = 4$. Then $C = \langle \mathbf{x}, \bar{\mathbf{x}}, \mathbf{y}, \bar{\mathbf{y}}, \mathbf{x} \rangle$ forms a 4-cycle of Q_n with $d_{Q_n}(\mathbf{x}, \mathbf{y}) = 2$. Suppose that $6 \leq k \leq 2^{n-1} + 2$. By induction, there is a $(k-2)$-cycle $C' = \langle \mathbf{x}, P, \bar{\mathbf{y}}, \mathbf{x} \rangle$ of Q_n^0 such that $d_P(\mathbf{x}, \bar{\mathbf{y}}) = k - 3$. Then $C = \langle \mathbf{x}, P, \bar{\mathbf{y}}, \mathbf{y}, \bar{\mathbf{x}}, \mathbf{x} \rangle$ forms a k-cycle of Q_n with $d_C(\mathbf{x}, \mathbf{y}) = 2$. Suppose that $k \geq 2^{n-1} + 4$. By induction, there is a 2^{n-1}-cycle C' of Q_n^0 with $d_{C'}(\mathbf{x}, \bar{\mathbf{y}}) = 1$. We write $C' = \langle \mathbf{x}, P, \mathbf{z}, \bar{\mathbf{y}}, \mathbf{x} \rangle$ with $d_P(\mathbf{x}, \mathbf{z}) = 2^{n-1} - 2$. By induction, there is a $(k - 2^{n-1})$-cycle C'' of Q_n^1 with $d_{C''}(\mathbf{y}, \bar{\mathbf{z}}) = 1$. We write $C'' = \langle \mathbf{y}, \bar{\mathbf{z}}, R, \mathbf{y} \rangle$ with $d_R(\mathbf{y}, \bar{\mathbf{z}}) = k - 2^{n-1} - 1$. Then $C = \langle \mathbf{x}, P, \mathbf{z}, \bar{\mathbf{z}}, R, \mathbf{y}, \bar{\mathbf{y}}, \mathbf{x} \rangle$ forms a k-cycle of Q_n with $d_C(\mathbf{x}, \mathbf{y}) = 2$.

Case 3. $d_{Q_n}(\mathbf{x}, \mathbf{y}) \geq 2$ and $l \geq 3$. Without loss of generality, we may assume that $\mathbf{x}$ is in Q_n^0 and $\mathbf{y}$ is in Q_n^1. Suppose that $k - l - d_{Q_n}(\mathbf{x}, \mathbf{y}) + 2 \leq 2^{n-1}$. By induction, there is a $(l + d_{Q_n}(\mathbf{x}, \mathbf{y}) - 2)$-cycle $C' = \langle \mathbf{x}, P, \bar{\mathbf{y}}, \mathbf{u}, R, \mathbf{x} \rangle$ of Q_n^0 such that $d_P(\mathbf{x}, \bar{\mathbf{y}}) = l - 1$ and $d_R(\mathbf{u}, \mathbf{x}) = d_{Q_n}(\mathbf{x}, \mathbf{y}) - 2$. By induction, there is a $(k - l - d_{Q_n}(\mathbf{x}, \mathbf{y}) + 2)$-cycle C'' of Q_n^1 with $d_{C''}(\mathbf{y}, \bar{\mathbf{u}}) = 1$. We write $C'' = \langle \mathbf{y}, S, \bar{\mathbf{u}}, \mathbf{y} \rangle$ with $d_S(\mathbf{y}, \bar{\mathbf{u}}) = k - l - d_{Q_n}(\mathbf{x}, \mathbf{y}) + 1$. Then $C = \langle \mathbf{x}, P, \bar{\mathbf{y}}, \mathbf{y}, S, \bar{\mathbf{u}}, \mathbf{u}, R, \mathbf{x} \rangle$ forms a k-cycle of Q_n with $d_C(\mathbf{x}, \mathbf{y}) = l$. Suppose that $k - l - d_{Q_n}(\mathbf{x}, \mathbf{y}) + 4 \geq 2^{n-1}$. By induction, there is a $(k - 2^{n-1})$-cycle $C' = \langle \mathbf{x}, P, \bar{\mathbf{y}}, \mathbf{u}, R, \mathbf{x} \rangle$ of Q_n^0 such that $d_P(\mathbf{x}, \bar{\mathbf{y}}) = l - 1$ and $d_R(\mathbf{u}, \mathbf{x}) = k - 2^{n-1} - l$. By induction, there is a 2^{n-1}-cycle C'' of Q_n^1 with $d_{C''}(\mathbf{y}, \bar{\mathbf{u}}) = 1$. We write $C'' = \langle \mathbf{y}, S, \bar{\mathbf{u}}, \mathbf{y} \rangle$ with $d_S(\mathbf{y}, \bar{\mathbf{u}}) = 2^{n-1} - 1$. Then $C = \langle \mathbf{x}, P, \bar{\mathbf{y}}, \mathbf{y}, S, \bar{\mathbf{u}}, \mathbf{u}, R, \mathbf{x} \rangle$ forms a k-cycle of Q_n with $d_C(\mathbf{x}, \mathbf{y}) = l$.

The theorem is proved. □

Thus, Q_n with $n \geq 2$ is bipanpositionable bipancyclic, bipanpositionable Hamiltonian, and bipanconnected. Yet there is a bipanconnected graph that is not bipanpositionable Hamiltonian. Let G be the bipartite graph shown in Figure 17.8. It is

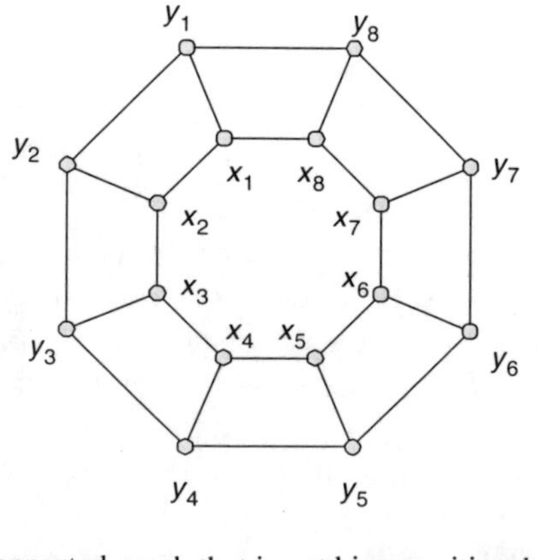

FIGURE 17.8 A bipanconnected graph that is not bipanpositionable Hamiltonian.

easy to check that G is bipanconnected. Moreover, there is no Hamiltonian cycle C of G such that $d_C(x_1,x_2)=5$. Thus, G is not bipanpositionable Hamiltonian. Since any bipanpositionable bipancyclic graph is bipanpositionable Hamiltonian, G is not bipanpositionable bipancyclic. However, we cannot find bipanpositionable Hamiltonian graphs that are not bipanconnected.

18 Mutually Independent Hamiltonian Cycles

18.1 INTRODUCTION

In Section 9.5, we discussed the concept of mutually independent Hamiltonian paths. Obviously, we can extend this concept into mutually independent Hamiltonian cycles. A Hamiltonian cycle C of graph G is described as $\langle u_1, u_2, \ldots, u_{n(G)}, u_1 \rangle$ to emphasize the order of vertices in C. Thus, u_1 is the beginning vertex and u_i is the ith vertex in C. Two Hamiltonian cycles of G beginning at a vertex x, $C_1 = \langle u_1, u_2, \ldots, u_{n(G)}, u_1 \rangle$ and $C_2 = \langle v_1, v_2, \ldots, v_{n(G)}, v_1 \rangle$, are *independent* if $x = u_1 = v_1$ and $u_i \neq v_i$ for every i, $2 \leq i \leq n(G)$. A set of Hamiltonian cycles $\{C_1, C_2, \ldots, C_k\}$ of G are *mutually independent* if any two different Hamiltonian cycles are independent. The *mutually independent Hamiltonianicity* of graph G, $IHC(G)$, is the maximum integer k such that for any vertex u of G there exist k mutually independent Hamiltonian cycles of G starting at u. Obviously, $IHC(G) \leq \delta(G)$ if G is a Hamiltonian graph.

Assume that K_5 is the complete graph with vertex set $\{0, 1, 2, 3, 4\}$. Let $C_1 = \langle 0, 1, 2, 3, 4, 0 \rangle$, $C_2 = \langle 0, 2, 3, 4, 1, 0 \rangle$, $C_3 = \langle 0, 3, 4, 1, 2, 0 \rangle$, and $C_4 = \langle 0, 4, 1, 2, 3, 0 \rangle$. Obviously, C_1, C_2, C_3, and C_4 are mutually independent. Thus, $IHC(K_5) = 4$. However, we rewrite C_1, C_2, C_3, and C_4 into the following Latin square:

1	2	3	4
2	3	4	1
3	4	1	2
4	1	2	3

A *Latin square* of order n is an $n \times n$ array made from the integers 1 to n with the property that any integer occurs once in each row and column. If we delete some rows from a Latin square, we will get a *Latin rectangle*. Obviously, a Latin square of order n can be viewed as n mutually independent Hamiltonian cycles with respect to the complete graph K_{n+1}. Thus, the concept of mutually independent Hamiltonian cycles can be interpreted as a Latin rectangle for graphs.

We can apply mutually independent Hamiltonian cycles in a lot of areas. Consider the following scenario. At Christmas, we have a holiday of ten days. A tour agency will organize a ten-day tour to Italy. Suppose that there will be a lot of people joining this tour. However, the maximum number of people staying in each local area is limited to say, 100 people, because of the hotel contract. One trivial solution is the first-come-first-serve approach. So only 100 people can attend this tour. (Note that we cannot schedule the tour in a pipelined manner, because the holiday period is fixed.)

Nonetheless, we observe that a tour is like a Hamiltonian cycle based on a graph, in which a vertex is denoted as a hotel and any two vertices are joined with an edge if the associated two hotels can be traveled in a reasonable time. Therefore, we can organize several subgroups; that is, each subgroup has its own tour. In this way, we do not allow two subgroups to stay in the same area during the same time period. In other words, any two different tours are indeed independent Hamiltonian cycles. Suppose that there are ten mutually independent Hamiltonian cycles. Then we may allow 1000 people to visit Italy on Christmas vacation. For this reason, we would like to find the maximum number of mutually independent Hamiltonian cycles. Such applications are useful for task scheduling and resource placement, and are also important for compiler optimization to exploit parallelism.

Recently, there have been some studies on the mutually independent Hamiltonianicity of some graphs and variations [163,207,209,236,290].

18.2 MUTUALLY INDEPENDENT HAMILTONIAN CYCLES ON SOME GRAPHS

The bound for $IHC(G)$ is studied in Ref. 236.

LEMMA 18.1 Assume that G is a Hamiltonian graph. Then $IHC(G) \leq \delta(G)$.

LEMMA 18.2 Let x be a vertex of a graph G such that $\deg_G(x) \geq n(G)/2$ and $G - \{x\}$ is Hamiltonian. Then there are $(2\deg_G(x) - n(G) + 1)$ mutually independent Hamiltonian cycles of G beginning at x.

Proof: Assume that $C = \langle v_1, v_2, \ldots, v_{n(G)-1}, v_1 \rangle$ is a Hamiltonian cycle of $G - \{x\}$.

Suppose that $\deg_G(x) = n(G) - 1$. We set $C_i = \langle x, v_i, v_{i+1}, \ldots, v_{n(G)-1}, v_1, v_2, \ldots, v_{i-1}, x \rangle$ for every $1 \leq i \leq n(G) - 1$. Then $\{C_1, C_2, \ldots, C_{n(G)-1}\}$ forms a set of $(n(G) - 1)$ mutually independent Hamiltonian cycles of G beginning at x. Note that $n(G) - 1 = 2\deg_G(x) - n(G) + 1$.

Suppose that $\deg_G(x) \leq n(G) - 2$. Without loss of generality, we may assume that $(x, v_1) \in E(G)$ and $(x, v_{n(G)-1}) \notin E(G)$. Let $S = \{v_i \mid (x, v_i) \in E(G)$ and $(x, v_{i+1}) \in E(G)$ for $1 \leq i \leq n(G) - 2\}$, and let $H = \{v_i \mid (x, v_i) \in E(G)$ and $(x, v_{i+1}) \notin E(G)$ for $1 \leq i \leq n(G) - 2\}$. We have $|H| = \deg_G(x) - |S|$.

Suppose that $|S| \leq 2\deg_G(x) - n(G)$. Then we have

$$\begin{aligned} n(G) &\geq |S| + 2|H| + 1 \\ &= |S| + 2(\deg_G(x) - |S|) + 1 \\ &= 2\deg_G(x) - |S| + 1 \\ &\geq 2\deg_G(x) - (2\deg_G(x) - n(G)) + 1 \\ &= n(G) + 1 \end{aligned}$$

We obtain a contradiction. Thus, $|S| \geq 2\deg_G(x) - n(G) + 1$.

We set $C_i = \langle x, v_i, v_{i+1}, \ldots, v_{n(G)-1}, v_1, v_2, \ldots, v_{i-1}, x \rangle$ for every $v_i \in S$. Then $\{C_i \mid$ for every $v_i \in S\}$ forms a set of $|S|$ mutually independent Hamiltonian cycles of G beginning at x. Therefore, we obtain at least $2\deg_G(x) - n(G) + 1$ mutually independent Hamiltonian cycles beginning at x. □

With Lemma 18.2, we have the following theorem.

THEOREM 18.1 Assume that G is a graph with $\delta(G) \geq n(G)/2$. Then $IHC(G) \geq 2\delta(G) - n(G) + 1$.

Proof: Since $\delta(G) \geq n(G)/2, n(G) \geq 3$.

Case 1. $n(G) = 3$. Then $G = K_3$. Obviously, $IHC(G) = 2 = 2\delta(G) - n(G) + 1$.

Case 2. $n(G) = 2k$ for some positive integer k with $k \geq 2$.

Suppose that $\delta(G) = n(G)/2$. By Dirac's Theorem, G is Hamiltonian. Thus, $IHC(G) \geq 1$.

Suppose that $\delta(G) \geq (n(G)/2) + 1$. We have $n(G) \geq 4$. Let x be an arbitrary vertex of G. Obviously, $\delta(G - \{x\}) \geq \delta(G) - 1 \geq n(G)/2 > n(G - \{x\})/2$ and $n(G - \{x\}) = n(G) - 1 \geq 3$. By Dirac's Theorem, $G - \{x\}$ is Hamiltonian. Then by Lemma 18.2, there exist $(2\deg_G(x) - n(G) + 1)$ mutually independent Hamiltonian cycles of G beginning at x. Since $\deg_G(x) \geq \delta(G)$, $IHC(G) \geq 2\delta(G) - n(G) + 1$.

Case 3. $n(G) = 2k + 1$ for some positive integer k with $k \geq 2$.

Obviously, $n(G) \geq 5$. Let x be an arbitrary vertex of G. Obviously, $\delta(G - \{x\}) \geq \delta(G) - 1 \geq k = n(G - \{x\})/2$ and $n(G - \{x\}) = n(G) - 1 \geq 4$. By Dirac's Theorem, $G - \{x\}$ is Hamiltonian. Then by Lemma 18.2, there exist $(2\deg_G(x) - n(G) + 1)$ mutually independent Hamiltonian cycles of G beginning at x. Since $\deg_G(x) \geq \delta(G)$, $IHC(G) \geq 2\delta(G) - n(G) + 1$.

The theorem is proved. □

In the following sections, we present some graphs that meet the bound mentioned previously.

THEOREM 18.2 $IHC(K_n) = n - 1$ if $n \geq 3$.

Proof: By Lemma 18.1, $IHC(K_n) \leq n - 1$. By Theorem 18.1, $IHC(K_n) \geq n - 1$. Thus, $IHC(K_n) = n - 1$. □

THEOREM 18.3 $IHC(G) = n(G) - 3$ if G is a graph with $\delta(G) = n(G) - 2 \geq 4$.

Proof: By Theorem 18.1, $IHC(G) \geq n(G) - 3$. Thus, we only need to show that $IHC(G) \leq n(G) - 3$.

Let x be any vertex of G with $\deg_G(x) = n(G) - 2$. Let $\{C_1, C_2, \ldots, C_r\}$ be a set of r mutually independent Hamiltonian cycles beginning at x. We may write $C_i = \langle x = v_1^i, v_2^i, \ldots, v_{n(G)}^i, x\rangle$ for every $1 \leq i \leq r$. Since $\deg_G(x) = n(G) - 2$, there is exactly one vertex y with $(x, y) \notin E(G)$. Let i be any index with $1 \leq i \leq r$. Obviously, $y \notin \{v_1^i, v_2^i, v_{n(G)}^i\}$. Thus, $y = v_i^{i(y)}$ for some $i(y)$ with $3 \leq i(y) < n(G)$. Since $i(y) \neq j(y)$ for any $1 \leq i < j \leq r$, $r \leq n(G) - 3$. Thus, $IHC(G) \leq n(G) - 3$.

The theorem is proved. □

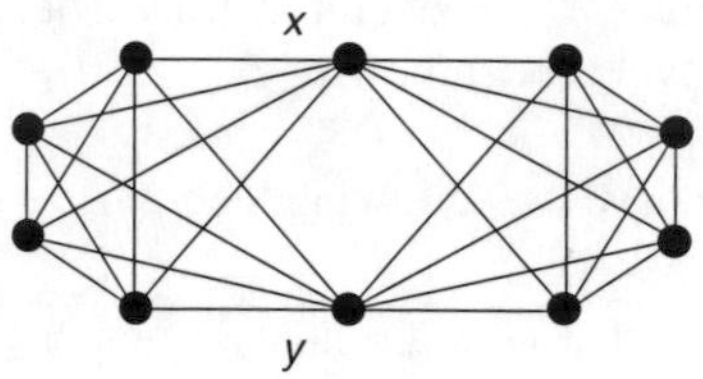

FIGURE 18.1 Illustration for $H(4)$.

Let G and H be two graphs. We use G^c to denote the complement of G, use $G+H$ to denote the disjoint union of G and H, and use $G \vee H$ to denote the graph obtained from $G+H$ by joining each vertex of G to each vertex of H.

Let m be a positive integer. We use $H(m)$ to denote the graph $K_2^c \vee (K_m + K_m)$. We illustrate $H(4)$ in Figure 18.1.

THEOREM 18.4 $IHC(H(m)) = 1$ for any positive integer m.

Proof: Obviously, $n(H(m)) = 2m+2$ and $\delta(H(m)) = m+1 = n(H(m))/2$. By Theorem 18.1, $IHC(H(m)) \geq 1$.

Let x and y be the vertices in $H(m)$ correspond to K_2^c. Let $C = \langle x = u_1, u_2, \ldots, u_{2m+2}, x \rangle$ be any Hamiltonian cycle of $H(m)$ beginning at x. It is easy to see that $u_{m+2} = y$. Thus, beginning at x, there does not exist any other Hamiltonian cycle of $H(m)$ independent with C. Thus, $IHC(H(m)) = 1$. □

18.3 MUTUALLY INDEPENDENT HAMILTONIAN CYCLES OF HYPERCUBES

In Ref. 291, Sun et al. study the mutually independent Hamiltonicity for hypercubes. We need the following lemmas.

LEMMA 18.3 Assume that $1 \leq n \leq 3$. Let $\{\mathbf{b_i} \mid 1 \leq i \leq n\}$ be any n-sets of black nodes of Q_n and $\{\mathbf{w_i} \mid 1 \leq i \leq n\}$ be any n-sets of white nodes of Q_n such that $(\mathbf{b_i}, \mathbf{w_i})$ is an edge of Q_n. Then there exist n Hamiltonian paths $P_1, \ldots, P_n$ of Q_n such that (1) P_i joins from $\mathbf{b_i}$ to $\mathbf{w_i}$ for $1 \leq i \leq n$ and (2) $P_i(k) \neq P_j(k)$ for $1 \leq k \leq 2^n$ and $1 \leq i < j \leq n$.

Proof: It is obvious that the theorem holds for $n = 1$. Assume that $n = 2$. Without loss of generality, we may assume that $(\mathbf{b_1}, \mathbf{w_1}) = (10, 00)$ and $(\mathbf{b_2}, \mathbf{w_2}) = (01, 11)$. Then $\langle 10, 11, 01, 00 \rangle$ and $\langle 01, 00, 10, 11 \rangle$ are the desired paths.

Assume that $n = 3$. Let $t_j = |\{(\mathbf{b_i}, \mathbf{w_i}) \mid (\mathbf{b_i}, \mathbf{w_i})$ is a j-dimensional edge with $1 \leq i \leq 3\}|$. Without loss of generality, we assume that $t_1 \geq t_2 \geq t_3$.

Suppose that $t_3 = 0$. Let $r_j = |\{\mathbf{b_i} \mid \mathbf{b_i} \in Q_3^j,\ 1 \leq i \leq 3\}|$ for $j = 0, 1$. Without loss of generality, we assume that $r_0 \geq r_1$. Obviously, $r_0 + r_1 = 3$. Since there are only four nodes in Q_2, we may assume that $\{(\mathbf{b_i}, \mathbf{w_i}) \mid 1 \leq i \leq 2\} \subset Q_3^0$ and $(\mathbf{b_3}, \mathbf{w_3}) \subset Q_3^1$. Since the theorem holds for $n = 2$, there exist two Hamiltonian paths P_1 and P_2 of Q_3^0 such that (1) P_i joins from $\mathbf{b_i}$ to $\mathbf{w_i}$ for $1 \leq i \leq 2$ and (2) $P_1(k) \neq P_2(k)$ for $1 \leq k \leq 2^2$.

By Theorem 13.1, there exists a Hamiltonian path P_3 of Q_3^1 from $\mathbf{b_3}$ to $\mathbf{w_3}$. Obviously, $\{P_i | 1 \leq i \leq 3\}$ forms the desired paths.

Now, we assume that $t_3 \neq 0$. Obviously, $t_1 = t_2 = t_3 = 1$. By the symmetric property of Q_3, we may assume that $\mathbf{b_1} = 100$ and $\mathbf{w_1} = 000$. Then $\mathbf{b_2} \in \{001, 111\}$.

Case 1. $\mathbf{b_2} = 001$. Then $\mathbf{w_2} = 011$. Moreover, $\mathbf{b_3} = 111$ and $\mathbf{w_3} = 110$. The required Hamiltonian paths are $\langle 100, 101, 111, 110, 010, 011, 001, 000 \rangle$, $\langle 001, 000, 100, 101, 111, 110, 010, 011 \rangle$, and $\langle 111, 011, 010, 000, 001, 101, 100, 110 \rangle$.

Case 2. $\mathbf{b_2} = 111$. Then $\mathbf{w_2} = 101$. Moreover, $\mathbf{b_3} = 010$ and $\mathbf{w_3} = 011$. The required Hamiltonian paths are $\langle 100, 101, 111, 110, 010, 011, 001, 000 \rangle$, $\langle 111, 110, 010, 011, 001, 000, 100, 101 \rangle$, and $\langle 010, 000, 001, 101, 100, 110, 111, 011 \rangle$. □

LEMMA 18.4 *$IHC(Q_n) = n - 1$ if $n \in \{1, 2, 3\}$.*

Proof: Let $n = 1$. Obviously, there is no Hamiltonian cycle in Q_1. Thus, $IHC(Q_1) = 0$.

Let $n = 2$. Since Q_2 is Hamiltonian, $IHC(Q_2) \geq 1$. Moreover, it is easy to see that there is no two mutually independent Hamiltonian cycles beginning at any node. Thus, $IHC(Q_2) = 1$.

Let $n = 3$. Obviously, Q_3 is Hamiltonian. Let $C = \langle 000 = \mathbf{v_1}, \mathbf{v_2}, \ldots, \mathbf{v_8}, \mathbf{v_1} \rangle$ be any Hamiltonian cycle beginning at 000. Obviously, 111 is either $\mathbf{v_4}$ or $\mathbf{v_6}$. Thus, $IHC(Q_3) \leq 2$. However, there are two mutually independent Hamiltonian cycles beginning at 000; namely $\langle 000, 100, 101, 111, 110, 010, 011, 001, 000 \rangle$ and $\langle 000, 010, 011, 001, 101, 111, 110, 100, 000 \rangle$. By the symmetric property of Q_n, $IHC(Q_3) \geq 2$. Therefore, $IHC(Q_3) = 2$. □

LEMMA 18.5 *$Q_4 - \{\mathbf{x}, \mathbf{y}\}$ is Hamiltonian laceable if $\mathbf{x}$ and $\mathbf{y}$ are any two nodes from different partite sets of Q_4.*

Proof: Without loss of generality, we can assume that $\mathbf{x} = 0000$. Then the Hamming weight of $\mathbf{y}$, $w(\mathbf{y})$, is either 1 or 3. Thus, we can assume that $\mathbf{y} = 1000$ or 1110.

Case 1. $\mathbf{y} = 1000$. Obviously, $\langle 0010, 1010, 1110, 1100, 0100, 0110, 0010 \rangle$ forms a Hamiltonian cycle C of $Q_4^0 - \{\mathbf{x}, \mathbf{y}\}$. Let $\mathbf{u}$ and $\mathbf{v}$ be any two nodes from different partite sets of $Q_4 - \{\mathbf{x}, \mathbf{y}\}$. Without loss of generality, we assume that $\mathbf{u}$ is a white node.

Case 1.1. $|\{\mathbf{u}, \mathbf{v}\} \cap Q_n^0| = 2$. Write C beginning at $\mathbf{u}$ as $\langle \mathbf{u} = \mathbf{x_1}, \mathbf{x_2}, \ldots, \mathbf{x_6}, \mathbf{x_1} \rangle$. Since C can be traversed backward and forward, $\mathbf{v}$ is either $\mathbf{x_2}$ or $\mathbf{x_4}$. By Theorem 13.1, there exists a Hamiltonian path P_1 of Q_n^1 joining $(\mathbf{x_3})^4$ to $(\mathbf{x_2})^4$, and there exists a Hamiltonian path P_2 of Q_n^1 joining $(\mathbf{x_3})^4$ to $(\mathbf{x_6})^4$ in Q_n^1. Then $\langle \mathbf{u} = \mathbf{x_1}, \mathbf{x_6}, \mathbf{x_5}, \mathbf{x_4}, \mathbf{x_3}, (\mathbf{x_3})^4, P_1, (\mathbf{x_2})^4, \mathbf{x_2} \rangle$ is a Hamiltonian path of $Q_4 - \{\mathbf{x}, \mathbf{y}\}$ from $\mathbf{u}$ to $\mathbf{x_2}$ and $\langle \mathbf{u} = \mathbf{x_1}, \mathbf{x_2}, \mathbf{x_3}, (\mathbf{x_3})^4, P_2, (\mathbf{x_6})^4, \mathbf{x_6}, \mathbf{x_5}, \mathbf{x_4} \rangle$ is a Hamiltonian path of $Q_4 - \{\mathbf{x}, \mathbf{y}\}$ from $\mathbf{u}$ to $\mathbf{x_4}$.

Case 1.2. $|\{\mathbf{u}, \mathbf{v}\} \cap Q_n^0| = 1$. Without loss of generality, we assume that $\mathbf{u} \in Q_n^0$. Write C as $\langle \mathbf{u} = \mathbf{x_1}, \mathbf{x_2}, \ldots, \mathbf{x_6}, \mathbf{x_1} \rangle$. By Theorem 13.1, there exists a Hamiltonian path P of Q_n^1 from $(\mathbf{x_6})^4$ to $\mathbf{v}$. Then $\langle \mathbf{u}, \mathbf{x_2}, \mathbf{x_3}, \ldots, \mathbf{x_6}, (\mathbf{x_6})^4, P, \mathbf{v} \rangle$ forms the desired path.

Case 1.3. $|\{\mathbf{u},\mathbf{v}\}\cap Q_n^0|=0$. Thus, **u** and **v** are in Q_n^1. Suppose that $\mathbf{u}\neq 1001$. Write C as $\langle \mathbf{u}^4=\mathbf{x}_1,\mathbf{x}_2,\dots,\mathbf{x}_6,\mathbf{x}_1\rangle$. Since C can be traversed backward and forward, we assume that $(\mathbf{x}_6)^4\neq \mathbf{v}$. By Theorem 13.1, there exists a Hamiltonian path P of $Q_4^1-\{\mathbf{u}\}$ joining $(\mathbf{x}_6)^4$ to **v**. Then $\langle \mathbf{u},(\mathbf{u})^4,\mathbf{x}_2,\dots,\mathbf{x}_6,(\mathbf{x}_6)^4,P,\mathbf{v}\rangle$ forms the desired path. Suppose that $\mathbf{v}\neq 0001$. Write C as $\langle \mathbf{v}^4=\mathbf{x}_1,\mathbf{x}_2,\dots,\mathbf{x}_6,\mathbf{x}_1\rangle$. Since C can be traversed backward and forward, we can assume that $(\mathbf{x}_6)^4\neq \mathbf{u}$. By Theorem 13.1, there exists a Hamiltonian path S of $Q_4^1-\{\mathbf{v}\}$ joining $(\mathbf{x}_6)^4$ to **u**. Then $\langle \mathbf{v},\mathbf{v}^4,\mathbf{x}_2,\dots,\mathbf{x}_6,(\mathbf{x}_6)^4,S,\mathbf{u}\rangle$ forms the desired path. Finally, we assume that $\mathbf{v}=0001$ and $\mathbf{u}=1001$. Then $\langle 0001,0101,1101,1111,0111,0011,0010,0110,0100,1100,1110,1010,1011,1001\rangle$ forms the desired path.

Case 2. $\mathbf{y}=1110$. Obviously, $\langle 0100,0110,0010,1010,1000,1100,0100\rangle$ forms a Hamiltonian cycle C of $Q_4^0-\{\mathbf{x},\mathbf{y}\}$. We can use the same technique as in Case 1 to prove that $Q_4^0-\{\mathbf{x},\mathbf{y}\}$ is Hamiltonian laceable. □

LEMMA 18.6 $Q_n-\{\mathbf{x},\mathbf{y}\}$ is Hamiltonian laceable if **x** and **y** are any two nodes from different partite sets of Q_n with $n\geq 4$.

Proof: We prove this lemma by induction. The induction basis is proved in Lemma 18.5. Assume that this lemma is true for Q_k with $4\leq k<n$. By the symmetric property of Q_n, we can assume that $\mathbf{x}=\mathbf{e}$. Thus, $w(\mathbf{y})$ is odd. Let **u** and **v** be any two nodes from different partite sets of $Q_n-\{\mathbf{x},\mathbf{y}\}$. Without loss of generality, we may assume that **u** is a white node.

Case 1. $w(\mathbf{y})<n$. Without loss of generality, we assume that $(\mathbf{y})_n=0$. Obviously, $\{\mathbf{x},\mathbf{y}\}\subset Q_n^0$.

Case 1.1. $|\{\mathbf{u},\mathbf{v}\}\cap Q_n^0|=2$. By induction, there exists a Hamiltonian path P of $Q_n-\{\mathbf{x},\mathbf{y}\}$ joining **u** to **v**. Write P as $\langle \mathbf{u}=\mathbf{y}_1,\mathbf{y}_2,\dots,\mathbf{y}_{2^{n-1}-2}=\mathbf{v}\rangle$. By Theorem 13.1, there exists a Hamiltonian path S joining $\mathbf{u}^n$ to $(\mathbf{y}_2)^n$. Then $\langle \mathbf{u},\mathbf{u}^n,S,(\mathbf{y}_2)^n,\mathbf{y}_2,\mathbf{y}_3,\dots,\mathbf{y}_{2^{n-1}-2}=\mathbf{v}\rangle$ forms the desired path.

Case 1.2. $|\{\mathbf{u},\mathbf{v}\}\cap Q_n^0|=1$. Without loss of generality, we assume that $\mathbf{u}\in Q_n^0$. Since there are 2^{n-2} black nodes in Q_n^0, we can choose a black node **r** of Q_n^0 such that $\mathbf{r}\neq\mathbf{y}$. By induction, there exists a Hamiltonian path P of $Q_n^0-\{\mathbf{x},\mathbf{y}\}$ joining **u** to **r**. Since every hypercube is Hamiltonian laceable, there exists a Hamiltonian path S of Q_n^1 joining $\mathbf{r}^n$ to **v**. Then $\langle \mathbf{u},P,\mathbf{r},\mathbf{r}^n,S,\mathbf{v}\rangle$ forms the desired path.

Case 1.3. $|\{\mathbf{u},\mathbf{v}\}\cap Q_n^0|=0$. Thus, **u** and **v** are in Q_n^1. Since every hypercube is Hamiltonian laceable, there exists a Hamiltonian path P of Q_n^1 between **u** and **v**. We write P as $\langle \mathbf{u}=\mathbf{x}_1,\mathbf{x}_2,\dots,\mathbf{x}_{2^{n-1}}=\mathbf{v}\rangle$. Obviously, there exists some index i with $1\leq i<2^{n-1}$ such that $\{(\mathbf{x}_i)^n,(\mathbf{x}_{i+1})^n\}\cap\{\mathbf{x},\mathbf{y}\}=\emptyset$. By induction, there exists a Hamiltonian path S of $Q_n^0-\{\mathbf{x},\mathbf{y}\}$ between $(\mathbf{x}_i)^n$ and $(\mathbf{x}_{i+1})^n$. Then $\langle \mathbf{u}=\mathbf{x}_1,\mathbf{x}_2,\dots,\mathbf{x}_i,(\mathbf{x}_i)^n,S,(\mathbf{x}_{i+1})^n,\mathbf{x}_{i+1},\mathbf{x}_{i+2},\dots,\mathbf{x}_{2^{n-1}-2}=\mathbf{v}\rangle$ forms the desired path.

Case 2. $w(\mathbf{y})=n$. Since $w(\mathbf{y})$ is odd, n is odd. Moreover, $w(\mathbf{v})<n$ because $\mathbf{y}\neq\mathbf{v}$. Without loss of generality, we assume that $(\mathbf{v})_n=0$. Therefore, $\mathbf{v}\in Q_n^0$ and $\mathbf{y}\in Q_n^1$.

Case 2.1. $\mathbf{u} \in Q_n^0$. Since every hypercube is hyper Hamiltonian laceable, there exists a Hamiltonian path P of $Q_n^0 - \{\mathbf{v}\}$ joining $\mathbf{u}$ to $\mathbf{e}$. Write P as $\langle \mathbf{u}, S, \mathbf{r}, \mathbf{e} \rangle$. Similarly, there exists a Hamiltonian path R of $Q_n^1 - \{\mathbf{y}\}$ joining $\mathbf{r}^n$ to $\mathbf{v}^n$. Then $\langle \mathbf{u}, S, \mathbf{r}, \mathbf{r}^n, R, \mathbf{v}^n, \mathbf{v} \rangle$ forms the desired path.

Case 2.2. $\mathbf{u} \notin Q_n^0$. Since there are 2^{n-2} black nodes in Q_n^0, we can choose a black node $\mathbf{r}$ of Q_n^0 such that $\mathbf{r} \neq \mathbf{v}$. Since every hypercube is hyper Hamiltonian laceable, there exists a Hamiltonian path P of $Q_n^0 - \{\mathbf{e}\}$ joining $\mathbf{r}$ to $\mathbf{v}$. Similarly, there exists a Hamiltonian path S of $Q_n^0 - \{\mathbf{y}\}$ joining $\mathbf{u}$ to $\mathbf{r}^n$. Then $\langle \mathbf{u}, S, \mathbf{r}^n, \mathbf{r}, P, \mathbf{v} \rangle$ forms the desired path. □

LEMMA 18.7 Let $\{\mathbf{b_i} \mid 1 \leq i \leq n-1\}$ be any $(n-1)$-sets of black nodes of Q_n and $\{\mathbf{w_i} \mid 1 \leq i \leq n-1\}$ be any $(n-1)$-sets of white nodes of Q_n such that $(\mathbf{b_i}, \mathbf{w_i})$ is an edge of Q_n. Then there exist $(n-1)$ Hamiltonian paths $P_1, \ldots, P_{n-1}$ of Q_n such that (1) P_i joins from $\mathbf{b_i}$ to $\mathbf{w_i}$ for $1 \leq i \leq n-1$ and (2) $P_i(k) \neq P_j(k)$ for $1 \leq k \leq 2^n$ and $1 \leq i < j \leq n-1$.

Proof: We prove this lemma by induction. The induction basis are $n = 1, 2, 3$, and 4. With Lemma 18.3, the lemma is true for $n = 1, 2$, and 3. Suppose that $n = 4$. Let $t_j = |\{(\mathbf{b_i}, \mathbf{w_i}) \mid (\mathbf{b_i}, \mathbf{w_i})$ be a j-dimensional edge with $1 \leq i \leq 3\}|$. Without loss of generality, we assume that $t_1 \geq t_2 \geq t_3 \geq t_4$. Obviously, $t_4 = 0$.

Let $r_j = |\{\mathbf{b_i} \mid \mathbf{b_i} \in Q_4^j,\ 1 \leq i \leq 3\}|$ for $j = 0, 1$. Without loss of generality, we may assume that $r_0 \geq r_1$. Obviously, $r_0 + r_1 = 3$. Suppose that $r_0 < 3$. Then $r_0 = 2$ and $r_1 = 1$. Without loss of generality, we may assume that $\{(\mathbf{b_i}, \mathbf{w_i}) \mid 1 \leq i \leq 2\} \subset Q_4^0$ and $(\mathbf{b_3}, \mathbf{w_3}) \subset Q_4^1$. By Lemma 18.3, there exist two Hamiltonian paths P_1 and P_2 of Q_4^0 such that (1) P_i joins from $\mathbf{b_i}$ to $\mathbf{w_i}$ for $1 \leq i \leq 2$ and (2) $P_1(k) \neq P_2(k)$ for $1 \leq k \leq 2^3$. By Theorem 13.1, there exists a Hamiltonian path P_3 of Q_4^1. Obviously, $\{\widetilde{P}_i \mid 1 \leq i \leq 3\}$ forms a set of the desired paths.

Suppose that $r_0 = 3$. Thus, $\{(\mathbf{b_i}, \mathbf{w_i}) \mid 1 \leq i \leq 3\} \subset Q_4^0$. By Lemma 18.5, there exist three Hamiltonian paths P_1, P_2, and P_3 of Q_4^0 such that (1) P_i joins from $\mathbf{b_i}$ to $\mathbf{w_i}$ for $1 \leq i \leq 3$ and (2) $P_i(k) \neq P_j(k)$ for $1 \leq i < j \leq 3$ and $1 \leq k \leq 2^3$. Then $\{\widetilde{P}_i \mid 1 \leq i \leq 3\}$ forms a set of the desired paths.

Suppose that the lemma is true for Q_k with $4 \leq k < n$. Let $t_j = |\{(\mathbf{b_i}, \mathbf{w_i}) \mid (\mathbf{b_i}, \mathbf{w_i})$ be a j-dimensional edge with $1 \leq i \leq n-1\}|$. Without loss of generality, we assume that $t_1 \geq t_2 \cdots \geq t_n$.

Obviously, $t_n = 0$. Let $r_j = |\{\mathbf{b_i} \mid \mathbf{b_i} \in Q_n^j,\ 1 \leq i \leq n-1\}|$ for $j = 0, 1$. Obviously, $r_0 + r_1 = n-1$. Without loss of generality, we may assume that $r_0 \geq r_1$. Suppose that $r_0 < n-1$. Without loss of generality, we may assume that $\{(\mathbf{b_i}, \mathbf{w_i}) \mid 1 \leq i \leq r_0\} \subset Q_n^0$ and $\{(\mathbf{b_i}, \mathbf{w_i}) \mid r_0 + 1 \leq i \leq n-1\} \subset Q_n^1$. By induction, there exist r_0 Hamiltonian paths $P_1, P_2, \ldots, P_{r_0}$ of Q_n^0 such that (1) P_i joins from $\mathbf{b_i}$ to $\mathbf{w_i}$ for $1 \leq i \leq r_0$ and (2) $P_i(k) \neq P_j(k)$ for $1 \leq i < j \leq r_0$ and $1 \leq k \leq 2^{n-1}$. Similarly, there exist $(n - r_0 - 1)$ Hamiltonian paths $P_{r_0+1}, P_{r_0+2}, \ldots, P_{n-1}$ of Q_n^1 such that (1) P_i joins from $\mathbf{b_i}$ to $\mathbf{w_i}$ for $r_0 + 1 \leq i \leq n-1$ and (2) $P_i(k) \neq P_j(k)$ for $r_0 + 1 \leq i < j \leq n-1$ and $1 \leq k \leq 2^{n-1}$. Obviously, $\{\widetilde{P}_i \mid 1 \leq i \leq n-1\}$ forms a set of desired paths.

Suppose that $r_0 = n-1$. Thus, $\{(\mathbf{b_i}, \mathbf{w_i}) \mid 1 \leq i \leq n-1\} \subset Q_n^0$. Note that the complete bipartite graph $K_{m,m}$ is not a subgraph of Q_m for $m \geq 3$. Thus, the subgraph

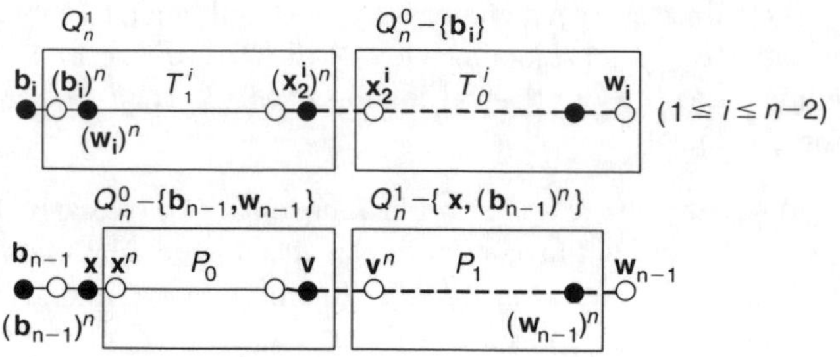

FIGURE 18.2 Illustration of Lemma 18.7.

induced by $\{(\mathbf{b_i})^n \mid 1 \le i \le n-1\} \cup \{(\mathbf{w_i})^n \mid 1 \le i \le n-1\}$ is not a complete bipartite graph. Therefore, there exists some index j and a neighbor $\mathbf{x}$ of $(\mathbf{b_j})^n$ such that $\mathbf{x}$ is not in $\{(\mathbf{w_i})^n \mid 1 \le i \le n-1\}$. Without loss of generality, we assume that $j = n-1$. By induction, there exist $(n-2)$ Hamiltonian paths $R_1, R_2, \ldots, R_{n-2}$ of Q_n^0 such that (1) P_i joins from $\mathbf{b_i}$ to $\mathbf{w_i}$ for $1 \le i \le n-2$ and (2) $R_i(k) \ne R_j(k)$ for $1 \le i < j \le n-2$ and $1 \le k \le 2^{n-1}$. Write R_i as $\langle \mathbf{b_i} = \mathbf{x_1^i}, \mathbf{x_2^i}, \ldots, \mathbf{x_{2^{n-1}}^i} = \mathbf{w_i} \rangle$. For $1 \le i \le n-2$, let $T_1^i = \langle (\mathbf{w_i})^n = (\mathbf{x_{2^{n-1}}^i})^n, (\mathbf{x_{2^{n-1}-1}^i})^n, \ldots, (\mathbf{x_2^i})^n \rangle$ and $T_0^i = \langle \mathbf{x_2^i}, \mathbf{x_3^i}, \ldots, \mathbf{x_{2^{n-1}}^i} = \mathbf{w_i} \rangle$.

For $1 \le i \le n-2$, we set S_i as $\langle \mathbf{b_i}, (\mathbf{b_i})^n, (\mathbf{w_i})^n, T_1^i, (\mathbf{x_2^i})^n, \mathbf{x_2^i}, T_0^i, \mathbf{w_i} \rangle$. Let $\mathbf{v}$ be any black node of Q_n^0 such that $\mathbf{v} \ne \mathbf{b_{n-1}}$. By Lemma 18.6, there exists a Hamiltonian path P_0 of $Q_n^0 - \{\mathbf{b_{n-1}}, \mathbf{w_{n-1}}\}$ from $\mathbf{x}^n$ to $\mathbf{v}$ and there exists a Hamiltonian path P_1 of $Q_n^1 - \{(\mathbf{b_{n-1}})^n, \mathbf{x}\}$ from $\mathbf{v}^n$ to $(\mathbf{w_{n-1}})^n$. Set S_{n-1} as $\langle \mathbf{b_{n-1}}, (\mathbf{b_{n-1}})^n, \mathbf{x}, \mathbf{x}^n, P_0, \mathbf{v}, \mathbf{v}^n, P_1, (\mathbf{w_{n-1}})^n, \mathbf{w_{n-1}} \rangle$. Then $\{S_i \mid 1 \le i \le n-1\}$ forms a set of desired paths. See Figure 18.2 for illustration. □

THEOREM 18.5 $IHC(Q_n) = n-1$ if $n \in \{1, 2, 3\}$ and $IHC(Q_n) = n$ if $n \ge 4$.

Proof: By Lemma 18.4, $IHC(Q_n) = n-1$ if $n \in \{1, 2, 3\}$. Now, we prove that $IHC(Q_n) = n$ for $n \ge 4$ by induction. We need to construct n mutually independent Hamiltonian cycles of Q_n beginning at any node $\mathbf{u}$. Without loss of generality, we assume that $\mathbf{u} = \mathbf{e}$.

Let $n = 4$. The corresponding mutually independent Hamiltonian cycles beginning at $\mathbf{e}$ are:

0000, 0010, 0110, 0100, 1100, 1000, 1010, 1110, 1111, 1101, 1001, 1011, 0011, 0111, 0101, 0001, 0000
0000, 0100, 0101, 0001, 0011, 0111, 0110, 0010, 1010, 1110, 1100, 1101, 1111, 1011, 1001, 1000, 0000
0000, 0001, 1001, 1011, 1111, 1101, 0101, 0111, 0011, 0010, 0110, 1110, 1010, 1000, 1100, 0100, 0000
0000, 1000, 1100, 1110, 1010, 1011, 1111, 1101, 1001, 0001, 0011, 0111, 0101, 0100, 0110, 0010, 0000

Let $n = 5$. The corresponding mutually independent Hamiltonian cycles beginning at $\mathbf{e}$ are:

00000, 00001, 01001, 01101, 01100, 01000, 01010, 01011, 01111, 01110, 11110, 11111, 11011, 11010, 11000, 11100, 11101, 11001, 10001, 10000, 10010, 10011, 10111, 10110, 10100, 10101, 00101, 00111, 00011, 00010, 00110, 00100
00000, 00010, 10010, 10110, 10100, 10000, 10001, 10011, 10111, 10101, 11101, 11001, 11000, 11100, 11110, 11010, 11011, 11111, 01111, 01101, 01001, 01011, 01010, 01000, 01100, 01110, 00110, 00100, 00101, 00111, 00011, 00001

00000, 00100, 00101, 00001, 00011, 00111, 00110, 01110, 01010, 01000, 01100, 01101, 01111, 01011, 01001, 11001, 11000, 11010, 11110, 11100, 11101, 11111, 11011, 10011, 10111, 10110, 10100, 10101, 10001, 10000, 10010, 00010
00000, 10000, 10001, 10101, 10111, 10011, 10010, 10110, 10100, 00100, 00101, 00111, 00110, 00010, 00011, 00001, 01001, 01011, 01010, 01110, 01111, 01101, 01100, 11100, 11101, 11111, 11110, 11010, 11011, 11001, 11000, 01000
00000, 01000, 11000, 11100, 11110, 11111, 11101, 11001, 11011, 11010, 10010, 10011, 10001, 10101, 10111, 10110, 00110, 00010, 00011, 00111, 00101, 00001, 01001, 01101, 01111, 01011, 01010, 01110, 01100, 00100, 10100, 10000

Thus, we assume that $IHC(Q_k) = n$ with $5 \leq k < n$. By induction, there exist $(n-2)$ mutually independent Hamiltonian cycles of Q_n^{00} beginning at $\mathbf{e}$. However, we pick any $(n-3)$ mutually independent Hamiltonian cycles $M_1, M_2, \ldots, M_{n-3}$ of Q_n^{00} beginning at $\mathbf{e}$. Write M_i as $\langle \mathbf{e}, P_{00}^i, \mathbf{a_i}, \mathbf{e}^i, \mathbf{e} \rangle$ for $1 \leq i \leq n-3$. By Lemma 18.7, there exist $(n-3)$ Hamiltonian paths $P_{01}^1, P_{01}^2, \ldots, P_{01}^{n-3}$ of Q_n^{01} such that (1) P_{01}^i joins from $(\mathbf{a_i})^n$ to $(\mathbf{e}^i)^n$ for $1 \leq i \leq n-3$ and (2) $P_{01}^i(k) \neq P_{01}^j(k)$ for $1 \leq i < j \leq n-3$ and $1 \leq k \leq 2^{n-2}$. Again, there exist $(n-3)$ Hamiltonian paths $\{P_{11}^1, P_{11}^2, \ldots, P_{11}^{n-3}\}$ of Q_n^{11} such that (1) P_{11}^i joins from $((\mathbf{e}^i)^n)^{n-1}$ to $((\mathbf{a_i})^n)^{n-1}$ for $1 \leq i \leq n-3$ and (2) $P_{11}^i(k) \neq P_{11}^j(k)$ for $1 \leq i < j \leq n-3$ and $1 \leq k \leq 2^{n-2}$. Moreover, there exist $(n-3)$ Hamiltonian paths $\{P_{10}^1, P_{10}^1, \ldots, P_{10}^{n-3}\}$ of Q_n^{10} such that (1) P_{10}^i joins from $(\mathbf{a_i})^{n-1}$ to $(\mathbf{e}^i)^{n-1}$ for $1 \leq i \leq n-3$ and (2) $P_{10}^i(k) \neq P_{10}^j(k)$ for $1 \leq i < j \leq n-3$ and $1 \leq k \leq 2^{n-2}$. For $1 \leq i \leq n-3$, we set C_ι as $\langle \mathbf{e}, P_{00}^i, \mathbf{a_i}, (\mathbf{a_i})^n, P_{01}^i, (\mathbf{e}^i)^n, ((\mathbf{e}^i)^n)^{n-1}, P_{11}^i, ((\mathbf{a_i})^n)^{n-1}, (\mathbf{a_i})^{n-1}, P_{10}^i, (\mathbf{e}^i)^{n-1}, \mathbf{e}^i, \mathbf{e} \rangle$.

Since there are 2^{n-2} nodes in Q_n^{01}, we can choose a path $\langle \mathbf{x_1}, \mathbf{y_1}, \mathbf{z_1} \rangle$ in Q_n^{01} such that (1) $\mathbf{y_1}$ is a black node not in $\{(\mathbf{a_i})^n \mid 1 \leq i \leq n-3\} \cup \{\mathbf{e}^n\}$ and (2) $\mathbf{z_1}$ is not in $\{P_{01}^i(2) \mid 1 \leq i \leq n-3\}$. Similarly, we can choose a path $\langle \mathbf{x_2}, \mathbf{y_2}, \mathbf{z_2} \rangle$ in Q_n^{11} such that (1) $\mathbf{y_2}$ is a black node not in $\{((\mathbf{e}^i)^n)^{n-1} \mid 1 \leq i \leq n-3\} \cup \{(\mathbf{z_1})^{n-1}\}$ and (2) $\mathbf{z_2}$ is not in $\{P_{11}^i(2) \mid 1 \leq i \leq n-3\}$. Moreover, we can choose a path $\langle \mathbf{x_3}, \mathbf{y_3}, \mathbf{z_3} \rangle$ in Q_n^{10} such that (1) $\mathbf{y_3}$ is a black node not in $\{(\mathbf{a_i})^{\mathbf{n-1}} \mid 1 \leq i \leq n-3\} \cup \{(\mathbf{z_2})^n\}$ and (2) $\mathbf{z_3}$ is not in $\{P_{10}^i(2) \mid 1 \leq i \leq n-3\} \cup \{(\mathbf{e}^{n-2})^{n-1}\}$. By Lemma 18.6, there exists a Hamiltonian path S_{01} of $Q_n^{01} - \{\mathbf{y_1}, \mathbf{z_1}\}$ from $\mathbf{e}^n$ to $\mathbf{x_1}$. Again, there exists a Hamiltonian path S_{11} of $Q_n^{11} - \{\mathbf{y_2}, \mathbf{z_2}\}$ from $(\mathbf{z_1})^{n-1}$ to $\mathbf{x_2}$. Moreover, there exists a Hamiltonian path S_{10} of $Q_n^{10} - \{\mathbf{y_3}, \mathbf{z_3}\}$ from $(\mathbf{z_2})^n$ to $\mathbf{x_3}$. By Theorem 13.1, there exists a Hamiltonian path S_{00} of $Q_n^{00} - \{\mathbf{e}\}$ from $(\mathbf{z_3})^{n-1}$ to $\mathbf{e}^{n-2}$. We set C_{n-2} as $\langle \mathbf{e}, \mathbf{e}^n, S_{01}, \mathbf{x_1}, \mathbf{y_1}, \mathbf{z_1}, (\mathbf{z_1})^{n-1}, S_{11}, \mathbf{x_2}, \mathbf{y_2}, \mathbf{z_2}, (\mathbf{z_2})^n, S_{10}, \mathbf{x_3}, \mathbf{y_3}, \mathbf{z_3}, (\mathbf{z_3})^{n-1}, S_{00}, \mathbf{e}^{n-2}, \mathbf{e} \rangle$.

Let $\mathbf{v}$ be the only node in $\{\mathbf{e}^i \mid 1 \leq i \leq n-2\} - \{P_{00}^i(2) \mid 1 \leq i \leq n-3\}$, let $\mathbf{r_1}$ be any white node of Q_n^{10} such that $\mathbf{r_1} \neq \mathbf{v}^{n-1}$, let $\mathbf{r_2}$ be any white node of Q_n^{00} such that $\mathbf{r_2} \neq \mathbf{e}$, and let $\mathbf{r_3}$ be any white node of Q_n^{01}. By Theorem 13.1, there exists a Hamiltonian path R_{10} of $Q_n^{10} - \{\mathbf{e}^{n-1}\}$ from $\mathbf{v}^{n-1}$ to $\mathbf{r_1}$. By Lemma 18.6, there exists a Hamiltonian path R_{00} of $Q_n^{00} - \{\mathbf{e}, \mathbf{v}\}$ from $(\mathbf{r_1})^{n-1}$ to $\mathbf{r_2}$. By Theorem 13.1, there exists a Hamiltonian path R_{01} of Q_n^{01} from $(\mathbf{r_2})^n$ to $\mathbf{r_3}$ and there exists a Hamiltonian path R_{11} of Q_n^{11} from $(\mathbf{r_3})^{n-1}$ to $(\mathbf{e}^{n-1})^n$. We set C_{n-1} as $\langle \mathbf{e}, \mathbf{v}, \mathbf{v}^{n-1}, R_{10}, \mathbf{r_1}, (\mathbf{r_1})^{n-1}, R_{00}, \mathbf{r_2}, (\mathbf{r_2})^n, R_{01}, \mathbf{r_3}, (\mathbf{r_3})^{n-1}, R_{11}, (\mathbf{e}^{n-1})^n, \mathbf{e}^{n-1}, \mathbf{e} \rangle$.

Let $\mathbf{q}$ be any black node of Q_n^{11} such that $\mathbf{q} \neq (\mathbf{z_1})^{n-1}$, let $\mathbf{s}$ be any white node of Q_n^{10} such that $\mathbf{s} \neq \mathbf{q}^n$, and let $\mathbf{w}$ be any black node of Q_n^{00} such that $\mathbf{w} \neq \mathbf{s}^{n-1}$. By Theorem 13.1, there exists a Hamiltonian path T_{11} of Q_n^{11} from $(\mathbf{e}^{n-1})^n$ to $\mathbf{q}$ and there exists a Hamiltonian path T_{01} of Q_n^{01} from $\mathbf{w}^n$ to $\mathbf{e}^n$. By Theorem 13.1, there exists a Hamiltonian path T_{10} of $Q_n^{10} - \{\mathbf{e}^{n-1}\}$ from $\mathbf{q}^n$ to $\mathbf{s}$

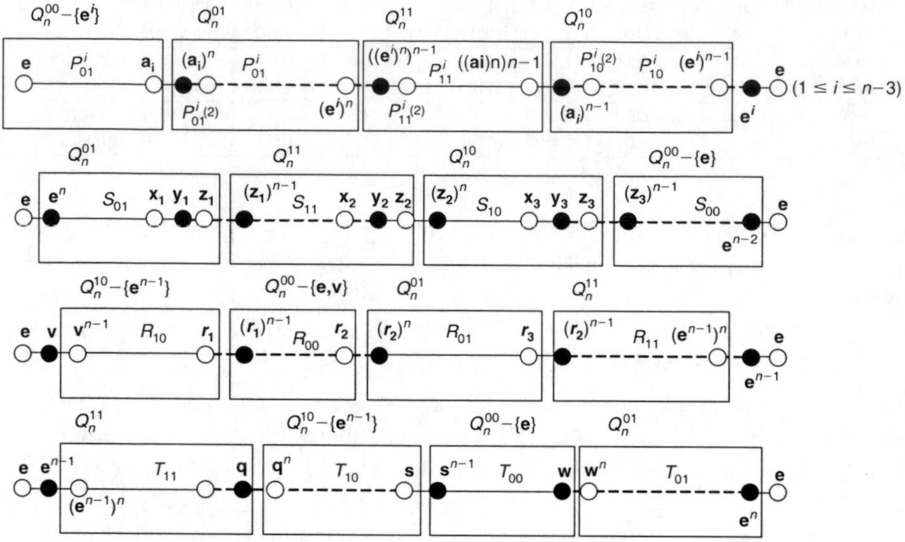

FIGURE 18.3 Illustration of Theorem 18.5.

and there exists a Hamiltonian path T_{00} of $Q_n^{00} - \{\mathbf{e}\}$ from $\mathbf{s}^{n-1}$ to $\mathbf{w}$. We set C_n as $\langle \mathbf{e}, \mathbf{e}^{n-1}, (\mathbf{e}^{n-1})^n, T_{11}, \mathbf{q}, \mathbf{q}^n, T_{10}, \mathbf{s}, \mathbf{s}^{n-1}, T_{00}, \mathbf{w}, \mathbf{w}^n, T_{01}, \mathbf{e}^n, \mathbf{e} \rangle$. Then $\{C_i \mid 1 \le i \le n\}$ forms a set of desired cycles. See Figure 18.3 for illustration. □

18.4 MUTUALLY INDEPENDENT HAMILTONIAN CYCLES OF PANCAKE GRAPHS

In Ref. 238, Lin et al. studied the mutually independent Hamiltonicity for pancake graphs.

LEMMA 18.8 Let $k \in \langle n \rangle$ with $n \ge 4$, and let $\mathbf{x}$ be a vertex of P_n. There is a Hamiltonian path P of $P_n - \{\mathbf{x}\}$ joining the vertex $(\mathbf{x})^n$ to some vertex $\mathbf{v}$ with $(\mathbf{v})_1 = k$.

Proof: Suppose that $n = 4$. Since P_4 is vertex-transitive, we may assume that $\mathbf{x} = 1234$. The required paths of $P_4 - \{1234\}$ are:

$k=1$	⟨4321, 3421, 2431, 4231, 1324, 3124, 2134, 4312, 1342, 2143, 4132, 2314, 3214, 4123, 2143, 3412, 1432, 2341, 3241, 1423, 2413, 4213, 1243⟩
$k=2$	⟨4321, 3421, 2431, 4231, 1324, 3124, 2134, 4312, 1342, 3142, 2413, 4213, 1243, 2143, 3412, 1432, 4132, 2314, 3214, 4123, 1423, 3241, 2341⟩
$k=3$	⟨4321, 3421, 2431, 4231, 1324, 3124, 2134, 4312, 1342, 3142, 4132, 2314, 3214, 4123, 1423, 2413, 4213, 1342, 2143, 3412, 1432, 2341, 3241⟩
$k=4$	⟨4321, 3421, 2431, 1342, 3142, 4132, 2314, 3214, 4123, 2143, 1243, 4213, 2413, 1423, 3241, 2341, 1432, 3412, 4312, 2134, 3124, 1324, 4231⟩

With Theorem 11.14, we can find the required Hamiltonian path in P_n for every n, $n \ge 5$. □

LEMMA 18.9 Let a and b be any two distinct elements in $\langle n \rangle$ with $n \geq 4$, and let $\mathbf{x}$ be a vertex of P_n. There is a Hamiltonian path P of $P_n - \{\mathbf{x}\}$ joining a vertex $\mathbf{u}$ with $(\mathbf{u})_1 = a$ to a vertex $\mathbf{v}$ with $(\mathbf{v})_1 = b$.

Proof: Suppose that $n = 4$. Since P_4 is vertex-transitive, we may assume that $\mathbf{x} = 1234$. Without loss of generality, we may assume that $a < b$. The required paths of $P_4 - \{1234\}$ are:

$a = 1$ and $b = 2$
⟨1423, 4123, 3214, 2314, 1324, 3124, 4213, 2413, 3142, 4132, 1432, 3412, 2143, 1243, 3421, 4321, 2341, 3241, 4231, 2431, 1342, 4312, 2134⟩
$a = 1$ and $b = 3$
⟨1423, 4123, 2143, 1243, 4213, 2413, 3142, 1342, 2431, 3421, 4321, 2341, 3241, 4231, 1324, 3124, 2134, 4312, 3412, 1432, 4132, 2314, 3214⟩
$a = 1$ and $b = 4$
⟨1423, 2413, 3142, 1342, 2431, 3421, 4321, 2341, 3241, 4231, 1324, 2314, 3214, 4123, 2143, 1243, 4213, 3124, 2134, 4312, 3412, 1432, 4132⟩
$a = 2$ and $b = 3$
⟨2134, 4312, 1342, 3142, 2413, 4213, 1243, 2143, 3412, 1432, 4132, 2314, 3214, 4123, 1423, 3241, 2341, 4321, 3421, 2431, 4231, 1324, 3124⟩
$a = 2$ and $b = 4$
⟨2134, 3124, 1324, 2314, 3214, 4123, 2143, 1243, 4213, 2413, 1423, 3241, 4231, 2431, 3421, 4321, 2341, 1432, 3412, 4312, 1342, 3142, 4132⟩
$a = 3$ and $b = 4$
⟨3214, 4123, 2143, 1243, 4213, 3124, 2134, 4312, 3412, 1432, 2341, 4321, 3421, 2431, 1342, 3142, 2413, 1423, 3241, 4231, 1324, 2314, 4132⟩

With Theorem 11.14, we can find the required Hamiltonian path on P_n for every n, $n \geq 5$. □

LEMMA 18.10 Let a and b be any two distinct elements in $\langle n \rangle$ with $n \geq 4$. Assume that $\mathbf{x}$ and $\mathbf{y}$ are two adjacent vertices of P_n. There is a Hamiltonian path P of $P_n - \{\mathbf{x}, \mathbf{y}\}$ joining a vertex $\mathbf{u}$ with $(\mathbf{u})_1 = a$ to a vertex $\mathbf{v}$ with $(\mathbf{v})_1 = b$.

Proof: Since P_n is vertex-transitive, we may assume that $\mathbf{x} = \mathbf{e}$ and $\mathbf{y} = (\mathbf{e})^i$ for some $i \in \{2, 3, \ldots, n\}$. Without loss of generality, we assume that $a < b$. Thus, $a \neq n$ and $b \neq 1$. We prove this statement by induction on n. For $n = 4$, the required paths of $P_4 - \{1234, (1234)^i\}$ are:

$\mathbf{y} = 2134$
$a = 1$ and $b = 2$
⟨1432, 2413, 3142, 4132, 1432, 3412, 4312, 1342, 2431, 3421, 4321, 2341, 3241, 4231, 1324, 3124, 4213, 1243, 2143, 4123, 3214, 2314⟩
$a = 1$ and $b = 3$
⟨1432, 4123, 2143, 1243, 3421, 4321, 2341, 3241, 4231, 2431, 1342, 4312, 3412, 1432, 4132, 3142, 2413, 4213, 3124, 1324, 2314, 3214⟩
$a = 1$ and $b = 4$
⟨1432, 4123, 3214, 2314, 1324, 3124, 4213, 2413, 3142, 4132, 1432, 2341, 3241, 4231, 2431, 1342, 4312, 3412, 2143, 1243, 3421, 4321⟩
$a = 2$ and $b = 3$
⟨2314, 3214, 4123, 2143, 1243, 4213, 3124, 1324, 4231, 2431, 1342, 4312, 3412, 1432, 4132, 3142, 2413, 1423, 3241, 2341, 4321, 3421⟩
$a = 2$ and $b = 4$
⟨2314, 3214, 4123, 2143, 3412, 4312, 1342, 2431, 3421, 1243, 4213, 3124, 1324, 4231, 3241, 1423, 2413, 3142, 4132, 1432, 2341, 4321⟩
$a = 3$ and $b = 4$
⟨3214, 4123, 2143, 1243, 3421, 2431, 4231, 3241, 1423, 2413, 4213, 3124, 1324, 2314, 4132, 3142, 1342, 4312, 3412, 1432, 2341, 4321⟩

y = 3214
$a = 1$ and $b = 2$
⟨1423, 4123, 2143, 1243, 3421, 4321, 2341, 3241, 4231, 2431, 1342, 3142, 2413, 4213, 3124, 1324, 2314, 4132, 1432, 3412, 4312, 2134⟩
$a = 1$ and $b = 3$
⟨1423, 4123, 2143, 1243, 4213, 2413, 3142, 1342, 2431, 3421, 4321, 2341, 3241, 4231, 1324, 2314, 4132, 1432, 3412, 4312, 2134, 3124⟩
$a = 1$ and $b = 4$
⟨1423, 4123, 2143, 1243, 3421, 2431, 1342, 3142, 2413, 4213, 3124, 2134, 4312, 3412, 1432, 4132, 2314, 1324, 4231, 3241, 2341, 4321⟩
$a = 2$ and $b = 3$
⟨2134, 4312, 1342, 2431, 4231, 3241, 1423, 4123, 2143, 3412, 1432, 2341, 4321, 3421, 1243, 4213, 2413, 3142, 4132, 2314, 1324, 3124⟩

$a = 2$ and $b = 4$
⟨2134, 3124, 4213, 2413, 3142, 1342, 4312, 3412, 1432, 4132, 2314, 1324, 4231, 2431, 3421, 1243, 2143, 4123, 1423, 3241, 2341, 4321⟩
$a = 3$ and $b = 4$
⟨3124, 2134, 4312, 1342, 3142, 2413, 4213, 1243, 3421, 2431, 4231, 1324, 2314, 4132, 1432, 3412, 2143, 4123, 1423, 3241, 2341, 4321⟩

y = 4321
$a = 1$ and $b = 2$
⟨1423, 4123, 3214, 2314, 4132, 3142, 2413, 4213, 3124, 1324, 4231, 3241, 2341, 1432, 3412, 2143, 1243, 3421, 2431, 1342, 4312, 2134⟩
$a = 1$ and $b = 3$
⟨1423, 4123, 2143, 3412, 1432, 2341, 3241, 4231, 1324, 3124, 2134, 4312, 1342, 2431, 2431, 1243, 4213, 2413, 3142, 4132, 2314, 3214⟩
$a = 1$ and $b = 4$
⟨1423, 2413, 4213, 3124, 2134, 4312, 3412, 2143, 1243, 3421, 2431, 1342, 3142, 4132, 1432, 2341, 3241, 4231, 1324, 2314, 3214, 4123⟩
$a = 2$ and $b = 3$
⟨2134, 4312, 1342, 3142, 4132, 2314, 3214, 4123, 2143, 3412, 1432, 2341, 3241, 1423, 2413, 4213, 1243, 3421, 2431, 4231, 1324, 3124⟩
$a = 2$ and $b = 4$
⟨2134, 3124, 4213, 2413, 1423, 3241, 2341, 1432, 4132, 3142, 1342, 4312, 3412, 2143, 1243, 3421, 2431, 4231, 1324, 2314, 3214, 4123⟩
$a = 3$ and $b = 4$
⟨3214, 2314, 1324, 4231, 3241, 2341, 1432, 4132, 3142, 1342, 2431, 3421, 1243, 2143, 3412, 4312, 2134, 3124, 4213, 2413, 1423, 4123⟩

Suppose that this statement holds for P_k for every k, $4 \leq k < n$. We have the following cases:

Case 1. $\mathbf{y} = (\mathbf{e})^i$ for some $i \neq 1$ and $i \neq n$, that is, $\mathbf{y} \in P_n^{\{n\}}$. Let $c \in \langle n-1 \rangle - \{a\}$. By induction, there is a Hamiltonian path R of $P_n^{\{n\}} - \{\mathbf{e}, (\mathbf{e})^i\}$ joining a vertex $\mathbf{u}$ with $(\mathbf{u})_1 = a$ to a vertex $\mathbf{z}$ with $(\mathbf{z})_1 = c$. We choose a vertex $\mathbf{v}$ in $P_n^{\langle n-1 \rangle - \{c\}}$ with $(\mathbf{v})_1 = b$. By Theorem 11.13, there is a Hamiltonian path H of $P_n^{\langle n-1 \rangle}$ joining the vertex $(\mathbf{z})^n$ to $\mathbf{v}$. Then $\langle \mathbf{u}, R, \mathbf{z}, (\mathbf{z})^n, H, \mathbf{v} \rangle$ is the desired path.

Case 2. $\mathbf{y} = (\mathbf{e})^n$, that is, $\mathbf{y} \in P_n^{\{1\}}$. Let $c \in \langle n-1 \rangle - \{1, a\}$ and $d \in \langle n-1 \rangle - \{1, b, c\}$. By Lemma 18.9, there is a Hamiltonian path R of $P_n^{\{n\}} - \{\mathbf{e}\}$ joining a vertex $\mathbf{u}$ with $(\mathbf{u})_1 = a$ to a vertex $\mathbf{w}$ with $(\mathbf{w})_1 = c$. Again, there is a Hamiltonian path H of $P_n^{\{1\}} - \{(\mathbf{e})^n\}$ joining a vertex $\mathbf{z}$ with $(\mathbf{z})_1 = d$ to a vertex $\mathbf{v}$ with $(\mathbf{v})_1 = b$. By Theorem 11.13, there is a Hamiltonian path Q of $P_n^{\langle n-1 \rangle - \{1\}}$ joining the vertex $(\mathbf{w})^n$ to the vertex $(\mathbf{z})^n$. Then $\langle \mathbf{u}, R, \mathbf{w}, (\mathbf{w})^n, Q, (\mathbf{z})^n, \mathbf{z}, H, \mathbf{v} \rangle$ is the desired path. □

LEMMA 18.11 Let a and b be any two distinct elements in $\langle n \rangle$ with $n \geq 4$. Let $\mathbf{x}$ be any vertex of P_n. Assume that $\mathbf{x_1}$ and $\mathbf{x_2}$ are two distinct neighbors of $\mathbf{x}$. There is a Hamiltonian path P of $P_n - \{\mathbf{x}, \mathbf{x_1}, \mathbf{x_2}\}$ joining a vertex $\mathbf{u}$ with $(\mathbf{u})_1 = a$ to a vertex $\mathbf{v}$ with $(\mathbf{v})_1 = b$.

Proof: Since P_n is vertex-transitive, we may assume that $\mathbf{x} = \mathbf{e}$. Moreover, we assume that $\mathbf{x_1} = (\mathbf{e})^i$ and $\mathbf{x_2} = (\mathbf{e})^j$ for some $\{i, j\} \subset \langle n \rangle - \{1\}$ with $i < j$. Without loss of generality, we assume that $a < b$. Thus, $a \neq n$ and $b \neq 1$. We prove this lemma by induction on n. For $n = 4$, the required paths of $P_4 - \{1234, (1234)^i, (1234)^j\}$ are:

$\mathbf{x_1} = 2134$ and $\mathbf{x_2} = 3214$
$a = 1$ and $b = 2$
⟨1423, 4123, 2143, 1243, 3421, 4321, 2341, 3241, 4231, 2431, 1342, 4312, 3412, 1432, 4132, 3142, 2413, 4213, 3124, 1324, 2314⟩
$a = 1$ and $b = 3$
⟨1423, 4123, 2143, 1243, 3421, 4321, 2341, 3241, 4231, 2431, 1342, 4312, 3412, 1432, 4132, 2314, 1324, 3124, 4213, 2413, 3142⟩
$a = 1$ and $b = 4$
⟨1423, 4123, 2143, 1243, 3421, 4321, 2341, 3241, 4231, 2431, 1342, 3142, 2413, 4213, 3124, 1324, 2314, 4132, 1432, 3412, 4312⟩
$a = 2$ and $b = 3$
⟨2143, 4123, 1423, 3241, 4231, 2431, 1342, 4312, 3412, 1432, 2341, 4321, 3421, 1243, 4213, 2413, 3142, 4132, 2314, 1324, 3124⟩
$a = 2$ and $b = 4$
⟨2143, 4123, 1423, 2413, 3142, 1342, 4312, 3412, 1432, 4132, 2314, 1324, 3124, 4213, 1243, 3421, 2431, 4231, 3241, 2341, 4321⟩
$a = 3$ and $b = 4$
⟨3124, 4213, 2413, 3142, 1342, 4312, 3412, 1432, 4132, 2314, 1324, 4231, 2431, 3421, 1243, 2143, 4123, 1423, 3241, 2341, 4321⟩

$\mathbf{x_1} = 2134$ and $\mathbf{x_2} = 4321$
$a = 1$ and $b = 2$
⟨1423, 2413, 3142, 4132, 1432, 2341, 3241, 4231, 1324, 3124, 4213, 1243, 3421, 2431, 1342, 4312, 3412, 2143, 4123, 3214, 2314⟩
$a = 1$ and $b = 3$
⟨1423, 4123, 2143, 1243, 3421, 2431, 1342, 4312, 3412, 1432, 2341, 3241, 4231, 1342, 3124, 4213, 2413, 3142, 4132, 2314, 3214⟩
$a = 1$ and $b = 4$
⟨1423, 4123, 3214, 2314, 1324, 3124, 4213, 2413, 3142, 1342, 4312, 3412, 2143, 1243, 3421, 2431, 4231, 3241, 2341, 1432, 4132⟩
$a = 2$ and $b = 3$
⟨2314, 3214, 4123, 2143, 3412, 4312, 1342, 3142, 4132, 1432, 2341, 3241, 1423, 2413, 4213, 1243, 3421, 2431, 4231, 1324, 3124⟩
$a = 2$ and $b = 4$
⟨2314, 3214, 4123, 2143, 1243, 3421, 2431, 4231, 1324, 3124, 4213, 2413, 1423, 3241, 2341, 1432, 3412, 4312, 1342, 3142, 4132⟩
$a = 3$ and $b = 4$
⟨3214, 2314, 4132, 3142, 1342, 4312, 3412, 1432, 2341, 3241, 1423, 2413, 4213, 3124, 1324, 4231, 2431, 3421, 1243, 2143, 4123⟩

$\mathbf{x_1} = 3214$ and $\mathbf{x_2} = 4321$
$a = 1$ and $b = 2$
⟨1423, 4123, 2143, 1243, 3421, 2431, 1342, 4312, 3412, 1432, 2341, 3241, 4231, 1324, 2314, 4132, 3142, 2413, 4213, 3124, 2134⟩
$a = 1$ and $b = 3$
⟨1423, 4123, 2143, 3412, 1432, 2341, 3241, 4231, 1324, 2314, 4132, 3142, 2413, 4213, 1243, 3421, 2431, 1342, 4312, 2134, 3124⟩
$a = 1$ and $b = 4$
⟨1423, 2413, 4213, 3124, 2134, 4312, 3412, 1432, 2341, 3241, 4231, 1324, 2314, 4132, 3142, 1342, 2431, 3421, 1243, 2143, 4123⟩
$a = 2$ and $b = 3$
⟨2134, 4312, 3412, 1432, 2341, 3241, 4231, 1324, 2314, 4132, 3142, 1341, 2431, 3421, 1243, 2143, 4123, 1423, 2413, 4213, 3124⟩
$a = 2$ and $b = 4$
⟨2134, 3124, 4213, 2413, 3142, 1342, 4312, 3412, 1432, 2341, 3241, 1423, 4123, 2143, 1243, 3421, 2431, 4231, 1324, 2314, 4132⟩
$a = 3$ and $b = 4$
⟨3124, 2134, 4312, 3412, 1432, 2341, 3241, 4231, 1324, 2314, 4132, 3142, 1342, 2431, 3421, 1243, 2143, 4123, 1423, 2413, 4213⟩

Suppose that this statement holds for P_k for every k, $4 \leq k < n$. We have the following cases.

Case 1. $j \neq n$; that is, $\mathbf{x_1} \in P_n^{\{n\}}$ and $\mathbf{x_2} \in P_n^{\{n\}}$. Let $c \in \langle n-1 \rangle - \{1, a\}$. By induction, there is a Hamiltonian path R of $P_n^{\{n\}} - \{\mathbf{e}, \mathbf{x_1}, \mathbf{x_2}\}$ joining a vertex $\mathbf{u}$ with $(\mathbf{u})_1 = a$ to a vertex $\mathbf{z}$ with $(\mathbf{z})_1 = c$. We choose a vertex $\mathbf{v}$ in $P_n^{\{1\}}$ with $(\mathbf{v})_1 = b$. By Theorem 11.13, there is a Hamiltonian path H of $P_n^{\langle n-1 \rangle}$ joining the vertex $(\mathbf{z})^n$ to $\mathbf{v}$. We set $P = \langle \mathbf{u}, R, \mathbf{z}, (\mathbf{z})^n, H, \mathbf{v} \rangle$. Then P is the desired path.

Case 2. $j = n$; that is, $\mathbf{x_1} \in P_n^{\{n\}}$ and $\mathbf{x_2} \in P_n^{\{1\}}$. Let $c \in \langle n-1 \rangle - \{1, a\}$ and $d \in \langle n-1 \rangle - \{1, b, c\}$. By Lemma 18.10, there is a Hamiltonian path R of $P_n^{\{n\}} - \{\mathbf{e}, \mathbf{x_1}\}$ joining a vertex $\mathbf{u}$ with $(\mathbf{u})_1 = a$ to a vertex $\mathbf{z}$ with $(\mathbf{z})_1 = c$. By Lemma 18.9, there is a Hamiltonian path H of $P_n^{\{1\}} - \{\mathbf{x_2}\}$ joining a vertex $\mathbf{w}$ with $(\mathbf{w})_1 = d$ to a vertex $\mathbf{v}$ with $(\mathbf{v})_1 = b$. By Theorem 11.13, there is a Hamiltonian path Q of $P_n^{\langle n-1 \rangle - \{1\}}$ joining the vertex $(\mathbf{z})^n$ to the vertex $(\mathbf{w})^n$. We set $P = \langle \mathbf{u}, R, \mathbf{z}, (\mathbf{z})^n, Q, (\mathbf{w})^n, \mathbf{w}, H, \mathbf{v} \rangle$. Then P is the desired path. □

We then have the following theorem:

THEOREM 18.6 $IHC(P_3) = 1$ and $IHC(P_n) = n - 1$ if $n \geq 4$.

Proof: It is easy to see that P_3 is isomorphic to a cycle with six vertices. Thus, $IHC(P_3) = 1$. Since P_n is an $(n-1)$-regular graph, it is clear that $IHC(P_n) \leq n - 1$. Since P_n is vertex-transitive, we need to show only that there exist $(n-1)$ mutually independent Hamiltonian cycles of P_n starting from the vertex $\mathbf{e}$. For $n = 4$, we prove that $IHC(P_4) \geq 3$ by listing the required Hamiltonian cycles as following:

$C_1 = \langle 1234, 2134, 4312, 3412, 2143, 1243, 4213, 3124, 1324, 4231, 3241, 2341, 1432, 4132, 2314, 3214, 4123, 1423, 2413, 3142, 1342, 2431, 3421, 4321, 1234 \rangle$
$C_2 = \langle 1234, 3214, 2314, 1324, 3124, 4213, 2413, 1423, 4123, 2143, 1243, 3421, 4321, 2341, 3241, 4231, 2431, 1342, 3142, 4132, 1432, 3412, 4312, 2134, 1234 \rangle$
$C_3 = \langle 1234, 4321, 2341, 1432, 4132, 2314, 1324, 4231, 3241, 1423, 2413, 3142, 1342, 2431, 3421, 1243, 4213, 3124, 2134, 4312, 3412, 2143, 4123, 3214, 1234 \rangle$

Let $n \geq 5$. Let B be the $(n-1) \times n$ matrix with

$$b_{i,j} = \begin{cases} i + j - 1 & \text{if } i + j - 1 \leq n \\ i + j - n + 1 & \text{if } n \geq i + j \end{cases}$$

More precisely,

$$B = \begin{bmatrix} 1 & 2 & 3 & 4 & \cdots & n-1 & n \\ 2 & 3 & 4 & 5 & \cdots & n & 1 \\ \vdots & \vdots & \vdots & \vdots & \ddots & \vdots & \vdots \\ n-1 & n & 1 & 2 & \cdots & n-3 & n-2 \end{bmatrix}.$$

It is not hard to see that $b_{i,1} b_{i,2} \ldots b_{i,n}$ forms a permutation of $\{1, 2, \ldots, n\}$ for every i with $1 \leq i \leq n - 1$. Moreover, $b_{i,j} \neq b_{i',j}$ for any $1 \leq i < i' \leq n - 1$ and $1 \leq j \leq n$. In other words, B forms a Latin rectangle with entries in $\{1, 2, \ldots, n\}$.

We construct $\{C_1, C_2, \ldots, C_{n-1}\}$ as follows:

(1) $k = 1$. By Lemma 18.8, there is a Hamiltonian path H_1 of $P_n^{\{b_{1,n}\}} - \{\mathbf{e}\}$ joining a vertex $\mathbf{x}$ with $\mathbf{x} \neq (\mathbf{e})^{n-1}$ and $(\mathbf{x})_1 = n-1$ to the vertex $(\mathbf{e})^{n-1}$. By Theorem 11.13, there is a Hamiltonian path H_2 of $\cup_{t=1}^{n-1} P_n^{\{b_{1,t}\}}$ joining the vertex $(\mathbf{e})^n$ to the vertex $(\mathbf{x})^n$ with $H_2(i+(j-1)(n-1)!) \in P_n^{\{b_{1,j}\}}$ for every $i \in \langle (n-1)! \rangle$ and for every $j \in \langle n-1 \rangle$. We set $C_1 = \langle \mathbf{e}, (\mathbf{e})^n, H_2, (\mathbf{x})^n, \mathbf{x}, H_1, (\mathbf{e})^{n-1}, \mathbf{e} \rangle$.

(2) $k = 2$. By Lemma 18.10, there is a Hamiltonian path Q_1 of $P_n^{\{b_{2,n-1}\}} - \{\mathbf{e}, (\mathbf{e})^2\}$ joining a vertex $\mathbf{y}$ with $(\mathbf{y})_1 = n-1$ to a vertex $\mathbf{z}$ with $(\mathbf{z})_1 = 1$. By Theorem 11.13, there is a Hamiltonian Q_2 of $\cup_{t=1}^{n-2} P_n^{\{b_{2,t}\}}$ joining the vertex $((\mathbf{e})^2)^n$ to the vertex $(\mathbf{y})^n$ such that $Q_2(i+(j-1)(n-1)!) \in P_n^{\{b_{2,j}\}}$ for every $i \in \langle (n-1)! \rangle$ and for every $j \in \langle n-2 \rangle$. By Theorem 11.14, there is a Hamiltonian path Q_3 of $P_n^{\{b_{2,n}\}}$ joining the vertex $(\mathbf{z})^n$ to the vertex $(\mathbf{e})^n$. We set $C_2 = \langle \mathbf{e}, (\mathbf{e})^2, ((\mathbf{e})^2)^n, Q_2, (\mathbf{y})^n, \mathbf{y}, Q_1, \mathbf{z}, (\mathbf{z})^n, Q_3, (\mathbf{e})^n, \mathbf{e} \rangle$.

(3) $3 \leq k \leq n-1$. By Lemma 18.11, there is a Hamiltonian path R_1^k of $P_n^{\{b_{k,n-k+1}\}} - \{\mathbf{e}, (\mathbf{e})^{k-1}, (\mathbf{e})^k\}$ joining a vertex $\mathbf{w_k}$ with $(\mathbf{w_k})_1 = n-1$ to a vertex $\mathbf{v_k}$ with $(\mathbf{v_k})_1 = 1$. By Theorem 11.13, there is a Hamiltonian path R_2^k of $\cup_{t=1}^{n-k} P_n^{\{b_{k,t}\}}$ joining the vertex $((\mathbf{e})^k)^n$ to the vertex $(\mathbf{w_k})^n$ such that $R_2^k(i+(j-1)(n-1)!) \in P_n^{\{b_{k,j}\}}$ for every $i \in \langle (n-1)! \rangle$ and for every $j \in \langle n-k \rangle$. Again, there is a Hamiltonian path R_3^k of $\cup_{t=n-k+2}^{n} P_n^{\{b_{k,t}\}}$ joining the vertex $(\mathbf{v_k})^n$ to the vertex $((\mathbf{e})^{k-1})^n$ such that $R_3^k(i+(j-1)(n-1)!) \in P_n^{\{b_{k,n-k+j+1}\}}$ for every $i \in \langle (n-1)! \rangle$ and for every $j \in \langle k-1 \rangle$. We set $C_k = \langle \mathbf{e}, (\mathbf{e})^k, ((\mathbf{e})^k)^n, R_2^k, (\mathbf{w_k})^n, \mathbf{w_k}, R_1^k, \mathbf{v_k}, (\mathbf{v_k})^n, R_3^k, ((\mathbf{e})^{k-1})^n, (\mathbf{e})^{k-1}, \mathbf{e} \rangle$.

Then $\{C_1, C_2, \ldots, C_{n-1}\}$ forms a set of $(n-1)$ mutually independent Hamiltonian cycles of P_n starting from the vertex $\mathbf{e}$. □

Example 18.1

We illustrate the proof of Theorem 18.6 with $n = 5$ as follows.

We set

$$B = \begin{bmatrix} 1 & 2 & 3 & 4 & 5 \\ 2 & 3 & 4 & 5 & 1 \\ 3 & 4 & 5 & 1 & 2 \\ 4 & 5 & 1 & 2 & 3 \end{bmatrix}$$

Then we construct $\{C_1, C_2, C_3, C_4\}$ as follows:

(1) $k = 1$. By Lemma 18.8, there is a Hamiltonian path H_1 of $P_5^{\{b_{1,5}\}} - \{\mathbf{e}\}$ joining a vertex $\mathbf{x}$ with $\mathbf{x} \neq (\mathbf{e})^4$ and $(\mathbf{x})_1 = 4$ to the vertex $(\mathbf{e})^4$. By Theorem 11.13, there is a Hamiltonian path H_2 of $\cup_{t=1}^{4} P_5^{\{b_{1,t}\}}$ joining the vertex $(\mathbf{e})^5$ to the vertex $(\mathbf{x})^5$ with $H_2(i+24(j-1)) \in P_5^{\{b_{1,j}\}}$ for every $i \in \langle 24 \rangle$ and for every $j \in \langle 4 \rangle$. We set $C_1 = \langle \mathbf{e}, (\mathbf{e})^5, H_2, (\mathbf{x})^5, \mathbf{x}, H_1, (\mathbf{e})^4, \mathbf{e} \rangle$.

(2) $k = 2$. By Lemma 18.10, there is a Hamiltonian path Q_1 of $P_5^{\{b_{2,4}\}} - \{\mathbf{e}, (\mathbf{e})^2\}$ joining a vertex $\mathbf{y}$ with $(\mathbf{y})_1 = 4$ to a vertex $\mathbf{z}$ with $(\mathbf{z})_1 = 1$. By Theorem 11.13, there is

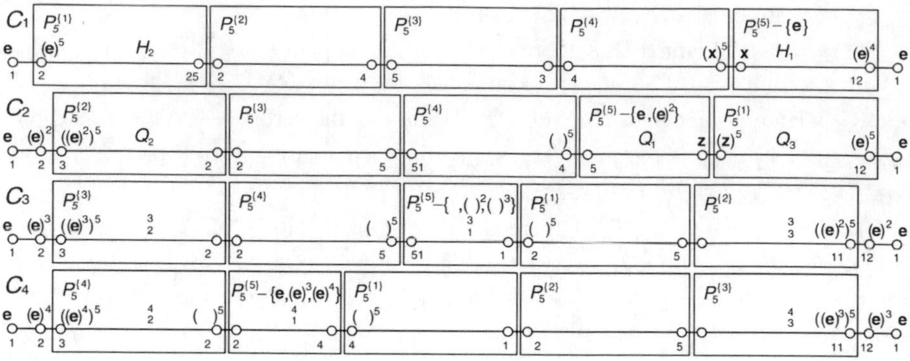

FIGURE 18.4 Illustration for Theorem 18.6 on P_5.

a Hamiltonian path Q_2 of $\cup_{t=1}^{3} P_5^{\{b_{2,t}\}}$ joining the vertex $((\mathbf{e})^2)^5$ to the vertex $(\mathbf{y})^5$ such that $Q_2(i+24(j-1)) \in P_5^{\{b_{2,j}\}}$ for every $i \in \langle 24 \rangle$ and for every $j \in \langle 3 \rangle$. By Theorem 11.14, there is a Hamiltonian path Q_3 of $P_5^{\{b_{2,5}\}}$ joining the vertex $(\mathbf{z})^5$ to the vertex $(\mathbf{e})^5$. We set $C_2 = \langle \mathbf{e}, (\mathbf{e})^2, ((\mathbf{e})^2)^5, Q_2, (\mathbf{y})^5, \mathbf{y}, Q_1, \mathbf{z}, (\mathbf{z})^5, Q_3, (\mathbf{e})^5, \mathbf{e} \rangle$.

(3) $3 \leq k \leq 4$. By Lemma 18.11, there is a Hamiltonian path R_1^k of $P_5^{\{b_{k,6-k}\}} - \{\mathbf{e}, (\mathbf{e})^{k-1}, (\mathbf{e})^k\}$ joining a vertex $\mathbf{w_k}$ with $(\mathbf{w_k})_1 = 4$ to a vertex $\mathbf{v_k}$ with $(\mathbf{v_k})_1 = 1$. By Theorem 11.13, there is a Hamiltonian path R_2^k of $\cup_{t=1}^{5-k} P_5^{\{b_{k,t}\}}$ joining the vertex $((\mathbf{e})^k)^5$ to the vertex $(\mathbf{w_k})^5$ such that $R_2^k(i+24(j-1)) \in P_5^{\{b_{k,j}\}}$ for every $i \in \langle 24 \rangle$ and for every $j \in \langle 5-k \rangle$. Again, there is a Hamiltonian path R_3^k of $\cup_{t=7-k}^{5} P_5^{\{b_{k,t}\}}$ joining the vertex $(\mathbf{v_k})^5$ to the vertex $((\mathbf{e})^{k-1})^5$ such that $R_3^k(i+24(j-1)) \in P_5^{\{b_{k,6-k+j}\}}$ for every $i \in \langle 24 \rangle$ and for every $j \in \langle k-1 \rangle$. We set $C_k = \langle \mathbf{e}, (\mathbf{e})^k, ((\mathbf{e})^k)^5, R_2^k, (\mathbf{w_k})^5, \mathbf{w_k}, R_1^k, \mathbf{v_k}, (\mathbf{v_k})^5, R_3^k, ((\mathbf{e})^{k-1})^5, (\mathbf{e})^{k-1}, \mathbf{e} \rangle$.

Then $\{C_1, C_2, C_3, C_4\}$ forms a set of four mutually independent Hamiltonian cycles of P_5 starting from the vertex $\mathbf{e}$. See Figure 18.4 for illustration.

18.5 MUTUALLY INDEPENDENT HAMILTONIAN CYCLES OF STAR GRAPHS

In Ref. 238, Lin et al. also studied the mutually independent Hamiltonicity for star graphs.

LEMMA 18.12 Let a and b be any two distinct elements in $\langle n \rangle$ with $n \geq 4$. Assume that $\mathbf{x}$ is a white vertex of S_n and assume that $\mathbf{x_1}$ and $\mathbf{x_2}$ are two distinct neighbors of $\mathbf{x}$. Then there is a Hamiltonian path P of $S_n - \{\mathbf{x}, \mathbf{x_1}, \mathbf{x_2}\}$ joining a white vertex $\mathbf{u}$ with $(\mathbf{u})_1 = a$ to a white vertex $\mathbf{v}$ with $(\mathbf{v})_1 = b$.

Proof: Since S_n is vertex-transitive and edge-transitive, we may assume that $\mathbf{x} = \mathbf{e}$, $\mathbf{x_1} = (\mathbf{e})^2$, and $\mathbf{x_2} = (\mathbf{e})^3$. Without loss of generality, we may also assume that $a < b$.

We have $a \neq n$ and $b \neq 1$. We prove this statement by induction on n. For $n=4$, the required paths of $S_4 - \{1234, 2134, 3214\}$ are:

$a=1$ and $b=2$	$\langle 1324, 3142, 4132, 1432, 3412, 4312, 2314, 1324, 3124, 4123, 2143, 1243, 4213, 2413, 1423, 3421, 4321, 2341, 3241, 4231, 2431\rangle$
$a=1$ and $b=3$	$\langle 1423, 2413, 4213, 1243, 2143, 4123, 3124, 1324, 2314, 4312, 3412, 1432, 4132, 3142, 1342, 2341, 4321, 3421, 2431, 4231, 3241\rangle$
$a=1$ and $b=4$	$\langle 1324, 3142, 4132, 1432, 3412, 4312, 2314, 1324, 3124, 4123, 2143, 1243, 4213, 2413, 1423, 3421, 2431, 4231, 3241, 2341, 4321\rangle$
$a=2$ and $b=3$	$\langle 2314, 1324, 3124, 4123, 2143, 1243, 4213, 2413, 1423, 3421, 4321, 2341, 3241, 4231, 2431, 1432, 4132, 3142, 1342, 4312, 3412\rangle$
$a=2$ and $b=4$	$\langle 2314, 1324, 3124, 4123, 2143, 1243, 4213, 2413, 1423, 3421, 4321, 2341, 3241, 4231, 2431, 1432, 3412, 4312, 1342, 3142, 4132\rangle$
$a=3$ and $b=4$	$\langle 3124, 1324, 2314, 4312, 3412, 1432, 4132, 3142, 1342, 2341, 4321, 3421, 2431, 4231, 3241, 1243, 2143, 4123, 1423, 2413, 4213\rangle$

Suppose that this statement holds for S_k for every $k, 4 \leq k \leq n-1$. Let c be any element in $\langle n-1\rangle - \{1, a\}$. By induction, there is a Hamiltonian path H of $S_n^{\{n\}} - \{\mathbf{e}, (\mathbf{e})^2, (\mathbf{e})^3\}$ joining a white vertex $\mathbf{u}$ with $(\mathbf{u})_1 = a$ to a white vertex $\mathbf{z}$ with $(\mathbf{z})_1 = c$. We choose a white vertex $\mathbf{v}$ in $S_n^{\{1\}}$ with $(\mathbf{v})_1 = b$. By Theorem 14.33, there is a Hamiltonian path R of $S_n^{\langle n-1\rangle}$ joining the black vertex $(\mathbf{z})^n$ to $\mathbf{v}$. Then $\langle \mathbf{u}, H, \mathbf{z}, (\mathbf{z})^n, R, \mathbf{v}\rangle$ is the desired path of $S_n - \{\mathbf{e}, (\mathbf{e})^2, (\mathbf{e})^3\}$. □

The following theorem is our main result for the star graph S_n:

THEOREM 18.7 $IHC(S_3) = 1$, $IHC(S_4) = 2$, and $IHC(S_n) = n-1$ if $n \geq 5$.

Proof: It is easy to see that S_3 is isomorphic to a cycle with six vertices. Thus, $IHC(S_3) = 1$. Using a computer, we have $IHC(S_4) = 2$ by brute force checking. Thus, we assume that $n \geq 5$. We know that S_n is an $(n-1)$-regular graph; it is clear that $IHC(S_n) \leq n-1$. Since S_n is vertex-transitive, we need to show only that there are $(n-1)$ mutually independent Hamiltonian cycles of S_n starting from $\mathbf{e}$. Let B be the $(n-1) \times n$ matrix with

$$b_{i,j} = \begin{cases} i+j-1 & \text{if } i+j-1 \leq n \\ i+j-n+1 & \text{if } n < i+j-1 \end{cases}$$

We construct $\{C_1, C_2, \ldots, C_{n-1}\}$ as follows:

(1) $k=1$. We choose a black vertex $\mathbf{x}$ in $S_n^{\{b_{1,n}\}} - \{(\mathbf{e})^{n-1}\}$ with $(\mathbf{x})_1 = n-1$. By Theorem 13.2, there is a Hamiltonian path H_1 of $S_n^{\{b_{1,n}\}} - \{\mathbf{e}\}$ joining $\mathbf{x}$ to the black vertex $(\mathbf{e})^{n-1}$. By Theorem 14.33, there is a Hamiltonian path H_2 of $\cup_{t=1}^{n-1} S_n^{\{b_{1,t}\}}$ joining the black vertex $(\mathbf{e})^n$ to the white vertex $(\mathbf{x})^n$ with $H_2(i + (j-1)(n-1)!) \in S_n^{\{b_{1,j}\}}$ for every $i \in \langle (n-1)!\rangle$ and for every $j \in \langle n-1\rangle$. We set $C_1 = \langle \mathbf{e}, (\mathbf{e})^n, H_2, (\mathbf{x})^n, \mathbf{x}, H_1, (\mathbf{e})^{n-1}, \mathbf{e}\rangle$.

(2) $k=2$. We choose a white vertex $\mathbf{y}$ in $S_n^{\{b_{2,n-1}\}} - \{\mathbf{e}, (\mathbf{e})^2\}$ with $(\mathbf{y})_1 = n-1$. By Lemma 13.35, there is a Hamiltonian path Q_1 of $S_n^{\{b_{2,j}\}} - \{\mathbf{e}, (\mathbf{e})^2\}$ joining $\mathbf{y}$ to a black vertex $\mathbf{z}$ with $(\mathbf{z})_1 = 1$. By Theorem 14.33, there is a Hamiltonian path

Q_2 of $\cup_{t=1}^{n-2} S_n^{\{b_{2,t}\}}$ joining the white vertex $((\mathbf{e})^2)^n$ to the black vertex $(\mathbf{y})^n$ such that $Q_2(i+(j-1)(n-1)!) \in S_n^{\{b_{2,j}\}}$ for every $i \in \langle (n-1)! \rangle$ and for every $j \in \langle n-2 \rangle$. Again, there is a Hamiltonian path Q_3 of $S_n^{\{b_{2,n}\}}$ joining the white vertex $(\mathbf{z})^n$ to the black vertex $(\mathbf{e})^n$. We set $C_2 = \langle \mathbf{e}, (\mathbf{e})^2, ((\mathbf{e})^2)^n, Q_2, (\mathbf{y})^n, \mathbf{y}, Q_1, \mathbf{z}, (\mathbf{z})^n, Q_3, (\mathbf{e})^n, \mathbf{e} \rangle$.

(3) $3 \le k \le n-1$. By Lemma 18.12, there is a Hamiltonian path R_1^k of $S_n^{\{b_{k,n-k+1}\}} - \{\mathbf{e}, (\mathbf{e})^{k-1}, (\mathbf{e})^k\}$ joining a white vertex $\mathbf{w_k}$ with $(\mathbf{w_k})_1 = n-1$ to a white vertex $\mathbf{v_k}$ with $(\mathbf{v_k})_1 = 1$. By Theorem 14.33, there is a Hamiltonian path R_2^k of $\cup_{t=1}^{n-k} S_n^{\{b_{k,t}\}}$ joining the white vertex $((\mathbf{e})^k)^n$ to the black vertex $(\mathbf{w_k})^n$ such that $R_2^k(i+(j-1)(n-1)!) \in S_n^{\{b_{k,j}\}}$ for every $i \in \langle (n-1)! \rangle$ and for every $j \in \langle n-k-1 \rangle$. Again, there is a Hamiltonian path R_3^k of $\cup_{t=n-k+2}^{n} S_n^{\{b_{k,t}\}}$ joining the black vertex $(\mathbf{v_k})^n$ to the black vertex $((\mathbf{e})^{k-1})^n$ such that $R_3^k(i+(j-1)(n-1)!) \in S_n^{\{b_{k,n-k+j+1}\}}$ for every $i \in \langle (n-1)! \rangle$ and for every $j \in \langle k-1 \rangle$. We set $C_k = \langle \mathbf{e}, (\mathbf{e})^k, ((\mathbf{e})^k)^n, R_2^k, (\mathbf{w_k})^n, \mathbf{w_k}, R_1^k, \mathbf{v_k}, (\mathbf{v_k})^n, R_3^k, ((\mathbf{e})^{k-1})^n, (\mathbf{e})^{k-1}, \mathbf{e} \rangle$.

Then $\{C_1, C_2, \ldots, C_{n-1}\}$ forms a set of $(n-1)$ mutually independent Hamiltonian cycles of S_n starting form the vertex $\mathbf{e}$. □

18.6 FAULT-FREE MUTUALLY INDEPENDENT HAMILTONIAN CYCLES IN A FAULTY HYPERCUBE

We can also study the mutually independent Hamiltonicity of a faulty graph. In Refs 163 and 209, the mutually independent Hamiltonicity of hypercubes with edge fault are discussed. Then we study the mutually independent Hamiltonicity of star graphs in the following section.

Let F be the set of faulty edges of Q_n. Suppose that Q_n is partitioned over dimension i into $\{Q_n^0, Q_n^1\}$ for some $1 \le i \le n$ and E_c is the set of crossing edges between Q_n^0 and Q_n^1. Then we define $F_0 = F \cap E(Q_n^0)$, $F_1 = F \cap E(Q_n^1)$, and $F_c = F \cap E_c$. Moreover, we set $\delta = n-1-|F|$ in the remainder of this paper. To tolerate faulty edges in hypercube networks, we have the following results.

LEMMA 18.13 Let $F \subseteq E(Q_n)$ be a set of at most $n-2$ faulty edges for $n \ge 3$. Suppose that $A = \{(\mathbf{w}_i, \mathbf{b}_i) \in E(Q_n) \mid \mathbf{w}_i \in V_0(Q_n), \mathbf{b}_i \in V_1(Q_n), 1 \le i \le \delta\}$ consists of δ distinct edges with no shared endpoints. Then $Q_n - F$ contains δ mutually fully independent Hamiltonian paths $P_1, P_2, \ldots, P_\delta$ where P_i is joining $\mathbf{w}_i$ to $\mathbf{b}_i$ for every $i \in \{1, 2, \ldots, \delta\}$.

Proof: By Lemma 18.7, this lemma holds for $|F| = 0$. Suppose that $|F| = n-2$. Then $\delta = n-1-(n-2) = 1$. By Theorem 13.1, there is a Hamiltonian path of $Q_n - F$ between any two vertices from different partite sets. Hence, we need to consider only the case where $1 \le |F| \le n-3$. The proof is by induction on n. One can see that the statement holds for Q_3, as the induction basis. In what follows, we assume that $n \ge 4$. Suppose that the statement holds for Q_{n-1}. Since $\delta + |F| = n-1 < n$, there must exist a dimension d such that no edge of $A \cup F$ is a d-dimensional edge. Since Q_n is edge-transitive, we can assume, without loss of generality, that $d = n$. Then we partition Q_n into $\{Q_n^0, Q_n^1\}$ over dimension $d = n$.

Thus, each $(\mathbf{w}_i, \mathbf{b}_i)$ is either in Q_n^0 or in Q_n^1. Let $r_0 = |\{(\mathbf{w}_i, \mathbf{b}_i) \in E(Q_n^0) \mid 1 \le i \le \delta\}|$ and $r_1 = |\{(\mathbf{w}_i, \mathbf{b}_i) \in E(Q_n^1) \mid 1 \le i \le \delta\}|$. Clearly, $r_0 + r_1 = \delta$. Without loss of generality, we assume that $\{(\mathbf{w}_1, \mathbf{b}_1), \ldots, (\mathbf{w}_{r_0}, \mathbf{b}_{r_0})\} \subset E(Q_n^0)$ and $\{(\mathbf{w}_{r_0+1}, \mathbf{b}_{r_0+1}), \ldots, (\mathbf{w}_\delta, b_\delta)\} \subset E(Q_n^1)$. Moreover, we have that $r_0 + |F_0| \le \delta + |F| = n-1$ and $r_1 + |F_1| \le \delta + |F| = n-1$. With symmetry, we assume that $r_0 + |F_0| \ge r_1 + |F_1|$. Then we have to take the following cases into account.

Case 1. Suppose that $r_i + |F_j| \le n-2$ for all $i, j \in \{0, 1\}$. Since $r_0 + |F_0| \le n-2$, $r_0 \le n-2-|F_0| = (n-1)-1-|F_0|$. By the induction hypothesis, there exist r_0 mutually fully independent Hamiltonian paths $H_i = \langle \mathbf{w}_i, H_i', \mathbf{u}_i, \mathbf{b}_i \rangle$ joining $\mathbf{w}_i$ to $\mathbf{b}_i$, $1 \le i \le r_0$, with some vertex $\mathbf{u}_i$ adjacent to $\mathbf{b}_i$ in $Q_n^0 - F_0$. Similarly, there exist r_1 mutually fully independent Hamiltonian paths $H_i = \langle \mathbf{w}_i, H_i', \mathbf{u}_i, \mathbf{b}_i \rangle$ joining $\mathbf{w}_i$ to $\mathbf{b}_i$, $r_0 + 1 \le i \le \delta$, with some $\mathbf{u}_i$ adjacent to $\mathbf{b}_i$ in $Q_n^1 - F_1$.

Next, we construct r_0 paths in $Q_n^1 - F_1$ to incorporate the previously established r_0 paths of $Q_n^0 - F_0$. Since $r_0 + |F_1| \le n-2$, we have $r_0 \le n-2-|F_1|$. By the induction hypothesis, $Q_n^1 - F_1$ also contains r_0 mutually fully independent Hamiltonian paths $R_1, R_2, \ldots, R_{r_0}$ where R_i joining $(\mathbf{u}_i)^n$ to $(\mathbf{b}_i)^n$ for every $i \in \langle\{1, 2, \ldots, r_0\}$. Similarly, $Q_n^0 - F_0$ also contains r_1 mutually fully independent Hamiltonian paths $R_{r_0+1}, R_{r_0+2}, \ldots, R_\delta$ where R_i joining $(\mathbf{u}_i)^n$ to $(\mathbf{b}_i)^n$ for every $i \in \langle\{r_0+1, r_0+2, \ldots, \delta\}$. Accordingly, we set $P_i = \langle \mathbf{w}_i, H_i', \mathbf{u}_i, (\mathbf{u}_i)^n, R_i, (\mathbf{b}_i)^n, \mathbf{b}_i \rangle$ for every $1 \le i \le \delta$. Thus, $\{P_1, P_2, \ldots, P_\delta\}$ forms a set of δ mutually fully independent Hamiltonian paths in Q_n. See Figure 18.5a for illustration.

Case 2. Suppose that $r_i + |F_i| = n-1$ for some $i \in \{0, 1\}$. Without loss of generality, we assume that $r_0 + |F_0| = n-1$. Since $r_0 = n-1-|F_0| \ge n-1-|F| = \delta$, we must have $r_0 = \delta$ and $|F_0| = |F| \le n-3$. By Theorem 13.1, there is a Hamiltonian path $H_1 = \langle \mathbf{w}_1, \mathbf{u}_1, H_1', (\mathbf{b}_1)^j, \mathbf{b}_1 \rangle$ joining $\mathbf{w}_1$ to $\mathbf{b}_1$ with some $1 \le j \le n-1$ and with some $\mathbf{u}_1$ adjacent to $\mathbf{w}_1$ in $Q_n^0 - F_0$. Note that $r_0 - 1 = \delta - 1 = n-2-|F| = (n-1)-1-|F_0|$. By induction hypothesis, there are $(r_0 - 1)$ mutually fully independent Hamiltonian paths $H_i = \langle \mathbf{w}_i, H_i', \mathbf{u}_i, \mathbf{b}_i \rangle$ joining $\mathbf{w}_i$ to $\mathbf{b}_i$, $2 \le i \le r_0$, with some $\mathbf{u}_i$ adjacent to $\mathbf{b}_i$ in $Q_n^0 - F_0$.

Case 2.1. Suppose that $n = 4$. Let $X_1 = \{((\mathbf{u}_1)^4, \mathbf{v}) \in E(Q_4^1) \mid \mathbf{v} = (\mathbf{u}_2)^4\}$ and $X_2 = \{((\mathbf{u}_2)^4, \mathbf{v}) \in E(Q_4^1) \mid \mathbf{v} = (\mathbf{u}_1)^4\}$. Obviously, $|X_0| \le 1$ and $|X_1| \le 1$. By Theorem 13.1, there is a Hamiltonian path R_1 joining $(\mathbf{w}_1)^4$ to $(\mathbf{u}_1)^4$ of $Q_4^1 - X_1$ and there is a Hamiltonian path R_2 joining $(\mathbf{u}_2)^4$ to $(\mathbf{b}_2)^4$ of $Q_4^1 - X_2$. One can see that $R_1(7) \ne R_2(1)$ and $R_1(8) \ne R_2(2)$. Then we set $P_1 = \langle \mathbf{w}_1, (\mathbf{w}_1)^4, R_1, (\mathbf{u}_1)^4, \mathbf{u}_1, H_1', (\mathbf{b}_1)^j, \mathbf{b}_1 \rangle$ and $P_2 = \langle \mathbf{w}_2, H_2', \mathbf{u}_2, (\mathbf{u}_2)^4, R_2, (\mathbf{b}_2)^4, \mathbf{b}_2 \rangle$. Consequently, $\{P_1, P_2\}$ forms a set of two mutually fully independent Hamiltonian paths in Q_4. See Figure 18.5b for illustration.

Case 2.2. Suppose that $n \ge 5$. By Lemma 18.6, there is a Hamiltonian path R_1 joining $(\mathbf{w}_1)^n$ to $(\mathbf{u}_1)^n$ in $Q_n^1 - \{(\mathbf{b}_1)^n, ((\mathbf{b}_1)^j)^n\}$. Then we set $P_1 = \langle \mathbf{w}_1, (\mathbf{w}_1)^n, R_1, (\mathbf{u}_1)^n, \mathbf{u}_1, H_1', (\mathbf{b}_1)^j, ((\mathbf{b}_1)^j)^n, (\mathbf{b}_1)^n, \mathbf{b}_1 \rangle$. Since $r_0 - 1 = \delta - 1 = n-2-|F| < n-2$, there exist $(r_0 - 1)$ mutually fully independent Hamiltonian paths R_i joining $(\mathbf{u}_i)^n$ to $(\mathbf{b}_i)^n$, $2 \le i \le r_0$ in $Q_n^1 - \{((\mathbf{b}_i)^n, ((\mathbf{b}_1)^j)^n)\}$. One can see that $R_i(2^{n-1} - 1) \ne ((\mathbf{b}_1)^j)^n$ for $2 \le i \le r_0$. Thus, we set $P_i = \langle \mathbf{w}_i, H_i', \mathbf{u}_i, (\mathbf{u}_i)^n, R_i, (\mathbf{b}_i)^n, \mathbf{b}_i \rangle$

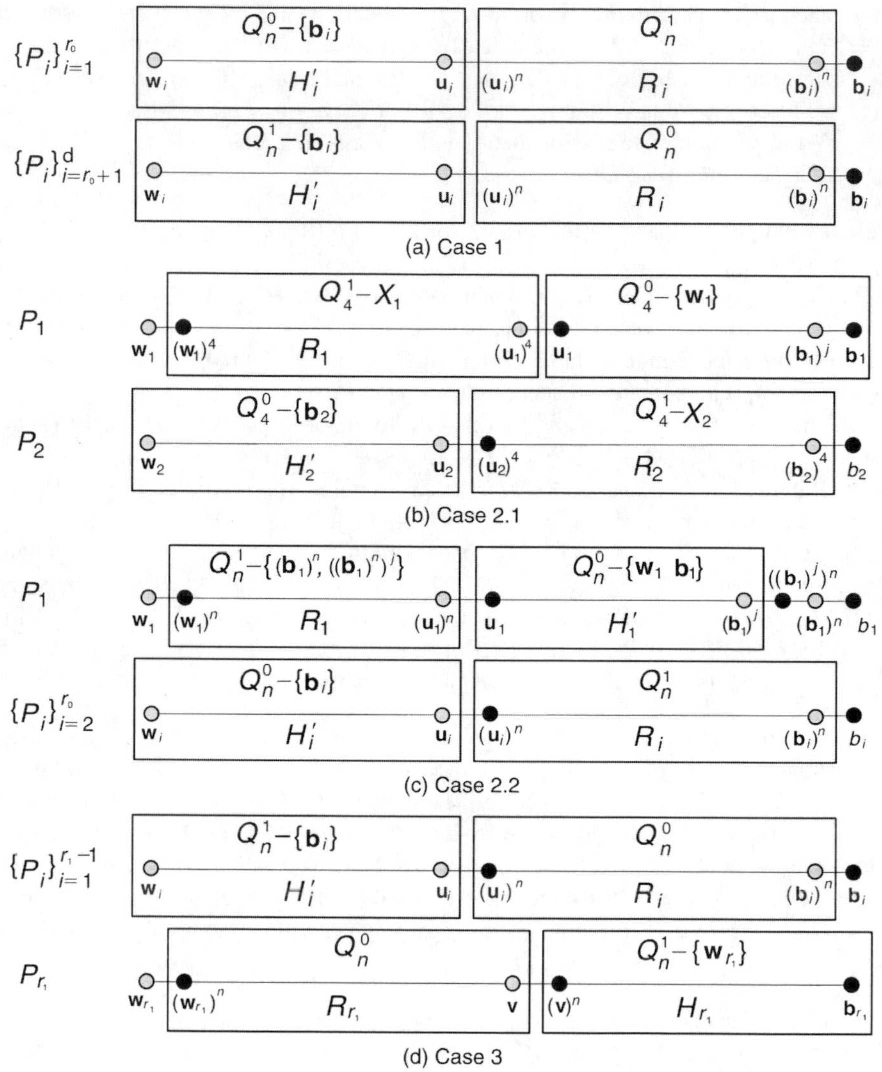

FIGURE 18.5 Illustration for the proof of Lemma 18.13.

for $2 \le i \le r_0$. Consequently, $\{P_1, \ldots, P_{r_0}\}$ forms a set of r_0 mutually fully independent Hamiltonian paths in Q_n. See Figure 18.5c for illustration.

Case 3. Suppose that $r_i + |F_{1-i}| = n - 1$ for some $i \in \{0, 1\}$. Without loss of generality, we assume that $r_1 + |F_0| = n - 1$. Since $r_1 = n - 1 - |F_0| \ge n - 1 - |F| = \delta$, we must have $r_1 = \delta$ and $|F_0| = |F|$. By induction hypothesis, there are $(r_1 - 1)$ mutually fully independent Hamiltonian paths $H_i = \langle \mathbf{w}_i, H_i', \mathbf{u}_i, \mathbf{b}_i \rangle$ joining $\mathbf{w}_i$ to $\mathbf{b}_i$, $1 \le i \le r_1 - 1$, with some $\mathbf{u}_i$ adjacent to $\mathbf{b}_i$ in Q_n^1. Since $r_1 - 1 = \delta - 1 = n - 2 - |F| = (n - 1) - 1 - |F_0|$, there exist $(r_1 - 1)$ mutually fully independent Hamiltonian paths R_i joining $(\mathbf{u}_i)^n$ to $(\mathbf{b}_i)^n$, $1 \le i \le r_1 - 1$, in $Q_n^0 - F_0$.

For $1 \leq i \leq r_1 - 1$, we set $P_i = \langle \mathbf{w}_i, H_i', \mathbf{u}_i, (\mathbf{u}_i)^n, R_i, (\mathbf{b}_i)^n, \mathbf{b}_i \rangle$. Next, we have to choose a vertex $\mathbf{v}$ of $V_0(Q_n^0)$ and construct a Hamiltonian path R_{r_1} joining $(\mathbf{w}_{r_1})^n$ to $\mathbf{v}$ in $Q_n^0 - F_0$ such that $\mathbf{v} \neq R_i(2)$ and $R_{r_1}(2^{n-1} - 1) \neq (\mathbf{u}_i)^n$ for every $1 \leq i \leq r_1 - 1$. This leads to the following cases.

Case 3.1. Suppose that $n \neq 5$ or $|F| > 1$. One can see that $(\mathbf{u}_1)^n, (\mathbf{u}_2)^n, \ldots, (\mathbf{u}_{r_1-1})^n$ have at most $(r_1 - 1)(n - 1)$ neighbors in $V_0(Q_n^0)$. Since $2^{n-2} > (r_1 - 1)(n - 1) = (n - 2 - |F|)(n - 1)$ for all $n \geq 3$, we can choose $\mathbf{v}$ other than all neighbors of $(\mathbf{u}_1)^n, (\mathbf{u}_2)^n, \ldots, (\mathbf{u}_{r_1-1})^n$. Obviously, $\mathbf{v} \neq R_i(2)$ for $1 \leq i \leq r_1 - 1$. By Theorem 13.1, there is a Hamiltonian path R_{r_1} joining $(\mathbf{w}_{r_1})^n$ to $\mathbf{v}$ of $Q_n^0 - F_0$ such that $R_{r_1}(2^{n-1} - 1) \neq (\mathbf{u}_i)^n$ for every $1 \leq i \leq r_1 - 1$. By Theorem 13.1, there is a Hamiltonian path H_{r_1} joining $(\mathbf{v})^n$ to $\mathbf{b}_{r_1}$ in $Q_n^1 - \{\mathbf{w}_{r_1}\}$. Then we set $P_{r_1} = \langle \mathbf{w}_{r_1}, (\mathbf{w}_{r_1})^n, R_{r_1}, \mathbf{v}, (\mathbf{v})^n, H_{r_1}, \mathbf{b}_{r_1} \rangle$. Consequently, $\{P_1, P_2, \ldots, P_{r_1}\}$ forms a set of r_1 mutually fully independent Hamiltonian paths in Q_n. See Figure 18.5d for illustration.

Case 3.2. Suppose that $n = 5$, $|F| = 1$, and $(\mathbf{u}_1)^n, (\mathbf{u})_2)^n, \ldots, (\mathbf{u}_{r_1-1})^n$ have at least one common neighbor. Thus, $r_1 = 3$. Since $2^{n-2} = 8 > 7 = (r_1 - 1)(n - 1) - 1$, we still can choose $\mathbf{v}$ other than all neighbors of $(\mathbf{u}_1)^n, (\mathbf{u}_2)^n, \ldots, (\mathbf{u}_{r_1-1})^n$. Obviously, $\mathbf{v} \neq R_i(2)$ for $1 \leq i \leq r_1 - 1$. By Theorem 13.1, there is a Hamiltonian path R_{r_1} joining $(\mathbf{w}_{r_1})^n$ to $\mathbf{v}$ of $Q_n^0 - F_0$ such that $R_{r_1}(2^{n-1} - 1) \neq (\mathbf{u}_i)^n$ for every $1 \leq i \leq r_1 - 1$. By Theorem 13.1, there is a Hamiltonian path H_{r_1} joining $(\mathbf{v})^n$ to $\mathbf{b}_{r_1}$ in $Q_n^1 - \{\mathbf{w}_{r_1}\}$. Similarly, we set $P_{r_1} = \langle \mathbf{w}_{r_1}, (\mathbf{w}_{r_1})^n, R_{r_1}, \mathbf{v}, (\mathbf{v})^n, H_{r_1}, \mathbf{b}_{r_1} \rangle$. Then $\{P_1, P_2, \ldots, P_{r_1}\}$ forms a set of r_1 mutually fully independent Hamiltonian paths in Q_n.

Case 3.3. Suppose that $n = 5$, $|F| = 1$, and $(\mathbf{u}_1)^n, (\mathbf{u}_2)^n, \ldots, (\mathbf{u}_{r_1-1})^n$ have no common neighbors. Then we choose $\mathbf{v}$ as the vertex that is adjacent to $(\mathbf{u}_1)^n$ but is not identical to $R_1(2)$. Obviously, $\mathbf{v} \neq R_i(2)$ for $1 \leq i \leq r_1 - 1$. By Theorem 13.1, $Q_n^0 - (F_0 \cup \{(\mathbf{v}, (\mathbf{u}_1)^n)\})$ remains Hamiltonian laceable. Thus, there is a Hamiltonian path R_{r_1} joining $(\mathbf{w}_{r_1})^n$ to $\mathbf{v}$ of $Q_n^0 - (F_0 \cup \{(\mathbf{v}, (\mathbf{u}_1)^n)\})$ such that $R_{r_1}(2^{n-1} - 1) \neq (\mathbf{u}_i)^n$ for every $1 \leq i \leq r_1 - 1$. By Theorem 13.1, there is a Hamiltonian path H_{r_1} joining $(\mathbf{v})^n$ to $\mathbf{b}_{r_1}$ in $Q_n^1 - \{\mathbf{w}_{r_1}\}$. Similarly, we set $P_{r_1} = \langle \mathbf{w}_{r_1}, (\mathbf{w}_{r_1})^n, R_{r_1}, \mathbf{v}, (\mathbf{v})^n, H_{r_1}, \mathbf{b}_{r_1} \rangle$. Then $\{P_1, P_2, \ldots, P_{r_1}\}$ forms a set of r_1 mutually fully independent Hamiltonian paths in Q_n. □

THEOREM 18.8 Let $n \geq 3$. Suppose that $F \subseteq E(Q_n)$ consists of at most $n - 2$ faulty edges. Then $Q_n - F$ contains $(n - 1 - |F|)$ mutually independent Hamiltonian cycles beginning from any vertex.

Proof: Since Q_n is vertex-transitive, we need to construct only δ mutually independent Hamiltonian cycles beginning from $e = 0^n$. Suppose that $|F| = 0$. By Theorem 18.5, the statement holds. Thus, we only consider the situation that F is nonempty. The proof is by induction on n. It is trivial that the statement holds for Q_3, as the induction basis. For $n \geq 4$, we assume that the statement holds for Q_{n-1}. Now we consider how to build δ mutually independent Hamiltonian cycles in $Q_n - F$. By selecting an arbitrary faulty edge of dimension d, $d \in \{1, 2, \ldots, n\}$, one can partition Q_n into $\{Q_n^0, Q_n^1\}$ over dimension d such that $|F_0|, |F_1| \leq n - 3$.

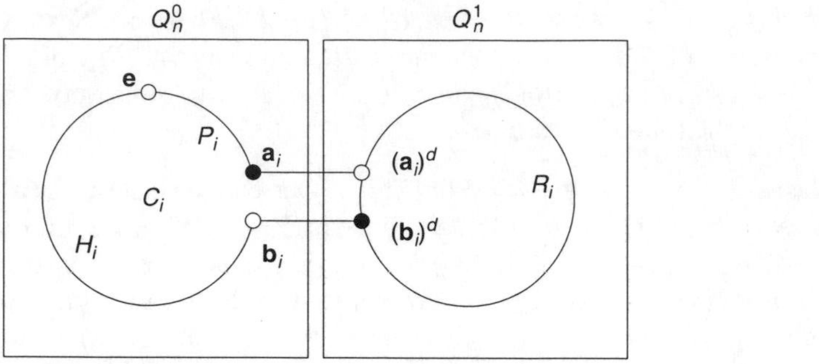

FIGURE 18.6 Illustration for the proof of Theorem 18.8.

By the induction hypothesis, $Q_n^0 - F_0$ contains δ mutually independent Hamiltonian cycles $C_1, C_2, \ldots, C_\delta$ beginning from $\mathbf{e} = 00\ldots0$, because of $|F_0| \leq |F| - 1 \leq n-3$ and $(n-1) - 1 - |F_0| \geq (n-1) - 1 - (|F| - 1) = n - 1 - |F| = \delta$. For convenience, we assume that the vertices of each cycle are ranked clockwise from 0 to $2^{n-1} - 1$; that is, the beginning vertex $\mathbf{e}$ has index 0. Next, we claim that there must exist an integer t, $0 \leq t \leq 2^{n-2} - 1$, so that the crossing edges $(C_i(t), (C_i(t))^d)$ and $(C_i(t + 1(\text{mod } 2^{n-1})), (C_i(t + 1(\text{mod } 2^{n-1})))^d)$ are both fault-free for all $1 \leq i \leq \delta$. If such edges do not exist, then $|F| \geq |F_c| \geq 2^{n-2}/\delta > |F|$ for $n \geq 3$, which leads to an immediate contradiction. Let $\mathbf{a}_i = C_i(t)$ and $\mathbf{b}_i = C_i(t + 1(\text{mod } 2^{n-1}))$. Thus, C_i can be represented as $\langle \mathbf{e}, P_i, \mathbf{a}_i, \mathbf{b}_i, H_i, \mathbf{e}\rangle$, $1 \leq i \leq \delta$. See Figure 18.6 for illustration. Note that $(\mathbf{a}_i)^d$ and $(\mathbf{b}_i)^d$ are adjacent in Q_n^1. By Lemma 18.13, $Q_n^1 - F_1$ contains δ mutually fully independent Hamiltonian paths $R_1, R_2, \ldots, R_\delta$ where R_i joining $(\mathbf{a}_i)^d$ to $(\mathbf{b}_i)^d$ for every $1 \leq i \leq \delta$. Therefore, $\{\langle \mathbf{e}, P_i, \mathbf{a}_i, (\mathbf{a}_i)^d, R_i, (\mathbf{b}_i)^d, \mathbf{b}_i, H_i, \mathbf{e}\rangle \mid 1 \leq i \leq \delta\}$ forms a set of δ mutually independent Hamiltonian cycles beginning from $\mathbf{e}$. □

18.7 FAULT-FREE MUTUALLY INDEPENDENT HAMILTONIAN CYCLES IN FAULTY STAR GRAPHS

Kueng et al. [211] study the Hamiltonicity of star graphs with edge faults. By Theorem 13.2, $S_n - F$ remains Hamiltonian for any $F \subset E(S_n)$ with $|F| \leq n-3$. Kueng et al. prove that $IHC(S_n - F) \geq n - 1 - f$ for any $F \subset E(S_n)$ with $f = |F| \leq n-3$.

LEMMA 18.14 Let $n \geq 5$. Assume that $F \subset E(S_n)$ with $|F| \leq n-4$ and assume that $I = \{a_1, \ldots, a_r\}$ is a subset of r elements of $\langle n\rangle$ for some $r \in \langle n\rangle$. Suppose that $\mathbf{u} \in V_0(S_n^{\{a_1\}})$ and $\mathbf{v} \in V_1(S_n^{\{a_r\}})$. Then there is a Hamiltonian path $H = \langle \mathbf{u} = \mathbf{x}_1, P_1, \mathbf{y}_1, \mathbf{x}_2, P_2, \mathbf{y}_2, \ldots, \mathbf{x}_r, P_r, \mathbf{y}_r = \mathbf{v}\rangle$ of $S_n^I - F$ joining $\mathbf{u}$ to $\mathbf{v}$ such that $\mathbf{x}_1 = \mathbf{u}$, $\mathbf{y}_r = \mathbf{v}$, and P_i is a Hamiltonian path of $S_n^{\{a_i\}} - F$ joining $\mathbf{x}_i$ to $\mathbf{y}_i$ for every $1 \leq i \leq r$.

Proof: By Theorem 13.2, this statement holds for $r = 1$. Thus, suppose that $r \geq 2$ and we set $\mathbf{x}_1 = \mathbf{u}$ and $\mathbf{y}_r = \mathbf{v}$. By Lemma 13.7, there are $(n-2)!/2 > n-4$ edges joining vertices of $V_1(S_n^{\{a_i\}})$ to vertices of $V_0(S_n^{\{a_{i+1}\}})$ for every $i \in \langle r-1 \rangle$. Therefore, we choose $(\mathbf{y}_i, \mathbf{x}_{i+1}) \in E^{a_i,a_{i+1}} - F$ with $\mathbf{y}_i \in V_1(S_n^{\{a_i\}})$ and $\mathbf{x}_{i+1} \in V_0(S_n^{\{a_{i+1}\}})$ for $i \in \langle r-1 \rangle$. By Theorem 13.2, there is a Hamiltonian path P_i of $S_n^{\{a_i\}} - F$ joining $\mathbf{x}_i$ to $\mathbf{y}_i$ for every $i \in \langle r \rangle$. As a result, $\langle \mathbf{u} = \mathbf{x}_1, P_1, \mathbf{y}_1, \mathbf{x}_2, P_2, \mathbf{y}_2, \ldots, \mathbf{x}_r, P_r, \mathbf{y}_r = \mathbf{v} \rangle$ forms the desired Hamiltonian path of $S_n^I - F$ joining $\mathbf{u}$ to $\mathbf{v}$. □

LEMMA 18.15 Let $n \geq 5$. Assume that $F \subset E(S_n)$ with $|F| \leq n-4$ and $|F \cap S_n^{\{i\}}| \leq n-5$ for every $i \in \langle n \rangle$. Moreover, assume that $I = \{a_1, \ldots, a_r\}$ is a subset of r elements of $\langle n \rangle$ for some $2 \leq r \leq n$. Suppose that $\mathbf{u} \in V_0(S_n^{\{a_1\}})$, $\mathbf{w} \in V_1(S_n^{\{a_1\}})$, and $\mathbf{v} \in V_0(S_n^{\{a_r\}})$. Then there is a Hamiltonian path H of $S_n^I - (F \cup \{\mathbf{w}\})$ joining $\mathbf{u}$ to $\mathbf{v}$.

Proof: By Lemma 13.7, there are $(n-2)!/2 > n-3$ edges joining vertices of $V_0(S_n^{\{a_1\}})$ to vertices of $V_1(S_n^{\{a_2\}})$. Thus, we choose a vertex $\mathbf{x}$ of $V_0(S_n^{\{a_1\}}) - \{\mathbf{u}\}$ with $(\mathbf{x})_1 = a_2$ and $(\mathbf{x}, (\mathbf{x})^n) \notin F$. By Theorem 13.2, there is a Hamiltonian path P of $S_n^{\{a_1\}} - (F \cup \{\mathbf{w}\})$ joining $\mathbf{u}$ to $\mathbf{x}$. By Lemma 18.14, there is a Hamiltonian path Q of $S_n^{I-\{a_1\}} - F$ joining $(\mathbf{x})^n$ to $\mathbf{v}$. As a result, $\langle \mathbf{u}, P, \mathbf{x}, (\mathbf{x})^n, Q, \mathbf{v} \rangle$ forms the desired Hamiltonian path. □

LEMMA 18.16 Let $i \in \langle n \rangle$ and $F \subset E(S_n)$ with $|F| \leq n-4$ for $n \geq 4$. Suppose that $\mathbf{w}$ and $\mathbf{b}$ are two adjacent vertices of S_n and $\mathbf{u} \in V_0(S_n) - \{\mathbf{w}, \mathbf{b}\}$. Then there is a Hamiltonian path of $S_n - (F \cup \{\mathbf{w}, \mathbf{b}\})$ joining $\mathbf{u}$ to some vertex $\mathbf{v}$ of $V_1(S_n) - \{\mathbf{w}, \mathbf{b}\}$ with $(\mathbf{v})_1 = i$.

Proof: Since S_n is vertex-transitive and edge-transitive, we assume that $\mathbf{w} = \mathbf{e}$ and $\mathbf{b} = (\mathbf{e})^j$ with some $j \in \langle n \rangle - \{1\}$. We set $F_k = F \cap E(S_n^{\{k\}})$ for every $k \in \langle n \rangle$. Then we prove this lemma by induction on n. The induction bases depend upon Lemma 13.35. Suppose that this statement holds on S_{n-1} with $n \geq 5$. We consider the dimensions of all edges of $F \cup \{(\mathbf{e}, (\mathbf{e})^j)\}$. If there is an edge of F whose dimension, say j', is different from j, we can partition S_n over dimension j'. Otherwise, every edge of F has the same dimension as j.

Case 1. There is an edge of F whose dimension, say j', is different from j. Since S_n is edge-transitive, we assume that $j' = n$. Thus, $(\mathbf{e}, (\mathbf{e})^j) \in E(S_n^{\{n\}})$ and $|F_k| \leq n-5$ for every $k \in \langle n \rangle$.

Case 1.1. Suppose that $\mathbf{u} \in V_0(S_n^{\{n\}})$. Since $|F| \leq n-4$, we can choose an integer $r \in \langle n-1 \rangle$ such that $|F \cap E^{r,n}| = 0$. By induction hypothesis, there is a Hamiltonian path P of $S_n^{\{n\}} - (F_n \cup \{\mathbf{e}, (\mathbf{e})^j\})$ joining $\mathbf{u}$ to a vertex $\mathbf{x} \in V_1(S_n^{\{n\}})$ with $(\mathbf{x})_1 = r$. We choose a vertex $\mathbf{v}$ in $V_1(S_n^{\langle n-1 \rangle - \{r\}})$ with $(\mathbf{v})_1 = i$. By Lemma 18.14, there is a Hamiltonian path Q of $S_n^{\langle n-1 \rangle} - F$ joining $(\mathbf{x})^n$ to $\mathbf{v}$. Then $\langle \mathbf{u}, P, \mathbf{x}, (\mathbf{x})^n, Q, \mathbf{v} \rangle$ is the desired path.

Case 1.2. Suppose that $\mathbf{u} \in V_0(S_n^{\{k\}})$ for some $k \in \langle n-1 \rangle$. By Lemma 13.7, there are $(n-2)!/2 > n-3$ edges joining vertices of $V_1(S_n^{\{k\}})$ to vertices of $V_0(S_n^{\{n\}})$. We choose a vertex $\mathbf{y}$ of $V_1(S_n^{\{k\}})$ such that $(\mathbf{y})^n \in V_0(S_n^{\{n\}}) - \{\mathbf{e}\}$ and $(\mathbf{y}, (\mathbf{y})^n) \notin F$. By

Theorem 13.2, there is a Hamiltonian path H of $S_n^{\{k\}} - F_k$ joining $\mathbf{u}$ to $\mathbf{y}$. We choose an integer r of $\langle n-1\rangle - \{k\}$ such that $|F \cap E^{r,n}| = 0$. By induction hypothesis, there is a Hamiltonian path P of $S_n^{\{n\}} - (F_n \cup \{\mathbf{e}, (\mathbf{e})^j\})$ joining $(\mathbf{y})^n$ to a vertex $\mathbf{x}$ of $V_1(S_n^{\{n\}}) - \{(\mathbf{e})^j\}$ with $(\mathbf{x})_1 = r$. Besides, we choose a vertex $\mathbf{v}$ of $V_1(S_n^{\langle n-1\rangle - \{k,r\}})$ with $(\mathbf{v})_1 = i$. By Lemma 18.14, there is a Hamiltonian path Q of $S_n^{\langle n-1\rangle - \{k\}} - F$ joining $(\mathbf{x})^n$ to $\mathbf{v}$. Then $\langle \mathbf{u}, H, \mathbf{y}, (\mathbf{y})^n, P, \mathbf{x}, (\mathbf{x})^n, Q, \mathbf{v}\rangle$ forms the desired path.

Case 2. Every edge of F has the same dimension j. Without loss of generality, we may assume that $j = n$. Thus, $|F_t| = 0$ for every $t \in \langle n\rangle$.

Case 2.1. Suppose that $\mathbf{u} \in V_0(S_n^{\{k\}})$ for some $k \in \langle n-1\rangle - \{1\}$. By Lemma 13.7, there are $(n-2)!/2 > n-4$ edges joining vertices of $V_1(S_n^{\{k\}})$ to vertices of $V_0(S_n^{\{1\}})$. Thus, we can choose a vertex $\mathbf{x}$ of $V_1(S_n^{\{k\}})$ with $(\mathbf{x})_1 = 1$ and $(\mathbf{x}, (\mathbf{x})^n) \notin F$. By Theorem 13.2, there is a Hamiltonian path H of $S_n^{\{k\}}$ joining $\mathbf{u}$ to $\mathbf{x}$. Similarly, we can choose a vertex $\mathbf{y}$ of $V_0(S_n^{\{1\}})$ with $(\mathbf{y})_1 = n$ and $(\mathbf{y}, (\mathbf{y})^n) \notin F$. By Theorem 13.2, there is a Hamiltonian path P of $S_n^{\{1\}} - \{(\mathbf{e})^n\}$ joining $(\mathbf{x})^n$ to $\mathbf{y}$. Let $\mathbf{v}$ be a vertex in $V_1(S_n^{\langle n-1\rangle - \{1,k\}})$ with $(\mathbf{v})_1 = i$. By Lemma 18.15, there is a Hamiltonian path Q of $S_n^{\langle n\rangle - \{1,k\}} - (F \cup \{\mathbf{e}\})$ joining $(\mathbf{y})^n$ to $\mathbf{v}$. Then $\langle \mathbf{u}, H, \mathbf{x}, (\mathbf{x})^n, P, \mathbf{y}, (\mathbf{y})^n, Q, \mathbf{v}\rangle$ forms the desired path.

Case 2.2. Suppose that $\mathbf{u} \in V_0(S_n^{\{1\}})$. By Lemma 13.7, there are $(n-2)!/2 > n-4$ edges joining vertices of $V_0(S_n^{\{1\}})$ to vertices of $V_1(S_n^{\{n\}})$. Thus, we can choose a vertex $\mathbf{x}$ of $V_0(S_n^{\{1\}}) - \{\mathbf{u}\}$ with $(\mathbf{x})_1 = n$ and $(\mathbf{x}, (\mathbf{x})^n) \notin F$. By Theorem 13.2, there is a Hamiltonian path H of $S_n^{\{1\}} - \{(\mathbf{e})^n\}$ joining $\mathbf{u}$ to $\mathbf{x}$. We choose a vertex $\mathbf{v}$ of $V_1(S_n^{\langle n-1\rangle - \{1\}})$ with $(\mathbf{v})_1 = i$. By Lemma 18.15, there is a Hamiltonian path Q of $S_n^{\langle n\rangle - \{1\}} - (F \cup \{\mathbf{e}\})$ joining $(\mathbf{x})^n$ to $\mathbf{v}$. Then $\langle \mathbf{u}, H, \mathbf{x}, (\mathbf{x})^n, Q, \mathbf{v}\rangle$ forms the desired path.

Case 2.3. Suppose that $\mathbf{u} \in V_0(S_n^{\{n\}})$. Since $|F| \le n-4$, we can choose two integers k_1 and k_2 in $\langle n-1\rangle - \{1\}$ such that $((\mathbf{e})^{k_1}, ((\mathbf{e})^{k_1})^n) \notin F$ and $((\mathbf{e})^{k_2}, ((\mathbf{e})^{k_2})^n) \notin F$. Let $X = \{(\mathbf{e}, (\mathbf{e})^t) \mid t \in \langle n-1\rangle - \{1, k_1, k_2\}\}$. Obviously, $|X| = n-4$. Moreover, we can choose a vertex $\mathbf{x} \in V_1(S_n^{\{n\}})$ such that $(\mathbf{x})_1 \in \langle n-1\rangle - \{1, k_1, k_2\}$ and $(\mathbf{x}, (\mathbf{x})^n) \notin F$. Since $(\mathbf{x})_1 \neq k_1$ and $(\mathbf{x})_1 \neq k_2$, we have $\mathbf{x} \neq (\mathbf{e})^{k_1}$ and $\mathbf{x} \neq (\mathbf{e})^{k_2}$. By Theorem 13.2, there is a Hamiltonian path $H = \langle \mathbf{u}, H_1, (\mathbf{e})^{k_1}, \mathbf{e}, (\mathbf{e})^{k_2}, H_2, \mathbf{x}\rangle$ of $S_n^{\{n\}} - X$ joining $\mathbf{u}$ to $\mathbf{x}$. Let $\mathbf{y} = (\mathbf{e})^{k_2}$. Since $(\mathbf{y})_1 \neq (\mathbf{x})_1$, we have $i \neq (\mathbf{x})_1$ or $i \neq (\mathbf{y})_1$.

Case 2.3.1. Suppose that $i \neq (\mathbf{x})_1$. Let $k_3 = (\mathbf{x})_1$. We choose a vertex $\mathbf{v}$ of $V_1(S_n^{\{k_3\}})$ with $(\mathbf{v})_1 = i$. By Lemma 13.7, there are $(n-2)!/2 > n-4$ edges joining vertices of $V_1(S_n^{\{k_1\}})$ to vertices of $V_0(S_n^{\{1\}})$. Thus, we can choose a vertex $\mathbf{z}$ of $V_1(S_n^{\{k_1\}})$ with $(\mathbf{z})_1 = 1$ and $(\mathbf{z}, (\mathbf{z})^n) \notin F$. By Theorem 13.2, there is a Hamiltonian path T of $S_n^{\{k_3\}}$ joining $(\mathbf{x})^n$ to $\mathbf{v}$. Similarly, there is a Hamiltonian path P of $S_n^{\{k_1\}}$ joining $((\mathbf{e})^{k_1})^n$ to $\mathbf{z}$. By Lemma 18.15, there exists a Hamiltonian path Q of $S_n^{\langle n-1\rangle - \{k_1, k_3\}} - (F \cup \{(\mathbf{e})^n\})$ joining $(\mathbf{z})^n$ to $(\mathbf{y})^n$. Then $\langle \mathbf{u}, H_1, (\mathbf{e})^{k_1}, ((\mathbf{e})^{k_1})^n, P, \mathbf{z}, (\mathbf{z})^n, Q, (\mathbf{y})^n, \mathbf{y}, H_2, \mathbf{x}, (\mathbf{x})^n, T, \mathbf{v}\rangle$ forms the desired path.

Case 2.3.2. Suppose that $i \neq (\mathbf{y})_1$. Let $k_3 = (\mathbf{y})_1$. Then the proof of this case is similar to that of Case 2.3.1. □

TABLE 18.1
The Required Hamiltonian Path of $S_4 - \{1234, 2134, 3214\}$

$a=1$ and $b=2$	$\langle 1324, 3142, 4132, 1432, 3412, 4312, 2314, 1324, 3124, 4123, 2143, 1243, 4213, 2413, 1423, 3421, 4321, 2341, 3241, 4231, 2431\rangle$
$a=1$ and $b=3$	$\langle 1423, 2413, 4213, 1243, 2143, 4123, 3124, 1324, 2314, 4312, 3412, 1432, 4132, 3142, 1342, 2341, 4321, 3421, 2431, 4231, 3241\rangle$
$a=1$ and $b=4$	$\langle 1324, 3142, 4132, 1432, 3412, 4312, 2314, 1324, 3124, 4123, 2143, 1243, 4213, 2413, 1423, 3421, 2431, 4231, 3241, 2341, 4321\rangle$
$a=2$ and $b=3$	$\langle 2314, 1324, 3124, 4123, 2143, 1243, 4213, 2413, 1423, 3421, 4321, 2341, 3241, 4231, 2431, 1432, 4132, 3142, 1342, 4312, 3412\rangle$
$a=2$ and $b=4$	$\langle 2314, 1324, 3124, 4123, 2143, 1243, 4213, 2413, 1423, 3421, 4321, 2341, 3241, 4231, 2431, 1432, 3412, 4312, 1342, 3142, 4132\rangle$
$a=3$ and $b=4$	$\langle 3124, 1324, 2314, 4312, 3412, 1432, 4132, 3142, 1342, 2341, 4321, 3421, 2431, 4231, 3241, 1243, 2143, 4123, 1423, 2413, 4213\rangle$

LEMMA 18.17 Let $\{a,b\} \subset \langle n\rangle$ with $a<b$ and let $F \subset E(S_n)$ with $|F| \leq n-4$ for $n \geq 4$. Suppose that $\mathbf{x} \in V_0(S_n)$, and assume that $\mathbf{x_1}$ and $\mathbf{x_2}$ are two distinct neighbors of $\mathbf{x}$. Then there is a Hamiltonian path of $S_n - (F \cup \{\mathbf{x}, \mathbf{x_1}, \mathbf{x_2}\})$ between two vertices $\mathbf{u}$ and $\mathbf{v}$ in $V_0(S_n) - \{\mathbf{x}\}$ such that $(\mathbf{u})_1 = a$ and $(\mathbf{v})_1 = b$.

Proof: Since S_n is vertex-transitive and edge-transitive, we assume that $\mathbf{x} = \mathbf{e}$, $\mathbf{x_1} = (\mathbf{e})^{i_1}$, and $\mathbf{x_2} = (\mathbf{e})^{i_2}$ with some $\{i_1, i_2\} \subset \{2, 3, \ldots, n\}$. We prove this lemma by induction on n.

Suppose that $n=4$. Thus, we have $|F|=0$. Since S_4 is edge-transitive, we assume that $\mathbf{x_1} = (\mathbf{e})^2 = 2134$ and $\mathbf{x_2} = (\mathbf{e})^3 = 3214$. The required paths of $S_4 - \{1234, 2134, 3214\}$ are listed in Table 18.1.

Suppose that the statement holds for S_{n-1} with $n \geq 5$. Let $F_k = F \cap E(S_n^{\{k\}})$ for every $k \in \langle n\rangle$. Without loss of generality, suppose that there is at least one edge of F in dimension n. Thus, $|F_k| \leq n-5$ for every $k \in \langle n\rangle$. Because $a<b$, we have $a \neq n$ and $b \neq 1$. Since $|F| \leq n-4$, we can choose an integer c in $\langle n-1\rangle - \{1, a\}$ such that $|F \cap E^{c,n}| = 0$. Moreover, we choose a vertex $\mathbf{v}$ of $V_0(S_n^{\{1\}})$ with $(\mathbf{v})_1 = b$.

Case 1. Suppose that $i_1 \neq n$ and $i_2 \neq n$. By induction hypothesis, there is a Hamiltonian path H of $S_n^{\{n\}} - (F_n \cup \{\mathbf{e}, (\mathbf{e})^{i_1}, (\mathbf{e})^{i_2}\})$ joining a vertex $\mathbf{u}$ of $V_0(S_n^{\{n\}})$ with $(\mathbf{u})_1 = a$ to a vertex $\mathbf{y}$ of $V_0(S_n^{\{n\}})$ with $(\mathbf{y})_1 = c$. By Lemma 18.14, there is a Hamiltonian path R of $S_n^{\langle n-1\rangle} - F$ joining $(\mathbf{y})^n$ to $\mathbf{v}$. As a result, $\langle \mathbf{u}, H, \mathbf{y}, (\mathbf{y})^n, R, \mathbf{v}\rangle$ forms the desired path in $S_n - (F \cup \{\mathbf{e}, (\mathbf{e})^{i_1}, (\mathbf{e})^{i_2}\})$.

Case 2. Either $i_1 = n$ or $i_2 = n$. Without loss of generality, we assume that $i_2 = n$. We choose a vertex $\mathbf{u} \in V_0(S_n^{\{n\}})$ with $(\mathbf{u})_1 = a$. By Lemma 18.16, there exists a Hamiltonian path H of $S_n^{\{n\}} - (F_n \cup \{\mathbf{e}, (\mathbf{e})^{i_1}\})$ joining a vertex $\mathbf{u}$ to some vertex $\mathbf{y}$ of $V_1(S_n^{\{n\}})$ with $(\mathbf{y})_1 = c$. By Lemma 18.15, there is a Hamiltonian path Q of $S_n^{\langle n-1\rangle} - (F \cup \{(\mathbf{e})^n\})$ joining $(\mathbf{y})^n$ to $\mathbf{v}$. As a result, $\langle \mathbf{u}, H, \mathbf{y}, (\mathbf{y})^n, Q, \mathbf{v}\rangle$ forms the desired path. □

TABLE 18.2
All Hamiltonian Cycles of $S_4 - \{(1234, 4231)\}$, Beginning from 1234

⟨1234, 2134, 3124, 1324, 2314, 4312, 3412, 1432, 4132, 3142, 1342, 2341, 4321, 3421, 2431, 4231, 3241, 1243, 2143, 4123, 1423, 2413, 4213, 3214, 1234⟩
⟨1234, 2134, 3124, 1324, 4321, 2341, 3241, 4231, 2431, 3421, 1423, 4123, 2143, 1243, 4213, 2413, 3412, 1432, 4132, 3142, 1342, 4312, 2314, 3214, 1234⟩
⟨1234, 2134, 3124, 4123, 1423, 2413, 4213, 1243, 2143, 3142, 4132, 1432, 3412, 4312, 1342, 2341, 3241, 4231, 2431, 3421, 4321, 1324, 2314, 3214, 1234⟩
⟨1234, 2134, 4132, 1432, 2431, 4231, 3241, 1243, 2143, 3142, 1342, 2341, 4321, 3421, 1423, 4123, 3124, 1324, 2314, 4312, 3412, 2413, 4213, 3214, 1234⟩
⟨1234, 2134, 4132, 3142, 1342, 4312, 3412, 1432, 2431, 4231, 3241, 2341, 4321, 3421, 1423, 2413, 4213, 1243, 2143, 4123, 3124, 1324, 2314, 3214, 1234⟩
⟨1234, 2134, 4132, 3142, 2143, 4123, 3124, 1324, 2314, 4312, 1342, 2341, 4321, 3421, 1423, 2413, 3412, 1432, 2431, 4231, 3241, 1243, 4213, 3214, 1234⟩
⟨1234, 3214, 2314, 1324, 3124, 4123, 2143, 1243, 4213, 2413, 1423, 3421, 4321, 2341, 3241, 4231, 2431, 1432, 3412, 4312, 1342, 3142, 4132, 2134, 1234⟩
⟨1234, 3214, 2314, 1324, 4321, 3421, 2431, 4231, 3241, 2341, 1342, 4312, 3412, 1432, 4132, 3142, 2143, 1243, 4213, 2413, 1423, 4123, 3124, 2134, 1234⟩
⟨1234, 3214, 2314, 4312, 1342, 3142, 4132, 1432, 3412, 2413, 4213, 1243, 2143, 4123, 1423, 3421, 2431, 4231, 3241, 2341, 4321, 1324, 3124, 2134, 1234⟩
⟨1234, 3214, 4213, 2413, 1423, 4123, 2143, 1243, 3241, 4231, 2431, 3421, 4321, 2341, 1342, 3142, 4132, 1432, 3412, 4312, 2314, 1324, 3124, 2134, 1234⟩
⟨1234, 3214, 4213, 2413, 3412, 4312, 2314, 1324, 3124, 4123, 1423, 3421, 4321, 2341, 1342, 3142, 2143, 1243, 3241, 4231, 2431, 1432, 4132, 2134, 1234⟩
⟨1234, 3214, 4213, 1243, 3241, 4231, 2431, 1432, 3412, 2413, 1423, 3421, 4321, 2341, 1342, 4312, 2314, 1324, 3124, 4123, 2143, 3142, 4132, 2134, 1234⟩

LEMMA 18.18 Let $f \in E(S_4)$. Then $\mathcal{IHC}(S_4 - \{f\}) = 1$.

Proof: Since S_4 is vertex-transitive, we only consider the mutually independent Hamiltonian cycles of $S_4 - \{f\}$ beginning from 1234. Suppose that $f = (1234, 4231)$. We list all Hamiltonian cycles of $S_4 - \{(1234, 4231)\}$, beginning from 1234, in Table 18.2. By brute force, there do not exist two mutually independent Hamiltonian cycles of $S_4 - \{(1234, 4231)\}$ beginning from 1234. Thus, $\mathcal{IHC}(S_4 - \{(1234, 4231)\}) \le 1$. By Theorem 13.2, there exists a Hamiltonian cycle in $S_4 - \{(1234, 4231)\}$. Hence, $\mathcal{IHC}(S_4 - \{f\}) = 1$. □

LEMMA 18.19 Suppose that $n \ge 5$ and $F \subset E(S_n)$ with $|F| = n - 3$. Let $\mathbf{u} \in V(S_n)$. Then there exist two mutually independent Hamiltonian cycles of $S_n - F$ beginning from $\mathbf{u}$.

Proof: Since S_n is vertex-transitive and edge-transitive, we assume that $\mathbf{u} = \mathbf{e}$ and also that F contains at least one edge in dimension n. Let $F_k = F \cap E(S_n^{\{k\}})$ for every $k \in \langle n \rangle$. As a result, $|F_k| \le n - 4$ for every $k \in \langle n \rangle$.

Case 1. Suppose that $(\mathbf{e}, (\mathbf{e})^n) \notin F$. Let $B = (b_{i,j})$ be the $2 \times n$ matrix with

$$b_{i,j} = \begin{cases} j & \text{if } i = 1 \\ n & \text{if } i = 2 \text{ and } j = 1 \\ j+1 & \text{if } i = 2 \text{ and } 2 \leq j \leq n-2 \\ 2 & \text{if } i = 2 \text{ and } j = n-1 \\ 1 & \text{if } i = 2 \text{ and } j = n \end{cases}$$

By Lemma 18.14, there is a Hamiltonian path P of $\left(\bigcup_{j=1}^{n} S_n^{\{b_{1,j}\}}\right) - F$ joining $(\mathbf{e})^n$ to $\mathbf{e}$. Similarly, there is a Hamiltonian path H of $\left(\bigcup_{j=1}^{n} S_n^{\{b_{2,j}\}}\right) - F$ joining $\mathbf{e}$ to $(\mathbf{e})^n$. Then we set $C_1 = \langle \mathbf{e}, (\mathbf{e})^n, P, \mathbf{e}\rangle$ and $C_2 = \langle \mathbf{e}, H, (\mathbf{e})^n, \mathbf{e}\rangle$. Obviously, $\{C_1, C_2\}$ forms a set of two mutually independent Hamiltonian cycles of $S_n - F$ beginning from $\mathbf{e}$. See Figure 18.7a for illustration.

Case 2. Suppose that $(\mathbf{e}, (\mathbf{e})^n) \in F$ and $|F_n| = n-4$. Obviously, $|F_k| = 0$ for every $k \in \langle n-1\rangle$. By Theorem 13.2, there is a Hamiltonian cycle $H = \langle \mathbf{e}, R, \mathbf{q}, \mathbf{p}, \mathbf{e}\rangle$ of $S_n^{\{n\}} - F_n$. Accordingly, we have $(\mathbf{p}, (\mathbf{p})^n) \notin F$ and $(\mathbf{q}, (\mathbf{q})^n) \notin F$. By Lemma 13.8, $(\mathbf{p})_1 \neq (\mathbf{q})_1$. We set $(\mathbf{p})_1 = i_{n-1}$ and $(\mathbf{q})_1 = i_1$. Let $i_2 i_3 \ldots i_{n-2}$ be an arbitrary permutation of $\langle n-1\rangle - \{i_1, i_{n-1}\}$.

For $1 \leq k \leq n-2$, let $\mathbf{x}_k$ be a vertex of $V_0(S_n^{\{i_k\}})$ such that $(\mathbf{x}_k)_1 = i_{k+1}$ and $(\mathbf{x}_k, (\mathbf{x}_k)^n) \notin F$. By Theorem 13.2, there is a Hamiltonian path P_1 of $S_n^{\{i_1\}}$ joining $(\mathbf{q})^n$ to $\mathbf{x}_1$. Similarly, there is a Hamiltonian path P_k of $S_n^{\{i_k\}}$ joining $(\mathbf{x}_{k-1})^n$ to $\mathbf{x}_k$ for $2 \leq k \leq n-2$ and there is a Hamiltonian path P_{n-1} of $S_n^{\{i_{n-1}\}}$ joining $(\mathbf{x}_{n-2})^n$ to $(\mathbf{p})^n$. Then we set $C_1 = \langle \mathbf{e}, R, \mathbf{q}, (\mathbf{q})^n, P_1, \mathbf{x}_1, (\mathbf{x}_1)^n, P_2, \mathbf{x}_2, (\mathbf{x}_2)^n, \ldots, \mathbf{x}_{n-2}, (\mathbf{x}_{n-2})^n, P_{n-1}, (\mathbf{p})^n, \mathbf{p}, \mathbf{e}\rangle$.

Obviously, we can choose a vertex $\mathbf{y}_{n-1}$ of $V_1(S_n^{\{i_{n-1}\}})$ such that $(\mathbf{y}_{n-1})_1 = i_2$ and $(\mathbf{y}_{n-1}, (\mathbf{y}_{n-1})^n) \notin F$. For $2 \leq k \leq n-3$, $|\{\mathbf{u} \in V_1(S_n^{\{i_k\}}) | (\mathbf{u})_1 = i_{k+1}$ and $d(\mathbf{u}, (\mathbf{x}_{k-1})^n) = 2\}| = n-3 < (n-2)!/2$ if $n \geq 5$. Thus, we choose a vertex $\mathbf{y}_k$ of $V_1(S_n^{\{i_k\}})$ such that $d(\mathbf{y}_k, (\mathbf{x}_{k-1})^n) > 2$, $(\mathbf{y}_k)_1 = i_{k+1}$, and $(\mathbf{y}_k, (\mathbf{y}_k)^n) \notin F$ for $2 \leq k \leq n-3$. Since $|\{\mathbf{u} \in V_1(S_n^{\{i_{n-2}\}}) | (\mathbf{u})_1 = i_1$ and $d(\mathbf{u}, (\mathbf{x}_{n-3})^n) = 2\}| = n-3 < (n-2)!/2$ if $n \geq 5$, we choose a vertex $\mathbf{y}_{n-2}$ of $V_1(S_n^{\{i_{n-2}\}})$ such that $d(\mathbf{y}_{n-2}, (\mathbf{x}_{n-3})^n) > 2$, $(\mathbf{y}_{n-2})_1 = i_1$, and $(\mathbf{y}_{n-2}, (\mathbf{y}_{n-2})^n) \notin F$. By Theorem 13.2, there exists a Hamiltonian path Q_1 of $S_n^{\{i_1\}}$ joining $(\mathbf{y}_{n-2})^n$ to $(\mathbf{q})^n$. Again, there is a Hamiltonian path Q_2 of $S_n^{\{i_2\}}$ joining $(\mathbf{y}_{n-1})^n$ to $\mathbf{y}_2$, there is a Hamiltonian path Q_{n-1} of $S_n^{\{i_{n-1}\}}$ joining $(\mathbf{p})^n$ to $\mathbf{y}_{n-1}$, and there is a Hamiltonian path Q_k of $S_n^{\{i_k\}}$ joining $(\mathbf{y}_{k-1})^n$ to $\mathbf{y}_k$ for $3 \leq k \leq n-2$. Then we set $C_2 = \langle \mathbf{e}, \mathbf{p}, (\mathbf{p})^n, Q_{n-1}, \mathbf{y}_{n-1}, (\mathbf{y}_{n-1})^n, Q_2, \mathbf{y}_2, (\mathbf{y}_2)^n, Q_3, \mathbf{y}_3, (\mathbf{y}_3)^n, \ldots, (\mathbf{y}_{n-2})^n, Q_1, (\mathbf{q})^n, \mathbf{q}, R^{-1}, \mathbf{e}\rangle$.

In summary, $\{C_1, C_2\}$ forms a set of two mutually independent Hamiltonian cycles of $S_n - F$ beginning from $\mathbf{e}$. Figure 18.7b illustrates C_1 and C_2 in S_5.

Case 3. Suppose that $(\mathbf{e}, (\mathbf{e})^n) \in F$ and $|F_n| \leq n-5$. Since $|F| = n-3$, there must exist an integer i_{n-1} of $\langle n-1\rangle - \{1\}$ such that $|F \cap E^{i_{n-1},n}| = 0$. Assume that

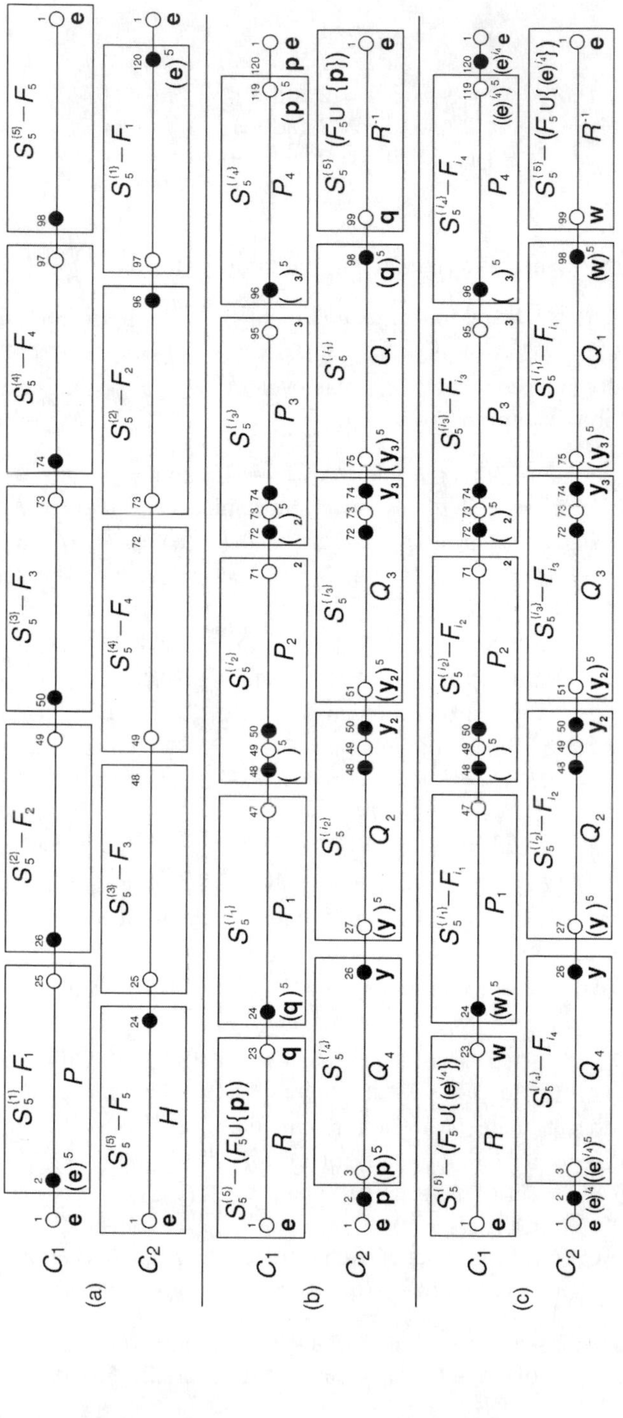

FIGURE 18.7 The two mutually independent Hamiltonian cycles in $S_5 - F$ for Lemma 18.19.

i_1 and i_2 are two integers of $\langle n-1\rangle - \{i_{n-1}\}$ such that $|F \cap E^{i_1,i_2}| = \max\{|F \cap E^{s,t}|\,|s,t \in \langle n-1\rangle - \{i_{n-1}\}\}$. Moreover, let $i_3 i_4 \ldots i_{n-2}$ be an arbitrary permutation of $\langle n-1\rangle - \{i_1, i_2, i_{n-1}\}$. Since $(\mathbf{e}, (\mathbf{e})^n) \in F$, we have $|F \cap E^{i_1,i_2}| \leq n-4$. Thus, $|F \cap E^{i_{n-2},i_1}| \leq n-5$ and $|F \cap E^{i_k,i_{k+1}}| \leq n-5$ for $2 \leq k \leq n-3$.

By Lemma 13.7, there are $(n-2)!/2 > n-3$ edges joining vertices of $V_0(S_n^{\{n\}})$ to vertices of $V_1(S_n^{\{i_1\}})$. Thus, we can choose a vertex $\mathbf{w} \in V_0(S_n^{\{n\}}) - \{\mathbf{e}\}$ such that $(\mathbf{w})_1 = i_1$ and $(\mathbf{w}, (\mathbf{w})^n) \notin F$. By Theorem 13.2, there exists a Hamiltonian path R of $S_n^{\{n\}} - (F_n \cup \{(\mathbf{e})^{i_{n-1}}\})$ joining $\mathbf{e}$ to $\mathbf{w}$. For $1 \leq k \leq n-2$, let $\mathbf{x}_k$ be a vertex of $V_0(S_n^{\{i_k\}})$ such that $(\mathbf{x}_k)_1 = i_{k+1}$ and $(\mathbf{x}_k, (\mathbf{x}_k)^n) \notin F$. By Theorem 13.2, there exists a Hamiltonian path P_1 of $S_n^{\{i_1\}} - F_{i_1}$ joining $(\mathbf{w})^n$ to $\mathbf{x}_1$. Similarly, there is a Hamiltonian path P_k of $S_n^{\{i_k\}} - F_{i_k}$ joining $(\mathbf{x}_{k-1})^n$ to $\mathbf{x}_k$ for $2 \leq k \leq n-2$ and there is a Hamiltonian path P_{n-1} of $S_n^{\{i_{n-1}\}} - F_{i_{n-1}}$ joining $(\mathbf{x}_{n-2})^n$ to $((\mathbf{e})^{i_{n-1}})^n$. Then we set $C_1 = \langle \mathbf{e}, R, \mathbf{w}, (\mathbf{w})^n, P_1, \mathbf{x}_1, (\mathbf{x}_1)^n, P_2, \mathbf{x}_2, (\mathbf{x}_2)^n, \ldots, (\mathbf{x}_{n-2})^n, P_{n-1}, ((\mathbf{e})^{i_{n-1}})^n, (\mathbf{e})^{i_{n-1}}, \mathbf{e}\rangle$.

Obviously, we can choose a vertex $\mathbf{y}_{n-1}$ of $V_1(S_n^{\{i_{n-1}\}})$ such that $(\mathbf{y}_{n-1})_1 = i_2$ and $(\mathbf{y}_{n-1}, (\mathbf{y}_{n-1})^n) \notin F$. For $2 \leq k \leq n-3$, $|\{\mathbf{u} \in V_1(S_n^{\{i_k\}}) | (\mathbf{u})_1 = i_{k+1}$ and $d(\mathbf{u}, (\mathbf{x}_{k-1})^n) = 2\}| = n-3$. By Lemma 13.7, there are $(n-2)!/2$ edges joining vertices of $V_1(S_n^{\{i_k\}})$ to vertices of $V_0(S_n^{\{i_{k+1}\}})$. We emphasize that $(n-2)!/2 > (n-3) + (n-5) = 2n-8$ if $n \geq 5$. Thus, we choose a vertex $\mathbf{y}_k$ of $V_1(S_n^{\{i_k\}})$ such that $d(\mathbf{y}_k, (\mathbf{x}_{k-1})^n) > 2$, $(\mathbf{y}_k)_1 = i_{k+1}$, and $(\mathbf{y}_k, (\mathbf{y}_k)^n) \notin F$ for $2 \leq k \leq n-3$. Since $(n-2)!/2 > |\{\mathbf{u} \in V_1(S_n^{\{i_{n-2}\}}) | (\mathbf{u})_1 = i_1$ and $d(\mathbf{u}, (\mathbf{x}_{n-3})^n) = 2\}| + (n-5) = (n-3) + (n-5) = 2n-8$ if $n \geq 5$, we choose a vertex $\mathbf{y}_{n-2}$ of $V_1(S_n^{\{i_{n-2}\}})$ such that $d(\mathbf{y}_{n-2}, (\mathbf{x}_{n-3})^n) > 2$, $(\mathbf{y}_{n-2})_1 = i_1$, and $(\mathbf{y}_{n-2}, (\mathbf{y}_{n-2})^n) \notin F$. Again, there exists a Hamiltonian path Q_1 of $S_n^{\{i_1\}} - F_{i_1}$ joining $(\mathbf{y}_{n-2})^n$ to $(\mathbf{w})^n$. Again, there is a Hamiltonian path Q_2 of $S_n^{\{i_2\}} - F_{i_2}$ joining $(\mathbf{y}_{n-1})^n$ to $\mathbf{y}_2$, there is a Hamiltonian path Q_{n-1} of $S_n^{\{i_{n-1}\}} - F_{i_{n-1}}$ joining $((\mathbf{e})^{i_{n-1}})^n$ to $\mathbf{y}_{n-1}$, and there is a Hamiltonian path Q_k of $S_n^{\{i_k\}} - F_{i_k}$ joining $(\mathbf{y}_{k-1})^n$ to $\mathbf{y}_k$ for $3 \leq k \leq n-2$. We set $C_2 = \langle \mathbf{e}, (\mathbf{e})^{i_{n-1}}, ((\mathbf{e})^{i_{n-1}})^n, Q_{n-1}, \mathbf{y}_{n-1}, (\mathbf{y}_{n-1})^n, Q_2, \mathbf{y}_2, (\mathbf{y}_2)^n, Q_3, \mathbf{y}_3, (\mathbf{y}_3)^n, \ldots, (\mathbf{y}_{n-2})^n, Q_1, (\mathbf{w})^n, \mathbf{w}, R^{-1}, \mathbf{e}\rangle$.

As a result, $\{C_1, C_2\}$ forms a set of two mutually independent Hamiltonian cycles of $S_n - F$ beginning from $\mathbf{e}$. Figure 18.7c illustrates C_1 and C_2 in S_5. □

LEMMA 18.20 Let f be any integer of $\langle n-4\rangle$ with $n \geq 5$. Suppose that $F \subset E(S_n)$ with $|F| = f$. Let $\mathbf{u} \in V(S_n)$. Then there exist $(n-1-f)$ mutually independent Hamiltonian cycles of $S_n - F$ beginning from $\mathbf{u}$.

Proof: Since S_n is vertex-transitive and edge-transitive, we assume that $\mathbf{u} = \mathbf{e}$ and also that F contains at least one edge in dimension n. Let $F_k = F \cap E(S_n^{\{k\}})$ for every $k \in \langle n\rangle$. Thus, $|F_k| \leq n-5$ for every $k \in \langle n\rangle$. Moreover, let $A_1 = E^{1,n} - \{(\mathbf{e}, (\mathbf{e})^n)\}$ and let $A_i = E^{i,n} \cup \{(\mathbf{e}, (\mathbf{e})^i)\}$ for $2 \leq i \leq n-1$.

Case 1. Suppose that $(\mathbf{e}, (\mathbf{e})^n) \in F$. We emphasize that there are at least $n-1-f$ elements of $\{|F \cap A_2|, |F \cap A_3|, \ldots, |F \cap A_{n-1}|\}$ equal to 0. Without loss of generality, we assume that $|F \cap (\cup_{i=f+1}^{n-1} A_i)| = 0$. Thus, at least one of $\{|F \cap A_1|, \ldots, |F \cap A_f|\}$ equals to 0.

Case 1.1. Suppose that $|F \cap A_1| = 0$. Let $B = (b_{i,j})$ be the $(n-1-f) \times n$ matrix with

$$b_{i,j} = \begin{cases} f+i+j & \text{if } f+i+j \le n \\ f+i+j-n & \text{otherwise} \end{cases}$$

Note that $b_{i,n-f-i} = n$ for every $1 \le i \le n-1-f$. Then we construct $(n-1-f)$ mutually independent Hamiltonian cycles $\{C_1, C_2, \ldots, C_{n-1-f}\}$ of $S_n - F$ beginning from $\mathbf{e}$ as follows.

Let $i \in \langle n-2-f \rangle$. We set $t_i = n-f-i$. By Lemma 18.17, there is a Hamiltonian path Q_i of $S_n^{\{b_{i,t_i}\}} - (F_{b_{i,t_i}} \cup \{\mathbf{e}, (\mathbf{e})^{b_{i,1}}, (\mathbf{e})^{b_{i,n}}\})$ joining two vertices $\mathbf{x}_i$ and $\mathbf{y}_i$ in $V_0(S_n^{\{b_{i,t_i}\}}) - \{\mathbf{e}\}$ such that $(\mathbf{x}_i)_1 = b_{i,t_i-1}$ and $(\mathbf{y}_i)_1 = b_{i,t_i+1}$. By Lemma 18.14, there is a Hamiltonian path P_i of $(\cup_{j=1}^{t_i-1} S_n^{\{b_{i,j}\}}) - F$ joining $((\mathbf{e})^{b_{i,1}})^n$ to $(\mathbf{x}_i)^n$. Similarly, there is a Hamiltonian path R_i of $(\cup_{j=t_i+1}^{n} S_n^{\{b_{i,j}\}}) - F$ joining $(\mathbf{y}_i)^n$ to $((\mathbf{e})^{b_{i,n}})^n$. Then we set $C_i = \langle \mathbf{e}, (\mathbf{e})^{b_{i,1}}, ((\mathbf{e})^{b_{i,1}})^n, P_i, (\mathbf{x}_i)^n, \mathbf{x}_i, Q_i, \mathbf{y}_i, (\mathbf{y}_i)^n, R_i, ((\mathbf{e})^{b_{i,n}})^n, (\mathbf{e})^{b_{i,n}}, \mathbf{e} \rangle$.

By Lemma 18.16, there is a Hamiltonian path T of $S_n^{\{b_{n-1-f,1}\}} - (F_{b_{n-1-f,n}} \cup \{\mathbf{e}, (\mathbf{e})^{b_{n-1-f,n}}\})$ joining $(\mathbf{e})^{b_{1,n}}$ to a vertex $\mathbf{z}$ of $V_0(S_n^{\{b_{n-1-f,1}\}}) - \{\mathbf{e}\}$ with $(\mathbf{z})_1 = b_{n-1-f,2}$. By Lemma 18.14, there is a Hamiltonian path W of $(\cup_{j=2}^{n} S_n^{\{b_{n-1-f,j}\}}) - F$ joining $(\mathbf{z})^n$ to $((\mathbf{e})^{b_{n-1-f,n}})^n$. Then we set $C_{n-1-f} = \langle \mathbf{e}, (\mathbf{e})^{b_{1,n}}, T, \mathbf{z}, (\mathbf{z})^n, W, ((\mathbf{e})^{b_{n-1-f,n}})^n, (\mathbf{e})^{b_{n-1-f,n}}, \mathbf{e} \rangle$.

As a result, $\{C_1, \ldots, C_{n-2-f}, C_{n-1-f}\}$ forms a set of $(n-1-f)$ mutually independent Hamiltonian cycles of $S_n - F$ beginning from $\mathbf{e}$. Figure 18.8 illustrates $\{C_1, C_2, C_3, C_4\}$ in $S_6 - F$ with $|F| = f = 1$.

Case 1.2. Suppose that $|F \cap A_1| > 0$. We emphasize that $f \ge 2$ in this subcase. Thus, at least one of $\{|F \cap A_2|, \ldots, |F \cap A_f|\}$ equals to 0. Without loss of generality, we assume that $|F \cap A_2| = 0$. Let $B = (b_{i,j})$ be the $(n-1-f) \times n$ matrix with

$$b_{i,j} = \begin{cases} f+i+j & \text{if } f+i+j \le n \\ 2 & \text{if } f+i+j = n+1 \\ 1 & \text{if } f+i+j = n+2 \\ f+i+j-n & \text{otherwise} \end{cases}$$

Then we build $(n-1-f)$ mutually independent Hamiltonian cycles $\{C_1, C_2, \ldots, C_{n-1-f}\}$ of $S_n - F$ beginning from $\mathbf{e}$ in the same manner as that of Case 1.1.

Case 2. Suppose that $(\mathbf{e}, (\mathbf{e})^n) \notin F$. We emphasize that there are at least $n-2-f$ elements of $\{|F \cap A_2|, |F \cap A_3| \ldots, |F \cap A_{n-1}|\}$ equaling to 0. Without loss of generality, we assume that $|F \cap (\cup_{i=f+2}^{n-1} A_i)| = 0$. Thus, at least one of $\{|F \cap A_1|, \ldots, |F \cap A_{f+1}|\}$ is 0.

Case 2.1. Suppose that $|F \cap A_1| = 0$. Let $B_n = (b_{i,j})$ be the $(n-1-f) \times n$ matrix with

$$B_5 = \begin{bmatrix} 1 & 2 & 3 & 4 & 5 \\ 4 & 5 & 1 & 2 & 3 \\ 5 & 4 & 2 & 3 & 1 \end{bmatrix}$$

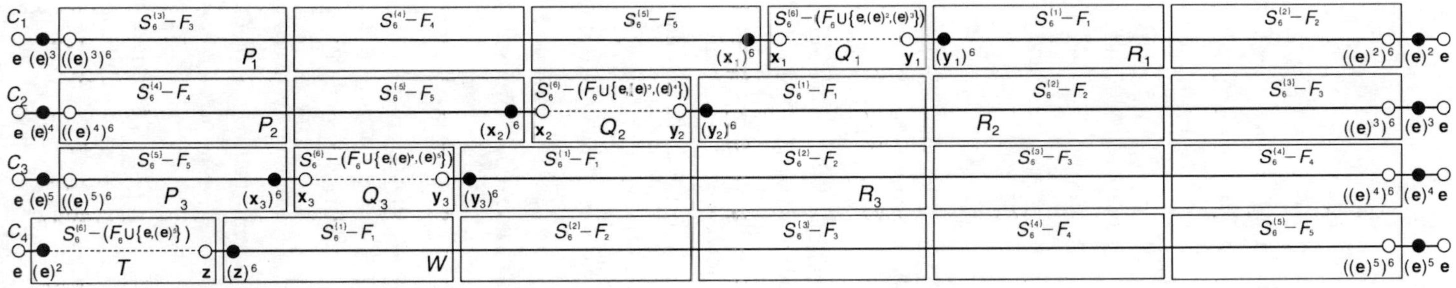

FIGURE 18.8 Mutually independent Hamiltonian cycles in $S_6 - F$ with $|F| = 1$ for Case 1.1 of Lemma 18.20.

and for $n \geq 6$,

$$b_{i,j} = \begin{cases} j & \text{if } i = 1 \\ f+i+j & \text{if } 2 \leq i \leq n-2-f \text{ and } f+i+j \leq n \\ f+i+j-n & \text{if } 2 \leq i \leq n-2-f \text{ and } f+i+j > n \\ n & \text{if } i = n-1-f \text{ and } j = 1 \\ 3 & \text{if } i = n-1-f \text{ and } j = 2 \\ 2 & \text{if } i = n-1-f \text{ and } j = 3 \\ n-1 & \text{if } i = n-1-f \text{ and } j = 4 \\ j-1 & \text{if } i = n-1-f \text{ and } 5 \leq j \leq n-1 \\ 1 & \text{if } i = n-1-f \text{ and } j = n \end{cases}$$

Then we build $(n-1-f)$ mutually independent Hamiltonian cycles $\{C_1, C_2, \ldots, C_{n-1-f}\}$ of $S_n - F$ beginning from $\mathbf{e}$ as follows.

We choose a vertex $\mathbf{v}$ of $V_1(S_n^{\{b_{1,n}\}}) - \{(\mathbf{e})^{b_{n-2-f,1}}\}$ with $(\mathbf{v})_1 = b_{1,n-1}$. By Theorem 13.2, there is a Hamiltonian path W of $S_n^{\{b_{1,n}\}} - (F_{b_{1,n}} \cup \{\mathbf{e}\})$ joining $\mathbf{v}$ to $(\mathbf{e})^{b_{n-2-f,1}}$. By Lemma 18.14, there is a Hamiltonian path D of $(\cup_{j=1}^{n-1} S_n^{\{b_{1,j}\}}) - F$ joining $(\mathbf{e})^n$ to $(\mathbf{v})^n$. We set $C_1 = \langle \mathbf{e}, (\mathbf{e})^n, D, (\mathbf{v})^n, \mathbf{v}, W, (\mathbf{e})^{b_{n-2-f,1}}, \mathbf{e} \rangle$.

Let $i \in \langle n-2-f \rangle - \{1\}$. We set $t_i = n - f - i$. By Lemma 18.17, there is a Hamiltonian path Q_i of $S_n^{\{b_{i,t_i}\}} - (F_{b_{i,t_i}} \cup \{\mathbf{e}, (\mathbf{e})^{b_{i,1}}, (\mathbf{e})^{b_{i,n}}\})$ joining two vertices $\mathbf{x}_i$ and $\mathbf{y}_i$ in $V_0(S_n^{\{b_{i,t_i}\}}) - \{\mathbf{e}\}$ such that $(\mathbf{x}_i)_1 = b_{i,t_i-1}$ and $(\mathbf{y}_i)_1 = b_{i,t_i+1}$. By Lemma 18.14, there is a Hamiltonian path P_i of $(\cup_{j=1}^{t_i-1} S_n^{\{b_{i,j}\}}) - F$ joining $((\mathbf{e})^{b_{i,1}})^n$ to $(\mathbf{x}_i)^n$. Similarly, there is a Hamiltonian path R_i of $(\cup_{j=t_i+1}^{n} S_n^{\{b_{i,j}\}}) - F$ joining $(\mathbf{y}_i)^n$ to $((\mathbf{e})^{b_{i,n}})^n$. Then we set $C_i = \langle \mathbf{e}, (\mathbf{e})^{b_{i,1}}, ((\mathbf{e})^{b_{i,1}})^n, P_i, (\mathbf{x}_i)^n, \mathbf{x}_i, Q_i, \mathbf{y}_i, (\mathbf{y}_i)^n, R_i, ((\mathbf{e})^{b_{i,n}})^n, (\mathbf{e})^{b_{i,n}}, \mathbf{e} \rangle$.

By Lemma 13.7, there are $(n-2)!/2 > n-3$ edges joining vertices of $V_0(S_n^{\{b_{n-1-f,k}\}})$ to vertices of $V_1(S_n^{\{b_{n-1-f,k-1}\}})$ for $3 \leq k \leq n-1$. Thus, we choose a vertex $\mathbf{z}_k$ of $V_0(S_n^{\{b_{n-1-f,k}\}})$ such that $(\mathbf{z}_k)_1 = b_{n-1-f,k-1}$, $(\mathbf{z}_k, (\mathbf{z}_k)^n) \notin F$, and $\mathbf{z}_k \neq C_1((k-1)(n-1)!+1)$. By Lemma 18.15, there is a Hamiltonian path T of $(\cup_{j=1}^{2} S_n^{\{b_{n-1-f,j}\}}) - (F \cup \{\mathbf{e}\})$ joining $(\mathbf{e})^{b_{2,n}}$ to $(\mathbf{z}_3)^n$. Again, there is a Hamiltonian path H_k of $S_n^{\{b_{n-1-f,k}\}} - F_{b_{n-1-f,k}}$ joining $\mathbf{z}_k$ to $(\mathbf{z}_{k+1})^n$ for $3 \leq k \leq n-2$. By Lemma 18.14, there is a Hamiltonian path H_{n-1} of $(\cup_{j=n-1}^{n} S_n^{\{b_{n-1-f,j}\}}) - F$ joining $\mathbf{z}_{n-1}$ to $(\mathbf{e})^n$. Then we set $C_{n-1-f} = \langle \mathbf{e}, (\mathbf{e})^{b_{2,n}}, T, (\mathbf{z}_3)^n, \mathbf{z}_3, H_3, (\mathbf{z}_4)^n, \ldots, \mathbf{z}_{n-2}, H_{n-2}, (\mathbf{z}_{n-1})^n, \mathbf{z}_{n-1}, H_{n-1}, (\mathbf{e})^n, \mathbf{e} \rangle$.

Consequently, $\{C_1, C_2, \ldots, C_{n-2-f}, C_{n-1-f}\}$ forms a set of $(n-1-f)$ mutually independent Hamiltonian cycles of $S_n - F$ beginning from $\mathbf{e}$. Figure 18.9a illustrates $\{C_1, C_2, C_3, C_4\}$ in $S_6 - F$ with $|F| = f = 1$.

Case 2.2. Suppose that $|F \cap A_1| > 0$. Thus, at least one of $\{|F \cap A_2|, \ldots, |F \cap A_{f+1}|\}$ equals to 0. Without loss of generality, we assume that $|F \cap A_2| = 0$. Let $B_n = (b_{i,j})$

be the $(n-1-f)\times n$ matrix with

$$b_{i,j}=\begin{cases} n & \text{if } i=1 \text{ and } j=1\\ j+1 & \text{if } i=1 \text{ and } 2\le j\le n-2\\ 2 & \text{if } i=1 \text{ and } j=n-1\\ 1 & \text{if } i=1 \text{ and } j=n\\ f+i+j & \text{if } 2\le i\le n-2-f \text{ and } f+i+j\le n\\ 2 & \text{if } 2\le i\le n-2-f \text{ and } f+i+j=n+1\\ 1 & \text{if } 2\le i\le n-2-f \text{ and } f+i+j=n+2\\ f+i+j-n & \text{if } 2\le i\le n-2-f \text{ and } f+i+j\ge n+3\\ j & \text{if } i=n-1-f \end{cases}$$

By Lemma 13.7, there are $(n-2)!/2>n-3$ edges joining vertices of $V_0(S_n^{\{b_{1,2}\}})$ to vertices of $V_1(S_n^{\{b_{1,1}\}})$. Thus, we choose a vertex $\mathbf{z}$ of $V_0(S_n^{\{b_{1,2}\}})$ such that $(\mathbf{z})_1=b_{1,1}$, $(\mathbf{z},(\mathbf{z})^n)\notin F$, and $(\mathbf{z})^n\neq(\mathbf{e})^{b_{2,n}}$. Again, there is a Hamiltonian path T of $S_n^{\{b_{1,1}\}}-(F_{b_{1,1}}\cup\{\mathbf{e}\})$ joining $(\mathbf{e})^{b_{2,n}}$ to $(\mathbf{z})^n$. By Lemma 18.14, there is a Hamiltonian path H of $(\cup_{j=2}^{n}S_n^{\{b_{1,j}\}})-F$ joining $\mathbf{z}$ to $(\mathbf{e})^n$. Then we set $C_1=\langle\mathbf{e},(\mathbf{e})^{b_{2,n}},T,(\mathbf{z})^n,\mathbf{z},H,(\mathbf{e})^n,\mathbf{e}\rangle$.

Let $i\in\langle n-2-f\rangle-\{1\}$. We set $t_i=n-f-i$. By Lemma 18.17, there is a Hamiltonian path Q_i of $S_n^{\{b_{i,t_i}\}}-(F_{b_{i,t_i}}\cup\{\mathbf{e},(\mathbf{e})^{b_{i,1}},(\mathbf{e})^{b_{i,n}}\})$ joining two vertices $\mathbf{x}_i$ and $\mathbf{y}_i$ in $V_0(S_n^{\{b_{i,t_i}\}})-\{\mathbf{e}\}$ such that $(\mathbf{x}_i)_1=b_{i,t_i-1}$ and $(\mathbf{y}_i)_1=b_{i,t_i+1}$. By Lemma 18.14, there is a Hamiltonian path P_i of $(\cup_{j=1}^{t_i-1}S_n^{\{b_{i,j}\}})-F$ joining $((\mathbf{e})^{b_{i,1}})^n$ to $(\mathbf{x}_i)^n$. Similarly, there is a Hamiltonian path R_i of $(\cup_{j=t_i+1}^{n}S_n^{\{b_{i,j}\}})-F$ joining $(\mathbf{y}_i)^n$ to $((\mathbf{e})^{b_{i,n}})^n$. Then we set $C_i=\langle\mathbf{e},(\mathbf{e})^{b_{i,1}},((\mathbf{e})^{b_{i,1}})^n,P_i,(\mathbf{x}_i)^n,\mathbf{x}_i,Q_i,\mathbf{y}_i,(\mathbf{y}_i)^n,R_i,((\mathbf{e})^{b_{i,n}})^n,(\mathbf{e})^{b_{i,n}},\mathbf{e}\rangle$.

By Lemma 13.7, there are $(n-2)!/2>n-3$ edges joining vertices of $V_0(S_n^{\{b_{n-1-f,2}\}})$ to vertices of $V_1(S_n^{\{b_{n-1-f,3}\}})$. Thus, we choose a vertex $\mathbf{w}$ of $V_0(S_n^{\{b_{n-1-f,2}\}})$ such that $(\mathbf{w})_1=b_{n-1-f,3}$, $(\mathbf{w},(\mathbf{w})^n)\notin F$, and $d(\mathbf{w},(\mathbf{y}_{n-2-f})^n)>1$. Moreover, we choose a vertex $\mathbf{v}$ of $V_1(S_n^{\{b_{n-1-f,n}\}})$ such that $(\mathbf{v})_1=b_{n-1-f,n-1}$ and $(\mathbf{v},(\mathbf{v})^n)\notin F$. By Lemma 18.14, there is a Hamiltonian path D_1 of $(\cup_{j=1}^{2}S_n^{\{b_{n-1-f,j}\}})-F$ joining $(\mathbf{e})^n$ to $\mathbf{w}$. Similarly, there is a Hamiltonian path D_2 of $(\cup_{j=3}^{n-1}S_n^{\{b_{n-1-f,j}\}})-F$ joining $(\mathbf{w})^n$ to $(\mathbf{v})^n$. By Theorem 15.2, there is a Hamiltonian path W of $S_n^{\{b_{n-1-f,n}\}}-(F_{b_{n-1-f,n}}\cup\{\mathbf{e}\})$ joining $\mathbf{v}$ to $(\mathbf{e})^{b_{n-2-f,1}}$. Then we set $C_{n-1-f}=\langle\mathbf{e},(\mathbf{e})^n,D_1,\mathbf{w},(\mathbf{w})^n,D_2,(\mathbf{v})^n,\mathbf{v},W,(\mathbf{e})^{b_{n-2-f,1}},\mathbf{e}\rangle$.

Hence, $\{C_1,C_2,\ldots,C_{n-2-f},C_{n-1-f}\}$ forms a set of $(n-1-f)$ mutually independent Hamiltonian cycles of S_n-F beginning from $\mathbf{e}$. Figure 18.9b illustrates $\{C_1,C_2,C_3,C_4\}$ in S_6-F with $|F|=f=1$. □

According to Lemmas 18.18 through 18.20, and the result in Ref. 211, we summarize the main result as follows.

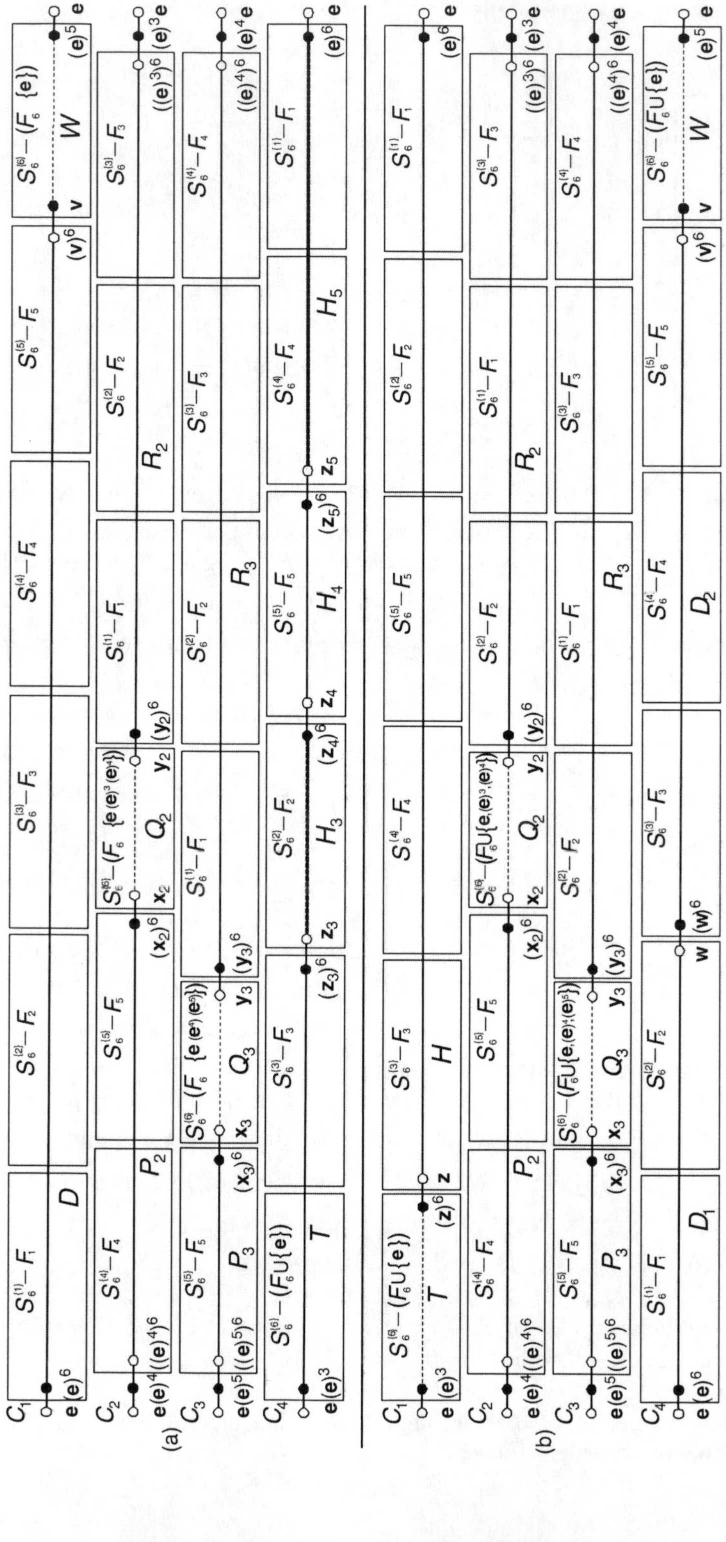

FIGURE 18.9 Mutually independent Hamiltonian cycles in $S_6 - F$ with $|F| = 1$ for Case 2 of Lemma 18.20.

THEOREM 18.9 Let $F \subset E(S_n)$ with $|F| \leq n-3$ for $n \geq 3$ and let $\mathbf{u} \in V(S_n)$. Then there exist $(n-2-|F|)$ mutually independent Hamiltonian cycles of $S_n - F$ beginning from $\mathbf{u}$ if $n \in \{3,4\}$ and also that there exist $(n-1-|F|)$ mutually independent Hamiltonian cycles of $S_n - F$ beginning from $\mathbf{u}$ if $n \geq 5$.

We believe that our result can be further refined. We conjecture that $\mathcal{IHC}(S_n - F) = \delta(S_n - F)$ where $\delta(S_n - F)$ denotes the minimum degree of $S_n - F$.

18.8 ORTHOGONALITY FOR SETS OF MUTUALLY INDEPENDENT HAMILTONIAN CYCLES

The concept of mutually independent Hamiltonian cycles can be viewed as a generalization of Latin rectangles. Perhaps one of the most interesting topics related to the Latin square is the orthogonal Latin square. Two Latin squares of order n are *orthogonal* if the n-squared pairs formed by juxtaposing the two arrays are all distinct. Similarly, two Latin rectangles of order $n \times m$ are *orthogonal* if the $n \times m$ pairs formed by juxtaposing the two arrays are all distinct. With this in mind, let G be a Hamiltonian graph and C_1 and C_2 be two sets of mutually independent Hamiltonian cycles of G from a given vertex x. We say that C_1 and C_2 are *orthogonal* if their corresponding Latin rectangles are orthogonal. For example, we know that $IHC(P_4) = 3$. The following Latin rectangle represents three mutually independent Hamiltonian cycles beginning at 1234:

2134,4312,1342,2431,3421,1243,4213,3124,1324,4231,3241,1423,2413,3142,4132,2314,3214,4123,2143,3412,1432,2341,4321
3214,2314,4132,1432,3412,4312,1342,3142,2413,4213,1243,2143,4123,1423,3241,2341,4321,3421,2431,4231,1324,3124,2134
4321,2341,1432,3412,2143,4123,1423,3241,4231,1324,3124,2134,4312,1342,2431,3421,1243,4213,2413,3142,4132,2314,3214

Yet the following Latin rectangle also represents three mutually independent Hamiltonian cycles beginning at 1234:

2134,3124,4213,1243,2143,4123,1423,2413,3142,4132,1432,3412,4312,1342,2431,3421,4321,2341,3241,4231,1324,2314,3214
3214,2314,4132,3142,2413,4213,1243,3421,2431,1342,4312,2134,3124,1324,4231,3241,1423,4123,2143,3412,1432,2341,4321
4321,3421,1243,2143,3412,4312,1342,2431,4231,1324,2314,3214,4123,1423,3241,2341,1432,4132,3142,2413,4213,3124,2134

We can confirm that these two Latin rectangles are orthogonal. Thus, we have two sets of three mutually independent Hamiltonian cycles that are orthogonal. With this example in mind, we can consider the following problem. Let G be any Hamiltonian graph. We can define $MOMH(G)$ as the largest integer k such that there exist k sets of mutually independent Hamiltonian cycles of G beginning from any vertex x such that each set contains exactly $IHC(G)$ Hamiltonian cycles and any two different sets are orthogonal. It would be interesting to study the value of $MOMH(G)$ for some Hamiltonian graph G.

Recently, people have become interested in a mathematical puzzle called Sudoku [341]. Sudoku can be viewed as a 9×9 Latin square with some constraints. There are several Sudoku variations that are presented. Mutually independent Hamiltonian cycles can also be considered a variation of Sudoku.

19 Mutually Independent Hamiltonian Paths

19.1 INTRODUCTION

In Chapter 18, we discussed the mutually independent hamiltonicity of graph G. Similarly, we can discuss the mutually independent Hamiltonian connectivity of graph G. A graph is k *mutually independent Hamiltonian connected* if there exist k mutually independent Hamiltonian paths between any two distinct vertices. Moreover, the *mutually independent Hamiltonian connectivity* of a graph G, $IHP(G)$, is the maximum integer k such that G is k mutually independent Hamiltonian connected.

THEOREM 19.1 (1) $IHP(K_n)=1$ if $n\in\{1,2\}$ and $IHP(K_n)=n-2$ if $n\geq 3$. (2) $IHP(G)\leq\delta(G)-1$. (3) Suppose that r is a positive integer. Then $IHP(G)\geq r$ for any graph G such that $\deg_G(x)+\deg_G(y)\geq n+r$ for any two distinct vertices x and y.

Proof: Statement (1) follows from Theorem 9.20; let u and v be two adjacent vertices in G such that $\deg(u)=\delta(G)$. Obviously, there are at most $\deg(u)-1$ mutually independent Hamiltonian paths joining u to v. Hence, Statement (2) follows. By Corollary 9.1, we obtain Statement (3). The theorem is proved. □

Obviously, the concept of mutually independent Hamiltonian connectivity is not suitable for bipartite graphs. For this reason, we say that a bipartite graph is *k mutually independent Hamiltonian laceable* if there exist k mutually independent Hamiltonian paths between any two vertices from distinct partite sets. Moreover, the *mutually independent Hamiltonian laceability* of a bipartite graph G, $IHP_L(G)$, is the maximum integer k such that G is k mutually independent Hamiltonian laceable. Let $K_{n,n}$ be the complete bipartite graph with n vertices in each partite set. It can be confirmed that $IHP_L(K_{n,n})=1$ if $n=1$ and $IHP_L(K_{n,n})=n-1$ if $n\geq 2$. Again, we have $IHP_L(G)\leq\delta(G)-1$ for any bipartite graph with at least three vertices.

19.2 MUTUALLY INDEPENDENT HAMILTONIAN LACEABILITY FOR HYPERCUBES

Hypercubes are one of the most popular interconnection networks. In Ref. 290, Sun et al. study the mutually independent Hamiltonian laceability for hypercubes. We need the following lemmas.

LEMMA 19.1 Suppose that $P_3 = \langle \mathbf{x}_1, \mathbf{x}_2, \mathbf{x}_3 \rangle$ is a path of Q_4 joining two white nodes $\mathbf{x}_1$ and $\mathbf{x}_3$. There exists a Hamiltonian path of $Q_4 - \{\mathbf{x}_1, \mathbf{x}_2, \mathbf{x}_3\}$ between any two different black nodes $\mathbf{u}$ and $\mathbf{v}$.

Proof: By the symmetric property of Q_4, we may assume that $P_3 = \langle 0000, 1000, 1100 \rangle$.

Case 1. $\mathbf{u}, \mathbf{v} \in Q_4^0$. Thus, $\mathbf{u}, \mathbf{v} \in \{1110, 0100, 0010\}$. Hence,

$$\langle 1110, 1010, 0010, 0110, 0111, 1111, 1011, 0011, 0001, 1001, 1101, 0101, 0100 \rangle$$
$$\langle 1110, 0110, 0100, 0101, 1101, 1111, 0111, 0011, 0001, 1001, 1011, 1010, 0010 \rangle$$
$$\langle 0100, 0110, 1110, 1010, 1011, 1111, 1101, 1001, 0001, 0101, 0111, 0011, 0010 \rangle$$

form the desired paths.

Case 2. $\mathbf{u} \in Q_4^0$ and $\mathbf{v} \in Q_4^1$. Then $\mathbf{u} \in \{1110, 0010, 0100\}$.

Suppose that $\mathbf{u} \in \{1110, 0010\}$. Set $\mathbf{r} = 0100$. Again, we set $P = \langle \mathbf{u}, 1010, 0010, 0110, \mathbf{r} \rangle$ if $\mathbf{u} = 1110$ and $P = \langle \mathbf{u}, 1010, 1110, 0110, \mathbf{r} \rangle$ if $\mathbf{u} = 0010$. By Theorem 13.1, there exists a Hamiltonian path R of Q_n^1 between $\mathbf{r}^4$ and $\mathbf{v}$. Obviously, $\langle \mathbf{u}, P, \mathbf{r}, \mathbf{r}^4, R, \mathbf{v} \rangle$ forms a Hamiltonian path of $Q_4 - \{\mathbf{x}_1, \mathbf{x}_2, \mathbf{x}_3\}$.

Suppose that $\mathbf{u} = 0100$. We set $\mathbf{r} = 0010$ and $P = \langle \mathbf{u}, 0110, 1110, 1010, \mathbf{r} \rangle$. By Theorem 13.1, there exists a Hamiltonian path S of Q_n^1 between $\mathbf{r}^4$ and $\mathbf{v}$. Obviously, $\langle \mathbf{u}, P, \mathbf{r}, \mathbf{r}^4, S, \mathbf{v} \rangle$ forms a Hamiltonian path of $Q_4 - \{\mathbf{x}_1, \mathbf{x}_2, \mathbf{x}_3\}$.

Case 3. $\mathbf{u}, \mathbf{v} \in Q_4^1$. The corresponding Hamiltonian paths of $Q_4 - P_3$ between $\mathbf{u}$ and $\mathbf{v}$ are:

0001, 1001, 1101, 0101, 0100, 0110, 1110, 1010, 0010, 0011, 0111, 1111, 1011
0001, 1001, 1101, 0101, 0100, 0110, 1110, 1010, 0010, 0011, 1011, 1111, 0111
0001, 1001, 1011, 0011, 0111, 0101, 0100, 0110, 0010, 1010, 1110, 1111, 1101
1011, 0011, 0010, 1010, 1110, 0110, 0100, 0101, 0001, 1001, 1101, 1111, 0111
1011, 1111, 1110, 1010, 0010, 0110, 0100, 0101, 0111, 0011, 0001, 1001, 1101
0111, 0011, 0010, 1010, 1110, 0110, 0100, 0101, 0001, 1001, 1011, 1111, 1101

The lemma is proved. □

LEMMA 19.2 Suppose that $P_3 = \langle \mathbf{x}_1, \mathbf{x}_2, \mathbf{x}_3 \rangle$ is a path of Q_n joining two white vertices $\mathbf{x}_1$ and $\mathbf{x}_3$. There exists a Hamiltonian path of $Q_n - \{\mathbf{x}_1, \mathbf{x}_2, \mathbf{x}_3\}$ between any two black vertices $\mathbf{u}$ and $\mathbf{v}$ for $n \geq 4$.

Proof: We prove this lemma by induction. The induction basis is proved by Lemma 19.1. Assume that this lemma is true for Q_k with $4 \leq k < n$. By the symmetric property of Q_n, we may assume that $P_3 \subset Q_n^0$.

Case 1. $\mathbf{u}, \mathbf{v} \in Q_n^0$. By induction, there exists a Hamiltonian path R of $Q_n - \{\mathbf{x}_1, \mathbf{x}_2, \mathbf{x}_3\}$ joining $\mathbf{u}$ to $\mathbf{v}$. Write R as $\langle \mathbf{u} = \mathbf{y}_1, \mathbf{y}_2, \ldots, \mathbf{y}_{2^{n-1}-3} = \mathbf{v} \rangle$. Since every hypercube is Hamiltonian laceable, there exists a Hamiltonian path S of Q_n^1 joining $\mathbf{u}^n$ to $(\mathbf{y}_2)^n$. Then $\langle \mathbf{u} = \mathbf{y}_1, \mathbf{u}^n, S, (\mathbf{y}_2)^n, \mathbf{y}_2, \mathbf{y}_3, \ldots, \mathbf{y}_{2^{n-1}-3} = \mathbf{v} \rangle$ forms the desired path.

Case 2. $\mathbf{u} \in Q_n^0$ and $\mathbf{v} \in Q_n^1$. Let $\mathbf{y}$ be any black vertex of Q_n^0 not in $\{\mathbf{x}_2, \mathbf{u}\}$. By induction, there exists a Hamiltonian path R of $Q_n^0 - \{\mathbf{x}_1, \mathbf{x}_2, \mathbf{x}_3\}$ from $\mathbf{u}$ to $\mathbf{y}$. Since every hypercube is Hamiltonian laceable, there exists a Hamiltonian path S of Q_n^1 from $\mathbf{y}^n$ to $\mathbf{v}$. Then $\langle \mathbf{u}, R, \mathbf{y}, \mathbf{y}^n, S, \mathbf{v} \rangle$ forms the desired path.

Case 3. $\mathbf{u}, \mathbf{v} \in Q_n^1$. Since $\deg_{Q_n^1}(\mathbf{u}) \geq 4$, there exists a neighbor $\mathbf{r}$ of $\mathbf{u}$ in Q_n^1 such that $\mathbf{r}^n \neq \mathbf{x}_2$. Let $\mathbf{t}$ be a black vertex of Q_n^0 not in $\{\mathbf{x}_2, \mathbf{r}^n\}$. By induction, there exists a Hamiltonian path S of Q_n^0 from $\mathbf{r}^n$ to $\mathbf{t}$. By Lemma 18.6, there exists a Hamiltonian path R of $Q_n^1 - \{\mathbf{u}, \mathbf{r}\}$ from $\mathbf{t}^n$ to $\mathbf{v}$. Then $\langle \mathbf{u}, \mathbf{r}, \mathbf{r}^n, S, \mathbf{t}, \mathbf{t}^n, R, \mathbf{v} \rangle$ forms the desired path. □

In the following, we use $\tilde{\mathbf{e}}$ to denote the vertex in Q_n which $(\tilde{\mathbf{e}})_i = 1$ for every $1 \leq i \leq n$.

LEMMA 19.3 Assume that n is even and $n \geq 4$. There exists a Hamiltonian path $P_1, P_2, \ldots, P_n$ of Q_n such that (1) P_i joins from $\mathbf{e}$ to $(\tilde{\mathbf{e}})^i$ and (2) $P_i(k) \neq P_j(k)$ for $1 \leq i < j \leq n$ and $2 \leq k \leq 2^n$.

Proof: We prove this lemma by induction. Suppose that $n = 4$. The required Hamiltonian paths joining $\mathbf{e}$ to $(\tilde{\mathbf{e}})^i$ in Q_4 are:

0000, 0100, 0110, 0010, 0011, 0001, 0101, 0111, 1111, 1101, 1001, 1011, 1010, 1000, 1100, 1110
0000, 0001, 0101, 0100, 0110, 0111, 0011, 0010, 1010, 1000, 1100, 1110, 1111, 1101, 1001, 1011
0000, 0010, 0011, 0001, 1001, 1011, 1010, 1000, 1100, 1110, 1111, 1101, 0101, 0100, 0110, 0111
0000, 1000, 1100, 1110, 1010, 0010, 0110, 0100, 0101, 0111, 0011, 0001, 1001, 1011, 1111, 1101

Thus, we assume that the lemma is true for Q_k with $4 \leq k < n$. By induction, there exist $(n-2)$ Hamiltonian paths $P_{00}^1, P_{00}^2, \ldots, P_{00}^{n-2}$ of Q_n^{00} such that (1) P_{00}^i joins from $\mathbf{e}$ to $(\tilde{\mathbf{e}})^i$ for $1 \leq i \leq n-2$ and (2) $P_{00}^i(k) \neq P_{00}^j(k)$ for $1 \leq i < j \leq n-2$ and $2 \leq k \leq 2^{n-2}$. However, we pick any $(n-3)$ Hamiltonian paths $P_{00}^1, P_{00}^2, \ldots, P_{00}^{n-3}$ of Q_n^{00}. We write P_{00}^i as $\langle \mathbf{e}, M_{00}^i, \mathbf{a}_i, (((\tilde{\mathbf{e}})^i)^{n-1})^n \rangle$ for $1 \leq i \leq n-3$. By Lemma 18.7, there exist $(n-3)$ Hamiltonian paths $\{P_{01}^1, P_{01}^2, \ldots, P_{01}^{n-3}\}$ of Q_n^{01} such that (1) P_{01}^i joins from $((\tilde{\mathbf{e}})^i)^{n-1}$ to $(\mathbf{a}_i)^n$ for $1 \leq i \leq n-3$ and (2) $P_{01}^i(k) \neq P_{01}^j(k)$ for $1 \leq i < j \leq n-3$ and $1 \leq k \leq 2^{n-2}$. Again, there exist $(n-3)$ Hamiltonian paths $\{Y_{11}^1, Y_{11}^2, \ldots, Y_{11}^{n-3}\}$ of Q_n^{11} such that (1) Y_{11}^i joins from $((\mathbf{a}_i)^n)^{n-1}$ to $(\tilde{\mathbf{e}})^i$ for $1 \leq i \leq n-3$ and (2) $Y_{11}^i(k) \neq Y_{11}^j(k)$ for $1 \leq i < j \leq n-3$ and $1 \leq k \leq 2^{n-2}$. Write Y_{11}^i as $\langle ((\mathbf{a}_i)^n)^{n-1}, P_{11}^i, \mathbf{b}_i, (\tilde{\mathbf{e}})^i \rangle$ for $1 \leq i \leq n-3$. By Lemma 18.7, there exist $(n-3)$ Hamiltonian paths $\{P_{10}^1, P_{10}^2, \ldots, P_{10}^{n-3}\}$ of Q_n^{10} such that (1) P_{10}^i joins from $(\mathbf{b}_i)^n$ to $((\tilde{\mathbf{e}})^i)^n$ for $1 \leq i \leq n-3$ and (2) $P_{10}^i(k) \neq P_{10}^j(k)$ for $1 \leq i < j \leq n-3$ and $1 \leq k \leq 2^{n-2}$. Set P_i as $\langle \mathbf{e}, P_{00}^i, (((\tilde{\mathbf{e}})^i)^{n-1})^n, ((\tilde{\mathbf{e}})^i)^{n-1}, P_{01}^i, (\mathbf{a}_i)^n, ((\mathbf{a}_i)^n)^{n-1}, P_{11}^i, \mathbf{b}_i, (\mathbf{b}_i)^n, P_{10}^i, ((\tilde{\mathbf{e}})^i)^n, (\tilde{\mathbf{e}})^i \rangle$ for $1 \leq i \leq n-3$.

Since there are 2^{n-2} vertices in Q_n^{01}, we can choose an edge $(\mathbf{y}_1, \mathbf{z}_1) \in E(Q_n^{01})$ such that (1) $\mathbf{z}_1$ is a white vertex not in $\{((\tilde{\mathbf{e}})^i)^{n-1} | 1 \leq i \leq n-3\}$ and (2) $\mathbf{y}_1 \neq \mathbf{e}^n$. Again, we can choose an edge $(\mathbf{y}_2, \mathbf{z}_2) \in E(Q_n^{11})$ such that (1) $\mathbf{z}_2$ is a white vertex not in $\{((\mathbf{a}_i)^n)^{n-1} \mid 1 \leq i \leq n-3\} \cup \{\tilde{\mathbf{e}}\}$ and (2) $\mathbf{y}_2 \neq (\mathbf{z}_1)^{n-1}$. Let $\mathbf{b}$ be any black vertex of Q_n^{00} such that $\mathbf{b} \neq (((\tilde{\mathbf{e}})^{n-2})^n)^{n-1}$. Since every hypercube is hyper Hamiltonian laceable, there exists a Hamiltonian path S_{01} of $Q_n^{01} - \{\mathbf{z}_1\}$ from $\mathbf{e}^n$ to $\mathbf{y}_1$. Again, there exists a Hamiltonian path S_{11} of $Q_n^{11} - \{\mathbf{z}_2\}$ from $(\mathbf{z}_1)^{n-1}$ to $\mathbf{y}_2$. Moreover, there exists a Hamiltonian path S_{00} of $Q_n^{00} - \{\mathbf{e}\}$ from $\mathbf{b}$ to $(((\tilde{\mathbf{e}})^{n-2})^n)^{n-1}$. By Lemma 18.6, there exists a Hamiltonian path S_{10} of $Q_n^{10} - \{((\tilde{\mathbf{e}})^{n-2})^n, (\tilde{\mathbf{e}})^n\}$ from $(\mathbf{z}_2)^n$ to $\mathbf{b}^{n-1}$. We set P_{n-2} as $\langle \mathbf{e}, \mathbf{e}^n, S_{01}, \mathbf{y}_1, \mathbf{z}_1, (\mathbf{z}_1)^{n-1}, S_{11}, \mathbf{y}_2, \mathbf{z}_2, (\mathbf{z}_2)^n, S_{10}, \mathbf{b}^{n-1}, \mathbf{b}, S_{00}, (((\tilde{\mathbf{e}})^{n-2})^n)^{n-1}, ((\tilde{\mathbf{e}})^{n-2})^n), (\tilde{\mathbf{e}})^n \rangle$.

Let $\mathbf{x}$ be the only vertex in $\{\mathbf{e}^i \mid 1 \leq i \leq n-2\} - \{P_{00}^i(2) \mid 1 \leq i \leq n-3\}$, let $\mathbf{r}_1$ be any black vertex of Q_n^{10} such that $\mathbf{r}_1 \neq \mathbf{e}^{n-1}$, and let $\mathbf{r}_2$ be any black vertex of Q_n^{00} such that $\mathbf{r}_2 \neq \mathbf{x}$. Since every hypercube is Hamiltonian laceable, there exists a Hamiltonian path R_{10} of Q_n^{10} from $\mathbf{x}^{n-1}$ to $\mathbf{r}_1$. Again, there exists a Hamiltonian path R_{01} of Q_n^{01} from $(\mathbf{r}_2)^n$ to $((\mathbf{r}_1)^{n-1})^n$. Moreover, there exists a Hamiltonian path R_{11} of Q_n^{11} from $(\mathbf{r}_1)^n$ to $(\tilde{\mathbf{e}})^{n-2}$. By Lemma 18.6, there exists a Hamiltonian path R_{00} of $Q_n^{00} - \{\mathbf{e}, \mathbf{x}\}$ from $(\mathbf{r}_1)^{n-1}$ to $\mathbf{r}_2$. Set P_{n-1} as $\langle \mathbf{e}, \mathbf{x}, \mathbf{x}^{n-1}, R_{10}, \mathbf{r}_1, (\mathbf{r}_1)^{n-1}, R_{00}, \mathbf{r}_2, (\mathbf{r}_2)^n, R_{01}, ((\mathbf{r}_1)^{n-1})^n, (\mathbf{r}_1)^n, R_{11}, (\tilde{\mathbf{e}})^{n-2} \rangle$.

Since there exist 2^{n-2} vertices in Q_n^{11}, we can choose an edge $(\mathbf{u}, \mathbf{v})$ of Q_n^{11} such that $\mathbf{u}$ is a white vertex, $\mathbf{u} \neq (\mathbf{e}^{n-1})^n$ and $\mathbf{v} \neq (\mathbf{z}_1)^{n-1}$. Let $\mathbf{w}$ be any white vertex of Q_n^{10} such that $\mathbf{w} \neq \mathbf{v}^n$ and let $\mathbf{t}$ be any black vertex of Q_n^{00} not in $\{\mathbf{b}, \mathbf{w}^{n-1}\}$. Since every hypercube is hyper Hamiltonian laceable, there exists a Hamiltonian path T_{11} of $Q_n^{11} - \{\mathbf{v}\}$ from $(\mathbf{e}^{n-1})^n$ to $\mathbf{u}$. Again, there exists a Hamiltonian path T_{10} of $Q_n^{10} - \{\mathbf{e}^{n-1}\}$ from $\mathbf{v}^n$ to $\mathbf{w}$. Moreover, there exists a Hamiltonian path T_{00} of $Q_n^{00} - \{\mathbf{e}\}$ from $\mathbf{w}^{n-1}$ to $\mathbf{t}$. Since every hypercube is Hamiltonian laceable, there exists a Hamiltonian path T_{01} of Q_n^{01} from $\mathbf{t}^n$ to $(\tilde{\mathbf{e}})^{n-1}$. Set P_n as $\langle \mathbf{e}, \mathbf{e}^{n-1}, (\mathbf{e}^{n-1})^n, T_{11}, \mathbf{u}, \mathbf{v}, \mathbf{v}^n, T_{10}, \mathbf{w}, \mathbf{w}^{n-1}, T_{00}, \mathbf{t}, \mathbf{t}^n, T_{01}, (\tilde{\mathbf{e}})^{n-1} \rangle$. Then $\{P_i \mid 1 \leq i \leq n-1\}$ forms a set of desired paths. See Figure 19.1 for illustration. □

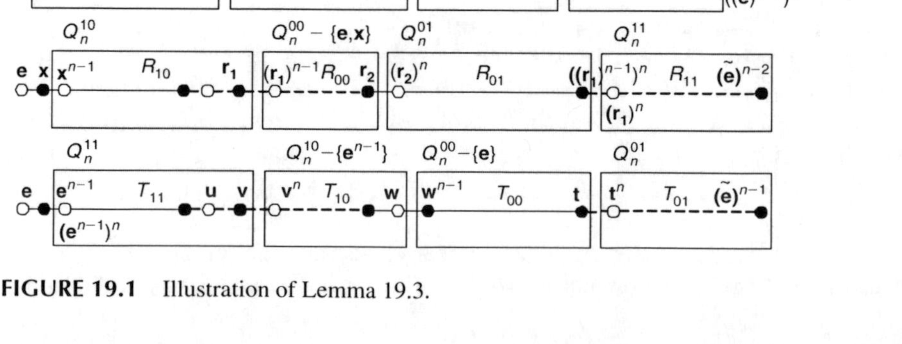

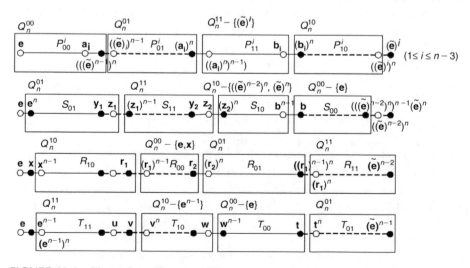

FIGURE 19.1 Illustration of Lemma 19.3.

THEOREM 19.2 $IHP_L(Q_n) = 1$ if $n \in \{1, 2, 3\}$ and $IHP_L(Q_n) = n - 1$ if $n \geq 4$.

Proof: By Lemma 19.9, $IHP_L(Q_n) = 1$ if $n \in \{1, 2, 3\}$. Now, we want to prove that $IHP_L(Q_n) = n - 1$ if $n \geq 4$ by induction. We need to construct $(n - 1)$ mutually independent Hamiltonian paths of Q_n from any white vertex $\mathbf{u}$ to any black vertex $\mathbf{v}$. Without loss of generality, we can assume that $\mathbf{u} = \mathbf{e}$.

Suppose that $n = 4$. Then the Hamming weight of $\mathbf{v}$, $w(\mathbf{v})$, is either 1 or 3. Thus, we can assume that $\mathbf{v} = 1000$ or 1110. The corresponding mutually independent Hamiltonian paths from $\mathbf{u}$ to $\mathbf{v}$ in Q_4 are:

$\mathbf{v} = 1000$	0000, 0010, 0110, 0100, 1100, 1110, 1010, 1011, 0011, 0001, 0101, 0111, 1111, 1101, 1001, 1000
	0000, 0100, 1100, 1110, 0110, 0010, 0011, 0001, 0101, 0111, 1111, 1101, 1001, 1011, 1010, 1000
	0000, 0001, 0011, 1011, 1001, 1101, 0101, 0111, 1111, 1110, 1010, 0010, 0110, 0100, 1100, 1000
$\mathbf{v} = 1110$	0000, 1000, 1100, 0100, 0110, 0010, 1010, 1011, 1001, 1101, 0101, 0001, 0011, 0111, 1111, 1110
	0000, 0010, 1010, 1000, 1100, 1101, 1001, 0001, 0011, 1011, 1111, 0111, 0101, 0100, 0110, 1110
	0000, 0001, 1001, 1011, 0011, 0111, 1111, 1101, 0101, 0100, 0110, 0010, 1010, 1000, 1100, 1110

Suppose that the theorem holds for Q_k with $4 \leq k < n$. We need to construct $(n - 1)$ mutually independent Hamiltonian paths of Q_n from $\mathbf{e}$ to $\mathbf{v}$. We have the following cases.

Case 1. $w(\mathbf{v}) < n$. Without loss of generality, we assume that $(\mathbf{v})_n = 0$. By induction, there exist $(n-2)$ mutually independent Hamiltonian paths $R_1, R_2, \ldots, R_{n-2}$ of Q_n^0 from $\mathbf{e}$ to $\mathbf{v}$. By Lemma 18.7, there exist $(n-2)$ Hamiltonian paths $S_1, S_2, \ldots, S_{n-2}$ of Q_n^1 such that (1) S_i joins from $(R_i(3))^n$ to $(R_i(4))^n$ and (2) $S_i(k) \neq S_j(k)$ for $1 \leq i < j \leq n-2$ and $1 \leq k \leq 2^{n-1}$. Set P_i as $\langle \mathbf{e}, R_i(2), R_i(3), (R_i(3))^n, S_i, (R_i(4))^n, R_i(4), R_i(5), \ldots, R_i(2^{n-1} - 1), \mathbf{v} \rangle$ for $1 \leq i \leq n-2$.

Let $X = \{\mathbf{x} | d_{Q_n^1}(\mathbf{x}, \mathbf{e}^n) = 2, \mathbf{x} \in Q_n^1\}$. Obviously, $|X| = \binom{n-1}{2}$. We can choose $\mathbf{r_2}$ in $X - (\{\mathbf{e}^n\} \cup \{S_i(1) | 1 \leq i \leq n-2\})$ such that path $R = \langle \mathbf{e}^n, \mathbf{r_1}, \mathbf{r_2} \rangle$ of Q_n^1. Since there are 2^{n-2} white vertices in Q_n^1, we can choose a white vertex $\mathbf{z}$ of Q_n^1 not in $\{(R_i(4))^n | 1 \leq i \leq n-2\} \cup \{\mathbf{r_1}, \mathbf{v}^n\}$. By Lemma 18.6, there exists a Hamiltonian path T_0 of $Q_n^0 - \{\mathbf{e}, \mathbf{v}\}$ from $(\mathbf{r_2})^n$ to $(\mathbf{z})^n$. By Lemma 19.2, there exists a Hamiltonian path T_1 of $Q_n^1 - \{\mathbf{e}^n, \mathbf{r_1}, \mathbf{r_2}\}$ from $\mathbf{z}$ to $\mathbf{v}^n$. Set P_{n-1} as $\langle \mathbf{e}, \mathbf{e}^n, \mathbf{r_1}, \mathbf{r_2}, (\mathbf{r_2})^n, T_0, (\mathbf{z})^n, \mathbf{z}, T_1, \mathbf{v}^n, \mathbf{v} \rangle$.

Then $\{P_i | 1 \leq i \leq n - 1\}$ forms a set of desired paths. See Figure 19.2a for illustration.

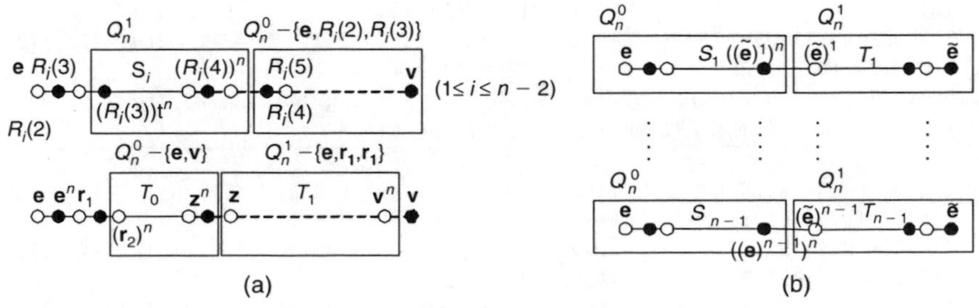

FIGURE 19.2 Illustration of Theorem 19.2.

Case 2. $w(\mathbf{v}) = n$. Since $w(\mathbf{v})$ is odd, n is odd. Obviously, $\mathbf{v} = \tilde{\mathbf{e}}$. Since Q_n^1 is isomorphic to Q_{n-1}, by Theorem 18.5, there exist $(n-1)$ mutually independent Hamiltonian cycles $C_1, C_2, \ldots, C_{n-1}$ of Q_n^1 beginning at $\tilde{\mathbf{e}}$. Write $C_i = \langle \tilde{\mathbf{e}}, T_i, (\tilde{\mathbf{e}})^i, \tilde{\mathbf{e}} \rangle$ for $1 \leq i \leq n-1$. Since Q_n^0 is isomorphic to Q_{n-1}, by Lemma 19.3, there exist $(n-1)$ Hamiltonian paths $S_1, S_2, \ldots, S_{n-1}$ of Q_n^0 such that (1) S_i joins from $((\tilde{\mathbf{e}})^i)^n$ to $\mathbf{e}$ and (2) $S_i(k) \neq S_j(k)$ for $1 \leq i < j \leq n-1$ and $1 \leq k < 2^{n-1}$. Set P_i as $\langle \mathbf{e}, S_i, ((\tilde{\mathbf{e}})^i)^n, (\tilde{\mathbf{e}})^i, T_i, \tilde{\mathbf{e}} \rangle$ for $1 \leq i \leq n-1$. Obviously, $\{P_1, P_2, \ldots, P_{n-1}\}$ forms a set of desired paths. See Figure 19.2b for illustration.

The theorem is proved. □

19.3 MUTUALLY INDEPENDENT HAMILTONIAN LACEABILITY FOR STAR GRAPHS

Similarly, we should study the corresponding problem for star graphs. Lin et al. [231] study the mutually independent Hamiltonian laceability for star graphs. We need the following properties for star graphs.

THEOREM 19.3 Let $\mathbf{u}$ be any white vertex of S_n with $n \geq 4$ and $\{a_1, a_2, \ldots, a_{n-1}\}$ be an $n-1$ subset of $\langle n \rangle$. Then there exist Hamiltonian paths $P_1, P_2, \ldots, P_{n-1}$ of S_n such that P_i is a path joining $\mathbf{u}$ to a black vertex $\mathbf{z}_i$ with

1. $\mathbf{z}_i = P_i(n!)$ and $(\mathbf{z}_i)_1 = a_i$ for $1 \leq i \leq n-1$
2. $|\{P_1(i), P_2(i), \ldots, P_{n-1}(i)\}| = n-1$ for $2 \leq i \leq n!$

Proof: Since S_n is vertex-transitive, we may assume that $\mathbf{u} = \mathbf{e}$. We prove that this theorem holds for S_4 by exhibiting the three required Hamiltonian paths, as follows:

$\{a_1, a_2, a_3\} = \{1, 2, 3\}$
(1234)(2134)(3124)(1324)(2314)(3214)(4213)(1243)(3241)(4231)(2431)(3421)(4321)(2341)(1342)(4312)(3412)(2413)(1423)(4123)(2143)(3142)(4132)(1432)
(1234)(3214)(2314)(4312)(1342)(3142)(4132)(1432)(3412)(2413)(4213)(1243)(2143)(4123)(1423)(3421)(2431)(4231)(3241)(2341)(4321)(1324)(3124)(2134)
(1234)(4231)(2431)(1432)(3412)(4312)(1342)(3142)(4132)(2134)(3124)(1324)(2314)(3214)(4213)(2413)(1423)(4123)(2143)(1243)(3241)(2341)(4321)(3421)
$\{a_1, a_2, a_3\} = \{1, 2, 4\}$
(1234)(4231)(2431)(3421)(4321)(2341)(3241)(1243)(2143)(3142)(1342)(4312)(3412)(1432)(4132)(2134)(3124)(4123)(1423)(2413)(4213)(3214)(2314)(1324)
(1234)(3214)(4213)(1243)(3241)(4231)(2431)(1432)(3412)(2413)(1423)(3421)(4321)(2341)(1342)(4312)(2314)(1324)(3124)(4123)(2143)(3142)(4132)(2143)
(1234)(2134)(3124)(1324)(2314)(3214)(4213)(2413)(1423)(4123)(2143)(1243)(3241)(4231)(2431)(3421)(4321)(2341)(1342)(3142)(4132)(1432)(3412)(4312)
$\{a_1, a_2, a_3\} = \{1, 3, 4\}$
(1234)(4231)(3241)(1243)(4213)(3214)(2314)(4312)(1342)(2341)(4321)(3421)(2431)(1432)(3412)(2413)(1423)(4123)(2143)(3142)(4132)(2134)(3124)(1324)
(1234)(3214)(4213)(2413)(1423)(3421)(2431)(4231)(3241)(1243)(2143)(4123)(3124)(2134)(4132)(1432)(3412)(4312)(2314)(1324)(4321)(2341)(1342)(3142)
(1234)(2134)(4132)(1432)(3412)(2413)(4213)(3214)(2314)(4312)(1342)(3142)(2143)(1243)(3241)(2341)(4321)(1324)(3124)(4123)(1423)(3421)(2431)(4231)

$\{a_1, a_2, a_3\} = \{2, 3, 4\}$
(1234)(3214)(4213)(1243)(3241)(4231)(2431)(1432)(3412)(2413)(1423)(3421)(4321)(2341)(1342)(4312)(2314)(1324)(3124)(4123)(2143)(3142)(4132)(2143)
(1234)(4231)(2431)(1432)(3412)(4312)(1342)(3142)(4132)(2134)(3124)(1324)(2314)(3214)(4213)(2413)(1423)(4123)(2143)(1243)(3241)(2341)(4321)(3421)
(1234)(2134)(3124)(1324)(2314)(3214)(4213)(2413)(1423)(4123)(2143)(1243)(3241)(4231)(2431)(3421)(4321)(2341)(1342)(3142)(4132)(1432)(3412)(4312)

Assume that the theorem holds on S_k for every $4 \le k < n$. Without loss of generality, we suppose that $a_1 < a_2 < \cdots < a_{n-3} < a_{n-1} < a_{n-2}$. Obviously, $a_i \ne i+2$ for $1 \le i \le n-3$, $a_{n-2} \ne 2$, and $a_{n-1} \ne n$. By the induction hypothesis, there exist Hamiltonian paths $H_1, H_2, \ldots, H_{n-2}$ of S_n^n such that H_i is a path joining $\mathbf{e}$ to a black vertex $\mathbf{v}_i$ with

1. $(\mathbf{v}_i)_1 = (H_i((n-1)!))_1 = i+3$ for $1 \le i \le n-4$, $(\mathbf{v}_{n-3})_1 = (H_{n-3}((n-1)!))_1 = 1$, $(\mathbf{v}_{n-2})_1 = (H_{n-2}((n-1)!))_1 = 3$
2. $|\{H_1(i), H_2(i), \ldots, H_{n-2}(i)\}| = n-2$ for $2 \le i \le (n-1)!$

For any $i, j \in \langle n \rangle$ with $i \ne j$, $|\{\mathbf{v} \in S_n^i | \mathbf{v}$ is a black vertex with $(\mathbf{v})_1 = j\}| = \frac{(n-2)!}{2} \ge 2$. We choose a black vertex $\mathbf{z}_i \in S_n^{i+2}$ with $(\mathbf{z}_i)_1 = a_i$ for $1 \le i \le n-3$, a black vertex $\mathbf{z}_{n-2} \in S_n^2$ with $(\mathbf{z}_{n-2})_1 = a_{n-2}$, and a black vertex $\mathbf{z}_{n-1} \in S_n^n$ with $(\mathbf{z}_{n-1})_1 = a_{n-1}$. We let B be the $(n-2) \times (n-1)$ matrix with

$$b_{i,j} = \begin{cases} i+j+2 & \text{if } i \le n-3 \text{ and } i+j+2 \le n-1 \\ i+j-n+3 & \text{if } i \le n-3 \text{ and } n-1 < i+j+2 \\ j+2 & \text{if } i = n-2 \text{ and } j \le n-3 \\ j-n+3 & \text{if } i = n-2 \text{ and } n-2 \le j \le n-1 \end{cases}$$

More precisely,

$$B = \begin{bmatrix} 4 & 5 & \cdots & n-2 & n-1 & 1 & 2 & 3 \\ 5 & 6 & \cdots & n-1 & 1 & 2 & 3 & 4 \\ \vdots & \vdots & \ddots & \vdots & \vdots & \vdots & \vdots & \vdots \\ 1 & 2 & \cdots & n-5 & n-4 & n-3 & n-2 & n-1 \\ 3 & 4 & \cdots & n-3 & n-2 & n-1 & 1 & 2 \end{bmatrix}$$

Obviously, $b_{i,1}, b_{i,2}, \ldots, b_{i,n-1}$ forms a permutation of $\{1, 2, \ldots, n-1\}$ for every i with $1 \le i \le n-2$. Moreover, $b_{i,j} \ne b_{i',j}$ for any $1 \le i < i' \le n-2$ and $1 \le j \le n-2$. In other words, B forms a rectangular Latin square with entries in $\{1, 2, \ldots, n-1\}$.

Let i be any index with $1 \le i \le n-2$. By Theorem 13.9, there exists a path $W_i = \langle \mathbf{x}_1^i, T_1^i, \mathbf{y}_1^i, \mathbf{x}_2^i, T_2^i, \mathbf{y}_2^i, \ldots, \mathbf{x}_{n-1}^i, T_{n-1}^i, \mathbf{y}_{n-1}^i \rangle$ joining the white vertex $(\mathbf{v}_i)^n$ to $\mathbf{z}_i$ such that $\mathbf{x}_1^i = (\mathbf{v}_i)^n$, $\mathbf{y}_{n-1}^i = \mathbf{z}_i$, and T_j^i is a Hamiltonian path of $S_n^{b_{ij}}$ joining $\mathbf{x}_j^i$ to $\mathbf{y}_j^i$ for every $1 \le j \le n-1$. Moreover, there exists a path $W = \langle \mathbf{x}_1^{n-1}, T_1^{n-1}, \mathbf{y}_1^{n-1}, \mathbf{x}_2^{n-1}, T_2^{n-1}, \mathbf{y}_2^{n-1}, \ldots, \mathbf{x}_{n-1}^{n-1}, T_{n-1}^{n-1}, \mathbf{y}_{n-1}^{n-1} \rangle$ joining the black vertex $(\mathbf{e})^n$ to the white vertex $((\mathbf{e})^{n-1})^n$ such that $\mathbf{x}_1^{n-1} = (\mathbf{e})^n$, $\mathbf{y}_{n-1}^{n-1} = ((\mathbf{e})^{n-1})^n$, and T_i^{n-1} is a Hamiltonian path of S_n^i joining

$\mathbf{x}_i^{n-1}$ to $\mathbf{y}_i^{n-1}$ for $1 \leq i \leq n-1$. Since S_n^n is isomorphic to S_{n-1}, by Theorem 13.2, there exists a Hamiltonian path R of $S_n^n - \{\mathbf{e}\}$ joining the black vertex $(\mathbf{e})^{n-1}$ to $\mathbf{z}_{n-1}$.

We set $P_i = \langle \mathbf{e}, H_i, \mathbf{v}_i, (\mathbf{v}_i)^n, W_i, \mathbf{z}_i \rangle$ for every $1 \leq i \leq n-2$ and $P_{n-1} = \langle \mathbf{e}, W, ((\mathbf{e})^{n-1})^n, (\mathbf{e})^{n-1}, R, \mathbf{z}_{n-1} \rangle$. Then $\{P_1, P_2, \ldots, P_{n-1}\}$ form the desired paths.

Hence, the theorem is proved. □

EXAMPLE 19.1

We illustrate the proof of Theorem 19.3 with $n = 5$ as follows.

Since $a_1 < a_2 < a_4 < a_3$, $a_1 \neq 3$, $a_2 \neq 4$, $a_3 \neq 2$, and $a_4 \neq 5$. By the induction hypothesis, there exist Hamiltonian paths H_1, H_2, and H_3 of S_5^5 such that H_i is a path joining $\mathbf{e}$ to a black vertex $\mathbf{v}_i$ with $(\mathbf{v}_1)_1 = (H_1(24))_1 = 4$, $(\mathbf{v}_2)_1 = (H_2(24))_1 = 1$, $(\mathbf{v}_3)_1 = (H_3(24))_1 = 3$ and $|\{H_1(i), H_2(i), H_3(i)\}| = 3$ for $2 \leq i \leq 24$.

We choose a black vertex $\mathbf{z}_1$ in S_5^3 with $(\mathbf{z}_1)_1 = a_1$, a black vertex $\mathbf{z}_2$ in S_5^4 with $(\mathbf{z}_2)_1 = a_2$, a black vertex $\mathbf{z}_3$ in S_5^2 with $(\mathbf{z}_3)_1 = a_3$, and a black vertex $\mathbf{z}_4$ in S_5^2 with $(\mathbf{z}_4)_1 = a_4$. We set

$$B = \begin{bmatrix} 4 & 1 & 2 & 3 \\ 1 & 2 & 3 & 4 \\ 3 & 4 & 1 & 2 \end{bmatrix}$$

By Theorem 13.9, there exists a path $W_1 = \langle \mathbf{x}_1^1, T_4^1, \mathbf{y}_1^1, \mathbf{x}_2^1, T_1^1, \mathbf{y}_2^1, \mathbf{x}_3^1, T_2^1, \mathbf{y}_3^1, \mathbf{x}_4^1, T_3^1, \mathbf{y}_4^1 \rangle$ joining the white vertex $(\mathbf{v}_1)^5$ to the black vertex $\mathbf{z}_1$ such that $\mathbf{x}_1^1 = (\mathbf{v}_1)^5$, $\mathbf{y}_4^1 = \mathbf{z}_1$, and T_j^1 is a Hamiltonian path of $S_5^{b_{1j}}$ for every $1 \leq j \leq 4$. Similarly, there exists a path $W_2 = \langle \mathbf{x}_1^2, T_1^2, \mathbf{y}_1^2, \mathbf{x}_2^2, T_2^2, \mathbf{y}_2^2, \mathbf{x}_3^2, T_3^2, \mathbf{y}_3^2, \mathbf{x}_4^2, T_4^2, \mathbf{y}_4^2 \rangle$ joining the white vertex $(\mathbf{v}_2)^5$ to the black vertex $\mathbf{z}_2$ such that $\mathbf{x}_1^2 = (\mathbf{v}_2)^5$, $\mathbf{y}_4^2 = \mathbf{z}_2$, and T_j^2 is a Hamiltonian path of $S_5^{b_{2j}}$ for every $1 \leq j \leq 4$. Furthermore, there exists a path $W_3 = \langle \mathbf{x}_1^3, T_3^3, \mathbf{y}_1^3, \mathbf{x}_2^3, T_4^3, \mathbf{y}_2^3, \mathbf{x}_3^3, T_1^3, \mathbf{y}_3^3, \mathbf{x}_4^3, T_2^3, \mathbf{y}_4^3 \rangle$ joining the white vertex $(\mathbf{v}_3)^5$ to the black vertex $\mathbf{z}_3$ such that $\mathbf{x}_1^3 = (\mathbf{v}_3)^5$, $\mathbf{y}_4^3 = \mathbf{z}_3$, and T_j^3 is a Hamiltonian path of $S_5^{b_{3j}}$ for every $1 \leq j \leq 4$. Moreover, there exists a path $W = \langle \mathbf{x}_1^4, T_1^4, \mathbf{y}_1^4, \mathbf{x}_2^4, T_2^4, \mathbf{y}_2^4, \mathbf{x}_3^4, T_3^4, \mathbf{y}_3^4, \mathbf{x}_4^4, T_4^4, \mathbf{y}_4^4 \rangle$ joining the black vertex $(\mathbf{e})^5$ to the white vertex $((\mathbf{e})^4)^5$ such that $\mathbf{x}_1^4 = (\mathbf{e})^5$, $\mathbf{y}_4^4 = ((\mathbf{e})^4)^5$, and T_i^4 is a Hamiltonian path of S_5^i for $1 \leq i \leq 4$. Since S_5^5 is isomorphic to S_4, by Theorem 13.2, there exists a Hamiltonian path R of $S_5^5 - \{\mathbf{e}\}$ joining the black vertex $(\mathbf{e})^4$ to $\mathbf{z}_4$.

Let $P_i = \langle \mathbf{e}, H_i, \mathbf{v}_i, (\mathbf{v}_i)^n, W_i, \mathbf{z}_i \rangle$ for every $1 \leq i \leq 3$ and $P_4 = \langle \mathbf{e}, W, ((\mathbf{e})^4)^5, (\mathbf{e})^4, R, \mathbf{z}_4 \rangle$. Thus, $\{P_1, P_2, P_3, P_4\}$ form the desired paths. See Figure 19.3 for illustration.

THEOREM 19.4 $IHP_L(S_2) = 1$, $IHP_L(S_3) = 0$, and $IHP_L(S_n) = n-2$ if $n \geq 4$.

Proof: Suppose that $n = 2$. Since S_2 is isomorphic to $K_{1,1}$, it is easy to see that $IHP_L(S_2) = 1$. Suppose that $n = 3$. Obviously, S_3 is isomorphic to the cycle graph with six vertices. It is easy to see that S_3 is not Hamiltonian laceable. Thus, $IPH_L(S_3) = 0$. Now, we assume that $n \geq 4$. Since $\delta(S_n) = n-1$, $IHP_L(S_n) \leq n-2$. To prove our theorem, we need to construct $(n-2)$ mutually independent Hamiltonian paths of S_n between any white vertex $\mathbf{u}$ and any black vertex $\mathbf{v}$. We prove this theorem by induction.

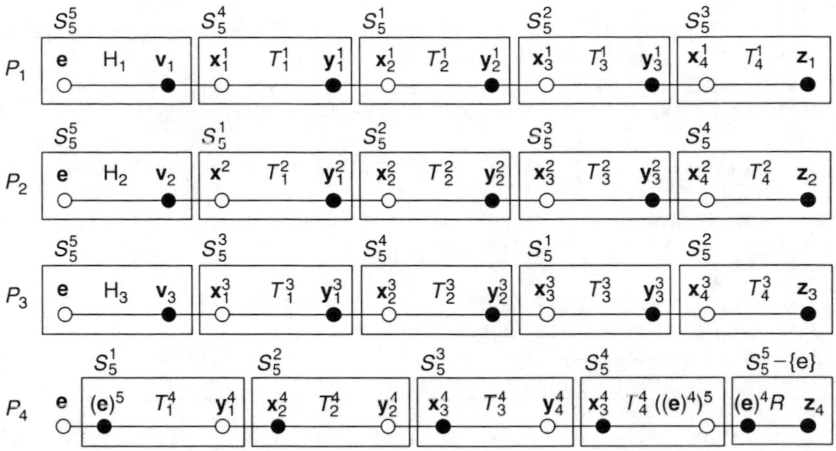

FIGURE 19.3 Illustration of Theorem 19.8 on S_5.

Since S_4 is vertex-transitive, we assume that $\mathbf{u} = 1234$. The required Hamiltonian paths of S_4 are:

v = (2134)
(1234)(4231)(2431)(1432)(4132)(3142)(2143)(1243)(3241)(2341)(1342)(4312)(3412)(2413)(4213)(3214)(2314)(1324)(4321)(3421)(1423)(4123)(3124)(2134)
(1234)(3214)(4213)(2413)(3412)(4312)(2314)(1324)(3124)(4123)(1423)(3421)(4321)(2341)(1342)(3142)(2143)(1243)(3241)(4231)(2431)(1432)(4132)(2134)

v = (3214)
(1234)(2134)(3124)(1324)(2314)(4312)(3412)(1432)(4132)(3142)(1342)(2341)(4321)(3421)(2431)(4231)(3241)(1243)(2143)(4123)(1423)(2413)(4213)(3214)
(1234)(4231)(2431)(3421)(4321)(2341)(3241)(1243)(4213)(2413)(1423)(4123)(2143)(3142)(1342)(4312)(3412)(1432)(4132)(2134)(3124)(1324)(2314)(3214)

v = (4231)
(1234)(2134)(4132)(1432)(2431)(3421)(4321)(1324)(3124)(4123)(1423)(2413)(3412)(4312)(2314)(3214)(4213)(1243)(2143)(3142)(1342)(2341)(3241)(4231)
(1234)(3214)(2314)(4312)(3412)(2413)(4213)(1243)(3241)(2341)(1342)(3142)(2143)(4123)(1423)(3421)(4321)(1324)(3124)(2134)(4132)(1432)(2431)(4231)

v = (1432)
(1234)(2134)(4132)(3142)(1342)(2341)(4321)(1324)(3124)(4123)(2143)(1243)(3241)(4231)(2431)(3421)(1423)(2413)(4213)(3214)(2314)(4312)(3412)(1432)
(1234)(3214)(2314)(4312)(3412)(2413)(4213)(1243)(3241)(4231)(2431)(3421)(1423)(4123)(2143)(3142)(1342)(2341)(4321)(1324)(3124)(2134)(4132)(1432)

v = (1243)
(1234)(4231)(3241)(2341)(1342)(4312)(3412)(1432)(2431)(3421)(4321)(1324)(2314)(3214)(4213)(2413)(1423)(4123)(3124)(2134)(4132)(3142)(2143)(1243)
(1234)(2134)(4132)(3142)(2143)(4123)(3124)(1324)(2314)(3214)(4213)(2413)(1423)(3421)(4321)(2341)(1342)(4312)(3412)(1432)(2431)(4231)(3241)(1243)

v = (1324)
(1234)(2134)(3124)(4123)(1423)(2413)(3412)(1432)(4132)(3142)(2143)(1243)(4213)(3214)(2314)(4312)(1342)(2341)(3241)(4231)(2431)(3421)(4321)(1324)
(1234)(4231)(2431)(3421)(4321)(2341)(3241)(1243)(4213)(3214)(2314)(4312)(1342)(3142)(2143)(4123)(1423)(2413)(3412)(1432)(4132)(2134)(3124)(1324)

v = (2341)
(1234)(2134)(4132)(3142)(1342)(4312)(2314)(3214)(4213)(1243)(2143)(4123)(3124)(1324)(4321)(3421)(1423)(2413)(3412)(1432)(2431)(4231)(3241)(2341)
(1234)(3214)(4213)(1243)(3241)(4231)(2431)(3421)(1423)(2413)(3412)(1432)(4132)(2134)(3124)(4123)(2143)(3142)(1342)(4312)(2314)(1324)(4321)(2341)
v = (3142)
(1234)(3214)(2314)(4312)(1342)(2341)(3241)(4231)(2431)(1432)(3412)(2413)(4213)(1243)(2143)(4123)(1423)(3421)(4321)(1324)(3124)(2134)(4132)(3142)
(1234)(4231)(2431)(1432)(4132)(2134)(3124)(4123)(1423)(3421)(4321)(1324)(2314)(3214)(4213)(2413)(3412)(4312)(1342)(2341)(3241)(1243)(2143)(3142)
v = (4123)
(1234)(4231)(2431)(3421)(1423)(2413)(4213)(3214)(2314)(1324)(4321)(2341)(3241)(1243)(2143)(3142)(1342)(4312)(3412)(1432)(4132)(2134)(3124)(4123)
(1234)(3214)(2314)(1324)(3124)(2134)(4132)(3142)(1342)(4312)(3412)(1432)(2431)(4231)(3241)(2341)(4321)(3421)(1423)(2413)(4213)(1243)(2143)(4123)
v = (4312)
(1234)(4231)(3241)(2341)(1342)(3142)(4132)(2134)(3124)(1324)(4321)(3421)(2431)(1432)(3412)(2413)(1423)(4123)(2143)(1243)(4213)(3214)(2314)(4312)
(1234)(2134)(3124)(1324)(2314)(3214)(4213)(2413)(1423)(4123)(2143)(1243)(3241)(4231)(2431)(3421)(4321)(2341)(1342)(3142)(4132)(1432)(3412)(4312)
v = (2413)
(1234)(2134)(3124)(4123)(1423)(3421)(2431)(4231)(3241)(1243)(2143)(3142)(4132)(1432)(3412)(4312)(1342)(2341)(4321)(1324)(2314)(3214)(4213)(2413)
(1234)(4231)(3241)(1243)(4213)(3214)(2314)(4312)(1342)(2341)(4321)(1324)(3124)(2134)(4132)(3142)(2143)(4123)(1423)(3421)(2431)(1432)(3412)(2413)
v = (3421)
(1234)(3214)(4213)(2413)(1423)(4123)(3124)(2134)(4132)(1432)(3412)(4312)(2314)(1324)(4321)(2341)(1342)(3142)(2143)(1243)(3241)(4231)(2431)(3421)
(1234)(2134)(4132)(1432)(2431)(4231)(3241)(2341)(1342)(3142)(2143)(1243)(4213)(3214)(2314)(4312)(3412)(2413)(1423)(4123)(3124)(1324)(4321)(3421)

Suppose that the theorem holds on S_k for every $4 \le k < n$. Without loss of generality, we assume that $(\mathbf{u})_n = n$ and $(\mathbf{v})_n = n-1$. We let C be the $(n-2) \times (n-2)$ matrix with

$$c_{ij} = \begin{cases} i+j-1 & \text{if } i+j-1 \le n-2 \\ i+j-n+1 & \text{if } n-2 < i+j-1 \end{cases}$$

More precisely,

$$C = \begin{bmatrix} 1 & 2 & 3 & \dots & n-2 \\ 2 & 3 & 4 & \dots & 1 \\ \vdots & \vdots & \vdots & \ddots & \vdots \\ n-2 & 1 & 2 & \dots & n-3 \end{bmatrix}$$

Obviously, $c_{i,1}, c_{i,2}, \dots, c_{i,n-2}$ forms a permutation of $\{1, 2, \dots, n-2\}$ for every i with $1 \le i \le n-2$. Moreover, $c_{i,j} \ne c_{i',j}$ for any $1 \le i < i' \le n-2$ and $1 \le j \le n-2$. In other words, C forms a Latin square with entries in $\{1, 2, \dots, n-2\}$.

It follows from Theorem 19.8 that there exist Hamiltonian paths $Q_1, Q_2, \dots, Q_{n-2}$ of S_n^n such that Q_i is a path joining $\mathbf{u}$ to a black vertex $\mathbf{s}_i$ with (1)

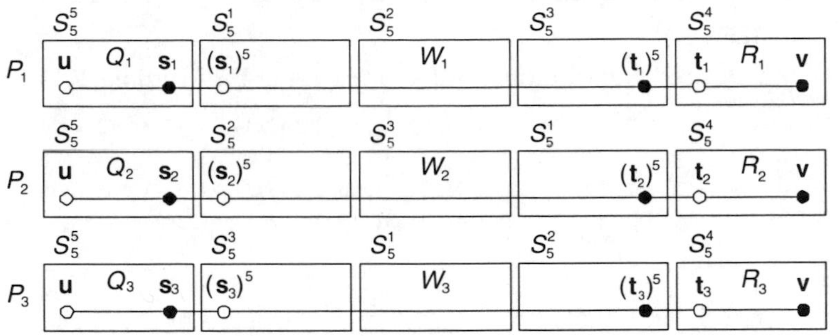

FIGURE 19.4 Illustration for three mutually independent Hamiltonian paths on S_5.

$(\mathbf{s}_i)_1 = c_{i1}$ for every $1 \le i \le n-2$ and (2) $|\{Q_1(i), Q_2(i), \ldots, Q_{n-2}(i)\}| = n-2$ for every $2 \le i \le (n-1)!$. Again, there exist Hamiltonian paths $R_1, R_2, \ldots, R_{n-2}$ of S_n^{n-1} such that R_i is a path joining $\mathbf{v}$ to a white vertex $\mathbf{t}_i$ with (1) $(\mathbf{t}_i)_1 = c_{i(n-2)}$ for every $1 \le i \le n-2$ and (2) $|\{R_1(i), R_2(i), \ldots, R_{n-2}(i)\}| = n-2$ for every $2 \le i \le (n-1)!$. For every $1 \le i \le n-2$ by Theorem 13.9, there exists a path $W_i = \langle \mathbf{x}_1^i, T_1^i, \mathbf{y}_1^i, \mathbf{x}_2^i, T_2^i, \mathbf{y}_2^i, \ldots, \mathbf{x}_{n-2}^i, T_{n-2}^i, \mathbf{y}_{n-2}^i \rangle$ joining the white vertex $\mathbf{x}_1^i$ to the black vertex $\mathbf{y}_{n-2}^n$ such that $\mathbf{x}_1^i = (\mathbf{s}_i)^n$, $\mathbf{y}_{n-2}^i = (\mathbf{t}_i)^n$, and T_j^i is a Hamiltonian path of $S_n^{b_{i+j-1}}$ joining $\mathbf{x}_j^i$ to $\mathbf{y}_j^i$ for every $1 \le j \le n-1$. Hence, $\{P_1, P_2, \ldots, P_{n-2}\}$ form the desired paths. See Figure 19.4 for illustration of the case $n = 5$.

Hence, the theorem is proved. □

19.4 MUTUALLY INDEPENDENT HAMILTONIAN CONNECTIVITY FOR (*n, k*)-STAR GRAPHS

Here, we recall the definition of (n, k)-star graphs. Let n and k be two positive integers with $n > k$. We use $\langle n \rangle$ to denote the set $\{1, 2, \ldots, n\}$. The (n, k)-star graph, $S_{n,k}$, is a graph with the node set $V(S_{n,k}) = \{u_1 u_2 \ldots u_k | u_i \in \langle n \rangle \text{ and } u_i \ne u_j \text{ for } i \ne j\}$. Adjacency is defined as follows: a node $u_1 u_2 \ldots u_i \ldots u_k$ is adjacent to (1) the node $u_i u_2 u_3 \ldots u_{i-1} u_1 u_{i+1} \ldots u_k$, where $2 \le i \le k$ (i.e., swap u_i with u_1) and (2) the node $x u_2 u_3 \ldots u_k$, where $x \in \langle n \rangle - \{u_i | 1 \le i \le k\}$. The edges of type (1) are referred to as an *i-edge* and the edges of type (2) are referred to as 1*-edges*.

By definition, $S_{n,k}$ is an $(n-1)$-regular graph with $n!/(n-k)!$ nodes. Moreover, it is node-transitive [60]. We use boldface to denote nodes in $S_{n,k}$. Hence $\mathbf{u}_1, \mathbf{u}_2, \ldots, \mathbf{u}_m$ denotes a sequence of nodes in $S_{n,k}$. Let $\mathbf{u} = u_1 u_2 \ldots u_k$ be any node of $S_{n,k}$. We say that u_i is the ith coordinate of $\mathbf{u}$, denoted by $(\mathbf{u})_i$, for $1 \le i \le k$. By the definition of $S_{n,k}$, there is exactly one neighbor $\mathbf{v}$ of $\mathbf{u}$ such that $\mathbf{u}$ and $\mathbf{v}$ are adjacent through an i-edge with $2 \le i \le k$. For this reason, we use $(\mathbf{u})^i$ to denote the unique i-neighbor of $\mathbf{u}$. Obviously, $((\mathbf{u})^i)^i = \mathbf{u}$. For $1 \le i \le n$, let $S_{n,k}^{\{i\}}$ denote the subgraph of $S_{n,k}$ induced by those nodes $\mathbf{u}$ with $(\mathbf{u})_k = i$. In Ref. 60, it is shown that $S_{n,k}$ can be decomposed into n subgraphs $S_{n,k}^{\{i\}}$, $1 \le i \le n$, such that each subgraph $S_{n,k}^{\{i\}}$ is isomorphic to $S_{n-1,k-1}$. Thus, the (n, k)-star graph can be constructed recursively. Obviously, $\mathbf{u} \in S_{n,k}^{\{(\mathbf{u})_k\}}$. Let

TABLE 19.1
The Desired Hamiltonian Paths for $S_{4,2} - \{14\}$ of Lemma 19.4

$\{i=1\}$	$\langle \mathbf{x}=41, 21, 31, 13, 23, 43, 34, 24, 42, 32, 12=\mathbf{y}\rangle$
$\{i=2\}$	$\langle \mathbf{x}=41, 21, 31, 13, 43, 34, 24, 42, 12, 32, 23=\mathbf{y}\rangle$
$\{i=3\}$	$\langle \mathbf{x}=41, 21, 31, 13, 23, 43, 34, 24, 42, 12, 32=\mathbf{y}\rangle$
$\{i=4\}$	$\langle \mathbf{x}=41, 21, 31, 13, 23, 32, 12, 42, 24, 34, 43=\mathbf{y}\rangle$

$I \subseteq \langle n \rangle$. We use $S^I_{n,k}$ to denote the subgraph of $S_{n,k}$ induced by those nodes $\mathbf{u}$ with $(\mathbf{u})_k \in I$. For $1 \le i \le n$ and $1 \le j \le n$ with $i \ne j$, we use $E^{i,j}_{n,k}$ to denote the set of edges between $S^{\{i\}}_{n,k}$ and $S^{\{j\}}_{n,k}$. $S_{3,1}$, $S_{4,1}$, $S_{4,2}$, and $S_{5,2}$ are shown in Figure 16.15.

Now, we need some properties concerning the (n, k)-star graphs.

LEMMA 19.4 Let n and k be any two integers with $k \ge 2$ and $n - k \ge 2$. Let $\mathbf{u}$ be any vertex of $S_{n,k}$ and let i be any positive integer with $1 \le i \le n$. Then there is a Hamiltonian path P of $S_{n,k} - \{\mathbf{u}\}$ joining a vertex $\mathbf{x}$ to a vertex $\mathbf{y}$ where $(\mathbf{x})_1 = (\mathbf{u})_k$ and $(\mathbf{y})_1 = i$.

Proof: By Theorem 16.6, this statement holds for $n \ge 5$. Thus, we assume that $n = 4$. Therefore, $k = 2$. Since $S_{4,2}$ is vertex-transitive, we assume that $\mathbf{u} = 14$. Note that $(\mathbf{u})_2 = 4$. We list all of the desired $\mathbf{x}$, $\mathbf{y}$, and Hamiltonian paths of $S_{4,2} - \{\mathbf{u} = 14\}$ in Table 19.1. Thus, this statement is proved. □

LEMMA 19.5 Let n and k be any two integers with $n \ge 5$, $k \ge 2$, and $n - k \ge 2$. Let $\mathbf{u}$ and $\mathbf{v}$ be any two distinct vertices of $S_{n,k}$ with $(\mathbf{u})_t = (\mathbf{v})_t$ for every $2 \le t \le k$, and let i and j be two integers in $\langle n \rangle$ with $i < j$. Then there is a Hamiltonian path P of $S_{n,k} - \{\mathbf{u}, \mathbf{v}\}$ joining a vertex $\mathbf{x}$ to a vertex $\mathbf{y}$ with $(\mathbf{x})_1 = i$ and $(\mathbf{y})_1 = j$.

Proof: Without loss of generality, we assume that $(\mathbf{u})_1 = 1$, $(\mathbf{v})_1 = 2$, and $(\mathbf{u})_t = (\mathbf{v})_t = n - k + t$ for every $2 \le t \le k$. Since $i < j$, $i \ne n$, and $j \ne 1$. We have the following cases.

Case 1. $n = 5$ and $k = 2$. Since $(\mathbf{u})_1 = 1$, $(\mathbf{v})_1 = 2$, and $(\mathbf{u})_2 = (\mathbf{v})_2 = 5$, we have $\mathbf{u} = 15$ and $\mathbf{v} = 25$. We list all of the desired Hamiltonian paths of $S_{5,2} - \{\mathbf{u}, \mathbf{v}\}$ in Table 19.2.

Case 2. $n = 5$ and $k = 3$. Since $(\mathbf{u})_1 = 1$, $(\mathbf{v})_1 = 2$, and $(\mathbf{u})_t = (\mathbf{v})_t = t + 2$ for every $2 \le t \le 3$, we have $\mathbf{u} = 145$ and $\mathbf{v} = 245$. Let $\mathbf{x}_1 = 135$, $\mathbf{x}_2 = 235$, $\mathbf{x}_3 = 325$, $\mathbf{x}_4 = 425$, and $\mathbf{p}_0 = 345$. Depending on the value i, we set $\mathbf{x}$ and R as

$$\mathbf{x} = \mathbf{x}_1 \text{ and } R = \langle 135, 315, 415, 215, 125, 425, 325, 235, 435, 345 = \mathbf{p}_0\rangle \text{ if } i = 1$$
$$\mathbf{x} = \mathbf{x}_2 \text{ and } R = \langle 235, 325, 425, 125, 215, 415, 315, 135, 435, 345 = \mathbf{p}_0\rangle \text{ if } i = 2$$
$$\mathbf{x} = \mathbf{x}_3 \text{ and } R = \langle 325, 425, 125, 215, 415, 315, 135, 235, 435, 345 = \mathbf{p}_0\rangle \text{ if } i = 3$$

and

$$\mathbf{x} = \mathbf{x}_4 \text{ and } R = \langle 425, 325, 125, 215, 415, 315, 135, 235, 435, 345 = \mathbf{p}_0\rangle \text{ if } i = 4$$

TABLE 19.2
The Desired Hamiltonian Paths for Case 1 of Lemma 19.5

$\{i=1, j=2\}$	$\langle 12,42,52,32,23,13,43,53,35,45,54,34,24,14,41,31,51,21\rangle$
$\{i=1, j=3\}$	$\langle 12,42,52,32,23,13,43,53,35,45,54,34,24,14,41,21,51,31\rangle$
$\{i=1, j=4\}$	$\langle 12,52,32,42,24,34,54,14,41,21,51,31,13,23,43,53,35,45\rangle$
$\{i=1, j=5\}$	$\langle 12,42,52,32,23,13,43,53,35,45,54,34,24,14,41,31,21,51\rangle$
$\{i=2, j=3\}$	$\langle 23,13,43,53,35,45,54,34,24,14,41,31,51,21,12,42,52,32\rangle$
$\{i=2, j=4\}$	$\langle 23,13,43,53,35,45,54,34,24,14,41,31,51,21,12,32,52,42\rangle$
$\{i=2, j=5\}$	$\langle 23,13,43,53,35,45,54,34,24,14,41,31,51,21,12,42,32,52\rangle$
$\{i=3, j=4\}$	$\langle 35,45,54,14,24,34,43,53,23,13,31,41,51,21,12,32,52,42\rangle$
$\{i=3, j=5\}$	$\langle 35,45,54,14,24,34,43,53,23,13,31,41,51,21,12,32,42,52\rangle$
$\{i=4, j=5\}$	$\langle 45,35,53,23,13,43,34,24,54,14,41,31,51,21,12,32,42,52\rangle$

Note that R is a Hamiltonian path of $S_{5,3}^{\{5\}} - \{\mathbf{u}, \mathbf{v}\}$ joining $\mathbf{x}$ to $\mathbf{p}_0$. We set $\mathbf{p}_1 = 453$, $\mathbf{p}_2 = 254$, and $\mathbf{p}_3 = 152$. By Lemma 16.17, $|\{\mathbf{z}|\mathbf{z} \in S_{5,3}^{\{1\}}$ and $(\mathbf{z})_1 = j\}| = |E_{5,3}^{1,j}| = 3$ for $2 \le j \le 5$. We can choose a vertex $\mathbf{p}_4$ in $S_{5,3}^{\{1\}} - \{(\mathbf{p}_3)^3\}$ with $(\mathbf{p}_4)_1 = j$. By Theorem 16.6, there is a Hamiltonian path Q_t of $S_{5,3}^{\{(\mathbf{p}_t)_3\}}$ joining $(\mathbf{p}_{t-1})^3$ to $\mathbf{p}_t$ for every $1 \le t \le 4$. We set $\mathbf{y} = \mathbf{p}_4$ and $P = \langle \mathbf{x}, R, \mathbf{p}_0, (\mathbf{p}_0)^3, Q_1, \mathbf{p}_1, (\mathbf{p}_1)^3, Q_2, \mathbf{p}_2, \ldots, (\mathbf{p}_3)^3, Q_4, \mathbf{p}_4 = \mathbf{y}\rangle$. Then P forms a desired path.

Case 3. $n \ge 6$, $k \ge 2$, and $n-k \ge 2$. Let $\mathbf{x}$ and $\mathbf{y}$ be any two vertices of $S_{n,k} - \{\mathbf{u}, \mathbf{v}\}$ with $(\mathbf{x})_1 = i$ and $(\mathbf{y})_1 = j$. By Theorem 16.6, there is a Hamiltonian path P of $S_{n,k} - \{\mathbf{u}, \mathbf{v}\}$ joining $\mathbf{x}$ to $\mathbf{y}$.

Thus, this statement is proved. □

We divided this section into three subsections, depending on $k=1$, $k=2$, and $k \ge 3$.

THEOREM 19.5 $IHP(S_{n,1}) = n-2$ if $n \ge 3$.

Proof: Since $IHP(G) \le \delta(G) - 1$ for every graph G with at least three vertices, $IHP(S_{n,1}) \le n-2$ for every $n \ge 3$. To prove this statement, we need to construct $(n-2)$ mutually independent Hamiltonian paths of $S_{n,1}$ between any two distinct vertices. Let $\mathbf{u}$ and $\mathbf{v}$ be any two distinct vertices of $S_{n,1}$. Since $S_{n,1}$ is isomorphic to complete graph K_n with order n vertices, we assume that $\mathbf{u} = n$ and $\mathbf{v} = n-1$. For every $1 \le i \le n-2$ and $1 \le j \le n-2$, we set $\mathbf{z}_j^i$ as

$$\mathbf{z}_j^i = \begin{cases} i+j-1 & \text{if } i+j \le n-1 \\ i+j-n+1 & \text{if } n \le i+j \end{cases}$$

We set $P_i = \langle \mathbf{u}, \mathbf{z}_1^i, \mathbf{z}_2^i, \ldots, \mathbf{z}_{n-2}^i, \mathbf{v}\rangle$ for every $1 \le i \le n-2$. Then $\{P_1, P_2, \ldots, P_{n-2}\}$ forms a set of $(n-2)$ mutually independent Hamiltonian paths of $S_{n,1}$ from $\mathbf{u}$ to $\mathbf{v}$. □

TABLE 19.3
All Hamiltonian Paths of $S_{4,2}$ between Vertices 12 and 14

$P_1 = \langle 12, 32, 42, 24, 34, 43, 23, 13, 31, 21, 41, 14\rangle$
$P_2 = \langle 12, 21, 41, 31, 13, 43, 23, 32, 42, 24, 34, 14\rangle$

THEOREM 19.6 $IHP(S_{4,2}) = 1$.

Proof: By Theorem 16.6, $S_{4,2}$ is Hamiltonian-connected. Thus, $IHP(S_{4,2}) \geq 1$. Using depth-first search, we list all Hamiltonian paths of $S_{4,2}$ from 12 to 14 in Table 19.3. It is easy to see that $P_1(6) = P_2(6) = 43$. Thus, $IHP(S_{4,2}) < 2$. Hence, $IHP(S_{4,2}) = 1$. □

LEMMA 19.6 Let $\mathbf{u}$ be any vertex of $S_{n,1}$, and let $\{i_1, i_2, \ldots, i_m\}$ be any nonempty subset of $\langle n\rangle - \{(\mathbf{u})_1\}$ with $n \geq 2$. Then, there are m Hamiltonian paths $P_1, P_2, \ldots, P_m$ of $S_{n,1}$ such that (1) P_j joins $\mathbf{u}$ to some vertex $\mathbf{x}_j$ with $\mathbf{x}_j = (\mathbf{x}_j)_1 = i_j$ for every $1 \leq j \leq m$ and (2)$|\{P_1(j), P_2(j), \ldots, P_m(j)\}| = m$ for every $2 \leq j \leq n$.

Proof: Without loss of generality, we assume that $\mathbf{u} = (\mathbf{u})_1 = n$ and $\mathbf{x}_j = (\mathbf{x}_j)_1 = i_j = j$ for every $1 \leq j \leq n-1$. For every $1 \leq j \leq n-1$ and for every $1 \leq l \leq n-1$, we set $\mathbf{z}_l^j$ as

$$\mathbf{z}_l^j = \begin{cases} j+l-1 & \text{if } j+l \leq n \\ j+l-n & \text{if } n+1 \leq j+l \end{cases}$$

We set $\mathbf{x}_j = \mathbf{z}_{n-1}^j$ and $P_j = \langle \mathbf{u}, \mathbf{z}_1^j, \mathbf{z}_2^j, \ldots, \mathbf{z}_{n-1}^j = \mathbf{x}_j\rangle$ for every $1 \leq j \leq n-1$. Then $P_{i_1}, P_{i_2}, \ldots, P_{i_m}$ form desired paths. □

THEOREM 19.7 $IHP(S_{n,2}) = n-2$ if $n \geq 5$.

Proof: Since $IHP(G) \leq \delta(G) - 1$ for every graph G with at least three vertices, $IHP(S_{n,2}) \leq n-2$ if $n \geq 5$. Let $\mathbf{u}$ and $\mathbf{v}$ be any two distinct vertices of $S_{n,2}$. We need to construct $(n-2)$ mutually independent Hamiltonian paths of $S_{n,2}$ from $\mathbf{u}$ to $\mathbf{v}$. We have the following cases.

Case 1. $(\mathbf{u})_2 = (\mathbf{v})_1$ and $(\mathbf{u})_1 = (\mathbf{v})_2$. Without loss of generality, we assume that $\mathbf{u} = 12$ and $\mathbf{v} = 21$. By Lemma 19.6, there are $(n-2)$ Hamiltonian paths $H_1, H_2, \ldots, H_{n-2}$ of $S_{n,2}^{\{2\}}$ such that (1) H_i joins $\mathbf{u}$ to some vertex $\mathbf{x}_i$ with $(\mathbf{x}_i)_1 = i+2$ for every $1 \leq i \leq n-2$ and (2) $|\{H_1(j), H_2(j), \ldots, H_{n-2}(j)\}| = n-2$ for every $2 \leq j \leq n-1$. Again, there are $(n-2)$ Hamiltonian paths $Q_1, Q_2, \ldots, Q_{n-2}$ of $S_{n,2}^{\{1\}}$ such that (1) Q_1 joins $\mathbf{v}$ to some vertex $\mathbf{y}_1$ with $(\mathbf{y}_1)_1 = n$ and Q_i joins $\mathbf{v}$ to some vertex $\mathbf{y}_i$ with $(\mathbf{y}_i)_1 = i+1$ for every $2 \leq i \leq n-2$, and (2) $|\{Q_1(j), Q_2(j), \ldots, Q_{n-2}(j)\}| = n-2$ for every $2 \leq j \leq n-1$. Let A be an $(n-2) \times (n-2)$ matrix defined by

$$a_{i,j} = \begin{cases} i+j+1 & \text{if } i+j \leq n-1 \\ i+j-n+3 & \text{if } n \leq i+j \end{cases}$$

More precisely,

$$A = \begin{bmatrix} 3 & 4 & \cdots & n-2 & n-1 & n \\ 4 & 5 & \cdots & n-1 & n & 3 \\ \vdots & \vdots & \ddots & \vdots & \vdots & \vdots \\ n & 3 & \cdots & n-3 & n-2 & n-1 \end{bmatrix}$$

Obviously, $a_{j,1}a_{j,2}\ldots a_{j,n-1}$ forms a permutation of $\{3,4,\ldots,n\}$ for every j with $1 \le j \le n-2$. Moreover, $a_{j,l} \ne a_{j',l}$ for any $1 \le j < j' \le n-2$ and $1 \le l \le n-2$. In other words, A forms a rectangular Latin square with entries in $\{3,4,\ldots,n\}$.

We set $\mathbf{p}_0^i = \mathbf{x}_i$ and $\mathbf{p}_{n-2}^i = (\mathbf{y}_i)^2$ for every $1 \le i \le n-2$. For every $1 \le i \le n-2$ and $1 \le j \le n-3$, we set $\mathbf{p}_j^i = a_{i,j+1}a_{i,j}$. By Theorem 16.6, there is a Hamiltonian path T_j^i of $S_{n,2}^{\{a_{i,j}\}}$ joining $(\mathbf{p}_{j-1}^i)^2$ to $\mathbf{p}_j^i$ for every $1 \le i \le n-2$ and $1 \le j \le n-2$. We set $P_i = \langle \mathbf{u}, H_i, \mathbf{x}_i = \mathbf{p}_0^i, (\mathbf{p}_0^i)^2, T_1^i, \mathbf{p}_1^i, (\mathbf{p}_1^i)^2, T_2^i, \mathbf{p}_2^i, \ldots, (\mathbf{p}_{n-3}^i)^2, T_{n-2}^i, \mathbf{p}_{n-2}^i = (\mathbf{y}_i)^3, \mathbf{y}_i, Q_i^{-1}, \mathbf{v}\rangle$ for every $1 \le i \le n-2$. Then $\{P_1, P_2, \ldots, P_{n-2}\}$ forms a set of $(n-2)$ mutually independent Hamiltonian paths of $S_{n,2}$ from $\mathbf{u}$ to $\mathbf{v}$. See Figure 19.5 for an illustration.

Case 2. Either $(\mathbf{u})_2 = (\mathbf{v})_1$ and $(\mathbf{u})_1 \ne (\mathbf{v})_2$ or $(\mathbf{u})_1 = (\mathbf{v})_2$ and $(\mathbf{u})_2 \ne (\mathbf{v})_1$. Without loss of generality, we assume that $(\mathbf{u})_2 = (\mathbf{v})_1$ and $(\mathbf{u})_1 \ne (\mathbf{v})_2$. Moreover, we assume that $\mathbf{u} = 12$ and $\mathbf{v} = 23$. By Lemma 19.6, there are $(n-3)$ Hamiltonian path $H_1, H_2, \ldots, H_{n-3}$ of $S_{n,2}^{\{2\}}$ such that (1) H_i joins $\mathbf{u}$ to some vertex $\mathbf{x}_i$ with $(\mathbf{x}_i)_1 = i+3$ for every $1 \le i \le n-3$ and (2) $|\{H_1(j), H_2(j), \ldots, H_{n-3}(j)\}| = n-3$ for every $2 \le j \le n-1$. Again, there are $(n-3)$ Hamiltonian path $Q_1, Q_2, \ldots, Q_{n-3}$ of $S_{n,2}^{\{3\}}$ such that (1) Q_1 joins $\mathbf{v}$ to some vertex $\mathbf{y}_1$ with $(\mathbf{y}_1)_1 = 1$ and Q_i joins $\mathbf{v}$ to some vertex $\mathbf{y}_i$ with $(\mathbf{y}_i)_1 = i+2$ for every $2 \le i \le n-3$, and (2) $|\{H_1(j), H_2(j), \ldots, H_{n-3}(j)\}| = n-3$ for every $2 \le j \le n-1$. Let A be an $(n-3) \times (n-2)$ matrix defined by

$$a_{i,j} = \begin{cases} i+j+2 & \text{if } i+j \le n-2 \\ 1 & \text{if } i+j = n-1 \\ i+j-n+4 & \text{if } n \le i+j \end{cases}$$

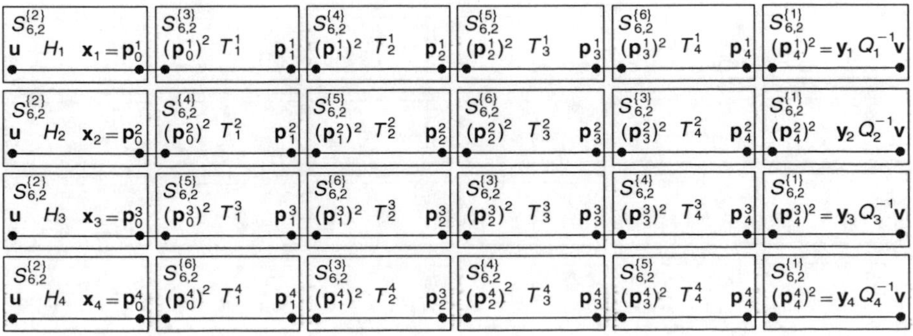

FIGURE 19.5 Illustration for Case 1 of Theorem 19.7 on $S_{6,2}$.

More precisely,

$$A = \begin{bmatrix} 4 & 5 & 6 & \cdots & n-1 & n & 1 \\ 5 & 6 & 7 & \cdots & n & 1 & 4 \\ \vdots & \vdots & \vdots & \ddots & \vdots & \vdots & \vdots \\ n & 1 & 4 & \cdots & n-3 & n-2 & n-1 \end{bmatrix}$$

We set $\mathbf{p}_0^i = \mathbf{x}_i$ and $\mathbf{p}_{n-2}^i = (\mathbf{y}_i)^2$ for every $1 \le i \le n-2$. For every $1 \le i \le n-3$ and $1 \le j \le n-3$, we set $\mathbf{p}_j^i = a_{i,j+1}a_{i,j}$. By Theorem 16.6, there is a Hamiltonian path T_j^i of $S_{n,2}^{\{a_{i,j}\}}$ joining $(\mathbf{p}_{j-1}^i)^2$ to $\mathbf{p}_j^i$ for every $1 \le i \le n-3$ and $1 \le j \le n-2$. We set $P_i = \langle \mathbf{u}, H_i, \mathbf{x}_i = \mathbf{p}_0^i, (\mathbf{p}_0^i)^2, T_1^i, \mathbf{p}_1^i, (\mathbf{p}_1^i)^2, T_2^i, \mathbf{p}_2^i, \ldots, (\mathbf{p}_{n-3}^i)^2, T_{n-2}^i, \mathbf{p}_{n-2}^i = (\mathbf{y}_i)^3, \mathbf{y}_i, Q_i^{-1}, \mathbf{v}\rangle$ for every $1 \le i \le n-3$.

Let $b_1 = 1$, $b_2 = 3$, $b_i = i+1$ for every $3 \le i \le n-1$, and $b_n = 2$. We set $\mathbf{w}_0 = \mathbf{u}$, $\mathbf{w}_i = b_i b_{i+1}$ for every $1 \le i \le n-1$, and $\mathbf{w}_n = (\mathbf{v})^2$. Note that $\{(\mathbf{w}_1)^2 = 13, \mathbf{w}_2 = 43\} \subset V(S_{n,2}^{\{3\}}) - \{\mathbf{v}\}$ and $\{(\mathbf{w}_{n-1})^2 = n2, \mathbf{w}_n = 32\} \subset V(S_{n,2}^{\{2\}}) - \{\mathbf{u}\}$. By Theorem 16.6, there is a Hamiltonian path W_i of $S_{n,2}^{\{b_i\}}$ joining $(\mathbf{w}_{i-1})^2$ to $\mathbf{w}_i$ for every $i \in \langle n-1\rangle - \{2\}$. By Lemma 16.18, there is a Hamiltonian path W_2 of $S_{n,2}^{\{3\}} - \{\mathbf{v}\}$ joining $(\mathbf{w}_1)^2$ to $\mathbf{w}_2$. Again, there is a Hamiltonian path W_n of $S_{n,2}^{\{2\}} - \{\mathbf{u}\}$ joining $(\mathbf{w}_{n-1})^2$ to $\mathbf{w}_n$. We set $P_{n-2} = \langle \mathbf{u} = \mathbf{w}_0, (\mathbf{w}_0)^2, W_1, \mathbf{w}_1, (\mathbf{w}_1)^2, W_2, \mathbf{w}_2, \ldots, (\mathbf{w}_{n-1})^2, W_n, \mathbf{w}_n = (\mathbf{v})^2, \mathbf{v}\rangle$.

Then $\{P_1, P_2, \ldots, P_{n-2}\}$ forms a set of $(n-2)$ mutually independent Hamiltonian paths of $S_{n,2}$ from $\mathbf{u}$ to $\mathbf{v}$. See Figure 19.6 for illustration.

Case 3. $(\mathbf{u})_2 = (\mathbf{v})_2$. Without loss of generality, we assume that $\mathbf{u} = 12$ and $\mathbf{v} = 32$. Suppose that $n = 5$. We show there are three mutually independent Hamiltonian paths on $S_{5,2}$ joining $\mathbf{u}$ to $\mathbf{v}$ by exhibiting the three required Hamiltonian paths in Table 19.4.

Thus, we assume that $n \ge 6$. For every $1 \le i \le n-3$ and $1 \le j \le n-3$, we set $\mathbf{z}_j^i$ as

$$\mathbf{z}_j^i = \begin{cases} (i+j+4)2 & \text{if } i+j \le n-4 \\ (i+j-n+7)2 & \text{if } n-3 \le i+j \le 2n-7 \\ 42 & \text{if } i+j = 2n-6 \end{cases}$$

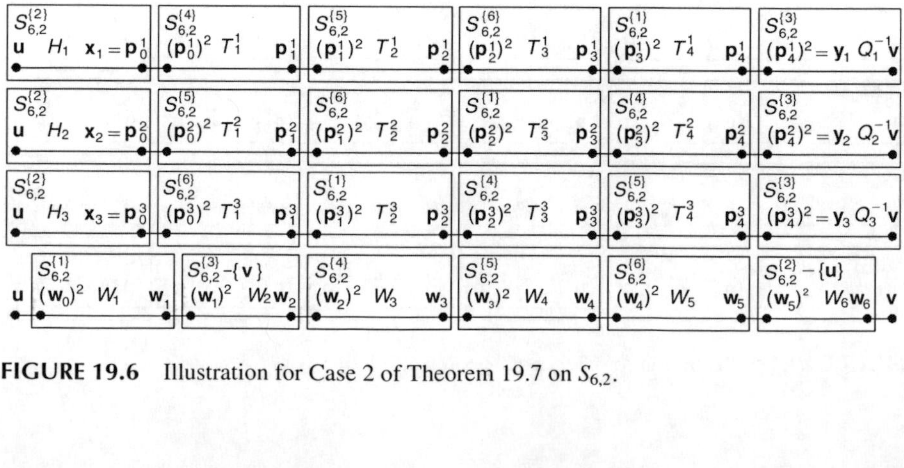

FIGURE 19.6 Illustration for Case 2 of Theorem 19.7 on $S_{6,2}$.

TABLE 19.4
The Three Mutually Independent Hamiltonian Paths of $S_{5,2}$ between 12 and 32

$P_1 = \langle 12, 42, 24, 34, 54, 14, 41, 51, 21, 31, 13, 43, 23, 53, 35, 45, 15, 25, 52, 32\rangle$
$P_2 = \langle 12, 52, 25, 15, 45, 35, 53, 43, 23, 13, 31, 21, 51, 41, 14, 34, 54, 24, 42, 32\rangle$
$P_3 = \langle 12, 21, 31, 41, 51, 15, 45, 35, 25, 52, 42, 24, 14, 54, 34, 43, 53, 13, 23, 32\rangle$

We set $H_i = \langle \mathbf{u}, \mathbf{z}_1^i, \mathbf{z}_2^i, \ldots, \mathbf{z}_{n-4}^i\rangle$ for every $1 \le i \le n-3$. Obviously, H_i is a Hamiltonian path of $S_{n,2}^{\{2\}} - \{\mathbf{v}, \mathbf{z}_{n-3}^i\}$ joining $\mathbf{u}$ to $\mathbf{z}_{n-4}^i$ for every $1 \le i \le n-4$, and $|\{H_1(j), H_2(j), \ldots, H_{n-3}(j)\}| = n-3$ for every $2 \le j \le n-3$. Let A be an $(n-3) \times (n-1)$ matrix defined by

$$a_{i,j} = \begin{cases} i-j+4 & \text{if } i \ne n-3 \text{ and } j \le i \\ 1 & \text{if } i \ne n-3 \text{ and } j-1 = i \\ 3 & \text{if } i \ne n-3 \text{ and } j-2 = i \\ n+i-j+3 & \text{if } i \ne n-3 \text{ and } j-3 \ge i \\ n & \text{if } i = n-3 \text{ and } j = 1 \\ 3 & \text{if } i = n-3 \text{ and } j = 2 \\ n-j+2 & \text{if } i = n-3 \text{ and } 3 \le j \le n-3 \\ 1 & \text{if } i = n-3 \text{ and } j = n-2 \\ 4 & \text{if } i = n-3 \text{ and } j = n-1 \end{cases}$$

More precisely,

$$A = \begin{bmatrix} 4 & 1 & 3 & n & n-1 & \cdots & 8 & 7 & 6 & 5 \\ 5 & 4 & 1 & 3 & n & \cdots & 9 & 8 & 7 & 6 \\ \vdots & \vdots & \vdots & \vdots & \vdots & \ddots & \vdots & \vdots & \vdots & \vdots \\ n-1 & n-2 & n-3 & n-4 & n-5 & \cdots & 4 & 1 & 3 & n \\ n & 3 & n-1 & n-2 & n-3 & \cdots & 6 & 5 & 1 & 4 \end{bmatrix}$$

Let $\mathbf{q}_1 = 24$, $\mathbf{q}_2 = 54$, $\mathbf{q}_3 = 34$, $\mathbf{q}_{n-1} = 14$, and $\mathbf{q}_i = (i+2)4$ for every $4 \le i \le n-2$. We set $Q_1 = \langle \mathbf{q}_1, \mathbf{q}_2, \ldots, \mathbf{q}_{n-1}\rangle$, $Q_2 = \langle \mathbf{q}_2, \mathbf{q}_1, \mathbf{q}_3, \mathbf{q}_4, \ldots, \mathbf{q}_{n-1}\rangle$, and $Q_3 = \langle \mathbf{q}_{n-1}, \mathbf{q}_2, \mathbf{q}_3, \ldots, \mathbf{q}_{n-2}, \mathbf{q}_1\rangle$. Obviously, $(\mathbf{z}_{n-4}^1)^2 = \mathbf{q}_1$, and Q_i is a Hamiltonian path of $S_{n,2}^{\{4\}}$ for every $1 \le i \le 3$.

We construct P_1 as following. Let $\mathbf{p}_1^1 = \mathbf{q}_{n-1}$, $\mathbf{p}_{n-1}^1 = (\mathbf{z}_{n-3}^1)^2$, and $\mathbf{p}_j^1 = a_{1,j+1}a_{1,j}$ for every $2 \le j \le n-2$. By Theorem 16.6, there is a Hamiltonian path T_j^1 of $S_{n,2}^{\{a_{1,j}\}}$ joining $(\mathbf{p}_{j-1}^1)^2$ to $\mathbf{p}_j^1$ for every $2 \le j \le n-1$. We set $P_1 = \langle \mathbf{u}, H_1, \mathbf{z}_{n-4}^1, \mathbf{q}_1, Q_1, \mathbf{q}_{n-1} = \mathbf{p}_1^1, (\mathbf{p}_1^1)^2, T_2^1, \mathbf{p}_2^1, \ldots, (\mathbf{p}_{n-2}^1)^2, T_{n-1}^1, \mathbf{p}_{n-1}^1 = (\mathbf{z}_{n-3}^1)^2, \mathbf{z}_{n-3}^1, \mathbf{v}\rangle$.

We construct P_2 as following. Let $\mathbf{p}_2^2 = \mathbf{q}_{n-1}$, $\mathbf{p}_{n-1}^2 = (\mathbf{z}_{n-3}^2)^2$, and $\mathbf{p}_j^2 = a_{2,j+1}a_{2,j}$ for every $3 \le j \le n-2$. By Theorem 16.6, there is a Hamiltonian path T_j^2 of $S_{n,2}^{\{a_{2,j}\}}$ joining $(\mathbf{p}_{j-1}^2)^2$ to $\mathbf{p}_j^2$ for every $3 \le j \le n-1$. Again, there is a Hamiltonian path H

of $S_{n,2}^{\{5\}}$ joining $(\mathbf{z}_{n-4}^2)^2$ to $(\mathbf{q}_2)^2$. We set $P_2 = \langle \mathbf{u}, H_2, \mathbf{z}_{n-4}^2, (\mathbf{z}_{n-4}^2)^2, H, (\mathbf{q}_2)^2, \mathbf{q}_2, Q_2, \mathbf{q}_{n-1} = \mathbf{p}_2^2, (\mathbf{p}_2^2)^2, T_3^2, \mathbf{p}_3^2, \ldots, (\mathbf{p}_{n-2}^2)^2, T_{n-1}^2, \mathbf{p}_{n-1}^2 = (\mathbf{z}_{n-3}^2)^2, \mathbf{z}_{n-3}^2, \mathbf{v}\rangle$.

We construct P_i for every $3 \le i \le n-3$ as following. Let $\mathbf{p}_0^i = \mathbf{z}_{n-4}^i$ and $\mathbf{p}_{n-1}^i = (\mathbf{z}_{n-3}^i)^2$ for every $3 \le i \le n-3$. We set $\mathbf{p}_j^i = a_{i,j+1}a_{i,j}$ for every $3 \le i \le n-3$ and $1 \le j \le n-2$. By Theorem 16.6, there is a Hamiltonian path T_j^i of $S_{n,2}^{\{a_{i,j}\}}$ joining $(\mathbf{p}_{j-1}^i)^2$ to $\mathbf{p}_j^i$ for every $3 \le i \le n-3$ and $1 \le j \le n-1$. We set $P_i = \langle \mathbf{u}, H_i, \mathbf{z}_{n-4}^i = \mathbf{p}_0^i, (\mathbf{p}_0^i)^2, T_1^i, \mathbf{p}_1^i, (\mathbf{p}_1^i)^2, T_2^i, \mathbf{p}_2^i, \ldots, (\mathbf{p}_{n-2}^i)^2, T_{n-1}^i, \mathbf{p}_{n-1}^i = (\mathbf{z}_{n-3}^i)^2, \mathbf{z}_{n-3}^i, \mathbf{v}\rangle$ for every $3 \le i \le n-3$.

We construct P_{n-2} as following. Let $\mathbf{x}_1 = n2$, $\mathbf{x}_{n-3} = 35$, and $\mathbf{x}_i = (n-i+1)(n-i+2)$ for every $2 \le i \le n-4$. By Lemma 16.18, there is a Hamiltonian path W_1 of $S_{n,2}^{\{2\}} - \{\mathbf{u}, \mathbf{v}\}$ joining $(\mathbf{q}_1)^2$ to $\mathbf{x}_1$. By Theorem 16.6, there is a Hamiltonian path W_0 of $S_{n,2}^{\{1\}}$ joining $(\mathbf{u})^2$ to $(\mathbf{q}_{n-1})^2$. Again, there is a Hamiltonian path W_i of $S_{n,2}^{\{n-i+2\}}$ joining $(\mathbf{x}_{i-1})^2$ to $\mathbf{x}_i$ for every $2 \le i \le n-3$. Moreover, there is a Hamiltonian path W_{n-2} of $S_{n,2}^{\{3\}}$ joining $(\mathbf{x}_{n-3})^2$ to $(\mathbf{v})^2$. We set $P_{n-2} = \langle \mathbf{u}, (\mathbf{u})^2, W_0, (\mathbf{q}_{n-1})^2, \mathbf{q}_{n-1}, Q_3, \mathbf{q}_1, (\mathbf{q}_1)^2, W_1, \mathbf{x}_1, (\mathbf{x}_1)^2, W_2, \mathbf{x}_2, \ldots, (\mathbf{x}_{n-4})^2, W_{n-3}, \mathbf{x}_{n-3}, (\mathbf{x}_{n-3})^2, W_{n-2}, (\mathbf{v})^2, \mathbf{v}\rangle$.

Then $\{P_1, P_2, \ldots, P_{n-2}\}$ forms a set of $(n-2)$ mutually independent Hamiltonian paths of $S_{n,2}$ from $\mathbf{u}$ to $\mathbf{v}$. See Figure 19.7 for illustration.

Case 4. $(\mathbf{u})_1 = (\mathbf{v})_1$. Without loss of generality, we assume that $\mathbf{u} = 12$ and $\mathbf{v} = 13$. By Lemma 19.6, there are $(n-2)$ Hamiltonian paths $H_1, H_2, \ldots, H_{n-2}$ of $S_{n,2}^{\{2\}}$ such that (1) H_i joins $\mathbf{u}$ to some vertex $\mathbf{x}_i$ with $(\mathbf{x}_i)_1 = i+4$ for every $1 \le i \le n-4$, $(\mathbf{x}_{n-3})_1 = 3$, and $(\mathbf{x}_{n-2})_1 = 4$, and (2) $|\{H_1(j), H_2(j), \ldots, H_{n-2}(j)\}| = n-2$ for every $2 \le j \le n-1$. Let A be an $(n-2) \times (n-1)$ matrix defined by

$$a_{i,j} = \begin{cases} i+j+3 & \text{if } i+j \le n-3 \\ i+j-n+5 & \text{if } n-2 \le i+j \le n-1 \\ 1 & \text{if } i+j = n \\ i+j-n+4 & \text{if } n+1 \le i+j \le 2n-4 \\ 3 & \text{if } i+j = 2n-3 \end{cases}$$

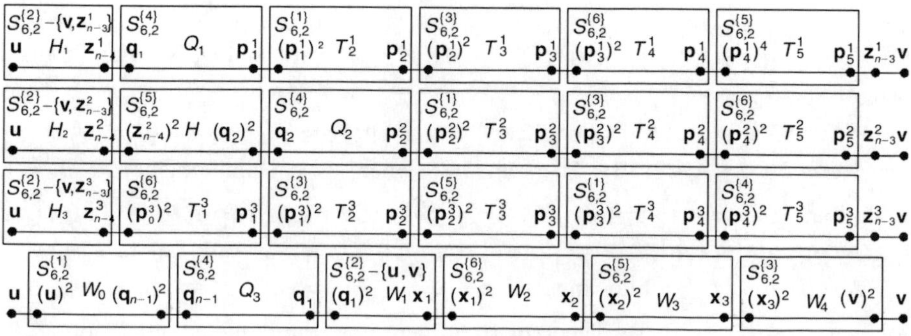

FIGURE 19.7 Illustration for Case 3 of Theorem 19.7 on $S_{6,2}$.

More precisely,

$$A=\begin{bmatrix} 5 & 6 & 7 & 8 & \cdots & n & 3 & 4 & 1 \\ 6 & 7 & 8 & 9 & \cdots & 3 & 4 & 1 & 5 \\ \vdots & \vdots & \vdots & \vdots & \ddots & \vdots & \vdots & \vdots & \vdots \\ n & 3 & 4 & 1 & \cdots & n-4 & n-3 & n-2 & n-1 \\ 3 & 4 & 1 & 5 & \cdots & n-3 & n-2 & n-1 & n \\ 4 & 1 & 5 & 6 & \cdots & n-2 & n-1 & n & 3 \end{bmatrix}$$

We set $\mathbf{p}_0^1=\mathbf{x}_1$, $\mathbf{p}_j^1=a_{1,j+1}a_{1,j}$ for every $1\le j\le n-2$, and $\mathbf{p}_{n-1}^1=(\mathbf{v})^2$. Obviously, $\{(\mathbf{p}_{n-4}^1)^2=n3, \mathbf{p}_{n-3}^1=43\}\subset V(S_{n,2}^{\{3\}})-\{\mathbf{v}\}$. By Lemma 16.18, there is a Hamiltonian path T_j^1 of $S_{n,2}^{\{a_{1,j}\}}$ joining $(\mathbf{p}_{j-1}^1)^2$ to $\mathbf{p}_j^1$ for every $j\in\langle n-1\rangle-\{n-3\}$ and there is a Hamiltonian path T_{n-3}^1 of $S_{n,2}^{\{3\}}-\{\mathbf{v}\}$ joining $(\mathbf{p}_{n-4}^1)^2$ to $\mathbf{p}_{n-3}^1$. We set $P_1=\langle \mathbf{u},H_1,\mathbf{x}_1=\mathbf{p}_0^1,(\mathbf{p}_0^1)^2,T_1^1,\mathbf{p}_1^1,(\mathbf{p}_1^1)^2,T_2^1,\mathbf{p}_2^1,\ldots,(\mathbf{p}_{n-2}^1)^2,T_{n-1}^1,\mathbf{p}_{n-1}^1=(\mathbf{v})^2,\mathbf{v}\rangle$.

Let $\mathbf{y}_i=a_{i+1,n-1}3$ for every $1\le i\le n-4$. For every $2\le i\le n-3$, we set $\mathbf{p}_0^i=\mathbf{x}_i$, $\mathbf{p}_j^i=a_{i,j+1}a_{i,j}$ for every $1\le j\le n-2$, and $\mathbf{p}_{n-1}^i=(\mathbf{y}_{i-1})^2$. Note that $\{(\mathbf{p}_{n-i-4}^{i+1})^2=n3,\ \mathbf{p}_{n-i-3}^{i+1}=43\}\subset V(S_{n,2}^{\{3\}})-\{\mathbf{v},\mathbf{y}_i\}$ for every $1\le i\le n-5$, and $\{(\mathbf{p}_0^{n-3})^2=23,\mathbf{p}_1^{n-3}=43\}\subset V(S_{n,2}^{\{3\}})-\{\mathbf{v},\mathbf{y}_{n-4}\}$. For every $2\le i\le n-3$, by Lemma 16.18, there is a Hamiltonian path T_j^i of $S_{n,2}^{\{a_{i,j}\}}$ joining the vertex $(\mathbf{p}_{j-1}^i)^2$ to $\mathbf{p}_j^i$ for every $j\in\langle n-1\rangle-\{n-i-2\}$ and there is a Hamiltonian path T_{n-i-2}^i of $S_{n,2}^{\{3\}}-\{\mathbf{v},\mathbf{y}_{i-1}\}$ joining $(\mathbf{p}_{n-i-3}^i)^2$ to $\mathbf{p}_{n-i-2}^i$. We set $P_i=\langle \mathbf{u},H_i,\mathbf{x}_i=\mathbf{p}_0^i,(\mathbf{p}_0^i)^2,T_1^i,\mathbf{p}_1^i,(\mathbf{p}_1^i)^2,T_2^i,\mathbf{p}_2^i,\ldots,(\mathbf{p}_{n-2}^i)^2,T_{n-1}^i,\mathbf{p}_{n-1}^i=(\mathbf{y}_{i-1})^2,\ \mathbf{y}_{i-1},\ \mathbf{v}\rangle$ for every $2\le i\le n-3$.

We set $\mathbf{p}_0^{n-2}=\mathbf{x}_{n-2}$, $\mathbf{p}_j^{n-2}=a_{n-2,j+1}a_{n-2,j}$ for every $1\le j\le n-3$, and $\mathbf{p}_{n-2}^{n-2}=4n$. By Theorem 16.6, there is a Hamiltonian path T_j^{n-2} of $S_{n,2}^{\{a_{n-2,j}\}}$ joining $(\mathbf{p}_{j-1}^{n-2})^2$ to $\mathbf{p}_j^{n-2}$ for every $j\in\langle n-2\rangle-\{1\}$. Let $\mathbf{w}=34$ and $\mathbf{z}=23$. By Lemma 16.18, there is a Hamiltonian path T_1^{n-2} of $S_{n,2}^{\{4\}}-\{\mathbf{w},(\mathbf{p}_{n-2}^{n-2})^2\}$ joining $(\mathbf{p}_0^{n-2})^2$ to $\mathbf{p}_1^{n-2}$. Again, there is a Hamiltonian path Q of $S_{n,2}^{\{3\}}-\{\mathbf{v}\}$ joining $(\mathbf{w})^2$ to $\mathbf{z}$. We set $P_{n-2}=\langle \mathbf{u},H_{n-2},\mathbf{x}_{n-2}=\mathbf{p}_0^{n-2},(\mathbf{p}_0^{n-2})^2,T_1^{n-2},\mathbf{p}_1^{n-2},(\mathbf{p}_1^{n-2})^2,T_2^{n-2},\mathbf{p}_2^{n-2},\ldots,(\mathbf{p}_{n-3}^{n-2})^2,T_{n-2}^{n-2},\mathbf{p}_{n-2}^{n-2},(\mathbf{p}_{n-2}^{n-2})^2,\mathbf{w},(\mathbf{w})^2,Q,\mathbf{z},\mathbf{v}\rangle$.

Then $\{P_1,P_2,\ldots,P_{n-2}\}$ forms a set of $(n-2)$ mutually independent Hamiltonian paths of $S_{n,2}$ from $\mathbf{u}$ to $\mathbf{v}$. See Figure 19.8 for illustration.

Case 5. $(\mathbf{u})_2\neq(\mathbf{v})_2$, $(\mathbf{u})_2\neq(\mathbf{v})_1$, $(\mathbf{u})_1\neq(\mathbf{v})_2$, and $(\mathbf{u})_1\neq(\mathbf{v})_1$. Without loss of generality, we assume that $\mathbf{u}=12$ and $\mathbf{v}=34$. By Lemma 19.6, there are $(n-3)$ Hamiltonian paths H_1, $H_2,\ldots,H_{n-3}$ of $S_{n,2}^{\{2\}}$ such that (1) H_i joins $\mathbf{u}$ to some vertex $\mathbf{x}_i$ with $(\mathbf{x}_i)_1=i+4$ for every $1\le i\le n-4$ and H_{n-3} joins $\mathbf{u}$ to some vertex $\mathbf{x}_{n-3}$ with $(\mathbf{x}_{n-3})_1=3$, and (2) $|\{H_1(j),H_2(j),\ldots,H_{n-3}(j)\}|=n-3$ for every $2\le j\le n-1$. Again, there are $(n-3)$ Hamiltonian paths Q_1, Q_2, $\ldots$, Q_{n-3} of $S_{n,2}^{\{4\}}$ such that (1) Q_1 joins $\mathbf{v}$ to some vertex $\mathbf{y}_1$ with $(\mathbf{y}_1)_1=1$ and Q_i joins $\mathbf{v}$ to some vertex $\mathbf{y}_i$ with

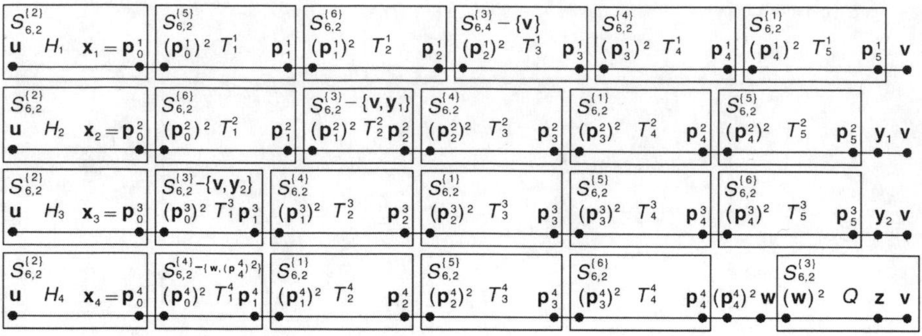

FIGURE 19.8 Illustration for Case 4 of Theorem 19.7 on $S_{6,2}$.

$(\mathbf{y}_i)_1 = i+3$ for every $2 \le i \le n-3$, and (2) $|\{Q_1(j), Q_2(j), \ldots, Q_{n-3}(j)\}| = n-3$ for every $2 \le j \le n-1$. Let A be an $(n-3) \times (n-2)$ matrix defined as

$$a_{i,j} = \begin{cases} i+j+3 & \text{if } i+j \le n-3 \\ 3 & \text{if } i+j = n-2 \\ 1 & \text{if } i+j = n-1 \\ i+j-n+5 & \text{if } n \le i+j \end{cases}$$

More precisely,

$$A = \begin{bmatrix} 5 & 6 & 7 & 8 & \cdots & n-1 & n & 3 & 1 \\ 6 & 7 & 8 & 9 & \cdots & n & 3 & 1 & 5 \\ \vdots & \vdots & \vdots & \vdots & \ddots & \vdots & \vdots & \vdots & \vdots \\ n & 3 & 1 & 5 & \cdots & n-4 & n-3 & n-2 & n-1 \\ 3 & 1 & 5 & 6 & \cdots & n-3 & n-2 & n-1 & n \end{bmatrix}$$

We set $\mathbf{p}_0^i = \mathbf{x}_i$ and $\mathbf{p}_{n-2}^i = (\mathbf{y}_i)^2$ for every $1 \le i \le n-3$. For every $1 \le i \le n-3$ and every $1 \le j \le n-3$, we set $\mathbf{p}_j^i = a_{i,j+1}a_{i,j}$. By Theorem 16.6, there is a Hamiltonian path T_j^i of $S_{n,2}^{\{a_{i,j}\}}$ joining the vertex $(\mathbf{p}_{j-1}^i)^2$ to $\mathbf{p}_j^i$ for every $1 \le i \le n-3$ and for every $1 \le j \le n-2$. We set $P_i = \langle \mathbf{u}, H_i, \mathbf{x}_i = \mathbf{p}_0^i, (\mathbf{p}_0^i)^2, T_1^i, \mathbf{p}_1^i, (\mathbf{p}_1^i)^2, T_2^i, \mathbf{p}_2^i, \ldots, (\mathbf{p}_{n-3}^i)^2, T_{n-2}^i, \mathbf{p}_{n-2}^i, (\mathbf{p}_{n-2}^i)^2 = \mathbf{y}_i, Q_i^{-1}, \mathbf{v} \rangle$ for every $1 \le i \le n-3$.

Let $b_1 = 1$, $b_2 = 4$, $b_i = i+2$ for every $3 \le i \le n-2$, $b_{n-1} = 2$, and $b_n = 3$. We set $\mathbf{w}_0 = \mathbf{u}$, $\mathbf{w}_i = b_{i+1}b_i$ for every $1 \le i \le n-1$ and $\mathbf{w}_n = (\mathbf{v})^2$. Note that $\{(\mathbf{w}_1)^2 = 14, \mathbf{w}_2 = 54\} \subset V(S_{n,2}^{\{4\}}) - \{\mathbf{v}\}$ and $\{(\mathbf{w}_{n-2})^2 = n2, \mathbf{w}_{n-1} = 32\} \subset V(S_{n,2}^{\{2\}}) - \{\mathbf{u}\}$. By Lemma 16.18, there is a Hamiltonian path W_i of $S_{n,2}^{\{b_i\}}$ joining the vertex $(\mathbf{w}_{i-1})^2$ to $\mathbf{w}_i$ for every $i \in \langle n \rangle - \{2, n-1\}$ and there is a Hamiltonian path W_2 of $S_{n,2}^{\{4\}} - \{\mathbf{v}\}$ joining the vertex $(\mathbf{w}_1)^2$ to $\mathbf{w}_2$. Again, there is a Hamiltonian path W_{n-1} of $S_{n,2}^{\{2\}} - \{\mathbf{u}\}$ joining the vertex $(\mathbf{w}_{n-2})^2$ to $\mathbf{w}_{n-1}$. We set $P_{n-2} = \langle \mathbf{u} = \mathbf{w}_0, (\mathbf{w}_0)^2, W_1, \mathbf{w}_1, (\mathbf{w}_1)^2, W_2, \mathbf{w}_2, \ldots, (\mathbf{w}_{n-1})^2, W_n, \mathbf{w}_n = (\mathbf{v})^2, \mathbf{v} \rangle$.

Then $\{P_1, P_2, \ldots, P_{n-2}\}$ forms a set of $(n-2)$ mutually independent Hamiltonian paths of $S_{n,2}$ from **u** to **v**. See Figure 19.9 as illustration.

Thus, this statement is proved. □

THEOREM 19.8 Let $\{i_1, i_2, \ldots, i_{n-1}\}$ be any subset of $\langle n\rangle$, and let **u** be any vertex of $S_{n,k}$ with $2 \le k \le n-2$. There are $(n-1)$ Hamiltonian paths $P_1, P_2, \ldots, P_{n-1}$ of $S_{n,k}$ such that (1) P_j joins **u** to some vertex $\mathbf{z}_j$ with $(\mathbf{z}_j)_1 = i_j$ for every $1 \le j \le n-1$ and (2) $|\{P_1(j), P_2(j), \ldots, P_{n-1}(j)\}| = n-1$ for every $2 \le j \le n!/(n-k)!$.

Proof: We prove this statement by induction on k. Suppose that $n=4$ and $k=2$. Since $S_{4,2}$ is vertex-transitive, we assume that $\mathbf{u} = 14$. We prove this statement holds on $S_{4,2}$ by exhibiting the three required Hamiltonian paths in Table 19.5.

Suppose that $n \ge 5$ and $k=2$. Since $S_{n,2}$ is vertex-transitive, we assume that $\mathbf{u} = 1n$. Without loss of generality, we suppose that $i_1 < i_2 < \cdots < i_{n-1}$. Obviously, $i_j = j$ or $j+1$ for $1 \le j \le n-1$. Hence, $i_j \notin \{j+2, j+3\}$ for every $1 \le j \le n-4$,

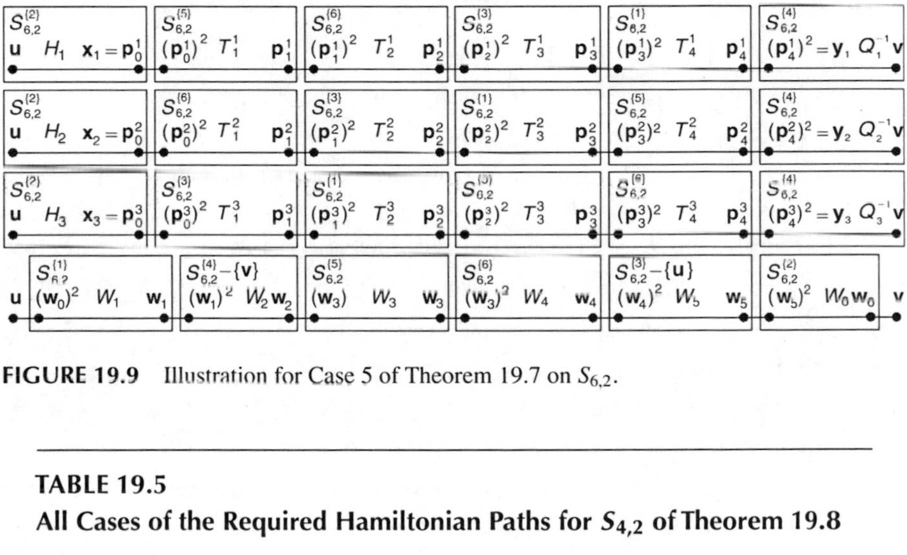

FIGURE 19.9 Illustration for Case 5 of Theorem 19.7 on $S_{6,2}$.

TABLE 19.5
All Cases of the Required Hamiltonian Paths for $S_{4,2}$ of Theorem 19.8

$\{i_1, i_2, i_3\} = \{1, 2, 3\}$	$P_1 = \langle 14, 34, 24, 42, 32, 23, 43, 13, 31, 41, 21, 12\rangle$
	$P_2 = \langle 14, 41, 21, 31, 13, 43, 34, 24, 42, 12, 32, 23\rangle$
	$P_3 = \langle 14, 24, 42, 32, 12, 21, 41, 31, 13, 23, 43, 34\rangle$
$\{i_1, i_2, i_3\} = \{1, 2, 4\}$	$P_1 = \langle 14, 34, 24, 42, 32, 23, 43, 13, 31, 41, 21, 12\rangle$
	$P_2 = \langle 14, 41, 31, 21, 12, 32, 42, 24, 34, 43, 13, 23\rangle$
	$P_3 = \langle 14, 24, 34, 43, 23, 13, 31, 41, 21, 12, 32, 42\rangle$
$\{i_1, i_2, i_3\} = \{1, 3, 4\}$	$P_1 = \langle 14, 24, 34, 43, 23, 32, 42, 12, 21, 41, 31, 13\rangle$
	$P_2 = \langle 14, 41, 21, 31, 13, 23, 43, 34, 24, 42, 12, 32\rangle$
	$P_3 = \langle 14, 34, 24, 42, 32, 12, 21, 41, 31, 13, 23, 43\rangle$
$\{i_1, i_2, i_3\} = \{2, 3, 4\}$	$P_1 = \langle 14, 24, 34, 43, 13, 31, 41, 21, 12, 42, 32, 23\rangle$
	$P_2 = \langle 14, 41, 21, 12, 32, 42, 24, 34, 43, 23, 13, 31\rangle$
	$P_3 = \langle 14, 34, 24, 42, 12, 32, 23, 43, 13, 31, 21, 41\rangle$

$i_{n-3} \notin \{1, n-1\}$, $i_{n-2} \notin \{1, 2\}$, and $i_{n-1} \notin \{2, 3\}$. By Lemma 19.6, there are $(n-2)$ Hamiltonian paths $H_1, H_2, \ldots, H_{n-2}$ of $S_{n,2}^{\{n\}}$ such that (1) H_j joins $\mathbf{u}$ to some vertex $\mathbf{x}_j$ with $(\mathbf{x}_j)_1 = j+1$ for every $1 \leq j \leq n-2$ and (2) $|\{H_1(j), H_2(j), \ldots, H_{n-2}(j)\}| = n-2$ for every $2 \leq j \leq n-1$. Since $k=2$, by Lemma 16.17, $|E_{n,2}^{j,l}| = (n-2)!/(n-2)! = 1$. We choose a vertex $\mathbf{z}_j \in S_{n,2}^{\{j+2\}}$ with $(\mathbf{z}_j)_1 = i_j$ for $1 \leq j \leq n-3$, a vertex $\mathbf{z}_{n-2} \in S_{n,2}^{\{1\}}$ with $(\mathbf{z}_{n-2})_1 = i_{n-2}$. Let A be an $(n-2) \times (n-1)$ matrix defined by

$$a_{j,l} = \begin{cases} j - l + 2 & \text{if } j \leq i+1 \\ j - l + n + 1 & \text{if } i + 2 \leq j \end{cases}$$

More precisely,

$$A = \begin{bmatrix} 2 & 1 & n-1 & n-2 & \cdots & 4 & 3 \\ 3 & 2 & 1 & n-1 & \cdots & 5 & 4 \\ \vdots & \vdots & \vdots & \vdots & \ddots & \vdots & \vdots \\ n-2 & n-3 & n-4 & n-5 & \cdots & 1 & n-1 \\ n-1 & n-2 & n-3 & n-4 & \cdots & 2 & 1 \end{bmatrix}$$

For every $j \in \langle n-2 \rangle - \{2\}$, we construct P_j as following. Let $\mathbf{p}_0^j = \mathbf{x}_j$ for every $j \in \langle n-2 \rangle - \{2\}$. Since $i_j \notin \{j+2, j+3\}$ for every $j \in \langle n-2 \rangle - \{2\}$, $i_{n-3} \notin \{1, n-1\}$, and $i_{n-2} \notin \{1, 2\}$, we can choose a vertex $\mathbf{p}_{n-1}^j = i_j a_{j,n-1} = \mathbf{z}_j$ in $S_{n,2}^{\{a_{j,n-2}\}}$. We set $\mathbf{p}_l^j = a_{j,l+1} a_{j,l}$ for every $1 \leq j \leq n-2$ and $1 \leq l \leq n-2$. Note that $(\mathbf{p}_l^j)^2 \in S_{n,2}^{\{a_{j,l+1}\}} - \{\mathbf{p}_{l+1}^j\}$ for every $j \in \langle n-2 \rangle - \{2\}$ and $0 \leq l \leq n-2$. By Lemma 16.18, there is a Hamiltonian path T_l^j in $S_{n,2}^{\{a_{j,l}\}}$ joining $(\mathbf{p}_{l-1}^j)^2$ to $\mathbf{p}_l^j$ for every $j \in \langle n-2 \rangle - \{2\}$ and $1 \leq l \leq n-1$. We set $P_j = \langle \mathbf{u}, H_j, \mathbf{x}_j = \mathbf{p}_0^j, (\mathbf{p}_0^j)^2, T_1^j, \mathbf{p}_1^j, (\mathbf{p}_1^j)^2, T_2^j, \mathbf{p}_2^j, \ldots, (\mathbf{p}_{n-2}^j)^2, T_{n-1}^j, \mathbf{p}_{n-1}^j = \mathbf{z}_j \rangle$ for every $j \in \langle n-2 \rangle - \{2\}$.

We construct P_2 as follows. Let $\mathbf{p}_0^2 = \mathbf{x}_2$. Since $i_2 \notin \{4, 5\}$, we can choose a vertex $\mathbf{p}_{n-1}^2 = i_2 a_{2,n-1} = \mathbf{z}_2$ in $S_{n,2}^{\{a_{2,n-2}\}}$. We set $\mathbf{p}_l^2 = a_{2,l+1} a_{2,l}$ for every $1 \leq l \leq n-2$. Note that $(\mathbf{p}_l^2)^2 \in S_{n,2}^{\{a_{2,l+1}\}} - \{\mathbf{p}_{l+1}^2\}$ for every $0 \leq l \leq n-2$. By Lemma 16.18, there is a Hamiltonian path T in $S_{n,2}^{\{3\}} - \{(\mathbf{p}_0^2)^2\}$ joining $\mathbf{y} = 43$ to $\mathbf{p}_1^2$. Again, there is a Hamiltonian path T_l^2 in $S_{n,2}^{\{a_{2,l}\}}$ joining $(\mathbf{p}_{l-1}^2)^2$ to $\mathbf{p}_l^2$ for every $2 \leq l \leq n-1$. We set $P_2 = \langle \mathbf{u}, H_2, \mathbf{x}_2 = \mathbf{p}_0^2, (\mathbf{p}_0^2)^2, \mathbf{y}, T, \mathbf{p}_1^2, (\mathbf{p}_1^2)^2, T_2^2, \mathbf{p}_2^2, (\mathbf{p}_2^2)^2, T_3^2, \mathbf{p}_3^2, \ldots, (\mathbf{p}_{n-2}^2)^2, T_{n-1}^2, \mathbf{p}_{n-1}^2 = \mathbf{z}_2 \rangle$.

We construct P_{n-1} as follows. Let $\mathbf{x} = 13$, $\mathbf{q}_0 = n3$, and $\mathbf{q}_i = (n-i)(n-i+1)$ for every $1 \leq i \leq n-2$. Since $i_{n-1} \notin \{2, 3\}$, we choose a vertex $\mathbf{z}$ in $S_{n,2}^{\{2\}}$ with $(\mathbf{z})_1 = i_{n-1}$ ($\mathbf{z} = i_{n-1}2$). By Theorem 16.6, there is a Hamiltonian path R of $S_{n,2}^{\{1\}}$ joining $(\mathbf{u})^2 = n1$ to $(\mathbf{x})^2 = 31$. Again, there is a Hamiltonian path Q_{n-1} of $S_{n,2}^{\{2\}}$ joining $(\mathbf{q}_{n-2})^2$ to $\mathbf{z}$. By Lemma 16.18, there is a Hamiltonian path Q_1 of $S_{n,2}^{\{n\}} - \{\mathbf{u}\}$ joining $(\mathbf{q}_0)^2$ to $\mathbf{q}_1$. And, there is a Hamiltonian path Q_i of $S_{n,2}^{\{n-i+1\}} - \{\mathbf{q}_i\}$ joining

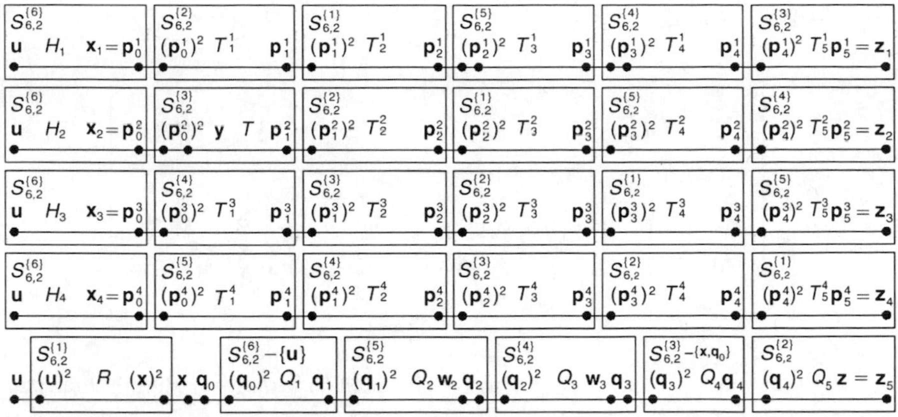

FIGURE 19.10 Illustration for Theorem 19.8 on $S_{6,2}$.

$(\mathbf{q}_1)^2$ to $\mathbf{w}_i = (n-i-1)(n-i+1)$ for $2 \le i \le n-3$. Moreover, there is a Hamiltonian path Q_{n-2} of $S_{n,2}^{\{3\}} - \{\mathbf{x}, \mathbf{q}_0\}$ joining $(\mathbf{q}_{n-3})^2$ to $\mathbf{q}_{n-2}$. We set $\mathbf{z}_{n-1} = \mathbf{z}$ and set $P_{n-1} = \langle \mathbf{u}, (\mathbf{u})^2, R, (\mathbf{x})^2, \mathbf{x}, \mathbf{q}_0, (\mathbf{q}_0)^2, Q_1, \mathbf{q}_1, (\mathbf{q}_1)^2, Q_2, \mathbf{w}_2, \mathbf{q}_2, (\mathbf{q}_2)^2, Q_3, \mathbf{w}_3, \mathbf{q}_3, (\mathbf{q}_3)^2, Q_4, \mathbf{w}_4, \mathbf{q}_4, \ldots, (\mathbf{q}_{n-3})^2, Q_{n-2}, \mathbf{q}_{n-2}, (\mathbf{q}_{n-2})^2, Q_{n-1}, \mathbf{z} = \mathbf{z}_{n-1} \rangle$. Then $P_1, P_2, \ldots, P_{n-1}$ form the desired paths of $S_{n,2}$. Note that $\mathbf{q}_i = (n-i)(n-i+1) = \mathbf{p}_{i+1}^1$; hence $\mathbf{q}_i$ does not equal to $T_{i+1}^1(2)$ for all $2 \le i \le n-3$. See Figure 19.10 as illustration.

Suppose that this statement holds for $S_{m,l}$ for every $4 \le m \le n-1$, $2 \le l \le k-1$, and $l \le m-2$. Let $\mathbf{u}$ be any vertex of $S_{n,k}$. Since $S_{n,k}$ is vertex-transitive, we assume that $\mathbf{u}$ is a vertex in $S_{n,k}$ with $(\mathbf{u})_1 = 1$ and $(\mathbf{u})_i = n-k+i$ for every $2 \le i \le k$. Note that $\mathbf{u} \in S_{n,k}^{\{n\}}$. Without loss of generality, we suppose that $i_1 < i_2 < \cdots < i_{n-3} < i_{n-1} < i_{n-2}$. Obviously, $i_j \ne j+2$ for every $1 \le j \le n-3$, $i_{n-2} \ne 2$, and $i_{n-1} \ne n$. By the induction hypothesis, there are $(n-2)$ Hamiltonian paths $H_1, H_2, \ldots, H_{n-2}$ of $S_{n,k}^{\{n\}}$ such that H_i is joining $\mathbf{u}$ to a vertex $\mathbf{v}_j$ such that (1) $(\mathbf{v}_j)_1 = (H_j((n-1)!/(n-k-1)!))_1 = j+3$ for every $1 \le j \le n-4$, $(\mathbf{v}_{n-3})_1 = (H_{n-3}((n-1)!/(n-k-1)!))_1 = 1$, $(\mathbf{v}_{n-2})_1 = (H_{n-2}((n-1)!/(n-k-1)!))_1 = 3$ and (2) $|\{H_1(j), H_2(j), \ldots, H_{n-2}(j)\}| = n-2$ for $2 \le j \le (n-1)!/(n-k-1)!$.

Since $n \ge 5$ and $k \ge 3$, by Lemma 16.17, $|E_{n,k}^{j,l}| = (n-2)!/(n-k)! \ge 3$. We choose a vertex $\mathbf{z}_j \in S_{n,k}^{\{j+2\}}$ with $(\mathbf{z}_j)_1 = i_j$ for $1 \le j \le n-3$, a vertex $\mathbf{z}_{n-2} \in S_{n,k}^{\{2\}}$ with $(\mathbf{z}_{n-2})_1 = i_{n-2}$, and a vertex $\mathbf{z}_{n-1} \in S_{n,k}^{\{n\}} - \{\mathbf{u}\}$ with $(\mathbf{z}_{n-1})_1 = i_{n-1}$. Let B be an $(n-2) \times (n-1)$ matrix defined as

$$b_{j,l} = \begin{cases} j+l+2 & \text{if } j \le n-3 \text{ and } j+l \le n-3 \\ j+l-n+3 & \text{if } j \le n-3 \text{ and } n-2 \le j+l \\ l+2 & \text{if } j = n-2 \text{ and } l \le n-3 \\ l-n+3 & \text{if } j = n-2 \text{ and } n-2 \le l \le n-1 \end{cases}$$

More precisely,

$$B = \begin{bmatrix} 4 & 5 & \cdots & n-2 & n-1 & 1 & 2 & 3 \\ 5 & 6 & \cdots & n-1 & 1 & 2 & 3 & 4 \\ \vdots & \vdots & \ddots & \vdots & \vdots & \vdots & \vdots & \vdots \\ n-1 & 1 & \cdots & n-6 & n-5 & n-4 & n-3 & n-2 \\ 1 & 2 & \cdots & n-5 & n-4 & n-3 & n-2 & n-1 \\ 3 & 4 & \cdots & n-3 & n-2 & n-1 & 1 & 2 \end{bmatrix}$$

Let j be any index with $1 \leq j \leq n-2$. By Theorem 16.7, there is a Hamiltonian path $W_j = \langle \mathbf{x}_1^j, T_1^j, \mathbf{y}_1^j, \mathbf{x}_2^j, T_2^j, \mathbf{y}_2^j, \ldots, \mathbf{x}_{n-1}^j, T_{n-1}^j, \mathbf{y}_{n-1}^j \rangle$ of $S_{n,k}^{\langle n-1 \rangle}$ joining the vertex $(\mathbf{v}_j)^k$ to $\mathbf{z}_j$ such that $\mathbf{x}_1^j = (\mathbf{v}_j)^k$, $\mathbf{y}_{n-1}^j = \mathbf{z}_j$, and T_l^j is a Hamiltonian path of $S_{n,k}^{\{b_{j,l}\}}$ joining $\mathbf{x}_l^j$ to $\mathbf{y}_l^j$ for every $1 \leq l \leq n-1$. We set $P_i = \langle \mathbf{u}, H_i, \mathbf{v}_i, (\mathbf{v}_i)^k = \mathbf{x}_1^i, W_i, \mathbf{y}_{n-1}^i = \mathbf{z}_i \rangle$ for every $1 \leq i \leq n-2$.

Since $S_{n,k}^{\{n\}}$ is isomorphic to $S_{n-1,k-1}$, by Lemma 19.4, there is a Hamiltonian path R of $S_{n,k}^n - \{\mathbf{u}\}$ joining a vertex $\mathbf{w}$ to a vertex $\mathbf{z}_{n-1}$ with $(\mathbf{w})_1 = (\mathbf{u})_{k-1} = n-1$ and $(\mathbf{z}_{n-1})_1 = i_{n-1}$. By Theorem 16.7, there is a Hamiltonian path $W = \langle \mathbf{x}_1^{n-1}, T_1^{n-1}, \mathbf{y}_1^{n-1}, \mathbf{x}_2^{n-1}, T_2^{n-1}, \mathbf{y}_2^{n-1}, \ldots, \mathbf{x}_{n-1}^{n-1}, T_{n-1}^{n-1}, \mathbf{y}_{n-1}^{n-1} \rangle$ of $S_{n,k}^{\langle n-1 \rangle}$ joining the vertex $(\mathbf{u})^k$ to the vertex $(\mathbf{w})^k$ such that $\mathbf{x}_1^{n-1} = (\mathbf{u})^k$, $\mathbf{y}_{n-1}^{n-1} = (\mathbf{w})^k$, and T_i^{n-1} is a Hamiltonian path of $S_{n,k}^{\{i\}}$ joining $\mathbf{x}_i^{n-1}$ to $\mathbf{y}_i^{n-1}$ for $1 \leq i \leq n-1$. We set $P_{n-1} = \langle \mathbf{u}, (\mathbf{u})^k, W, (\mathbf{w})^k, \mathbf{w}, R, \mathbf{z}_{n-1} \rangle$.

Then $\{P_1, P_2, \ldots, P_{n-1}\}$ forms a set of desired paths of $S_{n,k}$. Note that $(\mathbf{y}_1^{n-1})_1 = 2, \neq n = (\mathbf{x}_1^{n-3})_1$, and $(\mathbf{y}_i^{n-1})_1 = i+1 \neq i-1 = (\mathbf{x}_i^{n-3})_1$ for $2 \leq i \leq n-1$. See Figure 19.11 for an illustration.

Hence, this statement is proved. □

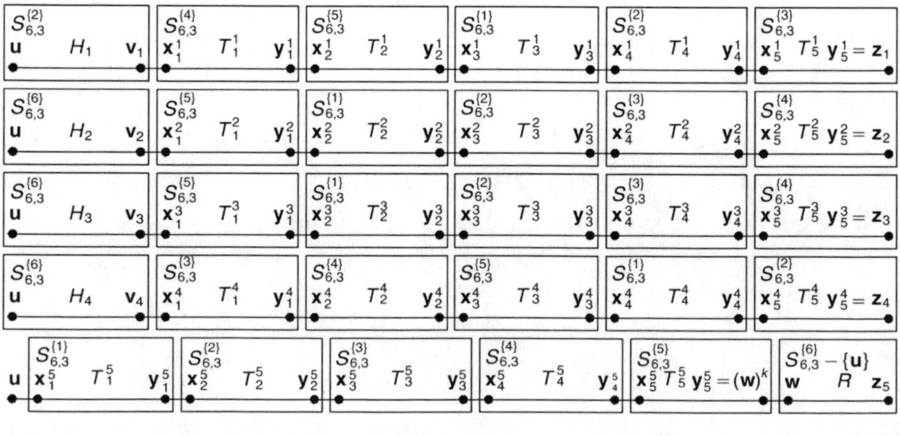

FIGURE 19.11 Illustration for Theorem 19.8 on $S_{6,3}$.

LEMMA 19.7 Let $\mathbf{u} = 1n$, $\mathbf{v} = 2n$, $\mathbf{y}_i = (i+2)n$ for every $1 \le i \le n-3$, and $\mathbf{y}_{n-2} = n2$ be vertices in $S_{n,2}$ with $n \ge 5$. Then there are $(n-2)$ paths $P_1, P_2, \ldots, P_{n-2}$ of $S_{n,2}$ such that (1) P_i is a Hamiltonian path of $S_{n,2} - \{\mathbf{v}, \mathbf{y}_i\}$ joining $\mathbf{u}$ to a vertex $\mathbf{x}_i$ with $(\mathbf{x}_i)_1 = i+1$ for every $1 \le i \le n-2$ and (2) $|\{P_1(j), P_2(j), \ldots, P_{n-2}(j)\}| = n-2$ for every $2 \le j \le n(n-1)-2$.

Proof: Obviously, $\{\mathbf{u}, \mathbf{v}, \mathbf{y}_1, \mathbf{y}_2, \ldots, \mathbf{y}_{n-3}\} = V(S_{n,2}^{\{n\}})$ and $\mathbf{y}_{n-2} \in S_{n,2}^{\{2\}}$. For every $1 \le i \le n-3$ and $1 \le j \le n-4$, let $\mathbf{z}_j^i$ be defined as

$$\mathbf{z}_j^i = \begin{cases} (i+j+2)n & \text{if } i+j \le n-3 \\ (i+j-n+5)n & \text{if } n-2 \le i+j \end{cases}$$

We set $R_i = \langle \mathbf{u}, \mathbf{z}_1^i, \mathbf{z}_2^i, \ldots, \mathbf{z}_{n-4}^i \rangle$ for every $1 \le i \le n-3$. Obviously, R_i is a Hamiltonian path of $S_{n,2}^{\{n\}} - \{\mathbf{v}, \mathbf{y}_i\}$ joining $\mathbf{u}$ to $\mathbf{z}_{n-4}^i$ for every $1 \le i \le n-3$, and $|\{R_1(j), R_2(j), \ldots, R_{n-3}(j)\}| = n-3$ for every $2 \le j \le n-3$. Moreover, $(\mathbf{z}_{n-4}^1)_1 = n-1$ and $(\mathbf{z}_{n-4}^i)_1 = i+1$ for every $2 \le i \le n-3$. Let C be an $(n-3) \times (n-1)$ matrix defined by

$$c_{i,j} = \begin{cases} n-1 & \text{if } i=1 \text{ and } j=1 \\ 5-j & \text{if } i=1 \text{ and } 2 \le j \le 4 \\ j-1 & \text{if } i=1 \text{ and } 5 \le j \\ 4-j & \text{if } i=2 \text{ and } 1 \le j \le 3 \\ j & \text{if } i=2 \text{ and } 4 \le j \\ i+j & \text{if } 3 \le i \text{ and } i+j \le n-1 \\ n-i-j+3 & \text{if } 3 \le i \text{ and } n \le i+j \le n+2 \\ i+j-n+1 & \text{if } 3 \le i \text{ and } n+3 \le i+j \end{cases}$$

More precisely,

$$C = \begin{bmatrix} n-1 & 3 & 2 & 1 & 4 & \ldots & n-5 & n-4 & n-3 & n-2 \\ 3 & 2 & 1 & 4 & 5 & \ldots & n-4 & n-3 & n-2 & n-1 \\ 4 & 5 & 6 & 7 & 8 & \ldots & n-1 & 3 & 2 & 1 \\ 5 & 6 & 7 & 8 & 9 & \ldots & 3 & 2 & 1 & 4 \\ \vdots & \vdots & \vdots & \vdots & \vdots & \ddots & \vdots & \vdots & \vdots & \vdots \\ n-2 & n-1 & 3 & 2 & 1 & \ldots & n-6 & n-5 & n-4 & n-3 \end{bmatrix}$$

For every $1 \le i \le n-3$, we set $\mathbf{p}_0^i = \mathbf{z}_{n-4}^i$, $\mathbf{p}_j^i = c_{i,j+1}c_{i,j}$ for every $1 \le j \le n-2$, and $\mathbf{p}_{n-1}^i = (i+1)c_{i,n-1}$. Note that $\mathbf{p}_j^i \in S_{n,2}^{\{c_{i,j}\}}$ and $(\mathbf{p}_j^i)^2 \in S_{n,2}^{\{c_{i,j+1}\}} - \{\mathbf{p}_{j+1}^i\}$ for every $1 \le i \le n-3$ and $0 \le j \le n-2$.

By Theorem 16.6, there is a Hamiltonian path T_j^2 of $S_{n,2}^{\{c_{2,j}\}}$ joining $(\mathbf{p}_{j-1}^2)^2$ to $\mathbf{p}_j^2$ for every $j \in \langle n-1 \rangle - \{2\}$ and there is a Hamiltonian path T of $S_{n,2}^{\{c_{2,2}=2\}} - \{(\mathbf{p}_1^2)^2\}$

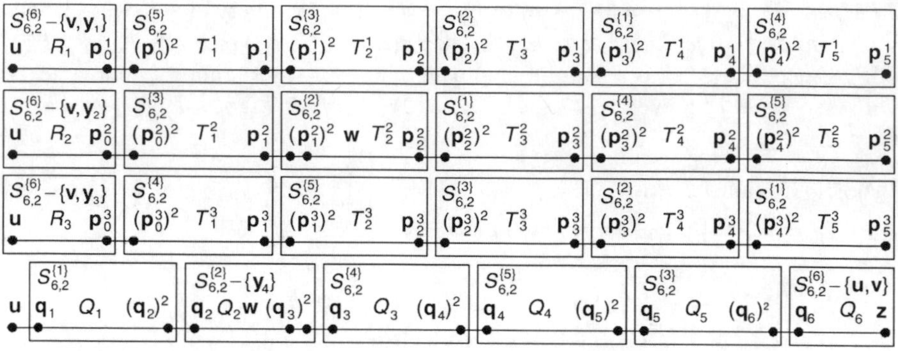

FIGURE 19.12 Illustration for Lemma 19.7 on $S_{6,2}$.

joining the vertex $\mathbf{w}=52$ to $\mathbf{p}_2^2$. We set $T_2^2=\langle(\mathbf{p}_1^2)^2,\mathbf{w},T,\mathbf{p}_2^2,\rangle$. By Theorem 16.6, there is a Hamiltonian path T_j^i of $S_{n,2}^{\{c_{i,j}\}}$ joining $(\mathbf{p}_{j-1}^i)^2$ to $\mathbf{p}_j^i$ for every $i\in\langle n-3\rangle-\{2\}$ and $1\le j\le n-1$. We set $P_i=\langle\mathbf{u},R_i,\mathbf{p}_0^i,(\mathbf{p}_0^i)^2,T_1^i,\mathbf{p}_1^i,(\mathbf{p}_1^i)^2,T_2^i,\mathbf{p}_2^i,\ldots,(\mathbf{p}_{n-2}^i)^2,T_{n-1}^i,\mathbf{p}_{n-1}^i\rangle$ for every $i\in\langle n-3\rangle$.

We set $d_1=1$, $d_2=2$, $d_i=i+1$ for every $3\le i\le n-2$, $d_{n-1}=3$, and $d_n=n$. We set $\mathbf{q}_1=(\mathbf{u})^2$ and $\mathbf{q}_i=d_{i-1}d_i$ for every $2\le i\le n$. Obviously, $\mathbf{q}_i\in S_{n,2}^{\{d_i\}}$ and $(\mathbf{q}_i)^2\in S_{n,2}^{\{d_{i-1}\}}-\{\mathbf{q}_{i-1}\}$ for every $2\le i\le n$. Moreover, $\mathbf{q}_2,(\mathbf{q}_3)^2\in S_{n,2}^{\{2\}}-\{\mathbf{y}_{n-2}=(\mathbf{v})^2\}$ and $\mathbf{q}_n\in S_{n,2}^{\{n\}}-\{\mathbf{u},\mathbf{v}\}$. By Lemma 16.18, and there is a Hamiltonian path Q_i of $S_{n,2}^{\{d_i\}}$ joining $\mathbf{q}_i$ to $(\mathbf{q}_{i+1})^2$ for every $i\in\langle n-1\rangle-\{2\}$. Again, there is a Hamiltonian path Q of $S_{n,2}^{\{d_2\}}-\{(\mathbf{q}_3)^2,\mathbf{y}_{n-2}\}$ joining $\mathbf{q}_2$ to $\mathbf{w}=52$. Moreover, there is a Hamiltonian path Q_n of $S_{n,2}^{\{d_n\}}-\{\mathbf{u},\mathbf{v}\}$ joining $\mathbf{q}_n=3n$ to the vertex $\mathbf{z}=(n-1)n$. We set $Q_2=\langle\mathbf{q}_2,Q,\mathbf{w},(\mathbf{q}_3)^2\rangle$. Then we set $P_{n-2}=\langle\mathbf{u},(\mathbf{u})^2=\mathbf{q}_1,Q_1,(\mathbf{q}_2)^2,\mathbf{q}_2,Q_2,(\mathbf{q}_3)^2,\ldots,\mathbf{q}_{n-1},Q_{n-1},(\mathbf{q}_n)^2,\mathbf{q}_n,Q_n,\mathbf{z}\rangle$.

Then $P_1,P_2,\ldots,P_{n-2}$ form desired paths. See Figure 19.12 for illustration. □

LEMMA 19.8 Let n and k be any two positive integers with $n\ge5$, $k\ge2$, and $n-k\ge2$. Let $\mathbf{u}$ and $\mathbf{v}$ be two vertices of $S_{n,k}$ with $(\mathbf{u})_1=1$, $(\mathbf{v})_1=2$, and $(\mathbf{u})_i=(\mathbf{v})_i=n-k+i$ for every $2\le i\le k$, and let $\mathbf{y}_i\in N_{S_{n,k}}(\mathbf{v})$ with $(\mathbf{y}_i)_1=i+2$ for every $1\le i\le n-2$. Then there are $(n-2)$ paths $P_1,P_2,\ldots,P_{n-2}$ of $S_{n,k}$ such that (1) P_i is a Hamiltonian path of $S_{n,k}-\{\mathbf{v},\mathbf{y}_i\}$ joining $\mathbf{u}$ to a vertex $\mathbf{x}_i$ with $(\mathbf{x}_i)_1=i+1$ for every $1\le i\le n-2$, and (2) $|\{P_1(j),P_2(j),\ldots,P_{n-2}(j)\}|=n-2$ for every $2\le j\le n!/(n-k)!-2$.

Proof: Note that when $k=2$, it is proved by Lemma 19.7. Obviously, $\{\mathbf{u},\mathbf{v},\mathbf{y}_1,\mathbf{y}_2,\ldots,\mathbf{y}_{n-3}\}\subset S_{n,k}^{\{n\}}$ and $\mathbf{y}_{n-2}\in S_{n,k}^{\{2\}}$. We have the following cases.

Case 1. $n=5$ and $k=3$. We have $\mathbf{u}=145$, $\mathbf{v}=245$, $\mathbf{y}_1=345$, $\mathbf{y}_2=425$, and $\mathbf{y}_3=542$. We prove that this statement holds on $S_{5,3}$ by listing every path in Table 19.6.

Case 2. $n\ge6$, $k\ge3$, and $n-k\ge2$. We prove this case by induction on n and k. By Lemma 19.7, this statement holds for every $n\ge5$ and $k=2$. By Case 1, this

TABLE 19.6
Desired Hamiltonian Paths of $S_{5,3} - \{245, \mathbf{y}_i\}$ for Lemma 19.8

$\mathbf{y}_1 = 345$	$\langle$145, 415, 315, 215, 125, 425, 325, 235, 135, 435, 534, 354, 154, 254, 524, 324, 124, 214, 514, 314, 134, 234, 432, 532, 352, 152, 452, 542, 342, 142, 412, 512, 312, 132, 231, 531, 431, 341, 541, 241, 421, 321, 521, 251, 451, 351, 153, 453, 543, 243, 143, 413, 513, 213, 123, 423, 523, 253$\rangle$
$\mathbf{y}_2 = 425$	$\langle$145, 345, 435, 235, 135, 315, 415, 215, 125, 325, 523, 423, 243, 143, 543, 453, 253, 153, 513, 413, 213, 123, 321, 421, 521, 251, 351, 451, 541, 241, 341, 431, 531, 231, 132, 532, 352, 452, 152, 512, 312, 412, 142, 542, 342, 432, 234, 324, 124, 524, 254, 154, 514, 214, 314, 134, 534, 354$\rangle$
$\mathbf{y}_3 = 524$	$\langle$145, 541, 241, 341, 431, 231, 321, 421, 521, 251, 451, 351, 531, 135, 315, 415, 215, 125, 425, 325, 235, 435, 345, 543, 143, 243, 423, 523, 123, 213, 413, 513, 153, 253, 453, 354, 154, 254, 524, 124, 324, 234, 534, 134, 314, 514, 214, 412, 142, 342, 432, 532, 132, 312, 512, 152, 352, 452$\rangle$

statement holds for $n=5$ and $k=3$. Thus, we assume that this statement holds on $S_{m,l}$ for every $5 \le m < n$ and for every $2 \le l < k$ with $m - l \ge 2$. Obviously, $S_{n,k}^{\{n\}}$ is isomorphic to $S_{n-1,k-1}$. By induction, there are $(n-3)$ paths $W_1, W_2, \ldots, W_{n-3}$ of $S_{n,k}^{\{n\}}$ such that (1) W_i is a Hamiltonian path of $S_{n,k}^{\{n\}} - \{\mathbf{v}, \mathbf{y}_i\}$ joining $\mathbf{u}$ to a vertex $\mathbf{w}_i$ with $(\mathbf{w}_i)_1 = i+1$ for every $1 \le i \le n-3$ and (2) $|\{W_1(j), W_2(j), \ldots, W_{n-3}(j)\}| = n-3$ for every $2 \le j \le (n-1)!/(n-k)! - 2$. Let C be an $(n-3) \times (n-1)$ matrix defined by

$$c_{i,j} = \begin{cases} i+j & \text{if } i+j \le n-2 \\ 1 & \text{if } i+j = n-1 \\ n-1 & \text{if } i+j = n \\ i+j-n+1 & \text{if } i+j \ge n+1 \end{cases}$$

More precisely,

$$C = \begin{bmatrix} 2 & 3 & 4 & 5 & \ldots & n-3 & n-2 & 1 & n-1 \\ 3 & 4 & 5 & 6 & \ldots & n-2 & 1 & n-1 & 2 \\ \vdots & \vdots & \vdots & \vdots & \ddots & \vdots & \vdots & \vdots & \vdots \\ n-2 & 1 & n-1 & 2 & \ldots & n-6 & n-5 & n-4 & n-3 \end{bmatrix}$$

We set $\mathbf{p}_0^i = \mathbf{w}_i$ for every $1 \le i \le n-3$. We choose a vertex $\mathbf{p}_j^i$ in $S_{n,k}^{\{c_{i,j}\}}$ with $(\mathbf{p}_j^i)_1 = c_{i,j+1}$ for every $1 \le i \le n-3$ and $1 \le j \le n-2$, and choose a vertex $\mathbf{p}_{n-1}^i$ in $S_{n,k}^{\{c_{i,n-1}\}}$ with $(\mathbf{p}_{n-1}^i)_1 = i+1$ for every $1 \le i \le n-3$. Note that $(\mathbf{p}_j^i)^k \in S_{n,k}^{\{c_{i,j+1}\}} - \{\mathbf{p}_{j+1}^i\}$ for every $1 \le i \le n-3$ and $0 \le j \le n-2$. By Theorem 16.6, there is a Hamiltonian path T_j^i of $S_{n,k}^{\{c_{i,j}\}}$ joining the vertex $(\mathbf{p}_{j-1}^i)^k$ to $\mathbf{p}_j^i$ for every $1 \le i \le n-3$ and $1 \le j \le n-1$. Let $\mathbf{x}_i = P_{n-1}^i$ for $1 \le i \le n-3$, then we set $P_i = \langle \mathbf{u}, W_i, \mathbf{w}_i = \mathbf{p}_0^i, (\mathbf{p}_0^i)^k, T_1^i, \mathbf{p}_1^i, (\mathbf{p}_1^i)^k, \mathbf{p}_2^i, \ldots, (\mathbf{p}_{n-2}^i)^k, T_{n-1}^i, \mathbf{p}_{n-1}^i = \mathbf{x}_i \rangle$ for every $1 \le i \le n-3$.

We set $d_1 = 1$, $d_2 = n$, $d_3 = n-1$, and $d_i = i-2$ for every $4 \le i \le n$. Since $n \ge 6$, $k \ge 3$, and $n-k \ge 2$, by Lemma 16.17, $|E_{n,k}^{d_i,d_{i+1}}| = (n-2)!/(n-k)! \ge 3$. By Lemma 19.5, there is a Hamiltonian path P of $S_{n,k}^{\{d_2=n\}} - \{\mathbf{u},\mathbf{v}\}$ joining a vertex $\mathbf{x}$ to a vertex $\mathbf{y}$ with $(\mathbf{x})_1 = 1$ and $(\mathbf{y})_1 = n-1$. We set $\mathbf{q}_1 = (\mathbf{u})^k$, $\mathbf{q}_2 = \mathbf{x}$, and $\mathbf{q}_3 = (\mathbf{y})^k$. We choose a vertex $\mathbf{q}_i$ in $S_{n,k}^{\{d_i\}}$ with $(\mathbf{q}_i)_1 = d_{i-1}$ for every $4 \le i \le n$. Obviously, $(\mathbf{q}_2)^k \in S_{n,k}^{\{d_1\}} - \{\mathbf{q}_1\}$ and $(\mathbf{q}_i)^k \in S_{n,k}^{\{d_{i-1}\}} - \{\mathbf{q}_{i-1}\}$ for every $3 \le i \le n$. Moreover, $\{\mathbf{q}_4, (\mathbf{q}_5)^k\} \subset V(S_{n,k}^{\{d_4\}}) - \{\mathbf{y}_{n-2}\}$. By Theorem 16.6, there is a Hamiltonian path Q_i of $S_{n,k}^{\{d_i\}}$ joining $\mathbf{q}_i$ to $(\mathbf{q}_{i+1})^k$ for every $i \in \langle n-1 \rangle - \{2,4\}$. Again, there is a Hamiltonian path Q_4 of $S_{n,k}^{\{d_4\}} - \{\mathbf{y}_{n-2}\}$ joining $\mathbf{q}_4$ to the vertex $(\mathbf{q}_5)^k$. Moreover, there is a Hamiltonian path Q_n of $S_{n,k}^{\{d_n\}}$ joining $\mathbf{q}_n$ to a vertex $\mathbf{z}$ with $(\mathbf{z})_1 = n-1$. We set $Q_2 = P$. Then we set $P_{n-2} = \langle \mathbf{u}, (\mathbf{u})^k = \mathbf{q}_1, Q_1, (\mathbf{q}_2)^k, \mathbf{q}_2, Q_2, (\mathbf{q}_3)^k, \ldots, \mathbf{q}_{n-1}, Q_{n-1}, (\mathbf{q}_n)^k, \mathbf{q}_n, Q_n, \mathbf{z} \rangle$.

Then $P_1, P_2, \ldots, P_{n-2}$ form the desired paths of $S_{n,k}$. Note that $((\mathbf{q}_4)^k)_1 = 2 \neq 1 = ((\mathbf{p}_2^{n-3})^k)_1$. See Figure 19.13 for illustration. □

THEOREM 19.9 $IHP(S_{n,k}) = n-2$ if $3 \le k \le n-2$.

Proof: Since $\delta(S_{n,k}) = n-1$, $IHP(S_{n,k}) \le n-2$. Let $\mathbf{u}$ and $\mathbf{v}$ be any two distinct vertices of $S_{n,k}$. To prove that this statement is true, we need to construct $(n-2)$ mutually independent Hamiltonian paths of $S_{n,k}$ between $\mathbf{u}$ and $\mathbf{v}$. We have the following cases.

Case 1. $(\mathbf{u})_i \neq (\mathbf{v})_i$ for some $2 \le i \le k$. Without loss of generality, we assume that $(\mathbf{u})_k \neq (\mathbf{v})_k$. Moreover, we assume that $(\mathbf{u})_k = n$ and $(\mathbf{v})_k = n-1$. Let C be an $(n-2) \times (n-2)$ matrix defined by

$$c_{i,j} = \begin{cases} i+j-1 & \text{if } i+j \le n-1 \\ i+j-n+1 & \text{if } n \le i+j \end{cases}$$

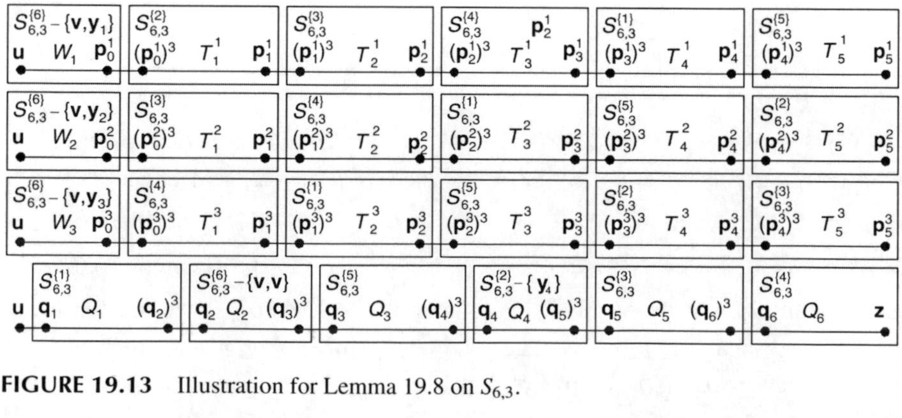

FIGURE 19.13 Illustration for Lemma 19.8 on $S_{6,3}$.

More precisely,

$$C = \begin{bmatrix} 1 & 2 & 3 & \dots & n-2 \\ 2 & 3 & 4 & \dots & 1 \\ \vdots & \vdots & \vdots & \ddots & \vdots \\ n-2 & 1 & 2 & \dots & n-3 \end{bmatrix}$$

By Theorem 19.8, there are Hamiltonian paths $Q_1, Q_2, \dots, Q_{n-2}$ of $S_{n,k}^{\{n\}}$ such that Q_i is a Hamiltonian path of $S_{n,k}^{\{n\}}$ joining $\mathbf{u}$ to a vertex $\mathbf{s}_i$ with (1) $(\mathbf{s}_i)_1 = c_{i,1}$ for every $1 \le i \le n-2$ and (2) $|\{Q_1(j), Q_2(j), \dots, Q_{n-2}(j)\}| = n-2$ for every $2 \le j \le (n-1)!/(n-k)!$. Again, there are Hamiltonian paths $R_1, R_2, \dots, R_{n-2}$ of $S_{n,k}^{\{n-1\}}$ such that R_i is a Hamiltonian path of $S_{n,k}^{\{n-1\}}$ joining $\mathbf{v}$ to a vertex $\mathbf{t}_i$ with (1) $(\mathbf{t}_i)_1 = c_{i,(n-2)}$ for every $1 \le i \le n-2$ and (2) $|\{R_1(j), R_2(j), \dots, R_{n-2}(j)\}| = n-2$ for every $2 \le j \le (n-1)!/(n-k)!$. For every $1 \le i \le n-2$, by Theorem 16.7, there exists a Hamiltonian path $W_i = \langle \mathbf{x}_1^i, T_1^i, \mathbf{y}_1^i, \mathbf{x}_2^i, T_2^i, \mathbf{y}_2^i, \dots, \mathbf{x}_{n-2}^i, T_{n-2}^i, \mathbf{y}_{n-2}^i \rangle$ of $S_{n,k}^{\langle n-2 \rangle}$ joining the vertex $\mathbf{x}_1^i$ to the vertex $\mathbf{y}_{n-2}^i$ such that $\mathbf{x}_1^i = (\mathbf{s}_i)^k$, $\mathbf{y}_{n-2}^i = (\mathbf{t}_i)^k$, and T_j^i is a Hamiltonian path of $S_{n,k}^{\{c_{i,j}\}}$ joining $\mathbf{x}_j^i$ to $\mathbf{y}_j^i$ for every $1 \le j \le n-2$. Let $P_i = \langle \mathbf{u}, Q_i, \mathbf{s}_i, (\mathbf{s}_i)^k = \mathbf{x}_1^i, W_i, \mathbf{y}_{n-2}^i = (\mathbf{t}_i)^k, \mathbf{t}_i, R_i^{-1}, \mathbf{v} \rangle$ for every $1 \le i \le n-2$. Hence, $\{P_1, P_2, \dots, P_{n-2}\}$ forms a set of $(n-2)$ mutually independent Hamiltonian paths of $S_{n,k}$ from $\mathbf{u}$ to $\mathbf{v}$. See Figure 19.14 for illustration.

Case 2. $(\mathbf{u})_i = (\mathbf{v})_i$ for every $2 \le i \le k$. Without loss of generality, we assume that $(\mathbf{u})_1 = 1$, $(\mathbf{v})_1 = 2$, and $(\mathbf{u})_i = (\mathbf{v})_i = n - k + i$ for every $2 \le i \le k$. Suppose that $n = 5$ and $k = 3$. We prove there are three mutually independent Hamiltonian paths on $S_{5,3}$ joining $\mathbf{u} = 145$ to $\mathbf{v} = 245$ by exhibiting the three required Hamiltonian paths in Table 19.7.

Thus, we assume that $n \ge 6$, $k \ge 3$, and $n - k \ge 2$. Let $\mathbf{y}_i$ be a vertex of $N_{S_{n,k}}(\mathbf{v})$ with $(\mathbf{y}_i)_1 = i + 2$ for every $1 \le i \le n-2$. Obviously, $\mathbf{u} \in S_{n,k}^{\{n\}}$, $\mathbf{v} \in S_{n,k}^{\{n\}}$,

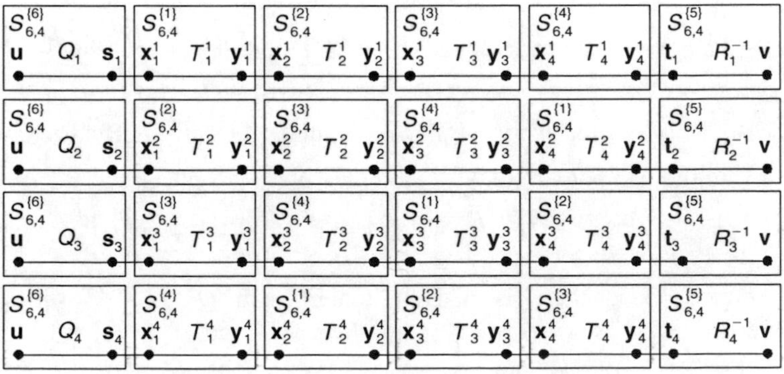

FIGURE 19.14 Illustration for Case 1 of Theorem 19.9 for $n = 6$ and $k = 4$.

TABLE 19.7
The Three Mutually Independent Hamiltonian Paths of $S_{5,3}$ between 145 and 245

$P_1 = \langle 145, 345, 435, 135, 235, 325, 523, 423, 243, 143, 543, 453, 253, 153, 513, 413, 213, 123, 321, 521, 421, 241, 341, 541, 451, 251, 351, 531, 431, 231, 132, 532, 352, 452, 152, 512, 312, 412, 142, 542, 342, 432, 234, 534, 134, 314, 214, 124, 324, 524, 254, 354, 154, 514, 415, 315, 215, 125, 425, 245\rangle$

$P_2 = \langle 145, 415, 315, 215, 125, 425, 524, 324, 124, 214, 514, 314, 134, 234, 534, 354, 154, 254, 452, 352, 532, 132, 432, 342, 542, 142, 412, 312, 512, 152, 251, 451, 541, 341, 241, 421, 521, 321, 231, 431, 531, 351, 153, 513, 413, 213, 123, 423, 243, 143, 543, 453, 253, 523, 325, 235, 135, 435, 345, 245\rangle$

$P_3 = \langle 145, 541, 341, 241, 421, 521, 321, 231, 431, 531, 351, 251, 451, 154, 354, 254, 524, 124, 324, 234, 534, 134, 314, 214, 514, 415, 315, 215, 125, 425, 325, 235, 135, 435, 345, 543, 143, 413, 213, 513, 153, 453, 253, 523, 123, 423, 243, 342, 142, 412, 512, 312, 132, 432, 532, 352, 152, 452, 542, 245\rangle$

$\mathbf{y}_i \in S_{n,k}^{\{n\}}$ for every $1 \le i \le n-3$, and $\mathbf{y}_{n-2} \in S_{n,k}^{\{2\}}$. By Lemma 19.8, there are $(n-3)$ paths $R_1, R_2, \ldots, R_{n-3}$ of $S_{n,k}^{\{n\}}$ such that (1) R_i is a Hamiltonian path of $S_{n,k}^{\{n\}} - \{\mathbf{v}, \mathbf{y}_i\}$ joining $\mathbf{u}$ to a vertex $\mathbf{x}_i$ with $(\mathbf{x}_i)_1 = i+1$ for every $1 \le i \le n-3$ and (2) $|\{R_1(j), R_2(j), \ldots, R_{n-3}(j)\}| = n-3$ for every $2 \le j \le (n-1)!/(n-k)! - 2$. Let C be an $(n-3) \times (n-1)$ matrix defined by

$$c_{i,j} = \begin{cases} i-j+2 & \text{if } j \le i+1 \\ i+n-j+1 & \text{if } i+2 \le j \end{cases}$$

More precisely,

$$C = \begin{bmatrix} 2 & 1 & n-1 & n-2 & \ldots & 4 & 3 \\ 3 & 2 & 1 & n-1 & \ldots & 5 & 4 \\ \vdots & \vdots & \vdots & \vdots & \ddots & \vdots & \vdots \\ n-2 & n-3 & n-4 & n-5 & \ldots & 1 & n-1 \end{bmatrix}$$

We set $\mathbf{p}_0^i = \mathbf{x}_i$ and $\mathbf{p}_{n-1}^i = (\mathbf{y}_i)^k$ for every $1 \le i \le n-3$. We choose a vertex $\mathbf{p}_j^i$ in $S_{n,k}^{\{c_{i,j}\}}$ with $(\mathbf{p}_j^i)_1 = c_{i,j+1}$ for every $1 \le i \le n-3$ and $1 \le j \le n-2$. Note that $(\mathbf{p}_j^i)^k \in S_{n,k}^{\{c_{i,j+1}\}} - \{\mathbf{p}_{j+1}^i\}$ for every $1 \le i \le n-3$ and $0 \le j \le n-2$. By Theorem 16.6, there is a Hamiltonian path T_j^i of $S_{n,k}^{\{c_{i,j}\}}$ joining $(\mathbf{p}_{j-1}^i)^k$ to $\mathbf{p}_j^i$ for every $1 \le i \le n-3$ and $1 \le j \le n-1$. We set $P_i = \langle \mathbf{u}, R_i, \mathbf{x}_i = \mathbf{p}_0^i, (\mathbf{p}_0^i)^k, T_1^i, \mathbf{p}_1^i, (\mathbf{p}_1^i)^k, T_2^i, \mathbf{p}_2^i, \ldots, (\mathbf{p}_{n-2}^i)^k, T_{n-1}^i, \mathbf{p}_{n-1}^i = (\mathbf{y}_i)^k, \mathbf{y}_i, \mathbf{v}\rangle$ for every $1 \le i \le n-3$.

By Lemma 19.5, there is a Hamiltonian path Q_2 of $S_{n,k}^{\{n\}} - \{\mathbf{u}, \mathbf{v}\}$ joining a vertex $\mathbf{x}$ to a vertex $\mathbf{y}$ with $(\mathbf{x})_1 = 1$ and $(\mathbf{y})_1 = n-1$. We set $\mathbf{q}_0 = \mathbf{u}$, $\mathbf{q}_1 = (\mathbf{x})^k$, $\mathbf{q}_2 = y$, and $\mathbf{q}_n = \mathbf{y}_{n-2}$. We choose a vertex $\mathbf{q}_i \in S_{n,k}^{\{n-i+2\}}$ for every $3 \le i \le n-1$ with $(\mathbf{q}_i)_1 = n-i+1$. Note that $(\mathbf{q}_0)^k \in S_{n,k}^{\{1\}} - \{\mathbf{q}_1\}$, and $(\mathbf{q}_i)^k \in S_{n,k}^{\{n-i+1\}} - \{\mathbf{q}_{i+1}\}$ for every $2 \le i \le n-1$. By Theorem 16.6, there is a

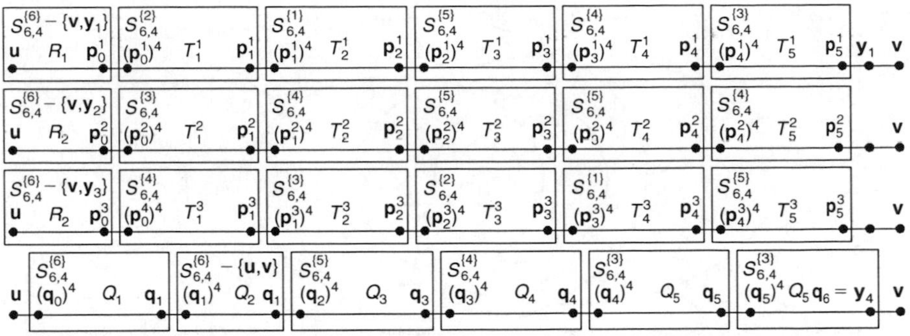

FIGURE 19.15 Illustration for Case 2 of Theorem 19.9 for $n=6$ and $k=4$.

Hamiltonian path Q_1 of $S_{n,k}^{\{1\}}$ joining $(\mathbf{q}_0)^k$ to $\mathbf{q}_1$. Again, there is a Hamiltonian path Q_i of $S_{n,k}^{\{n-i+2\}}$ joining $(\mathbf{q}_{i-1})^k$ to $\mathbf{q}_i$ for every $3 \le i \le n$. We set $P_{n-2} = \langle \mathbf{u} = \mathbf{q}_0, (\mathbf{q}_0)^k, Q_1, \mathbf{q}_1, (\mathbf{q}_1)^k = \mathbf{x}, Q_2, \mathbf{y} = \mathbf{q}_2, \ldots, (\mathbf{q}_{n-2})^k, Q_{n-1}, \mathbf{q}_{n-1}, (\mathbf{q}_{n-1})^k, Q_n, \mathbf{q}_n = \mathbf{y}_{n-2}, \mathbf{v}\rangle$. Then $\{P_1, P_2, \ldots, P_{n-2}\}$ forms a set of $(n-2)$ mutually independent Hamiltonian paths of $S_{n,k}$ joining $\mathbf{u}$ to $\mathbf{v}$. Note that $(\mathbf{q}_3)_1 = n-2 \neq 1 = ((\mathbf{p}_2^1)^k)_1$ and $(\mathbf{q}_i)_1 = n-i+1 \neq n-i+3 = ((\mathbf{p}_{i-1}^1)^k)_1$ for every $4 \le i \le n-1$. See Figure 19.15 for illustration.

Thus, this statement is proved. □

According to Theorems 19.5 through 19.7 and Theorem 19.9, we have the following result.

THEOREM 19.10 *$IHP(S_{n,k}) = n-2$ for any $1 \le k \le n-2$ except that $S_{4,2}$, and $IHP(S_{4,2}) = 1$.*

19.5 CUBIC 2-INDEPENDENT HAMILTONIAN-CONNECTED GRAPHS

It is interesting to point out that the set of all cubic graphs G with $IHP(G) = 2$ have more interesting properties. Ho et al. [150] observed that every cubic graph G with $IHP(G) = 2$ is a 1-fault-tolerant Hamiltonian graph.

THEOREM 19.11 Every cubic 2-independent Hamiltonian-connected graph is 1-fault-tolerant Hamiltonian.

Proof: Suppose that G is a cubic 2-independent Hamiltonian-connected graph. Let F be a subset of $V \cup E$ with $|F| = 1$.

Suppose that $e = (x, y)$ is an edge in F. Let $N(y) = \{x, u, v\}$. Since G is a cubic 2-independent Hamiltonian-connected graph, there are two independent Hamiltonian paths joining y and u. Apparently, these two Hamiltonian paths can be written

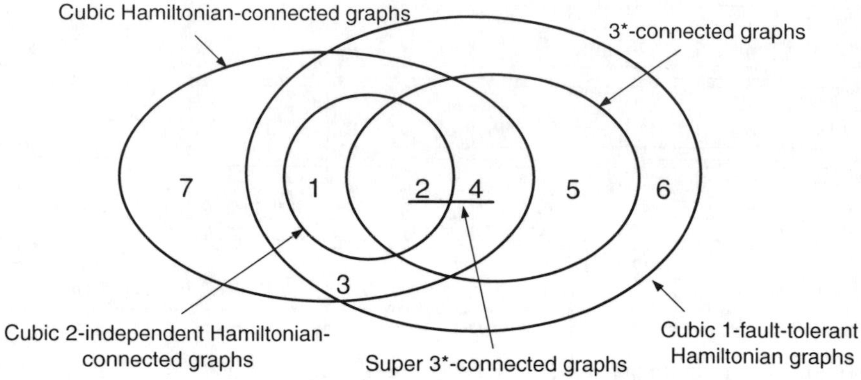

FIGURE 19.16 The relationship between cubic 1-fault-tolerant Hamiltonian graphs, 3*-connected graphs, super 3*-connected graphs, cubic Hamiltonian-connected graphs, and cubic 2-independent Hamiltonian-connected graphs.

as $\langle y, v, P_1, u\rangle$ and $\langle y, x, P_2, u\rangle$, where P_1 and P_2 are Hamiltonian paths in $G-y$. Obviously, $\langle y, v, P_1, u, y\rangle$ forms a Hamiltonian cycle of $G-e$.

Suppose that x is a vertex in F. Let $e=(x,y)$ be any edge incident with x and $N(y)=\{x, u, v\}$. Since G is cubic 2-independent Hamiltonian-connected graph, there are two independent Hamiltonian paths between x and v in G. Apparently, these two Hamiltonian paths can be written as $\langle x, y, u, Q_1, v\rangle$ and $\langle x, Q_2, u, y, v\rangle$, where Q_1 is a Hamiltonian path in $G-\{x, y\}$ and Q_2 is a Hamiltonian path in $G-\{y, v\}$. Obviously, $\langle y, u, Q_1, v, y\rangle$ forms a Hamiltonian cycle of $G-x$.

Thus, G is 1-fault-tolerant Hamiltonian. The theorem is proved. □

By Theorem 15.2, every cubic 3*-connected graph is also 1-fault-tolerant Hamiltonian. In Figure 19.16, we use a Venn diagram to illustrate the relation between cubic 1-fault-tolerant Hamiltonian graphs, 3*-connected graphs, super 3*-connected graphs, cubic Hamiltonian connected graphs, and cubic 2-independent Hamiltonian connected graphs. The set of cubic 2-independent Hamiltonian-connected graphs corresponds to regions 1 and 2; the set of cubic 1-fault-tolerant Hamiltonian graphs corresponds to regions 1, 2, 3, 4, 5, and 6; the set of 3*-connected graphs corresponds to regions 2, 4, and 5; the set of cubic Hamiltonian-connected graphs corresponds to regions 1, 2, 3, 4, and 7; the set of super 3*-connected graphs corresponds to regions 2 and 4.

We are wondering about the existence of graphs for all the possible regions. In Section 15.2, the existence of graphs in regions 5, 6, and 7 is proved. Ho et al. [150] present some examples of graphs in regions 1, 2, 3, and 4.

19.5.1 Examples of Cubic 2-Independent Hamiltonian-Connected Graphs That Are Super 3*-Connected

Here, we present a family of cubic 2-independent Hamiltonian-connected graphs that are super 3*-connected. In the following, we use $\oplus$ and $\ominus$ to denote addition and subtraction in integer modulo n, Z_n.

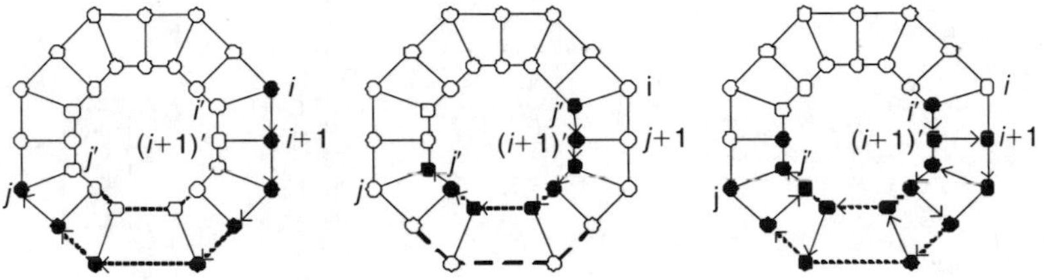

FIGURE 19.17 Illustrations for path patterns: (a) $P(i, j)$, (b) $Q(i',j')$, and (c) $M(i',j')$.

The *generalized Petersen graph* $P(n, 1)$ is the graph with vertex set $\{i \mid 0 \leq i < n\} \cup \{i' \mid 0 \leq i < n\}$ and edge set $\{(i, i \oplus 1) \mid 0 \leq i < n\} \cup \{(i', (i \oplus 1)') \mid 0 \leq i < n\} \cup \{(i, i') | 0 \leq i < n\}$.

THEOREM 19.12 $P(n, 1)$ is cubic 2-independent Hamiltonian-connected and super 3^*-connected if and only if n is odd.

Proof: It is proved in Ref. 194 that $P(n, 1)$ is super 3^*-connected if and only if n is odd. It is easy to see that $P(n, 1)$ is bipartite if and only if n is even. Thus, we consider the case where n is odd. We will prove that there exist two independent Hamiltonian paths between any two different vertices.

Let $n = 2k + 1$. To describe the required Hamiltonian paths, we use the following path patterns. Let $P(i, j)$ denote the path $\langle i, i \oplus 1, \ldots, j \ominus 1, j\rangle$, $Q(i',j')$ denote the path $\langle i', (i \oplus 1)', \ldots, (j \ominus 1)', j'\rangle$, and $M(i',j')$ denote the path $\langle i', (i \oplus 1)', i \oplus 1, i \oplus 2, (i \oplus 2)', (i \oplus 3)', \ldots, j \ominus 1, (j \ominus 1)', j'\rangle$. See Figure 19.17 for illustration. We also use $\overline{P}(j, i)$ to denote the path $(P(i, j))^{-1}$, $\overline{Q}(j', i')$ to denote the path $(Q(i',j'))^{-1}$, and $\overline{M}(j', i')$ to denote the path $(M(i',j'))^{-1}$.

By the symmetric property of $P(2k + 1, 1)$, we need to find only two independent Hamiltonian paths P_1 and P_2 between 0 to any vertex x in $\{1, 2, \ldots, k, 0', 1', \ldots, k'\}$. We have the following cases.

Case 1. $x = 0'$. Set $P_1 = \langle 0, \overline{P}(2k, 1), Q(1', (2k)'), 0'\rangle$ and $P_2 = \langle 0, P(1, 2k), \overline{Q}((2k)', 1'), 0'\rangle$. From Figure 19.18a, it is easy to see that P_1 and P_2 are independent.

Case 2. $x = 1$. Set $P_1 = \langle 0, \overline{P}(2k, 2), Q(2', 0'), 1', 1\rangle$ and $P_2 = \langle 0, 0', Q(1', (2k)'), \overline{P}(2k, 2), 1\rangle$. From Figure 19.18b, it is easy to see that P_1 and P_2 are independent.

Case 3. $x = k$ and k is odd. Set $P_1 = \langle 0, \overline{P}(2k, k + 1), Q((k + 1)', (2k)'), M(0', k'), k\rangle$ and $P_2 = \langle 0, M(0', k'), Q((k + 1)', (2k)'), \overline{P}(2k, k + 1), k\rangle$. From Figure 19.18c, it is easy to see that P_1 and P_2 are independent.

Case 4. $x = k$ and k is even. Set $P_1 = \langle 0, P(1, k - 1), \overline{Q}((k - 1)', 1'), \overline{M}(0', k'), k\rangle$ and $P_2 = \langle 0, \overline{M}(0', k'), \overline{Q}((k - 1)', 1'), P(1, k - 1), k\rangle$. From Figure 19.18d, it is easy to see that P_1 and P_2 are independent.

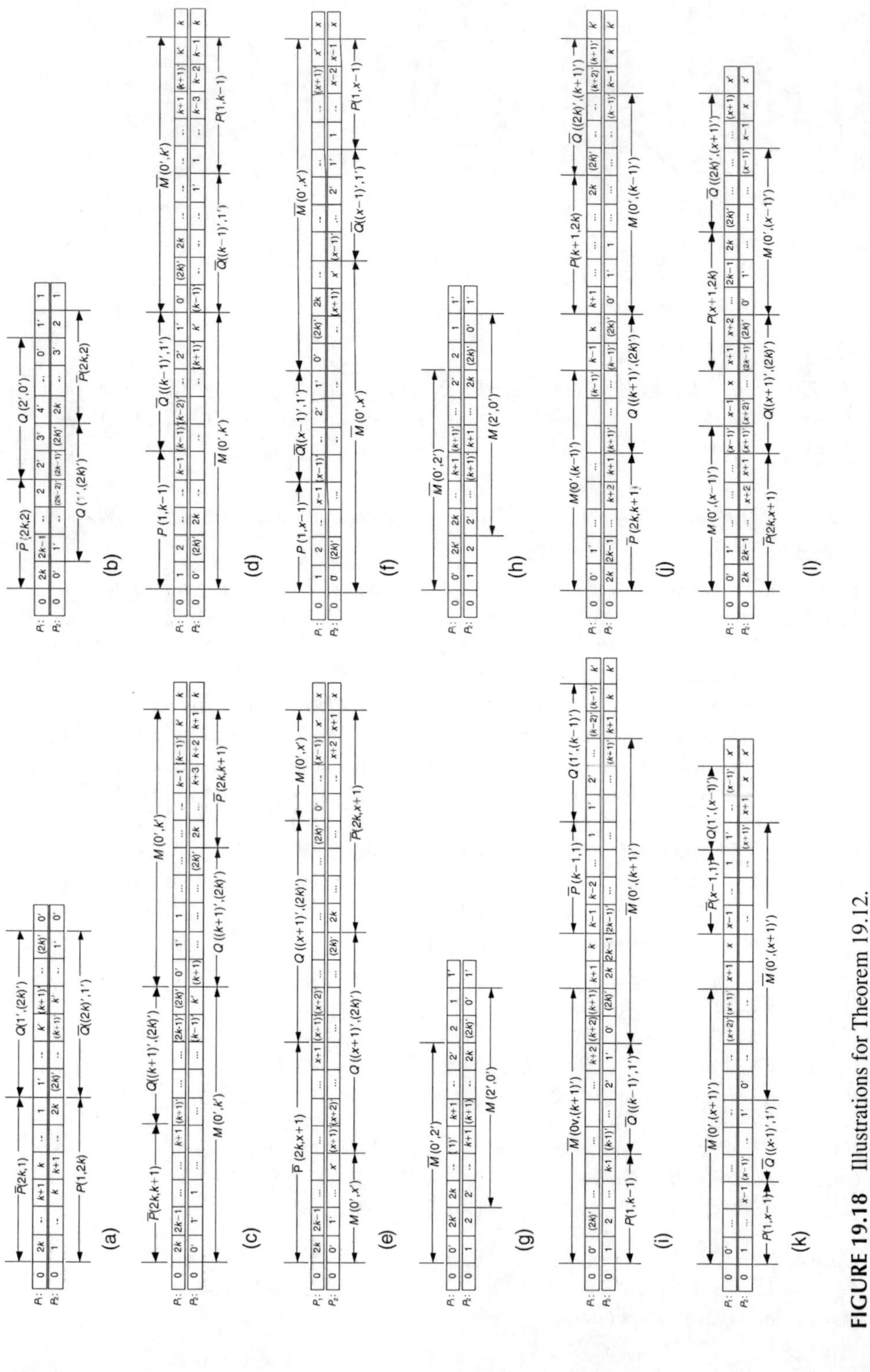

FIGURE 19.18 Illustrations for Theorem 19.12.

Case 5. $x = 2i + 1$ for some positive integer i. Set $P_1 = \langle 0, \overline{P}(2k, x+1), Q((x+1)', (2k)'), M(0', x'), x\rangle$ and $P_2 = \langle 0, M(0', x'), Q((x+1)', (2k)'), \overline{P}(2k, x+1), x\rangle$. From Figure 19.18e, it is easy to see that P_1 and P_2 are independent.

Case 6. $x = 2i$ for some positive integer i. Set $P_1 = \langle 0, P(1, x-1), \overline{Q}((x-1)', 1'), \overline{M}(0', x'), x\rangle$ and $P_2 = \langle 0, \overline{M}(0', x'), \overline{Q}((x-1)', 1'), P(1, x-1), x\rangle$. From Figure 19.18f, it is easy to see that P_1 and P_2 are independent.

Case 7. $x = 1'$ and k is odd. Set $P_1 = \langle 0, \overline{M}(0', 2'), 2, 1, 1'\rangle$ and $P_2 = \langle 0, 1, 2, \overline{M}(2', 0'), 1'\rangle$. From Figure 19.18g, it is easy to see that P_1 and P_2 are independent.

Case 8. $x = 1'$ and k is even. Set $P_1 = \langle 0, \overline{M}(0', 2'), 2, 1, 1'\rangle$ and $P_2 = \langle 0, 1, 2, \overline{M}(2', 0'), 1'\rangle$. From Figure 19.18h, it is easy to see that P_1 and P_2 are independent.

Case 9. $x = k'$ and k is odd. Set $P_1 = \langle 0, \overline{M}(0', (k+1)'), k+1, k, \overline{P}(k-1, 1), Q(1', (k-1)'), k'\rangle$ and $P_2 = \langle 0, P(1, k-1), \overline{Q}((k-1)', 1'), \overline{M}(0', (k+1)'), k+1, k, k'\rangle$. From Figure 19.18i, it is easy to see that P_1 and P_2 are independent.

Case 10. $x = k'$ and k is even. Set $P_1 = \langle 0, M(0', (k-1)'), k-1, k, P(k+1, 2k), \overline{Q}((2k)', (k+1)'), k'\rangle$ and $P_2 = \langle 0, \overline{P}(2k, k+1), Q((k+1)', (2k)'), M(0', (k-1)'), k-1, k, k'\rangle$. From Figure 19.18j, it is easy to see that P_1 and P_2 are independent.

Case 11. $x = (2i+1)'$ for some positive integer i. Set $P_1 = \langle 0, \overline{M}(0', (x+1)'), x+1, x, \overline{P}(x-1, 1), Q(1', (x-1)'), x'\rangle$ and $P_2 = \langle 0, P(1, x-1), \overline{Q}((x-1)', 1'), \overline{M}(0', (x+1)', x+1, x, x'\rangle$. From Figure 19.18k, it is easy to see that P_1 and P_2 are independent.

Case 12. $x = (2i)'$ for some positive integer i. Set $P_1 = \langle 0, M(0', (x-1)'), x-1, x, P(x+1, 2k), \overline{Q}((2k)', (x+1)'), x'\rangle$ and $P_2 = \langle 0, \overline{P}(2k, x+1), Q((x+1)', (2k)'), M(0', (x-1)'), x-1, x, x'\rangle$. From Figure 19.18l, it is easy to see that P_1 and P_2 are independent.

From the previous discussion, the theorem is proved. □

19.5.2 Examples of Super 3*-Connected Graphs That Are Not Cubic 2-Independent Hamiltonian-Connected

Now, we present a family of super 3*-connected graphs that are not cubic 2-independent Hamiltonian-connected. Again, we use $\oplus$ and $\ominus$ to denote addition and subtraction in integer modulo n, Z_n.

Assume that n is a positive even integers with $n \geq 4$. The graph $H(n)$ is the graph with vertex set $\{0, 1, \ldots, n-1\}$ and edge set $\{(i, n-i) \mid 1 \leq i < \frac{n}{2}\} \cup \{(i, i \oplus 1) \mid 0 \leq i \leq n-1\} \cup \{(0, \frac{n}{2})\}$.

THEOREM 19.13 $H(n)$ is super 3*-connected but not cubic 2-independent Hamiltonian-connected if n is even and $n \geq 4$.

Proof: By Theorem 15.3, $H(n)$ is Hamiltonian-connected. Hence, $H(n)$ is super 3*-connected. Now, we want to show that $H(n)$ is not cubic 2-independent Hamiltonian-connected if n is even and $n \geq 4$. We only prove that $L(4)$ is not cubic 2-independent

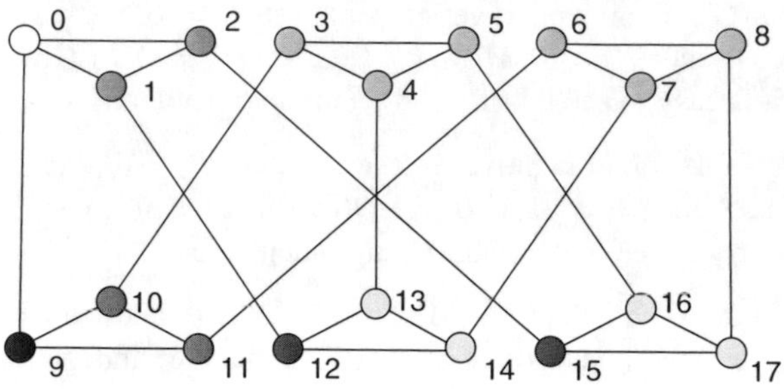

FIGURE 19.19 The graph N.

Hamiltonian-connected. Using the same technique, it is easy to see that $H(n)$ is not cubic 2-independent Hamiltonian-connected if n is even and $n \geq 4$.

By brute force, it is easy to see that there are only two Hamiltonian paths between 1 and 7; namely, $\langle 1, 2, 6, 5, 3, 4, 0, 7 \rangle$ and $\langle 1, 0, 4, 5, 3, 2, 6, 7 \rangle$. Apparently, these two Hamiltonian paths are not independent. □

19.5.3 Example of a Cubic 2-Independent Hamiltonian-Connected Graph That Is Not Super 3*-Connected

We will prove that the graph N in Figure 19.19 is 2-independent Hamiltonian-connected, but not super 3*-connected.

Obviously, N is obtained from $K_{3,3}$ by a sequence of 3-join with K_4. By Lemma 12.5, $K_{3,3}$ is not 1-fault-tolerant Hamiltonian. By Theorem 15.2, $K_{3,3}$ is not 3*-connected. Repeatedly using Theorem 15.1, N is not 3*-connected. Hence, N is not super 3*-connected. Now, we claim that N is cubic 2-independent Hamiltonian-connected. Since N is vertex-transitive, we need to find only two independent Hamiltonian paths between the vertex 0 to any other vertex of N. The corresponding cubic 2-independent Hamiltonian paths are:

$0 \to 1$	$\langle 0, 09, 11, 10, 03, 05, 04, 13, 12, 14, 07, 06, 08, 17, 16, 15, 02, 01 \rangle$
	$\langle 0, 02, 15, 16, 17, 08, 07, 06, 11, 09, 10, 03, 05, 04, 13, 14, 12, 01 \rangle$
$0 \to 2$	$\langle 0, 09, 10, 11, 06, 07, 08, 17, 15, 16, 05, 03, 04, 13, 14, 12, 01, 02 \rangle$
	$\langle 0, 01, 12, 14, 13, 04, 05, 03, 10, 09, 11, 06, 07, 08, 17, 16, 15, 02 \rangle$
$0 \to 3$	$\langle 0, 09, 10, 11, 06, 08, 07, 14, 13, 12, 01, 02, 15, 17, 16, 05, 04, 03 \rangle$
	$\langle 0, 02, 01, 12, 14, 13, 04, 05, 16, 15, 17, 08, 07, 06, 11, 09, 10, 03 \rangle$
$0 \to 4$	$\langle 0, 09, 10, 11, 06, 08, 07, 14, 13, 12, 01, 02, 15, 17, 16, 05, 03, 04 \rangle$
	$\langle 0, 01, 02, 15, 17, 16, 05, 03, 10, 09, 11, 06, 08, 07, 14, 12, 13, 04 \rangle$
$0 \to 5$	$\langle 0, 09, 10, 11, 06, 07, 08, 17, 16, 15, 02, 01, 12, 14, 13, 04, 03, 05 \rangle$
	$\langle 0, 02, 01, 12, 14, 13, 04, 03, 10, 09, 11, 06, 07, 08, 17, 15, 16, 05 \rangle$
$0 \to 6$	$\langle 0, 09, 11, 10, 03, 05, 04, 13, 14, 12, 01, 02, 15, 16, 17, 08, 07, 06 \rangle$
	$\langle 0, 01, 02, 15, 16, 17, 08, 07, 14, 12, 13, 04, 05, 03, 10, 09, 11, 06 \rangle$
$0 \to 7$	$\langle 0, 09, 11, 10, 03, 05, 04, 13, 14, 12, 01, 02, 15, 16, 17, 08, 06, 07 \rangle$
	$\langle 0, 01, 02, 15, 16, 17, 08, 06, 11, 09, 10, 03, 05, 04, 13, 12, 14, 07 \rangle$
$0 \to 8$	$\langle 0, 02, 01, 12, 13, 14, 07, 06, 11, 09, 10, 03, 04, 05, 16, 15, 17, 08 \rangle$
	$\langle 0, 09, 11, 10, 03, 04, 05, 16, 17, 15, 02, 01, 12, 13, 14, 07, 06, 08 \rangle$

$0 \to 9$	$\langle 0, 01, 02, 15, 17, 16, 05, 03, 04, 13, 12, 14, 07, 08, 06, 11, 10, 09\rangle$
	$\langle 0, 02, 01, 12, 13, 14, 07, 06, 08, 17, 15, 16, 05, 04, 03, 10, 11, 09\rangle$
$0 \to 10$	$\langle 0, 02, 01, 12, 14, 13, 04, 03, 05, 16, 15, 17, 08, 07, 06, 11, 09, 10\rangle$
	$\langle 0, 09, 11, 06, 08, 07, 14, 13, 12, 01, 02, 15, 17, 16, 05, 04, 03, 10\rangle$
$0 \to 11$	$\langle 0, 01, 02, 15, 16, 17, 08, 06, 07, 14, 12, 13, 04, 05, 03, 10, 09, 11\rangle$
	$\langle 0, 09, 10, 03, 05, 04, 13, 14, 12, 01, 02, 15, 16, 17, 08, 07, 06, 11\rangle$
$0 \to 12$	$\langle 0, 01, 02, 15, 17, 16, 05, 04, 03, 10, 09, 11, 06, 08, 07, 14, 13, 12\rangle$
	$\langle 0, 09, 11, 10, 03, 05, 04, 13, 14, 07, 06, 08, 17, 16, 15, 02, 01, 12\rangle$
$0 \to 13$	$\langle 0, 09, 10, 11, 06, 08, 07, 14, 12, 01, 02, 15, 17, 16, 05, 03, 04, 13\rangle$
	$\langle 0, 01, 02, 15, 17, 16, 05, 04, 03, 10, 09, 11, 06, 08, 07, 14, 12, 13\rangle$
$0 \to 14$	$\langle 0, 01, 02, 15, 16, 17, 08, 07, 16, 11, 09, 10, 03, 05, 04, 13, 12, 14\rangle$
	$\langle 0, 09, 11, 10, 03, 05, 04, 13, 12, 01, 02, 15, 16, 17, 08, 06, 07, 14\rangle$
$0 \to 15$	$\langle 0, 02, 01, 12, 14, 13, 04, 05, 03, 10, 09, 11, 06, 07, 08, 17, 16, 15\rangle$
	$\langle 0, 09, 11, 10, 03, 04, 05, 16, 17, 08, 06, 07, 14, 13, 12, 01, 02, 15\rangle$
$0 \to 16$	$\langle 0, 02, 01, 12, 14, 13, 04, 05, 02, 10, 09, 11, 06, 07, 08, 17, 15, 16\rangle$
	$\langle 0, 09, 10, 11, 06, 07, 08, 17, 15, 02, 01, 12, 14, 13, 04, 03, 05, 16\rangle$
$0 \to 17$	$\langle 0, 02, 01, 12, 13, 14, 07, 08, 06, 11, 09, 10, 03, 04, 05, 16, 15, 17\rangle$
	$\langle 0, 09, 11, 10, 03, 04, 05, 16, 15, 02, 01, 12, 13, 14, 07, 06, 08, 17\rangle$

Thus, we have the following theorem.

THEOREM 19.14 N is cubic 2-independent Hamiltonian-connected but not super 3^*-connected.

19.5.4 Example of a Cubic 1-Fault-Tolerant Hamiltonian Graph That Is Hamiltonian-Connected but Not Cubic 2-Independent Hamiltonian-Connected or Super 3*-Connected

We prove that the graph M in Figure 19.20a is 1-fault-tolerant Hamiltonian, and Hamiltonian-connected but not 2-independent Hamiltonian-connected or super 3^*-connected.

Obviously, M is obtained from the graph M_1, in Figure 19.20b, by a sequence of 3-joins with K_4. It is proved that M_1 is 1-fault-tolerant Hamiltonian [194]. By Theorem 12.1, M is 1-fault-tolerant Hamiltonian. Moreover, M is also obtained from

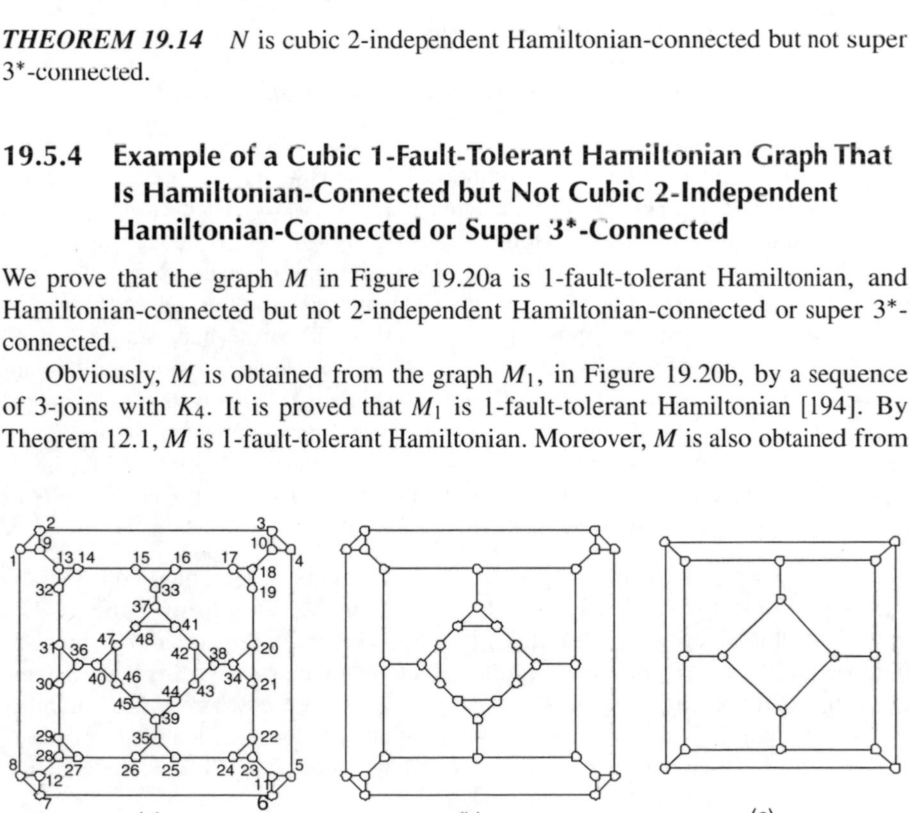

FIGURE 19.20 The graphs (a) M, (b) M_1, and (c) M_2.

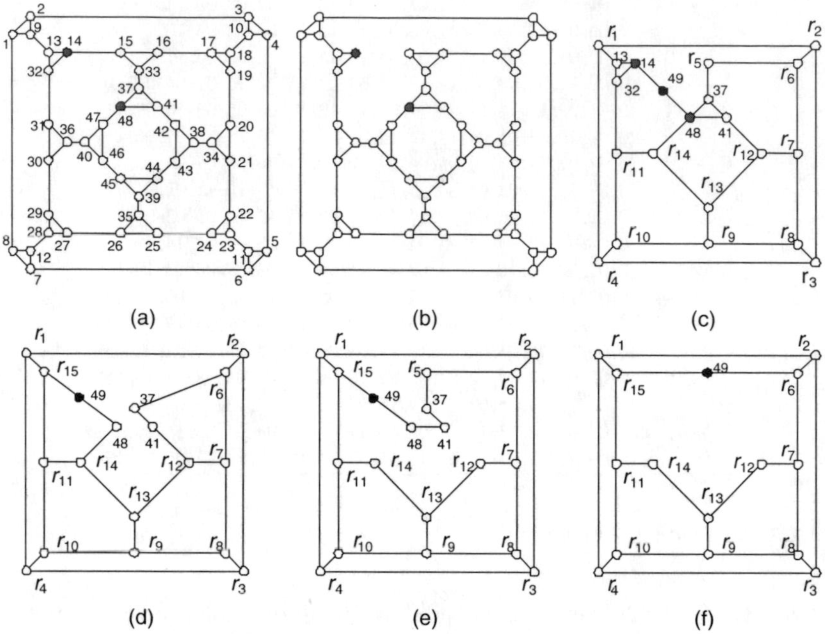

FIGURE 19.21 The graphs (a) M, (b) M_3, (c) M_4, (d) M_5, (e) M_6, and (f) M_7.

the graph M_2, in Figure 19.20c, by a sequence of 3-joins with K_4. It is easy to see that M_2 is isomorphic to the generalized Petersen graph $P(8,2)$. By Theorem 15.5, $P(8,2)$ is not 3*-connected. Repeatedly using Theorem 15.1, M is not 3*-connected.

By brute force, we prove that M is Hamiltonian-connected. Now, we want to show that M is not cubic 2-independent Hamiltonian-connected. Assume that $P = \langle v_1 = 14, v_2, \ldots, v_{48} = 48\rangle$ is any Hamiltonian path of M between 14 and 48 in M. We claim that v_2 must be 15. Then M is not 2-independent Hamiltonian-connected once our claim is proved. To prove this claim, it is equivalent to show that the graph M_3, shown in Figure 19.21b, is not Hamiltonian. Repeatedly using Theorem 14.9, it is sufficient to show that the graph M_4, shown in Figure 19.21c, is not Hamiltonian. Suppose there exists a Hamiltonian cycle H in M_4. We can write H as $H = \langle 49, 48, v_x, \ldots, v_z, 49\rangle$. Obviously, v_x is r_{14}, or 41.

Case 1. $v_x = r_{14}$. Then neither $(48,37)$ nor $(48,41)$ is in H. Thus, both $(37,41)$ and $(41,r_{12})$ are in H. This implies that the graph M_4, shown in Figure 19.21c, has a Hamiltonian cycle C'. And $deg_{M_4}(r_5) = 2$. Therefore, the graph M_5, shown in Figure 19.21d, has a Hamiltonian cycle C. Note that the graph M_5 is planar. By the Grinberg condition, $2(f'_4 - f''_4) + 3(f'_5 - f''_5) + 9(f'_{11} - f''_{11}) = 0$, where f'_i is the number of faces of length i inside C and f''_i is the number of faces of length i outside C for $i = 4, 5, 11$. Obviously, $2(f'_4 - f''_4) = 0 \pmod 3$. Since $|f'_4 - f''_4| = 1$, the equation cannot hold. We get a contradiction.

Case 2. $v_x = 41$. Neither $(48, r_{14})$ nor $(48,37)$ is in H. Thus, both $(48,41)$ and $(41,37)$ are in H. This implies that the graph M_6, shown in Figure 19.21e, has a

Hamiltonian cycle C'. Obviously, $deg_{M_6}(48) = deg_{M_6}(41) = deg_{M_6}(37) = deg_{M_6}(r_5) = 2$. Therefore, the graph M_7, shown in Figure 19.21f, has a Hamiltonian cycle C. Since the graph M_7 is planar, by the Grinberg condition, $2(f_4' - f_4'') + 3(f_5' - f_5'') + 6(f_8' - f_8'') = 0$, where f_i' is the number of faces of length i inside C and f_i'' is the number of faces of length i outside C for $i = 4, 5, 8$. Obviously, $2(f_4' - f_4'') = 0 \pmod 3$. Since $|f_4' - f_4''| = 1$, the equation cannot hold. We get a contradiction.

From the previous discussion, we have the following theorem.

THEOREM 19.15 M is cubic 1-fault-tolerant Hamiltonian, Hamiltonian-connected, but not cubic 2-independent Hamiltonian-connected or super 3*-connected.

20 Topological Properties of Butterfly Graphs

20.1 INTRODUCTION

The *wrapped around butterfly graph* $BF(n)$, introduced in Chapter 2, is an important family of interconnection networks, due to its nice topological properties, and thus has been widely discussed in Refs 24,50,132,185,220,300,303, and 343. However, $BF(n)$ is bipartite if and only if n is even. For this reason, it is difficult to discuss the corresponding topological properties in the previous chapters. Instead, we devote a chapter here to these properties.

Let $\mathbb{Z}_n = \{0, 1, \ldots, n-1\}$ denote the set of integers modulo n. The n-dimensional binary *wrapped butterfly graph* (or *butterfly graph* for short), denoted by $BF(n)$, is a graph with $\mathbb{Z}_n \times \mathbb{Z}_2^n$ as vertex set. Each vertex is labeled by a two-tuple $\langle \ell, a_0 \ldots a_{n-1} \rangle$ with a level $\ell \in \mathbb{Z}_n$ and an n-bit binary string $a_0 \ldots a_{n-1} \in \mathbb{Z}_2^n$. A level-$\ell$ vertex $\langle \ell, a_0 \ldots a_\ell \ldots a_{n-1} \rangle$ is adjacent to two vertices, $\langle (\ell+1)_{\mathbf{mod}\,n}, a_0 \ldots a_\ell \ldots a_{n-1} \rangle$ and $\langle (\ell-1)_{\mathbf{mod}\,n}, a_0 \ldots a_{\ell-1} \ldots a_{n-1} \rangle$, by *straight edges*, and is adjacent to another two vertices, $\langle (\ell+1)_{\mathbf{mod}\,n}, a_0 \ldots a_{\ell-1} \overline{a}_\ell a_{\ell+1} \ldots a_{n-1} \rangle$ and $\langle (\ell-1)_{\mathbf{mod}\,n}, a_0 \ldots a_{\ell-2} \overline{a}_{\ell-1} a_\ell \ldots a_{n-1} \rangle$, by *cross edges*. More formally, the edges of $BF(n)$ can be defined in terms of four generators g, g^{-1}, f, and f^{-1} as follows:

$$
\begin{aligned}
g(\langle \ell, a_0 \ldots a_\ell \ldots a_{n-1} \rangle) &= \langle (\ell+1)_{\mathbf{mod}\,n}, a_0 \ldots a_\ell \ldots a_{n-1} \rangle \\
f(\langle \ell, a_0 \ldots a_\ell \ldots a_{n-1} \rangle) &= \langle (\ell+1)_{\mathbf{mod}\,n}, a_0 \ldots a_{\ell-1} \overline{a}_\ell a_{\ell+1} \ldots a_{n-1} \rangle \\
g^{-1}(\langle \ell, a_0 \ldots a_\ell \ldots a_{n-1} \rangle) &= \langle (\ell-1)_{\mathbf{mod}\,n}, a_0 \ldots a_\ell \ldots a_{n-1} \rangle \\
f^{-1}(\langle \ell, a_0 \ldots a_{\ell-1} \ldots a_{n-1} \rangle) &= \langle (\ell-1)_{\mathbf{mod}\,n}, a_0 a_1 \ldots a_{\ell-2} \overline{a}_{\ell-1} a_\ell \ldots a_{n-1} \rangle
\end{aligned}
$$

where $\overline{a}_\ell \equiv a_\ell + 1 \ (\mathbf{mod}\ 2)$. Hence, the *g-edges*, $(u, g(u))$ or $(u, g^{-1}(u))$, and the *f-edges*, $(u, f(u))$ or $(u, f^{-1}(u))$ for any $u \in V(BF(n))$, represent the straight edges and cross edges, respectively. Consequently, we have Lemma 20.1.

LEMMA 20.1 $f^{-1}(g(u)) = g^{-1}(f(u))$ for any vertex u in $BF(n)$.

A level-ℓ edge of $BF(n)$ is an edge that joins a level-ℓ vertex to a level-$(\ell+1)_{\mathbf{mod}\,n}$ vertex. To avoid the degenerate case, we assume $n \geq 3$ in what follows. So $BF(n)$ is 4-regular. Moreover, $BF(n)$ is bipartite if and only if n is even. Figure 20.1a depicts the structure of $BF(3)$, and Figure 20.1b is the isomorphic structure of $BF(3)$ with level-0 replication for easy visualization.

For any $\ell \in \mathbb{Z}_n$ and $i \in \mathbb{Z}_2$, we use $BF_\ell^i(n)$ to denote the subgraph of $BF(n)$ induced by $\{\langle h, a_0 \ldots a_{n-1} \rangle \mid h \in \mathbb{Z}_n, a_\ell = i\}$. Obviously, $BF_{\ell_1}^i(n)$ is isomorphic to $BF_{\ell_2}^j(n)$ for

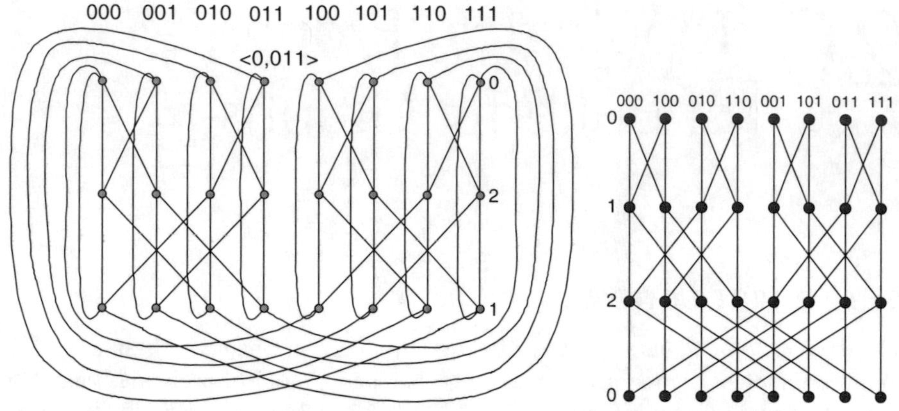

(a) (b)

FIGURE 20.1 (a) The structure of $BF(3)$ and (b) $BF(3)$ with level-0 replicated to ease visualization.

any $i,j \in \mathbb{Z}_2$ and $\ell_1, \ell_2 \in \mathbb{Z}_n$. Moreover, $\{BF_\ell^i(n) | i \in \mathbb{Z}_2\}$ forms a partition of $BF(n)$. With such observation, Wong [343] proposed a *stretching* operation to obtain $BF_\ell^i(n)$ from $BF(n-1)$. More precisely, the stretching operation can be described as follows. Assume that $\ell \in \mathbb{Z}_n$ and $i \in \mathbb{Z}_2$. Let $\Im_n$ be the set of all subgraphs of $BF(n)$ and let $G \in \Im_n$. We define the following subsets of $V(BF(n+1))$ and $E(BF(n+1))$:

$$V_1 = \{\langle h, a_0 \ldots a_{\ell-1} i a_\ell \ldots a_{n-1}\rangle | 0 \le h < \ell, \langle h, a_0 \ldots a_{\ell-1} a_\ell \ldots a_{n-1}\rangle \in V(G)\}$$

$$V_2 = \{\langle h+1, a_0 \ldots a_{\ell-1} i a_\ell \ldots a_{n-1}\rangle | \ell < h \le n-1, \langle h, a_0 \ldots a_{\ell-1} a_\ell \ldots a_{n-1}\rangle \in V(G)\}$$

$$V_3 = \{\langle \ell, a_0 \ldots a_{\ell-1} i a_\ell \ldots a_{n-1}\rangle | \langle \ell, a_0 \ldots a_{\ell-1} a_\ell \ldots a_{n-1}\rangle \text{ is incident with a level-}(\ell-1)_{\mathbf{mod}\, n} \text{ edge in } G\}$$

$$V_4 = \{\langle \ell+1, a_0 \ldots a_{\ell-1} i a_\ell \ldots a_{n-1}\rangle | \langle \ell, a_0 \ldots a_{\ell-1} a_\ell \ldots a_{n-1}\rangle \text{ is incident with a level-}\ell \text{ edge in } G\}$$

$$E_1 = \{(\langle h, a_0 \ldots a_{\ell-1} i a_\ell \ldots a_{n-1}\rangle, \langle h+1, b_0 \ldots b_{\ell-1} i b_\ell \ldots b_{n-1}\rangle) \,|\, 0 \le h < \ell, (\langle h, a_0 \ldots a_{\ell-1} a_\ell \ldots a_{n-1}\rangle, \langle h+1, b_0 \ldots b_{\ell-1} b_\ell \ldots b_{n-1}\rangle) \in E(G)\}$$

$$E_2 = \{(\langle h+1, a_0 \ldots a_{\ell-1} i a_\ell \ldots a_{n-1}\rangle, \langle (h+2)_{\mathbf{mod}\,(n+1)}, b_0 \ldots b_{\ell-1} i b_\ell \ldots b_{n-1}\rangle) \,|\, \ell \le h \le n-1, (\langle h, a_0 \ldots a_{\ell-1} a_\ell \ldots a_{n-1}\rangle, \langle (h+1)_{\mathbf{mod}\, n} b_0 \ldots b_{\ell-1} b_\ell \ldots b_{n-1}\rangle) \in E(G)\}$$

$$E_3 = \{(\langle \ell, a_0 \ldots a_{\ell-1} i a_\ell \ldots a_{n-1}\rangle, \langle \ell+1, a_0 \ldots a_{\ell-1} i a_\ell \ldots a_{n-1}\rangle) | \langle \ell, a_0 \ldots a_{\ell-1} a_\ell \ldots a_{n-1}\rangle \text{ is incident with at least one level-}(\ell-1)_{\mathbf{mod}\, n} \text{ edge and at least one level-}\ell \text{ edge in } G\}$$

Then we define the function γ_ℓ^i: $\cup_{n\ge 3}\Im_n \to \cup_{n \ge 4}\Im_n$ by assigning $\gamma_\ell^i(G)$ as the graph with the vertex set $V_1 \cup V_2 \cup V_3 \cup V_4$ and the edge set $E_1 \cup E_2 \cup E_3$. One may find that γ_ℓ^i is well-defined and one-to-one. Furthermore, $\gamma_\ell^i(G) \in \Im_{n+1}$ if $G \in \Im_n$. In

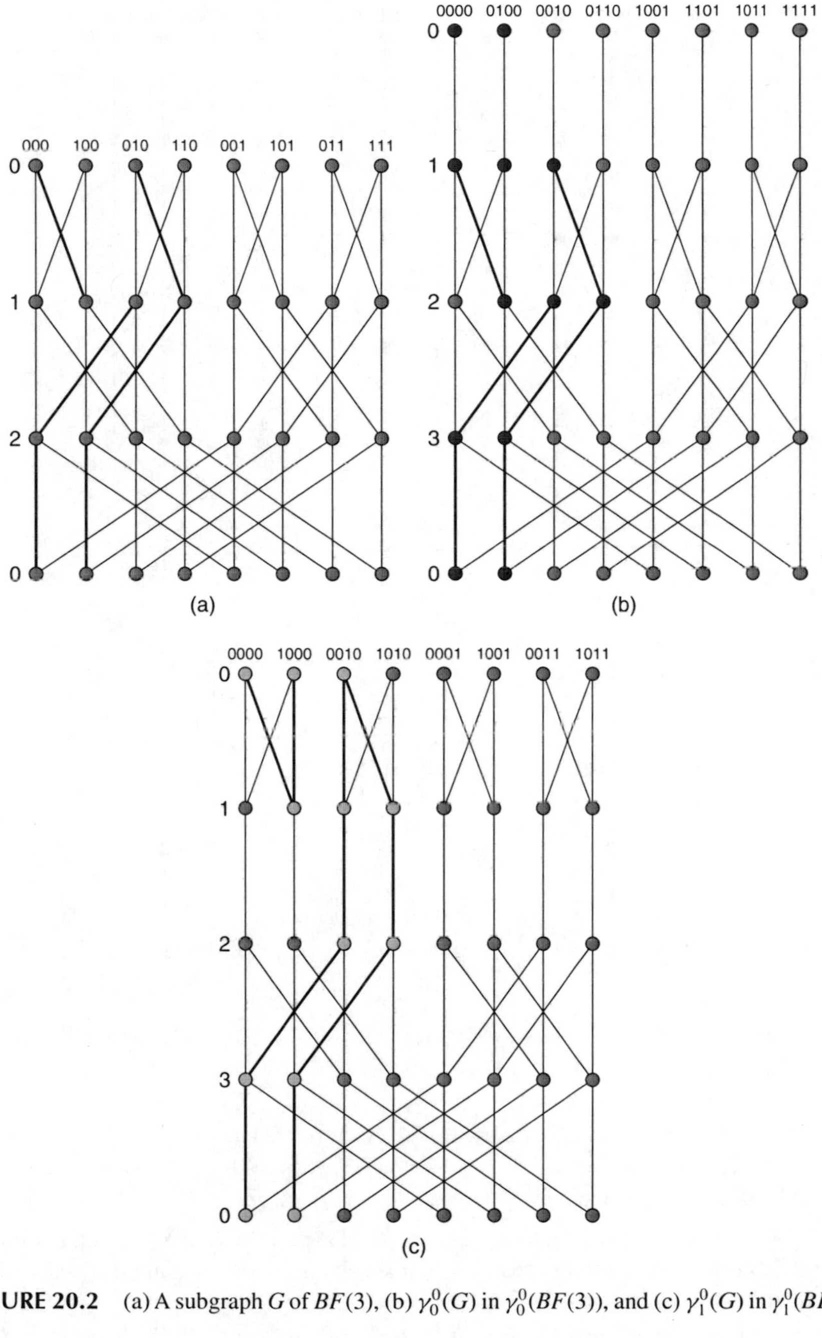

FIGURE 20.2 (a) A subgraph G of $BF(3)$, (b) $\gamma_0^0(G)$ in $\gamma_0^0(BF(3))$, and (c) $\gamma_1^0(G)$ in $\gamma_1^0(BF(3)$.

particular, $\gamma_\ell^i(BF(n)) = BF_\ell^i(n+1)$. Moreover, $\gamma_\ell^i(P)$ is a path in $BF(n+1)$ if P is a path in $BF(n)$. In Figure 20.2, we illustrate a subgraph G of $BF(3)$, $\gamma_0^0(G)$ in $\gamma_0^0(BF(3))$, and $\gamma_1^0(G)$ in $\gamma_1^0(BF(3))$. Obviously, $\gamma_{\ell_1}^i(BF(n))$ is isomorphic to $\gamma_{\ell_2}^j(BF(n))$ for any $\ell_1, \ell_2 \in \mathbb{Z}_n$, $i, j \in \mathbb{Z}_2$.

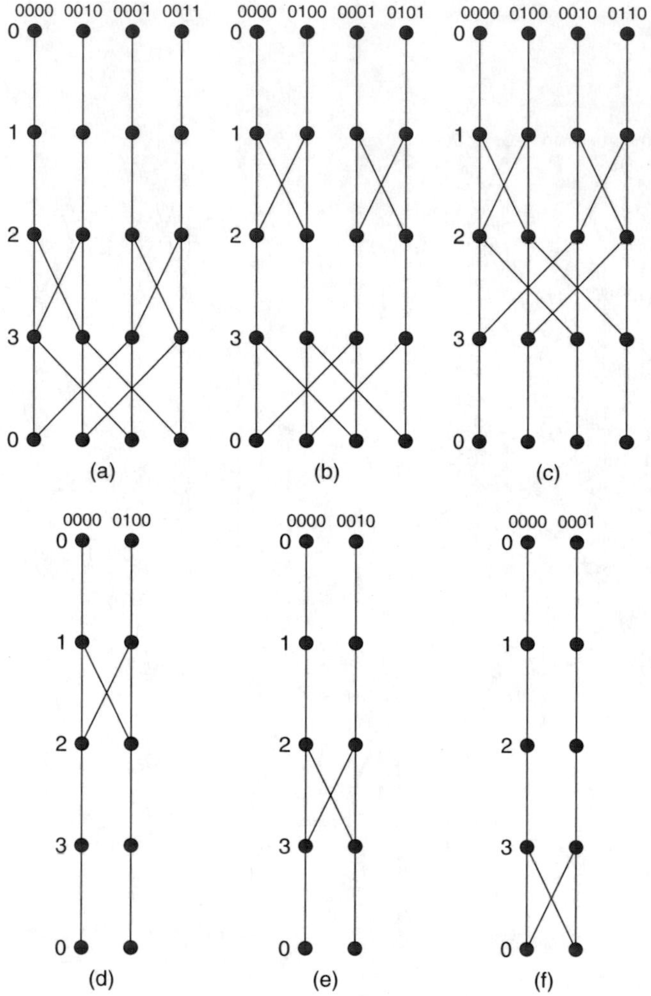

FIGURE 20.3 (a) $BF^{0,0}_{0,1}(4)$, (b) $BF^{0,0}_{0,2}(4)$, (c) $BF^{0,0}_{0,3}(4)$, (d) $BF^{0,0,0}_{0,1,3}(4)$, (e) $BF^{0,0,0}_{0,2,3}(4)$, and (f) $BF^{0,0,0}_{0,1,2}(4)$.

In fact, $BF(n)$ can be further partitioned. Assume that $1 \le m \le n$, $i_1, \ldots, i_m \in \mathbb{Z}_2$, and $\ell_1, \ldots, \ell_m \in \mathbb{Z}_n$ such that $\ell_1 < \cdots < \ell_m$. We use $BF^{i_1,\ldots,i_m}_{\ell_1,\ldots,\ell_m}(n)$ to denote the subgraph of $BF(n)$ induced by $\{\langle h, a_0 \ldots a_{n-1}\rangle | h \in \mathbb{Z}_n, a_{\ell_j} = i_j \text{ for } 1 \le j \le m\}$. So, $\{BF^{i_1,\ldots,i_m}_{\ell_1,\ldots,\ell_m}(n) | i_1, \ldots, i_m \in \mathbb{Z}_2, \ell_1, \ldots, \ell_m \in \mathbb{Z}_n, \ell_1 < \cdots < \ell_m\}$ forms a partition of $BF(n)$. To avoid the complicated case caused by modular arithmetic, we restrict our attention to $1 \le m \le n-1$, $0 \le \ell_1 < \cdots < \ell_m$, and $\ell_j < n - m + j - 1$ for each $1 \le j \le m$. Figure 20.3 illustrates $BF^{0,0}_{0,1}(4)$, $BF^{0,0}_{0,2}(4)$, $BF^{0,0}_{0,3}(4)$, $BF^{0,0,0}_{0,1,3}(4)$, $BF^{0,0,0}_{0,2,3}(4)$, and $BF^{0,0,0}_{0,1,2}(4)$. One can find that $BF^{0,0}_{0,1}(4)$ is isomorphic to $BF^{0,0}_{0,3}(4)$. Moreover, $BF^{0,0,0}_{0,1,3}(4)$, $BF^{0,0,0}_{0,2,3}(4)$, and $BF^{0,0,0}_{0,1,2}(4)$ are also isomorphic. However, $BF^{0,0}_{0,1}(4)$ is not isomorphic to $BF^{0,0}_{0,2}(4)$.

The following lemmas can be easily derived according to the stretching operation.

LEMMA 20.2 Assume that $n \geq 3$ and $i,j,k \in \mathbb{Z}_2$. Then $BF^{i,j}_{0,1}(n)$ is isomorphic to $BF^{i,j}_{0,n-1}(n)$; $BF^{i,j,k}_{0,1,2}(n)$, $BF^{i,j,k}_{0,1,n-1}(n)$, and $BF^{i,j,k}_{0,n-2,n-1}(n)$ are isomorphic.

LEMMA 20.3 Suppose that $i_1,\ldots,i_m \in \mathbb{Z}_2$ and $\ell_1,\ldots,\ell_m$ are integers such that $0 \leq \ell_1 < \cdots < \ell_m$ and $\ell_j < n-m+j-1$ for each $1 \leq j \leq m$. Then

$$BF^{i_1,\ldots,i_m}_{\ell_1,\ldots,\ell_m}(n) = \begin{cases} \gamma^{i_m}_{\ell_m} \circ \gamma^{i_{m-1}}_{\ell_{m-1}} \circ \ldots \circ \gamma^{i_3}_{\ell_3}\left(BF^{i_1,i_2}_{\ell_1,\ell_2}(3)\right) & \text{if } m = n-1 \\ \gamma^{i_m}_{\ell_m} \circ \gamma^{i_{m-1}}_{\ell_{m-1}} \circ \ldots \circ \gamma^{i_2}_{\ell_2}\left(BF^{i_1}_{\ell_1}(3)\right) & \text{if } m = n-2 \\ \gamma^{i_m}_{\ell_m} \circ \gamma^{i_{m-1}}_{\ell_{m-1}} \circ \ldots \circ \gamma^{i_1}_{\ell_1}(BF(n-m)) & \text{otherwise} \end{cases}$$

LEMMA 20.4 Let $n \geq 3$. Assume that $0 \leq \ell \leq n-1$, $\Gamma \in \Im_n$, and G is a connected spanning subgraph of Γ. For any $i,j \in \mathbb{Z}_2$, let

$$F_0 = \{\langle \ell, a_0 \ldots a_{n-1}\rangle = v \mid v \in V(G) \text{ is not incident with any level-}(\ell-1)_{\bmod n}, \text{ edge in } G\}$$

$$F_1 = \{\langle \ell, a_0 \ldots a_{n-1}\rangle = v \mid v \in V(G) \text{ is not incident with any level-}\ell \text{ edge in } G\}$$

$$\overline{F_0} = \bigcup_{\langle \ell, a_0 \ldots a_{\ell-1} a_\ell \ldots a_{n-1}\rangle \in F_0} \{\langle \ell, a_0 \ldots a_{\ell-1} ij a_\ell \ldots a_{n-1}\rangle, \langle \ell+1, a_0 \ldots a_{\ell-1} ij a_\ell \ldots a_{n-1}\rangle\}$$

$$\overline{F_1} = \bigcup_{\langle \ell, a_0 \ldots a_{\ell-1} a_\ell \ldots a_{n-1}\rangle \in F_1} \{\langle \ell+1, a_0 \ldots a_{\ell-1} ij a_\ell \ldots a_{n-1}\rangle, \langle \ell+2, a_0 \ldots a_{\ell-1} ij a_\ell \ldots a_{n-1}\rangle\}$$

$$X_0 = \bigcup_{\langle \ell, a_0 \ldots a_{\ell-1} a_\ell \ldots a_{n-1}\rangle \in F_0} \{(\langle \ell, a_0 \ldots a_{\ell-1} ij a_\ell \ldots a_{n-1}\rangle, \langle \ell+1, a_0 \ldots a_{\ell-1} ij a_\ell \ldots a_{n-1}\rangle)\}$$

$$X_1 = \bigcup_{\langle \ell, a_0 \ldots a_{\ell-1} a_\ell \ldots a_{n-1}\rangle \in F_1} \{(\langle \ell+1, a_0 \ldots a_{\ell-1} ij a_\ell \ldots a_{n-1}\rangle, \langle \ell+2, a_0 \ldots a_{\ell-1} ij a_\ell \ldots a_{n-1}\rangle)\}$$

$$M_0 = \bigcup_{\langle \ell, a_0 \ldots a_{n-1}\rangle \in G-(F_0 \cup F_1)} \{(\langle \ell, a_0 \ldots a_{\ell-1} ij a_\ell \ldots a_{n-1}\rangle, \langle \ell+1, a_0 \ldots a_{\ell-1} ij a_\ell \ldots a_{n-1}\rangle)\}$$

$$M_1 = \bigcup_{\langle \ell, a_0 \ldots a_{n-1}\rangle \in G-(F_0 \cup F_1)} \{(\langle \ell+1, a_0 \ldots a_{\ell-1} ij a_\ell \ldots a_{n-1}\rangle, \langle \ell+2, a_0 \ldots a_{\ell-1} ij a_\ell \ldots a_{n-1}\rangle)\}$$

Then $F_0 \cap F_1 = \emptyset$ and $\overline{F_0} \cap \overline{F_1} = \emptyset$. Thus, $\overline{F_0} \cup \overline{F_1} = V(\gamma^j_{\ell+1} \circ \gamma^i_\ell(\Gamma)) - V(\gamma^j_{\ell+1} \circ \gamma^i_\ell(G))$ can be represented as $\cup_{k=1}^m \{u_k, g(u_k) | u_k \in V(\gamma^j_{\ell+1} \circ \gamma^i_\ell(\Gamma))\}$ with some $m \geq 1$ such that

$\{u_k | 1 \le k \le m\} \cap \{g(u_k) | 1 \le k \le m\} = \emptyset$. Moreover, $M_0 \cup M_1 \subseteq E(\gamma^j_{\ell+1} \circ \gamma^i_\ell(G))$ and $X_0 \cup X_1 = \cup_{k=1}^m \{(u_k, g(u_k)) | u_k \in V(\gamma^j_{\ell+1} \circ \gamma^i_\ell(\Gamma))\}$ is a set of edges with no shared endpoints.

20.2 CYCLE EMBEDDING IN FAULTY BUTTERFLY GRAPHS

Embedding a cycle into a wrapped butterfly graph has attracted the efforts of many researchers in recent years [20,184,321,343]. Again, we would like to know the fault-tolerant Hamiltonicity of the butterfly graphs. Tsai [303] proved that the faulty wrapped butterfly graph contains a cycle of length $n2^n - 2$ if it has one vertex fault and one edge fault. Moreover, when n is an odd integer, we prove that the wrapped butterfly graph contains a Hamiltonian cycle if it has at most two faults and at least one of them is a vertex fault.

To get this goal, we need some properties concerning $BF(n)$. Let u be any vertex of $BF(n)$. We observe that $g^n(u) = u$. Moreover, $\langle u, g(u), g^2(u), \dots, g^n(u)\rangle$ forms a simple cycle C^u_g of length n. We call such cycle of $BF(n)$ a *g-cycle* at u. Hence, $C^v_g \simeq C^u_g$ if and only if $v \in V(C^u_g)$. As a result, all g-cycles form a partition of the straight edges of $BF(n)$. Meanwhile, any f-edge joins vertices from two different g-cycles. It can be seen that $(u,f(u))$ joins vertices from C^u_g and $C^{f(u)}_g$. Lemmas 20.5 and 20.6 were proved in Ref. 320.

LEMMA 20.5 **[320]** $(g(u), g^{-1}(f(u)))$ is an f-edge joining vertices of C^u_g and $C^{f(u)}_g$. Moreover, the path $\langle u, f(u), g^{-1}(f(u)), g(u), u\rangle$ forms a cycle of length 4.

Any C^u_g contains exactly one vertex at each level. In particular, C^u_g contains exactly one vertex at level 0, say $\langle 0, a_0 a_1 \dots a_{n-1}\rangle$. We use $C_g^{(a_0 a_1 \dots a_{n-1})}$ as the name for C^u_g. Now, we form a new graph $BF(n)^G$ with all the g-cycles of $BF(n)$ as vertices, where two different g-cycles are joined with an edge if and only if there exists at least one f-edge joining them. The vertex of $BF(n)^G$ corresponding to C^u_g is denoted by $\overline{C}^u_g$. We recall the definition of the hypercube as follows. An n-dimensional hypercube (abbreviated as n-cube) consists of 2^n vertices, which are labeled with the 2^n binary numbers from 0 to $2^n - 1$. Two vertices are connected by an edge if and only if their labels differ by exactly one bit.

LEMMA 20.6 **[320]** $BF(n)^G$ is isomorphic to an n-dimensional hypercube. Moreover, the set of vertices which are adjacent to $\overline{C}_g^{(a_0 a_1 \dots a_{n-1})}$ is $\{\overline{C}_g^{(\bar{a}_0 a_1 \dots a_{n-1})}, \overline{C}_g^{(a_0 \bar{a}_1 \dots a_{n-1})}, \dots, \overline{C}_g^{(a_0 a_1 \dots \bar{a}_{n-1})}\}$.

Let $h = (\overline{C}^u_g, \overline{C}^v_g)$ be any edge of $BF(n)^G$. We use $X(h)$ to denote the set of edges in $BF(n)$ joining vertices from C^u_g and C^v_g. Using standard counting techniques, we have the following two corollaries.

COROLLARY 20.1 **[184]** *If* $(u,f(u)) \in X(h)$ *then* $(g(u), f^{-1}(g(u))) \in X(h)$ *and* $|X(h)| = 2$. *Moreover,* $\{u, f(u), g(u), f^{-1}(g(u))\}$ *induces a 4-cycle in* $BF(n)$.

COROLLARY 20.2 **[320]** *There is a unique cycle C such that edges of* $BF(n)$ *joining vertices between* C_g^u *and C are exactly* $(u, f(u))$ *and* $(g(u), f^{-1}(g(u)))$ *and in that case, C is isomorphic to* $C_g^{f(u)}$.

According to Corollaries 20.1 and 20.2, any edge $h = (\overline{C}_g^u, \overline{C}_g^v)$ in $BF(n)^G$ induces a unique 4-cycle in $BF(n)$, with two f-edges and two g-edges. We use $X_f(C_g^u, C_g^v)$ to denote the set of f-edges in this 4-cycle, and $X_g(C_g^u, C_g^v)$ to denote the set of g-edges in this cycle.

LEMMA 20.7 Let T be any subtree of $BF(n)^G$ and let C_g^T be the subgraph of $BF(n)$ generated by

$$\left(\bigcup_{\overline{C}_g^u \in V(T)} E\left(C_g^u\right) \cup \bigcup_{\left(\overline{C}_g^u, \overline{C}_g^v\right) \in E(T)} X_f\left(C_g^u, C_g^v\right) \right) - \bigcup_{\left(\overline{C}_g^u, \overline{C}_g^v\right) \in E(T)} X_g\left(C_g^u, C_g^v\right)$$

Then C_g^T is a cycle of length $n \times |V(T)|$.

Proof: Let T be a subtree of $BF(n)^G$ and let $(\overline{C}_g^u, \overline{C}_g^v)$ be an edge of T. Hence, two cycles C_g^u and C_g^v in $BF(n)$ are joined by two f-edges $X_f(C_g^u, C_g^v)$. It is easy to see that $E(C_g^u) \bigcup E(C_g^v) \bigcup X_f(C_g^u, C_g^v) - X_g(C_g^u, C_g^v)$ forms a cycle of length $2n$ in $BF(n)$. Therefore, the cycle C_g^T of $BF(n)$ can be generated by

$$\left(\bigcup_{C_g^u \in V(T)} E\left(C_g^u\right) \cup \bigcup_{\left(\overline{C}_g^u, \overline{C}_g^v\right) \in E(T)} X_f\left(C_g^u, C_g^v\right) \right) - \bigcup_{\left(\overline{C}_g^u, \overline{C}_g^v\right) \in E(T)} X_g\left(C_g^u, C_g^v\right)$$

At the same time, the length of C_g^T is $n \times |V(T)|$. □

In Figure 20.4a, we have another layout of $BF(3)$. The graph $BF(3)^G$ is shown in Figure 20.4b. For example, $X_f(C_g^{(000)}, C_g^{(001)}) = \{(\langle 0, 000\rangle, \langle 2, 001\rangle), (\langle 2, 000\rangle, \langle 0, 001\rangle)\}$ and $X_g(C_g^{(000)}, C_g^{(001)}) = \{(\langle 0, 000\rangle, \langle 2, 000\rangle), (\langle 0, 001\rangle, \langle 2, 001\rangle)\}$. Let T be the tree indicated by bold lines in Figure 20.4b. The corresponding C_g^T is indicated by bold lines in Figure 20.4a.

Let $u = \langle k, a_0 a_1 \ldots a_{n-1}\rangle$ be any vertex of $BF(n)$ and let $\tilde{u}$ denote the vertex $\langle k, \overline{a}_0 \overline{a}_1 \ldots \overline{a}_{n-1}\rangle$. One can see that $f^n(u) = \tilde{u}$ and $f^{2n}(u) = u$. Moreover, $\langle u, f(u), f^2(u), \ldots, f^{2n}(u)\rangle$ forms a simple cycle of length $2n$, denoted by C_f^u. So all f-cycles form a partition of the cross edges of $BF(n)$. Meanwhile, any g-edge joins vertices from two different f-cycles. And then $(u, g(u))$ joins vertices from C_f^u and $C_f^{g(u)}$. Lemmas 20.8 and 20.9 were proved in Ref. 320.

LEMMA 20.8 **[320]** $(f(u), g^{-1}(f(u)))$, $(\tilde{u}, g(\tilde{u}))$, $(f(\tilde{u}), g^{-1}(f(\tilde{u})))$ are g-edges joining vertices of C_f^u and $C_f^{g(u)}$. Moreover, the paths $\langle u, f(u), g^{-1}(f(u)), g(u), u\rangle$ and $\langle \tilde{u}, f(\tilde{u}), g^{-1}(f(\tilde{u})), g(\tilde{u}), \tilde{u}\rangle$ form two 4-cycles in $BF(n)$.

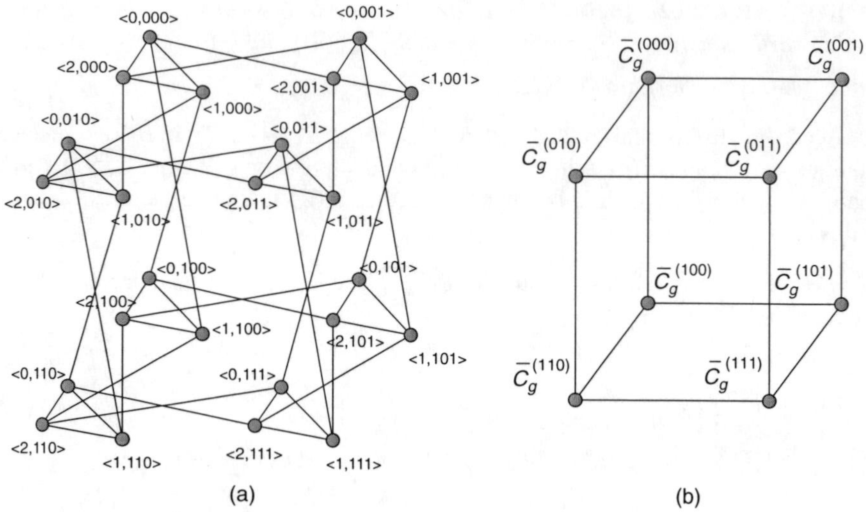

FIGURE 20.4 (a) Another layout of $BF(3)$ and (b) the graph $BF(3)^G$.

Any C_f^u contains exactly two vertices at each level. Suppose that u is one of the vertices in C_f^u at level i. Then the other vertex in C_f^u at level i is $\tilde{u}$. Thus, C_f^u contains exactly one vertex at level 0, say $\langle 0, a_0a_1 \ldots a_{n-1}\rangle$, with $a_{n-1} = 0$. We use $C_f^{(a_0a_1\ldots a_{n-2}0)}$ as the name for C_f^u. Now, we form a new graph $BF(n)^F$ with all the f-cycles of $BF(n)$ as vertices, where two different f-cycles are joined with an edge if and only if there exists a g-edge joining them. The vertex of $BF(n)^F$ corresponding to C_f^u is denoted by $\overline{C}_f^u$. We recall the definition of the folded hypercube as follows. An n-dimensional folded hypercube is basically an n-cube augmented with 2^{n-1} complement edges. Each complement edge connects two vertices whose labels are complements to each other.

LEMMA 20.9 **[320]** $BF(n)^F$ is isomorphic to the $(n-1)$-dimensional folded hypercube. Moreover, the set of vertices that are adjacent to $\overline{C}_f^{(a_0a_1\ldots a_{n-2}0)}$ is $\{\overline{C}_f^{(\bar{a}_0a_1\ldots a_{n-2}0)}, \overline{C}_f^{(a_0\bar{a}_1\ldots a_{n-2}0)}, \ldots, \overline{C}_f^{(a_0a_1\ldots \bar{a}_{n-2}0)}\} \cup \{\overline{C}_f^{(\bar{a}_0\bar{a}_1\ldots \bar{a}_{n-2}0)}\}$.

Let $h = (\overline{C}_f^u, \overline{C}_f^v)$ be any edge of $BF(n)^F$. We use $Y(h)$ to denote the set of edges of $BF(n)$ joining vertices from C_f^u and C_f^v. Using standard counting techniques, we have the following two corollaries.

COROLLARY 20.3 **[185]** *If* $(u, g(u)) \in Y(h)$ *then* $(f(u), g^{-1}(f(u)))$, $(\tilde{u}, g(\tilde{u}))$, *and* $(f(\tilde{u}), g^{-1}(f(\tilde{u})))$ *are in* $Y(h)$ *and* $|Y(h)| = 4$. *Moreover,* $\{u, f(u), g(u), g^{-1}(f(u))\}$ *and* $\{\tilde{u}, f(\tilde{u}), g(\tilde{u}), g^{-1}(f(\tilde{u}))\}$ *induce two 4-cycles in* $BF(n)$.

COROLLARY 20.4 **[320]** *There is a unique cycle* C *such that edges of* $BF(n)$ *joining vertices between* C_f^u *and* C *are exactly* $(u, g(u))$, $(f(u), g^{-1}(f(u)))$, $(\tilde{u}, g(\tilde{u}))$, *and* $(f(\tilde{u}), g^{-1}(f(\tilde{u})))$, *and in that case,* C *is isomorphic to* $C_f^{g(u)}$.

According to Corollaries 20.3 and 20.4, any edge $h=(\overline{C}_f^u, \overline{C}_f^v)$ induces two 4-cycles in $BF(n)$. Let α be an assignment of $(\overline{C}_f^u, \overline{C}_f^v) \in E(BF(n)^F)$ with one of the 4-cycles that it induced. We use $Y_f^\alpha(C_f^u, C_f^v)$ to denote the set of f-edges induced by $\alpha(h)$ and $Y_g^\alpha(C_f^u, C_f^v)$ to denote the set of g-edges induced by $\alpha(h)$. Hence, $|Y_f^\alpha(C_f^u, C_f^v)| = |Y_g^\alpha(C_f^u, C_f^v)| = 2$.

LEMMA 20.10 Let T be any subtree of $BF(n)^F$ and let $C_f^{T,\alpha}$ be the subgraph of $BF(n)$ generated by

$$\left(\bigcup_{\overline{C}_f^u \in V(T)} E\left(C_f^u\right) \cup \bigcup_{\left(\overline{C}_f^u, \overline{C}_f^v\right) \in E(T)} Y_g^\alpha\left(C_f^u, C_f^v\right) \right) - \bigcup_{\left(\overline{C}_f^u, \overline{C}_f^v\right) \in E(T)} Y_f^\alpha\left(C_f^u, C_f^v\right)$$

Then $C_f^{T,\alpha}$ is a cycle of $BF(n)$ of length $2n \times |V(T)|$.

Proof: Let T be a subtree of $BF(n)^F$ and let $(\overline{C}_f^u, \overline{C}_f^v)$ be an edge of T. Hence, $(\overline{C}_f^u, \overline{C}_f^v)$ induces two 4-cycles in $BF(n)$. Let α be an assignment of $(\overline{C}_f^u, \overline{C}_f^v) \in E(BF(n)^F)$ with one of the 4-cycles it induced. Consequently, two cycles C_f^u and C_f^v are joined by two g-edges in $Y_g^\alpha(C_f^u, C_f^v)$. It is easy to see that $E(C_f^u) \bigcup E(C_f^v) \bigcup Y_g^\alpha(C_f^u, C_f^v) - Y_f^\alpha(C_f^u, C_f^v)$ forms a cycle of length $4n$ in $BF(n)$. Therefore, the subgraph $C_f^{T,\alpha}$ of $BF(n)$ generated by

$$\left(\bigcup_{\overline{C}_f^u \in V(T)} E\left(C_f^u\right) \cup \bigcup_{\left(\overline{C}_f^u, \overline{C}_f^v\right) \in E(T)} Y_g^\alpha\left(C_f^u, C_f^v\right) \right) - \bigcup_{\left(\overline{C}_f^u, \overline{C}_f^v\right) \in E(T)} Y_f^\alpha\left(C_f^u, C_f^v\right)$$

is a cycle. Moreover, the length of $C_f^{T,\alpha}$ is $2n \times |V(T)|$. □

In Figure 20.5a, we have another layout of $BF(3)$. The graph $BF(3)^F$ is shown in Figure 20.5b. For example, $Y_g^\alpha(C_f^{(000)}, C_f^{(100)}) = \{(\langle 0,000\rangle, \langle 1,000\rangle), (\langle 0,100\rangle, \langle 1,100\rangle)\}$ and $Y_f^\alpha(C_f^{(000)}, C_f^{(100)}) = \{(\langle 0,000\rangle, \langle 1,100\rangle), (\langle 0,100\rangle, \langle 1,000\rangle)\}$. Let T be the tree indicated by bold lines in Figure 20.5b. The corresponding $C_f^{T,\alpha}$ is indicated by bold lines in Figure 20.5a.

Now, we propose three lemmas. Lemma 20.11 proves that the cycle of length $n2^n - 2$ can be embedded in $BF(n) - F$ for any $n \geq 3$ when F consists of one vertex and one edge. Lemma 20.12 shows that the maximum cycle length is $n2^n - 1$ if n is odd when F consists of one vertex and one edge. Lemma 20.13 verifies that the maximum cycle length is $n2^n - 2$ for odd $n \geq 3$ when F consists of two vertices. To prove these three lemmas, we use three fundamental cycles, denoted by $\mathcal{B}_1$, $\mathcal{B}_2$, and $\mathcal{B}_3(j)$, respectively, to construct a larger cycle in a faulty wrapped butterfly graph.

The cycle $\mathcal{B}_1$ shown in Figure 20.6a is constructed as follows. Let $a_1 = \langle 1, \underbrace{00\ldots0}_{n}\rangle$ and P_1 be the path $\langle a_1, g(a_1), g^2(a_1), \ldots, g^{n-2}(a_1)\rangle$. Hence,

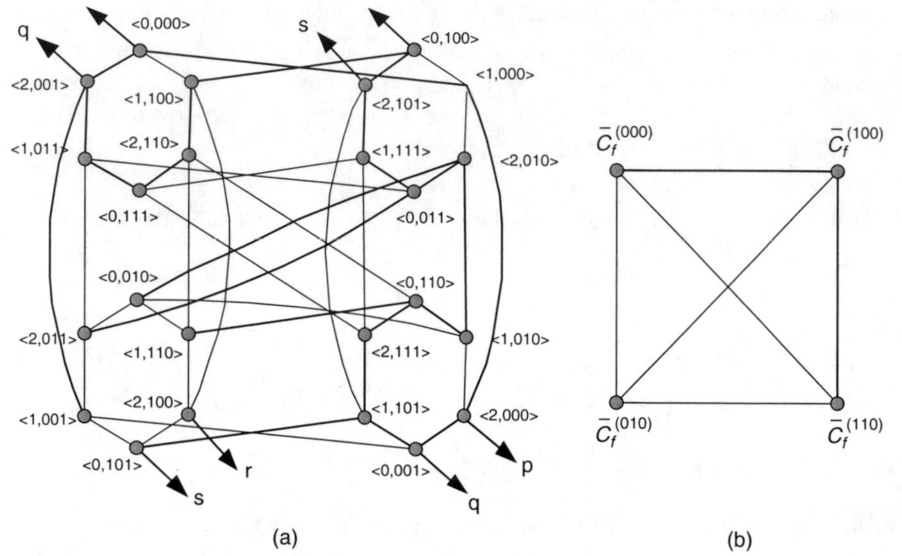

FIGURE 20.5 (a) Another layout of $BF(3)$ and (b) the graph $BF(3)^F$.

$g^{n-2}(a_1) = a_2 = \langle n-1, \underbrace{00\ldots0}_{n}\rangle$, $f(a_2) = \langle 0, \underbrace{00\ldots0}_{n-1}1\rangle = a_3$, and $f(a_3) = \langle 1, 1\underbrace{00\ldots0}_{n-2}1\rangle = a_4$. Let P_2 be the path $\langle a_4, g(a_4), g^2(a_4), \ldots, g^{n-1}(a_4)\rangle$. Consequently, $g^{n-1}(a_4) = a_5 = \langle 0, 1\underbrace{00\ldots0}_{n-2}1\rangle$ and $f^{-1}(a_5) = \langle n-1, 1\underbrace{00\ldots0}_{n-1}\rangle = a_6$. Let P_3 be the path $\langle a_6, g^{-1}(a_6), g^{-2}(a_6), \ldots, g^{-(n-1)}(a_6)\rangle$. Thus, $g^{-(n-1)}(a_6) = a_7 = \langle 0, 1\underbrace{00\ldots0}_{n-1}\rangle$ and $f(a_7) = a_1$. Let $\mathcal{B}_1$ be $\langle a_1, P_1, a_2, a_3, a_4, P_2, a_5, a_6, P_3, a_7, a_1\rangle$. Obviously, P_1 is a path in $C_g^{a_1}$, P_2 is a path in $C_g^{a_4}$, and P_3 is a path in $C_g^{a_6}$. Since $V(C_g^u) \cap V(C_g^v) = \emptyset$ for $u \neq v \in \{a_1, a_4, a_6\}$ and $n \geq 3$, P_1, P_2, and P_3 are disjoint paths. Therefore, $\mathcal{B}_1$ is a cycle of length $3n$.

Similarly, we can construct the basic cycle $\mathcal{B}_2$ shown in Figure 20.6b as follows. Let $\mathcal{B}_2 = \langle b_1, Q_1, b_2, b_3, Q_2, b_4, b_5, Q_3, b_6, b_7, Q_4, b_8, b_1\rangle$, where

$$b_1 = \left\langle 1, \underbrace{00\ldots0}_{n}\right\rangle, \quad b_2 = \left\langle n-1, \underbrace{00\ldots0}_{n}\right\rangle, \quad b_3 = \left\langle n-2, \underbrace{00\ldots0}_{n-2}10\right\rangle$$

$$b_4 = \left\langle 1, \underbrace{00\ldots0}_{n-2}10\right\rangle, \quad b_5 = \left\langle 0, 1\underbrace{00\ldots0}_{n-3}10\right\rangle, \quad b_6 = \left\langle n-1, 1\underbrace{00\ldots0}_{n-3}10\right\rangle$$

$$b_7 = \left\langle n-2, \underbrace{00\ldots0}_{n-1}\right\rangle, \quad b_8 = \left\langle 0, 1\underbrace{00\ldots0}_{n-1}\right\rangle$$

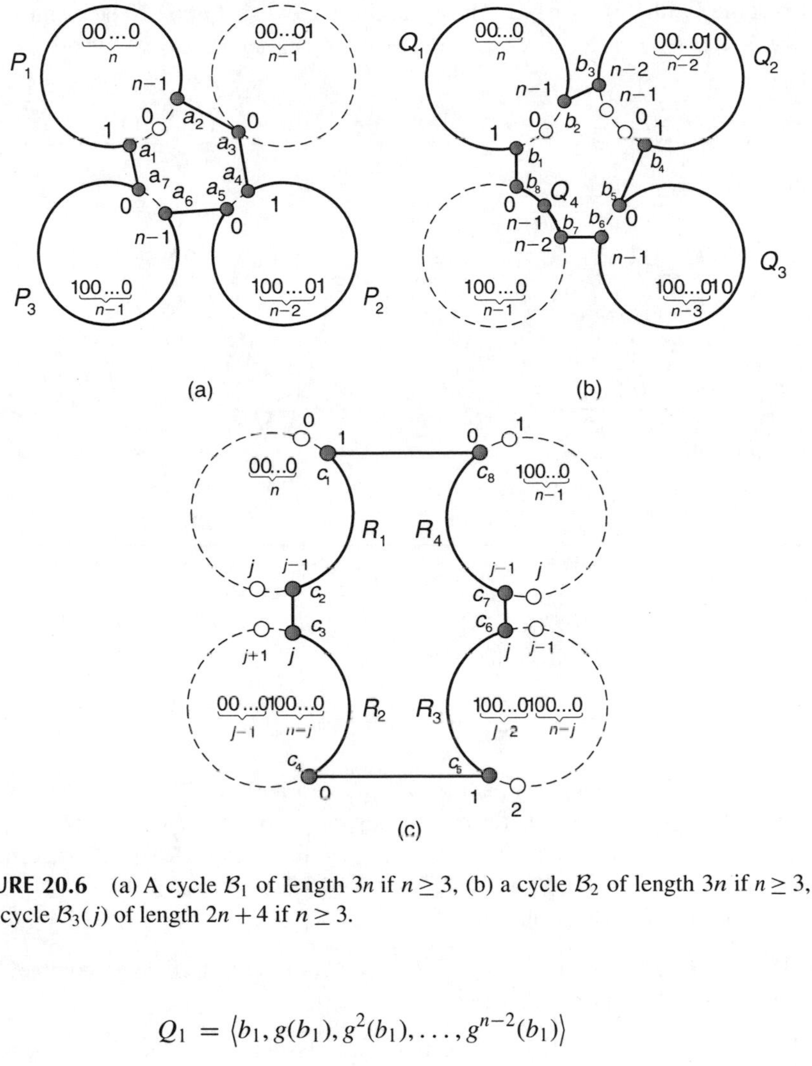

FIGURE 20.6 (a) A cycle $\mathcal{B}_1$ of length $3n$ if $n \geq 3$, (b) a cycle $\mathcal{B}_2$ of length $3n$ if $n \geq 3$, and (c) a cycle $\mathcal{B}_3(j)$ of length $2n+4$ if $n \geq 3$.

$$Q_1 = \langle b_1, g(b_1), g^2(b_1), \ldots, g^{n-2}(b_1) \rangle$$

$$Q_2 = \langle b_3, g^{-1}(b_3), g^{-2}(b_3), \ldots, g^{-(n-3)}(b_3) \rangle$$

$$Q_3 = \langle b_5, g(b_5), g^2(b_5), \ldots, g^{n-1}(b_5) \rangle$$

$$Q_4 = \langle b_7, g(b_7), g^2(b_7) \rangle$$

Q_1 is a path in $C_g^{b_1}$, Q_2 is a path in $C_g^{b_3}$, Q_3 is a path in $C_g^{b_5}$, and Q_4 is a path in $C_g^{b_7}$. Since $V(C_g^u) \cap V(C_g^v) = \emptyset$ for $u \neq v \in \{b_1, b_3, b_5, b_7\}$ and $n \geq 3$, Q_1, Q_2, Q_3, and Q_4 are disjoint paths. Consequently, $\mathcal{B}_2$ is a cycle of length $3n$.

Fix some j with $2 \le j \le n-1$. The cycle $\mathcal{B}_3(j)$ shown in Figure 20.6c is constructed as follows. Let $\mathcal{B}_3(j) = \langle c_1, R_1, c_2, c_3, R_2, c_4, c_5, R_3, c_6, c_7, R_4, c_8, c_1\rangle$, where

$$c_1 = \left\langle 1, \underbrace{00\ldots0}_{n} \right\rangle, \quad c_2 = \left\langle j-1, \underbrace{00\ldots0}_{n} \right\rangle, \quad c_3 = \left\langle j, \underbrace{00\ldots0}_{j-1} 1 \underbrace{00\ldots0}_{n-j} \right\rangle$$

$$c_4 = \left\langle 0, \underbrace{00\ldots0}_{j-1} 1 \underbrace{00\ldots0}_{n-j} \right\rangle, \quad c_5 = \left\langle 1, 1 \underbrace{00\ldots0}_{j-2} 1 \underbrace{00\ldots0}_{n-j} \right\rangle$$

$$c_6 = \left\langle j, 1 \underbrace{00\ldots0}_{j-2} 1 \underbrace{00\ldots0}_{n-j} \right\rangle, \quad c_7 = \left\langle j-1, 1 \underbrace{00\ldots0}_{n-1} \right\rangle, \quad c_8 = \left\langle 0, 1 \underbrace{00\ldots0}_{n-1} \right\rangle$$

$$R_1 = \langle c_1, g(c_1), g^2(c_1), \ldots, g^{j-2}(c_1)\rangle$$

$$R_2 = \langle c_3, g^{-1}(c_3), g^{-2}(c_3), \ldots, g^{-j}(c_3)\rangle$$

$$R_3 = \langle c_5, g^{-1}(c_5), g^{-2}(c_5), \ldots, g^{-(n-j+1)}(c_5)\rangle$$

$$R_4 = \langle c_7, g^{-1}(c_7), g^{-2}(c_7), \ldots, g^{-(n-j+1)}(c_7)\rangle$$

R_1 is a path in $C_g^{c_1}$, R_2 is a path in $C_g^{c_3}$, R_3 is a path in $C_g^{c_5}$, and R_4 is a path in $C_g^{c_7}$. Since $V(C_g^u) \cap V(C_g^v) = \emptyset$ for $u \ne v \in \{c_1, c_3, c_5, c_7\}$ and $n \ge 3$, R_1, R_2, R_3, and R_4 are disjoint paths. Consequently, $\mathcal{B}_3(j)$ is a cycle of length $2n+4$. We even have $b_3 = b_4$ and $c_1 = c_2$ if and only if $n = 3$. We have $V(\mathcal{B}_l) \cap V(\mathcal{B}_k) \ne \emptyset$ for $1 \le l \ne k \le 3$ and $n \ge 4$ because $P_1 = Q_1$ and $P_1 \cap R_1 \ne \emptyset$.

LEMMA 20.11 For any integer $n \ge 3$, $BF(n) - F$ contains a cycle of length $n2^n - 2$ if F consists of one vertex and one edge.

Proof: Since $BF(n)$ is vertex-transitive, we assume that the faulty vertex is $x = \langle 0, 00\ldots0\rangle$ and the faulty edge is e.

Case 1. e is a g-edge. Let $e = (u, g(u))$ for some $u \in V(BF(n))$. Since $n \ge 3$ and $BF(n)^F$ is isomorphic to an $(n-1)$-dimensional folded hypercube, $BF(n)^F - \{\overline{C}_f^x, (\overline{C}_f^u, \overline{C}_f^{g(u)})\}$ is connected. Let T be any spanning tree of $BF(n)^F - \{\overline{C}_f^x, (\overline{C}_f^u, \overline{C}_f^{g(u)})\}$. By Lemma 20.10, $C_f^{T,\alpha}$ for any α forms an $n2^n - 2n$ cycle. Let $S_g = \{(y, g(y)) \mid y = f^{2k}(x); 1 \le k \le n-1\}$, $S_g' = \{(f(y), g^{-1}(f(y))) \mid y = f^{2k}(x); 1 \le k \le n-1\}$, $S_f = \{(y, f(y)) \mid y = f^{2k}(x); 1 \le k \le n-1\}$, and $S_f' = \{(g(y), f^{-1}(g(y))) \mid y = f^{2k}(x); 1 \le k \le n-1\}$. Accordingly, $S_f \subset E(C_f^x)$ and $S_f' \subset E(C_f^{g(x)})$.

Case 1.1. $e \notin S_g \cup S_g'$ (see Figure 20.7). One can observe that $S_g \cup S_g' \cup S_f \cup S_f'$ forms $n-1$ disjoint 4-cycles in $BF(n)$. Meanwhile, $S_f' \subset E(C_f^{T,\alpha})$ and $S_f \cap E(C_f^{T,\alpha}) = \emptyset$.

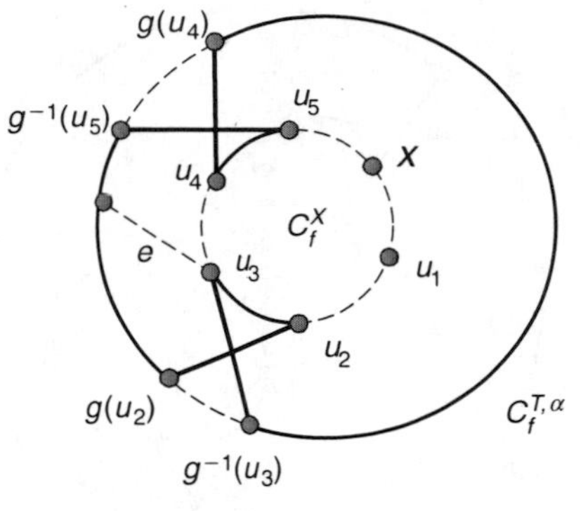

FIGURE 20.7 An illustration for Case 1.1 with $n = 3$, $S_g = \{(u_2, g(u_2)), (u_4, g(u_4))\}$, $S'_g = \{(u_3, g^{-1}(u_3)), (u_5, g^{-1}(u_5))\}$, $S_f = \{(u_2, u_3), (u_4, u_5)\}$, and $S'_f = \{(g(u_2), g^{-1}(u_3)), (g(u_4), g^{-1}(u_5))\}$.

Hence,

$$\left(E\left(C_f^{T,\alpha}\right) \cup S_g \cup S'_g \cup S_f\right) - S'_f$$

forms a cycle in $BF(n) - F$ and the cycle length is $n2^n - 2$.

Case 1.2. $e \in S_g \cup S'_g$. Constructing the cycle of length $n2^n - 2$ is similar to Case 1.1, except that $S_g = \{(y, g(y)) \mid y = f^{2k-1}(x); 1 \le k \le n-1\}$, $S'_g = \{(f(y), g^{-1}(f(y))) \mid y = f^{2k-1}(x); 1 \le k \le n-1\}$, $S_f = \{(y, f(y)) \mid y = f^{2k-1}(x); 1 \le k \le n-1\}$, and $S'_f = \{(g(y), f^{-1}(g(y))) \mid y = f^{2k-1}(x); 1 \le k \le n-1\}$.

Case 2. e is an f-edge. Let $e = (u, f(u))$ for some $u \in V(BF(n))$.

Case 2.1. n is an even integer. Since $n \ge 3$ and $BF(n)^G$ is isomorphic to the n-dimensional hypercube, $BF(n)^G - \{\overline{C}_g^x, (\overline{C}_g^u, \overline{C}_g^{f(u)})\}$ is connected. Let T be any spanning tree of $BF(n)^G - \{\overline{C}_g^x, (\overline{C}_g^u, \overline{C}_g^{f(u)})\}$. By Lemma 20.7, C_g^T forms a cycle spanning $V(BF(n)) - V(C_g^x)$. Since T does not contain the edge $(\overline{C}_g^u, \overline{C}_g^{f(u)})$ and $e \in X_f(C_g^u, C_g^{f(u)})$, C_g^T does not contain the faulty edge e. Let $S = \left\{X_f(C_g^y, C_g^{f(y)}) \mid y = g^{2k}(x); 1 \le k < n/2\right\}$.

Case 2.1.1. $e \notin S$. One can observe that $\bigcup_{(u,v)\in S} X_g(C_g^u, C_g^v) \cup S$ forms $n/2 - 1$ disjoint 4-cycles in $BF(n) - F$. Let $S_1 = \left\{(f(y), g^{-1}(f(y))) \mid y = g^{2k}(x); 1 \le k < n/2\right\}$. Hence, $S_1 \subset E(C_g^T)$ and $S_1 \subset \cup_{(u,v)\in S} X_g(C_g^u, C_g^v)$. Therefore,

$$\bigcup_{(u,v)\in S} X_g\left(C_g^u, C_g^v\right) \cup S \cup E\left(C_g^T\right) - S_1$$

forms a cycle of length $n2^n - 2$ in $BF(n) - F$.

Case 2.1.2. $e \in S$. The cycle can be constructed using the method of Case 2.1.1, except that $S = \left\{X_f(C_g^y, C_g^{f(y)}) \mid y = g^{2k-1}(x); 1 \le k < n/2\right\}$.

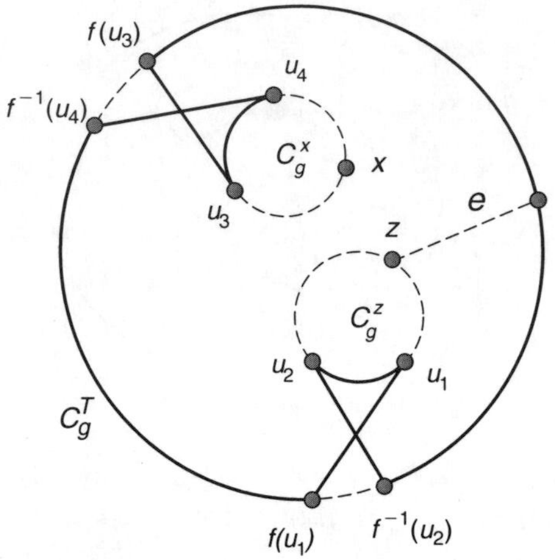

FIGURE 20.8 An illustration for Case 2.2.1 with $n=3$, $S_1=\{(u_3,f(u_3)),(u_4,f^{-1}(u_4))\}$, and $S_2=\{(u_1,f(u_1)),(u_2,f^{-1}(u_2))\}$.

Case 2.2. n is an odd integer. Let $S_1=\left\{X_f(C_g^y, C_g^{f(y)}) \mid y=g^{2k-1}(x);\ 1\le k\le (n-1)/2\right\}$.

Case 2.2.1. $e\notin S_1$ (see Figure 20.8). Since $\overline{C}_g^u\neq\overline{C}_g^{f(u)}$, we can choose $z\in\{u,f(u)\}$ according to the following rules:

1. If either $\overline{C}_g^u=\overline{C}_g^x$ or $\overline{C}_g^{f(u)}=\overline{C}_g^x$, then we choose z such that $\overline{C}_g^z\neq\overline{C}_g^x$.
2. Otherwise, we choose z such that $(\overline{C}_g^z,\overline{C}_g^x)\notin E(BF(n)^G)$.

Since $n\ge 3$ and $BF(n)^G$ is isomorphic to an n-dimensional hypercube, $BF(n)^G-\{\overline{C}_g^x,\overline{C}_g^z\}$ is connected. Let T be any spanning tree of $BF(n)^G-\{\overline{C}_g^x,\overline{C}_g^z\}$. By Lemma 20.7, C_g^T is a cycle spanning $V(BF(n))-V(C_g^x)-V(C_g^z)$ and it does not contain e. Let $S_2=\left\{X_f(C_g^y,C_g^{f(y)}) \mid y=g^{2k-1}(z);\ 1\le k\le (n-1)/2\right\}$. Hence, $S_1\cap S_2=\emptyset$ and $S_1\cup S_2$ is fault-free. We observe that $\bigcup_{(u,v)\in S_1} X_g(C_g^u,C_g^v)\cup S_1$ forms $(n-1)/2$ disjoint 4-cycles and $\cup_{(u,v)\in S_2}X_g(C_g^u,C_g^v)\cup S_2$ forms $(n-1)/2$ disjoint 4-cycles. Let $S_3=\left\{(f(y),g^{-1}(f(y))) \mid y=g^{2k-1}(x);\ 1\le k\le (n-1)/2\right\}$ and $S_4=\left\{(f(y),g^{-1}(f(y))) \mid y=g^{2k-1}(z);\ 1\le k\le (n-1)/2\right\}$. Hence, $S_3\cap S_4=\emptyset$ and $S_3\cup S_4\subset E(C_g^T)$. Meanwhile, $S_3\cup S_4\subset \cup_{(u,v)\in(S_1\cup S_2)}X_g(C_g^u,C_g^v)$. Therefore,

$$\bigcup_{(u,v)\in(S_1\cup S_2)} X_g\left(C_g^u,C_g^v\right)\cup S_1\cup S_2\cup E\left(C_g^T\right)-S_3-S_4$$

forms a cycle in $BF(n)-F$ and this cycle length is $n2^n-2$.

Case 2.2.2. $e\in S_1$ (see Figure 20.9). Since $n\ge 3$, there exists $y\in V(BF(n))$ such that $f(y)\in V(C_g^x)$. So both $(y,f(y))$ and $(g(y),f^{-1}(g(y)))$ join vertices of C_g^y and C_g^x. Let $W_1=V(C_g^{a_1})\cup V(C_g^{a_3})\cup V(C_g^{a_5})\cup V(C_g^{a_7})$ and $\overline{W}_1=\{\overline{C}_g^{a_1},\overline{C}_g^{a_3},\overline{C}_g^{a_5},\overline{C}_g^{a_7}\}$.

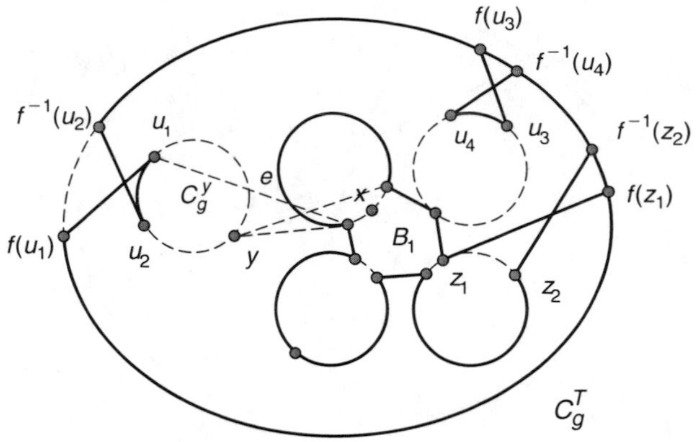

FIGURE 20.9 An illustration for Case 2.2.2 with $n=3$, $S_3=\{(u_3,f(u_3)),(u_4,f^{-1}(u_4))\}$, and $S_4=\{(u_1,f(u_1)),(u_2,f^{-1}(u_2))$.

Hence, $\overline{C}_g^y$ is not adjacent to any vertex in $\overline{W}_1-\{\overline{C}_g^x\}$. Since $n\geq 3$ and $BF(n)^G$ is isomorphic to an n-dimensional hypercube, $BF(n)^G-\overline{W}_1-\{\overline{C}_g^y\}$ is connected. Let T be any spanning tree of $BF(n)^G-\overline{W}_1-\{\overline{C}_g^y\}$. By Lemma 20.7, C_g^T is a cycle spanning $V(BF(n))-W_1-V(C_g^y)$. There exists $w\in V(\mathcal{B}_1)$ such that $X_f(C_g^w, C_g^{f(w)})$ is fault-free and $(w,f(w))$ joins some vertex in both $\mathcal{B}_1$ and C_g^T. Then

$$C_e=\left(E(\mathcal{B}_1)\bigcup E\left(C_g^T\right)\bigcup X_f\left(C_g^w,C_g^{f(w)}\right)\right)-X_g\left(C_g^w,C_g^{f(w)}\right)$$

forms a cycle of length $n2^n-2n$ spanning $(V(BF(n))-W_1-V(C_g^y))\cup V(\mathcal{B}_1)$.

Let $S_3=\left\{X_f(C_g^s,C_g^{f(s)})\,|\,s=g^{2k-1}(a_3);\,1\leq k\leq (n-1)/2\right\}$ and $S_4=\left\{X_f(C_g^t,C_g^{f(t)})\,|\,t=g^{2k-1}(y);\,1\leq k\leq (n-1)/2\right\}$. Hence, $S_3\cap S_4=\emptyset$ and $S_3\cup S_4$ is fault-free. We observe that $\cup_{(u,v)\in(S_3\cup S_4)}X_g(C_g^u,C_g^v))\cup S_3\cup S_4$ forms $n-1$ disjoint 4-cycles. Let $S_5=\{(f(s),g^{-1}(f(s)))\,|\,s=g^{2k-1}(a_3);\,1\leq k\leq (n-1)/2\}$ and $S_6=\{(f(t),g^{-1}(f(t)))\,|\,t=g^{2k-1}(y);\,1\leq k\leq (n-1)/2\}$. Hence, $S_5\cap S_6=\emptyset$, $S_5\cup S_6$ is a subset of both $E(C_e)$ and $\cup_{(u,v)\in(S_3\cup S_4)}X_g(C_g^u,C_g^v))$. Therefore,

$$\bigcup_{(u,v)\in(S_3\cup S_4)} X_g\left(C_g^u,C_g^v\right)\cup S_3\cup S_4\cup E(C_e)-S_5-S_6$$

forms a cycle of length $n2^n-2$ in $BF(n)-F$. □

LEMMA 20.12 For any odd integer $n\geq 3$, $BF(n)-F$ is Hamiltonian if F consists of one vertex and one edge.

Proof: Since $BF(n)$ is vertex-transitive, we may assume that the faulty vertex is $x=\langle 0,00\ldots 0\rangle$ and the faulty edge is e. Let $W_1=V(C_g^{a_1})\cup V(C_g^{a_3})\cup V(C_g^{a_5})\cup V(C_g^{a_7})$ and $\overline{W}_1=\{\overline{C}_g^{a_1},\overline{C}_g^{a_3},\overline{C}_g^{a_5},\overline{C}_g^{a_7}\}$.

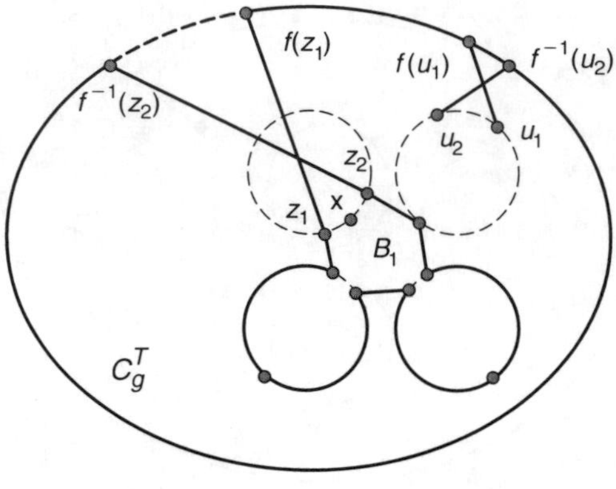

FIGURE 20.10 An illustration for Case 1.1 with $n=3$ and $S=\{(u_1, f(u_1)), (u_2, f^{-1}(u_2))\}$.

Case 1. e is an f-edge. Let $e=(u, f(u))$ for some $u \in V(BF(n))$.

Case 1.1. $\{u, f(u)\} \cap (V(C_g^{a_3}) \cup V(C_g^{a_7})) = \emptyset$ (see Figure 20.10). Since $n \geq 3$ and $BF(n)^G$ is isomorphic to an n-dimensional hypercube, $BF(n)^G - \overline{W}_1 - \{(\overline{C}_g^u, \overline{C}_g^{f(u)})\}$ is connected. Let T be any spanning tree of $BF(n)^G - \overline{W}_1 - \{(\overline{C}_g^u, \overline{C}_g^{f(u)})\}$. By Lemma 20.7, C_g^T is a cycle spanning $V(BF(n)) - W_1$ and it does not contain e.

Let $S = \left\{X_f(C_g^y, C_g^{f(y)}) \mid y = g^{2k-1}(a_3); 1 \leq k \leq (n-1)/2\right\}$. Hence, $\cup_{(u,v)\in S} X_g(C_g^u, C_g^v) \cup S$ forms $(n-1)/2$ disjoint 4-cycles. Let $S_1 = \left\{(f(y), g^{-1}(f(y))) \mid y = g^{2k-1}(a_3);\right.$ $1 \leq k \leq (n-1)/2\}$. Consequently, S_1 is a subset of both $\cup_{(u,v)\in S} X_g(C_g^u, C_g^v)$ and $E(C_g^T)$. Therefore,

$$C_e = \bigcup_{(u,v)\in S} X_g\left(C_g^u, C_g^v\right) \cup S \cup E\left(C_g^T\right) - S_1$$

forms a cycle of length $n2^n - 3n - 1$ spanning $V(BF(n)) - V(\mathcal{B}_1) - \{x\}$ and it does not contain e. Since the length of $\mathcal{B}_1$ is $3n$, there exists $z \in V(\mathcal{B}_1)$ such that $X_f(C_g^z, C_g^{f(z)})$ is fault-free and $(z, f(z))$ joins some vertex in both C_e and $\mathcal{B}_1$. Therefore,

$$\left(E(C_e) \bigcup E(\mathcal{B}_1) \bigcup X_f\left(C_g^z, C_g^{f(z)}\right)\right) - X_g\left(C_g^z, C_g^{f(z)}\right)$$

forms a Hamiltonian cycle of $BF(n) - F$.

Case 1.2. $\{u, f(u)\} \cap (V(C_g^{a_3}) \cup V(C_g^{a_7})) \neq \emptyset$. Let $S = \left\{X_f(C_g^y, C_g^{f(y)}) \mid y = g^{2k-1}(x);\right.$ $1 \leq k \leq (n-1)/2\}$ (see Figure 20.11). Hence, $e \notin S$. Since $n \geq 3$ and $BF(n)^G$ is isomorphic to an n-dimensional hypercube, $BF(n)^G - \{\overline{C}_g^x\}$ is connected. Let T be any spanning tree of $BF(n)^G - \{\overline{C}_g^x\}$. By Lemma 20.7, C_g^T is a cycle spanning

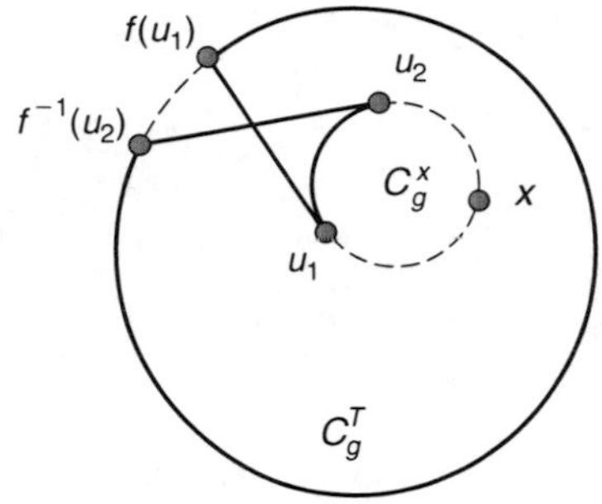

FIGURE 20.11 An illustration for Case 1.2 with $n=3$ and $S=\{(u_1, f(u_1)), (u_2, f^{-1}(u_2))\}$.

$V(BF(n)) - V(C_g^x)$. Since S is fault-free, $\cup_{(u,v)\in S} X_g(C_g^u, C_g^v) \cup S$ forms $(n-1)/2$ disjoint 4-cycles. Let $S_1 = \{(f(y), g^{-1}(f(y))) \mid y = g^{2k-1}(x); 1 \le k \le (n-1)/2\}$. Hence, S_1 is a subset of both $\cup_{(u,v)\in S} X_g(C_g^u, C_g^v)$ and $E(C_g^T)$. Therefore,

$$\bigcup_{(u,v)\in S} X_g\left(C_g^u, C_g^v\right) \cup S \cup E\left(C_g^T\right) - S_1$$

forms a Hamiltonian cycle of $BF(n) - F$.

Case 2. e is a g-edge. Let $e = (u, g(u))$ for some $u \in V(BF(n))$.

Case 2.1. $\overline{C}_g^u = \overline{C}_g^x$. Since $n \ge 3$ and $BF(n)^G$ is isomorphic to an n-dimensional hypercube, $BF(n)^G - \overline{W}_1$ is connected. Let T be any spanning tree of $BF(n)^G - \overline{W}_1$. Constructing a Hamiltonian cycle for this case is very similar to Case 1.1, except that the chosen vertex z is vertex u when $e \in E(\mathcal{B}_1)$.

Case 2.2. $\overline{C}_g^u \ne \overline{C}_g^x$ and $\overline{C}_g^u$ is not connected with $\overline{C}_g^x$ in $BF(n)^G$. Since $n \ge 3$ and $BF(n)^G$ is isomorphic to an n-dimensional hypercube, $BF(n)^G - \{\overline{C}_g^x\}$ is connected. Let T be a spanning tree of $BF(n)^G - \{\overline{C}_g^x\}$ such that $(\overline{C}_g^u, \overline{C}_g^{f(u)}) \in E(T)$. Then the Hamiltonian cycle of $BF(n) - F$ can be constructed by the same method used in Case 1.2.

Case 2.3. $(\overline{C}_g^u, \overline{C}_g^x) \in E(BF(n)^G)$ (that is, $\overline{C}_g^u$ and $\overline{C}_g^x$ are connected).

Case 2.3.1. $\overline{C}_g^u \ne \overline{C}_g^{a_3}$ and $\overline{C}_g^u \ne \overline{C}_g^{a_7}$ (see Figure 20.12). Since $n \ge 3$ and $BF(n)^G$ is isomorphic to an n-dimensional hypercube, $BF(n)^G - \overline{W}_1 - \{\overline{C}_g^u\}$ is connected. Let T be any spanning tree of $BF(n)^G - \overline{W}_1 - \{\overline{C}_g^u\}$. By Lemma 20.7, C_g^T is a cycle spanning $V(BF(n)) - W_1 - V(C_g^u)$. Let $S = \left\{X_f(C_g^y, C_g^{f(y)}) \mid y = g^{2k-1}(a_3); 1 \le k \le (n-1)/2\right\}$. Hence, $\cup_{(u,v)\in S} X_g(C_g^u, C_g^v) \cup S$ forms $(n-1)/2$ disjoint 4-cycles. Let $S_1 = \{(f(y), g^{-1}(f(y))) \mid y = g^{2k-1}(a_3); 1 \le k \le (n-1)/2\}$. Accordingly, S_1 is a subset of both

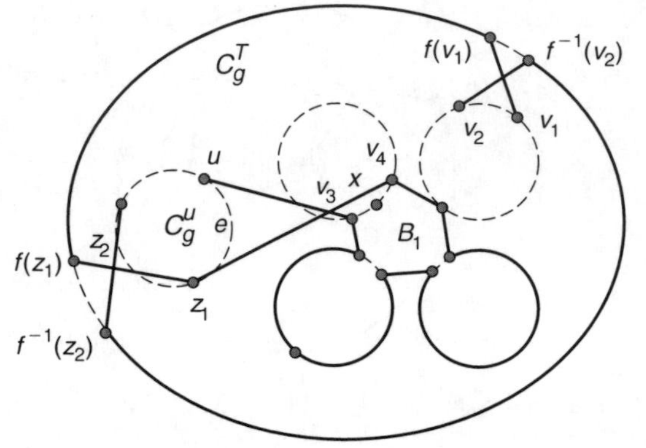

FIGURE 20.12 An illustration for Case 2.3.1 with $n=3$, $S=\{(v_1, f(v_1)), (v_2, f^{-1}(v_2))\}$, $X_f(C_g^u, C_g^{f(u)})=\{(u, v_3), (z_1, v_4)\}$, $X_f(C_g^z, C_g^{f(z)}))=\{(z_1, f(z_1)), (z_2, f^{-1}(z_2))\}$, $X_g(C_g^u, C_g^{f(u)})=\{(u, z_1), (v_3, v_4)\}$, and $X_g(C_g^z, C_g^{f(z)})=\{(z_1, z_2), (f(z_1), f^{-1}(z_2))\}$.

$\cup_{(u,v)\in S} X_g(C_g^u, C_g^v)$ and $E(C_g^T)$. Therefore,

$$C_e = \bigcup_{(u,v)\in S} X_g\left(C_g^u, C_g^v\right) \cup S \cup E\left(C_g^T\right) - S_1$$

forms a cycle spanning $V(BF(n)) - V(\mathcal{B}_1) - V(C_g^u) - \{x\}$. Since the length of $\mathcal{B}_1$ is $3n$, we can choose a vertex $z \in V(C_g^u)$ such that $f(z) \in V(C_e)$ if $f(u) \in V(\mathcal{B}_1)$ or $f(z) \in V(\mathcal{B}_1)$ if $f(u) \notin V(\mathcal{B}_1)$. Therefore,

$$\left(E(C_e) \cup E(\mathcal{B}_1) \cup E\left(C_g^u\right) \cup X_f\left(C_g^u, C_g^{f(u)}\right) \cup X_f\left(C_g^z, C_g^{f(z)}\right)\right)$$
$$- X_g\left(C_g^u, C_g^{f(u)}\right) - X_g\left(C_g^z, C_g^{f(z)}\right)$$

forms a Hamiltonian cycle of $BF(n) - F$.

Case 2.3.2. $\overline{C}_g^u = \overline{C}_g^{a_3}$ (see Figure 20.13). Let $S = \left\{X_g(C_g^y, C_g^{f(y)}) | y = g^{2k-1}(a_3); 1 \leq k \leq (n-1)/2\right\}$.

Assume that $e \notin S$. Since $n \geq 3$ and $BF(n)^G$ is isomorphic to an n-dimensional hypercube, $BF(n)^G - \overline{W}_1$ is connected. Let T be any spanning tree of $BF(n)^G - \overline{W}_1$. Then the Hamiltonian cycle of $BF(n) - F$ can be constructed by the same method used in Case 1.1.

Assume that $e \in S$. Let $W_2 = V(C_g^{b_1}) \cup V(C_g^{b_3}) \cup V(C_g^{b_5}) \cup V(C_g^{b_7})$ and $\overline{W}_2 = \{\overline{C}_g^{b_1}, \overline{C}_g^{b_3}, \overline{C}_g^{b_5}, \overline{C}_g^{b_7}\}$. Since $n \geq 3$ and $BF(n)^G$ is isomorphic to an n-dimensional hypercube, $BF(n)^G - \overline{W}_2$ is connected. Let T be a spanning tree of $BF(n)^G - \overline{W}_2$ such that $(\overline{C}_g^u, \overline{C}_g^{f(u)}) \in E(T)$. By Lemma 20.7, C_g^T is a cycle spanning $V(BF(n)) - W_2$. Let

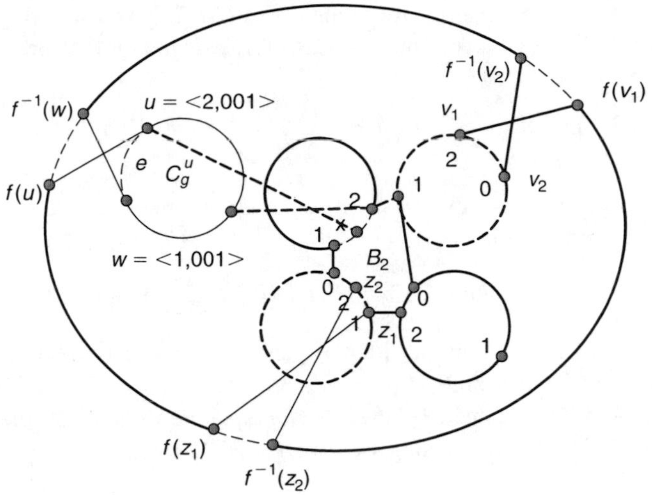

FIGURE 20.13 An illustration for the Case 2.3.2 with $n = 3$, $e = (\langle 2, 001\rangle, \langle 1, 001\rangle)$, and $S_2 = \emptyset$.

$S_2 = \left\{X_f(C_g^y, C_g^{f(y)}) \,\middle|\, y = g^{2k-1}(b_8);\ 1 \le k \le (n-3)/2\right\}$. Then

$$C_e = \left(E\left(C_g^T\right) \cup S_2 \cup \bigcup_{(u,v)\in S_2} X_g\left(C_g^u, C_g^v\right) \cup X_f\left(C_g^{g(b_3)}, C_g^{f(g(b_3))}\right)\right.$$
$$\left.\cup\, X_g\left(C_g^{g(b_3)}, C_g^{f(g(b_3))}\right)\right) - \left\{(f(y), g^{-1}(f(y))) \,\middle|\, y = g^{2k-1}(b_8);\ 1 \le k \le \frac{n-3}{2}\right\}$$
$$- \{(f(g(b_3)), f^{-1}(g^2(b_3)))\}$$

forms a cycle spanning $V(BF(n)) - V(\mathcal{B}_2) - \{x\}$ and it does not contain e. Since the length of $\mathcal{B}_2$ is $3n$, there exists $w \in V(\mathcal{B}_2)$ such that $(w, f(w))$ joins some vertex in both C_e and $\mathcal{B}_2$. Obviously, $X_f(C_g^w, C_g^{f(w)})$ is fault-free. Then

$$\left(E(C_e) \cup E(\mathcal{B}_2) \cup X_f\left(C_g^w, C_g^{f(w)}\right)\right) - X_g\left(C_g^w, C_g^{f(w)}\right)$$

forms a Hamiltonian cycle of $BF(n) - F$.

Case 2.3.3. $\overline{C}_g^u = \overline{C}_g^{a_7}$.
Suppose that $e \in X_g(C_g^x, C_g^{a_7})$. Hence, $e = (\langle 0, 10\ldots00\rangle, \langle 1, 100\ldots0\rangle)$. We observe that $e \in E(\mathcal{B}_1)$ and $e \notin E(\mathcal{B}_2)$. Using the same method used in the situation $e \in S$ of Case 2.3.2, the Hamiltonian cycle of $BF(n) - F$ can be constructed.

Suppose that $e \notin X_g(C_g^x, C_g^{a_7})$. Since $n \ge 3$ and $BF(n)^G$ is isomorphic to an n-dimensional hypercube, $BF(n)^G - \overline{W}_1$ is connected. Let T be any spanning tree of $BF(n)^G - \overline{W}_1$ and $e = (u, g(u))$ for some $u \in V(BF(n))$. Constructing a Hamiltonian cycle for this case is very similar to Case 1.1, except that the chosen vertex z is vertex u when $e \in E(\mathcal{B}_1)$. □

LEMMA 20.13 For any odd integer $n \geq 3$, $BF(n) - F$ is Hamiltonian if F consists of two vertices.

Proof: Since $BF(n)$ is vertex-transitive, we may assume that one faulty vertex is $x = \langle 0, 00\ldots0\rangle$ and the other is y. Let $\overline{W}_3 = \{\overline{C}_g^{c_1}, \overline{C}_g^{c_3}, \overline{C}_g^{c_5}, \overline{C}_g^{c_7}\}$.

Case 1. $\overline{C}_g^x = \overline{C}_g^y$ (see Figure 20.14). Let $y = \langle j, \underbrace{00\ldots0}_{n}\rangle$ for some $1 \leq j \leq n-1$. Since n is an odd integer, without loss of generality, we may assume that j is an even integer. Since $n \geq 3$ and $BF(n)^G$ is isomorphic to an n-dimensional hypercube, $BF(n)^G - \overline{W}_3$ is connected. Let T be any spanning tree of $BF(n)^G - \overline{W}_3$. By Lemma 20.7, C_g^T is a cycle spanning $V(BF(n)) - W_3$. Let

$$S_1 = \left\{X_f\left(C_g^y, C_g^{f(y)}\right) \middle| \, y = g^{2k}(c_2); 1 \leq k \leq \frac{n-j-1}{2}\right\}$$

$$S_2 = \left\{X_f\left(C_g^y, C_g^{f(y)}\right) \middle| \, y = g^{2k-1}(c_3); 1 \leq k \leq \frac{n-j-1}{2}\right\}$$

$$S_3 = \left\{X_f\left(C_g^y, C_g^{f(y)}\right) \middle| \, y = g^{2k-1}(c_5); 1 \leq k \leq \frac{j}{2} - 1\right\}$$

$$S_4 = \left\{X_f\left(C_g^y, C_g^{f(y)}\right) \middle| \, y = g^{2k-1}(c_8); 1 \leq k \leq \frac{j}{2} - 1\right\}$$

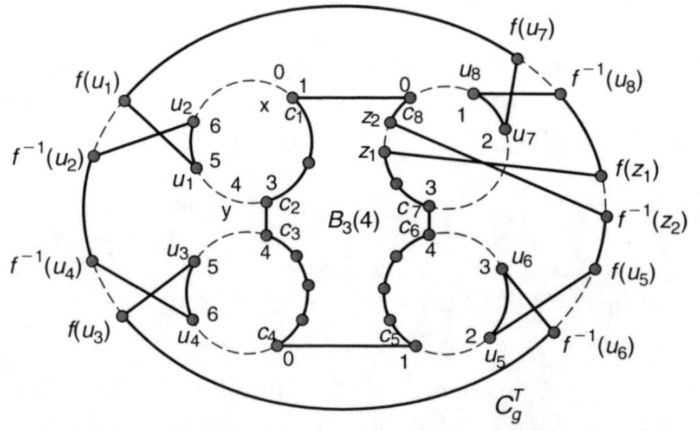

FIGURE 20.14 An illustration for Case 1 with $j = 4$, $n = 7$, $S_1 = \{(u_1, f(u_1)), (u_2, f^{-1}(u_2))\}$, $S_2 = \{(u_3, f(u_3)), (u_4, f^{-1}(u_4))\}$, $S_3 = \{(u_5, f(u_5)), (u_6, f^{-1}(u_6))\}$, and $S_4 = \{(u_7, f(u_7)), (u_8, f^{-1}(u_8))\}$.

One can observe that $S_i \cap S_j = \emptyset$ for $1 \le i \ne j \le 4$ and $\bigcup_{(u,v)\in S_1\cup S_2\cup S_3\cup S_4} X_g(C_g^u, C_g^v) \cup S_1 \cup S_2 \cup S_3 \cup S_4$ forms $n-3$ disjoint 4-cycles. Let

$$S_5 = \left\{ (f(y), g^{-1}(f(y))) \middle| y = g^{2k}(c_2); 1 \le k \le \frac{n-j-1}{2} \right\}$$

$$S_6 = \left\{ (f(y), g^{-1}(f(y))) \middle| y = g^{2k-1}(c_3); 1 \le k \le \frac{n-j-1}{2} \right\}$$

$$S_7 = \left\{ (f(y), g^{-1}(f(y))) \middle| y = g^{2k-1}(c_5); 1 \le k \le \frac{j}{2} - 1 \right\}$$

$$S_8 = \left\{ (f(y), g^{-1}(f(y))) \middle| y = g^{2k-1}(c_8); 1 \le k \le \frac{j}{2} - 1 \right\}$$

Consequently, $S_5 \cap S_6 \cap S_7 \cap S_8 = \emptyset$, and $S_5 \cup S_6 \cup S_7 \cup S_8$ is a subset of both $E(C_g^T)$ and $\cup_{(u,v)\in S_1\cup S_2\cup S_3\cup S_4} X_g(C_g^u, C_g^v)$. Therefore,

$$C_e = \bigcup_{(u,v)\in S_1\cup S_2\cup S_3\cup S_4} X_g\left(C_g^u, C_g^v\right) \cup S_1 \cup S_2 \cup S_3 \cup S_4 \cup E\left(C_g^T\right) - S_5 - S_6 - S_7 - S_8$$

forms a cycle of length $n2^n - 2n - 6$ spanning $V(BF(n)) - V(\mathcal{B}_3(j)) - \{x, y\}$. Since the length of $\mathcal{B}_3(j)$ is $2n+4$, there exists $z \in V(\mathcal{B}_3(j))$ such that $X_f(C_g^z, C_g^{f(z)})$ is fault-free and $(z, f(z))$ joins some vertex in both C_e and $\mathcal{B}_3(j)$. Then

$$\left(E(C_e) \cup E(\mathcal{B}_3(j)) \cup X_f\left(C_g^z, C_g^{f(z)}\right)\right) - X_g\left(C_g^z, C_g^{f(z)}\right)$$

forms a Hamiltonian cycle of $BF(n) - F$.

Case 2. $\overline{C}_g^x \ne \overline{C}_g^y$ and $\overline{C}_g^x$ is not connected with $\overline{C}_g^y$ in $BF(n)^G$ (see Figure 20.15). Since $n \ge 3$ and $BF(n)^G$ is isomorphic to an n-dimensional hypercube, $BF(n)^G - \{\overline{C}_g^x, \overline{C}_g^y\}$ is connected. Let T be any spanning tree of $BF(n)^G - \{\overline{C}_g^x, \overline{C}_g^y\}$. By Lemma 20.7, C_g^T is a cycle spanning $V(BF(n)) - V(C_g^x) - V(C_g^y)$. Let $S_1 = \left\{X_f(C_g^s, C_g^{f(s)}) \mid s = g^{2k-1}(x); 1 \le k \le (n-1)/2\right\}$ and $S_2 = \{X_f(C_g^t, C_g^{f(t)}) \mid t = g^{2k-1}(y); 1 \le k \le (n-1)/2\}$. Since $S_1 \cap S_2 = \emptyset$ and $S_1 \cup S_2$ is fault-free, $\cup_{(u,v)\in S_1\cup S_2} X_g(C_g^u, C_g^v) \cup S_1 \cup S_2$ forms $n-1$ disjoint 4-cycles. Let $S_3 = \left\{(f(s), g^{-1}(f(s))) \mid s = g^{2k-1}(x); 1 \le k \le (n-1)/2\right\}$ and $S_4 = \left\{(f(t), g^{-1}(f(t))) \mid t = g^{2k-1}(y); 1 \le k \le (n-1)/2\right\}$. Hence, $S_3 \cap S_4 = \emptyset$ and $S_3 \cup S_4 \subset E(C_g^T)$. At the same time, $S_3 \cup S_4 \subset \cup_{(u,v)\in(S_1\cup S_2)} X_g(C_g^u, C_g^v)$. Therefore,

$$\bigcup_{(u,v)\in(S_1\cup S_2)} X_g\left(C_g^u, C_g^v\right) \cup S_1 \cup S_2 \cup E\left(C_g^T\right) - S_3 - S_4$$

forms a Hamiltonian cycle of $BF(n) - F$.

Case 3. $(\overline{C}_g^x, \overline{C}_g^y) \in E(BF(n)^G)$ (that is, $\overline{C}_g^x$ and $\overline{C}_g^y$ are connected).

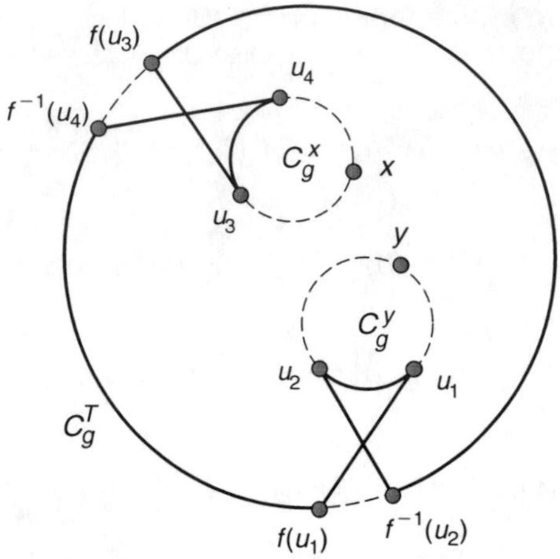

FIGURE 20.15 An illustration for Case 2 with $n=3$, $S_1=\{(u_3, f(u_3)), (u_4, f^{-1}(u_4))\}$, and $S_2=\{(u_1, f(u_1)), (u_2, f^{-1}(u_2))\}$.

Case 3.1. $\overline{C}_g^y \neq \overline{C}_g^{a_3}$ and $\overline{C}_g^y \neq \overline{C}_g^{a_7}$ (see Figure 20.9). Since $BF(n)^G$ is an n-dimensional hypercube with $n \geq 3$, $C_g^{a_3}$, $C_g^{a_5}$, and $C_g^{a_7}$ are fault-free. As a result, $\overline{C}_g^{a_3}$ is not adjacent to $\overline{C}_g^y$ in $BF(n)^G$. Since $BF(n)^G - \overline{W}_1 - \{\overline{C}_g^y\}$ is connected, let T be any spanning tree of $BF(n)^G - \overline{W}_1 - \{\overline{C}_g^y\}$. The Hamiltonian cycle of the graph $BF(n) - F$ can be constructed by the same method used in Case 2.2.2 of Lemma 20.11.

Case 3.2. $\overline{C}_g^y = \overline{C}_g^{a_3}$ or $\overline{C}_g^y = \overline{C}_g^{a_7}$. Let $S_1 = \left\{\langle 2k-1, \underbrace{00\ldots0}_{n-1}1\rangle \mid 1 \leq k \leq (n-1)/2\right\}$ and $S_2 = \left\{\langle 2k, 1\underbrace{00\ldots0}_{n-1}\rangle \mid 1 \leq k \leq (n-1)/2\right\}$. Therefore, there does not exist any edge in $X_f(C_g^u, C_g^{f(u)})$ that joins vertices of C_g^x and C_g^y for all $u \in S_1 \cup S_2$.

Case 3.2.1. $y \notin S_1 \cup S_2$ (see Figure 20.15). Since $n \geq 3$ and $BF(n)^G$ is isomorphic to an n-dimensional hypercube, $BF(n)^G - \{\overline{C}_g^x, \overline{C}_g^y\}$ is connected. Let T be any spanning tree of $BF(n)^G - \{\overline{C}_g^x, \overline{C}_g^y\}$. With the same method used in Case 2, the Hamiltonian cycle of the graph $BF(n) - F$ can be constructed.

Case 3.2.2. $y \in S_1 \cup S_2$.
Given an integer k with $0 \leq k < n$, the mapping σ_k from $V(BF(n))$ into $V(BF(n))$ can be defined by $\sigma_k(\langle l, a_0a_1\ldots a_{n-1}\rangle) = \langle (l-k) \textbf{ mod } n, a_ka_{k+1}\ldots a_{n-1}a_0a_1\ldots a_{k-1}\rangle$. Similarly, we can define the mapping φ_i from $V(BF(n))$ into $V(BF(n))$ as $\varphi_i(\langle l, a_0a_1\ldots, a_ia_{i+1},\ldots, a_{n-1}\rangle) = \langle l, a_0a_1\ldots \overline{a}_ia_{i+1}\ldots a_{n-1}\rangle$ where $0 \leq i < n$. Hence, σ_k and φ_i are two automorphisms of $BF(n)$.

Suppose that $y \in S_1$. Then $y = \langle l, \underbrace{00\ldots0}_{n-1}1\rangle$ for some odd l, $1 \le l \le n-2$. Accordingly, $\varphi_{n-1-l} \circ \sigma_l$ is an automorphism of $BF(n)$ such that $\varphi_{n-1-l} \circ \sigma_l(y) = x$ and $\varphi_{n-1-l} \circ \sigma_l(x) = \langle n-l, \underbrace{00\ldots0}_{n-1-l}1\underbrace{00\ldots0}_{l}\rangle = z$. Consequently, $z \notin S_1 \cup S_2$. Therefore, we can construct a Hamiltonian cycle C in $BF(n) - \{x, z\}$ by using the same method as in Case 3.2.1. Finally, $(\varphi_{n-1-l} \circ \sigma_l)^{-1}(C)$ also forms a Hamiltonian cycle of $BF(n) - \{x, y\}$.

Suppose that $y \in S_2$. Then $y = \langle l, 1\underbrace{00\ldots0}_{n-1}\rangle$ for some even l, $2 \le l \le n-1$. Hence, $\varphi_{n-l} \circ \sigma_l$ is an automorphism of $BF(n)$ such that $\varphi_{n-l} \circ \sigma_l(y) = x$ and $\varphi_{n-l} \circ \sigma_l(x) = \langle n-l, \underbrace{00\ldots0}_{n-l}1\underbrace{00\ldots0}_{l-1}\rangle = z$. Consequently, $z \notin S_1 \cup S_2$. With the same method used in Case 3.2.1, we can construct a Hamiltonian cycle C in $BF(n) - \{x, z\}$. Hence, $(\varphi_{n-l} \circ \sigma_l)^{-1}(C)$ also forms a Hamiltonian cycle of $BF(n) - \{x, y\}$. □

With a similar but easier proof, we have the following two lemmas.

LEMMA 20.14 **[185]** For any integer $n \ge 3$, $BF(n) - F$ is Hamiltonian if F consists of two edges.

LEMMA 20.15 **[321]** For any integer $n \ge 3$, $BF(n) - F$ contains a cycle of length $n2^n - 2$ if F consists of one vertex and a cycle of length $n2^n - 4$ if F consists of two vertices.

Combining Lemmas 20.11 through 20.15, we have the following theorem.

THEOREM 20.1 For any integer n with $n \ge 3$, let $F \subset V(BF(n)) \cup E(BF(n))$, $f_v = |F \cap V(BF(n))|$, and $|F| \le 2$. $BF(n) - F$ contains a cycle of length $n \times 2^n - 2f_v$. In addition, $BF(n) - F$ contains a Hamiltonian cycle if n is an odd integer.

20.3 SPANNING CONNECTIVITY FOR BUTTERFLY GRAPHS

Of course, we would like to know the spanning connectivity properties of a hypercube. Kueng et al. [207] proved that the wrapped butterfly graph $BF(n)$ is super spanning laceable if n is even and $BF(n)$ is super spanning connected if n is odd. We only prove that $BF(n)$ is 1^*-connected and 2^*-connected if n is odd, and is 1^*-laceable and 2^*-laceable if n is even. Yet we need more properties about $BF(n)$.

Let G be a subgraph of $BF(n)$ and let C be a cycle of G. Then C is an *ℓ-scheduled* cycle [343] with respect to G if every level-ℓ vertex of G is incident with a level-$(\ell-1)_{\mathbf{mod}\,n}$ edge and a level-ℓ edge in C. Furthermore, C is a *totally scheduled* cycle [343] of G if it is an ℓ-scheduled cycle of G for all $\ell \in \mathbb{Z}_n$. Obviously, $\gamma_\ell^i(C)$ with $i \in \{0, 1\}$ is a totally scheduled cycle of $\gamma_\ell^i(G)$ if C is a totally scheduled cycle of G.

LEMMA 20.16 Assume that $n \ge 3$ and $k \ge 1$. Suppose that $\{P_1, \ldots, P_k\}$ is a k-container of $BF(n)$ between two vertices x and y with the following conditions: (1) $V(BF(n)) - \bigcup_{i=1}^{k} V(P_i) = \bigcup_{i=1}^{m} \{u_i, g(u_i) | u_i \in V(BF(n))\}$ with some $m \ge 1$

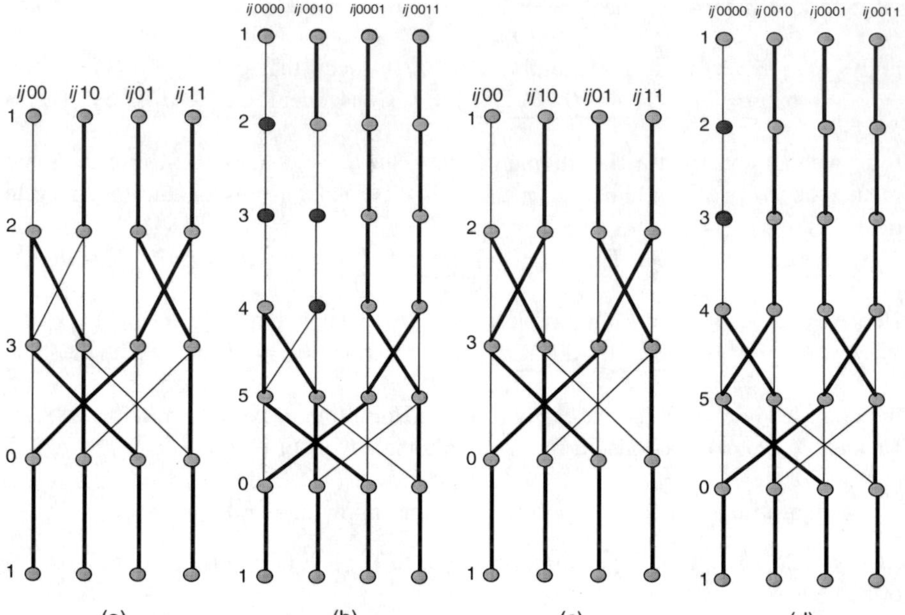

FIGURE 20.16 (a) A weakly 2-scheduled Hamiltonian path P_1 of $BF_{0,1}^{i,j}(4)$ joins $\langle 1, ij00\rangle$ to $\langle 2, ij10\rangle$; (b) $\gamma_3^0 \circ \gamma_2^0(P_1)$ in $BF_{0,1,2,3}^{i,j,0,0}(6) = \gamma_3^0 \circ \gamma_2^0(BF_{0,1}^{i,j}(4))$; (c) a weakly 2-scheduled Hamiltonian path P_2 of $BF_{0,1}^{i,j}(4)$ joins $\langle 1, ij00\rangle$ to $\langle 2, ij00\rangle$; and (d) $\gamma_3^0 \circ \gamma_2^0(P_2)$ in $BF_{0,1,2,3}^{i,j,0,0}(6)$.

such that $\{u_i | 1 \le i \le m\} \cap \{g(u_i) | 1 \le i \le m\} = \emptyset$ and (2) $\cup_{i=1}^m \{(f(u_i), g^{-1} \circ f(u_i))\} \subseteq \cup_{i=1}^k E(P_i)$. Then there exists a k^*-container of $BF(n)$ between x and y.

Proof: Let $A = \bigcup_{i=1}^m \{(u_i, g(u_i))\} \cup \bigcup_{i=1}^m \{(u_i, f(u_i))\} \cup \bigcup_{i=1}^m \{(g(u_i), f^{-1} \circ g(u_i))\}$ and $B = \bigcup_{i=1}^m \{(f(u_i), g^{-1} \circ f(u_i))\}$. Obviously, $A \cap (\bigcup_{i=1}^k E(P_i)) = \emptyset$. Then $((\bigcup_{i=1}^k E(P_i)) \cup A) - B$ forms a k^*-container of $BF(n)$ between x and y. □

A path P of $BF(n)$ is *ℓ-scheduled* if every level-ℓ vertex of $I(P)$ is incident to a level-$(\ell - 1)_{\mathbf{mod}\, n}$ edge and a level-ℓ edge on P. A path P of $BF(n)$ is *weakly ℓ-scheduled* if at least one level-ℓ vertex of $I(P)$ is incident to a level-$(\ell - 1)_{\mathbf{mod}\, n}$ edge and a level-ℓ edge on P. Figure 20.16 illustrates two weakly 2-scheduled Hamiltonian paths P_1 and P_2 of $BF_{0,1}^{i,j}(4)$ and their images $\gamma_3^0 \circ \gamma_2^0(P_1)$ and $\gamma_3^0 \circ \gamma_2^0(P_2)$ in $BF_{0,1,2,3}^{i,j,0,0}(6)$, respectively.

THEOREM 20.2 **[343]** Assume that $n \ge 3$. Every $BF(n)$ has a totally scheduled Hamiltonian cycle. Thus, $BF(n)$ is 2^*-connected.

Proof: We prove this theorem by induction on n. Obviously, $BF(3)$ contains a totally scheduled Hamiltonian cycle (See Figure 20.17). Assume that $BF(n-1)$ has a totally scheduled Hamiltonian cycle C for $n \ge 4$. Accordingly, $\gamma_0^0(C)$ and $\gamma_0^1(C)$ are totally scheduled Hamiltonian cycles of $\gamma_0^0(BF(n - 1)) = BF_0^0(n)$

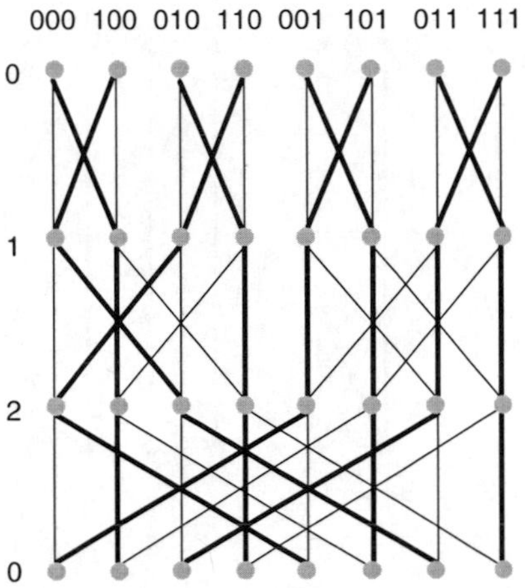

FIGURE 20.17 A totally scheduled Hamiltonian cycle of $BF(3)$.

and $\gamma_0^1(BF(n-1)) = BF_0^1(n)$, respectively. Since $BF(n)$ can be partitioned into $\{BF_0^0(n), BF_0^1(n)\}$, then the cycle generated by

$$(E(\gamma_0^0(C)) \cup E(\gamma_0^1(C)) \cup \{(\langle 0, 0^n\rangle, \langle 1, 10^{n-1}\rangle), (\langle 1, 0^n\rangle, \langle 0, 10^{n-1}\rangle)\})$$
$$- \{(\langle 0, 0^n\rangle, \langle 1, 0^n\rangle), (\langle 0, 10^{n-1}\rangle, \langle 1, 10^{n-1}\rangle)\}$$

is a totally scheduled Hamiltonian cycle of $BF(n)$. Thus, $BF(n)$ is 2^*-connected. □

By stretching operation, we have the following lemma and corollary.

LEMMA 20.17 Let $n \geq 3$. Assume that $0 \leq \ell \leq n-2$ and $i, j \in \mathbb{Z}_2$. Then there exists a totally scheduled Hamiltonian cycle of $BF_{\ell,\ell+1}^{i,j}(n)$.

COROLLARY 20.5 *Assume that $n \geq 4$ and $i, j, p, q \in \mathbb{Z}_2$. Then there exists a totally scheduled Hamiltonian cycle of $BF_{0,1,2,3}^{i,j,p,q}(n)$, including all straight edges of level 0, level 1, level 2, and level 3 in $BF_{0,1,2,3}^{i,j,p,q}(n)$.*

To prove that $BF(n)$ is 1^*-connected (respectively 1^*-laceable) for odd n (respectively even n), we need the following two lemmas.

LEMMA 20.18 Let $u = \langle 0, 000\rangle$ and let v be a level-ℓ vertex of $BF(3)$ for any $\ell \in \mathbb{Z}_3$. Then there exists a weakly ℓ-scheduled Hamiltonian path of $BF(3)$ joining u to v.

Proof: With symmetry, we assume that $\ell \in \{0, 2\}$. Let $v = \langle \ell, x_0x_1x_2\rangle$.

Case 1. Suppose that $\ell = 1$. For convenience, let $v_h^{ij} = \langle h, x_0ij\rangle$. In Figure 20.18a, we show a path P from u to v_2^{00} in $BF_{1,2}^{0,0}(3)$. By Lemma 20.17, there exists a totally scheduled Hamiltonian cycle C^{ij} in $BF_{1,2}^{i,j}(3)$.

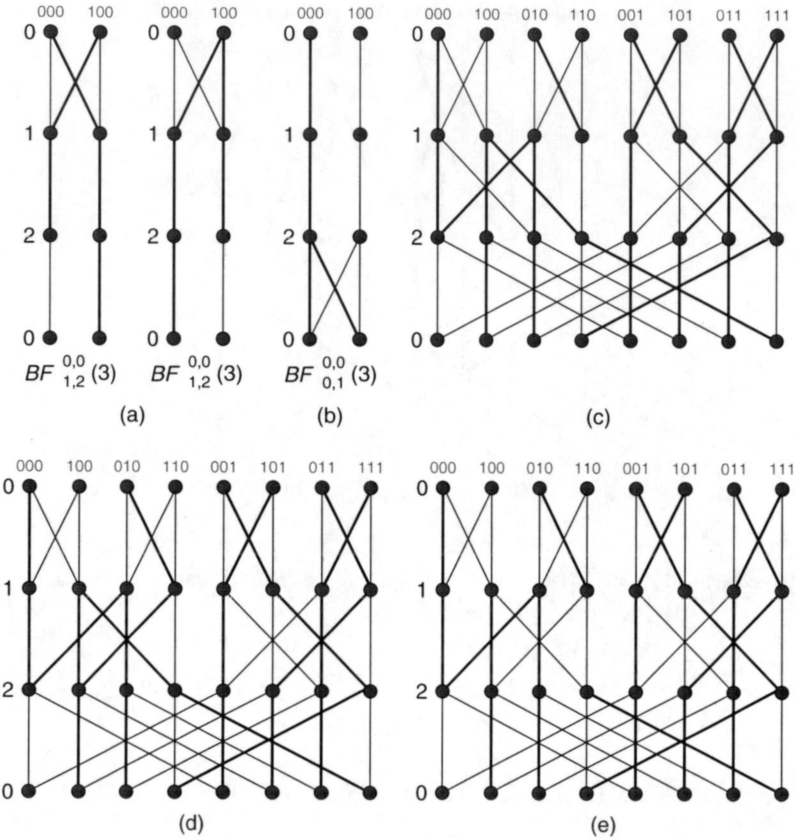

FIGURE 20.18 (a) Hamiltonian paths from $u = \langle 0, 000\rangle$ to every level-2 vertex in $BF^{0,0}_{1,2}(3)$; (b) a path from $u = \langle 0, 000\rangle$ to $\langle 0, 001\rangle$ in $BF^{0,0}_{0,1}(3)$; (c) a weakly 0-scheduled Hamiltonian path from $u = \langle 0, 000\rangle$ to $\langle 0, 100\rangle$; (d) a weakly 0-scheduled Hamiltonian path from $u = \langle 0, 000\rangle$ to $\langle 0, 010\rangle$; and (e) a weakly 0-scheduled Hamiltonian path from $u = \langle 0, 000\rangle$ to $\langle 0, 110\rangle$.

Case 1.1. Suppose that $x_1 = 0$ and $x_2 = 0$. Obviously, we have $v = v_2^{00}$. Since $BF(3)$ can be partitioned into $\{BF^{0,0}_{1,2}(3), BF^{1,0}_{1,2}(3), BF^{0,1}_{1,2}(3), BF^{1,1}_{1,2}(3)\}$, the path P' generated by

$$\Big(E(P) \cup E(C^{10}) \cup E(C^{01}) \cup E(C^{11})$$
$$\cup \big\{(v_1^{00}, v_2^{10}), (v_2^{00}, v_1^{10}), (v_2^{10}, v_0^{11}), (v_0^{10}, v_2^{11}), (v_1^{01}, v_2^{11}), (v_2^{01}, v_1^{11})\big\}\Big)$$
$$- \big\{(v_1^{00}, v_2^{00}), (v_1^{10}, v_2^{10}), (v_2^{10}, v_0^{10}), (v_2^{11}, v_0^{11}), (v_1^{01}, v_2^{01}), (v_1^{11}, v_2^{11})\big\}$$

is a weakly 2-scheduled Hamiltonian path of $BF(3)$ joining u to v.

Case 1.2. Suppose that $x_1 = 1$ and $x_2 = 0$. Obviously, the path P' generated by

$$\Big(E(P) \cup E(C^{10}) \cup E(C^{01}) \cup E(C^{11}) \cup \{(v_2^{00}, v_1^{10}), (v_2^{10}, v_0^{11}), (v_0^{10}, v_2^{11}), (v_1^{01}, v_2^{11}), (v_2^{01}, v_1^{11})\}\Big) - \{(v_1^{10}, v_2^{10}), (v_2^{10}, v_0^{10}), (v_2^{11}, v_0^{11}), (v_1^{01}, v_2^{01}), (v_1^{11}, v_2^{11})\}$$

is a weakly 2-scheduled Hamiltonian path of $BF(3)$ joining u to v.

Case 1.3. Suppose that $x_1 = 0$ and $x_2 = 1$. Obviously, the path P' generated by

$$\Big(E(P) \cup E(C^{10}) \cup E(C^{01}) \cup E(C^{11}) \cup \{(v_2^{00}, v_0^{01}), (v_1^{01}, v_2^{11}), (v_2^{01}, v_1^{11}), (v_2^{10}, v_0^{11}), (v_0^{10}, v_2^{11})\}\Big) - \{(v_2^{01}, v_0^{01}), (v_1^{01}, v_2^{01}), (v_1^{11}, v_2^{11}), (v_2^{10}, v_0^{10}), (v_2^{11}, v_0^{11})\}$$

is a weakly 2-scheduled Hamiltonian path of $BF(3)$ joining u to v.

Case 1.4. Suppose that $x_1 = 1$ and $x_2 = 1$. Obviously, the path P' generated by

$$\Big(E(P) \cup E(C^{10}) \cup E(C^{01}) \cup E(C^{11}) \cup \{(v_2^{00}, v_0^{01}), (v_2^{01}, v_1^{11}), (v_2^{10}, v_0^{11}), (v_0^{10}, v_2^{11})\}\Big) - \{(v_2^{01}, v_0^{01}), (v_1^{11}, v_2^{11}), (v_2^{10}, v_0^{10}), (v_2^{11}, v_0^{11})\}$$

is a weakly 2-scheduled Hamiltonian path of $BF(3)$ joining u to v.

Case 2. Suppose that $\ell = 0$. For convenience, let $v_h^{ij} = \langle h, ij1\rangle$ and $u_h^{ij} = \langle h, ij0\rangle$. In Figure 20.18b, we show a path P from u to v_0^{00} in $BF_{0,1}^{0,0}(3)$. By Lemma 20.17, there exists a totally scheduled Hamiltonian cycle C^{ij} in $BF_{0,1}^{i,j}(3)$.

Case 2.1. Suppose that $x_0 = 0, x_1 = 0$, and $x_2 = 1$. Obviously, we have $v = v_0^{00}$. Since $BF(3)$ can be partitioned into $\{BF_{0,1}^{0,0}(3), BF_{0,1}^{1,0}(3), BF_{0,1}^{0,1}(3), BF_{0,1}^{1,1}(3)\}$, the path P' generated by

$$\Big(E(P) \cup E(C^{10}) \cup E(C^{01}) \cup E(C^{11}) \cup \{(u_0^{00}, u_1^{10}), (u_1^{00}, u_0^{10}), (u_1^{00}, u_2^{01}), (u_2^{00}, u_1^{01}), (u_0^{01}, u_1^{11}), (u_1^{01}, u_0^{11})\} \cup \{(v_1^{00}, v_2^{00}), (v_1^{00}, v_2^{01}), (v_2^{00}, v_1^{01})\}\Big) - \{(u_0^{00}, u_1^{00}), (u_0^{10}, u_1^{10}), (u_1^{00}, u_2^{00}), (u_1^{01}, u_2^{01}), (u_0^{01}, u_1^{01}), (u_0^{11}, u_1^{11}), (v_1^{01}, v_2^{01})\}$$

is a weakly 0-scheduled Hamiltonian path of $BF(3)$ joining u to v.

Case 2.2. Suppose that $x_0 = 1, x_1 = 0$ and $x_2 = 1$. Obviously, the path P' generated by

$$\begin{aligned}&\Big(E(P) \cup E(C^{10}) \cup E(C^{01}) \cup E(C^{11})\\ &\cup \{(v_0^{00}, v_1^{10}), (u_1^{00}, u_2^{01}), (u_2^{00}, u_1^{01}), (u_0^{01}, u_1^{11}), (u_1^{01}, u_0^{11})\}\\ &\cup \{(v_1^{00}, v_2^{00}), (v_1^{00}, v_2^{01}), (v_2^{00}, v_1^{01})\}\Big)\\ &\quad - \{(v_0^{10}, v_1^{10}), (u_1^{00}, u_2^{00}), (u_1^{01}, u_2^{01}), (u_0^{01}, u_1^{01}), (u_0^{11}, u_1^{11}), (v_1^{01}, v_2^{01})\}\end{aligned}$$

is a weakly 0-scheduled Hamiltonian path of $BF(3)$ joining u to v.

Case 2.3. Suppose that $x_0 = 0, x_1 = 1$ and $x_2 = 1$. Obviously, the path P' generated by

$$\begin{aligned}&\Big(E(P) \cup E(C^{10}) \cup E(C^{01}) \cup E(C^{11})\\ &\cup \{(v_0^{00}, v_1^{10})1, (v_2^{10}, v_1^{11}), (v_0^{11}, v_1^{01})\} \cup \{(v_1^{00}, v_2^{00}), (v_1^{00}, v_2^{01}), (v_2^{00}, v_1^{01})\}\Big)\\ &\quad - \{(v_1^{10}, v_2^{10}), (v_0^{11}, v_1^{11}), (v_0^{01}, v_1^{01}), (v_1^{01}, v_2^{01})\}\end{aligned}$$

is a weakly 1-scheduled Hamiltonian path of $BF(3)$ joining u to v.

Case 2.4. Suppose that $x_0 = 1, x_1 = 1$ and $x_2 = 1$. Obviously, the path P' generated by

$$\begin{aligned}&\Big(E(P) \cup E(C^{10}) \cup E(C^{01}) \cup E(C^{11})\\ &\cup \{(v_0^{00}, v_1^{10}), (v_2^{10}, v_1^{11}), (u_0^{11}, u_1^{01}), (u_1^{11}, u_0^{01})\}\\ &\cup \{(v_1^{00}, v_2^{00}), (v_1^{00}, v_2^{01}), (v_2^{00}, v_1^{01})\}\Big)\\ &\quad - \{(v_1^{10}, v_2^{10}), (v_0^{11}, v_1^{11}), (u_0^{01}, u_1^{01}), (u_0^{11}, u_1^{11}), (v_1^{01}, v_2^{01})\}\end{aligned}$$

is a weakly 1-scheduled Hamiltonian path of $BF(3)$ joining u to v.

Case 2.5. For $x_2 = 0$, we illustrate the required paths in Figures 20.18c, through 20.18e. □

LEMMA 20.19 Let $u = \langle 0, 0000\rangle$ and let v be a level-ℓ vertex of $BF(4)$ for odd $\ell \in \mathbb{Z}_4$. Then there exists a weakly ℓ-scheduled Hamiltonian path of $BF(4)$ joining u to v.

Proof: First of all, we assume that $\ell = 1$. For convenience, let $v = \langle 1, x_0x_1x_2x_3\rangle$ and $v_h^{ij} = \langle h, x_0ijx_3\rangle$. In Figure 20.19, we show a weakly 1-scheduled Hamiltonian path P from u to v_1^{00} in $BF_{1,2}^{0,0}(4)$. By Lemma 20.17, there exists a totally scheduled Hamiltonian cycle C^{ij} in $BF_{1,2}^{i,j}(4)$. Since P is weakly ℓ-scheduled, there is at least one level-ℓ vertex $w = \langle \ell, y_000y_3\rangle$ of P incident to a level-0 edge and a level-1 edge on P. For convenience, let $w_h^{ij} = \langle h, y_0ijy_3\rangle$.

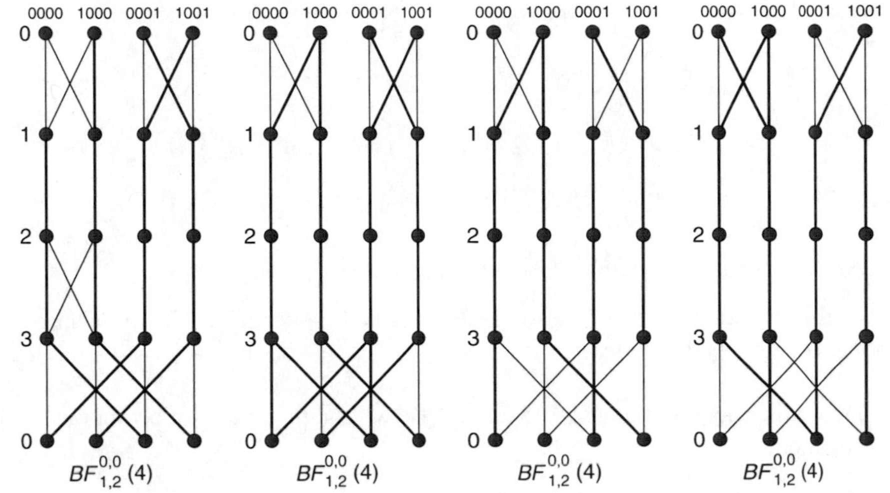

FIGURE 20.19 Weakly 1-scheduled Hamiltonian paths from $u = \langle 0, 0000\rangle$ to every level-1 vertex in $BF^{0,0}_{1,2}(4)$.

Case 1. Suppose that $x_1 = 0$ and $x_2 = 0$. Obviously, we have $v = v_1^{00}$. Since $BF(4)$ can be partitioned into $\{BF^{0,0}_{1,2}(4), BF^{1,0}_{1,2}(4), BF^{0,1}_{1,2}(4), BF^{1,1}_{1,2}(4)\}$, the path P' generated by

$$\Big(E(P) \cup E(C^{10}) \cup E(C^{01}) \cup E(C^{11}) \\ \cup \big\{\big(w_1^{00}, w_2^{10}\big), \big(w_2^{00}, w_1^{10}\big), \big(w_2^{00}, w_3^{01}\big), \big(w_3^{00}, w_2^{01}\big), \big(w_1^{01}, w_2^{11}\big), \big(w_2^{01}, w_1^{11}\big)\big\}\Big) \\ - \big\{\big(w_1^{00}, w_2^{00}\big), \big(w_1^{10}, w_2^{10}\big), \big(w_2^{00}, w_3^{00}\big), \big(w_2^{01}, w_3^{01}\big), \big(w_1^{01}, w_2^{01}\big), \big(w_1^{11}, w_2^{11}\big)\big\}$$

is a weakly 1-scheduled path of $BF(4)$ joining u to v.

Case 2. Suppose that $x_1 = 1$ and $x_2 = 0$. Obviously, the path P' generated by

$$\Big(E(P) \cup E(C^{10}) \cup E(C^{01}) \cup E(C^{11}) \\ \cup \big\{\big(v_1^{00}, v_2^{10}\big), \big(w_2^{00}, w_3^{01}\big), \big(w_3^{00}, w_2^{01}\big), \big(w_1^{01}, w_2^{11}\big), \big(w_2^{01}, w_1^{11}\big)\big\}\Big) \\ - \big\{\big(v_1^{10}, v_2^{10}\big), \big(w_2^{00}, w_3^{00}\big), \big(w_2^{01}, w_3^{01}\big), \big(w_1^{01}, w_2^{01}\big), \big(w_1^{11}, w_2^{11}\big)\big\}$$

is a weakly 1-scheduled path of $BF(4)$ joining u to v.

Case 3. Suppose that $x_1 = 0$ and $x_2 = 1$. Obviously, the path P' generated by

$$\Big(E(P) \cup E(C^{10}) \cup E(C^{01}) \cup E(C^{11}) \cup \big\{\big(v_1^{00}, v_2^{10}\big), \big(v_3^{10}, v_2^{11}\big), \big(v_1^{11}, v_2^{01}\big)\big\}\Big) \\ - \big\{\big(v_2^{10}, v_3^{10}\big), \big(v_1^{11}, v_2^{11}\big), \big(v_1^{01}, v_2^{01}\big)\big\}$$

is a weakly 1-scheduled path of $BF(4)$ joining u to v.

Case 4. Suppose that $x_1 = 1$ and $x_2 = 1$. Obviously, the path P' generated by

$$\Big(E(P) \cup E(C^{10}) \cup E(C^{01}) \cup E(C^{11}) \cup \big\{\big(v_1^{00}, v_2^{10}\big), \big(v_3^{10}, v_2^{11}\big), \big(w_1^{11}, w_2^{01}\big), \big(w_2^{11}, w_1^{01}\big)\big\}\Big) - \big\{\big(v_2^{10}, v_3^{10}\big), \big(v_1^{11}, v_2^{11}\big), \big(w_1^{01}, w_2^{01}\big), \big(w_1^{11}, w_2^{11}\big)\big\}$$

is a weakly 1-scheduled path of $BF(4)$ joining u to v.

With symmetry, we are able to create a weakly 3-scheduled Hamiltonian path from u to every level-3 vertex. □

Depending upon Lemmas 20.18 and 20.19, we can prove the following theorem by induction.

THEOREM 20.3 **[343]** Assume that n is an even integer greater than two. Let $u = \langle 0, 0^n \rangle$ and let v be any level-ℓ vertex of $BF(n)$ for odd $\ell \in \mathbb{Z}_n$. Then there exists a weakly ℓ-scheduled Hamiltonian path of $BF(n)$ joining u to v.

Proof: For convenience, let $v = \langle \ell, x_0 x_1 \ldots x_{n-1} \rangle$ and $v_h^{ij} = \langle h, x_0 \ldots x_{\ell-1} ij x_{\ell+2} \ldots x_{n-1} \rangle$. We proceed by induction on n. For $n = 4$, Lemma 20.19 confirms this condition. Now we assume that $BF(n-2)$ is Hamiltonian laceable for $n \geq 6$. By Theorem 20.2, there exists a totally scheduled Hamiltonian cycle C in $BF(n-2)$. By the inductive hypothesis, there also exists a weakly ℓ-scheduled Hamiltonian P of $BF(n-2)$ from u to any vertex on an odd level. Since P is weakly ℓ-scheduled, there is at least one level-ℓ vertex $w = \langle \ell, y_0 \ldots y_{\ell-1} y_{\ell+2} \ldots y_{n-1} \rangle$ of P incident to a level-(ℓ-1) edge and a level-ℓ edge on P. If the final edge of P is a level-ℓ edge, then we assume that $P' = \langle u, \gamma_{\ell+1}^0 \circ \gamma_\ell^0(P), v_{\ell+2}^{0,0}, v_{\ell+1}^{0,0}, v_\ell^{0,0} \rangle$. Otherwise, we assume that $P' = \gamma_{\ell+1}^0 \circ \gamma_\ell^0(P)$. In either case, P' is a weakly ℓ-scheduled path joining u to v_ℓ^{00}. For convenience, let $w_h^{ij} = \langle h, y_0 \ldots y_{\ell-1} ij y_{\ell+2} \ldots y_{n-1} \rangle$.

Case 1. Suppose that $x_\ell = 0$ and $x_{\ell+1} = 0$.

Since $BF(n)$ can be partitioned into $\{BF_{\ell,\ell+1}^{0,0}(n), BF_{\ell,\ell+1}^{1,0}(n), BF_{\ell,\ell+1}^{0,1}(n), BF_{\ell,\ell+1}^{1,1}(n)\}$, the path P'' generated by

$$\Big(E(P') \cup E\big(\gamma_{\ell,\ell+1}^{1,0}(C)\big) \cup E\big(\gamma_{\ell,\ell+1}^{0,1}(C)\big) \cup E\big(\gamma_{\ell,\ell+1}^{1,1}(C)\big) \cup \big\{\big(w_\ell^{00}, w_{\ell+1}^{10}\big), \big(w_{\ell+1}^{00}, w_\ell^{10}\big), \big(w_{\ell+1}^{00}, w_{\ell+2}^{01}\big), \big(w_{\ell+2}^{00}, w_{\ell+1}^{01}\big), \big(w_\ell^{01}, w_{\ell+1}^{11}\big), \big(w_{\ell+1}^{01}, w_\ell^{11}\big)\big\}\Big) - \big\{\big(w_\ell^{00}, w_{\ell+1}^{00}\big), \big(w_\ell^{10}, w_{\ell+1}^{10}\big), \big(w_{\ell+1}^{00}, w_{\ell+2}^{00}\big), \big(w_{\ell+1}^{01}, w_{\ell+2}^{01}\big), \big(w_\ell^{01}, w_{\ell+1}^{01}\big), \big(w_\ell^{11}, w_{\ell+1}^{11}\big)\big\}$$

is a weakly ℓ-scheduled path of $BF(n)$ joining u to v.

Case 2. Suppose that $x_\ell = 1$ and $x_{\ell+1} = 0$. Obviously, the path P'' generated by

$$\begin{aligned}&\Big(E(P') \cup E\left(\gamma^{1,0}_{\ell,\ell+1}(C)\right) \cup E\left(\gamma^{0,1}_{\ell,\ell+1}(C)\right) \cup E\left(\gamma^{1,1}_{\ell,\ell+1}(C)\right)\\ &\cup \{(v^{00}_\ell, v^{10}_{\ell+1}), (w^{00}_{\ell+1}, w^{01}_{\ell+2}), (w^{00}_{\ell+2}, w^{01}_{\ell+1}), (w^{01}_\ell, w^{11}_{\ell+1}), (w^{01}_{\ell+1}, w^{11}_\ell)\}\Big)\\ &- \{(v^{10}_\ell, v^{10}_{\ell+1}), (w^{00}_{\ell+1}, w^{00}_{\ell+2}), (w^{01}_{\ell+1}, w^{01}_{\ell+2}), (w^{01}_\ell, w^{01}_{\ell+1}), (w^{11}_\ell, w^{11}_{\ell+1})\}\end{aligned}$$

is a weakly ℓ-scheduled path of $BF(n)$ joining u to v.

Case 3. Suppose that $x_\ell = 0$ and $x_{\ell+1} = 1$. Obviously, the path P'' generated by

$$\begin{aligned}&\Big(E(P') \cup E\left(\gamma^{1,0}_{\ell,\ell+1}(C)\right) \cup E\left(\gamma^{0,1}_{\ell,\ell+1}(C)\right) \cup E\left(\gamma^{1,1}_{\ell,\ell+1}(C)\right)\\ &\cup \{(v^{00}_\ell, v^{10}_{\ell+1}), (v^{10}_{\ell+2}, v^{11}_{\ell+1}), (v^{11}_\ell, v^{01}_{\ell+1})\}\Big)\\ &- \{(v^{10}_{\ell+1}, v^{10}_{\ell+2}), (v^{11}_\ell, v^{11}_{\ell+1}), (v^{01}_\ell, v^{01}_{\ell+1})\}\end{aligned}$$

is a weakly ℓ-scheduled path of $BF(n)$ joining u to v.

Case 4. Suppose that $x_\ell = 1$ and $x_{\ell+1} = 1$. Obviously, the path P'' generated by

$$\begin{aligned}&\Big(E(P') \cup E\left(\gamma^{1,0}_{\ell,\ell+1}(C)\right) \cup E\left(\gamma^{0,1}_{\ell,\ell+1}(C)\right) \cup E\left(\gamma^{1,1}_{\ell,\ell+1}(C)\right)\\ &\cup \{(v^{00}_\ell, v^{10}_{\ell+1}), (v^{10}_{\ell+2}, v^{11}_{\ell+1}), (w^{11}_\ell, w^{01}_{\ell+1}), (w^{11}_{\ell+1}, w^{01}_\ell)\}\Big)\\ &- \{(v^{10}_{\ell+1}, v^{10}_{\ell+2}), (v^{11}_\ell, v^{11}_{\ell+1}), (w^{01}_\ell, w^{01}_{\ell+1}), (w^{11}_\ell, w^{11}_{\ell+1})\}\end{aligned}$$

is a weakly ℓ-scheduled path of $BF(n)$ joining u to v.

In either case, $\{P''\}$ is a 1-container of $BF(n)$ between u and v. By Lemma 20.4, we have $V(BF(n)) - V(P'') = V(BF^{0,0}_{\ell,\ell+1}(n)) - V(P') = \bigcup_{i=1}^{m} \{z_i, g(z_i)\}$ with some $m \geq 1$ such that $\{z_i \mid 1 \leq i \leq m\} \cap \{g(z_i) \mid 1 \leq i \leq m\} = \emptyset$. Moreover, we have $\{(f(z_i), g^{-1} \circ f(z_i)) \mid 1 \leq i \leq m\} \subseteq E(P'')$. Thus, by Lemma 20.16, there exists a weakly ℓ-scheduled Hamiltonian path $BF(n)$ joining u to v. □

Using an approach similar to that mentioned earlier, we have the following theorem.

THEOREM 20.4 **[343]** Assume that n is an odd integer greater than two. Let $u = \langle 0, 0^n \rangle$ and let v be any level-ℓ vertex of $BF(n)$ for any ℓ. Then there exists a weakly ℓ-scheduled Hamiltonian path of $BF(n)$ joining u to v.

Theorems 20.3 and 20.4 imply the following corollary.

COROLLARY 20.6 **[343]** *Assume that $n \geq 3$. Then $BF(n)$ is 1^*-laceable (respectively 1^*-connected) if n is even (respectively odd).*

Using a similar technique, Kueng et al. [208] prove that $BF(n)$ is i^*-laceable (respectively i^*-connected) if n is even (respectively odd) for $i \in \{3, 4\}$. Thus, we have the following theorem.

THEOREM 20.5 **[208]** Let $n \geq 3$. Then $BF(n)$ is super spanning connected if n is odd, and is super spanning laceable otherwise.

20.4 MUTUALLY INDEPENDENT HAMILTONICITY FOR BUTTERFLY GRAPHS

Kueug et al. [208] also study the mutually independent Hamiltonicity for butterfly graphs. It is proved that $\mathcal{IHC}(BF(n)) = 4$ for $n \geq 3$. By Theorem 20.2, there exists a totally scheduled Hamiltonian cycle in $BF(n)$. Using the aforementioned stretching operation, we have the following corollary.

COROLLARY 20.7 *Assume that $n \geq 3$ and $i, j, k \in \mathbb{Z}_2$. Then there exists a totally scheduled Hamiltonian cycle of $BF_{0,1,2}^{i,j,k}(n)$ including all straight edges of level 0, level 1, and level 2 in $BF_{0,1,2}^{i,j,k}(n)$.*

Suppose that $e_1 = (u_1, v_1)$ and $e_2 = (u_2, v_2)$ are either any two cross edges of $BF(n)$ or any two straight edges of $BF(n)$. It is known that there exists an isomorphism μ over $V(BF(n))$ such that $u_2 = \mu(u_1)$ and $v_2 = \mu(v_1)$. Clearly, every Hamiltonian cycle of $BF(n)$ contains at least one cross edge and at least one straight edge. Hence, we have the following lemma.

LEMMA 20.20 For any edge e of $BF(n)$ with $n \geq 3$, there exists a totally scheduled Hamiltonian cycle of $BF(n)$ including e.

LEMMA 20.21 Assume that $i, j, k \in \mathbb{Z}_2$ and e is a non-level-3 straight edge of $BF_{0,1,2}^{i,j,k}(4)$. There exists a totally scheduled Hamiltonian cycle of $BF_{0,1,2}^{i,j,k}(4)$ including e.

Proof: Obviously, $\langle\langle 0, ijk0\rangle, \langle 1, ijk0\rangle, \langle 2, ijk0\rangle, \langle 3, ijk0\rangle, \langle 0, ijk1\rangle, \langle 1, ijk1\rangle, \langle 2, ijk1\rangle, \langle 3, ijk1\rangle$, and $\langle 0, ijk0\rangle\rangle$ forms a desired Hamiltonian cycle of $BF_{0,1,2}^{i,j,k}(4)$. □

By stretching operation and Corollary 20.7, we have the following corollary.

COROLLARY 20.8 *Let e be any edge of $BF_{0,1,2}^{i,j,k}(n)$ for $n \geq 5$ and some $i, j, k \in \mathbb{Z}_2$. There exists a totally scheduled Hamiltonian cycle of $BF_{0,1,2}^{i,j,k}(n)$ including e.*

To prove the main result, we need the following lemmas.

LEMMA 20.22 Let $n \geq 3$. Assume that $i, j \in \mathbb{Z}_2$ and $0 \leq \ell \leq n-3$. Suppose that s is any level-$(\ell+1)$ vertex of $BF_{\ell,\ell+1}^{i,j}(n)$ and d is any level-$(\ell+2)$ vertex of $BF_{\ell,\ell+1}^{i,j}(n)$. When $n = 3$, there exists a 0-scheduled Hamiltonian path P_3 of $BF_{0,1}^{i,j}(3)$ joining s to d

with $P_3(2)=g^{-1}(s)$. When $n\geq 4$, there exists an ℓ-scheduled Hamiltonian path P_n of $BF^{i,j}_{\ell,\ell+1}(n)$ joining s to d such that P_n is weakly $(\ell+2)$-scheduled with $P_n(2)=g^{-1}(s)$. In particular, $P_n(n\times 2^{n-2}-2)=g^{-2}(d)$ and $P_n(n\times 2^{n-2}-1)=g^{-1}(d)$ for $n\geq 3$ if $d\neq g(s)$.

Proof: Without loss of generality, we assume $\ell=0$ such that $s=\langle 1, ij0^{n-2}\rangle$ and $d=\langle 2, ijz\rangle$ with some $z\in\mathbb{Z}_2^{n-2}$. The desired Hamiltonian paths of $BF^{i,j}_{0,1}(3)$ are

$$\langle\langle 1, ij0\rangle, \langle 0, ij0\rangle, \langle 2, ij1\rangle, \langle 1, ij1\rangle, \langle 0, ij1\rangle, \langle 2, ij0\rangle\rangle$$

$$\langle\langle 1, ij0\rangle, \langle 0, ij0\rangle, \langle 2, ij0\rangle, \langle 0, ij1\rangle, \langle 1, ij1\rangle, \langle 2, ij1\rangle\rangle$$

For $n\geq 4$, we further assume that $d=\langle 2, ijpqx\rangle$ with some $p,q\in\mathbb{Z}_2$ and $x\in\mathbb{Z}_2^{n-4}$. Then we construct the desired Hamiltonian path by induction on n.

The induction bases are listed in Tables 20.1 and 20.2. Then we assume that the statement holds for $BF^{i,j}_{0,1}(n-2)$ with $n\geq 6$ and partition $BF^{i,j}_{0,1}(n)$ into $\{BF^{i,j,h,k}_{0,1,2,3}(n)\mid h,k\in\mathbb{Z}_2\}$. By induction hypothesis, there exists a 0-scheduled Hamiltonian path P^{00} of $BF^{i,j}_{0,1}(n-2)$ such that P^{00} joins $\langle 1, ij0^{n-4}\rangle$ to $\langle 2, ijx\rangle$ and also that P^{00} is weakly 2-scheduled. Thus, there is at least one level-2 vertex $v=\langle 2, ijy\rangle$ with $y\neq x$ of $I(P^{00})$ incident to a level-1 edge and a level-2 edge on P^{00}. By Lemma 20.3, $BF^{i,j,0,0}_{0,1,2,3}(n)=\gamma_3^0\circ\gamma_2^0(BF^{i,j}_{0,1}(n-2))$. Thus, $\gamma_3^0\circ\gamma_2^0(P^{00})$ is obviously either a path joining s to $\langle 2, ij00x\rangle$ or a path joining s to $\langle 4, ij00x\rangle$. By Corollary 20.5, there is a totally scheduled Hamiltonian cycle C^{hk} of $BF^{i,j,h,k}_{0,1,2,3}(n)$, including all straight edges of levels 2 and 3, for every $hk\in\mathbb{Z}_2^2-\{00\}$.

Let $F_k=\{\langle 2, ijw\rangle\in V(P^{00})\mid \langle 2, ijw\rangle$ is not incident to any level-$(k+1)$ edge in $P^{00}\}$ with $k\in\{0,1\}$. By Lemma 20.4, $V(\gamma_3^0\circ\gamma_2^0(P^{00}))=V(BF^{i,j,0,0}_{0,1,2,3}(n))-(\overline{F_0}\cup\overline{F_1})$, where $\overline{F_0}=\{\langle 2, ij00w\rangle\mid\langle 2, ijw\rangle\in F_0\}\cup\{\langle 3, ij00w\rangle\mid\langle 2, ijw\rangle\in F_0\}$ and $\overline{F_1}=\{\langle 3, ij00w\rangle\mid\langle 2, ijw\rangle\in F_1\}\cup\{\langle 4, ij00w\rangle\mid\langle 2, ijw\rangle\in F_1\}$. If $\gamma_3^0\circ\gamma_2^0(P^{00})$ joins s to $\langle 2, ij00x\rangle$, then let $\overline{P^{00}}=\gamma_3^0\circ\gamma_2^0(P^{00})$ and $\widetilde{F_0}=\overline{F_0}$. Otherwise, let $\overline{P^{00}}=\langle s, \gamma_3^0\circ\gamma_2^0(P^{00}), \langle 4, ij00x\rangle, \langle 3, ij00x\rangle, \langle 2, ij00x\rangle\rangle$ and $\widetilde{F_0}=\overline{F_0}-\{\langle 2, ij00x\rangle, \langle 3, ij00x\rangle\}$. Then we consider the following cases.

TABLE 20.1
Hamiltonian Paths of $BF^{i,j}_{0,1}(4)$ as Induction Bases

$\langle\langle 1, ij00\rangle, \langle 0, ij00\rangle, \langle 3, ij01\rangle, \langle 2, ij11\rangle, \langle 1, ij11\rangle, \langle 0, ij11\rangle, \langle 3, ij11\rangle, \langle 2, ij01\rangle,$
$\langle 1, ij01\rangle, \langle 0, ij01\rangle, \langle 3, ij00\rangle, \langle 2, ij10\rangle, \langle 1, ij10\rangle, \langle 0, ij10\rangle, \langle 3, ij10\rangle, \langle 2, ij00\rangle\rangle$
$\langle\langle 1, ij00\rangle, \langle 0, ij00\rangle, \langle 3, ij00\rangle, \langle 2, ij00\rangle, \langle 3, ij10\rangle, \langle 0, ij11\rangle, \langle 1, ij11\rangle, \langle 2, ij11\rangle,$
$\langle 3, ij01\rangle, \langle 0, ij01\rangle, \langle 1, ij01\rangle, \langle 2, ij01\rangle, \langle 3, ij11\rangle, \langle 0, ij10\rangle, \langle 1, ij10\rangle, \langle 2, ij10\rangle\rangle$
$\langle\langle 1, ij00\rangle, \langle 0, ij00\rangle, \langle 3, ij00\rangle, \langle 2, ij00\rangle, \langle 3, ij10\rangle, \langle 2, ij10\rangle, \langle 1, ij10\rangle, \langle 0, ij10\rangle,$
$\langle 3, ij11\rangle, \langle 0, ij11\rangle, \langle 1, ij11\rangle, \langle 2, ij11\rangle, \langle 3, ij01\rangle, \langle 0, ij01\rangle, \langle 1, ij01\rangle, \langle 2, ij01\rangle\rangle$
$\langle\langle 1, ij00\rangle, \langle 0, ij00\rangle, \langle 3, ij01\rangle, \langle 2, ij01\rangle, \langle 1, ij01\rangle, \langle 0, ij01\rangle, \langle 3, ij00\rangle, \langle 2, ij00\rangle,$
$\langle 3, ij10\rangle, \langle 2, ij10\rangle, \langle 1, ij10\rangle, \langle 0, ij10\rangle, \langle 3, ij11\rangle, \langle 0, ij11\rangle, \langle 1, ij11\rangle, \langle 2, ij11\rangle\rangle$

TABLE 20.2
Hamiltonian Paths of $BF^{i,j}_{0,1}$ (5) as Induction Bases

$\langle\langle 1, ij000\rangle, \langle 0, ij000\rangle, \langle 4, ij001\rangle, \langle 3, ij011\rangle, \langle 2, ij111\rangle, \langle 1, ij111\rangle, \langle 0, ij111\rangle, \langle 4, ij111\rangle, \langle 3, ij111\rangle, \langle 2, ij011\rangle,$
$\langle 1, ij011\rangle, \langle 0, ij011\rangle, \langle 4, ij011\rangle, \langle 3, ij001\rangle, \langle 2, ij101\rangle, \langle 1, ij101\rangle, \langle 0, ij101\rangle, \langle 4, ij101\rangle, \langle 3, ij101\rangle, \langle 2, ij001\rangle,$
$\langle 1, ij001\rangle, \langle 0, ij001\rangle, \langle 4, ij000\rangle, \langle 3, ij010\rangle, \langle 2, ij110\rangle, \langle 1, ij110\rangle, \langle 0, ij110\rangle, \langle 4, ij110\rangle, \langle 3, ij110\rangle, \langle 2, ij010\rangle,$
$\langle 1, ij010\rangle, \langle 0, ij010\rangle, \langle 4, ij010\rangle, \langle 3, ij000\rangle, \langle 2, ij100\rangle, \langle 1, ij100\rangle, \langle 0, ij100\rangle, \langle 4, ij100\rangle, \langle 3, ij100\rangle, \langle 2, ij000\rangle\rangle$
$\langle\langle 1, ij000\rangle, \langle 0, ij000\rangle, \langle 4, ij001\rangle, \langle 3, ij011\rangle, \langle 2, ij111\rangle, \langle 1, ij111\rangle, \langle 0, ij111\rangle, \langle 4, ij111\rangle, \langle 3, ij111\rangle, \langle 2, ij011\rangle,$
$\langle 1, ij011\rangle, \langle 0, ij011\rangle, \langle 4, ij011\rangle, \langle 3, ij001\rangle, \langle 2, ij101\rangle, \langle 1, ij101\rangle, \langle 0, ij101\rangle, \langle 4, ij101\rangle, \langle 3, ij101\rangle, \langle 2, ij001\rangle,$
$\langle 1, ij001\rangle, \langle 0, ij001\rangle, \langle 4, ij000\rangle, \langle 3, ij010\rangle, \langle 2, ij110\rangle, \langle 1, ij110\rangle, \langle 0, ij110\rangle, \langle 4, ij110\rangle, \langle 3, ij110\rangle, \langle 2, ij010\rangle,$
$\langle 1, ij010\rangle, \langle 0, ij010\rangle, \langle 4, ij010\rangle, \langle 3, ij000\rangle, \langle 2, ij000\rangle, \langle 3, ij100\rangle, \langle 4, ij100\rangle, \langle 0, ij100\rangle, \langle 1, ij100\rangle, \langle 2, ij100\rangle\rangle$
$\langle\langle 1, ij000\rangle, \langle 0, ij000\rangle, \langle 4, ij001\rangle, \langle 3, ij011\rangle, \langle 2, ij111\rangle, \langle 1, ij111\rangle, \langle 0, ij111\rangle, \langle 4, ij111\rangle, \langle 3, ij111\rangle, \langle 2, ij011\rangle,$
$\langle 1, ij011\rangle, \langle 0, ij011\rangle, \langle 4, ij011\rangle, \langle 3, ij001\rangle, \langle 2, ij101\rangle, \langle 1, ij101\rangle, \langle 0, ij101\rangle, \langle 4, ij101\rangle, \langle 3, ij101\rangle, \langle 2, ij001\rangle,$
$\langle 1, ij001\rangle, \langle 0, ij001\rangle, \langle 4, ij000\rangle, \langle 3, ij000\rangle, \langle 2, ij000\rangle, \langle 3, ij100\rangle, \langle 2, ij100\rangle, \langle 1, ij100\rangle, \langle 0, ij100\rangle, \langle 4, ij100\rangle,$
$\langle 3, ij110\rangle, \langle 4, ij110\rangle, \langle 0, ij110\rangle, \langle 1, ij110\rangle, \langle 2, ij110\rangle, \langle 3, ij010\rangle, \langle 4, ij010\rangle, \langle 0, ij010\rangle, \langle 1, ij010\rangle, \langle 2, ij010\rangle\rangle$
$\langle\langle 1, ij000\rangle, \langle 0, ij000\rangle, \langle 4, ij001\rangle, \langle 3, ij011\rangle, \langle 2, ij111\rangle, \langle 1, ij111\rangle, \langle 0, ij111\rangle, \langle 4, ij111\rangle, \langle 3, ij111\rangle, \langle 2, ij011\rangle,$
$\langle 1, ij011\rangle, \langle 0, ij011\rangle, \langle 4, ij011\rangle, \langle 3, ij001\rangle, \langle 2, ij101\rangle, \langle 1, ij101\rangle, \langle 0, ij101\rangle, \langle 4, ij101\rangle, \langle 3, ij101\rangle, \langle 2, ij001\rangle,$
$\langle 1, ij001\rangle, \langle 0, ij001\rangle, \langle 4, ij000\rangle, \langle 3, ij010\rangle, \langle 2, ij010\rangle, \langle 1, ij010\rangle, \langle 0, ij010\rangle, \langle 4, ij010\rangle, \langle 3, ij000\rangle, \langle 2, ij000\rangle,$
$\langle 3, ij100\rangle, \langle 2, ij100\rangle, \langle 1, ij100\rangle, \langle 0, ij100\rangle, \langle 4, ij100\rangle, \langle 3, ij110\rangle, \langle 4, ij110\rangle, \langle 0, ij110\rangle, \langle 1, ij110\rangle, \langle 2, ij110\rangle\rangle$
$\langle\langle 1, ij000\rangle, \langle 0, ij000\rangle, \langle 4, ij000\rangle, \langle 3, ij000\rangle, \langle 2, ij000\rangle, \langle 3, ij100\rangle, \langle 2, ij100\rangle, \langle 1, ij100\rangle, \langle 0, ij100\rangle, \langle 4, ij100\rangle,$
$\langle 3, ij110\rangle, \langle 2, ij010\rangle, \langle 1, ij010\rangle, \langle 0, ij010\rangle, \langle 4, ij010\rangle, \langle 3, ij010\rangle, \langle 2, ij110\rangle, \langle 1, ij110\rangle, \langle 0, ij110\rangle, \langle 4, ij110\rangle,$
$\langle 0, ij111\rangle, \langle 1, ij111\rangle, \langle 2, ij111\rangle, \langle 3, ij011\rangle, \langle 4, ij011\rangle, \langle 0, ij011\rangle, \langle 1, ij011\rangle, \langle 2, ij011\rangle, \langle 3, ij111\rangle, \langle 4, ij111\rangle,$
$\langle 3, ij101\rangle, \langle 4, ij101\rangle, \langle 0, ij101\rangle, \langle 1, ij101\rangle, \langle 2, ij101\rangle, \langle 3, ij001\rangle, \langle 4, ij001\rangle, \langle 0, ij001\rangle, \langle 1, ij001\rangle, \langle 2, ij001\rangle\rangle$
$\langle\langle 1, ij000\rangle, \langle 0, ij000\rangle, \langle 4, ij000\rangle, \langle 3, ij000\rangle, \langle 2, ij000\rangle, \langle 3, ij100\rangle, \langle 2, ij100\rangle, \langle 1, ij100\rangle, \langle 0, ij100\rangle, \langle 4, ij100\rangle,$
$\langle 3, ij110\rangle, \langle 2, ij010\rangle, \langle 1, ij010\rangle, \langle 0, ij010\rangle, \langle 4, ij010\rangle, \langle 3, ij010\rangle, \langle 2, ij110\rangle, \langle 1, ij110\rangle, \langle 0, ij110\rangle, \langle 4, ij110\rangle,$
$\langle 0, ij111\rangle, \langle 1, ij111\rangle, \langle 2, ij111\rangle, \langle 3, ij011\rangle, \langle 4, ij001\rangle, \langle 0, ij001\rangle, \langle 1, ij001\rangle, \langle 2, ij001\rangle, \langle 3, ij001\rangle, \langle 4, ij011\rangle,$
$\langle 0, ij011\rangle, \langle 1, ij011\rangle, \langle 2, ij011\rangle, \langle 3, ij111\rangle, \langle 4, ij111\rangle, \langle 3, ij101\rangle, \langle 4, ij101\rangle, \langle 0, ij101\rangle, \langle 1, ij101\rangle, \langle 2, ij101\rangle\rangle$
$\langle\langle 1, ij000\rangle, \langle 0, ij000\rangle, \langle 4, ij000\rangle, \langle 3, ij000\rangle, \langle 2, ij000\rangle, \langle 3, ij100\rangle, \langle 2, ij100\rangle, \langle 1, ij100\rangle, \langle 0, ij100\rangle, \langle 4, ij100\rangle,$
$\langle 3, ij110\rangle, \langle 2, ij010\rangle, \langle 1, ij010\rangle, \langle 0, ij010\rangle, \langle 4, ij010\rangle, \langle 3, ij010\rangle, \langle 2, ij110\rangle, \langle 1, ij110\rangle, \langle 0, ij110\rangle, \langle 4, ij110\rangle,$
$\langle 0, ij111\rangle, \langle 1, ij111\rangle, \langle 2, ij111\rangle, \langle 3, ij111\rangle, \langle 4, ij111\rangle, \langle 3, ij101\rangle, \langle 4, ij101\rangle, \langle 0, ij101\rangle, \langle 1, ij101\rangle, \langle 2, ij101\rangle,$
$\langle 3, ij001\rangle, \langle 2, ij001\rangle, \langle 1, ij001\rangle, \langle 0, ij001\rangle, \langle 4, ij001\rangle, \langle 3, ij011\rangle, \langle 4, ij011\rangle, \langle 0, ij011\rangle, \langle 1, ij011\rangle, \langle 2, ij011\rangle\rangle$
$\langle\langle 1, ij000\rangle, \langle 0, ij000\rangle, \langle 4, ij000\rangle, \langle 3, ij010\rangle, \langle 2, ij110\rangle, \langle 1, ij110\rangle, \langle 0, ij110\rangle, \langle 4, ij110\rangle, \langle 3, ij110\rangle, \langle 2, ij010\rangle,$
$\langle 1, ij010\rangle, \langle 0, ij010\rangle, \langle 4, ij010\rangle, \langle 3, ij000\rangle, \langle 2, ij000\rangle, \langle 3, ij100\rangle, \langle 2, ij100\rangle, \langle 1, ij100\rangle, \langle 0, ij100\rangle, \langle 4, ij100\rangle,$
$\langle 0, ij101\rangle, \langle 1, ij101\rangle, \langle 2, ij101\rangle, \langle 3, ij001\rangle, \langle 4, ij011\rangle, \langle 0, ij011\rangle, \langle 1, ij011\rangle, \langle 2, ij011\rangle, \langle 3, ij011\rangle, \langle 4, ij001\rangle,$
$\langle 0, ij001\rangle, \langle 1, ij001\rangle, \langle 2, ij001\rangle, \langle 3, ij101\rangle, \langle 4, ij101\rangle, \langle 3, ij111\rangle, \langle 4, ij111\rangle, \langle 0, ij111\rangle, \langle 1, ij111\rangle, \langle 2, ij111\rangle\rangle$

Case 1. If $ij = 00$, $d = \langle 2, ij00x\rangle$. Let $A = \{(\langle 2, ij10y\rangle, \langle 3, ij00y\rangle), (\langle 2, ij00y\rangle, \langle 3, ij10y\rangle), (\langle 2, ij11y\rangle, \langle 3, ij01y\rangle), (\langle 2, ij01y\rangle, \langle 3, ij11y\rangle), (\langle 3, ij11y\rangle, \langle 4, ij10y\rangle), (\langle 3, ij10y\rangle, \langle 4, ij11y\rangle)\}$ and $B = \{(\langle 2, ij00y\rangle, \langle 3, ij00y\rangle), (\langle 2, ij10y\rangle, \langle 3, ij10y\rangle), (\langle 2, ij01y\rangle, \langle 3, ij01y\rangle), (\langle 2, ij11y\rangle, \langle 3, ij11y\rangle), (\langle 3, ij10y\rangle, \langle 4, ij10y\rangle), (\langle 3, ij11y\rangle, \langle 4, ij11y\rangle)\}$. Then the subgraph P, generated by $(E(\overline{P^{00}}) \cup E(C^{10}) \cup E(C^{01}) \cup E(C^{11}) \cup A) - B$, forms a weakly 2-scheduled path of $BF^{i,j}_{0,1}(n)$ joining s to d with $V(BF^{i,j}_{0,1}(n)) - V(P) = \widetilde{F_0} \cup \overline{F_1}$. Moreover, since $\overline{P^{00}}$, C^{10}, C^{01}, and C^{11} are all 0-scheduled, P is 0-scheduled.

For convenience, let $d^{ab}_h = \langle h, ijabx\rangle$ and $v^{ab}_h = \langle h, ijaby\rangle$ for any $a, b \in \mathbb{Z}_2$, $h \in \{2, 3, 4\}$.

Case 2. If $ij = 10$, $d = \langle 2, ij10x\rangle$. Let $A = \{(d_2^{00}, d_3^{10}), (v_2^{11}, v_3^{01}), (v_2^{01}, v_3^{11}), (v_3^{11}, v_4^{10}), (v_3^{10}, v_4^{11})\}$ and $B = \{(d_2^{10}, d_3^{10}), (v_2^{01}, v_3^{01}), (v_2^{11}, v_3^{11}), (v_3^{10}, v_4^{10}), (v_3^{11}, v_4^{11})\}$. Then the subgraph P, generated by $(E(\overline{P^{00}}) \cup E(C^{10}) \cup E(C^{01}) \cup E(C^{11}) \cup A) - B$, forms the desired weakly 2-scheduled path of $BF_{0,1}^{i,j}(n)$ joining s to d.

Case 3. If $ij = 01$, $d = \langle 2, ij01x\rangle$. Let $A = \{(d_2^{00}, d_3^{10}), (d_2^{11}, d_3^{01}), (d_3^{11}, d_4^{10})\}$ and $B = \{(d_2^{01}, d_3^{01}), (d_2^{11}, d_3^{11}), (d_3^{10}, d_4^{10})\}$. Then the subgraph P, generated by $(E(\overline{P^{00}}) \cup E(C^{10}) \cup E(C^{01}) \cup E(C^{11}) \cup A) - B$, forms the desired weakly 2-scheduled path of $BF_{0,1}^{i,j}(n)$ joining s to d.

Case 4. If $ij = 11$, $d = \langle 2, ij11x\rangle$. Let $A = \{(d_2^{00}, d_3^{10}), (d_3^{11}, d_4^{10}), (v_3^{01}, v_4^{00}), (v_3^{00}, v_4^{01})\}$ and $B = \{(d_3^{10}, d_4^{10}), (v_3^{00}, v_4^{00}), (v_3^{01}, v_4^{01}), (d_2^{11}, d_3^{11})\}$. Then the subgraph P, generated by $(E(\overline{P^{00}}) \cup E(C^{10}) \cup E(C^{01}) \cup E(C^{11}) \cup A) - B$, forms the desired weakly 2-scheduled path of $BF_{0,1}^{i,j}(n)$ joining s to d.

Note that P constructed previously is not a Hamiltonian path. For any $h, k \in \mathbb{Z}_2$, let

$$\begin{aligned}
X_0^{hk} &= \{(\langle 2, ijhkw\rangle, \langle 3, ijhkw\rangle) \mid \langle 2, ij00w\rangle \text{ and } \langle 3, ij00w\rangle \text{ are in } \widetilde{F_0}\} \\
Y_0^{hk} &= \{(\langle 2, ijhkw\rangle, \langle 3, ij\bar{h}kw\rangle) \mid \langle 2, ij00w\rangle \text{ and } \langle 3, ij00w\rangle \text{ are in } \widetilde{F_0}\} \\
X_1^{hk} &= \{(\langle 3, ijhkw\rangle, \langle 4, ijhkw\rangle) \mid \langle 3, ij00w\rangle \text{ and } \langle 4, ij00w\rangle \text{ are in } \overline{F_1}\} \\
Y_1^{hk} &= \{(\langle 3, ijhkw\rangle, \langle 4, ijh\bar{k}w\rangle) \mid \langle 3, ij00w\rangle \text{ and } \langle 4, ij00w\rangle \text{ are in } \overline{F_1}\}
\end{aligned}$$

Obviously, $X_0^{10} \subset E(C^{10})$ and $X_1^{01} \subset E(C^{01})$. Moreover, $(X_0^{10} \cup X_1^{01}) \cap B = \emptyset$. Therefore, $(X_0^{10} \cup X_1^{01}) \subset E(P)$. As a result, let P' be the subgraph generated by $(E(P) \cup (X_0^{00} \cup Y_0^{00} \cup Y_0^{10}) \cup (X_1^{00} \cup Y_1^{00} \cup Y_1^{01})) - (X_0^{10} \cup X_1^{01})$. Then P' is the desired Hamiltonian path of $BF_{0,1}^{i,j}(n)$ joining s to d. In Figure 20.20, we illustrate P' for **Case 1**. □

In terms of the symmetry of $BF(n)$, we have the following corollary.

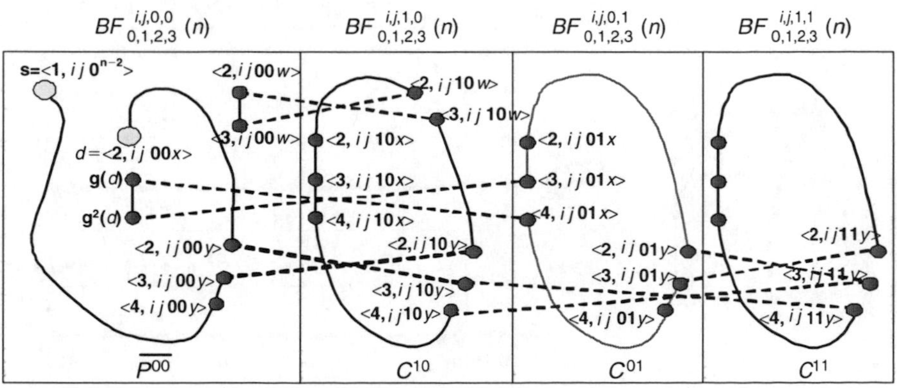

FIGURE 20.20 The path P' generated by $(E(P) \cup (X_0^{00} \cup Y_0^{00} \cup Y_0^{10}) \cup (X_1^{00} \cup Y_1^{00} \cup Y_1^{01})) - (X_0^{10} \cup X_1^{01})$.

COROLLARY 20.9 *Let $n \geq 3$. Assume $i,j \in \mathbb{Z}_2$ and $0 \leq \ell \leq n-3$. Suppose that s is any level-$(\ell+1)$ vertex of $BF^{i,j}_{\ell,\ell+1}(n)$ and d is any level-ℓ vertex of $BF^{i,j}_{\ell,\ell+1}(n)$. When $n=3$, there exists a 2-scheduled Hamiltonian path P_3 of $BF^{i,j}_{\ell,\ell+1}(n)$ joining s to d with $H(2)=g(s)$. When $n \geq 4$, there exists an $(\ell+2)$-scheduled Hamiltonian path P_n of $BF^{i,j}_{\ell,\ell+1}(n)$ joining s to d such that P_n is weakly ℓ-scheduled with $P_n(2)=g(s)$. In particular, $P_n(n \times 2^{n-2}-2)=g^2(d)$ and $P_n(n \times 2^{n-2}-1)=g(d)$ for $n \geq 3$ if $d \neq g^{-1}(s)$.*

LEMMA 20.23 Assume that $n \geq 4$. Let $s=\langle 1,0^n\rangle$, $d_1=\langle 2,0^2 10^{n-3}\rangle$, and $d_2=\langle 0,0^n\rangle$ be vertices of $BF^{0,0}_{0,1}(n)$. Then there exist two Hamiltonian paths H_1 and H_2 of $BF^{0,0}_{0,1}(n)$ such that (1) H_1 joins s to d_1, (2) H_2 joins s to d_2, and (3) $H_1(1)=H_2(1)=s$ and $H_1(t) \neq H_2(t)$ for all $2 \leq t \leq |V(BF^{0,0}_{0,1}(n))| = n \times 2^{n-2}$.

Proof: Let $u_1=g(s)=\langle 2,0^n\rangle$, $u_2=f(u_1)=g(d_1)=\langle 3,0^2 10^{n-3}\rangle$, $u_3=g^{-1}(d_1)=\langle 1,0^2 10^{n-3}\rangle$, $u_4=f(u_2)=\langle 4,0^2 110^{n-4}\rangle$, and $u_5=g(u_1)=f(d_1)=\langle 3,0^n\rangle$ be vertices of $BF^{0,0}_{0,1}(n)$. We partition $BF^{0,0}_{0,1}(n)$ into $\{BF^{0,0,0}_{0,1,2}(n), BF^{0,0,1}_{0,1,2}(n)\}$. By Corollary 20.7, there is a Hamiltonian cycle C_0 of $BF^{0,0,0}_{0,1,2}(n)$ including all straight edges of level 2. Thus, $(u_1,u_5) \in E(C_0)$. By Lemma 20.21 and Corollary 20.8, there is a Hamiltonian cycle C_1 of $BF^{0,0,1}_{0,1,2}(n)$ with $(u_2,u_4) \in E(C_1)$. Note that s and d_1 are vertices of degree 2 in $BF^{0,0,0}_{0,1,2}(n)$ and $BF^{0,0,1}_{0,1,2}(n)$, respectively. Therefore, we can write $C_0=\langle s,u_1,u_5,P_0,d_2,s\rangle$ and $C_1=\langle d_1,u_2,u_4,P_1,u_3,d_1\rangle$. As an example, Figure 20.21a depicts C_0 and C_1 on $BF^{0,0}_{0,1}(4)$. Besides, Figure 20.21b displays the

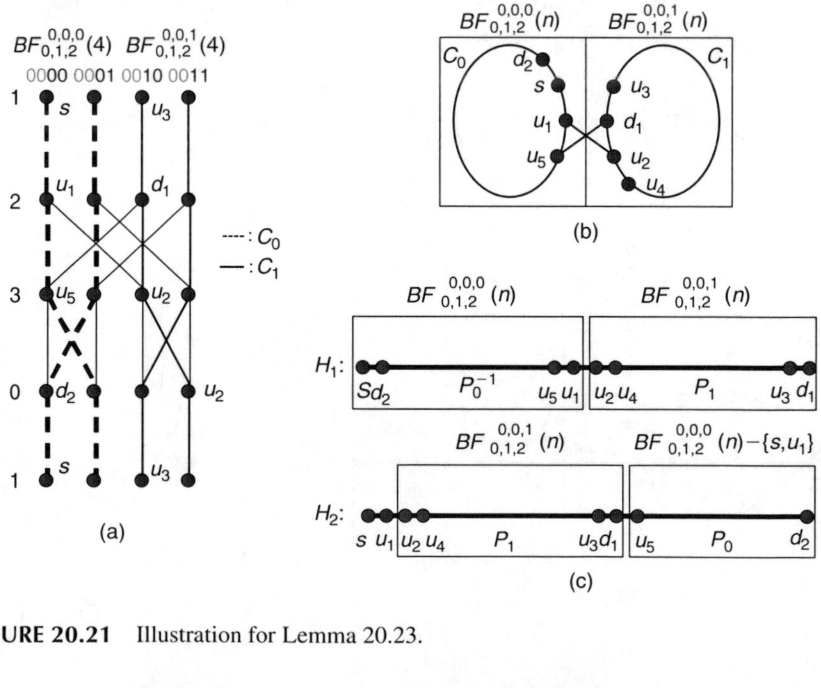

FIGURE 20.21 Illustration for Lemma 20.23.

abstraction of C_0 and C_1 for general n. Since $\{(u_1,u_2),(d_1,u_5)\} \subset E(BF_{0,1}^{0,0}(n))$, we set

$$H_1 = \langle s, d_2, P_0^{-1}, u_5, u_1, u_2, u_4, P_1, u_3, d_1 \rangle$$
$$H_2 = \langle s, u_1, u_2, u_4, P_1, u_3, d_1, u_5, P_0, d_2 \rangle$$

One can find, as shown in Figure 20.21c, that H_1 and H_2 satisfy the requirements. □

LEMMA 20.24 Given any $k \in \{0,1\}$ and $n \geq 4$, let (b_1, w_1) be a level-1 straight edge of $BF_{0,1,n-1}^{1,1,k}(n)$ and let (b_2, w_2) be a level-0 straight edge of $BF_{0,1,n-1}^{1,1,k}(n)$ such that w_1 and w_2 are two distinct level-1 vertices. There exist two Hamiltonian paths H_1 and H_2 of $BF_{0,1}^{1,1}(n)$ such that the following conditions are all satisfied:

1. $H_1(1) = b_1$ and $H_1(n \cdot 2^{n-2}) = w_1$
2. $H_2(1) = b_2$ and $H_2(n \cdot 2^{n-2}) = w_2$
3. $H_1(t) \neq H_2(t)$ for all $1 \leq t \leq n \times 2^{n-2}$

Proof: Without loss of generality, we assume that $k=0$. Let $u_1 = g^{n-3}(b_1)$, $u_2 = f(u_1)$, $u_3 = g(u_2)$, $u_4 = g(u_3)$, $u_5 = g^{n-3}(u_4) = g^{-1}(u_2)$, $u_6 = f(u_5) = g^{-1}(w_1)$, $v_1 = f^{-1}(b_2)$, $v_2 = g^{-n+3}(v_1)$, $v_3 = g^{-1}(v_2)$, $v_4 = g^{-1}(v_3) = g(v_1)$, $v_5 = f^{-1}(v_4) = g^{-1}(b_2)$, and $v_6 = g^{-n+3}(v_5) = g(w_2)$ be vertices of $BF_{0,1}^{1,1}(n)$. By Corollary 20.7 and Lemma 20.2, there exists a totally scheduled Hamiltonian cycle C_0 of $BF_{0,1,n-1}^{1,1,0}(n)$. Note that w_1 is adjacent to u_6. Moreover, w_1, u_6, b_2, and w_2 are all vertices of degree 2 in $BF_{0,1,n-1}^{1,1,0}(n)$. So C_0 can be written as $C_0 = \langle w_1, b_1, P_0, u_1, u_6, w_1 \rangle$, in which $P_0 = \langle b_1, P_{01}, v_5, b_2, w_2, v_6, P_{02}, u_1 \rangle$. Similarly, there exists a totally scheduled Hamiltonian cycle C_1 of $BF_{0,1,n-1}^{1,1,1}(n)$ with $(u_5, u_2) \in E(C_1)$. Since u_2, u_3, v_3, and v_4 are all vertices of degree 2 in $BF_{0,1,n-1}^{1,1,1}(n)$, we can write $C_1 = \langle u_3, u_4, P_1, u_5, u_2, u_3 \rangle$, in which $P_1 = \langle u_4, P_{11}, v_1, v_4, v_3, v_2, P_{12}, u_5 \rangle$.

As an example, Figure 20.22a depicts C_0 and C_1 of $BF_{0,1}^{1,1}(4)$. Figure 20.22b displays the abstraction of C_0 and C_1 for general n. We set

$$H_1 = \langle b_1, P_{01}, v_5, b_2, w_2, v_6, P_{02}, u_1, u_2, u_3, u_4, P_{11}, v_1, v_4, v_3, v_2, P_{12}, u_5, u_6, w_1 \rangle$$
$$H_2 = \langle b_2, v_1, P_{11}^{-1}, u_4, u_3, u_2, u_5, P_{12}^{-1}, v_2, v_3, v_4, v_5, P_{01}^{-1}, b_1, w_1, u_6, u_1, P_{02}^{-1}, v_6, w_2 \rangle$$

Since $w_1 \neq w_2$, $u_2 \neq v_2$, $u_3 \neq v_3$, $u_4 \neq v_4$, and $u_6 \neq v_6$, one can check, as shown in Figure 20.22c, that H_1 and H_2 satisfy the requirements. □

THEOREM 20.6 [207] For all $n \geq 3$, $\mathcal{IHC}(BF(n)) = 4$.

Proof: It is trivial that $\mathcal{IHC}(BF(n)) \leq \delta(BF(n)) = 4$. Suppose that $n = 3$. By the symmetric property of $BF(3)$, we have to find four mutually independent Hamiltonian cycles starting from vertex $\langle 0, 000 \rangle$. A set $\{C_1, C_2, C_3, C_4\}$ of four Hamiltonian cycles is listed in Table 20.3 and it is easy to check whether they are mutually independent.

For $n > 3$, we partition $BF(n)$ into $\{BF_{0,1}^{i,j}(n) | i,j \in \mathbb{Z}_2\}$. Without loss of generality, we assume that the beginning vertex is $s = \langle 1, 0^n \rangle$. Let $u_1 = \langle 2, 0^2 10^{n-3} \rangle$,

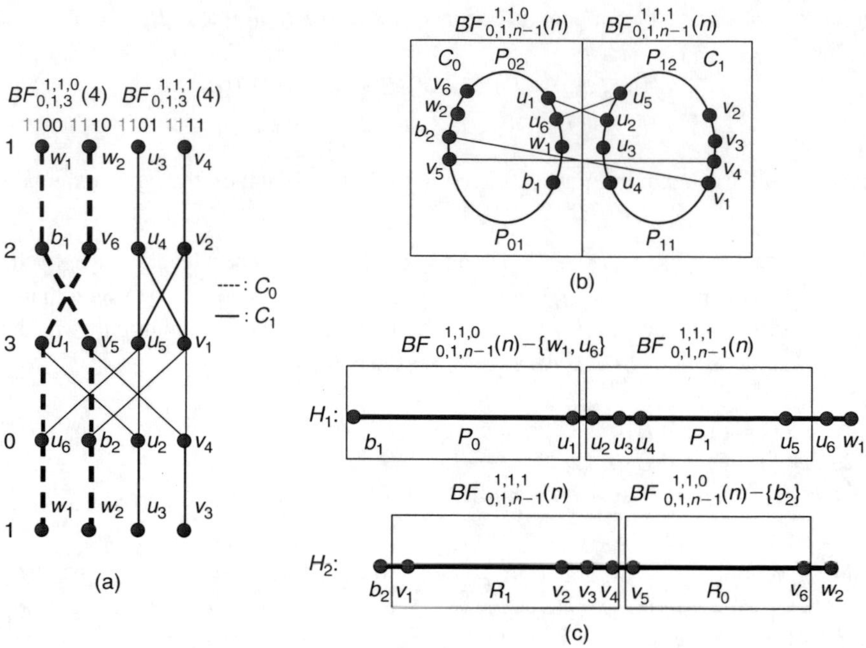

FIGURE 20.22 Illustration for Lemma 20.24. In (c), $R_1 = \langle v_1, P_{11}^{-1}, u_4, u_3, u_2, u_5, P_{12}^{-1}, v_2\rangle$ and $R_0 = \langle v_5, P_{01}^{-1}, b_1, w_1, u_6, u_1, P_{02}^{-1}, v_6\rangle$.

TABLE 20.3

Four Mutually Independent Hamiltonian Cycles C_1, C_2, C_3, C_4 of $BF(3)$ Starting from ⟨0,000⟩

C_1	⟨⟨0, 000⟩, ⟨2, 001⟩, ⟨0, 001⟩, ⟨1, 001⟩, ⟨2, 011⟩, ⟨0, 011⟩, ⟨1, 011⟩, ⟨0, 111⟩, ⟨2, 111⟩, ⟨1, 111⟩, ⟨2, 101⟩, ⟨1, 101⟩, ⟨0, 101⟩, ⟨2, 100⟩, ⟨0, 100⟩, ⟨1, 100⟩, ⟨2, 110⟩, ⟨0, 110⟩, ⟨1, 110⟩, ⟨0, 010⟩, ⟨2, 010⟩, ⟨1, 010⟩, ⟨2, 000⟩, ⟨1, 000⟩, ⟨0, 000⟩⟩
C_2	⟨⟨0, 000⟩, ⟨1, 000⟩, ⟨2, 000⟩, ⟨0, 001⟩, ⟨1, 001⟩, ⟨2, 011⟩, ⟨0, 011⟩, ⟨1, 111⟩, ⟨2, 101⟩, ⟨0, 101⟩, ⟨1, 101⟩, ⟨2, 111⟩, ⟨0, 110⟩, ⟨1, 010⟩, ⟨2, 010⟩, ⟨0, 010⟩, ⟨1, 110⟩, ⟨2, 100⟩, ⟨0, 100⟩, ⟨1, 100⟩, ⟨2, 110⟩, ⟨0, 111⟩, ⟨1, 011⟩, ⟨2, 001⟩, ⟨0, 000⟩⟩
C_3	⟨⟨0, 000⟩, ⟨1, 100⟩, ⟨2, 100⟩, ⟨0, 100⟩, ⟨1, 000⟩, ⟨2, 010⟩, ⟨0, 010⟩, ⟨1, 110⟩, ⟨2, 110⟩, ⟨0, 111⟩, ⟨1, 111⟩, ⟨0, 011⟩, ⟨2, 011⟩, ⟨1, 011⟩, ⟨2, 001⟩, ⟨0, 001⟩, ⟨1, 001⟩, ⟨0, 101⟩, ⟨2, 101⟩, ⟨1, 101⟩, ⟨2, 111⟩, ⟨0, 110⟩, ⟨1, 010⟩, ⟨2, 000⟩, ⟨0, 000⟩⟩
C_4	⟨⟨0, 000⟩, ⟨2, 000⟩, ⟨1, 000⟩, ⟨2, 010⟩, ⟨0, 010⟩, ⟨1, 010⟩, ⟨0, 110⟩, ⟨2, 110⟩, ⟨1, 110⟩, ⟨2, 100⟩, ⟨0, 101⟩, ⟨1, 001⟩, ⟨2, 001⟩, ⟨0, 001⟩, ⟨1, 101⟩, ⟨2, 111⟩, ⟨0, 111⟩, ⟨1, 011⟩, ⟨2, 011⟩, ⟨0, 011⟩, ⟨1, 111⟩, ⟨2, 101⟩, ⟨0, 100⟩, ⟨1, 100⟩, ⟨0, 000⟩⟩

$u_2 = f^{-1}(u_1) = \langle 1, 01^20^{n-3}\rangle$, $u_3 = g^{-1}(u_2) = \langle 0, 01^20^{n-3}\rangle$, $u_4 = f(u_3) = \langle 1, 1^30^{n-3}\rangle$, $u_5 = g(u_4) = \langle 2, 1^30^{n-3}\rangle$, $u_6 = f^{-1}(u_5) = \langle 1, 1010^{n-3}\rangle$, $u_7 = f^{-1}(s) = \langle 0, 10^{n-1}\rangle$, $v_1 = g^{-1}(s) = \langle 0, 0^n\rangle$, $v_2 = f(v_1) = \langle 1, 10^{n-1}\rangle$, $v_3 = g(v_2) = \langle 2, 10^{n-1}\rangle$, $v_4 = f^{-1}(v_3) = \langle 1, 1^20^{n-2}\rangle$, $v_5 = g^{-1}(v_4) = \langle 0, 1^20^{n-2}\rangle$, $v_6 = f(v_5) = \langle 1, 010^{n-2}\rangle$, and $v_7 = g(v_6) = f(s) = \langle 2, 010^{n-2}\rangle$ be vertices of $BF(n)$. Obviously, $\{u_1, u_2, u_3, u_4, u_5, u_6, u_7, v_1, v_2, v_3, v_4, v_5, v_6, v_7\}$ consists of 14 different vertices of $BF(n)$ such that all (u_1, u_2), (u_3, u_4), (u_5, u_6), (u_7, s), (v_1, v_2), (v_3, v_4), (v_5, v_6), and (v_7, s) are in $E(BF(n))$. By Lemma 20.23, there exist two Hamiltonian paths P_1 and P_2 of $BF_{0,1}^{0,0}(n)$ such that (1)

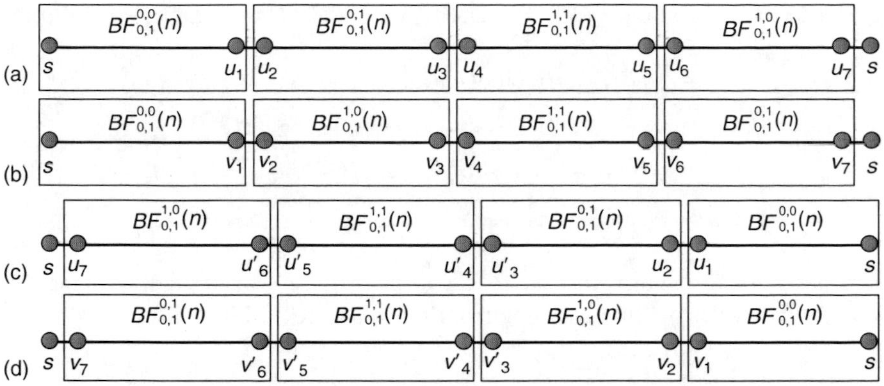

FIGURE 20.23 Illustration for Theorem 20.6. (a) C_1, (b) C_2, (c) C_3, and (d) C_4.

P_1 joins s to u_1, (2) P_2 joins s to v_1, and (3) $P_1(1) = P_2(1) = s$ and $P_1(t) \neq P_2(t)$ for all $2 \leq t \leq n \times 2^{n-2}$. By Lemma 20.22, there is a Hamiltonian path Q_1 of $BF^{0,1}_{0,1}(n)$ joining u_2 to u_3. Similarly, there is a Hamiltonian path Q_2 of $BF^{1,0}_{0,1}(n)$ joining v_2 to v_3, a Hamiltonian path R_1 of $BF^{1,0}_{0,1}(n)$ joining u_6 to u_7, and a Hamiltonian path R_2 of $BF^{0,1}_{0,1}(n)$ joining v_6 to v_7. Using Lemma 20.24, we can find two Hamiltonian paths S_1 and S_2 of $BF^{1,1}_{0,1}(n)$ such that (1) S_1 joins u_4 to u_5, (2) S_2 joins v_4 to v_5, and (3) $S_1(t) \neq S_2(t)$ for all $1 \leq t \leq n \times 2^{n-2}$. We set $C_1 = \langle s, P_1, u_1, u_2, Q_1, u_3, u_4, S_1, u_5, u_6, R_1, u_7, s \rangle$ and $C_2 = \langle s, P_2, v_1, v_2, Q_2, v_3, v_4, S_2, v_5, v_6, R_2, v_7, s \rangle$. Figures 20.23a and 20.23b illustrate C_1 and C_2, respectively. Obviously, C_1 and C_2 are both Hamiltonian cycles of $BF(n)$. We claim that C_1 and C_2 are independent in what follows. First, Lemma 20.23 guarantees that $C_1(t) \neq C_2(t)$ with $2 \leq t \leq n \times 2^{n-2}$. Next, $C_1(t) \neq C_2(t)$ if $n \times 2^{n-2} + 1 \leq t \leq n \times 2^{n-1}$ because C_1 and C_2 pass through different subgraphs separately. Moreover, Lemma 20.24 guarantees that $C_1(t) \neq C_2(t)$ with $n \times 2^{n-1} + 1 \leq t \leq 3n \times 2^{n-2}$. Finally, it is obvious that $C_1(t) \neq C_2(t)$ if $3n \times 2^{n-2} + 1 \leq t \leq n \times 2^n$. So C_1 and C_2 are independent.

Let $u'_3 = \langle 0, 01^2 0^{n-4} 1 \rangle$, $u'_4 = f(u'_3) = \langle 1, 1^3 0^{n-4} 1 \rangle$, $u'_5 = g(u'_4) = \langle 2, 1^3 0^{n-4} 1 \rangle$, $u'_6 = f^{-1}(u'_5) = \langle 1, 1010^{n-4} 1 \rangle$, $v'_3 = \langle 2, 10^{n-2} 1 \rangle$, $v'_4 = f^{-1}(v'_3) = \langle 1, 1^2 0^{n-3} 1 \rangle$, $v'_5 = g^{-1}(v'_4) = \langle 0, 1^2 0^{n-3} 1 \rangle$, and $v'_6 = f(v'_5) = \langle 1, 010^{n-3} 1 \rangle$ be vertices of $BF(n)$. Obviously, $u'_i \neq u_i$ and $v'_i \neq v_i$ for $3 \leq i \leq 6$. By Lemma 20.22, there is a Hamiltonian path Q_3 of $BF^{0,1}_{0,1}(n)$ joining u_2 to u'_3, a Hamiltonian path Q_4 of $BF^{1,0}_{0,1}(n)$ joining v_2 to v'_3, a Hamiltonian path R_3 of $BF^{1,0}_{0,1}(n)$ joining u'_6 to u_7, and a Hamiltonian path R_4 of $BF^{0,1}_{0,1}(n)$ joining v'_6 to v_7. Using Lemma 20.24, we can construct two Hamiltonian paths S_3 and S_4 of $BF^{1,1}_{0,1}(n)$ such that (1) S_3 joins u'_4 to u'_5, (2) S_4 joins v'_4 to v'_5, and (3) $S_3(t) \neq S_4(t)$ for all $1 \leq t \leq n \times 2^{n-2}$. We set $O_1 = \langle s, P_1, u_1, u_2, Q_3, u'_3, u'_4, S_3, u'_5, u'_6, R_3, u_7, s \rangle$ and $O_2 = \langle s, P_2, v_1, v_2, Q_4, v'_3, v'_4, S_4, v'_5, v'_6, R_4, v_7, s \rangle$. Similar to C_1 and C_2, O_1 and O_2 are independent in $BF(n)$.

Let $C_3 = O_1^{-1}$ and $C_4 = O_2^{-1}$. For clarity, we list C_1, C_2, C_3, and C_4 as follows:

$$\begin{aligned}
C_1 &= \langle s, P_1, u_1, u_2, Q_1, u_3, u_4, S_1, u_5, u_6, R_1, u_7, s\rangle \\
C_2 &= \langle s, P_2, v_1, v_2, Q_2, v_3, v_4, S_2, v_5, v_6, R_2, v_7, s\rangle \\
C_3 &= \langle s, u_7, R_3^{-1}, u_6', u_5', S_3^{-1}, u_4', u_3', Q_3^{-1}, u_2, u_1, P_1^{-1}, s\rangle \\
C_4 &= \langle s, v_7, R_4^{-1}, v_6', v_5', S_4^{-1}, v_4', v_3', Q_4^{-1}, v_2, v_1, P_2^{-1}, s\rangle
\end{aligned}$$

Then it is easy to confirm that C_1, C_2, C_3, and C_4 are four mutually independent Hamiltonian cycles of $BF(n)$ starting from s. See Figure 20.23 for illustration. □

21 Diagnosis of Multiprocessor Systems

21.1 INTRODUCTION

With the rapid development of technology, the need for high-speed parallel processing systems has been continuously increasing. The reliability of the processors in parallel computing systems is therefore becoming an important issue. In order to maintain the reliability of a system, whenever a processor is found to be faulty, it should be replaced by a fault-free processor. The process of identifying all the faulty vertices is called the *diagnosis* of the system. System-level diagnosis appears to be an alternative to circuit-level testing in a complex multiprocessor system.

Many terms for system-level diagnosis have been defined and various models have been proposed [19,115,248,273]. If all allowable fault sets can be diagnosed correctly and completely based on a single syndrome, which is defined in the following section, then the diagnosis is referred to as *one-step diagnosis* or *diagnosis without repairs*. A system is called *sequentially t-diagnosable*, if at least one faulty unit can be identified, provided that the number of faulty vertices does not exceed t. A system is said to be *t-diagnosable* if, for every syndrome, there is a unique set of faulty vertices that could produce the syndrome, as long as the number of faulty vertices does not exceed t. The maximum number of faulty vertices that the system can guarantee to identify is called the *diagnosability* of the system.

Another way of diagnosis is to allow a certain number of processors to be incorrectly diagnosed. Friedman [113] introduced the notation of t/s-diagnosable. A system is t/s-diagnosable if, given any syndrome, the faulty units can be isolated to within a set of at most s units, provided that the number of faulty units does not exceed t. Chwa and Hakimi [72] studied the t_1/t_1-diagnosable systems.

21.2 DIAGNOSIS MODELS

For the purpose of self-diagnosis of a given system, several different models have been proposed [245,247,273]. Preparata et al. [273] first introduced a model, called the PMC-model, for system-level diagnosis in multiprocessor systems. In this model, it is assumed that a processor can test the faulty or fault-free status of another processor. The *comparison model*, called the *MM model* and proposed by Maeng and Malek [245,247], is considered to be another practical approach for fault diagnosis in multiprocessor systems.

21.2.1 PMC Model

A multiprocessor system is modeled as an undirected graph $G=(V,E)$ whose vertices represent processors and edges represent communication links. Adjacent processors are capable of performing tests on each other. For adjacent vertices $u, v \in V$, the ordered pair (u, v) represents the test performed by u on v. In this situation, u is called the *tester* and v is called the *tested* vertex. The outcome of a test (u, v) is 1 (respectively 0) if u evaluates v as faulty (respectively fault-free).

A *test assignment* for a system $G=(V,E)$ is a collection of tests (u, v) for some adjacent pairs of vertices. Throughout this chapter, it can be modeled as a directed graph $T=(V,L)$, where $(u, v) \in L$ implies that u tests v in G. The collection of all test results for a test assignment T is called a *syndrome*. Formally, a syndrome is a function $\sigma: L \to \{0, 1\}$.

For a PMC model, some known results about the definition of a t-diagnosable system and related concepts are listed as follows. Some of these previous results are on directed graphs and others are on undirected graphs.

A system of n units is t-diagnosable if all faulty units can be identified without replacement, provided that the number of faults presented does not exceed t.

Let $F_1, F_2 \subset V$ be two distinct sets and let the symmetric difference $F_1 \Delta F_2 = (F_1 - F_2) \cup (F_2 - F_1)$. Dahbura and Masson [77] proposed a polynomial time algorithm to check whether a system is t-diagnosable.

LEMMA 21.1 **[77]** A system $G(V,E)$ is t-diagnosable under the PMC model if and only if for each pair $F_1, F_2 \subset V$ with $|F_1|, |F_2| \le t$ and $F_1 \neq F_2$, there is at least one test from $V - (F_1 \cup F_2)$ to $F_1 \,\Delta\, F_2$.

The following two results related to t-diagnosable systems are due to Hakimi and Amin [136], and Preparata et al. [273], respectively.

LEMMA 21.2 **[273]** Let $G(V,E)$ be the graph representation of a system G, with V representing the processors and E the interconnection among them. Let $|V|=n$. The following two conditions are necessary for G to be t-diagnosable under the PMC model:

1. $n \ge 2t+1$
2. Each processor is tested by at least t other processors

LEMMA 21.3 **[136]** The following two conditions are sufficient for a system G of n processors to be t-diagnosable under the PMC model:

1. $n \ge 2t+1$
2. $\kappa(G) \ge t$

For a directed graph G and vertex $v \in V(G)$, let $\Gamma(v)=\{v_i \mid (v, v_i) \in E\}$ and $\Gamma(X)=\bigcup_{v\in X}\Gamma(v)-X, X\subset V$. Hakimi and Amin presented the following necessary and sufficient condition for a system G to be t-diagnosable.

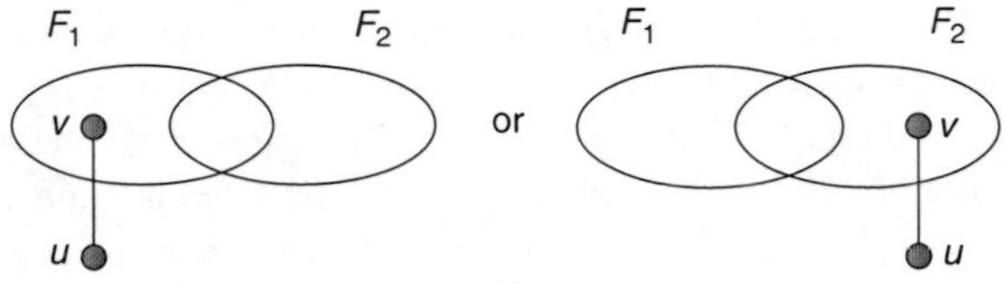

FIGURE 21.1 Illustration for a distinguishable pair (F_1, F_2).

THEOREM 21.1 **[136]** Let $G(V,E)$ be the directed graph of a system G with n units. Then G is t-diagnosable under the PMC model if and only if: (1) $n \geq 2t+1$, (2) $d_{in}(v) \geq t$ for all $v \in V$, and (3) for each integer p with $0 \leq p \leq t-1$, and each $X \subset V$ with $|X| = n - 2t + p$, $|\Gamma(X)| > p$.

Let $G(V,E)$ be an undirected graph of a system G. The following lemma flows directly from Lemma 21.1.

LEMMA 21.4 For any two distinct sets $F_1, F_2 \subset V$, (F_1, F_2) is a distinguishable pair under the PMC model if and only if $\exists u \in V - (F_1 \cup F_2)$ and $\exists v \in F_1 \,\Delta\, F_2$ such that $(u, v) \in E$ (see Figure 21.1).

It follows from Section 21.2.1 that the following lemma holds.

LEMMA 21.5 A system is t-diagnosable under the PMC model if and only if for each distinct pair of sets $F_1, F_2 \subset V$ with $|F_1| \leq t$ and $|F_2| \leq t$, F_1 and F_2 are distinguishable.

An equivalent way of stating the previous lemma is the following.

LEMMA 21.6 A system is t-diagnosable under the PMC model if and only if for each indistinguishable pair of sets $F_1, F_2 \subset V$, it implies that $|F_1| > t$ or $|F_2| > t$.

By Lemma 21.2, a similar result for undirected graph is stated as follows.

COROLLARY 21.1 **[273]** *Let $G(V,E)$ be an undirected graph. The following two conditions are necessary for G to be t-diagnosable under the PMC model:*

1. $n \geq 2t+1$
2. $\delta(G) \geq t$

For later discussion, an alternative characterization of a t-diagnosable system is given.

THEOREM 21.2 Let $G(V,E)$ be the graph of a system G. Then G is t-diagnosable under the PMC model if and only if for each vertex set $S \subset V$ with $|S| = p$, $0 \leq p \leq t-1$, every component C of $G - S$ satisfies $|V_C| \geq 2(t-p)+1$.

Proof: To check whether $|V_C| \geq 2(t-p)+1$ is necessary, we prove it by contradiction. Then there exists a set of vertices $S \subset V$ with $|S|=p, 0 \leq p \leq t-1$, such that one of the components $G-S$ has strictly less than $2(t-p)+1$ vertices. Let C be such a component with $|V_C| \leq 2(t-p)$. We then arbitrarily partition V_C into two disjoint subsets, $V_C = A_1 \cup A_2$ with $|A_1| \leq t-p$ and $|A_2| \leq t-p$. Let $F_1 = A_1 \cup S$ and $F_2 = A_2 \cup S$. Then $|F_1| \leq t$ and $|F_2| \leq t$. It is clear that there is no edge between $V-(F_1 \cup F_2)$ and $F_1 \Delta F_2$. By Lemma 21.4, F_1 and F_2 are indistinguishable. This contradicts the assumption that G is t-diagnosable.

To prove the sufficiency, suppose, that to the contrary, that G is not t-diagnosable; that is, there exists an indistinguishable pair (F_1, F_2) with $|F_i| \leq t, i=1,2$. By Lemma 21.4, there is no edge between $V-(F_1 \cup F_2)$ and $F_1 \Delta F_2$. Let $S = F_1 \cap F_2$. Thus, in $G-S$, $F_1 \Delta F_2$ is disconnected from other parts. We observe that $|F_1 \Delta F_2| = 2(t-p)$, where $|S|=p$ and $0 \leq p \leq t-1$. Therefore, there is at least one component C of $G-S$ with $|V_C| \leq 2(t-p)$, which is a contradiction. This completes the proof of the theorem. □

21.2.2 Comparison Model

In this approach, the diagnosis is carried out by assigning the same testing task to a pair $\{u, v\}$ of processors and comparing their responses. The comparison is performed by a third processor w that has direct communication links to both processors u and v. The third processor w is called a comparator of u and v.

If the comparator is fault-free, a disagreement between the two responses is an indication of the existence of a faulty processor. To gain as much knowledge as possible about the faulty status of the system, it is assumed that a comparison is performed by each processor for each pair of distinct neighbors with which it can communicate directly. This special case of the MM-model is referred to as the MM*-model. Sengupta and Dahbura [282] studied the MM-model and the MM*-model, gave a characterization of diagnosable systems under the comparison approach, and proposed a polynomial time algorithm to determine faulty processors under MM*-model. In this section, we study the diagnosabilities of an Matching Composition Network (MCN) (which will be defined subsequently) and Cartesian Product Networks under MM*-model.

In the study of multiprocessor systems, the topology of networks is usually represented by a graph $G=(V,E)$, where each vertex $v \in V$ represents a processor and each edge $(u,v) \in E$ represents a communication link. The diagnosis by comparison approach can be modeled by a labeled multigraph, called a comparison graph, $T=(V,L)$, where V is the set of all processors and L is the set of labeled edges. A labeled edge $(u,v)_w \in L$, with w being a label in the edge, connects u and v, which implies that processors u and v are being compared by w. Under the MM-model, processor w is a comparator for processors u and v only of $(w,u) \in E$ and $(w,v) \in E$. The MM*-model is a special case of the MM-model; it is assumed that each processor w such that $(w,u) \in E$ and $(w,v) \in E$ is a comparator for the pair of processors u and v. The comparison graph $T=(V,L)$ of a given system can be a multigraph, for the same pair of vertices may be compared by several different comparators.

For $(u, v)_w \in L$, the output of the comparison w of u and v is denoted by $r((u, v)_w)$, and a disagreement of the outputs is denoted by the comparison results $r((u, v)_w) = 1$, whereas an agreement is denoted by $r((u, v)_w) = 0$.

Therefore, if the comparator w is fault-free and $r((u, v)_w) = 0$, then u and v are both fault-free. If $r((u, v)_w) = 1$, then at least one of u, v, and w must be faulty. The set of all comparison results of a multicomputer system that are analyzed together to determine the faulty processors is called a *syndrome* of the system.

Next, we discuss the diagnosability under the comparison model. Given a graph G, let T be the comparison graph of G. For a vertex $v \in V(G)$, we define X_v to be the set of vertices $\{u|(v, u) \in E(G)\} \cup \{u|(v, u)_w \in E(T)$ for some $w\}$ and Y_v to be the set of edges $\{(u, w)|u, w \in X_v$ and $(v, u)_w \in E(T)\}$. In Ref. 282, the *order graph* of vertex v is defined as $G(v) = (X_v, Y_v)$ and the *order* of the vertex v, denoted by $order_G(v)$, is defined to be the cardinality of a minimum vertex cover of $G(v)$. Let $U \subset V(G)$, we use $\Gamma(G, U)$ to denote the set $\{v|(u, v)_w \in E(T)$ and $w, u \in U, v \in = \overline{U}\}$. We observe that $\Gamma(G, U) = N(\overline{U}, U)$ if $G[U]$ is connected and $|U| > 1$. This observation can be extended to the following lemma.

LEMMA 21.7 Let U be a subset of $V(G)$ and $G[U_i]$, $1 \leq i \leq k$, be the connected components of the subgraph $G[U]$ such that $U = \cup_{i=1}^{k} U_i$. Then $\Gamma(G, U) = \cup_{i=1}^{k} \{N(\overline{U}, U_i) \mid |U_i| > 1\}$.

In Figure 21.2, taking Q_3 as an example, we have $\Gamma(G, U) = \{4, 5, 6, 7\}$, where $U = \{0, 1, 2, 3\}$.

The next lemma follows directly from the definition of connectivity of G.

LEMMA 21.8 **[102]** Let G be a connected graph and U be a subset of $V(G)$. Then $|N(\overline{U}, U)| \geq \kappa(G)$ if $|\overline{U}| \geq \kappa(G)$ and $N(\overline{U}, U) = \overline{U}$ if $|\overline{U}| < \kappa(G)$.

Five lemmas and theorems presented by Sengupta and Dahbura [282] must be applied to characterize whether a system is t-diagnosable. The results of these theorems follow.

The first one is a necessary and sufficient condition for ensuring distinguishability.

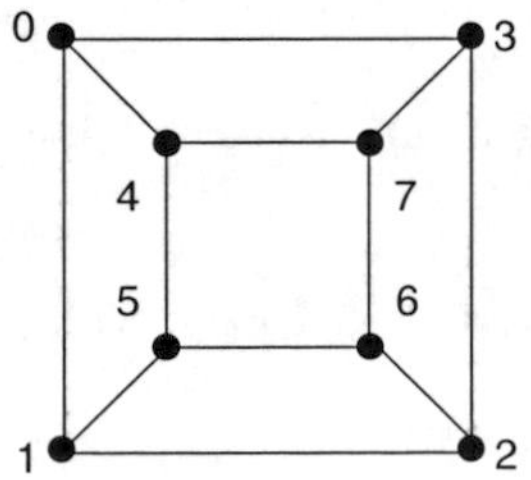

FIGURE 21.2 An example for $\Gamma(G, U)$ of Q_3.

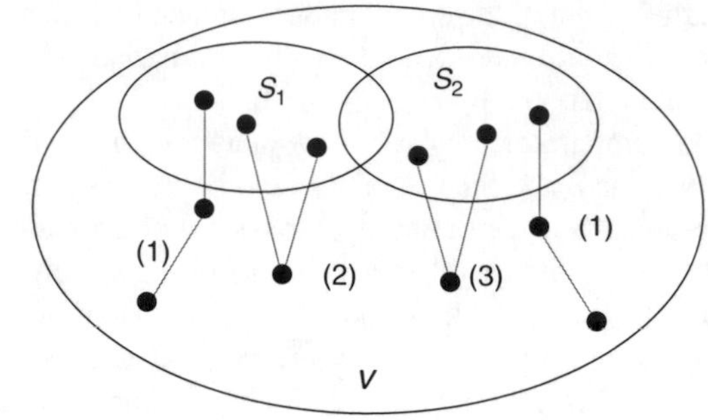

FIGURE 21.3 Description of distinguishability.

THEOREM 21.3 **[282]** For any S_1, S_2 where $S_1, S_2 \subset V$ and $S_1 \neq S_2$, (S_1, S_2) is a distinguishable pair under the comparison model if and only if at least one of the following conditions is satisfied (see Figure 21.3):

1. $\exists i$, there are $k \in V - S_1 - S_2$ and $\exists$ there are $j \in (S_1 - S_2) \cup (S_2 - S_1)$ such that $(i, j)_k \in C$
2. $\exists$ there are $i, j \in S_1 - S_2$ and $\exists$ there are $k \in V - S_1 - S_2$ such that $(i, j)_k \in C$
3. $\exists$ there are $i, j \in S_2 - S_1$ and $\exists$ there are $k \in V - S_1 - S_2$ such that $(i, j)_k \in C$

Two necessary conditions for confirming a system to be t-diagnosable are as follows.

LEMMA 21.9 **[282]** If a system with N vertices is t-diagnosable, then $N \geq 2t + 1$.

LEMMA 21.10 **[282]** If, in a system, each vertex has order at least t, then for each $S_1, S_2 \subset V$ such that $|S_1 \cup S_2| \leq t$, (S_1, S_2) is a distinguishable pair.

Another necessary and sufficient condition for ensuring distinguishability is the following theorem.

THEOREM 21.4 **[282]** A system is t-diagnosable under the comparison model if and only if each vertex has order at least t and for each distinct pair of sets $S_1, S_2 \subset V$ such that $|S_1| = |S_2| = t$, at least one of the conditions of Theorem 21.3 is satisfied.

The following theorem is a sufficient condition for verifying a system to be t-diagnosable.

THEOREM 21.5 **[282]** A system G with N vertices is t-diagnosable under the comparison model if:

1. $N \geq 2t + 1$
2. $order_G(v) \geq t$ for every vertex v in G

3. $|\Gamma(G,U)| > p$ for each $U \subset V(G)$ such that $|U| = N - 2t + p$ and $0 \le p \le t-1$

According to the Theorems 21.3 through 21.5, we observe that condition (3) of Theorem 21.5 restricts G, satisfying the first condition of Theorem 21.3, and ignores conditions (2) and (3). Hence, we present a hybrid theorem to test whether a system is t-diagnosable.

THEOREM 21.6 A system G with N vertices is t-diagnosable under the comparison model if:

1. $N \ge 2t+1$
2. $order_G(v) \ge t$ for every vertex v in G
3. For any two distinct subsets $S_1, S_2 \subset V(G)$ such that $|S_1| = |S_2| = t$ either (a) $|\Gamma(G,U)| > p$, where $U = V(G) - (S_1 \cup S_2)$, and $|S_1 \cap S_2| = p$; or (b) The pair (S_1, S_2) satisfies condition (2) or (3) of Theorem 21.3

Proof: Conditions (1) and (2) are the same as conditions (1) and (2) of Theorem 21.5. Consider condition (3a). S_1 and S_2 are two distinct subsets of $V(G)$ with $|S_1| = |S_2| = t$, $U = V(G) - (S_1 \cup S_2)$, and $|S_1 \cap S_2| = p$. Then $0 \le p \le t-1$ and $|U| = N - 2t + p$. If $|\Gamma(G,U)| > p$, it implies that the pair (S_1, S_2) satisfies condition (1) of Theorem 21.3. Combining conditions (3a) and (3b), by Theorems 21.3 and 21.4, this theorem follows. □

21.3 DIAGNOSABILITY OF THE MATCHING COMPOSITION NETWORKS

Now we define the MCN. Let G_1 and G_2 be two graphs with the same number of vertices. Let M be an arbitrary perfect matching between the vertices of G_1 and G_2; that is, M is a set of edges connecting the vertices of G_1 and G_2 in a one-to-one fashion, the resulting composition graph is called a *Matching Composition Network*. For convenience, G_1 and G_2 are called the M-components of the MCN. Formally, we use the notation $G_1 \oplus G_2$ to denote a MCN, which has vertex set $V(G_1 \oplus G_2) = V(G_1) \cup V(G_2)$ and edge set $E(G_1 \oplus G_2) = E(G_1) \cup E(G_2) \cup M$. See Figure 21.4.

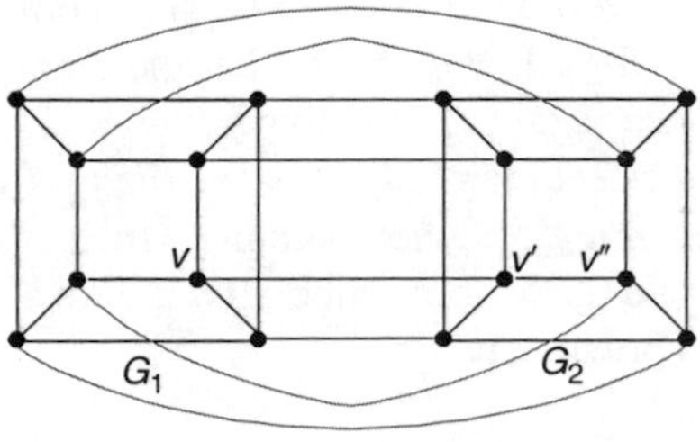

FIGURE 21.4 An example of MCN $G(G_1 \oplus G_2)$.

21.3.1 Diagnosability of the Matching Composition Networks under the PMC Model

THEOREM 21.7 Let $G_1(V_1, E_1)$ and $G_2(V_2, E_2)$ be two t-diagnosable systems with the same number of vertices, where $t \geq 1$. Then MCN $G = G_1 \oplus G_2$ is $(t+1)$-diagnosable.

Proof: We shall use Theorem 21.2 to prove this theorem. Let $G = G(V, E) = G_1 \oplus G_2$ and $S \subset V$ with $|S| = p$, $0 \leq p \leq t$. Let $S_1 = S \cap V_1$ and $S_2 = S \cap V_2$ with $|S_1| = p_1$ and $|S_2| = p_2$. In the following proof, we consider two cases: (1) $S_1 = \emptyset$ or $S_2 = \emptyset$ and (2) $S_1 \neq \emptyset$ and $S_2 \neq \emptyset$. We shall prove that $|V_C| \geq 2((t+1) - p) + 1$ for every component C of $G - S$ as $0 \leq p \leq t$.

Case 1. $S_1 = \emptyset$ or $S_2 = \emptyset$. Without loss of generality, we assume that $S_1 = \emptyset$ and $S_2 = S$. We know that each vertex of V_2 has an adjacent neighbor in V_1, so $G - S$ is connected. The only component C of $G - S$ is $G - S$ itself. Hence, $|V_C| = |V - S| = |V_1| + |V_2| - p$. G_i is t-diagnosable, $i = 1, 2$, and by Lemma 21.2, $|V_i| \geq 2t + 1$. So $|V_C| \geq 2(2t+1) - p \geq 2((t+1) - p) + 1$ for $t \geq 1$.

Case 2. $S_1 \neq \emptyset$ and $S_2 \neq \emptyset$. Since $S_1 \neq \emptyset$ and $S_2 \neq \emptyset$, $1 \leq p_1 \leq t - 1$ and $1 \leq p_2 \leq t - 1$. Let C_1 be a component of $G_1 - S_1$. G_1 is t-diagnosable, and by Theorem 21.2, $|V_{C_1}| \geq 2(t - p_1) + 1$. We claim that $2(t - p_1) + 1 \geq p_2 + 1$. Since $p = p_1 + p_2$, $2(t - p_1) + 1 = 2(t - (p - p_2)) + 1 = 2p_2 + 2(t - p) + 1$. Notice that $p \leq t$. Hence, $|V_{C_1}| \geq 2(t - p_1) + 1 \geq p_2 + 1$. That is, V_{C_1} has at least one adjacent neighbor $v \in V_2$ and $v \notin S_2$. G_2 is t-diagnosable. By Theorem 21.2, every component of $G_2 - S_2$ has at least $2(t - p_2) + 1$ vertices. Let C_2 be the component of $G_2 - S_2$ such that $v \in V_{C_2}$ and let C be the component of $G - S$ such that $V_{C_1} \cup V_{C_2} \subset V_C$. Then $|V_C| \geq |V_{C_1}| + |V_{C_2}| \geq (2(t - p_1) + 1) + (2(t - p_2) + 1) = 2(2t - p + 1) \geq 2((t+1) - p) + 1$ as $t \geq 1$.

So every component of $G - S$ has at least $2((t+1) - p) + 1$ vertices in this case. Consequently, the lemma follows. □

21.3.2 Diagnosability of the Matching Composition Networks under the Comparison Model

What we have in mind is the following. Let G_1 and G_2 be two t-connected networks with the same number of vertices and $order_{G_i}(v) \geq t$ for every vertex v in G_i, where $i = 1, 2$, and let M be an arbitrary perfect matching between the vertices of G_1 and G_2. Then the degree of any vertex v in $G(G_1 \oplus G_2)$ as compared with that of vertex v in G_i, $i = 1, 2$, is increased by 1. We expect that the diagnosability of $G(G_1 \oplus G_2)$ is also increased to $t + 1$. For example, the hypercube Q_{n+1} is constructed from two copies of Q_n by adding a perfect matching between the two, and the diagnosability is increased from n to $n + 1$ for $n \geq 5$. Other examples, such as the twisted cube TQ_{n+1}, the crossed cube CQ_{n+1}, and the Möbius cube MQ_{n+1}, are all constructed recursively using the same method as previously.

THEOREM 21.8 Let G_1 and G_2 be two networks with the same number of vertices, and t be a positive integer. Suppose that $order_{G_i}(v) \geq t$ for every vertex v in G_i, where $i = 1, 2$. Then $order_{G_1 \oplus G_2}(v) \geq t + 1$ for vertex v in $G_1 \oplus G_2$.

Proof: See Figure 21.4. Let v be a vertex of $G = G_1 \oplus G_2$. Without loss of generality, we assume that $v \in V(G_1)$, $v' \in V(G_2)$, and $(v, v') \in M$. Of course, vertex v' is connected to at least another vertex v'' in $V(G_2)$. Let $G_1(v)$ and $G(v)$ be the order graph of v in graph G_1 and G, respectively. We observe that $G_1(v)$ is a proper subgraph of G. Both v' and v'' are in the latter, none of them are in the former, and (v', v'') is an edge in $G(v)$. Therefore, every vertex cover of the order graph $G(v)$ contains a vertex cover of the order graph $G_1(v)$. Besides, any vertex cover of $G(v)$ has to include at least one of v' and v''. Thus, $order_{G_1 \oplus G_2}(v) \geq order_{G_i}(v) + 1$ for any vertex v in G_i, $i = 1, 2$. This completes the proof. □

We need the following lemma later in Theorem 21.9.

LEMMA 21.11 Let G be a t-connected network, $|V(G)| \geq t + 2$ and $order_G(v) \geq t$ for every vertex v in G, where $t \geq 2$. Suppose that U is a subset of vertices of $V(G)$ with $|\overline{U}| \leq t$. Then $\Gamma(G, U) = \overline{U}$.

Proof: By assumption $|\overline{U}| \leq t$ and $\kappa(G) \geq t$, we prove the lemma by two cases; the first for $|\overline{U}| < \kappa(G)$ and the second for $|\overline{U}| = \kappa(G)$.

If $|\overline{U}| < \kappa(G)$, the induced graph $G[U]$ is connected. By Lemma 21.7, $\Gamma(G, U) = N(\overline{U}, U)$. By Lemma 21.8, $N(\overline{U}, U) = \overline{U}$. This case holds.

Suppose that $|\overline{U}| = \kappa(G)$. We observe that, adding any vertex v of $\overline{U}$ to U, the induced subgraph $G[U \cup \{v\}]$ forms a connected graph. It implies that every vertex v of $\overline{U}$ is adjacent to every connected component of $G[U]$. We claim that the subgraph induced by U contains a connected component A with cardinality at least 2 (See Figure 21.5a). Then the connected component A is adjacent to all vertices in $\overline{U}$ and, so $\Gamma(G, U) = \overline{U}$.

Now, we prove the claim. Suppose to the contrary that every connected component of the subgraph induced by U is an isolated vertex. Let v be an arbitrary vertex in $\overline{U}$, and let $G(v) = (X_v, Y_v)$ be the order graph of v in G. Then $\overline{U} - \{v\}$ is a vertex cover of $G(v)$, because every connected component of $G[U]$ is an isolated vertex. Since $|\overline{U}| \leq t$, we have $|\overline{U} - \{v\}| \leq t - 1$. Therefore, even if the induced graph $G[\overline{U} - \{v\}]$ is a complete graph (see Figure 21.5b), the cardinality of a minimum vertex cover of the order graph $G(v)$ is at most $t - 1$. However, this contradicts the hypothesis of

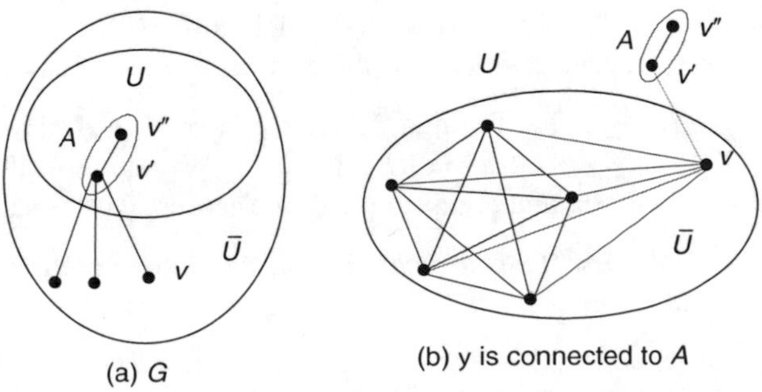

FIGURE 21.5 An example of $\Gamma(G, U)$ when $|U| = t$.

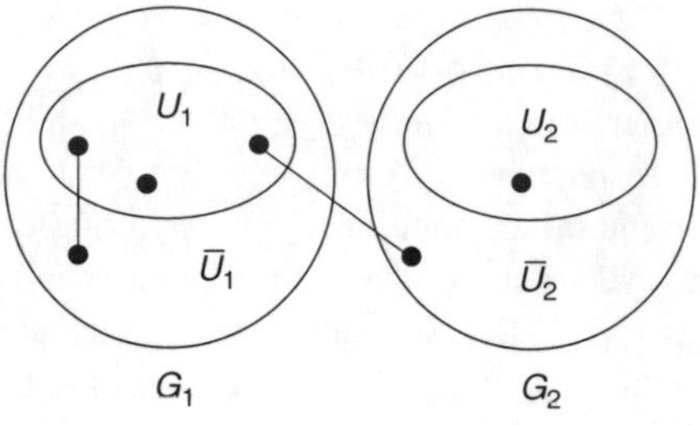

FIGURE 21.6 Illustration of Theorem 21.9 for an example of Case 2.1.

$order_G(v) \geq t$ for every vertex v in G. So $G[U]$ has a connected component A with cardinality at least 2. This proves the claim, and the lemma follows. □

We are now ready to state and prove the following theorem about the diagnosability of the MCN under the comparison model. As an illustration, the conditions of the following theorem are applicable to some well-known interconnection networks, such as Q_n, CQ_n, TQ_n, and MQ_n for $n = t \geq 3$.

THEOREM 21.9 For $t \geq 2$, let G_1 and G_2 be two graphs with the same number of vertices N, where $N \geq t+2$. Suppose that $order_{G_i}(v) \geq t$ for every vertex v in G_i and the connectivity $\kappa(G_i) \geq t$, where $i = 1, 2$. Then MCN $G_1 \oplus G_2$ is $(t+1)$-diagnosable under the comparison model.

Proof: Since $|V(G_1)| = |V(G_2)| = N$, $2N \geq 2(t+2) > 2(t+1)+1$. By Theorem 21.8, $order_{G_1 \oplus G_2}(v) \geq t+1$ for any vertex v in $G_1 \oplus G_2$. It remains to prove that $G_1 \oplus G_2$ satisfies condition (3) of Theorem 21.6.

Let F_1 and F_2 be two distinct subsets of $V(G)$ with the same number $t+1$ of vertices, and let $|F_1 \cap F_2| = p$, then $0 \leq p \leq t$. In order to prove this theorem, we will prove that F_1 and F_2 are distinguishable; that is, that this pair (F_1, F_2) satisfies either condition (3a) or (3b) of Theorem 21.6.

Let $G = G_1 \oplus G_2$ and $U = V(G) - (F_1 \cup F_2)$, then $|U| = 2N - 2(t+1) + p$. Let $U = U_1 \cup U_2$ with $U_i = U \cap V(G_i)$ and $\overline{U}_i = V(G_i) - U_i$, $i = 1, 2$. Without loss of generality, we assume that $|U_1| \geq |U_2|$. Let $|\overline{U}_1| = n_1$, $|\overline{U}_2| = n_2$, $n_1 + n_2 = 2(t+1) - p$, and $n_1 \leq n_2$. Since $0 \leq n_1 \leq (2(t+1)-p)/2$, the maximum value of n_1 is equal to $t+1$ when $p = 0$ and $n_2 = t+1$. According to different values of n_1 and n_2, we divide the proof into two cases. The first case $n_2 \leq t$, which implies $n_1 \leq t$. The second case $n_2 > t$, and this case is further divided into three subcases $n_1 < t$, $n_1 = t$, and $n_1 > t$.

Case 1. $n_1 \leq t$ and $n_2 \leq t$. By Lemma 21.11, we have $|\Gamma(G, U)| \geq |\Gamma(G_1, U_1)| + |\Gamma(G_2, U_2)| = |\overline{U}_1| + |\overline{U}_2| = n_1 + n_2 = 2(t+1) - p$. We know that $0 < p \leq t$, $|\Gamma(G, U)| \geq 2(t+1) - p > p$, and condition (3a) of Theorem 21.6 is satisfied.

Case 2. $n_2 > t$. We discuss the case according to the following three subcases: (1) $n_1 < t$, (2) $n_1 = t$, and (3) $n_1 > t$.

Case 2.1. $n_1 < t$. Since $\kappa(G_1) \geq t$ and $|\overline{U}_1| = n_1 < t$, $G[U_1]$ is connected. By Lemmas 21.7 and 21.8, $\Gamma(G_1, U_1) = N(\overline{U}_1, U_1) = n_1$. There are n_1 and n_2 vertices in $\overline{U}_1$ and $\overline{U}_2$, respectively, and $n_2 = 2t + 2 - p - n_1$ (see Figure 21.6).

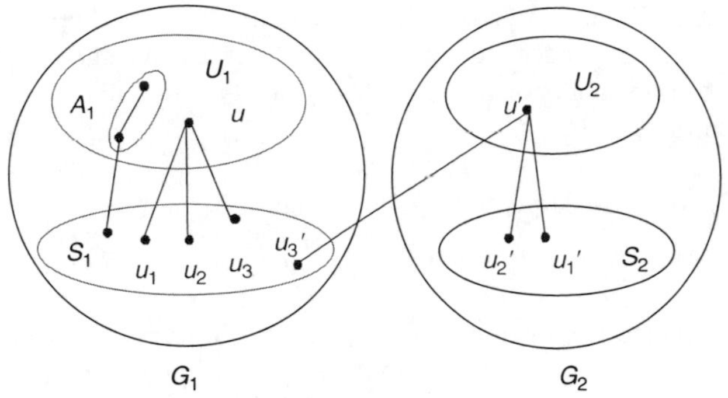

FIGURE 21.7 Illustration of Theorem 21.9 for an example of Case 2.3.

If all the vertices in $\overline{U}_1$ are adjacent to some n_1 vertices in $\overline{U}_2$, there are still at least $n_2 - n_1 = 2t + 2 - p - 2n_1$ vertices in $\overline{U}_2$ such that each of them is adjacent to some vertex in U_1 under the matching M. So, $|\Gamma(G, U)| \geq |\Gamma(G_1, U_1)| + (n_2 - n_1) = n_1 + (n_2 - n_1) = n_2$. Because $n_2 > t \geq p$, the proof of this subcase is complete.

Case 2.2. $n_1 = t$. We know that $n_1 + n_2 = 2(t + 1) - p$, $0 \leq p \leq t$, $n_2 > t$, and $n_1 = t$. The only two valid values for n_2 are $t + 1$ and $t + 2$. $n_2 = t + 1$ implies $p = 1$ and $n_2 = t + 2$ implies $p = 0$. By Lemma 21.11, $|\Gamma(G_1, U_1)| = |\overline{U}_1| = t \geq 2 > p$ for $p = 0$ or 1. Then the subcase holds.

Case 2.3. $n_1 > t$. Observe that $0 \leq n_1 \leq (2(t + 1) - p)/2$, where $0 \leq p \leq t$ and $n_2 \geq n_1 > t$. Therefore, $n_1 = n_2 = t + 1$. It also implies $p = 0$. Here, we will prove that the subcase satisfies either condition (3a) or (3b) of Theorem 21.6.

First, if the subgraph induced by U contains a connected component A_1 with cardinality at least 2 (see Figure 21.7), then it must be adjacent to some vertex in $\overline{U}$. Thus, we know that $|\Gamma(G, U)| > 0 = p$, and condition (3a) of Theorem 21.6 is satisfied.

Otherwise, every connected component of U contains a single vertex only. By Theorem 21.3, we know that F_1 and F_2 are distinguishable if there exists a path form u_1 to u_2 that contains a vertex u in U and $u_1, u_2 \in F_1 - F_2$, or $u_1, u_2 \in F_2 - F_1$. If $p = 0$, it implies $F_1 \cap F_2 = \phi$, any vertex u in $G[U]$ with degree more than 2 must be connected to at least two vertices in F_1 or F_2 (see Figure 21.7). By Theorem 21.8, $order_{G_1 \oplus G_2}(v) \geq t + 1$ for every vertex v in $G_1 \oplus G_2$; therefore, $deg(v) \geq t + 1$ for every vertex v in $G_1 \oplus G_2$. Since $t \geq 2$, condition (3b) of Theorem 21.6 is satisfied.

Hence, the subcase holds and the theorem follows. □

By Theorems 21.5 and 21.9, we have the following corollary:

COROLLARY 21.2 *Let G_1 and G_2 be two graphs with the same number of vertices N. Suppose that both G_1 and G_2 are t-diagnosable under the comparison model and have connectivity $\kappa(G_1) = \kappa(G_2) \geq t$, where $t \geq 2$. Then MCN $G_1 \oplus G_2$ is $(t + 1)$-diagnosable under the comparison model.*

In Ref. 330, D. Wang has proved that the diagnosability of hypercube-structured multiprocessor systems under the comparison model is n when $n \geq 5$. However, the diagnosability of Q_4 is not known to be 4. Using our Theorem 21.9, we can strengthen the result as follows.

THEOREM 21.10 The hypercube Q_n is n-diagnosable for $n \geq 4$.

Proof: We observe that Q_3 is 3-connected, $order_{Q_3}(v) = 3$ for every vertex v in Q_3, and the number of vertices of Q_3 is 8, $8 \geq t + 2 = 5$ for $t = 3$. It is well-known that Q_4 can be constructed from two copies of Q_3 by adding a perfect matching between these two copies. Therefore, by Theorem 21.9, Q_4 is 4-diagnosable.

Then the proof is by induction on n. We have shown that Q_4 is 4-diagnosable. Assume that it is true for $n = m - 1$. Considering $n = m$, Q_m is obtained from two copies G_1, G_2 of Q_{m-1} by adding a perfect matching joining corresponding vertices in G_1 and G_2. It is well-known that Q_{m-1} is $(m - 1)$-connected. By Corollary 21.2, Q_m is m-diagnosable. This completes the induction proof. □

However, Q_3 is not 3-diagnosable. In Figure 21.8, there is a Q_3, let $S_1 = \{0, 5, 7\}$ and $S_2 = \{2, 5, 7\}$. Then, by Theorem 21.3, S_1 and S_2 are not distinguishable, as shown in Figure 21.8.

As we observe that most of the related results on diagnosability of multiprocessors systems [101,330] are based on a sufficient theorem; namely, Theorem 21.5. Not satisfying this sufficient condition, such as in the case of Q_4, does not necessarily imply that the network is not 4-diagnosable. Therefore, we propose a hybrid condition, of (3a) and (3b) of Theorem 21.6, to check the diagnosability of multiprocessor systems under the comparison model, which is more powerful. Applying our Theorems 21.6 and 21.9, we show that the diagnosability of Q_4 is indeed 4.

It is known [47,103,212] that the crossed cube CQ_n [92], the twisted cube TQ_n [146], and the Möbius cube MQ_n [76] are all n-connected. By Theorem 21.8, we can prove that the order of each vertex in these two cubes is n. We observe that the two cubes are both constructed recursively using a similar way satisfying the requirements of Theorem 21.9 and Corollary 21.2. Therefore, we can prove that CQ_n, TQ_n, and MQ_n are all n-diagnosable for $n \geq 4$. Then we list the following three theorems.

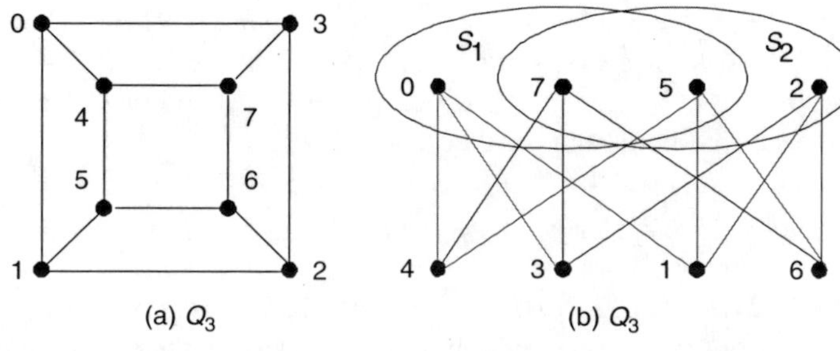

FIGURE 21.8 An example of an indistinguishable pair for Q_3.

THEOREM 21.11 [**101**] The crossed cube CQ_n is n-diagnosable for $n \geq 4$.

THEOREM 21.12 The twisted cube TQ_n is n-diagnosable for $n \geq 4$.

THEOREM 21.13 The Möbius cube MQ_n is n-diagnosable for $n \geq 4$.

21.4 DIAGNOSABILITY OF CARTESIAN PRODUCT NETWORKS

Many multiprocessor networks are constructed by the Cartesian product, such as grids, hypercubes, meshes, and tori. The product networks constitute very important classes for the interconnection networks. The diagnosability of the product networks under the PMC model was investigated in Ref. 16. In this section, we study the diagnosability of the product network of G_1 and G_2, where G_i is t_i-diagnosable or t_i-connected for $i = 1, 2$. Furthermore, we use different combinations of t_i-diagnosability and t_i-connectivity to study the diagnosability of the product networks. We show that the product network of $G_1, G_2, \ldots,$ and G_k is $(t_1 + t_2 + \cdots + t_k)$-diagnosable, where each G_i is either t_i-diagnosable or t_i-connected with regularity t_i for $1 \leq i \leq k$.

The Cartesian product $G = G_1 \times G_2$ of two graphs $G_1 = (V_1, E_1)$ and $G_2 = (V_2, E_2)$ is the graph $G = (V, E)$, where the set of vertices V and the set of edges E are given by:

1. $V = \{\langle x, y\rangle \mid x \in V_1 \text{ and } y \in V_2\}$
2. For $u = \langle x_u, y_u\rangle$ and $v = \langle x_v, y_v\rangle$ in V, $(u, v) \in E$ if and only if $(x_u, x_v) \in E_1$ and $y_u = y_v$, or $(y_u, y_v) \in E_2$ and $x_u = x_v$

Let y be a fixed vertex of G_2. The subgraph G_1^y-component of $G_1 \times G_2$ has vertex set $V_1^y = \{(x, y) | x \in V_1\}$ and edge set $E_1^y = \{(u, v) | u = <x_u, y>, v = <x_v, y>, (x_u, x_v) \in E_1\}$. Similarly, let x be a fixed vertex of G_1, the subgraph G_2^x-component of $G1 \times G2$ has vertex set $V_2^x = \{(x, y) | y \subset V_2\}$ and edge set $E_2^x = \{(u, v) | u = <x, y_u>, v = <x, y_v>, (y_u, y_v) \in E_2\}$. It is clear that the G_1^y-component (abbreviated as G_1^y) and the G_2^x-component (abbreviated as G_2^x) are isomorphic to G_1 and G_2, respectively (as illustrated in Figure 21.9). The following lemma lists a set of known results [80,81,90,97,353] related to the topological properties of the Cartesian product of $G_1 \times G_2$ of two graphs G_1 and G_2.

LEMMA 21.12 Let $u = \langle x_u, y_u\rangle$ and $v = \langle x_v, y_v\rangle$ be two vertices in $G_1 \times G_2$. The following properties hold:

1. $G_1 \times G_2$ is isomorphic to $G_2 \times G_1$
2. $|G_1 \times G_2| = |G_1| \cdot |G_2|$, where $|G|$ is the number of vertices in G
3. $deg_{G_1 \times G_2}(u) = deg_{G_1}(x_u) + deg_{G_2}(y_u)$
4. $dist_{G_1 \times G_2}(u, v) = dist_{G_1}(x_u, x_v) + dist_{G_2}(y_u, y_v)$, where $dist_G(u, v)$ is the distance between u and v in G
5. $D(G_1 \times G_2) = D(G_1) + D(G_2)$, where $D(G)$ is the diameter of G
6. $\kappa(G_1 \times G_2) \geq \kappa(G_1) + \kappa(G_2)$, where $\kappa(G)$ is the connectivity of G

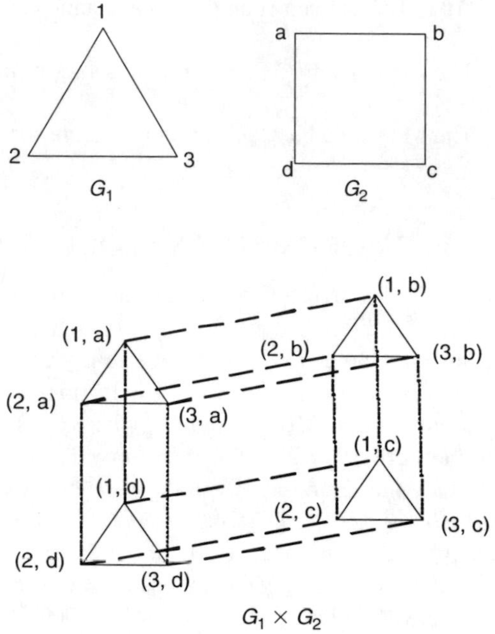

FIGURE 21.9 An example of product network $G_1 \times G_2$.

21.4.1 Diagnosability of Cartesian Product Networks under the PMC Model

Araki and Shibata [16] proposed some results for the t-diagnosability and t/t-diagnosability of Cartesian product systems under the PMC model, listed as follows.

THEOREM 21.14 **[16]** Let G_1 and G_2 be digraphs of t_1- and t_2-diagnosable systems, respectively. Then the system $G = G_1 \times G_2$ is $(t_1 + t_2)$-diagnosable.

LEMMA 21.13 **[16]** Let G be a digraph of a t-diagnosable system. Then $(G \times K_2)$ is a digraph of a $(t + 1)$-diagnosable system.

THEOREM 21.15 **[16]** Let G_1 and G_2 be digraphs of t_i/t_i-diagnosable system $(i = 1, 2)$. If $\delta_i \geq \left\lceil \frac{(t_i+1)}{2} \right\rceil$ for $i = 1, 2$. Then the system $G = G_1 \times G_2$ is $(t_1 + t_2)/(t_1 + t_2)$-diagnosable.

LEMMA 21.14 **[16]** Let G be a digraph of a t/t-diagnosable system such that $\delta_{in}(G) \geq \left\lceil \frac{t+1}{2} \right\rceil$. Then a system represented by $(G \times K_2)$ is $(t + 1)/(t + 1)$-diagnosable.

THEOREM 21.16 **[16]** Let G_1 and G_2 be digraphs of t_i/t_i-diagnosable system $(i = 1, 2)$. If $\delta_i \geq \left\lceil \frac{t_i}{2} \right\rceil + 1$ and $t_i \geq 2 (i = 1, 2)$. Then the system $G = G_1 \times G_2$ is $(t_1 + t_2 + 2)/(t_1 + t_2 + 2)$-diagnosable.

21.4.2 Diagnosability of Cartesian Product Networks under the Comparison Model

In this section, we distinguish the product networks into homogeneous product networks and heterogeneous product networks. By homogeneous product networks, we mean that every factor network of the product has the same properties of being t-diagnosable and t-regular (or being t-connected and t-regular, respectively), while heterogeneous product networks mean that one of the factor networks is t-diagnosable and another is t-connected. Before discussing homogeneous product networks and heterogeneous product networks, we consider the problem of whether a t-regular and t-connected interconnection network is t-diagnosable.

21.4.3 Diagnosability of t-Connected Networks

This section considers the problem of whether under suitable conditions, a t-regular and t-connected interconnection network is also t-diagnosable. A t-regular and t-connected interconnection network with at least $2t+3$ vertices is first proven also to be t-diagnosable. Moreover, the product network of G_1 and G_2 is shown to be (t_1+t_2) diagnosable, where G_i is t_i connected with regularity t_i for $i=1,2$.

LEMMA 21.15 Let G be a t-regular and t-connected network with $N \geq 2t+1$ vertices and $t > 2$. Then each vertex v of G has order t.

Proof: Let v be a vertex of G and let $G(v)$ be the order graph of v in G. Let $\chi(v)$ be a vertex cover of $G(v)$. Assume that vertex v has order $k < t$. Since G contains $N \geq 2t+1$ vertices and the order of v is $k < t$, there exists at least one vertex $y \in V, y \neq v, y \notin N(v)$ and $y \notin \chi(v)$. The distance between v and y is at least 2. Each edge of $G(v)$ has at least one endpoint in $\chi(v)$, so all paths from v to y in G must be from v via z, which is a vertex in $\chi(v)$. Deleting all the vertices of $\chi(v)$ in G ensures that no paths exist from v to y. However, exactly k vertices are deleted, contradicting the assumption that G is a t-connected network, so $k \geq t$. $N(v)$ is a vertex cover of $G(v)$, so the vertex v must have order $k=t$. □

Given a t-diagnosable system, by Lemma 21.9, the number of vertices must exceed or be equal to $2t+1$. However, a t-regular and t-connected network with $N=2t+1$ vertices is not necessarily t-diagnosable. The graph shown in Figure 21.10 is a 4-regular and 4-connected network with $N=9$ vertices, since any two arbitrarily

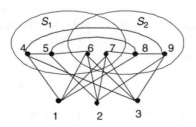

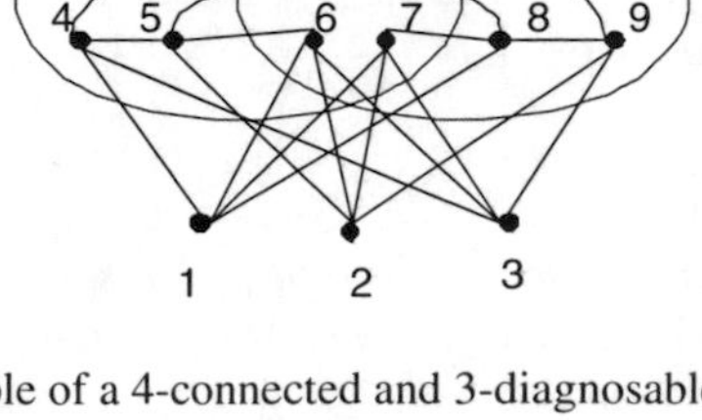

FIGURE 21.10 An example of a 4-connected and 3-diagnosable system.

distinct vertices in Figure 2.10 are contained in two disjoint cycles. For example, two distinct vertices 4 and 5 are present in cycles $\langle 4,9,8,5\rangle$ and $\langle 4,1,6,5,2,7,3\rangle$. This graph can be easily seen to be not 4-diagnosable, since $\{4,5,6,7\}$ and $\{6,7,8,9\}$ constitute an indistinguishable pair. With regard to $N=2t+2$, the three, dimensional crossed cube CQ_3 and the three-dimensional hypercube Q_3 are 3-regular, 3-connected networks and each vertex has order $t=3$. However, it was demonstrated [101,196] that CQ_3 and Q_3 are not 3-diagnosable under the comparison diagnosis model. The t-regular and t-connected network G with $N \geq 2t+3$ vertices is thus considered in the following theorem.

THEOREM 21.17 Let $G=(V,E)$ be a t-regular and t-connected network with N vertices and $t>2$. G is t-diagnosable if $N \geq 2t+3$.

Proof: Let S_1 and S_2 be two distinct subsets of V with $|S_1|=|S_2|=t$, $|S_1 \cap S_2|=p$ and $0 \leq p \leq t-1$. By Theorem 21.4 and Lemma 21.15, G can be shown to be t-diagnosable by showing that (S_1,S_2) is a distinguishable pair. Let $V''=S_1 \cup S_2$ and $V'=V-V''$. Then $|V''|=2t-p>t$. Notably, V' may not be connected. The case in which all connected components of the subgraph induced by V' are isolative vertices is considered first. For $0 \leq p \leq t-1$, the following cases are considered.

Case 1. $0 \leq p \leq t-3$. Since $0 \leq p \leq t-3$ and G is a t-regular graph, each vertex of V' has at least two neighbors in S_1-S_2 or S_2-S_1 for $t>2$. Thus, either condition (2) or (3) in Theorem 21.3 is satisfied.

Case 2. $p=t-2$. In this case, $|V''|=t+2, N \geq 2t+3$ and $|V'|=N-(t+2) \geq t+1$. Assume that the pair S_1,S_2 are indistinguishable. Therefore, conditions (2) and (3) in Theorem 21.3 cannot be satisfied, implying that each vertex of V' must be connected to $t-2$ vertices in $S_1 \cap S_2$: one vertex in S_1-S_2 and one vertex in S_2-S_1. Therefore, at most t vertices in V' satisfy this assumption, contradicting the condition $|V'| \geq t+1$. Hence, either condition (2) or (3) in Theorem 21.3 must be satisfied.

Case 3. $p=t-1$. $|V''|=t+1$ and $|V'|=N-t-1$. The subgraph induced by V' consists of isolative vertices and G is a t-regular graph, so $(N-t-1)t$ edges are adjacent to the vertices of V' and V''. However, G has exactly $Nt/2$ edges. For $N \geq 2t+3$, we have $(N-t-1)t > Nt/2$, which is a contradiction, so $p=t-1$ is impossible.

Now consider that the subgraph induced by V' contains a connected component R with cardinality of at least 2. Let $u \in R$ and $v \in (S_1-S_2) \cup (S_2-S_1)$. G is t-connected, so there exist t disjoint paths from u to v. However, at most p disjoint paths exist from u to v via the vertices of $S_1 \cap S_2$. Therefore, there exists at least one path from u to v such that no vertex of the path belongs to $S_1 \cap S_2$. Since u is a vertex in R, there exists another vertex w adjacent to u. Hence, the condition (1) in Theorem 21.3 is satisfied, completing the proof of the theorem. □

COROLLARY 21.3 *For $t_1,t_2>2$, let G_1 and G_2 be two t_1-connected and t_2-connected networks, with regularity t_1 and t_2, respectively. Let $G=(V,E)$ be the product network of G_1 and G_2. Then the product network $G=G_1 \times G_2$ is (t_1+t_2)-diagnosable with regularity t_1+t_2.*

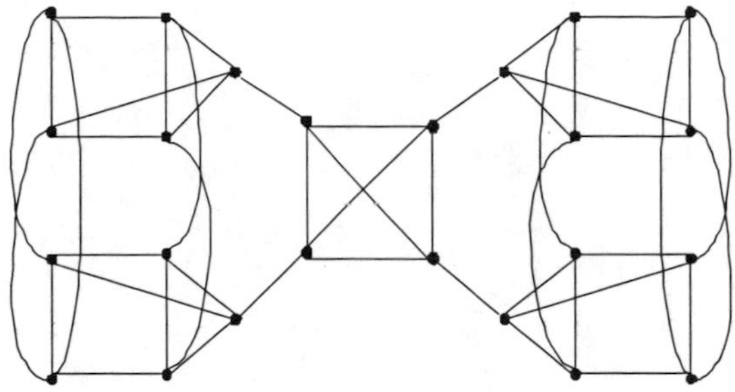

FIGURE 21.11 An example of a 2-connected and 4-diagnosable system.

Proof: G_1 is t_1-regular and t_1-connected, so at least $t_1 + 1$ vertices exist in G_1. Similarly, the number of vertices in G_2 is at least $t_2 + 1$. Therefore, G contains at least $(t_1 + 1)(t_2 + 1)$ vertices. Moreover, by Lemma 21.12, the degree of every vertex in G is $t_1 + t_2$ (regularity $t_1 + t_2$). $\delta(G)$ is used to denote the minimum degree of G. That $\kappa(G) \leq \delta(G)$ is well known [250]. However, by Lemma 21.12, $\kappa(G) \geq \kappa(G_1) + \kappa(G_2) = t_1 + t_2$. Since $t_1 + t_2 \leq \kappa(G) \leq \delta(G) = t_1 + t_2$, $\kappa(G) = t_1 + t_2$. Since $(t_1 + 1)(t_2 + 1) > 2(t_1 + t_2) + 3$ for $t_1, t_2 > 2$, Theorem 21.17 implies that G is $(t_1 + t_2)$-diagnosable. Therefore, the corollary follows. □

COROLLARY 21.4 *Let G be a product network of $G_1, G_2, \ldots,$ and G_k. Each G_i is t_i-regular, t_i-connected, and $t_i > 2$ for $1 \leq i \leq k$ where $k > 2$. Then, the product network G is $(t_1 + t_2 + \cdots + t_k)$-regular and $(t_1 + t_2 + \cdots + t_k)$-diagnosable.*

Theorem 21.17 indicates that a t-connected network with $N \geq 2t + 3$ vertices is also t-diagnosable. However, a t-diagnosable network is not necessarily a t-connected network (as depicted in Figure 21.11). The example shown in Figure 21.11 is 4-regular and 4-diagnosable, but not 4-connected. The t diagnosability and t connectivity are not equivalent terms, but these two concepts are closely related; Theorem 21.17 provides an example.

21.4.4 Diagnosability of Homogeneous Product Networks

By Corollary 21.3, the homogeneous product network $G_1 \times G_2$ is $(t_1 + t_2)$-diagnosable, where G_i is t_i-connected and t_i-regular, $t_i > 2$, $i = 1, 2$. The homogeneous product network $G_1 \times G_2$ is also $(t_1 + t_2)$-diagnosable, where G_i is t_i-diagnosable and t_i-regular, $t_i > 2$, $i = 1, 2$. Several lemmas must be proven first.

LEMMA 21.16 Let $G = (V, E)$ be a t-regular network with $N \geq 2t + 1$ vertices. Suppose that each vertex of G has order t, $t > 2$. If $V' \subset V$ and $|V - V'| \leq t$, then $\Gamma(G, V') = V - V'$.

Proof: Let v be an arbitrary vertex in $V - V'$, and let $G(v)$ be the order graph of v in G. The following two cases are considered.

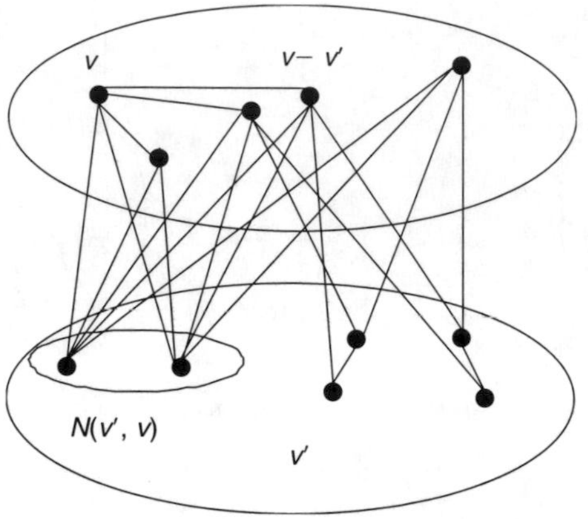

FIGURE 21.12 Illustration of Lemma 21.16 for an example of Case 2.

Case 1. $|V - V'| < t$. For $|V - V'| < t$, the degree of each vertex is t, so each vertex in V' has at least one neighbor in V'. Therefore, no isolated vertex exists in V'. Similarly, every vertex in $V - V'$ has at least one neighbor in V'. Hence, $\Gamma(G, V') = V - V'$.

Case 2. $|V - V'| = t$. For $|V - V'| = t$, each vertex in $V - V'$ has at least one neighbor in V'. $N(v, V')$ is used to denote the neighbor set of v in V'. Assume that no vertex in $N(v, V')$ is adjacent to any other vertices in V'. Then, every vertex in $N(v, V')$ is adjacent only to $V - V'$ (as shown in Figure 21.12). Thus, $V - V' - \{v\}$ is a vertex cover of $G(v)$, because every vertex in $N(v, V')$ is an isolated vertex in V'. The cardinality of a minimum vertex cover of the order graph $G(v)$ can be easily determined to be at most $t - 1$. However, this contradicts the hypothesis that each vertex has order t. Therefore, $N(v, V')$ contains at least one neighbor u of v such that the vertex u is adjacent to another vertex w in V'. Hence, $\Gamma(G, V') = V - V'$. □

LEMMA 21.17 Let H be a t-regular network, $t > 2$, and let K_2 be the complete network with two vertices. Suppose that the order of each vertex in H is t. Then, each vertex of the product network $G = H \times K_2$ has order $t + 1$.

Proof: Let G^0 and G^1 be two copies of H in G. $M = (V, C)$ represents the comparison scheme of G. Let v be a vertex of G and let $G(v)$ be the order graph of v in G. Without loss of generality, assume that v is a vertex in G^0, and that u is a neighbor of v in G^1. There exists at least one vertex w in G^1 such that $(v, w)_u \in C$. Then, let $G^0(v)$ be the order graph of v in G^0. Since $G^0(v)$ is a proper subgraph of $G(v)$, every vertex cover of $G(v)$ must contain a vertex cover of $G^0(v)$. However, (w, u) is an edge in $G(v)$ rather than in $G^0(v)$. Therefore, a vertex cover of $G(v)$ must include at least either u or w. The order of v in G, therefore, exceeds that of v in G^0 by one. Thus, the lemma is proven. □

THEOREM 21.18 For $t > 2$, let H be a t-regular and t-diagnosable network with N vertices. Then the product network $G = H \times K_2$ is $(t + 1)$-diagnosable.

Proof: Let $G^0 = (V^0, E^0)$ and $G^1 = (V^1, E^1)$ be two copies of H in $G = (V, E)$. Let S_1 and S_2 be two distinct subsets of V and let $V'' = S_1 \cup S_2$ with $|S_1| = |S_2| = t + 1$, $|S_1 \cap S_2| = p$ and $0 \leq p \leq t$. Then, let $V' = V - V''$ with $|V'| = 2N - 2(t+1) + p$. Since G has $2N$ vertices, $2N \geq 2(2t+1) > 2(t+1) + 1$. Lemma 21.17 implies that each vertex of G has order $t + 1$. Hence, the theorem is proven if one of the conditions of Theorem 21.3 is satisfied. Now, let $V^{0'} = V' \cap V^0$ and $V^{1'} = V' \cap V^1$. G^0 and G^1 are isomorphic to H, so without loss of generality, assume that $|V^{0'}| \geq |V^{1'}|$. Let $|V^0 - V^{0'}| = k$ and $|V^1 - V^{1'}| = 2(t + 1) - p - k$. Since $|V^{0'}| \geq |V^{1'}|$, $k \leq 2(t + 1) - p - k$. Thus, the proof is divided into the following cases.

Case 1. $2(t+1) - p - k \leq t$ and $k \leq t$. From Lemma 21.16, $|\Gamma(G, V')| \geq |\Gamma(G^0, V^{0'})| + |\Gamma(G^1, V^{1'})| = k + 2(t+1) - p - k = 2(t+1) - p$. Since $p \leq t$, $|\Gamma(G, V')| \geq 2(t + 1) - p > p$. By Theorem 21.5, this case holds.

Case 1.1. $2(t+1) - p - k > t$ and $k < t$. From Lemma 21.16, $|\Gamma(G^0, V^{0'})| = k$. Since $V^0 - V^{0'}$ contains $k < t$ vertices, each vertex in $V^{0'}$ has at least one neighbor in $V^{0'}$. Therefore, no isolated vertex is present in $V^{0'}$. Notably, at least $2(t+1) - p - 2k$ vertices in $V^1 - V^{1'}$ are adjacent to some $2(t+1) - p - 2k$ vertices in $V^{0'}$. Thus, $|\Gamma(G, V')| \geq |\Gamma(G^0, V^{0'})| + N(V^1 - V^{1'}, V^{0'}) \geq k + 2(t+1) - p - 2k = 2(t+1) - p - k$. Since $2(t+1) - p - k > t \geq p$, by Theorem 21.5, the case holds.

Case 1.2. $2(t+1) - p - k > t$ and $k = t$. Since $2(t+1) - p - k > t$ and $k = t$, $(t+2) - p > t$, implying $p < 2$. From Lemma 21.16, $|\Gamma(G, V')| \geq |\Gamma(G^0, V^{0'})| = t > 2 > p$. Then, the next case follows.

Case 1.3. $2(t+1) - p - k > t$ and $k > t$. Since $2(t+1) - p - k > t$ and $k > t$, the number of vertices in $V - V'$ is $2(t+1)$, indicating $p = 0$. Condition (1) in Theorem 21.3 is supposed to be satisfied in G^0 first. Then, the subgraph induced by $V^{0'}$ includes at least one connected component R with a cardinality of at least two. Given $|V^0 - V^{0'}| = t + 1$, Theorem 21.5 implies $|\Gamma(G^0, V^{0'})| \geq t > 2$ since G^0 is t-diagnosable. Therefore, $|\Gamma(G, V')| \geq |\Gamma(G^0, V^{0'})| > 2 > p$. This result implies that condition (1) in Theorem 21.3 is also satisfied in G.

Next, consider that the condition (1) in Theorem 21.3 is violated in G^0. Then, either condition (2) or (3) in Theorem 21.3 is satisfied in G^0. Since G^0 is t-regular and $t > 2$, one vertex v in $V^{0'}$ is adjacent to at least three vertices in $V^0 - V^{0'}$. Now, let u, w, and x be three vertices in $V^0 - V^{0'}$ such that $u, w \in S_1$, and $x \in S_2$. Since $u, w \in S_1 - S_2$, $v \in V - S_1 - S_2$ and $p = 0$, condition (2) in Theorem 21.3 is also satisfied in G. The theorem follows. □

Let G_i be a t_i-regular interconnection network $i = 1, 2$, and let $G = G_1 \times G_2$ be the product network of G_1 and G_2. Then the order of each vertex v in G is estimated from the following lemma.

LEMMA 21.18 Let $G_i = (V_i, E_i)$ be a t_i-regular network with $t_i > 2$. Suppose each vertex of G_i has order at least t_i, $i = 1, 2$. Then each vertex of the product network $G = G_1 \times G_2$ has order $t_1 + t_2$.

Proof: Let $v = \langle x, y\rangle$ be an arbitrary vertex of G and let $G(v)$ be the order graph of v in G. According to the definition of product networks, x is a vertex of V_1 and y is a vertex of V_2. Therefore, the order of x is at least t_1 and the order of y is at least t_2. Let $G_1(x)$ be the order graph of x in G_1 and let $G_2(y)$ be the order graph of y in G_2. $N(x)$ is a vertex cover of $G_1(x)$, so the order of vertex x is exactly t_1. Similarly, the order of vertex y is t_2. Let $G_1^y(v)$ be the order graph of v in the subgraph G_1^y of G and let $G_2^x(v)$ be the order graph of v in the subgraph G_2^x of G. Since $V_1^y \cap V_2^x = v$, $G_1^y(v) \cap G_2^x(v) = \emptyset$, where $V(G_1^y(v))$ and $V(G_2^x(v))$ are the vertex sets of $G_1^y(v)$ and $G_2^x(v)$, respectively. $G_1^y(v)$ and $G_2^x(v)$ are observed to be subgraphs of $G(v)$. Thus, every vertex cover of $G(v)$ must contain a vertex cover of both $G_1^y(v)$ and $G_2^x(v)$. Since the subgraphs G_1^y and G_2^x of G are isomorphic to G_1 and G_2, respectively, $G_1^y(v)$ is isomorphic to $G_1(x)$ and $G_2^x(v)$ is isomorphic to $G_2(y)$. Therefore, the order of v in $G_1^y(v)$ is t_1 and the order of v in $G_2^x(v)$ is t_2. Since $V(G_1^y(v)) \cap V(G_2^x(v)) = \emptyset$, the order of v in $G(v)$ is $t_1 + t_2$. Hence, the lemma follows. □

Corollary 21.3 was proven; it states that the product network $G_1 \times G_2$ is $(t_1 + t_2)$-diagnosable, in which G_i is t_i connected for $t_i > 2$, $i = 1, 2$. The previous section also established that a t_i-diagnosable network is not equivalent to a t_i-connected network. The following theorem states that the product network $G_1 \times G_2$ is $(t_1 + t_2)$-diagnosable, where G_i is t_i-diagnosable for $t_i > 2$, $i = 1, 2$. We present Theorem 21.19 and omit the related proof, which is completely illustrated in Ref. 42.

THEOREM 21.19 **[42]** For $t_i > 2$, let $G_i = (V_i, E_i)$ be a t_i-diagnosable and t_i-regular network with N_i vertices, $i = 1, 2$. Let $G = (V, E)$ be the product network of G_1 and G_2. Then the product network $G = G_1 \times G_2$ is $(t_1 + t_2)$-diagnosable under the comparison diagnosis model with regularity $t_1 + t_2$.

Notice that in Theorem 21.19 the number of vertices N_i is larger than or equal to $2t_i + 1$ for t_i-diagnosable, $i = 1, 2$. From Theorem 21.19 and by induction, the following corollary is obtained.

COROLLARY 21.5 *Let G be the product network of $G_1, G_2, \ldots,$ and G_k, where each G_i is t_i-diagnosable with regularity t_i and $t_i > 2$ for $1 \le i \le k$. Then the product network G is $(t_1 + t_2 + \cdots + t_k)$-diagnosable with regularity $(t_1 + t_2 + \cdots + t_k)$.*

21.4.5 Diagnosability of Heterogeneous Product Networks

This subsection considers different combinations of t_i-diagnosability and t_i-connectivity to study the diagnosability of the product networks. The diagnosability of the heterogeneous product network G of G_1 and G_2, is considered, in which G_1 is t_1-diagnosable and G_2 is t_2-connected. Although the heterogeneous product network differs from the homogeneous product network, a similar result is obtained to that obtained for the homogeneous product network. Lemmas 21.15 and 21.17 immediately yield the following lemma.

LEMMA 21.19 Let G_1 be a t_1-regular and t_1-diagnosable network with $t_1 > 2$ and let G_2 be a t_2-regular and t_2-connected network with $N_2 \geq 2t_2 + 1$ vertices and $t_2 > 2$. Then, each vertex of the product network $G = G_1 \times G_2$ has order $t_1 + t_2$.

Section 21.4.3 presents some examples to show that a t-diagnosable network is not equivalent to a t-connected network. Therefore, the following theorem is not implied by Theorem 21.19, but it can be proven by a similar technique.

THEOREM 21.20 **[42]** For $t_1, t_2 > 2$, let $G_1 = (V_1, E_1)$ be a t_1-regular and t_1-diagnosable network with N_1 vertices and let $G_2 = (V_2, E_2)$ be a t_2-regular and t_2-connected network with N_2 vertices where $N_2 \geq 2t_2 + 1$ vertices. Then the product network $G = G_1 \times G_2$ is $(t_1 + t_2)$-diagnosable under the comparison diagnosis model with regularity $t_1 + t_2$.

In the previous theorem, the factor network G_2 must have at least $2t_2 + 1$ vertices. Therefore, by Corollary 21.5 and Theorem 21.20, the following corollary holds.

COROLLARY 21.6 *Let G be the product network of $G_1, G_2, \ldots, G_k$. Suppose that G_1 is t_1-regular and t_1-connected with $N_1 \geq 2t_1 + 1$ vertices, and suppose that G_i is t_i-regular and t_i-diagnosable, $t_i > 2$ for $2 \leq i \leq k$. Then the product network G is $(t_1 + t_2 + \cdots + t_k)$-diagnosable with regularity $(t_1 + t_2 + \cdots + t_k)$.*

However, Corollaries 21.4 and 21.5 yield the following corollary.

COROLLARY 21.7 *Let G be the product network of $G_1, G_2, \ldots,$ and G_k. Suppose that G_i is t_i-regular and t_i-connected, $t_i > 2$ for $1 \leq i \leq m$ where $m > 2$, and suppose that G_j is t_j-regular and t_j-diagnosable, $t_j > 2$ for $m + 1 \leq j \leq k$. Then the product network G is $(t_1 + t_2 + \cdots + t_k)$-diagnosable under the comparison diagnosis model with regularity $(t_1 + t_2 + \cdots + t_k)$.*

21.5 STRONGLY *t*-DIAGNOSABLE SYSTEMS

The hypercube Q_n, the crossed cube CQ_n, the Möbius cube MQ_n, and the twisted cube TQ_n are all known to be n-connected but not $(n+1)$-connected. For each of these cubes, every vertex cut of size n has a particular structure, as stated in the following lemma.

LEMMA 21.20 Let $n \geq 2$ and let XQ_n represent any n-dimensional cube that belongs to the cube family. For each set of vertices $S \subset V(XQ_n)$ with $|S| = n$, if $XQ_n - S$ is disconnected, there exists a vertex $v \in V(XQ_n)$ such that $N(v) = S$.

Proof: We prove this lemma by induction on n. A two-dimensional cube XQ_2 is simply a cycle of length 4. Clearly, this lemma is true for XQ_2. Assume that it holds for some $n \geq 2$. We now show that it holds for $n + 1$.

Let $(n+1)$-dimensional cube XQ_{n+1} be obtained from two n-dimensional cubes XQ_n, denoted by XQ_n^L and XQ_n^R, by adding a perfect matching between them.

Let $S \subset V(XQ_{n+1})$, $|S| = n+1$, and $S_L = V(XQ_n^L) \cap S$ and $S_R = V(XQ_n^R) \cap S$. In the remainder of this proof, we show that XQ_{n+1} satisfies one of the two conditions: (1) $XQ_{n+1} - S$ is connected or (2) $XQ_{n+1} - S$ is disconnected and there is a vertex $v \in V(XQ_{n+1})$ such that $N(v) = S$.

We study three cases: (1) $|S_L| \leq n-1$ and $|S_R| \leq n-1$, (2) either $|S_L| = n$ or $|S_R| = n$, and (3) either $|S_L| = n+1$ or $|S_R| = n+1$.

Case 1. $|S_L| \leq n-1$ and $|S_R| \leq n-1$. Since XQ_n is n-connected, both $XQ_n^L - S_L$ and $XQ_n^R - S_R$ are connected. For $n \geq 2$, we know that $|V(XQ_n^L) - S_L| \geq 2^n - (n-1) > n - 1 \geq |S_R|$ and $|V(XQ_n^R) - S_R| \geq 2^n - (n-1) > n-1 \geq |S_L|$. So the subgraph $XQ_n^L - S_L$ is connected to the other subgraph $XQ_n^R - S_R$. Hence, $XQ_{n+1} - S$ is connected.

Case 2. Either $|S_L| = n$ or $|S_R| = n$. Without loss of generality, suppose that $|S_L| = n$ and $|S_R| = 1$. Suppose $XQ_n^L - S_L$ is connected. Using a similar argument used in Case 1, we can prove that $XQ_{n+1} - S$ is connected. Otherwise, $XQ_n^L - S_L$ is disconnected. By induction hypothesis, there exists a vertex $v \in V(XQ_n^L)$ such that $N(\{v\}, XQ_n^L) = S_L$. Now, consider XQ_n^R and consider the matching neighbor u of v in XQ_n^R. Note that $XQ_n^R - S_R$ is connected for $n \geq 2$ and every vertex in XQ_n^R has a matching neighbor in XQ_n^L. Thus, $XQ_{n+1} - S$ is connected if $S_R \neq \{u\}$. If $S_R = \{u\}$, $XQ_{n+1} - S$ is disconnected, and $S = N(v)$. This proves Case 2.

Case 3. Either $|S_L| = n+1$ or $|S_R| = n+1$. Without loss of generality, suppose that $|S_L| = n+1$ and $|S_R| = 0$. Since there is one corresponding matched vertex for each vertex $v \in V(XQ_n^L - S_L)$ in $V(XQ_n^R)$, $XQ_{n+1} - S$ is connected.

Consequently, this lemma holds. □

Let F_1 and F_2 be two distinct sets of vertices of XQ_n with $|F_i| \leq n+1, i = 1, 2$ and let $S = F_1 \cap F_2$. Then $|S| \leq n$. By the previous lemma, either $XQ_n - S$ is connected, or $XQ_n - S$ is disconnected and there is a vertex $v \in V(XQ_n)$ such that $S = N(v)$. If $XQ_n - S$ is connected, the two sets $V(XQ_n) - (F_1 \cup F_2)$ and $F_1 \,\Delta\, F_2$ both belong to the same component $XQ_n - S$. Thus, there exists one edge connecting $V(XQ_n) - (F_1 \cup F_2)$ and $F_1 \,\Delta\, F_2$. By Lemma 23.4, F_1 and F_2 are distinguishable. Therefore, if F_1 and F_2 are indistinguishable, $|F_i| \leq n+1, i = 1, 2$, $XQ_n - S$ is disconnected and there exists a vertex v such that $S = N(v)$. $S = F_1 \cap F_2$, so $N(v) \subseteq F_1$ and $N(v) \subseteq F_2$. We then propose the following concept.

A system G is strongly t-diagnosable if the following two conditions hold:

1. G is t-diagnosable
2. For any two distinct subsets $F_1, F_2 \subset V(G)$ with $|F_i| \leq t+1$, $i = 1, 2$, either (a) (F_1, F_2) is a distinguishable pair; or (b) (F_1, F_2) is an indistinguishable pair and there exists a vertex $v \in V$ such that $N(v) \subseteq F_1$ and $N(v) \subseteq F_2$.

A $(t+1)$-diagnosable system is stronger than a t-diagnosable system, and of course it is strongly t-diagnosable according to the previous definition. However, among all those strongly t-diagnosable systems, we are interested in the one that is t-diagnosable but not $(t+1)$-diagnosable.

Following Lemma 21.3 and the definition of strongly t-diagnosable, we propose a sufficient condition for verifying whether a system G is strongly t-diagnosable.

PROPOSITION 21.1 *A system $G(V, E)$ with n vertices is strongly t-diagnosable if the following three conditions hold:*

1. $n \geq 2(t+1)+1$
2. $\kappa(G) \geq t$
3. *For any vertex set $S \subset V$ with $|S| = t$, if $G - S$ is disconnected, there exists a vertex $v \in V$ such that $N(v) \subset S$*

Proof: With conditions (1) and (2), by Lemma 21.3, G is t-diagnosable. Now, we want to prove whether condition (2) the definition of strongly t-diagnosable holds. Let $F_1, F_2 \subset V$ be two distinct sets with $|F_i| \leq t+1, i = 1,2$ and $S = F_1 \cap F_2$. Suppose that $G - S$ is connected. Then there exists one edge connecting $V - (F_1 \cup F_2)$ and $F_1 \Delta F_2$. By Lemma 21.4, F_1 and F_2 are distinguishable. That is, condition (2a) of the definition of strongly t-diagnosable holds.

Otherwise, $G - S$ is disconnected. By condition (2), the connectivity of G is at least t, and $0 \leq |S| \leq t$, so $|S| = t$. Then by condition (3), there exists one vertex $v \in V$ such that $N(v) \subset S$. Therefore, $N(v) \subset F_1$ and $N(v) \subset F_2$. So condition (2b) of the definition of strongly t-diagnosable holds. This completes the proof of this proposition. □

Next, we present a necessary and sufficient condition for a system G to be strongly t-diagnosable.

LEMMA 21.21 A system $G(V, E)$ with $|V| = n$ is strongly t-diagnosable if and only if the following three conditions hold:

1. $n \geq 2(t+1)+1$
2. $\delta(G) \geq t$
3. For any two distinct subsets $F_1, F2 \subset V(G)$ with $|F_i| \leq t+1$, $i = 1,2$, the pair (F_1, F_2) satisfy condition (2a) or (2b) of the definition of strongly t-diagnosable

Proof: We first prove the necessity. To prove condition (1), we show that the assumption $n \leq 2(t+1)$ leads to a contradiction. Assume that $n \leq 2(t+1)$. We can partition V into two disjoint vertex sets V_1 and V_2, $V_1 \cap V_2 = \emptyset$ and $V = V_1 \cup V_2$, with $|V_i| \leq t+1$, $i = 1,2$. By Lemma 21.4, V_1 and V_2 are indistinguishable. Since G is strongly t-diagnosable, by the definition of strongly t-diagnosable, $N(v) \subset V_1$ and $N(v) \subset V_2$, for some vertex $v \in V$, contradicting the assumption $V_1 \cap V_2 = \emptyset$.

To prove condition (2), since G is strongly t-diagnosable, it is t-diagnosable by definition. Then by condition (2) of Corollary 21.1, $|N(v)| \geq t$ for each vertex $v \in V$. So condition (2) is necessary. Condition (3) of this lemma is the same as condition (2) of the definition of strongly t-diagnosable. This proves the necessity.

To prove the sufficiency of conditions (1), (2), and (3). We need to show only that G is t-diagnosable. Suppose not; then there exists an indistinguishable pair of sets $F_1, F_2 \subset V$, $F_1 \neq F_2$, and $|F_i| \leq t$, $i = 1,2$. By condition (2b) of the definition of strongly t-diagnosable, there exists a vertex $v \in V$ such that $N(v) \subset F_1$ and $N(v) \subset F_2$.

By condition (2), $|N(v)| \geq t$. However, $|F_1| \leq t$ and $|F_2| \leq t$. Hence, $F_1 = F_2 = N(v)$. This contradicts the fact that $F_1 \neq F_2$. The lemma follows. □

We now give another necessary and sufficient condition for checking whether a system is strongly t-diagnosable. The motivation of these conditions is as follows: Let $G(V,E)$ be a strongly t-diagnosable system. Suppose that G is $(t+1)$-diagnosable. Then by Theorem 21.2, for every set $S \subset V$, $0 \leq p \leq t$ where $|S| = p$, each component C of $G - S$ satisfies $|V_C| \geq 2((t+1) - p) + 1$. Otherwise, G is t-diagnosable but not $(t+1)$-diagnosable. Then there exists an indistinguishable pair (F_1, F_2), $F_1 \neq F_2$, with $|F_i| \leq t+1$, $i = 1, 2$. By condition (2b) of the definition of strongly t-diagnosable, there exists a vertex $v \in V$ such that $N(v) \subset F_1$ and $N(v) \subset F_2$. Note that $\delta(G) \geq t$, and therefore, $|N(v)| \geq t$. It means that $\{v\}$ is a trivial component of $G - (F_1 \cap F_2)$. Setting $S = F_1 \cap F_2$ and $|S| = t$, $G - S$ has a trivial component.

THEOREM 21.21 A system $G = (V,E)$ is strongly t-diagnosable if and only if for each vertex set $S \subset V$ with cardinality $|S| = p$, $0 \leq p \leq t$, the following two conditions are satisfied:

1. For $0 \leq p \leq t-1$, every component C of $G - S$ satisfies $|V_C| \geq 2((t+1) - p) + 1$
2. For $p = t$, either (a) every component C of $G - S$ satisfies $|V_C| \geq 3$; or else, (b) $G - S$ contains at least one trivial component. (Remark: $2((t+1) - p) + 1 = 3$ as $p = t$.)

Proof: We use Theorem 21.2 to prove the sufficiency of conditions (1) and (2). Let S be a set of vertices with $|S| = p$, $0 \leq p \leq t-1$. By condition (1), every component C of $G - S$ satisfies $|V_C| \geq 2((t+1) - p) + 1 \geq 2(t - p) + 1$. Then by Theorem 21.2, G is t-diagnosable.

To show that G is strongly t-diagnosable, we need to prove that condition (2) of the definition of strongly t-diagnosable holds. Suppose that conditions (1) and (2a) are both satisfied. Then by Theorem 21.2, G is $(t+1)$-diagnosable. Now consider the case that G is not $(t+1)$-diagnosable. Let (F_1, F_2) be an indistinguishable pair, $F_1 \neq F_2$, with $|F_1| \leq t+1$ and $|F_2| \leq t+1$. Let $S = F_1 \cap F_2$ and $X = V - (F_1 \cup F_2)$, then $0 \leq p \leq t$, where $|S| = p$. Since F_1 and F_2 are indistinguishable, by Lemma 21.4, there is no edge between X and $F_1 \Delta F_2$. Therefore, in $G - S$, $F_1 \Delta F_2$ is disconnected from the other components. Observe that $|F_1 \Delta F_2| \leq 2((t+1) - p)$, by condition (1), p cannot be in the range from 0 to $t-1$. So $p = t$ and $|F_1 \Delta F_2| \leq 2((t+1) - p) = 2((t+1) - t) = 2$. Then, by condition (2b), $G - S$ must have a trivial component $\{v\}$. So $N(v) \subset S$. G is t-diagnosable, by condition (2) of Corollary 21.1, $|N(v)| \geq t$. Hence, $S = N(v)$. Since $S = F_1 \cap F_2$, $N(v) \subset F_1$ and $N(v) \subset F_2$. Therefore, G is strongly t-diagnosable.

This proves the sufficiency. Next, we prove that the conditions (1) and (2) are also necessary.

To prove condition (1), suppose that to the contrary, that there exists a set of vertices $S \subset V$ with $|S| = p$, $0 \leq p \leq t-1$, such that $G - S$ has a component with strictly less than $2((t+1)-p)+1$ vertices. Let C be such a component with $|V_C| \leq 2((t+1)-p)$. We can partition V_C into two disjoint subsets A_1 and A_2, $A_1 \cup A_2 = V_C$ and $A_1 \cap A_2 = \emptyset$, with $|A_i| \leq (t+1)-p$, $i=1,2$. Let $F_1 = A_1 \cup S$ and $F_2 = A_2 \cup S$. Then $|F_i| \leq t+1$, $i=1,2$, and F_1 and F_2 are indistinguishable by Lemma 21.4. Since G is strongly t-diagnosable, by condition (2b) of the definition of strongly t-diagnosable, there exists a vertex v such that $N(v) \subset F_1$ and $N(v) \subset F_2$. G is t-diagnosable, by Corollary 21.1, each vertex of G has degree at least t. So $|N(v)| \geq t$. However, $N(v) \subset F_1 \cap F_2 = S$ and $|S| = p \leq t-1$; this is a contradiction. Thus, condition (1) is necessary.

Now, we prove that condition (2) is necessary. Let S be a set of vertices with $|S| = p$ and $p = t$. Suppose that G is $(t+1)$-diagnosable. By Theorem 21.2, for $p = t$, every component C of $G-S$ satisfies $|V_C| \geq 2((t+1)-t)+1 = 3$. That is, condition (2a) holds if G is $(t+1)$-diagnosable. Otherwise, G is not $(t+1)$-diagnosable and there exists a component C in $G-S$ with strictly fewer than three vertices, $|V_C| \leq 2$. We have to show that there is a trivial component in $G-S$. It holds on $|V_c| = 1$. Assume that $|V_C| = 2$; say, $V_C = \{v_1, v_2\}$. Let $F_1 = S \cup \{v_1\}$ and $F_2 = S \cup \{v_2\}$. Then $|F_1| = t+1$, $|F_2| = t+1$, and F_1 and F_2 are indistinguishable. Since G is strongly t-diagnosable, by condition (2b) of the definition of strongly t-diagnosable, there exists a vertex v such that $N(v) \subset F_1$ and $N(v) \subset F_2$. We have $S = F_1 \cap F_2$ and $N(v) \subset S$. Therefore, $\{v\}$ is a trivial component in $G-S$; this proves condition (2b).

Consequently, the theorem holds. □

The previous theorem again states that a strongly t-diagnosable system is almost $(t+1)$-diagnosable, if it is not so. The only case that stops it from being $(t+1)$-diagnosable occurs in the following situation: all the neighboring vertices $N(v)$ of some vertex v are faulty simultaneously.

21.5.1 Strongly t-Diagnosable Systems in the Matching Composition Networks

In previous studies, the diagnosability of many practical interconnection networks have been explored. Actually, some of them are not only n-diagnosable but also strongly n-diagnosable. For example, the hypercube Q_n, the crossed cube CQ_n, the Möbius cube MQ_n, and the twisted cube TQ_n are so. In the following, we shall prove that all members in the cube family are strongly n-diagnosable for $n \geq 4$.

Under the comparison model [245,247], it is proved that an MCN with two t-connected and t-diagnosable M-components is $(t+1)$-diagnosable in Ref. 215 and Chapter 1. In the following theorem, we shall show that an MCN with two t-diagnosable M-components is strongly $(t+1)$-diagnosable under the PMC model.

THEOREM 21.22 Let $G_1(V_1, E_1)$ and $G_2(V_2, E_2)$ be two t-diagnosable systems with the same number of vertices, where $t \geq 2$. Then MCN $G = G_1 \oplus G_2$ is strongly $(t+1)$-diagnosable.

Proof: We use Theorem 21.21 to prove it. Let $G = G(V, E) = G_1 \oplus G_2$ and $S \subset V$ with $|S| = p$, $0 \leq p \leq t+1$. Let $S_1 = S \cap V_1$, $S_2 = S \cap V_2$, $|S_1| = p_1$, and $|S_2| = p_2$. In the following proof, we consider two cases: (1) $S_1 = \emptyset$ or $S_2 = \emptyset$ and (2) $S_1 \neq \emptyset$ and $S_2 \neq \emptyset$. We shall prove that: (a) $|V_C| \geq 2((t+2)-p)+1$ for every component C of $G-S$ as $0 \leq p \leq t$ and (b) for $p = t+1$, either (i) every component C of $G-S$

satisfies $|V_C| \geq 3$, or else (ii) $G - S$ contains at least one trivial component. Then by Theorem 21.21, G is strongly $(t+1)$-diagnosable.

Case 1. $S_1 = \emptyset$ or $S_2 = \emptyset$. Without loss of generality, we assume that $S_1 = \emptyset$ and $S_2 = S$. We know that each vertex of V_2 has an adjacent neighbor in V_1, so $G - S$ is connected. The only component C of $G - S$ is $G - S$ itself. Hence, $|V_C| = |V - S| = |V_1| + |V_2| - p$. G_i is t-diagnosable, $i = 1, 2$; by Corollary 21.1, $|V_i| \geq 2t + 1$. So $|V_C| \geq 2(2t+1) - p \geq 2((t+2) - p) + 1$ for $t \geq 2$. That is, conditions (1) and (2a) of Theorem 21.21 are satisfied.

Case 2. $S_1 \neq \emptyset$ and $S_2 \neq \emptyset$. Since $S_1 \neq \emptyset$ and $S_2 \neq \emptyset$, $p_1 \geq 1$ and $p_2 \geq 1$. Then, we divide the case into two subcases: (1) both $p_1 \leq t - 1$ and $p_2 \leq t - 1$, and (2) either $p_1 = t$ or $p_2 = t$. Note that $0 \leq p \leq t + 1$ and $p = p_1 + p_2$. For subcase (1), $1 \leq p_1 \leq t - 1$ and $1 \leq p_2 \leq t - 1$, and for subcase (2), either $p_1 = t$ and $p_2 = 1$ or $p_2 = t$ and $p_1 = 1$.

Case 2.1. $1 \leq p_1 \leq t - 1$ and $1 \leq p_2 \leq t - 1$. Let C_1 be a component of $G_1 - S_1$. Note that G_1 is t-diagnosable. By Theorem 21.2, $|V_{C_1}| \geq 2(t - p_1) + 1$. We claim that $2(t - p_1) + 1 \geq p_2 + 1$. Since $p = p_1 + p_2$, $2(t - p_1) + 1 = 2(t - (p - p_2)) + 1 = 2p_2 + 2(t - p) + 1$. Suppose $p \leq t$, $|V_{C_1}| \geq 2p_2 + 1$. Otherwise, $p = t + 1$. Since $p_1 \leq t - 1$, $p_2 \geq 2$ and $2p_2 + 2(t - p) + 1 \geq p_2 + 1$. Hence, $|V_{C_1}| \geq 2(t - p_1) + 1 \geq p_2 + 1$. That is, V_{C_1} has at least one adjacent neighbor $v \in V_2$ and $v \notin S_2$. Note that G_2 is t-diagnosable. By Theorem 21.2, every component of $G_2 - S_2$ has at least $2(t - p_2) + 1$ vertices. Let C_2 be the component of $G_2 - S_2$ such that $v \in V_{C_2}$ and let C be the component of $G - S$ such that $V_{C_1} \cup V_{C_2} \subset V_C$. Then $|V_C| \geq |V_{C_1}| + |V_{C_2}| \geq (2(t - p_1) + 1) + (2(t - p_2) + 1) = 2(2t - p + 1) \geq 2((t+2) - p) + 1$ as $t \geq 2$. So every component of $G - S$ has at least $2((t+2) - p) + 1$ vertices in this subcase. It means that conditions (1) and (2a) of Theorem 21.21 are satisfied.

Case 2.2. Either $p_1 = t$ and $p_2 = 1$ or $p_2 = t$ and $p_1 = 1$. Without loss of generality, assume that $p_2 = t$ and $p_1 = 1$. Since $p = p_1 + p_2 = t + 1$, we need to prove only whether either condition (2a) or (2b) of Theorem 21.21 holds. Let C_1 be a component of $G_1 - S_1$. Note that G_1 is t-diagnosable. By Theorem 21.2, $|V_{C_1}| \geq 2(t - p_1) + 1 = 2(t - 1) + 1$. Since $t \geq 2$, $|V_{C_1}| \geq 2(t - 1) + 1 \geq 3$. So the component of $G - S$ containing the vertex set V_{C_1} has at least three vertices.

Let C_2 be a component of $G_2 - S_2$, $N(V_{C_2}, V_2) \subset S2$. If V_{C_2} has some adjacent neighbor $v_1 \in V_1$ and vertex v_1 belongs to some component C_1 of $G_1 - S_1$, then the component C containing the two vertex sets V_{C_1} and V_{C_2} has at least four vertices. Thus, condition (2a) of Theorem 21.21 holds. Otherwise, $N(V_{C_2}, V_1) \subset S_1$. Since $|S_1| = p_1 = 1$, $|N(V_{C_2}, V_1)| = 1$. That is, $|V_{C_2}| = 1$ and $N(V_{C_2}) \subset S_1 \cup S_2$. Hence, C_2 is a trivial component of $G - S$, and therefore, condition (2b) of Theorem 21.21 holds.

Consequently, the theorem follows. □

For $t = 1$, the previous result is not necessarily true; we give an example shown in Figure 21.13. Let G_1 and G_2 be two path graphs of length 4 with vertex sets $\{u_1, u_2, u_3, u_4, u_5\}$ and $\{v_1, v_2, v_3, v_4, v_5\}$, respectively. Let G be the MCN constructed by adding a perfect matching (the dash lines in Figure 21.13a) between G_1 and G_2. By Lemma 21.3, both G_1 and G_2 are 1-diagnosable and G is 2-diagnosable. See Figure 21.13b; let $F_1 = \{u_1, u_2, v_2\}$ and $F_2 = \{v_1, v_2, u_2\}$. By Lemma 21.4, F_1 and F_2

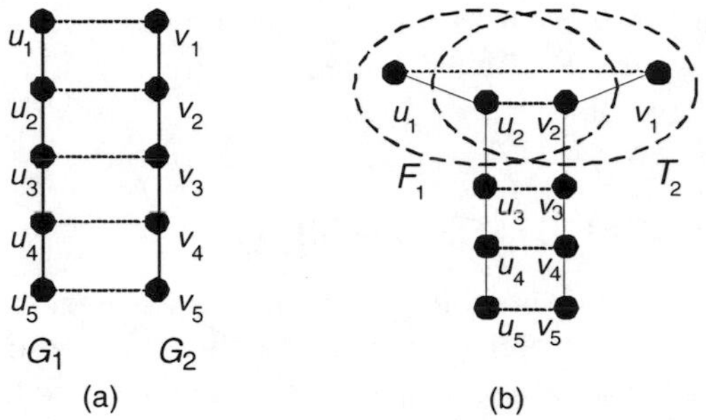

FIGURE 21.13 An example of nonstrongly $(t+1)$-diagnosable; $t=1$.

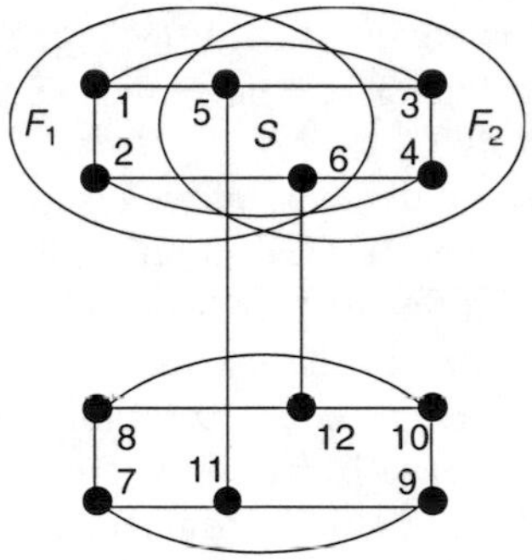

FIGURE 21.14 An example of a nonstrongly 3-diagnosable system.

are indistinguishable, but there does not exist any vertex $v \in V(G_i)$, $i=1,2$, such that $N(v) \subset F_1$ and $N(v) \subset F_2$. So G is not strongly 2-diagnosable.

Applying Theorem 21.22, all systems in the cube family are strongly $(t+1)$-diagnosable if their subcubes are t-diagnosable for $t \geq 2$. The hypercube Q_n, the crossed cube CQ_n, the twisted cube TQ_n, and the Möbius cube MQ_n are well-known members in the cube family. For $n=2$, these cubes are all isomorphic to the cycle of length 4; they are 1-diagnosable but not 2-diagnosable. For $n=3$, these cubes are all 3-connected. By Lemma 21.3, they are 3-diagnosable. So we have the following corollary.

COROLLARY 21.8 *The hypercube Q_n, the crossed cube CQ_n, the Möbius cube MQ_n, and the twisted cube TQ_n are all strongly n-diagnosable for $n \geq 4$.*

We now give some examples that are not strongly t-diagnosable. Consider the three-dimensional hypercube Q_3; it is 3-diagnosable but not strongly 3-diagnosable, due to the fact that $|V(Q_3)| = 8 \leq 2(t+1)+1$ as $t=3$, which contradicts condition (1) of Lemma 21.21. Let C_n be a cycle of length n, $n \geq 7$. By Lemma 21.3, C_n is 2-diagnosable, but it is not strongly 2-diagnosable. Another nontrivial example is presented in Figure 21.14. This graph G is 3-regular, 2-connected, and by Theorem 21.2, it is 3-diagnosable. As shown in Figure 21.14, $F_1 = \{1,2,5,6\}$ and $F_2 = \{3,4,5,6\}$.

(F_1, F_2) is an indistinguishable pair, but there does not exist any vertex v in $V(G)$ such that $N(v) \subset F_1$ and $N(v) \subset F_2$. By the definition of strongly t-diagnosable, the graph is not strongly 3-diagnosable.

21.6 CONDITIONAL DIAGNOSABILITY

Consider a system G with diagnosability $t(G) = t$; so G is t-diagnosable but not $(t+1)$-diagnosable. In previous researches on diagnosability, the investigated networks are often strongly t-diagnosable; for example, members in the cube family are so. Given a system G, suppose that it is strongly t-diagnosable but not $(t+1)$-diagnosable. As we mentioned before, the only thing that stops it from being $(t+1)$-diagnosable is that there exists a vertex v whose neighboring vertices are faulty simultaneously. We are, therefore, led to the following question: How large can the maximum value of t can be such that G remains t-diagnosable under the condition that every faulty set F satisfies $N(v) \subseteq F$ for each vertex $v \in V^{\text{a}}$

For classical measurement of diagnosability, it is usually assumed that processor failures are statically independent. It does not reflect the total number of processors in the system and the probabilities of processor failures. In Ref. 258, Najjar and Gaudiot proposed *fault resilience* as the maximum number of failures that can be sustained while the network remains connected with a reasonably high probability. For a hypercube, the fault resilience is shown as 25% for the four-dimensional cube Q_4 and it increases to 33% for the 10-dimensional cube Q_{10}. More particularly, for the 10-dimensional cube Q_{10}, 33% processors can fail and the network still remains connected with a probability of 99%. They also gave the conclusion that large-scale systems with a constant degree are more susceptible to failures by disconnection than smaller networks. With the observation of Lemma 21.4, a connected network gives higher probability to a diagnosis of faulty processors and has better ability to distinguish any two sets of processors.

Motivated by the deficiency of the classical measurement of diagnosability and the broadness of a system being strongly t-diagnosable, we introduce a measure of conditional diagnosability by claiming the property that any faulty set cannot contain all neighbors of any processor. We formally introduce some terms related to the conditional diagnosability. A faulty set $F \subset V$ is called a *conditional faulty set* if $N(v) \subseteq F$ for any vertex $v \in V$. A system $G(V, E)$ is *conditionally t-diagnosable* if F_1 and F_2 are distinguishable, for each pair of conditional faulty sets F_1, $F_2 \subset V$ and $F_1 \neq F_2$, with $|F_1| \leq t$ and $|F_2| \leq t$. The *conditional diagnosability* of a system G, written as $t_c(G)$, is defined to be the maximum value of t such that G is conditionally t-diagnosable. It is clear that $t_c(G) \geq t(G)$.

LEMMA 21.22 Let G be a network system. Then $t_c(G) \geq t(G)$.

Let F_1 and F_2 be two distinct subsets of V. We say that (F_1, F_2) is a *distinguishable conditional-pair* (an indistinguishable conditional pair, respectively) if F_1 and F_2 are conditional faulty sets and are distinguishable (indistinguishable, respectively).

It follows from the definition that a strongly t-diagnosable system is clearly conditionally $(t+1)$-diagnosable. However, the conditional diagnosability of some strongly

t-diagnosable systems can be far greater than $t+1$. This motivates us to study the conditional diagnosability of the hypercube.

LEMMA 21.23 Let G be a strongly t-diagnosable system. Then G is conditionally $(t+1)$-diagnosable.

21.6.1 CONDITIONAL DIAGNOSABILITY OF Q_n UNDER THE PMC MODEL

Before discussing conditional diagnosability, we have some observations as follows. Let (F_1, F_2) be an indistinguishable conditional pair. Let $X = V - (F_1 \cup F_2)$. Then there is no edge between X and $F_1 \triangle F_2$. So $N(F_1 \triangle F_2, X) = \emptyset$ and $N(X, F_1 \triangle F_2) = \emptyset$. Let vertex $v \in F_1 - F_2$ (or $v \in F_2 - F_1$). Then $N(v) \subset (F_1 \cup F_2)$. F_1 is a conditional faulty set, so $N(v) \subseteq F_1$ and $N(v) \cap (F_2 - F_1) \neq \emptyset$. Similarly, F_2 is a conditional faulty set, $N(v) \subseteq F_2$ and $N(v) \cap (F_1 - F_2) \neq \emptyset$. So $|N(v) \cap (F_1 - F_2)| \geq 1$ and $|N(v) \cap (F_2 - F_1)| \geq 1$ for every vertex $v \in F_1 \triangle F_2$. Now consider a vertex $u \in X = V - (F_1 \cup F_2)$. Since F_1 and F_2 are an indistinguishable conditional pair, $N(u) \cap (F_1 \triangle F_2) = \emptyset$, $N(u) \subseteq F_1$ and $N(u) \subseteq F_2$. So $N(u) \subseteq (F_1 \cup F_2)$. Therefore, every vertex $u \subset X$ has at least one neighbor in X (see Figure 21.15). We state this fact in the following lemma.

LEMMA 21.24 Let $G(V, E)$ be a system. Given an indistinguishable conditional pair (F_1, F_2), $F_1 \neq F_2$, the following two conditions hold:

1. $|N(u) \cap (V - (F_1 \cup F_2))| \geq 1$ for $u \in (V - (F_1 \cup F_2))$
2. $|N(v) \cap (F_1 - F_2)| \geq 1$ and $|N(v) \cap (F_2 - F_1)| \geq 1$ for $v \in F_1 \triangle F_2$

Let (F_1, F_2) be an indistinguishable conditional pair, and let $S = F_1 \cap F_2$. By the earlier observations, every component of $G - S$ is nontrivial. Moreover, for each component C_1 of $G - S$, if $V_{C_1} \cap (F_1 \triangle F_2) = \emptyset$, $deg_{C_1}(v) \geq 1$ for $v \in V_{C_1}$; for each component C_2 of $G - S$, if $V_{C_2} \cap (F_1 \triangle F_2) \neq \emptyset$, $deg_{C_2}(v) \geq 2$ for $v \in V_{C_2}$. To find the conditional diagnosability of the hypercube Q_n, we need to study the cardinality of the set S.

First, we give an example to show that the conditional diagnosability of the hypercube Q_n is no greater than $4(n-2)+1$. As shown in Figure 21.16, we take a cycle of length four in Q_n, let $\{v_1, v_2, v_3, v_4\}$ be the four consecutive vertices on this cycle and let $F_1 = N(\{v_1, v_2, v_3, v_4\}) \cup \{v_1, v_2\}$ and $F_2 = N(\{v_1, v_2, v_3, v_4\}) \cup \{v_3, v_4\}$. It is a simple matter to check whether (F_1, F_2) is an indistinguishable conditional-pair.

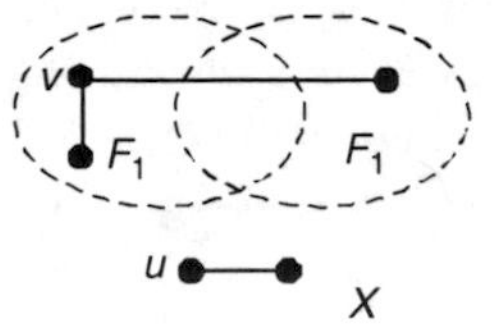

FIGURE 21.15 Illustration for Lemma 21.24.

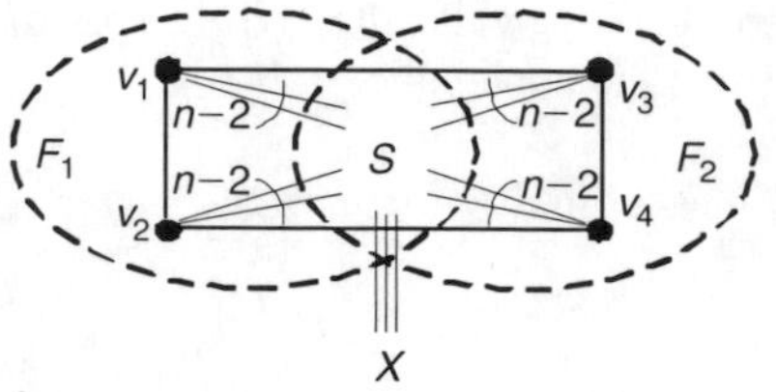

FIGURE 21.16 Illustration for an indistinguishable conditional pair (F_1, F_2), where $|F_1| = |F_2| = 4(n-2)+2$.

Note that the hypercube Q_n has no triangle and any two vertices have at most two common neighbors. It is obvious that $|F_1 - F_2| = |F_2 - F_1| = 2$ and $|F_1 \cap F_2| = 4(n-2)$. Hence, Q_n is not conditionally $(4(n-2)+2)$-diagnosable and $t_c(Q_n) \leq 4(n-2)+1$. Then, we shall show that Q_n is in fact conditionally t-diagnosable, where $t = 4(n-2)+1$.

LEMMA 21.25 $t_c(Q_n) \leq 4(n-2)+1$ for $n \geq 3$.

Let S be a set of vertices, $S \subset V(Q_n)$. Suppose that $Q_n - S$ is disconnected and C is a component of $Q_n - S$. We need some results on the cardinalities of S and V_C under some restricted conditions. The results are listed in Lemmas 21.26 and 21.27.

These two lemmas are both proved by dividing Q_n into two Q_{n-1}'s, denoted by Q_{n-1}^L and Q_{n-1}^R. To simplify the explanation, we define some symbols as follows: $V_L = V(Q_{n-1}^L)$, $V_R = V(Q_{n-1}^R)$, $C_L = Q_{n-1}^L \cap C$, $C_R = Q_{n-1}^R \cap C$, $V_{C_L} = V(C_L)$, $V_{C_R} = V(C_R)$, $S_L = V_L \cap S$, and $S_R = V_R \cap S$.

The following result is also implicit in Ref. 216.

LEMMA 21.26 Let Q_n be the n-dimensional hypercube, $n \geq 3$, and let S be a set of vertices $S \subset V(Q_n)$. Suppose that $Q_n - S$ is disconnected. Then the following two conditions hold:

1. $|S| \geq n$
2. If $n \leq |S| \leq 2(n-1)-1$, then $Q_n - S$ has exactly two components, one is trivial and the other is nontrivial. The nontrivial component of $Q_n - S$ contains $2^n - |S| - 1$ vertices

Proof: Since $\kappa(Q_n) = n$ [279], condition (1) holds. We need to prove that only condition (2) is true. Because $Q_n - S$ is disconnected, there are at least two components in $Q_n - S$. We consider three cases: (1) $Q_n - S$ contains at least two trivial components, (2) $Q_n - S$ has at least two nontrivial components, and (3) there are exactly one trivial component and one nontrivial component in $Q_n - S$. In Cases 1 and 2, we shall prove that $|S| \geq 2(n-1)$. Then $n \leq |S| \leq 2(n-1)-1$, which implies that $Q_n - S$ belongs to Case 3.

Case 1. $Q_n - S$ contains at least two trivial components. Let $\mathbf{v}_i \in V$, $i = 1, 2$, and $\{\mathbf{v}_1\}, \{\mathbf{v}_2\} \subset V(Q_n)$ be two trivial components of $Q_n - S$. It means that

$N(\mathbf{v}_1) \subset S$ and $N(\mathbf{v}_2) \subset S$. For Q_n, it is not difficult to see that any two vertices have at most two common neighbors. That is, $|N(\mathbf{v}_1) \cap N(\mathbf{v}_2)| \leq 2$. Hence, $|S| \geq |N(\mathbf{v}_1) \cup N(\mathbf{v}_2)| = |N(\mathbf{v}_1)| + |N(\mathbf{v}_2)| - |N(\mathbf{v}_1) \cap N(\mathbf{v}_2)| \geq 2n - 2 = 2(n-1)$.

Case 2. $Q_n - S$ has at least two nontrivial components. We prove, by induction on n, that $|S| \geq 2(n-1)$. For $n = 3$, suppose $n \leq |S| \leq 2(n-1) - 1$, it implies that $|S| = 3$. The connectivity of Q_3 is 3. By Lemma 21.20, the only vertex cut S with $|S| = 3$ in Q_3 is $S = N(\mathbf{v})$ for some vertex $\mathbf{v} \in V(Q_3)$. It follows that $Q_3 - S$ has exactly two components: one is trivial and the other is nontrivial. Therefore, if $Q_3 - S$ has at least two nontrivial components, $|S| \geq 2(n-1)$, where $n = 3$. Assume that the case holds for some $n-1$, $n-1 \geq 3$. We now show that it holds for n.

Let C and C' be two nontrivial component of $Q_n - S$. So $|V_C| \geq 2$. It is feasible to divide Q_n into the two disjoint Q_{n-1}'s, denoted by Q^L_{n-1} and Q^R_{n-1}, such that $|V_{C_L}| \geq 1$ and $|V_{C_R}| \geq 1$. There is another component C' of $Q_n - S$, so at least one of the two graphs $Q^L_{n-1} - S_L$ and $Q^R_{n-1} - S_R$ is disconnected.

Suppose that both $Q^L_{n-1} - S_L$ and $Q^R_{n-1} - S_R$ are disconnected. Since $\kappa(Q_{n-1}) = n-1$, $|S_L| \geq n-1$ and $|S_R| \geq n-1$. Then $|S| = |S_L| + |S_R| \geq 2(n-1)$. Otherwise, one of the two subgraphs $Q^L_{n-1} - S_L$ and $Q^R_{n-1} - S_R$ is connected. Without loss of generality, assume that $Q^L_{n-1} - S_L$ is connected and $Q^R_{n-1} - S_R$ is disconnected. Then $V_L = V_{C_L} \cup S_L$ and the other nontrivial component C' of $Q_n - S$ is completely contained in $Q^R_{n-1} - S_R$. Since $V_{C'}$ is disconnected from V_{C_L}, the corresponding matched vertices of $V_{C'}$ in Q^L_{n-1} are in S_L. That is, $N(V_{C'}, Q^L_{n-1}) \subseteq S_L$. Hence, $|S_L| \geq |V_{C'}| \geq 2$.

If $|S_R| \geq 2(n-2)$, then $|S| = |S_L| + |S_R| \geq 2 + 2(n-2) = 2(n-1)$. Otherwise, $n-1 \leq |S_R| \leq 2(n-2) - 1$. By induction hypothesis, $Q_R - S_R$ cannot have two nontrivial components, and by the result of Case 1, $Q_R - S_R$ has exactly two components; one is trivial and the other is nontrivial. We know that $Q^R_{n-1} - S_R$ has C_R and C' as its components, and C' is a nontrivial component. So C_R must be a trivial component of $Q^R_{n-1} - S_R$, and $|V_{C'}| = 2^{n-1} - |S_R| - 1$. Note that $N(V_{C'}, Q^L_{n-1}) \subseteq S_L$. Then $|S| = |S_L| + |S_R| \geq |V_{C'}| + |S_R| = 2^{n-1} - |S_R| - 1 + |S_R| = 2^{n-1} - 1 \geq 2(n-1)$ for $n \geq 4$.

Consequently, condition (2) is true and the lemma holds. □

Suppose that $Q_n - S$ is disconnected, every component of $Q_n - S$ is nontrivial, and there exists one component C of $Q_n - S$ such that $deg_C(\mathbf{v}) \geq 2$ for every vertex $\mathbf{v}$ in C. In view of the example given in Figure 21.15 and Lemma 21.24, we shall prove that either $|S|$ is sufficiently large or else $|V_C|$ is large, as stated in the following lemma.

LEMMA 21.27 Let Q_n be the n-dimensional hypercube and $n \geq 5$, and let S be a vertex set $S \subseteq V(Q_n)$. Suppose that $Q_n - S$ is disconnected and every component of $Q_n - S$ is nontrivial, and suppose that there exists one component C of $Q_n - S$ such that $deg_C(v) \geq 2$ for every vertex $\mathbf{v}$ in C. Then one of the following two conditions holds:

1. $|S| \geq 4(n-2)$
2. $|V_C| \geq 4(n-2) - 1$

Proof: Since $deg_C(\mathbf{v}) \geq 2$ for every vertex $\mathbf{v}$ in C, it is feasible to divide Q_n into two disjoint Q_{n-1}'s, denoted by Q^L_{n-1} and Q^R_{n-1}, such that $V(Q^L_{n-1} \cap C) \neq \emptyset$ and $V(Q^R_{n-1} \cap C) \neq \emptyset$. Let $C_L = Q^L_{n-1} \cap C$ and $C_R = Q^R_{n-1} \cap C$. For each vertex x in C_L ($\mathbf{y}$ in C_R, respectively), it has at most one neighbor in C_R (C_L, respectively). Hence, $deg_{C_L}(\mathbf{x}) \geq 1$ and $deg_{C_R}(\mathbf{y}) \geq 1$ for $\mathbf{x} \in V_{C_L}$ and $\mathbf{y} \in V_{C_R}$, respectively.

$Q_n - S$ is disconnected; there are at least two components in $Q_n - S$. Let $S_L = V_L \cap S$ and $S_R = V_R \cap S$. Note that both Q^L_{n-1} and Q^R_{n-1} contain some nonempty part of the component C. So at least one of the two subgraphs $Q^L_{n-1} - S_L$ and $Q^R_{n-1} - S_R$ is disconnected. In the following proof, we investigate two cases: (1) one of $Q^L_{n-1} - S_L$ and $Q^R_{n-1} - S_R$ is connected, (2) both $Q^L_{n-1} - S_L$ and $Q^R_{n-1} - S_R$ are disconnected.

Case 1. One of $Q^L_{n-1} - S_L$ and $Q^R_{n-1} - S_R$ is connected, and the other is disconnected. Without loss of generality, we assume that $Q^L_{n-1} - S_L$ is connected and $Q^R_{n-1} - S_R$ is disconnected. Let C' be another component of $Q_n - S$ other than C. Then $V_L = S_L \cup V_{C_L}$ and the component C' of $Q_n - S$ is in $Q^R_{n-1} - S_R - V_{C_R}$. Since C_R and C' are both nontrivial component, by Lemma 21.26, $|S_R| \geq 2(n-2)$. If $|S_L| \geq 2(n-2)$, then $|S| = |S_L| + |S_R| \geq 4(n-2)$ and condition (1) holds. Otherwise, $|S_L| \leq 2(n-2) - 1$. Then $|V_{C_L}| = 2^{n-1} - |S_L| \geq 2^{n-1} - 2(n-2) + 1$. That is, $|V_C| = |V_{C_L}| + |V_{C_R}| \geq (2^{n-1} - 2(n-2) + 1) + 2 = 2^{n-1} - 2(n-2) + 3 \geq 4(n-2) - 1$ for $n \geq 4$ and condition (2) holds.

Case 2. Both $Q^L_{n-1} - S_L$ and $Q^R_{n-1} - S_R$ are disconnected. By Lemma 21.26, we consider the following three subcases: (1) $|S_L| \geq 2(n-2)$ and $|S_R| \geq 2(n-2)$, (2) $n-1 \leq |S_L| \leq 2(n-2) - 1$ and $n-1 \leq |S_R| \leq 2(n-2) - 1$, and (3) either $|S_L| \geq 2(n-2)$, $n-1 \leq |S_R| \leq 2(n-2) - 1$; or $|S_R| \geq 2(n-2)$, $n-1 \leq |S_L| \leq 2(n-2) - 1$.

Case 2.1. $|S_L| \geq 2(n-2)$ and $|S_R| \geq 2(n-2)$. Since $|S_L| \geq 2(n-2)$ and $|S_R| \geq 2(n-2)$, $|S| = |S_L| + |S_R| \geq 4(n-2)$. Hence, condition (1) holds.

Case 2.2. $n-1 \leq |S_L| \leq 2(n-2) - 1$ and $n-1 \leq |S_R| \leq 2(n-2) - 1$. In this subcase, $|V_{C_L}| = 2^{n-1} - |S_L| - 1$ and $|V_{C_R}| = 2^{n-1} - |S_R| - 1$. So $|V_C| = |V_{C_L}| + |V_{C_R}| = 2^n - |S| - 2$. Suppose $|S| \geq 4(n-2)$. Then condition (1) holds. Otherwise, $|S| \leq 4(n-2) - 1$. Then $|V_C| = 2^n - |S| - 2 \geq 2^n - (4(n-2) - 1) - 2 = 2^n - 4(n-2) - 1 \geq 4(n-2) - 1$ for $n \geq 4$. Hence, condition (2) holds.

Case 2.3. Either $|S_L| \geq 2(n-2)$, $n-1 \leq |S_R| \leq 2(n-2) - 1$; or $|S_R| \geq 2(n-2)$, $n-1 \leq |S_L| \leq 2(n-2) - 1$. Without loss of generality, assume that $|S_L| \geq 2(n-2)$, $n-1 \leq |S_R| \leq 2(n-2) - 1$. Then $|V_{C_R}| = 2^{n-1} - |S_R| - 1 \geq 2^{n-1} - 2(n-2)$. Since $deg_{C_L}(x) \geq 1$ for each vertex $\mathbf{x} \in V_{C_L}$, we have $|V_{C_L}| \geq 2$. Thus, $|V_C| = |V_{C_L}| + |V_{C_R}| \geq 2 + (2^{n-1} - 2(n-2)) = 2^{n-1} - 2(n-2) + 2 \geq 4(n-2) - 1$ for $n \geq 5$.

This completes the proof of the lemma. □

We are now ready to prove that conditional diagnosability of Q_n is $4(n-2)+1$ for $n \geq 5$. Let (F_1, F_2) be an indistinguishable conditional pair, with $n \geq 5$. We shall prove our result by proving that either $|F_1| \geq 4(n-2)+2$ or $|F_2| \geq 4(n-2)+2$.

Let $S = F_1 \cap F_2$. We consider two cases: (1) $Q_n - S$ is connected and (2) $Q_n - S$ is disconnected.

LEMMA 21.28 Let Q_n be the n-dimensional hypercube, $n \geq 5$. Let $F_1, F_2 \subset V(Q_n)$, $F_1 \neq F_2$, be an indistinguishable conditional pair and $S = F_1 \cap F_2$. Then either $|F_1| \geq 4(n-2)+2$ or $|F_2| \geq 4(n-2)+2$.

Proof: Suppose that $Q_n - S$ is connected. Then $F_1 \Delta F_2 = V(Q_n - S)$ and $V(Q_n) = F_1 \cup F_2$. Suppose that on the contrary, that $|F_1| \leq 4(n-2)+1$ and $|F_2| \leq 4(n-2)+1$. Then $2^n = |F_1| + |F_2| - |F_1 \cap F_2| \leq (4(n-2)+1) + (4(n-2)+1) - 0 = 8(n-2)+2$. This contradicts the fact that $2^n > 8(n-2)+2$ for $n \geq 5$. Hence, the result holds, as $Q_n - S$ is connected.

Now we consider the case where $Q_n - S$ is disconnected, by Lemma 21.24, $Q_n - S$ has a component C with $deg_C(v) \geq 2$ for every vertex $\mathbf{v} \in V_C$. By Lemma 21.27, we have $|S| \geq 4(n-2)$ or $|V_C| \geq 4(n-2) - 1$.

Suppose that $|S| \geq 4(n-2)$. Since $deg_C(\mathbf{v}) \geq 2$ for every vertex v in C, and Q_n does not contain any cycle of length 3, so $|V_C| \geq 4$. With the observation that $V_C \subset F_1 \Delta F_2$, we conclude that either $(F_1 - F_2) \geq \left\lceil \frac{|V_C|}{2} \right\rceil \geq 2$ or $(F_2 - F_1) \geq \left\lceil \frac{|V_C|}{2} \right\rceil \geq 2$. Therefore, either $|F_1| = |S| + |F_1 - F_2| \geq 4(n-2)+2$ or $|F_2| = |S| + |F_2 - F_1| \geq 4(n-2)+2$.

Otherwise, $|V_C| \geq 4(n-2) - 1$. Then either $(F_1 - F_2) \geq \left\lceil \frac{|V_C|}{2} \right\rceil \geq 2(n-2)$ or $(F_2 - F_1) \geq \left\lceil \frac{|V_C|}{2} \right\rceil \geq 2(n-2)$. Because there are at least two nontrivial components in $Q_n - S$, by Lemma 21.26, $|S| \geq 2(n-1)$. Hence, $|F_1| = |S| + |F_1 - F_2| \geq 4(n-2)+2$ or $|F_2| = |S| + |F_2 - F_1| \geq 4(n-2)+2$.

Therefore, for any indistinguishable conditional pair $F_1, F_2 \subset V(Q_n)$, it implies that $|F_1| \geq 4(n-2)+2$ or $|F_2| \geq 4(n-2)+2$. This proves the lemma. □

By Lemma 21.25, $t_c(Q_n) \leq 4(n-2)+1$, and by Lemmas 21.6 and 21.28, Q_n is conditionally $(4(n-2)+1)$-diagnosable for $n \geq 5$. Hence, $t_c(Q_n) = 4(n-2)+1$ for $n \geq 5$. For Q_3 and Q_4, we observe that Q_3 is not conditionally 4-diagnosable and Q_4 is not conditionally 8-diagnosable, as shown in Figures 21.17a and 21.17b. So $t_c(Q_3) \leq 3$ and $t_c(Q_4) \leq 7$. Hence, the conditional diagnosabilities of Q_3 and Q_4 are both strictly less than $4(n-2)+1$.

The Q_3 is 3-diagnosable and is not conditionally 4-diagnosable. It follows from Lemma 21.22 that $t_c(Q_3) = 3$. For Q_4, we prove that $t_c(Q_4) = 7$ in the following lemma.

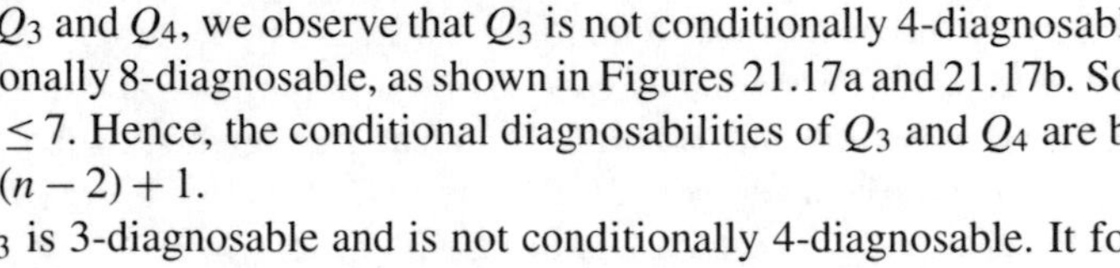

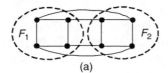

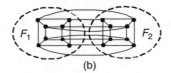

FIGURE 21.17 Illustration for two indistinguishable conditional pairs for Q_3 and Q_4.

LEMMA 21.29 $t_c(Q_4)=7$.

Proof: We already know that $t_c(Q_4)\leq 7$. Suppose that on the contrary that Q_4 is not conditionally 7-diagnosable. Let $F_1,F_2\subset V(Q_4)$ be an indistinguishable conditional pair with $|F_i|\leq 7$, $i=1,2$, and let $S=F_1\cap F_2$. It follows from Lemmas 21.24 and 21.26 that $|S|\geq 2(n-1)=6$ for $n=4$. Furthermore, $|F_1-F_2|\geq 2$ and $|F_2-F_1|\geq 2$. Then $|F_1|\geq 8$ and $|F_2|\geq 8$, which is a contradiction. So $t_c(Q_4)=7$. □

Finally, the conditional diagnosability of hypercube Q_n is stated as follows.

THEOREM 21.23 The conditional diagnosability of Q_n is $t_c(Q_n)=4(n-2)+1$ for $n\geq 5$, $t_c(Q_3)=3$ and $t_c(Q_4)=7$.

21.7 CONDITIONAL DIAGNOSABILITY OF Q_n UNDER THE COMPARISON MODEL

As we review some previous papers [102,103,215,330], the hypercube Q_n, the crossed cube CQ_n, the twisted cube TQ_n, and the Möbius cube MQ_n all have diagnosability n under the comparison model. In classical measures of system-level diagnosability for multiprocessor systems, if all the neighbors of some processor v are faulty simultaneously, it is not possible to determine whether processor v is fault-free or faulty. As a consequence, the diagnosability of a system is limited by its minimum degree. Hence, Lai et al. introduced a restricted diagnosability of multiprocessor systems called *conditional* diagnosability in Ref. 214. Lai et al. considered a measure by restricting the event that for each processor v in a system, all the processors, which are directly connected to v, do not fail at the same time. Under this condition, Lai et al. proved that the conditional diagnosability of an n-dimensional hypercube Q_n is $4(n-2)+1$ under the PMC model.

In this section, we study the conditional diagnosability of Q_n under the comparison model and show that it is $3(n-2)+1$ for $n\geq 5$. The conditional diagnosability of Q_n is about three times larger than that of the classical diagnosability of Q_n.

Before studying the conditional diagnosability of the hypercube, we need some definitions for further discussion. Let $G(V,E)$ be a graph. For any set of vertices $U\subseteq V(G)$, $G[U]$ denotes the subgraph of G induced by the vertex subset U. Let H be a subgraph of G and v be a vertex in H. We use $V(H;3)=\{v\in V(H)\mid deg_H(v)\geq 3\}$ to represent the set of vertices which has degree 3 or more in H. Let F_1 and F_2 be two distinct sets of $V(G)$ and $S=F_1\cap F_2$. We use $C_{F_1\,\Delta\,F_2,S}$ to denote the subgraph induced by the vertex subset $(F_1\,\Delta\,F_2)\cup\{u\mid \text{there exists a vertex } v\in F_1\,\Delta\,F_2 \text{ such that } u \text{ and } v \text{ are connected in } G-S\}$. The following result is a useful sufficient condition for checking whether (F_1,F_2) is a distinguishable pair.

THEOREM 21.24 Let $G(V,E)$ be a graph. For any two distinct sets $F_1,F_2\subset V$ with $|F_i|\leq t$, $i=1,2$, and $S=F_1\cap F_2$. (F_1,F_2) is distinguishable if the subgraph $C_{F_1\Delta F_2,S}$ of $G-S$ contains at least $2(t-|S|)+1$ vertices having degree 3 or more.

Proof: Given any pair of distinct sets of vertices F_1 and F_2 of V with $|F_i|\leq t$, $i=1,2$. Let $S=F_1\cap F_2$. Then $0\leq |S|\leq t-1$, and $|F_1\,\Delta\,F_2|\leq 2(t-|S|)$. Consider the subgraph $C_{F_1\Delta F_2,S}$, the number of vertices having degree 3 or more is at least

$2(t - |S|) + 1$ in $C_{F_1 \Delta F_2,S}$, the subgraph $C_{F_1 \Delta F_2,S}$ contains at least $2(t - |S|) + 1$ vertices. There is at least one vertex with degree 3 or more lying in $C_{F_1 \Delta F_2,S} - F_1 \Delta F_2$. Let u be one of such vertices with degree 3 or more. Let i, j, and k be three distinct vertices linked to u. If one of i, j, and k lies in $C_{F_1 \Delta F_2,S} - F_1 \Delta F_2$, condition (1) of Theorem 21.3 holds, obviously. Suppose that all these three vertices belong to $F_1 \Delta F_2$. Without loss of generality, assume that i lies in $F_1 - F_2$; one of the two cases will happen: (1) if j lies in $F_1 - F_2$, condition (2) of Theorem 21.3 holds; or (2) if j lies in $F_2 - F_1$, wherever k lies in $F_1 - F_2$ or $F_2 - F_1$, condition (2) or (3) of Theorem 21.3 holds. So (F_1, F_2) is a distinguishable pair and the proof is complete. □

By Theorem 21.24, we now propose a sufficient condition to verify whether a system is t-diagnosable under the comparison diagnosis model.

COROLLARY 21.9 *Let $G(V, E)$ be a graph. G is t-diagnosable if, for each set of vertices $S \subset V$ with $|S| = p$, $0 \leq p \leq t - 1$, every connected component C of $G - S$ contains at least $2(t - p) + 1$ vertices having degree at least 3. More precisely, $|V(C; 3)| \geq 2(t - p) + 1$.*

Let (F_1, F_2) be an indistinguishable conditional pair. Let $X = V - (F_1 \cup F_2)$. Since F_1 and F_2 are an indistinguishable conditional pair, none of the three conditions of Theorem 21.3 holds and every vertex has at least one fault-free neighbor. Let vertex $u \in X$. If $N(u) \cap X \neq \emptyset$, then $N(u) \cap (F_1 \Delta F_2) = \emptyset$ (see Figure 21.18a); otherwise, $|N(u) \cap (F_1 - F_2)| = 1$ and $|N(u) \cap (F_2 - F_1)| = 1$ (see Figure 21.18b). Let vertex $v \in F_1 \Delta F_2$. If $N(v) \cap X = \emptyset$, then $|N(v) \cap (F_1 - F_2)| \geq 1$ and $|N(v) \cap (F_2 - F_1)| \geq 1$ (see Figure 21.18c). We state this fact in the following lemma.

LEMMA 21.30 Let $G(V, E)$ be a graph and let $F_1, F_2 \subset V$ be an indistinguishable conditional pair. Let $X = V - (F_1 \cup F_2)$. The following three conditions hold:

1. $|N(u) \cap (F_1 \Delta F_2)| = 0$ for $u \in X$ and $N(u) \cap X \neq \emptyset$
2. $|N(u) \cap (F_1 - F_2)| = 1$ and $|N(u) \cap (F_2 - F_1)| = 1$ for $u \in X$ and $N(u) \cap X = \emptyset$
3. $|N(v) \cap (F_1 - F_2)| \geq 1$ and $|N(v) \cap (F_2 - F_1)| \geq 1$ for $v \in F_1 \Delta F_2$ and $N(v) \cap X = \emptyset$

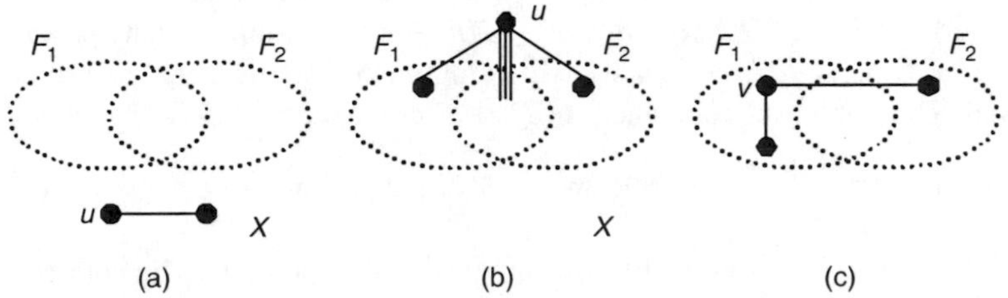

FIGURE 21.18 An indistinguishable conditional pair (F_1, F_2).

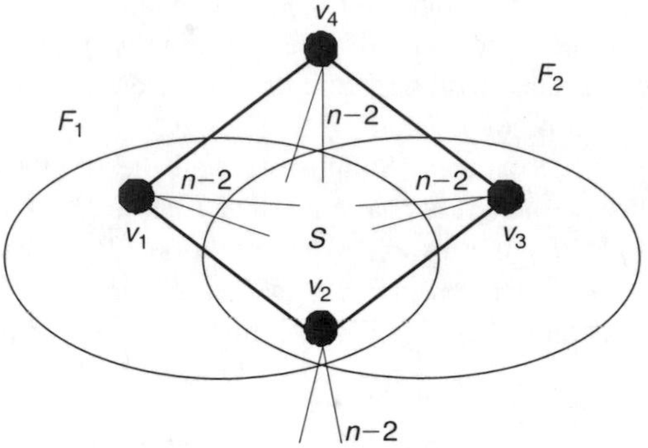

FIGURE 21.19 An indistinguishable conditional pair (F_1, F_2), where $|F_1| = |F_2| = 3(n-2)+2$.

Now, we give an example to show that the conditional diagnosability of the hypercube Q_n is no greater than $3(n-2)+2$, $n \geq 5$. As shown in Figure 21.19, we take a cycle of length four in Q_n. Let $\{\mathbf{v}_1, \mathbf{v}_2, v_3, v_4\}$ be the four consecutive vertices on this cycle, and let $F_1 = N(\{v_1, v_3, v_4\}) \cup \{v_1\}$ and $F_2 = N(\{v_1, v_3, v_4\}) \cup \{v_3\}$, then $|F_1| = |F_2| = 3(n-2)+2$. It is straightforward to check whether F_1 and F_2 are two conditional faulty sets, and F_1 and F_2 are indistinguishable by Theorem 21.3. Note that the hypercube Q_n has no cycle of length 3 and any two vertices have at most two common neighbors. Obviously, $|F_1 - F_2| = |F_2 - F_1| = 1$ and $|F_1 \cap F_2| = 3(n-2)+1$. Therefore, Q_n is not conditionally $(3(n-2)+2)$-diagnosable and $t_c(Q_n) \leq 3(n-2)+1$, $n \geq 3$. Then, we shall show that Q_n is conditionally t-diagnosable, where $t = 3(n-2)+1$.

LEMMA 21.31 $t_c(Q_n) \leq 3(n-2)+1$ for $n \geq 3$.

Let F be a set of vertices $F \subset V(Q_n)$ and C be a connected component of $Q_n - F$. We need some results on the cardinalities of F and $V(C)$ under some restricted conditions. The results are listed in Lemmas 21.32 and 21.36. In Lemma 21.32, Lai et al. [214] proved that deleting at most $2(n-1)-1$ vertices from Q_n, the incomplete hypercube Q_n has one connected component containing at least $2^n - |F| - 1$ vertices. We expand this result further. In Lemma 21.36, we show that deleting at most $3n-6$ vertices from Q_n, the incomplete hypercube Q_n has one connected component containing at least $2^n - |F| - 2$ vertices.

LEMMA 21.32 **[214]** Let Q_n be an n-dimensional hypercube, $n \geq 3$, and let F be a set of vertices $F \subset V(Q_n)$ with $n \leq |F| \leq 2(n-1)-1$. Suppose that $Q_n - F$ is disconnected. Then $Q_n - F$ has exactly two components: one is trivial and the other is nontrivial. The nontrivial component of $Q_n - F$ contains $2^n - |F| - 1$ vertices.

In order to prove Lemma 21.36, we need some preliminary results, as follows.

LEMMA 21.33 **[279]** Let Q_n be an n-dimensional hypercube. The connectivity of Q_n is $\kappa(Q_n) = n$.

LEMMA 21.34 For any three vertices $\mathbf{x}$, $\mathbf{y}$, and $\mathbf{z}$ in Q_4, $|N(\{\mathbf{x},\mathbf{y},\mathbf{z}\})| \geq 7$.

Proof: A four-dimensional hypercube Q_4 can be divided into two Q_3's, denoted by Q_3^L and Q_3^R. Any two vertices in the Q_n have at most two common neighbors. If these three vertices x, y, and z all fall in Q_3^L, then $\mathbf{x}$, $\mathbf{y}$, and $\mathbf{z}$ have at least four neighboring vertices, all in Q_3^L. Besides, $\mathbf{x}$, $\mathbf{y}$, and $\mathbf{z}$ have three more neighboring vertices in Q_3^R. Therefore, $|N(\{\mathbf{x},\mathbf{y},\mathbf{z}\})| \geq 4+3=7$. Suppose that now $\mathbf{x}$ and $\mathbf{y}$ fall in Q_3^L, z falls in Q_3^R. Vertex $\mathbf{x}$ and $\mathbf{y}$ have at least 4 neighboring vertices, all in Q_3^L. Vertex $\mathbf{z}$ will bring in at least three neighboring vertices in Q_3^R. Therefore, $|N(\{\mathbf{x},\mathbf{y},\mathbf{z}\})| \geq 4+3=7$. □

We are going to prove Lemma 21.36 by induction on n, and we need a base case to start with. As we observed, for $n=4$, we found a counter example that the result of Lemma 21.36 does not hold. So we have to start with $n=5$.

LEMMA 21.35 Let Q_5 be a five-dimensional hypercube, and let F be a set of vertices $F \subset V(Q_5)$ with $|F| \leq 3n-6=9$. Then $Q_5 - F$ has a connected component containing at least $2^n - |F| - 2 = 30 - |F|$ vertices.

Proof: A five-dimensional hypercube Q_5 can be divided into two Q_4's, denoted by Q_4^L and Q_4^R. Let $F_L = F \cap V(Q_4^L), 0 \leq |F_L| \leq 9$ and $F_R = F \cap V(Q_4^R), 0 \leq |F_R| \leq 9$. Then $|F| = |F_L| + |F_R|$. Without loss of generality, we may assume that $|F_L| \geq |F_R|$. In the following proof, we consider three cases by the size of F_R: (1) $0 \leq |F_R| \leq 2$, (2) $|F_R| = 3$, and (3) $|F_R| = 4$.

Case 1. $0 \leq |F_R| \leq 2$. Since $\kappa(Q_4)=4$, $Q_4^R - F_R$ is connected and $|V(Q_4^R - F_R)| = 2^4 - |F_R|$. Let $F_R^{(L)} \subset V(Q_4^L)$ be the set of vertices, that has neighboring vertices in F_R. For each vertex $v \in Q_4^L - F_L - F_R^{(L)}$, there is exactly one vertex $v^{(R)}$ in $Q_4^R - F_R$, such that $(v, v^{(R)}) \in E(Q_5)$. Besides, $|V(Q_4^L - F_L - F_R^{(L)})| \geq 2^4 - |F_L| - |F_R|$. Hence, $Q_5 - F$ has a connected component that contains at least $[2^4 - |F_R|] + [2^4 - |F_L| - |F_R|] = 32 - |F| - |F_R| \geq 30 - |F|$ vertices.

Case 2. $|F_R| = 3$. Since $\kappa(Q_4)=4$, $Q_4^R - F_R$ is connected and $|V(Q_4^R - F_R)| = 2^4 - |F_R|$. Let $F_R = \{x,y,z\}$ and $F_R^{(L)} = \{\mathbf{x}^{(L)}, \mathbf{y}^{(L)}, \mathbf{z}^{(L)}\} \subset V(Q_4^L)$, where $(\mathbf{x},\mathbf{x}^{(L)})$, $(\mathbf{y},\mathbf{y}^{(L)})$, $(\mathbf{z},\mathbf{z}^{(L)}) \in E(Q_5)$. For each vertex $\mathbf{v} \in Q_4^L - F_L - F_R^{(L)}$, there is exactly one vertex $\mathbf{v}^{(R)}$ in $Q_4^R - F_R$, such that $(\mathbf{v}, \mathbf{v}^{(R)}) \in E(Q_5)$. If at least one of the three vertices $\mathbf{x}^{(L)}$, $\mathbf{y}^{(L)}$, and $\mathbf{z}^{(L)}$ belongs to F_L, then $|V(Q_4^L - F_L - F_R^{(L)})| \geq 2^4 - |F_L| - 2$. Hence, $Q_5 - F$ has a connected component that contains at least $[2^4 - |F_R|] + [2^4 - |F_L| - 2] = 30 - |F|$ vertices; otherwise, $|V(Q_4^L - F_L - F_R^{(L)})| \geq 2^4 - |F_L| - 3$. Since $|F_L| \leq 6$, by Lemma 21.34, $\mathbf{x}^{(L)}$, $\mathbf{y}^{(L)}$, and $\mathbf{z}^{(L)}$ have at least one neighboring vertex in $Q_4^L - F_L - F_R^{(L)}$. Hence, $Q_5 - F$ has a connected component that contains at least $[2^4 - |F_R|] + [2^4 - |F_L| - 3] + 1 = 30 - |F|$ vertices.

Case 3. $|F_R| = 4$. Since $|F_R| = 4$ and $|F_L| \leq 5$, by Lemma 21.32, $Q_4^L - F_L$ ($Q_4^R - F_R$, respectively) has a connected component C_L (C_R, respectively) that contains at least $2^4 - |F_L| - 1$ ($2^4 - |F_R| - 1$, respectively) vertices. Since $|V(C_L)| \geq |F_R| + 1$, there exists a vertex $u \in C_L$ and a vertex $v \in C_R$ such that

$(u, v) \in E(Q_5)$. Hence, $Q_5 - F$ has a connected component that contains at least $[2^4 - |F_L| - 1] + [2^4 - |F_R| - 1] = 30 - |F|$ vertices.

Consequently, the lemma holds.

We now prove Lemma 21.36. □

LEMMA 21.36 Let Q_n be an n-dimensional hypercube, $n \geq 5$, and let F be a set of vertices $F \subset V(Q_n)$ with $|F| \leq 3n - 6$. Then $Q_n - F$ has a connected component containing at least $2^n - |F| - 2$ vertices.

Proof: We prove the lemma by induction on n. By Lemma 21.35, the lemma holds for $n = 5$. As the inductive hypothesis, we assume that the result is true for Q_{n-1}, for $|F| \leq 3(n-1) - 6$, and for some $n \geq 6$. Now we consider Q_n, $|F| \leq 3n - 6$. An n-dimensional hypercube Q_n can be divided into two Q_{n-1}'s, denoted by Q_{n-1}^L and Q_{n-1}^R. Let $F_L = F \cap V(Q_{n-1}^L), 0 \leq |F_L| \leq 3n - 6$ and $F_R = F \cap V(Q_{n-1}^R), 0 \leq |F_R| \leq 3n - 6$. Then $|F| = |F_L| + |F_R|$. Without loss of generality, we may assume that $|F_L| \geq |F_R|$. In the following proof, we consider two cases by the size of F_R: (1) $0 \leq |F_R| \leq 2$ and (2) $|F_R| \geq 3$.

Case 1. $0 \leq |F_R| \leq 2$. Since $0 \leq |F_R| \leq 2$, $Q_{n-1}^R - F_R$ is connected and $|V(Q_{n-1}^R - F_R)| = 2^{n-1} - |F_R|$. Let $F_R^{(L)} \subset V(Q_{n-1}^L)$ be the set of vertices that has neighboring vertices in F_R. For each vertex $v \in Q_{n-1}^L - F_L - F_R^{(L)}$, there is exactly one vertex $v^{(R)}$ in $Q_{n-1}^R - F_R$, such that $(\mathbf{v}, \mathbf{v}^{(R)}) \in E(Q_n)$. Besides, $|V(Q_{n-1}^L - F_L - F_R^{(L)})| \geq 2^{n-1} - |F_L| - |F_R|$. Hence $Q_n - F$ has a connected component that contains at least $[2^{n-1} - |F_R|] + [2^{n-1} - |F_L| - |F_R|] = 2^n - |F| - |F_R| \geq 2^n - |F| - 2$ vertices.

Case 2. $|F_R| \geq 3$. Since $|F_R| \geq 3, 3 \leq |F_L| \leq 3(n-1) - 6$ and $3 \leq |F_R| \leq 3(n-1) - 6$. By the inductive hypothesis, $Q_{n-1}^L - F_L$ ($Q_{n-1}^R - F_R$, respectively) has a connected component C_L (C_R, respectively) that contains at least $2^{n-1} - |F_L| - 2$ ($2^{n-1} - |F_R| - 2$, respectively) vertices. Next, we divide the case into three subcases: (1) $|V(C_L)| = 2^{n-1} - |F_L| - 2$ and $Q_{n-1}^R - F_R$ is disconnected, (2) $|V(C_L)| = 2^{n-1} - |F_L| - 2$ and $Q_{n-1}^R - F_R$ is connected, and (3) $|V(C_L)| \geq 2^{n-1} - |F_L| - 1$ and $|V(C_R)| \geq 2^{n-1} - |F_R| - 1$.

Case 2.1. $|V(C_L)| = 2^{n-1} - |F_L| - 2$ and $Q_{n-1}^R - F_R$ is disconnected. This is an impossible case. Since $\kappa(Q_{n-1}) = n - 1$, $|F_R| \geq n - 1$. By Lemma 21.32, $|F_L| \geq 2((n-1) - 1)$. Then the total number of faulty vertices is at least $(n-1) + 2((n-1) - 1) = 3n - 5$, which is greater than $3n - 6$, a contradiction.

Case 2.2. $|V(C_L)| = 2^{n-1} - |F_L| - 2$ and $Q_{n-1}^R - F_R$ is connected. Since $Q_{n-1}^R - F_R$ is connected, $|V(Q_{n-1}^R - F_R)| = 2^{n-1} - |F_R|$. Since $|V(C_L)| \geq |F_R| + 1$, there exists a vertex $u \in C_L$ and a vertex $v \in C_R$ such that $(u, v) \in E(Q_n)$. Hence, $Q_n - F$ has a connected component that contains at least $[2^{n-1} - |F_R|] + [2^{n-1} - |F_L| - 2] = 2^n - |F| - 2$ vertices.

Case 2.3. $|V(C_L)| \geq 2^{n-1} - |F_L| - 1$ and $|V(C_R)| \geq 2^{n-1} - |F_R| - 1$. Since $|V(C_L)| \geq |F_R| + 1$, there exists a vertex $u \in C_L$ and a vertex $v \in C_R$ such that

$(u, v) \in E(Q_n)$. Hence, $Q_n - F$ has a connected component that contains at least $[2^{n-1} - |F_L| - 1] + [2^{n-1} - |F_R| - 1] = 2^n - |F| - 2$ vertices.

This completes the proof of the lemma. □

By Lemma 21.36, we have the following corollary.

COROLLARY 21.10 *Let Q_n be an n-dimensional hypercube, $n \geq 5$, and let F be a set of vertices $F \subset V(Q_n)$ with $|F| \leq 3n - 6$. Then $Q_n - F$ satisfies one of the following conditions:*

1. *$Q_n - F$ is connected.*
2. *$Q_n - F$ has two components, one of which is K_1, and the other one has $2^n - |F| - 1$ vertices.*
3. *$Q_n - F$ has two components, one of which is K_2, and the other one has $2^n - |F| - 2$ vertices.*
4. *$Q_n - F$ has three components, two of which are K_1, and the third one has $2^n - |F| - 2$ vertices.*

Let $G(V, E)$ be a graph. A subset M of $E(G)$ is called a *matching* in G if its elements are links and no two are adjacent in G; the two ends of an edge in M are said to be matched under M. A vertex cover of G is a subset $\mathcal{K}$ of $V(G)$ such that every edge of G has at least one end in $\mathcal{K}$. A subset I of $V(G)$ is called an independent set of G if no two vertices of I are adjacent in G. To prove the conditional diagnosability of the hypercube, we need the following classical results.

THEOREM 21.25 **[339]** Let $G(V, E)$ be a bipartite graph. The maximum size of a matching in G equals the minimum size of a vertex cover of G.

PROPOSITION 21.2 **[339]** *Let $G(V, E)$ be a bipartite graph. The set $I \subset V(G)$ is a maximum independent set of G if and only if $V - I$ is a minimum vertex cover of G.*

The hypercube can be described as follows. Let Q_n denote an n-dimensional hypercube. Q_1 is a complete graph with two vertices labeled with 0 and 1, respectively. For $n \geq 2$, each Q_n consists of two Q_{n-1}'s, denoted by Q^0_{n-1} and Q^1_{n-1}, with a perfect matching M between them. That is, M is a set of edges connecting the vertices of Q^0_{n-1} and the vertices of Q^1_{n-1} in a one-to-one manner. It is easy to see that there are 2^{n-1} edges between Q^0_{n-1} and Q^1_{n-1}. The hypercube is a bipartite graph with 2^n vertices. Hence, we have the following lemma.

LEMMA 21.37 Let Q_n be an n-dimensional hypercube. In hypercube Q_n, the maximum size of a matching, the minimum size of a vertex cover and the maximum size of an independent set are all 2^{n-1}.

We are now ready to show that the conditional diagnosability of Q_n is $3(n-2)+1$ for $n \geq 5$. Let $F_1, F_2 \subset V(Q_n)$ be two conditional faulty sets with $F_1 \leq 3(n-2)+1$

and $F_2 \leq 3(n-2)+1$, $n \geq 5$. We shall show our result by proving that (F_1, F_2) is a distinguishable conditional pair under the comparison diagnosis model.

LEMMA 21.38 Let Q_n be an n-dimensional hypercube with $n \geq 5$. For any two conditional faulty sets $F_1, F_2 \subset V(Q_n)$, and $F_1 \neq F_2$, with $F_1 \leq 3(n-2)+1$ and $F_2 \leq 3(n-2)+1$. Then (F_1, F_2) is a distinguishable conditional pair under the comparison diagnosis model.

Proof: We use Theorem 21.24 to prove this result. Let $S = F_1 \cap F_2$, then $0 \leq |S| \leq 3(n-2)$. We will show that, deleting S from Q_n, the subgraph $C_{F_1 \Delta F_2, S}$ containing $F_1 \Delta F_2$ has *many* vertices having degree 3 or more. More precisely, we are going to prove that, in the subgraph $C_{F_1 \Delta F_2, S}$, the number of vertices having degree 3 or more is at least $2[3(n-2)+1-|S|]+1 = 6n-2|S|-9$. In the following proof, we consider three cases by the size of S: (1) $0 \leq |S| \leq n-1$, (2) $|S| = n$, and (3) $n+1 \leq |S| \leq 3(n-2)$.

Case 1. $0 \leq |S| \leq n-1$. Since the connectivity of Q_n is n, $Q_n - S$ is connected and the subgraph $C_{F_1 \Delta F_2, S}$ is the only component in $Q_n - S$. Since the hypercube Q_n has no cycle of length 3 and any two vertices have at most two common neighbors, it is straightforward, though tedious, to confirm that the number of vertices that have degree 2 or 1 is at most 2 in $C_{F_1 \Delta F_2, S}$. Hence, the number of vertices having degree 3 or more is at least $2^n - |S| - 2$, which is greater than $6n - 2|S| - 9$, for $n \geq 5$. By Theorem 21.24, (F_1, F_2) is a distinguishable conditional pair under the comparison diagnosis model.

Case 2. $|S| = n$. If $Q_n - S$ is disconnected, by Lemma 21.32, $Q_n - S$ has one trivial component $\{v\}$ such that $N(v) \subset F_1$ and $N(v) \subset F_2$. Since F_1 and F_2 are two conditional faulty sets, this is an impossible case. So $Q_n - S$ is connected, and the subgraph $C_{F_1 \Delta F_2, S}$ is the only component in $Q_n - S$. Let $U = Q_n - (F_1 \cup F_2)$. If there exist two vertices u and v in $V(U)$ such that u is adjacent to v, then condition (1) of Theorem 21.3 holds and therefore (F_1, F_2) is a distinguishable conditional pair; otherwise, $V(U)$ is an independent set. Since $|S| = n$ and $|F_1 \Delta F_2| \leq 2(2n-5)$, $|V(U)| \geq 2^n - 2(2n-5) - n = 2^n - 5n + 10$. By Lemma 21.37, the maximum size of an independent set is 2^{n-1} in Q_n. Comparing the lower bound $2^n - 5n + 10$ and the upper bound 2^{n-1}, we have $2^n - 5n + 10 > 2^{n-1}$ for $n \geq 5$—a contradiction.

Case 3. $n+1 \leq |S| \leq 3(n-2)$. By Corollary 21.10, there are four cases in $Q_n - S$ that we need to consider. For Case 1 of Corollary 21.10, $Q_n - S$ is connected, the proof is exactly the same as that of Case 2; hence, the detail is omitted. For Case 2 and Case 4 of Corollary 21.10, $Q_n - S$ has at least one trivial component $\{\mathbf{v}\}$ such that $N(\mathbf{v}) \subset F_1$ and $N(\mathbf{v}) \subset F_2$. Since F_1 and F_2 are two conditional faulty sets, the two cases are disregarded. Therefore, we need to consider only that $Q_n - S$ has two components: one of which is K_2, and the other one has $2^n - |S| - 2$ vertices. Let $(\mathbf{x}, \mathbf{y})$ be the component with only one edge. Since $N(\{\mathbf{x}, \mathbf{y}\}) \subseteq S$ and F_1 and F_2 do not contain all the neighbors of any vertex, vertex x and y cannot belong to $F_1 \Delta F_2$. So the subgraph $C_{F_1 \Delta F_2, S}$ is the other large connected component of $Q_n - S$. Let $U = Q_n - (F_1 \cup F_2) - \{\mathbf{x}, \mathbf{y}\}$. If no two vertices of $V(U)$ are adjacent, then $V(U)$ is an independent set and $|V(U)| \geq 2^n - 6n + |S| + 8$. By Lemma 21.37, the maximum size of a matching is $2^{n-1} - 1$ in $Q_n - \{\mathbf{x}, \mathbf{y}\}$. By Theorem 21.25 and Proposition 21.2, the maximum size of an independent set is $2^{n-1} - 1$ in $Q_n - \{\mathbf{x}, \mathbf{y}\}$. Comparing the lower bound $2^n - 6n + |S| + 8$ with the upper bound $2^{n-1} - 1$, we have $2^n - 6n + |S| + 8 > 2^{n-1} - 1$ for $n \geq 5$, $n + 1 \leq |S| \leq 3(n-2)$, which is a contradiction. Hence, there exist two vertices $\mathbf{u}$ and $\mathbf{v}$ in $V(U)$ such that $\mathbf{u}$ is adjacent to $\mathbf{v}$, then condition 1 of Theorem 21.3 is satisfied, and therefore, (F_1, F_2) is a distinguishable conditional pair.

In Case 1, we prove that at least one of the conditions of Theorem 21.3 is satisfied in subgraph $C_{F_1 \Delta F_2, S}$. In Case 2 and Case 3, condition (1) of Theorem 21.3 holds in subgraph $C_{F_1 \Delta F_2, S}$. Therefore, (F_1, F_2) is a distinguishable conditional pair under the comparison diagnosis model. □

By Lemma 21.31, $t_c(Q_n) \leq 3(n-2) + 1$, and by Lemma 21.38, Q_n is conditionally $(3(n-2)+1)$–diagnosable for $n \geq 5$. Hence, $t_c(Q_n) = 3(n-2) + 1$ for $n \geq 5$. For Q_3 and Q_4, we observe that Q_3 is not conditionally 4-diagnosable and Q_4 is not conditionally 6-diagnosable, as shown in Figure 21.20. So $t_c(Q_3) \leq 3$ and $t_c(Q_4) \leq 5$. Hence, the conditional diagnosabilities of Q_3 and Q_4 are both strictly less than $3(n-2) + 1$.

For the three-dimensional hypercube Q_3, Q_3 is 3-diagnosable and is not conditionally 4-diagnosable. It follows from Lemma 21.22 that $t_c(Q_3) = 3$. For the four-dimensional hypercube Q_4, we can use a similar technique to that used in proving Lemma 21.38 to prove that for any two conditional faulty subsets F_1 and F_2 of $V(Q_4)$, and $F_1 \neq F_2$, with $|F_1| \leq 5$ and $|F_2| \leq 5$, then (F_1, F_2) is a distinguishable conditional pair under the comparison diagnosis model. Hence, the conditional diagnosability of Q_4 is 5. In summary, the conditional diagnosability of Q_n is stated as follows.

THEOREM 21.26 The conditional diagnosability of Q_n is $t_c(Q_n) = 3(n-2) + 1$ for $n \geq 5$, $t_c(Q_3) = 3$ and $t_c(Q_4) = 5$.

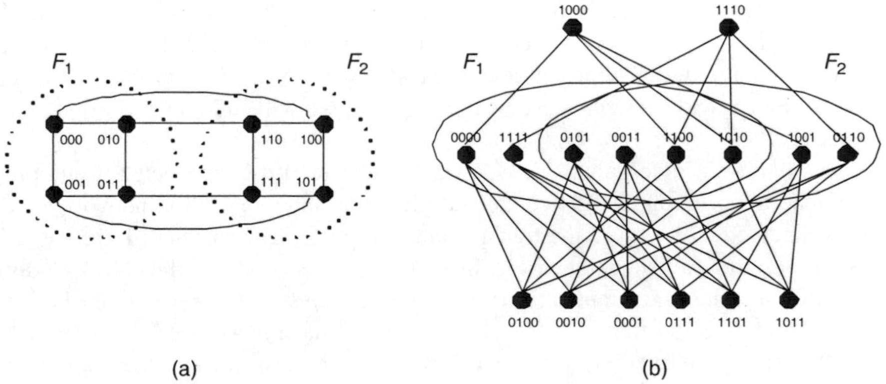

FIGURE 21.20 Two indistinguishable conditional pairs for Q_3 and Q_4.

21.7.1 Conditional Diagnosability of Cayley Graphs Generated by Transposition Trees under the Comparison Diagnosis Model

In this section, we study the conditional diagnosability of the star graph S_n and a class of graphs that arise as a generalization of the star graph. These graphs are Cayley graphs generated by transposition trees.

We consider the comparison model and show that the conditional diagnosability of these graphs is $3n - 8$ for $n \geq 4$, except for the n-dimensional star graph, for which it is $3n - 7$.

21.7.2 Cayley Graphs Generated by Transposition Trees

In this section, we summarize the connectivity properties of Cayley graphs generated by transposition trees. These graphs arise naturally as a common generalization of star graphs and bubble-sort graphs. Some papers studying these graphs include Refs. 15, 56–58,296.

Let Γ be a finite group and S be a set of elements of Γ such that the identity of the group does not belong to S. The *Cayley graph* $\Gamma(S)$ is the directed graph whose vertex set is Γ, and there is an arc from u to v if and only if there is an $s \in S$ such that $u = vs$. The graph $\Gamma(S)$ is connected if and only if S is a generating set for Γ. A Cayley graph is always vertex-transitive, so it is maximally arc-connected if it is connected; however, its vertex connectivity may be low.

In this paper, we choose the finite group to be Γ_n, the symmetric group on $\{1, 2, \ldots, n\}$, and the generating set S to be a set of transpositions. The vertices of the corresponding Cayley graph are permutations, and since S has only transpositions, there is an arc from vertex u to vertex v if and only if there is an arc from v to u. Hence we can regard these Cayley graphs as undirected graphs by replacing every pair of arcs between two vertices with an edge; let the resulting graph be $\Gamma_n(S)$. A simple way to depict S is via a graph $G(S)$ with vertex set $\{1, 2, \ldots, n\}$, where there is an edge between i and j if and only if the transposition (ij) belongs to S. This graph is called the *transposition generating graph* of $\Gamma_n(S)$ or simply *transposition (generating) graph* if it is clear from the context. In fact, the star graph S_n was introduced via the generating graph $K_{1,n-1}$, where the center is 1 and the leaves are $2, 3, \ldots, n$. Notice, that if we change the label of the center, we still get a graph isomorphic to the star graph S_n; hence with a slight abuse of terminology, we will call all these graphs *star graphs*.

Note that the Cayley graph $\Gamma_n(S)$ is $|S|$-regular, and it is connected if and only if the generating graph $G(S)$ is connected. Since an interconnection network needs to be connected, we require the transposition graph to be connected. Here we will consider only the fundamental case, when $G(S)$ is a tree, and call the corresponding transposition generating graph a *transposition tree*. Thus the Cayley graphs obtained by these transposition trees are $(n - 1)$-regular and have $n!$ vertices. In addition to the star graph mentioned earlier, these Cayley graphs also include the bubble-sort graph whose transposition tree is a path. Figure 21.21 shows the bubble-sort graph for $n = 4$.

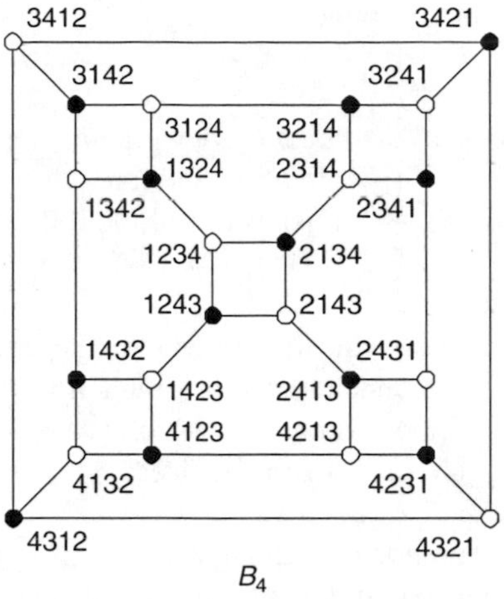

FIGURE 21.21 The bubble-sort graph for $n = 4$.

Let $\Gamma_n(S)$ be a Cayley graph generated by a transposition tree S. To help us describe the structure of the Cayley graph $\Gamma_n(S)$ when $G(S)$ is a tree, without loss of generality we may assume that a leaf of the transposition tree is n. We use boldface letters to denote vertices in $\Gamma_n(S)$. Hence, $\mathbf{u}_1, \mathbf{u}_2, \ldots, \mathbf{u}_n$ is a sequence of n vertices in $\Gamma_n(S)$. It is known that the connectivity of $\Gamma_n(S)$ is $n-1$. Clearly $\Gamma_n(S)$ is a bipartite graph with one partite set containing the vertices corresponding to odd permutations and the other partite set containing the vertices corresponding to even permutations. Let $\mathbf{u} = u_1 u_2 \ldots u_n$ be any vertex of the Cayley graph $\Gamma_n(S)$. We say that u_i is the ith coordinate of $\mathbf{u}$, denoted by $(\mathbf{u})_i$, for $1 \le i \le n$. For $1 \le i \le n$, let $\Gamma_n^{\{i\}}$ denote the subgraph of $\Gamma_n(S)$ induced by those vertices $\mathbf{u}$ with $(\mathbf{u})_n = i$.

Since n is a leaf in the generating tree, it is easy to see that the Cayley graph $\Gamma_n(S)$ has the following properties:

1. $\Gamma_n(S)$ consists of n vertex-disjoint subgraphs: $\Gamma_n^{\{1\}}, \Gamma_n^{\{2\}}, \ldots, \Gamma_n^{\{n\}}$; each isomorphic to another Cayley graph $\Gamma_{n-1}(S')$ with $S' = S \setminus \{\pi\}$ where π is the transposition corresponding to the edge incident to the leaf n.
2. $\Gamma_n^{\{i\}}$ has $(n-1)!$ vertices, and it is $(n-2)$-regular for all i.
3. For all i, each vertex in $\Gamma_n^{\{i\}}$ has a unique neighbor outside $\Gamma_n^{\{i\}}$, and these outside neighbors are all different. There are exactly $(n-2)!$ independent edges between $\Gamma_n^{\{i\}}$ and $\Gamma_n^{\{j\}}$ for all $i \ne j$.

These properties are illustrated in Figures 7 of Chapter 3.5 and 21.21, as (for example) S_4 and the bubble-sort graph contain four copies of a smaller Cayley graph, the 6-cycle. Note that the 6-cycle is the shortest cycle in star graphs, whereas in other Cayley graphs we also have 4-cycles.

Cayley graphs generated by transposition trees have strong connectivity properties. Roughly speaking, after deleting a large number of vertices from these graphs, they will still contain a large connected component, as shown by the following theorem:

THEOREM 21.27 **[58]** Let $\Gamma_n(S)$ be a Cayley graph obtained from a transposition generating tree S on $\{1, 2, \ldots, n\}$ with $n \geq 4$, and let T be a set of vertices of G such that $|T| \leq 3n - 8$. Then $\Gamma_n(S) - T$ satisfies one of the following conditions:

1. $\Gamma_n(S) - T$ is connected.
2. $\Gamma_n(S) - T$ has two components, one of which is K_1 or K_2.
3. $\Gamma_n(S) - T$ has three components, two of which are singletons.
4. $\Gamma_n(S) - T$ has two components, one of which is a path of length 3, and T is the union of the neighbor sets of the vertices on the path except the vertices of the path itself with $|T| = 3n - 8$.
5. $\Gamma_n(S) - T$ has four components, three of which are singletons, and T is the union of the neighbor sets of the singletons with $|T| = 3n - 8$.
6. $\Gamma_n(S) - T$ has two components, one of which is a 4-cycle, $n = 4$ and $|T| = 4$.

Note: Cases 4, 5, and 6 can occur only when $\Gamma_n(S)$ is not a star graph, because each require a 4-cycle in the graph.

21.7.3 Conditional Diagnosability

In classical measures of system-level diagnosability for multiprocessor systems, if all the neighbors of some processor v are faulty simultaneously, it is not possible to determine whether processor v is fault-free or faulty. So the diagnosability of a system is limited by its minimum vertex degree. In particular, as we have mentioned before, the star graph S_n has diagnosability $n - 1$ (see [355]). The same result can be proven easily for Cayley graphs generated by transposition trees as well, whose proof we omit.

THEOREM 21.28 Let $\Gamma_n(S)$ be a Cayley graph obtained from a transposition generating tree S on $\{1, 2, \ldots, n\}$ with $n \geq 4$. Then $t(\Gamma_n(S)) = n - 1$.

A Cayley graph $\Gamma_n(S)$ has $\binom{n!}{n-1}$ vertex subsets of size $n - 1$, among which there are only $n!$ vertex subsets which contain all the neighbors of some vertex. Since the ratio $n!/\binom{n!}{n-1}$ is very small for large n, in case of independent failures the probability of a faulty set containing all the neighbors of any vertex is very low.

Now we give an example in the Cayley graph $\Gamma_n(S)$ to get a bound on the conditional diagnosability. As shown in Figure 21.22, we take a path of length two in $\Gamma_n(S)$. Let $\langle \mathbf{u}_1, \mathbf{u}_2, \mathbf{u}_3 \rangle$ be a path with length two. We set $A = N_{\Gamma_n(S)}(\mathbf{u}_1) \cup N_{\Gamma_n(S)}(\mathbf{u}_2) \cup N_{\Gamma_n(S)}(\mathbf{u}_3)$, $F_1 = A - \{\mathbf{u}_2, \mathbf{u}_3\}$ and $F_2 = A - \{\mathbf{u}_1, \mathbf{u}_2\}$. It is straightforward to confirm that F_1 and F_2 are two conditional faulty sets, and F_1 and F_2 are indistinguishable by Theorem 21.3. When $\Gamma_n(S)$ is a star graph, it has no cycles with length less than six; hence the vertices in $N_{\Gamma_n(S)}(\mathbf{u}_1)$, $N_{\Gamma_n(S)}(\mathbf{u}_2)$, and $N_{\Gamma_n(S)}(\mathbf{u}_3)$ are all different; thus $|F_1| = |F_2| = 3n - 6$. On the other hand, if $\Gamma_n(S)$ is not a star graph, it contains 4-cycles, so some of those neighbors may be the same. However, it

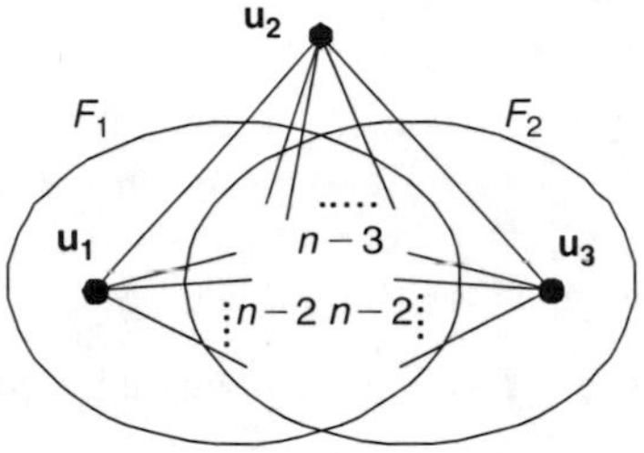

FIGURE 21.22 An indistinguishable conditional pair (F_1, F_2).

is easy to see that any two vertices in $\Gamma_n(S)$ can have at most two common neighbors. Thus when the path $\langle \mathbf{u}_1, \mathbf{u}_2, \mathbf{u}_3 \rangle$ is part of a 4-cycle, we get $|F_1| = |F_2| = 3n - 7$. In both cases we have $|F_1 - F_2| = |F_2 - F_1| = 1$, therefore when $\Gamma_n(S)$ is a star graph, it is not conditionally $(3n-6)$-diagnosable; otherwise, $\Gamma_n(S)$ is not conditionally $(3n-7)$-diagnosable. Hence we have the following result.

PROPOSITION 21.3 *For $n \geq 4$, $t_c(\Gamma_n(S)) \leq 3n - 7$ when $\Gamma_n(S)$ is a star graph; otherwise, $t_c(\Gamma_n(S)) \leq 3n - 8$.*

The following two lemmas will be needed to show our result on the conditional diagnosability of $\Gamma_n(S)$ for $n \geq 4$.

LEMMA 21.39 For $n \geq 4$, let F_1 and F_2 be any two distinct conditional faulty subsets of $V(\Gamma_n(S))$ with $|F_1| \leq 3n - 7$ and $|F_2| \leq 3n - 7$ if $\Gamma_n(S)$ is a star graph, and $|F_1| \leq 3n - 8$ and $|F_2| \leq 3n - 8$ otherwise. Denote by H the maximum component of $\Gamma_n(S) - (F_1 \cap F_2)$. Then for every vertex $\mathbf{u}$ in $F_1 \triangle F_2$, $\mathbf{u}$ is in H.

Proof: Without loss of generality, we assume that $\mathbf{u}$ is in $F_1 - F_2$. Since F_2 is a conditional faulty set, there is vertex $\mathbf{v}$ in $(V(\Gamma_n(S)) - F_2) - \{\mathbf{u}\}$ such that $(\mathbf{u}, \mathbf{v}) \in E(\Gamma_n(S))$. Suppose that $\mathbf{u}$ is not a vertex of H. Then $\mathbf{v}$ is not in $V(H)$, so $\mathbf{u}$ and $\mathbf{v}$ are part of a small component in $\Gamma_n(S) - (F_1 \cap F_2)$. Since F_1 and F_2 are distinct, we have $|F_1 \cap F_2| \leq 3n - 8$ when $\Gamma_n(S)$ is a star graph and $|F_1 \cap F_2| \leq 3n - 9$ otherwise. Thus in Theorem 21.27, Cases 4 through 6 cannot occur, hence $\{\mathbf{u}, \mathbf{v}\}$ forms a component K_2 of $\Gamma_n(S) - (F_1 \cap F_2)$, that is, $\mathbf{u}$ is the unique neighbor of $\mathbf{v}$ in $\Gamma_n(S) - (F_1 \cap F_2)$. This is a contradiction since F_1 is a conditional faulty set, but all the neighbors of $\mathbf{v}$ are faulty in $\Gamma_n(S) - F_1$. □

LEMMA 21.40 Let G be a graph with $\delta(G) \geq 2$, and let F_1 and F_2 be any two distinct conditional faulty subsets of $V(G)$ with $F_2 \subset F_1$. Then (F_1, F_2) is a distinguishable conditional pair under the comparison diagnosis model.

Proof: Let u be any vertex of $F_1 - F_2$. Since F_1 is a conditional faulty subset of $V(G)$, there is a vertex v of $V(G) - F_1$ such that $(u, v) \in E(G)$ and there is a vertex w of $V(G) - F_1$ such that $(v, w) \in E(G)$. Since $F_2 \subset F_1$, neither v nor w is in F_2. By Theorem 21.3, (F_1, F_2) is a distinguishable pair. □

Now we can prove our main results.

THEOREM 21.29 For $n \geq 4$, let F_1 and F_2 be two distinct conditional faulty subsets of $V(\Gamma_n(S))$. Assume that $|F_1| \leq 3n-7$ and $|F_2| \leq 3n-7$ when $\Gamma_n(S)$ is a star graph, and $|F_1| \leq 3n-8$ and $|F_2| \leq 3n-8$ otherwise. Then (F_1, F_2) is a distinguishable conditional pair under the comparison diagnosis model.

Proof: By Lemma 21.40, (F_1, F_2) is a distinguishable pair if $F_1 \subset F_2$ or $F_2 \subset F_1$. Thus we assume that $|F_1 - F_2| \geq 1$ and $|F_2 - F_1| \geq 1$. Let $A = F_1 \cap F_2$. Then we have $|A| \leq 3n-8$ when $\Gamma_n(S)$ is a star graph, and $|A| \leq 3n-9$ otherwise. Let H be the maximum component of $\Gamma_n(S) - A$. By Lemma 21.39, every vertex in $F_1 \Delta F_2$ is in H.

We claim that H has a vertex $\mathbf{v}$ outside $F_1 \cup F_2$ that has no neighbor in A. Since every vertex has degree $n-1$, vertices in A can have at most $|A|(n-1)$ neighbors in H. There are at most $2(3n-7) - |A|$ vertices in $F_1 \cup F_2$, and at most two vertices of $\Gamma_n(S) - A$ may not belong to H by Theorem 21.27. Since $|A| \leq 3n-8$, we have $n! - |A|(n-2) - 2(3n-7) - 2 \geq n! - (3n-8)(n-2) - 2(3n-7) - 2 \geq 4$ when $n \geq 4$. Thus there must be vertices of H outside $F_1 \cup F_2$ having no neighbor in A; let $\mathbf{v}$ be such a vertex.

If $\mathbf{v}$ has no neighbor in $F_1 \cup F_2$, then we can find a path of length at least 2 within H to a vertex $\mathbf{p}$ in $F_1 \Delta F_2$. We may assume that $\mathbf{p}$ is the first vertex of $F_1 \Delta F_2$ on this path, and let $\mathbf{q}$ and $\mathbf{w}$ be the two vertices on this path immediately before $\mathbf{p}$ (we may have $\mathbf{v} = \mathbf{q}$), so $\mathbf{q}$ and $\mathbf{w}$ are not in $F_1 \cup F_2$. Then the edges $(\mathbf{q}, \mathbf{w})$ and $(\mathbf{w}, \mathbf{p})$ show that (F_1, F_2) is a distinguishable conditional pair. Now assume that $\mathbf{v}$ has a neighbor in $F_1 \Delta F_2$. Then since the degree of $\mathbf{v}$ is at least 3, and $\mathbf{v}$ has no neighbor in A, there are three possibilities:

1. $\mathbf{v}$ has two neighbors in $F_1 - F_2$
2. $\mathbf{v}$ has two neighbors in $F_2 - F_1$
3. $\mathbf{v}$ has at least one neighbor outside $F_1 \cup F_2$

In each case, Theorem 21.3 implies that (F_1, F_2) is a distinguishable conditional pair of $\Gamma_n(S)$ under the comparison diagnosis model, finishing the proof. □

To summarize, with Proposition 21.3 and Theorem 21.29, we have the following result.

THEOREM 21.30 For $n \geq 4$, $t_c(\Gamma_n(S)) = 3n-7$ when $\Gamma_n(S)$ is a star graph, and $t_c(\Gamma_n(S)) = 3n-8$ otherwise.

REMARK: Theorem 21.28 can be proved similarly, indeed much simpler, using that its connectivity is $n-1$, proved in Ref. 56.

21.8 LOCAL DIAGNOSABILITY

We observe that the diagnosability of a system discussed in previous sections are all in a global sense, but not in a local sense. A system is t-diagnosable if all the faulty processors can be uniquely identified, provided that the number of faulty processors does not exceed t. However, it is possible to correctly indicate all the faulty processors in a t-diagnosable system when the number of faulty processors is greater than t. For example, consider a multiprocessor system generated by integrating two arbitrary subsystems with a few communication links in some way, where the two subsystems are m-diagnosable and n-diagnosable, respectively, and $m >> n$. The diagnosability of this system is limited by n, but it is possible to correctly point out all the faulty processors even if the number of the faulty ones is between m and n. Therefore, if only considering the global faulty/fault-free status, we lose some local systematic details. For this reason, Hsu et al. [175] introduced a new measure of diagnosability called *local diagnosability* under the PMC model. The aim of local diagnosis is to study the local status of each vertex instead of the whole system. Given a single vertex, we need to identify only the status of this particular processor correctly. We now propose the following concept.

Let $G(V,E)$ be a graph and $v \in V$ be a vertex. The G is locally t-diagnosable at vertex v if, given a syndrome σ_F produced by a set of faulty vertices $F \subseteq V$ containing vertex v with $|F| \leq t$, every set of faulty vertices F' compatible with σ_F and $|F'| \leq t$, must also contain vertex v.

Let $G(V,E)$ be a graph and $v \in V$ be a vertex. The local diagnosability of vertex v, written as $t_l(v)$, is defined to be the maximum value of t such that G is locally t-diagnosable at vertex v.

The following result is another point of view for checking whether a vertex is locally t-diagnosable.

LEMMA 21.41 Let $G(V,E)$ be a graph and $v \in V$ be a vertex. The G is locally t-diagnosable at vertex v if and only if, for any two distinct sets of vertices $F_1, F_2 \subset V$, $|F_1| \leq t$, $|F_2| \leq t$ and $v \in F_1 \Delta F_2$, (F_1, F_2) is a distinguishable pair.

In the following, we study some properties of a system being locally t-diagnosable at a given vertex and its relationship between a system that is t-diagnosable.

PROPOSITION 21.4 *Let $G(V,E)$ be a graph and $v \in V(G)$ be a vertex. The G is locally t-diagnosable at vertex v, then $|V(G)| \geq 2t+1$.*

Proof: We show this by contradiction. Assume that $|V(G)| \leq 2t$. We partition $V(G)$ into two disjoint subsets F_1, F_2 with $|F_1| \leq t$, $|F_2| \leq t$. The vertex v is either in F_1 or in F_2. Since $V-(F_1 \cup F_2) = \emptyset$, there is no edge between $V-(F_1 \cup F_2)$ and $F_1 \Delta F_2$. By Lemma 21.4, (F_1, F_2) is an indistinguishable pair; this contradicts the assumption that G is locally t-diagnosable at vertex v. So the result follows. □

PROPOSITION 21.5 *Let $G(V,E)$ be a graph and $v \in V$ be a vertex with $deg(v) = n$. The local diagnosability of vertex v is at most n.*

Proof: Let F_1 be the set of vertices adjacent to vertex v, $F_1 = N_G(v)$ and $|F_1| = n$. Let $F_2 = F_1 \cup \{v\}$ with $|F_2| = n+1$. It is a simple matter to check whether there is no edge between $V-(F_1 \cup F_2)$ and $F_1 \Delta F_2$. By Lemma 21.4, (F_1, F_2) is an indistinguishable pair. Thus, G is not locally $(n+1)$-diagnosable at vertex v, so $t_l(v) \leq n = deg(v)$. We have the stated result. □

PROPOSITION 21.6 *Let $G(V,E)$ be a graph. G is t-diagnosable if and only if G is locally t-diagnosable at every vertex.*

Proof: To prove the necessity, we assume that G is t-diagnosable. If the result is not true, there exists a vertex $v \in V$ such that G is not locally t-diagnosable at vertex v. By Lemma 21.41, there exists a distinct pair of sets $F_1, F_2 \subset V$ with $|F_1| \leq t$, $|F_2| \leq t$ and $v \in F_1 \Delta F_2$, (F_1, F_2) is an indistinguishable pair. By Lemma 21.1, G is not t-diagnosable. This contradicts the assumption; hence, the necessary condition follows.

To prove the sufficiency, suppose that on the contrary that G is not t-diagnosable; there exists a distinct pair of sets $F_1, F_2 \subset V$ with $|F_1| \leq t$, $|F_2| \leq t$, and (F_1, F_2) is an indistinguishable pair. Being distinct, the set $F_1 \Delta F_2 \neq \emptyset$, we can find a vertex $v \in F_1 \Delta F_2$. By Lemma 21.41, G is not locally t-diagnosable at vertex v, which is a contradiction. This completes the proof. □

By the definition of local diagnosability and Proposition 21.6, we know that the diagnosability of a multiprocessor system is equal to the minimum local diagnosability of all vertices of the system. Thus, we have the following theorem.

THEOREM 21.31 Let $G(V,E)$ be a multiprocessor system. The diagnosability of G is t if and only if $min\{t_l(v) \mid \text{for every } v \in V\} = t$.

From Theorem 21.31, we can identify the diagnosability of a system by computing the local diagnosability of each vertex. Because many well-known systems are vertex-symmetric, the diagnosability of these system can be easily identified by this effective method.

Before studying the local diagnosability of a vertex, we need some definitions for further discussion. Let S be a set of vertices and v be a vertex not in S. After deleting the vertices in S from G, we use C_v to denote the connected component, which vertex v belongs to. Now, we propose a necessary and sufficient condition for verifying whether a system is locally t-diagnosable at a given vertex v.

THEOREM 21.32 Let $G(V,E)$ be a graph and $v \in V$ be a vertex. The G is locally t-diagnosable at vertex v if and only if, for each set of vertices $S \subset V$ with $|S| = p$, $0 \leq p \leq t-1$ and $v \notin S$, the connected component, which v belongs to in $G - S$, has at least $2(t-p)+1$ vertices.

Proof: To prove the necessity, we assume that G is locally t-diagnosable at vertex v. If the result does not hold, there exists a set of vertices $S \subset V$ with $|S| = p, 0 \leq p \leq t-1$, $v \notin S$ such that the connected component C_v has strictly less than $2(t-p)+1$ vertices, $|V(C_v)| \leq 2(t-p)$. We then arbitrarily partition $V(C_v)$ into two disjoint subsets, $V(C_v) = S_1 \cup S_2$ with $|S_1| \leq t-p$, $|S_2| \leq t-p$. Let $F_1 = S_1 \cup S$ and $F_2 = S_2 \cup S$. It is clear that $|F_1| \leq (t-p)+p = t$, $|F_2| \leq (t-p)+p = t$, the vertex $v \in F_1 \Delta F_2$ and there is no edge between $V - (F_1 \cup F_2)$ and $F_1 \Delta F_2$. By Lemma 21.41, (F_1, F_2) is an

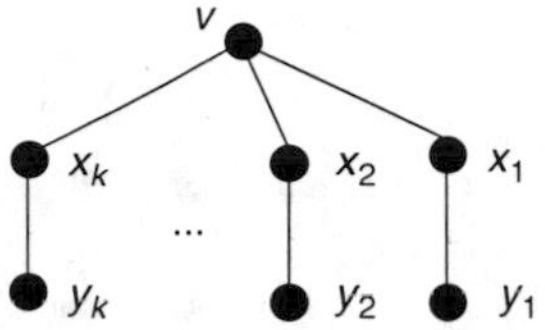

FIGURE 21.23 A Type I structure $T_1(v;k)$ consists of $2k+1$ vertices and $2k$ edges.

indistinguishable pair. This contradicts the assumption that G is locally t-diagnosable at vertex v.

We now prove the sufficiency by contradiction. Suppose G is not locally t-diagnosable at vertex v, then, there exists an indistinguishable pair (F_1, F_2) with $|F_1| \leq t$, $|F_2| \leq t$, and $v \in F_1 \,\Delta\, F_2$. By Lemma 21.4, there is no edge between $V-(F_1 \cup F_2)$ and $F_1 \,\Delta\, F_2$. Let $S = F_1 \cap F_2$ with $|S| = p$, $0 \leq p \leq t-1$ and $v \notin S$. $F_1 \,\Delta\, F_2$ is disconnected from other parts after removing all the vertices in S from G. We observe that $|F_1 \,\Delta\, F_2| \leq 2(t-p)$. Thus, the connected component C_v has at most $2(t-p)$ vertices and $|V(C_v)| \leq 2(t-p)$. This contradicts the assumption that the connected component C_v has to satisfy $|V(C_v)| \geq 2(t-p)+1$. Hence, the theorem holds. □

We now propose two special subgraphs called the Type I structure and the Type II structure. They provide us with an efficient and simple method to identify the local diagnosability of each vertex of a system under the PMC diagnosis model.

Let $G(V,E)$ be a graph, $v \in V$ be a vertex and k be an integer, $k \geq 1$; a Type I structure $T_1(v;k)$ of order k at vertex v is defined to be the following graph:

$$T_1(v;k) = [V(v;k), E(v;k)]$$

which is composed of $2k+1$ vertices and of $2k$ edges, as illustrated in Figure 21.23, where $V(v;k) = \{v\} \cup \{x_i, y_i \mid 1 \leq i \leq k\}$ and $E(v;k) = \{(v,x_i), (x_i,y_i) | 1 \leq i \leq k\}$.

Following Theorem 21.32 and the definition of local diagnosability, we propose a sufficient condition for verifying whether it is locally t-diagnosable at a given processor in a system.

THEOREM 21.33 Let $G(V,E)$ be a graph and $v \in V$ be a vertex. The G is locally t-diagnosable at vertex v if G contains a Type I structure $T_1(v;t)$ of order t at vertex v as a subgraph.

Proof: We use Theorem 21.32 to prove this result. Assume that G contains a subgraph $T_1(v;t)$ at vertex v. Let $e_i = (x_i, y_i)$ be the edge for each i, $1 \leq i \leq t$, with respect to $T_1(v;t)$. The number of vertices of the connected component including vertex v is at least $2t+1$. Let $S \subset V(G)$ be a set of vertices with $|S| = p, 0 \leq p \leq t-1$ and $v \notin S$. After deleting S from $V(G)$, there are at least $(t-p)$ complete e_i's still remaining in $T_1(v;t)$. Therefore, the number of vertices of the connected component C_v is at least $2(t-p)+1$. By Theorem 21.32, G is locally t-diagnosable at vertex v. The proof is complete. □

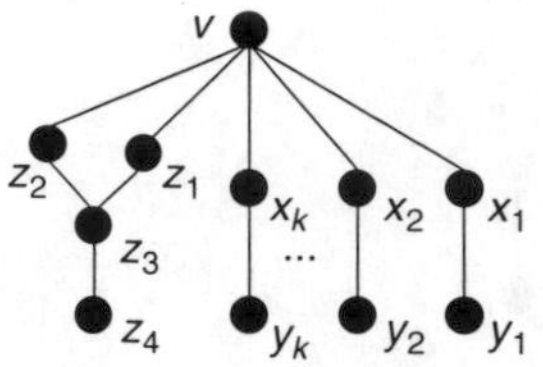

FIGURE 21.24 A Type II structure $T_2(v; k, 2)$ consists of $2k+5$ vertices and $2k+5$ edges.

A Type II structure $T_2(v; k, 2)$ at a vertex v is defined as follows.

Let $G(V, E)$ be a graph, $v \in V$ be a vertex and k be an integer, $k \geq 1$; a Type II structure $T_2(v; k, 2)$ of order $k+2$ at vertex v is defined to be the following graph, $T_2(v; k, 2) = [V(v; k, 2), E(v; k, 2)]$, which is composed of $2k+5$ vertices and of $2k+5$ edges, as illustrated in Figure 21.24, where

- $V(v; k, 2) = \{v\} \cup \{x_i, y_i \mid 1 \leq i \leq k\} \cup \{z_1, z_2, z_3, z_4\}$
- $E(v; k, 2) = \{(v, x_i), (x_i, y_i) | 1 \leq i \leq k\} \cup \{(v, z_1), (v, z_2), (z_1, z_3), (z_2, z_3), (z_3, z_4)\}$

In the following discussion, we propose another sufficient condition for verifying whether a graph it is locally t-diagnosable at a given processor in a system.

THEOREM 21.34 Let $G(V, E)$ be a graph and $v \in V$ be a vertex. The G is locally t-diagnosable at vertex v if G contains a Type II structure $T_2(v; k, 2)$ of order $k+2$ at vertex v as a subgraph, where $t = k+2$.

Proof: We use Theorem 21.32 to prove this result. Assume that G contains a subgraph $T_2(v; k, 2)$ of order $t = k+2$ at vertex v. The number of vertices of the connected component including vertex v is at least $2k+5 = 2t+1$. Let $S \subset V$ be a set of vertices with $|S| = p$, $0 \leq p \leq t-1$ and $v \notin S$, the number of vertices of C_v is at least $(2k+5) - 2*1$ after removing one vertex in S, the number of vertices of C_v is at least $(2k+5) - 2*2$ after removing two vertices in S, and so on. Thus, the connected component C_v satisfies $|V(C_v)| \geq (2k+5) - 2p = 2(t-p)+1$. By Theorem 21.32, G is locally t-diagnosable at vertex v. This proves the theorem. □

In the following, we give some examples.

EXAMPLE 21.1

Let us consider a cycle of length four as shown in Figure 21.25a. We can find a Type I structure $T_1(v; 1)$ of order 1 at vertex v, as shown in Figure 21.25b. Hence, vertex v is locally 1-diagnosable.

EXAMPLE 21.2

Consider examples as shown in Figures 21.26a, 21.26b, and 21.26c. It is a routine matter to check whether there is a subgraph $T_1(v_1; 2)$, $T_1(v_2; 2)$, and $T_2(v_3; 1, 2)$ at vertex v_1, v_2,

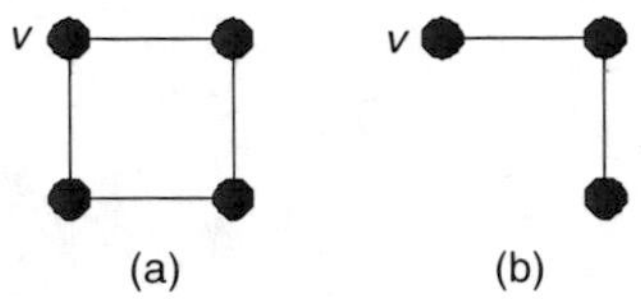

FIGURE 21.25 A cycle of length four and a Type I structure $T_1(v; 1)$ of order 1 at v.

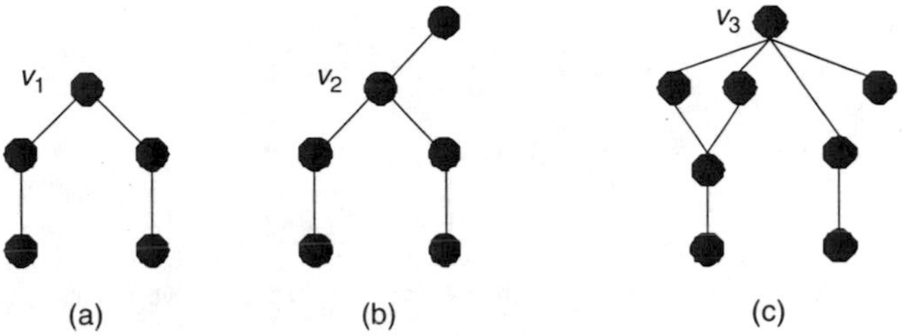

FIGURE 21.26 Some examples of local diagnosability.

and v_3, respectively. Hence it is locally 2-diagnosable, 2-diagnosable, and 3-diagnosable at vertex v_1, v_2, and v_3, respectively.

By Theorems 21.33 and 21.34, we have the following result.

THEOREM 21.35 Let $G(V, E)$ be a graph and $v \in V$ be a vertex with $deg(v) = n$. The local diagnosability of vertex v is n if G contains a subgraph, which is either a Type I structure $T_1(v; n)$ of order n or a Type II structure $T_2(v; n-2, 2)$ of order n, at vertex v.

21.8.1 Strongly Local Diagnosable Property

We use the hypercube and the star graph as two examples to introduce the concept of the strongly local diagnosable property.

In previous section, we presented two sufficient conditions for identifying the local diagnosability of a vertex. It seems that identifying the local diagnosability of a vertex is the same as counting its degree. We give an example to show that this is not true in general. As shown in Figure 21.27, we take a vertex v in the two-dimensional hypercube Q_2, let $F_1 = \{v, 1\}$ and $F_2 = \{2, 3\}$ with $|F_1| = 2$ and $|F_2| = 2$. It is a simple matter to confirm that (F_1, F_2) is an indistinguishable pair. Hence, $t_l(v) \neq deg(v) = 2$. We then propose the following two concepts.

Let $G(V, E)$ be a graph and $v \in V$ be a vertex. Vertex v has the strongly local diagnosability property if the local diagnosability of vertex v is equal to its degree.

Let $G(V, E)$ be a graph. G has the strongly local diagnosability property if, every vertex in the graph G has the strongly local diagnosability property.

According the definitions about strong local diagnosability above, we have the following theorems.

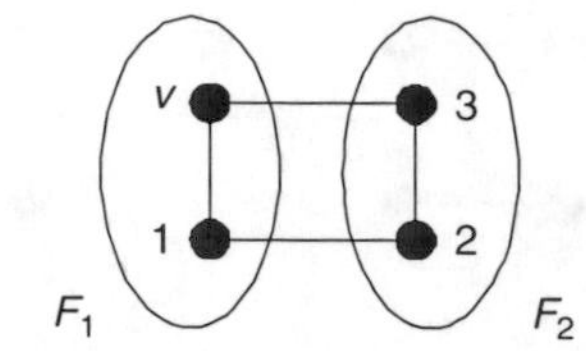

FIGURE 21.27 An indistinguishable pair (F_1, F_2) in Q_2.

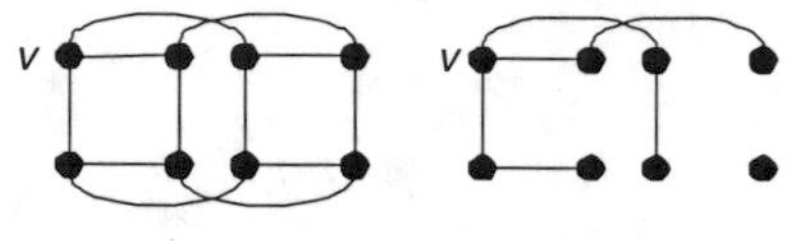

FIGURE 21.28 A Q_3 and a Type I structure $T_1(\mathbf{v}; 3)$ of order 3 at vertex $\mathbf{v}$.

THEOREM 21.36 Let Q_n be an n-dimensional hypercube, $n \geq 3$. Q_n has the strongly local diagnosability property.

Proof: We use Theorem 21.35 to prove this result, and we shall construct a Type I structure of order n at each vertex for $n \geq 3$. We prove this by induction on n. Since an n-dimensional hypercube Q_n is vertex-symmetric, we can concentrate on the construction of Type I structure at a given vertex v. For $n = 3$, $deg(\mathbf{v}) = 3$ and it is clear that Q_3 contains a Type I structure $T_1(\mathbf{v}; 3)$ of order 3 at vertex $\mathbf{v}$ (see Figure 21.28). Using the inductive hypothesis, we assume that Q_{n-1} contains a Type I structure $T_1(\mathbf{v}; n-1)$ of order $n - 1$ at each vertex, for some $n \geq 4$. Now we consider Q_n; Q_n can be decomposed into two subcubes Q^0_{n-1} and Q^1_{n-1} by some dimension. Without loss of generality, we may assume that the vertex $\mathbf{v} \in Q^0_{n-1}$. By the inductive hypothesis, Q^0_{n-1} contains a Type I structure $T_1(\mathbf{v}; n-1)$ of order $n - 1$ at vertex $\mathbf{v}$. Consider the vertex $(\mathbf{v})^n$ in Q^1_{n-1}. Vertex $(\mathbf{v})^n$ has an adjacent neighbor that is in Q^1_{n-1} due to $deg((\mathbf{v})^n) = n$, where $n \geq 3$. Thus, Q_n contains a Type I structure $T_1(\mathbf{v}; n)$ of order n at vertex $\mathbf{v}$. By Theorem 21.35, the definition of local diagnosability and the definition of strongly local diagnosability, Q_n has the strongly local diagnosability property. See Figure 21.28. □

THEOREM 21.37 Let S_n be an n-dimensional star graph, $n \geq 3$. S_n has the strongly local diagnosable property.

Proof: We shall construct a Type I structure of order $n-1$ at each vertex, for $n \geq 3$. We prove this by induction on n. Since an n-dimensional star graph S_n is vertex-symmetric, we can concentrate on an arbitrary vertex $\mathbf{v} = v_1 v_2 \ldots v_n$. For $n = 3$, $deg(\mathbf{v}) = 2$ and it is clear that S_3 contains a Type I structure $T_1(\mathbf{v}; 2)$ of order 2 at vertex $\mathbf{v}$. By the inductive hypothesis, we assume that S_{n-1} contains a Type I structure $T_1(\mathbf{v}; n-2)$ of order $n - 2$ at each vertex, for some $n \geq 4$. Now we consider S_n. By the definition of star graphs, S_n can be decomposed into n subgraphs $S_n^{\{v_1\}}$, $S_n^{\{v_2\}}$, ..., and $S_n^{\{v_n\}}$. So $\mathbf{v} \in S_n^{\{v_n\}}$. By the inductive hypothesis, $S_n^{\{v_n\}}$ contains a Type I structure $T_1(\mathbf{v}; n-2)$

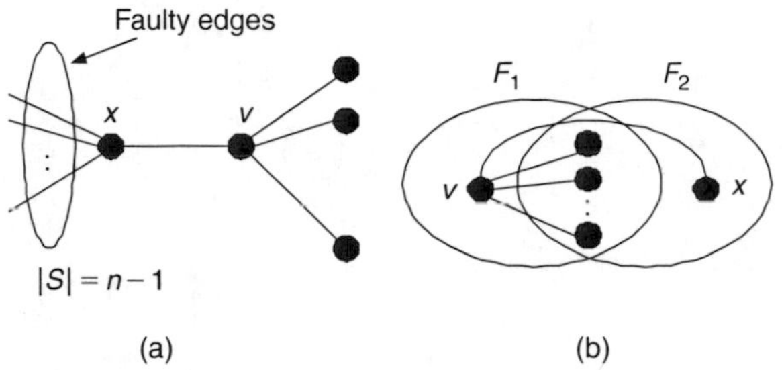

FIGURE 21.29 An indistinguishable pair (F_1, F_2), where $|F_1| = |F_2| = n$.

of order $n-2$ at vertex $\mathbf{v}$. Consider the vertex $(\mathbf{v})^n$ in $S_n^{\{v_1\}}$. Vertex $(\mathbf{v})^n$ has at least one adjacent neighbor in $S_n^{\{v_1\}}$ due to $deg((\mathbf{v})^n) = n-1$, where $n \geq 3$. Thus, S_n contains a Type I structure $T_1(\mathbf{v}; n-1)$ of order $n-1$ at vertex $\mathbf{v}$. By Theorem 21.35, the local diagnosability of each vertex in S_n is $n-1$, for $n \geq 3$. □

We now consider a system that is not vertex-symmetric. Let $G(V, E)$ be a graph and $S \subset E(G)$ be a set of edges. Removing the edges in S from G, the degree of each vertex in the resulting graph $G - S$ is called the remaining degree of v, and is denoted by $deg_{G-S}(v)$. We consider a faulty hypercube Q_n with a faulty set $S \subset E(Q_n)$, $n \geq 3$. We shall prove that Q_n has the strong local diagnosability property even if it has up to $(n-2)$ faulty edges. The number $n-2$ is optimal, in the sense that a faulty hypercube Q_n cannot be guaranteed to have this strong property if there are $n-1$ faulty edges. As shown in Figure 21.29, we take a vertex $v \in V(Q_n)$ and a vertex x, which is an adjacent neighbor of v. Let $S = \{(y, x) \in E(Q_n) \mid \text{vertex } y \text{ is directly adjacent to } x\} - \{(v, x)\}$, then $|S| = n-1$ and the remaining degree of v in $Q_n - S$ is n. Let $F_1 = (N_{Q_n-S}(v) - \{x\}) \cup \{v\}$ and $F_2 = N_{Q_n-S}(v)$, then $|F_1| = |F_2| = n$ and $v \in F_1 \Delta F_2$. It is clear that there is no edge between $V - (F_1 \cup F_2)$ and $F_1 \Delta F_2$. By Lemma 21.4, (F_1, F_2) is an indistinguishable pair; hence, $t_l(\mathbf{v}) \neq deg_{Q_n-S}(\mathbf{v}) = n$. Therefore, $Q_n - S$ may not have this strong property, if $|S| \geq n-1$.

THEOREM 21.38 Let Q_n be an n-dimensional hypercube with $n \geq 3$, and $S \subset E(Q_n)$ be a set of edges, $0 \leq |S| \leq n-2$. Removing all the edges in S from Q_n, the local diagnosability of each vertex is still equal to its remaining degree.

Proof: We use Theorem 21.35 to prove this result, and we shall construct a Type I structure at each vertex. We prove this by induction on n. For $n = 3$, $0 \leq |S| \leq 1$, if $|S| = 0$, it is clear that Q_3 contains a Type I structure $T_1(\mathbf{v}; 3)$ of order 3 at every vertex. If $|S| = 1$, a three-dimensional hypercube Q_3 with one missing edge is shown in Figure 21.30. It is a routine task to see that every vertex has a Type I structure $T_1(\mathbf{v}; k)$ of order k at it, where k is the remaining degree of the vertex. By the inductive hypothesis, we assume that the result is true for Q_{n-1}, $0 \leq |S| \leq (n-1) - 2$, for some $n \geq 4$. Now we consider Q_n, $0 \leq |S| \leq n-2$. If $|S| = 0$, we refer to the proof of Theorem 21.36, Q_n contains a Type I structure $T_1(\mathbf{v}; n)$ of order n at every vertex. If $1 \leq |S| \leq n-2$, we choose an edge in S. The edge is in some dimension that decomposed Q_n into

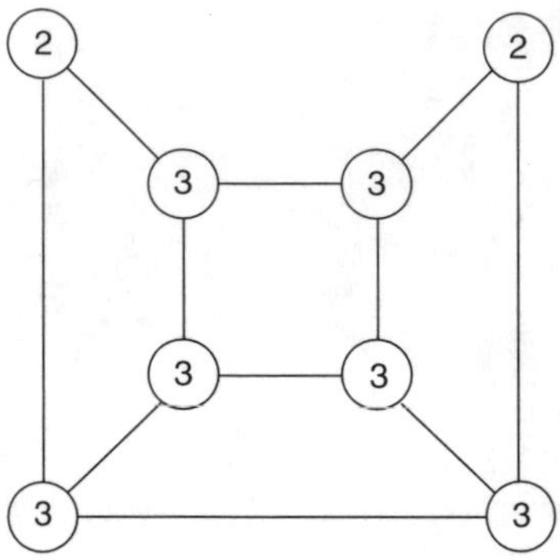

FIGURE 21.30 Q_3 with one missing edge. The number labeled on each vertex represents its local diagnosability.

two subcubes Q_{n-1}^0 and Q_{n-1}^1 by this dimension, such that the edge is a crossing edge. Let v be any vertex in $V(Q_n)$. Let $S_0 = S \cap E(Q_{n-1}^0)$, $0 \leq |S_0| \leq (n-3)$ and $S_1 = S \cap E(Q_{n-1}^1)$, $0 \leq |S_1| \leq (n-3)$. Without loss of generality, we may assume that the vertex v is in Q_{n-1}^0 and $deg_{Q_{n-1}^0 - S_0}(\mathbf{v}) = k$. By the inductive hypothesis, $Q_{n-1}^0 - S_0$ contains a Type I structure $T_1(\mathbf{v}; k)$ at v. Consider the crossing edge $(\mathbf{v}, (\mathbf{v})^n)$. If $(\mathbf{v}, (\mathbf{v})^n) \in S$, $Q_n - S$ contains a Type I structure $T_1(\mathbf{v}; k)$ of order k at vertex v. If $(\mathbf{v}, (\mathbf{v})^n) \notin S$, the remaining degree of v in $Q_n - S$ is $k+1$ and the vertex $(\mathbf{v})^n$ has at least an adjacent neighbor in Q_{n-1}^1 due to $0 \leq |S_1| \leq (n-1) - 2$. Therefore, $Q_n - S$ contains a Type I structure $T_1(v; k+1)$ of order $k+1$ at vertex $\mathbf{v}$. By Theorem 21.35, removing all the edges in S from Q_n, the local diagnosability of each vertex is still equal to its remaining degree. □

We have the following corollary.

COROLLARY 21.11 *Let Q_n be an n-dimensional hypercube with $n \geq 3$, and $S \subset E(Q_n)$ be a set of edges, $0 \leq |S| \leq n-2$. Then, $Q_n - S$ has the strong local diagnosability property.*

We now prove that S_n has the strongly local diagnosable property, even if it has up to $(n-3)$ faulty edges. The number $(n-3)$ is optimal, in the sense that a faulty S_n cannot be guaranteed to have this strong property if there are $(n-2)$ faulty edges.

THEOREM 21.39 Let S_n be an n-dimensional star graph with $n \geq 3$, and $F \subset E(S_n)$ be a set of edges, $0 \leq |F| \leq n-3$. Removing all the edges in F from S_n, the local diagnosability of each vertex is still equal to its remaining degree.

Proof: We prove this result by constructing a Type I structure at each vertex. We prove this by induction on n. For $n = 3$, $|F| = 0$, it is clear that S_3 contains a Type I structure $T_1(\mathbf{v}; 2)$ of order 2 at every vertex. By the inductive hypothesis, we assume that the result is true for S_{n-1}, $0 \leq |F| \leq (n-1) - 3$, for some $n \geq 4$. Now we consider

S_n, $0 \le |F| \le n-3$. If $|F|=0$, we refer to the proof of Theorem 21.37, S_n contains a Type I structure $T_1(\mathbf{v}; n-1)$ of order $n-1$ at every vertex. If $1 \le |F| \le n-3$, we choose an edge $e \in F$ in some dimension. The star graph can be decomposed into n subgraphs $S_n^{\{1\}}$, $S_n^{\{2\}}, \ldots,$ and $S_n^{\{n\}}$. By the symmetric property of S_n, we may assume that e is a crossing edge between $S_n^{\{1\}}$ and $S_n^{\{2\}}$. Consider a vertex $\mathbf{v} \in V(S_n)$. Let $F_i = F \cap E(S_n^{\{i\}})$, $0 \le |F_i| \le (n-4)$ for all $1 \le i \le n$. Without loss of generality, we may assume that vertex $\mathbf{v}$ is in $S_n^{\{1\}}$ and $deg_{S_n^{\{1\}}-F_1}(\mathbf{v}) = k$. By the inductive hypothesis, $S_n^{\{1\}} - F_1$ contains a Type I structure $T_1(\mathbf{v}; k)$ at $\mathbf{v}$. Consider the crossing edge $(\mathbf{v}, (\mathbf{v})^n)$. If $(\mathbf{v}, (\mathbf{v})^n) \in F$, $S_n - F$ contains a Type I structure $T_1(\mathbf{v}; k)$ of order k at vertex $\mathbf{v}$. If $(\mathbf{v}, (\mathbf{v})^n) \notin F$, the remaining degree of $\mathbf{v}$ in $S_n - F$ is $k+1$ and the vertex $(\mathbf{v})^n$ has at least one adjacent neighbor in $S_n^{\{(\mathbf{v})_1\}}$ due to $0 \le |F_{\{(\mathbf{v})_1\}}| \le (n-1)-3$. Therefore, $S_n - F$ contains a Type I structure $T_1(\mathbf{v}; k+1)$ of order $k+1$ at vertex $\mathbf{v}$. By Theorem 21.35, removing all the edges in F from S_n, the local diagnosability of each vertex is still equal to its remaining degree. □

With Theorem 21.39, we have the following corollary.

COROLLARY 21.12 *Let S_n be an n-dimensional star graph with $n \ge 3$, and $S \subset E(S_n)$ be a set of edges, $0 \le |S| \le n-3$. Then, $S_n - S$ has the strongly local diagnosable property.*

21.8.2 Conditional Fault Local Diagnosability

In the previous section, we know that Q_n does not have the strongly local diagnosability property; if there are $n-1$ faulty edges, all these faulty edges are incident to a single vertex, and this vertex is incident to only one fault-free edge. Therefore, we are led to the following question: How many edges can be removed from Q_n such that Q_n keeps the strong local diagnosability property under the condition that each vertex of the faulty hypercube Q_n is incident to at least two fault-free edges? First, we give an example to show that a faulty hypercube Q_n with $3(n-2)$ faulty edges may not have the strong local diagnosability property even if each vertex of the faulty hypercube Q_n is incident to at least two fault-free edges. As shown in Figure 21.31a, we take a cycle of length four in Q_n, $n \ge 3$. Let $\{v, a, b, c\}$ be the four consecutive vertices on this cycle, and $S \subset E(Q_n)$ be a set of edges, $S = S_1 \cup S_2 \cup S_3$, where S_1 is the set of all edges incident to $\mathbf{a}$ except $(\mathbf{v}, \mathbf{a})$ and $(\mathbf{b}, \mathbf{a})$, S_2 is the set of all edges incident to b except $(\mathbf{a}, \mathbf{b})$ and $(\mathbf{c}, \mathbf{b})$, and S_3 is the set of all edges incident to c except (v, c) and $(\mathbf{b}, \mathbf{c})$, then $|S_1| = |S_2| = |S_3| = n-2$. The remaining degree of vertex $\mathbf{v}$ in $Q_n - S$ is n, $deg_{Q_n-S}(\mathbf{v}) = n$. As shown in Figure 21.31b, let $F_1 = (N_{Q_n-S}(v) - \{\mathbf{c}\}) \cup \{\mathbf{v}\}$ and $F_2 = (N_{Q_n-S}(v) - \{\mathbf{a}\}) \cup \{\mathbf{b}\}$, then $|F_1| = |F_2| = n$ and $v \in F_1 \,\Delta\, F_2$. It is clear that there is no edge between $V(Q_n) - (F_1 \cup F_2)$ and $F_1 \,\Delta\, F_2$. By Lemma 21.4, (F_1, F_2) is an indistinguishable pair; hence, $t_l(\mathbf{v}) \ne deg_{Q_n-S}(\mathbf{v}) = n$. So some vertex of $Q_n - S$ may not have this strong property, if $|S| \ge 3(n-2)$. Then, we shall show that $Q_n - S$ has the strong local diagnosability property, if each vertex of $Q_n - S$ is incident to at least two fault-free edges and $|S| \le 3(n-2) - 1$. We need the following results to construct a Type I structure or a Type II structure at a vertex of a faulty hypercube.

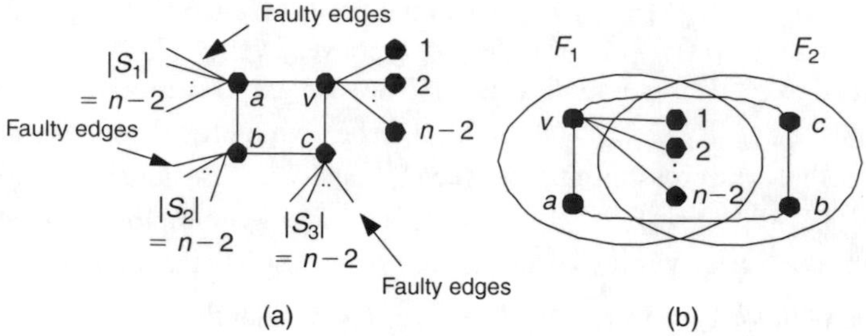

FIGURE 21.31 An indistinguishable pair (F_1, F_2), where $|F_1| = |F_2| = n$.

THEOREM 21.40 **[339]** Let $G(V, E)$ be a bipartite graph with bipartition (X, Y). Then G has a matching that saturates every vertex in X if and only if $|N(S)| \geq |S|$, for all $S \subseteq X$.

THEOREM 21.41 **[339]** Let $G(V, E)$ be a bipartite graph. The maximum size of a matching in G equals the minimum size of a vertex cover of G.

LEMMA 21.42 An n-dimensional hypercube Q_n has no cycle of length 3 and any two vertices have at most two common neighbors.

For our following discussion, we need some definitions. Let Q_n be an n-dimensional hypercube and $S \subseteq E(Q_n)$ be a set of edges. Removing the edges in S from Q_n, for a vertex v in the resulting graph $Q_n - S$, we define $BG(\mathbf{v}) = (L_1(\mathbf{v}) \cup L_2(\mathbf{v}), E)$ to be the bipartite graph under $\mathbf{v}$ with bipartition $(L_1(\mathbf{v}), L_2(\mathbf{v}))$, where $L_1(\mathbf{v}) = \{\mathbf{x} \in V(Q_n) \mid \text{vertex } \mathbf{x} \text{ is adjacent to vertex } \mathbf{v} \text{ in } Q_n - S\}$, $L_2(\mathbf{v}) = \{\mathbf{y} \in V(Q_n) \mid \text{there exists a vertex } \mathbf{x} \in L_1(\mathbf{v}) \text{ such that } (\mathbf{x}, \mathbf{y}) \in E(Q_n) \text{ in } Q_n - S\} - \{\mathbf{v}\}$, and $E(BG(\mathbf{v})) = \{(\mathbf{x}, \mathbf{y}) \in E(Q_n) \mid \text{vertex } \mathbf{x} \in L_1(\mathbf{v}) \text{ and vertex } \mathbf{y} \in L_2(\mathbf{v})\}$. $L_1(\mathbf{v})$ ($L_2(\mathbf{v})$, respectively) is called the level 1 (level 2, respectively) vertex under $\mathbf{v}$ (see Figure 21.32).

THEOREM 21.42 Let Q_n be an n-dimensional hypercube with $n \geq 3$, and $S \subset E(Q_n)$ be a set of edges, $0 \leq |S| \leq 3(n-2) - 1$. Assume that each vertex of $Q_n - S$ is incident to at least two fault-free edges. Removing all the edges in S from Q_n, the local diagnosability of each vertex is still equal to its remaining degree.

Proof: According to Theorem 21.35, we can concentrate on the construction of a Type I structure or Type II structure at each vertex. Consider a vertex $\mathbf{v}$ in $Q_n - S$ with $deg_{Q_n - S}(\mathbf{v}) = k$. As shown in Figure 21.32, let $BG(\mathbf{v}) = (L_1(\mathbf{v}) \cup L_2(\mathbf{v}), E)$ be the bipartite graph under $\mathbf{v}$. Let $M \subset E(BG(\mathbf{v}))$ be a maximum matching from $L_1(\mathbf{v})$ to $L_2(\mathbf{v})$. In the following proof, we consider three cases by the size of M: (1) $|M| = k$, (2) $|M| = k - 1$, and (3) $|M| \leq k - 2$.

Case 1. $|M| = k$. Since $|M| = k$ and $|L_1(\mathbf{v})| = k$, there exists a Type I structure $T_1(\mathbf{v}; k)$ of order k at vertex $\mathbf{v}$. By Theorem 21.35, the local diagnosability of vertex $\mathbf{v}$ is equal to k.

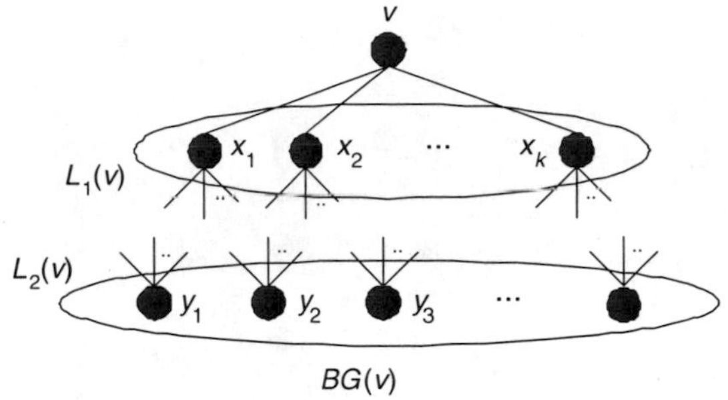

FIGURE 21.32 The bipartite graph $BG(v)$.

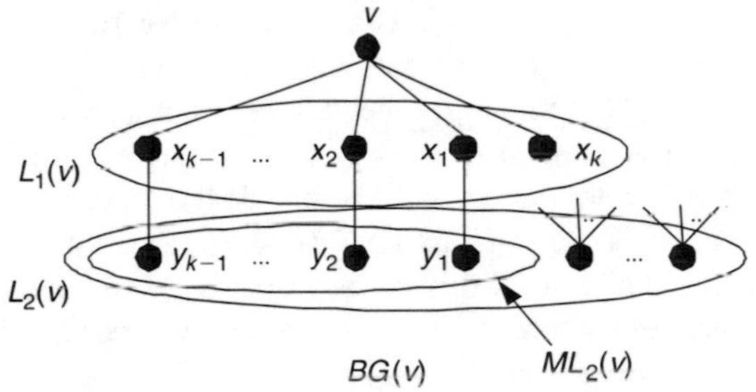

FIGURE 21.33 Illustration for the Case 2 of Theorems 21.42 and 21.44.

Case 2. $|M| = k - 1$. We shall show that there is a Type II structure of order k at vertex v. As shown in Figure 21.33, let $L_1(\mathbf{v}) = \{\mathbf{x}_1, \mathbf{x}_2, \ldots, \mathbf{x}_k\}$ and let $ML_2(\mathbf{v}) \subset L_2(\mathbf{v})$ be the set of vertices matched under M, $ML_2(\mathbf{v}) = \{\mathbf{y} \in L_2(\mathbf{v}) \mid$ there exists a vertex $\mathbf{x} \in L_1(\mathbf{v})$ such that $(\mathbf{x}, \mathbf{y}) \in M\}$. So $|ML_2(\mathbf{v})| = k - 1$. Let $ML_2(\mathbf{v}) = \{\mathbf{y}_1, \mathbf{y}_2, \ldots, \mathbf{y}_{k-1}\}$ and assume that vertex $\mathbf{x}_i$ is matched with vertex $\mathbf{y}_i$ for each i, $1 \le i \le k - 1$. Then there exists a vertex $\mathbf{x}_k \in L_1(\mathbf{v})$, $\mathbf{x}_k$ is unmatched by M. Since each vertex of $Q_n - S$ is incident to at least two fault-free edges, there exists a vertex $\mathbf{y}_i \in ML_2(\mathbf{v})$, $i \in \{1, 2, \ldots, k - 1\}$, such that $(\mathbf{x}_k, \mathbf{y}_i) \in E(BG(\mathbf{v}))$. Without loss of generality, let $(\mathbf{x}_k, \mathbf{y}_1) \in E(BG(\mathbf{v}))$. If the remaining degree of $\mathbf{y}_1$ is at least three, as shown in Figure 21.34, there exists a Type II structure $T_2(\mathbf{v}; k - 2, 2)$ of order k at vertex $\mathbf{v}$. By Theorem 21.35, the local diagnosability of vertex $\mathbf{v}$ is equal to k and the result follows. If the remaining degree of $\mathbf{y}_1$ is two, the number of faulty edges incident to $\mathbf{y}_1$ is $n - 2$. Next, we divide the case into two subcases: (1) both $\mathbf{x}_k$ and $\mathbf{x}_1$ have remaining degree two and (2) one of $\mathbf{x}_k$ and $\mathbf{x}_1$ has remaining degree at least three and the other has at least two.

Case 2.1. Both $\mathbf{x}_k$ and $\mathbf{x}_1$ have remaining degree 2. This is an impossible case. Since the number of faulty edges incident to $\mathbf{x}_k$ and $\mathbf{x}_1$ is $2(n - 2)$, the total number of faulty edges is at least $3(n - 2)$ which is greater than $3(n - 2) - 1$, which is a contradiction.

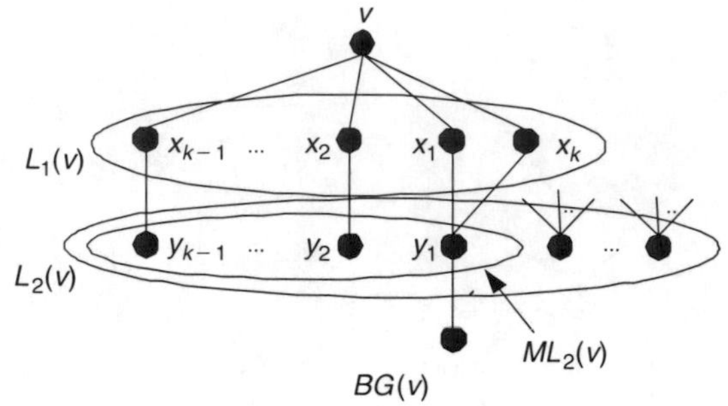

FIGURE 21.34 A Type II structure $T_2(v; k-2, 2)$ of order k at vertex v.

Case 2.2. One of $\mathbf{x}_k$ and $\mathbf{x}_1$ has remaining degree at least three and the other has at least two. Without loss of generality, assume that $\mathbf{x}_k$ has remaining degree at least three and $\mathbf{x}_1$ has remaining degree at least two. Since $deg_{Q_n-S}(\mathbf{x}_k) \geq 3$, there exist at least two vertices in $ML_2(\mathbf{v})$ that are the neighbors of vertex $\mathbf{x}_k$. Then, we can find a vertex $\mathbf{y}_i \in ML_2(\mathbf{v})$ and $\mathbf{y}_i \neq \mathbf{y}_1$, $i \in \{2, 3, \ldots, k-1\}$, such that $(\mathbf{x}_k, \mathbf{y}_i) \in E(BG(\mathbf{v}))$. Without loss of generality, let $(\mathbf{x}_k, \mathbf{y}_2) \in E(BG(\mathbf{v}))$. If the remaining degree of $\mathbf{y}_2$ is at least three, there exists a Type II structure $T_2(\mathbf{v}; k-2, 2)$ of order k at vertex $\mathbf{v}$. By Theorem 21.35, the local diagnosability of vertex $\mathbf{v}$ is equal to k and the result follows. If the remaining degree of $\mathbf{y}_2$ is two, the number of faulty edges incident to $\mathbf{y}_2$ is $n-2$. We then consider two further cases.

Case 2.2.1. $\mathbf{x}_1$ has remaining degree 2. This is an impossible case. Since the number of faulty edges incident to $\mathbf{x}_1$ is $n-2$, the total number of faulty edges is at least $3(n-2)$, which is greater than $3(n-2)-1$. A contradiction holds.

Case 2.2.2. $\mathbf{x}_1$ has remaining degree at least three. Since $deg_{Q_n-S}(\mathbf{x}_1) \geq 3$, there exist at least two vertices in $ML_2(\mathbf{v})$ that are the neighbors of vertex $\mathbf{x}_1$. By Lemma 21.42, any two vertices of Q_n have at most two common neighbors. We can find a vertex $\mathbf{y}_i \in ML_2(v)$, $\mathbf{y}_i \neq \mathbf{y}_1$ and $\mathbf{y}_i \neq \mathbf{y}_2$, $i \in \{3, 4, \ldots, k-1\}$, such that $(\mathbf{x}_1, \mathbf{y}_i) \in E(BG(\mathbf{v}))$. Without loss of generality, let $(\mathbf{x}_1, \mathbf{y}_3) \in E(BG(\mathbf{v}))$. If the remaining degree of $\mathbf{y}_3$ is at least three, there exists a Type II structure $T_2(\mathbf{v}; k-2, 2)$ of order k at vertex $\mathbf{v}$. By Theorem 21.35, the local diagnosability of vertex $\mathbf{v}$ is equal to k and the result follows. If the remaining degree of $\mathbf{y}_3$ is two, then the number of faulty edges incident to $\mathbf{y}_3$ is $n-2$, and the total number of faulty edges is at least $3(n-2)$, which is greater than $3(n-2)-1$, a contradiction.

Case 3. $|M| \leq k-2$. We shall see that this is an impossible case. By Theorem 21.41, the minimum size of a vertex cover of the bipartite graph $BG(v)$ is no greater than $k-2$. We take a vertex cover with the minimum size, and let $VCL_1(\mathbf{v}) \subset L_1(\mathbf{v})$, $VCL_2(\mathbf{v}) \subset L_2(\mathbf{v})$, and $VCL_1(\mathbf{v}) \cup VCL_2(\mathbf{v})$ be the vertex cover, as shown in Figure 21.35. $VCL_1(\mathbf{v})$ and $VCL_2(\mathbf{v})$ can cover all the edges of $BG(\mathbf{v})$. Let $NVCL_1(\mathbf{v}) = L_1(\mathbf{v}) - VCL_1(\mathbf{v})$. We claim that the total number of faulty edges is at least $(n-1)|NVCL_1(\mathbf{v})| - 2|VCL_2(\mathbf{v})|$, and this number is greater than $3(n-2)$, which is a contradiction. With this claim, the case is impossible.

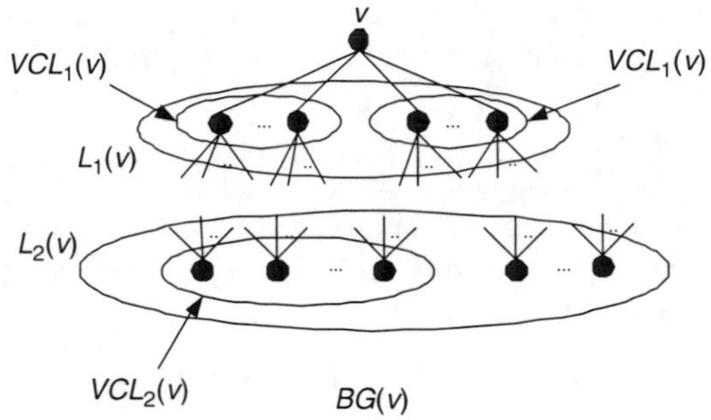

FIGURE 21.35 Illustration for Case 3 of Theorems 21.42 and 21.44.

Now we prove the claim. First, for each vertex $\mathbf{x} \in NVCL_1(\mathbf{v})$, the edges connecting $\mathbf{x}$ except $(\mathbf{x}, \mathbf{v})$ must be incident to the vertices in $VCL_2(\mathbf{v})$. For each vertex $\mathbf{y} \in VCL_2(\mathbf{v})$, by Lemma 21.42, at most two edges connecting $\mathbf{y}$ are incident to the vertices in $NVCL_1(\mathbf{v})$. Then, the total number of faulty edges is at least $(n-1)|NVCL_1(\mathbf{v})| - 2|VCL_2(\mathbf{v})|$. Since $VCL_1(\mathbf{v}) \cup VCL_2(\mathbf{v})$ is a minimum vertex cover, $|VCL_1(\mathbf{v})| + |VCL_2(\mathbf{v})| \leq k-2$. Since $|L_1(\mathbf{v})| = k$ and each vertex of $Q_n - S$ is incident to at least two fault-free edges, there exists a vertex in $L_1(\mathbf{v}) - VCL_1(\mathbf{v})$ such that the vertex has at least one neighbor in $VCL_2(\mathbf{v})$. Thus, $|VCL_2(\mathbf{v})| \geq 1$. Now, we show that the number $(n-1)|NVCL_1(\mathbf{v})| - 2|VCL_2(\mathbf{v})|$ is greater than $3(n-2)$. With $|VCL_1(\mathbf{v})| + |VCL_2(\mathbf{v})| \leq k-2$ and $|VCL_2(\mathbf{v})| \geq 1$, we have the following:

$$\begin{aligned}
&[(n-1)|NVCL_1(\mathbf{v})| - 2|VCL_2(\mathbf{v})|] - [3(n-2)] \\
&\quad = [(n-1)(k - |VCL_1(\mathbf{v})|) - 2|VCL_2(\mathbf{v})|] - [3(n-2)] \\
&\quad \geq [(n-1)(|VCL_2(\mathbf{v})| + 2) - 2|VCL_2(\mathbf{v})|] - [3(n-2)] \\
&\quad = (|VCL_2(\mathbf{v})| - 1)(n-3) + 1 \\
&\quad > 0, \quad \text{for all } n \geq 3
\end{aligned}$$

□

Thus, our claim holds.

In summary, aside from those impossible cases, we show that $Q_n - S$ contains either a Type I structure $T_1(\mathbf{v}; k)$ or a Type II structure $T_2(\mathbf{v}; k-2, 2)$ of order k at vertex $\mathbf{v}$. By Theorem 21.35, removing all the edges in S from Q_n, the local diagnosability of each vertex is still equal to its remaining degree.

By Theorem 21.42, we have the following corollary.

COROLLARY 21.13 *Let Q_n be an n-dimensional hypercube with $n \geq 3$, and $S \subset E(Q_n)$ be a set of edges, $0 \leq |S| \leq 3(n-2) - 1$. $Q_n - S$ has the strongly local diagnosability property, provided that each vertex of $Q_n - S$ is incident with at least two fault-free edges.*

Next, we shall show that S_n keeps the strongly local diagnosable property, no matter how many edges are faulty, provided that each vertex of $S_n - F$ is incident to at least two fault-free edges.

THEOREM 21.43 Let S_n be an n-dimensional star graph with $n \geq 3$, and $F \subset E(S_n)$ be a set of edges. Assume that each vertex of $S_n - F$ is incident with at least two fault-free edges. Removing all the edges in F from S_n, the local diagnosability of each vertex is still equal to its remaining degree.

Proof: According to Theorem 21.35, we can concentrate on the construction of a Type I structure at each vertex. Consider a vertex $\mathbf{v}$ in $S_n - F$ with $deg_{S_n - F}(\mathbf{v}) = k$. Let $N_{S_n - F}(v) = \{\mathbf{x}_1, \mathbf{x}_2, \ldots, \mathbf{x}_k\}$ be the neighborhood of $\mathbf{v}$. Let $L_2(\mathbf{v}) = \{\mathbf{y} \in V(S_n) \mid$ there exists a vertex $\mathbf{x} \in N_{S_n - F}(\mathbf{v})$ such that $(\mathbf{x}, \mathbf{y}) \in E(S_n)\} - \{\mathbf{v}\}$. Since each vertex of $S_n - F$ is incident with at least two fault-free edges and S_n has no cycle of length less than six, the maximum size of a matching from $N_{S_n - F}(\mathbf{v})$ to $L_2(\mathbf{v})$ is equal to k. As a result, there must exists a Type I structure $T_1(\mathbf{v}; k)$ of order k at vertex $\mathbf{v}$. By Corollary 21.31, removing all the edges in F from S_n, the local diagnosability of each vertex is still equal to its remaining degree. □

By Theorem 21.43, the following corollary holds.

COROLLARY 21.14 *Let S_n be an n-dimensional star graph with $n \geq 3$, and $F \subset E(S_n)$ be a set of edges. S_n keeps the strongly local diagnosable property no matter how many edges are faulty, provided that each vertex of $S_n - F$ is incident with at least two fault-free edges.*

Finally, we consider another condition: each vertex of a faulty hypercube Q_n is incident to at least three fault-free edges. Based on this condition, we prove that Q_n keeps the strong local diagnosability property, no matter how many edges are faulty.

THEOREM 21.44 Let Q_n be an n-dimensional hypercube with $n \geq 3$, and $S \subset E(Q_n)$ be a set of edges. Assume that each vertex of $Q_n - S$ is incident with at least three fault-free edges. Removing all the edges in S from Q_n, the local diagnosability of each vertex is still equal to its remaining degree.

Proof: According to Theorem 21.35, we can concentrate on the construction of a Type I structure or Type II structure at each vertex. Consider a vertex $\mathbf{v}$ in $Q_n - S$ with $deg_{Q_n - S}(\mathbf{v}) = k$. Let $BG(\mathbf{v}) = (L_1(\mathbf{v}) \cup L_2(\mathbf{v}), E)$ be the bipartite graph under $\mathbf{v}$. Then, $|L_1(\mathbf{v})| = k$. Let $M \subset E(BG(\mathbf{v}))$ be a maximum matching from $L_1(\mathbf{v})$ to $L_2(\mathbf{v})$. In the following proof, we consider three cases by the size of M: (1) $|M| = k$, (2) $|M| = k - 1$, and (3) $|M| \leq k - 2$.

Case 1. $|M| = k$. Since $|M| = k$ and $|L_1(\mathbf{v})| = k$, there exists a Type I structure $T_1(\mathbf{v}; k)$ of order k at vertex $\mathbf{v}$. By Theorem 21.35, the local diagnosability of vertex $\mathbf{v}$ is equal to k.

Case 2. $|M| = k - 1$. We will show that there is a Type II structure of order k at vertex v. As shown in Figure 21.33, let $L_1(\mathbf{v}) = \{\mathbf{x}_1, \mathbf{x}_2, \ldots, \mathbf{x}_k\}$ and let $ML_2(\mathbf{v}) \subset L_2(\mathbf{v})$

be the set of vertices matched under M, $ML_2(\mathbf{v}) = \{\mathbf{y} \in L_2(\mathbf{v}) \mid$ there exists a vertex $\mathbf{x} \in L_1(\mathbf{v})$ such that $(\mathbf{x}, \mathbf{y}) \in M\}$. So $|ML_2(\mathbf{v})| = k - 1$. Let $ML_2(\mathbf{v}) = \{\mathbf{y}_1, \mathbf{y}_2, \ldots, \mathbf{y}_{k-1}\}$ and assume vertex $\mathbf{x}_i$ is matched with vertex $\mathbf{y}_i$ for each i, $1 \le i \le k-1$. Then there exists a vertex $\mathbf{x}_k \in L_1(\mathbf{v})$, $\mathbf{x}_k$ is unmatched by M. Since each vertex of $Q_n - S$ is incident with at least three fault-free edges, there exists a vertex $\mathbf{y}_i \in ML_2(\mathbf{v})$, $i, \in \{1, 2, \ldots, k-1\}$, such that $(\mathbf{x}_k, \mathbf{y}_i) \in E(BG(\mathbf{v}))$. Without loss of generality, let $(\mathbf{x}_k, \mathbf{y}_1) \in E(BG(\mathbf{v}))$. Since the remaining degree of $\mathbf{y}_1$ is at least three, as shown in Figure 21.34, there exists a Type II structure $T_2(\mathbf{v}; k-2, 2)$ of order k at vertex $\mathbf{v}$. By Theorem 21.35, the local diagnosability of vertex $\mathbf{v}$ is equal to k, and the result follows.

Case 3. $|M| \le k-2$. We will see that this is an impossible case. By Theorem 21.41, the minimum size of a vertex cover of the bipartite graph $BG(\mathbf{v})$ is no greater than $k-2$. However, we claim that any $k-2$ vertices of $BG(\mathbf{v})$ cannot cover all the edges of $BG(\mathbf{v})$. With this claim, the case is impossible.

Now we prove this claim. Suppose to the contrary, that we take a vertex cover with the minimum size, and let $VCL_1(\mathbf{v}) \subset L_1(\mathbf{v})$, $VCL_2(\mathbf{v}) \subset L_2(\mathbf{v})$, and $VCL_1(\mathbf{v}) \cup VCL_2(\mathbf{v})$ be the vertex cover, as shown in Figure 21.35. $VCL_1(\mathbf{v})$ and $VCL_2(\mathbf{v})$ can cover all the edges of $BG(\mathbf{v})$. Since $|VCL_1(\mathbf{v})| + |VCL_2(\mathbf{v})| \le k-2$, we rewrite this inequality into the following equivalent form: $2(k - |VCL_1(\mathbf{v})|) \ge 2(|VCL_2(\mathbf{v})| + 2)$. Let $NVCL_1(\mathbf{v}) = L_1(\mathbf{v}) - VCL_1(\mathbf{v})$. Since each vertex of $Q_n - S$ is incident with at least three fault-free edges, for each vertex $\mathbf{x} \in NVCL_1(\mathbf{v})$, aside from the edge $(\mathbf{x}, \mathbf{v})$, at least two edges connecting $\mathbf{x}$ must be incident to the vertices in $VCL_2(\mathbf{v})$. So the total number of edges incident to the vertices in $VCL_2(\mathbf{v})$ is at least $2|NVCL_1(\mathbf{v})|$. For each vertex $\mathbf{y} \in VCL_2(\mathbf{v})$, by Lemma 21.42, at most two edges connecting $\mathbf{y}$ are incident to the vertices in $NVCL_1(\mathbf{v})$. So the total number of edges incident to the vertices in $NVCL_1(\mathbf{v})$ is at most $2|VCL_2(\mathbf{v})|$. Compare the lower bound $2|NVCL_1(\mathbf{v})|$ and the upper bound $2|VCL_2(\mathbf{v})|$. We have the following inequality:

$$2|NVCL_1(\mathbf{v})| = 2(k - |VCL_1(\mathbf{v})|) \ge 2(|VCL_2(\mathbf{v})| + 2) > 2|VCL_2(\mathbf{v})|.$$

The lower bound $2|NVCL_1(\mathbf{v})|$ is greater than the upper bound $2|VCL_2(\mathbf{v})|$. It means that some edges are not covered by $VCL_1(\mathbf{v})$ or $VCL_2(\mathbf{v})$ in $BG(\mathbf{v})$. Thus, our claim follows.

In Case 1, $Q_n - S$ contains a Type I structure $T_1(\mathbf{v}; k)$ of order k at vertex $\mathbf{v}$. In Case 2, $Q_n - S$ contains a Type II structure $T_2(\mathbf{v}; k-2, 2)$ of order k at vertex $\mathbf{v}$. We also prove that Case 3 is impossible. By Theorem 21.35, removing all the edges in S from Q_n, the local diagnosability of each vertex is still equal to its remaining degree. □

By Theorem 21.44, the following corollary holds.

COROLLARY 21.15 *Let Q_n be an n-dimensional hypercube with $n \ge 3$, and $S \subset E(Q_n)$ be a set of edges. Q_n keeps the strongly local diagnosability property no matter how many edges are faulty provided that each vertex of $Q_n - S$ is incident to at least three fault-free edges.*

References

1. Abraham, S. and Padmanabhan, K., The twisted cube topology for multiprocessors: a study in network asymmetry, *Journal of Parallel and Distributed Computing*, 13, 104, 1991.
2. Akers, S.B., Harel, D., and Krishnamurthy, B., The star graph: an attractive alternative to the n-cube, *Proceedings of the International Conference on Parallel Processing*, 1987, 393.
3. Akers, S. and Krishnameurthy, B., A group-theoretic model for symmetric interconnection networks, *IEEE Transactions Computers*, 38, 555, 1989.
4. Akers, S. and Krishnameurthy, B., The star graph: an attractive alternative to n-cube, *Proceedings of the International Conference on Parallel Processing*, 1987, 393.
5. Alavi, Y. and Williamson, J.E., Panconnected graphs, *Studia Scientiarum Mathematicarum Hungarica*, 10, 1, 1975.
6. Al-Ayyouba, A.E. and Day, K., Comparative study of product networks, *Journal of Parallel and Distributed Computing*, 62, 1, 2002.
7. Albert, M., Aldread, R.E.L., Holton, D., and Sheehan, J., On 3*-connected graphs, *Australia Journal of Combinatorics*, 24, 193, 2001.
8. Alon, N., Transmitting in the n-dimensional cube, *Discrete Applied Mathematics*, 37/38, 9, 1992.
9. Alspach, B., The classification of Hamiltonian generalized Petersen graphs, *Journal of Combinatorial Theory, Series B*, 34, 293, 1983.
10. Alspach, B. and Hare, D., Edge-pancyclic block-intersection graphs, *Discrete Mathematics*, 97, 17, 1991.
11. Alspach, B., Robinson, P.J., and Rosenfeld, M., A result on Hamiltonian cycles in generalized Petersen graphs, *Journal of Combinatorial Theory, Series B*, 31, 225, 1981.
12. Appel, K. and Haken, W., Every planar map is four colorable, *Bulletin of the American Mathematical Society*, 82, 711, 1976.
13. Appel, K. and Haken, W., Every planar map is four colorable, Part I: discharging, *Illinois Journal of Mathematics*, 21, 429, 1977.
14. Appel, K. and Haken, W., The four color proof suffices, *The Mathematical Intelligencer*, 8, 10, 1986.
15. Araki, T., Hyper Hamiltonian laceability of cayley graphs generated by transpositions, *Networks*, 48, 121, 2006.
16. Araki, T. and Shibata, Y., Diagnosability of networks by the Cartesian product, *IEICE Transactions Fundamentals*, E38-A, 465, 2000.
17. Bagherzadeh, N., Dowd, M., and Nassif, N., Embedding an arbitrary tree into the star graph, *IEEE Transactions on Computers*, 45, 475, 1996.
18. Bannai, K., Hamiltonian cycles in generalized Petersen graphs, *Journal of Combinatorial Theory, Series B*, 24, 181, 1978.
19. Barsi, F., Grandoni, F., and Maestrini, P., A theory of diagnosability without repairs, *IEEE Transactions on Computers*, C-25, 585, 1976.
20. Barth, D. and Raspaud, A., Two edge-disjoint Hamiltonian cycles in the butterfly graph, *Information Processing Letters*, 51, 175, 1994.

21. Berge, C., Sur le couplage maximum d'un graphe, *C.R. Acad. Sci. Paris*, 247, 258, 1958.
22. Berge, C., Two theorems in graph theory, *Proceedings of the National Academy of Sciences U.S.A.*, 1957, 842.
23. Bermond, J.C., Comellas, F., and Hsu, D.F., Distributed loop computer networks: a survey, *Journal of Parallel and Distributed Computing*, 24, 2, 1995.
24. Bermond, J.C., Darrot, E., Delmas, O., and Perennes, S., Hamilton cycle decomposition of the butterfly networks, *Parallel Processing Letters*, 8, 371, 1998.
25. Bermond, J.C., Delorme, C., and Quisquater, J.J., Strategies for interconnection networks, some methods from graph theory, *Journal of Parallel and Distributed Computing*, 3, 433, 1986.
26. Birkhoff, G.D., A determinant formula for the number of ways of coloring a map, *The Annals of Mathematics*, 14, 42, 1912.
27. Boesch, F.T. and Bogdanowicz, Z.R., The number of spanning trees in a prism, *International Journal of Computer Mathematics*, 21, 229, 1987.
28. Bondy, J.A., Pancyclic graphs, I, *Journal of Combinatorial Theory, Series B*, 11, 80, 1971.
29. Bondy, J.A., Variations on the Hamiltonian theme, *Canadian Mathematical Bulletin*, 15, 57, 1972.
30. Bondy, J.A. and Chvátal, V., A method in graph theory, *Discrete Mathematics*, 15, 111, 1976.
31. Booth, K.S. and Lueker, G.S., Testing for the consecutive ones property, interval graphs, and graph planarity using *PQ*-tree algorithms, *Journal of Computer and System Sciences*, 13, 335, 1976.
32. Bouabdallah, A., Heydemann, M.C., Opatrny, J., and Sotteau, D., Embedding complete binary trees into star and pancake graphs, *Theory of Computing Systems*, 31, 279, 1998.
33. Brooks, R.L., On coloring the nodes of a network, *Proceedings of the Cambridge Philosophical Society*, 1941, 194.
34. Bruck, J., Cypher, R., and Ho, C.-T., Wildcard dimensions, coding theory and fault-tolerant meshes and hypercubes, *IEEE Transactions on Computers*, 44, 150, 1995.
35. Carlson, B.S., Chen, C.-Y., and Meliketian, D.S., Daul Eulerian properties of plane multigraphs, *SIAM Journal on Discrete Mathematics*, 8, 33, 1995.
36. Catlin, P.A., Hajós' graph-coloring conjecture: variations and counterexamples, *Journal of Combinatorial Theory, Series B*, 26, 604, 1979.
37. Cayley, A., A theorem on trees, *The Quarterly Journal of Mathematics*, 23, 276, 1889.
38. Chan, M.-Y. and Lee, S.-J., On the existence of Hamiltonian circuits in faulty hypercubes, *SIAM Journal on Discrete Mathematics*, 4, 511, 1991.
39. Chang, C.-H., Chuang, Y.-C., and Hsu, L.-H., Additive multiplicative increasing functions on nonnegative square matrices and multidigraphs, *Discrete Mathematics*, 240, 21, 2001.
40. Chang, S.-P., Juan, J. S.-T., Lin, C.-K., Tan, J.J.M., and Hsu, L.-H., Mutually independent Hamiltonian connectivity of (n, k)-star graphs, to appear.
41. Chang, J.H. and Kim, J., Ring embedding in faulty (n, k)-star graphs, *Proceedings of the Eighth International Conference on Parallel and Distributed Systems*, 2001, 99.
42. Chang, C.-P., Lai, P.-L., Tan, J.J.M., and Hsu, L.-H., The diagnosability of t-connected networks and product networks under the comparison diagnosis model, *IEEE Transactions on Computers*, 53, 1582, 2004.
43. Chang, C.-H., Lin, C.-K., Huang, H.-M., and Hsu, L.-H., The super laceability of the hypercubes, *Information Processing Letters*, 92, 15, 2004.
44. Chang, C.-H., Lin, C.-K., Huang, H.-M., Tan, J.J.M., and Hsu, L.-H., The super spanning connectivity and super spanning laceability of the enhanced hypercubes, submitted.

45. Chang, C.-H., Sun, C.-M., Huang, H.-M., and Hsu, L.-H., On the equitable k^*-laceability of hypercubes, *Journal of Combinatorial Optimization*, 14, 349, 2007.
46. Chang, C.-P., Sung, T.-Y., and Hsu, L.-H., Edge congestion and topological properties of crossed cubes, *IEEE Transaction on Parallel and Distributive Systems*, 11, 64, 2000.
47. Chang, C.-P., Wang, J.-N., and Hsu, L.-H., Topological properties of twisted cubes, *Information Sciences*, 113, 147, 1999.
48. Chang, J.-M., Yang, J.-S., Wang, Y.-L., and Cheng, Y., Panconnectivity, fault-tolerant Hamiltonicity and Hamiltonian connectivity in alternating group graphs, *Networks*, 44, 302, 2004.
49. Chen, C.-Y., Jean, E.-Y., and Hsu, L.-H., A class of additive multiplicative graph functions, *Discrete Mathematics*, 65, 53, 1987.
50. Chen, G. and Lau, F.C.M., Comments on a new family of Cayley graph interconnection networks of constant degree four, *IEEE Transactions on Parallel and Distributed Systems*, 8, 1299, 1997.
51. Chen, M.-S. and Shin, K.-G., Processor allocation in an N-cube multiprocessor using gray codes, *IEEE Transaction on Computer*, C-36, 1396, 1987.
52. Chen, Y.-C., Tan, J.J.M., Hsu, L.-H., and Kao, S.-S., Super-connectivity and super-edge-connectivity for some interconnection networks, *Applied Mathematics and Computation*, 140, 245, 2003.
53. Chen, Y.-C., Tsai, C.-H., Hsu, L.-H., and Tan, J.J.M., A recursively construction scheme for super fault-tolerant Hamiltonian graphs, *Applied Mathematics and Computation*, 177, 465, 2006.
54. Chen, Y.-C., Tsai, C.-H., Hsu, L.-H., and Tan, J.J.M., On some super fault-tolerant Hamiltonian graphs, *Applied Mathematics and Computation*, 148, 729, 2004.
55. Cheng, C.-J., Maximizing the total number of spanning trees in a graph: two related problems in graph theory and optimal design theory, *Journal of Combinatorial Theory, Series B*, 31, 240, 1981.
56. Cheng, E. and Lipták, L., Fault resiliency of Cayley graphs generated by transpositions, *International Journal of Foundations of Computer Science*, 18, 1005, 2007.
57. Cheng, E. and Lipták, L., Linearly many faults in Cayley graphs generated by transposition trees, *Information Sciences*, 177, 4877, 2007.
58. Cheng, E. and Lipták, L., Structural properties of Cayley graphs generated by transposition trees, *Congressus Numerantium*, 180, 81, 2006.
59. Cheng, E., Lipman, M.J., and Park, H., Super connectivity of star graphs, alternating group graphs and split-satrs, *ARS Combinatoria*, 59, 107, 2001.
60. Chiang, W.-K. and Chen, R.-J., The (n,k)-star graph: a generalized star graph, *Information Processing Letters*, 56, 259, 1995.
61. Chiue, W.-S. and Shieh, B.-S., On connectivity of the cartesian product of two graphs, *Applied Mathematics and Computation*, 102, 129, 1999.
62. Cho, H.-J. and Hsu, L.-Y., Generalized honeycomb torus, *Information Processing Letters*, 86, 185, 2003.
63. Cho, H.-J. and Hsu, L.-Y., Ring embedding in faulty honeycomb rectangular torus, *Information Processing Letters*, 84, 277, 2002.
64. Chou, R.-S. and Hsu, L.-H., 1-Edge fault-tolerant design for meshes, *Parallel Processing Letters*, 4, 385, 1994.
65. Choudum, S.A. and Nahdini, U., Complete binary trees in folded and enhanced cubes, *Networks*, 43, 266, 2004.
66. Choudum, S.A. and Sunitha, V., Augmented cubes, *Networks*, 40, 71, 2002.
67. Chuang, Y.-C., Chang, C.-H., and Hsu, L.-H., Optimal 1-edge fault-tolerant designs for ladders, *Information Processing Letters*, 84, 87, 2000.

68. Chung, F.R.K., Diameters of graphs: old problems and new results, *Proceedings of the 18th Southeastern Conference Combinatorics, Graph Theory, and Computing, Congressus Numerantium* 1987, 298.
69. Chvátal, V., Flip-flops in hypohamiltonian graphs, *Canadian Mathematical Bulletin*, 16, 33, 1973.
70. Chvátal, V., On Hamilton's ideal, *Journal of Combinatorial Theory, Series B*, 12, 163, 1972.
71. Chvátal, V. and Erdös, P., A note on Hamiltonian circuits, *Discrete Mathematics*, 2, 111, 1972.
72. Chwa, K.Y. and Hakimi, S.L., On fault identification in diagnosable systems, *IEEE Transactions on Computers*, C-30, 414, 1981.
73. Cohen, D.S. and Blum, M., On the problem of sorting burnt pancakes, *Discrete Applied Mathematics*, 61, 105, 1995.
74. Colbourn, C.J., *The Combinatorics of Network Reliability*, Oxford University Press, 1987.
75. Collier, J.B. and Schmeichel, E.F., New flip-flop constructions for hypohamiltonian graphs, *Discrete Mathematics*, 18, 193, 1977.
76. Cull, P. and Larson, S.M., The Möbius cubes, *IEEE Transactions Computers*, 44, 647, 1995.
77. Dahbura, A.T. and Masson, G.M., An $O(n^{2.5})$ faulty identification algorithm for diagnosable systems, *IEEE Transactions on Computers*, 33, 486, 1984.
78. Dandamudi, S.P. and Eager, D.L., Hierarchical interconnection networks for multicomputer systems, *IEEE Transactions on Computers*, 39, 786, 1990.
79. Das, K. and Banerjee, A., Hyper Petersen network: yet another hypercube-like topology, *Proceedings of the Second Symposium on the Frontiers of Massively Parallel Computation*, 1992, 270.
80. Day, K. and Al-Ayyoub, A.-E., Minimal fault diameter for highly resilient product networks, *IEEE Transactions on Parallel and Distributed Systems*, 11, 926, 2000.
81. Day, K. and Al-Ayyoub, A.-E., The cross product of interconnection networks, *IEEE Transactions on Parallel and Distributed Systems*, 8, 109, 1991.
82. Day, K. and Al-Ayyoub, A.E., Topological properties of OTIS-networks, *IEEE Transactions on Parallel and Distributed Systems*, 13, 359, 2002.
83. Day, K. and Tripathi, A., Arrangement graphs: a class of generalized star graphs, *Information Processing Letters*, 42, 235, 1992.
84. Day, K. and Tripathi, A., Embedding of cycles in arrangement graphs, *IEEE Transactions on Computers*, 12, 1002, 1993.
85. de Bruijn, N.G., A combinatorial problem, *Proceedings of the Koninklijke Nederlandsche Akademie van Wetenschappe*, 1946, 758.
86. Dirac, G.A., A property of 4-chromatic graphs and some remarks on critical graphs, *Journal of the London Mathematical Society*, 27, 85, 1952.
87. Dirac, G.A., In abstrakten graphen vorhandene vollständige 4-graphen und ihre unterteilungen, *Mathematische Nachrichten*, 22, 61, 1960.
88. Dirac, G.A., Some theorems on abstract graphs, *Proceedings of the London Mathematical Society*, 1952, 69.
89. Dirac, G.A., The structure of k-chromatic graphs, *Fundamenta Mathematicae*, 40, 42, 1953.
90. Dudeney, H.E., *Amusements in Mathematics*, Nelson, 1917.
91. Edmonds, J. and Johnson, E., Matching, Euler tours, and Chinese postman, *Mathematical Programming*, 5, 147, 1973.

92. Efe, K., A variation on the hypercube with lower diameter, *IEEE Transactions Computers*, 40, 1312, 1991.
93. Efe, K., The crossed cube architecture for parallel computing, *IEEE Transactions Parallel and Distruibuted Systems*, 3, 513, 1992.
94. Efe, K., Blackwell, P.K., Sloubh, W., and Shiau, T., Topological properties of the crossed cube architecture, *Parallel Computing*, 20, 1763, 1994.
95. Egerváry, E., On combinatorial properties of matrices, *Matematikai Lapok*, 38, 16, 1931.
96. El-Amawy, A. and Latifi, S., Properties and performance of folded hypercubes, *IEEE Transactions Parallel and Distributed Systems*, 2, 31, 1991.
97. El-Ghazawi, T. and Youssef, A., A generalized framework for developing adaptive fault-tolerant routing algorithms, *IEEE Transactions on Reliability*, 42, 250, 1993.
98. Erdös, P. and Gallai, T., On maximal paths and circuits of graphs, *Acta Mathematica Hungarica*, 10, 337, 1959.
99. Euler, L., Demonstratio nonullarum insignium proprietatum quibus solida hedris planis inclusa sunt praedita, *Novi Comm. Acad. Sci. Imp. Petropol*, 4, 140, 1758.
100. Euler, L., Solutio problematis ad geometriam situs pertinentis, *Commetarii Academiae Scientiarum Imperialis Petropolitanae*, 8, 128, 1736.
101. Fan, J., Diagnosability of crossed cubes under the comparison diagnosis model, *IEEE Transactions on Parallel and Distributed Systems*, 13, 687, 2002.
102. Fan, J., Diagnosability of crossed cubes under the two strategies, *Chinese Journal of Computers*, 21, 456, 1998.
103. Fan, J., Diagnosability of the Möbius cubes, *IEEE Transactions on Parallel and Distributed Systems*, 9, 923, 1998.
104. Fan, J., Jia, X., and Lin, X., Complete path embeddings in crossed cubes, *Information Science*, 176, 3332, 2006.
105. Fan, J., Jia, X., and Lin, X., Optimal embeddings of paths with various lengths in twisted cubes, *IEEE Transactions on Parallel and Distributed Systems*, 18, 511, 2007.
106. Fan, J. X., Lin, X. L., and Jia, X. H., Optimal path embedding in crossed cubes, *IEEE Transactions on Parallel and Distributed Systems*, 16, 1190, 2005.
107. Fáry, I., On the straight line representations of planar graphs, *Acta Scientiarum Mathematicarum*, 11, 229, 1948.
108. Fernandes, R., Friesen, D.K., and Kanevsky, A., Embedding rings in recursive networks, *Proceeding of the 6th Symposium Parallel and Distributed Processing*, 1994, 273.
109. Fiduccia, C.M. and Hedrick, P.J., Edge congestion of shortest path systems for all-to-all communication, *IEEE Transactions Parallel and Distributed Systems*, 8, 1043, 1997.
110. Ford, L.R. and Fulkerson, D.R., *Flows in Networks*, Princeton, Princeton University Press, 1962.
111. Fragopoulou, P. and Akl, S.G., Edge-disjoint spanning trees on the star networks with applications to fault tolerance, *IEEE Transactions on Computers*, 45, 174, 1996.
112. Fragopoulou, P. and Akl, S.G., Optimal communication algorithms on the star graphs using spanning tree constructions, *Journal of Parallel and Distributed Computing*, 24, 55, 1995.
113. Friedman, A.D., A new measure of digital system diagnosis, *Proceedings of the 5th International Symposium Fault-Tolerant Computing*, 1975, 167.
114. Friedman, A.D. and Simoncini, L., System level fault diagnosis, *Computer Magazine*, 13, 47, 1980.
115. Fu, J.-S., Conditional fault-tolerant Hamiltonicity of star graphs, *Parallel Computing*, 33, 488, 2007.
116. Fu, J.-S., Hamiltonicity of the *WK*-recursive network with and without faulty nodes, *IEEE Transaction on Parallel and Distributed Systems*, 16, 853, 2005.

117. Fu, J.-S., Longest fault-free paths in hypercubes with vertex faults, *Information Sciences*, 176, 756, 2006.
118. Fu, A.-W. and Chau, S.-C., Cyclic-cubes: a new family of interconnection networks of even fixed-degrees, *IEEE Transactions on Parallel and Distributed Systems*, 9, 1253, 1998.
119. Gallai, T., On directed paths and circuits, *Theory of Graphs, Proceedings of the Colloquium Tihnay*, 1968, 115.
120. Galli, T., Über extreme punkt-und kantenmengen, *Annales Univ. Sci. Budapest*, 20, 133, 1959.
121. Gates, W.H. and Papadimitriou, C.H., Bounds for sorting by prefix reversal, *Discrete Mathematics*, 27, 47, 1979.
122. Georges, J.P., Non-Hamiltonian bicubic graphs, *Journal of Combinatorial Theory, Series B*, 46, 121, 1989.
123. Germa, A., Heydemann, M.C., and Sotteau, D., Cycles in cube-connected cycles graph, *Discrete Applied Mathematics*, 83, 135, 1998.
124. Ghouila-Houri, A., Une condition suffisante d'existence d'un circuit Hamiltonien, *C.R. Adac. Sci. Paris*, 156, 161, 1960.
125. Ghose, K. and Desai, K.R., Hierarchical cubic networks, *IEEE Transaction on Parallel and Distributed Systems*, 6, 427, 1995.
126. Gibbons, A., *Algorithmic Graph Theory*, Cambridge, Cambridge University Press, 1985.
127. Good, I.J., Normal recurring decimals, *Journal London Mathematical Society*, 21, 167, 1946.
128. Graham, R., Rothschild, B., and Spencer, J.H., *Ramsey Theory*, New York, John Wiley and Sons, 1980.
129. Grinberg, E.J., Plane homogeneous graphs of degree three without Hamiltonian circuits, *Latvian Mathematical Yearbook*, 5, 51, 1968.
130. Gu, Q.P. and Peng, S., Node-to-node cluster fault tolerant routing in star graphs, *Information Processing Letters*, 56, 29, 1995.
131. Guan, M., Graphic programming using odd and even points, *Chinese Mathematics*, 1, 273, 1962.
132. Gupta, A.-K. and Hambrusch, S.-E., Embedding complete binary trees into butterfly networks, *IEEE Transactions on Computers*, 40, 853, 1991.
133. Hadwiger, H., Über eine Klassifikation der Strekenkomplexe, Zürich, *Vierteljschr. Naturforch. Ges.*, 88, 133, 1943.
134. Hajós, G., Über eine Konstruktion nicht *n*-fäbbarer Graphen, *Wiss. Z. Martinluther-Univ. Halle-Wittenberg Math. -Nat. Reihe*, 10, 116, 1961.
135. Hakimi, S.L., On the realizability of a set of integers as degrees of the vertices of a graph, *SIAM Journal on Applied Mathematics*, 10, 496, 1962.
136. Hakimi, S.L. and Amin, A.T., Characterization of connection assignment of diagnosable systems, *IEEE Transactions on Computers*, 23, 86, 1974.
137. Hall, P., On representation of subsets, *Journal of the London Mathematical Society*, 10, 26, 1935.
138. Harary, F., The maximum connectivity of a graph, *Proceedings of the National Academy of Sciences U.S.A.*, 1962, 1141.
139. Harary, F. and Hayes, J.P., Edge fault tolerance in graphs, *Networks*, 23, 135, 1993.
140. Harary, F. and Hayes, J.P., Node fault tolerance in graphs, *Networks*, 27, 19, 1996.
141. Havel, V., A remark on the existence of finite graphs, *Casopis Pro Pestován í Matematiky*, 80, 477, 1955.
142. Heawood, P.J., Map-colour theorem, *Quarterly Journal of Mathematics*, 24, 332, 1890.

143. Hedetniemi, S.T., *Homomorphisms of graphs and automata*, doctoral dissertation, University of Michigan, Ann Arbor, 1966.
144. Heydari, M.H. and Sudborough, I.H., On the diameter of the pancake networks, *Journal of Algorithms*, 25, 67, 1997.
145. Hierholzer, C. Über die Möglichkeit, einen Linienzug ohne Wiederholung und ohne Unterbrechnung zu umfahren, *Mathematische Annalen*, 6, 30, 1873.
146. Hilbers, P.A.J., Koopman, M.R.J., and van de Snepscheut, J.L.A., The twisted cube, *Parallel Architectures and Languages Europe, Lecture Notes in Computer Science*, 1, 152, 1987.
147. Ho, T.-Y. and Hsu, L.-H., A note on the minimum cut cover of graphs, *Operations Research Letters*, 15, 193, 1994.
148. Ho, T.-Y. and Lin, C.-K., Fault-tolerant Hamiltonian connectivity and fault-tolerant Hamiltonicity of the fully connected cubic networks, submitted.
149. Ho, T.-Y., Hsu, L.-H., and Sung, T.-Y., Transmitting on various network topologies, *Networks*, 27, 145, 1996.
150. Ho, T.-Y., Hung, C.-N., and Hsu, L.-H., On cubic 2-independent Hamiltonian connected graphs, *Journal of Combinatorial Optimization*, 14, 275, 2007.
151. Ho, T.-Y., Lin, C.-K., Tan, J.J.M., and Hsu, L.-H., On the mutually orthogonal Hamiltonian connected graphs, submitted.
152. Ho, T.-Y., Lin, C.-K., Tan, J.J.M., and Hsu, L.-H., The super spanning connectivity of augmented cube, to appear.
153. Ho, T.-Y., Sung, T.-Y., and Hsu, L.-H., A note on the edge fault-tolerance with respect to hypercubes, *Applied Mathematics Letters*, 18, 1125, 2005.
154. Ho, T.-Y., Sung, T.-Y., Hsu, L.-H., Tsai, C.-H., and Hwang, J.-Y. The recognition of double Euler trail for series-parallel networks, *Journal of Algorithms*, 28, 216, 1998.
155. Hobbs, A., The square of a block is vertex pancyclic, *Journal of Combinatorial Theory, Series B*, 20, 1, 1976.
156. Holten, D.A. and Sheehan, J., *The Petersen graph*, Cambridge, Cambridge University Press, 1993.
157. Hopcroft, J. and Tarjan, R.E., Efficient planarity testing, *Journal of the Association for Computing Machinery*, 21, 549, 1974.
158. Horton, J.D., On two-factors of bipartite regular graphs, *Discrete Mathematics*, 41, 35, 1982.
159. Hsieh, S.-Y., Embedding longest fault-free paths onto star graphs with more vertex faults, *Theoretical Computer Science*, 337, 370, 2005.
160. Hsieh, S.-Y., Chen, G.-H., and Ho, C.-W., Hamiltonian-laceability of star graphs, *Networks*, 36, 225, 2000.
161. Hsieh, S.-Y., Chen, G.-H., and Ho, C.-W., Longest fault-free paths in star graphs with edge faults, *IEEE Transactions on Computers*, 50, 960, 2001.
162. Hsieh, S.-Y. Chen, G.-H., and Ho, C.-W., Longest fault-free paths in star graphs with vertex faults, *Theoretical Computer Science*, 262, 215, 2001.
163. Hsieh, S.-Y. and Yu, P.-Y., Fault-free mutually independent Hamiltonian cycle in hypercubes with faulty edges, *Journal of Combinatorial Optimization*, 13, 153, 2007.
164. Hsu, D.F., On container width and length in graphs, groups and networks, *IEICE Transactions on Fundamentals of Electronics, Communications and Computer Sciences*, E77-A, 668, 1994.
165. Hsu, L.-H., A classification of graph capacity functions, *Journal of Graph Theory*, 21, 251, 1996.
166. Hsu, L.-H., A note on the ultimate categorical matching in a graph, *Discrete Mathematics*, 256, 487, 2002.

167. Hsu, L.-H. Generalized homomorphism graph functions, *Discrete Mathematics*, 84, 31, 1990.
168. Hsu, L.-H., *Monotone multiplicative graph functions and dependence of tree copy functions*, doctoral dissertation, State University of New York at Stony Brook, 1981.
169. Hsu, L.-H., On a multiplicative graph function conjecture, *Discrete Mathematics*, 45, 245, 1983.
170. Hsu, L.-H., On a strongly multiplicative graph function conjecture, *Chinese Journal of Mathematics*, 13, 103, 1985.
171. Hsu, H.-C., Chiang, L.-C., Tan, J.J.M., and Hsu, L.-H., Fault Hamiltonicity of augmented cubes, *Parallel Computing*, 31, 131, 2005.
172. Hsu, H.-C., Hsieh, Y.-L., Tan, J.J.M., and Hsu, L.-H., Fault Hamiltonicity and fault Hamiltonian connectivity of the (n, k)-star graphs, *Networks*, 42, 189, 2003.
173. Hsu, H.-C., Li, T.-K., Tan, J.J.M., and Hsu, L.-H., Fault Hamiltonicity and fault Hamiltonian connectivity of the arrangement graphs, *IEEE Transactions on Computers*, 53, 39, 2004.
174. Hsu, L.-H., Liu, S.-C., and Yeh, Y.-N., Hamiltonicity of hypercubes with constraint of required and faulty edges, *Journal of Combinatorial Optimization*, 14, 197, 2007.
175. Hsu, G.-H. and Tan, J.J.M., A local diagnosability measure for multiprocessor systems, *IEEE Transactions on Parallel and Distributed Systems*, 18, 598, 2007.
176. Hu, K.-S., Yeoh, S.-S., Hsu, L.-H., and Chen, C.Y., Node-pancyclicity and edge-pancyclicity of hypercube variants, *Information Processing Letters*, 102, 1, 2007.
177. Huang, H.-L. and Chen, G.-H., Combinatorial properties of two-level hypernet networks, *IEEE Transactions on Parallel and Distributed Systems*, 10, 1192, 1999.
178. Huang, W.-T., Chuang, Y.-C., Hsu, L.-H., and Tan, J.J.M., On the fault-tolerant Hamiltonicity of crossed cubes, *IEICE Transaction on Fundamentals of Electronics, Communications and Computer Sciences*, E85-A, 1359, 2002.
179. Huang, W.-T., Chuang, Y.-C., Tan, J.J.M., and Hsu, L.-H., Fault-free Hamiltonian cycle in faulty Möbius cubes, *Computación y Sistemas*, 4, 106, 2000.
180. Huang, W.-T., Tan, J.J.M., Hung, C.-N., and Hsu, L.-H., Fault-tolerant Hamiltonicity of twisted cubes, *Journal of Parallel and Distributed Computing*, 62, 591, 2002.
181. Hung, C.-N., Hsu, H.-C., Liang, K.-Y., and Hsu, L.-H., Ring embedding in faulty pancake graphs, *Information Processing Letters*, 86, 271, 2003.
182. Hung, C.-N., Hsu, L.-H., and Sung, T.-Y., Christmas tree: a versatile 1-fault-tolerant design for token rings, *Information Processing Letters*, 72, 55, 1999.
183. Hung, C.-N., Hsu, L.-H., and Sung, T.-Y., On the construction of combined k-fault-tolerant Hamiltonian graphs, *Networks*, 37, 165, 2001.
184. Hung, C.-N., Wang, J.-J., Sung, T.-Y., and Hsu, L.-H., On the isomorphism between cyclic-cubes and wrapped butterfly networks, *IEEE Transaction on Parallel and Distributive Systems*, 11, 864, 2000.
185. Hwang, S.-C. and Chen, G.-H., Cycles in butterfly graphs, *Networks*, 35, 161, 2000.
186. INMOS Limited, *Transputer Reference Manual*, Prentice-Hall, 1988.
187. Jordan, C., Sur les assemblages de lignes, *Journal für die Reine und Angewandte Mathematik*, 70, 185, 1896.
188. Jwo, J.S., Lakshmivarahan, S., and Dhall, S.K., Embedding of cycles and grids in star graphs, *Journal of Circuits, Systems, and Computers*, 1, 43, 1991.
189. Kao, S.-S. and Hsu, L.-H., Brother trees: a family of optimal 1_p-Hamiltonian and 1-edge Hamiltonian graphs, *Information Processing Letters*, 86, 263, 2003.
190. Kao, S.-S. and Hsu, L.-H., Spider web networks: a family of optimal, fault tolerant, Hamiltonian bipartite graphs, *Applied Mathematics and Computation*, 160, 269, 2005.

191. Kao, S.-S., and Hsu, L.-H., The globally bi-3* and the hyper bi-3* connectedness of the spider web networks, *Applied Mathematics and Computation*, 170, 597, 2005.
192. Kao, S.-S., Hsu, K.-M., and Hsu, L.-H., Cubic Hamiltonian graphs of various types, *Discrete Mathematics*, 306, 1364, 2006.
193. Kao, S.-S., Hsu, H.-C., and Hsu, L.-H., Globally bi-3*-connected graphs, submitted.
194. Kao, S.-S., Huang, H.-M., Hsu, K.-M., and Hsu, L.-H., Cubic 1-fault-tolerant Hamiltonian graphs, globally 3*-connected graphs, and super 3-spanning connected graphs, submitted.
195. Kautz, W.H., Bounds on directed (d, k) graphs, *Theory of Cellular Logic Networks and Machine*, AFCRL-68-0668, 20, 1968.
196. Kavianpour, A. and Kim, K.H., Diagnosability of hypercube under the pessimistic one-step diagnosis strategy, *IEEE Transactions on Computers*, 40, 232, 1991.
197. Kel'mans, A.K., Comparison of graphs by their number of trees, *Discrete Mathematics*, 16, 241, 1976.
198. Kel'mans, A.K., Connectivity of probabilistic networks, *Automation and Remote Control*, 16, 197, 1974.
199. Kel'mans, A.K. and Chelenkov, V.M., A certain ploynomial of a graph and graphs with an extremal numbers of trees, *Journal of Combinatorial Theory, Series B*, 16, 197, 1974.
200. Kempe, A.B., On the geographical problem of four colours, *American Journal of Mathematics*, 2, 193, 1879.
201. Kirchoff, G., Über die auflösung der gleichungen, auf welche man bei der untersuchung der linearen verteilung galvanischer ströme geführt wird, *Ann. Phys. Chem.*, 72, 497, 1847.
202. Kobeissi, M. and Mollard, M., Disjoint cycles, and spanning graphs of hypercubes, *Discrete Mathematics*, 288, 73, 2004.
203. König, D., Graphen and matrizen, *Matematikai Lapok*, 38, 116, 1931.
204. König, D., Über grappen und ihre anwendung auf determinatentheorie und mengenlehre, *Mathematische Annalen*, 38, 453, 1916.
205. Kounoike, Y., Kaneko, K., and Shinano, Y., Computing the diameters of 14- and 15-pancake graphs, *Proceedings of the 8th International Symposium on Parallel Architectures, Algorithms, and Networks*, 2005, 490.
206. Krishnamoorthy, M.S. and Krishnamurthy, B., Fault diameter of interconnection networks, *Computers and Mathematics with Applications*, 13, 577, 1987.
207. Kueng, T.-L., Liang, T., and Hsu, L.-H., Mutually independent Hamiltonian cycles of the binary wrap-around butterfly networks, submitted.
208. Kueng, T.-L., Liang, T., and Hsu, L.-H., Super spanning properties of the binary wrapped butterfly networks, submitted.
209. Kueng, T.-L., Lin, C.-K., Liang, T., Tan, J.J.M., and Hsu, L.-H., A note on fault-free mutually independent Hamiltonian cycles in hypercubes with faulty edges, forthcoming.
210. Kueng, T.-L., Lin, C.-K., Liang, T., Tan, J.J.M., and Hsu, L.-H., Fault-tolerant Hamiltonian connectedness of cycle composition networks, forthcoming.
211. Kueng, T.-L., Lin, C.-K., Liang, T., Tan, J.J.M., and Hsu, L.-H., Fault-free mutually independent Hamiltonian cycles of faulty star networks, submitted.
212. Kulasinghe, P., Connectivity of the crossed cube, *Information Processing Letters*, 61, 221, 1997.
213. Kuratowski, K., Sur le probléme des courbes gauches en topologie, *Fundamenta Mathematicae*, 15, 271, 1930.
214. Lai, P.-L., Tan, J.J.M., Chang, C.-P., and Hsu, L.-H., Conditional diagnosability measures for large multiprocessor systems, *IEEE Transactions on Computers*, 54, 165, 2007.

215. Lai, P.-L., Tan, J.J.M., Tsai, C.-H., and Hsu, L.-H., The diagnosability of matching composition network under the comparison diagnosis model, *IEEE Transactions on Computers*, 53, 1064, 2004.
216. Latifi, S., Combinatorial analysis of the fault-diameter of the n-cube, *IEEE Transactions on Computers*, 42, 27, 1993.
217. Latifi, S., On the fault-diameter of the star graph, *Information Processing Letters*, 46, 143, 1993.
218. Lee, C.-M., Tan, J.J.M., and Hsu, L.-H., Embedding Hamiltonian paths in hypercubes with a required vertex in a fixed position, submitted.
219. Lee, C.-M., Teng, Y.-H., Tan, J.J.M., and Hsu, L.-H., Embedding Hamiltonian paths in augmented cubes with a required vertex in a fixed position, submitted.
220. Leighton, F.T., *Introduction to Parallel Algorithms and Architecture: Arrays·Trees·Hypercubes*, San Mateo, Morgan Kaufmann, 1992.
221. Lewinter, M. and Widulski, W., Hyper-Hamilton laceable and caterpillar-spannable product graphs, *Computers and Mathematics with Applications*, 34, 99, 1997.
222. Li, T.-K., Tan, J.J.M., and Hsu, L.-H., Hyper Hamiltonian laceability on the edge fault star graph, *Information Sciences*, 165, 59, 2004.
223. Li, T.-K., Tan, J.J.M., Hsu, L.-H., and Sung, T.-Y., Optimum congested routing strategy on twisted cubes, *Journal of Interconnection Networks*, 1, 115, 2000.
224. Li, T.-K., Tan, J.J.M., Hsu, L.-H., and Sung, T.-Y., The shuffle-cubes and their generalization, *Information Processing Letters*, 77, 35, 2001.
225. Li, T.-K., Tsai, C.-H., Tan, J.J.M., and Hsu, L.-H., Bipanconnected and edge-fault-tolerant bipancyclic of hypercubes, *Information Processing Letters*, 87, 107, 2003.
226. Li, T.-K., Tsai, C.-H., Tan, J.J.M., Hsu, L.-H., and Hung, C.-N., Cells for optimal 1-Hamiltonian regular graphs, submitted.
227. Lin, C.-K., Chang, C.-P., Ho, T.-Y., Tan, J.J.M., and Hsu, L.-H., A new isomorphic definition of the crossed cube and its spanning connectivity, submitted.
228. Lin, C.-K., Ho, T.-Y., Tan, J.J.M., and Hsu, L.-H., Fault-tolerant Hamiltonicity and fault-tolerant Hamiltonian connectivity of the folded Petersen cube networks, submitted.
229. Lin, C. K., Hsu, H.-C., Huang, H.-M., Tan, J.J.M., and Hsu, L.-H., Fault tolerance spanning connectivity of pancake networks and fault tolerance spanning laceability of star networks, submitted.
230. Lin, C.-K., Huang, H.-M., and Hsu, L.-H., On the spanning connectivity of graphs, *Discrete Mathematics*, 307, 285, 2007.
231. Lin, C.-K., Huang, H.-M., Hsu, L.-H., and Bau, S., Mutually independent Hamiltonian paths in star networks, *Networks*, 46, 110, 2005.
232. Lin, C.-K., Huang, H.-M., Tan, J.J.M., and Hsu, L.-H., On spanning connected graphs, forthcoming.
233. Lin, C.-K., Huang, H.-M., and Hsu, L.-H., The super connectivity of the pancake graphs and the super laceability of the star graphs, *Theoretical Computer Science*, 339, 257, 2005.
234. Lin, C.-K., Huang, H.-M., Hsu, D.F., and Hsu, L.-H., The spanning diameter of the star graphs, *Proceedings of the 2004 International Symposium on Parallel Architectures, Algorithms and Networks (ISPAN'04), IEEE Computer Society Press*, 551, 2004.
235. Lin, C.-K., Huang, H.-M., Hsu, D.F., and Hsu, L.-H., The spanning diameter of the star graph, *Networks*, 48, 235, 2006.
236. Lin, C.-K., Shih, Y.-K., Tan, J.J.M., and Hsu, L.-H., Mutually independent Hamitonian cycles in some graphs, forthcoming.

237. Lin, C.-K., Tan, J.J.M., Hsu, L.-H., Cheng, E., and Lipták, L., Conditional fault Hamiltoncity and conditional Hamiltonian laceability of the star graph, submitted.
238. Lin, C.-K., Tan, J.J.M., Huang, H.-M., and Hsu, L.-H., Mutually independent Hamiltonian cycles for the pancake graphs and the star graphs, submitted.
239. Lin, C.-K., Tan, J.J.M., Hsu, D.F., and Hsu, L.H., On the spanning connectivity and spanning laccability of hypercube-like networks, *Theoretical Computer Science*, 381, 218, 2007.
240. Lin, C.-K., Tan, J.J.M., Hsu, D.F., and Hsu, L.-H., On the spanning fan-connectivity of graphs, submitted.
241. Loulou, R., Minimal cut cover of a graph with an application to the testing of electronic boards, *Operations Research Letters*, 12, 301, 1992.
242. Lovász, L., Three short proofs in graph theory, *Journal of Combinatorial Theory, Series B*, 19, 269, 1975.
243. Ma, M., Liu, G., and Xu, J.-M., Panconnectivity and edge-fault-tolerant pancyclicity of augmented cubes, *Parallel Computing*, 33, 35, 2007.
244. Ma, M. and Xu, J.-M., Panconnectivity of locally twisted cubes, *Applied Mathematics Letters*, 19, 673, 2006.
245. Maeng, J. and Malek, M., A comparison connection assignment for self-diagnosis of multiprocessors systems, *Proceedings of the 11th International Symposium Fault-Tolerant Computering*, 1981, 173.
246. Mai, T.-C., Wang, J.-J., and Hsu, L.-H., Hyper Hamiltonian generalized Petersen graphs, submitted.
247. Malek, M., A comparison connection assignment for diagnosis of multiprocessors systems, *Proceedings of the 7th annual symposium on Computer Architecture*, 1980, 31.
248. Mallela, S. and Masson, G.M., Diagnosable system for intermittent faults, *IEEE Transactions on Computers*, C-27, 461, 1978.
249. Maziasz, R. and Hayes, J., Layout optimization of static CMOS functional cells, *IEEE Transactions on Computer-Aided Design of Integrated Circuits and Systems*, 9, 708, 1990.
250. Mchugh, J.A., *Algotithmic Graph Theorey*, London, Prentice Hall, 1990.
251. Megson, G.M., Yang, X., and Liu, X., Honeycomb tori are Hamiltonian, *Information Processing Letters*, 72, 99, 1999.
252. Menger, K., Zur allgemeinen Kurventheorie, *Fundamenta Mathematicae*, 10, 95, 1927.
253. Miller, Z., Pritikin, D., and Sudborough, I.H., Near embeddings of hypercubes into Cayley graphs on the symmetric group, *IEEE Transactions on Computers*, 43, 13, 1994.
254. Mitchem, J. and Schmeichel, E., Pancyclic and bipancyclic graphs—a survey, *Graphs and Applications*, 271, 1982.
255. Moon, J.W., On a problem of Ore, *Mathematical Gazette*, 49, 40, 1965.
256. Mukhopadhyaya, K. and Sinha, B.P., Hamiltonian graphs with minimum number of edges for fault-tolerant topologies, *Information Processing Letters*, 44, 95, 1992.
257. Mycielski, J., Sur le coloriage des graphes, *Colloquim Mathematiques*, 3, 161, 1955.
258. Najjar, W. and Gaudiot, J.L., Network resilience: a measure of network fault tolerance, *IEEE Transactions on Computers*, 39, 174, 1990.
259. Nash-Williams, C.St.J.A., On orientations, connectivity and odd-vertex pairings in finite graphs, *Canadian Journal of Mathematics*, 12, 555, 1960.
260. Öhring, S. and Das, S., Folded Petersen cube networks: new competitors for the hypercubes, *IEEE Transactions on Parallel and Distributed Systems*, 7, 151, 1996.
261. Ore, O., Coverings of graphs, *Annali di Matematica Pura ed Applicata*, 55, 315, 1961.
262. Ore, O., Note on Hamilton circuit, *American Mathematical Monthly*, 67, 625, 1960.

263. Paoli, M., Wong, W.-W., and Wong, C.-K., Minimum k-Hamiltonian graphs II, *Journal of Graph Theory*, 10, 79, 1986.
264. Parhami, B., Swapped interconnection networks: topological, performance, and robustness attributes, *Journal of Parallel and Distributed Computing*, 65, 1443, 2005.
265. Parhami, B., The Hamiltonicity of swapped (OTIS) networks built of Hamiltonian component networks, *Information Processing Letters*, 95, 441, 2005.
266. Parhami, B. and Kwai, D.-M., A unified formulation of honeycomb and diamond networks, *IEEE Transaction on Parallel and Distributive Systems*, 12, 74, 2001.
267. Park, J.H., One-to-many disjoint path covers in a graph with faulty elements, *Proceedings of the 10th Annual International Conference*, 2004, 392.
268. Park, C.D. and Chwa, K.Y., Hamiltonian properties on the class of hypercube-like networks, *Information Processing Letters*, 91, 11, 2004.
269. Park, J.H. and Chwa, K.Y., Recursive circulant: a new topology for multicomputer networks, *Proceedings of the International Symposium Parallel Architectures, Algorithms and Networks*, 1994, 73.
270. Park, J.H. and Kim, H.C., Longest paths and cycles in faulty star graphs, *Journal of Parallel and Distributed Computing*, 64, 1286, 2004.
271. Park, J.H., Kim, H.C., and Lim, H.S., Many-to-many disjoint path covers in hypercube-like interconnection networks with faulty elements, *IEEE Transactions on Parallel and Distributed Systems*, 17, 227, 2006.
272. Petersen, J., Die Theorie der reguären graphen, *Acta Math.*, 15, 193, 1891.
273. Preparata, F.P., Metze, G., and Chien, R.T., On the connection assignment problem of diagnosis systems, *IEEE Transactions on Electronic Computers*, 16, 848, 1967.
274. Prüfer, H., Beweis eines statzs über permutationen, *Arch. Math. Phys.*, 27, 742, 1918.
275. Robbins, H.E., A theorem on the graphs, with an application to the problem in traffic control, *American Mathematical Monthly*, 46, 281, 1839.
276. Roberts, F.S., *Graph Theory and its Applications to the Problems of Society*, CBMS-NSF Monograph 29, SIAM Publication, 1978.
277. Robertson, G.N., *Graphs, under girth, valency, and connectivity constraints*, doctoral thesis, University of Waterloo, 1968.
278. Roy, B., Nombre chromatique et plus longs chemins d'un graphe, *Rev. Française Automat. Informat. Recherche Opérationelle sér Rounge*, 1, 127, 1967.
279. Saad, Y. and Shultz, M.H., Topological properties of hypercube, *IEEE Transactions on Computers*, 37, 867, 1988.
280. Saxena, P.C., Gupta, S., and Rai, J., A delay optimal coterie on the k-dimensional folded Petersen graph, *Journal of Parallel and Distributed Computing*, 63, 1026, 2003.
281. Sengupta, A., On ring embedding in hypercubes with faulty nodes and links, *Information Processing Letters*, 68, 207, 1998.
282. Sengupta, A. and Dahbura, A., On self-diagnosable multiprocessor systems: diagnosis by the comparison approach, *IEEE Transactions on Computers*, 41, 1386, 1992.
283. Shannon, C.E., A theorem on coloring the lines of a network, *Journal of Mathematical Physics*, 28, 148, 1949.
284. Shih, C.-J. and Batcher, K.E., Adding multiple-fault tolerance to generalized hypercube networks, *IEEE Transaction on Parallel and Distributive Systems*, 5, 785, 1994.
285. Simmons, G.J., Almost all n-dimensional rectangular lattices are Hamilton-laceable, *Proceedings of the Ninth Southeastern Conference on Combinatorics, Graph Theory, and Computing*, 1978, 649.
286. Stanley, R.P., Acyclic orientations of graphs, *Discrete Mathematics*, 5, 171, 1973.
287. Stein, S.K., Convex maps, *Proceedings of the American Mathematical Society*, 1951, 464.

288. Stojmenovic, I., Honeycomb networks: topological properties and communication algorithms, *IEEE Transaction on Parallel and Distributive Systems*, 8, 1036, 1997.
289. Sun, C.-M., Hung, C.-N., Huang, H.-M., Hsu, L.-H., and Jou, Y.-D., Hamiltonian laceability of faulty hypercube, *Journal of Interconnection Networks*, 8, 133, 2007.
290. Sun, C.-M., Lin, C.-K., Huang, H.-M., and Hsu, L.-H., Mutually independent Hamiltonian paths and cycles in hypercubes, *Journal of Interconnection Networks*, 7, 235, 2006.
291. Sung, T.-Y., Ho, T.-Y., and Hsu, L.-H., Optimal k-fault-tolerant networks for token rings, *Journal of Information Science and Engineering*, 16, 381, 2000.
292. Sung, T.-Y., Lin, C.-Y., Chuang, Y.-C., and Hsu, L.-H., Fault-tolerant token ring embedding in double loop networks, *Information Processing Letters*, 66, 201, 1998.
293. Sung, T.-Y., Lin, M.-Y., and Ho, T.-Y., Multiple-edge-fault tolerance with respect to hypercubes, *IEEE Transaction on Parallel and Distributive Systems*, 8, 187, 1997.
294. Szekeres, G. and Wilf, H.S., An inequality for chromatic number of a graph, *Journal of Combinatorial Theory*, 4, 1, 1968.
295. Tait, P.G., On the colouring of maps, *Proceedings of the Royal Society of Edinburgh*, 1880, 501.
296. Tchuente, M., Generation of permutations by graphical exchanges, *Ars Combinatoria*, 14, 115, 1982.
297. Teng, Y.-H., Tan, J.J.M., Ho, T.-Y., and Hsu, L.-H., On mutually independent Hamiltonian paths, *Applied Mathematics Letters*, 19, 345, 2006.
298. Teng, Y.-H., Tan, J.J.M., and Hsu, L.-H., Honeycomb rectangular disks, *Parallel Computing*, 31, 371, 2005.
299. Teng, Y.-H., Tan, J.J.M., and Hsu, L.-H., The globally bi-3*-connected property of the honeycomb rectangular torus, forthcoming.
300. Touzene, A., Edges-disjoint spanning trees on the binary wrapped butterfly network with applications to fault tolerance, *Parallel Computing*, 28, 649, 2002.
301. Tsai, C.-H., Linear array and ring embeddings in conditional faulty hypercubes, *Theoretical Computer Science*, 314, 431, 2004.
302. Tsai, C.-H., Chuang, Y.-C., Tan, J.J.M., and Hsu, L.-H., Hamiltonian properties of faulty recursive circulant graphs, *Journal of Interconnection Networks*, 3, 273, 2002.
303. Tsai, C.-H., Liang, T., Hsu, L.-H., and Lin, M.-Y., Cycle embedding in faulty wrapped butterfly graphs, *Networks*, 42, 85, 2003.
304. Tsai, C.-H., Tan, J.J.M., and Hsu, L.-H., The super connected property of recursive circulant graphs, *Information Processing Letters*, 91, 293, 2004.
305. Tsai, C.-H., Tan, J.J.M., Liang, T., and Hsu, L.-H., Fault-tolerant Hamiltonian laceability of hypercubes, *Information Processing Letters*, 83, 301, 2002.
306. Tsen, F.-S., Sung, T.-Y., Lin, M.-Y., Hsu, L.-H., and Myrvold, W., Finding the most vital edges with respect to the number of spanning trees, *IEEE Transactions on Reliability*, 43, 600, 1994.
307. Tseng, Y.-C., Chang, S.-H., and Sheu, J.-P., Fault-tolerant ring embedding in a star graph with both link and node failures, *IEEE Transactions on Parallel and Distributed Systems*, 8, 1185, 1997.
308. Tseng, S.-S. and Wang, L.-R., Maximizing the number of spanning trees of networks based on cycle basis representation, *International Journal of Computer Mathematics*, 28, 47, 1989.
309. Tucker, A.C. and Bodin, L., A model for municipal street-sweeping operations, *Case Studies in Applied Mathematics (CUPM), The Mathematical Association of America*, 1976.
310. Tutte, W.T., Convex representations of graphs, *Proceedings of the London Mathematical Society*, 1960, 304.

311. Tutte, W.T., How to draw a graph, *Proceedings of the London Mathematical Society*, 1963, 743.
312. Tutte, W.T., On the 2-factors of bicubic graphs, *Discrete Mathematics*, 1, 203, 1971.
313. Tutte, W.T., On Hamiltonian circuits, *Journal of the London Mathematical Society*, 21, 107, 1946.
314. Tutte, W.T., The dissection of equilateral triangles into equilateral triangles, *Proceedings of the Cambridge Philosophical Society*, 44, 463, 1948.
315. Tutte, W.T., The factorization of linear graphs, *Journal of the London Mathematical Society*, 22, 107, 1947.
316. Tutte, W.T. and Smith, C.A.B., On universal paths in a network of degree 4, *American Mathematical Monthly*, 48, 233, 1941.
317. Tzeng, N.-F. and Wei, S., Enhanced hypercubes, *IEEE Transactions on Computers*, 40, 284, 1991.
318. Uehara, T. and vanCleemput, W.M., Optimal layout of CMOS functional arrays, *IEEE Transactions on Computers*, 30, 305, 1981.
319. Ueno, S., Bagchi, A., Hakimi, S.L., and Schmeichel, E.F., On minimum fault-tolerant networks, *SIAM Journal on Discrete Mathematics*, 6, 565, 1993.
320. Vadapalli, P. and Srimani, P.K., A new family of Cayley graph interconnection networks of constant degree four, *IEEE Transactions on Parallel and Distributed Systems*, 7, 26, 1996.
321. Vadapalli, P. and Srimani, P.K., Fault tolerant ring embedding in tetravalent Cayley network graphs, *Journal of Circuits, Systems, and Computers*, 6, 527, 1996.
322. Vaidya, A.S., Rao, P.S.N., and Shankar, S.R., A class of hypercube-like networks, *Proceedings of the 50th IEEE Symposium on Parallel and Distributed Processing*, 1993, 800.
323. van Aardenne-Ehrenfest, T. and de Bruijn, N.G., Circuits and trees in oriented linear graphs, *Simon Stevin*, 28, 203, 1951.
324. Vecchia, G.D. and Sanges, C., Recursively scalable networks for message passing architectures, *Parallel Processing and Applications*, 33, 1988.
325. Vitaver, L.M., Determination of minimal coloring of vertices of a graph by means of boolean powers of the incidence matrix (Russian), *Doklady Akademii Nauk SSSR*, 147, 758, 1962.
326. Vizing, V.G., Critical graphs with given chromatic class, *Diskret. Analiz.*, 5, 9, 1965.
327. Vizing, V.G., On an estimate of the chromatic class of a p-graph, *Diskret. Analiz.*, 3, 25, 1964.
328. Wagner, K. Bemerkungen zum vierfarbenproblem, *Jahresbericht Deutsch Math. Verein*, 46, 21, 1936.
329. Wagner, K., Úber eine eigenschaft der ebenen komplexe, *Mathematische Annalen*, 114, 570, 1937.
330. Wang, D., Diagnosability of hypercubes and enhanced hypercubes under the comparison diagnosis model, *IEEE Transactions on Computers*, 48, 1369, 1999.
331. Wang, H.-Y. and Chang, R.K.C., Fully connected cubic network: a highly recursive interconnection network, *Proceedings of the 11th Conference on Parallel and Distributed Computing Systems*, 1998, 250.
332. Wang, S.-Y., Hsu, L.-H., and Sung, T.-Y., Faithful 1-edge fault tolerant graphs, *Information Processing Letters*, 61, 173, 1997.
333. Wang, J.-J., Hung, C.-N., and Hsu, L.-H., Optimal 1-Hamiltonian graphs, *Information Processing Letters*, 65, 157, 1998.
334. Wang, J.-J., Hung, C.-N., Tan, J.J.M., Hsu, L.-H., and Sung, T.-Y., Construction schemes for fault-tolerant Hamiltonian graphs, *Networks*, 35, 233, 2000.

335. Wang, J.-J., Sung, T.-Y., Hsu, L.-H., and Lin, M.-Y., A new family of optimal 1-Hamiltonian graphs with small diameter, *Proceedings of the 4th Annual International Computing and Combinatorics Conference, Lecture Notes in Computer Science*, 1998, 269.
336. Wang, C.-F. and Sahni, S., Matrix multiplication on the OTIS-mesh optoelectronic computer, *IEEE Transactions on Computers*, 50, 635, 2001.
337. Welsh, A.J.A. and Powell, M.B., An upper bound for the chromatic number of a graph and its application to timetabling problems, *The Computer Journal*, 10, 85, 1967.
338. Wang, J.-F. and Yang, C.-S., On the number of trees of circulant graphs, *International Journal of Computer Mathematics*, 16, 229, 1984.
339. West, D.B., *Introduction to Graph Theory*, Upper Saddle River, NJ, Prentice Hall, 2001.
340. Whitney, H., Congruent graphs and the connectivity of graphs, *American Journal of Mathematics*, 54, 150, 1932.
341. Wikipedia, The free encyclopedia, http://wikipedia.org/wiki/Sudoku. Online; accesssed July 15, 2005.
342. Wilson, R.J., An Eulerian trail through Königberg, *Journal of Graph Theory*, 10, 265, 1986.
343. Wong, S.-A., Hamilton cycles and paths in butterfly graphs, *Networks*, 26, 145, 1995.
344. Wong, C.-K. and Coppersmith, D., A combinatorial problem related to multimode memory organizations, *Journal of the Association for Computing Machinery*, 21, 392, 1974.
345. Wong, W.-W. and Wong, C.-K., Minimum k-Hamiltonian graphs, *Journal of Graph Theory*, 8, 155, 1984.
346. Xu, M. and Xu, J.-M., The forwarding indices of augmented cubes, *Information Processing Letters*, 101, 185, 2007.
347. Yamada, T., Yamamoto, K., and Ueno, S., Fault-tolerant graphs for hypercubes and tori, *IEICE Transactions on Information and Systems*, E79–D, 1147, 1996.
348. Yang, X., Megson, G.M., and Evans, D.J., An oblivious shortest-path routing algorithm for fully connected cubic networks, *Journal of Parallel and Distributed Computing*, 66, 1294, 2006.
349. Yang, M.-C., Tan, J.J.M., and Hsu, L.-H., Hamiltonian circuit and linear array embeddings in faulty k-ary n-cubes, *Journal of Parallel and Distributed Computing*, 67, 362, 2007.
350. Yang, M.-C., Tan, J.J.M., and Hsu, L.-H., Highly fault-tolerant cycle embeddings of hypercubes, *Journal of Systems Architecture*, 53, 227, 2007.
351. Yang, C.-S., Wang, J.-F., Lee, J.-Y., and Boesch, F.T., Graph theoretic analysis for the boolean n cube networks, *IEEE Transactions on Circuits and Systems*, 35, 1175, 1988.
352. Yeh, C.H. and Varvarigos, E.A. Macro-star networks: efficient low-degree alternatives to star graphs, *IEEE Transactions on Parallel and Distributed Systems*, 9, 987, 1998.
353. Youssef, A., Design and analysis of product networks, *Proceedings of the Fifth Symposium on the Frontiers of Massively Parallel Computation*, 1995, 521.
354. Yun, S.K. and Park, K.H., Hierarchical hypercube networks (HHN) for massively parallel computers, *Journal of Parallel and Distributed Computing*, 37, 194, 1996.
355. Zheng, J., Latifi, S., Regentova, E., Luo, K., and Wu, X., Diagnosability of star graphs under the comparison diagnosis model, *Information Processing Letters*, 93, 29, 2005.

Index